# Innovations and Advances in Computer Sciences and Engineering

Tarek Sobh
Editor

# Innovations and Advances in Computer Sciences and Engineering

Springer

*Editor*
Dr. Tarek Sobh
University of Bridgeport
School of Engineering
221 University Avenue
Bridgeport CT 06604
USA
sobh@bridgeport.edu

ISBN 978-90-481-3657-5 e-ISBN 978-90-481-3658-2
DOI 10.1007/978-90-481-3658-2
Springer Dordrecht Heidelberg London New York

Library of Congress Control Number: 2009942992



Printed on acid-free paper

Springer is part of Springer Science+Business Media (www.springer.com)

## Dedication

*To Nihal, Omar, Haya, Sami and Adam*

# Preface

This book includes Volume I of the proceedings of the 2008 International Conference on Systems, Computing Sciences and Software Engineering (SCSS). SCSS is part of the International Joint Conferences on Computer, Information, and Systems Sciences, and Engineering (CISSE 08). The proceedings are a set of rigorously reviewed world-class manuscripts presenting the state of international practice in Innovations and Advanced Techniques in Computer and Information Sciences and Engineering.

SCSS 08 was a high-caliber research conference that was conducted online. CISSE 08 received 948 paper submissions and the final program included 390 accepted papers from more than 80 countries, representing the six continents. Each paper received at least two reviews, and authors were required to address review comments prior to presentation and publication.

Conducting SCSS 08 online presented a number of unique advantages, as follows:

- All communications between the authors, reviewers, and conference organizing committee were done on line, which permitted a short six week period from the paper submission deadline to the beginning of the conference.
- PowerPoint presentations, final paper manuscripts were available to registrants for three weeks prior to the start of the conference
- The conference platform allowed live presentations by several presenters from different locations, with the audio and PowerPoint transmitted to attendees throughout the internet, even on dial up connections. Attendees were able to ask both audio and written questions in a chat room format, and presenters could mark up their slides as they deem fit
- The live audio presentations were also recorded and distributed to participants along with the power points presentations and paper manuscripts within the conference DVD.

The conference organizers and I are confident that you will find the papers included in this volume interesting and useful. We believe that technology will continue to infuse education thus enriching the educational experience of both students and teachers.

Tarek M. Sobh, Ph.D., PE
Bridgeport, Connecticut
June 2009

# Table of Contents

# Acknowledgements

The 2008 International Conference on Systems, Computing Sciences and Software Engineering (SCSS) and the resulting proceedings could not have been organized without the assistance of a large number of individuals. SCSS is part of the International Joint Conferences on Computer, Information, and Systems Sciences, and Engineering (CISSE). CISSE was founded by Professor Khaled Elleithy and myself in 2005, and we set up mechanisms that put it into action. Andrew Rosca wrote the software that allowed conference management and interaction between the authors and reviewers online. Mr. Tudor Rosca managed the online conference presentation system and was instrumental in ensuring that the event met the highest professional standards. I also want to acknowledge the roles played by Sarosh Patel and Ms. Susan Kristie, our technical and administrative support team.

The technical co-sponsorship provided by the Institute of Electrical and Electronics Engineers (IEEE) and the University of Bridgeport is gratefully appreciated. I would like to express my thanks to Prof. Toshio Fukuda, Chair of the International Advisory Committee and the members of the SCSS Technical Program Committee including: Abdelaziz AlMulhem, Alex A. Aravind, Anna M. Madueira, Hamid Mcheick, Hani Hagras, Julius Dichter, Low K.S., Marian P. Kazmierkowski, Michael Lemmon, Mohamed Dekhil, Mostafa Aref, Natalia Romalis, Raya Al-Qutaish, Rodney G. Roberts, Sanjiv Rai, Shivakumar Sastry ,Tommaso Mazza, Samir Shah, and Mohammed Younis.

The excellent contributions of the authors made this world-class document possible. Each paper received two to four reviews. The reviewers worked tirelessly under a tight schedule and their important work is gratefully appreciated. In particular, I want to acknowledge the contributions of all the reviewers. A complete list of reviewers is given on page XXVII.

Tarek M. Sobh, Ph.D., PE
Bridgeport, Connecticut
June 2009

# Reviewers List

Aamir, Wali
Aaron Don, Africa
Abd El-Nasser, Ghareeb
Abdelsalam, Maatuk, 387
Adam, Piorkowski, 373
Adrian, Runceanu
Adriano, Albuquerque, 287, 293, 305, 361
Ahmad Sofian, Shminan, 171
Ahmad, Saifan
Ahmed, Zobaa
Alcides de Jesús, Cañola
Aleksandras Vytautas, Rutkauskas
Alexander, Vaninsky
Alexei, Barbosa de Aguiar
Alice, Arnoldi, 229
Alionte, Cristian Gabriel
Amala V. S., Rajan
Ana María, Moreno, 409
Anna, Derezinska
Antal, Tiberiu Alexandru
Anton, Moiseenko
Anu, Gupta
Asma, Paracha, 465
Atif, Mohammad, 129, 135, 141
Aubrey, Jaffer, 97
Baba Ahmed, Eddine
Biju, Issac, 333
Brana Liliana, Samoila
Buket, Barkana
Cameron, Cooper
Cameron, Hughes
Cecilia, Chan
chetankumar, Patel
Chwen Jen, Chen
Cornelis, Pieters
Craig, Caulfield
Curila, Sorin
Daniel G., Schwartz
Daniela, López De Luise, 421
David, Wyld
Denis, Berthier
Dierk, Langbein
Dil, Hussain
Dmitry, Kuvshinov, 263
D'Nita, Andrews-Graham
Ecilamar, Lima, 593
Edith, Lecourt
Emmanuel Ajayi, Olajubu
Erki, Eessaar
Ernesto, Ocampo, 273
Fernando, Torres
Gennady, Abramov
Ghulam, Rasool, 321
Gururajan, Erode
Hadi, Zahedi, 345
He, xing-hua
Hector, Barbosa Leon
Houming, FAN
Igor, Aguilar Alonso, 431
Ilias, Karasavvidis
Jaakko, Kuusela
James, Feher
Jan, GENCI
Janett, Williams
Jian-Bo, Chen, 147
Jonathan, White, 241
José L., Fuertes, 223, 505
Jozef, Simuth
József, Berke, 251
Juan, Garcia, 339
junqi, liu
Jussi, Koskinen
Jyri, Naarmala
Kenneth, Faller II
Khaled, Elleithy
Krystyna Maria, Noga
Kuderna-Iulian, Benta, 165
Laura, Vallone, 205
Lei, Jiasu
Leszek, Rudak
Leticia, Flores
Liang, Xia, 193
madjid, khalilian
Mandhapati, Raju, 529
Margareth, Stoll
Maria, Pollo Cattaneo
Marina, Müller
Marius, Marcu
Marius-Daniel, Marcu
Martina, Hedvicakova, 447
Md. Abdul, Based, 397
Miao, Song, 553
Mircea, Popa
Mohammad Abu, Naser
Morteza, Sargolzaei Javan, 159
Muthu, Ramachandran
Nagm, Mohamed
Nazir, Zafar, 327
Neander, Silva, 593
Nilay, Yajnik
Nita, Sarang
Nova, Ovidiu
Olga, Ormandjieva
Owen, Foley, 491
Paola, Ferrari, 593
Paul, David and Chompu, Nuangjamnong
Peter, Nabende, 511
Petr, Silhavy
PIIA, TINT
Radek, Silhavy
Richard, Barnett
S. R., Kodituwakku
S. Shervin, Ostadzadeh
Sajad, Shirali-Shahreza, 443
Salvador, Bueno, 55, 247
Samir Chandra, Das
Santiago, de Pablo
Šarunas, Packevicius
Seibu, Mary Jacob
Sergiy, Popov, 357
Serguei, Mokhov
shalini, batra
Sherif, Tawfik
Shinichi, Sobue
shukor sanim, m. fauzi, 85
Siew Yung, Lau
Soly Mathew, Biju
Somesh, Dewangan
Sridhar, Chandran
Sunil Kumar, Kopparapu, 101
sushil, chandra, 119
Svetlana, Baigozina
Syed Sajjad, Rizvi
Tariq, Abdullah
Thierry, Simonnet
Thomas, Nitsche, 533
Thuan, Nguyen Dinh, 7
Tibor, Csizmadia, 79
Timothy, Ryan
Tobias, Haubold
Tomas, Sochor, 437
Umer, Rashid
Ushasri, anilkumar
Vaddadi, Chandu
Valeriy, Cherkashyn, 459
Veselina, Jecheva
Vikram, Kapila
Xinqi, Zheng
Yaohui, Bai, 377
Yet Chin, Phung, 37
Youming, Li
Young, Lee, 403
Yuval, Cohen
Zeeshan-ul-hassan, Usmani, 19, 25, 517
Zsolt Tibor, Kosztyá, 79

# Possibilities of Computer Simulation in Power Engineering and Environmental Engineering

Lucie Nohacova, Karel Nohac

University of West Bohemia, Pilsen, 30614, Czech Republic

*Abstract* — **This article concerns on overview of today software products for ordinary personal computers and small workstations, gives list of important features relevant to solution of environmental foresighting. For this branch of problems are important not only usual advantages of software packages like computing power, simplicity of use, number of prepared models, but in the first place multidisciplinarity and reliability of gained output. This time there are many software packages prepared for instant use in many branches, this paper describes only few of them and concerns on most known and accessible systems for simulation with partial focus on systems for power engineering problems, which are one of most complex and important problems to be solved.**

**Especially in environmental foresighting there is very high level of complexity of possible systems influence with direct and indirect way. For that reasons there are in this branch many chances to use computer simulations of systems to gain exact overview of future conditions of systems depending of various factors.**

*Index Terms* — **Power engineering, Engineering, Computer graphics software, Computer science education**

## I. Introduction

For successful decision making in all branches with respect to environmental foresighting, there is necessary to know behavior of usually very complex systems, not only in the near future but also in long time distance. Therefore prediction of states of systems is simultaneously important and difficult. Especially in environmental foresighting there is very high level of complexity of possible systems influence with direct and indirect way. For that reasons there are in this branch many chances to use computer simulations of systems to gain exact overview of future conditions of systems depending of various factors. [1],[2].

## II. Simulation Tools Overview

### 1. *Automation Studio, Famic Technologies Inc.*

Automation Studio is an innovative schematic capture, simulation and project documentation software package for automation, pneumatic and hydraulic systems, electrical controls systems and fluid power systems design, but it is intended to be used by engineers in a wide variety of related fields.

Automation Studio will play a useful role in design, simulation, training, documentation and presentation making, service and troubleshooting supporting.

Automation Studio's simulation features were designed for ease of use and quick setup. Users do not need to develop models, the models are included for all supplied components and users only need to determine a limited number of parameters. This makes the software useful for a broader range of users and eliminates the need for advanced training and support.

Automation Studio works on Microsoft Windows NT, 2000, or XP.

Automation Studio is a fully commercial software package. [5]

Other information at www.automationstudio.com

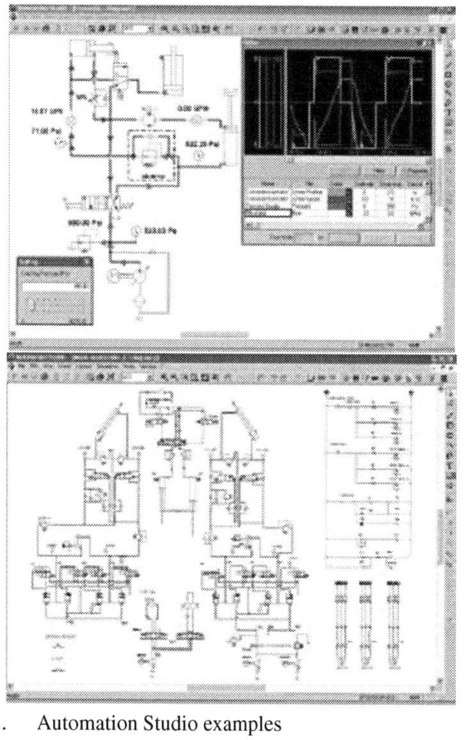

Fig. 1. Automation Studio examples

### 2. *20-sim, University of Twente*

20-sim is a modeling and simulation program. With 20-sim you can model and simulate the behavior of dynamic systems, such as electrical, mechanical and hydraulic systems or any combination of these.

Simulation results in 20-sim can be shown as animations using a 3D Animation Editor. Animations

T. Sobh (ed.), *Innovations and Advances in Computer Sciences and Engineering*,
DOI 10.1007/978-90-481-3658-2_1, © Springer Science+Business Media B.V. 2010

are composed of predefined objects like cubes, spheres, lines, squares, cameras and lights. Complex objects can be imported from CAD packages using common exchange formats. Animations are fully linked to the Simulator in 20-sim. While a plot is being drawn, simultaneously the animation will be shown! This link is kept after a simulation has finished.

20-sim Pro can generate ANSI C-Code as well as MATLAB code for any 20-sim model. The Matlab template is used to generate C-code for Matlab / Simulink S-functions. These S-functions can be used in the Matlab real-time Workshop. The real-time Linux template is new to 20-sim. This template coordinates the C-code generation for the Real-Time Linux software of Syscon GMBH. This software is based on RTAI. This makes the template easily adaptable for other version of real-time Linux and can fully cooperate with hardware model in real time.

20-sim works on Windows 9X/NT/2000/XP. 20sim is a fully commercial software package. Other information at www.20sim.com [5]

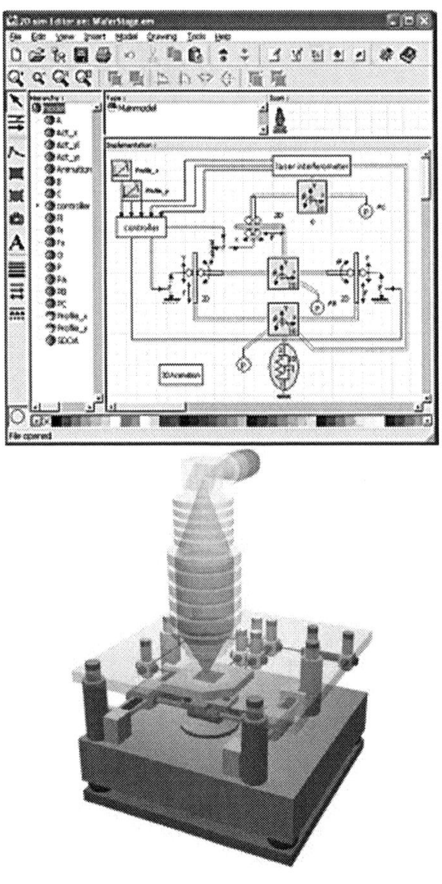

Fig. 2. 20-sim examples

*3. Dynast - Dynamic and Static problems, Icosym*

Developed from the end of seventies at the Czech Technical University in Prague, Dynast admits system models characterized by a set of algebro-differential equations, by a block diagram, by a physical model in the form of a multipole diagram or by a combination of the above approaches. The multipole diagrams allow for portraying the actual physical structure of the modeled real systems without deriving any equations or constructing any graphs. Block and multipole diagrams can be submitted to DYNAST in a graphical form using a web-based Java applet or a more sophisticated OrCAD schematic capture editor. This implies Dynast to be very powerful and universal simulation tool. [6].

Using either DYNCAD or DYNSHELL, the plant model can be easily set up in a graphical form. DYNAST can be then used to simulate the plant and to validate its open-loop model. If the model is nonlinear, DYNAST is capable of linearizing it. DYNAST then can compute the required plant transfer-function poles and zeros, and export them to MATLAB in an M-file. After designing the digital control in the MATLAB environment, the resulting control structure can be implemented in Simulink while the controlled plant model remains in DYNAST. Simulink installed on the client computer can then communicate with the remote DYNAST at each time step across the Internet exploiting the Simulink S-function. [3] ],[4], [6].

Dynast works on Windows 9X/NT/2000/XP and computing core on some UNIX system. Dynast is a fully commercial software package, but offers free simulation demo through internet server. Other information at www.icosym.cvut.cz [5]

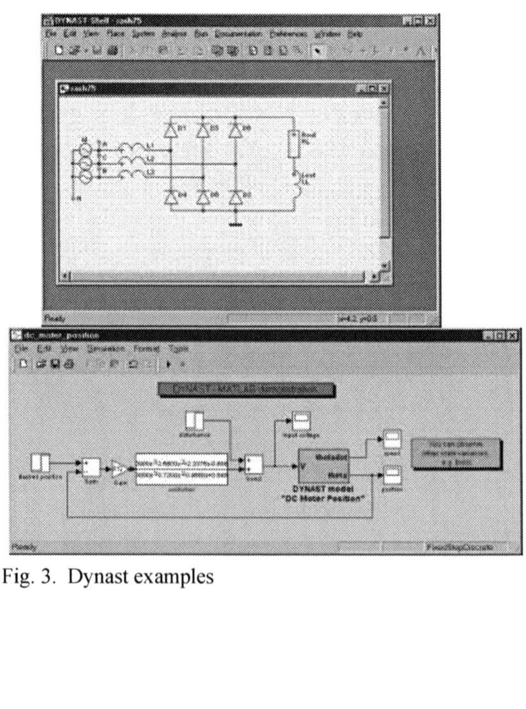

Fig. 3. Dynast examples

4. *Octave , Octave Consortium*

GNU Octave is a high-level language, primarily intended for numerical computations. It provides a convenient command line interface for solving linear and nonlinear problems numerically, and for performing other numerical experiments using a language that is mostly compatible with Matlab. It may also be used as a batch-oriented language.

Octave has extensive tools for solving common numerical linear algebra problems, finding the roots of nonlinear equations, integrating ordinary functions, manipulating polynomials, and integrating ordinary differential and differential-algebraic equations. It is easily extensible and customizable via user-defined functions written in Octave's own language, or using dynamically loaded modules written in C++, C, Fortran, or other languages. Because of Octave is fully functional Mathlab Clone, there is possible to create and utilize Simuling features in Octave, which makes Octave very powerful and universal simulation tool.

Octave works on Windows 9X/NT/2000/XP. Octave is completely free software. Other information at www.octave.org [5]

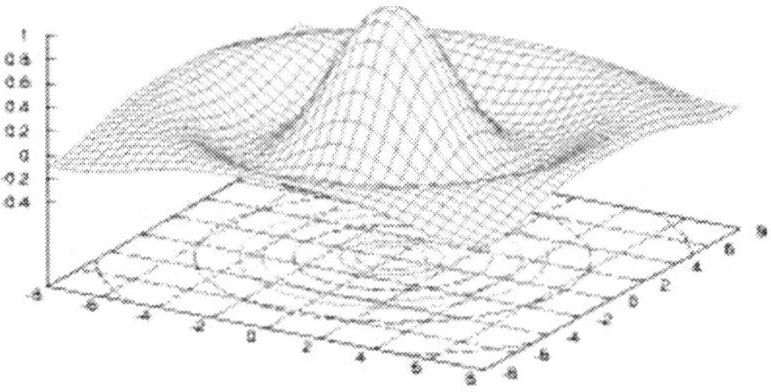

Fig. 4. Octave example

5. *Scilab, Scilab Consortium*

Scilab is another Matlab like computing and simulation program, which language is not so close to Matlab, but includes further features for simulation and visualization of results. Scilab is a scientific software package for numerical computations providing a powerful open computing environment for engineering and scientific applications. Powerful feature of Scilab is a SIMULINK like tool for interactive dynamic system modeling Scicos.

A number of other toolboxes are available with the system:

- 2-D and 3-D graphics, animation
- Linear algebra, sparse matrices
- Polynomials and rational functions
- Simulation: ODE solver and DAE solver
- Differentiable and non-differentiable optimization
- Signal processing and many others.

Scilab works on most Unix systems including GNU/Linux and on Windows 9X/NT/2000/XP. Scilab is completely free software. Other information at www.scilabsoft.inria.fr [5]

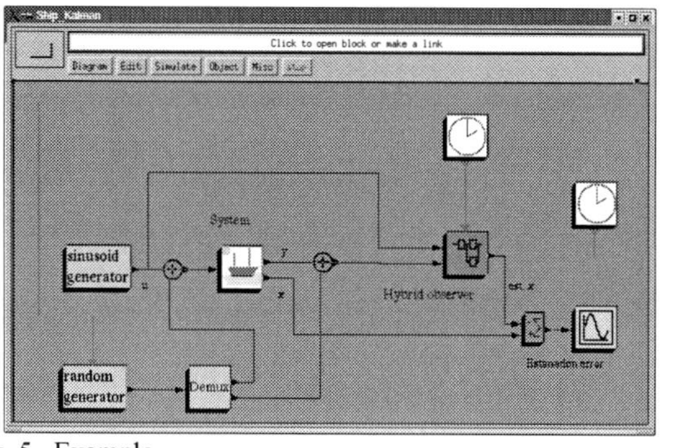


Fig. 5. Example

## III. Other Software simulation tools

There are many software packages to solve simulation problems. In this paper cannot be described all of them. So we maybe can very shortly introduce some other examples:

- CAMP-G (Software package that helps engineers and scientists design Mechatronics and Dynamic Systems using as input physical models described by the method of Bond Graphs, www.bondgraph.com ) [5]
- ACSL - Advanced Continuous Simulation Language (first commercially available, modeling and simulation language designed for simulating continuous systems, www.acslsim.com) [5]
- AnyLogic (tool with very various modeling possibilities for the description of structure, behavior and data of the system being modeled, www.xjtek.com) [5]
- ITI-SIM and SimulationX (Can easy combine formal model description and use of explicit differential equations and programming statements, www.iti.de) [5]
- ATP-EMTP (universal program system for digital simulation of transient phenomena of power systems, electromagnetic as well as electromechanical nature, www.emtp.org) [5]
- Multisim 7, Electronic Workbench (powerful electronic and control simulation tool, www.interactiv.com) [5]

In branch of simulations for building energy analysis we should remark EnergyPlus (www.eere.energy.gov), Energy-10 (www.nrel.gov) and Ener-Win (www.cox-internet.com). [5]

Except that, there is a important branch for power engineering and environmental engineering, namely tools for lighting designers and architects by predicting the light levels and appearance of a space prior to construction. There are also many programs, for example Radiance (radsite.lbl.gov/radiance).

- DOE-2, DOE2Parm,VisualDOE is classic program for building energy use analysis provides computer simulation program including weather conditions, can quickly determine the choice of building parameters which improve energy efficiency. DOE-2 and VisualDOE are commercial software packages. DOE2Parm is completely free software. Other information at gundog.lbl.gov, www.deringergroup.com, www.archenergy.com [5]
- Simplorer, Ansoft Corporation (best usable in branch is automotive, aerospace, power electronics, and electric drive systems. Simplorer is a fully commercial software package. Other information at www.ansoft.com) [5]
- Dymola - Dynamic Modeling Laboratory, Dynasim AB Dymola is a new technology for multi-engineering modeling and simulation.,. Dymola allows simulation of the dynamic behavior and complex interactions between, multiple systems of different kind using new methodology based on object orientation and equations (automatic formula manipulation, model libraries from many engineering domains, object oriented modeling language with support of hierarchical structuring, interface to Simulink) Best usable in branch: is automotive, aerospace, robotics, process. Dymola is a fully commercial software package. Works on Microsoft Windows 9X/NT/2000/XP Other info at www.dynasim.se [5]
- AMESim, Imagine s. a. is graphical system for modeling, simulation and analysis of dynamic engineering systems, is based on a large set of validated libraries issued from different physical domains. AMESim allows you to access rapidly the ultimate goal of modeling without code writing. (strong numerical capabilities, wide library of models, advanced tools to study the static/dynamic behavior, sensitivity analysis, parametric study analysis, open concept with interface to CAE software, Matlab, Simulink, Flux or other user code) Best usable in branch: system design in automotive and aerospace. Works on Microsoft Windows, most Unix systems including GNU/Linux AMESim is a fully commercial software package. Other information at www.amesim.com [5]

## IV. SUMMARY AND CONCLUSION

In this paper is given overview of following software packages features, which are applicable for computer simulation in branch Power engineering and environmental engineering

- Automation Studio, Famic Technologies Inc.
- 20-sim, University of Twente
- Dynast - Dynamic and Static problems, Icosym
- Octave, Octave Consortium
- Scilab, Scilab Consortium

Not many of them are fully suitable for power engineering and environmental engineering. Some tools are universal enough, but they are commercial (Automation Studio, 20-sim,), others offer usually less pre-build components, although they have strong special tools to create them quickly (firstly Dynast). Except that there are some freeware clones of Matlab, which have no support, but they are powerful enough to solve very wide range of problems for free (Octave and Scilab). Finally some packages suitable only for specific branch in power engineering and environmental engineering, because they focus on some engineering problem solving, Electronic Workbench and other electronic tools, other power engineering tools, Radiance and other lighting design tools, etc.

## V. ACKNOWLEDGEMENT

The authors wish to thank to the assistance and support of Prof. Ing. Jan Mühlbacher, CSc. – head of our common Department of Power Engineering and Ecology on University of West Bohema in Pilsen.

This paper was written under solving science project 2A-TP/051.

## REFERENCES

*Books:*

[1] J. Mühlbacher, M. Kolcun and J. Haler, "Mathematical analysis of electrical networks", vol. I. Czech Republic: 2004.

*Papers from Conference Proceedings (Published):*

[2] J. Mühlbacher, K. Nohá and L. Nohá ová, "Distributed power systems," in Proc. 2003 Maribor. Power Engineering 2003. Slovenia Republic., *International Conference "Power Engineering"*, Maribor, 12th International Expert Meeting, pp. 1-4

[3] K. Nohá , M. B lík and M. Tesa ová., "Simulation of asynchronous engine of main pump in nuclear power plant for monitoring purposes," in Proc. 2006 Chang Won. CMD 2006 *"International Conference on Condition Monitoring and Diagnosis 2006"*. Republic of Korea , pp. 1-8

[4] L. Nohá ová, K. Nohá , "Some cases of distributed resources connected to the distribution network," in Proc. 2004 Maribor. Power Engineering 2004. Slovenia Republic., *International Conference "Power Engineering"*, Maribor, 13th International Expert Meeting, pp. 1-6

[5] Internet adresses: www.automationstudio.com www.20sim.com www.icosym.cvut.cz www.octave.org www.scilabsoft.inria.fr

www.bondgraph.com www.acslsim.com www.xjtek.com www.iti.de www.emtp.org www.interactiv.com www.eere.energy.gov www.nrel.gov www.cox-internet.com gundog.lbl.gov, www.deringergroup.com, www.archenergy.com www.ansoft.com www.dynasim.se www.amesim.com

[6] L. Nohá ová, K. Nohá , "New Possible aproach to Modelling in Power Engineering", in Proc. 2008 Pernink. Environmental Impacts of Power Industry 2008. Czech Republic., *International Conference*, Pernink, 5th International Expert Meeting, pp. 1-6

# Mining Time Pattern Association Rules in Temporal Database

Nguyen Dinh Thuan
Faculty of Information Technology
NhaTrang University, Viet Nam

*Abstract* – **The discovery of association rules in large databases is considered an interesting and important research problem. Recently, different aspects of the problem have been studied, and several algorithms have been presented in the literature, among others in [3,8,9]. A time pattern association rule is an association rule that holds a specific time interval. For example, bread and coffee are frequently sold together in morning hours, or mooncake, lantern and candle are often sold before Mid-autumn Festival. This paper extends the a priori algorithm and develops the optimization technique for mining time pattern association rules.**

**Index Terms: time pattern, association rules, temporal data mining.**

## I. INTRODUCTION

Data mining, also known as knowledge discovery in databases, has been recognized as a new area for database research. The problem of discovering association rules the first was introduced in [1][2]. Given a set of transactions, where each transaction is a set of items, an association rule is an expression of the form X→Y, where X and Y are sets of items. For example, given a database of orders (transactions) placed in a restaurant, we may have an association rule of form:

egg → milk (support: 10%, confidence: 70%)

which means that 10% of all transactions contain both egg and milk, and 70% of transaction that have egg also have milk in them. The two percentage parameters above are commonly referred to as support and confidence, respectively.

One important extension to association rules is to include a temporal dimension. For example, eggs and coffee may be ordered together primarily between 7AM and 10AM. Therefore, we may find that the above association rule has a support as high as 40% for transactions that happen between 7AM and 10AM and a support as low as 0.05% for other transactions.

This paper studies temporal association rules during time intervals that follow some user-given time schemas. An example of time schema is (day, month, year) and the notation (10, *, 2006), which corresponds to the time interval consisting of all the 10th days of all months in year 2006. We refer to this association rule as cyclic if the rule has the minimum confidence and support at regular time intervals. Such a rule need not hold for entire transactional databases, but rather only for transactional data in a particular periodic time interval.

The remainder of this paper is organized as follows. In the next section, we give a formal definition of time pattern association rules and the Apriori algorithm. In Section 3, discuss the time pattern association rules algorithm with its correctness proved. Finally, we present our conclusions and future work.

## II. THE APRIORI ALGORITHM

### *A. Problem Description [1][2]*

Definition 1: Let $I = \{I_1, I_2, .., I_m\}$ be a set of m distinct attributes, also called literals. Let a database D be a set of transactions, where each transaction T is a set of items such that $T \subseteq I$ An association rule is an implication of the form $X \rightarrow Y$, where $X, Y \subset I$, are sets of items called itemsets, and $X \cap Y = \emptyset$. Here, X is called antecedent, and Y consequent.

Two important measures for association rules, support (s) and confidence (γ), can be defined as follows.

Definition 2: The support (s) of an association rule is the ratio (in percent) of the records that contain $X \cup Y$ to the total number of records in the database.

Therefore, if we say that the support of a rule is 5% then it means that 5% of the total records contain $X \cup Y$.

Let $N_{XY}$ be the number of records that contain $X \cup Y$

$$Supp(X \rightarrow Y) = \frac{N_{XY}}{|D|}$$

Definition 3: For a given number of records, confidence (γ) is the ratio (in percent) of the number of records that contain $X \cup Y$ to the number of records that contain X.

$$Conf(X \rightarrow Y) = \frac{Supp(X \cup Y)}{Supp(X)}$$

Thus, if we say that a rule has a confidence of 85%, it means that 85% of the records containing X also contain Y. The confidence of a rule indicates the degree of correlation in the dataset between X and Y. Confidence is a measure of a rule's strength. Often a large confidence is required for association rules.

Two problems in mining of association rules:

- Find all sets of items L, which occur with a frequency that is greater than or equal to the user-specified threshold support, Min_Supp = s0.

- Generate the desired rules using the large itemsets, which have user-specified threshold confidence, Min_Conf = γ0.

T. Sobh (ed.), *Innovations and Advances in Computer Sciences and Engineering*,
DOI 10.1007/978-90-481-3658-2_2, 

The first step finds large or frequent itemsets. Itemsets other than those are referred as small itemsets. Here an itemset is a subset of the total set of items of interest from the database.

Example 1: Consider a small database with six items I={A, B, C, D, E, F} and twelve transactions as shown in Table 1.
We have association rule as follows:
Supp (A→D) = 5/12= 41.66%, Conh (A→D) = 5/6 = 83.33%
Supp (B → D) = 25% and Conf (B → D)= 42.85%
Supp (D → F) = 16.7% and Conf (D → F) = 25%
Supp (F → D) = 16.7% and Conf (F → D) =100%

TABLE 1

| $T_{ID}$ | Items | | | | | |
|---|---|---|---|---|---|---|
| $t_1$ | | B | | D | | |
| $t_2$ | | B | C | | | |
| $t_3$ | A | B | | D | | |
| $t_4$ | A | | | D | | |
| $t_5$ | | B | | | E | |
| $t_6$ | A | | C | D | | |
| $t_7$ | A | | | D | E | |
| $t_8$ | A | B | | | E | |
| $t_9$ | | B | C | | | |
| $t_{10}$ | | B | | D | | |
| $t_{11}$ | A | | | D | E | F |
| $t_{12}$ | | | | D | | F |

*B. A Priority Algorithm [3][4]*

The Apriori algorithm was developed by Agrawal (1993) is a great achievement in the history of mining association rules (Cheung1996c) . It is by far the most well-known association rule algorithm. Basically, an a priori approach is based on the heuristic: if any itemset of length k is not frequent in the database, its length (k + 1) super-itemset will never be frequent. As we know that a superset of one large itemset and a small itemset will result in a small itemset, these techniques generate too many candidate itemsets which turn out to be small. The a priori algorithm addresses this important issue. The a priori generates the candidate itemsets by joining the large itemsets of the previous pass and deleting those subsets, which are small in the previous pass without considering the transactions in the database. By only considering large itemsets of the previous pass, the number of candidate large itemsets is significantly reduced.

In the first pass, the itemsets with only one item are counted. The discovered large itemsets of the first pass are used to generate the candidate sets of the second pass using the a priori_gen() Procedure. Once the candidate itemsets are found, their supports are counted to discover the large itemsets of size two by scanning the database. In the third pass, the large itemsets of the second pass are considered as the candidate sets to discover large itemsets of this pass. This iterative process terminates when no new large itemsets are found. Each pass i of the algorithm scans the database once and determines large itemsets of size i. Li denotes large itemsets of size i, while Ci is candidates of size i.

**Procedure 1: finds all *large* or *frequent itemsets***
**Input:** I, D, s
**Output:** L: Large Itemsets

**Method:**
begin
  $L_1$ = {large 1-itemsets}
  **For** (k = 2; $L_{k-1} \neq \emptyset$; k++) **do**
  **begin**
    $C_k$ = Apriori-gen($L_k$-1) // New candidates
    For all transactions T ∈ D do
      Subset($C_k$,T)//c.count++ if c∈ $C_k$ is contained in T
      $L_k$ = {c∈ $C_k$ | c.Count >=Min_Sup*|D|}
  **end**
  L= ∪$L_k$
end.

**Procedure 2: A*priori*_gen()**

The candidates' generation is made by means of the Procedure ***apriori***-gen(). It takes as argument $L_{k-1}$, the set of all frequent (k-1)-itemsets, and returns a superset of the frequent k-itemsets. This procedure has been organized into a *Join Step and* a *Pruning Step.*

1. Join Step:
In SQL
*Insert into* $C_k$
  *select* p.item$_1$, p.item$_2$, ..., p.item$_{k-1}$, q.item$_{k-1}$
  *from* $L_{k-1}$ p, $L_{k-1}$ q
  *where* p.item$_1$ = q.item$_1$ *and* ... *and* p.item$_{k-2}$ = q.item$_{k-2}$ *and p.item$_{k-1}$ < q.item$_{k-1}$;*

In the next step (*pruning*) all the candidate itemsets c ∈ $C_k$ such that any subset of c with k-1 items is not in $L_{k-1}$ are deleted.

2. Pruning Step:
**for each** itemset c ∈ $C_k$ **do**
  **for each** (k-1) subsets s in c **do**
    **if** (s ∉ $L_{k-1}$) **then delete** c from $C_k$;

**Procedure 3: Find Association Rules Given Large Itemsets**:

**Input**: I, D, s, γ, L

**Output**: Association rules satisfying s and γ

**Method:**

- Find all nonempty subsets, X, of each large itemset, l∈L
- For every subset, obtain a rule of the form x→ (*l*-x) if the ratio of the frequency of occurrence of *l* to that of x is greater than or equal to the threshold confidence.

Example 2: Let be the data in example 1, $c_0$=60% and $s_0$=40%
The set of large itemsets:L = {{A}, {B},{D}, {E}, {AD}}
Confidence and Support of some association rules:

TABLE 2

| Rule | Confidence | Support |
|---|---|---|
| A→D | 83.33% | 41.66% |
| D→A | 62.5% | 41.66% |

## III. MINING TIME PATTERN ASSOCIATION RULES

*A. Related Work*

Since the concept of association rule was first introduced in Agrawal, R., et al, 1993[1][2]. The concept of association rules was also extended in several ways, including generalized rules and multi-level rules[4] constraint-based rules [5], cyclic association rules[6]. The problems of mining cyclic association rules are association rules with perfect periodicity in the sense that each rule holds in every cycle with no exception. The perfection in periodicity plays a significant role in designing efficient algorithms for cyclic association rules: If we know that a rule does not hold at a particular time instant, then the rule will not hold in any cycle which includes this time instant.

In [7], the work of [8] was extended, approximately discover user-defined temporal pattern in association rules. The work of [9] is more flexible and perhaps more practical than that of [7]. However, it requires user's prior knowledge about what exact temporal patterns are to be discovered. In some cases, users lack such priori knowledge.

*B. Time schema*

*Definition 4:* A time schema is a set $R = \{d_i \in DT_i \mid i=1,.., m\}$, where: $d_i$ : time element name likes hour, day, month, year, ...

$DT_i = Dom(d_i) \cup \{*\}$ : domain of di or the wild card symbol '*'

Example 3:

$R_1 = \{day \in [1..31], month \in [1..12], year \in [1980 .. 2006\}$

$R_2 = \{day \in [1..7], week \in [1..52]\}$

*Definition 5:* We say a time pattern on time schema R is a tuple $t \in R$ and t-i*, which i is number of symbols '*' in the time pattern.

Example 4: given the time schema $R_1$ in Ex3

t1 = (25,2,2006) is t-0*

t2 = (25,*,2006) is t-1*

*Definition 6:* Given $t=(t_1, t_2, ..., t_m)$, $t'=(t'_1, t'_2, ..., t'_m)$ are two time patterns on the same time schema R.

We say t-*cover t' : t>* t' if either $t_i=t_i'$ or $t_i$='*' $\forall i=1,..,m$

Example 5: We have (25,*,*) >* (25,*,2006) and (25,*,2006) >* (25,2,2006)

*Definition 7:* Given $t=(t_1, t_2, ..., t_m)$ is a time pattern on time schema R and a database D be a set of transactions. The partition of D over time pattern t, is $D = \bigcup_{k=1}^{n} D_{t*k}$ and $D_{t*i} \cap D_{t*j} \neq \emptyset$ $\forall i \neq j$, where $D_{t*k}$ is corresponding the values of $t_i \neq$'*'.

Example 6: Consider the time schema R1={day, month, year}

+ if t=(day, *, *) then D is divided into 31 partitions: $D_{t*1}, D_{t*2}, ..., D_{t*31}$, where $D_{t*i}$ corresponds to the time interval consisting of all the ith days of all months in all year.

+ if t=(day, month, *) then D is divided into: $\{D_{t*ij} \mid i=1,..,31, j=1,..,12\}$

Let: $D = \bigcup_{k=1}^{n} D_{t*k}$

$|D_{t*k}|$ be the number of transaction in Dt*k

$N_{Dt*k}(X)$ be the number of transaction in $D_{t*k}$ that contain itemset X

In this section, we propose the algorithm for mining of time pattern association rules (abbreviated as MTP), denoted by <(X→$_{t*}$Y), t*>, where Supp (X→$_{t*}$Y)>Min_Supp and Conf (X →$_{t*}$ Y) > Min_Conf.

In addition, these rules correspond with time pattern t* and the concept of support and confidence are produced by newly defined.

*Definition 8:* The support value of an itemset X is:

$$Supp_{D_{t*k}}(X) = \frac{N_{D_{t*k}}(X)}{|D_{t*k}|}$$

*Definition 9:* The support value of an rule (X→$_{t*}$Y), corresponding time pattern t* is

$$Supp(X \rightarrow_{t*} Y) = \frac{N_{D_{t*k}}(XY)}{|D_{t*k}|}$$

*Definition 10:* The confidence value of rule (X→$_{t*}$Y), corresponding time pattern t* is

$$Conf(X \rightarrow_{t*} Y) = \frac{Supp_{D_{t*k}}(XY)}{Supp_{D_{t*k}}(X)}$$

*C. Algorithm MTP:*

*Procedure 4:* Find every set of 2-itemsets that is cyclic and frequent.

```
Begin
Ret = Ø
L2*0={frequent 2-itemsets with t-0*}
For each value k in time pattern t-*cover t
   For each X2 ∈ L2*0
      if (X2 ∉ Ret) then
            X2.Count = 1
            X2.Start= k
            Ret = Ret ∪ X2
            end if
            if (X2 ∈ Ret) then
                  X2.Count = X2.Count +1
      if (X2.Count + (|Dt*k|-k) < Min_Supp*|Dt*k| ) then
            Ret = Ret - X2
Return <Ret, t*>
end
```

*Procedure 5:* Find every set of k-itemsets, which is cyclic and frequent.

```
begin
L = ∅
Find every set of 2-itemsets, which is cyclic and frequent.
m = 2
while (Cm ≠∅) do
      Cm = Cm-1* Cm-1 //Join
      m = m+1
end
For each value k in time pattern t-*cover t do
      For each itemset X ∈ Cm do
            X.Count = X.Count + 1
      For each itemset X∈ Cm do
      if (X2.Count + (|Dt*k|-k) >= Min_Supp*|Dt*k| ) then
            L = L ∪ X;
Return <L, t*>;
end.
```

*D. An example of MTP:*

Given a time schema R={day, week}, Min_Supp=50% and Min_Conf=70% and database D be a set of transactions as

shown in Table 3.
Under Apriori algorithm, we have:

Supp(A→ D) = 5/35= 14.38% <Min_supp

Conf(A→ D) =5/8 = 62.5% <Min_Conf

TABLE 3

| | Day 1 | Day 2 | Day 3 | Day 4 | Day 5 | Day 6 | Day 7 |
|---|---|---|---|---|---|---|---|
| Week 1 | ABD | ABC | | AD | | | BCF |
| Week 2 | BDF | BC | CD | D | | EF | CF |
| Week 3 | AD | B | AF | BD | BC | | |
| Week 4 | ACDF | BD | | DE | | | ACEF |
| Week 5 | ADF | BE | | | DEF | | DEF |

These rules do not satisfy Min_supp and Min_Conf requirement because the total number of transactions in database D contain both A and D do not exist through all time of the database. It may be the case that A and D occur in Day1 of all weeks.

Our proposed algorithm, with time pattern t=(day 1,*), we have some values: In this case $|D_{t*k}|$=5, Min_Supp=50%

TABLE 4

| $L_2(1,*)$, k=1 | | |
|---|---|---|
| $X_2$ | Start | Count |
| AB | 1 | 1 |
| AD | 1 | 1 |
| BD | 1 | 1 |

TABLE 5

| $L_2(1,*)$, k=2 | | |
|---|---|---|
| $X_2$ | Start | Count |
| AB | 1 | 1 |
| AD | 1 | 1 |
| BD | 1 | 2 |
| BF | 2 | 1 |
| DF | 2 | 1 |

After 1st scan database D, we have candidate 2-itemsets as follows: {AB, AD, BD, BF, DF}

TABLE 6

| $L_2(1,*)$, k=3 | | |
|---|---|---|
| $X_2$ | Start | Count |
| AB | 1 | 1 |
| AD | 1 | 2 |
| BD | 1 | 2 |
| BF | 2 | 1 |
| DF | 2 | 1 |

TABLE 7

| $L_2(1,*)$, k=4 | | |
|---|---|---|
| $X_2$ | Start | Count |
| AB | 1 | 1* |
| AC | 4 | 1* |
| AD | 1 | 3 |
| AF | 4 | 1* |
| BD | 1 | 2 |
| BF | 2 | 1* |
| CD | 4 | 1* |
| CF | 4 | 1* |
| DF | 2 | 2 |

After ith scan database D, we drop itemset $X_2$ if its counter $X_2$.count does not statisfy $X_2$.Count + ($|D_{t*k}|$-k) > Min_Supp*$|D_{t*k}|$ . The itemsets marked with '*' do not satisfy this condition.

TABLE 8

| $L_2(1,*)$, k=5 | | |
|---|---|---|
| $X_2$ | Start | Count |
| AD | 1 | 4 |
| AF | 5 | 1* |
| BD | 1 | 2* |
| DF | 2 | 3 |

TABLE 9

| $L_2(1,*)$, k=5 | | |
|---|---|---|
| $X_2$ | Start | Count |
| AD | 1 | 4 |
| DF | 2 | 3 |

TABLE 10

| Rules | Supp | Conf |
|---|---|---|
| ($A \to_{t*} D$) | 80% | 100% |
| ($D \to_{t*} A$) | 80% | 80% |
| ($D \to_{t*} F$) | 60% | 60% |
| ($F \to_{t*} D$) | 60% | 100% |

Pruning with Min_Conf=70%:

TABLE 11

| Rules | Supp | Conf |
|---|---|---|
| ($A \to_{t*} D$) | 80% | 100% |
| ($D \to_{t*} A$) | 80% | 80% |
| ($F \to_{t*} D$) | 60% | 100% |

*E. Correctness of MTP*

*Lemma:* An itemset T remains in <L, t*> after the processing of kth time pattern t-*cover t if satisfy this condition: X.Count + ($|D_{t*k}|$-k) > Min_Supp*$|D_{t*k}|$

*Proof:*

It can be seen that the proof follows directly from the Algorithm MTP. Suppose there are totally $|D_{t*k}|$ be the number of transaction in $D_{t*k}$. It is easy see that at the end of algorithm X.Count is updated at least Min_Supp*$|D_{t*k}|$ times.

At the processing of k-time pattern t-*cover t, the number of times that we can update X.Count is <= ($|D_{t*k}|$-k). Therefore, after the processing of kth, we have

X.Count + ($|D_{t*k}|$-k) > Min_Supp*$|D_{t*k}|$

*Theorem:* At the end of Algorithm MTP, we have <L, t*> contains all and only the k-itemsets that are frequent for at least Min_Supp of the t0* time pattern covered by t*.

*Proof:*

Consider any time pattern t*. Suppose $|D_{t*k}|$ basic time intervals are covered by t*. For each itemset $X \in L_k$, its count X.Count >= Min_Supp*$|D_{t*k}|$ since it is not drop in the last update. Following from Lemma 1, for each itemset Y dropped in the k-th update of $L_k$, its counter:

Y.Count <= Y.Count + ($|D_{t*k}|$-k) <= Min_Supp*$|D_{t*k}|$

i.e., $Supp_{D_{t*k}}$ (X)<=Min_Supp of the basic time pattern covered by t*. Since all frequent itemsets are processed, for all time patterns, t*, L contains all and only the itemsets that their are large for at least Min_Supp of the basic time pattern covered by t*.

## IV. CONCLUSION AND FUTURE WORK

In this paper, we studied the discovery of association rules in term time schemas. We developed algorithm MTP to generate the temporal cyclic association rules. The algorithm is easier to use. For example, given a time schema (day, month, year), we discover all rules that repeat themselves daily, monthly, yearly, i.e., temporal association rules with time schemas.

The future work includes two directions. First, we would like to consider temporal patterns in other data mining problems such as clustering. Second, we would like to study the computational complexity of mining time pattern association rules.

## REFERENCES

[1] Agrawal. R., Imielinski. T., Swami. A., Mining Associations between Sets of Items in Massive Databases. In Proc.of the 1993 ACM-SIGMOD Int'l Conf. on Management of Data, 207-216.

[2] Agrawal. R., Mannila. H., Srikant. R., Toivonen. H., Verkamo. A. I., Fast Discovery of Association Rules. In Advances in Knowledge Discovery and Data Mining, AAAI Press, 1996, p307-328.

[3] Ali. K., Manganaris. S., Srikant. R., Partial Classification using Association Rules. In Proc. of the 3rd Int'l Conference on Knowledge Discovery in Databases and Data Mining, 1997, p115-118.

[4] Chang-Hung Lee, Cheng-Ru Lin And Ming-Syan Chen. On Mining General Temporal Association Rules in a Publication DataBase. In ICDM01, pages 337-344, 2001.

[5] Li. Y, Ning. P., Wang. X. S., And Jajodia. S., Discovering calendar-based temporal association rules. Manuscript. http://www.ise.gmu.edu/_pning/tdm.ps, Nov. 2000.

[6] Ozden. B., Ramaswamy. S., And Silberschatz., A Cyclic association rules. In Proc. of the 14th Int'l Conf. on Data Engineering, pages 412–421, 1998.

[7] Nguyen Xuan Huy, Nguyen Dinh Thuan. Mining Association Rules With Transaction-Weight in Temporal Database. In Proc. of the First Symposium "Fundamental and Applied Information Technology Research" (FAIR) Ha Noi, 10/2003. p137-p147

[8] Nguyen Dinh Ngoc, Nguyen Xuan Huy, Nguyen Dinh Thuan. Mining cyclic association rules in temporal database. The journal Science & Technology Development, Vietnam National University – Ho Chi Minh City (VNU-HCM). Vol 7, N8, 2004 p12-19

[9] Nguyen Dinh Thuan. Algorithm for incremental data mining in temporal database. Journal of Computer science and Cybernetics. Vol 20, N1, 2004. p80-90.

# Domain-Based Intelligent Tutoring System

Dawod Kseibat, Ali Mansour, Osei Adjei

*Abstract* - **In this paper we present a new framework for predicting the proper instructional strategy for a given teaching material based on its attributes. The framework is domain-based in the sense that it is based on the qualitative observation of the teaching materials' attributes stored by the system. The prediction process is based on a machine learning approach using feed forward artificial neural network to generate a model that both fit the input data attributes and predict the proper instructional strategy by extracting knowledge implicit in these attributes. The framework was adapted in an Intelligent Tutoring System (ITS) to teach Modern Standard Arabic language to adult English-speaking learners with no pre-knowledge of Arabic language is required. The learning process will be through the Internet since the online education is better suited to mature individuals who are self-motivated and have a good sense of purpose.**

## I. Introduction

Many researchers of ITSs combined expert systems and machine learning techniques in the design of the student modelling and the system's decision making process. The basic principle is to produce evaluation of the student knowledge based on his/her observable behaviors. These techniques include artificial neural networks (ANN), fuzzy logic, and decision trees. Nevertheless, the focus was on applying fuzzy logic and ANNs separately or combined (neuro-fuzzy) in the development of these systems. The use of ANN to predict the student's behaviour and the number of errors the student will make was presented by Wang and Mitovic (2002). Neuro-fuzzy systems which are based on fuzzy systems trained by a learning algorithm derived from neural network theory were used in different approaches for student modelling. Stathacopoulou *et. al.* (2004; 2007) developed neuro-fuzzy approaches to detect the student motivation and monitoring of student's actions. Sevarac (2006) presented a neuro-fuzzy system that enables the classification of the students based on qualitative observation of their characteristics. The system has 100% successful rate for the test data. The majority of the previously discussed systems presented methods that applied the use of different machine learning techniques in the development of ITSs. However, most of these systems directed towards student modelling with little attention given to the expert model (domain model) and other ITSs components. This had led to an architecture that focused on representing the learner's knowledge (student model) not the knowledge to be learned (domain model).

## II. Model Design

The constructed framework will be applied to a sequence of cases, in which each new case must be assigned to one of a set of pre-defined classes (outputs) on the basis of observed attributes or features (inputs). Each case can be characterized by a tuple (**x**, y), where **x** is attributes set and **y** is a predefined class label (instructional strategy). The input attributes' set includes properties from learning materials such as lessons and questions. The prediction process is the task to learn a target function ***f*** that maps each attributes' set **x** to one of the predefined labels **y**. Since the ***Course manger*** is the part of the system which responsible for controlling the learning process, the goal is to provide the course manger with expertise on the proper strategy of how to teach ***certain*** materials. The prediction process is based on ***Back propagation*** feed forward ANN (see Figure 1).

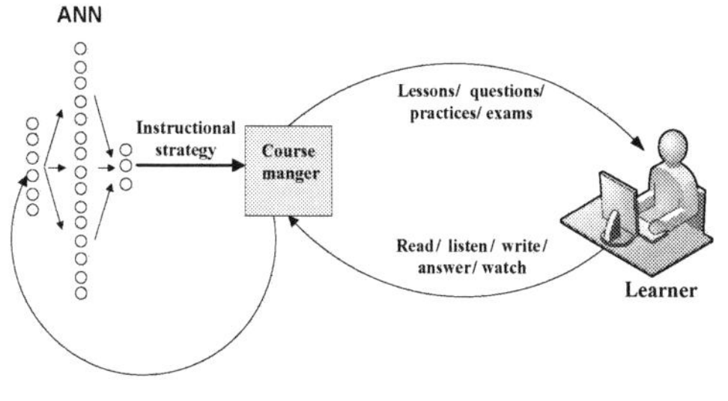


Fig. 1. The prediction process of the instructional strategy based on ANN

## III. Inputs Attributes

Input data of the problem domain is required by the ANN in the learning process. From this data we must establish what is necessary and what is redundant for the training process which can be difficult to determine. Therefore, all relevant attributes with their possible values were included in the ANN training process. The inputs to the network correspond to the learning materials' attributes extracted from the system's database include: learning level, difficulty, time, category (i.e. lesson, question), type (i.e. listening, reading), and history (i.e. studied, never studied). On the basis of these inputs, the network predicts the proper learning strategy (i.e. strategy 1, strategy 2,...). Since all input attributes are verbal attributes representing different properties, hence, each attribute discretiszied into equal intervals using set of linguistic terms where each term represents a value (category) of that attribute. Each value is coded into a discrete number from 0 to 1 as represented in Table I. 288 different combinations of all

T. Sobh (ed.), *Innovations and Advances in Computer Sciences and Engineering*,
DOI 10.1007/978-90-481-3658-2_3, 

attributes' values were generated to be used as inputs to the ANN.

TABLE I
Learning material attributes

| learning material attributes | | | |
|---|---|---|---|
| | Attributes | Discrete attributes values | Coded data |
| 1 | learning levels | Beginner | 0.0 |
| | | Intermediate | 0.5 |
| | | advance | 1.0 |
| 2 | difficulty | Easy | 0.0 |
| | | Medium | 0.5 |
| | | Hard | 1.0 |
| 3 | category | Lesson | 0 |
| | | Question | 1 |
| 4 | type | Listening | 0.2 |
| | | Reading | 0.4 |
| | | Writing | 0.6 |
| | | Grammar | 0.8 |
| 5 | history | Studied | 0 |
| | | Never Studied | 1 |
| 6 | time | Short (less than 15 minutes) | 0 |
| | | Full (more than and equal 15 minutes) | 1 |

## IV. OUTPUT ATTRIBUTES

Using the Multiple Intelligence concept (Gardner, 2006), two groups of different instructional strategies were developed. Each group contains different strategies and each strategy is pre-defined by the partition of the input space, i.e. of the attributes themselves. The first group is concerned with providing the proper instructional strategy for certain lesson. The second group is concerned with providing the proper instructional strategy for certain question. For the first group, four factors which govern the instructional strategy of lessons were proposed as foundations for the design of each strategy. The complete list of these lessons' instructional strategies is presented in Table II and their values are presented in Table III. These factors can be described as follows:

1. **Practice navigation (PN)**. The process of a learner going through all practices related to a certain lesson.
2. **Lesson prerequisites (LP)**. The process of completing all the prerequisites related to a certain lesson by the learner. Prerequisites are different lessons that precede the given lesson in the learning level-class.
3. **Time (T)**. The time dedicated for a learner to study a certain lesson. This time is assigned to each lesson at the design time.
4. **Extra tutoring (ET)**. This is a new concept introduced in this work which is a mechanism for providing extra explanations and descriptions on certain lessons to specific learners. Extra tutoring is best suitable for tutoring material which is in a "Beginner" learning level or its difficulty is "Hard".

Table 2 provides five different modules of instructional strategies ($S_i$, i∈{1,2,3,4,5}) for lessons. These modules will be used according to the combinations of attributes been detected form the presented materials. For example, strategy 1 supports the discovery teaching which operates by providing the learner with the freedom to work on an unconstrained domain (Veermans and Joolingen, 2004). This strategy is suitable for the advanced learner who is self motivated and has a good sense of purpose. On the other hand, strategy 5 which supports the coaching teaching method in which the system takes full control of the learning process by allowing the learner to have enough control so that mistakes can be made without permitting the learning of incorrect information (United Kingdom Department for Education and Skills, 2005). This strategy is best suited for a beginner who needs more supervision and support. Strategies 2, 3 and 4 include principles from the above stated strategies. Based on expertise, each combination of these attributes is mapped into certain instructional strategy. Each strategy is represented as three digits binary numbers (ANN outputs). Table III describes the factors for the lessons' strategies and their values.

TABLE II

| | $S_i$ | PN | LP | ET | T | First output | Second output | Third output |
|---|---|---|---|---|---|---|---|---|
| 1 | $S_1$ | ✓ | ✓ | × | ✓ | 0 | 0 | 0 |
| 2 | $S_2$ | ✓ | ✓ | × | ✓ | 0 | 0 | 1 |
| 3 | $S_3$ | ✓ | × | ✓ | × | 0 | 1 | 0 |
| 4 | $S_4$ | × | ✓ | × | ✓ | 1 | 0 | 0 |
| 5 | $S_5$ | × | × | ✓ | × | 1 | 1 | 1 |

Lesson's strategies

TABLE III
Factors values in lessons strategies

| Feature | Values | Description |
|---|---|---|
| Practices navigation | ✓ | Learner does not have to complete all lessons' practices before start new lesson. |
| | × | Learner must complete all lessons' practices before start new lesson. |
| Lesson's prerequisites | × | Learner must complete all lessons' prerequisites before start this lesson. |
| | ✓ | Learner does not have to complete all lessons' prerequisites before start this lesson. |
| Tutoring | ✓ | Apply tutoring to the learner. |
| | × | Don't apply tutoring to the learner. |
| Time | ✓ | Learner has unlimited amount of time in each lesson. |
| | × | Learner must spend at most a specified amount of time in each lesson. |

The second group is concerned with providing the proper instructional strategy for questioning the learner ($S_j$, j∈{6,7,8}). Four factors which administrate the questioning process are proposed as foundations for the design of each strategy. Each strategy is represented as three digits binary numbers (ANN outputs). The complete list of these questions' instructional strategies is presented in Table IV and their values are presented in Table V. These factors can be described as follows:

1. **Lesson navigation (LN)**. The process of a learner going through all the lessons related to certain given question.
2. **Time (T).** The time dedicated for a learner to go through certain question. This time is assigned to each question at design time.
3. **Feedback (F).** One or more statements presented to the learner according to a specific strategy to provide some help in answering a certain question.
4. **Repeat content (RC).** Repeating the content of the question to the learner.

TABLE IV
Questions strategies

| | | LN | T | F | RC | First Output | Second output | Third output |
|---|---|---|---|---|---|---|---|---|
| 1 | $S_6$ | ✓ | × | x | x | 0 | 1 | 1 |
| 2 | $S_7$ | ✓ | × | ✓ | × | 1 | 0 | 1 |
| 3 | $S_8$ | × | ✓ | ✓ | ✓ | 1 | 1 | 0 |

TABLE V
Factors values in questions strategies

| Feature | Values | Description |
|---|---|---|
| Lessons' navigation | ✓ | Learner does not have to complete all lessons related to the given question. |
| | × | Learner must complete all lessons related to the given question. |
| Time | ✓ | Learner has unlimited time to answer the given question. |
| | × | Learner must answer the given question in a certain time. |
| Feedbacks | ✓ | system provide feedbacks to the learner. |
| | × | No feedbacks provided to the learner. |
| Repeat contents | ✓ | Learner can repeat the content of the question. |
| | × | Learner can't repeat the content of the question. |

A total of 288 patterns (input/output), each with 6 inputs and 3 outputs, were generated to be used in the ANN training and testing process.

## V. THE DESIGN OF THE NEURAL NETWORK

A back-propagation of error is simply a gradient descent method to minimise the total root mean square error of the output computed by the ANN. The training of an ANN by back-propagation involves three stages: the feedforward of the input training pattern, the calculation and back-propagation of the associated errors, and the adjustment of the weights. After training, application of ANN involves only the computation of the feedforward phase, a testing phase (Rumelhart, 1986). The learning process is described by equation 1.

$$g_k(x)=f\left(\sum_{j=1}^{n}w_{kj}f\left(\sum_{i-1}^{d}w_{ji}x_i+w_{j0}\right)+w_{k0}\right) \quad (1)$$

Where $g_k(x)$ is the $k^{th}$ output, n is the number of the hidden layers, d is the number of inputs. $W_{kj}$ represents the weights form the hidden layer to the output layer, $w_{ji}$ represents the weights form the input layer to the hidden layer, and $w_{j0}$ and $w_{k0}$ represent the biases to different layers. The binary sigmoid function which is the most common ANN activation function was used in this study (Fausett, 1994) to map the real numbers into the interval (0 to 1). The sigmoid function ***f(t)*** used in the training is defined by equation 2.

$$f(t)=\frac{1}{1+e^{-t}} \quad (2)$$

To properly train the ANN, a training set containing input and output data was constructed as described in the previous sections. The ANN creates connections and learning patterns based on these inputs and outputs. Each pattern creates a unique configuration of network structure with a unique set of connection strengths or weights. The decision of the proper ANN structure used is complex as it depends on the specific problem being solved. With too few units, the network may not be powerful enough for a given learning task. With a large number of units, computation is too lengthy. Sometimes the ANN may ***memorize*** the input training sample; such a network tends to perform poorly on the new test samples. To reach the best generalization the data (288 patterns) was divided into two sets. The first set consists of 80% (230 patterns) of the data for learning. The other 20% (58 patterns) of the data were randomly chosen and used for testing since typically between 1/3 and 1/10 of the data set held out for testing (Mitchell 1997; Witten and Frank 2000). The learning rate was 0.5 which is an average value of learning rate (Haykin, 1994), and the error tolerance was 0.0001. The number of learning runs ranged from 10,000 to 40,000 runs. The number of hidden units in each hidden layer was varied from 5 to 20. The Error is calculated according to equation 3.

$$\text{Error (E)} = |\text{Actual output} - \text{Neural output}| \quad (3)$$

In testing mode, the test data, containing only input patterns, were provided to the trained ANN for testing since the prediction on the test data gives an unbiased estimate of the error rate of the of the prediction process. The test data were applied to the trained network without changing the weights and the network size which were used to train the network. Outputs from the ANN for all the testing input patterns were generated. The network went through one cycle of operation in this mode covering all the patterns in the test data. ANN software was compiled with C++ based on an ANN source code provided by Rao (1995). In each learning cycle, the ANN goes through all the training patterns in the training mode and all the testing patterns in the testing mode.

## VI. EXPERMINTAL RESULTS

### *A. Training results*

The results of training errors for 3-layer networks with different number of hidden unites presented in this section. In the training mode, the learning cycles were fixed to 10,000 while the number of units in the hidden layer was varied from

5 to 20. The learning rate was chosen to be an average number of 0.5. The final minimum average errors last cycle in the training set are presented in Table VI.

TABLE VI
Errors from 15 hidden unites

| Total number of cycles | 10,000 |
|---|---|
| Number of hidden units in the hidden layer | 15 |
| Error last cycle | 0.0013 |

The ANN showed excellent training results and the network converge to a minimum error with 15 unites in the hidden layer. In the next stage, the network structure was fixed to 15 unites in the hidden layer and the learning cycles were varied from 10,000 to 60,000. The average error last cycle in the training set is presented in Table VII.

TABLE VII
Errors from 50,000 learning cycles

| Total number of cycles | 50,000 |
|---|---|
| Number of hidden units in the hidden layer | 15 |
| Error last cycle | 0.00021 |

From the last results, it is clear that the neural network with sixteen hidden unites and 50,000 number of cycles was converge to a minimum error, hence, the weights from that network were saved to be used later in the testing process. The optimal ANN's architecture is presented in Figure 2.

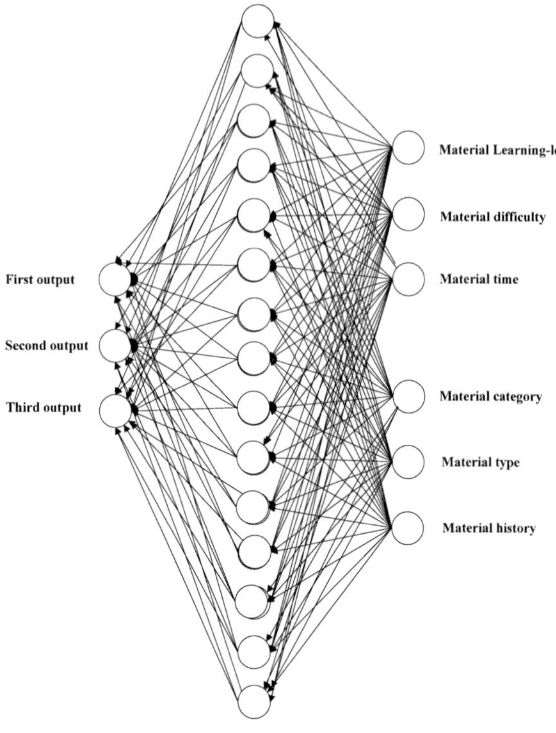


Fig. 2. The optimal ANN's architecture

*B. Testing results*

In the testing mode, the ANN went trough one learning cycle in all testing patterns, 58 patterns. The generated outputs were drawn and the average error between the ANN outputs and the real output were calculated. The correlation between each real output and the ANN output was calculated by the function correl() using Microsoft office Excel. The correlation indicates the strength and direction of a linear relationship between the two outputs. The testing process showed excellent results where the average error in the prediction of each output was less than 0.001%. The correlation results are presented in Table VIII.

TABLE VIII
Correlation results

| | First output | Second output | Third output |
|---|---|---|---|
| Correlation(r) | 0.99 | 0.99 | 1 |

CONCLUSION

In this paper a new framework for predicting the proper instructional' strategy for the given teaching material based on its attributes (domain based) was presented. Eight different instructional' strategy were developed, five for lessons and 3 for questions. The implemented approach is flexible since it accommodates different styles of instructional strategy. The prediction process is based on a multilayer feed forward back propagation network. The average errors indicate excellent prediction by the ANN for the three outputs. The correlation results between each real output and the ANN output indicate strong correlation between real outputs and the ANN outputs which imply that both outputs are almost identical. For the present problem, the ANN showed good convergence based on the proposed topology of six inputs, fifteen hidden unites and 3 outputs (6-15-3) with learning rate 0.5 and total number of learning cycles is 50,000.

FUTURE WORK

Instructional strategy is closely associated with student assessment in which both influence the tutor's next action hence we hope we can combine ANN results with student assessment in order to improve the tutor system performance.

## REFERENCES


[1] Fausett, L., (1994). "Fundamentals of Neural Networks. Architectures Algorithm and Applications", Prentice Hall, Englewood Cliffs, NJ, USA.

[2] Gardner, H., (2006), "Multiple Intelligences: New Horizons in Theory and Practice", Basic books, USA.

[3] Koedinger, K., Anderson, J., Hadley, W., Mark, M., (1997), "Intelligent tutoring goes to school in the big city", Journal of Artificial intelligence in Education, 8(1), pp. 30-43.

[4] Haykin, Simon, (1994), "Neural networks a comprehensive foundation", Macmillan publishing company, USA.

[5] Mitchell, T. (1997), "Machine Learning", New York: McGraw-Hill.

[6] Rao, V. B. and Rao, H. V. (1995). "Neural Networks and Fuzzy logic". MIS Press, New York, USA.

[7] Rumelhart, D., and McClelland, J. (1986). "Parallel Distributed Processing". M.I.T. press. Cambridge, MA.

[8] Sevarac, Z, (2006), "Neuro Fuzzy Reasoner for student modelling", Proceeding of the sixth international conference on advance learning technologies (ICALT'06).

[9] Smith, P., and Ragan, T. (1999), "Instructional design", second edition, John Wiley & sons, Inc, NY, USA.

[10] Stathacopoulou, R., Samarakou, M., R., and Magoulas, G. (2004), "A Neuro-Fuzzy Approach to Detect Student's Motivation", ICALT.

[11] Stathacopoulou, R., Grigoriadou, M., Samarakou, M., and Mitropoulos, D., (2007), "Monitoring students' actions and using teachers' expertise in implementing and evaluating the neural network-based fuzzy diagnostic model", Expert Syst. Appl. 32(4).

[12] United Kingdom Department for Education and Skills, (2005), "Working together: coaching and assessment for learning", Crown copyright 2005, UK.

[13] Veermans, K., and Joolingen, W. (2004), "Combining heuristics and formal methods in a tool for supporting simulation-based discovery learning", Proceedings of the Workshop on Intelligent Tutoring Systems for Ill-Defined Domains, pages 217–226. Jhongli, Taiwan.

[14] Wang, T., and Mitovic, A. (2002), "Using neural networks to predict students performance", In Proceeding of the international conference on computer education (ICCE'02), IEEE.

[15] Witten, I., and Frank, E. (2000), "Data Mining: Practical Machine Learning Tools and Techniques with Java Implementations", San Diego, CA: Morgan Kaufmann.

# I Know What You Did This Summer – Users' Behavior on Internet

Zeeshan-ul-hassan Usmani[1], Fawzi Abdulkhaliq Alghamdi[2], Amina Tariq[3] and Talal Naveed Puri[4]
[1]*Department of Computer Science, Florida Institute of Technology, Melbourne, FL, USA*
[2] *Department of Electrical Engineering, Florida Institute of Technology, Melbourne, FL, USA*
[3]*National University of Computer and Emerging Sciences, Islamabad, Pakistan*
[4]*GE Infrastructure Transportation, Florida Institute of Technology, Melbourne, FL, USA*
*zusmani@fit.edu, falghamd@fit.edu, amina.tariq@nu.edu.pk, talal.puri@ge.com*

**Abstract – Human age is surrounded by assumed set of rules and behaviors imposed by local culture and the society they live in. This paper introduces software that counts the presence of a person on the Internet and examines the activities he/she conducts online. The paper answers questions such as how "old" are you on the Internet? How soon will a newbie be exposed to adult websites? How long will it take for a new Internet user to know about social networking sites? And how many years a user has to surf online to celebrate his/her first "birthday" of Internet presence? Paper findings from a database of 105 school and university students containing their every click of first 24 hours of Internet usage are presented. The findings provide valuable insights for Internet Marketing, ethics, Internet business and the mapping of Internet life with real life. Privacy and ethical issues related to the study have been discussed at the end.**

## I. Introduction

"How old are you?" is a common question followed by the simple answer of the age of a person being asked. However the answer portrays much more complex assumptions and preset rules imposed by the local culture and the settings [5]. For example, if the answer is 21 years old, it is assumed that the fellow is allowed to drink, should be studying somewhere in college, can drive and should be able to decide what is right and wrong for him/her with no or limited supervision [6]. Likewise, when we discuss Internet usage and the time we spend online, it would be useful to make assumptions and categorize the user. These categorizations may be beneficial to differentiate between a novice and a veteran Internet user. With this central theme in mind we start with the problem of counting and recording Internet life of the users to get valuable insights. To the authors' knowledge, the software utility discussed in the following sections is unique. There are several related attempts like calculating the worth of users' FaceBook profile by Weblo Digital Assets [2], and by calculating the worth of a human being on his/her physical characteristics by Humans for Sale [4]. Time Tracker is another interesting Plug-In for FireFox which can only track the time you spend using that web-browser [3]. The presented software – Internet Age, is unique in a way that it not only counts the time user spends online but also records every website he/she visits. In Internet Age time is not only counted for particular web browser, it can be anything from email to web-surfing, from watching a video on YouTube to chatting on MSN messenger. Internet Age can answer questions like: How old are you on the Internet? How soon is a novice exposed to adult websites? How long will it take for a new Internet user to know about social networking sites? And how many years a user has to surf to celebrate his/her first birthday of the Internet presence? The results are revealing and surprising for a few of these questions. There are dozens of applications for this software from calculating Internet credit scores to determining the trust level of the user, from feedback to Internet marketing, from filtering of inappropriate websites to recommending some.

The paper is organized as follows: next section presents an overview of the Internet Age software and its salient features. Section 3 describes the data sets used for testing the software and some relevant assumptions from the data perspective. Section 4 presents the results obtained from the test data sets and provides a detailed analysis of the significance of those results. The paper concludes by describing some limitations and proposed future work in the same direction.

## II. Internet Age

Internet Age® is a research utility to collect users' Internet usage data. The beta version can be downloaded from http://my.fit.edu/~zusmani/IAge.html. The software was developed using Microsoft.Net platform version 2.0 and Visual C#. SQL server has been used as a database on server side. The software is using BHO (Browser Helper Object) – using COM interfaces, .NET Remote Server (Message Listener), Shell Integration – for taskbar services, and SMTP services for sending mails (sessions reports). Figure 1 shows the application architecture of the software.

BHO (Browser Helper Object) is a special component which extends the client's browser functionalities. It uses the COM interfaces to interact with the operating system. BHO also intercepts all Internet Explorers (Browser) requests, and

T. Sobh (ed.), *Innovations and Advances in Computer Sciences and Engineering*,
DOI 10.1007/978-90-481-3658-2_4, © Springer Science+Business Media B.V. 2010

sends it on HTTP Channel port 9098 down to the Internet Age that is running on the same machine.

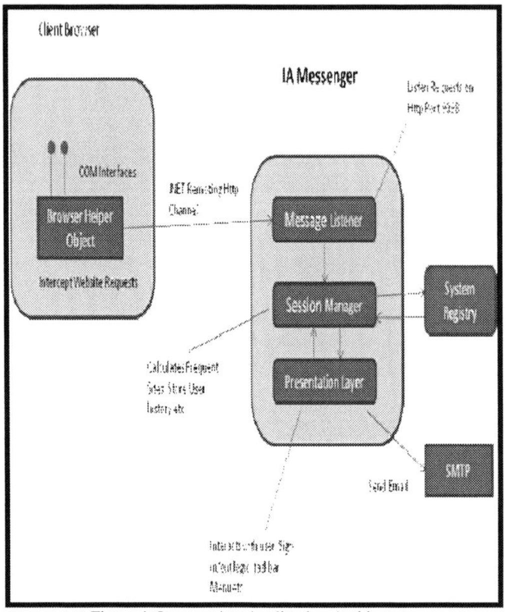


Figure 1: Internet Age Application Architecture

The Message Listener is a .NET remote server which runs under Internet Age. It listens to messages (i.e. visited URLs) from one or more BHOs. The Message Listener listens asynchronously and is capable of processing multiple requests at the same time. Once a message is received, the Message Listener passes this message to the Session Manager.

Internet Age stores various kinds of information in the Session Manager, such as Visited Websites, Frequent Websites, Login ID, Time Stamps, and most importantly the User's Age. In order to use the information between subsequent sessions, the software stores the information in System Registry.

Shell Integration interacts with Microsoft Windows Shell, i.e. Taskbars, NotifyIcons, Class libraries etc. In order to run as background application the software use NotifyIcon component of .NET framework 2.0. NotifyIcon component is responsible for putting the application's icon in the extreme right corner of the taskbar and provides various events (like click, right click etc) for the server application to interact with the user. The software use *System.Net.Mail.SmtpClient* class for sending emails. The *SmtpClient* class implements standard email functionality of SMTP protocol use for sending the email messages.

As soon as a user installs the software, it will record the first login as the time and date of birth (the first recorded login) of that particular user on the Internet. The software will then automatically generate a unique *User-ID* for the user which will be used in all future communications of session reports from user machine to admin server. *User-ID* is generated by users' age, gender and the number of installation. For example, if the user is a male of 27 years of age and this is the 30th copy of the software being installed, then his id will look like 27M0030. The only purpose of the *User-ID* is to keep track of the session reports generated by unique users. There is no password required to login into the software. The user have the choice to indicate in the installation process if they want the software to use for their own personal use and do not want to share the session reports with the server, in that case, software will not send any sessions reports or user's activities summary to the server. User can also choose to install the software under their windows profile – in that case software can distinguish various users on the same machine. Figure 2 shows the main screen of the software. This beta version of the software only works with Internet Explorer 6 and above. The next version will be enhanced for Firefox and Safari browsers and for non-window machines.

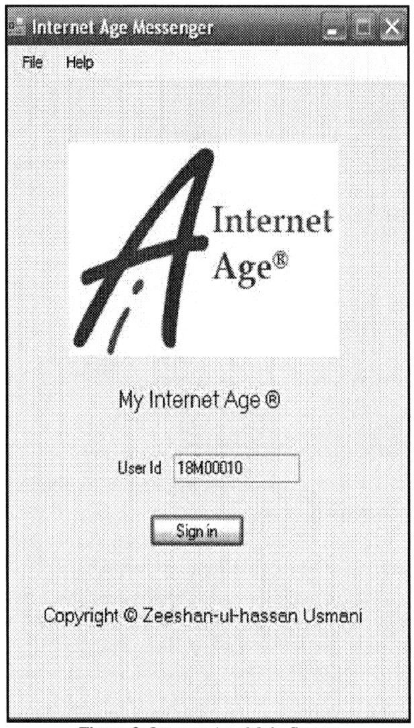


Figure 2: Internet Age Main Screen

There are three sections in the software status screen as shown in Figure 3: Frequent Websites, First Recorded Login and Age.

*Frequent Websites:* Frequent websites were calculated according to users' browsing history. If a user is visiting a particular website more than his/her average number of visits per website, the website and number of visits will be included on the frequent websites column of the screen. The panel can only show a maximum of ten websites.

*First Recorded Login:* This is the time and date stamp of the user's first login with the software. We consider it as the user being 'Born' on the Internet. There is no general model to estimate the user's Internet age, however, few estimators can be used. For example first email creation date, Google search history etc. Most of these estimators usually supplied by the user and are hard to validate.

*Age:* This is the total time a user has spent on the Internet since he/she had installed the software. This time includes all of users' activities like email, browsing (through Internet Explorer), chatting, watching videos etc. If the user becomes idle for ten minutes, the software will automatically session-out and will send the session report summary to the server using SMTP.

Figure 3: Internet Age Status Screen

The session report, as shown in Figure 4, contains the following information:

*Serial Numbers – serial numbers* of the websites (in the same order they were visited).

*Time Stamp* – first time the user went to that site.

*Site Visited* – URL's of the websites

*Number of Times* – frequency of visits on each website.

The session report also contains the information about *User-ID*, total time spent during that session with starting and ending time-stamps and session id for that particular machine/user. When recording the websites, the sub-domains are considered as separate websites; for example, www.yahoo.com is different from finance.yahoo.com, the former is in the category of search engines while the later is in the finance category. The discussion on website categories is presented later in this section.

| User ID: 18M0010 | | |
|---|---|---|
| S# | Time Stamp (Age) | Site Visited |
| 1 | 12 seconds | www.google.com |
| 2 | 1 min 1 sec | www.cnn.com |
| 3 | 1 day, 1 hour, 11 minutes and 0 seconds | www.amazon.com |
| 4 | 7 days, 2 hours, 2 min and 4 sec | www.porn.com |
| 5 | 1 month, 3 days, 2 hours, 0 min and 11 sec | www.youtube.com |
| Session ID | | 001 |
| Session Starts | | 14:23:01 EST |
| Session Ends | | 15:25:11 EST |
| Total Session Time | | 01 Hour, 2 Minutes, |

Figure 4: Sample Session Report

Figure 5 shows the sample cumulative summary of the server. The report has User-ID, Age, Gender, Time-Stamp (The first visit to the website in that category), Category of the website, Time Spent on Category per I-Day – Total time a user has spent on that category in 24 hours of Internet use (Internet-day=24 hours of Internet Use), Total Time Spent Per R-Day – total time a user has spent per regular day on the Internet; for example, if a user spends 2 hours per day on the Internet, he will be 1 day old (I-Age=1 day) after 12 days of Internet use. Number of unique categories per I-Day is the unique number of categories a user has visited in twenty four hours of Internet surfing.

| User ID | Age | Gender | Time Stamp | Category | Time Spend on Category per I-Day | Total Time Spend per R-Day | No of Categories per I-Day User |
|---|---|---|---|---|---|---|---|
| 1 | 8 | M | 0:02:12 | 11 | 3:12 | 2:19 | 5 |
| | | | 0:02:32 | 8 | 6:09 | | |
| | | | 0:12:03 | 5 | 4:05 | | |
| | | | 0:17:01 | 3 | 3:11 | | |
| | | | 0:21:12 | 4 | 7:23 | | |
| 2 | 8 | M | 0:00:03 | 11 | 1:12 | 2:11 | 5 |
| | | | 0:01:14 | 3 | 4:34 | | |
| | | | 0:02:30 | 2 | 6:09 | | |
| | | | 0:05:46 | 1 | 10:09 | | |
| | | | 0:10:10 | 6 | 1:56 | | |
| 3 | 10 | M | 0:00:03 | 11 | 0:10 | 1:45 | 11 |
| | | | 0:01:14 | 3 | 0:13 | | |
| | | | 0:02:30 | 2 | 0:34 | | |
| | | | 0:05:46 | 1 | 4:34 | | |
| | | | 0:10:10 | 6 | 1:11 | | |
| | | | 0:17:13 | 7 | 2:23 | | |
| | | | 0:33:00 | 8 | 4:33 | | |
| | | | 2:30:11 | 4 | 5:19 | | |
| | | | 4:32:11 | 5 | 0:13 | | |
| | | | 4:31:09 | 9 | 2:13 | | |
| | | | 8:01:14 | 10 | 2:37 | | |

Figure 5: Cumulative Server Summary

### III. DATA AND ASSUMPTIONS

The overall evaluation of the software has been carried out based on two data sets collected from different regions of the world. The first data set was collected from an elementary school in Karachi, Pakistan. There were 62 students from grade 3 to grade 5 between the ages of 8 to 10. The software was installed in their computer labs. The students had no prior experience with the Internet. It makes the data more important for marketing and users' behavior studies. Table 1 presents the age and gender wise distribution of the students in data set 1.

Table 1 : Distribution for Data Set 1

| Age | No of Students | Male | Female |
|---|---|---|---|
| 8-10 | 62 | 33 | 29 |

The second data set was collected from students of a university in UAE. There were 43 students between the ages of 17 to 27. The software was installed in their computer labs. Table 2 presents the age and gender wise distribution of the students in data set 2.

Table 2: Distribution for Data Set 2

| Age | No of Students | Male | Female |
|---|---|---|---|
| 17-27 | 43 | 24 | 19 |

There were no adult or child filters installed on the machines and students were in uncontrolled environment without the teacher. For the collection of first data set, the teacher simply signed-up the students and then the students were on their own on what to search and what to read. Parents of the minor students and school authorities have given the required consent to perform the research study and data will not be shared with any third party for any commercial or other interests. It should also be noted that the students had previous experience working on computers, so they knew how to use the mouse, word processing tools, and other mundane tasks of computer applications.

For the sake of consistency, reports were generated on the basis of one Internet Day, that is 24 hours of Internet surfing. In real life, it could interpret differently for different students. For example, the data of a student who spends only 2 hours per day on average on the Internet will comprise of 12 days of Internet browsing, while for other students it might be just 4 days of Internet browsing with 6 hours per day use. To cater the broad variety of websites visited by students as part of the analysis, we have created a generic list of eleven different categories of websites as presented in Table 3 along with frequent examples for understanding. After reviewing session summaries, websites were manually classified into different categories. This makes it easier to compare and analyze the trends among different sets of people, for example, an adult sitting in California might access www.bankofamerica.com on daily basis, while another adult in middle-east might access www.standardchartered.com for the very same purpose on a daily basis. In our case both were classified as "Finance".

Table 3. Categories and Examples

| S# | Category | Examples |
|---|---|---|
| 1 | Books | Libraries, Amazon, Barnes and |
| 2 | Retailers | WalMart, BestBuy, Circuit City, |
| 3 | Entertainment | TV, Movies, Cinemas, Theater |
| 4 | Finance | Yahoo Finance, Banks, Stock M |
| 5 | Sex | Pronography, Adult Movies, 18+ |
| 6 | Educational | Universities, Schools, Scholarsh |
| 7 | Government | Government departments, Pass |
| 8 | News | CNN, BBC, News Papers, TV N |
| 9 | Videos | Youtube, MetaCare |
| 10 | Social Networks | Orkut, Facebook, Myspace, Lin |
| 11 | Search Engines | Google, Yahoo, MSN, alltheweb |

### IV. RESULTS

This section presents the overall results obtained from the analysis of the collected data sets. Findings from the data sets are presented, supported by analysis, to give an insight on the worth of the developed tool in identifying the complex patterns in internet usage.

The results obtained from the data sets indicate that male and female students have almost similar internet usage behavior in terms of exploring the number of web site categories, with male students slightly dominating as shown in Figure 6. The male users tend to explore on average 6 website categories per I-day, while females are not far behind, exploring 5 website categories per I-day. In the future, we will be collecting data from all age groups to see if there is a relationship between age groups and number of categories, popular literature suggests it is linear [6].

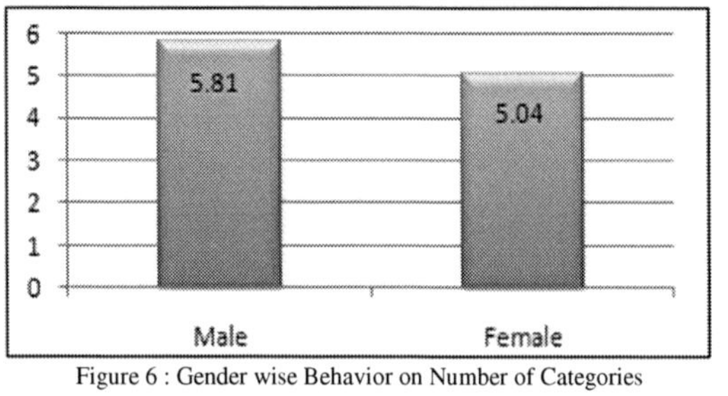


Figure 6 : Gender wise Behavior on Number of Categories

The average time each gender spends on the internet is approximately the same and the difference is statistically insignificant as shown in Figure 7. For the data sets, based on average per day use, if we can consider 3.14 hours of Internet usage per day including weekends, a user will be able to celebrate his first year of Internet presence in 7.6 actual years.

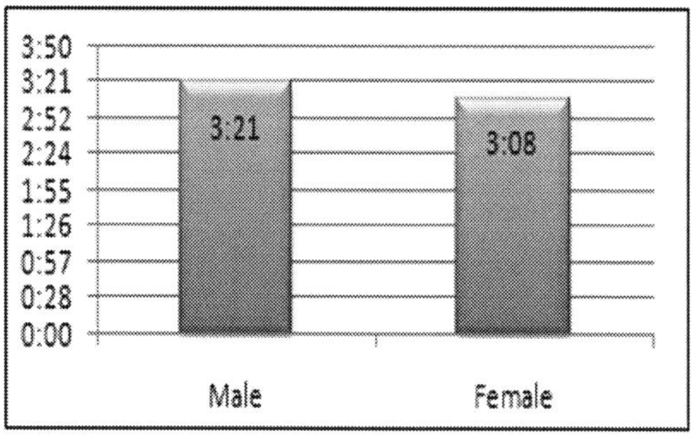


Figure 7 : Gender wise Behavior on Average Time Spend Per Regular Day (in hours)

Further analysis of the gathered data indicates that the entertainment is one of the prime reasons for which users spent their time online as presented in Figure 8. It clearly shows that average time users spent on entertainment category (category no 3) of the websites is around 7 hours per I-day, which is followed by the social networking, search engines and retailing websites (category number 10, 11 and 2 respectively). All of these categories have almost similar average user time which is around 5 hours. The results also identify government and news website categories as least visited websites by internet users in our study. Another very interesting observation from the data is that users spend almost an equal amount of time on sex-oriented websites and on educational resources, which is around 3.5 hours per I-day.

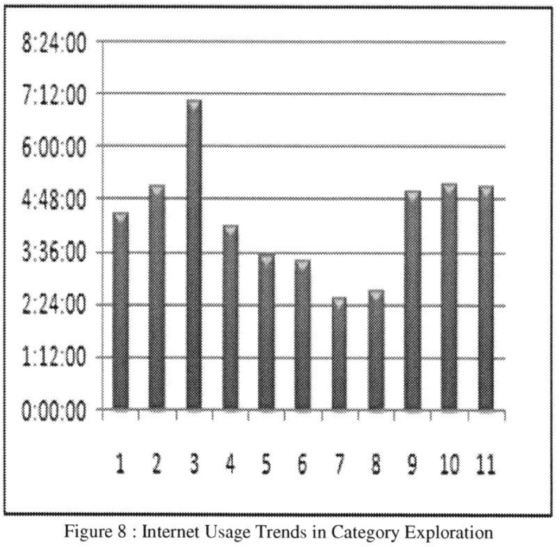


Figure 8 : Internet Usage Trends in Category Exploration

The average time spent by the females on the entertainment and educational websites is almost twice to that spent by the males as shown in Table 5.

Table 5: Gender wise Category Exploration

| Category | Avg. Time Male on a Specific Category per I-Day | Avg. Time Female on a Specific Category per I-Day |
|---|---|---|
| 1 | 4:35:04 | 4:20:50 |
| 2 | 5:14:41 | 4:52:16 |
| 3 | 5:18:34 | 8:45:46 |
| 4 | 3:36:29 | 4:45:46 |
| 5 | 3:22:42 | 3:38:13 |
| 6 | 2:38:06 | 4:10:05 |
| 7 | 3:06:27 | 1:56:51 |
| 8 | 2:44:18 | 2:39:26 |
| 9 | 4:43:10 | 5:15:27 |
| 10 | 4:45:13 | 5:28:16 |
| 11 | 4:52:52 | 5:16:18 |

Another interesting observation is that male tend to spend almost double average time on government websites in contrast to the females as indicated by Figure 9 and 10. Both genders enjoy almost similar level of interests in the remaining website categories.

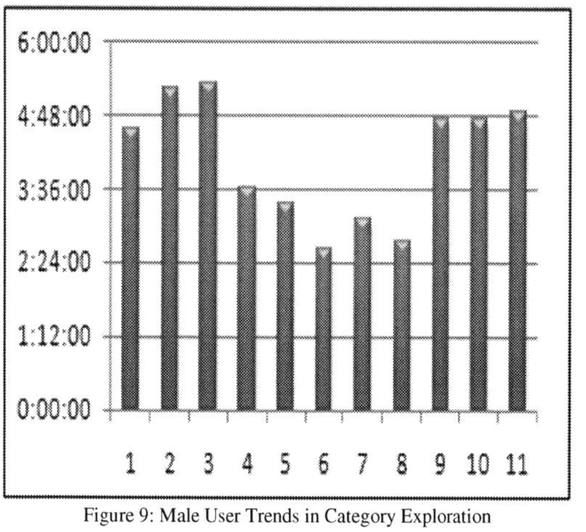


Figure 9: Male User Trends in Category Exploration

These obsevations exploring the internet usage behaviour of genders can be very effectively applied to the research in applied psycology [8,9 ]and can further be fruitful for different organizations to have know how of their target audience.

The data obtained was further utilized to explore another dimension of the internet usage behaviour which is basically how quickly user reaches a website of specific category. These observations can help the organizations to analyse the effectivity of their internet marketing strategies and evaluate the accesability of their websites. As depicted by Figure 11 the average time to visit a specific category is least for the entertainment category which is in complete alignment with the earlier results of entertainment being the most popular category.

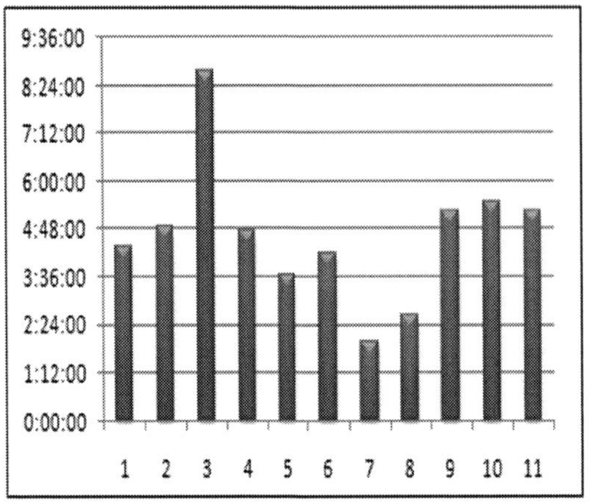


Figure 10: Female User Trends in Category Exploration

The second most easily visited category is that of retailers followed by education and books categories which are conceptually corelated as well. The categories that took the maximum average time to be visited by the users include the news, videos and finance websites, which can be justified by the not so popular nature of these websites as described earlier within our group of studied subjects.

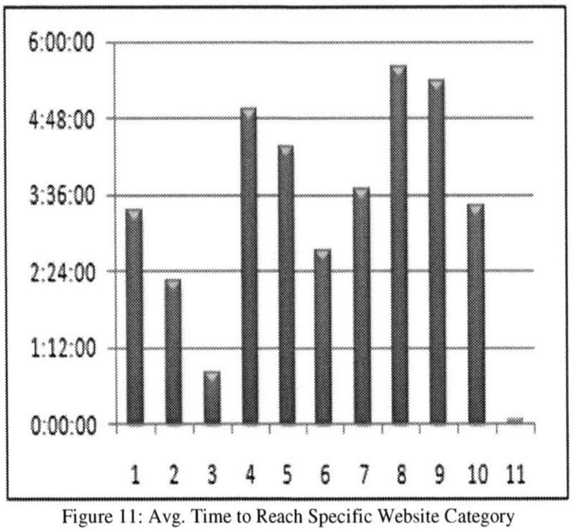


Figure 11: Avg. Time to Reach Specific Website Category

Based on the usage behavior, the students in the study will be celebrating their first year of Internet presence in approximately eight actual years.

## V. Conclusion

This paper has introduced a new tool, Internet Age, to keep record of the amount of time a user spends on the Internet and also tracks the sites he/she visits. The paper offers revealing insights on accessibility of different websites and the time required to being exposed to different categories of websites. It has been found, based on the database of 105 school and university students, that average users' 7.6 years of life would be equal to one year of Internet presence. The findings, although preliminary, can be used for classifying customers; calculating credit scores and marketing the websites for given group of people. We will be extending the work with the new release of the software on multiple platforms and browsers, and a comparative study of Internet usage trends in different regions of the world. There are a few privacy issues associated with the software that has to be considered before any commercial use of this application. For example, by collecting data, we might get the users' data who involve in illegal activities like gambling, child pornography, illegal movies and software downloads and hacking. When and how to release this information to law agencies would require the understanding of a complex set of laws and international rules. Building trust with the users to encourage them to be the part of the research study is another dimension we will be working on in the future.

## References

[1] Jupiter Research Report, "Time Spent Online", PEW Internet and American Life Series, 2006

[2] http://www.weblo.com/digital/

[3] https://addons.mozilla.org/en-US/firefox/addon/1887

[4] http://www.humanforsale.com/

[5] L. Leung and P.S.N. Lee , "Multiple determinants of life quality: the roles of Internet activities, use of new media, social support, and leisure activities", Telematics and Informatics, Elsevier 2006

[6] Katelyn Y. A. McKenna and John Bargh, "Consequences of the Internet for Self and Society: Is Social Life Being Transformed, Journal of Social Issues, Wiley-Blackwell, March 15, 2002

[7] A. Contarello and M. Sarrica, "ICTs, social thinking and subjective well-being - The Internet and its representations in everyday life", Computers in Human Behavior, Elsevier, March 1, 2007

[8] Subrahmanyam, K., Greenfield, P. M., & Tynes, B. "Constructing sexuality and identity in an online teen chatroom" Journal of Applied Developmental Psychology, 25, 651–666,2004

[9] Kraut, R., Lundmark, V., Patterson, M., Kiesler, S., Mukopadhyay, T., & Scherlis, W." Internet paradox: A social technology that reduces social involvement and psychological well-being?" American Psychologist, 53, 1017-1031, 1998

# Relative Ranking – A Biased Rating

Zeeshan-ul-Hassan Usmani[1], Fawzi Abdulkhaliq Alghamdi[2], Amina Tariq[3] and Talal Naveed Puri[4]
[1]*Dept. of Computer Science, Florida Institute of Technology, Melbourne, FL, USA*
[2]*Dept. of Electrical and Computer Engineering, Florida Institute of Technology, Melbourne, FL, USA*
[3]*National University of Computer and Emerging Sciences, Islamabad, Pakistan*
[4]*General Electronics Transportation, Melbourne, FL, USA*
*zusmani@fit.edu, falghamd@fit.edu,amina.tariq@nu.edu.pk, talal.puri@ge.com*

**Abstract: Reviewers' ratings have become one of the most influential parameters when making a decision to purchase or rent the products or services from the online vendors. Star Rating system is the de-facto standard for rating a product. It is regarded as one of the most visually appealing rating systems that directly interact with the consumers; helping them find products they will like to purchase as well as register their views on the product. It offers visual advantage to pick the popular or most rated product. Any system that is not as appealing as star system will have a chance of rejection by online business community. This paper argues that, the visual advantage is not enough to declare star rating system as a triumphant, the success of a ranking system should be measured by how effectively the system helps customers make decisions that they, retrospectively, consider correct. This paper argues and suggests a novel approach of Relative Ranking within the boundaries of star rating system to overcome a few inherent disadvantages the former system comes with.**

## I. INTRODUCTION

The movement toward E-commerce in the virtual space has produced business strategies that could never exist in the physical world. These novel strategies have been originated to address the traditional consumer behavioral needs which have not changed as such for instance reliance of people on the opinions of others for "experience goods" – products whose quality is difficult to observe or test adequately before purchase. Experience goods include a fairly broad range of areas including books, movies, and even various kinds of advice, such as medical and financial [20].

This trend has resulted in the advent of very dominant e-business strategies which is based on the utilization of customer comments and ratings by business web sites to supplement their credibility and create a greater sense of community. Reviewers are likely to visit the site each time they consume a product since they enjoy sharing their opinions and comment readers may come to depend on reviews to help guide their purchases [18]. A rating scale is a technique used to classify a product or service in terms of its effectiveness or use [1, 2]. Rating scale can be of any type from a numerical scale of 1 to 100 to a simple thumbs-up and thumbs-down system, from a descriptive dialogue box to a letter grade system of A, B, C, D and F (commonly used in academia). Using stars in praise of something is the oldest form of rating known to human being. And it is still the most commonly used form of product/service ratings.

People tend to look for star rating before making their decisions, whether it is selecting a hotel for vacations or picking an airline for business travel, buying a book online from Amazon [16] or renting a movie from Netflix, selecting a seller on e-Bay [17] or locating a good restaurant in the neighborhood. Star rating becomes the de-facto standard for the quality of goods and services. There are usually five stars to present the quality of a product or service as shown in Figure 1.

Figure 1: Visual Depiction of Star Rating System [19]

The common interpretation of the product ranking is taken by the number of stars a product has been assigned, one star can interpret as product/service being poor, two stars for average, and three for good, four for very good, and five is the perfect score for excellent product or services. Further, the e-commerce businesses tend to customize this rating according to the type of the product being bought by the customer. For instance at Amazon customers are asked to rate books they have read on a 5-point legend from "hated it" to "loved it." After rating a sample of books, customers may request recommendations for books that they might like [16, 18]. Epinion.com's product rating ranks the products. All of the products in a category or subcategory are ranked based on their overall star rating. The overall star rating is calculated based on the various factors including, overall product rating (with extra weight given to high quality reviews), and number of reviews about the product and recency of reviews about the

T. Sobh (ed.), *Innovations and Advances in Computer Sciences and Engineering*,
DOI 10.1007/978-90-481-3658-2_5, 

product [19]. There are few exceptional examples as well which go beyond the common five-star ranking standard. For instance some hotels claims to be six stars and Burj-ul-arab in Dubai, UAE had claimed to be the only seven star hotel in the world [6].

Overall one glimpse of the star rating enables the consumers to perceive the product/service's quality indication. And this visual strength of star rating system makes it almost impossible to be replaced by some other system.

When users submit their rating for products or services, rating interface usually comes with few other information, statistics and links. That additional information can help the user to make the rating and provide the opportunity to further explore more items and to give more ratings based on current selection of his/her interest [8]. Figure 2 summarizes the general architecture of participation for a given instance of rating.

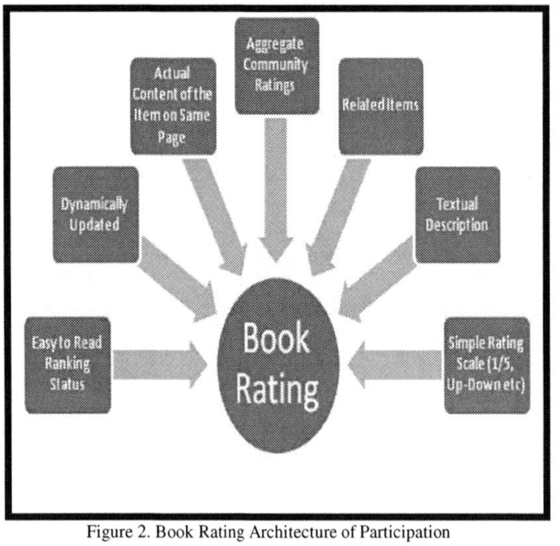


Figure 2. Book Rating Architecture of Participation

Some e-business websites have tried visual variations in the star rating as well for instance Bizrate.com allows its visitors to rank the Merchant or Stores on the four level scales, which are represented by Smiley scale. This Smiley Scale helps the visitors in finding the stores with the level of quality required by the consumer [19].

Some businesses have replaced stars with other characters, for example, spoon and forks in the restaurant's menu, CDs in movie shops and a clapping person sitting on the chair in case of a San Francisco Chronicles movie critics [7]. YouTube also gives the opportunity to submit comments and video response and shows views statistics and allows sharing of the link on social networking sites [10]. Digg.com works on simple thumbs-up and thumbs-down approach and shows the number of diggers for particular site or item [11]. Internet Movie Database (IMDB) also shows the demographic information of the raters along with number of raters and the percentage of raters on each scale [13]. Netflix shows the brief description of the movie with member reviews [12]. Figure 3 shows the snapshots of these websites ranking criteria.

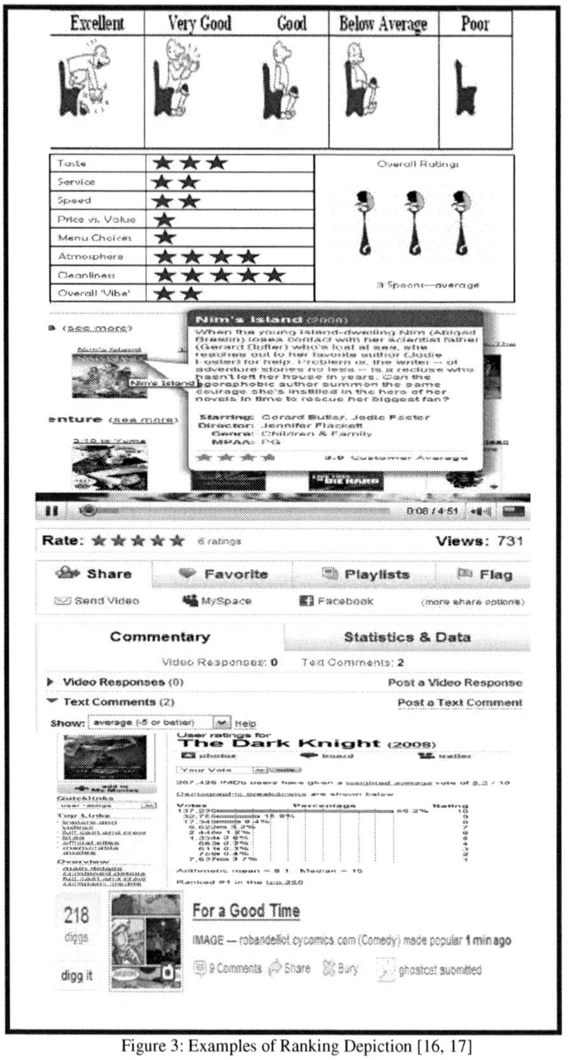


Figure 3: Examples of Ranking Depiction [16, 17]

Based on this initial study of the different types of star rankings introduced by various e-businesses this paper aims at analyzing the problems and limitations of this ranking system and identifying the alternative solutions to minimize the impact of these limitation. Next section presents the analysis of the issues with the current star-ranking system. Section 3 proposes the relative ranking based solution that can be used in combination with the current star-ranking system to minimize the impact of the limitations of the prevailing system. The paper concludes by proposing some future work that is required to further enhance and validate the proposed solution.

## II. PROBLEM ANALYSIS

Like any other system Star rating system has few disadvantages as well. Consider the profiles of Book 1 and Book 2 in the Table 1. Book 1 has eighty points and twenty reviewers/raters, out of those twenty, half of them gave it 5 stars and another 4 of them gave it 4 star rating. But, still Book 2 with almost half of the points of the Book 1 and half of the number of reviewers will take the lead with 4.4 stars over 4 stars of Book 1. This can be quite misleading.

Another view to look at this is to compare a book in hand with some other book. For example, there is a new novel by some infamous author that receives 5 stars rating with its only reviewer (who might be the author by self), but now how good this novel is compare to Harry Potter or any other best seller in that category? The bestseller might have 4.3 star rating with 2,000 reviewers while the new novel will have 5 star perfect rating with only one reviewer.

Table 1. Book Rating Profiles

| Book 1 | | | |
|---|---|---|---|
| | No of Reviewers | Stars | Points |
| | 10 | 5 | 50 |
| | 4 | 4 | 16 |
| | 3 | 3 | 9 |
| | 2 | 2 | 4 |
| | 1 | 1 | 1 |
| Total | 20 | | 80 |
| Avg Stars Rating | | | 4 |
| | | | |
| Book 2 | | | |
| | No of Reviewers | Stars | Points |
| | 5 | 5 | 25 |
| | 4 | 4 | 16 |
| | 1 | 3 | 3 |
| | 0 | 2 | 0 |
| | 0 | 1 | 0 |
| Total | 10 | | 44 |
| Avg Stars Rating | | | 4.4 |

Amazon, Netflix, Barnes and Nobel, E-bay and other sites have taken some steps to overcome this problem by providing the total number of reviewers in each star as shown in Figure 4, but the extra information required the user to scroll down and look for detail descriptions instead of giving some measure or goodness criteria in a glimpse [3], [5]. This violates the overall motivation of using star rating in contrast to displaying complex information in any other viable form. Further providing additional information might be a good step but is not helping the user to compare the book with some other book in the same category.

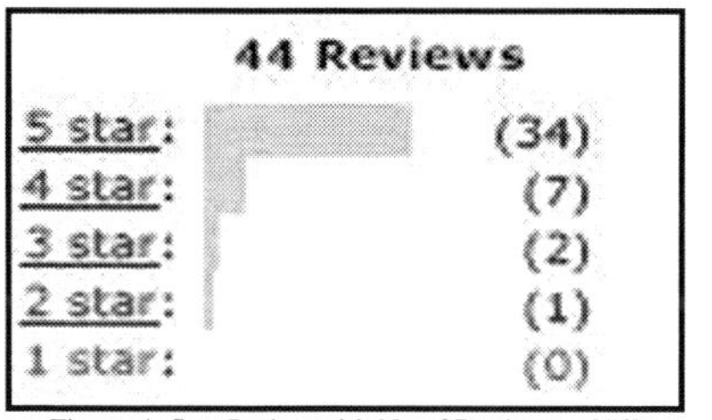


Figure 4. Star Rating with No of Reviewers [16]

Researchers have introduced different equations and efficient methods to calculate the ratings and to compare the books with variable number of reviewers by incorporating independent t and effect sizes, using degree of freedom and advance stochastic models [5] [4] [1]. It should be considered that this work is not proposing a recommendation method like item-to-item collaboration as used by Amazon [9] or advance statistical method of comparison of independent sets, like Wilcoxon signed-rank test [15] or Kruskal-Wallis one-way analysis of variance [14]. We are introducing here a simple yet powerful approach to solve this problem within the boundaries of star rating system with minimal apriori knowledge from the website or service provider. Next Section presents the details of the proposed relative ranking system and describes its validity by applying the similar book ranking scenario.

## III. RELATIVE RANKING

As the name suggests Relative Ranking is a measure of comparing a given book with some other book. Relative Ranking requires two book profiles to work. Book 1 is the benchmark, Book 1 even does not have to exist, it could be the average of all of the books ever published and rated, or average ratings of all of the books in artificial intelligence in Computer Science or based on any other category, subject, author or region specific criteria. The application of relative ranking is basically the comparison of Book 2 with Book 1 based on certain rules and criteria.

Book 2 Relative Ranking cannot be more than the average star rating of Book 1 until it has the same number or more number of points. It means if Book 1 has the profile as presented in Table 1, Book 2 with the same profile as shown in Table 1 cannot have the Relative Ranking of 4 or more until it has the same or more number of points as Book 1. Once, Book 2 has the same or greater points, we do not have to calculate the Relative Ranking, because with more number of points Book 2 has the capability to become the benchmark. For example, if two million copies of new novel have been sold, we do not have to compare it with a novel with five thousand sold copies. Table 2 demonstrates the stepwise approach taken to evaluate the relative ranking of the product, further it provides working example with the data given in Table 1 and mathematical notations.

Table 2. Pseudo Code, Mathematical Notations and Application Scenario for Relative Ranking Calculation

| | Description | Example | Mathematical Notation |
|---|---|---|---|
| Step 1 | Calculate point influence per star scale. Points in the respective scale divided by total points in all scales. | Therefore, 50/80 = 0.6250; 16/80 = 0.2000; 9/80 = 0.1125; 4/80 = 0.0500; 1/80 = 0.0125; | $i = [1,2,3,4,5]$<br>$X_i$ = No. of reviewers in the $i$ scale<br>T = Total number of points<br>$D_i = X_i * i$<br>$J_i = \frac{D_i}{T}$ |
| Step 2 | Calculate the contribution of each point in the average score. Multiply point influence by the average. | Therefore: 0.6250 * 4.00 = 2.5; 0.20 * 4 = 0.8; 0.1125 * 4 = 0.45; 0.0500 * 4 = 0.2; 0.0125 * 4 = 0.05; | A = Average Overall Rating<br>$C_i = J_i * A$ |
| Step 3 | Calculate influence per person. Divide the contribution obtained in Step 2 by the number of reviewers in the respective star scale. | Therefore, 2.5/10 = 0.25; 0.8/4 = 0.2; 0.45/3 = 0.15; 0.2/2 = 0.1; 0.05/1 = 0.05; | $I_i = \frac{C_i}{X_i}$ |
| Final Step | $Y_i$ = No. of reviewers in the $i$ scale (Book2)<br>$R_i = I_i * Y_i$<br>$RR$ = Relative Ranking = $\sum_{i=1}^{5} R_i$ | | |

Table 3 presents the profile of Book 1 when considered as the benchmark. Table 4 shows the Profile of Book 2 with relative ranking along with its average star rating.

Table 3. Benchmark (Book 1) Profile

| Book 1 | | | |
|---|---|---|---|
| | No of Reviewers | Stars | Points |
| | 10 | 5 | 50 |
| | 4 | 4 | 16 |
| | 3 | 3 | 9 |
| | 2 | 2 | 4 |
| | 1 | 1 | 1 |
| Total | 20 | | 80 |
| Avg Stars Rating | | | 4 |

Table 4. Book 2 Profile with Relative Ranking

| Book 2 | | | |
|---|---|---|---|
| | No of Reviewers | Stars | Points |
| | 5 | 5 | 25 |
| | 4 | 4 | 16 |
| | 1 | 3 | 3 |
| | 0 | 2 | 0 |
| | 0 | 1 | 0 |
| Total | 10 | | 44 |
| Avg Stars Rating | | | 4.4 |
| Relative Ranking | | | 2.2 |

We recommend that the relative ranking should be depicted on the similar lines as the star rating. Figure 5 shows the proposed visual drawing with Relative Ranking.

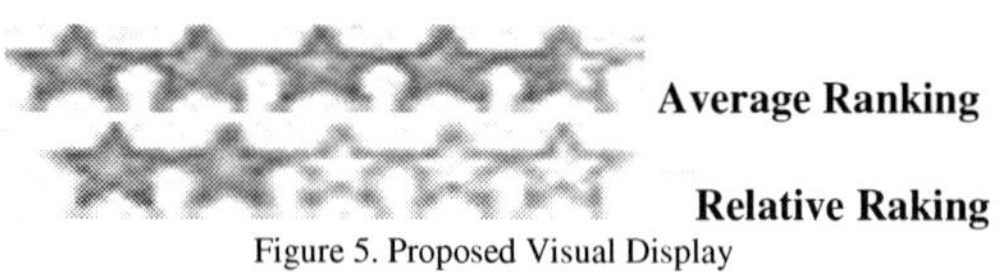


Figure 5. Proposed Visual Display

Some people might argue about the significance of number of raters and their influence on final star rating. For example, reviewer one may be the most knowledgeable person in the subject area of the book and reviewer two may be a high school student, who is just killing time by writing meaningless reviews about books. Both should not be the same, but we do not have that information to influence our results and it would be as controversial as the current approach on how to rate the reviewers.

Another thing we should be concerned about is to select the base book (benchmark), it might not be appropriate to have the general benchmark or base case for all comparisons. A book of Artificial Intelligence in Computer Science category should not be compared with a famous novel from Science Fiction category.

## IV. CONCLUSION AND FUTURE WORK

Star rating has become the standard of user reviews and ratings over the Internet. It has few limitations when it comes to averaging the ratings and comparing the books with different number of raters and variable reviews. Proposed Relative Ranking approach is one step further to make it more realistic and give more confidence to users of this system. This approach should not be confused with recommender system; it simply provides another way to look at the aggregate community data biased by a benchmark. We would like to extend this work with a comparative study of contemporary rating scales and novel applications in online marketing, trust and feedback and would also like to define the novel methods for selecting a benchmark for books, movies, electronics and other items. Moreover, some improvements in the visual depiction of the relative ranking in-contrast to the average star scale can be made to increase the applicability and understanding of this scale by the target audience.

## REFERENCES

[1] Hanna Yehuda and Jennifer McGinn. "Coming to Terms: Comparing and Combining the Results of Multiple Evaluators Performing Heuristic Evaluation"; CHI 2007, April 28-May 3, 2007, San Jose, California, USA

[2] Muzaffer Ozakca and Youn-Kyung Lim. "A Study of Reviews and Ratings on the Internet"; CHI 2006, April 22-27, 2006, Montreal, Quebec, Canada

[3] Juha Leino and Kari-Jouko Raiha. "Case Amazon: Ratings and Reviews as Part of Recommendations"; RecSys '07, October 19-20, 2007, Minneapolis, Minnesota, USA

[4] Ba-Quy Vuong, Ee-Peng Lim, Aixin Sun, Minh-Tam Le, Hady Wirawan Lauw and Kuiyu Chang. "On Ranking Controversies in Wikipedia: Models and Evaluation"; WSDM '08, February 11-12, 2008, Palo Alto, California, USA

[5] Tzu-Kuo Huang, Chih-Jen Lin and Ruby C. Weng. "Ranking Individuals by Group Comparisons"; 23rd International Conference on Machine Learning, 2006, Pittsburgh, PA, USA.

[6] http://www.burj-al-arab.com/, accessed on Friday, May 14th 2008

[7] www.sfgate.com/eguide/movies/reviews/, accessed on Friday, May 14th 2008

[8] Turnbull, Don, "Rating, Voting and Ranking: Designing for Collaboration and Consensus", CHI 2007, San Jose, California, USA

[9] Linden Greg, Smith Brent, and York Jeremy, "Amazon.com Recommendations – item-to-item Collaborative Filtering", IEEE Intelligent Computing, Jan-Feb 2003

[10] www.YouTube.com

[11] www.Digg.com

[12] www.Netflix.com

[13] www.IMDB.com

[14] William H. Kruskal and W. Allen Wallis. Use of ranks in one-criterion variance analysis. Journal of the American Statistical Association 47 (260): 583–621, December 1952

[15] http://en.wikipedia.org/wiki/Wilcoxon_signed-rank_test, accessed on August 1st 2008

[16] www.amazon.com

[17] www.ebay.com

[18] J. Ben Schafer, Joseph A. Konstan and John Riedl, "E-commerce Recommender Applications", Data Mining and Knowledge Discovery, Vol. 5, pp. 115–153, 2001

[19] Chang, Elizabeth and Hussain, Farookh Khadeer and Dillon, Tharam S. 2005. : Trustworthiness Measure for e-Service, Third Annual Conference on Privacy, Security and Trust, 12-14 Oct 2005. St. Andrews, New Brunswick, Canada: University of New Brunswick, Canada.

[20] Dhar, Vasant,Does Chatter Matter? The Impact of User-Generated Content on Music Sales(February 2008). CeDER Working Paper No. 07-06. Available at SSRN: http://ssrn.com/abstract=1113536

# Resource Redundancy – A Staffing Factor using SFIA

C. Nuangjamnong, S. P. Maj, D. Veal
Edith Cowan University
2 Bradford Street
Western Australia

*Abstract* - **There are many technologies and associated protocols that can provide different levels of redundancy in order to ensure continuous twenty-four-seven network connectivity and associated corporate services. The basic design principle of redundancy is to ensure there is no single point of failure hence the use of additional resources (active/active; active/passive or active/standby) to take over when the active component fails. It is important to address all aspects of the IT infrastructure such as utilities, power supplies, switching fabric etc. Configurations with no single point of failure are able to offer an annual down time of 0.5 hours or less. A detailed analysis of a large corporate network clearly demonstrated that it conformed to 'world's best' practice for a highly redundant network design. However, an analysis of their network staffing profiles clearly indicated the lack of a systematic approach to the provision of skill set redundancy. The authors recommend the use of the Skills Framework for an Information Age (SFIA) as a simple, quantifiable reference model for the more effective management of human assets.**

## I. INTRODUCTION

Corporate networks, both large and small, are expected to provide continuous network connectivity on a twenty-four-seven basis with annual down times of 0.5 hours – or less. This is only possible if the network infrastructure is designed to have no single point of failure. For the purposes of this paper, the network itself will be logically partitioned into three main sectors – access network, enterprise network and corporate WAN [1]. The access network connects the enterprises private network to the Internet Service Provider (ISP); the enterprise network represents the enterprise's internal network typically partitioned from the external network; the corporate WAN provides long distance connectivity.

In order to determine network failure adequate information from both physical and logical systems is required. One possible method is used the Information Technology (IT) audit. This process involves collecting and evaluating evidence of an organization's network systems, practices, and operations. Three major components for an IT audit in network infrastructure are [2]:

- Availability – will the organization's network systems be available for the business at the all times when required?
- Confidentiality – will the information in the network systems be disclosed only to authorize users?
- Integrity – will information provided by the network systems always be accurate, reliable, and timely?

However, tasks related with Information and Communication Technology (ICT) are particularly vulnerable to human error. An example of human error in network security can occur in the management and implementation of appropriate security policies to address the particular needs of a business organization. Even through network administrators may fully understand how to apply security policy within their networks, there are some several reasons why this task is difficult such as configuration language, vendor proprietary, and firewall rule-bases [3]. This makes the network administrators' task more difficult in terms of time, pressure and an increase in stressful situations that may well be a factor contributing to skill shortages within this area. ICT professionals were a decline compared to the previous year's 82 percent of vacancy filling rate [4].

Another threat to networks is reliability and availability, Mean Time Between Failures (MTBF) is the mean (average) time between failures of a system, and is often attributed to the "useful life" of the device. It does not include 'infant mortality' or 'end of life'. Calculations of MTBF assume that a system is "renewed", after failure and again made operational. The average time between failing and being returned to service is termed Mean Down Time (MDT) or Mean Time To Repair (MTTR). MTBF figures may be reduced by parallelism. In effect serial devices reduce availability and parallelism increases availability. Therefore, the network topology has a direct impact on system MTBF [5]. Duplicated devices typically may operate in one of three different modes – active/active, active/passive or active/standby. In order to prevent network failure, each data centre needs to have network resources backup and network planned.

There are various vendor neutral and vendor specific protocols operating at different OSI levels of the seven-layer model. At the OSI layer 2 Link Aggregation Control Protocol – (LACP) distributes traffic over multiple links. Spanning Tree Protocol (STP) eliminates potential loops resulting from redundant links. The OSI layer 3 routing redundancy (Hot Standby Routing Protocol (HSRP) or Virtual Router Redundancy Protocol (VRRP)) removes single point of router failure. There are a number of different protocols and technologies for the different network functions. The major vendors typically provide 'white paper' reference guides and associated configuration guides [6]. These serve as various deployment scenarios, typically based on specific products, in

T. Sobh (ed.), *Innovations and Advances in Computer Sciences and Engineering*,
DOI 10.1007/978-90-481-3658-2_6, 

a tested and proven architecture for all aspects of a network – access, enterprise, corporate etc.

However, it is also important to consider the entire network infrastructure. For example, all elements of the electrical systems, including backup systems, should be fully duplicated with critical services connected to both 'A-side' and 'B-side' power feeds hence providing N+1 redundancy. The Gartner Group defines a resilient business as one that can 'bounce back from any kind of setback, whether a natural disaster, an economic change, a competitive onslaught, cyber-espionage, or a terrorist attack' [7]. According to a 2008 Gartner report cited by Cisco Systems only about one third of enterprises surveyed had plans to cope with the complete loss of physical assets and employee space [7].

To apply the Skills Framework for the Information Age (SFIA), the authors used a Data Centre in a public university in Australasia as a case study. The goals of this study are to present how to match IT professional skills for redundant responsibilities and accountabilities in IT management department. Better network resource redundancy in staff issue may be employed and created due to IT professional skills shortage.

## II. SKILLS FRAMEWORK FOR THE INFORMATION AGE (SFIA)

In order to keep network reliability and performance, not only network resources need backup and redundancy, but also IT staff redundancy is essential for every data centre. Most IT managers clearly indicated the need for a more systematic approach to skill set auditing – especially within the field of network support and management [8]. The SFIA is a UK government backed initiative designed to more explicitly define a competency framework and the associated skill sets [9]. The SFIA provides a common reference model for the identification of the skills needed to develop effective information systems (IS) making use of information technology. The model is a simple and logical two dimensional framework consisting of areas of work on axis and levels of responsibility and accountability (TABLE I) [10]. The overall purpose of the SFIA is to assist organizations employing IT staff to [10]:

- reduce IT project risk
- maintain staff
- make recruitment effective
- enhance the effectiveness and efficiency of the IT function

Furthermore, according to the SFIA standard,

*"SFIA enables employers of IT professionals to carry out a range of HR activities against a common framework of reference - including skill audit, planning future skill requirements, development programmes, standardization of job titles and functions, and resource allocation* [11]*."*

*'The aim of SFIA was to clarify job roles and the levels at which individuals operate, and to provide standardized terminology that all fits into a single framework. This gives a clear demarcation between one person and the next, and gives them an incentive to improve their skills and strive to reach the next level, not just in operations, but also in other areas such as project management* [12]*.'*

TABLE I
SFIA LEVELS OF RESPONSIBILITY AND ACCOUNTABILITY

| SFIA Levels of responsibility | Autonomy | Influence | Complexity |
|---|---|---|---|
| **Level 7:** set strategy, inspire, mobilize | Responsibility for all aspects of a significant area of work, including policy formation and application | Decisions critical to organizational success<br>Develops long-term strategic relationships with customers and industry leaders | Work involves application of highest level management and leadership skills<br>Has deep understanding of information systems industry and emerging technologies and implications for the wider business environment |
| **Level 6:** initiate and influence | Responsibility for significant area of work, including technical financial and quality aspects<br>Establishes organizational objectives and delegates assignment | Influences policy formation on contribution of specialization to business objectives | Highly complex work activities covering technical, financial and quality aspects and contributing to formulation of IS strategy |
| **Level 5:** ensure and advise | Works under broad direction<br>Full accountability for own technical work or project/supervisory responsibility | Influence organization, customers, suppliers and peers within industry on contribution of specialization | Challenging range and variety of complex technical or professional work activities |
| **Level 4:** enable | Works under general direction within a clear framework of accountability | Influences team, and specialist peers internally<br>Participates in external activities related to specialization | Broad range of complex technical or professional work activities, in a variety contexts |
| **Level 3:** apply | Resolving complex problems and assignments<br>Specific instruction is usually given and work is reviewed at frequent milestones | Frequent external contact with customers and suppliers | Broad range of work, sometimes complex and non routine, in variety of environments |
| **Level 2:** assist | Works under routine supervision | May have some external contact with customers, suppliers, and own domain | Performs range of varied work activities in variety of structured environments |
| **Level 1:**follow | Expected to seek guidance in unexpected situation | Interacts with department | Requires assistance in resolving unexpected problems |

Source: SFIA Version 3.0

## III. CASE STUDY

This study concentrated on quantitative research methodology because it directly relates to the applying of SFIA via a Data Centre in public university within Australasia. The name of the university will remained anonymous for network security purposes. In this scenario, a large corporate network was analyzed. This network provides services for over 30,000 users with about 4,000 Internet users at peak time. There are 15,000 Unshielded Twisted Pair (UTP) points distributed over six locations. Two of the locations are geographically remote (hundreds of kilometers). Within the main site there are 400 layer 2 switches, over 30 layer 3 routers along with 40 virtual firewalls, and various load balancers, cache engines and Internet accounting engines. Standard operating procedures are commensurate with best practices. For example new equipment must undergo a 'burn in' period to test for faults and identify any MTBF 'infant mortality'.

Network and server infrastructure is hosted in a specialized room, which is referred to as the "Data Center" room. This room is equipped to provide the necessary infrastructure redundancy including: backup power generation, redundant Uninterruptable Power Supplies (UPS), fire alarms, smoke detectors, cooling backup and also electrical power fed from two separate substations. Each data centre hosts all of the primary equipment e.g. Core routers, VPNs, firewalls, application, email servers etc. Adjacent to the Data Centre is a similarly equipped Disaster Recovery room in order to provide both physical and logical redundancy in case of major equipment failures such as routing, hard disc etc. In addition to the Disaster Recovery room a remotely located Disaster Recovery location is currently under construction.

The core campus network infrastructure is designed to provide a 95% uptime. To address such a high requirement of uptime, many servers and networking devices in the university provide levels of redundancy that can help to provide the additional availability for the systems. However, the uptime is also dependent upon the applications running on and the design of the server infrastructure hosting those applications. In most cases the network infrastructure is capable of providing up to 100% redundancy but the applications themselves might not be capable of providing that level of redundancy. The case study network is clearly a highly redundant design providing twenty-four-seven services with an annual downtime less than 5 %.

## IV. STAFFING SKILL SET REDUNDANCY

This corporate network is managed by thirty staff – consisting of a variety of network engineers and systems engineers variously responsible for operating systems, switching, routing, firewalls, DNS, DHCP etc. Staff are organized in different groups – each with a clearly defined role. For example the layer 2 switching group consists of four network engineers. The Storage Area Network (SAN) group consists of ten staff. In keeping with good management practices some of the network functions are managed by staff from more than one group. In effect for some tasks responsibility for some network maintenance is shared between groups (Figure 1).

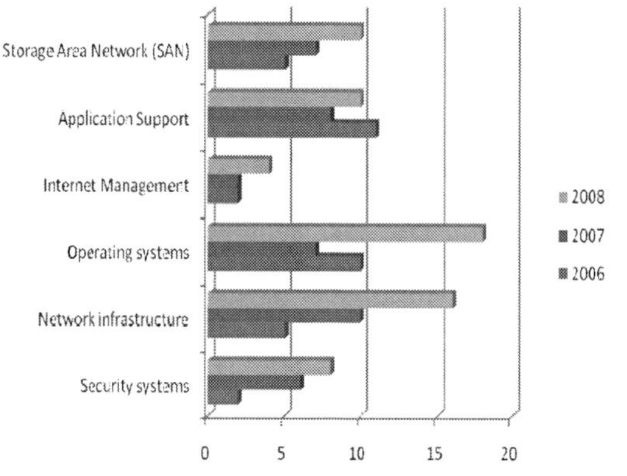


Figure 1. The combination of IT staff functions in Data Centre between 2006 and 2008

Regardless of the how the network groups are organized the number of staff in any group is only a coarse metric for evaluating skill set overlap and redundancy. What must also be considered is the skill set of each staff member in a given group. This is important because this may be the determining factor in how well a group can respond to system problems and failure.

In order to determine staffing skill set redundancy, a more detailed analysis of staff in each of the eleven functional groups was conducted. This consisted of an evaluation of every staff member according to a number of different criteria that included:

- knowledge and experience of any new technologies introduced
- ability to troubleshoot

## V. RESULTS AND DISCUSSION

The results for each staff member was aggregated and combined into a final score (0 low, 5 high) for each group (TABLE II). This 'backup scale' was an evaluation, based upon the above criteria, for the ability of a given group to cope with system failure should there be sudden and unexpected changes in staff availability e.g. illness, holiday, resignations. In order to prevent unpredicted situations such as resignations and illness of IT staff in the data centre, multi-skilled IT professional and overlapped responsibility are required at least one group from each staff. This approach can assist the data centre when system failure occurs during peak period and lack of replacement staff. It also promotes intergroup collaboration such as exchanging technical knowledge, improving high performance existing systems and maintaining staff in the data centre.

It can be clearly seen that some functional groups, such as switching, provide a high level of staffing redundancy. However, there are functional groups that provide low or even non-existent skill set redundancies. Even though there are ten staffs in the SAN group the skill set profiles result in a very low aggregated score. The situation is even worse for the physical infrastructure group (UPS, cooling etc).

TABLE II
STAFF SKILL SET REDUNDANCY

| Group/Number in Group | Role | Responsibility | Backup (Scale 0 – 5) |
|---|---|---|---|
| Windows System Group (6) | Windows System/ Unix Support | Email, File and Print | 4 |
| Unix System Group (6) | Windows / Unix Support | Operating Systems management, Database Administration | 4 |
| VmWare System Group (6) | Windows System Support | ESX (VMWARE infrastructure) | 2 |
| Switching Group (4) | Network System Support | Switching (Layer 2) | 5 |
| Routing Group (4) | Network System Support | Routing (Layer 3) | 3 |
| Security Group (4) | Network System Support Engineers | Firewall | 3 |
| Security Group (4) | Network System Support | VPN | 2 |
| SAN Group (10) | System/Network System Support | SAN | 1 |
| Internet Management Group (4) | Network System Support | Internet Accounting and Traffic Management | 2 |
| Application Support Group (10) | System/Network System Support | DNS and DHCP | 4 |
| Network infrastructure Group (4) | Network System Support | Physical Infrastructure, UPS, Cooling | 0 |

The SFIA framework is therefore designed to represent a consistent and systematic approach to skill set audits, job function standardization and hence staff resource planning. As a generic tool is it not limited to any specialist IT department.

A preliminary analysis of the five of the groups was conducted using a simplified SFIA matrix. Three levels of responsibility and accountability were used:

Level 1: first line of support e.g. help desk
Level 2: second line of support e.g. technical support
Level 3: expert knowledge

TABLE III
LEVEL OF EXPERTISE

| Group | Average years of Experience | Qualifications | Level of Support |
|---|---|---|---|
| Windows System Group | 3 | No Vendor qualifications, Bachelors and TAFE qualifications | Level 2 and level 3 support provided |
| Routing Group | 3 | CCNA, CCNP, Bachelors and TAFE qualifications | Level2 and level 3 support provided |
| SAN Group | 2 | No Vendor qualification, Bachelors and TAFE qualifications | Level 2 support provided. Level 3 support is provided by outside consultant |
| Unix System Group | 4 | Oracle certification, Bachelors and TAFE qualifications | Level 2 support provided, Some Level 3 support is provided by outside consultant |
| Network infrastructure Group | 5 | Engineering and TAFE qualifications | Level 1 support provided. Level 2 and Level 3 provided by consultant |

In terms of SFIAplus, a value added product administered by the British Computer Society (BCS), is a three dimensional matrix that adds further detail including skill training and development [13], [14]. Criteria used to determine the appropriate level included: experience, qualifications and training along with the aggregated level of responsibility and accountability (TABLE III).

From the results it can be seen that both the SAN group and Unix System group depend on outside consultants for their level 3 technical support. Even more significantly, the Network Infrastructure Group provide only level 1 support and hence are completely dependent for level 2 and 3 support from external consultants. This has both advantages and disadvantages. Hiring staff with specialized skills in infrastructure systems is relatively expensive. The converse argument is that there is a dependency on external consultants. Regardless of the relative merits of each case the results clearly identified a functional group requiring a more detailed analysis as the basis of recommendations to senior management.

The average years of experience in all groups were surprisingly small. It was found that as staff become experienced they have a tendency to move to different groups in order to enhance their skill set – and hence employability prospects. It is suggested that this degree of staff migration may result in a lower level of technical support in some of the functional groups. In effect technical skill sets may be lost to a group unless there are clearly defined procedures for sharing tasks between groups. However, a more detailed analysis is currently being conducted and the results will be used as a report to senior management.

This exercise, in particular the definition of levels of technical support, highlighted the need for routine testing and failover exercises. Whilst there are defined standard operating procedures, currently there is no 'live' failover testing. It is suggested that 'live' failover testing should be conducted in order to evaluate the standard operating procedures and associated staff preparedness. Given the critical nature of this type of exercise participating staff must have a level 3 skill set standard. Accordingly three of the groups analyzed (SAN, Unix System and Network Infrastructure) would be unable to conduct routine exercises of this nature without the assistance of external contractors and or consultants.

The SFIA skill set is also applicable to more senior levels of network support i.e. management. The results, even in this preliminary study, are such that this work is currently being undertaken.

It was found that the SFIA was a useful template to evaluate staff skills and more significantly, the staff preparedness for major system failures.

## VI. CONCLUSIONS AND RECOMMENDATIONS

A system is only as strong as the weakest link. This work has clearly demonstrated that even though a large corporate network, designed to internationally defined best practices for redundancy, may experience problems due to staff skill set profiles. Staffing redundancy and backup is important and this often needs to be quantified to make the needs more explicit. SFIA helps to assist in systematizing such processes. Redundancy and backup are both an important system requirement and both staff as well as equipment need to be considered.

It would be interesting to consider the SFIA process within a larger organization and to note if a greater degree of specialization with the increased numbers of staff would have led to more or less overlap of skill sets. However, further work is currently being undertaken and it is intended to investigate a range both smaller and larger organizations to determine if the variation in the amount of staff backup provided across a range of situations.

## REFERENCES

[1] Sun Microsystems, "Enterprise network design patterns: high availability," December 2003, http://www.sun.com/blueprints/1203/8174683.pdf

[2] A. Kadam, "A career as Information Systems Auditor," December 2003, http://www.networkmagazineindia.com/200312/securedview01.shtml

[3] A. Mayer, A. Wool, and E. Ziskind, "Offline firewall analysis," in Springer – Verlag, Int. J. Inf. Secur., pp. 1 – 20, June 2005, http://www.eng.tau.ac.il/~yash/ijis05.pdf

[4] State Government of Western Australia, "Information and Communication Technology (ICT) skill shortages," May 2007, http://www.skillsinfo.gov.au/NR/rdonlyres/ACE15B26-7E56-4FF8-AF61-D7342B810447/0/ICTreportNT.doc

[5] Vicor Corporation, "Reliability and MTBF overview," December 2007, http://www.vicorpower.com/documents/quality/Rel_MTBF.pdf

[6] Cisco Systems, "Capacity and performance management: best practices white paper," October 2005, http://www.cisco.com/en/US/tech/tk869/tk769/technologies_white_paper09186a008011fde2.shtml

[7] Cisco Systems, "Network availability: how much do you need? how do you get it," 2004, http://www.cisco.com/en/US/netsol/ns206/networking_solutions_white_paper09186a008015829c.shtml

[8] P. Dixit, M.A. Vouk, D.L. Bitzer, and C. Alix, "Reliability and availability of a wide area network-based education system," *The Seventh International Symposium on Software Reliability Engineering (ISSRE '96)*, p. 213, 1996

[9] SFIA Foundation Ltd, "SFIA as the IT skills standard," 2007, http://www.sfia.org.uk/cgi-bin/wms.pl/927

[10] SFIA Foundation Ltd, "Skills framework for information age: version 3.0," 2008, http://www.sfia.org.uk/cdv3/SFIA3webrefTOC.html

[11] J. Slater, "SFIA: Skills Framework for the Information Age," January 2008, http://www.jasonslater.co.uk/2008/01/08/sfia-skills-framework-for-the-information-age/

[12] C. Everett, "A local council is investing in an SFIA scheme to identify the IT skills gap and broaden expertise in the area," April 2007, http://www.computing.co.uk/computing/analysis/2187528/case-study-north-cornwall

[13] British Computer Society (BCS), "Introduction to SFIA/SFIAplus for employers," 2007, http://www.bcs.org/server.php?search_word=SFIA&show=nav.7015&pp=10

[14] British Computer Society (BCS), "Embracing the challenge, exploiting the opportunities: building a world class IT profession in the era of global sourcing," 2004, http://www.bcs.org/server.php?search_word=SFIA&show=nav.7015&pp=10

# Integrating Text Mining and Genetic Algorithm for Subject Selection

Y.C. Phung*, S. Phon-Amnuaisuk** and R. Komiya**

* Monash University Sunway Campus/School of Information Technology, Malaysia

** Multimedia University/Faculty of Information Technology, Malaysia

***Abstract*—Advances in science and technology have brought about a growing number of study areas and disciplines. This in turn, results in the increase of subjects or units being offered via modular programmes or courses in universities or institutes of higher educations. The modular method of selecting subjects for completion of a course can be likened to the process of selecting events to complete a timetable, albeit a possibly less constrained variation. This paper combines the possibility of text mining the web and applying timetabling strategies using genetic algorithm (GA) into the subject selection problem. It aims to provide a basis for a system that provides advisory selections for students of higher educations. A subject selection system integrating text mining and specific GA mechanisms is implemented. Experiments and test runs with randomly generated excess population of chromosomes indicated a fair degree of success in this implementation.**

## I. Introduction

The advances in science and technology brings along an increasing number of disciplines; this, in turn, is increasing the number of subjects or units that can be offered in institutions of higher education, or at the very least, an increase in the areas of studies. The institutions of higher education referred to in this paper is not necessarily restricted to only universities but also to any institutions of educations that offer modular courses or programmes at tertiary level. A modular course is where enrolled students are required to undertake a certain number of unit modules to complete a course. These unit modules are usually measured using credit points. Each course would have its own specific set of credit point constraints that need to be satisfied for its completion. For example, one course might require students to acquire specific amounts of credit points from several different schools (or faculties), while another might simply limit the number of credit points that can be taken from one particular school.

Such an arrangement provides the students with more flexibility in the sense that the students are free to select any subject; as long as they satisfy constraints, including course, subject, and availability constraints. Reactions to the given selection flexibilities could be varied, possibly ranging from enthusiasm to confusion. Students already sure of their interests or their future career path will no doubt select subjects with confidence, based on their own individual preferences. Students who are unsure or indecisive might resort to random selection or by imitating their peers, which could lead to unavoidable selection changes or poor academic performance.

*Paper Outline*

The rest of this paper is organized as follows: Section II details subject selection problem further while Section III explores in detail similarities of timetabling and subject selection. Section IV showcases actual implementation framework whereby it capitalizes on leveraging text mining and genetic algorithm (GA) mechanisms and the overall implementations. Section V explores results and analysis while Section VI concludes and provides some information on future works.

## II. Problem Statement

Looking at subject selection, in a way, it is about fitting a number of subjects into credit point slots such that they satisfy all or most of the constraints. Abstracting from this view, both the timetabling problem and the subject selection problem can be viewed as problems of fitting in a number of items into a number of slots in such a way that it generates the best possible effect. This is, therefore, an optimization problem as dictated in [9] that seeks to produce an optimal placement or selection of items given a number of slots and constraints. GA has apparently met with the most successes in optimization problems [2, 9]. It has been widely applied to the timetabling problem in a variety of ways, as seen in [1, 3, 4, 5, 6, 8] and their references. In these and most others, GA appeared to have mostly met with successes, at varying degrees, at producing optimal or near-optimal timetabling solutions.

With the noted similarities of the timetabling problem and the subject selection problem, the possibility of leveraging text mining from web sources and adopting GA timetabling strategies for the subject selection problem is explored.

## III. Ttimetabling Vs Subject Selection

Reference [8] has defined a timetable as $n$ event assignments, each event assignment has four parts $(e, t, l, p)$ from four different sets $(E, T, L, P)$ where $e \in E$, $t \in T$, $l \in L$, and $p \in P$. $E$ is the set of events that require an assignment of time and place. This set includes lectures, tutorials, and laboratories sessions, and so on per week. $T$ is a set of usable times and $L$ is a set of possible locations. $P$ is then the set of persons that have a role to play in an event.

T. Sobh (ed.), *Innovations and Advances in Computer Sciences and Engineering*, DOI 10.1007/978-90-481-3658-2_7, © Springer Science+Business Media B.V. 2010

So in their own words, an event assignment is therefore interpreted as "event *e* starts at time *t* in place *l* and looked after by person *p.*"

Subject selection can be seen from a similar perspective. A subject selection is defined as n subject enrolments, where each subject enrolment, s, is simply one subject from the set S, which is a set of all available subjects, constrained by availability and course/subject requirements. This appears to be a generalization of a timetable. The main part of an event assignment is the event itself. Time, location and involved persons are extra, but realistic constraints. If a very large number of rooms are available, together with numerous qualified staff, the timetable problem can be reduced to a much simpler problem of assigning different events to different times to avoid clashes. This simplified view can be carried forward to subject selection; a problem of enrolling a student in subjects constrained by semester and year while at the same time avoiding "clashes" of not meeting subject or course requirements.

## IV. IMPLEMENTATION FRAMEWORK

Monash University Malaysia's (Monash) official website is used to test this implementation. Several test cases using its undergraduate degrees are experimented. The complete list of available subjects for all courses is sourced from its online handbook for undergraduate provided at http://www.monash.edu.au/pubs/2004handbooks/undergrad/.

To prepare for the implementation, a priority list of both hard and soft constraints for the proposed system has been identified and is given in Table I. This identification is based on the approach adopted from the work of Ross [8] and Terashima et al [5], whereby each constraint is associated with a penalty, based on perceived influence or importance. Given that the number of hard constraints closely matches the timetabling problem, the complexity of producing an optimal subject selection may likely be similar, if not equal.

For the purpose of clarity and classification for implementation, only the first two of the constraints listed are soft constraints, the rest are hard constraints.

TABLE I
HARD AND SOFT CONSTRAINTS

| Constraints | Descriptions |
|---|---|
| Spread | A subject selection must be able to spread out the different types of subjects such that compulsory subjects and electives are alternately selected. |
| Bias | Addresses the factor of human subjectivity referring to current trends and demands (i.e. the current "in" or "hot" skills or knowledge). |
| Compulsory | The requirement where a set of subjects must be taken for the successful completion of a course. |
| Uniqueness | The requirement that a subject can only be taken once for each subject selection. |
| Prerequisite | Another subject or subjects that determine the legality of its selection. In other words, a subject cannot be selected if its prerequisites were not selected earlier. |
| Prohibition | Just the opposite of a prerequisite, where a prerequisite must be present, and a prohibition must not. |
| Load | Refers to the acceptable number of subjects that come from a particular set, and not to be taken in the sense of reduced or extra workload (underload and/or overload) |
| Availability | Actual period of time (year and semester) a subject will be offered. |

### A. Text Mining and GA Mechanisms

The text mining mechanism used is as shown in Fig. 1. It adopts KEA's approach and has been explored by the author which is detailed in Phung [11]. Raw material in the form of html format from website source will be processed and cleaned to obtain the appropriate feed which satisfies the priorities and constraints set forth.

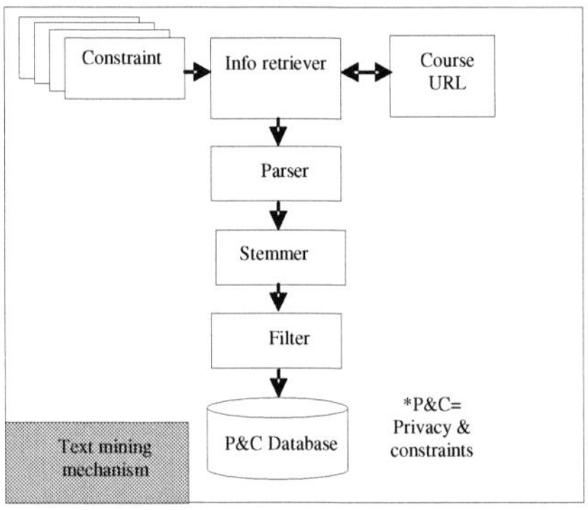


Fig. 1. Text mining mechanism

The GA mechanisms on the other hand is in truth a hybrid GA in the sense that it utilizes an external function to maintain chromosome feasibilities by the use of bounds, rather than relying fully on the genetic operators and the GA process. This is necessary, as evidenced in many of the implementations in literature, as the subject selection problem is a highly constrained problem, though likely not as constrained as the timetabling problem. The basic structure of such mechanism is shown in Fig. 2.

In this implementation, a one-dimensional array is used to represent chromosome, i.e. an individual (candidate solution) of the population. The array will have array positions corresponding to selection slots as given in Table II; and the array elements corresponding to selected subjects. The basic crossover-mutation pair of genetic operators are then implemented and used for the population of subject selection. The overall system architecture is an integration of text mining and GA, where the cleaned P&C database obtained from the text mining process is used as the feed to the GA mechanism; which in turn must satisfy the constraints set forth.

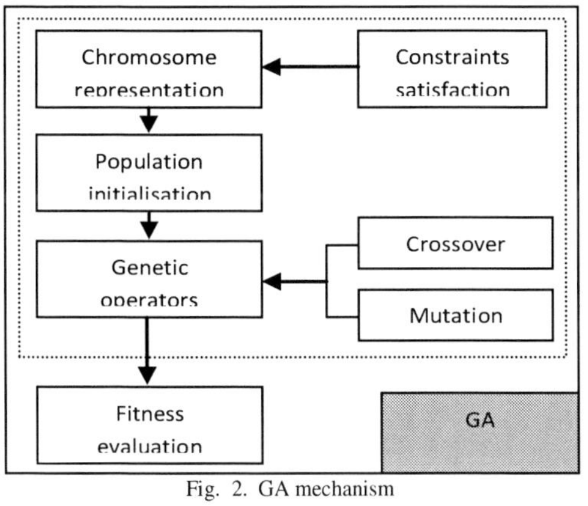


Fig. 2. GA mechanism

### B. *Overall Implementation*

For the context of this implementation, subject selection is viewed as the problem of selecting a certain number of subjects, each with a credit point value, and using them to fill a fixed number of selection slots measured in credit points. Graphically this is viewed as given in Table II; a total of 24 selection slots are available, where each selection slot represents a 6 credit points.

TABLE II
SUBJECT SELECTION SLOTS

| Year, Semester | Selection Slots | | | |
|---|---|---|---|---|
| Y1, S1 | 1 | 2 | 3 | 4 |
| Y1, S2 | 5 | 6 | 7 | 8 |
| Y2, S1 | 9 | 10 | 11 | 12 |
| Y2, S2 | 13 | 14 | 15 | 16 |
| Y3, S1 | 17 | 18 | 19 | 20 |

The implementation is designed to handle 24 selection slots instead of 144 credit-point-slots, although mechanisms were left in place to facilitate changes to incorporate such. The numbers in each selection slot also represents the position of the selection slot within the one-dimensional chromosome's representation array, which doubles up as its reference number.

## V. RESULTS AND ANALYSIS

### A. *Constraints Penalties*

Chromosomes that violate constraints are penalized according to their perceived level of important. The rationale is that poor chromosomes will remain at the bottom end of the population, and will be more likely to be replaced by new offspring [2], rather than relying on repair functions. The quality of the chromosomes will fully rely on the GA itself. A reference table for violation of constraints along with their penalties is given in table III.

TABLE III
CONSTRAINTS PENALTIES

| Constraint Type | Penalty | Constraint Type | Penalty |
|---|---|---|---|
| Compulsory | 800 | Staff bias | 10 |
| Unique | 1200 | Preferred faculty bias | 8 |
| Load | 600 | Discouraged faculty bias | 8 |
| Available | 1000 | Preferred discipline bias | 6 |
| Prerequisite | 400 | Discouraged discipline bias | 6 |
| Prohibition | 400 | Primary user preference | 14 |
| Spread | 75 | Secondary user preference | 8 |

The penalty value is used to penalize a chromosome the moment the value's corresponding constraint type is violated. Multiple violations incur multiple penalties. The values have been fine-tuned up to the present values. They have been selected based on the definition that constraints must be sufficiently fulfilled. These numbers are spaced apart with large intervals to allow the GA a greater distinction between good and bad chromosomes. Spread constraint is strongly stressed by academic counselors; therefore a relatively high penalty is given as compared to biases and preferences to reflect the emphasis.

### B. *Implementation Test Runs*

An initial population of 100 chromosomes is randomly generated. A hundred may be in excess but in reality, overall performance was slightly poorer for populations of 50 and 75. The chromosomes of the population are guaranteed only partial feasibility. These will go into cycles of reproduction and re-population. The maximum number of generations is set at 5000, although the GA will converge earlier if no change to the best chromosome's cost is detected after 1000 generations.

At each generation, elitism is practiced, whereby the best chromosome of the current generation will be preserved unchanged to the next generation. A cull rate of 10 percent is placed on the population, which means that 10 percent of the poorest (highest cost) chromosomes are discarded and new offspring generated to replace them (steady-state re-population [10]). The parents of the offspring are randomly selected from the population after culling, with a slight bias towards fitter chromosomes. Mutations occur at a rate of a high 15 percent at each generation. The high rate is to slow down GA convergence as overall best fitness tends to remain constant very early.

Two forms of gene crossover operators were implemented and tested: subject crossover (SubX) (4-point and 8-point), and semester crossover (SemX) (2-point and 3-point). Gene crossovers were used to prevent invalid gene values from occurring. The mutation operators consist of a simple mutation operator (SM), and a biased mutation operator (BM). All possible combinations of these operators will be tested, with each being executed 10 times for a better overview and easier examination of the quality of the results.

TABLE IV
CHROMOSOMES FITNESS

| Genetic Operators | Fitness | | | Complete Success | Total Success |
|---|---|---|---|---|---|
| | Best | Worst | Average | | |
| SemX2Pt, SM | 216 | 1278 | 402.4 | 80% | 80% |
| SemX3Pt, SM | 204 | 284 | 258.4 | 100% | 100% |
| SemX2Pt, BM | 200 | 282 | 243.4 | 100% | 100% |
| SemX3Pt, BM | 204 | 898 | 311.4 | 90% | 90% |
| SubX4Pt, SM | 200 | 268 | 230.6 | 100% | 100% |
| SubX8Pt, SM | 204 | 1020 | 322.4 | 90% | 90% |
| SubX4Pt, BM | 218 | 1018 | 397.8 | 70% | 70% |
| SubX8Pt, BM | 218 | 2690 | 655.0 | 70% | 70% |

Table IV above summarizes the fitness of the chromosomes after the full GA run. The first column indicates the combination of crossover and mutation operators used. The next three columns respectively indicate best fitness, worst fitness, and average fitness of the best chromosome of all the runs. Complete success indicates the percentage of the chromosomes containing a feasible solution that does not violate any of the constraints. Total success on the other hand indicates the percentage of all feasible solutions (including those violating the spread constraint).

On the whole, the semester crossover operators did better than the subject crossovers, with 2-point and 3-point semester crossover holding little differences, though 4-point subject crossover generally did better than its 8-point counterpart. Biased mutation did better than simple mutation, though both are aimed at the similar task of attempting to improve a chromosome.

Specifically, the results are only somewhat satisfactory. This test run produced mostly complete successes, with only a few chances of bad generations out of the total runs, given the tendency of the average towards the best.

Table V below shows the summed total of all constraint violations of the 10 best chromosomes produced from the 10 runs. Note that the violations are mainly due to the infeasible solutions. The number of chromosomes that actually incurred violations is noted in the second column. The first column is as before, the others correspond to the constraint types as detailed and given in Table III earlier. Each value below the columns indicates how many such violations have been incurred by all the produced solutions.

TABLE V
TOTAL CONSTRAINT VIOLATIONS

| Genetic Operators | Chrom | Co | Un | Lo | Av | Pr | Pro | Sp |
|---|---|---|---|---|---|---|---|---|
| SemX2Pt, SM | 2 | – | – | 1 | – | 1 | 1 | – |
| SemX3Pt, SM | 0 | – | – | – | – | – | – | – |
| SemX2Pt, BM | 0 | – | – | – | – | – | – | – |
| SemX3Pt, BM | 1 | – | – | 1 | – | – | – | – |
| SubX4Pt, SM | 0 | – | – | – | – | – | – | – |
| SubX8Pt, SM | 1 | – | – | – | – | 2 | – | – |
| SubX4Pt, BM | 3 | – | – | – | – | 2 | 2 | – |
| SubX8Pt, BM | 3 | 4 | – | – | – | 2 | – | – |

The important area of discussion would be the comparisons of a student's actual subject selection with this implementation. Table VI and VII show the comparisons, with the subjects arranged in order from the first semester of the first year to the second semester of the third year.

TABLE VI
REAL SUBJECT SELECTION

| 1 | 2 | 3 | 4 |
|---|---|---|---|
| CSE1301 | MAT1841 | CSE1308 | BUS1060 |
| CSE1303 | MAT1830 | CSE1434 | MGW1010 |
| CSE2303 | CSE2304 | CSE2324 | BUS2011 |
| CSE2302 | CSE2305 | CSE2316 | BEW1601 |
| CSE3305 | CSE3308 | CSE3313 | CSE3318 |
| CSE3302 | CSE3322 | CSE3309 | CSE3301 |

TABLE VII
BEST PRODUCED SUBJECT SELECTION

| 1 | 2 | 3 | 4 |
|---|---|---|---|
| CSE1301 | MAT1841 | CSE1308 | BUS1010 |
| CSE1303 | MAT1830 | CSE1434 | MKW1120 |
| CSE2303 | CSE2304 | CSE2324 | CSE2318 |
| CSE2302 | CSE2305 | CSE2316 | BUS1042 |
| CSE3305 | CSE3308 | CSE3313 | BUS2011 |
| CSE3302 | CSE3322 | CSE2309 | CSE3301 |

The above results indicate that only 2 subjects truly differ: BEW1601 and MGW1010. BUS1060 was a subject taken by the student prior to the amendment that prohibits it for those also taking CSE1301. CSE2318/CSE3318 and CSE2309/CSE3309 is essentially the same subject; hence do not differ in content in any way. In short, the produced selection appears to closely match what a student would have selected given the limited amount of preference, and that the missing subjects had been selected simply to fill in the credit points. The test run served as an indication of a fair degree of success of the implementation.

## VI. Future Work And Conclusion

Much work is still needed to provide a full integration of both the text mining and GA mechanisms to envisage a more complete analysis of applications of GA to the timetabling problem, specifically to identify and categorize the different approaches. This should be conducted to maintain the proposed structure of subject selection to take advantage of the proven structure of timetables.

There is also a need to formalize the investigation into the application of a proposed subject selection system and how it is to fit into the current administrative functions. Integration should also be considered, together with possible events on staff and students. Reactions or side effects such as over-dependence or inappropriate usage that could reduce student academic performance need to be addressed.

Yet as an overall view, subject selection appears to have a place in most institutions of higher education; our subject selection could become an advisory tool for students, or at the very least, reduce the need for selection changes, thereby reducing the hassle associated with it for both students and staff.

## References


[1] A. Colorni, M. Dorigo, and V. Maniezzo, "Genetic Algorithms and Highly Constrained Problems: the Time-Table Case," *Proceedings: First International Workshop on Parallel Problem Solving from Nature*, Berlin, Springer-Verlag, Germany, pp. 55–59, 1991.

[2] D.E. Goldberg, "Genetic Algorithms," *ACM Computing Surveys*, New York, ACM Press, Vol28 (1):pp. 77–80, 1999.

[3] E. K. Burke, D.G. Elliman, and R.F. Weare, "A Genetic Algorithm Based University Timetabling System," *Proc. 2nd East-West Conference on Computer Technologies in Education*, Crimea, Ukraine, Vol1:pp. 35-40, 1994.

[4] G.R. Filho, and L.A.N. Lorena, "A Constructive Evolutionary Approach to School Timetabling," *EvoWorkshops*, pp. 130-139, 2001.

[5] H. Terashima-Marin, P. Ross, M. Valenzuela-RENDÓN, "Evolution of Constraint Satisfaction Strategies in Examination Timetabling," *Proc. Genetic and Evolutionary Computation Conference (GECC-99)*, pp. 635-642, Morgan Kaufmann Publishers, 1999.

[6] J.P. Caldeira, and C.R. Agostinho, "School Timetabling Using Genetic Search, *Practice and Theory of Automated Timetabling*," Toronto, http://citeseer.nj.nec.com/caldeira97school.html, 1997.

[7] P. Darbyshire, and A. Wenn, "A Matter of Necessity: Implementing Web-Based Subject Administration, *Managing Web-Enabled Technologies in Organizations: A Global Perspective*, Idea Group Publishing, Hershey, USA. pp. 162-190, 2000.

[8] P. Ross, D. Corne, and H.-L. Fang, "Successful Lecture Timetabling with Evolutionary Algorithms," *Proc European Conference on Artificial Intelligence (ECAI) '94 Workshop W17: Applied Genetic and other Evolutionary Algorithms*, Amsterdam, Holland, Springer-Verlag, 1994.

[9] R.L. Haupt, and S.E. Haupt, "*Practical Genetic Algorithms*," Reading, USA, John Wiley & Sons, Inc., 1998.

[10] T. Back, F. Hoffmeister, and H-P. Schwefel, "A Survey of Evolution Strategies," *Proc. 4th International Conference on Genetic Algorithms*, San Diego, pp. 2–9, Morgan Kaufmann Publishers, 1991.

[11] Y.-C. Phung, "Text mining for stock movement predictions: a Malaysian perspective," sixth international conference on data mining, Data Mining VI: data mining, text mining and their business applications, WIT Press, pp. 103-111, 2005.

# "DESIGN, DEVELOPMENT & IMPLEMENTATION OF E-LEARNING TOOLS"

**Bagale G. S.**
girishbagale_08@rediffmail.com

**Naik Sawankumar**
naiksa1@gmail.com

**Deshmukh A. J.**
ashu_nmims@rediffmail.com

**Patil R. Y.**
raj_fmht@yahoo.com

NMIMS(U), Mukesh Patel School of Technology Management and Engineering(MPSTME) ,Mumbai-400056

*ABSTRACT*

*The ICT (Information & Communication Technology) can be used as a resource; a resource that provides access point for information as well as real world requirements. This paper shows way how to use ICT for effective & efficient teaching-learning process. The paper throws light on need and use of ICT in Teaching, Learning and Laboratory work, contribution by teachers, students and others to make ICT a reality in technical education. Also paper presents benefits, constraints and steps required to be taken for use effective use of ICT.*

*This paper gives information about some free software which helps to create online courses & helps in teaching-learning process. This paper also explains "BITS Virtual University-A Case Study"*

## I.INTRODUCTION

Education is the backbone of any economy. Technical Education in particular, as it provides skilled manpower is essential for the growth of industries and service organizations. In the context of globalization, there is an urgent need to look at revamping our education system, as there have been rapid changes and developments happening in all areas including science and technology. One such technology, which has revolutionized the way we live in the last few years including education, is the 'Information & Communication Technology' (ICT).We are now experiencing an ICT revolution, just like industrial revolution in the 19th century. The advancements in computers, communication technology /networking and the advent of Internet / Web have opened up new avenues in all fields and education is no exception to it.

## II. ROLE OF ICT IN EDUCATION

The impact of ICT on education has been so tremendous that there has been lot of efforts through research to study the impact of the same. The inclusion of ICT in education has caused a paradigm shift from a 'teacher centered' to a 'learner centered' orientation. The developments in ICT and learning technology has led to a number of trends on the future of educational systems and activities like - E-Learning, Web-based ITS, distance learning, virtual university, virtual school etc.[4] .

Educators and policy makers agree that ICTs are of paramount importance to the future of education. ICT can help education in various ways like:

• Increasing access through distance learning – ICTs can provide new and innovative means to bring educational opportunities to greater numbers of people of all ages, especially those who have historically been excluded, such as populations in rural areas, women facing social barriers and students with disabilities.

• Enabling a knowledge network of students – Effective use of ICTs can

T. Sobh (ed.), *Innovations and Advances in Computer Sciences and Engineering*,
DOI 10.1007/978-90-481-3658-2_8, © Springer Science+Business Media B.V. 2010

contribute to the timely transmission of information and knowledge, thereby helping education systems meet this challenge.

• Training teachers – The use of ICTs can help in meeting teacher training targets. Moreover, ICTs provide opportunities to complement on the job training and continuing education for teachers.

## III. ICT TOOL FOR TEACHING

In the 21st century technical education will require teachers to adapt and adopt a different set of pedagogic practices. The force to pedagogic change in technical education is the new modes of enquiry offered by computer based tools and resources, known as ICT. This technology offers easy access to a vast array of internet resources and extends opportunities for the students both inside and outside the classroom. The main forms of ICT activity include: tools for data capture, processing and interpretation (multimedia software and information system's), publishing and presentation tools (i.e. computer projection technology), data handling tools (word processing), presentation tools (computer controlled microscope, digital still and video cameras).

**Video-conferencing**

The paradigm shift in distance learning has created a lot of wonders in distance learning. The present day technological innovations made everything very simple and one can attend conference from Mumbai to Delhi or to any corner of the globe via Videoconferencing upsetting the geographical constraints. It connects everything through satellite and internet which gives a real experience to the learners and it is a boon from heaven to the learners.

**Tele-conferencing & Chatting**

Tele conferencing is a powerful tool for distance learners. It helps learner to interact with other person over a telephone or via a modem. In present day scenario, mobile technology reduces our work to the maximum extent and is reachable even to an ordinary man and mobile interactions is very easy that makes distance learning a fruitful experience.

**Use of Internet**

Internet is a great mantra of 20th century which is leading human life towards more comfortable zone. Everything is being digitalized in the world, from student-filled, single teacher directed classrooms to "teacher-less", boundary-less and time-optimised learning kiosks. The new trend of distance learning is being utilized successfully by the ICT revolution.

**Virtual classroom**

Some teachers and university professors have adopted virtual worlds for educational purposes. Educators create an online community that students can log into and interact with. Within these educational virtual worlds, students will use their avatar to learn about new assignments and to create projects that are viewable within the virtual world.

**Moodle Architecture**

The word Moodle is an acronym for Modular Object-Oriented Dynamic Learning Environment. Moodle is a student centered course management system designed to help educators who want to create quality online courses. Such e-learning systems are called as Learning Management Systems (LMS) or Virtual

Learning Environments (VLE). One of the main advantages of Moodle over other systems is a strong grounding in social constructionist pedagogy.

## IV. RELEASES FROM LAB & ASSIGNMENT WORK

Mechanical aspects of laboratory process includes setting up experiments, taking complex measurements, tabulating data, drawing graphs and executing multiple or difficult calculations. All these process in most of the cases are laborious and time consuming. Preparation of set up, actual execution and computations are difficult to perform in stipulated time (time table). This increases burden on teaching staff as well as students. Using software tools linked to data in different ways with a less time and effort, interactive computer simulation can help to avoid 'bogged down' with the mechanics of simply setting up equipment The interactive and dynamic nature of tools such as simulation, data analysis software and graphing technologies can be influential in: allowing students to visualize process more clearly and give qualitative or numeric relationships between variables; focusing attention on overarching issues; increasing salience of underlying features of situations and abstracts concepts, helping students to access ideas more quickly and easily. Using ICT to support practical investigation can help to experience the entire process as holistic and cyclic, whereas time constraint on conventional laboratory work tend to isolate and obscure the links between- planning, practical work, writing up and evaluation. .

Well documented motivational effects of ICT, which seems to be intrinsically more interesting and exciting to students than using other resources, ICT offers the opportunity to greatly enhance the quality of presentation by using movement, light, sound and colour rather than static text and images, which is attractive and more authentic.

Above all, due to ICT an increase persistence and participation of students through enhancing laboratory activity, but by providing immediate accurate results and reducing laboriousness of work.[6] [8]

## V. ROLE OF TEACHERS, STUDENTS

Teachers have to recognize the potential benefits of computer supported technical teaching and learning and its specific role in meeting their classroom aspirations. Most of the experienced teachers are not fully aware of ICT and its use in their teaching. The advantage to the teachers that they can relate their course material with real time examples present on Internet. Well-integrated and effective classroom use of ICT is currently rare. Teachers tend to use ICT largely to support, enhance and complement existing classroom practice rather than reshaping subject content, goals and pedagogies. Teacher should make aware and encourage students to use the technology in effective manner to support their learning to ask lots of questions about data collected and to find their own solutions and interpretations.

Students can collect more relative information and have interactive classroom and lab sessions. Students might be encouraged to compare sets of data; they can look at each other's solutions and discuss the differences and similarities or compare it with standard sample data. They might take a broader view of what constitutes relevant and useful information. [6] [5]

## VI. RETURN ON INVESTMENT

**1. Real-time learning.**

E-Learning offers real-time learning and application of critical knowledge. Knowledge will no longer need to be taken from the shelf of the training department, brushed off, and reviewed. E-Learning is immediate and provides up-to-date information.

**2. Learner-centric training.**

E-Learning changes the focus of training from traditional instructor-centric to learner-centric training. This is how training and learning should be done. E-Learning is tailored to the learners' professional responsibilities and capabilities, creating relevant application to their immediate and future needs. A learner and his needs should be the sole focus and goal of any training or educational program.

**3. Attract, train and retain.**

The most important asset in an organization is its' knowledge workers. The shortage of skilled workers is global. Research shows that the number one reason for loss of key employees is that they feel their organization hasn't invested sufficient resources for their professional development. E-Learning not only addresses the workers' need to develop new knowledge and skills, but also provides learning-on-demand.

**4. Ownership and Empowerment.**

E-Learners are responsible for their own learning. E-Learning empowers them to manage and implement their own learning and development plans. Ownership of learning is crucial for individual growth and retention of employees. Empowerment creates learner ownership and direction – leading to powerful learning and growth potential. E-Learning gives the E-learner the ability to measure their progress and assess their 'gap' in desired skills.

**5. Simulation.**

We learn by seeing and doing. E-Learning introduces a truly innovative way of simulating each learning experience or event with content and ideas provided by some of the leading professionals in the world. Simulation also introduces the required, 'interactive' part of learning – interaction and participation with a local or global audience.

**6. Anytime and anywhere.**

One difficult and costly process of traditional training is coordinating travel, resources, materials, classroom settings, or seminar training for a global workforce. No longer is it necessary to dedicate critical resources to plan, coordinate and manage travel, reservations, rentals, and equipment for each learner and event. The reality of training in a virtual information classroom, across continents, is now possible – anytime, anywhere.

**7. Cost effective.**

Costs can be applied to each learner, and results can be measured against the incurred costs. More importantly,E-Learning is less intrusive to the daily work duties and schedule of the organization and learner, saving time and money through less interruption of the learner's regularly scheduled duties. This cost effective training is tangible ROI – immediately recognized by the organization.

## VII. ORGANIZATIONAL HURDLES

Organizations face following hurdles in managing and understanding their knowledge workforce: knowledge isn't static and neither is the knowledge workforce.

-Knowledge, such as 'best practices' can and will become obsolete – some knowledge has little or no shelf life at all.
-Knowledge is not only Internet mobile, but mobile with each employee.
- Initial investment for ICT is high for which management of the technical institutes is not supportive
- Resistance to change from workforce.
-Lack of time to gain confidence and experience with technology;
- Limited access to reliable resources;
- Technical curriculum overloaded with content;
-Assessment that requires no use of the technology and lack of subject specific guidance
for using ICT to support learning. [8]

## VIII. BITS VIRTUAL UNIVERSITY-A CASE STUDY

One of the best models, which is already implemented, is by BITS Virtual University. To make its model of distance learning scalable, reachable to wider audience and leverage the benefits of emerging technologies, BITS conceived and designed the BITS Virtual University (VU).This VU project envisages the design and development of multimedia course that is web-enabled that goes towards the curriculum of a full-fledged degree program. This enables people who are off-campus to avail of the facilities offered to a normal on-campus student registered under the same program. Since these courses are web-enabled, the student can work in his own comfortable environment and is not restricted to the classroom. Moreover he can work at a convenient pace. All the students registered for a semester are given a login and a password. Using this student can access course materials. Prescribed textbooks and materials which will be useful to the student are made available separately. Multimedia based Soft-Teachers are deployed for explaining concepts. Some innovative methods of using Java based "concept applets" for educational resource development have been used. An attempt has been made to simulate classroom teaching so that it is easier for the student to understand the course. To give the student a feel of lab environment, an introductive virtual lab framework has been designed, which can be reused for certain categories of practice-intensive courses. Desktop IP based Video-Conferencing, Scheduled Video over IP and Video-on-demand over IP facilities are available as integral components of this learning support system.

Although any concept offers lots of advantages it is unlikely that it will replace the traditional learning system. After seeing this model few questions still remain unanswered.

> Most of the models outlined emphasize meeting immediate market demands for course work as well as treating students primarily as 'customers'. One concern is that the pure 'student as consumer model' rests on questionable assumption that they know what they want when they enroll in an educational program and confidently decide what they want.

> Although E-learning gives more advantages to the students' along with knowledge there is requirement of communication to express their thoughts and ideas to the teacher. Therefore, there has to be synchronization in the time of interaction for exchange of views.

> E-learning infrastructure with proper security (cyber laws) issue is the minimum requirement for E-learning movement

among the people of sub-urban, rural areas of the country, which demands for capital.
> To design any new system one of the important factors, which cannot be neglected, is proper management. Therefore more and more management courses are required to be adopted in UG level program. CCIR is one of the institutes making management revolution through 'Information Integrity Program'. IIT- Delhi, SNDT–Mumbai and now Pune University is making an attempt to introduce this concept. [1] [2]

## IX. CONCLUSION

This paradigm shift from student-filled, single teacher-directed classrooms to "teacherless", boundary-less, time-optimised learning or schooling is causing a degree of stress to major stake holders in the educational community. Thus the advent of virtual classrooms, blogs, video classrooms and teleconferencing had simplified learning irrespective of the age group, competence and education level of learners. For effective implementation, and therefore acceptance of the use of technologies, educators might view this shift to be like all educational changes of value which require new skills, behaviours, and beliefs or understandings. The emerging technologies of ICT can have a positive effect on the educational system if we recognize that change is a journey, not a blueprint and that the development of new skills, behaviors, and beliefs is a complex process that must embrace the problems inherent in change. We must keep ourselves ready for change.

## REFERENCES

[1] M.P.Dale, K.H.Munde, "Potential role of ICT in supporting technical education", paper presented at *XXXVI ISTE Annual Convention & National Seminar 14-16 December 2006*

[2] S.N.Dharwadkar, M.I.Agnani ,"Vision for education-the new era ", paper presented at *XXXVI ISTE Annual Convention & National Seminar 14-16 December 2006*

[3] N.Thamil Selvi, "Emerging learning services & architecture", paper presented at *XXXVI ISTE Annual Convention & National Seminar 14-16 December 2006*

[4] Dr.Srinivasa Pai P.,Dr.Niranjan N.Chiplunkar, "Role of information technology in enhancing the effectiveness of Continuing education ", paper presented at *XXXVI ISTE Annual Convention & National Seminar 14-16 December 2006.*

[5]A.Subramaniam,D.Kothandaraman,R.Sabitha,M.Suresh Babu, "Emerging learning system e-learning ",paper presented at *XXXVI ISTE Annual Convention & National Seminar 14-16 December 2006*

[6]V.Rajasekaran,A.Palaniappan, "Evolving technologies in distance learning ", paper presented at *XXXVI ISTE Annual Convention & National Seminar 14-16 December 2006*

[7] Q.S.Zakiuddin, "Modern learning Environment ", paper presented at *XXXVI ISTE Annual Convention & National Seminar 14-16 December 2006*

[8] Dr.K.Sukumaran,Dr. M.V.Srinath, "Mobile learning: an innovative learning method of life long learning ", paper presented at *XXXVI ISTE Annual Convention & National Seminar 14-16 December 2006*

[9] www.Moodle.org

[10] www.blogger.com

[11] www.nicenet.org

# Search for equilibrium state flight

Jaroslav Tupy
Ivan Zelinka

Department of Applied Informatics
Tomas Bata University in Zlin
Nad Stranemi 4511, Zlin 760 05, Czech Republic
jtupy@fai.utb.cz, zelinka@fai.utb.cz

***Abstract* - In this paper, the calculation of aircraft steady state flight optima is discussed using a unique combination of global optimization theory and the direct computer simulation. The methods of artificial intelligence and heuristic algorithms handling with the equations of motion are presented in this paper. The main aim was to apply them in actuating the flying airplane into equilibrium state upon setting the control elements of various kinds to the needed position. New approach of SOMA (Self Organizing Migrating Algorithm) and DE (Differential Evolution) has been used here to find the vehicle stable state. The method can be practically utilized in construction of airplanes, unmanned aircraft as well as the flight simulators design.**

## I. INTRODUCTION

This study of searching for the global optima in aircraft dynamics is concerned with how the evolutionary algorithms may bring forward new visions of modern engineering in the aircraft industry. Aviation, with its core responsibility for passengers' lives and traveling comfort, laid the same requirement of devotion and robustness on all tools utilized in flying vehicle design and manufacture.

This paper describes the procedure of how the heuristic approaches in minimization of nonlinear differential equation model (of the 7D+ complexity) can successfully compete with the classical numerical optimization methods. Those are SOMA (Self Organizing Migrating Algorithm) and DE (Differential Evolution) from the class of EA (Evolutionary Algorithms). The other party represents the simplex method or highest gradient method.

Many aircraft designers were centuries longing for a rigid optimization method capable to answer the greatest challenge in their business – to design the lightest possible airplane, which would be sufficiently firm and secure to fly. Studying the nature of flight dynamics (Etkin, 1995) was crucial for construction of the airplane control elements efficient in driving the vehicle by human forces. No flying object can stop its movement immediately, high above the ground, without fatal consequences. In addition the aircraft flies through the airspace in three axes, so in vertical, horizontal and lateral directions. And there are always forces and moments of inertia acting and influencing its 3D flight pathway. Here the optimization method should find the steady state flight stability for a set of external conditions and all available control elements positions to keep the aircraft stable in maneuvers. The flight situations are necessary to be investigated within whole envelope of permissible altitudes, speeds of flight and changing body masses and completed with disturbed situations of uneven engines thrust (different revolves, one engine stall) and/or the flight under side wind conditions.

Zelinka, Lampinen (1999) made first research studies comparing the overall global optima search algorithms, including the genetic algorithms, with evolutionary algorithms, stating no absolute priority to any of them, but showing a systematic contrast, confrontation and comparison between them.

Very thorough analysis of evolutionary algorithms potential in global minima search was described by Zelinka (2002), with conclusion of suitability of SOMA and DE, as an heuristic approach, to start winning over deterministic ones with their robustness and high likelihood of local extreme hang-up problem outmaneuvering, finishing in the required global minima.

Neumaier (2004), Hamacher (2006) discovered that modified approach of Stochastic tunneling can provide also sufficiently reliable results in continuous minima search, however due to FAI UTB intensive research in evolutionary algorithms field, we realized to perform the aircraft trim analysis with SOMA, which has got also improved modifications of adaptive search strategies.

The overwhelming number of aircraft designers and manufacturers utilize, maybe from traditional reasons, the methods based on direct numerical calculations.

This approach consists in a process of systematical iterations through the whole space of potential solutions, which, however, can cause problems: very long and sometimes even interminable computation time, substantial simplification by the model linearization or omission of "minor" relations and thus not full matching the real situation, need of the aircraft/method specialist who changes the data and even the algorithm "en route", differences in algorithms "made-to-measure" for different airplane models. The worst of all, those methods often end in the local extreme without even touching the area where the global extreme lies.

The present research we perform at FAI aims to prove that the heuristic algorithms can prove to have considerable

T. Sobh (ed.), *Innovations and Advances in Computer Sciences and Engineering*,
DOI 10.1007/978-90-481-3658-2_9, 

advantages in global optimization. The biggest difference from the classic approach is utilizing the random element in the scanned through pathways. SOMA offers to find the functional hyper minimum always and within the reasonable time.

What are the undisputable and by our research supported advantages of such approach:

The algorithm from the point of view of service is easy to run. The specialist deals „only"with three elements:

preparation of the data – wide complex database - exactly describing the model

suggestion of the cost function properly fitting the extreme valuation

appreciation of results – determination the constraints and interpretation of the numbers for the design, simulation and alteration the airplane behavior.

It handles full-fledged nonlinear system of the equations of motion including perturbations and noises, however for higher Mach numbers needed for jet planes analysis these nonlinearities play substantial role and thus it is important to reflect them.

Only the ruling parameters of the algorithm can change. The algorithm itself remains in a form of an independent routine unchanged.

Behavior of the cost function as well as the precise position of the extreme is easy to study on subspaces of the hyper space of possible solutions and as a result facilitate the verification of conformity in behavior of the model and the airplane.

## II. METHODS

In order to investigate the aircraft stability - with the view of finding the well satisfying function describing the aircraft handling, we examined its occurrence in specialized papers, professional books and on the internet. The main topic description appearing in the AIAA („American Institute for Aeronautics and Astronautics") Education series books were utilized namely:

- Performance, Stability, Dynamics & Control of Airplanes by Bandu N. Pamadi (NASA),
- Flight Dynamics Principles by Michael V. Cook, and
- Aerodynamics for Engineers by John J. Bertin.

Each occurrence of the topic, moreover divided to static and dynamic stability, was identified, highlighted and then saved in a computer thesis textual database, including explanatory pictures and graphics. Using the software Mathematica 6.0 (by Wolfram Research) we developed a stability simulation program – Trim algorithm - capable to make the optimization analysis based on SOMA and DE algorithms.

### *Trim Method*

The non-linear state equation describing the general rigid-body dynamics is:

$$x' = f(\boldsymbol{x}(t), \boldsymbol{u}(t), \boldsymbol{v}(t), t)$$

with state vector $\boldsymbol{x}$, input vector $\boldsymbol{u}$, and external disturbance vector $\boldsymbol{v}$.

The *singular point* or *equilibrium point* of a time-invariant system with no external control inputs is defined as:

$$f(x', x, \boldsymbol{u}) = 0, \text{ with } \boldsymbol{x}' = 0 \text{ and } \boldsymbol{u} = 0 \text{ or constant}$$

The system is "at rest" when all of the time-derivatives are identically zero.

**Steady-state flight** can be determined as a condition in which all of the motion variables are constant or zero and all acceleration components are zero. The definition allows finding

steady wings-level flight

steady turning flight

wings-level climb

a climbing turn

steady pull-up or push-over, and

steady roll.

So steady-state flight can be defined in terms of the state variables of the flat-Earth equations:

$V', \alpha', \beta', p', q', r', = 0, \ \boldsymbol{u} = constant$

which are limited by the following constraints:

steady wings-level flight: $\varphi, \varphi', \theta', \psi' = 0$ *(i.e. p, q, r = 0)*

steady turning flight: $\varphi', \theta' = 0, \quad \psi' = turn\ rate$

steady pull-up: $\varphi, \varphi', \psi' = 0, \quad \theta' = pull\text{-}up\ rate$

steady roll: $\theta', \psi' = 0, \quad \varphi' = roll\ rate$

The conditions p', q', r' = 0 require the angular rates - and therefore also the **aerodynamic** and **thrust moments** - to be zero or constant. The conditions $V', \alpha', \beta' = 0$ require the aerodynamic forces to be zero or constant. For this reason, the steady-state pull-up / push-over and steady roll conditions can only exist instantaneously.

To find a steady-state flight condition, a set of non-linear simultaneous differential equations, derived from the state model, must be solved. Due to the very complex functional dependence of the aerodynamic data, it is in general not possible to solve these equations analytically. Instead, a numerical algorithm must be used to iteratively adjust the independent variables until a solution criterion is met. The solution will be interpolated, but can be made arbitrarily close to the exact solution by tightening up the criteria.

The trim algorithm presented here will deal with the aircraft model only through its input and output signals. It does not have to work within the model to balance forces and moments separately, which makes the **trim-routine generally applicable**. Hence, any aircraft model using the same concept of input and state vectors can be trimmed with the same program, the internal structure of the aircraft model does not matter.

The structure of the equations of motions is:

$$\begin{bmatrix} \dot{p} \\ \dot{v} \\ \dot{\omega} \\ \dot{q} \end{bmatrix} = \begin{bmatrix} \Omega_E & B^T & 0 & 0 \\ -B\Omega_E^2 & -(\Omega_B + B\Omega_E B^T) & 0 & 0 \\ 0 & 0 & -J^{-1}\Omega_B J & 0 \\ 0 & 0 & 0 & -\frac{1}{2} \end{bmatrix} * \begin{bmatrix} p \\ v \\ \omega \\ q \end{bmatrix} + \begin{bmatrix} 0 \\ Bg(p) + \frac{F_B}{m} \\ J^{-1}M_B \\ 0 \end{bmatrix}$$

Where $\boldsymbol{\Omega}_E$, $\boldsymbol{\Omega}_B$, and $\boldsymbol{\Omega}_q$ – are coefficient matrixes of angular speeds, $\boldsymbol{\varphi}$, $\boldsymbol{\theta}$, $\boldsymbol{\psi}$ = roll angle, pitch angle, and yaw angle, **p**, **q**, **r** = roll rate, pitch rate and yaw rate, **B** – transformation matrix of ECI into airplane coordinate system, **J** – inertial momentum matrix and $\mathbf{F}_B$ – vector of forces – sum of aerodynamic, engine and undercarriage forces acting on the airplane.

As for the method itself the Nondeterministic Polynomial Time – Complete Problems (NP-C) are a set of tasks which are extremely difficult to solve. The time which is needed increases exponentially or with factorial in accordance with the dimension of the potential solutions space. Very sophisticated method how to challenge such problems is based on evolutionary principles – genetic algorithms (GA), stochastic (randomized) algorithms (SA), evolutionary algorithms (EA) and similar. Evolutionary algorithms are heuristic optimization algorithms that were inspired by the Darwinian Evolution Theory of Species. In these algorithms a candidate for the best optima is an individual in a population, and the cost function determines the environmental conditions criteria under which the individual should live and survive. Each of iteration of the EA involves a competitive selection that rejects the poorest individuals out and adopts individuals with better skills in the population. The candidates with highest fitness (appraised by the cost function value) are recombined with other individuals by mutual swapping their parts (sub chains). Such method is called Differential Evolution (DE). The resultant chains are additionally subject to mutations by a small change in their elements. Recombination and mutation are the evolutionary drivers for creation of the new populations that are biased towards areas of the hyper-space, for which better fitness values have already been seen.

Additional example of EA algorithm is SOMA (Self-Organizing Migrating Algorithm) – a highly effective method of hyper-functional optimization. Its set of hypothetical solutions compose a group of individuals (population) traveling through the N-dimensional hyper-space of potential solutions aiming to find the best optima. During the migration moves the individuals tend to „organize"itself within the extreme area making the final solution more accurate.

## III. RESULTS

Investigating the aircraft stability and designing an algorithm capable to provide required minima location was a complex task. We made such work for the Czech aircraft industry submitter to prove newly designed optimization methods (Zelinka, Soma & DE 2005) and show their efficiency and viability.

We performed a deep mesh analysis on the data envelope given by the outer limits of the aircraft flight speed, flight altitude and the side wind, analyzed for varying masses of the aircraft body.

After the program intense run we compared the obtained results with other (classical) methods results obtained from our submitter by his own computational tools using the traditional convex gradient method.

The initial mesh was stated for 3 variants of mass (5000, 5500 and 6000 kg), 3 variants of flight speed (250, 400 and 800 km/h) and 3 variants of height level (1500, 4000 and 10000 m). The table of results is as follows:

TABLE 1
INVESTIGATION MESH OF INPUT PARAMETERS AND REACHED AIRCRAFT TRIMS

| Aircraft Trim | TAS Speed [m/s] **70** | | | Side wind [m/s] **0** | | | | | |
|---|---|---|---|---|---|---|---|---|---|
| Altitude | 1 500 m | | | 4 000 m | | | 10 000 m | | |
| Mass [kg] | 5000 | 5500 | 6000 | 5000 | 5500 | 6000 | 5000 | 5500 | 6000 |
| Equilibrium state SOMA | 1.72e-3 | 2.65e-2 | 3.41e-1 | 4.72e-3 | 3.95e-2 | 7.12e-1 | 2.24e-3 | 5.48e-2 | 7.25e-1 |
| SIMPLX | 0.196 | - | 12.345 | 1.531 | - | 8.962 | 1.983 | - | 12.567 |

Objective function used:

$$J = A * |\dot{v}| + B * |\dot{\alpha}| + C * |\dot{\beta}| + D * |\dot{p}| + E * |\dot{q}| + F * |\dot{r}|$$

| Mass | Speed | Altitude |
|---|---|---|
| 4700 | 150 | 5000 |

| Side wind velocity [m/s] | Cost Value | |
|---|---|---|
| | SOMA | Trim |
| 0 | 2.39469e-32 | 2.0891497596641907e-16 |
| 10 | 1.66781e-22 | 5.620165516943417e-16 |
| 20 | 2.27642e-11 | 86.3993 |
| 30 | 5.18040e-11 | 193.88 |
| 50* | 2.91439e-2 | 549.341 |

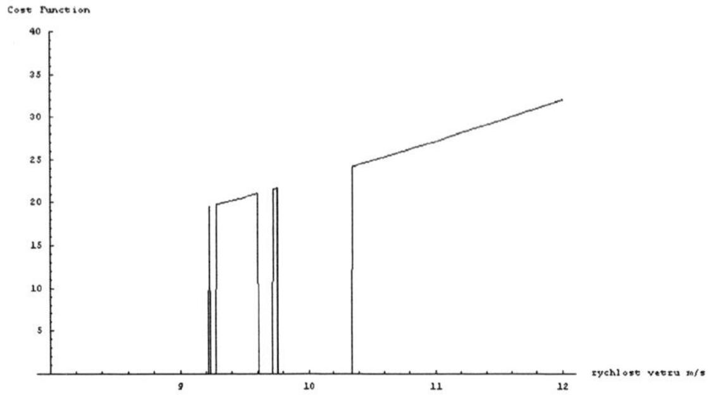


Fig. 1 - Cost Value of the Trim algorithm based on the Side Wind Velocity

| Mass | Speed | Altitude |
|---|---|---|
| 5000 | 100 | 500 |

| Side wind Velocity | Cost Value | |
|---|---|---|
| [m/s] | SOMA | Trim |
| 0 | 8.37814e-85 | 3.115737821596512e-16 |
| 10 | 1.67390e-18 | 22.2359 |
| 20 | 1.54968e-21 | 88.2112 |
| 30 | 2.36506e-2 | 201.284 |
| 40 | 2.11508e-1 | 366.72 |
| 70 | 1.23821e2 | 1300.62 |

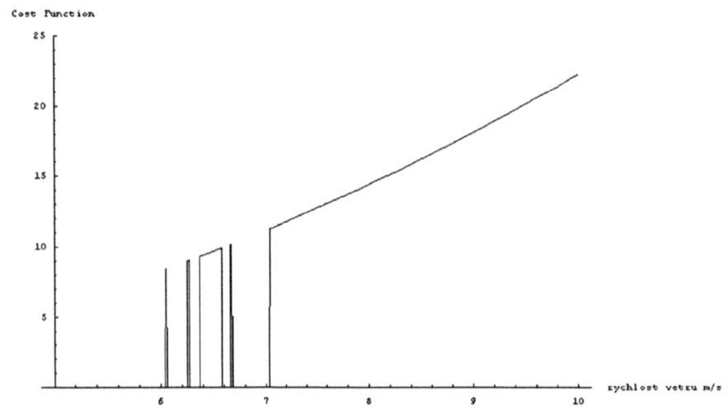


Fig. 2 - Cost Value of the Trim algorithm based on the Side Wind Velocity

| Mass | Speed | Altitude |
|---|---|---|
| 5000 | 200 | 10000 |

| Side wind Velocity | Cost Value | |
|---|---|---|
| [m/s] | SOMA | Trim |
| 0 | 3.10946e-56 | 5.124230669479612e-16 |
| 15 | 4.20329e-10 | 3.4924852535559254e-16 |
| 20 | 3.23727e-8 | 50.1555 |
| 30 | 1.47137e-6 | 111.266 |
| 40 | 1.39449e-6 | 197.701 |
| 50 | 6.35855e-9 | 310.385 |
| 80 | 2.95552e-2 | 820.55 |
| 100 | 1.65702e-1 | 1327.38 |

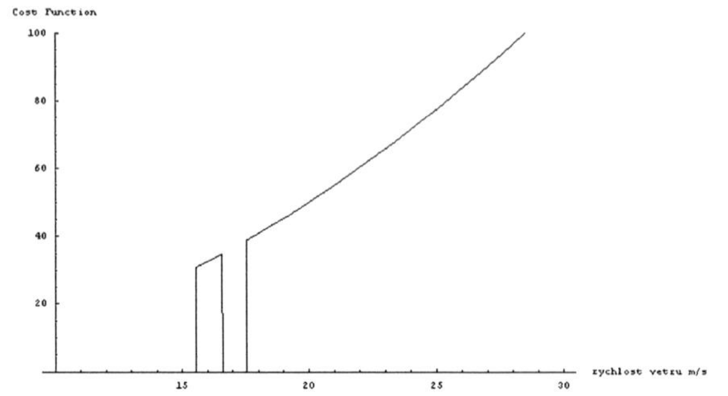


Fig. 3 - Cost Value of the Trim algorithm based on the Side Wind Velocity

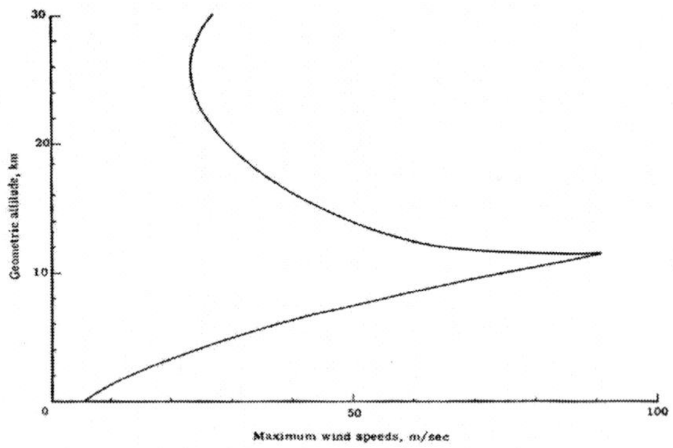


Fig. 4 - Typical statistic maximum Wind Speed envelope.

Used control parameters of SOMA algorithm – version All_To_All_Adaptive:

| PopSize = 20 | Migrations = 35 | Step = 0.4 |
|---|---|---|
| PathLength = 3.0 | PRT = 0.3 | MinDiv = 0.1 |

## *Graphs of relation of the State Vector items and the Side Wind Velocity*

Airplane flight conditions:

| Mass | Speed | Altitude |
|---|---|---|
| [kg] | [m/s] | [m] |
| 4700 | 150 | 5000 |

### Legend:

V' – TAS speed derivative [m/s$^2$]
α' – Angle of attack derivative [rad/s]
β' – Angle of sideslip derivative [rad/s]
p' – Roll rate derivative [rad/s$^2$]
q' – Pitch rate derivative [rad/s$^2$]
r' – Yaw rate derivative [rad/s$^2$]

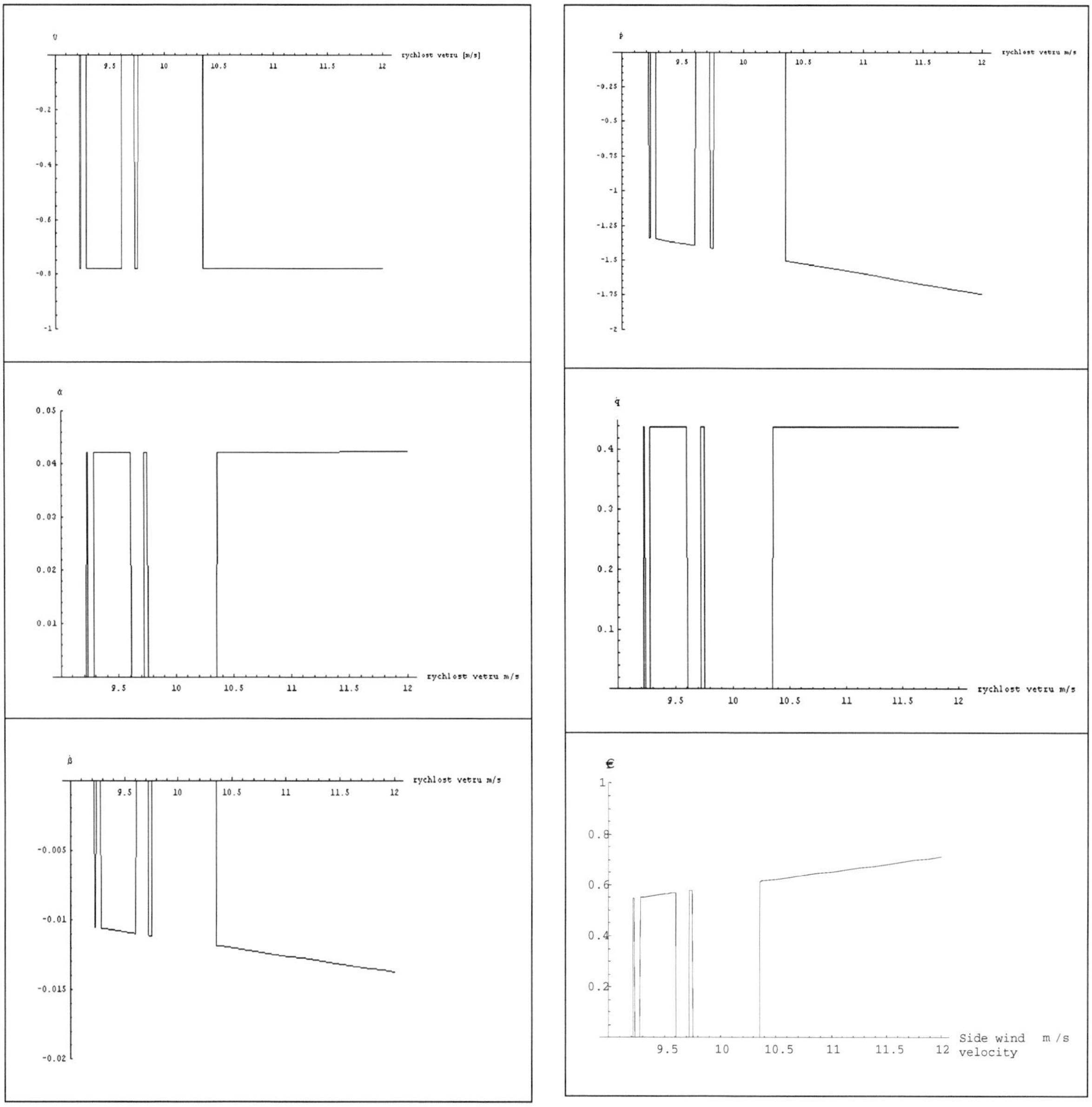


## IV. CONCLUSION

The designed Cost Function $J = A * |\dot{v}| + B * |\dot{\alpha}| + C * |\dot{\beta}| + D * |\dot{p}| + E * |\dot{q}| + F * |\dot{r}|$

works with SOMA properly also under disturbances caused by the side wind, and SOMA in version <u>All to All</u>

Adaptive gives the right results of the airplane steady state flight.

All reached results showed better SOMA performance than the SIMPLEX algorithm. At the classical method several values were even not achievable due to the excessive time demand (2 – 3 days of procedure run), and labor consumption at our submitter. The main criterion of steady state flight has been reached with well accepted accuracy and reliability offering much universal use. This is valid both for number of input variables and more exact approach in adapting the physical rules without simplifying considerations of nonlinear modeling.

The objective of the study was to prove finding of airplane equilibrium state at different flight conditions through exploitation of the heuristic algorithms using the embedded model of equations of motion. We found out that the adaptive versions of SOMA and DE supply effective results in a sufficiently robust way.

Analysis did not prove the All_to_One and All_to_All to be fast and reliable enough for appropriate final results, however, those are convenient for initial calculations to see immediate impact of adjusted ruling parameters. While All_to_All_Adaptive gives very good results in all modeled regimes and deep airplane dynamic equilibrium analysis.

DE was in several cases slightly better than SOMA; however both methods always finished safely in global extreme.

In cooperation with the HGS company (Brno) we performed the computational simulations using their aircraft model data. Results showing the conformity with the real aircraft were presented in a manner of submitting the cost function, numerical data and final graphs. The problem complexity would be more deeply studied for a real aircraft in the near future.

We proved that our Trim algorithm with embedded nonlinear differential equation model and optimization solver SOMA is a sufficiently robust tool, however it's belonging to evolutionary algorithms and thus bringing in the element of coincidence could repel some physicists or old school aircraft experts. That is our task to continue in performing further computational groundwork (as i.e. Hamacher (2006) or Tempo (2007) published lately) and try to continue improving the proposed optima search methods to support the aircraft industry in designing competitive products.

The practical usage of aircraft flight trim solution could help also the Czech aircraft industry as well as small private companies in designing and testing new airplanes as well as building jet fighter flight simulators.

REFERENCES

[1] Lampinen Jouni, Zelinka Ivan, New Ideas in Optimization – Mechanical Engineering Design Optimization by Differential Evolution. Volume 1. London: McGraw-Hill, 1999.20 p, ISBN 007-709506-5

[2] Siddal James N, *Optimal engineering design: principles and application.* Mechanical engineering series / 14. Marcel Dekker Inc. ISBN 0-8247-1633-7

[3] Zelinka Ivan, Vladimir Vasek, Jouni Lampinen, New Algorithms of Global Optimization, Journal of Automation, Czech Ed., 10/01, 628-634, ISSN 0005-125X

[4] Zelinka Ivan, Lampinen Jouni, SOMA - Self-Organizing Migrating Algorithm, Nostradamus 2000,
3rd International Conference on Prediction and Nonlinear Dynamic, Zlin, Czech Republic

[5] R. Storn and K. Price, „Differential Evolution - a simple and efficient heuristic for global optimization over continuous spaces", Journal of Global Optimization, 11(4):341-359, December 1997, Kulwer Academic Publishers

[6] A. Neumaier, Complete Search in Continuous Global Optimization and Constraint Satisfaction, pp. 271-369 in: Acta Numerica 2004 (A. Iserles, ed.), Cambridge University Press 2004.

[7] K. Hamacher. Adaptation in Stochastic Tunneling Global Optimization of Complex Potential Energy Landscapes, Europhys. Lett. 74(6):944, 2006.

[8] Bernard ETKIN, Lloyd Duff Reid. Dynamics of flight – Stability and Control. John Wiley & Sons, Inc, 1996, ISBN: 978-0-471-03418-6

[9] Tempo Roberto. Randomized Algorithms for Systems and Control: Theory and Applications, IEIIT-CNR, Politecnico di Torino, presentation: CTU Prague 2007

# Evaluating acceptance of OSS-ERP based on user perceptions


Salvador Bueno
University of Pablo de Olavide
Ctra. Utrera, km.1
Seville, 41013 Spain

M. Dolores Gallego
University of Pablo de Olavide
Ctra. Utrera, km.1
Seville, 41013 Spain



***Abstract*-Organizations implement Enterprise Resource Planning (ERP) systems with the objective of reaching operational efficiency and the incorporation to new markets through the information flow control on time of the entire organization. However, ERP systems are complex tools, mainly for the small and medium size enterprises (SMEs). For these reason, new ERP configurations have arisen for SMEs such as Open Source Software-ERP (OSS-ERP). OSS-ERP is a research topic barely analyzed by the literature. Specifically, this paper's aim is to focus on the OSS-ERP users' acceptance and use. The authors have developed a research model based on the Technology Acceptance Model (TAM) for testing the users' behavior toward OSS-ERP.**


## I. Introduction

Enterprise Resource Planning (ERP) can be defined as that Information System (IS) which integrates relative organization tools, data and information flows by means of a data base [1] [2] [3]. The actual competitive environment has impelled that many organizations implement ERP systems. Furthermore, organizations implement ERP systems with the objective of reaching operational efficiency and the incorporation to new markets through the information flow control on time of the entire organization.

In spite of these ERP advantages for organizations, the main commercial ERP systems (i.e. SAP, Peoplesoft, Oracle, etc.) have relevant implementation obstacles: (1) high implementation complexity, (2) high average implementation time (3) high software, maintenance and consultation cost and (4) low adaptation flexibility [4]. Due to these characteristics, ERP systems are reasonable mainly for large organizations, although the small and medium size enterprises (SMEs) have the same necessities with respect to the information management and control. For these reason, new ERP configurations have arisen for SMEs such as adapted ERP for SMEs, ERP through application service providers (ASP) or Open Source Software-ERP (OSS-ERP).

OSS was born as an answer to today's shaping of the software market and it has spread out through the main areas related to IS. OSS offers several organizational advantages, such as saving costs related to IS or the capacity to adapt to the changing enterprise requirements. These relevant advantages increase the attraction of technological solutions based on OSS as compared to proprietary technologies.

In spite of the increasing penetration of market of OSS-ERP, scientific studies do not exist on OSS-ERP. In particular, there aren't works that analyze the diffusion and acceptation of OSS-ERP. This research is focused in the OSS-ERP acceptance topic. Our purpose is to define a research model in order to observe the intention of use on OSS-ERP by users. With this objective, we have applied the Technological Acceptance Model (TAM) as the theoretical frameworks which have been formulated the research hypotheses.

The rest of the paper is organized as follows. In the section two, we expose the research context. In the section three and four, the hypotheses, our research model and design are explained. Finally, the findings and discussions are gathered.

## II. Background on OSS-ERP

ERP is an old research field. Since the beginning of the decade of the 70s, the number of studies about ERP has been growing progressively around two research topics: (1) technical factors and (2) organizational impact. Although organizations perceive that ERP are complex tools, ERP are considered a strategic resource [5] and they can provide organizations a high level of competitiveness by means of acquiring a strong market position [6].

From a market view, ERP have covered the old necessity to improve the information control of an entire organization. Recently, ERP systems implementations have been growing strongly [7] [8] [9] [10] [11]. Some authors, such as [12], affirm that the ERP market will reach one trillion of dollar in the year 2010. However, this growth would not be possible without the opening towards SMEs. This kind of companies considers to traditional ERP systems as expensive and complex tools with a high impact in organizational structures, process and culture [13]. For that reason, ERP complexity management can be considered a crucial activity [14] [15].

Based on this point of view, OSS-ERP reduce the disadvantages of ERP for the SMEs. OSS was born as an answer to today's shaping of the software market [16]. The strict translation of open source makes reference to the possibility to have access to source, whether the software is free (of charge) or not [17]. Actually, we have to indicate that the basis for the consolidation of OSS as an alternative to proprietary software is three-fold [18] [19].

Besides, OSS-ERP has three relevant advantages to organizations: (1) increased adaptability, (2) decreased reliance on a single supplier, (3) reduced costs [20]. Generally, OSS-ERP include the necessary functions to manage integrally all the activities of a company. Due to their high flexibility, these tools can be adapted the client's specific needs. Furthermore, as OSS-ERP are based on open software technologies, organizations are not put under the payment


T. Sobh (ed.), *Innovations and Advances in Computer Sciences and Engineering*,
DOI 10.1007/978-90-481-3658-2_10, © Springer Science+Business Media B.V. 2010

of licenses or exclusive contracts. With the collaboration of partners, OSS-ERP vendors receive benefits for the support services.

In the website Sourceforge.net, we can identify 2058 projects about OSS-ERP, although all has not had an impact in ERP market. This data is an indicative of the increasing relevance of this enterprise solution for organizations. In this moment, OSS-ERP vendor with greater diffusion are Compiere, Openbravo, Abanq (before FacturaLux), ERP5, Tiny ERP, Fisterra, OFBiz, SQL-Ledger and WebERP.

One of the key factors for fomenting the diffusion of OSS-ERP is the users' acceptance. OSS-ERP can produce relevant change in organizations during the implementation stage. In this sense, one of the main efforts of the change management is the promotion of actions that increases OSS-ERP users' acceptance. OSS-ERP success or failure will be largely determined by users' acceptance degree. One after another, it depends on the complexity of the transformations that OSS-ERP incorporate in organizations when they are implemented.

## III. TAM METHODOLOGY AND HYPOTHESIS

The TAM model developed by [21] has been widely applied with the purpose of understanding the conduct and motivational factors that influence IS adoption and use. TAM is based on the Theory of Reasoned Action (TRA) proposed by [22]. TAM tests the users behavior toward IS, based on the perceived usefulness (PU), perceived ease of use (PEU), attitude toward use (ATU) and behavioral intention to use (BIU). PU is "the degree to which a person believes that using a particular system would enhance his or her job performance", and PEU as "the degree to which a person believes that using a particular system would be free of effort" [23].

Additionally, TAM postulates that BIU depends of IU of an IS [21]. ATU are based on PU and PEU, although [21] [23] doesn't detail what factors are exogenous variables. In relation to this aspect, [24] indicates three large groups of external variables: (1) regarding the user, (2) regarding the organization and (3) regarding to IS. On the other hand, TAM was designed to improving the measures for prediction and explanation of IS use [25].

Based on the TAM model proposed by [23], we formulate the first working hypotheses of this research which make reference to the intrinsic constructs of the model. These hypotheses are stated in the following way:

H1. PEU for OSS-ERP systems has a positive effect on PU.

H2. PU for OSS-ERP systems has a positive effect on ATU.

H3. PEU for OSS-ERP systems has a positive effect on ATU.

H4. PU for OSS-ERP systems has a positive effect on BIU.

H5. ATU for OSS-ERP systems has a positive effect on BIU.

## IV. RESEARCH MODEL AND DESIGN

The proposed research model, which is graphically displayed in Fig. 1, summarizes all the aforementioned formulated hypotheses. With such a model, we propose to uncover which factors influence the acceptance of OSS-ERP by users. In order to proceed in confirming the hypotheses, we have designed a field study as the necessary tool to obtain the information that would allow us to carry out this test. The election process of the sample and the instrument validity are detailed below.

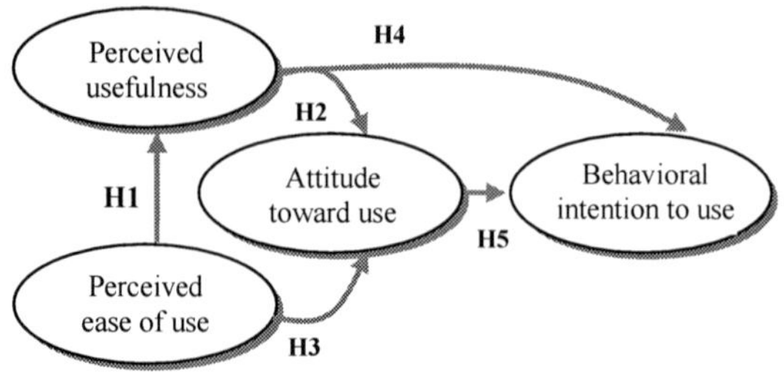


Fig. 1. Research model and hypotheses.

### A. Sample

For our research, we have taken OSS-ERP users as the individuals that will form the sample for our study. We turned to Sourceforge website (http://sourceforge.net/index.php), where we were able to access to 703 contact information of the users that were registered in the OSS-ERP projects. Also, we have contacted with registered users in the forums and mailing list of the vendors Tiny ERP, ERP 5 and JFire. We sent an invitation letter by e-mail with the URL address where the questionnaire was published. In the end, we received 160 survey responses. Of those, 128 were complete and valid for our study. This number represents a response rate of 18.06%. The Table I shows the sample information.

### B. Survey design

In order to measure each one the variables included in the TAM model developed for our study (Fig. 1), we carried out a review of the literature that allowed us to identify items for each one of the constructs. We finally included 15 items in the survey (see Table II).

TABLE II
ITEMS

| Items | Source |
|---|---|
| **PU1**-Using the system in my job enabled to accomplish tasks more quickly. | [25] |
| **PU2**-Using the OSS-ERP improves my performance in my job. | [46] |
| **PU3**-Using the OSS-ERP in my job increases my productivity. | [47] |
| **PU4**-Using the OSS-ERP enhances my effectiveness in my job. | |
| **PEU1**-My interaction with the OSS-ERP systems is clear and understandable. | |
| **PEU2**-Interacting with the OSS-ERP does not require a lot of my mental effort. | [25] [46] |
| **PEU3**-I find the OSS-ERP to be easy to use. | |
| **PEU4**-I find it easy to get the systems to do what I want it to do. | |
| **ATU1**-The OSS-ERP will provide access to more data. | |
| **ATU2**-The OSS-ERP will make data analysis easier. | |
| **ATU3**-The OSS-ERP will be better than the old system. | [25] [44] [46] |
| **ATU4**-The OSS-ERP will provide accurate information. | |
| **ATU5**-The OSS-ERP will provide integrated, timely and reliable information. | |
| **BIU1**-I expect to use the new system. | [25] [44] [46] |
| **BIU2**-I expect the information from the new system to be useful. | |

A seven point Likert-type scale was used in the questionnaire from (1) "strongly disagree" to (7) "strongly agree". This broad scale allows users a wide range of possible answers to correctly express their opinion and has been used in other studies similar to ours, for example [26] utilized TAM to study from a user acceptance perspective about web-based negotiation support systems; [27] examine the adoption of WebCT using TAM; [28] investigates how customers perceive and adopt Internet banking developing a theoretical model based on TAM; or [29] developed a study of consumer acceptance of the Internet as a channel of distribution in Taiwan using TAM.

TABLE I
SAMPLE INFORMATION

| | N | % |
|---|---|---|
| **Age** | | |
| < 25 | 10 | 7.8125 |
| 25-30 | 35 | 27.34375 |
| 31-40 | 39 | 30.46875 |
| 41-50 | 27 | 21.09375 |
| 51-60 | 13 | 10.15625 |
| >60 | 4 | 3.125 |
| Total | 128 | 100 |
| **Gender** | | |
| Male | 125 | 97.65625 |
| Female | 3 | 2.34375 |
| Total | 128 | 100 |
| **Education** | | |
| Less than high school | 2 | 1.5625 |
| High school graduate | 20 | 15.625 |
| College/university graduate | 51 | 39.84375 |
| Master's | 49 | 38.28125 |
| Ph.D. | 6 | 4.6875 |
| Total | 128 | 100 |

### C. *Instrument validity*

Before conducting the main survey, we carried out a pre-test and a pilot test to validate the questionnaire. For the pre-test, various research experts in the application of TAM methodology revised the survey structure as well as its explanatory capability. They examined aspects such as the study invitation letter, the appropriateness of the questions and answers, the indications in each part of the survey and the survey length. Afterwards, a pilot test took place which included twenty-five OSS-ERP users.

Once the explanatory capability of the survey was verified, it was necessary to test that the selected items were capable of explaining the associated constructs. For this, Cronbach´s alpha test was applied to the groups of items for the constructs of the model, based on the data obtained in the pre-test (see Table III). In our particular case, all the variables reached very satisfactory levels, going above the recommended level of 0.7 [30]. This circumstance demonstrates that the items are capable of measuring the variables for which they were selected. These results were expected, given the fact that the selected items were taken from similar studies whose predictive capability had been demonstrated.

TABLE III
RELIABILITY COEFFICIENTS

| Construct | Cronbach's α |
|---|---|
| Perceived usefulness (PU) | 0.983 |
| Perceived ease of use (PEU) | 0.887 |
| Attitude toward use (ATU) | 0.705 |
| Behavioral intention to use (BIU) | 0.918 |

## V. PRELIMINARY ANALYSIS

Before proceeding to evaluate the validity of the scales proposed, we consider it appropriate to present a few of indexes used within descriptive statistics, such as (N), referring to the number of responses, mean and standard deviation (S.D.). Each one of the items conforms to the constructs in our study (see Table IV). This analysis has allowed us to reveal the averages for each one of the items included for each constructs in this study.

TABLE IV
SUMMARY OF MEASUREMENT SCALES

| Items | N | Mean | S.D. |
|---|---|---|---|
| **PEU1** | 128 | 2.102 | 1.222 |
| **PEU2** | 127 | 2.811 | 1.632 |
| **PEU3** | 128 | 2.438 | 1.338 |
| **PEU4** | 128 | 2.625 | 1.490 |
| **PU1** | 127 | 1.945 | 1.274 |
| **PU2** | 127 | 1.953 | 1.126 |
| **PU3** | 126 | 2.024 | 1.249 |
| **PU4** | 127 | 1.976 | 1.185 |
| **BIU1** | 128 | 1.688 | 1.209 |
| **BIU2** | 128 | 1.570 | 1.284 |
| **ATU1** | 126 | 1.778 | 1.226 |
| **ATU2** | 126 | 1.857 | 1.319 |
| **ATU3** | 126 | 1.746 | 1.213 |
| **ATU4** | 125 | 1.824 | 1.115 |
| **ATU5** | 126 | 1.817 | 1.176 |

From the Table IV we can appreciate the high response rate for all the items included in the study, all close to one-hundred percent. This information allows us to believe that the survey is easy to understand and that the participating users didn't have any difficulty in responding to the majority of the consulted variables. Likewise, we can say that the typical deviations are acceptable for all the variables considered.

## VI. ANALYSIS AND FINDINGS

In order to agree upon all the hypotheses collected in the investigational model, an exploratory factor analysis (EFA), a confirmatory factor analysis (CFA) and finally, a causal analysis have been developed. The statistical analysis of the causal model will be carried out with the software Liserl 8.51 (2001).

### A. *Exploratory factorial analysis.*

The EFA was done to identify those items that had a strong impact on its constructs [31] as well as to reduce the number of items to be used in later analyses. This EFA was done separately for each one of the nine constructs included in our model and enabled us to make an analysis of the converging validity of these variables and its dimension.

The EFA was calculated using the SPSS 15.0 statistical software. The extraction method selected was Principal Axis Factoring, as this is the most appropriate method for identifying the constructs [31]. We extracted those variables whose self-values were greater than one. Thus, we eliminated the factors which had a low factorial charge and whose exclusion from the study allowed us to obtain a Cronbach's alpha greater than the recommended minimum of 0.7 [30]. Likewise, the EFA (see Table V) carried out for each one of the constructs of the model has been done by means of the Kaiser's Varimax Rotation in order to determine the unidimensionality of the scales [32], [33].

The EFA was able to demonstrate the unidimensionality of all the constructs in the study. Likewise, we can observe that the constructs perceived ease of use, perceived usefulness, behavioral intention to use, attitude toward use (see Table V) far surpass the recommended minimum value with respect to Cronbach´s alpha of 0.7. Also, all items have a good factorial charge and for this reason we recommend taking them for testing the model.

TABLE V
EXPLORATORY FACTOR ANALYSIS

| Construct | Items | Loading | Cronbach's α |
|---|---|---|---|
| PU | PU1 | 0.924 | 0.965 |
| | PU2 | 0.943 | |
| | PU3 | 0.956 | |
| | PU4 | 0.918 | |
| PEU | PEU1 | 0.780 | 0.881 |
| | PEU2 | 0.728 | |
| | PEU3 | 0.927 | |
| | PEU4 | 0.824 | |
| BIU | BIU1 | 0.845 | 0.811 |
| | BIU2 | 0.845 | |
| ATU | ATU1 | 0.857 | 0.914 |
| | ATU2 | 0.884 | |
| | ATU3 | 0.755 | |
| | ATU4 | 0.868 | |
| | ATU5 | 0.769 | |

## B. *Confirmatory factor analysis.*

The CFA was carried out using the Lisrel structural equations software [34]. This software has been used in other studies similar to ours, such as [26] [35] [36] [37] [38] [39].

The CFA was carried out using maximum likelihood robust statistical method. The evaluation of the adjustment is first carried out for the measurement model, verifying the statistical significance of each charge obtained among the indicator, the construct and the measurement reliability of each one of the constructs included in our model. The measurement analysis shows clearly that all standardized lambda coefficients are superior to the minimum value allowed of 0.5.

Regarding the discriminatory validity of the scales, our objective was to determine that each factor represents a separate dimension. This was done by means of the standardized linear or covariance correlations among the constructs. The results show discriminatory validity indexes among the different analyzed dimensions as they take values far from one [40]. Once the covariance correlations are squared, the variance quantity extracted became less and thus, we were then able to guarantee the discriminatory validity of the constructs. In order to study in depth this validity, we made sure that the correlation confidence interval between each pair of constructs didn't have a value of one, demonstrating that these factors represent notably different concepts [41].

Once the discriminatory validity of the scales was demonstrated, and before proceeding to the CFA results interpretation, it was necessary to determine the adjustment fits of the estimated model, using the indexes indicated in Table VI. The indicators used went above the maximum limits established by [31].

TABLE VI
OVERALL FITS OF MODELS

| Fit index | Results | Recommended value |
|---|---|---|
| Chi-square/grade of freedom | 1.47 | ≤3.00 |
| Normed fit index (NFI) | 0.917 | ≥0.90 |
| Non-normed fit index (NNFI) | 0.965 | ≥0.90 |
| Comparative fit Index (CFI) | 0.971 | ≥0.90 |
| Adjusted goodness-of-fit index (AGFI) | 0.834 | ≥0.80 |
| Root mean square error of approximation (RMSEA) | 0.048 | ≤0.05 |
| Goodness-of-fit index (GFI) | 0.903 | ≥0.90 |
| Incremental fit index (IFI) | 0.972 | ≥0.90 |
| Parsimony Normed Fit Index (PNFI) | 0.742 | >0.5 |

Upon verifying the discriminatory validity of the scales and the model adjustment fit, the CFA of the whole model was carried out showing an adequate specification of the proposed factorial structure in the results (see Table VII).

The proposed reliability of the measurement scales was evaluated from the Cronbach´s alpha and Composite Reliability (Cr) coefficients [42]. We declare that a model possesses internal consistency when the composite reliability reaches values greater than 0.7 for all the constructs defined in the model.

In order to estimate the discriminatory validity of the model, we have calculated the average variance extracted (AVE) proponed by [43]. On one hand, the standardized factorial charges are statistically significant, around 0.7 and with individual reliabilities at or above 50%. This demonstrates the converging validity of the measurement scales. This information, together with the strong Cronbach alpha, provides sufficient evidence for the internal consistency of the measurements [31]. Besides, all constructs surpass the recommended AVE value of 0.5 (see Table VII) and all AVE square roots surpass the correlations with the rest of the constructs.

TABLE VII
SUMMARY OF MEASUREMENT SCALES

| Construct | Lambda stand. | $R^2$ | Composite reliability | AVE | Cronbach's α |
|---|---|---|---|---|---|
| **PEU** | | | | | |
| PEU1 | 0.790 | 0.624 | 0.893 | 0.676 | 0.899 |
| PEU2 | 0.773 | 0.598 | | | |
| PEU3 | 0.859 | 0.738 | | | |
| PEU4 | 0.841 | 0.743 | | | |
| **PU** | | | | | |
| PU1 | 0.881 | 0.776 | 0.952 | 0.832 | 0.966 |
| PU2 | 0.968 | 0.937 | | | |
| PU3 | 0.934 | 0.872 | | | |
| PU4 | 0.862 | 0.743 | | | |
| **BIU** | | | | | |
| BIU1 | 0.923 | 0.852 | 0.861 | 0.756 | 0.873 |
| BIU2 | 0.813 | 0.661 | | | |
| **ATU** | | | | | |
| ATU1 | 0.829 | 0.687 | 0.901 | 0.645 | 0.914 |
| ATU2 | 0.831 | 0.691 | | | |
| ATU3 | 0.773 | 0.598 | | | |
| ATU4 | 0.803 | 0.645 | | | |
| ATU5 | 0.779 | 0.607 | | | |

# VII. MODEL HYPOTHESES CONTRAST

The research models were tested by structural equation modeling (SEM) using Lisrel 8.51 with maximum-likelihood estimation. The parameters that represent the regression coefficients among the constructs are indicated with the symbol β.

$$PU = \beta_1 PEU + \varepsilon_2$$

$$ATU = \beta_2 PEU + \beta_3 PU + \varepsilon_3$$

$$BIU = \beta_4 ATU + \beta_5 PU + \varepsilon_4$$

For the Lisrel application the t-student statistics were reached which allowed support for each one of the formulated hypothesis in the study (Fig. 2). In addition, this software permitted us to quantitatively define the strength of each one of the relationships among the constructs defined in the model, which ever one they corresponded to.

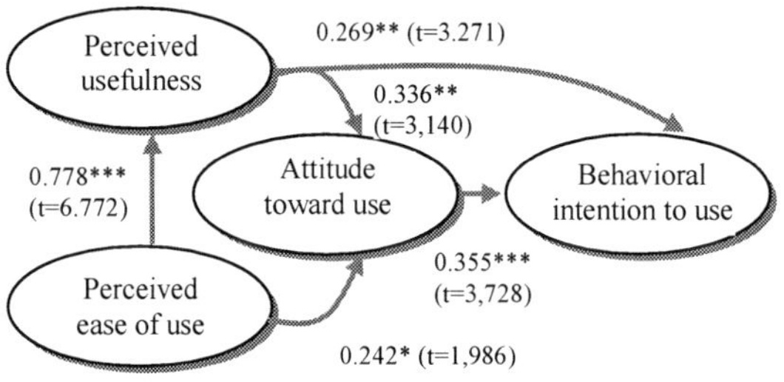


Fig. 2. Lisrel test.

Regarding the significance level of each relationship among the constructs of the TAM model, its t-student statistic, we can claim that this relationship has been significantly confirmed in all cases. All hypotheses were tested significantly for $p < 0.05^*$, $p < 0.01^{**}$, $p < 0.001^{***}$ (based on t(499); t(0.05;499) = 1.967007242; t(0.01; 499) = 2.590452926; t(0.001;499) = 3.319543035) and the findings support the causal relationships proposed in the research model [41].

In this sense, the high significance of the relationship posed in hypothesis H1 between the perceived ease of use and the perceived usefulness ($\beta$=0.778, $p<0.001$) is fulfilled. Likewise, hypothesis H2, dealing with the perceived usefulness and the Attitude toward use, has been confirmed with a high level of significance ($\beta$=0.336, $p<0.01$). Turning to hypothesis H3, we can observe the significant support of the perceived ease of use with that of the Attitude toward use ($\beta$=0.242, $p<0.05$). In hypothesis H4 we are able to state the significance of the effect that the perceived usefulness has on the behavioral intention to use of the OSS-ERP ($\beta$=0.269, $p<0.01$). Finally, the effect that the Attitude toward use an OSS-ERP has on the behavioral intention to use has been verified in hypothesis H5 ($\beta$=0.355, $p<0.001$).

Moreover, we were able to prove how the model very adequately explains the variance in perceived usefulness ($R^2$=0.415), Attitude toward use ($R^2$=0.293) and behavioral intention to use ($R^2$=0.378). Based on these results, we can state the research model defined satisfactorily explains the intentions of the OSS-ERP use as far as the final users are concerned. The variability reached for the attitude toward use the OSS-ERP (28.1%) can be considered significant compared to other studies that apply the TAM model in the IS field. This can be seen, for example, in the studies done by [44] with a $R^2$ of 0.288 for the intentions to use an ERP system of SAP and by [45] with a $R^2$ of 0.26 related to the intention to use IS for e-learning.

## VIII. DISCUSSIONS

From a theoretical point of view, these findings confirm the applicability of TAM to explain the users' acceptance of OSS-ERP systems. All the relationships proposed by TAM have been tested satisfactorily. In this sense, this study contributes evidence about OSS-ERP systems acceptance.

From a practical point of view, we can observe some relevant implications for organizations and practitioners that could highlight the utility of management resources for OSS-ERP acceptance. First, organizations would have to involve potential users in the main stages of the OSS-ERP implantation project with the intention of reaching a successful implementation. Second, in order to stimulate an adequate use of OSS-ERP, organizations and users would have to select OSS-ERP which is useful and easy to use. Third, based on our findings, OSS-ERP solutions for organizations seem to be a viable alternative front to ERP proprietary software, especially for SMEs.

Finally, our findings highlight the necessity of continuing with research about this topic. Specifically, we could start future work based on these findings. First, we are interested in analyzing how affect the behavioral intention to use an OSS-ERP system factors such as technological complexity, training, organizational communication and top management support. Second, we consider that it would be interesting to develop future research to analyze OSS-ERP acceptation after completing an implementation in a particular organization.

## REFERENCES

[1] T.H Davenport, "Putting the enterprise into the enterprise system", Harvard Business Review Vol. 76 (4), pp. 121-13, 1998.
[2] T.H Davenport, "The Future of Enterprise System-Enabled Organizations". Information Systems Frontiers Vol. 2 (2), 163-174, 2000.
[3] F.R. Jacobs and D.C. Whybark. "Why ERP? A primer on SAP implementation". Irwin McGraw-Hill. New York, 2000.
[4] S. Bueno and J.L. Salmeron, "Fuzzy modeling Enterprise Resource Planning tool selection", Computer Standards & Interfaces 30 (3), pp. 137-147, 2008.
[5] H.R. Yen and C. Sheu, "Aligning ERP implementation with competitive priorities of manufacturing firms: An exploratory study". International Journal of Production Economics Vol. 92 (3), pp. 207-220, 2004.
[6] A. Tchokogue, C. Bareil, C.R. Duguay, C.R., "Key lessons from the implementation of an ERP at Pratt & Whitney Canada". International Journal of Production Economics Vol. 95 (2), pp. 151–163, 2005.
[7] B. Lea, M.C. Gupta, W. Yu, "A prototype multi-agent ERP system: an integrated architecture and a conceptual framework", Technov. Vol. 25 (4), pp. 433-441, 2005.
[8] V.A. Mabert, A. Soni and M.A. Venkataramanan. "Enterprise Resource Planning: Common Myths versus evolving reality", Business Horizons Vol. 44 (3), pp. 69-76, 2001.
[9] L. Wu, C. Ong and Y. Hsu, "Active ERP implementation management: A Real Options perspective", Journal of Systems and Software Vol. 81 (6), pp. 1039-1050, 2008.
[10] J.W. Ross and M.R. Vitale, "The ERP revolution: surviving vs. Thriving", Information Systems Frontiers Vol. 2 (2), pp. 233-241, 2000.
[11] E. Bendoly and F. Kaefer, "Business technology complementarities: impacts of the presence and strategic timing of ERP on B2B e-commerce technology efficiencies". Omega 32 (5), pp. 395-405, 2004.
[12] P. Seddon, G. Shanks and L. Willcocks. Introduction: ERP-The Quiet Revolution? In Second-Wave Enterprise Resource Planning Systems. Implementing for Effectiveness. Edited by Shanks, G, Seddon, P.B. and Willcocks, L. P. Cambridge University Press, Cambridge, 2003.
[13] C.P. Holland and B. Light. "A critical success factors model for ERP Implementation", Software IEEE Vol. 16 (3), pp.30-36, 1999.

[14] F. Fui-Hoon, S. Faja, T. Cata. "Characteristics of ERP software maintenance: a multiple case study", Journal of software maintenance and evolution: research and practice Vol. 13 (6), pp. 399-414, 2001.
[15] S. Abdinnour-Helm, M.L. Lengnick-Hall and C.A. Lengnick-Hall, "Pre-implementation attitudes and organizational readiness for implementing an Enterprise Resource Planning system", European Journal of Operational Research Vol. 146 (2), pp. 258-273, 2003.
[16] E. Morgan, "Possibilities for open source software in libraries", Information Technology and Libraries Vol. 21 (1), pp. 12–15, 2002.
[17] M.D. Gallego, P. Luna, and S. Bueno, "Designing a forecasting analysis to understand the diffusion of open source software in the year 2010", Technological Forecasting and Social Change Vol. 75 (5), pp. 672-686, 2008.
[18] J. Lerner, and J. Tirole, "Some simple economics of open source", Journal of Industrial Economics Vol. 50 (2), pp. 197–234, 2002.
[19] A. Fuggetta, "Open source software—an evaluation". Journal of Systems and Software Vol. 66 (1), pp. 77–90, 2003
[20] N. Serrano and J.M. Sarriegi, "Open Source Software ERPs: A New Alternative for an Old Need". IEEE Software Vol. 23 (3), pp. 94-97, 2006.
[21] F.D. Davis, "A technology acceptance for empirically testing new end user information systems: theory and results", Doctoral dissertation, Sloan School of Management, Massachusetts Institute of Technology, 1986.
[22] M. Fishbein and I. Ajzen, "Belief, Attitude, Intention, and Behavior: An Introduction to Theory and Research". Addison-Wesley. New York, 1985.
[23] F.D. Davis, "Perceived usefulness, perceived ease of use, and user acceptance of Information Technology". MIS Quarterly Vol. 13 (3), pp. 319-340, 1989.
[24] A. Shirani, M. Aiken and B. Reithel, 1994. A model of user information satisfaction. Data Base 25 (4), pp. 17–23, 1994.
[25] Davis, F.D., Bagozzi, R.P., Warshaw, P.R., 1989. User acceptance of computer technology: a comparison of two theoretical models, Management Science 35 (8), 982–1003.
[26] Lee, K.C., Kang, I., & Kim, J.S., 2007. Exploring the user interface of negotiation support systems from the user acceptance perspective. Computers in Human Behavior 23 (1), pp. 220–239.
[27] E.W.T. Ngai, J.K.L. Poon and Y.H.C. Chan, "Empirical examination of the adoption of WebCT using TAM". Computers and Education Vol. 48 (2), pp. 250–267, 2007.
[28] T.C.E. Cheng, D.Y.C. Lam and A.C.L. Yeung, "Adoption of internet banking: An empirical study in Hong Kong". Decision Support Systems Vol. 42 (3), pp. 1558–1572, 2006.
[29] J.M.S. Cheng, G.J. Sheen and G.C. Lou, "Consumer acceptance of the internet as a channel of distribution in Taiwan – A channel function perspective". Technovation Vol. 26 (7), pp. 856–864, 2006.
[30] J. Nunally, I. Bernstein, "Psychometric theory", third ed. McGraw-Hill, New York, 1995.
[31] J.F. Hair, R.E. Anderson, R.L. Tathan, W.C. Black, "Multivariate Analysis", Prentice Hall. New York, 2000.
[32] H. Kaiser, "A second generation little jiffy". Psychometrika Vol. 35, pp. 401–415, 1970.
[33] H.F. Kaiser and J. Rice, Little Jiffy, Mark IV. Educational and Psychological Measurement Vol. 34, pp. 111–117, 1974.
[34] K.G. Joreskog and D. Sorbom, "Lisrel VIII manual". Chicago: Scientific Software International, 1993.
[35] K.A. Pituch and Y.K. Lee, "The influence of system characteristics on e-learning use". Computers and Education Vol. 47 (2), pp. 222–244, 2006.
[36] M.Y. Yi, J. Jackson, J. Park, and J. Probst, "Understanding information technology acceptance by individual professionals: Toward an integrative view". Information and Management Vol. 43 (3), pp. 350–363, 2006.
[37] M. Lee, C. Cheung and Z. Chen, "Acceptance of Internet-based learning medium: The role of extrinsic and intrinsic motivation". Information and Management Vol. 42 (8), pp. 1095–1104, 2005.
[38] P. Luarn and H.H. Lin, "Toward an understanding of the behavioral intention to use mobile banking". Computers in Human Behavior Vol. 21 (6), pp. 873–891, 2005.
[39] H. Selim, "An empirical investigation of student acceptance of course websites". Computers and Education Vol. 40(4), pp. 343–360, 2003.
[40] R. Bagozzi, "Structural equation model in marketing research. Basic principles, principles of marketing research". Oxford: Blackwell Publishers, 1994.
[41] J.C. Anderson and D.W. Gerbing, "Structural equation modelling in practice: A review and recommended two-step approach". Psychological Bulletin Vol. 103 (3), pp. 411–423, 1988.
[42] R. Bagozzi and Y. Yi, "On the evaluation of structural equation models", Academy of Marketing Science1Vol. 6 (1), pp. 74–94, 1988.
[43] C. Fornell and D. Larcker, "Evaluating structural equation models with unobservable variables and measurement error". Journal of Marketing Research Vol. 18 (1), pp. 39-50, 1981.
[44] K. Amoako-Gyampah and A.F. Salam, "An extension of the technology acceptance model in an ERP implementation environment". Information & Management Vol. 41 (6), pp. 731-745, 2004.
[45] R: Saade´ and B. Bahli, "The impact of cognitive absorption on perceived usefulness and perceived ease of use in on-line learning: An extension of the technology acceptance model". Information and Management Vol. 42 (2), pp. 317–327, 2005.
[46] V. Venkatesh and F.D. Davis, "A theoretical extension of the technology acceptance model: four longitudinal field studies". Management Science Vol. 46 (2), pp. 186-204, 2000.
[47] F. Calisir and F. Calisir, "The relation of interface usability characteristics, perceived usefulness, and perceived ease of use to end-user satisfaction with enterprise resource planning (ERP) systems". Computers in Human Behavior 20 (4), pp. 505-515, 2004.

# Enhanced Progressive Vector Data Transmission For Mobile Geographic Information Systems (MGIS)

AHMED ABDEL HAMID
Faculty of Computers and Information, Helwan University, Egypt
ahmedabdel_hamed@yahoo.com

MAHMOUD AHMED
Post Doctoral Fellow
University of Waterloo, ON, Canada
and Assistant Prof., Faculty of Computers and Information,
Helwan University, Egypt
mfouad@engmail.uwaterloo.ca, mfouad4@gmail.com

YEHIA HELMY
Professor and Dean
of Faculty of Computers and Information,
Helwan University, Egypt
ymhelmy@yahoo.com

*Abstract-* **Dramatic increases in mobile applications are evolving; accordingly, faster spatial data transmission is required. Despite the transmission of Raster and TIN data types had significant enhancements; the vector data, which constitutes the main bulk of any GIS, is lacking the speed suitable for practical solutions. Vector data in Mobile GIS faces several challenges, e.g., volume of data, limited bandwidth, expenses of wireless services, and mobile device limited capabilities, so, Mobile GIS user suffers long response time. Progressive vector data transmission has been investigated to overcome above obstacles. In this research, enhanced progressive data transmission is reported, new modifications of another approach published by different author are investigated; the proposed modifications minimized the number of required topological checks and reduced the response time. A prototype based on client-server architecture is developed using Java technology, real datasets are examined using Nokia 6300 mobile handset, and the visualization at client side uses Scalable Vector Graphics format.**

## I. Introduction

The wide spread of mobile devices, e.g., PDA, cell phones, pocket PC,.,,, etc in addition to their decreasing prices, enabled a lot of services to users, in Egypt, like many other countries, there is an increasing market demand towards providing location based services and geo spatial information to users through wireless services, the recent analysis of Egyptian Geospatial Data Infrastructure (GSDI) as in [1] indicates that there is a great potential for providing more geographic and Location Based Services (LBS) for a large number of users.

Providing geographic information services through the mobile technology is known as Mobile GIS (MGIS). MGIS is an interdisciplinary branch which integrates GPS, mobile communication (Such as GSM, GPRS, and CDMA etc), wireless internet, LBS, and other technologies [2].

Despite the spread of using MGIS and advances in mobile technology; MGIS still suffers from the limitations of mobile devices such as power, screen size and colors, networking, etc. These limitations affect the volume of data to displayed and processed and affect the applications capabilities. Mobile device limitations complicate the design of MGIS applications. The low bandwidth networks and high volumes of spatial data lead to longer response time. Response time is the period from the user submitting of the request until he can interact with the response [3]. In MGIS, response time is very long because of two reasons. First, the mobile devices use wireless networks, these networks are not fast enough to give acceptable performance. Second, geo-spatial data is high in volume by nature. These reasons lead to long wait time until the user can interact with the data. Progressive transmission can be used as solution for this problem. The progressive transmission provides the user with sample data to be transmitted faster and the user does not wait long and begin interact with this portion of data in parallel to downloading the reminder of the data and displaying what is downloaded. Progressive transmission is used only for downloading the large amount of data through slow networks [4].

## II. Related Work

Progressive transmission was successfully used to deliver raster and TIN data on the internet.

But with the vector data, little progress was achieved. Ref. [4] defined a framework of progressive vector transmission and discussed the need to progressive vector transmission and its challenges, only the topological operators are used, no metrics or semantic operators were used for creation of multiple representations. Ref. [5] presented the design and implementation of an algorithm that implement the progressive transmission. This implementation was not so realistic, that is it had been tested only on single polyline packets in small dataset. Ref. [6] presented real implementation of progressive Vector transmission which embedded into the GEOSERVNET (GSN) software and addressed many implementations issues such as spatial index, line simplification and multiple representation and encoding algorithms, definition and interpretation of data request and response communication protocol, data consistency, interoperability and openness, client and server sides buffer management and presentation issues, however the applied generalization was offline. Ref. [7] presented implementation

T. Sobh (ed.), *Innovations and Advances in Computer Sciences and Engineering*,
DOI 10.1007/978-90-481-3658-2_11, © Springer Science+Business Media B.V. 2010

for client/server progressive transmission of spatial vector data to a small mobile device, starting with coarse overview information and progressively streaming more detailed information to the client, however, the generalization method was also done offline. Most previous work in progressive vector transmission was based on pre-generated multiple resolutions. Ref. [8] presented the prototype for progressive vector transmission based on on-fly generalization. He proposed vertex decimation method based on some constraint rules to preserve the topology and prevent self intersections in multiple representations for polygons. The author in [3] extended the work in [8] by modifying the used constraint rules; this is to preserve the topology, prevent self intersections in multiple representations for polygons and to improve the performance. The additional improvements in [9] enhanced the performance of the work in [8], moreover, minimized the size of coding of the removed vertices and added some rules to preserve the topology, however, the above research did not deal with point features, also the defined rules do not guarantee that there are no self intersections in generalized data. Moreover, he used large number of topological checks that increase the generalization time.

## III. Progressive Transmission for Spatial Data

Progressive transmission is known as streaming also, but the term streaming was related with multimedia on the web to provide the user with some data as stream and the delivered data is stored in the buffer until the size becomes sufficient to be played on the client machine [10]. In the spatial domain one can implement the progressive transmission by creating multiple resolutions, providing the user with lowest level of detail (LOD), rendering it, and allowing the user to interact with available LOD while downloading the increments that increase the details (Fig. 1). These increments are downloaded until full data completed or until the user interrupts this process when she/he feels that the downloaded data is enough. Because there is no robust generalization method that maintains topology and time constraints, the progressive vector transmission is still challenging. Because vector data consists of a set of lines, polygons, points, and a set of spatial relations, the progressive transmission of vector data is not simple task as with raster data by adding and deleting pixels [4]. In progressive vector transmission, the user must have the control to select resolution and abort the downloading process of increments when he/she is satisfied by the downloaded data [5]. Progressive vector transmission is interdisciplinary field which encompasses computer graphics, network communication, spatial data structure and models, plus cartographic generalization and multiple representations. One of the main challenges in progressive vector transmission is combining all these disciplines together [6]. From implementation view, progressive vector transmission technique consists of two sides (server and client) [4]. Server side will generate online or pre-construct the multi representations of the original dataset, process client request, transfer the coarser level of detail at first time, then transfer the increments to increase details for individual clients. Client side will send request to the server, download the lower level of detail, draw this level, and download the increments and draw them to increase the details.

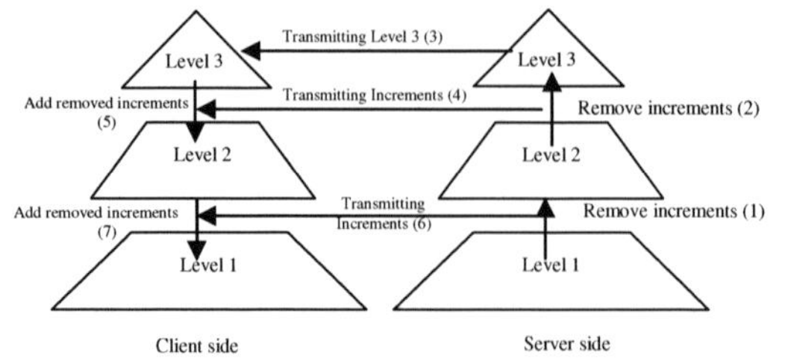


Fig. 1. Concept of progressive transmission in spatial domain

## IV. Proposed Generalization Algorithm

This research can be considered as an extension of the work reported in [9].

In this research the map is considered as collections of line, polygon, and point features. The line is an open chain of vertices ($V_0$, $V_1$,......, $V_n$) and the polygon is a closed chain of vertices ($V_0$, $V_1$,......, $V_n$) but $V_0$ is connected to $V_n$. These features must be simple according to OpenGIS simple features specification for SQL Revision 1.1 [11] and not collections. The point feature is a pair of coordinates. All layers have been given the same importance. To simplify the line and area features; the simplification algorithm in [12] is applied. This algorithm ranks the vertices of the linear features (line or polygon) for removal. So that the vertices that have the small effective area will be removed firstly. The effective area of vertex ($V_i$) is calculated by the area of the triangle formed with their two adjacent vertices ($V_{i-1}$, $V_i$, $V_{i+1}$). Despite the applied algorithm found to be able to extract low level of detail, it does not preserve the topology in different levels of detail. Hence, it was necessary to refine its performance by developing a number of rules to maintain the consistency of topology and guarantee no self intersections will occur after removing any vertex during simplification.

### A. *Proposed Rules*

1. If the feature is isolated and does not have unique attributes, it can be removed completely. The line can be removed if its length is smaller than given threshold. The polygon can be removed if its area is smaller than given threshold.
2. If the vertex ($V_i$) is shared by more than two features, it is irremovable (Fig. 2a).
3. For vertex that is shared by only two features, it is irremovable if the two segments $V_{i-1}V_i$ and $V_iV_{i+1}$ are not shared by two features or the vertex is a start or an end of the line feature (Fig. 2b).
4. For the vertex $V_i$, it is irremovable if there are other vertices from the same feature intersect the triangle ($V_{i-1}$, $V_i$ , $V_{i+1}$). This will prevent the self intersections (Fig. 2c).

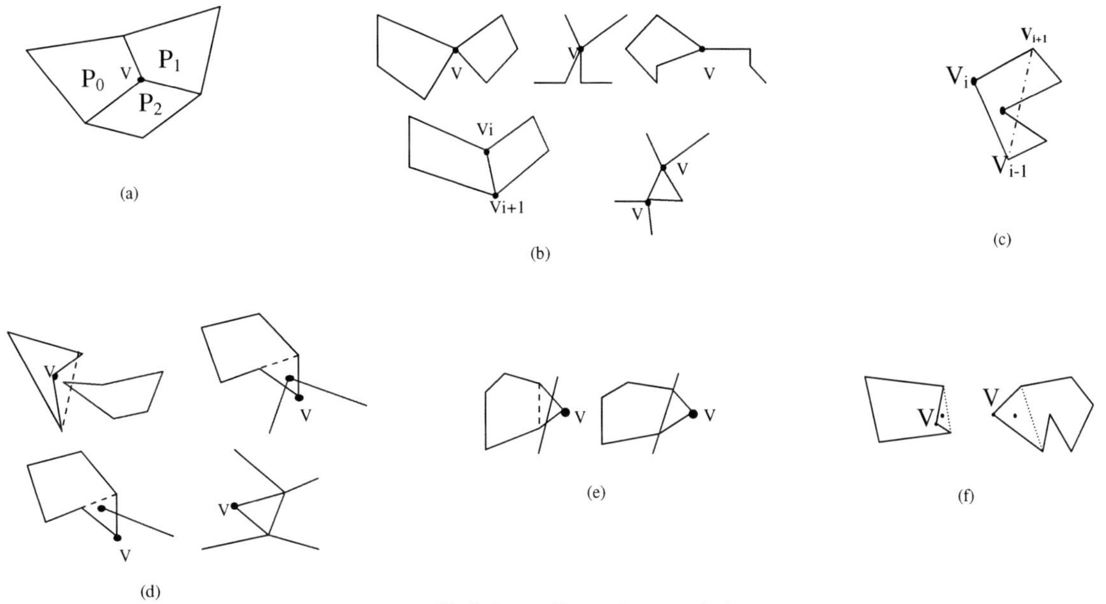


Fig. 2. Irremovable cases by proposed rules

5. For the vertex $V_i$, $V_i$ is irremovable if there are other vertices from other features (same or different layer) intersect the triangle ($V_{i-1}$, $V_i$ , $V_{i+1}$) or the segment $V_{i-1}V_{i+1}$ is segment from this feature. (Fig. 2d).
6. If there are features from the different layers intersect the two segments $V_{i-1}V_i$ and $V_iV_{i+1}$, and not intersect the segment $V_{i-1}V_{i+1}$or this segment is part of this feature, vertex ($V_i$) is irremovable. (Fig. 2e).
7. The vertex ($V_i$) is irremovable if there is any point feature is identical to vertex ($V_i$) or exist in the triangle ($V_{i-1}$, $V_i$, $V_{i+1}$), except the points that are identical to one of vertices ($V_{i-1}$, $V_{i+1}$) (Fig. 2f).

Rule 1 as proposed by [9] is used for implementation of selection operator to minimize the number of features. The features are removed completely and then will be transmitted. The elimination process can be done, if this feature satisfies three conditions. First, the feature is isolated i.e. it does not have any shared portion with another feature from any layer. Second, if this feature is small (in length for lines and area for polygons). This shape condition will be determined by given threshold. The threshold for length of lines will be 5 pixels, and for the area of polygons will be 3 X 3 pixels. The third condition is that the feature does not have unique attribute. Rule 2 and rule 3 preserve the topology. Rule 2 is also proposed by [9] to prevent the removal of the vertex that is shared by more than two features. Rule 3 is modified rule. This rule deals with the vertex that is shared by two features whether these two features are from the same layer or not. If two features share line, after generalization this line will not contracted to vertex. So this will keep the visual impression of the original topology. Also this rule will prevent the inconsistent topology in the line features, for example if two lines share two vertices but are not adjacent in one or both features. Rule 4 is used to prevent self intersection. The removal of vertex ($V_i$) will cause self intersection, if there is any other vertex in the triangle ($V_{i-1}$, $V_i$, $V_{i+1}$) [13]. The other rules are used to preserve the topology. If we apply rule 5 on the features from the same layer, it will prevent any new intersections so it will preserve the topology for the features from the same layers. Also this rule can be applied on the features from different layers to minimize the topological checks that perform in the next rules. The time used in these points checks is largely smaller than time for topological checks. Rule 6 is applied after rule 5 in the simplification algorithm to check the topology with the features that are not have vertices intersect the triangle ($V_{i-1}$, $V_i$ , $V_{i+1}$). This rule is used only to check the topology in different layers and these layers are intersected and the intersection point is not vertex in the vertices of the features. Also this rule will not used if the features from different layers are disjoint. This rule is performed at steps. First if there are no features intersect the two segments $V_{i-1}V_i$ and $V_iV_{i+1}$, we can remove vertex ($V_i$). But if this feature intersects those segments and does not intersect the new segment or this feature has line spatially equal to new segment, vertex ($V_i$) is irremovable to avoid the wrong topologies. All these checks are performed to the features that intersect Minimum Bounding Box of old segments ($V_{i-1}V_i$ and $V_iV_{i+1}$). Rule 7 deals with point features in the map. It is not a rule to manage the selection of point features but to preserve the topology of linear features with point features. The vertex ($V_i$) is irremovable if there is any point feature in the triangle ($V_{i-1}$, $V_i$ , $V_{i+1}$) except the points that are identical to the two adjacent vertices ($V_{i-1}$, $V_{i+1}$). If point feature is outside the polygon, after generalization it will not be within the polygon and the points that are within the polygon will not be outside the polygon. Also if one point is at the left of the line it will not be at the right side after generalization of the line. Also this rule will keep the points that are on the interior of lines and boundary of the polygons in their locations without topology change.

### B. *Operations for removal*

Progressive transmission does not need generalization algorithm only; it needs refinement or reconstruction algorithm too. The generalization algorithm is done on server side whereas the reconstruction algorithm is done on client side. So each removal in generalization method must be retrieved in the reconstruction method. The proposed generalization algorithm can do two removal operations. It can remove complete features and vertices. The removed feature will be represented by its ID, the layer ID that the feature belongs to, and the geometry of this feature. The representation of the removed vertices is important because it must define the topology and geometry aspects. These two aspects have important role during the reconstruction on the client side. We will use the representation of [9]. If the feature (Line or polygon) is represented by a series of vertices, so

Line feature = $V_0, V_1, \ldots V_n$

Polygon feature = $V_0, V_1, \ldots, V_n$

After removing one vertex (V) from feature (F) and this vertex belongs to this feature only, the feature will be

$$Fg = F - V \tag{1}$$

And the removed vertex (V) can be expressed as

$$V = X, Y, FID, order \tag{2}$$

Where X, Y are coordinates of the removed vertex. FID is the ID of this feature. Order is the position of the removed vertex in the vertices series of feature. This removal is called simple removal.

But the proposed rules give the possibility of removing the vertex that is shared between two features. In this case the removal is called complex removal. The representation of the removed vertex in this case must not store the removed vertex only but also the topology of this vertex. The generalized features can be expressed as the following equation after removing the shared vertex.

$$Fg1 = F1 - V \tag{3}$$
$$Fg2 = F2 - V \tag{4}$$

And the removed vertex is

$$V = X, Y, FID1, FID2, Order1, Order2 \tag{5}$$

Where X and Y are the vertex coordinates and FID1, FID2 are the identifiers of the feature that the vertex belongs to. Order1 and Order2 are positions of the vertex in the features vertices.

By continuous removing of vertices, the lower level of detail can be obtained. To reconstruct the original data set, the process is done in reverse order (last removed, first reconstructed). By storing the removed vertices in stack structure; we can transmit the last removed vertex to client first. If the feature has relation with any feature or its length or area greater than specific threshold, the feature can be expressed as producing three levels

$$F = F_{low} + V_{i3} + V_{i2} + V_{i1} \tag{6}$$

Where $F_{low}$ is the feature at the lowest level. $V_{i3}$, $V_{i2}$, and $V_{i1}$ are the removed vertices from this feature in each level (level 3, level 2, and level 1) respectively.

### C. *Transmission Model*

The spatial data and the increments are transmitted in binary format rather than textual format because binary format reduces the data volume transmitted [9]. As the client side is J2ME, there are some extra data must be transmitted to the client side to be able to extract the geometry data.

The Lowest level of detail consists of collection of generalized features; each feature has ID and geometry data. Also it corresponds to specific layer. Each feature will be transmitted in binary format and is decoded at the client side then is added to its layer.

Increments consist of two types (removed features and removed vertices). The removed features are transmitted in the same manner as transmission of features of the lowest level of detail plus transmitting the layer ID of each feature. These features added easily to the map in their layer at the client side. In case of removed vertices, we must transmit data that is used to reconstruct the geometry and the topology of the feature.

For each vertex, we will send this format:

X coordinate: double
Y coordinate: double
FID1: string
Order1: integer
Type: byte
FID2: string
Order2: integer

Where X and Y are the coordinates of this removed vertex, FID1 is the feature ID of the first feature, and Order1 is the position of this vertex in the first feature. Type is assigned the value 1 (for complex removal) or 0 (for simple removal). FID2 and Order2 will not be transmitted if the vertex belongs to one feature. FID2 is the feature of the second feature. Order2 is the position of the vertex in the second feature. Type is transmitted to inform the client side the type of removal and if there is second feature or no.

## V. IMPLEMENTATION

To investigate the proposed approach, a prototype is

developed based on client-server architecture (Fig. 3) which consists of client side and server side.

Client side consists of Configuration builder that enables user to enter the parameters to retrieve the data from the server side, Connection handler that handles the connection with server to obtain the lowest level of detail and also to get the increments from the server, and SVG Image Manager that creates and edits the Scalable Vector Graphics (SVG) image that represent map. Also it executes the actions of the user including zoom in, zoom out, pan, show, or not show specific layer.

The server side has Application Tier for executing the application functions and logic and Data Tier to manage the maps. The Application Tier consists of Query Processor to manage the connection between user and the server, Map Generalization for extracting multiple resolutions from the master dataset by applying the simplification algorithm and selection operator in the generalization method, and Data Reader which is used to retrieve the data from the source.

The client side was developed in java 2 micro edition (J2ME). The client side uses Scalable 2D Vector Graphics (JSR-226), or M2G, API. This API gives the classes and interfaces for providing the scalable vector graphics features in J2ME. The connection is separated in different thread to allow the user to interact with the map during downloading the increments. The server side was developed in java and JTS version 1.8 (Java Topology Suite) was chosen as environment for generalization in the server side.

## VI. EXPERIMENTS AND RESULTS

The prototype was tested with topological dataset contains building, roads, and points that represent the artworks. Fig. 4 shows different phases on the user screen for this dataset. The features from different layers can not intersect i.e. the artworks do not intersect with building nor roads, and the buildings do not intersect with the roads. So rule 6 will not be used. The original dataset was 742 KB. The server is desktop computer with 3 GHZ Processor, 1 G RAM and use dial up connection (33.6 Kbps). The client is Nokia 6300 mobile.

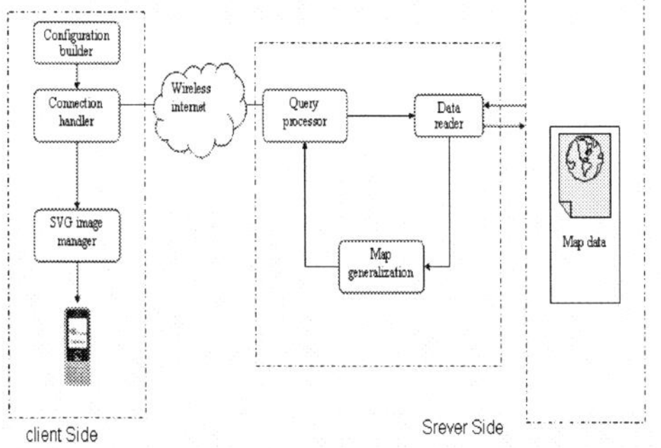


Fig. 3. Prototype architecture for progressive vector transmission

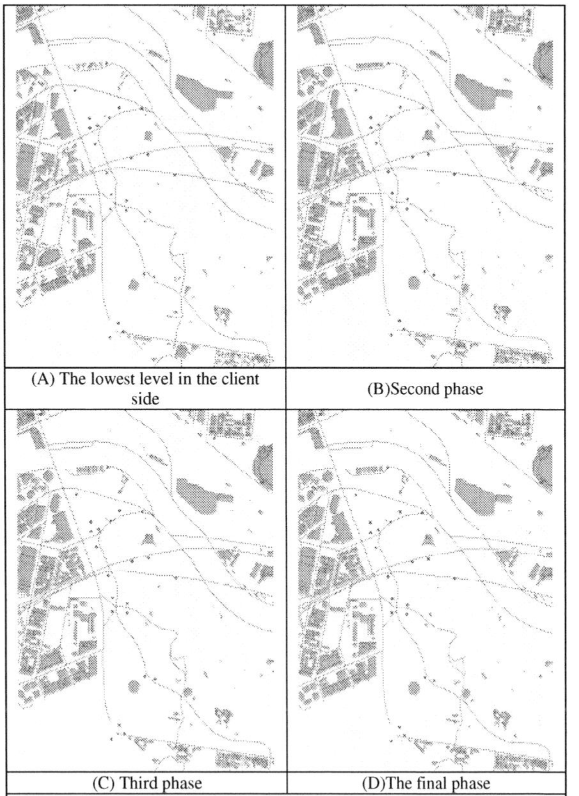

(A) The lowest level in the client side | (B)Second phase | (C) Third phase | (D)The final phase

Fig. 4. Different phases in the user devices (Data from ArcPad samples)

The client downloaded the area with (150, 100,200, and 250) as (Minimum X, Minimum Y, Width, and Height). The selected area is 198 KB and has 561 buildings, 84 roads, and 24 artworks. Three lower LODs from original dataset were created on the server side to be transmitted progressively to the client. These LODs were tested with ARCGIS 9.1 for intersection from different layers, overlapping in polygons and self intersections and there are no violations. Table 1 shows the sizes in each level and the performance of the generalization algorithm.

Table1 shows that the response time using any level produced by proposed algorithm is less than using one step transmission. Table 2 shows the time need to transmit the increments in case of transmitting the dataset in four stages (level 3 at Table 1).

TABLE I
DIFFERENT LEVELS AND THEIR GENERALIZATION AND RESPONSE TIME

| Level | Data Size | Generalization Time (ms) | Response Time (seconds) |
|---|---|---|---|
| Level 3 | 76.5 KB | 390 | 21.13 |
| Level 2 | 93.7 KB | 328 | 23.78 |
| Level 1 | 120 KB | 234 | 28.53 |
| Full Level | 198 KB | ----- | 38.85 |

TABLE 2
TRANSMISSION TIME OF INCREMENTS TO DOWNLOAD THE FULL DATA IN FOUR STAGES (LEVEL 3 AT TABLE 1)

| | Transmission time (seconds) |
|---|---|
| Increments of level 3 | 6.92 |
| Increments of level 2 | 7.12 |
| Increments of level 1 | 25.09 |

## VII. CONCLUSION AND FUTURE WORK

In this paper, some constraint rules are proposed and tested to overcome the disadvantages of current approaches of progressive data transmission. These new rules are expected to represent a new step towards faster geospatial data transmission, hence, contribute to bridge the gap between the huge demands for faster geo-spatial data transmission and current status of data transmission in mobile GIS applications and services. To investigate the proposed rules, a complete prototype was developed and heavily tested using real world datasets. Experiments proved the efficiency of the applied new rules and shown a number of advantages e.g., they minimize the required number of topological checks reported in previous research while they do not need topological checks at all in that cases where the dataset has disjoint layers. Moreover, the experiments and results show that the seven rules constrained algorithm can reduce the overall response time, perform real time spatial data generalization, preserve the topology and prevent the self intersections at different resolutions. Also, the minor changes due to data generalization is not affecting visual perception of map context which maintain the quality of provided mobile services so that any resolution can be used for analysis. The future work will concentrate on applying this algorithm more dynamic and movable clients so that the progressive transmission will take into account the predicted position of the user when the data is visualized at his client device screen.

## REFERENCES


[1] Mahmoud Ahmed, "Situation analysis of National Geo-Spatial Data Infrastructure (NGSDI) in Egypt," Proceedings of The 11th World Multi-Conference on, Systemics, Cybernetics and Informatics: WMSCI. Orlando, Florida, USA, 2007.

[2] Li Luqun and Li Minglu. "A research on development of Mobile GIS architecture," ISEIS - International Society for Environmental Information Sciences, Volume 2, 2004, Pp 920-926.

[3] B. Yang. "A multi-resolution model of vector map data for rapid transmission over the Internet," Computers & Geosciences 31, 2005, Pp 569–578.

[4] Michela Bertolotto and Max J. Egenhofer. "Progressive Vector Transmission," Proceedings of the 7th International Symposium on Advances in GIS, November 1999, Pp. 152-157.

[5] Barbara P. Buttenfield, "Transmitting vector geospatial data across the Internet," Proceedings GIScience, September 2002.

[6] H. HAN, V. TAO, and H. WU, "Progressive vector data transmission," In Proceedings of the 6th AGILE, Lyon, France, 2003, pp. 103–113.

[7] M. Sester and C. Brenner, "Continuous generalization for visualization on small Mobile devices," 11th International Symposium on Spatial Data Handling, 2004, pp. 469–480.

[8] B. Yang, R. S. Purves and R. Weibel, "Implementation of progressive transmission algorithms for vector map data in web-based visualization," International Archives of the Photogrammetry, Remote Sensing and Spatial Information Sciences, Vol. 34, part XXX, 2004.

[9] B. Yang, R. Purves and R. Weibel, "Efficient transmission of vector data over the internet," International Journal of Geographical Information Science, Vol.21, No.2, February 2007, pp. 215–237.

[10] Haiyang Han, "Progressive vector data transmission," M.Sc Thesis, York University, Toronto, Ontario, December 2003.

[11] OpenGIS Simple Features Specification for SQL Revision 1.1, Open GIS Consortium, Inc. OpenGIS project Document 99-049. http://www.opengeospatial.org/standards/sfs

[12] M. Visvalingam and J.D. Whyatt, "Line generalisation by repeated elimination of points," Cartographic Journal, Vol. 30, 1993, pp. 46-51.

[13] L. Kulik, M. Duckham, and M. Egenhofer, "Ontology-driven map generalization", Journal of visual language and computing, 16(2), 2005.

# Mining Inter-transaction Data Dependencies for Database Intrusion Detection*

Yi Hu
Department of Computer Science
Northern Kentucky University
Highland Heights, KY 41099
Email:huy1@nku.edu

Brajendra Panda
Computer Science and Computer Engineering Department
University of Arkansas
Fayetteville, AR 72701
bpanda@uark.edu

***Abstract*** ***-Existing database security mechanisms are not sufficient for detecting malicious activities targeted at corrupting data. With the increase of attacks toward database-centered applications, an effective intrusion detection system is essential for application security. Although someresearches havebeen done on the database intrusion detection, methods for detecting anomalous activitiesin databases haveonly recently been explored in detail. In this paper, we present an approach employing inter-transaction data dependency mining fordetecting well-crafted attacks thatconsists a group of seemingly harmless database transactions. Our experiments illustrated the advantage of this new approach and validated the effectiveness of the model proposed.***

## I. INTRODUCTION

Host-based and network-basedintrusion detection is a long-standing research domain. Many different approacheshave beenemployed tosafeguardcritical information in today's data centeredenvironment.One way to make data less vulnerable is to deploy Intrusion Detection System (IDS) in critical computer systems.In recent years, researchers have proposed a variety of approaches for increasing the intrusion detection efficiency and accuracy.However, none of these approaches is fail proof.PGP Corporation's 2006 annual study on the cost of data breaches showed that the average total cost of a data breach per reporting company was $4.8 million per breach, and could reach up to $22 million [1].Methods for detecting intrusionin databases haveonly recently been explored in detail [2,3, 4,5,6,7,8,9, 10, 11, 12].The subtlety of identifying suspiciousqueries in a database iscritical for protecting important data in enterprise environment.

The existing approaches for identifying database intrusion can be classified based on the intrusion classification mechanisms. The intrusion classification procedure can use signature based or non-signature based database intrusion detection methods. The signature based database intrusion detection system stores the signatures ofonline attacks such as the SQL injections and identifies future attacks that match these signatures. The non-signature based database intrusion detection system profiles the patterns of normal user transactions and uses these patterns for identifying intrusive behavior. If the usertransactions donot conform to the patterns of "good transactions", they are identified as the anomalous.

This paper proposed a non-signature based data miningapproachthat employs inter-transaction data dependencies for identifying malicioustransactions. The data dependencies generated will beusedas the profile of normal user transactions.Transactionsnot compliant with thesedata dependency rules are recognized as malicious transactions.

## II. LITERATURE REVIEW AND MOTIVATIONS

Researchers strive to designIDSs that can detect more varieties ofdatabaseintrusions in a timely fashion with lower false alarms.The Hidden Markov Model has been proposed in [6] to detect malicious data corruption. Lee et al. [8]have used time signatures in discovering database intrusions. Their approach is to tag the time-signature to data items. A security alarm is raised when a transaction attempts to write a temporal data object that has already been updatedwithina certain period. Another method presented by Chung et al. [7] identifies data items frequently referenced together and saves this information for later comparison.

Hu et al. [10] offered a database intrusion detection model that uses data dependency relationships observed in user transactions by analyzingapplication codeusing a semantic analyzer. Another model proposed by Hu et al. [9] for detecting malicious transactions uses an approach concentrating on mining intra-transaction data dependencies among data items in the database. By intra-transaction data dependency it refers to the data access correlations between two or more data items accessed in a database transaction. The transactions that do not conform to the intra-transaction data dependencies are identified as malicious transactions.The model proposed by Srivastava et al. [11] uses weighted intra-transactional rule mining for database intrusion detection.The approach offered byFonseca et al. [12]consists of a comprehensive representation of user database utilization profiles to perform concurrent intrusion detection.

* This work was supported in part by a 2008 NKU Faculty Summer Fellowship and US AFOSR under grant FA 9550-04-1-0429

T. Sobh (ed.), *Innovations and Advances in Computer Sciences and Engineering*,
DOI 10.1007/978-90-481-3658-2_12, © Springer Science+Business Media B.V. 2010

Even with theseapproaches, there are still some shortcomings. The malicious transactions launched by the attacker can be very well crafted.Anattacker may launch a group of malicious transactions each of whichappears as a normal transaction. This makes the adversary's activities very difficult to be detected. This research developednew methodologies for detecting these cyber attacks toward corrupting data by using anovelapproach that profiles the normal database access patterns across different transactions. Data access operations that are not compliant to these intrinsicinter-transactiondatadependenciesare identified as anomalous activities.

## III. METHODOLOGY

In a database system, there can be many concurrent transactions executed at any given time. Also, many transactions can be submitted sequentially for achieving one complex user task. In order to analyze transactions to profile the normal database access patterns across transactions, it is critical to develop an algorithm that clusters user transactions into user tasks. Otherwise, simplyperformingsequential pattern miningalgorithm [13]across database transactions for profiling legitimate access patterns is not useful for our purpose. Our approach for detecting well-crafted malicious transactions is divided into following steps.First, the database log used for discovering data dependencies across transactionsis clustered into user tasks. Then the inter-transaction data dependencies represented by*Inter-transaction Write Dependency*and*Inter-transaction Write Sequence*are identified.The details are explained in the following sub-sections.

### *A.ClusteringUser Transactions into User Tasks*

A user task is a group of transactions that are semantically related to perform a user task. In this step, we use a data mining method to cluster the user transactions based on the time series data ofdatabasetransactions. Each user transaction is tagged with the submission timeandthe ID of the user who submitted the transaction in the database log.

First, the database log containing user transaction informationis pre-processed based on the IDof the user submitting thetransactions.Since transactions or user tasks submitted by different users are not semantically related, the database log is first clustered by the user ID. That is, all transactions submitted by the same user are clustered into one group inside which the relative executionorder of transactions iskept. This step is trivial. Then the transactions submitted by the same user are clustered again so that transactions belonging to the same user task are grouped into the same cluster. To achieve this, we propose to use a data mining method to cluster the user transactions based on submission times of the transactions.

Beforedescribingour methodfor clustering user transactions, let us consideran example of user transactions and user tasks. Figure 1 shows 3 sample user tasks thatcontain9 transactions. Tasks 1, 2, and 3 have 3, 2, and 4 transactions respectively. All of these transactions are inside one cluster that representsthetransactions submitted by the same user.Figure1also shows thetimegapsbetween transactions andbetweenuser tasks.

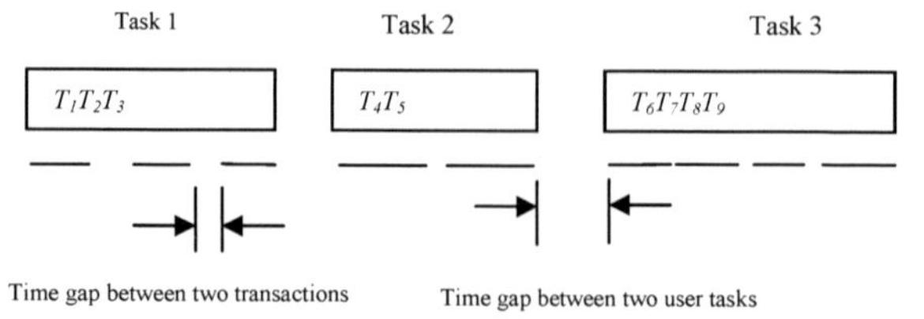


Figure 1: Time Gaps between Transactions and between User Tasks

Our data mining method for clustering user transactions into user tasks is based on the following observation. Normally thetimegap between twoadjacenttransactions inside one user task is significantly greater that between twoadjacentuser tasks. The procedure for clustering user transactions is divided into three steps.

First, the mean absolute deviation between adjacent transactionsis computed as:

$$d = \frac{1}{j-i}(| x_{i,i+1} - m | + | x_{i+1,i+2} - m | + ... + | x_{j-1,j} - m |) \quad (1)$$

where$x_{i,i+1}$, ..., $x_{j-1,j}$are$j-i$measurements of thetimegapsbetween transactions, and$m$is the mean value of thetimegap between transactions, i.e.,$m = \frac{1}{j-i}(x_{i,i+1} + x_{i+1,i+2} + ... + x_{j-1,j})$.

Second, the*z-score*is calculatedfor thetimegap between transactions.

$$z_{i,i+1} = \frac{x_{i,i+1} - m}{c} \quad (2)$$

where $c$ is the standard deviation,$x_{i,i+1}$is thetimegap between transaction$i$and transaction$i$+1,$m$is the mean of thetimegap between transactions.

Third, whenz-score$z_{i,i+1}$is greater than a predefined threshold$t$, atimegap between user tasks is detected. This can be explained by using an example case involving banking transactions. Normally, each client in a bank needs to wait for the previous clientto complete his/her tasks in order to get served; moreover,before severing a client the bank clerk or the operator of a certain banking application needs to interact with the client to find theparticular needs of the client. Thus the*z-score*for thetimegap betweentheuser tasks is significantlylargerthan that of two transactions in a user task. In order to find a desired threshold$t$, the mean*z-score*between two adjacentuser tasksduring the busiest business-hour of a particular operator can be calculated and used as the threshold. It is desirable to use differentthresholdsfor different operators.

### *B. Identification of Inter-transaction Data Dependencies*

We propose to identify the malicious transactions based on the followingobservation.In many cases, when data is updated in a transaction, a sequence of other data itemsisupdated in several transactions of the sameusertask. Although the transaction code decides the exact the update sequence of these data items,oftendata updated in one user taskreflects the data dependencies across transactions, i.e.,*inter-transaction data dependencies*. For example,assume a clerk in a bank assistsa customerinhis banking needs.First, the customerwantstoknow the balances of his accounts;this is the first transaction. After that, the customer decidesto transfer money fromhissaving account tohischecking account in order to keep the balance of the checking account above the minimum requiredbalance. This isthe secondtransaction. Then, the customer withdraws some money from his saving accounts. This is the third transaction. After that, there might be a separate transaction executedby the systemfor auditing or logging purpose.Thus, one task may involve multiple transactions.

***Definition1***: The***Inter-transaction Write Dependency***rulespecifies that afteradata item, say$x$,is updated, a group of write operationsmustbe performed onsome otherdata items, say$d_1,d_2, \ldots, d_n$,within auser task.We use the notation$w(x) \rightarrow w(d_1,d_2, \ldots, d_n)$ to denote such a rule.It must be noted thatthe sequence for updating dataitems$d_1,d_2, \ldots, d_n$is irrelevanthere.Several inter-transaction write dependencies may existfor any data item.Each of them may have different support and confidence.The*Inter-transaction Write Dependency Set*$ITWD(x)$ denotes the set containing allwrite dependenciesof data item$x$.

The inter-transaction writedependency does notconsiderthe exact updateorder of the data items$d_1,d_2, \ldots, d_n$. Rather the orderis dependent on the intrinsic data dependencies and the programming logic for user task(s) that is (are) used for updating data item$x$. If data items$d_1,d_2, \ldots, d_n$are always updated in certain sequenceafter data item$x$is updated, thisupdating sequence isconsideredausefulpropertyfor detecting malicious database transactions. Thus,we define*Inter-transaction Write Sequence*for this purpose.

***Definition2***: The***Inter-transaction Write Sequence***is an operation sequence with the format$w(x) \rightarrow w<d_1,d_2, \ldots, d_n>$. The sequence specifies that after a transaction updates data item$x$, the transaction needstoperform write operations on$d_1,d_2, \ldots, d_n$in sequence within auser task.Several inter-transaction write sequences may existfor any data item$x$.Each of them may have different support and confidence.The inter-transactionwritesequenceset$ITWS(x)$ denotes the set containing all dataupdatingsequences of data item$x$.

It can be seen that inter-transaction write sequencescan more precisely profile the data dependencies across transactions than inter-transactionwritedependenciessincethe formerillustrates theexactupdateorder of data itemsinvolved in a user task. However, due to the possible conditional execution of transactions involved in the user tasks, the exact execution order of these transactionsand the number of data items updated may not always be the same.

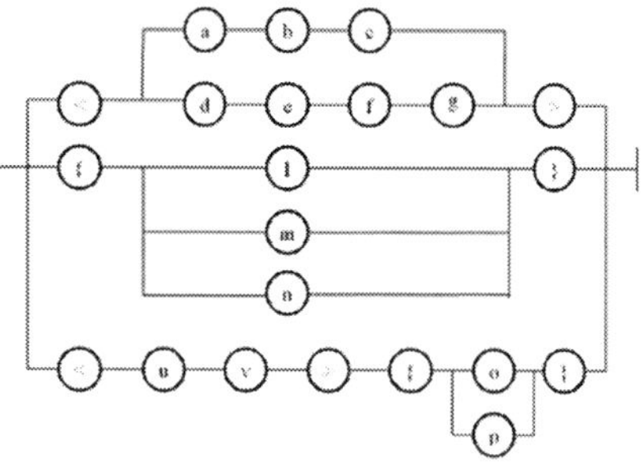


Figure 2. An example of Inter-transaction Write Dependency and Write Sequence

Figure 2 illustrates an example of inter-transaction write dependency andinter-transactionwrite sequence.In this figure, left and right angle brackets indicate the beginning and end ofone or more inter-transaction write sequencesrespectively. While the left and right curly brackets indicate the beginning and end of one inter-transaction write dependency respectively. Figure 2shows following four inter-transaction write sequences, i.e., $w(x) \rightarrow <w(a),w(b),w(c)>$,$w(x) \rightarrow <w(d),w(e),w(f),w(g)>$,$w(x) \rightarrow <w(u),w(v),w(o),w(p)>$,and $w(x) \rightarrow <w(u),w(v),w(p),w(o)>$. There is one inter-transaction write dependency $w(x) \rightarrow w(l,m,n)$.Since the inter-transaction write dependency does not care about data updating sequence, thus we draw data items on different parallel lines to indicate this category of data dependency. It must be noted in the graph that we use the concate nation of an inter-transaction write sequence and an inter-transaction write dependency to represent two inter-transaction write sequences for simplifying the graphical representation of two inter-transaction write sequences $w(x) \rightarrow <w(u),w(v),w(o),w(p)>$ and $w(x) \rightarrow <w(u),w(v),w(p),w(o)>$.Figure 2 shows that after data item $x$ is updated, some data items have to be updated based on *at least one* inter-transaction data dependency rule specified.

Followingexampleillustrates how the inter-transaction data dependencieshelp detect malicious transactions. Supposean attacker operating a remote terminal submitsfivemalicious databasetransactionseach of which satisfies*intra-transaction*data dependencies and thus appears as legitimate user transaction. These transactions arepresented in Table 1. Also assume the attacker does not have the knowledge ofinter-transaction data dependencies specified in Figure 2.Let us consider two scenarios. The first scenario is that the attacker submitted these transactions slowly, thus each of them is considered by our intrusion detection system as a*separate*user task. In transaction 1, data item$x$is updated. The onlyotherdata items updated intheuser task consisting of transaction 1 is data item$d$and$f$.

Theinter-transaction data dependencyrule consisting of data item*d*and*f*is$w(x) \rightarrow <w(d),w(e),w(f),w(g)>$ which specifies data item*d,e,f,* and*g*have to be updated in sequencein a user taskafter*x*is updated. So the user task does not comply with the data dependencies specified and transaction 1 is considered as a malicious transaction. The second scenario is that the attacker submitted these transactionsin a very short time window;thus,all of these are considered as a part of a user task. Although in this case, data item*d,e,f,* and*g*are updated in the user taskby transaction 1, 3, and 5, theupdating sequence$d \rightarrow f \rightarrow g \rightarrow e$is out of order.Therefore,this user task is identified as malicious.

Table 1. Example Malicious Transactions

| Trans. ID | Transaction Operations |
|---|---|
| 1 | r(*b*), w(*x*), r(*y*), w(*d*), w(*f*); |
| 2 | r(*m*), r(*n*); |
| 3 | r(*a*), w(*g*), w(*y*), r(*b*), w(*j*); |
| 4 | w(*t*), w(*s*); |
| 5 | r(*z*), w(*e*), w(*q*), w(*i*), r(j), w(j), r(p); |

Next, we illustrate our steps for mining inter-transaction write dependencies and inter-transaction write sequences across transactions. First, the transactions in each user task need to be preprocessed. This is because the transactions in the database log not only contain write operations but may also have read operations. These read operations are not of our interest since we only consider inter-transaction data dependencies based on the updating operations. Also information about aborted transactions also needs to be filtered. Any transactions submitted by system administrators are also removed since the system maintenance activities are not considered as a part of "normal" system operations. For example, the system administrator may manually manipulate database tables for fixing or cleaning up inconsistent database tables due to logic errors of applications. After this data preprocessing phase, a filtered user task only contains write operations of each transaction in the task.

### *C. Algorithms for Generating Inter-transaction Data Dependency Rules*

It must be noted that although transactions are clusteredinto user tasks based on user IDs,when generating the inter-transaction write dependenciesor inter-transaction write sequences,the proposedapproachdoes not profile inter-transaction data dependencies for each individual user. Rather, it tries to find the intrinsic data correlations across all user tasks inside the database. Since profiling individual user's activities is prone to the change of users' usage patterns and involves a lot of overhead for storing and matching classification rules for detecting malicious activities against each user, our approachdoes have notable advantagefrom this point of view.

Since inter-transactionwritedependencies specifydata dependencies across user transactions regardless of dataupdating sequence, discovering this kind of data correlation is intrinsically similar to the procedure of association rule mining. Although in this case, we are trying to find associationrules across different transactions, the user tasks clustered are considered as atomic units that are analyzed for association rule discovery. The inter-transactionwritesequencesillustrate data update sequences;thus,mining this category of data relationships is similar tothesequentialpatternmining process. Following is the algorithms for discovering theinter-transactiondata dependencies.

***TheAlgorithm:***

1. Initialize the inter-transaction write dependency set$ITWD$= {},
Initialize the inter-transaction write sequence set$ITWS$= {},
Initialize a temporary sequence set$T$={}.
2. Generate association rules$R= \{r_i |$ support $(r_i) > \tau_1$,confidence$(r_i) > \tau_2\}$ by using existing association rule mining algorithm on write operations of user tasks, where$\tau_1$and$\tau_2$are the minimum support and confidence respectively and$r_i$representsan instance of inter-transaction write dependency such as$w(x_1) \rightarrow w(a,b,c)$.
3. Generate sequential patterns$P= \{p_i|$ support$(p_i) > \tau_1\}$ by using existing sequential pattern mining algorithmon write operations of user tasks, where$\tau_1$is the minimum support and$p_i$represent an instance ofwrite sequence pattern such as$<w(d),w(e),w(f)>$.
4. **for**each sequential pattern$p_i$where $|p_i| > 1$
**for**each write operation$w(d_i)$insidethe sequential pattern
 **if**$<w(d_i),w(d_{j1}),w(d_{j2}),w(d_{j3}),\ldots,w(d_{jk})> \notin P$and
 $<w(d_{j1}),w(d_{j2}),w(d_{j3}),\ldots,w(d_{jk})> \neq <\emptyset>$
 $T=T \cup \{<w(d_i),w(d_{j1}),w(d_{j2}),w(d_{j3}),\ldots,w(d_{jk})>\}$, where$w(d_{j1}),w(d_{j2}),w(d_{j3}),\ldots,w(d_{jk})$ are all sequenced write operations after$w(d_i)$
**for**each sequence in$T$
**if**support($<w(d_i),w(d_{j1}),w(d_{j2}),\ldots,w(d_{jk})>$)/ support($<w(d_i)>$) $>\tau_2$
 $ITWS=ITWS \cup \{w(d_i) \rightarrow w<d_{j1},d_{j2},\ldots,d_{jk}>\}$
5. **for**each association rule$r_i$:$w(d_i) \rightarrow w(d_{j1},d_{j2},d_{j3},\ldots,d_{jk}) \in R$
**if**there exists a more restrictive inter-transaction write sequence rule that contains the same data items updated
 $R=R- \{w(d_i) \rightarrow w(d_{j1},d_{j2},d_{j3},\ldots,d_{jk})\}$
6. Assignthe resulting associationrule set$R$to$ITWD$, i.e.$ITWD=R$

To clarify the above algorithm, here we offer some explanations. Step 3 generates the association rules based on write operations of user tasks. But the result set is not directly used as inter-transaction write dependency rules. The reason is that there could be a more restrictive inter-transaction write sequence rule that contain all of the data items. For example, suppose there is an association rule$w(x) \rightarrow w\ (a,b,c)$ and another write sequence rule$w(x) \rightarrow <w(a),w(b),w(c)>$,thenit is more desirable to use the rule$w(x) \rightarrow <w(a),w(b),w(c)>$for detectingmalicious database transactions. This is becausethemalicious database transactionshave to update these four data items in exact sequence without being detected. Thus,in Step 5 of the algorithm, we eliminate these redundant association rules from the set*R*.

## IV. Experiments and Analysis

In order to evaluate the performance of the proposed approach and to examine how this approach affects the true positive rates and false positive rates, we designed two separate experiments. One experiment used only intra-transaction data dependency approach proposed by Hu et al. [9] for detecting malicious transactions. Another experiment used both intra-transaction data dependency and the proposed inter-transaction data dependency for detecting malicious database transactions. Two database logs were employed in these experiments. The first one was the log that consisted of legitimate user transactions that were used by both experiments for mining data dependencies. The second log contained synthetic transactions that were randomly generated and were treated as malicious transactions.

Figure 3and Figure 4illustrates the false and true positive rates respectively in detecting malicious transactions in user tasks. The true positive rates were generated using our proposed approach on the database log containing malicious transactions. It can be seen that by employing both inter-transaction and intra-transaction data dependencies for detecting malicious database transactions, the true positive rate increased greatly than that by only employing intra-transaction data dependencies. When the average number of write operations in a transaction was low, the increase was more significant. For example, when there was one write operation in each transaction, employing both intra-transaction and inter-transaction data dependencies for detecting malicious transactions made the true positive rate increase by 16%. The false positive rate was generated using the database log containing normal user transactions. It can be seen that this new approach did not affect the false positive rates very much. The average increase of false positive rate was about 0.6%.

Figure5and Figure6presents the effectiveness of the algorithm for detecting user tasks consisting of different number of transactions. It shows that the more number transactions in legitimate user tasks for profiling inter-transaction data dependency, the more effective the inter-transaction data dependencies in detecting malicious database transactions. It was observed that the true positive rate increased by 25% when the average number of transactions in legitimate user tasks increased from 2 to 5 (This was observed when the average number of write operations in a transaction was 1). Also from the false positive rate shown in this figure, it is noted that the false positive rate only varied from 1% to 2% when the average number of transactions increased from 2 to 5 in a user task.

## V. Conclusions

In this paper, we presented an approach for studying inter-transaction data dependencies and mining them for identifying malicious database transactions. The developed method proved to be capable of detecting attacks carried out by a set of seemingly harmless database transactions. We first proposed a data mining method for clustering legitimate user transactions into user tasks for facilitating discovery of inter-transaction data dependencies. Then we illustrated the methodology for mining inter-transaction write dependencies and inter-transaction write sequences. The experiments confirmed the advantage of mining inter-transaction data dependency for detecting malicious database transactions. As part of our future work, we will develop a model for identifying malicious transactions in clustered database environment where data fragmentation and parallel queries pose new challenge to database intrusion detection. We also plan to develop a database intrusion detection framework for the Service-Oriented Architec

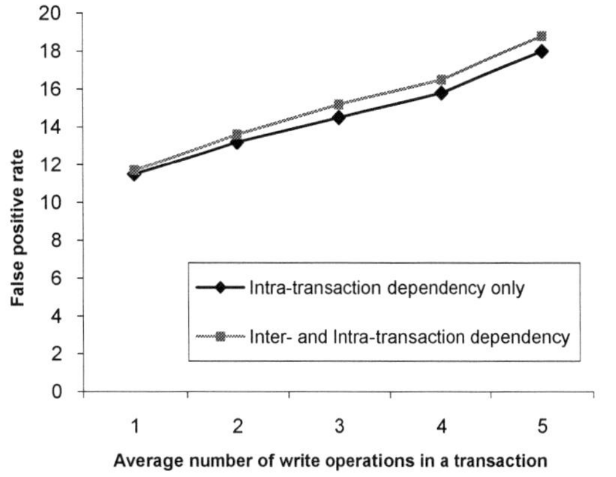


Figure 3. False Positive Rates in Detecting Malicious Transactions

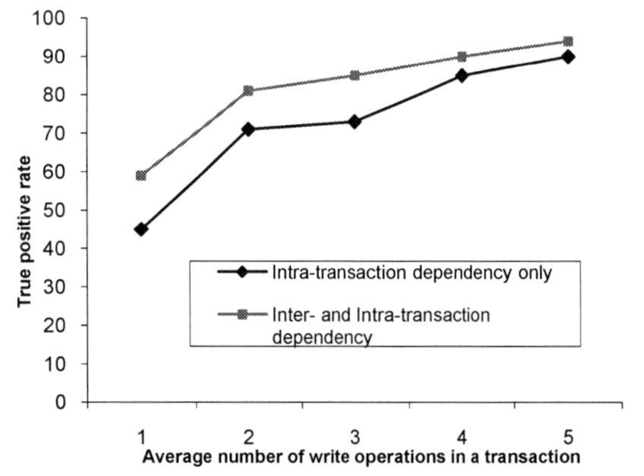


Figure 4. True and False Positive Rates in Detecting Malicious Transactions

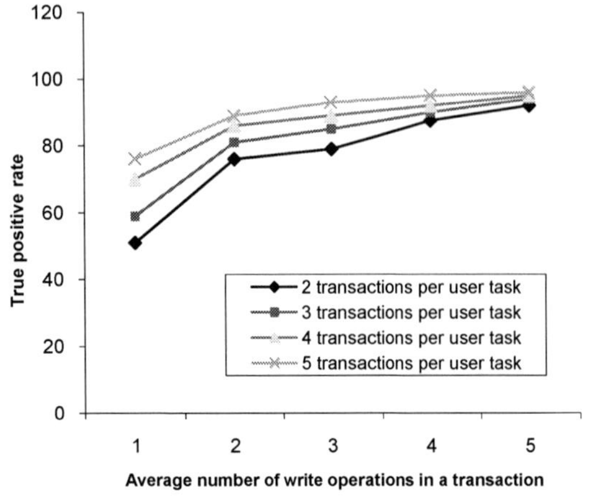


Figure5.True positive rate related to number of transactions per user task

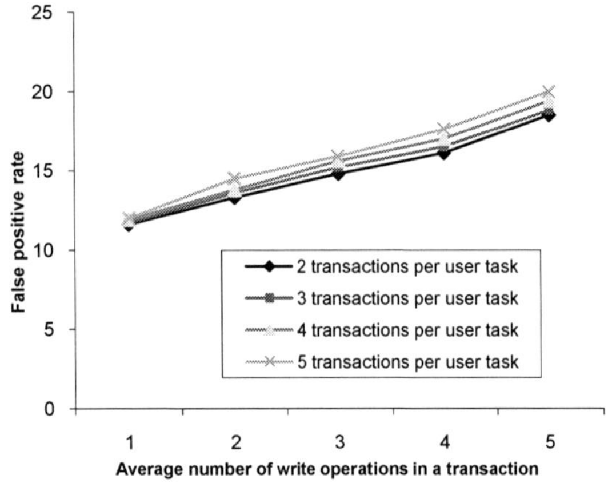


Figure 6. False positive rate related to number of transactions per user task

ACKNOWLEDGEMENT

Research of Brajendra Panda has been supported in part by US AFOSR under grant FA 9550-04-1-0429. We are thankful to Dr. Robert. L. Herklotz for his support, which made this work possible.

REFERENCES

[1] PGP Corporation's 2006 annual study on the cost of data breaches. http://download.pgp.com/pdfs/Ponemon2-Breach-Survey_061020_F.pdf

[2] Rietta, F.S. Application Layer Intrusion Detection for SQL Injection. In ACM Southeast Regional Conference, 2006.

[3] Ramasubramanian, P., Kannan, A. Intelligent multi-agent based database hybrid intrusion prevention system. In ADBIS 2004, LNCS, vol. 3255, pp. 393-408. Springer, Heidelberg, 2004.

[4] Valeur, F., Mutz, D., Vigna, G. A learning-based approach to the detection of SQL attacks. In DIMVA 2005, LNCS, vol. 3538, pp.123-140. Springer, Heidelberg, 2005.

[5] Bertino, E., Kamra, A., Terzi, E., Vakali, A. Intrusion Detecton in RBAC-administered databases. In ACSAC, pp. 170-182. IEEE Computer Society Press, Los Alamitos, 2005.

[6] Barbara, D., Goel, R., and Jajodia, S. Mining Malicious Data Corruption with Hidden Markov Models. In Proceedings of the 16th Annual IFIP WG 11.3 Working Conference on Data and Application Security, Cambridge, England, July 2002.

[7] Chung, C., Gertz M., and Levitt, K. DEMIDS: A Misuse Detection System for Database Systems. In Third Annual IFIP TC-11 WG 11.5 Working Conference on Integrity and Internal Control in Information Systems, Kluwer Academic Publishers, pages 159-178, November 1999.

[8] Lee, V. C.S., Stankovic, J. A., Son, S. H. Intrusion Detection in Real-time Database Systems Via Time Signatures. In Proceedings of the Sixth IEEE Real Time Technology and Applications Symposium, 2000.

[9] Hu, Y.and PandaB.A Data Mining Approach for Database Intrusion Detection.In Proceedings of the 19th ACM Symposium on Applied Computing, Nicosia, Cyprus, Mar. 2004.

[10] Hu, Y.and PandaB.Design and Analysis of Techniques for Detection of MaliciousActivities in Database Systems.Journal of Network and Systems Management, Vol. 13, No. 3, Sep. 2005.

[11] Srivastava, A.,Sural, S.,and Majumdar,A.Weighted Intra-transactional Rule Mining for Database Intrusion Detection. Advances in Knowledge Discovery and Data Mining. Vol. 3918, pp. 611-620, 2006.

[12]Fonseca, J., Vieira, M., and Madeira H. Monitoring Database Application Behavior for Intrusion Detection. In Proceedings of the 12th Pacific Rim International Symposium on Dependable Computing, 2006.

[13]Agrawal, R. and Srikant, R. Mining Sequential Patterns. In Proceedings of the 1995 Int. Conf. Data Engineering, Taipei, Taiwan, March 1995. Pages 3 -14.

# Energy-Security Adaptation Scheme of Block Cipher Mode of Operations

Amit K. Beeputh, M. Razvi Doomun, Padaruth Dookee
Faculty of Engineering
University of Mauritius
Reduit, Mauritius
amit.beeputh@gmail.com, r.doomun@uom.ac.mu, dookeepadaruth@gmail.com

*Abstract*— **With rapid growth in wireless network technologies, there is high need for secure communication in such as highly heterogeneous environment. Since mobile wireless devices have limited battery capacity, wireless network security should make optimal use of this resource. Implementing security is challenging under this condition. This paper aims to investigate optimal use of energy under the block cipher AES (Advanced Encryption Standard) algorithm by adapting to the energy available at the network nodes. Besides, the error rate parameter is also considered in this adaptation framework. Different modes of operation electronic codebook (ECB), Cipher Block Chaining (CBC), and Counter (CTR), have different energy performance and error resilience.**

***Keywords:*** *wireless security, block cipher, mode of operation, AES*

## I. INTRODUCTION

For the past few years, wireless technology has been an innovative field of research. Wireless access technologies [6] have become more visible and affordable and recently, industry has made significant progress in resolving some constraints to the widespread adoption of wireless technologies, such as disparate standards, low bandwidth, and high infrastructure and service cost [5]. However, failure to implement security in the wireless network, gives rise to several problems like service violations, bandwidth shortages, abuse by malicious users, monitoring of the user activity, direct attack on user's device [1]. Thus, in this energy constraint environment, to provide a secure mechanism is a serious field of research in the wireless sensor network (WSN) [4][16][17]. There are many uses for WSN such monitoring, tracking, and controlling. Some of the specific applications are habitat monitoring, object tracking, nuclear reactor controlling, fire detection, traffic monitoring, etc. There are quite a few systems which adapt dynamically to the system [2][3][9][14]. For example the REAR [13] (reliable energy aware routing) has been developed which is a distributed, on-demand, reactive routing protocol that is intended to provide a reliable transmission environment for data packet delivery.

Since the sensors are geographically distributed, there is a need to make optimal use of energy usage so as to increase the battery lifespan of the sensors. This needs to be because it may happen that some of the sensors cannot be physically replaced easily, for example if a sensor is in the enemies' territory. The security requirements for WSN are similar to traditional network but more challenging to implement. Security for WSN needs to be a compromise between adequacy of security and energy availability [7]. The encryption used for this paper is the AES.

## II. ADVANCE ENCRYPTION STANDARD

After DES, the introduction of Advanced Encryption Standard (AES) [11] was adopted by the US government as the encryption standard in the start millennium after about five years of standardization process in 2006. It is now one of the most important cryptographic algorithms today and has received a lot of attention from researchers due to its ease to deploy, little memory usage and easy implementation.

### *A. Architecture of AES*

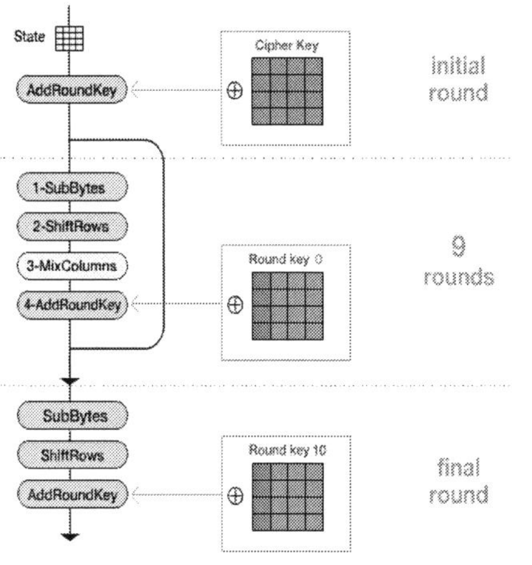


Figure 1. Architecture of the AES

During encryption input block is exclusive-or (XOR) with a round-key in the pre-processing step and the output is iteratively transformed Nr times using a round function where Nr denotes the total number of rounds and each iteration uses a different round key. Usually for the 128 AES, there are 10 rounds which are carried out as shown in **the figure 1**; for 192 bit key AES there are 12 rounds, and for 256 bit key AES there are 14 rounds. The final round does not have the MixColumns stage.

T. Sobh (ed.), *Innovations and Advances in Computer Sciences and Engineering*,
DOI 10.1007/978-90-481-3658-2_13, 

AES operates on a 4×4 array of bytes, referred as the state. For other versions of Rijndael, with a larger block size, additional columns will be present in the state. For encryption, each round (called the round function) of AES (except the last round which is different) consists of four stages: SubBytes, ShiftRows, MixColumns and AddRoundkey.

### B. *SubBytes*

This transformation is a non-linear byte substitution that operates independently on each byte of the State using a substitution table (S-box). The latter is appreciated for its non-linearity properties and is derived from the multiplicative inverse over GF ($2^8$). In this stage, each byte in the state is replaced with its corresponding entry in a fixed lookup table.

### C. *ShiftRows*

A permutation operation which makes each row of the state shifts cyclically a certain number of steps. The first row is not shifted whereas each byte in the second row is shifted one to the left. The third row is shifted two to the left, fourth row is shifted three to the left. This operation moves the bytes to "lower" positions in the row, while the "lowest" bytes wrap around into the "top" of the row.

### D. *MixColumns*

In MixColumn, the columns of the State are considered as polynomials over GF(28) and multiplied modulo x4 + 1 with a fixed polynomial c( x ), given by c(x) = '03' x3 + '01' x2 + '01' x + '02' . This polynomial is coprime to x4 + 1 and therefore invertible. The application of this operation on all columns of the State is denoted by MixColumn (State). The inverse of MixColumn is similar to MixColumn. Every column is transformed by multiplying it with a specific multiplication polynomial d(x), defined by ('03' x3 + '01' x2 + '01' x + '02') X d(x) = '01'.

It is given by: d(x) = '0B' x3 + '0D' x2 + '09' x + '0E'

### E. *AddRoundKey*

In this step, the subkey is combined with the state. For each round, there is a subkey which is derived from the main key using Rijndael's key schedule; each subkey is the same size as the state. The subkey is added by combining each byte of the state with the corresponding byte of the subkey using bitwise XOR.In most cases, increasing the number of encryption rounds improves security at the cost of system throughput and vice-versa. The proposal for Rijndael provides details on the implementation of AES for a 32-bit processor. Accordingly different steps of the round can be combined in a single set of table lookups, allowing for very fast implementation. Each cell of the state can be separated and treated differently.

## III. MODE OF OPERATION

### A. *The Electronic Codebook Mode*

It is the simplest mode of operation where the plaintext is divided into blocks of required size and each block is encrypted separately [8]. The resulting disadvantage is that the same plaintext will yield same ciphertext. There is no feedback associated with this mode of operation. For decryption, the blocks of ciphertext are independently to the inverse cipher function. The resulting sequence produces back the plaintext. Since the encryption of each block is done independently, there can be parallel processing of the plaintext to ciphertext and vice versa.

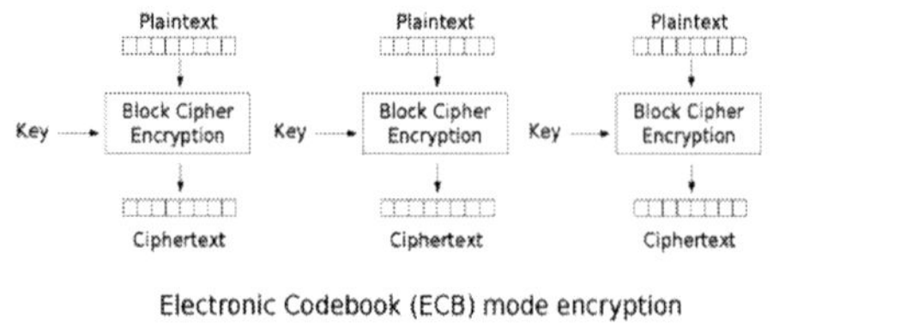


Figure 2. ECB Encryption

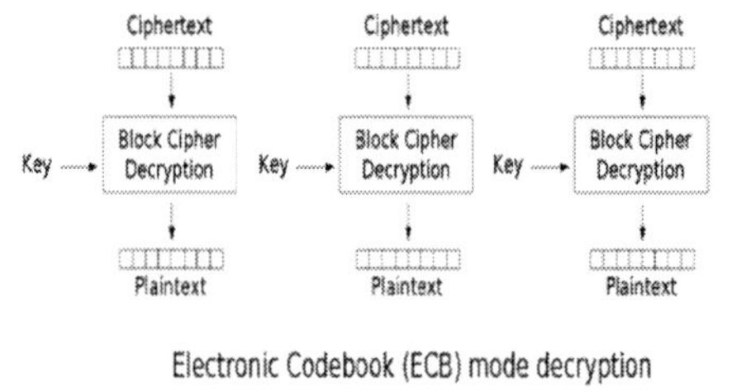


Figure 3. ECB Decryption

In the ECB, the pattern is not hidden efficiently. Thus attacker can make use of ciphertext/plaintext pair to cause damage. This, in term, can cause serious threat to the message confidentiality. In ECB encryption and ECB decryption, multiple forward cipher functions and inverse cipher functions can be computed in parallel. However, errors are not propagated in this mode. That is if ever an error occurs in a block, it will not be propagated in other blocks.

### B. *Cipher Block Chaining Mode*

The first plaintext block combines with an initialisation vector (IV) in the Cipher Block Chaining (CBC) mode [15]. The IV is not secret but unpredictable. The integrity of the IV should also be protected. The CBC mode is defined as follows:

CBC-Encryption:

$C_i = E_k\ (P_i\ XOR\ C_{i-1}),\ C_0 = IV$

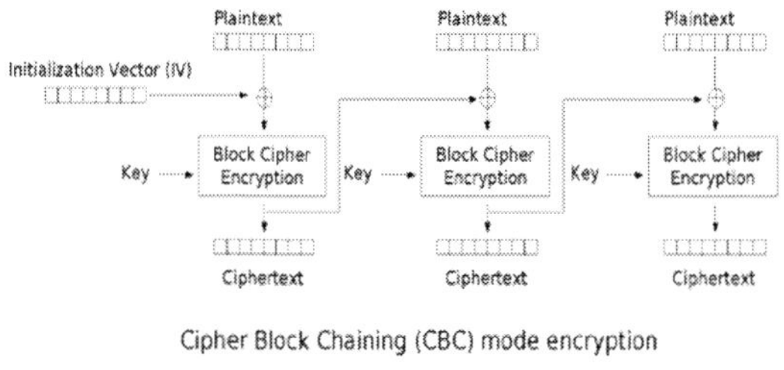


Figure 4. CBC Encryption

Once the XOR of the first plaintext and IV are done, this input block is applied to the cipher function along with the key. The resulting output is the first block of ciphertext. The latter is also XORed with the second plaintext block to manufacture the second output block as shown in the diagram above. The second output block is XORed again with the third plaintext block to produce the next input block. Thus each successive plaintext block is XORed with the previous ciphertext bloc in order to produce the new input block for the cipher function. This process is repeated till the last plaintext block has been processed.

CBC-Decryption:

$P_i = D_K(C_i) \text{ XOR } C_{i-1}, C_0 = IV$

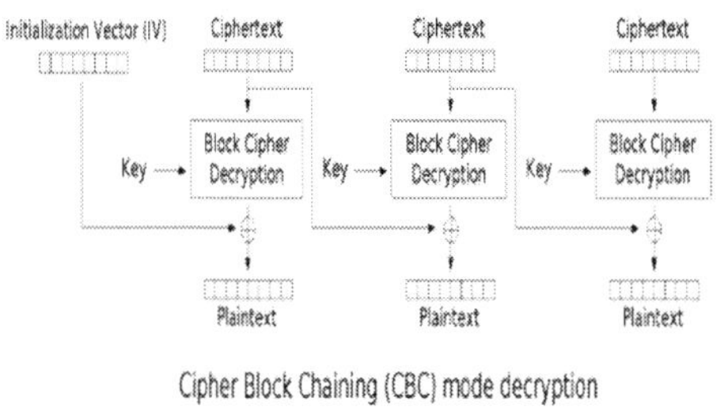


Figure 5. CBC Decryption

For the decryption in CBC, the first ciphertext block is decrypted using the inverse cipher function. The resulting output block is then XORed with the IV to produce the first plaintext block. For the second ciphertext block, it is decrypted using the key and the result is XORed with the first ciphertext block to recover the second plaintext block. This process continues till the last plaintext has been recovered.

Since there is dependency on previously generated blocks in the encryption process, this mode cannot be run in parallel. In CBC decryption, however, the input blocks for the inverse cipher function, i.e., the ciphertext blocks, are immediately available, so that multiple inverse cipher operations can be performed in parallel. Moreover, if ever there is a one-bit change in a plaintext, this affects all following ciphertext blocks, and a plaintext can be recovered from just two adjacent blocks of ciphertext.

### C. *Counter mode (CTR)*

The counter mode (CTR) [18] makes a block cipher into a stream cipher. The keystream is XORed with the plaintext block to produce the ciphertext block. The next keystream block is generated by the encryption of successive values of a counter. The latter is a simple operation such as an integer increment modulo 2w, where w is a convenient register width and the sequence is guaranteed not to repeat for a long time.

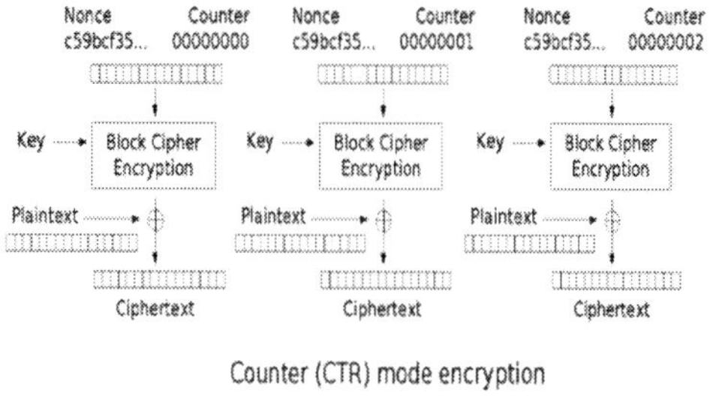


Figure 6. CTR Encryption

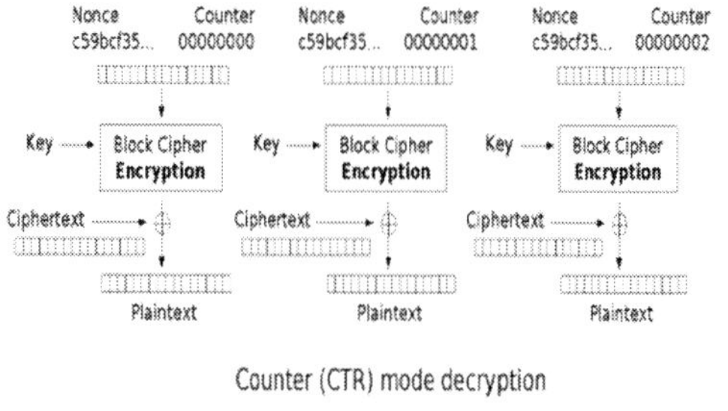


Figure 7. CTR Decryption

## IV. Error Propagation

### A. *ECB*

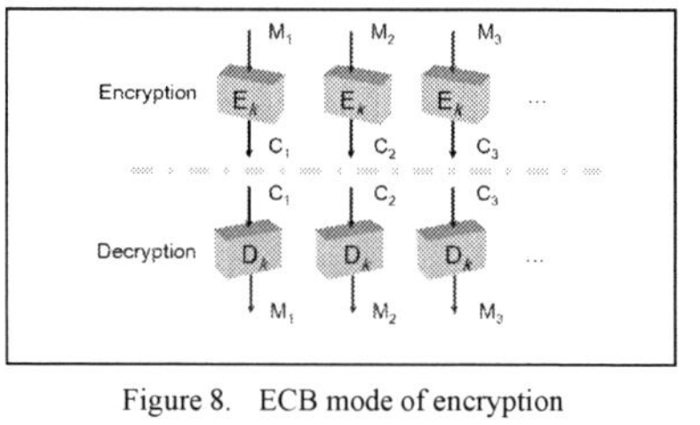


Figure 8. ECB mode of encryption

Electronic codebook (ECB) being the simplest of the cipher modes with the message divided into blocks and each block is encrypted separately. The immediate drawback of this mode is that identical plaintext blocks are encrypted into identical cipher blocks and thus not hiding data patterns well [12].

The encryption formula is given by:

$Ci = Ek\ (Pi)$

The decryption formula is given by:

$Pi = Dk\ (Ci)$

ECB does not have error-propagation characteristics since no feedback technique is employed. The following formula represents the error propagation index of EBC mode.

***No of Corrupted deciphered ciphertext = No of corrupted ciphertext***

$P_{corrupt} = C_{corrupt}$

***Total Number of Blocks*** $= T_B$

***Error Index*** $= P_{corrupt} / T_B$

B. *CBC*

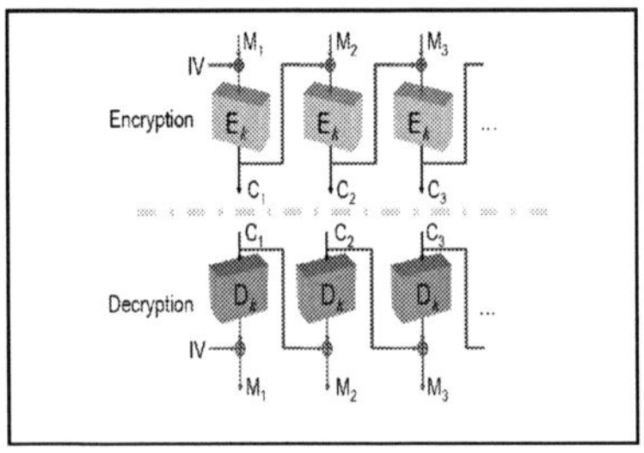


Figure 9. CBC mode of encryption

In cipher-block-chaining (CBC) mode, each block of plaintext is XORed with the previous ciphertext block before being encrypted. This way, each ciphertext block is dependent on all plaintext blocks processed up to that point. Moreover, to make each message unique, an initialization vector must be used in the first block.

The encryption formula is given by:

$Ci = Ek\ (Pi\ XOR\ Ci\text{-}1)$

The decryption formula is given by:

$Pi = Dk\ (Ci)\ XOR\ Ci\text{-}1$

CBC has limited error-propagation during decryption. An error in one block propagates only to the decryption of the next block. The following formula represents the error propagation index characteristic of CBC mode.

***No. of Corrupted deciphered ciphertext = No of corrupted ciphertext + 1***

$P_{corrupt} = C_{corrupt\ +1}$

***Total Number of Blocks*** $= T_B$

***Error Index*** $= P_{corrupt} / T_B$

CBC error propagation is independent on the number of corrupted bits within each block.

C. *CTR mode*

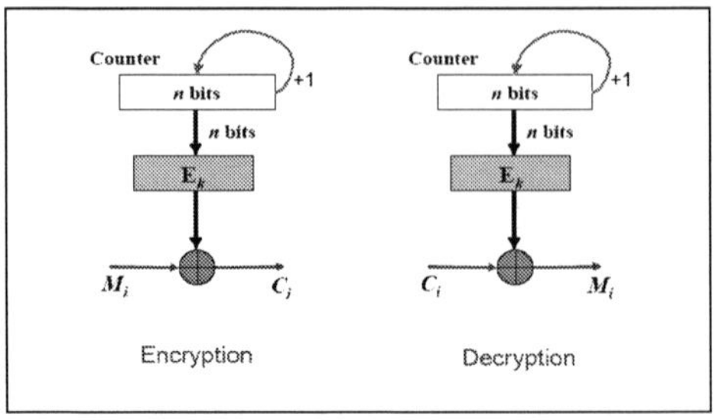


Figure 10. CTR mode of encryption

Counter mode turns a block cipher into a stream cipher. It generates the next keystream block by encrypting successive values of a 'counter'. The counter can be any simple function which produces a sequence which is guaranteed not to repeat for a long time.

CTR mode does not have error propagation and thus have the same error propagation model as ECB. The following formula represents the error propagation index characteristic of CTR mode.

***No of Corrupted deciphered ciphertext = No of corrupted ciphertext***

$P_{corrupt} = C_{corrupt}$

***Total Number of Blocks in ciphertext*** $= T_B$

***Error Index*** $= P_{corrupt} / T_B$

## V. SYSTEM OVERVIEW

Each of the mode of operation have their own rating to the criteria such as robustness, confidentiality, scalability, error propagation, complexity, energy consumption, extent of parallelization and flexibility.

TABLE I. RATING OF THE BLOCK CIPHER MODE

| | ECB | CBC | CTR |
|---|---|---|---|
| Robustness | 2 | 3 | 4 |
| Confidentiality | 1 | 3 | 5 |
| Scalability | 3 | 3 | 3 |
| Error propagation | 1 | 2 | 1 |
| Complexity | 1 | 2 | 4 |
| Energy consumption | 2 | 3 | 5 |
| Parallel | 5 | 2 | 4 |
| Flexibility | 3 | 3 | 4 |

1-Bad 2-Poor 3-Fair 4-Good 5-Excellent

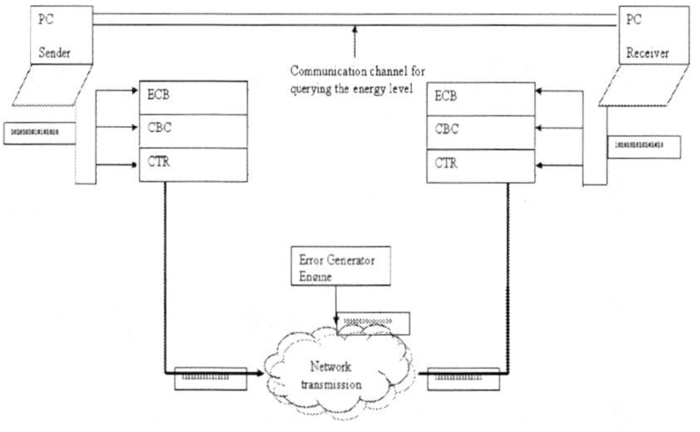


Figure 11. System overview

The sender will query the energy level at the receiver after an interval of 30 seconds. This interval is optimum because energy depletion takes place after every cycle. For our purpose, this interval is optimum. After acquiring the energy level of the receiver, the sender gets the minimum of the energy level at the two ends. For example if the energy level at the sender side is A% and energy level at the receiver side is B% and A<B, then the energy level to determine the encryption mode is A.

The query of the energy level is done via an independent communication channel. To avoid excess traffic in the channel, no security is used. In fact, the sender, X, will send a probe to the receiver, Y. Upon the reception of that probe the receiver Y will immediately calculate its energy level and send it to the sender. At the sender side, the minimum energy level is determined and that determines which block encryption method is to be used. For example, if the energy level at the sender side is 10% and that of the receiver is 100%, then the least secure encryption, which is the ECB, is used for encryption. The mode of encryption used is also sent via the network in order for the receiver to know which decryption method is to be used.

### A. *Client Server Simulation*

The client server simulation exists for this application. The energy level will be queried. Ideally, the client should send a probe to the receiver via the network. Besides, the error engine should reside at the receiver side. The stream of packets received, are passed through an error generator engine in order to generate some errors in the packets. The error engine can create an error at particular random position depending on the percentage error to be injected.

### B. *Energy adaptation Framework*

The following table displays the energy level that will be used for the swapping of the mode in the system.

TABLE II. ENERGY LEVEL SWAPPING RATE.

| | Total Rating | Battery Level Percentage |
|---|---|---|
| ECB | 18 | <=40% |
| CBC | 21 | >40% & <=70% |
| CTR | 30 | >70% |

### C. *Experimental platform*

The experimental platform needs to be simple and feasible with respect to the available resources for this project. Thus some of the highly viable measurement techniques analyzed can not be employed due to budget constraints. Alternatively an alternative but satisfactory experimental framework has been designed. The specifications of the target platform are shown in the table below.

TABLE III. EXPERIMENTAL PLATFORM

| | |
|---|---|
| **Processor** | Intel Pentium 3 (1.60 MHz) |
| **RAM** | 512 MB |
| **OS** | Windows XP Professional (SP2) |

The codes are compiled and executed using Microsoft Visual C++ 6.0 © Enterprise Edition with optimizations enabled.

## VI. SIMULATION & RESULTS

The energy level can be used to categorize the energy level of the mode of operation. ECB will be a low energy level, CBC a medium energy level and CTR as high energy level.

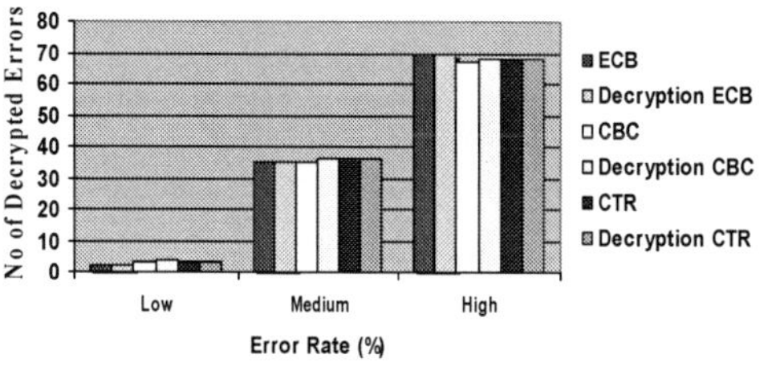


Figure 12. Error Propagation results

From the graph, it can be observed that for every error, irrespective of the level of errors, is injected in the network, the same number of block errors is decrypted at the decryption side in the ECB mode. Similarly, for every error that the CTR mode encounters, the same number of errors obtained while decryption. However, the CBC mode is different. If X number of error is injected in the network, X+1 number of block errors are decrypted at the decryption side. The energy adaptation framework will provide a means to automatically change the mode of operation so that the energy is not depleted quickly.

Energy is optimally used and battery life prolonged when energy adaptation system (EAS) is used. The blue line highlight the energy level on the device when there is swapping of mode of operation and the other line represent the counter mode being used throughout an encryption process of a 20 MB of text file. Results obtained prove that there is about 3% energy gain while using EAS. Error propagation index has also been used as a parameter to the algorithm. Depending on the error index the appropriate encryption mode is employed.

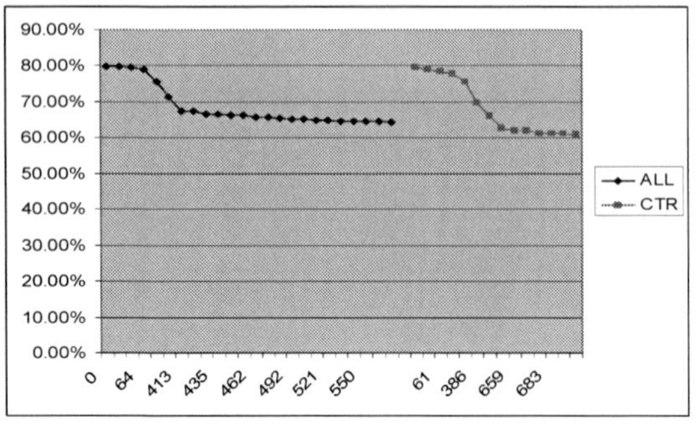


Figure 13. Energy adaptation results

### A. *Constraint*

Since there were no sensor devices to test this system, only a laptop has been used for testing and implementation of the system. To monitor the battery level, software called BatteryMon was used which required license for used/ however, only the trial version has been used.

## VII. CONCLUSION & FUTURE WORKS

### A. *Conclusion*

Energy aware security framework prolongs battery life. Thus the availability of the network is high as the nodes will be "alive" for longer lapses of time. If all important parameters affecting processing are factored in the EAS, then a highly viable and reliable system for WSN can be developed. In this project, only error propagation has been taken into consideration. Errors in a network are inevitable. The network is erroneous; the most secure algorithm having the least error propagation can be used depending on the energy available.

### B. *Future works*

AES consideration for data type (Video, audio, pictures) will be another related field of research. Several types of files can potentially have different energy adaptation framework. This may additionally enhance the lifetime of the device. Variation of key size can be another potentially candidate for the optimization of battery. Additionally, key size directly influences energy consumption. Thus it can be an important parameter to EAS. The framework can be enhanced to cater for different encryption algorithms. E.g RC5 and serpent are among algorithms that are viable for WSN and have been subject of extensive research. A comparison of the results obtained with hardware implementation of EAS will determine which suits best. Different types of hardware devices may reveal different responses to this system.

## REFERENCES

[1] Tom Karygiannis, Les Owens, "Wireless Network Security 802.11 Bluetooth and Handheld Devices," National Institute Of Standards and Technology Special Publication 800-48, Nov 2002.

[2] Hans Van Antwerpen, Nkhil Dutta, Rajest Gupta, Shivajit Mohapatra, Cristiano Pereira, Nalini Venkatasubramanian, Raph von Vignau., "Energy aware System Design for wireless multimedia," Feb 2004.

[3] G V Merrett, B M Al-Hashimi, N M White and N R Harris, "Resource aware sensor nodes in wireless sensor network," Electronic System Design Group, School of Electronics and Computer Science, University of Southampton, SO17 1BJ, UK, 2005.

[4] F. L. LEWIS, "Wireless Sensor Networks," D.J.Cook and S.K. Das, editors, Smart Environments: Technologies, Protocols, and Applications, John Wiley, New York, 2004.

[5] 25 M. Weiser, "The Computer for the 21st Century," Scientific Am., Sept. 1991, pp. 94-104; reprinted inIEEE Pervasive Computing, Jan.-Mar. 2002, pp. 19-25.

[6] Kay Romer and Friedemann Mattern, "The Design Space of Wireless Sensor Networks," IEEE Wireless Communications, pp. 54- 61, Dec. 2004.

[7] Hemant Kumar, "Energy aware security services for sensor network.," Electrical and Computer Engineering, University of Massachusetts Amherst, April 2005.

[8] Yang, F. - Heys, H, "Comparison of Two Self-Synchronizing Cipher Modes," In Proceedings of Queen's 22nd Biennial Symposium on Communications, Kingston, Ontario, Jun. 2004.

[9] S. Bandyopadhyay, et. al., "An Energy-Efficient Hierarchical Clustering Algorithm for Wireless Sensor Networks," IEEE INFOCOM'03.

[10] Y.W. Law, S. Dulman, S. Etalle and P. Havinga, "Assessing Security-Critical Energy-Efficient Sensor Networks", Department of Computer Science, University of Twente, Technical Report TR-CTIT-02-18, Jul 2002.

[11] B. Gladman, "Input and Output Block Conventions for AES Encryption Algorithms," AES Round 2 public comment, June 6, 1999.

[12] A. Menezes, P. van Oorschot, S. Vanstone, "Handbook of Applied Cryptography," CRC Press, 1997.

[13] Rahul C. Shah and Jan M. Rabaey, "Energy aware routing for low energy ad hoc sensor networks," In Proc. IEEE Wireless Communications and Networking Conference (WCNC), volume 1, pages 350--355, Orlando, FL, March, 17-21 2002.

[14] P.J.M. Havinga and G.J.M. Smit, "Energy-Efficient Wireless Networking for Multimedia Applications," Wireless Communications and Mobile Computing, Wiley, 2001, pp. 165-184.

[15] D.I Manfred Lindner, "L93 – Secret-Key Cryptography," Institute Of Computer Technology – Vienna University Of Technology. 2005.

[16] Fernando C. Colon Osorio, Emmanuel Agu, And Kerry Mckay, "Tradeoffs Between Energy And Security In Wireless Networks.," Wireless System Security Research Laboratory. 27 September 2005.

[17] Hasan Çam, Suat Özdemir, Prashant Nair, "Energy-Efficient Security Protocol For Wireless Sensor Networks," IEEE Vehicular Technology Conference, 2003.

[18] Knudsen, L. "Block Chaining Modes of Operation.," Technical Report, Department of Informatics, University of Bergen, 2000.

# Treating measurement uncertainty in complete conformity control system

ZS. T. KOSZTYÁN, T. CSIZMADIA, CS. HEGEDŰS, Z. KOVÁCS
Department of Management, University of Pannonia, Veszprém, Hungary
kzst@vision.vein.hu, csizi@gtk.uni-pannon.hu, hegeduscs@gtk.uni-pannon.hu

*Abstract* – **The effective management and control of technological processes as a source of competitive advantage is of vital importance for many organizations. In process control system, SPC is a widely used optimization philosophy, which uses a collection of (statistical) tools for data and process analysis, making inferences about process behavior and decision-making. Concerning decision-making, however, the traditional SPC methods do not take the measurement uncertainty into account even if it is known. We developed a method, which takes the measurement uncertainty, the costs coming up in connection with the production process and the revenues from their sold into account. Simulations show that taking measurement uncertainty into account as a priori information is suitable because Type II decision error can be decreased considerably and profit can be increased significantly, which are key factors in process control and in enhancing competitiveness.**

## I. INTRODUCTION

There are two seemingly different theories dealing with measurement and conformity. Traditional SPC tools are mainly used in production [1]. The goal of SPC is to eliminate variability in the process [2-5]. of detecting assignable causes of variation, hence at reducing the overall variability of the control parameters [6]. In process control, the product is considered appropriate depending on whether it lies within a tolerance field. The measurement is completed with measuring equipment, which can introduce major deficiencies. Even if the measurement uncertainty is known, it is not taken into account in decision-making. To tackle this problem, a protocol to treat measurement uncertainty was developed in 1993 [7]. However, this ISO-GUM method [8] is now only applied in laboratories and still does not answer the question of whether we should accept a particular product when we know that uncertainty exists [9-10]. The paper attempts to identify how the advantages of the two different theories can be combined and unified and how the uncertainty of the measuring equipment can be taken into account in decision-making.

The essential novelty of the proposal is that we use measuring intervals in our model instead of measuring points. The length of the interval is a product that is equal to $k$ multiplied the measurement uncertainty. We investigate how the cost and profit of a company are affected depending on how they treat measurement uncertainty.

Measurement uncertainty should even be taken into account in the course of optimisation if the error introduced by measuring equipment is much smaller than the tolerance value concerning the product's conformity. This article does not address the problems that the measures are often indirect and the "weighted with" quantities also increase measurement uncertainty.

## II. DETERMINING MEASUREMENT UNCERTAINTY

The ISO-GUM [8]. assumes that the measured quantity can be expressed with a model function (e.g. a resistance can be expressed as a function of length and specific resistance). The standard deviation of the parameters of this function (e.g. length and specific resistance in this case) derived from the measurements and its systematic failure are called standard uncertainty. According to the ISO-GUM method, Type-A uncertainty is estimated statistically whereas Type-B uncertainty is evaluated by other means. A Type-B uncertainty should commonly be based on a pool of comparatively reliable information. If the model function and standard uncertainties are known or they can be determined, the combined standard uncertainty ($u_c$) of the measured quantity (e.g. resistance in this case) can be determined with Monte Carlo simulation or analytically with Taylor series development (according to the ISO-GUM method). Definitions of the measurement uncertainties according to ISO-GUM [8]:

**Definition**: "***Standard uncertainty*** is the uncertainty of a measurement expressed as a standard deviation."

**Note:** Calculation of Type A standard uncertainty, in case of $x_i = \overline{X}_i$ $u(x_i) = s(X_i) = \sqrt{\frac{1}{n-1}\sum_{k=1}^{n}\left(X_{k,i} - \overline{X}_i\right)^2}$, where $\overline{x}_i$ is the mean of the $i^{th}$ parameter of $n$ measured values, $X_{k,i}$ is the $i^{th}$ parameter of $k^{th}$ measured value.

**Definition:** "***Combined standard uncertainty*** is the standard uncertainty of a measurement when that result is obtained from the values of a number of other quantities, equal to the positive square root of a sum of terms, the terms being the variances or covariances of these other quantities weighted according to how the measurement results varies with changes in these quantities."

T. Sobh (ed.), *Innovations and Advances in Computer Sciences and Engineering*,
DOI 10.1007/978-90-481-3658-2_14, 

Calculation: in case of $y = f(x_1, x_2, ..., x_N)$

$$u_c^2(y) = \sum_{i=1}^{N}\sum_{j=1}^{N} \frac{\partial f}{\partial x_i}\frac{\partial f}{\partial x_j} u(x_i, x_j) = \\ = \sum_{i=1}^{N}\left(\frac{\partial f}{\partial x_i}\right)^2 u(x_i) + 2\sum_{i=1}^{N-1}\sum_{j=i+1}^{N} \frac{\partial f}{\partial x_i}\frac{\partial f}{\partial x_j} u(x_i)u(x_j)r(x_i, x_j),$$

where $r(x_i, x_j)$ is the correlation between measured values $x_i$ and $x_j$.

If we describe combined uncertainty with a number instead of the density function of distribution then we can do this to different distributions. For instance, in case of combined standard uncertainty following $N(0,\sigma)$ standard deviation the value of $u_c$ is $\sigma$. In addition, the ISO-GUM method includes a third uncertainty concept; this is extended uncertainty, which is an interval. $U=ku_c$. The value $k$ depends on e.g. the number of measurements, the type of standard uncertainty, and the distribution of standard uncertainty.

**Definition:** "***Expanded uncertainty*** is a quantity defining an interval about the result of a measurement that may be expected to encompass a large fraction of the distribution of values that could reasonably be attributed to the measurand.

**Note:** 1.) The fraction may be viewed as the coverage probability or level of confidence of interval. 2.) To associate a specific level of confidence with the interval defined by the expanded uncertainty requires explicit or implicit assumptions regarding the probability distribution characterised by the measurement result and its combined standard uncertainty. The level of confidence that may be attributed to this interval can be known only to the extent to which such assumptions may be justified."

We can say over the course of a calibration that the measured value of a component (e.g. the resistance in this case) is $x \pm U$ (e.g. 12,04 kΩ ± 102 Ω ). Larger companies can also calibrate their own equipment, and do not need to bring in standardisation specialists. However, these companies are not fully familiar with the exact methodology of determining measurement uncertainty. The paper does not wish to introduce this methodology. We would like to point out that knowing combined standard uncertainty (referred to as combined uncertainty in this article) is *a priori* information that should be determined not only because of the inherent value of the measurement, but also because it can significantly increase profit. It should be noted that our method has been worked out for those cases when all products are measured and the measurement uncertainty is smaller with at least one magnitude compared to the prescribed tolerance field.

## III. CONSIDERING MEASUREMENT UNCERTAINTY

### A. *Analytical treatment of measurement uncertainty*

Many companies have tried to manage measurement uncertainty in industrial processes. Among these we find, for instance, the OSRAM's method [11], in which during the traditional conformity control, not only is a product's acceptability decided, but also if the total value of the extended measurement uncertainty added to (or subtracted from) the measurement point was outside the specification limits. If this occurs, the measurement is repeated. Actual intervals were used here, but the 'optimal' length of the interval was not calculated [11].

The main goal of this section is to answer the question of how the knowledge of measurement uncertainty of the measuring equipment alters our decision-making concerning the conformity of the product. First, let us assume that both the actual process and measurement uncertainty can be modelled.

Let us assume that a product is manufactured through one production stage. If there are more stages, the half-finished product from an operation place is then considered to be "sold" for the next phase, or the total product realisation process is still considered to be one stage (black box). Let a lower LSL and an upper USL specification limit (we take into account those cases where the nominal value of the quality index is the best to be reached) and a T specified value. Let us assume that the distribution $F(x(t))$, the expected value $E(x(t))$ and the standard deviation $D(x(t))$ of the actual process are all known. In addition, we know the distribution of the uncertainty of measuring equipment $G(m(t))$ and the expected value of the measuring fault of measuring equipment $E(m(t))$ and its standard deviation $D(m(t))$. We mark the value received in place $y(t)$ of the density function of the measuring fault of measuring equipment: $g_m(y(t), \mu_m, c_m)$, where $\mu_m = E(m(t))$, $c_m = D(m(t))$. Let the $t^{th}$ value of $y$ measured with measuring equipment $m$ — $y(t)=x(t)+m(t)$ — which can be characterised as the sum of the real value and the deviation of measuring equipment. If the value $x(t)$ were known, then we could clearly decide to accept ($LSL \le x(t) \le USL$) the value $x(t)$ or refuse it ($x(t) < LSL$, or $USL < x(t)$). However, we do not know the value $x(t)$ in practice, and we can only measure a value $y(t)$. The question is whether the abovementioned decision-making rule can also be applied at this point or whether it is worth taking the uncertainty of measuring equipment into account. In the simulation in this paper the value $x(t)$ is also postulated to be known so that we can compare the result of our decision-making with the case when the measuring uncertainty is not taken into account. In order to check the importance of treating measurement uncertainty we used the model in Fig. 1. The uncertainty of measuring equipment and its distribution can be modelled. At the output side the value $y(t)$ and for the sake of comparability $x(t)$ can be produced.

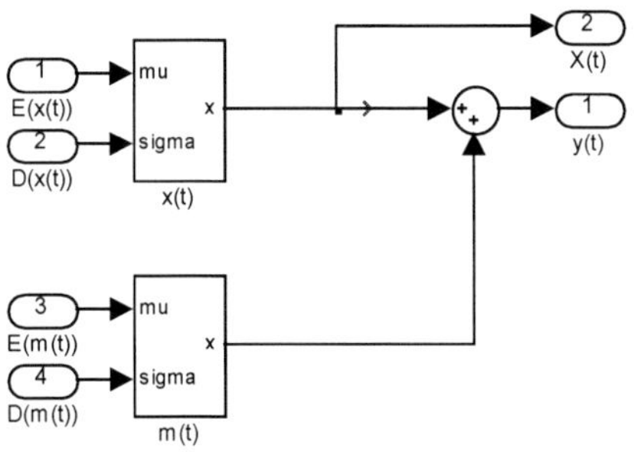


Fig. 1. The simulation of measuring failure

We do not know the value $x(t)$ in practice therefore we also have to take into account the costs that are derived from bad decision-making and on the probability of making a bad decision.

Marks: be $\mu_y = E(y(t))$, $\mu_m = E(m(t))$, the expected value of measured $y(t)$ process, and the expected value of the measurement uncertainty of measuring equipment $m(t)$. Let the $u_c(m(t))$ combined standard measuring uncertainty of measuring instrument (in short measuring uncertainty) be described by $c_m$ standard deviation and let us assume that the $\mu_m$ expected value is zero.

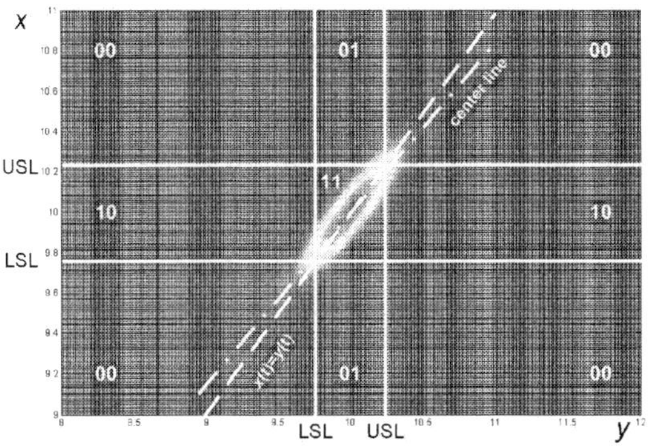


Fig. 2. The 2D probability power distribution of real and measured values

The combination of the real ($x(t)$) and measured values ($y(t)$) are located in Fig. 2. and 3. On the oblique these values correspond to each other ($x(t)=y(t)$), the measuring error is zero ($m(t)=0$). Diverging from the oblique the measuring error and the possibility of decision error are increasing.

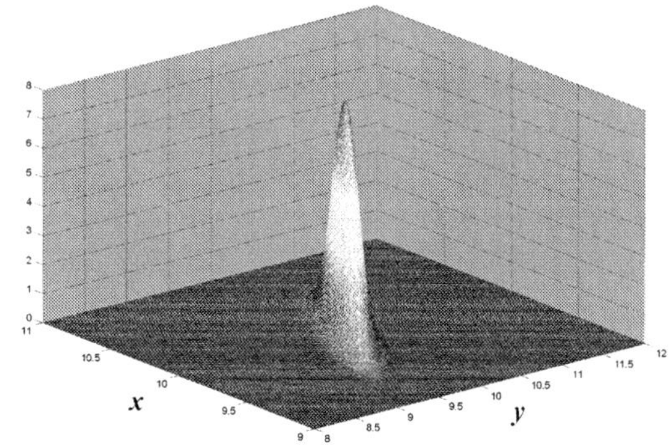


Fig. 3. The 3D probability density function of the real and measured values

The probabilities of the combination of the real and measured values are:

$$p(x(t), y(t)) = f(x(t), \mu_x, c_x) \cdot f(y(t) - x(t), \mu_m, c_m) \tag{1}$$

The $f(x(t), \mu_x, c_x)$ probability density function belongs to the actual process and the $f(y(t) - x(t), \mu_m, c_m)$ function belongs to the deviation of the measurement. The two probability distributions are independent from each other therefore, their product is the resultant probability. The fields of the different results of decision have been numbered as 00, 01, 10, 11, and these numbers will be used further on.

The conformity of the products can correctly be judged in two cases:

(**11**) The conforming product is accepted i.e. $y(t) \in$ [LSL, USL] and $x(t) \in$ [LSL, USL] with the probability:

$$p_{11} = \int_{LSL}^{USL} \int_{LSL}^{USL} p(x(t), y(t))\,dx\,dy \tag{2}$$

(**00**) The non-conforming product is rejected i.e. $y(t) \notin$ [LSL, USL] and $x(t) \notin$ [LSL, USL] with the probability:

$$p_{00} = \int_{-\infty}^{LSL} \int_{-\infty}^{LSL} p(x(t), y(t))\,dx\,dy + \int_{USL}^{\infty} \int_{-\infty}^{LSL} p(x(t), y(t))\,dx\,dy + \int_{-\infty}^{LSL} \int_{USL}^{\infty} p(x(t), y(t))\,dx\,dy + \int_{USL}^{\infty} \int_{USL}^{\infty} p(x(t), y(t))\,dx\,dy \tag{3}$$

Two kinds of errors are distinguished: Types I and II errors [12]:

(**10**) In case of Type I decision error, the product is rejected because the measured value is not in the area of specification limits ($y(t) \notin$ [LSL, USL]), but the real value is the element of these specification limits ($x(t) \in$ [LSL, USL]). At this time the probability function is:

$$p_{10} = \int_{-\infty}^{LSL}\int_{LSL}^{USL} p(x(t), y(t))dxdy + \int_{USL}^{\infty}\int_{LSL}^{USL} p(x(t), y(t))dxdy \quad (4)$$

(**01**) In case of Type II decision error, the product are accepted, despite the fact that the real $x(t)$ value is not the element of the specification limits. At this time the probability function is:

$$p_{01} = \int_{LSL}^{USL}\int_{-\infty}^{LSL} p(x(t), y(t))dxdy + \int_{LSL}^{USL}\int_{USL}^{\infty} p(x(t), y(t))dxdy \quad (5)$$

*B. Calculating profits*

One of the most important goals of the organisations is to maximise profit, which can be reached by aiming both for maximising income and minimising cost.

Income is determined by the decisions of an organisation. If the measure shows that a product fails to conform, it is discarded independently of its conformity. We do not have any income. If we sell the non-conforming product then some of the income derived from its sale will have to be used for repair of the non-conforming product (see the cases in the following sections).

If there is no measurement uncertainty, then $x(t)=y(t)$. Since the real value of $x(t)$ is unknown, losses must be considered in cases of rejecting a conform product and of accepting the non-conform one. Let us see which kind of costs come up in connection with the production process. The production cost and the cost of conformity inspection already occur before decision-making; therefore, they are independent of it and do not influence the optimum place, only its value.

We consider the following scenarios in relation to the decision on product conformity:

- $c_{11}$: production and inspection cost
- $c_{10}$: production and inspection cost + discarding cost
- $c_{01}$: production and inspection cost + cost of non-conformity
- $c_{00}$: production and inspection cost + discarding cost

If we sell non-conforming products ($c_{01}$), the cost of the non-conformity depends on the number of products sold.

The cost $c_{01}$ can be derived from the losses of the cheaper sale or the costs of the expenses for repair of the returned product. If a large number of the non-conform products are returned, the costs can increase further because they need to be repaired at the supplier's costs or they can be refused. In case of serious problems, the supplier could potentially have their contract cancelled.

TABLE 1
THE TENDENCY OF PROFIT AS A FUNCTION OF DECISIONS AND THE FACTS

| Profits | | Decision: Accepting | Decision: Discarding |
|---|---|---|---|
| Fact | Conform | $\pi_{11} = r_{11} - c_{11}$ | $\pi_{10} = r_{10} - c_{10}$ |
| | Non-conform | $\pi_{01} = r_{01} - c_{01}$ | $\pi_{00} = r_{00} - c_{00}$ |

The total profit (Table 1) equals the sum product of the specific profits and its probabilities:

$$\Pi = p_{00}\cdot\pi_{00} + p_{10}\cdot\pi_{10} + p_{01}\cdot\pi_{01} + p_{11}\cdot\pi_{11} \quad (6)$$

*C. Modifying the criteria of decision-making taking the measurement uncertainty into account*

Let us assume that the measurement uncertainty of measuring instrument is known. This time the following rule of decision will be used:

If $[y(t) - k_{LSL}\cdot c_m] > LSL$ and $[y(t) + k_{USL}\cdot c_m] < USL$, where $k_{LSL}, k_{USL} \in \mathbf{R}$ are coverage factors depend on the costs and revenues, then we accept the product, else ($[y(t) - k_{LSL}\cdot c_m] \le LSL$ or $[y(t) + k_{USL}\cdot c_m] \ge USL$) we reject it. The question is the value of the two constants $k_{LSL}$, $k_{USL}$. It can be seen in Fig. 4. that this way instead of measuring points, measuring intervals are taken into consideration.

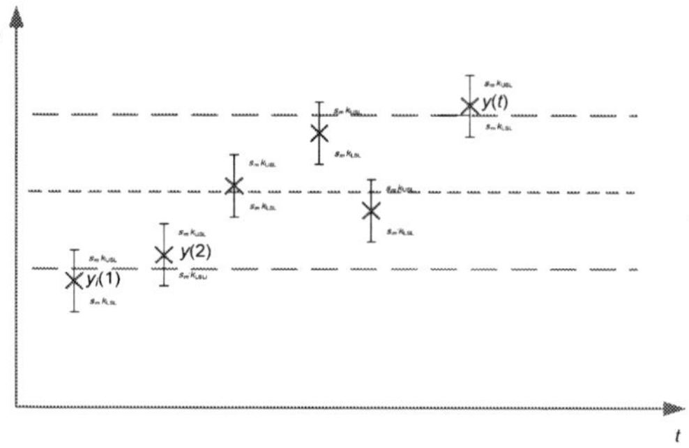


Fig. 4. Decision-making of adequacy based on measuring intervals instead of measuring points

(**11**) The real value and the measuring intervals are element of the specification limits, $LSL \le x(t) \le USL$, and $LSL \le y(t) - k_{LSL}c_m$ and $y(t) + k_{USL}c_m \le USL$, i.e. $LSL + k_{LSL}c_m \le y(t) \le USL - k_{USL}c_m$

$$p_{11}^k = p_{11}(k_{LSL}, k_{USL}) = \int_{LSL+k_{LSL}\sigma_m}^{USL-k_{USL}\sigma_m}\int_{LSL}^{USL} p(x(t), y(t))dxdy \quad (7)$$

(**00**) Neither the real value, nor the measuring intervals are element of the specification limit ($LSL \ge x(t)$ or $x(t) \ge USL$ and $y(t) \ge USL - k_{USL}c_m$ or $LSL + k_{LSL}c_m \ge y(t)$):

$$p_{00}^k = p_{00}(k_{LSL}, k_{USL}) =$$

$$= \int_{-\infty}^{LSL+k_{LSL}\sigma_m}\int_{-\infty}^{LSL} p(x(t), y(t))dxdy + \int_{USL-k_{USL}\sigma_m}^{\infty}\int_{-\infty}^{LSL} p(x(t), y(t))dxdy + \quad (8)$$

$$+ \int_{-\infty}^{LSL+k_{LSL}\sigma_m}\int_{USL}^{\infty} p(x(t), y(t))dxdy + \int_{USL-k_{USL}\sigma_m}^{\infty}\int_{USL}^{\infty} p(x(t), y(t))dxdy$$

(**10**) Type I decision error: the conforming product ( $LSL \le x(t) \le USL$ ) is rejected, because the measuring interval does not lie within the tolerance field ( $LSL \le y(t) - k_{LSL}\sigma_m$ and $y(t) + k_{USL}\sigma_m \le USL$ ).

$$p_{10}^{k} = p_{10}(k_{LSL}, k_{USL}) = \int_{-\infty}^{LSL+k_{LSL}\sigma_m} \int_{LSL}^{USL} p(x(t), y(t))dxdy + \int_{USL-k_{USL}\sigma_m}^{\infty} \int_{LSL}^{USL} p(x(t), y(t))dxdy \qquad (9)$$

(**01**) Type II decision error: the non-conforming product ( $LSL \geq x(t)$ or $x(t) \geq USL$ ) is accepted, because the measuring interval is element of the tolerance field ( $LSL + k_{LSL}\sigma_m \le y(t) \le USL - k_{USL}\sigma_m$ ).

$$p_{01}^{k} = p_{01}(k_{LSL}, k_{USL}) = \int_{LSL+k_{LSL}\sigma_m}^{USL-k_{USL}\sigma_m} \int_{-\infty}^{LSL} p(x(t), y(t))dxdy + \int_{LSL+k_{LSL}\sigma_m}^{USL-k_{USL}\sigma_m} \int_{USL}^{\infty} p(x(t), y(t))dxdy \qquad (10)$$

Let us assume that the measurement uncertainty is given and described by $\sigma_m$. This time the following target functions can be determined as the function of *k*:

1. Maximising the profit. (unconstrained optimisation)

$$\Pi_k = \Pi(k_{LSL}, k_{USL}) = \pi_{00} p_{00}(k_{LSL}, k_{USL}) + \pi_{11} p_{11}(k_{LSL}, k_{USL}) + \pi_{10} p_{10}(k_{LSL}, k_{USL}) + \pi_{00} p_{00}(k_{LSL}, k_{USL}) \rightarrow \max \qquad (11)$$

2. Profit will be maximised with the condition that the number of accepted non-conforming product does not exceed a certain value. (constrained optimisation)

$$\Pi_k = \Pi(k_{LSL}, k_{USL}) = \pi_{00} p_{00}(k_{LSL}, k_{USL}) + \pi_{11} p_{11}(k_{LSL}, k_{USL}) + \pi_{10} p_{10}(k_{LSL}, k_{USL}) + \pi_{00} p_{00}(k_{LSL}, k_{USL}) \rightarrow \max \qquad (12)$$

Subject to

$p_{01}(k_{LSL}, k_{USL}) < \varepsilon$, where $\varepsilon \in R^{+}$

The first target function (without taking into consideration any kind of constraints) is a simple profit maximisation function (11). The second problem is how to maximise profit taking the probability of acceptance of a non-conforming product that is lower than $\varepsilon \in R^{+}$ into account (12).

If the power density functions of the real process and the measurement uncertainty is known the optimal *k* value can be analytically calculated. Unfortunately the parameters of the real process can only be estimated by a measured process and measurement uncertainties. Therefore the optimal *k* value can only be estimated by a stochastic optimisation method, or can be estimated by simulations. Fig. 5. and 6. show the results of simulations.

*D. Treating measurement uncertainty supported by simulation methods*

It is worth dealing with measurement uncertainty as *a priori* information, and simulation tools can be used in order to determine the value of *k*. The advantage of simulation is its use of almost any distribution with any parameters. We used the Matlab simulation software to determine significant differences.

In order to compare the two cases—the measurement uncertainty was either taken into consideration or not—let us assume that we know the actual process and so we try to define the value of *k* for particular distributions. We assumed normal distribution both for the process and the measuring equipment. A test configuration can also be constructed for other distributions.

We created Fig. 5. using Matlab, running one million test simulations in the process of $k_{USL}$, $k_{LSL}$. Fig. 5. shows that the distribution of profit is largely symmetrical. The value of the two *k*s is symmetrical if the distribution of the process and measurement uncertainty is also symmetrical, and the expected value of the process is in the middle of the tolerance field. The profit can be increased taking parameter $k=k_{LSL}=k_{US}$ into account.

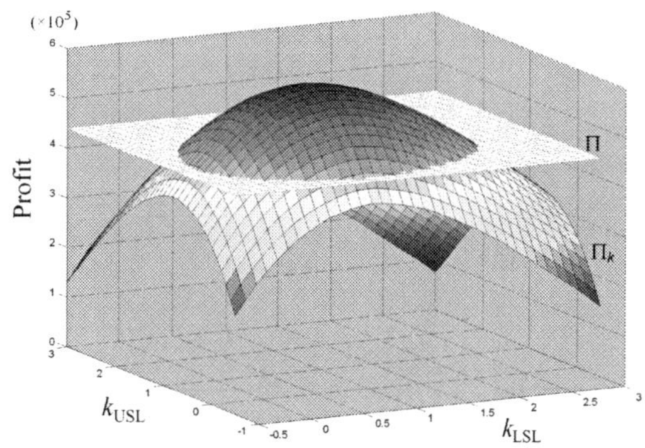


Fig. 5. The tendency of the total profit as a function of taking measurement uncertainty into account

Let us now take a closer look at the extent to which $k=k_{LSL}=k_{USL}$, where we take the measurement uncertainty of the measuring equipment into account. We produced Fig. 6. inferring normal distribution.

The profit was decreased in an approximately linear way with the rise of measurement uncertainty. Increasing the value for *k*, there will be a greater chance that the measuring area will be outside of the tolerance field that increases the number of rejected but conforming products and decreases the number of accepted but non-conform products at once. Decreasing the value of *k* gives the opposite result. The value of $k$ belonging to maximum profit is determined by the costs of Type I and II errors. If the non-conforming product can be sold at a reduced price or can be repaired cheaply, then we can increase the total profit while decreasing the value of $k$.

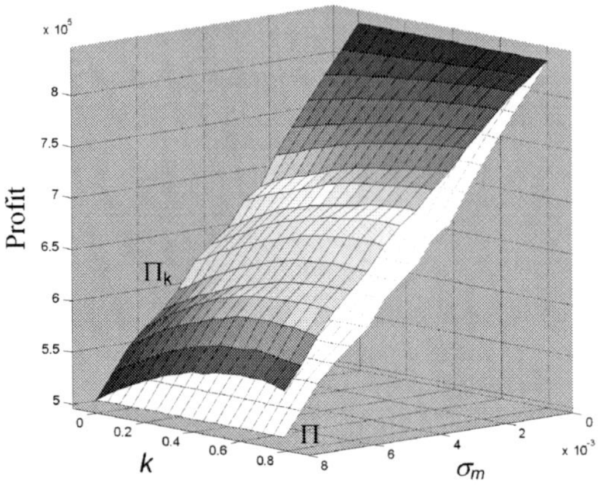


Fig. 6. The tendency of the total profit as a function of the measurement uncertainty of measuring equipment and value $k$.

## IV. Summary, Future Challanges

The results reported in this paper show that it is useful to extend the tools of SPC to measurement uncertainty. The consideration of measurement uncertainty can significantly increase the profit of organisations. The value of $k$ can be developed by simulation, which shows the extent to which measurement uncertainty should be taken into account. Further research needs to be done, however, to analyse the correlation between more measured parameters.

In addition, this method can be applied in its simple form (stationer cases, not sampling control) in a number of fields. For example, it is important in the automotive industry that measurement uncertainty be under 10% in comparison with the change of the production process. The method can also help to decrease the cost while taking advantage of the possibilities of 'producing on the margin' [13]. We hope that this paper develops a new approach that will help to stimulate work in theoretical and empirical research.

## References

[1] Galicinski, L. M. *The SPC Book*. Qualitran Professional Services Inc. 1999

[2] Montgomery, D.C. *Introduction to Statistical Quality Control*. New York, Wiley 2001

[3] Küppers, S. A validation concept for purity determination and assay in the pharmaceutical industry using measurement uncertainty and statistical process control. *Accred Qual Assur*, pp.338–341. 1997

[4] Schippers, W.A.J. Applicability of statistical process control techniques. *International Journal of Production Economics*, pp. 525–535. 1998

[5] Chen, S.H., Yang, C.C., Lin, W.T. & Yeh, T.M. Performance evaluation for introducing statistical process control to the liquid crystal display industry. *International Journal of Production Economics*, pp. 80–92, 2008

[6] Morgan, C. & Dewhurst, A. Multiple retailer supplier performance: An exploratory investigation into using SPC techniques. *International Journal of Production Economics*, pp. 13–26, 2008

[7] Kessel, W. (2002). Measurement uncertainty according to ISO/BIPM-GUM. *Thermochimica Acta*, pp. 1-16, 2002

[8] International Organisation for Standardisation (ISO), *Guide to the Expression of Uncertainty in Measurement*, International Organisation for Standardisation (ISO), Geneva, Switzerland, 1993 (corrected and reprinted 1995).

[9] Kosztyan, Zs.T., Csizmadia T. & Hegedűs, Cs. Mérési bizonytalanság figyelembe vétele mintavételes megfelelőségvizsgálatnál. (Treating measurement uncertainty in sampling conformity control). *XX. Nemzetközi Karbantartási Konferencia*, pp. 152-163, 2008

[10] Kosztyán, Zs.T., Kovács, Z, & Csizmadia T. Mérési bizonytalanság kezelése statisztikai folyamatirányításnál. (Treating measurement uncertainty in SPC). *XIX. Nemzetközi Karbantartási Konferencia*, pp. 1-19, 2007

[11] Jordan, W. Measurement, test and calibration: industrial tasks with respect to DIN EN ISO/IEC 17025 and general quality management, *2nd CIE Expert Symposium*, 12-13 June, pp. 181-189. 2006

[12] Duncan, A.J. *Quality Control and Industrial Statistics*. Irwin, Homewood, IL, 1974

[13] Phadke, M.S. *Quality Engineering Using Roboust Design*. Prentice-Hall 1989

# Software Process Improvement Models Implementation in Malaysia

[1]Shukor Sanim M.Fauzi, [2]Nuraminah R. and [3]M. Hairul Nizam M. Nasir
[1]Faculty of Information Technology and Quantitative Sciences
Universiti Teknologi Mara, Perlis Campus, 02600, Arau, Perlis, Malaysia
Tel: +60-04-987 4319 Fax: +60-04-987 4225
E-mail: shukorsanim@perlis.uitm.edu.my

[2]Faculty of Information and Communication Technology
Universiti Pendidikan Sultan Idris, 35900 Tg Malim, Perak, Malaysia
*Tel: +60-05-450 5076,*
E-mail: nuraminah@ftmk.upsi.edu.my

[3]Department of Software Engineering,
Faculty of Computer Science and Information Technology
University of Malaya 50603 Kuala Lumpur, Malaysia
Tel: +60-03-7967 6435 Fax: +60-03-2178 4965
E-mail: hairulnizam@um.edu.my

*Abstract*— **The growing focus on software development has highlighted the importance of software process. Over the last few years, many large foreign organizations like Motorola, Thales, Lockheed Martin and other organizations have adopted the software process improvement model to provide effective software process management and control of the software they developed. What about the IT Company in Malaysia? This research attempts to identify and analyze the extent of Capability Maturity Model Integration (CMMI) and other associated process model or frameworks in Malaysian organizations. Surveys in Malaysian software organizations which have an experience in initiating and conducting software process improvement initiative have been conducted. There were 20 respondents involved in this research. Most of the respondents were professionals who are directly involved in SPI initiative. This paper will present a survey analysis on organization's initiative in CMMI, IT Service Management and Other framework. From the survey, most of the organizations do not plan to implement IT service management framework such as AS8018 ICT Service Management, HP ITSM (IT service Management Reference Model), MOF (Microsoft Operations Framework), IBM SMSL (Systems Manage Solution Lifecycle, and CobiT (Control Objectives for Information and related Technology). Most of the organizations have implemented CMM/CMMI in their organizations. Other than that, they also implemented ISO 9001, Balance Scorecard and also PMBOK. From the result, it shows that most of the organizations in Malaysia have already shifted to the new paradigm by enforcing software process improvement in their organization.**

## I. INTRODUCTION

Software Process Improvement (SPI) is a relatively new area, developed in the last twenty years [1]. Several methodologies and models have been developed in recent years. Perhaps the most well known and common of these is the Capability Maturity Model (CMM) from the Software Engineering Institute In a recent time, Software Engineering Institute has established new model commonly known as Capability Maturity Model Integration (CMMI) [2]. Examples of other models and methodologies are Six Sigma, Lean Development, ISO Standard, Agile Methodologies, Balbridge [3] [4] [5] [6] [7] [8] [9]. The description of each model will be discussed in section III in this paper. Most ideas in SPI were adopted from the theories and methodologies for quality in manufacturing systems developed in the last few decades by *Shewhart*, Deming, Crosby and Juran [10] [11] [12] [13].

There is a number of studies have been conducted on software process improvement by using above models and methodologies in other continents such as Europe [14], Australia, New Zealand [15] and North America [16]. However, there is still lack of published studies on software process improvement or software best practices in Southeast Asia, especially Malaysia itself.

The purpose of this paper is to discuss about the extent of other associated model or framework in Malaysian organizations. This paper also will touch on several of SPI models which have been used in real environment. Next section, we will discuss about the result and discussion on the data collected on current initiatives and progress in other associated process model or framework.

## II. SOFTWARE PROCESS IMPROVEMENT STANDARDS

There are various software process improvement standards in place. There are:

T. Sobh (ed.), *Innovations and Advances in Computer Sciences and Engineering*,
DOI 10.1007/978-90-481-3658-2_15, © Springer Science+Business Media B.V. 2010

**CMM** - According to Paulk *et* al. in [17] reported that CMM guides software groups on how to gain control of their processes for developing and maintaining software and how to evolve toward a culture of software engineering and excellence of management. It provides a framework for organizing these evolutionary steps into five maturity levels to act as a successive foundation for continuous process improvement. The five levels as defined by are initial, repeatable, defined, managed and optimized as mention by Herbsleb *et.* al in [18]

As described by Herbsleb *et.* al [18] each level in CMM needs to reach the Key Process Area (KPA) which consists of key practices that contribute to satisfying its goals. Each Key Area must implement the pre-established goals, which are, activities to be developed, or necessary infrastructure to the goals satisfaction. The key practices are aggregated can either be the implementation or Institutionalization. Each maturity level establishes a different component in the software process, resulting in the capability increase of the organization process.

**CMMI** - Software Engineering Institute (SEI) describes that the Capability Maturity Model Integration (CMM-I) as an SPI process which provide a guidance for improving on each organization's processes and the capability to control and manage the development, achievement and maintenance of products or services during software process [19] It is a staged representation, organizes process areas in five maturity levels which is same as CMM. The different is that, CMM-I is an integrated approach which establish a framework to integrate current and future models or build an initial set of integrated models.

In order to achieve the CMMI, the organization must select the models by choosing either continuous or staged representation [19]. Continuous refers to an organization to select the order of improvement that best meets the organization's business objectives and mitigates the organization's areas of risk. Meanwhile representation staged helps by providing a proven sequence of improvements, starting with basic management practices and progressing through a predefined and proven path of successive levels, each serving as a foundation for the next.

**SIX SIGMA** - Six Sigma is quality program pioneered by Motorola invented by Bill Smith who is senior engineer at Motorola whereby the main objective is to reach a quality goal on their products [20]. It emphasizes on a good financial results that can be achieved through the virtual elimination of product and process defects [21]. Six Sigma follows a sequence of process which involving 5-steps, known as DMAIC (Define, Measure, Analyze, Improve and Control) as mention in which each steps improves to satisfy the software process [21].

Two implementation methods involve in Six Sigma as mentioned by Knuth *et.* al are the most well-defined if a problem with an unknown solution existing products, processes or services which method is DMAIC [22].( Define, Measure, Analyze, Improve and Control) The latest method, which still in the developing stage is called Design for Six Sigma (DFSS) as described by Kermani [20]. The goal of DFSS is to develop a new product, process or service that is eliminate the defects. However, a few of consulting companies have invented roadmap for DFSS like IDOV (Identify, Design, Optimize and Validate) as mention by Hurber and Mazur in [11] and DMADV (Design, Measure, Analyze, Design and Verify) by Knuth *et.* al [22]. A paper described by Ajit in [23] introduce "DFSS" (Design for Six Sigma) methodology to boost product quality which can improve software process.

**ISO-9000** - It is a set of standards, adopted by the International Organization for Standardization (ISO) based in Geneva, Switzerland which defines the requirements for an organization's Quality Management System (QMS) as mentiond by Galin [24]. Paulk describes ISO 9000 as a quality management and assurance standards which provides guideline for selection and use; clarifies the dissimilarities and interrelationships between quality concepts and provides guidelines for the selection and use of a series of international standards [25]. ISO-9000 can be divided into three main components which is the first one is a framework that describes the overall requirements for the QMS. Secondly, a life-cycle activity which covers requirements for all phases of the development process until software maintenance and finally supports activities which identifies requirements for sustaining all the actions.

According to Galin [24],for any organization that is requesting certification, initially the organization need to prepare planning processes which require them to have a proper development of organization that fulfills the SQA system. The certifying organization will undergoing to the certification audits which include review the quality manual and SQA procedure developed by the organization .If those criteria comply with ISO 9000 ,the performance audit of management system take place. If its not, the organization need to refine back the organization's SQA system. Meanwhile, the implementation of organization's SQA system takes place. Followed by the performance of SQA management system is checked whether its comply or not with ISO 9000.If all the criteria is fulfill ,the ISO-9000 is issue and if its not the organization need to carry out the performance improvements of organization of SQA management system again.

## III. MATERIALS AND METHOD

The survey has been conducted from March 2008 until April 2008. The focus of this research will be on identifying the extent of Malaysian organizations in implementing SPI initiatives such as CMMI and other associated process model or framework. Essentially, we perform the initial literature review on SPI, looking at the broader context of SPI. In order to acquire the overall picture of SPI models and frameworks, some subjects related to the software process aspects, software qualities and SPI standards itself are reviewed. All the information above is collected using on-line search via the internet specifically on the online databases namely ACM, IEEE, technical reports published by Software Engineering

Institute (SEI), academic textbooks, magazines, online articles and others.

Secondly, we construct the questionnaires to acquire the objective of the research. The questionnaire consists of three parts. The first part investigates the profile of respondent and organization. Second part investigates on organization's current initiatives and progress on different framework. Third part investigates on CMMI initiatives and progress of the respondent's organization. The surveys were conducted using paper survey and also online survey due to the different risk or constraint of cost, time, and distance can be overcome.

In the data analysis, the respondents are characterized according to the attribution of weight to their answers. Hence, the final results take into consideration the respondent's organization status.

## IV. RESULT AND DISCUSSION

The Formula 1 shows the calculation to attribute weight to a respondent, and tables I, II, III, IV, and V are scoring table used in the calculation base on the question 1 to 5 in section A of the questionnaires and the CMMI level achieved of the respondent's organization.

**Formula 1:**

$$TS = Q1(i) + Q2(i) + Q3(i) + Q4(i) + Q5(i) + CMMI(i) \quad (1)$$

Where:

*TS* is the total score attributed to respondent. It is the sum of the respondent characteristic. *Q1(i), Q2(i), Q3(i), Q4(i),* and *Q5(i)* is the respondent's score due to their answer in section A question number 1 until question number 5. Different answer given different score based on the table below. *CMMI (i)* is the CMMI level of the organization of respondent i.

TABLE I: Q1(I) ON RESPONDENT POSITION

| | |
|---|---|
| Manager | 5 |
| Project or Team Leader | 4 |
| Technical Manager | 4 |
| Engineering Process Group (EPG) Member | 3 |
| Other | 1 |

TABLE II: Q2(I) ON RESPONDENT'S ORGANIZATION OWNERSHIP

| | |
|---|---|
| Wholly Malaysian owned | 2 |
| Partially Malaysian and foreign owned | 3 |
| Wholly funded by foreign capital | 4 |
| Don't know | 1 |

TABLE III: Q3(I) ON RESPONDENT'S ORGANIZATION BUDGET

| | |
|---|---|
| < RM5 million | 2 |
| RM5million – RM9million | 3 |
| RM10million – RM49million | 4 |
| RM50million – RM150million | 5 |
| > RM150million | 6 |
| Don't know | 1 |

TABLE IV: Q4(I) ON RESPONDENT'S ORGANIZATION SIZE

| | |
|---|---|
| <25 full time staff or equivalent | 1 |
| 25 – 49 full time staff | 2 |
| 50 – 99 full time staff | 3 |
| 100 – 499 full time staff | 4 |
| 500 – 2000 full time staff | 5 |
| > 2000 full time staff | 6 |

TABLE V: Q5(I) ON RESPONDENT'S ORGANIZATION OPERATION YEARS

| | |
|---|---|
| < 5 years | 1 |
| 5 – 10 years | 2 |
| 11 – 15 years | 3 |
| > 15 years | 4 |

TABLE VI: TOTAL SCORE OF THE 20 RESPONDENTS

| R(i) | A1 | A2 | A3 | A4 | A5 | CMMI level | TS |
|---|---|---|---|---|---|---|---|
| R1 | 1 | 3 | 1 | 2 | 3 | 5 | 15 |
| R2 | 4 | 3 | 1 | 2 | 3 | 5 | 18 |
| R3 | 1 | 3 | 1 | 2 | 3 | 5 | 15 |
| R4 | 1 | 3 | 1 | 2 | 3 | 5 | 15 |
| R5 | 5 | 2 | 3 | 2 | 3 | 3 | 18 |
| R6 | 1 | 2 | 3 | 2 | 3 | 3 | 14 |
| R7 | 4 | 2 | 3 | 2 | 3 | 3 | 17 |
| R8 | 1 | 2 | 3 | 2 | 3 | 3 | 14 |
| R9 | 1 | 2 | 3 | 2 | 3 | 3 | 14 |
| R10 | 1 | 2 | 3 | 2 | 3 | 3 | 14 |
| R11 | 5 | 4 | 3 | 2 | 1 | 5 | 20 |
| R12 | 5 | 4 | 3 | 2 | 1 | 5 | 20 |
| R13 | 4 | 4 | 3 | 2 | 1 | 5 | 19 |
| R14 | 1 | 4 | 3 | 2 | 1 | 5 | 16 |
| R15 | 1 | 4 | 3 | 2 | 1 | 5 | 16 |
| R16 | 4 | 4 | 5 | 4 | 4 | 5 | 26 |
| R17 | 3 | 4 | 5 | 4 | 4 | 5 | 25 |
| R18 | 1 | 4 | 5 | 4 | 4 | 5 | 23 |
| R19 | 1 | 4 | 5 | 4 | 4 | 5 | 23 |
| R20 | 1 | 4 | 5 | 4 | 4 | 5 | 23 |

**Legend:**

| | |
|---|---|
| *R(i)* | Respondent |
| *A1* | Answer for Question 1 (respondent position) |
| *A2* | Answer for Question 2 (respondent's organization ownership) |
| *A3* | Answer for Question 3 (respondent's organization budget) |
| *A4* | Answer for Question 4 (respondent's organization size) |
| *A5* | Answer for question 5 (respondent's organization operation years) |
| *CMMI Level* | CMMI Level for each respondents |
| *TS* | Total score (calculated based on Formula 1) |

Table VI shows the *TS* based on the questions answered by the respondents. In demographic information section, the respondents have been asked about their position, the respondent's organization ownership, and respondent's organization budget. Respondents also were asked about their organization size, organization operation years and also their achievement in CMMI. Based on the results, it demonstrates that the respondents have a good and sufficient knowledge of this research and also can provide reliable input to this survey.

## V. SURVEY ANALYSIS ON IMPLEMENTATION OF SERVICE MANAGEMENT FRAMEWORK

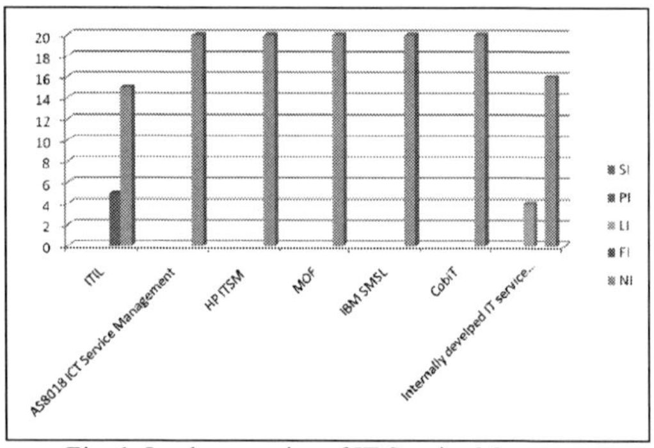


Fig. 1. Implementation of IT Service Management Framework

Figure 1 shows the implementation of IT service management framework by the respondent's organization. From the survey, most of the organizations do not plan to implement IT service management framework: Information Technology Infrastructure Library (ITIL), AS8018 ICT Service Management, HP ITSM (IT service Management Reference Model), MOF (Microsoft Operations Framework), IBM SMSL (Systems Manage Solution Lifecycle, CobiT (Control Objectives for Information and related Technology), Internally developed IT service management Framework. However, there are 5 respondents where their organization have fully implement the Information Technology Infrastructure Library (ITIL) framework and 4 respondents where their organization largely implemented developed IT service management Framework.

## VI. SURVEY ANALYSIS ON OTHER FRAMEWORK

From the survey, CMM/CMMI framework is largely and fully implemented in the respondent's organization. Besides, some other frameworks are also implemented in the respondent's organization:

According to the survey, there are 4 respondents where ISO 9001 is largely implemented in their organization, 6 and 10 respondents where Balanced Scorecard is partially implemented and fully implemented respectively in their organization. There are about 5 respondents where their organization partially implemented PMBOK (project Management Body of Knowledge) framework and six sigma, another popular framework which is fully implemented in some organization according to 10 respondents, other framework such as Project Professional by Project Management Institute also fully implemented according to 6 respondents. Lastly, Prince 2 (Project Management Methodology) which is the least favorite framework among the other frameworks above.

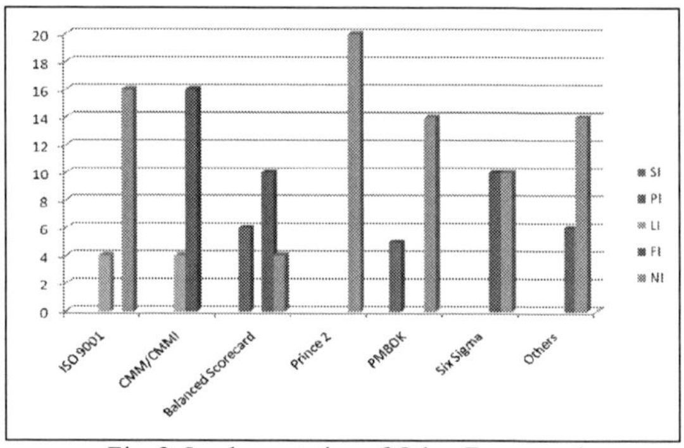


Fig. 2. Implementation of Other Framework

## VII. CMMI INITIATIVE AND PROGRESS

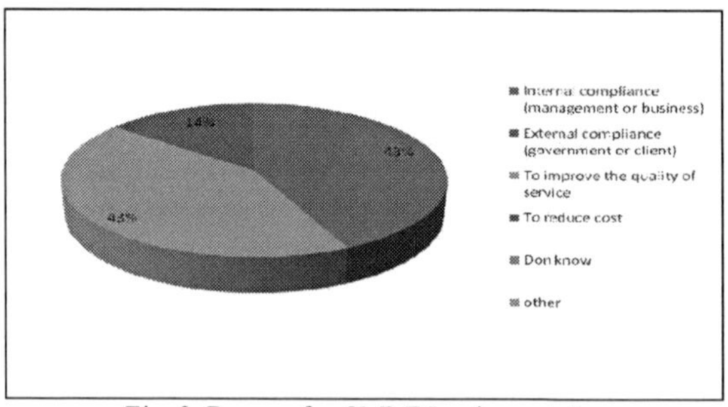


Fig. 3. Reason for CMMI Implementation

Figure 3 shows the reason for CMMI implementation. From the Figure 3, the main reasons of the CMMI implementation in the organization are internal compliance (management or business) and to improve the quality of service which is 43% equally and respectively according to the respondents. To reduce cost is also one of the reasons why the organization implemented the CMMI framework, according to the 14% of the respondents.

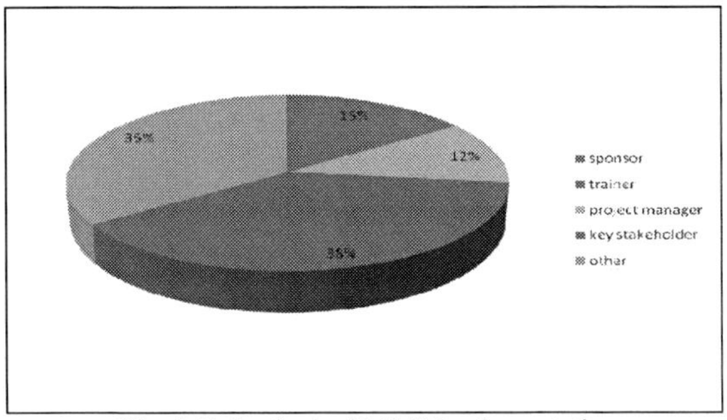


Fig. 4. Roles in the CMMI Implementation

From Figure 4, it shows that the role of respondents in the CMMI implementation mainly is the key stakeholder which contributes to 38% of the respondents. There are 15% of the respondents whose are playing the trainer role, where as 12% of the respondents are project manager. There are a big number of respondents that contribute to other roles especially user which is about 35% of the respondents.

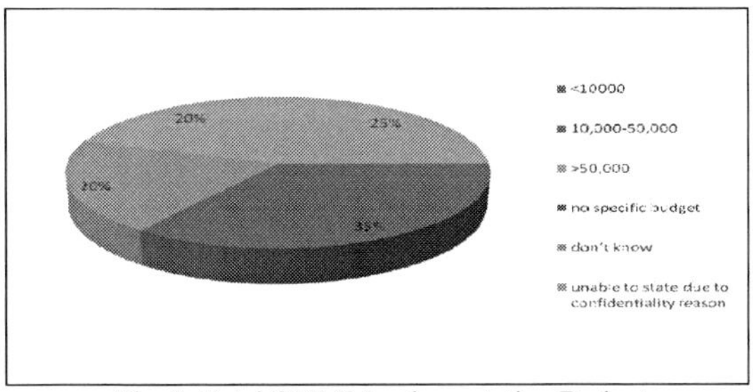


Fig. 5. Total CMMI Implementation Budget

There are 35% respondents state that their organization does not have any specific budget of the CMMI implementation where as 25% of the respondents where their organizations spend >RM 50,000 for the total CMMI implementation budget. It can be seen in Figure 5, where there are 20% of the respondents who does not know their organization's CMMI implementation budget and another 20% of the respondents unable to state the organization's budget for the CMMI implementation due to confidentiality reason.

Figure 6 shows that there are 3 variety of CMMI budget is spent based on the respondents' organization.

According to Figure 6, 50% of the respondents, the percentages of the budget in their organization are spent as below:

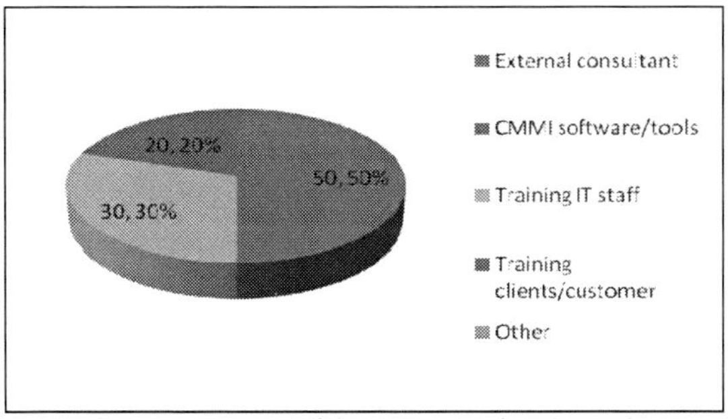


Fig. 6. Percentage of the CMMI Budget Spent On

50% of the budget are allocated for the CMMI software and tools, 20% of the budget spent on Training IT staff where as another 30% spent on Training clients/customer.

Whereas, according to another 25% of respondents whose state that the percentages of the CMMI budget in their organization is spent as in Figure 7 below:

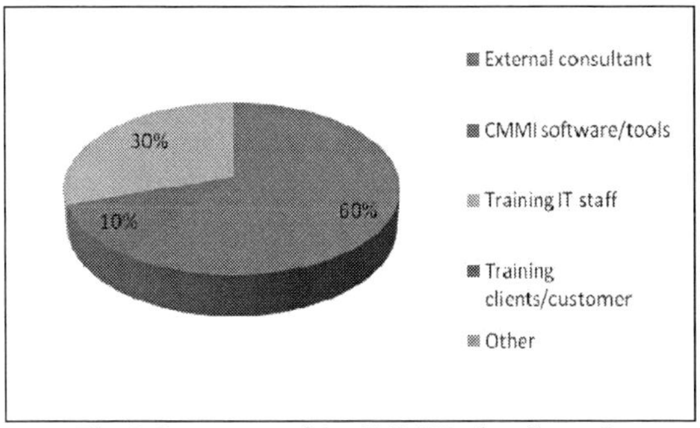


Fig. 7. Percentage of the CMMI Budget Spent On

60% of the CMMI budget spent for the external consultants and another 30% are allocated for the Training IT staff and another 10% are allocated for the CMMI software and tools.

While, 25% of the respondents state that the percentages of the CMMI budget in their organization are spent as in Figure 8 below:

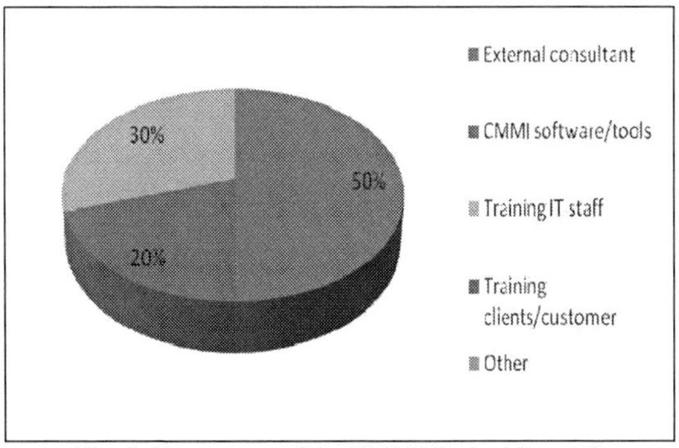


Fig. 8. Percentage of the CMMI Budget Spent On

According to the respondents, 50% of the CMMI budget spent for the external consultants, 30% are allocated for the Training IT staff and 20% are spent for the CMMI software and tools.

## VIII. Conclusion and Future Works

In software development process, it is believed that a high quality software development process is a key success to develop a high quality product. From the finding, it shows that most of the Malaysian organizations have already implemented SPI initiatives. From the result, we hope that it can motivate other software companies to take part in SPI initiative. From now on, most of the project involves in information technology and software development requires the organization to implement SPI initiative. This is to ensure that the product produced is more quality.

Our next action for SPI initiative research in Malaysia will be 1) to identify what are the factors that contribute to the success of SPI initiative. This study will help other Malaysian companies to learn the necessary actions before implementing SPI initiatives 2) to identify the attributes that can help the implementation of SPI 3) to identify what are the critical success factors for SPI.

## IX. Acknowledgement

We would like to acknowledge Universiti Teknologi Mara, for providing the Short-Term Grant to support this project.

## References

[1] Miguel A.S., *State of the Art and Future of Research in Software Process Improvement*, COMPSAC'04.

[2] CMMI Product Development Team (2000). *CMMI for Systems Engineering/Software Engineering,* Version 1.02, . CMU/SEI-2000-TR-018. Pittsburgh, PA:Software Engineering Institute.

[3] ISO/IECJTC1/WG10, "SPICE Products," Technical Report, Type 2, June 1995.

[4] Dorling. A. , (1993) SPICE: Software Process Improvement and Capability dEtermination. *Software Quality Journal* 2, 209-224.

[5] Harry, M. and R. Schroeder, *Six Sigma*, Douleday, 2000.

[6] Mary and Tom Poppendieck, *Lean Software Development*, Addison Wesley, 2003.

[7] International Organization for Standardization, *ISO Standard 9001: Quality Management Systems*, 2000.

[8] Tayntor, C., *Six Sigma Development Software* , CRC Press, 2003.

[9] Cockburn, A., *Agile Software Development,*Addison Wesley, 2002.

[10] Shewart, W., *Economical Control of Quality of Manufactured Product*, Van Nostrand, 1931.

[11] Deming, E., *Out of the Crisis*, MIT Press, 1986.

[12] Crosby, P., *Quality is Free*, McGraw-Hill, 1979.

[13] Juran, J. ,*Juran on Planning for Quality*, MacMillan, 1988.

[14] Dutta,S., Lec, M. & Van Wassenhove, L. Software Engineering in Europe: A study of best practices. *IEEE Software*, vol 16, no 3, May 1999, pp.82-90.

[15] Groves, L., Nickson, R., Reeve, G., Reeves, S., & Utting, M., A survey of software requirements specification practices in the New Zealand software industry. *Proc Australian Software Engineering Conference 2000*. pp. 189-201.

[16] McConnell, S. Professional Software Development, *2nd Ed Boston*, Mass,: Addison-Wesley, 2004.

[17] Paulk, M. C. , A Comparison of ISO 9001 and the Capability Maturity Model for Software, Software Capability Maturity Model Project, 1994, pp. 1-71.

[18] Knuth, K., Niebuhr, G., Lind, M. amd Mauer, B. Six Sigma – Making Corporations and Stockholders,*Industrial Engineering 361*, 2002.

[19] Clarke, L A. and Osterweil, L. J.,Continuous Self-Evaluation for the Self-Improvement of Software, International workshop on Self-adaptive software, 2000, pp. 27- 39.

[20] Abrahamsson, P. Commitment Development in Software Process Improvement: Critical Misconceptions, *Software Engineering*, 2001, pp.71-80.

[21] Carnegie Mellon, Software Engineering Institute, "*Capability Maturity Model Integration (*CMMI)", Version 1.1, 2002.

[22] Kumar, G. P. Software Process Improvement –TRIZ and Six Sigma, (Using Contradiction Matrix and 40 Principles), 2005.

[23] Ashok Shenvi A., Design for six sigma: software product quality, *Proceedings of the 1st conference on India software engineering conference,*2008 , pp. 97-106.

[24] Galin D., *Sotware Quality Assurance :from theory to implementation, Addison* Wesly, United States, 2003.

# Neural Network and Social Network to enhance the customer loyalty process

Carlos Andre Reis Pinheiro
School of Computing
Dublin City University
cpinheiro@computing.dcu.ie

Markus Helfert
School of Computing
Dublin City University
markus.helfert@computing.dcu.ie

*Abstract* – **Due to the increased competition in the telecommunications, customer relation and churn management is one of the most crucial aspects for companies in this sector. Over the last decades, researchers have proposed many approaches to detect and model historical events of churn. Traditional approaches, like neural networks, aim to identify behavioral pattern related to the customers. This kind of supervised learned model is suitable to establish likelihood assigned to churn. Although these models can be effective in terms of predictions, they just present the isolated likelihood about the event. However these models do not consider the influence among the customers. Based on the churn score, companies are able to perform an efficient process to retain different types of customer, according to their value in any corporate aspects. Social network analysis can be used to enhance the knowledge related to the customers' influence in an internal community. The approach we propose in this paper combines traditional predictive model with social network analysis. This new proposition to valuate the customers can arise distinguishes aspects about the virtual communities inside the telecommunications' networks, allowing companies to establish a action plan more effective to enhance the customer loyalty process. Combined scores from predictive modeling and social network analysis can create a new customer centric view, based on individual pattern recognition and community overview understanding. The combination of scores provided by the predictive model and the social network analysis can optimize the offerings to retain the customer, increasing the profit and decreasing the cost assigned to the marketing campaigns.**

## I. TELECOMMUNICATIONS ENVIRONMENT AND THE NEW CHALLENGES

The telecommunications market is characterized by an increased competitive environment. In order to maintain the customer base growing, or at least stable, all telecommunications companies should provide valuable offers, suitable product plans and a range of pricing discounts. Most often, these actions are based on the customer's behavior and also their values in a corporate perspective. Those aggressive offers can be quite expensive to the companies. According to those high expenditures and aiming to keep the companies in a profit operation, it is quite important to establish an effective and efficient customer loyalty program. This program, when well performed, can become a real competitive advantage in such arduous market.

Establishes a likelihood assigned to the possibility of churn is not enough. Defines a customer value according personal information or billing behavior is not suitable anymore. In order to optimize the customer loyalty operation it is necessary to discern the differences between the customers and highlight those characteristics more relevant based on the corporation perspective. The fast market has changed the more dynamically should be the company to adapt itself in these new scenarios. And one of the most important changes certainly is concern about the customer behavior and the way how telecommunications has been used for them. New ways of communications and novel patterns of usage have been established new business issues and hence new threats and opportunities.

Realizing these changes and creating new approaches to address them is one of the most suitable methodologies to transform those changes in opportunities. These brand new approaches should be based on the creation of new predictive models due distinct variables, new methodology to valuate the customers according distinguish perspective of usage and so on. Ii is possible to use the same traditional techniques, or combined them in order to achieve this main goal. The key factor of success here is realize that the customer behavior has been changing quickly, and therefore the models to recognize their pattern and to assess their corporate values should change as well.

An effective customer loyalty program passes through by the right understanding of the customer's behavior and by the establishment of an accurate corporate value for the customers. The aforesaid models can address these issues creating a graph related to the customers interactions and connections a churn prediction likelihood, respectively assigned to the social network analysis and to the artificial neural networks.

Telecommunication's customers left an understandable track about their behaviors, showing to the companies how they use the services and products, how the call each other, how they pay their bills, how they acquire and consume new offers and how often they complain about something. This track is quite important to realize the customer's needs and to understand the churn's behavior so this kind of event could be predicted. Monitoring and following these tracks is a suitable way to prevent the churn events from the customers.

All customers create naturally a community of relationship. Among others, one indicator about the relationship may be the frequency of calls between their members. Those communities can be viewed as social networks, where each individual have

T. Sobh (ed.), *Innovations and Advances in Computer Sciences and Engineering*,
DOI 10.1007/978-90-481-3658-2_16, © Springer Science+Business Media B.V. 2010

connections, stronger or weaker, with the other members of the community. In that way any individual can exert some influence over the other members, and particularly, over the events which happens in a social network.

Some events within the network can be influenced by activities of other customers. In the example of churn, word of mouth, rumors, commentaries and mostly activities of churn of other customers may create a chain process. This chain process can be started by a strong node of the social network, even by a less revenue value customers but high influent one, which can causes severe impacts in the customer loyalty's process.

Telecom companies should be able to recognize the high value customers, not in terms of revenue, usage or billing, but in terms of influence. The influence represents the impact which some customers can exert over the others within a virtual community or a social network. All customers identified as central or strong nodes in a social network can be offered distinguish loyalty promotions, not based on what they are, but based on what they represent, how they can influence several distinct customers to follow them in a chain's event like churn.

In general, the telecommunications market is very stable in terms of customer base, and due the quite expensive process to acquire new customers, an effective customer loyalty program could be a considerable competitive advantage, allowing companies to retain their best customers and also their best virtual communities. The retention process is a new concept in telecommunications which means in corporate terms not just retain the best or profitable customers but also retain the relations and connections among them, and hence, keeping the service usage and the revenue assigned to this.

Traditional predictive modeling based on artificial neural networks and new pattern recognition method like social network analysis can be used in conjunction in order to create a distinguish approach to manage the customer relationship and the loyalty

## II. TRADITIONAL PREDICTIVE MODELS BASED ON ARTIFICIAL NEURAL NETWORKS

Artificial Neural Networks can be used as a supervised learning model [4] to predict the churn events which usually happens in the telecommunications environment.

Predictive modeling can establish a very good approach to assigned likelihoods to the customer's events, indicating for each customer the propensity that they will terminate their business relation.

Neural network is a data basis modeling, which means that is necessary a specific timeframe of observation and a target class definition. The class in this case will be the event of churn, and the time frame will be used to train the neural network. It will be composed by historic calling information plus demographic and corporate data about the customers.

That model will recognize the pattern assigned to the customers which leave the company, and also to those ones to keep on it. Using this target class the predictive model mighty prevent future events, establishing for each customer a likelihood related to the possibilities that they have to quite the company or to keep on them.

Learning with the past and creating rules which can be applied to current customers, inferring the chances that they have to leave, make the companies able to anticipate some events and prevent it.

Neural network model provides a likelihood assigning to each class event occurrence. In that case, the occurrences are the customers and the predicted class is the churn event. Therefore the likelihood provided by the model is an indication about the distance of the occurrence to that the predicted class. This distance will establish how close the customers are to make churn or to stay in the company.

That score shall be used to create specifics approaches to retain customers, allowing establish distinct offerings according to the risk of churn assigned to them. Each customer, or a group of them based on a range of scores, can be managed through different ways, maximizing the retention rate and minimizing the cost of operation. The offering will be based on the risk of churn instead a generic overview.

In spite of the predictive model can make the companies able to identify the more propensity customers to leave and then proceed focused actions to try to retain them, there is an additional way to enhance that approach. By using the churn score companies can direct the effort toward to the high likelihood customers to churn. But what happens when two distinct customers have the same likelihood? How to differentiate them in terms of value and hence rank them in an actions perspective? Valuate the customers is the best way to rank them in a massive retention actions. The question is which the best wise to value them is.

However, customers who tend to leave the company may have distinct values. Therefore, in order to optimize the retention's process, it is mandatory to know the customers' value and cross this information with the propensity of churn. Aware about those differences, companies are able to offer retention plans adjustable for the customer value and the churn propensity at the same time, increasing the retention rate and reducing the operational cost.

## III. SOCIAL NETWORKS IN THE TELECOMMUNICATIONS ENVIRONMENT

The main feature of any community is the relationship events between their members [1,3,6,8]. Any kind of community is established and maintained based on this kind of relation events. According these brand new technologies and communication devices available currently, this type of characteristic have been increasing in relevance and becoming more evident in telecommunications scenery. From novel options for communications and relationships, like smart phones and personal digital assistants, with flexibility and mobility, those communities gained new boundaries, creating new means of relationships and social networks. Based on different ways to communicate, people are ready to create new types of virtual networks. Due the dynamism and cohesion of these communities the people's influence can be significantly more important and relevant to the companies.

Additionally, some kind of events can be faced as a chain process [7], which means that some specific points of the network can trigger a sequence of similar events due its influence and weight. The linkages between the customers can describe the way that the events will run and the weight of the linkages can represent the impact of the chain process in the network.

Understanding the way of these relationships occurs and recognizing the influences of some specific points inside the network could be a great competitive advantage, especially with respect to the event which could be performed in a chain process like churn.

Due to the unique relationship between the telephones and the individuals, and the influence which some person could exert over the others members of the community, the recognition of the characteristics of the social network within the telecommunications' environment is quite relevant. Discovering the central and strong nodes of those social networks and understanding how they behavior could be one of the most effective way to establish a new customer value, and to predict the impact of the churn, but in chain process overview. Realizing the customers connections and identifying the central nodes of each chain inside the social network mighty be the best way to prevent a massive event of churn and hence a huge leakage of revenue.

## IV. SOCIAL NETWORK ANALYSIS TO ESTABLISH A NEW METHOD TO VALUE THE CUSTOMERS

Social network analysis can reveal the possible correlations among the churn's events inside the virtual community [2], proving the stronger influence and impact when these events are triggered by a core node within the social network. Similarly, if the event is triggered by a boundary node the impact over the others members should be weaker. The influence means the number of customers which should follow the initial churn's event in a chain process.

Monitoring and analyzing the social network, particularly those which are interconnected, can allow companies to assess the revenue impact assigned to the churn's events in a chain process. Based on the likelihood assigned to the churn's event and the level of influence related to the customer, companies can perform different actions to avoid the chain's process initiation.

Social network analysis in telecommunications can help companies to recognize the customer's behavior and then predict the strength of links between the customers and the impact of the events among them. In this particular scenery it is more important to retain an influencer customer than a valuable one. In fact, the length of influence is important, thus representing the span of the triggered chain process.

As described by figure 1, a central node in a social network can be referred to a strategic customer. The point highlighted in the figure can be assigned to a customer more relevant to the company due his influence score rather than his traditional value. The other nine customers linked to this core node can represent more value than itself in an isolated way. This kind of valuation in an environment highly related with communities like telecommunications should be view as a competitive advantage.

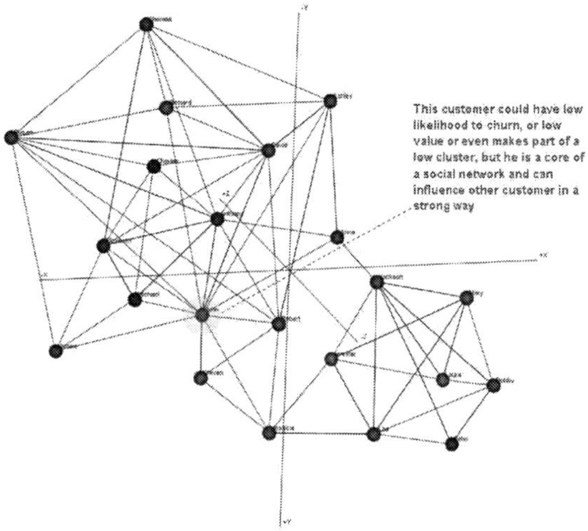


Fig. 1. Representation of the core node in a social network and the high influence assigned to it. Regardless the corporate value assigned to this customers, he presents a high influence over the virtual community which can represent the impact of any activity from him.

Considering the hypothesis that there are strong correlation between the churn's events [5] when they occurred inside a social network, one of the yields of this research is to evaluate the impact of each node in the social network, which means each valued customer in a specific community. Doing that, the telecoms' operators will be able to define and perform distinguish actions to retain customers based on their influence inside the community rather than the regular measures like billing and usage. In particular cases, the customer's influence can be more relevant to the company than the regular billing evaluation. Perhaps, the sum of the billing accounts assigned to the customers who can follow a core node in churn event chain can be greater than the billing related to a single boundary node, even considering a high billing, but with low influence in the social network. In the first case several distinct billing accounts should be considered rather than a single bill in the second scenario.

## V. A PROPOSITION OF USAGE A COMBINED MODELS APPROACH

Traditional supervised learned models based on neural artificial networks can establish a distinguish knowledge about the customers' behavior. This type of model is fitted to specific goals as to predict the churn events. In order to assess the correlation among the churn events it is important to analyze the events in a perspective of a chain. Are the events related to each other? Are the correlation assigned to the events related to any customers' attribute? Are some correlations about the strength of the links and the churn events?

Those questions can be addressed using social network analysis rather than the traditional models. However, binding both models, the traditional one based on artificial neural networks to predictive modeling, and the social network analysis to arise the customer's behavior based on a chain process enable companies to achieve a more effective and accurate methodology to understand and prevent churn.

Due to the huge complexity related to relationships between humans, the main challenge involving social network analysis is the capacity to recognize the patterns of behaviors among the individuals and the impact of each event can keep in terms of individual's influences.

The knowledge about the length of the customers' influence can be used to define a new value and allow companies to establish distinguish loyalty campaigns. This new perspective change substantially the way companies can manage their customers, evaluating them based on their influence instead based on their revenue generating. Particularly in the telecommunications market this new approach can be quite relevant due the natural social networks hidden inside the data.

In order to identify the valuable customers compared to the regular or less valuable ones, we propose to apply social network analysis. A distinguish differentiation among the customers can be raised by this technique, identifying the customers' influence and hence the kind of retention action which should be performed. The influence factor can reveal the customers which are able to trigger churn events in a chain perspective. This kind of knowledge is more relevant in a telecommunication's scenery than traditional attributes and thus enables companies able to select more effective actions for customer retention.

In order to compare a set of customers we calculate an influence factor, representing the value assigned to each individual customers plus the sum of their satellites' nodes values. The value of the satellites' nodes is weighted due the strength of the links between the nodes, which means the strength of the customer relationship. In this way, the customer's value is established based on their linkages with other customers, the weights, the frequency and the recency of it, plus personal demographic and behavioral attributes. The value depends on more the importance of the customer within the social network than his isolated score. This is quite different considering to other approaches, where the customer's value is based on isolated characteristics. In our approach, the customer's value is based on the relationship characteristics more than the isolated ones, making the value more assigned to the influence factors than the personal behavior. The customer's value can be calculated based on the relative values assigned to the linkages and the value related to the other customers. In a chain event like churn it is important to understand all aspects related.

In this way companies will be able to keep not just the higher value nodes but also the adjacent nodes within the virtual social networks. In practical terms, companies will be able to retain the customers and their relations rather than just the customers.

Based on the influence's measure companies are able to consider not only the customers, but also the existing relationships in their network. Retain the relationships within the social network means to keep the customer related using the telecommunications' services and products. Consequently, this usage maintenance represents revenue which keeps coming in to the company.

Combining the knowledge arose by the both models; the predictive one based on artificial neural network and the pattern recognition based on social network analysis; it is possible to establish a different type of retention action, as showed in table 1.

TABLE I
THE MODELS' SOCRES COMBINED TO DEFINE A MORE EFFECTIVE ACTION PLAN.

| Propensity to churn | Customer's influence | | |
|---|---|---|---|
| | Level 1 | Level 2 | Level 3 |
| 0-30 | no action | no action | no action |
| 30-50 | strong shield campaign | mid shield campaign | weak shield campaign |
| 50-70 | strong anti-attrition action | mid anti-attrition action | weak anti-attrition action |
| 70-100 | srong retention promotion | mid retention promotion | weak retention promotion |

The traditional predictive modeling based on neural network can establish likelihood for each customer, defining the propensity of each one to terminate the business relation. Due to the complexity to create several distinct marketing campaigns, it is quite usual to set up a range of prediction's likelihood. In that case, the propensities will be categorized into different limits like from 0 to 30% of churn in one group, from 30 to 50% in other group, from 50 to 70% in another one and finally greater than 70% in the last group. That kind of summarization can be understood as all customers up to 30% of propensity to leave should be considered as loyal customers with low probability to churn, implying no action for retention is required. Analogous strategy can be established for the other groups, 30-50% be considered as low chance to leave, 50-70% be considered as medium and greater than 70% as high propensity to churn.

However, two distinct customers can have the same propensity to leave company, but they can represent different values to the corporation. Therefore must be a way to differentiate those customers in order to offer to them the right package of actions for retention, minimizing thus the cost of operation and increasing the hit rate of loyalty.

The traditional approach to select the good customers based on their billing or revenue history is very mature and usual performed. However these approaches just consider the individual's characteristics. According to the virtual communities existing within the telecommunications' networks, some specific customers may have more importance or relevance for the company because their connections and relations rather than their isolated value.

A novel approach to arise the customer's influence will be established using social network analysis, recognizing the core nodes into the communities.

Based on the churn probability and the level of influence we can develop a matrix of possible actions for the retention process, as presented in the Table 1.

Assuming level 1 as the higher influence and level 5 as the lowest, and the likelihood as a percentage of churn's propensity, the cells in the table describes the possible actions to be taken by the companies to improve the customer loyalty process. In this way, the customers who have more influence over the others and have the highest likelihood to churn will receive the strongest retention promotion. Analogously, the customers who have the lowest influence and the smallest probability to leave will receive a weak shield campaign to retain them. Additionally, the customers who have no propensity to churn don't need to receive any kind of action primarily.

## VI. Evaluation and Conclusion

Based on the main goals established previously there are several business issues to be addressed using the combined model's approach. Due to the high predictive capacity related to the artificial neural networks it is possible to assign distinct likelihoods to each customer in terms of churn prediction. Also, due the strong pattern recognition capability related to the social network analysis it is possible to assign for each customer a level of influence and impact which can be exerted into the community. Both churn prediction and customer influence scores can be combined in order to achieve best results in corporate environments, especially in a customer loyalty process.

The proposed approach to use a combination of predictive and pattern recognition models, referred in the table 2 as CM, emphasizes the best features of each one. As described in that table, artificial neural networks, referred as ANN, is a very good model to predict particular events, like churn. Also it is adaptable to changing in database or data flows, something very useful in telecommunications. Neural networks also can be replicable to be used in different scenarios and it able to be deployed in a production environment. As a supervised learning model, neural networks can use previous premises to direct the model learning. Specific characteristics of telecommunications can be highlighted in such learning process. Finally, neural network is based on historical data which can be very effective in a kind of event that vary in time like churn.

Social network analysis, referred in the table as SNA, is a good technique to recognize patterns of behaviors in large amount of data, typical for the telecom industry. As an unsupervised model it does not require previous premise, which is relevant in order to understand the customer behavior in a virtual network. It is able to establish an overall overview of relationships, which can provide a distinguish knowledge about the customers' behavior. That model defines a comprehensive view about the community's behavior. This allows companies to use that kind of models as a decision support approach in many different business issues. Also, social network analysis is an interpretable model where their rules can be described in simple sentences, understandable by the business analysts, and deployed in a production environment. This characteristic is important so that the business analysts can understand the rules behind the knowledge and create the actions to apply this knowledge in terms of actions.

TABLE II
BENEFITS BY THE APPROACH BASED ON THE COMBINATION OF THE MODELS.

| Features | Models | | |
|---|---|---|---|
| | ANN | SNA | CM |
| Capacity to predict events | X | | X |
| Capacity to recognize patterns | | X | X |
| Historical basis | X | | X |
| Snapshot basis | | X | X |
| Delivery overall overview | | X | X |
| Interpretable | | X | X |
| Able to be replicated | X | | X |
| Implementable | X | X | X |
| Able to be deployed | X | X | X |
| Learning adaptive | X | X | X |
| Supervised learning | X | | X |
| Unsupervised learning | | X | X |
| Able to score databases | X | X | X |

Telecommunications market is quite dynamic, and some particulars scenarios change very often. When the market changes the data related to it changes as well. Both models are adaptable to different changes which happen in data, pursuing the customer's behavior in terms of actions, usage, consuming and relationships. The adaptable feature is fundamental in a of market segment characterized by high competition, as the telecommunications. In this market the conditions assigned to the customer's behavior can change rapidly. The dynamical characteristic assigned to the customers should be reflected on data and hence the model basis on data is able to recognize this kind of change.

Either model can score the customers' database in order to direct an action plan definition. Using neural networks and social network analysis companies are able to establish a distinguish methodology to improve the customer loyalty program, taking the unique benefits from both effective analytical models.

In this approach based on combined models, it is possible to use the predictive score of churn to focus the retention actions just for customers which have high propensity to terminate the business relation. Simultaneously, it is possible to use the social network's knowledge to recognize the customer's influence and thus rank the retention actions for the most important customers, the ones which have more influence over the community. In this way, it is possible to increase substantially the rate of accuracy assigned to the loyalty process and also to reduce the cost of the operation.

Ranking the customers not just based on the likelihood of churn but according to their corporate value, considering their influence in some virtual community or social network, make possible to establish and to perform actions in a straightforward way to retain the distinguish customers, especially the ones who have more possibilities to exert impact over the others.

The innovation factor here is indeed the approach to establish the value assigned to the customers, considering their influence's attributes and their relations' weights more than their isolated characteristics. In an event chain like churn this approach may represent a completely different way to accomplish the customer loyalty program.

The traditional way to evaluate the customers is usually according to their billing, demographic information or even based on their behavior, using clustering or segmentation models. However, due the virtual communities created within the telecommunications' networks, it is mandatory to establish a distinct manner to value the customers, considering now their importance and influence in those communities. Following this approach companies will be able to retain more than just the high value customers but instead, they will maintain the relations inside the networks which means more products and services usage. Understand and keep the customer's relations are save the revenue assigned to them, meaning the sustainability of the corporate performance and the business profitability.

## VII. Summary and Further Research

Predictive modeling using artificial neural networks can discover the knowledge behind the huge amount of data, very often in the telecommunications' segment, and establish an accurate way to assign for each customer a specific likelihood to the churn's event. These probabilities can be used to define and perform different actions in order to retain the customers, specially the profitable and influent ones.

Social network analysis is a suitable methodology to establish a new customer's value based on their influence and impact among the others customers rather than based on isolated attributes. Considering the existing virtual communities within the telecommunications' network, this type of approach may succeed more than the traditional methodologies which consider just the personal attributes to assess the customer's value.

This research intends to combine the benefits from both predictive and pattern recognition models in order to create an effective methodology to improve and enhance the customer loyalty program.

However, features and capabilities assigned to simulation are required to evaluate the best options in terms of actions and campaigns to be performed and shall be focus of this research onward. In order to create a suitable environment to support business decisions, besides the knowledge arose by the models, a simulation process would be required to improve the evaluation of customers, the impacts of churn events, how the process will perform in a chain's perspective and which suppose to be the best action to retain the central customers.

The simulation environment should address business questions based on assessments of the network chain's event, evaluating the different ways of impacts according to the customers churns. Due some specific event of churn how the social network will behavior afterward? How the simulation of events could influence the customers' value and then update their importance and relevance according to the ongoing situations? This kind of simulation process should create a huge improvement of accuracy in the analytical decision support environment. Corporate decisions would be defined due simulation assessments rather than just historical numbers and statistical analysis.

## References


[1] Carrington, Peter J., Scott, John, Wasserman, Stanley. Models and Methods in Social Network Analysis. *Cambridge University Press*, 2005.

[2] Dasgupta, Koustuv, Singh, Rabul, Viswanathan, Balaji, Chakraborty, Dipanjan, Mukherjea, Sougata, Navati, Amit A. Social ties and their relevance to churn in mobile telecom networks. *EDBT'08*, 2008.

[3] Degenne, Alain, Forse, Michel. Introducing Social Networks. *Sage Publications*, 1999.

[4] Haykin, Simon. Neural Networks: A Comprehensive Foudation, *Prentice Hall*, 1998.

[5] Kempe, David, Kleinberg, Jon, Tardos, Eva. Maximizing the spread of influence through a social network. *SIGKDD'03*, 2003.

[6] Knoke, David, Yank, Song. Social Network Analysis. *Sage Publications*, 2007.

[7] Seshadri, Mukund, Machiraju, Sridhar, Sridharan, Ashwin. Mobile call graphs: beyond power-law and longnormal distributions. *KDD'08*, 2008.

[8] Wasserman, Stanley; Faust, Katherine. Social Network Analysis: methods and applications. *Cambridge University Press*, 1994.

# Using B-trees to Implement Water: a Portable, High Performance, High-Level Language

A. Jaffer, M. Plusch, and R. Nilsson
Clear Methods, Inc.
One Broadway, 14th Floor
Cambridge, Massachusetts 02142 USA

*Abstract* - **To achieve high performance, the next generation of high-level programming languages should incorporate databases as core technology. Presented here are the design considerations for the Water language leading to the use of a (B-tree) Indexed Sequential Access Method database at its core.**

## INTRODUCTION

As increasing CPU speed outpaces growth in memory bandwidth, programming languages must deal with aggregations of data in order to keep pace both with execution speed and programmer time. To scale with CPU speed, lower level language constructs require more and more sophistication from the (high-level) compiler. Higher level language constructs need only be optimized by the implementation language - provided they are the right constructs.

An example of this is the ***for_each*** construct of the Water language. This language's sole iteration construct maps over elements of vectors, maps over database keys and values, or maps over integers, and can collect results in a vector, in a database, or reduce them with a given function. The input keys can be arbitrarily filtered.

This is an example using ***for_each*** to iterate over a vector containing two strings.

```
<vector "zero" "one"/>.<for_each combiner=join>
  <join key "=" value "; "/>
</>
```

Returns: `"0=zero; 1=one; "`

The implementation of ***for_each*** has code optimized for the possible combinations of input type, map function, and filter. It also has code optimizing the common cases where primitive methods are given for filter, map, or reduce. Because ***for_each*** usually iterates multiple times per invocation, the time to dispatch to the correct code is negligible when amortized over the iterations.

The Water language had it origins in 1998 when Christopher Fry and Mike Plusch, the founders of Clear Methods, recognized both the potential and the limitations of XML and Web services. Water has since become a platform enabling businesses to make use of Web services and XML without the inherent limitations and complexity of traditional Web services development.

The first version of Water was released in 2001 to run on Java virtual machines. This first implementation suffers from slow operation and long startup times. In late 2006, a new higher-performance implementation was needed in order to achieve the performance level appropriate for a lightweight browser plug-in. In addition to achieving high performance, it was critical that the language be compatible with multiple operating systems and platforms.

The Water language is object-oriented. Object-key pairs are associated with methods and other values. Water has lexically-scoped environments; environment-variable pairs are associated with values. Water is reflexive; in the Java implementation, code is stored as objects. The top-level environment and root of the class hierarchy can grow to have a large number of associations.

The Java implementation of Water spends considerable time loading library code into the runtime image. Startup would be much faster if binary code objects could be saved and restored from a file.

Embedded platforms do not all support virtual memory. And many platforms (embedded or not) perform poorly as memory use by applications or plugins grows. To improve performance it is important to control RAM use by applications and plugins.

So we are looking for a core technology that:

- stores associations (in databases)
- has fast access times for both large and small databases
- can be saved to and restored from binary files
- has a bounded RAM footprint

B-trees [1] [2] [3] [4] have all of these properties. Such use of B-trees is not without precedent; created in 1966, the MUMPS (Massachusetts General Hospital Utility Multi-Programming System) and its derivatives are based on an Indexed Sequential Access Method (ISAM) database, most often B-trees.

We have adapted the ***WB*** B-tree library [5] for Water's use. It has the additional benefit of being thread-safe; critical update operations are protected by distributed locks; inter-thread communication is supported through mutual-exclusion operations provided in the application programmer interface. Thus ***WB*** can be used to support multiple sessions in a server or in a browser's multiple frames.

## MULTILINGUAL PROGRAMMING

***WB*** is written in a subset of the Algorithmic Language Scheme which can be translated to C by the ***Schlep*** compiler

T. Sobh (ed.), *Innovations and Advances in Computer Sciences and Engineering*,
DOI 10.1007/978-90-481-3658-2_17, © Springer Science+Business Media B.V. 2010

[6] which is included in the WB distribution. At Clear Methods, Aubrey Jaffer and Ravi kiran Gorrepati adapted ***Schelp*** to create ***scm2java*** and ***scm2cs***, producing completely compatible implementations of ***WB*** in Java-1.5 and C#. This same translation technology is used for translating the Scheme sources for the Water compiler and runtime engine into C, Java, and C#.

The use of these translators means that compiler and engine development (and releases) can proceed in parallel with Water code development using any of Water's compatible platforms. The Scheme implementation (***SCM*** [7]) used for development does comprehensive bounds and type checking, eliminating the need for writing many program-logic checks into the source code of the Water compiler and engine.

Another mechanical translation is done by a simple bespoke Scheme program processing the data-representation design document, extracting the version, byte opcode, and (numerical) type assignments and producing source files which are included or otherwise used in the builds and runtime.

Java and C# provide garbage-collection. In C, the new Water implementation uses the Boehm conservative garbage collector [8] for temporary allocations.

## PRIMITIVE TYPES

The keys and their associated values in ***WB*** are from 0 to 255 bytes in length. The 250 bytes are more than enough to host all the codepoints, identifiers, and numbers including bignums that users (other than number-theorists) need. Integers are from 1 byte to 249 bytes in length and are stored big-endian with a length prefix so they sort correctly as keys in B-trees. Water also encodes strings and binary objects smaller than 253 bytes as ***immediate*** objects.

Although there are techniques for extending B-tree keys and values in length, at some point it becomes burdensome for the runtime infrastructure to allocate and store large primitive types in the runtime image; doing so also can exceed the bounded RAM footprint. So the new Water implementation picks as its boundary 253 bytes. A string or binary object larger than this is given a unique identifier and its data is stored under numerical keys appended to its identifier. Whether a string or binary object is represented as a single immediate or as associations in a B-tree is not discernable to the user.

The index used for each string chunk is the UTF-8 codepoint offset of the end of the chunk from the beginning of the first chunk. Strings thus have $O(\log N)$ access time even though their UTF-8 representation has variable numbers of bytes per codepoint.

## OBJECT ENCODING

The straightforward embedding of Water object structures into B-trees is that every record instance (classes are also record instances) has a unique identifier; and every slot corresponds to an association of the slot-value with that identifier combined with the slot-name. A slot-name is a non-negative integer or immediate string. A slot-value is either an immediate datum or an identifier for a (long string or) record.

A straightforward embedding of Water program expressions into B-trees builds on the object representation. Each expression is represented as a record. The ***_subject*** field (the object that gets the method call), if present, contains the literal datum or the identifier of the subject expression. The ***_method*** field contains the method-name string or the identifier of the method expression. Named arguments associate their keys (appended to the expression identifier) with their values or value expressions. Unnamed arguments associate their numerical order (0 ...) with their literal values or value expressions.

Variables and certain other strings used as keys or symbols are assigned unique identifiers; the forward and backward identification with strings being stored in B-trees. These identifiers are one to five bytes in length.

Although independent from other representation decisions, lexical bindings are also convenient to store in B-trees. Each environment has an identifier; and each variable (combined with the environment-identifier) is associated with its value. An environment's associations are deleted just before the environment is reused.

To support good error reporting, it is desirable to link every program expression to its location in a source file. This can be done simply in a B-tree while presenting no bloat or overhead to the code itself. A dedicated B-tree associates the identifier of each expression with its filename and offsets.

In ***WB***, a B-tree ***handle*** caches the last block visited, bypassing would-be full traversals from the B-tree's root for nearby references. To take advantage of this, the Water implementation uses six ***WB*** handles: string-to-variable, variable-to-string, bindings, records, program, and program-annotations.

## SECURITY

The six ***WB*** handles, along with directories to which a session has access, are the set of capabilities passed to routines in the Water compiler and runtime engine. They cannot be accessed or modified from a Water program. They are not stored in B-trees. Pointed to only from the call stack, they provide a measure of protection and isolation from other threads, which have separate call stacks.

## EXECUTION

Modern CPUs execute dozens of instructions in the time it takes to fetch one cache-line from main memory. Few applications today tend to be CPU-bound; most are memory- or cache-bound. (CPU-bound programs tend to be overrepresented in benchmark suites.) For all their benefits, access to small datums through B-trees does incur significant overhead. But for a runtime interpreter, multiple fetches from the straightforward embedding of program expressions precede each data access. Thus, the most productive area to optimize for overall performance is to reduce the bandwidth of program B-tree fetches.

Toward this end, we would like to aggregate a program expression into single B-tree value. But ***WB*** values are limited in length. So the aggregate expression format should also be space efficient. And the format should provide for overflow by being able to indicate that an expression is continued in the next B-tree association's value (***WB*** supports ISAM).

The aggregate expression format is a byte encoding with several types of codes. All the primitive methods are assigned single byte codes, as are prefix-markers. Identifiers, of which there are eight types (including symbol, long-string, method, and expression), have byte codes, the bottom two bits of which indicate the number of bytes following: 1, 2, 3, or 4. (Those identifiers are then between two and five bytes in size.)

For expressions there are markers delimiting the boundary between keyed and unkeyed arguments and the end of arguments. For the method definition, there are 24 codes indicating whether the following parameter is keyed or unkeyed, evaluated or unevaluated, required or optional, whether a default expression follows, and whether a type specifier follows.

## SYSTEM STATE

As described here, all the state of a Water session except for the call stack is contained in its B-trees. ***WB*** being disk-backed, those B-trees are stored in a file on disk or other mass storage device. The time to run the 230kB Water executable, resume a 285kB saved session, compile and execute a trivial program, and exit takes about 6ms (3ms user + 3ms system) on a 3.0GHz Pentium-4 running Fedora-7 Linux. This time doesn't increase no matter how large the saved session is because ***WB*** reads only the blocks it needs from disk.

The ability to save program and data together into a format that runs on all platforms opens intriguing possibilities. Database files can contain their report generators, accessible with one click. Documents can adjust their formatting to suit the platform they are opened on.

## ABOUT THE WATER LANGUAGE

Water is a secure, dynamic object-oriented language and database. It is an all-purpose language (and meta-language) that runs Web applications off-line in the browser or server-side [9]. The language is compatible with .NET, Java, and C on Windows, Linux and Mac OS X systems. Water handles all aspects of software including UI, logic, and database.

Water programs can store persistent data locally with Water's embedded object/XML database. The small footprint (<500kB) and instant-startup are well suited for running programs in the browser. HTML, CSS and JavaScript are naturally part of Water programs. The same Water program can be flexibly deployed to run either locally in the browser or on the server. Programs install automatically with one click.

The simplest ***Hello World*** program in Water is:

```
"Hello Water!"
```

It displays the text *Hello Water!* in a browser window.

The following Water program uses a model-view-controller (MVC) design pattern.
(http://waterlanguage.org/examples/model_view_controller.h2o)

```
<class model_view_controller
       model_data=<v "sample string"/> />

<method model_view_controller.htm>
   <form action=<w .<controller_method/> />
    .model_data.<for_each combiner=insert>
      <div value/>
    </>
    <input name="an_input"
           value=.model_data.<last/> />
    <input type="submit" value="Submit"/>
   />
</>

<method model_view_controller.controller_method
        an_input=req>
   .model_data.<insert an_input/>
   _subject
</>

model_view_controller
```

Opening the URL runs the Water program in a browser and displays the screen shown in Fig 1.

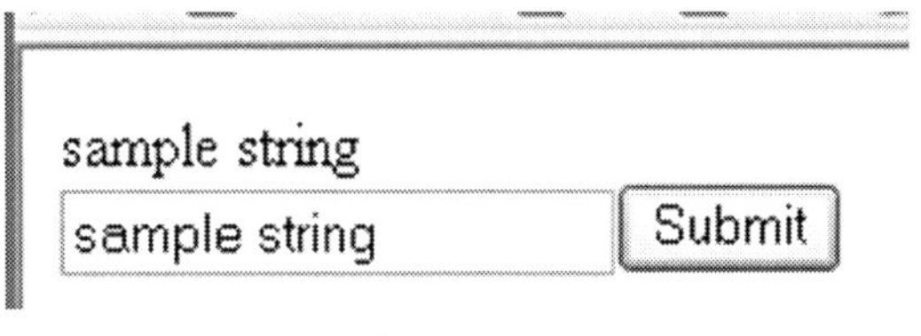


Fig. 1 Screen shot 1

This display of the application was created from a view implemented with the htm method. The model data is displayed in div tags using:

```
.model_data.<for_each combiner=insert>
            <div value/>
          </>
```

The input box displaying the last value in model data is created by:

```
<input name="an_input" value=.model_data.<last/> />
```

The submit button is created by:

```
<input type="submit" value="Submit"/>
```

If the user replaces ***sample string*** with ***Water*** in the input field, the program displays the screen shown in Fig 2.

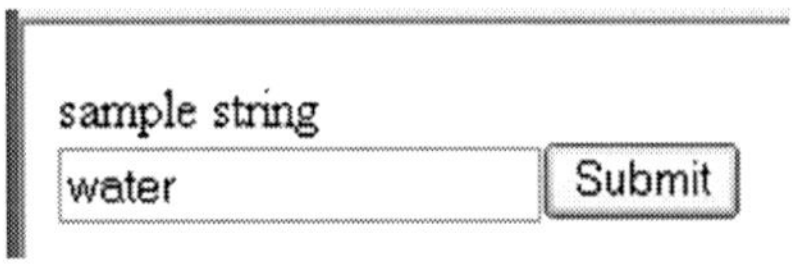


Fig. 2 Screen shot 2

Clicking ***Submit*** will call the controller_method and pass in the argument an_input with value ***Water***.

```
model_view_controller.<controller_method
                            an_input="Water"/>
```

w

When the controller method is called, it inserts an_input argument into the model data:

```
<method model_view_controller.controller_method
         an_input=req>
   .model_data.<insert an_input/>
   _subject
</>
```

This causes the application's presentation to refresh showing the value added to model data as in Fig. 3.

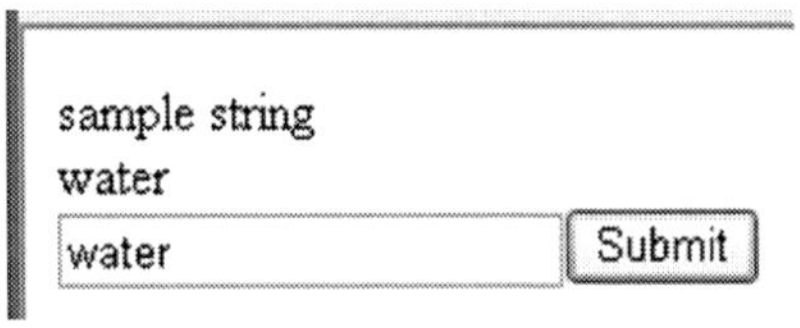


Fig. 3 Screen shot 3

REFERENCES

[1] Michael Ley, *B-Tree, Computer Science Bibliography* [Online], dblp.uni-trier.de, Universität Trier, Available: http://www.informatik.uni-trier.de/~ley/db/access/btree.html

[2] R. Bayer and E. McCreight, "Organization and maintenance of large ordered indexes," Acta Informatica, 1:173-189, 1972.

[3] Yehoshua Sagiv, "Concurrent Operations on B*-trees with Overtaking," JCSS 33(2): 275-296 (1986).

[4] W.E. Weihl and P. Wang, "Multi-version memory: Software cache management for concurrent B-trees," in *Proc. 2nd IEEE Symp. Parallel and Distributed Processing*, 1990, pp 650-655.

[5] R. Zito-Wolf, J. Finger, and A. Jaffer, *WB B-tree Library Reference Manual (2a2)* [Online], February 2008. Available: http://people.csail.mit.edu/jaffer/WB

[6] A. Jaffer, *Schlep: Scheme to C translator for a subset of Scheme* [Online], Available: http://people.csail.mit.edu/jaffer/Docupage/schlep.html

[7] A. Jaffer, *SCM Scheme Implementation Reference Manual (5e5)* [Online], February 2008, Available: http://people.csail.mit.edu/jaffer/SCM

[8] H. Boehm, A. Demers, and M. Weiser, *A garbage collector for C and C++* [Online], Available: http://www.hpl.hp.com/personal/Hans_Boehm/gc

[9] Plusch, Mike, *Water: Simplified Web Services and XML Programming* [Online], Available: http://waterlanguage.org/water_book_2002/index.htm

# VOICE BASED SELF HELP SYSTEM: USER EXPERIENCE VS ACCURACY

*Sunil Kumar Kopparapu*

TCS Innovation Lab - Mumbai
Tata Consultancy Services, Yantra Park, Thane (West), Maharastra, India.
Email: SunilKumar.Kopparapu@TCS.Com

## ABSTRACT

In general, self help systems are being increasingly deployed by *service based* industries because they are capable of delivering better customer service and increasingly the switch is to voice based self help systems because they provide a *natural* interface for a human to interact with a machine. A speech based self help system ideally needs a speech recognition engine to convert spoken speech to text and in addition a language processing engine to take care of any misrecognitions by the speech recognition engine. Any off-the-shelf speech recognition engine is generally a combination of acoustic processing and speech grammar. While this is the norm, we believe that ideally a speech recognition application should have in addition to a speech recognition engine a *separate* language processing engine to give the system better performance. In this paper, we discuss ways in which the speech recognition engine and the language processing engine can be combined to give a better user experience.

***Index Terms***— User Interface; Self help solution; Spoken Language System, Speech Recognition, Natural Language

## 1. INTRODUCTION

Self help solutions are being increasingly deployed by many service oriented industries essentially to serve their customer-base any time, any where. Technology based on artificial intelligence are being used to develop self help solutions. Typically self help solutions are web based and voice based. Oflate use of voice based self help solutions are gaining popularity because of the ease with which they can be used fueled by the significant development in the area of speech technology. Speech recognition engines are being increasingly used in several applications with varying degree of success. The reason businesses are investing in speech are several. Significant reasons among them are the return on investment (RoI) which for speech recognition solutions is typically $9 - 12$ months. In many cases it is as less as 3 months [1]. In addition, speech solutions are economical and effective in improving customer satisfaction, operations and workforce productivity. Password resetting using speech [2], airline enquiry, talking yellow pages [3] and more recently in the area of contact centers are some of the areas where speech solutions have been demonstrated and used.

The increased performance of the speech solution can be primarily attributed to several factors. For example, the work in the area of dialog design, language processing have contributed to the performance enhancement of the speech solution making them deployable in addition to the fact that people have become more comfortable using voice as an interface to transact now. The performance of a speech engine is primarily based on two aspects, namely, (a) the acoustic performance and (b) the non-acoustic performance. While the change in the acoustic performance of the speech engine has increased moderately the mature use of non-acoustic aspects have made the speech engine usable in applications; a combination of this in total enables good user experience.

Any speech based solution requires a spoken speech signal to be converted into text and this text is further processed to derive some form of information from an electronic database (in most practical systems). The process of converting the spoken speech into text is broadly the speech recognition engine domain while the later, converting the text into a meaningful text string to enable a machine to process it is the domain of natural language (NL) processing. In literature, these two have a very thin line demarcating them and is usually fuzzy, because the language processing is also done in the form of speech grammar in a speech recognition engine environment. This has been noted by Pieraccini et al [4], where they talk about the complexity involved in integration language constraints into a large vocabulary speech recognition system. They propose to have a limited language capability in the speech recognizer and transfer the complete language capability to a post processing unit. Young et al [5] speak of a system which combines natural language processing with speech understanding in the context of a problem solving dialog while Dirk et al [6] suggest the use language model to integrate speech recognition with semantic analysis. The MIT Voyager speech understanding system [7] interacts with the user through spoken dialog and the authors describe their attempts at the integration between the speech recognition and natural language components. [8] talks of combining statistical and knowledge based spoken language to enhance speech based solution.

T. Sobh (ed.), *Innovations and Advances in Computer Sciences and Engineering*,
DOI 10.1007/978-90-481-3658-2_18, © Springer Science+Business Media B.V. 2010

In this paper, we describe a non-dialog based self help system, meaning, a query by the user is responded by a single answer by the system; there is no interactive session between the machine and the user (as in a dialog based system). The idea being that the user queries and the system responds with an answer assuming that the query is complete in the sense that an answer is *fetchable*. In the event the query is incomplete the natural language processing (NL) engine responds with a close answer by making a few assumptions, when required [9]. The NL engine in addition also *corrects* any possible speech engine mis-recognition. In this paper, we make no effort to distinguish the differences in the way language is processed in the speech recognition module and the natural language processing module. We argue that it is optimal (in the sense of performance of the speech solution plus user experience) to use as a combination of language model in the speech recognition and natural language processing modules. We first show (Section 2) that language processing has to be distributed and can not be limited to either the speech recognition or the natural language processing engine. We then show how using a readily available SR engine and a NL engine [9] how the distribution of language processing helps in proving better user experience. We describe a speech based self self help system in Section 3 and describe the user experiences in Section 4. We conclude in Section 5.

## 2. BACKGROUND

For any speech based solution to work in field there are two important parameters, namely, the accuracy of the speech recognition engine and the overall user experience. While both of these are not entirely independent it is useful to consider them as being independent to be able to understand the performance of the speech solution and the user experience associated with the solution. User experience is measured by the freedom the system gives the user in terms of (a) who can speak (speaker independent), (b) what can be spoken (large vocabulary) and (c) how to speak (restricted or free speech) while the speech recognition accuracies are measured as the ability of the speech engine to convert the spoken speech into *exact* text.

Let $\mathbb{F}$ represent speech recognition engine and let $\mathbb{E}$ be the natural language processing engine. Observe that $\mathbb{F}$ converts the acoustic signal or a time sequence into a string sequence (string of words)

$$\mathbb{F} : \text{time sequence} \rightarrow \text{string sequence}$$

while $\mathbb{E}$ processes a string sequence to generate another string sequence.

$$\mathbb{E} : \text{string sequence} \rightarrow \text{string sequence}$$

Let $q_t$ represents the spoken query corresponding to, say, the string query $q_s$ (it can be considered the read version of the written string of words $q_s$). Then the operations of the speech and the natural language processing engines can be represented as

$$\begin{aligned} \mathbb{F}(q_t) &= q_{s'} \text{ (speech engine)} \\ \mathbb{E}(q_{s'}) &= q_{s''} \text{ (NL processing)} \end{aligned} \quad (1)$$

Clearly, the speech recognition engine $\mathbb{F}$ uses acoustic models (usually hidden Markov Model based) and language grammar which are tightly coupled to convert $q_t$ to $q_{s'}$ while the natural language engine $\mathbb{E}$ operates on $q_{s'}$ and uses *only* statistical or knowledge based language grammar to convert it into $q_{s''}$. It is clear that the language processing happens both in $\mathbb{F}$ and $\mathbb{E}$ the only difference being that the language processing in $\mathbb{F}$ is tightly coupled with the overall functioning of the speech recognition engine unlike in $\mathbb{E}$. Language processing or grammar used in $\mathbb{F}$ is tightly coupled with the acoustic models and hence the degree of configurability is very limited (speech to text). At the same time language processing is necessary to perform *reasonable* recognition (speech recognition performance). While there is a relatively high degree of configurability possible in $\mathbb{E} : q_s \rightarrow q_s$ (text to text). The idea of any speech based solution is to build $\mathbb{F}$ and $\mathbb{E}$ such that their combined effort, namely, $\mathbb{E}(\mathbb{F}(q_t)) = q_{s''}$ is such that $q_{s''} \approx q_s$. Do we need language processing in both $\mathbb{F}$ and $\mathbb{E}$ or is it sufficient to (a) isolate $\mathbb{F}$ and $\mathbb{E}$; and have language processing only in $\mathbb{E}$ or (b) combine all language processing into $\mathbb{F}$ and do away with $\mathbb{E}$ completely. Probably there is an optimal combination of $\mathbb{F}$ and $\mathbb{E}$ which produces a usable speech based solution.

An ideal speech recognition system should be able to convert $q_t$ into the exact query string $q_s$. Assume that there are three different types of speech recognition engines. Let the speech recognition engine $\mathbb{F}_1$ allow any user to speak anything (speaker independent dictation system); $\mathbb{F}_2$ be such that it is $\mathbb{F}_1$ but the performance is tuned to a particular person (person dependent) and $\mathbb{F}_3$ is such that it is $\mathbb{F}_2$ additionally constrained in the sense that it allows the user to speak from within a restricted grammar. Clearly the user experience is best for $\mathbb{F}_1(x_t) = x^1_{s'}$ (user experience: $\uparrow$) and worst for $\mathbb{F}_3(x_t) = x^3_{s'}$ (user experience: $\downarrow$) and it between experience is provided by $\mathbb{F}_1(x_t) = x^2_{s'}$ (user experience: $\leftrightarrow$).

Let $d(x_s, y_s)$ be the distance between the string $x_s$ and $y_s$. Clearly, $d(x^1_{s'}, x_s) > d(x^2_{s'}, x_s) > d(x^3_{s'}, x_s)$, the performance of the speech engine is best for $\mathbb{F}_3$ followed by $\mathbb{F}_2$ followed by $\mathbb{F}_1$. Observe that in terms of user experience it is the reverse. For the overall speech system to perform *well* the contribution of $\mathbb{E}$ would vary, namely $\mathbb{E}$ should be able to generate $q^1_{s''}$, $q^2_{s''}$ and $q^3_{s''}$ using $q^1_{s'}$, $q^2_{s'}$ and $q^3_{s'}$ respectively, so that $d(q^1_{s''}, q_s) \approx d(q^2_{s''}, q_s) \approx d(q^3_{s''}, q_s) \approx 0$. The performance of $\mathbb{E}$ has to be better to compensate for the *poor* performance of $\mathbb{F}$; for example the performance of $\mathbb{E}_1$ has to be better than the performance of $\mathbb{E}_3$ to compensate for the poor performance of $\mathbb{F}_1$ compared to $\mathbb{F}_3$.

Typically, a $\mathbb{F}_1$ (ideal user experience) speech recognition would be categorized by (a) Open Speech (free speech - speak without constraints), (b) Speaker independent (different accents, dialects, age, gender) and (c) Environment independent (office, public telephone). While (a) greatly depends on the language model used in the speech recognition system, both (b) and (c) depend on the acoustic models in the SR. For $\mathbb{F}_1$ type of system, the user experience is good but speech recognition engine accuracies are poor. On the other hand, a typical $\mathbb{F}_3$ (bad on user experience) would be categorized by limiting the domain of operation and the system would be tuned (in other words constrained) to make use of prior information on expected type of queries.

In the next section we describe a voice based self help system which enables us to tune the language grammar and hence control the performance of the speech recognition engine.

### 3. VOICE BASED SELF HELP SYSTEM

Voice based self help system is a speech enabled solution which enables human users to interact with a machine using their speech to carry out a transaction. To better understand the role of $\mathbb{F}$ and $\mathbb{E}$ in a speech solution we actually built a voice based self help system. The self help system was built using the Speech Recognition ($\mathbb{F}$) engine of Microsoft using Microsoft SAPI SDK [10] and the Language Processing ($\mathbb{E}$) module was developed in-house [9].

In general, insurance agents act as intermediaries between the insurance company (service providing company) and their clients (actual insurance seekers). Usually, the insurance agents keep track of information of their clients (policy status, maturity status, change of address request among other things) by being in touch with the insurance company. In the absence of a self help system, the insurance agents got information by speaking to live agents at a call center run by the insurance company. The reason for building a self help system was to enable the insurance company to lower the use of call center usage and additionally providing dynamic information needed by agents; both this together provide better customer service. The automated self help system, enabled answering queries of an insurance agent. Figure 1 shows a high level functional block representation of the self help system. The user (represented as a mobile phone in Figure 1) calls a predetermined number and speaks his queries to get information. The speech recognition engine converts the spoken query (speech signal) into text; this text is operated upon by the natural language processing block. This processed string is then used to fetch an answer from the database. The response to this query, which is a text string, is (a) spoken out to the user using a text to speech engine and (b) alternatively is sent to the user as a SMS. We used Kannel, an open source WAP and SMS gateway to send the answer string as SMS [11].

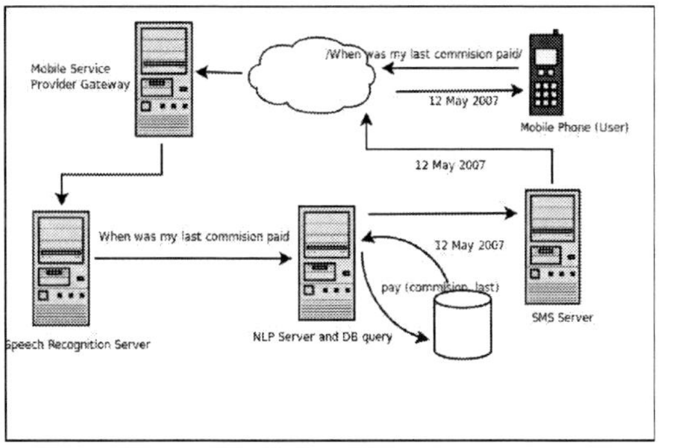


**Fig. 1**. Block Diagram of a Self Help System

The self help system was designed to cater to

1. different kinds of information sought by insurance agent
   (a) on behalf of their clients (example, *What is the maturity value of the policy TRS1027465*) and
   (b) themselves (example, *When was my last commission paid?*).
2. different accents,
3. handle different complexity of queries and
4. Additionally the system should be able to accept natural English query.
5. different ways in which same queries can be asked, Examples:
   (a) Surrender value of policy TRS1027465?
   (b) What is the surrender value of policy TRS1027465?
   (c) Can you tell me surrender value of policy TRS1027465?
   (d) Please let me know the surrender value of policy TRS1027465?
   (e) Please tell me surrender value of policy TRS1027465?
   (f) Tell me surrender value of policy TRS1027465?
   (g) My policy is TRS1027465. What is its surrender value?

   should all be understood as being queried for the

   ... *surrender_value* of the *policy TRS1027465*....

Note that the performance of the speech recognition engine is controlled by the speech grammar (see Figures 2, 3, 4 for examples of speech grammar) that drive the speech recognition engine. The speech grammar is used by the speech engine before converting the spoken acoustic signal into a text string. In Section 4 we show how the performance of the speech engine can be controlled by varying the speech grammar.

## 4. EXPERIMENTAL RESULTS

We built three versions of the self help system with varying degrees of processing distributed in $\mathbb{F}$ and $\mathbb{E}$, namely,

1. $\mathbb{F}_1$ has no grammar (Figure 2), giving a very high degree of freedom to the user as to what they can ask, giving them scope to ask invalid queries.
2. $\mathbb{F}_2$ (Figure 3) has liberal grammar; more processing in $\mathbb{E}$ and
3. $\mathbb{F}_3$ has a constrained grammar (see Figure 4) which constraints the flexibility of what the user can say

For example, $\mathbb{F}_1$ grammar would validate even an out of domain query like *What does this system do?* in one dimension and an incorrect query like *What is last paid commission address change?*. On the other extreme a $\mathbb{F}_3$ grammar would only recognize queries like *What is the surrender value of Policy number* or *Can you please tell me the maturity value of Policy number* and so on. Note that the constrained grammar $\mathbb{F}_3$ gives a very accurate speech recognition because the speaker speaks what the speech engine expects this in turn puts very less load in terms of processing on $\mathbb{E}$.

For the experimental setup, the $\mathbb{F}_1$ grammar generated a total of 27 possible queries of which only 3 were not responded by the $\mathbb{F}_1$, $\mathbb{E}$ combined system. On the other hand for a grammar of type $\mathbb{F}_3$ a total of 357 different queries that the user could ask possible (very high degree of flexibility to the user). Of these only a total of 212 queries were valid in the sense that they were meaningful and could be answered by the $\mathbb{F}_3$, $\mathbb{E}$ system the rest, 145, were processed by $\mathbb{E}$ but were not meaningful and hence an answer was not provided. The performance of $\mathbb{F}_2$ grammar was in between these two cases producing a total of 76 possible queries that the user could ask, of which 20 were invalided by the $\mathbb{F}_2$, $\mathbb{E}$ combine.

```
<GRAMMAR>
 <RULE NAME="F_1" TOPLEVEL="ACTIVE">
  <RULEREF NAME="DonotCare"/>
 </RULE>
</GRAMMAR>
```

**Fig. 2**. $\mathbb{F}_1$: No grammar; the speaker can speak anything.

```
<GRAMMAR>
 <RULE NAME="F_2" TOPLEVEL="ACTIVE">
  <RULEREF NAME="DonotCare"/>
  <RULEREF NAME="KeyConcept"/>
  <RULEREF NAME="DonotCare"/>
  <RULEREF NAME="KeyWord"/>
  <RULEREF NAME="DonotCare"/>
 </RULE>
 <RULE NAME="KeyConcept">
  <P> Surrender Value </P>
  <P> Maturity Value </P>
  <P> ... </P>
  <P> Address Change </P>
 </RULE>
 <RULE NAME="KeyWord">
  <P> Policy Number </P>
  <P> ... </P>
  <P> ... </P>
 </RULE>
</GRAMMAR>
```

**Fig. 3**. $\mathbb{F}_2$: Liberal grammar: Some restriction on the user.

## 5. CONCLUSIONS

The performance of a voice based self help solution has two components; user experience and the performance of the speech engine in converting the spoken speech into text. It was shown that $\mathbb{F}$ and $\mathbb{E}$ can be used jointly to come up with types of self help solutions which have varying effect on the user experience and performance of the speech engine. Further, we showed that on one hand by controlling the language grammar one could provide better user experience but the performance of the speech recognition became poor while on the other hand when the grammar was such that the performance of speech engine was good the user experience became poor. This shows that there is a balance between the speech recognition accuracy and user experience that is to be maintained by people who design voiced based self help systems so that both the speech recognition accuracy is good without sacrificing the user experience.

## 6. REFERENCES

[1] Daniel Hong, "An introductory guide to speech recognition solutions," Industry white paper by Datamonitor, 2006.

[2] Microsoft Research, "Microsoft speech - solutions: Password reset," http://www.microsoft.com/ speech/ solutions/ pword/ default.mspx, 2007.

[3] Tellme, "Every day info," http://www.tellme.com/ products/ TellmeByVoice, 2007.

[4] Roberto Pieraccini and Chin-Hui Lee, "Factorization of language constraints in speech recognition," in *Proceedings of the 29th annual meeting on Association for Computational Linguistics*, Morristown, NJ, USA, 1991, pp. 299–306, Association for Computational Linguistics.

[5] S. L. Young, A. G. Hauptmann, W. H. Ward, E. T. Smith, and P. Werner, "High level knowledge sources in usable

```
<GRAMMAR>
 <RULE NAME="F_3" TOPLEVEL="ACTIVE">
  <o> <RULEREF NAME="StartTag"/> </o>
  <RULEREF NAME="KeyConcept"/>
  <o> of <o> the </o> </o>
  <o> in <o> the </o> </o>
  <RULEREF NAME="KeyWord"/>
  <o> <RULEREF NAME="EndTag"/> </o>
 </RULE>
 <RULE NAME="StartTag">
  <P> What is the </P>
  <P> Please send me </P>
  <P> Can you please send me</P>
  <P> Can you tell me </P>
 </RULE>
 <RULE NAME="KeyConcept">
  <P> Surrender Value </P>
  <P> Maturity Value </P>
  <P> ... </P>
  <P> Address Change </P>
 </RULE>
 <RULE NAME="KeyWord">
  <P> Policy Number </P>
  <P> ... </P>
  <P> ... </P>
 </RULE>
 <RULE NAME="EndTag">
  <P> Thank You </P>
  <P> ... </P>
 </RULE>
</GRAMMAR>
```

**Fig. 4**. $\mathbb{F}_3$: Constrained grammar - speaker is highly constrained in what he can speak.

speech recognition systems," *Commun. ACM*, vol. 32, no. 2, pp. 183–194, 1989.

[6] Dirk Buhler, Wolfgang Minker, and Artha Elciyanti, "Using language modelling to integrate speech recognition with a flat semantic analysis," in *6th SIGdial Workshop on Discourse and Dialogue*, Lisbon, Portugal, September 2005.

[7] Victor W. Zue, James Glass, David Goodine, Hong Leung, Michael Phillips, Joseph Polifroni, and Stephanie Seneff, "Integration of Speech Recognition and Natural Language Processing in the MIT Voyager System," in *Proc. ICASSP*, 1991, vol. 1, pp. 713–716.

[8] Ye-Yi Wang, Alex Acero, Milind Mahajan, and John Lee, "Combining statistical and knowledge-based spoken language understanding in conditional models," in *Proceedings of the COLING/ACL on Main conference poster sessions*, Morristown, NJ, USA, 2006, pp. 882–889, Association for Computational Linguistics.

[9] Sunil Kumar Kopparapu, Akhlesh Srivastava, and P. V. S. Rao, "Minimal parsing key concept based question answering system," in *HCI (3)*, Julie A. Jacko, Ed. 2007, vol. 4552 of *Lecture Notes in Computer Science*, pp. 104–113, Springer.

[10] Microsoft, "Microsoft Speech API," *http://msdn.microsoft.com/en-us/library/ms723627(VS.85).aspx*, Accessed Nov 2008.

[11] Open Source, "Kannel: Open source WAP and SMS gateway," *http://www.kannel.org/*, Accessed Nov 2008.

# Using GIS to produce Vulnerability Maps of Non-Gauged Watersheds area

## (Qena Valley– Pilot area)

Eng. Amal Ahmed Abd-Ellatif Yousef

*Lecturer in Hail University – Saudi Arabia*

**Abstract-** Qena area located at the western side of the Red Sea Hills. The area is embedded within a network of active watersheds, which are subjected to recurrent flash flooding. This report is directed towards developing a methodology for flood risk assessment and relative vulnerability classification for watersheds associated with wadi systems in arid regions. Geographic Information System (GIS) is used as the main analysis tool for modeling Qena area. Watersheds in the study area have been identified, and geomorphology parameters and watershed characteristics have been estimated. Different parameters, which mostly affect the runoff, have been calculated. The HBV rainfall runoff model has been calibrated and used to define the flash flood vulnerable areas at the catchment. The results of flood risk classification compared well with the estimated runoff peak discharge. GIS has proved, as expected, to be an easy and efficient toll for watersheds flood risk assessment.

## I. INTRODUCTION

More than ninety percent of the Egyptian territories are classified as arid and hyper arid desert regions. In many locations the desert is characterized by the presence of an intense wadi system, which is subjected to harsh climatic conditions, and extreme water scarcity. Nevertheless, many of such wadies experience extreme precipitation events in the form of flash floods, where a considerable amount of rainfall occurs, suddenly, for a short duration, and with a long period of recurrence. Efforts are therefore directed to serve two objectives: (1) Making use of the available water during rare precipitation events, and (2) Protection against potential damage associated with flash floods. A methodology for flood predictions, risk assessment, and vulnerability estimation is seen to be in evitable. In the last few years, a considerable amount of attention has been devoted to the development of the Eastern Desert of Egypt, especially Wadi Qena area as shown in Fig.1.

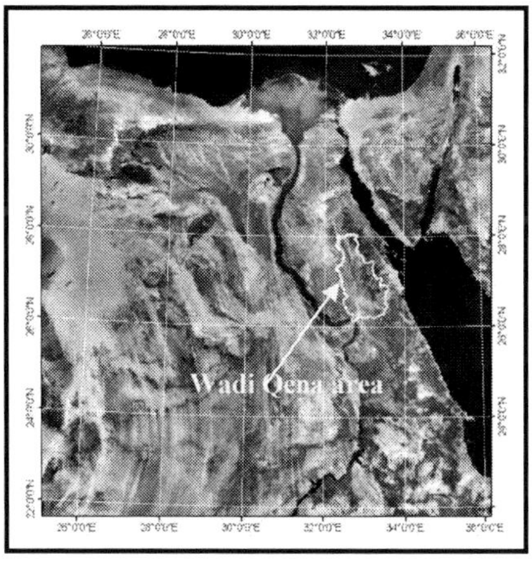


Fig.1. General Layout for Wadi Qena Area

Wadi Qena is embedded within a network of active watersheds, which are subjected to recurrent flash flooding. In some cases these flash floods cause damage to roads and infrastructure. On the other hand the area is classified as arid zone which, in general, is limited in water resources. To ensure the sustainable development in this area, an assessment for flash flooding potential, and vulnerability of the watersheds located in this area should be carried out. Limited watershed data are available from digital elevation model (DEM) and rainfall records. Therefore a suitable approach that depends on such limited data should be applied to study those watersheds.

## II. THE STUDY AREA

Wadi Qena is located at the western side of the Red Sea Hills and joins the Nile with right angle north of Qena and collects the water of many effluents which join it all along its 270 km course from the east and south east, It is a unique geomorphic feature in the Egyptian Eastern Desert . It extends in a nearly north-south direction (parallel to the Nile River) and joins the Nile Valley at Qena Town. The Wadi Qena basin covers approximately 18.000 sq. km and is bounded by Latitudes 26° 00' 00"- 28° 15' N and Longitudes 32° 30'- 33° 00'E.

## III. DATA

The Digital Elevation Model (using contour lines) has been developed and Land use/cover maps (vegetation, domestic, etc.) Are generated using Satellite Image (Landsat 7 band), in addition to Soil classification maps and Location of climatic stations around Qena Valley. The Hydrological parameters (drainage network, rainfall, evaporation and evapo-transpiration, temperature, infrastructure locations, etc.) And Geological classification and ground water aquifers. Fig.1 shows the general layout of Qena Wadi area.

The only available 2 meteorological stations representing the observed rainfall data for Wadi-Qena are used in this analysis. The location of the rain gauge stations is represented in fig.2.

T. Sobh (ed.), *Innovations and Advances in Computer Sciences and Engineering*,
DOI 10.1007/978-90-481-3658-2_19, 

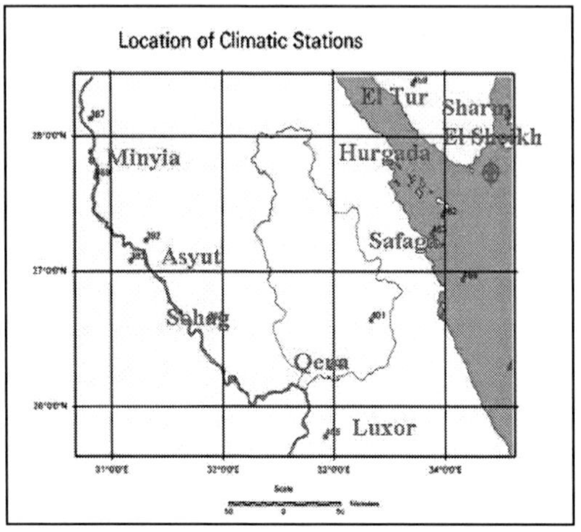


**Fig.2.** The Location of climatic stations around the Qena Vally

The main storms events are used together with daily-observed data are used for calibrating the hydrologic model.

## IV. GEOMORPHOLOGIC CHARACTERISTICS

The geomorphologic characterize of Wadi Qena basin have been extracted using the available data and modern GIS tools. It could be classified into the following main landforms:

*1. Limestone Plateau*

This plateau is dissected and consists mainly of hard, jointed and fractured limestone beds.These beds are horizontal to slightly dipping to the north. Dendritic and sub-parallel drainage patterns are dominant at Gabal Aras and Gabal Ras El-Jisr. On Landsat images, they show rough drainage texture. A flat-topped surface at Gabal Aras (523 m. a.s.l.) represents a hard, massive, structurally-controlled landform and provides a suitable catchment area.

*2.Nubia Sandstone Plateau*

This plateau is composed mainly of hard, massive sandstone beds forming dissected patches. These patches comprise some beds of iron oxides and clays, that highly affect the ground water conditions and quality. The plateau is characterized by a dendritic drainage pattern and fine texture as in the cases of Gabal Abu Had and Assuray, whereas at Gebel Qurayaah the drainage texture appears coarse on the Landsat images. This sandstone plateau is cut by a few main faults.

*3. Tors*

This geomorphologic unit is represented by a small part at the northeast corner of the investigated area and represents exposures of Precambrian basement rocks. The rocks are hard, massive and form dissected isolated hills of medium to high topographic relief. Also, they are characterized by high weathering and represent a part of the groundwater aquifers catchment's areas.

*4. Fault Scarps*

The area is affected by structural disturbances that created major fault scarps with steep slopes (38° - 75°), which are well developed in the sandstone terrain. These scarps moderate to trend NW-SE and N-S.

*5. Alluvial Fans*

Alluvial fans are dispersed in the investigated area due to the presence of fault scarps inducing topographic difference between the plateaus and the wadis. These fans are composed mainly of sands, clay and gravels. Most of them are adjacent to Wadi Qena.

*6. Flood Plain*

The flood plain surrounds the River Nile and is composed mainly of mud, silt and clay with some sands. It belongs to the Pre-Nile and is of Quaternary age (Said, 1981). This flood plain is nearly flat and completely cultivated. All Geomorphologic Units of Wadi Qena shown in Fig.3.

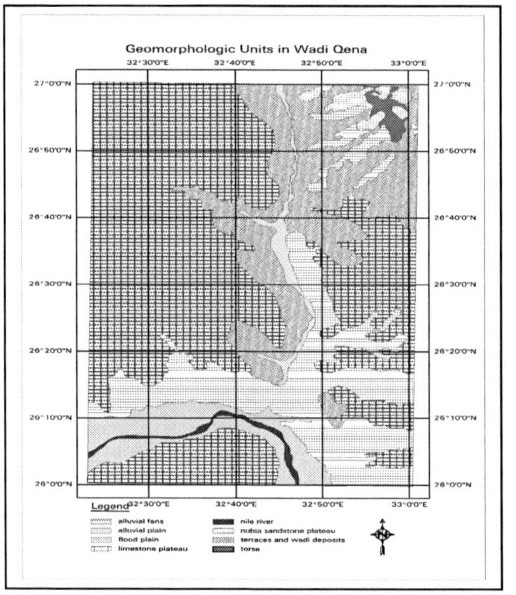


Fig.3.Geomorphologic Units of Wadi Qena

## V. MORPHOLOGICAL CHARACTERISTICS

The morphological parameters as described in TABLE 1 for the watersheds in the study area are computed using the generated DEMfor the study area using the GIS tools.

TABLE 1

Morphological Parameters for the Watersheds in the Study Area

| Parameter | Value | Description |
|---|---|---|
| Area (A) | 16000000000 | Area of watershed |
| Perimeter (P) | 730000 | Perimeter of watershed |
| Basin length ($L_B$) | 215542 | The maximum distance from the outlet point to the watershed divide. |
| Valley length ($V_L$) | 351374 | The distance measured along the main channel from the watershed outlet to the basin divide |
| Stream frequency (F) | 1.28275E-06 | $F=\frac{SNu}{A}$ |
| Drainage density (D) | 0.001185457 | $D=\frac{SLu}{A}$ |
| Length of overland flow ($L_o$) | 421.7784012 | $L_o=\frac{1}{2D}$ |
| Stream highest order (U) | 16 | Strahler order |
| Sum of stream number ($SN_u$) | 20524 | Sum of stream numbers |
| Sum of stream length ($SL_u$) | 18967306 | Sum of stream lengthes |
| Bifurcation ratio ($R_B$) | 3.7 | $R_B=\frac{N_{\omega-1}}{N_\omega}$ |
| Length ratio ($R_L$) | 1.354120593 | $R_L=\frac{\overline{L}_\omega}{\overline{L}_{\omega-1}}$ |
| Shape index ($I_{sh}$) | 0.038130981 | $I_{sh}=1.27\frac{A}{P^2}$ |
| Circularity ratio ($R_c$) | 0.038212866 | $R_c=\frac{A}{A_0}$ |
| Elongation ratio ($R_e$) | 0.662147873 | $R_e=\frac{2}{L_B}\left(\frac{A}{\pi}\right)^{0.5}$ |
| Relief ( R) | 1508 | the elevation difference between watershed outlet and the highest point on the watershed perimeter. |
| Internal relief (E) | 327 | the difference between elevation of 10% of the watershed length from the source and 15% of watershed length from the outlet. |
| Relief ratio ($R_r$) | 0.001517106 | $R_r=\frac{R}{L_B}$ |
| Sinuosity (Si) | 1.630188084 | $S_i=\frac{V_l}{L_B}$ |
| Slope index (SI) | 0.001240843 | $S_l=\frac{E}{0.75\,VL}$ |
| Ruggedness number (Rn) | 0.387644316 | $R_n=R.D$ |
| Texture ratio (Rt) | 0.028115068 | $R_t=\frac{\sum_{\omega=1}^{\omega=n}N_\omega}{P}$ |

## VI. METHODOLOGY

A methodology for flood risk assessment and relative vulnerability classification for watersheds associated with wadi systems in arid regions is herein proposed and verified for Wadi Qena study area. The flood risk assessment methodology comprises ranking the studied watershed according to its flood risk using morphological parameters as in TABLE 1. The study uses the advantages of modern technologies including remote sensing and Geographic Information System, as recommended by several researchers; DeVantire et al (1993), Garbrecht et al (1995), Maidment (1993) (1994), and Elbadawy (2003). Fig. 4 shows the methodology flowchart.

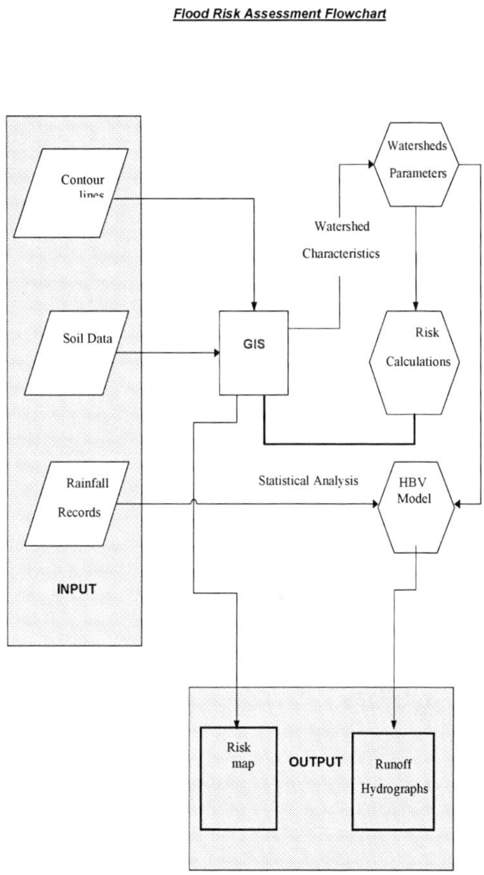


**Fig. 4 .** GIS Analysis Flowchart

## VII. THE HYDROLOGIC MODEL

The catchment is divided into a number of grid cells. For each of the cells individually, daily runoff is computed through the application of the standard version of the HBV model, as distributed by Killingtveit and

Saelthun (1995). The use of the grid cells offers the possibility to turn the HBV modeling concept, which is originally lumped, into a distributed model. Figure (8) shows a schematic view of the HBV hydrologic model concept. The land-phase of the hydrological cycle is represented by three different components: a snow routine (neglected) a soil routine and a runoff response routine.

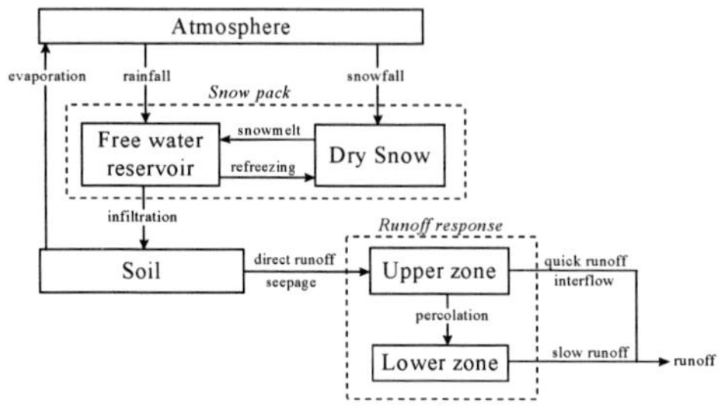


**Figure (8 ):** Schematic view of the relevant components of the HBV model

***The Soil Routine***

The incoming water from the rainfall (snow routine), $S_{in}$, is available for infiltration in the soil routine. The soil layer has a limited capacity, $F_c$, to hold soil water, which means if $F_c$ is exceeded the abundant water cannot infiltrate and, consequently, becomes directly available for runoff.

$$S_{dr} = \max\{(SM + S_{in} - F_c);0\} \qquad (1)$$

Where $S_{dr}$ is the abundant soil water (also referred to as direct runoff) and $SM$ is the soil moisture content. Consequently, the net amount of water that infiltrates into the soil, $I_{net}$, equals:

$$I_{net} = S_{in} - S_{dr} \qquad (2)$$

Part of the infiltrating water, $I_{net}$, wills runoff through the soil layer (seepage). This runoff volume, $SP$, is related to the soil moisture content, $SM$, through the following power relation:

$$SP = \left(\frac{SM}{F_c}\right)^{\beta} I_{net} \qquad (3)$$

where $\beta$ is an empirically based parameter. Application of equation (3) implies that the amount of seepage water increases with increasing soil moisture content. The fraction of the infiltrating water which doesn't runoff, $I_{net} - SP$, is added to the available amount of soil moisture, $SM$.

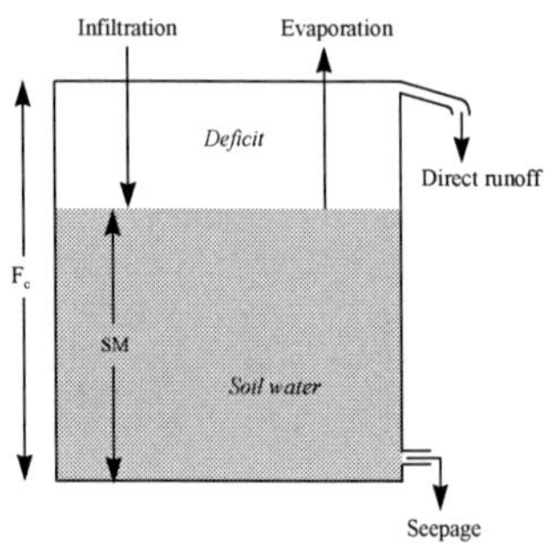


**Fig. 5.** Schematic view of the soil moisture routine

A percentage of the soil moisture will evaporate. This percentage is related to the measured potential evaporation and the available amount of soil moisture:

$$E_a = \frac{SM}{T_m} E_p ; SM < T_m$$

$$E_a = E_p \; ; SM \geq T_m \qquad (4)$$

where $E_a$ is the actual evaporation, $E_p$ is the potential evaporation and $T_m$ ($\leq F_c$) is a user defined threshold, above which the actual evaporation equals the potential evaporation.

***The runoff response routine***

The volume of water which becomes available for runoff, $S_{dr} + SP$, is transferred to the runoff response routine. In this routine the runoff delay is simulated through the use of a number of linear reservoirs. Three types of runoff are distinguished:

1. Quick runoff
2. Interflow
3. Slow runoff (baseflow)

Two linear reservoirs are defined to simulate these three different processes: the *upper zone* (generating quick runoff and interflow) and the *lower zone* (generating slow runoff). The available runoff water from the soil routine (i.e. direct runoff, $S_{dr}$, and seepage, $SP$) in principle ends up in the lower zone, unless the percolation threshold, $P_m$, is exceeded, in which case the redundant water ends up in the upper zone: **(2)**

$$\Delta V_{LZ} = \min\{P_m;(S_{dr} + SP)\}$$

$$\Delta V_{UZ} = \max\{0;(S_{dr} + SP - P_m)\} \qquad (5)$$

**(3)**

where $V_{UZ}$ is the content of the upper zone, $V_{LZ}$ is the content of the lower zone and $\Delta$ means "increase of".

The lower zone is a linear reservoir, which means the rate of slow runoff, $Q_{LZ}$, which leaves this zone during one time step equals:

$$Q_{LZ} = K_{LZ} * V_{LZ} \qquad \textbf{(4)} \qquad (6)$$

where $K_{LZ}$ is the reservoir constant.

The upper zone is also a linear reservoir, but it is slightly more complicated than the lower zone because it is divided into two zones: A lower part in which interflow is generated and an upper part in which quick flow is generated.

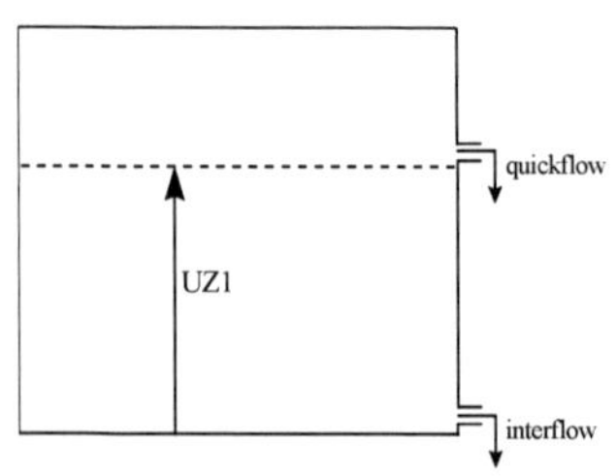


**Fig. 6.** Schematic view of the Upper zone

If the total water content of the upper zone, $V_{UZ}$, is lower than a threshold $UZ1$, the upper zone only generates interflow. On the other hand, if $V_{UZ}$ exceeds $UZ1$, part of the upper zone water wills runoff as quick flow:

$$Q_i = K_i * \min\{UZ1; V_{UZ}\}$$

$$Q_q = K_q * \max\{(V_{UZ} - UZ1); 0\} \qquad (7)$$

Where $Q_i$ is the amount of generated interflow in one time step, $Q_q$ is the amount of generated quick flow in one time step and $K_i$ and $K_q$ are reservoir constants for interflow and quick flow respectively.

The total runoff rate, $Q$, is equal to the sum of the three different runoff components:

$$Q = Q_{LZ} + Q_i + Q_q \qquad (8)$$

The runoff behaviour in the runoff response routine is controlled by two threshold values $P_m$ and UZ1 in combination with three reservoir parameters, $K_{LZ}$, $K_i$ and $K_q$. In order to represent the differences in delay times between the three runoff components, the reservoir constants have to meet the following requirement:

$$K_{LZ} < K_i < K_q \qquad (9)$$

## VIII. RISK ANALYSIS & OUTPUTS

Fig. 7 Shows the longitudinal cross section along the 5 basin digital elevation model in the direction of moving the water after coming the flash flood .Fig. 8. Elevation & volume relationship for all basins. , Fig. 9 show the different cross section along the 5 basins which token in Digital Elevation Model (DEM).

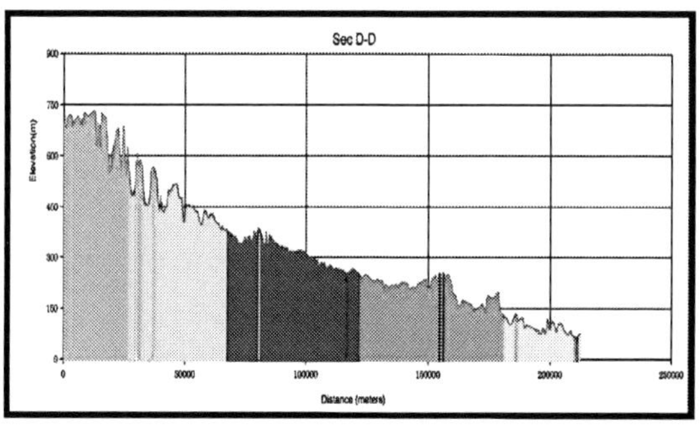


**Fig. 7.** Longitudinal cross section (D-D)

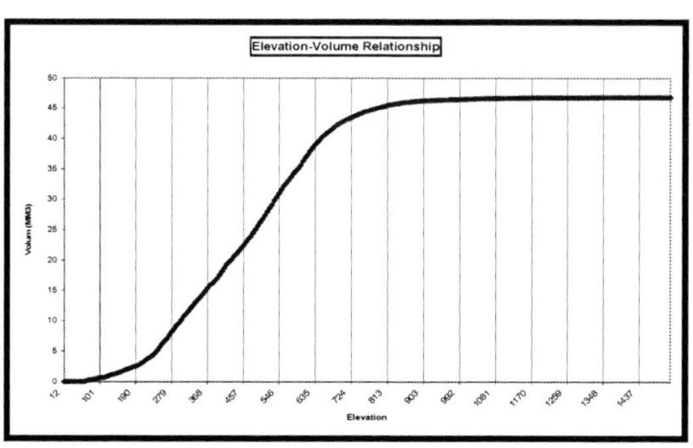


**Fig. 8.** Elevation & volume relationship for all basins

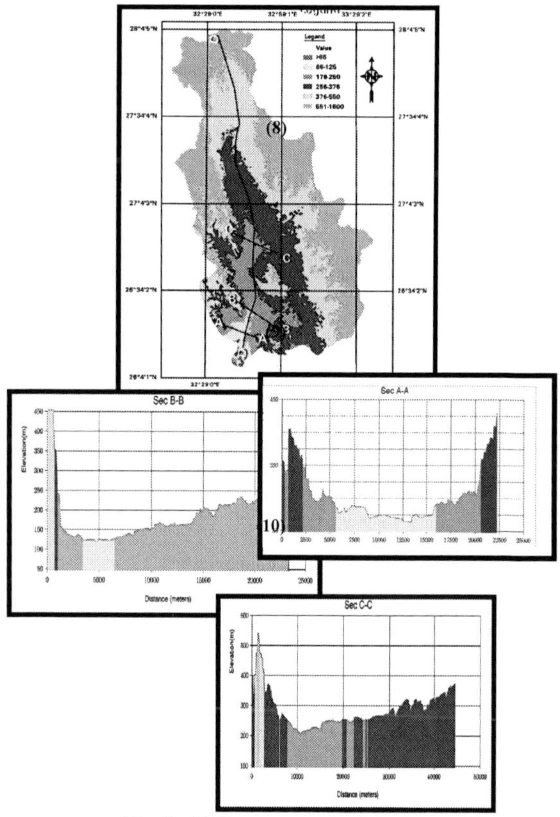


**Fig. 9.** Different cross sections

Basin 1 has minimum elevation and consider the first risk area after coming flood. Cross sections (a-a) shows the risk map for the study area after analyses the results of HBV model and apply it in the GIS layers ,We found that the effected area will be under elevation 70 meters as shown in Fig. 10, Fig. 11.

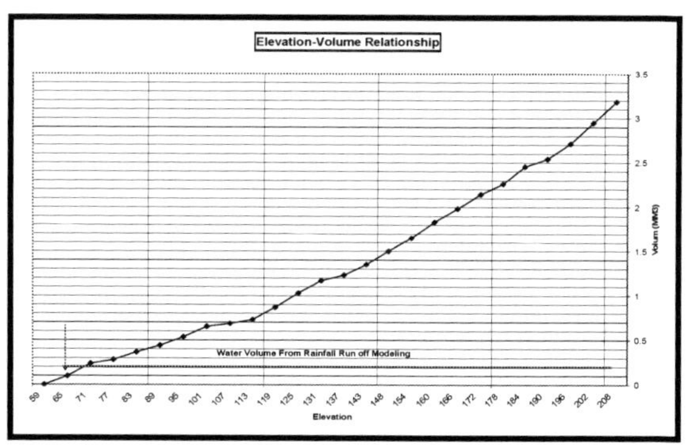


**Fig.10.** Elevation and Volume relationship at basin outlet

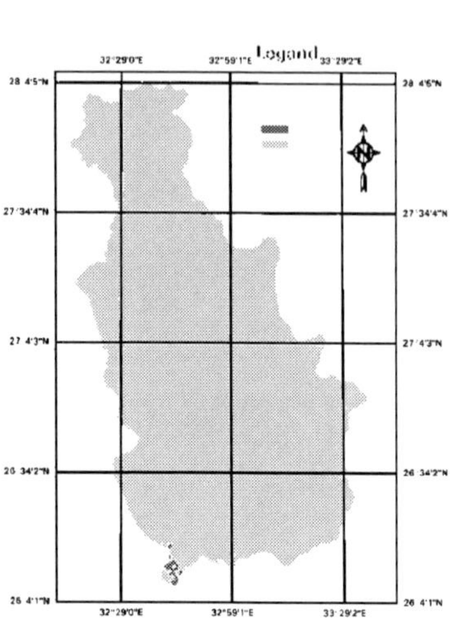

**Fig. 11.** horizontal projection for the flooded elevation

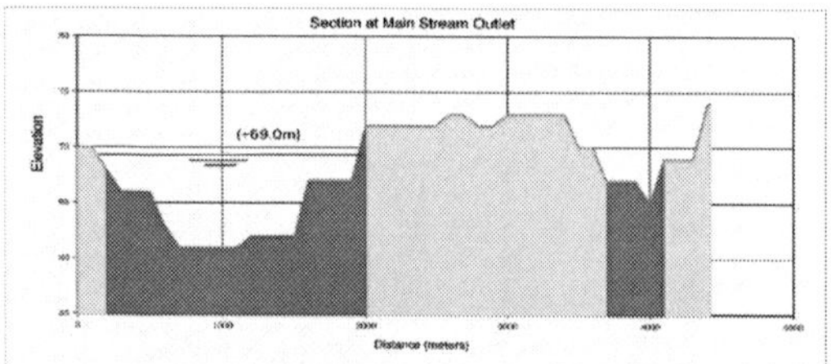


By calculation the amount of runoff at the basin outlet and relate it to the basin elevation from Fig. 10 The areas below the elevation (69 meter) will be flooded and by projecting that elevation on the DEM the area on the plane that will be damaged is in red color as shown in Fig.11.

While Fig. 12 shows the risk map for the study area after analyses the results of HBV model and apply it in the GIS layers . We found there are many cities will be effected by the flash flood and it should be take a protect policy for this zone to safe people, buildings, roads, cultivated land .... Etc. the pink circle consider the most dangerous zone all the development plan should avoid this buffer.

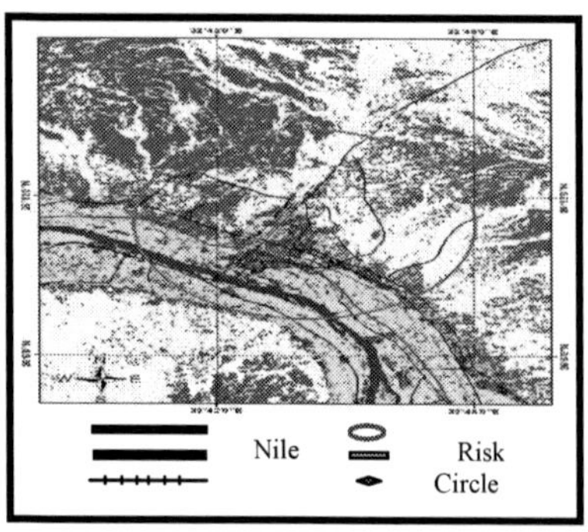


**Fig. 12.** Risk Map for the Study Area

## IX. CONCLUSIONS

The following major conclusions could be drawn from this study:

- The different characteristics of watersheds are estimated using Geographic Information System (GIS) techniques. This means that GIS is a useful tool for delineating the characteristics of water sheds.
- The proposed flood risk assessment and relative vulnerability methodology is proved to be suitable for small wadi systems especially when detailed data is lacking.
- The selected geomorphological parameters resulted in risk assessment, which is well matched, with the results from estimated runoff hydrograph when both peak discharge are considered
- The risk classification presented provides a general prioritisation scheme for flood control and flood protection programmes.
- Assessing the risk potential on the sub-catchment level is crucial to eliminate mis-interpretation of results based on full-scale watershed analysis.
- The study presents an integrated approach for flood risk assessment for Wadi Qena Valley area.
- The methodology used represents an appropriate way for evaluating the vulnerability to risks and can be applied in different catchment areas.

## X. REFERENCES

[1] Aerts J.C.J.H., Kriek M., and Schepel M., 1999, "STREAM (Spatial Tools for River Basins and Environment and Analysis of Management Options): 'Set Up and Requirements'", Phys. Chem. Earth (B), Vol. 24, No. 6, pp. 591-595.

[2] Bergstrom, S., 1995, "The HBV model", In:V.P. Singh (Editor), Computer Models of Watershed Hydolrogy, Water Resources Publications, Highlands Ranch, Colorado, PP. 443-476.

[3] DeVantire, B.A. and Feldman, A.D., 1993, "Review of GIS Applications in Hydrologic Modeling", 119(2), pp246-261.

[4] Horton, R. E. 1932, "Drainage basin characteristics, Trans. Am. Geophys. Union, vol. 13, pp. 350-361.

[5] Rodriguez-Iturbe (a), L.M. Gonzales Sanabria, and R.L. Bras 1982, "A Geomorphoclimatic Theory of the Instantaneous Unit Hydrograph." Water Resour. Res. v. 18(4), p. 877-886.

[6] Wagdy, Ahmed, (2004), "Flood Risk Assessment for Non-Gauged Watersheds", Proceedings of the Arab Water conference, Cairo, Egypt.

# Arabic Character Recognition Using Gabor Filters

Hamdi A. Al-Jamimi and Sabri A. Mahmoud
Information and Computer Science, King Fahd University of Petroleum and Minerals, Dhahran, Saudi Arabia.
{ g200505810, smasaad @kfupm.edu.sa}

**Abstract- A technique for the Automatic recognition of Arabic characters using Gabor filters is presented. K-Nearest Neighbor (KNN) is used for classification. Although KNN is a simple classifier, the achieved recognition rates proved that Gabor filters are effective in the classification of Arabic characters. Different number of orientations and scales, resulting in 30 and 48 feature vector sizes, are used and the recognition rates compared. A high recognition rate of 99.57%.is achieved using 30 features (3 scales and 5 orientations). The results are compared with two previously published techniques using Modified Fourier Spectrum and Fourier descriptors using the same data. This technique has 2.6% and 4% higher recognitions rate than Fourier descriptors and Modified Fourier Spectrum descriptors, respectively.**

**Key words:** Arabic Character Recognition, K-Nearest Neighbour, Gabor filters, OCR

## I. Introduction

Optical character recognition systems can contribute to the advancement of office automation. It can improve human-computer interaction in several applications. Office automation, automatic mail routing, check verification, invoice and shipping receipt processing, subscription collections, machine processing of forms, are few examples [1].

In this work, Gabor features are used following the success of Gabor filters in a wide range of applications in image processing, and pattern recognition. Gabor features have been successfully used in face recognition, character recognition, browsing and retrieving of image data, to name a few applications [2],[3],[4],[5],and [6].

Character recognition has been receiving significant interest in recent decades. Optical recognition of Latin, Chinese and Japanese characters achieved high recognition rates. However, Arabic character recognition is still lagging. The advancement in Arabic character recognition is expected to lead to the advancement of other languages that use the alphabet of Arabic language like, Urdu, Persian, and Jawi.

Arabic text is written from right to left; Arabic has 28 basic characters, of which 16 have from one to three dots. The dots differentiate between similar characters. Additionally, three characters can have a zigzag like stroke (Hamza). The dots and Hamza are called secondaries and they may be located above or below the character primary part. Written Arabic text is cursive in both machine-printed and hand-written text. The shape of an Arabic character depends on its position in the word; a character might have up to four different shapes depending on its position. The character size varies according to its position in the word. Table (1) shows 78 different shapes of Arabic characters used in this work. We used characters of traditional Arabic. Since the paper is addressing characters, diacritics are normally attached with Arabic text. Isolated characters are not diacritised. Hence, diacritics are not addressed here.

This paper is organized as follows; Section II addresses related work; Gabor filters theory is presented in Section III; Section IV details Gabor filter-based feature extraction method for Arabic character recognition; classification of Arabic characters is presented in Section V; the experimental results are detailed in Section VI; and finally, the conclusions are stated in Section VII.

## II. Related Work

A system for Arabic character recognition based on Modified Fourier Spectrum (MFS) descriptors is presented in [7]. Ten MFS descriptors are used to classify the Arabic character models. A model represents all characters with the same primary part and different number and location of dots. For example, characters with numbers 3,6,9,64,and 74 have the same primary part with different number and location of dots (from 1 to 3 dots and the dots are above or below the character primary part). After classification, the particular character is identified using the number and location of dots. Al-Yousefi and Udpa [5] introduced a statistical approach for Arabic character recognition. The character is segmented to primary and secondary parts. Secondary parts are then isolated and identified separately and moments of horizontal and vertical projections of the primary parts are computed. Finally a statistical classification approach is used to classify the characters. Sari et al. [8] introduced a new character segmentation algorithm (ACSA) of Arabic scripts. Their algorithm is based on morphological rules, which are constructed at the feature extraction phase. This segmentation algorithm may be used along with the presented technique for the classification of Arabic cursive text.

For almost three decades, features based on Gabor filters are used for their useful properties in image processing. The most important properties are related to invariance to illumination, rotation, scale, and translation. Furthermore, these properties are based on the fact that they are all parameters of Gabor filters themselves [9]. It is evident that Gabor filters have several advantageous in feature extraction. Gabor feature vectors can be used directly as input to a classification or segmentation operator or they can first be transformed into feature vectors which are then used as input to another stage [10].

T. Sobh (ed.), *Innovations and Advances in Computer Sciences and Engineering*,
DOI 10.1007/978-90-481-3658-2_20, © Springer Science+Business Media B.V. 2010

TABLE 1
ARABIC CHARACTERS

| | | | | | | | | | | | |
|---|---|---|---|---|---|---|---|---|---|---|---|
| 1 | ـا | 14 | حـ | 27 | س | 40 | ـظ | 53 | ـقـ | 66 | ن |
| 2 | ا | 15 | ح | 28 | شـ | 41 | عـ | 54 | ق | 67 | هـ |
| 3 | بـ | 16 | خـ | 29 | ـشـ | 42 | ـعـ | 55 | كـ | 68 | ـهـ |
| 4 | ـبـ | 17 | خ | 30 | ش | 43 | ـع | 56 | ـك | 69 | ـه |
| 5 | ب | 18 | ـد | 31 | صـ | 44 | ع | 57 | ك | 70 | ـو |
| 6 | تـ | 19 | د | 32 | ـصـ | 45 | غـ | 58 | لـ | 71 | و |
| 7 | ـتـ | 20 | ـذ | 33 | ص | 46 | ـغـ | 59 | ـلـ | 72 | ـي |
| 8 | ت | 21 | ذ | 34 | ضـ | 47 | ـغ | 60 | ل | 73 | ى |
| 9 | ثـ | 22 | ـر | 35 | ـضـ | 48 | غ | 61 | مـ | 74 | يـ |
| 10 | ـثـ | 23 | ر | 36 | ض | 49 | فـ | 62 | ـمـ | 75 | ـيـ |
| 11 | ث | 24 | ـز | 37 | ـط | 50 | ـفـ | 63 | م | 76 | ي |
| 12 | ـجـ | 25 | ز | 38 | ط | 51 | ف | 64 | نـ | 77 | ـة |
| 13 | ج | 26 | ـسـ | 39 | ظ | 52 | قـ | 65 | ـنـ | 78 | لا |

In [4] and [11] Huo et. al. used Gabor filters to construct high performance Chinese OCR engine for machine printed documents. Three key techniques contributing to the high recognition rates are highlighted (viz. the use of Gabor features, the use of discriminative feature extraction, and the use of minimum classification error as a criterion for model training). Hamamoto et al.[6] proposed a method for hand printed Chinese character recognition using Gabor filters. Their results showed that Gabor features seem to be robust to noisy patterns. In [12] Gabor filter-based method is applied to handwritten numeral recognition using k-NN, Fisher's linear, quadratic and Euclidean distance classifiers. Experimental results showed that the Gabor filter-based method is an effective means in classifying Chinese numerals as well as Chinese characters.

Deng et al.[2] introduced a robust discriminant analysis of Gabor Feature (RGF) method for face recognition. In [13] a new method using Gabor filters for character recognition in gray-scale images is proposed, where the features are extracted directly from gray-scale character images by Gabor filters.

Manjunathi in [14] proposed the use of Gabor wavelet features for texture analysis and provided a comprehensive experimental evaluation of the Gabor features compared with other multi-resolution texture features using the Brodatz texture database. The results indicate that Gabor features provide the best pattern retrieval accuracy.

The effectiveness of Gabor features have been demonstrated in several OCR publications (e.g. the recognition of both machine-printed and handwritten alphanumeric characters, and small to medium vocabulary Chinese characters). However, it is interesting to notice that in OCR area Gabor features have not become as popular as they are in face and Iris pattern recognition [11].

In this paper we showed that by proper selection of Gabor number of scales and orientations we were able to achieve recognition rates that are higher than previously published work using the same data. An improvement of 2.6% to 4% was achieved using our selected Gabor features. Since the suitable number of scales and orientations are problem dependent, many experiments were carried out to select the proper number of scales and orientations for printed Arabic character recognition. The authors are not aware of any published work that addressed Arabic character recognition using Gabor features.

## III. GABOR FILTERS

Gabor filter is a linear filter whose impulse response is defined by a harmonic function multiplied by a Gaussian function. That is, Gabor filter can be viewed as a sinusoidal plane of particular frequency and orientation, modulated by a Gaussian envelope as shown in (1).

$$g(x,y) = s(x,y)\, h(x,y), \qquad (1)$$

where s(x,y) is a complex sinusoid, known as a carrier, and h(x, y) is a 2-D Gaussian shaped function, known as envelope. The complex sinusoid is defined as follows,

$$s(x,y) = e^{-j2\pi(u_0 x + v_0 y)} \qquad (2)$$

The 2-D Gaussian function is defined, in (3), as follows,

$$h(x,y) = \frac{1}{\sqrt{2\pi}\,\sigma} e^{-\frac{1}{2}\left(\frac{x^2}{\sigma_x^2} + \frac{y^2}{\sigma_y^2}\right)} \qquad (3)$$

Thus the 2-D Gabor filter can be written as:

$$g(x,y) = \frac{1}{\sqrt{2\pi}\sigma} e^{-\frac{1}{2}\left(\frac{x^2}{\sigma_x^2} + \frac{y^2}{\sigma_y^2}\right)} e^{-j2\pi(u_0 x + v_0 y)} \qquad (4)$$

In general, a filter bank consisting of Gabor filters with various scales and rotations is created. The filters are convolved with the signal, resulting in a so-called Gabor space. This process is closely related to processes in the primary visual cortex [14]. Relations between activations for a specific spatial location are very distinctive between objects in an image. Furthermore, important activations can be extracted from the Gabor space in order to create a sparse object representation.

## IV. FEATURE EXTRACTION

Arabic character image features are extracted using Gabor filters which can be written as a two dimensional Gabor function g(x, y) as given in Equation (5). Its Fourier transform G(u, v) is given in Equation(6) [14]:

$$g(x,y) = \frac{1}{2\pi\sigma_x\sigma_y} \exp\left[-\frac{1}{2}\left(\frac{x^2}{\sigma_x^2} + \frac{y^2}{\sigma_y^2}\right) + 2\pi jWx\right] \qquad (5)$$

$$G(u,v) = \exp\left[-\frac{1}{2}\left(\frac{(u-W)^2}{\sigma_u^2} + \frac{v^2}{\sigma_v^2}\right)\right] \qquad (6)$$

Where $\sigma_x$, $\sigma_y$ are the variances of x and y along the x, y axis, respectively; $\sigma_u = ½ \pi\sigma_x$ and $\sigma_v = ½ \pi \sigma_y$. Gabor functions form a complete but non-orthogonal basis set. Expanding a signal using this basis set provides localized frequency description of the signal.

Let g(x, y) be the mother Gabor wavelet, then this self-similar filter dictionary can be obtained by appropriate dilations and rotations of g(x, y) through the generating function of (7) [14]:

$$g_{mn}(x,y) = a^{-m}G(x',y'),$$
$$x' = a^{-m}(x\cos\theta + y\sin\theta),\ and$$
$$y' = a^{-m}(-x\sin\theta + y\cos\theta), \qquad (7)$$

Where $a > 1$, $m,n$ are integers, $\theta = n\pi/K$ and $K$ is the total number of orientations. The scale factor $a^{-m}$ is to ensure that the energy is independent of m.

After filtering the given input image, statistical features such as the mean and the variance of the image are computed. The extracted feature vector is constructed from the means and variances of all filtered images.

Filtering in the frequency domain requires taking the Fourier transform of an image, processing it, and then computing the inverse transform to obtain the result. Thus, given a digital image, i(x, y), the basic filtering equation in which we are interested has the form:

$$g(x,y)=\Im^{-1}[H(u,v)\, I(u,v)]$$

Where $\Im^{-1}$ is the Inverse Fourier Transform (IFT), $I(u, v)$ is the FT of the input image, $i(x, y)$. $H(u, v)$ is a filter function (i.e. the Gabor filter), and $g(x, y)$ is the filtered (output) image.

The FFT is computed for the Arabic characters' images. The Gabor filters are applied using the different orientations and scales. Then the inverse Fourier transform of the filtered images is computed. The mean μ and the standard deviation σ for each filtered image are then computed to form the character feature vector. Figure (1) shows the feature extraction process.

In general, Scales and Orientations determine the number of extracted features according to the scales and orientations using the mean and variance as features. Different number for scales and orientations are tested. Large number of experiments was conducted to estimate the suitable number of scales and orientations that produces the highest recognition rates.

## V. Classification

The classification phase consists of two phases, training (modeling) and testing.

### 1. Training

In this work, the data base of the isolated Arabic character of [7] is used. The data was captured using a scanner with a resolution of 300 pixels per inch. The scanned document images are transformed into binary images. The data consists of 50 samples of each character of the Arabic characters shown in Table (1). 70 % of the data was used for training and 30 % for testing. This was necessary to compare our results with the published work using the same data [7].

The K-NN is used in the classification stage. Hence, in the training phase, the features of the Arabic characters (of the training data) are extracted and saved as reference models of the respective classes. Each class will have 30 reference models that will be used by the K-NN classifier to identify the label of the sample under test.

### 2. Recognition

Whereas the intent of feature extraction is to map input patterns onto points in a feature space, the goal of classification is to assign a class label to the unknown sample based on its similarity with the reference character models. In this work, the K-Nearest Neighbor (KNN) is used for classification. It is a simple classifier and hence the results will indicate how effective the features are. The magnitude (city block) distance measure is used to measure the similarity/dissimilarity between the input sample and the character reference models. The feature vector ($V$) for the unknown character is computed and then compared with the feature vectors of all the characters' reference models. The magnitude (city block) distance is computed using a simple formula given by:

$$E_i = \sum_{j=1}^{r} \left| M_{ij} - V_j \right| \qquad (8)$$

where $E_i$ is the distance between the input character and the reference model $i$ (i.e. sum of the absolute difference between the features of the input character and those of model $i$), r is the total number of parameters in the feature vector (i.e. 48 and 30 in our case), $M_{ij}$ is the $j^{th}$ feature of model $i$, and $V_j$ is feature $j$ of the input character feature vector.

The distances ($E_i$) between the new sample and all reference models are found. The classification decision is based on the k-nearest neighbor using the top (i.e. least magnitude distance) 1, 3 and 5 as presented in the tables below. The nearest distance is computed using Equation (8). An object is classified by a majority vote of its neighbours, with the object being assigned to the class most common amongst its k-nearest neighbours. If k = 1, then the object is simply assigned to the class of its nearest neighbour.

## VI. Experimental Results

In this section the experimental results using different number of scales and orientations are presented. These scales and orientations resulted in feature vectors of 48 and 30 features per character. Each of these cases is detailed below; comparison of the recognition rates using 1-NN, 3-NN, 5-NN of each case is given. Misclassifications are analyzed and comparison of the two cases is presented.

1. Using 48 features

Four scales with (2, 4, 6, 8) frequencies and 6 orientations (viz. 30°, 60°, 90°, 120°, 150°, 180°) are used to extract 48 features. For each scale 6 filtered images are produced. The mean $\mu_{mn}$ and standard deviation $\sigma_{mn}$ are, then, computed; where m refers to the scale number and n the orientation number. The feature vectors of all used scales are concatenated to form the feature vector of the Arabic character, $V = [\mu_{11}, \sigma_{11}, \mu_{12}\ \sigma_{12}, .., \mu_{48}, \sigma_{48}]$. The recognition rates using 1-, 3-, and 5-nearest neighbours is shown in Table (2). The highest recognition rate of 99.49% is achieved with 1-NN.

TABLE 2
THE RECOGNITION RATES USING 48 FEATURES

| K- Nearest Neighbors | Recognition Rate % |
|---|---|
| 1-NN | 99. 49% |
| 3-NN | 98.80% |
| 5-NN | 97.94% |

The misrecognized test samples are shown in Table (3). The image of the sample, the image of the recognized sample, and number of misclassified samples are shown in the table. Analyzing the erroneous samples indicate that 5 out of the six errors are between characters with same main character part but with different number of dots (i.e. all rows excluding 3rd row). All cases differ by one dot. The 3rd row shows Ghain character misclassified as DhaD.

TABLE 3
THE MISCLASSIFIED CHARACTERS USING 48 FEATURES WITH 1-NN

| Original Character | Recognized as | Number of errors | Row number |
|---|---|---|---|
|  |  | 1 | 1 |
|  |  | 1 | 2 |
|  |  | 1 | 3 |
|  |  | 1 | 4 |
|  |  | 2 | 5 |

2. Using 30 features

Three scales with (2, 4, 6) frequencies and 5 orientations (viz. 30°, 60°, 90°, 120°, 150°) are used to extract 30 features. For each scale, 5 filtered images are produced. The mean and standard deviation are computed. The feature vectors of all used scales are concatenated to form the feature vector of the Arabic character, $V = [\mu_{11}, \sigma_{11}, \mu_{12}\ \sigma_{12}, \dots, \mu_{30}, \sigma_{30}]$. The recognition rates using 1-,3-, and 5-nearest neighbors is shown in Table (4). The highest recognition rate of 99.57% is achieved with 1-NN.

TABLE 4
THE RECOGNITION RATES USING 30 FEATURES

| K- Nearest Neighbors | Recognition Rate % |
|---|---|
| Top 1 | 99. 57% |
| Top 3 | 98.71% |
| Top 5 | 98.69% |

The misrecognized test samples are shown in Table (5). The image of the sample, the image of the recognized sample, and number of misclassified samples are shown in the table. Analyzing the erroneous samples indicate that all the errors are between characters with same main character part but with different number of dots. All cases differ by one dot. The number of erroneous samples is one less than the previous case.

TABLE 5
THE MISCLASSIFIED CHARACTERS USING 30 FEATURES WITH 1-NN

| Original Character | Recognized as | Number of errors | Row Number |
|---|---|---|---|
|  |  | 1 | 1 |
|  |  | 1 | 2 |
|  |  | 1 | 3 |
|  |  | 1 | 4 |
|  |  | 1 | 5 |

Figure (2) shows the recognition rates using 30 and 48 features with 1-NN, 3-NN, 5-NN, 7-NN. On the average, using 30 features is better than using 48 features (i.e. it results in higher recognition rate). 1-NN is achieving the highest recognition rate.

Comparing the results of the presented technique to those of reference [7] using the same data, the average recognition rate using the Modified Fourier Transform (MFS), Fourier descriptors, and Gabor-based descriptors are 95.6%, 97%, and 99.57%, respectively. It is clear that Gabor-based descriptors are superior.

## VII. CONCLUSIONS AND FUTURE WORK

Gabor features have been used successfully in computer vision tasks. In this work, a technique based on Gabor filters for automatic Arabic character recognition is presented.

Gabor-based features are extracted using a number of scales and orientations.

The average recognition rate using Gabor features and KNN classifier is 99.57%. Only 5 samples were misrecognized (out of 1170 samples) using 30 Gabor-based features. The misclassification is between characters having the same primary part and only differs by the number of the character dots. The dots problem is reported in previously published work where two techniques for addressing the dots problem are proposed [15]. It requires that the dots be extracted before classification. In this case, the character models are used, hence, reducing the number of classes from 78 to 45. This is expected to improve the recognition rate in two ways. It addresses the dot problem and reducing the number of classes improves classification results.

The average recognition rates using the same data and the Modified Fourier Transform (MFS, Fourier descriptors, and Gabor-based descriptors are 95.6%, 97%, and 99.57%, respectively. A 2.57% to 4% improvement in recognition rates over Fourier descriptors and Modified Fourier Spectrum descriptors, respectively.

The presented technique could be used for automatic character recognition for certain applications (car plate recognition is an example). The recognition of Arabic numbers (as they are non-cursive) is another application. The researchers are currently investigating the use of this technique with Hidden Markov Models (HMM) using the sliding window technique as HMM has the capability to recognize cursive text without segmentation.

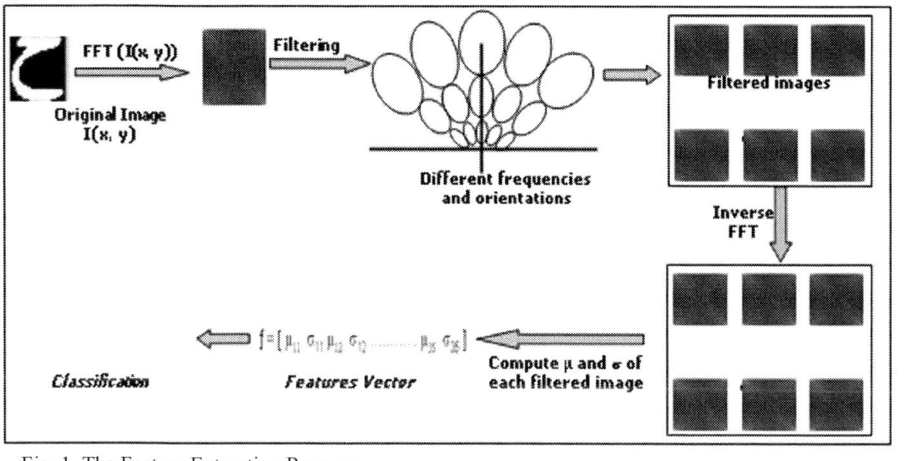


Fig. 1. The Feature Extraction Process

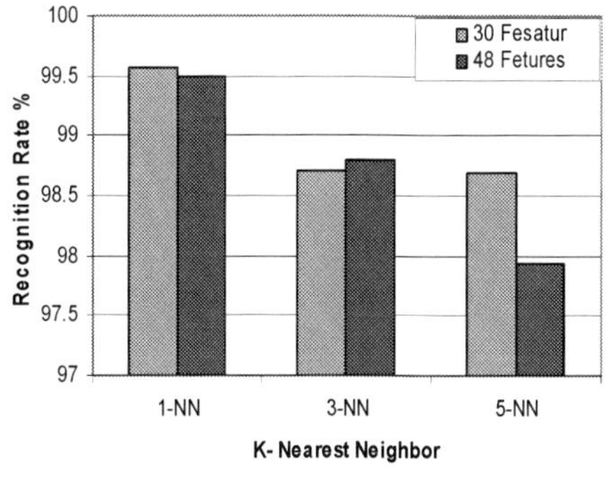


Fig.2. The recognition rates using 1-NN, 3NN, 5-NN and 30 & 48 features

ACKNOWLEDGMENT

The authors would like to thank King Fahd University of Petroleum and Minerals for supporting this research work. This work is partially supported by KFUPM Internal Project number IN060337.

REFERENCES

[1] A. Amin, "Off-line Arabic character recognition: the state of the art," *Pattern recognition.*, vol. 31, 1998, p. 517.

[2] Deng, Weihong et al., "Robust Discriminant Analysis of Gabor Feature for Face Recognition," *Fourth International Conference on Fuzzy Systems and Knowledge Discovery,* vol. Volume 3, Aug. 2007, pp. 248 - 252 .

[3] C.J. Lee and S. Wang, "Fingerprint feature reduction by principal Gabor basis function," *Pattern recognition.*, vol. 34, 2001, pp. 2245-2248.

[4] Q. Huo, , Z. Feng, and Y. Ge, "A study on the use of Gabor features for Chinese OCR," *Proceedings of 2001 international Symposium on Intelligent Multimedia, Video and Speech Processing, Hong Kong,* May 2001.

[5] X. Wang, "Gabor Filters Based Feature Extraction for Robust Chinese Character Recognition," *ACTA ELECTRONICA SINICA*, vol. 30, 2002, pp. 1317-1322.

[6] Y. Hamamoto et al., "Recognition of Hand-printed Chinese Characters Using Gabor Features," *Proceedings of the Third International Conference on Document*

*Analysis and Recognition* , vol. Volume 2, Aug. 1995, pp.819 - 823 .

[7] S. Mahmoud and Mahmoud, Ashraf, "Arabic Character Recognition using Modified Fourier Spectrum (MFS)," *Proceeding of the International Conference Geometric Modelling & Imaging-New Trends, London, UK,* , Jul. 2006, pp. 155 - 159.

[8] Sari, T, Souici, L, and Sellami, M, "Off-line handwritten Arabic character segmentation algorithm: ACSA," *Proceedings. Eighth International Workshop on Frontiers in Handwriting Recognition, 2002.*, Aug. 2002, pp. 452 - 457.

[9] J. Kamarainen, V. Kyrki, and H. Kalviainen, "Invariance Properties of Gabor Filter-Based Features-Overview and Applications," *IEEE Transactions on Image Processing,* vol. 15, 2006, pp. 1088-1099.

[10] S.E. Grigorescu, N. Petkov, and P. Kruizinga, "Comparison of Texture Features Based on Gabor Filters," *IEEE Transactions on Image Processing,* vol. 11, 2002, pp. 1160-1167.

[11] Q. Huo, Y. Ge, and Z. Feng, "High Performance Chinese OCR based on Gabor Features, Discriminative Feature Extraction and Model Training," *IEEE International Conference on Acoustic Speech and Signal Processing,* 2001, pp. pp. III-1517.

[12] Y. Hamamoto et al., "Recognition of Handwritten Numerals Using Gabor Features," *Proceedings /,* vol. 3, 1996, pp. 250-253.

[13] X. Wang, "Gabor Filters-Based Feature Extraction for Character Recognition," *Pattern recognition.,* vol. 38, 2005, pp. 369-380.

[14] B.S. Manjunath and W.K. Ma, "Texture Features for Browsing and Retrieval of Image Data," *IEEE Transactions on Pattern Analysis and Machine Intelligence,* vol. 18, 1996, pp. 837-841.

[15] S.A. Mahmoud, "Arabic character recognition using fourier descriptors and character contour encoding," *Pattern Recognition,* vol. 27, Jun. 1994, pp. 815-824.

# Entropy, Autocorrelation and Fourier analysis of HIV-1 Genome

**Sushil Chandra**

sushil.chandra@gmail.com
Institute of Nuclear Medicine and Allied Sector
DRDO, Delhi
India

**Ahsan Zaigam Rizvi**

ahsan.zaigam@gmail.com
Department of Electronics and Communication Engineering
Tezpur University, Assam, India

***Abstract*** **- Genome Signal Processing (G.S.P.) is a new emerging field, in which we are trying develop new concepts and devices. In this paper we are applying the theoretical fundamentals of Electronics and Bioinformatics on HIV-1 genome. We report the applications of entropy, autocorrelation and frequency domain analysis on large genomic data. This entropy, autocorrelation and frequency domain analysis will help to solve long standing problems, like hidden truths and patterns of HIV-1 genome.**

## I. INTRODUCTION

As we know that DNA and RNA is made up of four bases namely – A, T/U, C, and G. In this paper we are converting nucleotide string into digital signal and trying to solve complex problems with Digital Signal Processing techniques. This leads to more accurate computational advantages [1]. The frequency analysis of genome opens another door for Computational Biology as well as Bioelectronics. It can be use to understand the coding regions of genome, different repeat and complex regions in genome. Fourier transformation can use to identify ORFs, and use to understand the functional and structural characteristics of nucleotide as well as amino acid strings.

## II. FOURIER TRANSFORM ANALYSIS

In HIV-1 genome there are many missing sites and gene sequence are not complete [1]. The statistical methods for sequence analysis are less useful.

Fourier transform give a computational advantages and give more accuracy in order to examine complex genomes [2]. Complex genomes can be analyzed with the help of Fourier transform. It is more accurate and having many computational advantages. With the help of Fourier transform, we can identify protein coding regions and functional complexities of genome. It is also useful to identify different patterns present in genome.

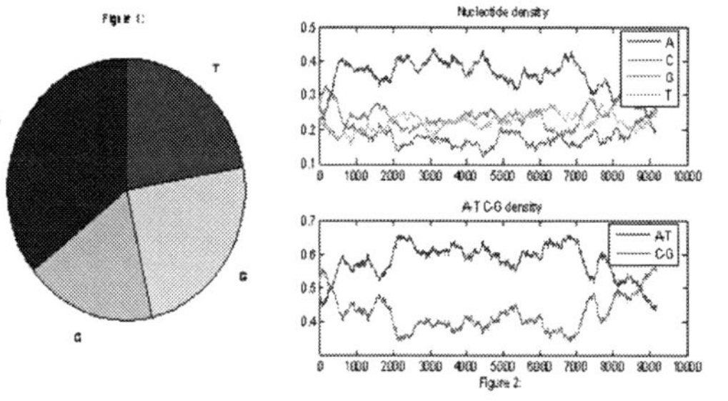


Figure 1: Pie Chart of Bases present in HIV-1 Genome
Figure 2: Nucleotide density of HIV-1 genome

Discrete Fourier Transform (DFT) is one of the specific forms of Fourier analysis. It transforms time domain function into frequency domain function. DFT requires a discrete input function which having non-zero values and under finite durations. DFT provides the frequency contents at 'k', which correspond to period on N/k samples.
Let us assign the random complex number 'l' to character A, random complex number 'm' to character T, random complex number 'n' to character C and random complex number 'o' to character G. So, sequence X[n] becomes:

$$X[n] = l \times b_A[n] + m \times b_T[n] + n \times b_C[n] + o \times b_G[n], \quad n = 1,2,3 \ldots N$$

Where, $b_A[n]$, $b_T[n]$, bC[n], $b_G[n]$ are the binary indicators and N is length of genome.

Let us take Discrete Fourier Transformation (DFT) of sequence x[n] of length L is:

$$X[k] = \sum_{n=1}^{N} X[n] \times e^{-(j.2\pi.Kn)/N}, \quad k = 1,2,3\ldots\ldots N$$

Where, 'j' is imaginary unit.

The resulting sequences $U_A[k]$, $U_T[k]$, $U_C[k]$, $U_G[k]$ are the DFT of binary sequences $b_A[n]$, $b_T[n]$, $b_C[n]$, $b_G[n]$.
Let us take, a, b, c, d are random numeric values, then X[k] becomes:

T. Sobh (ed.), *Innovations and Advances in Computer Sciences and Engineering*,
DOI 10.1007/978-90-481-3658-2_21, 

$$X[k]=a\times U_A[k]+b\times U_T[k]+c\times U_C[k]+d\times U_G[k],\quad k=1,2,3..N$$

Total power spectra can be defined as, squire of absolute value of DFT of individual characters. If $U_A[k]$, $U_T[k]$, $U_C[k]$, $U_G[k]$ are the DFT of sequences then total power spectra P[k] can be define as:

$$P[k]=|U_A[k]|^2+|U_T[k]|^2+|U_C[k]|^2+|UG[k]|^2,\quad k=1,2,3....N$$

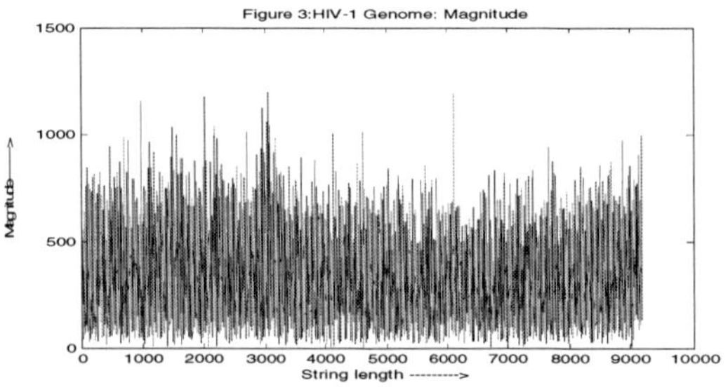


Figure 3: Magnitude plot of HIV-1 genome

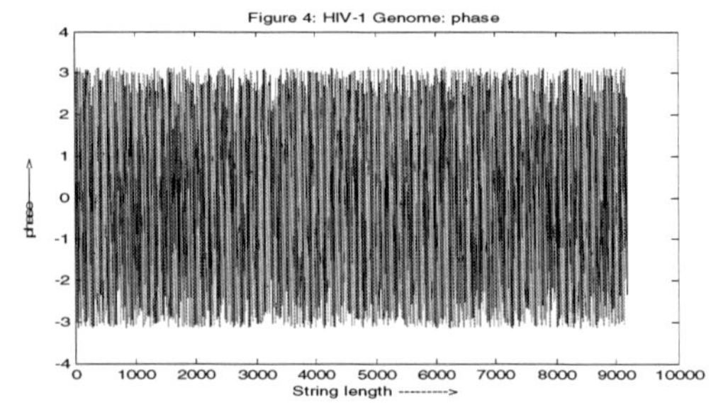


Figure 4: Phase plot of HIV-1 genome

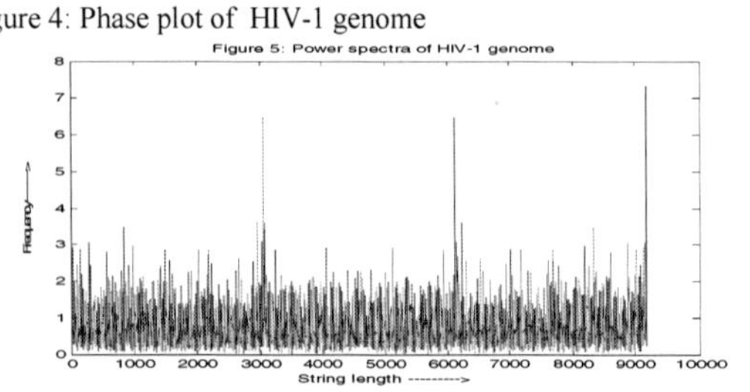


Figure 5: Power spectra of HIV-1 genome

## III. AUTOCORRELATION ANALYSIS

Autocorrelation analysis can be used for determining patterns present in genome. It predicts dependence of character with rest of characters present in genome. These autocorrelation indices for DNA analysis (AIDAs) can be applied to RFLP and genome sequence data. The resulting set of autocorrelation coefficients measure, whether and to what extent, individual DNA sequences resemble the sequence sampled. It is useful tool to explore the genetic structure of a genome and to suggest hypotheses on the evolutionary processes. To determine correlation sequence 'Cr' of a nucleotide sequence x[n] is:

$$Cr\,[n]=\sum_{k=1}^{N} X\,[k]\times x[n+k],\quad k=1,2,3....N$$

We can determine the complexities by indices of autocorrelation. The regions of probable genes or structural parts of genome may be more complex then rest sequence. Figure [6] shows autocorrelation analysis of HIV-1 genome. We simply adjust the parameters and can also do cross-correlation of genomic data.

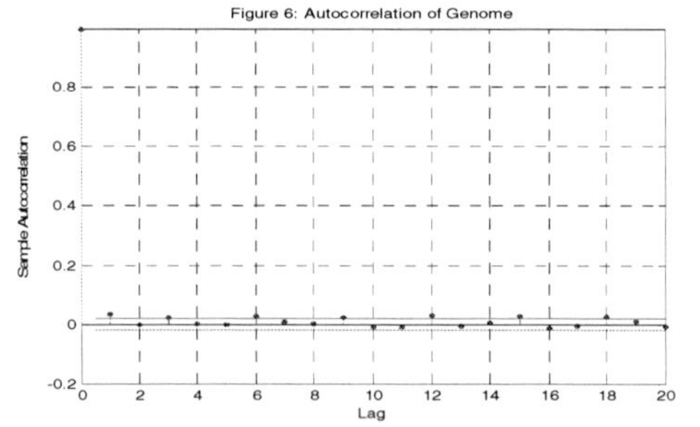


Figure 6: Autocorrelation analysis of HIV-1 Genome

## IV. ENTROPY ANALYSIS

The Shannon entropy or information entropy is use to understand uncertainty associated with random variables. The Shannon entropy is a measure of the average information content. The uncertainty of ribosomal finger after binding is:

$$H(X)=E(I(X))=\ -\sum_{i=1}^{N} p(x_i)\times\log_2\times p(x_i)$$

Where, I(X) is the information content of X. p $(x_i)$ is the probability mass function of X.

For extracting information, we slide the window on nucleotide string. Take the frequency content and measure its probability mass function. The negative summation of log value gives the information about ribosomal binding sites. The window size is directly influence the result obtains. Decision of window size depends upon knowledge and there is no any exact method to give window size. Window length should be lower than the size of genome. More complex genome parts have more information. The structural parts of genome have more randomneness. Repetitive regions of genome have lower complexities.

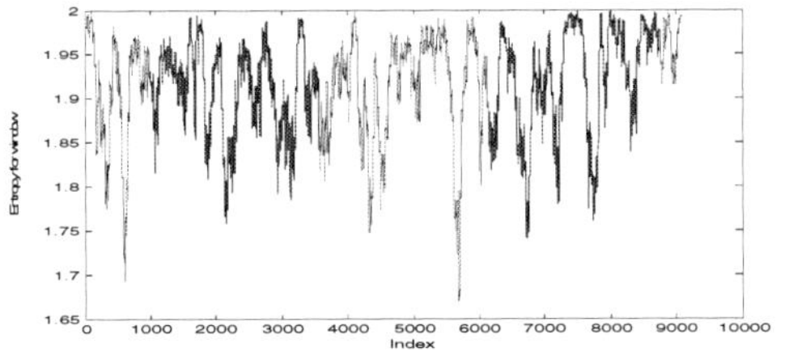

Figure 6: Entropy plot of sliding windows of HIV-1 genome.

## V. CONCLUSION

The frequency domain analysis, autocorrelation analysis and entropy analysis give idea about patterns present with in HIV-1 genome; it is also provide feasible methods for functionality analysis, optimization, prototyping etc. Results obtain by frequency domain analysis is useful for (1) Homology search and gene detection. (2) Robust Biological data analysis. (3) Pathway analysis. (4) To detect correlation between multiple biological databases etc.

## REFERENCES

[1] Sergey Edward Lyshevski and Frank A. Krueger, "Nanoengineering Bioinformatics: Fourier Transform and Entropy Analysis", Proceedings American Control Conference, Boston, pp-317-322, 2004.

[2] Sergey Edward Lyshevski, Frank A. Krueger and Elias Theodorou, "Nanoengineering Bioinformatics: Nanotechnology Paradigm and its Applications" Proceedings IEEE conference on Nanotechnology", San Francisco, CA, pp. 896-899, 2003.

[3] Donald C. Benson, "Fourier methods of Biosequence Analysis", Nucleic Acid Research, Vol. 18, No-21, pp-6305-6310, 1990.

[4] Donald C. Benson, "Digital Signal Processing methods for biosequence comparison", Nucleic Acids Research, Vol. 18, No. 10, pp-3001-3006, April 1990.

[5] T.A. Brown, *Genomes*, John Wiley & Sons, Inc., New York, 2002.

[6] D. T. Mount, *Bioinformatics Sequence and Genome Analysis*. Cold Spring Harbor Laboratory Press, 2001.

[7] D. Anastassiou, "Genomic Signal Processing", IEEE Signal Processing Mag., pp- 8-20, July 2001.

[8] P.P. Vaidyanathan, Multirate system and filter banks, Englewood Cliffs, JN: Prentice Hall, 1993.

[9] P.P. Vaidyanathan and B. Yoon, "Digital filter for gene prediction application", Proc. 36th Asilomer Conference on Signal, System, and Computers, Monterey, CA, November 2002.

[10] S. Tiwari, S. Ramachandran, A. Bhattacharya, S. Bhattacharya, and R. Ramaswamy, " Prediction of probable genes by Fourier Analysis of genomic sequences", CABIOS, vol. 13, No. 3, pp – 263-270, 1997.

[11] NCBI website- http://www.ncbi.nlm.nih.gov gi|9629357|ref|NC_001802.1|Human immunodeficiency virus 1, complete genome

# Contextual Data Rule Generation For Autonomous Vehicle Control

Kevin McCarty
University of Idaho
1776 Science Center Dr.
Idaho Falls, ID 83402 USA
kmccarty@ieee.org

Milos Manic
University of Idaho
1776 Science Center Dr.
Idaho Falls, ID 83402 USA
misko@ieee.org

Sergiu-Dan Stan
Tech. University of Cluj-Napoca,
C. Daicoviciu no. 15, 400020
Cluj-Napoca, Romania
sergiustan@ieee.org

*Abstract*— **Autonomous vehicles are often called upon to deal with complex and varied situations. This requires analyzing input from sensor arrays to get as accurate a description of the environment as possible. These ad-hoc descriptions are then compared against existing rule sets generated from decision trees that decide upon a course of action. However, with so many environmental conditions it is often difficult to create decision trees that can account for every possible situation, so techniques to limit the size of the decision tree are used. Unfortunately, this can obscure data which is sparse, but also important to the decision process. This paper presents an algorithm to analyze a decision tree and develops a set of metrics to determine whether or not sparse data is relevant and should be include. An example demonstrating the use of this technique is shown.**

## I. Introduction

A typical autonomous vehicle must manage a large array of sensors from traction, to temperature, to motion detection. Each sensor represents an input, or dimension, in a knowledge base. This knowledge base was probably derived from a large dataset generated through numerous tests and case studies of sensory data, gathered into a data warehouse and analyzed using a data mining technique such as a Decision Tree.

Effective data mining requires the ability to quickly and meaningfully sift through mountains of data, such as that provided by an array of sensors, and extract meaningful kernels of knowledge [1], [2]. From this new knowledge rules for intelligent systems from e-commerce to intelligent controllers are extracted [3], [4], [5]. One of the more popular techniques is the Decision Tree shown in Fig. 1.

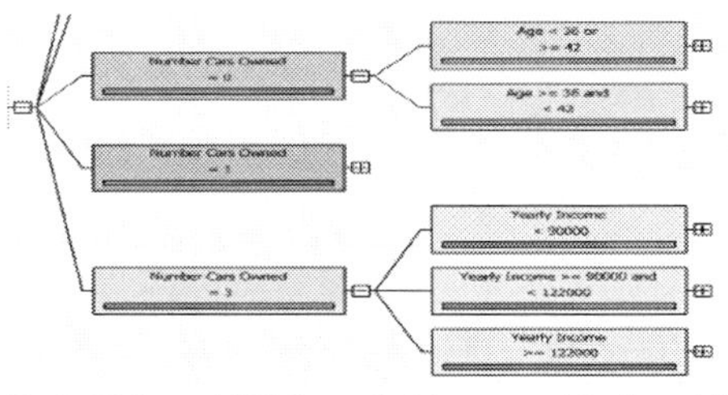

Fig. 1. A Microsoft SQL Server decision tree used for data mining

In addition to the Decision Tree there other techniques, including Bayesian, artificial neural network (ANN), fuzzy and distance classifiers [6], [7], and others whose purpose is to derive meaningful associations from data [1], [8], [9].

Decision Trees are built using techniques such as Iterative Dichotomiser 3 (ID3) and Classification 4.5 (C4.5) [1], [10]. These decision tree induction algorithms grow the tree, starting from a single parent node which contains a set of data, by selecting an attribute from among a set of candidate attributes at each node as demonstrated in Fig. 2. ID3 uses a simpler notion of "information content" while C4.5 attempts to overcome the bias of uneven sampling by normalizing across attributes.

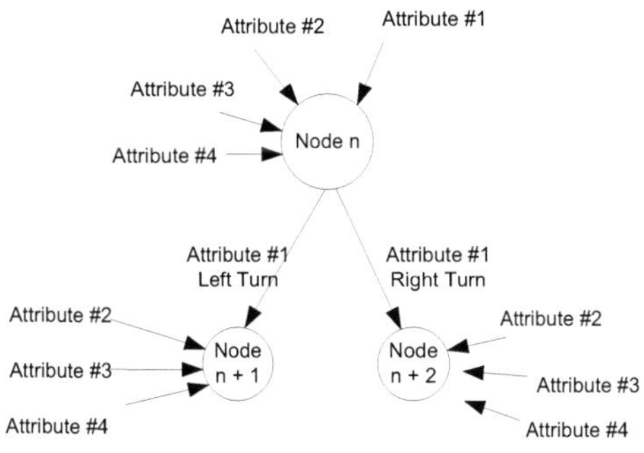



Fig. 2. A decision tree node generation node with attributes

The tree is grown in the following steps:

1. Determine appropriate information "threshold", designed to yield optimal "information content".

2. Choose attribute from among set of attributes with maximum "information gain"

3. If information gain from attribute exceeds threshold create child nodes by splitting attribute accordingly [1], [10], [11].

ID3/C4.5 determines the maximum gain by choosing the attribute which will yield the most "information content" or clear differentiation of the data with a minimum of noise or randomness. If the gain is above a predetermined threshold, i.e. there is sufficient differentiation that new knowledge is likely then the node will produce one or more leaf offspring, with each leaf containing a subset of the parent node data partitioned along the attribute. As a simple example of this technique, consider the node *n* in Fig. 2 as representing a data sample with 100 turns, 50 left and 50 right. Now consider the attribute #1 as *Turn*. *Turn* achieves maximum gain because it affects every data point and partitions the data into subsets of

T. Sobh (ed.), *Innovations and Advances in Computer Sciences and Engineering*,
DOI 10.1007/978-90-481-3658-2_22, 

equal size. In contrast, a sample of 99 right turns and 1 left turn generates little gain.

As Fig 2. shows, by applying the decision tree algorithm to node $n$, 2 new nodes are generated in the tree along the *Turn* attribute, one for left and one for right.

This process continues recursively for each child node until no new nodes can be produced.

Once a Decision Tree is constructed, the next step is rule generation. Each leaf node $nl_i$ of a Decision Tree represents a condition described by the path to the node.

$$nl_i = \cap A_p \quad (1)$$

where $A_p$ is an attribute along the path from the root to the leaf. Each leaf node also consists of a set of data points that are the result of applying those attributes as filters to the original dataset. The combination of attributes and resulting data become a set of probabilities for a given result, or event, based upon the attributes. With this information, rules can be constructed.

DEFINE RULE *rule_name* (2)
ON *event*
IF *condition1*
*AND condition2*
...
*AND condition*
DO *action*

Decision Trees are a very effective tool for data mining [1], [12] but when the number of dimensions is large, the number of nodes on the resulting decision tree may be too large to process. The number of nodes in a Decision Tree with $n$ dimensions is determined by the cross product of the number of elements $e$ of each dimension $d_i$ used to branch:

$$\textit{Total number of nodes in Decision Tree} = \prod_{i=1}^{n} d_i \quad (3)$$

An autonomous vehicle can have well over 100 sensors. Even limiting each sensor to a small range of values could still generate a decision tree with trillions of nodes.

One proposed method to solve this problem is the Contextual Fuzzy Type-2 Hierarchies for Decision Trees (CoFuH-DT) method [12]. The CoFuH-DT is fuzzification of the decision tree followed by application a fuzzy type-2 context. CoFuH-DT can construct a fuzzy Decision Tree that accentuates the importance of certain criteria, even if the data is sparse by applying a fuzzy type-2 context. Under CoFuH-DT, decision trees can be pruned quickly via fuzzy set operators and understood in the context of polymorphic sets of rules.

This paper demonstrates application of several advanced data mining techniques to a Decision Tree generated from a sample data set and how the resulting contexts are available for use by CoFuH-DT. This paper is organized as follows: Section II introduces an autonomous vehicle problem. Section III describes the various steps on applying the ADMT to the decision tree to generate contexts. Section IV applies the algorithm to a sample data set to generate useful contexts. Section V presents the conclusions and future work.

## II. Problem Statement

Consider an autonomous vehicle with an array of sensors. As it moves through its environment, it will encounter a variety of situations that will require decisive action. It also has a Decision Tree and database of rules so that for any given combination of sensor inputs, it can find an appropriate rule. The problem is that it only has a limited amount of storage space and computing capacity so the Decision Tree cannot be overly large. One technique to limit the growth of a Decision Tree is to increase the information threshold. But raising the information threshold can obscure important information that happens to be sparse. Suppose, for example, an autonomous vehicle happens across a small patch of ice on one side while climbing a hill. One wheel begins to slip. There may be data related to slippage of a particular wheel going up a hill; but if there is not enough of it, the Decision Tree will not generate a node reflecting that condition so there can be no rule. The problem lies in being able to account for important but sparse conditions without having to burden the autonomous vehicle with an overly large Decision Tree or complex set of rules.

CoFuH-DT can reduce the need for a large tree by accentuating sparse data using Fuzzy Type 2 contexts. However, in order for CoFuH-DT to be effective there must exist contextual information that can be applied. Simply running ID3 or C4.5 over the data is unlikely to produce anything but a more or less detailed tree; so a different, hybrid technique is required. Advanced Data Mining Techniques (ADMT) such as Artificial Neural Networks (ANNs) are an effective means of generating classifications and learning about patterns that may contain sparse or noisy data [12], [13]. As such they are an effective tool for generating a range of candidates for a Fuzzy Type 2 context.

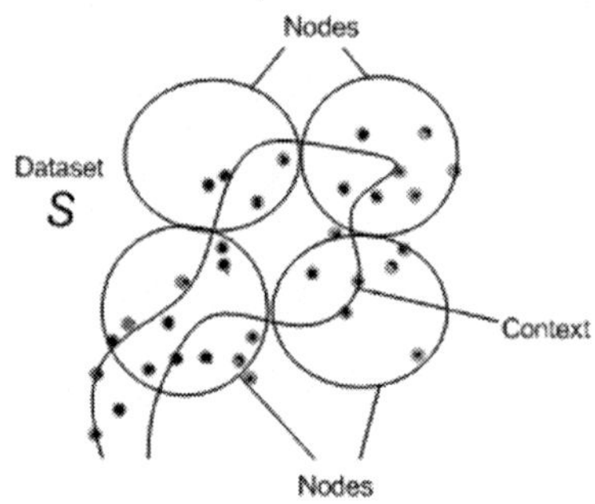


Fig. 3. Context spanning several nodes

Consider for the autonomous vehicle a small array of sensors. Some control power to each wheel, others detect the presence of obstacles while others monitor speed, angle of incline, etc. Using this array, the vehicle wants to generate a profile of various hazards using a Decision Tree. The attributes of the data could end up looking like that of Table 1.

TABLE I. ATTRIBUTES OF ENVIRONMENT

| Attribute | Potential Values |
|---|---|
| Speed | >10 |
| Road Conditions | 6 |
| Incline | >10 |
| Weather | 6 |

| Lateral Forces | >10 |
|---|---|
| Obstacle Types | 5 |
| Power Consumption | >10 |
| Wheel Traction FL | 4 |
| Wheel Traction FR | 4 |
| Wheel Traction RL | 4 |
| Wheel Traction RR | 4 |

By limiting the number of distinct ranges of values of *Speed*, *Incline*, *Lateral Forces* and *Power Consumption* to just 10, a decision tree could still have over 400 million potential nodes. Making things even more difficult is that some values, like *Speed* and *Incline*, have varying weights in lieu of other factors, such as the severity of the weather and steepness of the incline. Other values, such as *Power Consumption* appear to have little relevance at all but may actually be very important in accurately assessing risk, as in the case of a low power situation.

An expert wanting to create rules using the resulting Decision Tree is faced with a dilemma. He must choose between analyzing a huge tree in the hope of gaining the necessary insight, or setting the information gain threshold high enough to reduce the tree to a manageable number of nodes. In the first case, resources and time required in order to process and analyze a huge base of nodes can be substantial [14]. In the second case, by increasing the threshold for the decision tree algorithm, the resulting tree may be much smaller and more manageable but a lot of information could be lost in the process, potentially leaving the vehicle facing hazards to which it cannot adequately respond.

CoFuH-DT presents a better alternative by combining the efficiency of the decision tree with the power of fuzzy type-2 contexts. Generating the contexts can be a difficult task but is made easier through use of ADMT such as an artificial neural network (ANN) [6], [15]. This is accomplished by applying the ANN to process the resulting datasets representing the nodes of the decision tree and generating a series of classifications or contexts. These contexts can be reapplied to the fuzzified decision tree using CoFuH-DT. The final decision tree is smaller, more semantically concise and appropriate to the situation but without the loss associated with traditional methods.

## III. CoT-DT Algorithm

"Interestingness" for CoT-DT takes into account the concept of node distance. Consider the dataset *S*. Applying ID3 or C4.5 or other algorithm to generate a decision tree produces a number of nodes *M* with *N* leaf nodes. Each leaf node $n_i$ of the set of all leaf nodes *N* contains a subset $s_i$ of the dataset *S*.

$$\forall n_i \in N, f(n_i) = \{s_i \subset S\}, \cup s_i = S, i = 1,..,N \quad (4)$$

where $f(n_i)$ is a filter applying all the attributes of $n_i$ against *S*. Let the distance $f_d(n_i, n_j)$ between any two nodes $n_i$, $n_j$ be the number of intermediate nodes that must be traversed when traveling from $n_i$ to $n_j$ as demonstrated in Fig. 4.

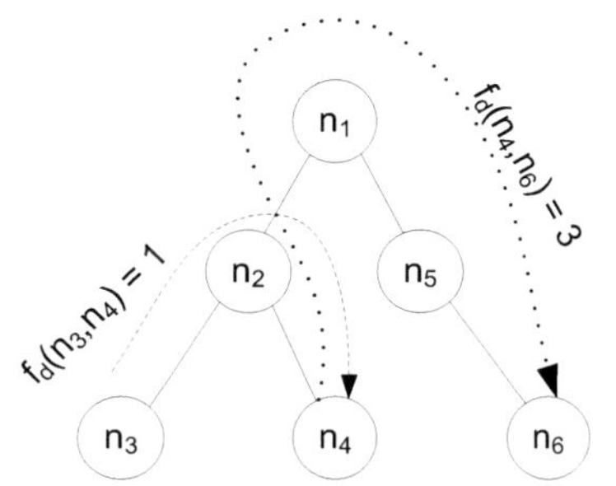


Fig. 4. Calculating distance between nodes

Unlike traditional classification using ANNs or other Advanced Data Mining Technique (ADMT) which seeks to create clusters of data based upon some measure of "closeness", context generation seeks to discover relationships that exist between sets of data within a given set of nodes. This is accomplished by examining the intersection of a particular classification across a set of nodes. "Interestingness" is a function of the node and data characteristics for that classification.

Pseudocode for the Contextual Derivation From Decision Trees (CoT-DT) algorithms is as follows:

```
CoT-DT program

  Starting with a dataset
  Step 1: Generate Decision Tree nodes using process such as ID3, C4.5, CART, etc.
  Step 2: Create subsets from nodes/attributes
  For each node in Decision Tree
    If node is leaf
      Add to subset
  Step 3: Create clusters using advanced data mining clustering technique over dataset, generating series of data clusters
  For each cluster generated
      Step 4: Evaluate cluster span to determine interestingness value
      If Interestingness of cluster span > Interestingness Threshold
        Then Add cluster span to context list
  Return context list
end of program
```

In order to elaborate the details of the proposed CoT-DT algorithm, consider a set of data *S* residing in a data warehouse or large relational database. The goal is to derive a Decision Tree and then to extract data clusters using some clustering algorithm such as an ANN. Through the process of examining the characteristics of the clusters in relation to the Decision Tree, useful contexts may emerge. The steps of CoT-DT proceed as follows:

**Step 1. Decision tree generation** – using ID3, C4.5, etc.

**Step 2. Node Selection.** Look at the decision tree from the point of view of a set-based operator. Each leaf $n_i$ of the tree encompasses a subset $s_i \in S$ demonstrated in Fig 5.

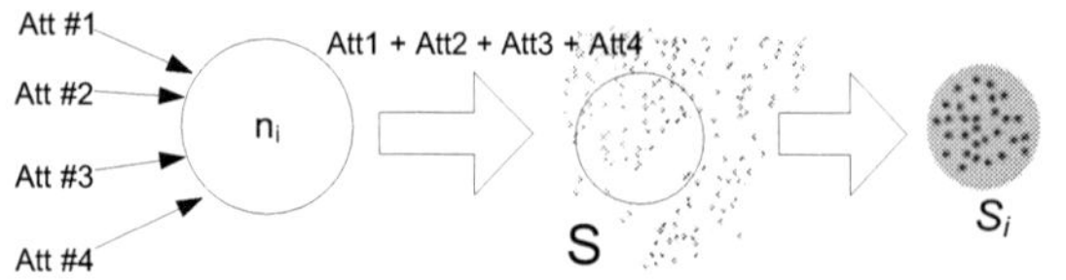

Fig. 5. Nodes of Decision Tree produce subset $s_i$ of original set $S$

Fig 5. shows how the collection of attributes $A$ of the leaf combine to create a filter that when applied to $S$, produces the data set $s_i$ of the leaf.

$$\forall n_i \epsilon N,\ A_{n_i}(S) = s_i, i = 1,..,N \tag{5}$$

Node selection then combines $s_i$ into subsets of $S$ for analysis in Step 3.

**Step 3. ADMT classification.** From the $s_i$ created in Step 2, use an artificial neural network (ANN) to create a set of data clusters $C$ as shown in Fig 6.

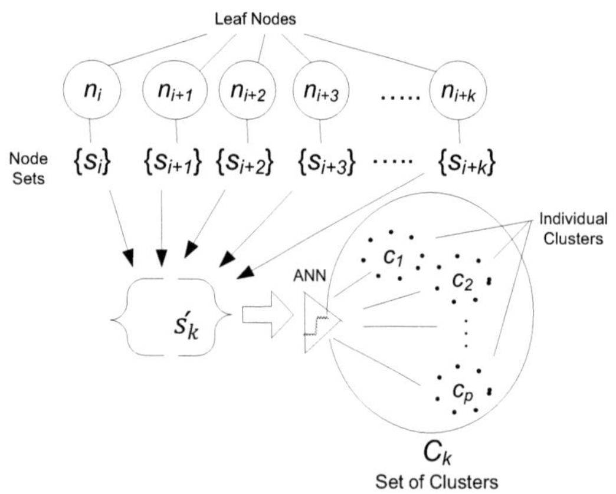

Fig. 6. ANN classifier applied to leaf node sets produces clusters

Each resulting cluster $c_p$ in the set of generated clusters $C_k$ represents a degree of "closeness" between a series of data points $s'_k$ which is a combination of leaf node $s_i$ created in Step 2 and is a subset of $S$.

$$s'_k \subset S,\ g(s'_k) = \{c_p \mid \cup c_p = C_k\ p = 1,..,k\} \tag{6}$$

where $g(s'_k)$ is an ADMT such as an ANN that when applied to $s'_k$ produces the set of clusters $C_k$.

Fig. 7 demonstrates how cluster creation using an ANN combines subsets of a node set into one or more unique clusters.

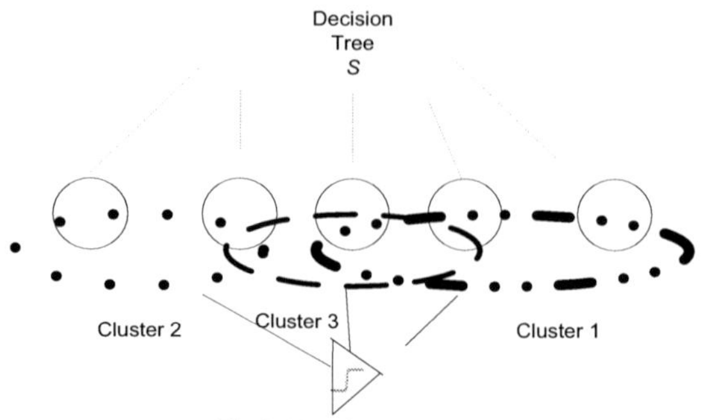

Fig. 7. ANN cluster generation

**Step 4. Context Evaluation and Creation.** Compare each cluster $c_p \in C_k$ to each node $n_i$. Denote the non-empty intersection of $c_p$ with each $s_i$ in $n_i$ as the element $e_j$.

$$e_j = s_i \cap c_p,\ e_j \neq \emptyset \tag{7}$$

The union of the node elements $e_j$ over all or some subset of the leaf nodes $N$ is called a cluster-span as shown in Fig 8.

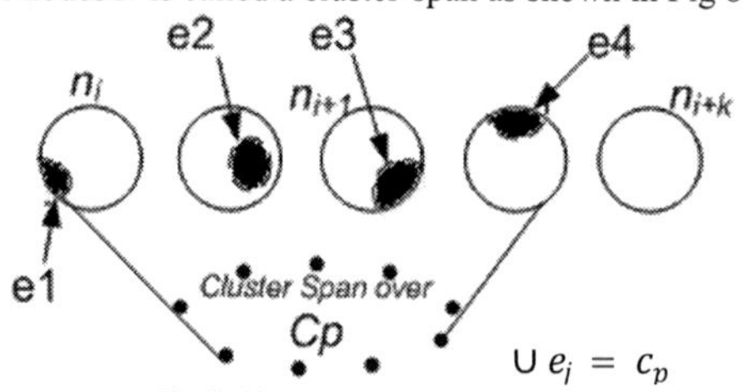

Fig. 8. Cluster span over several nodes

Each single node element $e_j$ of the cluster span consists of a "coverage". Let $f_{dp}(e_j)$ represent the number of data points in $e_j$, and let $f_{dp}(s_i)$ represent the total number of data points in the node's corresponding data set $s_i$. The coverage $f_{cvg}(e_j)$ is the ratio of the number of data points in $e_j$ to the number of data points in $s_i$.

$$f_{cvg}(e_j) = \frac{f_{dp}(e_j)}{f_{dp}(s_i)} \tag{8}$$

Let $f_d(e_i,e_j)$ be the distance between the corresponding nodes for $e_i$ and $e_j$ as illustrated in Fig. 4. Let $f_{dm}(e_j)$ represent the greatest distance between the node containing the element $e_j$ and any other node in the cluster-span.

$$f_{dm}(e_i) = max(f_d(e_i, e_j)), \forall e_j \subset c_p, \quad i = 1,..,n,\ j = 1,..,n,\ p = 1,..,k \tag{9}$$

Let "interestingness" of an element $f_{int}(e_j)$ be a function of its coverage multiplied by its distance function.

$$f_{int}(e_j) = f_{cvg}(e_j) * f_{dm}(e_j) \tag{10}$$

In addition any cluster-span containing some non-empty set of elements $e_1..e_j$ also creates a "context" $CT_i$.

$$CT_i = \cup e_j \tag{11}$$

Note that if a given context $CT_i$ has only one element, the distance function for that element equals 0 as does the measure of interestingness for the context. The context may be particularly interesting but belonging to a single node it adds no new information to the decision tree. Hence for any given context $CT_i$ to be "interesting" its corresponding cluster-span must have at least 2 elements. Interestingness of a entire context, $F_{int}$ is the weighted) sum of the interestingness of its corresponding elements.

$$F_{int}(CT_i) = \sum_j w_j f_{int}(e_j),\ e_j \subset c_p \in C_k,$$
$$j = 1,..,p,\ i=1,..,k \quad (12)$$

where $w_j$ represents a given weight assigned to the corresponding $e_j$. Weights are a means to take into account the relative size or relevance of a node or to reduce the impact of noisy data.

As an example consider the following basic decision tree with four leaf nodes as shown in Fig 9. Each leaf node contains exactly 100 elements.

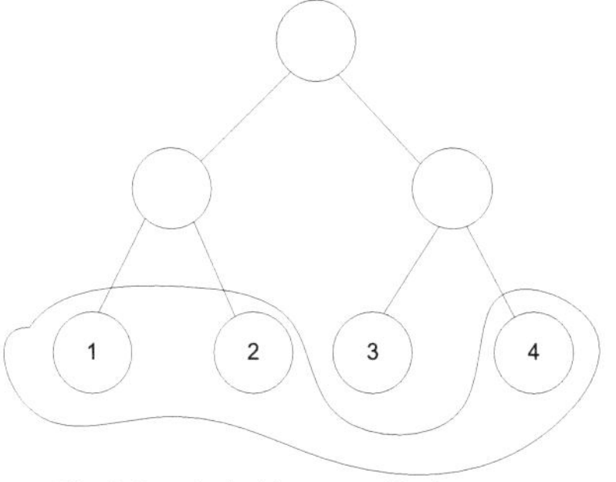


Fig. 9. Sample decision tree with cluster-span

Now consider a cluster-span which contains 50 elements from nodes 1 and 2 and another 25 elements from node 4. Assuming all nodes are weighted equally, by Eq. 9, its corresponding context "interestingness" is calculated as follows:

$$f_{int}(e_1) = 3 \text{ x } .5 = 1.5,$$
$$f_{int}(e_2) = 3 \text{ x } .5 = 1.5,$$
$$f_{int}(e_3) = 3 \text{ x} .25 = .75$$

$$F_{int}(CT_i) = 1.5 + 1.5 + .75 = 3.75 \quad (13)$$

Contexts with sufficient interestingness can now be employed with CoFuH-DT to perform fuzzy-set operations, classification and context pruning.

## IV. TEST EXAMPLE

A sample database provided for users of Microsoft's SQL Server 2005 Analysis Services contains approximately 60,000 purchase records. The dataset contains 12 relevant attributes, each with 2-70 possible values. The total potential size of the decision tree is $2\text{x}10^{10}$ nodes. The Microsoft Decision Tree induction algorithm is described as a proprietary hybrid algorithm based on a C4.5 algorithm combined with elements of CART (Classification And Regression Trees). Using the Microsoft Decision Tree induction algorithm to construct the decision tree resulted in 187 actual nodes.

A standard error back propagation artificial neural network (EBP-NN) was used for the classification described in Step 3. The EBP-ANN was used for several reasons:

1. The EBP-ANN is a well-recognized and commonly used technique.
2. The EBP-ANN provides a standard baseline against which other ANN's can be measured against.
3. The EBP-ANN was provided with the commercial software.

The CoT-DT process was performed on a 2.8 GHz, Pentium-4 Desktop and took approximately 2 minutes to complete. The EBP-ANN produced approximately 50 clusters which were evaluated as potential contexts. Some of the more interesting contexts were based upon the customer's age and income. Creating fuzzy regions for both by breaking the span of ages into the fuzzy sets, YOUNG, MIDDLE-AGED, and OLD and the span of income into the fuzzy sets POOR, LOWER-CLASS, MIDDLE-CLASS, UPPER-CLASS, RICH generated a new series of classifications.

From these classifications two contexts in particular emerged with a high degree of "interestingness": RICH_AND_YOUNG and RICH_AND_OLD. The data were too sparse to meet the information threshold for the decision tree algorithm but covered a number of distant nodes and thus were still quite interesting.

Each of the two contexts mentioned shows a very high correlation between membership in the corresponding fuzzy region and high volume and high dollar purchases. Other cases, for example RICH_AND_MIDDLE-AGED, had a much lower correlation.

Two other ADMT were also applied, a k-means clustering algorithm and a Bayesian network. These also generated contexts. For the Bayesian network, there was a focus on marital status and no kids while the k-means added the dimension of home ownership. Contexts would be described as MARRIED_NO_CHILDREN (M-NC) and MARRIED_HOMEOWNER_NO_CHILDREN (M-HNC).

Customers who were members of the contexts described all showed significantly higher predispositions to make more and/or higher value purchases than average.

Applying the contexts all reduced the number of potential nodes on the original decision tree. This reduction tends to be very dramatic due to the specificity of the context which makes irrelevant or "out of context" all other attributes. However, as occurred with the commercial algorithm in this case, contexts can often get lost or obfuscated in a decision tree induction because of the sparseness of the data, or overlooked because the relationship is non-intuitive.

A reasonable interpretation of the aforementioned contexts might be that younger buyers are more impulsive while older buyers more secure in their finances than members in the middle group and hence more likely to take on greater and more premium discretionary purchases. Whatever the reason, a sales manager now has a collection of FLoST-based, semantically simple, yet powerful contexts with which to frame and generate rules for particular customers.

Finally, rule generation is made much simpler. In traditional rule generation, rules define an action for the set of conditions represented by a node [12], [16] shown in Eq. 2.

The situation described above would necessarily involve a great number of conditionals to accurately represent the large number of affected attributes and sub-conditions. However, generating a context-base rule is much simpler because the many disparate products and customers now belong to a single contextual category and can be dealt with as such:

DEFINE RULE *RECOMMEND_PURCHASE* (14)
ON CustomerPurchase
IF Customer IS RICH_AND_YOUNG
DO Recommend purchase PREMIUM_PRODUCT

For an autonomous vehicle with an array of sensors, use of CoT-DT, CoFuH-DT enables it to determine how to respond to its environment using Decision Trees that are smaller, yet more suitable with rules that are more appropriate. A robotic land rover attempts to navigate a landscape with a myriad of environmental and physical obstacles and hazards. The faster it moves, the more quickly it must process all the various attributes and come to a good decision. However, there are times when certain factors become so overwhelming that a good decision only needs to take those into account while ignoring the others. Take the case where the land rover has to navigate a steep slope. Turning to the right or left greatly increases the possibility of a roll-over so virtually any decision which would involve such a turn is not a good one. It makes no sense to contemplate turning decisions or pursue decision branches which might be considered irrelevant when making a turn. At other times, outliers in behavior or actions which would in most cases be considered abnormal, suddenly become "normal" or preferred within a given context. For example suppose under low battery conditions the rover has an overriding need to seek a power source and may have to engage in any number of aberrant moves and behaviors to meet that goal. CoFuH-DT/CoT-DT allow the rover to frame those actions in a meaningful context as well as more quickly prune its decision tree in order to generate a more understandable set of rules.

Comparisons of decision trees using the aforementioned derived contexts are shown in Table 2.

TABLE II. NODE COMPARISONS USING VARIOUS CONTEXTS

| | Nodes | Avg. # Purch | Avg. $ Purch |
|---|---|---|---|
| Org DT | $2x10^{10}$ | 3.27 | 1588 |
| MS SQL HDT | 187 | 3.27 | 1588 |
| EBP Cond 1 | 17 | 4.07 | 3343 |
| EBP Cond 2 | 13 | 4.0 | 1537 |
| Bayes | 43 | 3.46 | 1839 |
| K-Means | 24 | 3.51 | 2000 |

## V. CONCLUSION

This paper demonstrates two significant benefits to Contextual Derivation from Decision Trees (CoT-DT) algorithm using Advanced Data Mining Techniques (ADMT):

The first benefit is that ADMT under CoT-DT can derive new contextual information from a fully-formed decision tree for use by Contextual Fuzzy Type-2 Hierarchies for Decision Trees (CoFuH-DT) rule generation.

The second benefit of the CoT-DT approach is that it can be used to measure and validate the overall effectiveness of a decision tree induction algorithm. The more accurate or complete an algorithm, the fewer and less interesting contexts that are likely derivable. By the same token, CoT-DT can compensate for an ineffective algorithm by providing useful contexts for appropriate rule generation.

As demonstrated by experimental results of this paper, CoT-DT approach produced new and meaningful contexts. Viewing a decision tree within the narrow frame of a context reduced the in-context decision tree by many orders of magnitude over what was theoretically possible. Even after applying a commercial algorithm, CoT was still able to achieve an additional contextual reduction of over 90%.

Future work involving the effectiveness of other neural network strategies, such as Kohonen Self-Organizing Maps and other related techniques such as Support Vector Machines could to be done to improve the effectiveness of the context generation algorithm. Additional work could be done to extend this approach to include other advanced data mining techniques beyond the decision tree.

## REFERENCES


[1] J. Han, M. Kamber; *Data Mining Concepts and Techniques*, 2nd Ed, Morgan Kaufmann Publishers, 2006 pp 291-310

[2] I.S.Y. Kwan; *A mental cognitive model of Web semantic for e-customer profile*; 13th International Workshop on Database and Expert Systems Applications, Sept. 2002

[3] N. Li, Y. Xiao-Wei, Z. Cheng-Qi, Z. Shi-Chao; *Product hierarchy-based customer profiles for electronic commerce recommendation*; International Conference on Machine Learning and Cybernetics, Nov. 2002

[4] C. Wang, G. Chen, S. Ge, D. Hill; *Smart Neural Control of Pure-Feedback Systems*; International Conference on Neural Information Processing; Nov. 2002

[5] M. Dai, Y. Huang, *Data Mining Used in Rule Design for Active Database Systems*; 4th International Conference on Fuzzy Systems and Knowledge Discovery, June 2007

[6] A. Tatuzov, N. Kurenkov, W. Lee; *Neural Network Data Clustering on the Basic of Scale Invariant Entropy*; IEEE International Conference on Neural Networks, July 2006

[7] Z.P. Wang, S. S. Ge, T. H. Lee, X. C. Lai; *Adaptive Smart Neural Network Tracking Control of Wheeled Mobile Robots*; International Conference on Control, Automation, Robotics and Vision, Dec. 2006

[8] Adomavicius, G.; Tuzhilin, A.; *Using data mining methods to build customer profiles*; Computer, Volume 34, Issue 2, Feb 2001

[9] X. Wang, Z. Shi, C. Wu, W. Wang; *An Improved Algorithm for Decision-Tree-Based SVM*; Proceedings of the 6th World Congress on Intelligent Control and Automation, June 2006

[10] J. Zhao, Z. Chang; *Neuro-Fuzzy Decision Tree by Fuzzy ID3 Algorithm and Its Application to Anti-Dumping Early-Warning System*; IEEE International Conference on Information Acquisition, August 2006

[11] J. Sun, X. Wang; *An Initial Comparison on Noise Resisting Between Crisp and Fuzzy Decision Trees*; 4th International Conference on Machine Learning and Cybernetics, August 2005

[12] K. McCarty, M. Manic; *Contextual Fuzzy Type-2 Hierarchies for Decision Trees (CoFuH-DT) – An Accelerated Data Mining Technique*; IEEE International Conference on Human System Interaction, May 2008

[13] T. Fuering, J. Buckley, Y. Hayashi; *Fuzzy Neural Nets Can Solve the Overfitting Problem;* International Joint Conference on Neural Networks, July 1999

[14] S. Russell, P. Norvig; *Artificial Intelligence – A Modern Approach, 2nd Ed.*; Prentice Hall 2003, pp 95-114

[15] L. Yu, S. Wang, K.K. Lai; *An Integrated Data Preparation Scheme for Neural Network Data Analysis*; IEEE Transactions on Knowledge and Data Engineering, Feb. 2006

[16] C. Hsu, H. Huang, D. Schuschel; *The ANNIGMA-Wrapper Approach to Fast Feature Selection for Neural Nets*; International Conference on Systems, Man, and Cybernetics, April 2002

# A New Perspective in Scientific Software Development

Atif Farid Mohammad
School of Computing, Queen's University, Kingston, ON
atif@cs.queensu.ca

**Abstract:** *Scientific software development is a process whereby software is created to assist scientists in achieving their target solutions. This process lacks harmonious communication between scientists and software engineers and thus, a gap needs to be breached between the scientific community and the computing world. This vital issue can be resolved by utilizing a new perspective in scientific software development, using well-established practices of software engineering. This new paradigm is discussed in a case study with several scientists, who confirm its effectiveness for developing scientific software if it can be adapted to their environment.*

**Keywords**
Science, software, design, documentation, techniques, tools, methodologies

## 1. Introduction

Computing is a branch of mathematics, and human-computer interaction is the main area providing a bridge between the computing and scientific communities. Software engineering is one of the pillars of this bridge. This paper presents an understanding of the field of scientific software development by software engineers.

Scientific software development has played a key role in the information age - a role that is expected to gain more advancement in the future. On the basis of a detailed case study, this paper will describe the process of identifying requirements of the objects and their relevant processes classification; as per the design theorist Horst Rittel [1] "*a statement of a problem is a statement of the solution.*" Even though this is a relatively uncomplicated observation, it offers a fine model often ignored by software solution providers: requirements classification for scientific software is in a premature stage due to the closed nature of the scientific world of research.

An elementary requirement of any scientific and/or engineering field is that its rational underpinnings be original and well thought out. These originations of ideas or algorithms provide initial objects and related states of these objects, which are alterable due to associated processes. This methodology is called (OPM) Object-Process Methodology, which is a combination of associated language and diagrams to depict a software model in a generalized way for everyone to understand easily.

This paper also brings forward the best practices for software engineers to adapt while working with scientists. A list has been compiled after careful review of the software engineering practices utilized in general for software development:

Table 1

| | |
|---|---|
| A | Initial functionality configuration (Requirements) |
| B | Establishment of a relationship with point A |
| C | Establishment of a common testing ground |
| D | Daily log maintenance |
| E | Modules development tracking |
| F | Module inspections |
| G | Disaster recovery plan |
| H | Improvement of solution safety and security |
| I | Reliability |
| J | Code quality check |

These practices arose as a result of trying to organize and categorize the interactions of the scientists [10] and software engineers' [11] work ethics. This paper is arranged in ten sections and contains detailed discussion on these practices and a case study with empirical analysis of the new paradigm.

## 2. Scientific Software Development

Scientific software development can be divided into five important factors: **Ideas, Concepts, Structures, Architectures** and **Operations**. Galileo had discovered that the sun was the centre of our known universe. This idea was established by Galileo on the basis of his study of Copernicus's work. The discoveries of Newton offered mathematical explanations as to why planets orbited the sun and also suggested that the whole scientific world could be explained by fundamental laws of nature. Scientific R&D contains specific relevance to accuracy and performance of research in progress [3]. Scientific R&D involves the following phases:

- Planning of research
- Organization of research ideas in logical sequence and System Study of control points
- System Development
- Individual module inspections
- Field trials and experiments

These phases require that research scientists offer guidance to software engineers involved in the development of scientific solutions. Meyersdorf, Dori et al. [3] used Object-Process Analysis to establish a sound methodology for R&D performance evaluation. Every system has a structured architecture. Scientific software architecture [5] is the embodiment of

T. Sobh (ed.), *Innovations and Advances in Computer Sciences and Engineering*,
DOI 10.1007/978-90-481-3658-2_23, 

scientifically established concepts. For example, a change of voltage is proportional to an electric current, or a change in voltage can also be proportional to a charge to a rechargeable battery. As per IEEE 1471-2000, *software architecture is the fundamental organization of a system, embodied in its components, their relationships to each other and the environment, and the principles governing its design and evolution.* Scientific software is either built by a scientist or most efficiently by a team of scientists and software engineers. To develop a scientific software solution requirement, one needs the following:

- Conceptual modeling
- Well-defined process
- Power tools (Computer language(s))

Scientists perform several sorts of tests on their daily experiments, and extract data. This type of data processing is in fact a conceptual modeling [5] [6] performed by scientists. *In a few systems, the conceptual view plays a primary role. The module, execution, and code views are defined indirectly via rules or conventions, and the conceptual view is used to specify the software architecture of a system, perhaps with some attributes to guide the mapping of other views* [9]. It is important that a software engineer use the mix of software engineering practices to prepare the software for delivery to a scientist with an eye to upcoming changes.

# 3. Best Practices

**3.1** An **initial functionality configuration** with an individual scientist's perspective needs to be created to communicate a shared understanding of the new product among all project stakeholders, such as scientists, software designers, and developers, etc.

**3.2** A **relationship should be established** among all requirements provided by the scientists, according to the mentioned functionalities in an analysis to achieve results with an operation's prime designer's authority to authorize any changes if needed at the time of and after solution development.

**3.3 A common testing ground** needs to be established by all stakeholders on a modular basis to achieve an error-free solution. The solution testing should have classifications for each module in the desired software as per usage priority.

**3.4** A **daily log** should be maintained by software engineers to have a back track to the actual system development as well as alterations, additions, and updates in the requirements document's deviations needed by the scientists.

**3.5** All **modules development tracking** should be maintained in a development log, to manage the relationships established in the raw detail manual showing all stakeholders' mentioned requirements in the document of initial configurations.

**3.6** Software engineers need to ensure that **inspections are performed by the scientists** at the completion of each module to determine the correct version on the basis of the first four points.

**3.7** A solution **development disaster recovery** plan is to be injected, ensuring that versions of the solution are secured off site until the actual solution is ready to replace those in the backup repositories.

**3.8** At the time the software modules are compiled, software engineers should plan to **improve safety and security** of the software through the integration of sound software engineering principles and standards to effectively meet the objectives of the solution.

**3.9 Reliability** is a key factor to be measured by scientists in their peer-to-peer work with software engineers, as per software reliability standards documented in NASA-STD-8739.8. "*Trending reliability tracks the failure data produced by the software system to develop a reliability operational profile of the system over a specified time.*"

**3.10** Independent module **code quality checking** of the software is needed at every step of scientific software development to ensure that the modules are readable with step-by-step technical documentation to facilitate future enhancements. Software module failures may be due to some of the following factors:

- Incorrect logic
- Incorrect statements or incorrect input data

Andrew Tanenbaum has said [15]: *"The nice thing about standards is that you have so many to choose from; further, if you do not like any of them, you can just wait for next year's model."* There can be at least three major directions concluded after going through the best practices mentioned above:

I. Perspective of the scientist
II. Perspective of the user
III. Perspective of the "interface designer"

# 4. Design and implementation

Scientific software architecture needs to be designed by software engineers and developers such that scientists can take advantage of emerging computational environments such as the Web and information grids, and pattern identifier applications on the available data with user-friendly front-end inputs. By defining patterns in scientific software, authors [12] aspired to the development of software by the use of object components only. Whereas scientists mostly tend to use programming languages, such as FORTRAN, operations organized as toolkits to provide resultant desired solutions and association rule mining capabilities can be produced using other languages as given in [12 & 13], such as C/C++,

Python and most recently C Sharp and Java. Authors [14] mentioned the non-availability of modern aspects of software design in Fortran 90 without discussing the aspects of a certain generation of language.

## 5. Case Study

A case study has been prepared and conducted from actual interviews with scientists to analyze the new paradigm's affective impacts on scientific software development. Following is the questionnaire that resulted from actual interviews of scientists.

**Q. 1:** What will you do if you need software to help you as an integral part in achieving your goal? Please select your priorities by renumbering from selections given in Section 1 from A to J.

**Q. 2:** Please rate your choices in importance from 1 to 7 of the above given choices, 1 being the least important and 7 being of maximum importance.

Table 2

| A | | B | | C | | D | | E | |
|---|---|---|---|---|---|---|---|---|---|
| F | | G | | H | | I | | J | |

**Q. 3:** Do you agree or disagree with the following statement: if a software engineer and an operational user are involved in helping you achieve your goal, (please add remarks in a couple of sentences, if you like):

It is most essential for the software engineer to write down requirements of the scientist in some accepted, structured format specified by the scientist verbatim, to gather and analyze them in order to get a solution provision.

**Q. 4:** Will you want to inspect your required solution at each step of verification and validation by yourself or would you prefer the software code be inspected by outside consultants of the same trade as your own?

**Q. 5:** Please rate software security as being critical to your desired solution on the same scale of 1 to 7, where 1 is least important and 7 is of maximum importance, and 9 is used for not applicable. Please explain your rating in a few words.

**Q. 6:** In your opinion, is the following perception correct: that scientific knowledge has to be implemented in mathematical form? As the pen became the routine tool, so does the computer, as our technology today, demand an even stronger definition of scientific knowledge. This knowledge becomes useful only when represented by a software application. You can answer this on the scale of 1 to 7, with 1 as agreeing least and 7 as agreeing most.

**Q. 7:** If you ever have developed software, what problems did you face in development, and how did the process you use help you in a unique way?

**Q. 8:** If you involve a software engineer to get your scientific software developed, is there a unique reason? Do you prefer using the sequence of methodologies you selected in Question 1?

## 6. Lessons Learned

The interviews with our scientists resulted in a particularly rich resource of information about scientific software. The case study of scientific software development reported herein may not be representative of all scientists.

It is possible that scientists have provided information in the interviews regarding what they require to get their software job done, and what steps they actually would like to take, if given a chance to get it developed by themselves. Table 1 contains coding of prescribed practices in Section 1 for the interviewees, of a software engineer with scientists involved in scientific software development.

Table 3 contains coding of best practices. These codes are utilized in pictorial diagrams in bar charts.

Table 3: Participant Scientists

| 1 | BPY | 2 | BKK | 3 | BSF | 4 | BZL | 5 | BJY |
|---|---|---|---|---|---|---|---|---|---|
| 6 | ZAN | 7 | BKB | 8 | MP1 | 9 | MP2 | 10 | SP1 |
| 11 | SP2 | | | | | | | | |

Data has been categorized as shown in Tables 3 and 4 for the purpose of analysis. All tables contain actual data by participants. The section of lessons learned on analysis is presented to prove the effectiveness of prescribed software engineering practices toward scientific software development.

**Analysis of Question 1:** The selection made by participants on average for all practices generated results displayed in Chart 1:

Chart 1: Average selection by participant

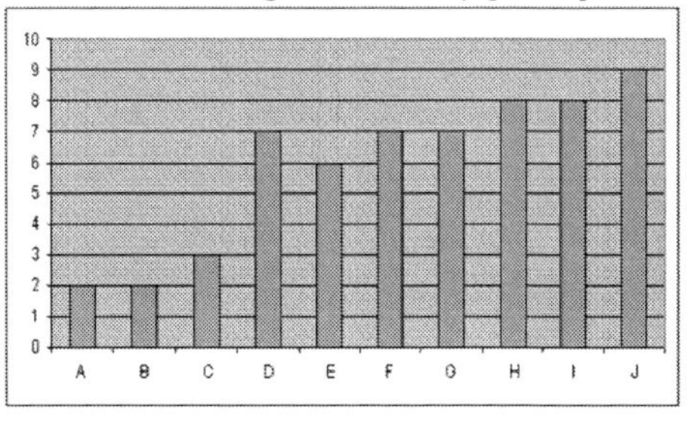


The result of Question 1 clearly demonstrates the interest of scientists in the prescribed best practices and gives a new sequence shown below:

Table 4: Scientist-selected sequence of work

| | |
|---|---|
| A | Initial functionality configuration |
| B | Relationship among requirements |
| C | Testing ground |
| E | Module development tracking |
| F | Module inspection |
| G | Disaster recovery plan |
| D | Daily log |
| H | Safety |
| I | Reliability checking |
| J | Code quality check |

As per our participant BKK of the Department of Biology at Queen's University, the actual sequence is in perfect logical order. Selection results do not demonstrate a significant difference in the sequence.

**Analysis of Question 2:** This analysis brought us an equal importance in response from the scientists of the steps in the sequence from most important to least important.

Chart 2: Pictorial representation of importance

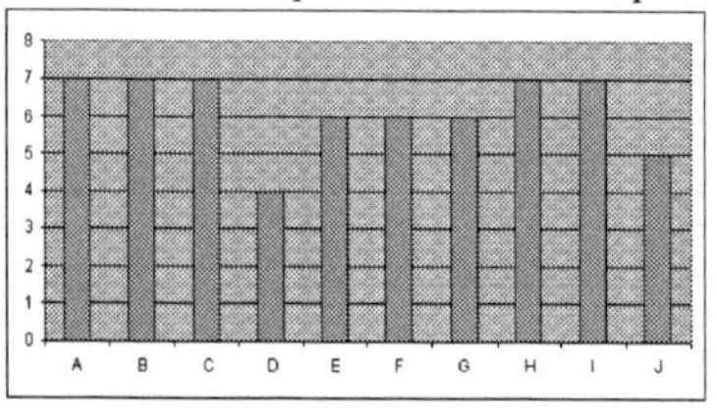


Table 5.1: Selections by participants

| | |
|---|---|
| A | Initial functionality configuration – Most Imp. |
| B | Relationship among requirements – Most Imp. |
| C | Testing ground – Most Imp. |
| H | Safety – 2nd Most Imp. |
| I | Reliability checking – 2nd Most Imp. |
| E | Modules development tracking – 3rd Most Imp. |
| F | Module inspection – 3rd Most Imp. |
| G | Disaster recovery plan – 3rd Most Imp. |
| J | Code quality check – 4th |
| D | Daily log – 5th |

**Analysis of Question 3:** In response to this question, all participants agreed with the statement. In the opinion of our expert scientist BPY: "*It is important that all stakeholders (Scientists, software engineers and operational users) are to be at one platform. This agreement will start a scientific software development process to get the objective achieved to work without any software bursts in between scientific experiments, data input and analysis.* "

Our second expert BKK said: "*Verbatim documentation might not be possible by software engineers, as scientists might come up with different ways to solve one issue. It is important to make a daily log of discussions by all participants.*"

**Analysis of Question 4:** Except for BKK, SP1 of York University and SP2 of Toronto College of Technology, all participants were interested in inspecting software development by themselves.

In the opinion of BKK, it will give him an edge to inspect software after getting a second opinion from industry experts of his trade to garner more comments that will help him rethink the way he finds target software for his test results.

In the opinion of SP1 of York University, outside inspectors will take up his time asking for details, which will not be that fruitful. As he is the expert, it is better that he by himself do the inspection. SP2's thoughts about software inspection, as well as software safety and security, are given below:

"*Software inspection is the progression used to help categorize the correctness, completeness, security, and value of developed software. Inspection is a procedure of technical investigation, performed on behalf of stakeholders by either outside consultants or software initializing scientists, which is intended to reveal quality-related information about the desired software with respect to the context in which it is intended to operate.*"

**Analysis of Question 5:** The result came as six participants out of 11 thought the safety and security to their desired software is most critical as shown in Chart 3.

Chart 3: Security criticality rating graph

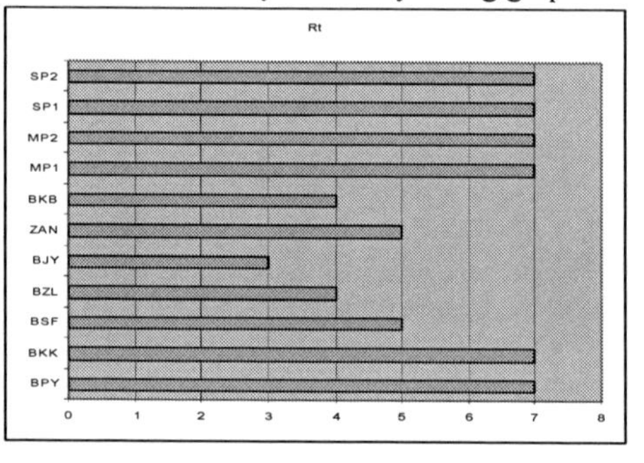


As per a few participants, the comments are given below:

Participant BPY: "*I think scientific software developed for a scientist does not need any significant security. Scientist's software is a part of the whole experimentation process. This process might need software to only calculate some parameters, for example, to get the change in strength of a chemical's density only.*"

Participant BSF: "*Software is highly valuable; it must be protected from unauthorized use. The importance of software safety and piracy continues to increase globally as our research work is now done around the world.*"

Participant BJY: *"If software is developed for any of my research work, I do not think that needs any security."*
Participant BKK: *"In any case, security is the most vital aspect to our software. It does not matter if it is commonly available in the market or if it is been developed by a software engineer."*

**Analysis of Question 6:**

Chart 4: Perception correctness rating graph

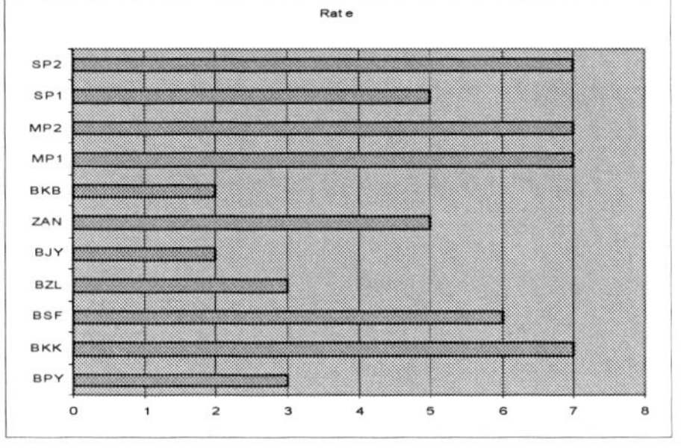


**Analysis of Question 7:** Except SP1 and SP2, all other participants have used available software in their trades. The first two mentioned participants did develop some of their desired software and commonly used the following path:

1. Conceived an idea of scientific experiment.
2. Noted all possible solutions.
3. Devised result calculation algorithms.
4. Designed a software model in UML/BPEL.
5. Coded software in C++/Java.
6. Did software testing module by module.
7. Packaged the solution for future use.

**Analysis of Question 8:** BPY and BKK involved software engineers and have given their thoughts as following: BPY: *I have worked with School of Computing students and faculty members to get my required software developed. Scientific software development can be considered difficult work for a software developer with little or no knowledge of the scientific curriculum. We have to explain all requirements to software engineers step by step.*
BKK: *I like to work on an idea to convert it to a mathematical model first. If I have to involve a software developer, I always try to explain in detail my devised models. I also like to show the results I get on paper after the dry run of my own algorithms. This gives an understanding to the software developer, as to what I need in actuality.*

# 7. Use of Methodologies

Our statistical case study and analysis shows that scientists think, innovate, and build a scientific solution to answer a specific need or to test some of their critical/innovative ideas. The case study also shows that the need for desired scientific software can usually be presented in the requirements document. There are several software development methodologies available, such as Waterfall, Spiral, Capability Maturity Model Integration (CMMI), and Object-Process Methodology (OPM). This paper will look at OPM as a formal standard to software engineering like Unified Modeling Language (UML) [02]. In reality, when we start designing a system, in general, we usually do not start from scratch. This is inevitable in the case of scientific software development, because we might have to design and develop systems of greater complexity.

# 8. Expanding Knowledge Base

As per our discussions on Questions 7 and 8 in the case study, the requirements that evolve in the development of a scientific solution from a possible knowledge base necessitates an understanding of ideas in a compact and easy way. In order to better comprehend these requirements, elicitation techniques are used to work with scientists. To explain why these techniques are important for the success of the products, a brief review is given below [7, 8].
**Interviewing** [8] is a simple, direct technique used in order to bring to the surface new information or to uncover conflicts or complexities. **Brainstorming** can be done by software engineers and scientists to define issues and questions regarding the scientific solution. **Prototyping** is an initial versioning of a system, which is available early in the development phase. **Questionnaires** are the collections of closed and open-ended questions to be provided to scientists; their responses are analyzed to understand the general trends and their opinions. **Observation** is an ethnographic technique in which software engineers can observe their design's dry-run activities as performed by scientists. **Protocol Analysis** involves the users engaging in a task and discussing their thoughts with the software engineer.

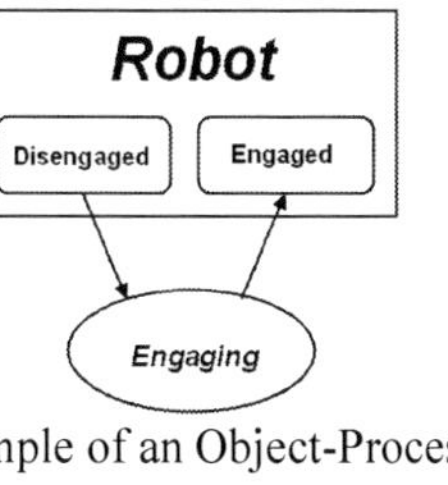


Figure 1: An example of an Object-Process diagram.

Hence, we need a visual modeling language that will allow us to compactly understand systems, incorporate parts of previously designed systems in our new designs and validate that those parts were used appropriately. OPM [2] provides a medium to view a solution's model in a generalized way. OPM is a combination of objects, and processes that can change the states of these objects.

Objects are things in existence; process is a thing that transforms an object, and state is the situation of an object. OPM encompasses the third major factor of modeling named "State" [4]. It is a situation that an object can be in at some point during its lifecycle. At any point in time, an object is in exactly one state. A state change can happen only as a result of an occurrence of a process. A simple example toward understanding OPM is mentioned in Figure 1. A robot can have two states interchangeable due to the process of engaging.

## 9. Conclusion

Scientific software development is a field currently lacking an established curriculum like project management or software engineering's well-defined architectures possess. The finding in this article clearly put an emphasis on the scientists, their needs and their goals, and the prescribed way of software engineering practices to achieve workable scientific software. Scientists have ideas with the capability to convert into well thought out and proven concepts. The need to achieve their goals does exist in their minds. This requirement is primarily outside of the software engineering activity and is expressed often in fuzzy or unclear terms. Object-Process Methodology is an approach that as yet has not been properly explored in the software engineering or scientific world of research and development. OPM provides an excellent view of a system model. It can be quite suitable for scientific software modeling and development.

## Acknowledgements

I like to thank Dr. Diane Kelly of Royal Military College of Canada for her wise guidance during this research work. I also like to thank Dr. Daniel Amyot, Dr. Craig Kuziemsky and Dr. Liam Peyton of University of Ottawa, as well as Dr. Paul Young and Dr. Kenton Ko of Queen's University for their invaluable input in this research work.

## References

[01] Rittel, Horst and Melvin Webber (1973) "Dilemmas in a General Theory of Planning," Policy Sciences 4, Elsevier Scientific Publishing, pp. 155-159.

[02] Dov Dori, Object-Process Analysis of Computer Integrated Manufacturing Documentation and Inspection. International Journal of Computer Integrated Manufacturing, 9, 5, pp. 339-353, 1996.

[03] Mor Peleg and Dov Dori, The Model Multiplicity Problem: Experimenting with Real-Time Specification Methods. IEEE Transaction on Software Engineering, 26, 8, pp. 742-759, 2000.

[04] Pnina Soffer, Boaz Golany, Dov Dori and Yair Wand, Modeling Off-the-Shelf Information Systems Requirements: An Ontological Approach. Requirements Engineering, 6, pp. 183-199, 2001.

[05] Liu, H., and Gluch, D. 2004. Conceptual modeling with the object-process methodology in software architecture. J. of Computing in Small Colleges, 19 (3), 10-21.

[06] Dori D, Choder M (2007) Conceptual Modeling in Systems Biology Fosters Empirical Findings: The mRNA Lifecycle. PLoS ONE 2(9): e872. doi:10.1371/journal.pone.0000872

[07] Hickey, A. Davis, and D. Kaiser, "Requirements Elicitation Techniques: Analyzing the Gap between Technology Availability and Technology Use," Comparative Technology Transfer and Society Journal (CTTS), 1 (3), pp. 279-302, 2003.

[08] Davis, O. Dieste, A. Hickey, N. Juristo, and A. Moreno, "Systematic Review of the Effectiveness of Requirements Elicitation Techniques," Proceedings of the Fourteenth International Requirements Engineering Conference (RE06), September 2006.

[09] C. Hofmeister, R. Nord and D. Soni, Applied Software Architecture, Addison-Wesley, 2000

[10] Dorian Arnold and Jack Dongarra, "Developing an Architecture to Support the Implementation and Development of Scientific Computing Applications," in The Architecture of Scientific Software, (IFIP TC2/WG2.5), Ottawa, Canada, October 2000.

[11] Ian Foster, Carl Kesselman, "Scaling System-Level Science: Scientific Exploration and IT Implications", November 2006.

[12] Charles Blilie, "Patterns in Scientific Software: An Introduction", Computing in Science and Engineering, May/June 2002, pp. 48-53

[13] Viktor K. Decyke, Charles D. Norton, Hanry J. Gardner, "Why Fortran?" Computing in Science and Engineering, July/August 2007, pp. 68-71

[14] Charles D. Norton, Viktor K. Decyke, Boleslaw Szymanski, Hanry J. Gardner, "The Transition and Adoption of Modern Programming Concepts for Scientific Computing in Fortran", Scientific Programming, Vol. 15, no. 1, spring 2007, 27 pages

[15] Tanenbaum, A.S.: Distributed Operating Systems, Prentice Hall, Upper Saddle River, NJ U.S.: Prentice Hall, 614 pages, 1995.

# Supply Chain Requirements Engineering: A Simulated Reality Check

Atif Farid Mohammad[1], Dustin E.R. Freeman[2]
[1]School of Computing Queen's University, Kingston, ON, [2]University of Toronto
[1]atif@cs.queensu.ca, [2] dustin@cs.toronto.edu

Abstract- *This paper presents a realistic understanding of the field of software requirement engineering of automotive spare parts dealers' distribution and maintenance workshops services and spare parts supply chain management information systems. It attempts to elaborate elicitation techniques to get to actual requirements used by system analysts. These requirements establish needs of customers and users associated to the automotive parts, which can lead to desired software goal achievement of an organization. The magnitudes are also characterized by process interdependencies, interpersonal and inter-organizational conflicts and information uncertainties, and their interrelations. Problems are also described in this paper that occur in implementing a major organizational change initiative where various employee groups have different understandings of the rationale of the project and strategies intended to achieve its goals.*

**Keywords:** Requirements Engineering, Self-managing, Cost reduction, conflict resolution

## 1. Introduction

An inventory control of automotive spare parts warehouses' software project requirements engineering (RE) procedures to achieve a target supply chain management system is discussed on an actual automotive parts distributor/dealers chain. This paper will describe the process of identifying stakeholders and their requirements as per the design theorist Horst Rittel [3]: "a statement of a problem is a statement of the solution." Even though this is a relatively uncomplicated observation, it holds a fine model often ignored by software solution providers: requirements classification is just a premature stage in software design. If it is even assumed, the requirements collection and classification and software design as well as development are interrelated. The stakeholders, which are described in the following case study, were several and had distributed tasks. Their objectives were different and had various conflicts. This paper presents a synopsis of current research in RE in relation to supply chain management in the terms of major activities that comprise the RE field. Even though these activities are described in parallel and in a meticulous order, there is a strong relationship they all have in common, and it did cover the entire software project development life cycle. Section 2 contains the case study of a successfully completed software project. Section 3 sketches out the controls that build the fundamentals [2] for successful elicitation techniques:

- elicitation of the requirements and
- design and analysis [1]

Section 4 illustrates the conflicts and conflict resolution techniques needed in order to begin the RE process to achieve all stakeholders' agreement to get the software project completed:

- agreement on requirements and
- evolution

This conflict resolution [6] process is also characterized by process interdependencies, interpersonal and inter-organizational conflicts and the solutions associated. Section 5 discusses supply chain management's model designing process. Section 6 provides a simulated case study of our work. Finally, the paper is concluded with section 8, a summary of the work done using standard RE practices while following Waterfall [1] methodology. This paper does offer important focal point of view of the key challenges for future RE research.

## 2. Case Study: Inventory control of Automotive Spare Parts

The case study is of automotive spare parts inventory control for stocks needed to assist repair and maintenance of automotive vehicles for both independent customers as well as garage owners. The main subject under study was to convert a manual system to manage the inventory control to a customized in-house developed software system.

**2.1. History of Physical Inventory Control System:** ABC Motors did not have computerized system in 1993; the system used consisted of physical cards called the CARDEX SYSTEM. These cards were of four types;

T. Sobh (ed.), *Innovations and Advances in Computer Sciences and Engineering*,
DOI 10.1007/978-90-481-3658-2_24, © Springer Science+Business Media B.V. 2010

**Order, Transaction, Sales** and **Audit**

All physical cards were to be filled with a blue ink pen, and a tick or correction in red ink to be audited by the inventory physical audit department, after the physical counting of parts on shelves in the warehouse every week. At the sales counter, the staff used to receive both customers as well as garage owners to place orders to get the spare parts needed. A sales order form contained the following columns to be filled in by the customer. Customer used to fill in the form with blue/black ink, where as "Price, Net Cost" used to be filled in by card-viewing staff with red ink, and a manual total with sales tax was also mentioned at the bottom of sales order form. This department also had a sales return form.

An invoice used to be prepared on the basis of sales order form, and included a copy for the customer, one copy for the inventory transaction update staff, and one copy for the accounts department for their ledgers. The parts delivered to either a garage or an individual customer used to be subtracted in the Transaction Card, and if it was equal or less than the "Order Level" on the card, an order form was to be completed simultaneously to place an order to refill the stock by placing the order of required spare part with regard to Order Card to be filled with above mentioned information on the physical card as well.

There were six (6) directly and six (6) indirectly related stakeholders in this warehouse situation: Inventory Management Staff, Order Placement Staff, Audit, Sales, Management Tier, Invoicing Department, Customers, General Supplier, Parts Manufacturer, Garage, After Market Manufacturers/Refurbishers, Transportation and Delivery Actors. There was another system associated with this functionality, and it was the "Workshop (Garage) Management System", as this service provider had both spare parts warehouses as well as attached garages to provide Refurbishment Services, and it contained the following main modules:

(i) Service Level Provisions,
(ii) Work Order Management,
(iii) Spare Parts Ordering Module and
(iv) Invoicing

**2.2. Customized Software Development: The Goal:** This study will focus on the warehouse inventory supply chain related system's customized software development, which was the goal to be achieved. There had been sub-goals identified during this study which will be elaborated. These sub-goals related to quite a few factors involved in developing this computerized system, which are as follows:

- System study start point and initial conflicts resolution and the role of stakeholders
- Financial viability and related conflicts
- Information normalization
- Database design with self-managing characteristics
- Reports structure and related conflicts
- Input screens and related conflicts
- Live testing and Training of staff
- Post implementation support

The waterfall model [1] has been adapted for the completion of desired software project.

**2.2.1: System study start point and conflicts:** As mentioned earlier, there was no utilization of computers in the original system of organization. It was relatively difficult to draw a system study start point. After having brief discussions on an individual department basis with management and collectively with related departments, the conflicts of interest to be resolved were drawn up first. Management required starting work on their reporting system. The inventory control staff, order department, the audit department, sales department and the invoicing department wanted to start printing invoices to get started with.

**2.2.2: Role of stakeholders:** At the time of finding the start point, interviews and group discussions had been initiated with six (6) directly related stakeholders. The findings were as follows:

- **Inventory management staff group discussions:** There were eight (8) people to fill in the cards in CARDEX, for both order and transactions as well as to verify the quantity of a sales order form requested by sales department.
- **Order Placement Staff:** There has been staff of six (6) to process orders of required parts.
- **Audit:** This department had a staff of ten (10), which had been following registers called "Audit Ledgers".
- **Sales:** The sales department had a staff of four (4) to collect clients' requirements and fill in customer order forms.
- **Management Tier:** There were five (5) managers involved in this tier. Their job was to manage the day-to-day departmental activities and analyze reports, and conduct meetings to discuss any issues.
- **2.2.3: Financial viability and related conflicts:** The management had issues with the financial viability of the computers software and hardware as well as the recruitment of technical staff to work on the newly developed system.

There was a fear among staff of the probable loss of their jobs, which was interfering with their cooperation. The development of the systems was done in a way that all the staff was advised to learn typing with a tutor in their scheduled work shifts.

**2.2.4: Information normalization:** This was the phase where all collected information was studied to remove redundant information.

**2.2.5: Database design:** The starts of database designing happened after all redundancies of the data field were removed.

**2.2.6: Reports structure development and related conflicts:** On the basis of the older manual system reports, quite a few more reports designs were generated, giving the management trends of sales, damage and lost items info to resolve the issues they were having between the order, audit and physical inventory control staff.

**2.2.7: User friendly input screen generation and related conflicts:** The designed layouts for the various departments encountered opposition from most of the departments, the formats of the physical system were followed as much as possible to get a prototype system on the computer for the ease of use for the staff and management.

**2.2.8: Live testing:** At the time of the supply chain management system's first phase completion, it was than tested on live data with the simultaneous use of the physical system to debug all errors. The **second phase** was to put up the "Sales Display" module test. In the **third phase** was to test and replace the "Invoice Print" module to get computerized invoices. The **fourth phase** was to test and replace the "Order Management" module. It was automated as per the set order level of spare parts in "Inventory Control Transaction Management" module used to generate automatic orders and print those out to be sent to the manufacturer to provide the required spare parts. In the **fifth** and last phase the "Management Required Reports" module was tested and handed over to the managers.

**2.2.9: Training of staff:** During the live testing of these modules the staff was given hands-on training to use the different screens and print out required reports. During the phase of system study and designing, staff had been trained to use computers by the use of typing tutorial software.

**2.2.10: Post implementation support:** Two developers having the necessary network maintenance knowledge were hired for this section of the organization to manage day-to-day activities and debugging, if required, and to maintain the network as well as the applications running in all departments mentioned.

# 3. Elicitation of the requirements

Let us have a look at elicitation techniques. There is a very concise definition given below: [2] "The choice of elicitation technique depends on the time and resources available to the requirements engineer, and of course, the kind of information that needs to be elicited". There are a number of well-known elicitation techniques:

**3.1. Data Collection:** Both top-down and reverse look up on the reports helped in the above-mentioned case to gather requirements, even before any other technique utilized.

**3.2. Questionnaires, surveys** and both team and individual interview techniques also were utilized.

**3.3. Group elicitation techniques** helped to attain stakeholders' agreement on conflicting issues. That included brainstorming and focus groups [2].

**3.4. Iterative Feedback of Stakeholders:** After completion of one module of the software, it was beta tested on the live test data, in front of the user, to make corrections, if required, until the user was satisfied and had no more objections or requirements to get embedded.

After completing the requirements collection phase, the most important phase, "the design and analysis", should be initiated. In Dr. Royce's words [1] "*some reason what a software design is going to do is subject to wide interpretation even after previous agreement. It is important to involve the customer in a formal way so that he has committed himself at earlier points before final delivery. To give the contractor free rein between requirement definition and operation is inviting trouble.*" The concept of Agile Methodology is relatively very new. Waterfall does contain iterations and the Agile Methodology follows most of this model's features to get to solutions. Both of the design and analysis phase described in both Waterfall and Agile modeling require iterations of continuous visit to the requirements, by embracing the change make sure that the requirements of the user are fulfilled, and the door of change should always be open for the in-house developers to make the changes as and when needed.

# 4. Conflicts & resolution techniques

This is the most important aspect of any project. It does not matter whether it is a building construction project or a software development process: there are always differences of opinion between the stakeholders, which are eliminated by getting the agreement of all stakeholders, involved in that certain project. [2] As requirements are elicited and modeled, maintaining agreement with all stakeholders can be a problem, especially where stakeholders have divergent goals. Recall that validation is the process of establishing that the requirements and models elicited provide an accurate account of stakeholder requirements. Explicitly describing the requirements is a necessary precondition not only for validating requirements, but also for resolving conflicts between stakeholders. The conflicts can be between requirements [6]. In the project mentioned in section 2, the conflicts had been inspected to find resolutions for all stakeholders, particularly in dealings with users, managers and developers participating in the project development. Stakeholders' conflict resolution behaviors alleviate the unconstructive effects of project complexity. In 2.2.1 there had been quite a few conflicts, which were resolved in the following manner:

- Prioritization of information as per department.
- Beginning of training computer skills for all the departments at once.

To resolve the conflicts between the management groups, detailed interviews of individuals were conducted, and in a detailed group meeting with managerial staff, agreement upon all the adjusted requirements was achieved.

## 5. Supply chain model design

Automotive spare parts inventory models differ significantly from regular inventory models [07]. With respect to automotive spare parts inventory, the customer's sole interest [08] is that his vehicle is not operational due to a lack of spare parts. For the service provider, which can be a garage owner with spare parts supplying side business, or an independent spare parts store, this creates the opportunity for smart inventory management, where on average the goal to provide the required part is met. An automated system which keeps track of orders over a history should have some success in predicting future orders.

Different automotive vehicles have parts in common, if they are of a make such as GM, FORD, MITSUBISHI, etc. In general, a repair garage services various types of make and model vehicles, and it is good practice that per vehicle make and model at a customer a goal-oriented repair, tune-up and maintenance level is set, since a customer may want to have a guaranteed repair for the vehicle. However, it can be expected that collection of the inventory for various automotive brands can channel to cost savings. In such models, the frequently used parts made by one manufacturer such as MOPER [9] contribute to achieve goals of satisfying various one brand owner/customer.

Supply chain [10, 11, 12, 13, 14, 15, 16, 17, 18] delivery management model as shown in Figure 1, basically is the transshipment of spare parts from a main to a certain area general suppliers/stores, can be done in a regular way, but also in a faster, perhaps more costly way, in an urgent requirement situation. It is observed that if a certain area supplier is out of stock and no consignment is feasible, the customer is not going to wait until an ordinary replacement from the main supplier has been done. Instead, a faster and urgent delivery method will be used. The system's database is to be designed in a way as mentioned in section 2.2.5 that the queries are to generate orders and send these orders to the main/sub suppliers. The system should get an acknowledgement of the receiving of the order by the part sender(s). This type of behaviour can be called a self-managing system to provide the users' capability to know the date of order submission and the users can make a timely decision to serve the customer. This idea is also called autonomic computing [8].

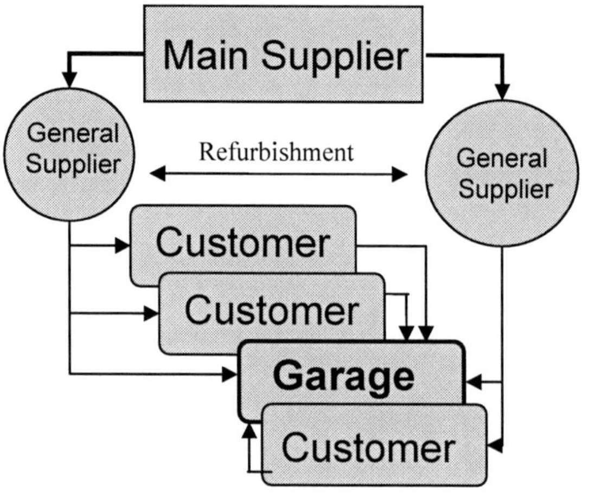


Figure 1: Supply Chain Cross Station Delivery

Let's introduce some terminology to express some of these ideas. The main components of the computer system are **parts** and **locations**. Both parts and locations are related by **shipping** and **labour**. Parts have both **parents** and **children**, where the children of a part are the parts it is composed of, and the parent of the part is what the part makes when it is put together with other parts. For example, the children of a wheel would be the hubcap, brake assembly, etcetera. The parent of a wheel could be the axle. Parts may be shipped back and forth from different locations, at a certain time and cost, which depends on the part, and on the locations from and to which it is shipped. In fact, **time** and **cost** will be the most important ways to measure effectiveness of this system. Shipping relationships are shown in Figure 2.

Finally, labour may be applied to parts to install children into parents, uninstall children from parents, or to service certain parts. A particular piece of labour will have a specific time and cost. An example of a labour procedure is shown below.

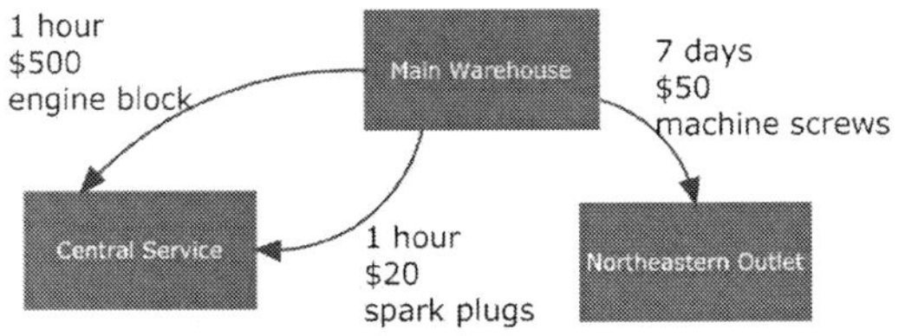


Figure 2: Shipping cost and times example

Figure 3: Servicing Pistons. The acts of Labour are the arrows between the ovals, which represent "states" of the system.

All of the above are simplified versions of day-to-day work in the automotive maintenance industry. The specific data, such as the time and cost it takes to install Part # 12345 into its parent, Part # 67890, will have to be determined by experience. The quality of the system depends entirely on the quality of this data, and should improve over time as the system becomes more of an "expert" and workers get better at using it.

## 6. System Simulation

Here we present a simulation of some aspects of a hypothetical system. The requirements engineering techniques used in the paper allow for beneficial control over the supply chain of an automotive company, and this section will attempt to determine potential benefits. As mentioned in section 5, the major performance metrics of the new versus the old system are **time** and **cost**. Cost here will specifically be cost to the company, which we will assume correlates with cost to the customer.

Time, by itself, is not a factor to the company but to the customer the less time they are without a working car the better, regardless of what is happening with the car's repair. Cost, from both the company's and customer's perspective, should always be at a minimum. In a real-life situation, a balance should be found between these two, but that is beyond the scope of this paper.

We shall look at a case study to determine how the knowledge in this system could aid an automotive company to improve their time and cost of repairs. While this case is very simple, the same ideas could be applied to more complex systems.

The case we shall examine is that of a camshaft that has seized up because of old and worn-out bearings. There are two possible ways of repair illustrated in Figure 4.

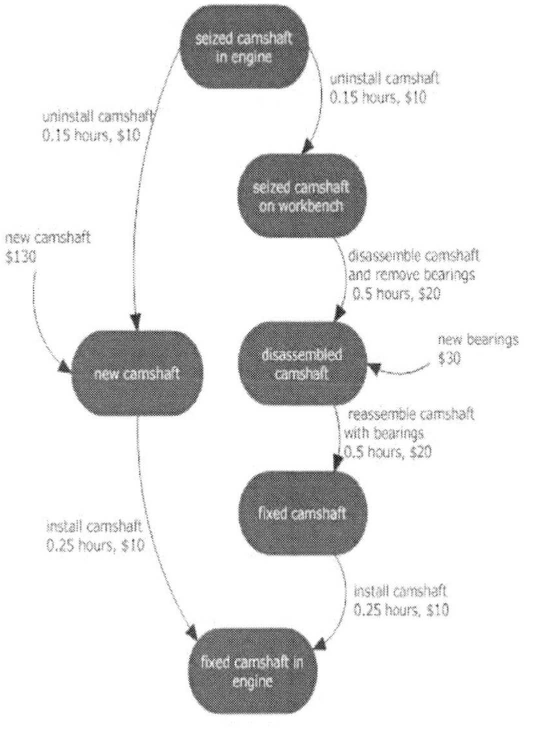


Figure 4: Comparison of methods to repair a seized camshaft

Should we disassemble the camshaft to repair the children parts, or simply replace the parent? The customer and company only care about the end result. By "traveling" along the two alternative paths, we can add up the total time and cost each path would take. The path with the least time and cost should be the preferred choice, and what the computer system would recommend. In Figure 4, the path of replacing the camshaft completely takes 0.4 hours and $150; the path of replacing only the bearings takes 1.4 hours and $90. So, the desired path is dependent on the priorities of the company in this case.

Using this system, certain non-intuitive revelations may occur, such as determining it is more cost- and time- effective to replace an entire engine when enough cheap, tiny parts are broken. In the camshaft example, it may also be the case that the garage doesn't have either camshafts or ball bearings in stock. This would mean that the items "new camshaft" or "new bearings" in Figure 4 would have a time amount attached to them, anywhere from a few hours to over a week. Then it may make more sense to choose the other path, depending on the time and cost of ordering parts. The above example may be extended to much more complex systems, given input from workers and computing power.

## 10. Conclusion

In the section 2.2, the Waterfall [1] has been utilized, and further in 2.2.8, the iterative approach, which is now known as Agile Methodology [5] has been adapted in the years of 1993-1994. As with Waterfall [1], the idea of iteration has existed since 1970. It was not an easy task to make an environment automated by using computers and printers, where the fear of losing jobs or demotion, transfers to a department, which might have some entirely new challenges for the existing staff. In section 5, the model designing is discussed, which can be implemented in any environment; the environment can either be a computer or electronics, cell phone accessories, a chain of stores containing any and all of the items, etc. The area in RE where we need more research is to establish a certain guideline containing a defined sequence of requirement elicitation techniques on the basis of experienced consultants' opinions as well as successful projects to follow, following the resolution techniques of embedded conflicts of any kind mentioned earlier. The examination schema for the future recommends giving weight to the reprocess of requirements model, should be designed and iteratively tested during the entire software solution development lifecycle. The software solution provision should happen by the utilization of requirements revelation. Additionally, using a software system instead of the traditional system means that the system could give suggestions on how to improve the efficiency of the business. Most of the ideas for this are inspired from [19] Project Management practices.

## 11. References

[01] Royce, Winston (1970), "Managing the Development of Large Software Systems", Proceedings of IEEE WESCON 26 (August): 1-9

[02] Bashar Nuseibeh, Steve Easterbrook, "Requirements engineering: a roadmap", May 2000 Proceedings of the Conference on The Future of Software Engineering ICSE '00 Publisher: ACM Press

[03] Rittel, Horst and Melvin Webber (1973) "Dilemmas in a General Theory of Planning," Policy Sciences 4, Elsevier Scientific Publishing, Amsterdam, pp. 155-159.

[04] Lawrence Peters "Relating software requirements and design", January 1978 ACM SIGSOFT Software Engineering Notes, Volume 3 , 7 Issue 5 , 3-4. Publisher: ACM Press

[05] Scott W. Ambler, http://www.agilemodeling.com/

[06]Gamel O. Wiredu, "A Framework for the Analysis of Coordination in Global Software Development", May 2006. GSD '06 Publisher: ACM Press

[07]Madeiro, Salomao S.; de Lima Neto, Fernando B.;Intelligent optimization for an integrated inventory model with multi-product and multi-storehouse Automation and Logistics, 2008. ICAL 2008. IEEE Conference on 1-3 Sept. 2008 Page(s):682 - 687

[08] Wells, J.D.; Sarker, Sa.; Urbaczewski, A.; Sarker, Su.; Studying customer evaluations of electronic commerce applications: a review and adaptation of the task-technology fit perspective System Sciences. Proceedings of the 36th Annual Hawaii International Conference on 6-9 Jan 2003 Page(s):10

[09] http://www.mopar.ca/en/: Accessed on 10/17/2008

[10] Scott J. Mason, P. Mauricio Ribera, Jennifer A.Farris, "Integrating the warehousing and transportation functions of the supply chain", Transportation Research Part E 39, 141-159,2003

[11]Young Hae Lee, Sook Han Kim,"Production-distribution planning in supply chain considering capacity constraints", Computers & Industrial Engineering 43, 169-190,2002

[12]Beamon, B. M., "Supply chain design and analysis: models and methods", International Journal of Production Economics, 55, 281-294, 1998

[13]Marco Perona, etc., "The integrated management of logistic chains in the white goods industry-- A field research in Italy", Int. J Production Economics 69, 227-238, 2001

[14]Laoucine Kerbachea, James MacGregor Smith,"Queueing networks and the topological design of supply chain systems", International Journal of Production Economics 91, 251-272, 2004

[15]Andrew P., Robert M., "The evolution towards an integrated steel supply chain: A case study from the UK", Int. J. Production Economics 89, 207-216, 2004

[16] Hokey Min and Gengui Zhou, "Supply chain modeling: past, present and future", Computers & Industrial Engineering, 43, 23 1-249, 2002

[17]Keah Choon Tan, "A framework of supply chain management literature", European Journal of Purchasing & Supply Management 7, 39-48, 2001

[18] Jeremy Shapiro, "Modeling the Supply Chain",Thomson Learning Inc., 2001

[19]Heizer, Jay and Render, Barry. "Operations Management" 7th Ed.

# A Path from a Legacy System to GUI System

Atif Farid Mohammad
*School of Computing, Queens University, Kingston, ON*
*atif@cs.queensu.ca*

**Abstract-** *There has been an assessment done in this paper of a large credit card company's software reengineering project. This organization is currently involved in its MIS migration from a legacy system built in late 60s on mini-frame using DB2 database on both COBOL and RPG to Microsoft Dot-Net technology. The new GUI system is to be built with more user friendly access to get more information on one screen rather jumping in between various black and white screens and waist the important time of both the staff and the card users as well as the merchants using the merchant services provided. There has been a practical experimentation given in this paper to provide the credit card service provider organization to convert legacy system to an efficient GUI system.*

**Keywords**: Legacy System, migration, business reengineering

## 1. Introduction

Software Reengineering [1] activities in general require techniques and tools that help to dig out valuable knowledge about the structures and inner mechanism of software systems from the source code using reverse engineering, since other information sources such as system documentation and developer's knowledge are often not available or consistent. A key dilemma during reverse engineering tasks is that the information extracted from the source code has to be condensed and abstracted towards an abstraction level which can support software engineers to perform essential analyses or modifications of the system and preventing them from an information overflow that makes software reengineering [2] goals complicated. These activities have to be supported by tools. An important consideration in building tool support for reengineering is what information must be provided, and how this information is modeled.

Therefore, a basic assignment is to develop a brief model as a basis for such tools, which enables to characterize program artifacts needed during the reengineering process and to build high-level abstractions from low level legacy program elements. Problems to be addressed in this project are as follows:

A. Specify a model to represent design information of a mini-frame system.
B. Specify operations to query the design information stored in the model.
C. Specify operations to abstract low level program entities into more abstract high-level entities using ERD and DFD mechanisms.

Assessment of the implementation of the prototype GUI of the model and suitable operations carried over in ASP.NET using C Sharp language.

## 2. Related Work

This work relates to software reengineering assessment of a real time running reengineered system [3] and an understanding of its legacy parent system as well as the database understanding.

**2.1 An understanding of database utilization.** There is a main database repository that contains information, mainly "Credit card numbers, Line of credits, Approvals, Denials, Transaction time stamp, Merchant No and more", by utilizing this data at the backend following information in the form of reports and screens are generated e.g. Statements, Modifications, Cancellations, Sales, Authorizations, Balance Transfers, Letter Generations, etc. Figure 1 explains the database utilization by various departments using this new web UI system.

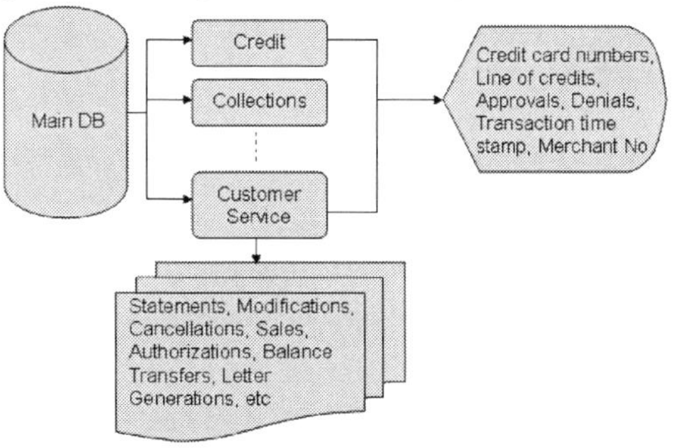


Figure 1: An understanding of database utilization

**2.2 Individual legacy module utilization sequence.** User needed to login after every 20 minutes, if idle.

T. Sobh (ed.), *Innovations and Advances in Computer Sciences and Engineering*,
DOI 10.1007/978-90-481-3658-2_25, 

User had to jump in between various interfaces individually such as:

- Approvals
- Authorizations
- Transaction Displays e.g. Merchant detail, Amount etc.
- Dispute initiations, status detail, add a comment etc.
- Fraud application (s).
- Additional Card issuance
- Sales of various features e.g. Insurance, Protections etc.

Each department had its own interfaces, none of the other department could be a backup in case of a department's shutdown due to any unanticipated reason e.g. storm, blackout etc.

**Why software reengineering?** Software reengineering is adapted to improve system's maintainability, evolvability, reusability and improve one's understanding of the system. To get a better understanding 60% of reverse engineering [4] is done by gathering knowledge of design recovery as well as system level abstraction and the most important factor is re-documentation of the legacy system following standard reformatting and 40% forward engineering [5] is done by the application knowledge, restructuring, redesign and retargeting the new platform.

## 3. Business process reengineering

Business Process Reengineering [6] wants to thus obtain by the fundamental reorganization of business processes as core of the processes salient improvements of the business achievements. A goal of the change strategy is thus the increase of economy (efficiency) by simultaneous reaching of **quality service improvements, cost lowering and time savings** as well.

Chandler and Liang shaped the set of "Structure follows strategy" [7]. Business Process Reengineering inserts an intermediate step: Directly from the strategy the cores of the processes are derived, those necessary are too realized around the strategy. The definition of the core of the processes makes it possible to find organizational solutions which support the strategy.

i. Determination of the strategy,
ii. Regulation the core of processes (3 - 5 regulations per enterprise),
iii. Derivative of the structure

The dominance of the processes over the structure, evident from it, leads to the extension of the sentence from Chandler to "Structure follows process follows strategy". Figure 2 explains the relationship and connectivity between the constraints involved in BPR.

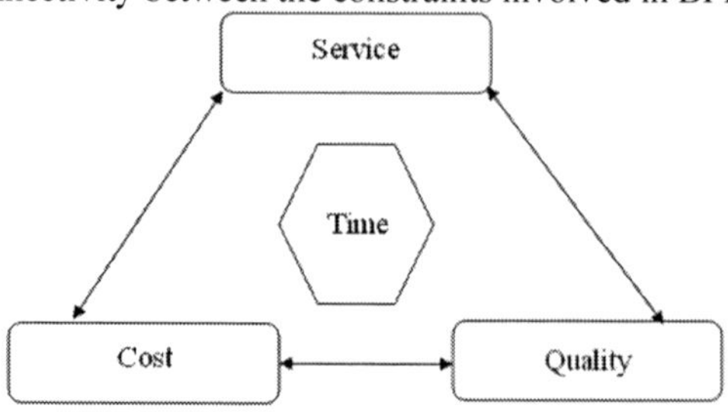


Figure 2 BPR constraints and their connectivity

### 3.1 Classifications

*Enterprise processes*: This concerns those activities, which create a value for the customer. As there are two aspects:

***Horizontal aspect:***

The enterprise understands itself as a bundle about core of the processes.

***Vertical aspect:***

The enterprise understands itself as an organizational hierarchical system.

***Core of the processes***: Linkage of coherent activities, decisions, information and flow of material, which constitute together the competition advantage of an enterprise with the following characteristics:

- Perceptible customer utilization
- Enterprise specificity
- Non-Imitating
- Non-Replaceable

### 3.2 Principles of BPR

Fundamental considering:

***Thought negation***:

As we operate our business?

***Challenging thought:***

Why do we make that at all?

**Alternative thinking**: Traditional rules and traditional acceptance (**"that we made however"** - set of symptoms always in such a way) are to be forgotten.

Drastic changes:

***Thought negation***:

As we became what we make, no matter its better or faster?

***Challenging thought:***

As we'd make it, if begin from front?

Improvements around orders of magnitude

***Thought negation***:

Gradual improvements

***Challenging thought:***
As we would make it, if we could begin from the front?

# 4. Identification of the Requirements

Before a segmenting of the processes (Triage) to take place can be identified, must be segmented in the individual processes. After an analysis of the processes these are visualized. This result in a review with the cooperating elements becomes subsequently, evaluated. It is to be paid attention on the following factors:

- Process chains,
- Linkages between the processes,
- Process interruptions,
- Interfaces of the processes

As well as quantitative data such as run and downtimes, frequent repetition of the processes, process costs and probabilities e.g. Process interruptions.

**4.1 A Call Center Environment**

The very rapid growth of call centers [8] has been one of the most striking features of work life in many industrialized countries over the past few years. Their development has changed the nature of white-collar work for the many people who spend their working days handling telephone calls in these purpose-built units. Call centers have already played a considerable part in the restructuring of the banking and financial services sector, in particular by facilitating the development of telephone-based 'direct' banking and insurance. Call centers are of course not the first or only structural change in the financial industry made possible by technology: the extensive relocation of back office functions which began in the 1960s and 1970s also made use of new technology to restructure banking and insurance operations. However call centers alter the nature of working life in a much more radical way than these previous changes.

As IVR [9] has taken over this sort of routine enquiry handling, so the work of call centre staff has shifted noticeably from a customer service role to a customer sales role. Efficient call centre staff [10] are needed for adding value – or in other words, for using the opportunity to develop the relationship with a customer by trying to sell additional services. This suggests that handling of inbound calls may increasingly be automated, but that banks and insurers may want to further develop their use of outbound telephone calling (cold-calling) of customers.

The development of the internet, and in particular of on-line banking and insurance services, reinforces this trend. In many countries a customer can now not only check their bank balance over the Web, but can also compare car or household insurance quotes, arrange bank loans or even take out a house mortgage loan.

| Reason category | %Call Reason |
|---|---|
| *Calls that could be automated* | *62%* |
| Default readout information | 16% |
| Available balance/credit | 17% |
| Make Payment | 10% |
| Recent transactions/charges | 8% |
| Get payment mailing address | 1% |
| New automation (not in current IVR) | 10% |
| *Calls handled by default agent* | *36%* |
| Billing question | 8% |
| Other payment-related questions | 7% |
| Misc. account maintenance | 6% |
| Lost/Stolen Card | 3% |
| Cancel card | 2% |
| Other | 6% |
| *Calls that need to be transferred* | *2%* |
| Spanish | 2% |
| Info on rewards program | 0% |
| | 100% |

Table 1: Various aspects of calls received

The take-up of Internet banking [11] has been rapid in several countries, and the introduction in the next two years of Web-enabled third-generation mobile phones using Wireless application protocol (WAP) [12] and of digital television will extend the opportunities for direct banking transactions by customers. Another likely development, the introduction of electronic purses based on smart card technology which can be loaded with e-cash from home computers, also puts the individual customer more in direct control of their finances. In each case, the implication is that call centers may become less utilized. It is clear that the Internet is and will be an important means of communication with customers. Already some corporate web sites are equipped with 'call me' buttons, enabling customers who are visiting web sites to request that a telephone call is made to them at a time of their choice.

# 5. Analysis of Legacy System

The software re-engineering has three steps:

I. Analyze the legacy system,
II. Specify the characteristics of the target system,
III. Create a standard testbed or validation suite to validate the correct transfer of functionality.

The analysis step [13] did begin by locating all available code and documentation, including user manuals, design documents and requirement specifications. Once all of the information on the legacy system was collected, it was analyzed to identify its unique aspects such as the current condition of the existing system, its maintainability and operability.

Once the legacy system and its quality characteristics of legacy code [14] were specified, the step of stating the desired characteristics and specification of the target system initiated. The characteristics of the existing system that must be changed are specified, such as its operating system, hardware platform, design structure, and language.

Finally, a standard testbed and validation suite has been created. These were used to prove the new system is functionally equivalent to the legacy system and to demonstrate functionality has remained unchanged after re-engineering [15]. The testbed for the target system implemented incrementally.

**5.1 Legacy System**

- Requirement used to be fixed
- Applications were isolated
- No shared substrate
- Architecture and code separation
- Implicit architecture
- Design info and trade-off used to be discarded
- Designed was followed by maintenance
- Technology dependant

**5.2 Reengineered System**

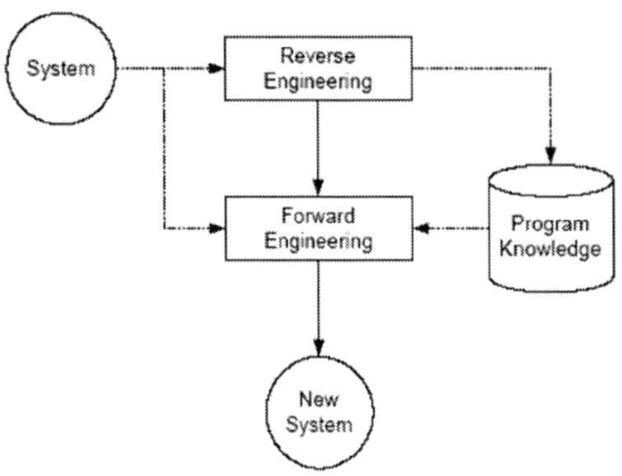


Figure 3: System reengineering [18]

- Requirements may change
- Application use is global via web
- Evolving substrate
- Architecture and code integrated and evolved together
- Explicit architecture
- Design info and trade-off is preserved to guide evolution
- Design and maintenance is a single activity Technology independent

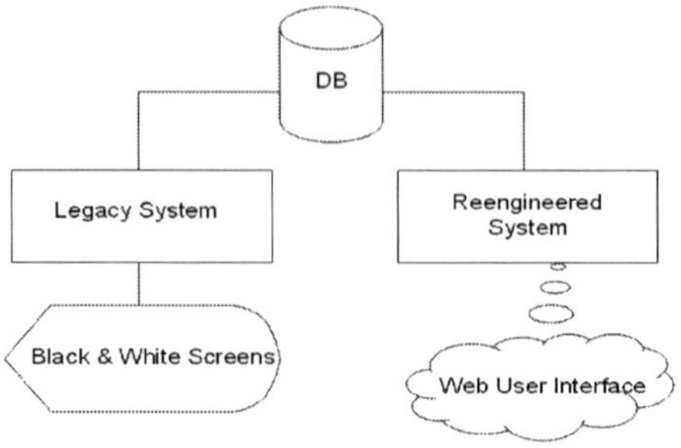


Figure 4: Legacy system to a Web UI application

## 6. Process Bridge Development

To better understand the software reengineering process bridging, pictorial diagram given in figure 4, explains the new GUI based web application generation process:

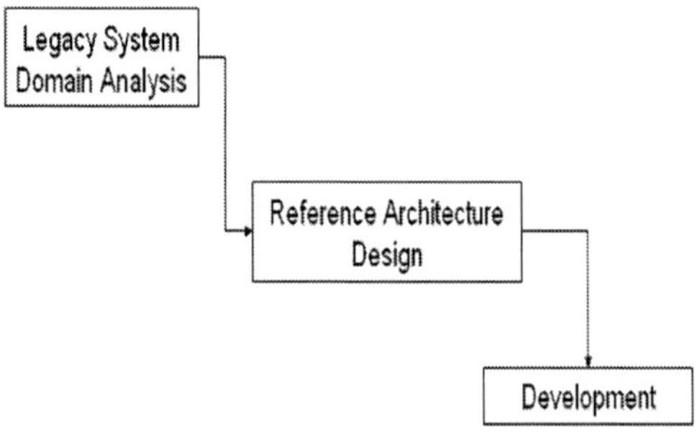


Figure 5: Legacy system bridge development

- **Legacy System Domain analysis:**
  - requirements description
  - domain definition (entities, attributes, relations, constraints, ...)
  - system analysis
  - legacy code description (architecture and traceability with requirements)
- **Reference architecture design:**
  - reference requirements
  - reference architecture
  - traceability
- **Development:**
  - library of reusable components
  - exploitation environment

**Reengineering Process:** The goal is to reuse the work-products of domain (software) reengineering [17] in order to produce a new application satisfying new specific requirements.

It entails the following activities:

- Identify specific requirements

- Identify architectural modifications
- Modify/adapt/generate components
- Build the global application

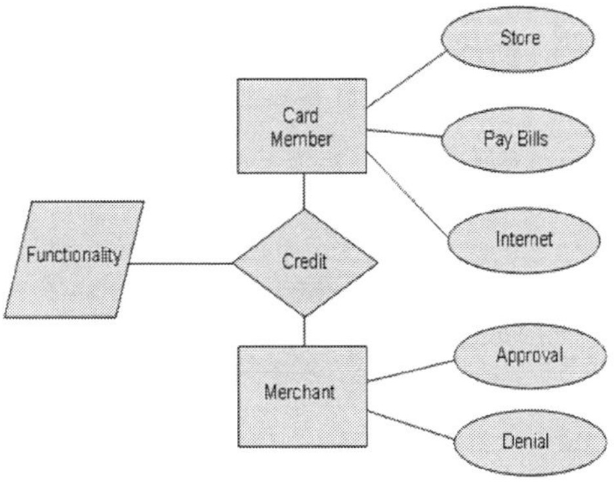

Figure 6: A simplified entity relationship diagram

**Lessons learned:** As companies apply re-engineering techniques [17], valuable lessons can be learned. Some general lessons learned are below:

- Re-engineering is not a solution and can never substitute for lack of planning many small gains can eventually add up to one large gain.
- Identify motivations and what is to be achieved by re-engineering. Direct, visible benefits to the customer are vital for continued support.
- Cost of planning, training, execution, testing and internal marketing are real.
- Re-engineering requires a highly trained staff that has experience in the current and target system, the automated tools, and the specific programming languages.
- Evaluate the system functionality with the intent of discovering what is worth retaining for future use and what is not.
- While design recovery is difficult, time-consuming, and essentially a manual process, it is vital for recovering lost information and information transfer.
- Evaluate the code, documentation, maintenance history and appropriate metrics to determine the current condition of the system.

# 7. Conclusion

This research paper presented an assessment is done on the real life migration of a legacy system to a web user interface. It provides a solid idea of re-structuring or re-writing part or all of a legacy system without changing its functionality. This change is applicable where some but not all sub-systems of a larger system requires costly and time consuming maintenance. Software reengineering involves adding effort to make them easier to maintain these systems, which may be re-structured and re-documented as per the business process may also be reengineered simultaneously. Entity Relationship Diagram given in figure 6 elaborates new service platform for "Customer Service Representatives" (CSRs). This new system is a fully functional backup for CSRs' department in case of any department shutdown and/or to reduce the load of that department.

# Appendix

To achieve the results the legacy functionality of the credit card business has been coded in Foxpro, which was utilized to code applications during 1992 to 1997 era. The physical legacy and GUI examples with screenshots are given to better understand the work done on this software reengineering research work.

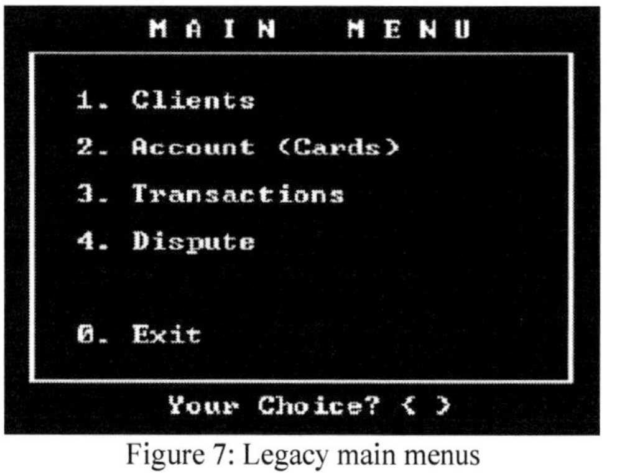

Figure 7: Legacy main menus

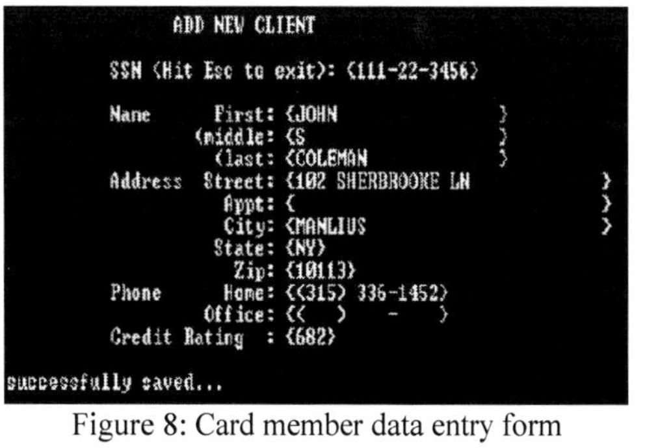

Figure 8: Card member data entry form

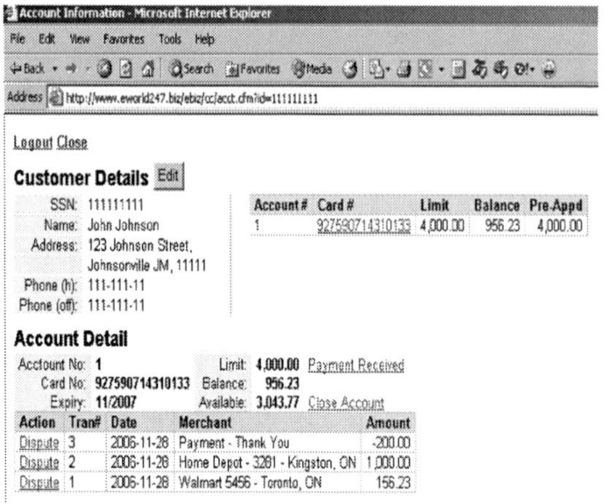

Figure 9: Web UI converted from Legacy System

## Acknowledgement

I like to thank Dr. Ying Zou of ECE at Queen's University for her wise guidance during this research work.

## References

[01] Chia-Chu Chiang; Leveraging software reengineering systems for heterogeneous distributed computing environments. Software Maintenance, 2000. Proceedings. International Conference on 11-14 Oct. 2000 Page(s):254 - 261

[02] Zhou, S.; Yang, H.; Luker, P.; He, X.; A useful approach to developing reverse engineering metrics. Computer Software and Applications Conference, 1999. COMPSAC '99. Proceedings. The Twenty-Third Annual International 27-29 Oct. 1999 Page(s):320 - 321

[03] Swafford, D.; Elman, D.; Aiken, P.; Merhout, J.;Experiences reverse engineering manually. Software Maintenance and Reengineering, 2000. Proceedings of the Fourth European 29 Feb.-3 March 2000 Page(s):189 - 197

[04] Estadale, J. and Zuylen, H.J. 1993. Views, Representations and Development Methods. The REDO Compendium: Reverse Engineering for Sofrware Maintenance, 93- 109.

[05] Baxter, I.D.; Mehlich, M.; Reverse engineering is reverse forward engineering. Reverse Engineering, 1997. Proceedings of the Fourth Working Conference on 6-8 Oct. 1997 Page(s):104 - 113

[06] J. T. C. Teng, V. Grover, and K. D. Fiedler, "Business process reengineering: Charting a strategic path for the information age," California Management Review, Spring, pp. 9-31, 1994.

[07] Alfred Chandler, Ting-Peng Liang, Developing Expert Systems for Business Applications, February 1995, Book.

[08] Saltzman, R. and V. Mehrotra. 2001. A call center uses simulation to drive strategic change. Inferfaces 3137-101.

[09] He Shu-guang; Li Li; Qi Er-shi; Study on the Continuous Quality Improvement of Telecommunication Call Centers Based on Data Mining Service Systems and Service Management, 2007 International Conference on 9-11 June 2007 Page(s):1 - 5

[10] Green, L.V., P.J. Kolesar, and W. Whitt, Coping with Time-Varying Demand when Setting Staffing Requirements for a Service System. 2005, Columbia University. p. 58.

[11] K. Furst, W. W. Lang, and D. E. Nolle, "Internet banking: Developments and prospects," in Economic and Policy Analysis—Working Paper 2000-9, 2000.

[12] S.P. Savino, "WAP: Wireless Application Protocol—Wireless Wave of the Future," Proc. Portland Int'l Conf. Management of Eng. and Technology, pp. 178-179, 2001.

[13] A. Cimitile, A. De Lucia, G. A. Di Lucca, and A. R. Fasolino. Identifying objects in legacy systems using design metrics. Journal of Systems and Software, 44(3):199–211, 1999.

[14] Yijun Yu, Yiqiao Wang, John Mylopoulos, Sotirios Liaskos, Alexei Lapouchnian, Julio Cesar Sampaio do Prado Leite: Reverse Engineering Goal Models from Legacy Code, RE 05: Proceedings of the 13th IEEE International Conference on Requirements Engineering

[15] Chikofsky, E.J.: Learning From Past Trial and Error: Some History of Reverse Engineering to Requirements, presentation slide in RETR Workshop at WCRE,05

[16] Zigman, Franklin and Mark Wilson, "Integrating Reengineering, Reuse and Specificaton Tool Environments to Enable Reverse Engineering", Proceedings of the Second Working Conference on Reverse Engineering, Toronto, Canada, July 1995, pp. 78-84.

[17] R. S. Arnold, Software Reengineering, IEEE Computer Society, 1994.

[18] Chia-Chu Chiang; Software Stability in Software Reengineering. Information Reuse and Integration, 2007. IRI 2007. IEEE International Conference on 13-15 Aug. 2007 Page(s):719 - 723

# Features Based Approach to Identify the P2P File Sharing

Jian-Bo Chen
Department of Information & Telecommunications Engineering
Ming Chuan University
Taoyuan, Taiwan

*Abstract*-**With the improvement of computer networks and the growth of bandwidth, the types of Internet traffic have changed. The P2P file sharing application became the significant source of network traffic. This paper examines the features for P2P file sharing application according to the packets of OSI layer 3 and layer 4. Four features are defined in this paper, including quantity of packet count, percentage of TCP packet count, percentage of specific size of packet count, and percentage of duplicate destination port numbers. Based on these features, the hosts running P2P file sharing application can be identified. The specific signatures in the payload of OSI layer 7 are used to verify which P2P file sharing application the host is running.**

## I. INTRODUCTION

With the growth of the Internet, the network usage is increasing rapidly. Many applications are processed by the Internet. At the end of 1990, the function of end user devices was diversification, and the bandwidth of the Internet enlarged. The P2P (peer to peer) transmission is the most popular application on the Internet [1-3]. The idea of P2P is to alleviate the heavy traffic of a single server. The peers can act as both client and server which provide the contents to other peers. There are many types of P2P applications and architectures, but the most appropriate application is the P2P file sharing application. The P2P file sharing application can indeed improve the file transfer performance, but most problems of P2P file sharing application are network congestion and intellectual property rights [4-5].

This paper defines the features of P2P file sharing application according to the layer-3 and layer-4 information of OSI reference model [6-7]. The layer-3 is network layer, which contains the IP addresses of both source and destination. The layer-4 is transport layer, which contains the port number of each transmission site. In addition to this information, we also collect the number of packets and the size of each packet. Based on this information, four features for P2P file sharing application are defined, including *quantity of packet count*, *percentage of TCP packet count*, *percentage of specific size of packet count*, and *duplication of destination port number*.

Based on these features, it can be guessed which IP address seems to be running P2P file sharing application. In order to avoid an identification error, if an IP address conforms to any three features of the four, it will be the assumed one is running P2P file sharing application. Afterwards, the payload for this IP address is captured. The information of layer-7, which is the application layer, can be used to compare the known signature of each P2P file sharing application [8-9]. When the layer-7 payload matches some specific keywords, it can be clearly identified which P2P file sharing application this IP address is running.

The remainder of this paper is organized as follows. In section 2, different types of P2P communications are addressed. In section 3, the features of P2P file sharing application are defined. The experimental results are given in section 4. Finally, in section 5, the conclusion is given.

## II. P2P FILE SHARING

### A. eDonkey/Mule

eDonkey/eMule is a decentralized architecture which does not rely on a central server to search and find files [10]. Its characteristic is fast searching, and it can search any file globally. It also allows peers to transmit any kind of file and provides the function to change to other peers for sharing the same file. Peers can download the same file from different peers in order to improve the performance of file sharing. When connected with another peer, the source peer will announce which other peers contain the same file. Consequently, the peer can download this file from other peers simultaneously.

### B. Foxy

Foxy is the most popular P2P file sharing application in Taiwan [11]. Its architecture is like Gnutella, but its sharing unit is based on a folder. Peers share the files when they are on the shared folder. There are no bandwidth limitations for uploading and downloading for Foxy; it also does not need to find the seed to search for the shared files. This application can find the files and peers automatically. In order to enhance the searching performance, even if peer has no file to share, it will also respond to the source peer, which will waste lots of bandwidth. The advantage of Foxy is that it can upload and download files simultaneously. Therefore, the more other peers downloading, the more speed it can achieve.

T. Sobh (ed.), *Innovations and Advances in Computer Sciences and Engineering*,
DOI 10.1007/978-90-481-3658-2_26, © Springer Science+Business Media B.V. 2010

*C. BitTorrent*

BitTorrent, sometimes call BT, cannot search shared files directly from other peers [12-13]. The peers must first find the seed (torrent file). The Tracker is contained inside the seed and records the network position, original source address, file hash values, and so on. BitTorrent downloads any shared file according the seed. The architecture of BitTorrent consists of the Tracker, seed, and peers. The Tracker is very important in BitTorrent architecture. It contains the information about which files are owned by which peers, so the peer can download the shared file from these peers. The seed always comes from some forum or exchange by peers. When a peer joins to BitTorrent, all he needs to do is to find the seed, then the peer can start to download any file.

## III. FEATURES ANALYSIS

*A. Quantity of packet count*

When the P2P file sharing application is running, it will issue lots of packets in order to communicate with other peers. According to the observation of packet count for P2P file sharing, we can find that the amount of packet count is increasing. Because the peer must check both the availability of peers and the availability of files, it must send out many query packets to get the information. Normally, any other computer hosts, except servers, will not issue too many packets in the same time. Thus, we define our first feature according to the quantity of packet count. Firstly, we determine the quantity of packet count for normal hosts and hosts running P2P file sharing application over a set period of time. We define the threshold, when the packet count for one host is larger than this threshold, this host may be running P2P file sharing application.

$$T_a > \text{threshold} \qquad (1)$$

where $T_a$ is the total packet count for the specific host.

*B. Percentage of TCP packet count*

From the observation of layer-4 packet types, the UDP packet count for normal users is always very small. Before this experiment, we collected the packet for a host running browser, running e-mail, connected to BBS, using instant messaging, and so on for several hours. The ratio of UDP packet count is very small which it approaches zero percent. This means that the TCP packet count is one hundred percent. Statistically, in this experiment, the percentage of TCP packet count with hosts running P2P file sharing application is always between 40% and 77%. Hence, the host running P2P file sharing application will decrease the percentage of TCP packet count. Here, the feature for percentage of TCP packet count is defined below.

$$T = \frac{T_t}{T_a} \qquad (2)$$

where $T_a$ is total packet count for the specific host, $T_t$ is the TCP packet count for this host, and $T$ is the percentage of TCP packet count.

*C. Percentage of specific size of packet*

During the period of P2P file sharing, it can be observed that the packet sizes for TCP packets are almost 1500 bytes. The percentage of packet size larger than 1400 bytes is nearly 30%. In comparison with normal users, the percentage of packet size larger than 1400 bytes is far less than 30%. Thus, the number of large packet size is another important feature for P2P file sharing detection.

$$P = \frac{T_s - T_n}{T_t} \qquad (3)$$

where $T_t$ is total TCP packet count for the specific host, $T_s$ is the number of TCP packets with size between 1400 to 1500 bytes, $T_n$ is number of TCP packet with size between 1400 to 1500 bytes and the port number is well-known, and $P$ is the percentage of specific packet size.

*D. Percentage of duplicate destination port number*

After the handshakes between P2P file sharing peers, the host starts to share files with other peers. In order to improve the download performance, one host may download the same file from other peers. That is, the same source IP address will communicate with other destination IP addresses and different destination port numbers. Thus, we use the packet count for duplicate destination port number as the numerator, and the packet count for different IP address as the denominator. The value we calculate is another feature for P2P file sharing application.

$$D = \frac{T_{dp}}{T_{ip}} \qquad (4)$$

where $T_{ip}$ is the packet count for different IP address, $T_{dp}$ is the packet count for duplicate destination port number, and $D$ is the percentage of duplicate destination port number.

*E. Signatures of P2P file sharing*

In the transmission process of P2P file sharing, every P2P file sharing application has its unique signature in its payload, or in its application header. The signature can be clearly identified which P2P file sharing application is being used. But the most difficult problem is the difficulty to capture the payload for all the traffic. Table 1 lists some signatures for different P2P file sharing applications.

- eDonkey/eMule

eDonkey and eMule use the same protocol. The data transmission is based on TCP. Whereas the control signals are based on both TCP and UDP. Either the data transmission or the control signals are started with a fixed value. The eDonkey starts with "0xe3" and the eMule starts with "0xe5". The successor is followed by a four byte field which represents the packet length. According to the first five bytes, eDonkey/eMule protocol can be identified. After that, the eMule has an advanced version for compressing the data in order to save the bandwidth which starts with "0xd4".

- BitTorrent

BitTorrent also starts with fixed format. The first byte is "0x13" and is followed by a fixed 19 byte string contained with "BitTorrent protocol". Thus, the 20 byte in the payload can be the signature for BitTorrent protocol. There are"Info_hash20……get_peers1" string in UDP packets, and "GET/announce?info_hash=%" string in the TCP packets.

- Foxy

The operation for Foxy is similar to Gnutella protocol and its signature is "GNUTELLA". When a host issues a request for downloading a multimedia file, the signature "GET /uri-res/N2R?..." will appear in the TCP packet; when responding to the host, the signature "HTTP/1.1 206 Partial Content…" will appear in TCP packet.

TABLE I
KEYWORDS FOR P2P APPLICATIONS

| P2P application | Keywords |
|---|---|
| eDonkey/eMule | 0xe3<br>0xe5<br>0xd4 |
| BitTorrent | 0x13BitTorrent<br>Info_hash20……get_peers1<br>GET/announce?info_hash=% |
| Foxy | Gnutella<br>GET/uri-res/N2R?<br>HTTP/1.1 206 Partial Content… |

## IV. EXPERIMENTAL RESULTS

In this section, the experimental architecture and flows are introduced. The experimental results and analysis for each feature are also described.

### *A. Experimental architecture*

In network environments, NetFlow is probably the most useful standard tool for network traffic accounting. In the implementation described here, both a NetFlow probe (nProbe) and a collector to monitor the inbound and outbound flows were used. The architecture of traffic collection is shown in Fig. 1. When the nProbe is activated, it will collect traffic data and emit flows in NetFlow format towards the specified collector. A set of packet of the same transaction (the same source IP, source port, destination IP, destination port, and protocol) is called a flow. Every flow, even a very long-standing ISO CD image download, has a limited lifetime. This is because the flow collector should periodically receive flow chunks to account for traffic precisely. The collector is used to receive the NetFlow data and store all the information into a database.

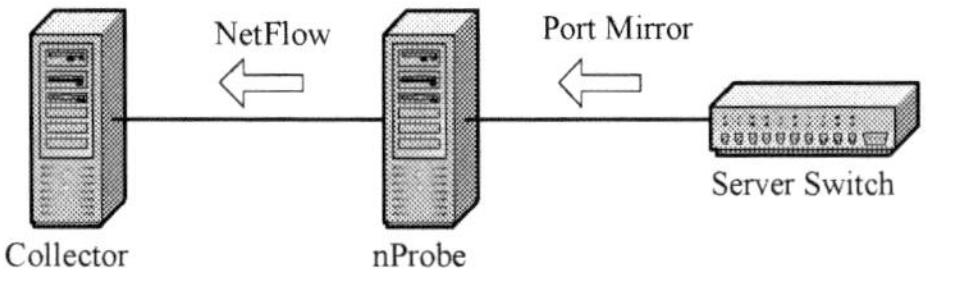


Fig. 1. NetFlow traffic collection

Ethereal is one of the most useful tools for network administrators to capture and analyze the packets [14]. It is an open source software that can be used to check the payload for an application. In the environment described in this paper, the machine that runs the nProbe can also run the Ethereal software to capture the packets for a specified IP address. A sample command for Ethereal to capture the payload is

```
/usr/sbin/tethereal -f ip host 192.168.1.100
                    -a duration:300
                    -i eth1
                    -w cap_file
```

In this example, the IP address being tracked is 192.168.1.100. The duration:300 parameter means packets for 300 seconds for that IP address are captured. The eth1 is the NIC interface that connects to the mirrored port of the switch. The payloads in the cap_file are stored for future analysis.

### *B. The analysis of features*

In the experimental environment, the packet count for the P2P file sharing applications are larger than 500, as shown in Fig. 2. Thus, we can define the threshold as 500 packet count per second. When comparing the percentage of TCP packet counts, the hosts running P2P file sharing are always less than 77% as shown in Fig. 3. The percentage of specific size of packet is shown in Fig. 4, where all the percentages of packet size larger than 1400 bytes are greater than 14%. Finally, the percentage of duplicate destination port number is shown in Fig. 5, where all the percentages are greater than 16%.

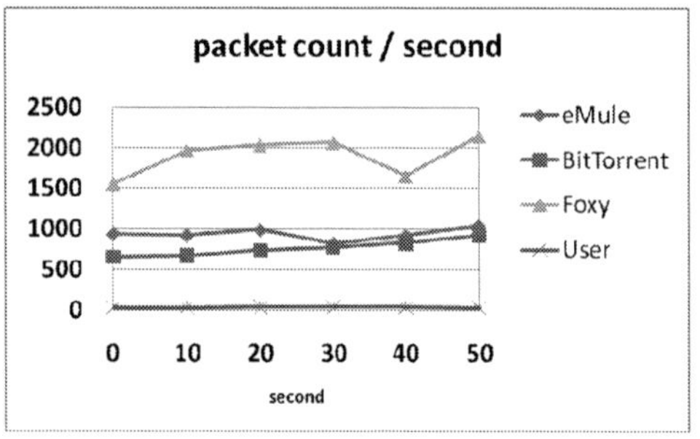


Fig. 2. The comparison for packet count

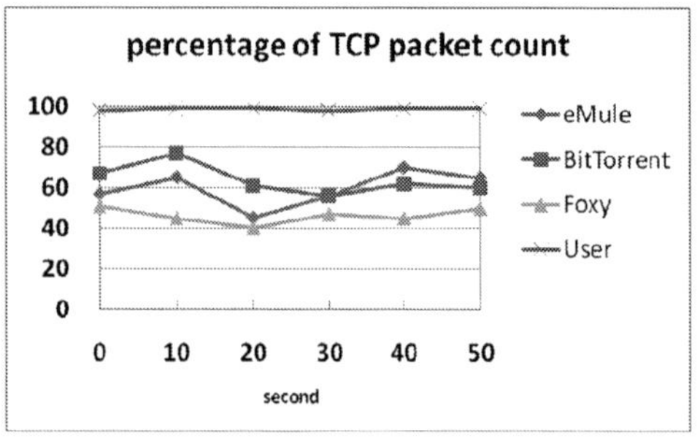


Fig. 3. The comparison for percentage of TCP packet count

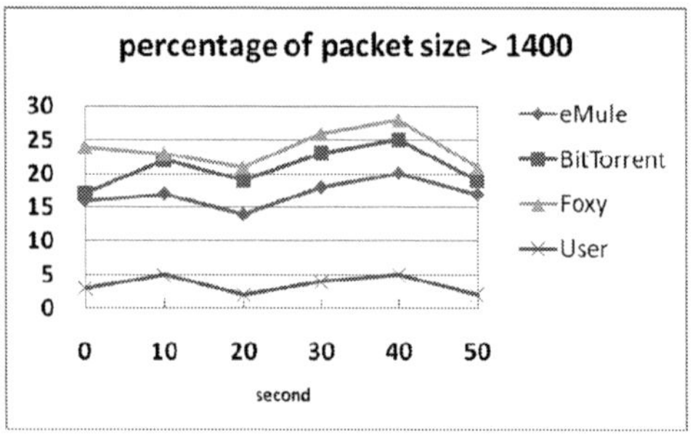


Fig. 4. The comparison for percentage of specific size of packet count

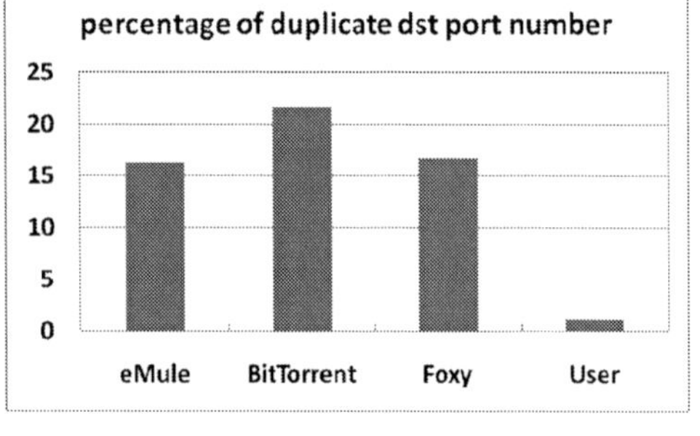


Fig. 5. The comparison for percentage of duplicate destination port number

*C. The analysis of payload*

In the above experiment, the thresholds for these four features are defined. But in some cases, the host not running P2P file sharing application may be identified as running P2P file sharing application because it matches two or more of the thresholds. The number of features used will determine the error rate as described in Table II. If only one feature is used to identify P2P file sharing application, the average error rate, no matter which feature is used, is about 72%. If two of these four features are used, the average error rate decreases to about 37%. The average error rate for three and four features are 11% and 9% respectively.

TABLE II
THE ERROR RATE FOR P2P IDENTIFICATION

| Number of Feature | Error Rate |
|---|---|
| 1 | 72% |
| 2 | 37% |
| 3 | 11% |
| 4 | 9% |

In order to find out which kind of P2P file sharing application is running, the payload needs to be analyzed. Capturing and analyzing the payload for all packets is time consuming and inefficient, so only the payload of specific hosts which match three or four of the above features are captured and analyzed. For the captured payload, comparing with the signatures in Table I, the usage percentage of different P2P file sharing application can be determine, as shown in Fig. 6.

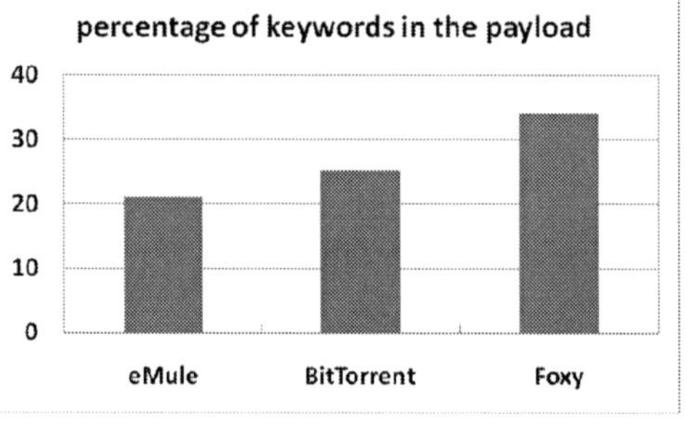


Fig. 6. The percentages of different P2P file sharing application

## V. CONCLUSION

In this paper, four features for P2P file sharing application are defined. According to these experiments, four thresholds for these features are also defined. If the thresholds for specific host match three or four of these four, it can be said that the percentage of this host running P2P file sharing application is 89%. Only the payload for this host is captured and compared with the payload of pre-defined signatures to find out what P2P file sharing application it is running.

## REFERENCES


[1] S. Sen, and Jia Wang, "Analyzing Peer-To-Peer Traffic Across Large Networks," *Trans. IEEE/ACM Networking*, Vol. 12, No. 2, pp. 219-232, April, 2004

[2] Hamada, T., Chujo, K., Chujo, T., and Yang, X., "Peer-to-Peer Traffic in Metro Networks: Analysis, Modeling, and Policies," *Proc. Network Operations and Management Symposium*, Vol. 1, pp. 425-438, April, 2004

[3] S. Saroiu, P. Krishna Gummadi, and Steven D. Gribble, "A Measurement Study of Peer-to-Peer File Sharing Systems," *Proc. Multimedia Computing and Networking*, January, 2002

[4] Spognardi, Alessandro Lucarelli, and Roberto Di Pietro, "A Methodology for P2P File-Sharing Traffic Detection," *Proc. the Second International Workshop on Hot Topics in Peer-to-Peer Systems*, pp. 52-61, July, 2005

[5] Matsuda T., Nakamura F., Wakahara, and Y. Tanaka, "Traffic Features Fit for P2P Discrimination," *Proc. 6th Asia-Pacific Symposium on Information and Telecommunication Technologies*, pp. 230- 235, 2005

[6] T. Karagiannis, A. Broido, M. Faloutsos, and Kc claffy, "Transport Layer Identification of P2P Traffic," *Proc. 4th ACM SIGCOMM Conference on Internet Measurement*, pp. 121-134, New York, NY, USA, 2004

[7] Li Juan Zhou, Zhi Tong Li, and Bin Liu, " P2P Traffic Identification by TCP Flow Analysis," *Proc. International Workshop on Networking, Architecture, and Storages*, pp. 47-50, 2006

[8] Wang Jin Song, Zhang Yan, Wu Qing, and Wu Gong Yi, "Connection Pattern-Based P2P Application Identification Characteristic," *Proc. Network and Parallel Computing Workshops*, pp. 437-441, September, 2007

[9] S Sen, O Spatscheck, and D. Wang, "Accurate, Scalable In-Network Identification of P2P Traffic Using Application Signatures," *Proc.13th International conference on World Wide Web*, 2004

[10] The eMule protocol, http://www.emule-project.net/

[11] The Foxy protocol, http://tw.gofoxy.net/

[12] The BitTorrent protocol, http://www.bittorrent.com/

[13] Mong-Fong Horng, Chun-Wei Chen, Chin-Shun Chuang, and Cheng-Yu Lin, "Identification and Analysis of P2P Traffic – An Example of BitTorrent," *Proc. First International Conference on Innovative Computing, Information and Control*, pp. 266-269, 2006

[14] The Ethereal, http://www.ethereal.com/

# Evaluating the Performance of 3D Face Reconstruction Algorithms

Andreas Lanitis
Dept. of Multimedia and Graphic Arts,
Cyprus University of Technology,
P.O. Box 50329, 3036, Lemesos, Cyprus
andreas.lanitis@cut.ac.cy

Georgios Stylianou
Dept. of Computer Science,
European University Cyprus,
P.O. Box 22006, 1516, Nicosia, Cyprus,
g.stylianou@euc.ac.cy

*Abstract*—**The use of 3D data in face image processing applications received increased attention during the recent years. However, the development of efficient 3D face processing applications (i.e face recognition) relies on the availability of appropriate 3D scanning technology that enables real time, accurate, non-invasive and low cost 3D data acquisition. 3D scanners currently available do not fulfill the necessary criteria. As an alternative to using 3D scanners, 3D face reconstruction techniques, capable of generating a 3D face from a single or multiple face images, can be used. Although numerous 3D reconstruction techniques were reported in the literature so far the performance of such algorithms was not evaluated in a systematic approach. In this paper we describe a 3D face reconstruction performance evaluation framework that can be used for assessing the performance of 3D face reconstruction techniques. This approach is based on the projection of a set of existing 3D faces into 2D using different orientation parameters, and the subsequent reconstruction of those faces. The comparison of the actual and reconstructed faces enables the definition of the reconstruction accuracy and the investigation of the sensitivity of an algorithm to different conditions. The use of the proposed framework is demonstrated in evaluating the performance of two 3D reconstruction techniques.**

## I. Introduction

During the last ten years, 3D face reconstruction has been receiving a continuously increased scientific attention. During this period of time, researchers have developed different methodologies for 3D face reconstruction including reconstruction from a single image, stereo based reconstruction and video based reconstruction [15]. Even though direct 3D face reconstruction via the use of 3D digitization equipment (i.e. 3D scanners) provides the best quality, this technology is still quite expensive, not portable, requires a considerable amount of post-processing time and most importantly requires the cooperation of the subject to be scanned during the scanning process. In addition most 3D scanners available in the market require controlled lighting conditions in order to produce accurate results.

Therefore the 3D digitization process limits the breadth of applications that 3D faces can be used for. These applications include surveillance and security, entertainment, virtual reality, realistic video conferencing and cultural heritage reconstruction [8].

As an alternative to using 3D scanners, it is possible to reconstruct the 3D structure of a face using single or multiple images captured using ordinary imaging equipment such as cameras and camcorders. Even though numerous 3D face reconstruction methods have been developed, in the relevant papers the focus is primarily on describing the reconstruction methods instead of the performance evaluation aspect. Most of the authors provided an extensive description of their proposed method but limited the evaluation procedure to a visual evaluation of the results while sometimes provided a minimal quantitative evaluation of the quality of the 3D reconstruction. Being able to judge on the quality of a method individually and/or compare different methods is very important as it allows the isolation of 3D reconstruction deficiencies for a single method. In addition a comprehensive performance evaluation test will allow the definition of the most promising methods from a pool of methods.

The contribution of this paper is the development of a framework for performance evaluation and the provision of an extensive quantitative performance evaluation of 3D face reconstruction. To our knowledge this is the first paper that deals in detail specifically with this issue. The proposed performance evaluation framework is composed of 3D face projection to 2D under different conditions and the assessment of 3D reconstruction quality by comparing the original face and the reconstructed face leading to statistics of error for the different conditions and comparison techniques. The use of the evaluation methodology is demonstrated by comparing the performance of two 3D face reconstruction techniques: A PCA-Based method [14] and a Shape Approximation-Based Method [8].

In the remainder of paper we give an overview of the related 3D face reconstruction literature, describe the performance evaluation framework and briefly describe the two reconstruction methods to be evaluated. Experimental comparative results, conclusions related to the performance of two techniques under investigation and plans for future work are also presented.

Work supported by a Cyprus Research Promotion Foundation research grant (EPYAN/0205/08).

T. Sobh (ed.), *Innovations and Advances in Computer Sciences and Engineering*,
DOI 10.1007/978-90-481-3658-2_27, 

## II. Literature Review

In this section we provide a brief overview of 3D face reconstruction with emphasis on 3D face reconstruction performance evaluation. More comprehensive reviews related to 3D face reconstruction [15] and 3D face processing appear elsewhere [4, 17].

### *A. 3D Face Reconstruction Methods*

Image-based 3D face reconstruction methods can be classified as single-image, stereo and video-based methods. Single-image methods operate on a single input image, stereo methods operate on a pair of images and video-based methods operate on multiple input frames. In this section we provide a brief overview of single-image 3D face reconstruction as in its current form the proposed evaluation framework is applicable to evaluation of single image reconstruction techniques.

Due to the fact that in a single image there are always regions of the face that are not visible, single image reconstruction methods rely on the analysis of a large number of training samples, to learn the structure and typical deformations associated with 3D faces. This knowledge enables the implementation of top-down approaches that can estimate the overall 3D face structure based on the visible facial parts.

Vetter and Blanz [2] use a Principal Component Analysis (PCA) based model to generate a 3D morphable model that is used to achieve 3D face reconstruction by computing the optimum weights of the model shape and texture eigenvectors, in order to approximate the 3D structure for a novel 2D face image. During the reconstruction process they also estimate 22 rendering parameters such as pose, focal length of the camera, light intensity, color and direction. The reconstruction is done by minimizing the distance between the intensity of a novel 2D face image and the intensity of a 2D projection of a 3D model instance. Several variations of these techniques were reported by a number of researchers [1, 7, 11, 13].

Reiter et al. [12] use canonical correlation analysis as a basis for generating 3D depth maps of faces from color frontal images. This method requires a training set of face images and their corresponding depth maps so that Canonical Correlation analysis is applied in order to define pairs of RGB images and depth maps with maximum correlation. Based on this approach the relationship between 3D depth maps and face images is learned and used for predicting the 3D appearance of a face, given a single image. This method is only applicable when dealing with frontal images captured under controlled illumination.

### *B. Performance Evaluation*

In the literature only a small number of researchers report some kind of performance evaluation method in relation with 3D face reconstruction.

Blanz et al [3] divide the MPI database into two sets of 100 3D models and use each part of the database for training and testing. 3D faces from the test set are projected to 2D using an orthographic projection in frontal orientation. A morhable-based reconstruction technique [2] is then applied and the resulting 3D faces are compared against the ground truth. The evaluation is based on the following quantities:

- The image plane matching error for all feature points (in units of pixels in a 300x300 image).
- The Mahalanobis distance from the average 3D face among the training set.
- The per-vertex average of Euclidean distances in 3D space between reconstruction and original, computed over the entire set of vertices.

Lee et al [10] assess their silhouette-based 3D reconstruction method using a test set with 100 images. 50 images in the test set are random faces generated using a morphable 3D model. The remaining test faces are 3D models from the USF database. The reconstruction accuracy is assessed based on the Hausdorff and Euclidean distance between actual and reconstructed vertices.

Wang et al [16] generate 14 2D instances of the test faces with different poses and attempt to reconstruct the original 3D geometry using the 14 2D projected faces. The shape and texture reconstruction errors are computed using the Euclidean distance between the recovered and the original synthesized models. In their results they provide the minimum, maximum, mean and standard deviation of the shape and texture errors.

Leclercq et al [9] assess the 3D face reconstruction ability of nine stereo-based 3D reconstruction methodologies. Quantitative results are reported for few stereo pairs of certain subjects and show that in the task of 3D face reconstruction most stereo-based reconstruction methods exhibit similar performance.

The performance evaluation framework reported in this paper, shares some basic components from previous efforts. However, with our work we propose the evaluation of 3D reconstruction techniques in relation to specific sources of variation that may be encountered in faces to be reconstructed, enabling in that way the definition of the sensitivity of different methods to different types of variation. This type of analysis can be of utmost importance for improving certain aspects that make a certain 3D reconstruction technique more vulnerable to reconstruction errors.

## III. Performance Evaluation Framework

The evaluation framework adopted, involves the projection of existing 3D faces into 2D and the subsequent reconstruction of the 2D projected faces, so that it is feasible to compare reconstructed 3D faces against the ground truth. An important aspect is the evaluation of reconstruction accuracy in relation to different conditions, so that it is possible to obtain comprehensive conclusions related to the robustness of a 3D reconstruction method for different test scenarios. In particular, the sensitivity of face reconstruction methods to rotation around the x and y axis (yaw and pitch angles) and sensitivity to the accuracy of point location on faces, are investigated. The most important aspects of the evaluation methodology are discussed in the following sections.

### A. *Image Datasets*

During the tests we use 100 3D faces separated into two groups of 50 faces. During the experiments we use the first group for training and the second for testing and vice versa. It should be noted that the two groups contain 3D faces from different subjects. Typical examples of 3D faces used in our experiments are shown in figure1. About 70000 vertices and the corresponding RGB intensities at each vertex are used for representing the shape and texture of 3D faces used in our experiments.

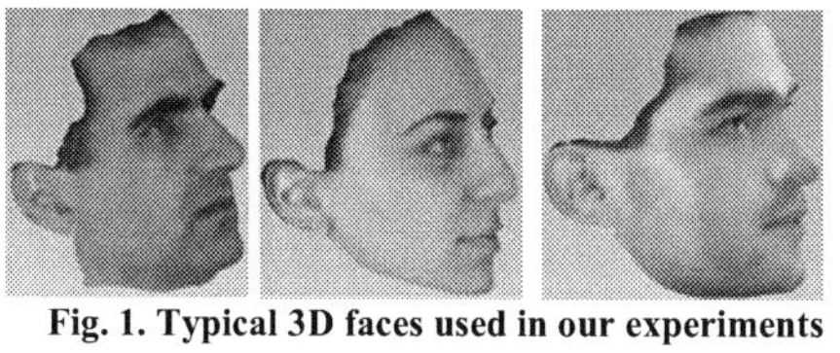

**Fig. 1. Typical 3D faces used in our experiments**

3D reconstruction methods often require the location of a number of landmarks on faces to be reconstructed. For the evaluation experiments described in this paper, we assume that the location of the 68 landmarks shown in figure 2 are provided. In order to define the location of the 68 landmarks on all training and test samples (both 3D and 2D projected faces), we establish the correspondence between the 68 2D landmarks and the vertices from a 3D sample face. Once the correspondence is established, whenever we project a 3D face to the 2D plane we define automatically the positions of the 68 landmarks on 2D projected samples.

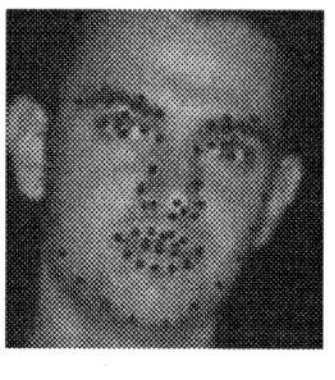

**Fig. 2. Location of the 68 points on a face to be reconstructed**

### B. *Test Conditions*

An important aspect of the evaluation framework is the assessment of the ability of a 3D reconstruction method to deal with difficult cases, such as non-frontal faces. Within this framework we test the following cases:

*i. Rotation around the x axis ( Yaw):* 3D faces from the test set are projected to 2D faces with x-rotation angles ranging from -1 to +1 radians (-60 to 60 degrees). In our experiments we vary the x rotation angle in steps of 0.1 within the minimum and maximum limits stated above. For each rotation angle we reconstruct the corresponding 3D faces so that it is feasible to collect statistics related to the sensitivity of the reconstruction in the presence of rotation of faces around the x axis.

*ii. Rotation around the y axis (Pitch):* The ability to reconstruct faces rotated around the y-axis with angles ranging from -1 to +1 radians, is assessed using a similar framework as the one adopted in relation with the x-axis rotation test.

*iii. Simultaneous rotation around the x and y axis (Yaw and Pitch):* As part of this experiment we rotate 3D faces both around the x and y-axis with angles ranging from -1 to +1 radians so that the reconstruction accuracy as a function of combined x and y face rotation is established.

*iv. Point Displacement:* 3D reconstruction methods often rely on the location of a number of landmarks on the faces to be reconstructed. For the experiments described in this evaluation, we assume that the location of 68 points is already given (see figure 2), hence the results refer to the case that the points are accurately located. Because in real applications, it is possible that some of the points are not located accurately, we include in our evaluation an experiment that aims to assess the sensitivity of 3D reconstruction techniques to the landmark location accuracy. In this context we apply at each landmark a randomly selected x and y displacement prior to the reconstruction process. By varying the maximum displacement allowed it is possible to assess the sensitivity of a reconstruction method to misallocation of facial landmarks. In our experiments we vary the point location error from 0 to 60 pixels.

Visual examples of the scenarios tested in the proposed evaluation framework are shown in figure 3.

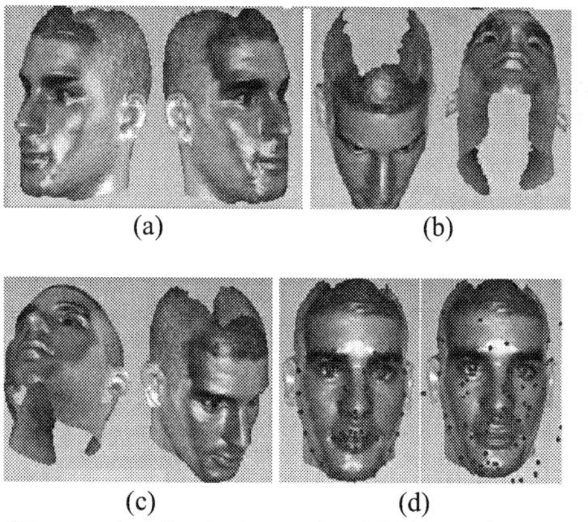

(a) (b) (c) (d)

**Fig. 3. 3D reconstruction test scenarios. (a) Range of yaw angles (b) Range of pitch angles, (c) Range of combined yaw and pitch angles (d) Minimum and maximum random point displacement.**

### C. Similarity Metrics

In our preliminary investigation, reconstructed faces are compared with the original 3D faces using two metrics: A shape-distance (shape_d) metric that assesses the accuracy of the reconstructed geometry and a texture-distance (texture_d) metric that shows the overall difference between the texture of reconstructed and original faces. Since the correspondences between vertices of reconstructed and real 3D faces are known, both the shape and texture distance measures are defined as Euclidean distances calculated using:

$$shape\_d = \frac{1}{M}\sum_{i=1}^{M}\sqrt{[(x_i - x_i')^2 + (y_i - y_i')^2 + (z_i - z_i')^2]} \quad (1)$$

$$texture\_d = \frac{1}{M}\sum_{i=1}^{M}\sqrt{[(r_i - r_i')^2 + (g_i - g_i')^2 + (b_i - b_i')^2]} \quad (2)$$

Where M is the number of vertices, $x_i$ $y_i$ $z_i$ are the coordinates of the $i^{th}$ vertex of the actual 3D face and $x'_i$ $y'_i$ $z'_i$ are the coordinates of the $i^{th}$ vertex of the reconstructed 3D face. Similarly $r_i$ $g_i$ $b_i$ and $r'_i$ $g'_i$ $b'_i$ are the RGB components of the actual and reconstructed texture at the $i^{th}$ vertex.

Prior to the calculation of the similarity measures the two 3D point clouds to be compared are aligned by using rigid transformations (translation, scaling, rotation) so that erroneous results due to the global positioning and orientation of the two set of vertices, are eliminated. Similarly in the case of texture we normalize the reconstructed and actual texture so that both samples have the same mean RGB intensity among all vertices.

During the evaluation process for each test condition we reconstruct all 2D projected faces from the test set and the mean values of similarity metrics among the test set are calculated.

## IV. 3D RECONSTRUCTION METHODOLOGIES

In this section we briefly describe the two 3D face reconstruction techniques to be evaluated using the proposed evaluation framework: The PCA-Based method [14] and the Shape Approximation Method [8]. Both methods utilize (in a different way) a statistical 3D face model [2, 5] as the basis of the reconstruction process.

### A. PCA-Based Reconstruction

PCA-based reconstruction relies on the observation that a 2D and a 3D representation of the same face are correlated [12]. In this approach we first use samples from the training set for training statistical 2D and 3D appearance models [2,5]. We then represent all 3D training samples and the corresponding 2D projected faces into parameters of the 3D and 2D statistical models respectively. Based on this representation we train a PCA-based model that is used as the basis for reconstructing 3D faces. More details of the method are provided in subsequent sections.

*Building a 3D Face Appearance Model:* 3D faces from the training set are used for training a statistical appearance model that models both shape and intensity variations within the training set. The model training procedure involves the alignment of all training samples so that there is one-to-one correspondence between vertices of all samples in the training set [2] and the estimation of the mean 3D face shape and mean texture among the training set. A PCA-based model training procedure similar to the one used by Cootes et al [5] and Blanz and Vetter [2] is used for building a face shape model and an intensity model. Both the shape and intensity models are then used as the basis for generating a combined shape-intensity model based on the Active Appearance Model (AAM) formulation [5]. Within this framework 3D faces can be represented by a small number of 3D model parameters enabling in this way the compact parameterization of 3D faces. Since this representation is reversible it is also possible to generate novel 3D faces consistent with the training set by setting different values to the model parameters.

*Building a 2D Face Model:* All samples from the training set are projected to 2D. All 2D projected faces are then used for training a 2D statistical appearance model [5] using a similar approach as the one used for training a 3D statistical face model. In this framework 2D faces from the training set can be parameterized using a number of 2D model parameters.

*3D Reconstruction Training Phase:* 3D and 2D faces from the training set are represented using the parameters of the 3D and 2D models respectively. A vector (**X**) containing both the 3D (**X3D**) and 2D parameters (**X2D**) of each pair of samples of the same individual are formulated (see equation 3).

$$\mathbf{X}_i = [\mathbf{X}_{3D_i}, \mathbf{X}_{2D_i}] \quad (3)$$

Where $\mathbf{X}_i$ is the combined 3D and 2D model representation of the ith sample. Based on this representation PCA is applied for training a statistical model that learns the correlations between 2D and 3D model parameters.

*3D Reconstruction:* Given a 2D face to be reconstructed, the 2D model representation is estimated and the initial values of the corresponding 3D model parameters are set to zero. A vector (**X'**) containing the current estimate of the 3D model parameters (zero vector initially) and the 2D model parameters is formulated (see equation 3). An optimisation method is applied in order to define an optimum estimate of the parameters of the 3D model, enabling in that way the reconstruction of the 3D face. The cost function minimized by the optimiser is calculated by considering the reconstruction error when a candidate solution **X'** is projected to the PCA space and afterwards reconstructed. In this respect minimum reconstruction error indicates the compatibility between the candidate 3D model parameters and the actual 2D model parameters. Once the optimum 3D model parameters are defined, the shape and texture of the resulting 3D face can be generated.

### *B. Shape Approximation - Based Reconstruction*

The Shape Approximation-Based method requires the existence of a statistical 3D shape model, similar to the one used for the PCA-based method described in section IV.A. The aim of this method is to deform the shape of a statistical 3D face shape model in order to approximate the shape of a face that appears in a face image as defined by the 68 landmarks shown in figure 2. The main steps involved in the reconstruction process are:

*Alignment:* During this step we align the 3D model vertices with the 68 landmarks located on the face to be reconstructed. During this process the 68 vertices that correspond to the locations of the landmarks located on the face are identified and projected to 2D. The overall distance (dis) between the two sets of points is calculated using:

$$dis = \sum_{i=1}^{N} \sqrt{[(\mathbf{X2D}_i - \mathbf{X3D}_i)^2 + (\mathbf{Y2D}_i - \mathbf{Y3D}_i)^2]} \qquad (4)$$

Where N is the number of selected landmarks (i.e 68), **X2D**, **Y2D** are the x and y coordinates of the 68 landmarks located on the face to be reconstructed and **X3D**, **Y3D** are the x and y coordinates of the selected 68 shape model vertices projected to 2D.

The alignment process involves rigid transformations based on the Generalized Procroustes Analysis alignment method [6].

*Model Deformation:* Once the alignment is completed the Yaw and Pitch angles of the face in the given image are estimated by defining the x and y rotation angles required for rotating the vertices of the mean 3D shape, so that the distance *(dis)* between the 68 landmarks and the corresponding 2D projected vertices is minimized. The next step involves the use of an optimization algorithm for estimating optimum values of the 3D shape model that minimize the distance (*dis*). During the process the set of shape model parameters that best approximate the shape of the face in the given image are defined.

*Texture Mapping:* The texture mapping step involves the projection of the reconstructed 3D vertices to the source face image and the retrieval of the texture for each vertex through a sampling process.

Figure 4 shows the results of applying the face reconstruction algorithms under evaluation, to a previously unseen face image.

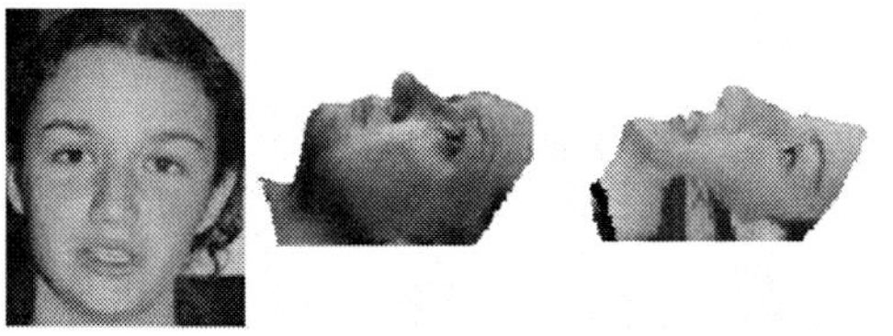

**Fig. 4. Example of 3D face reconstruction. The raw face (left) is reconstructed using the PCA-Based (centre) and the Shape Approximation method (right).**

## V. EXPERIMENTAL EVALUATION

The performance of the two methods is evaluated using the evaluation method described in section III. In particular experiments that assess the sensitivity of the two 3D reconstruction methods in relation with variation in the yaw angle, pitch angle and accuracy of point location were carried out. Figures 5 and 6 show the results of the experiments.

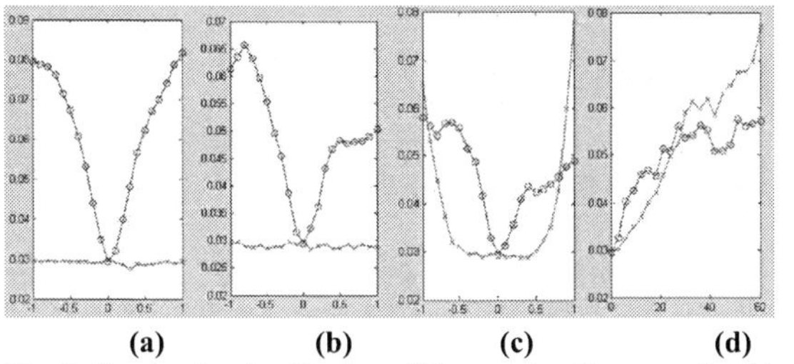

**(a) (b) (c) (d)**

**Fig. 5. Graphs showing the shape 3D reconstruction error for (a) Different yaw angles, (b) Different pitch angles, (c) Combined variation of pitch and yaw angles and (d) Accuracy of point location. The results for the PCA-Based method are shown in blue and results for the Shape Approximation-Based method are shown in red.**

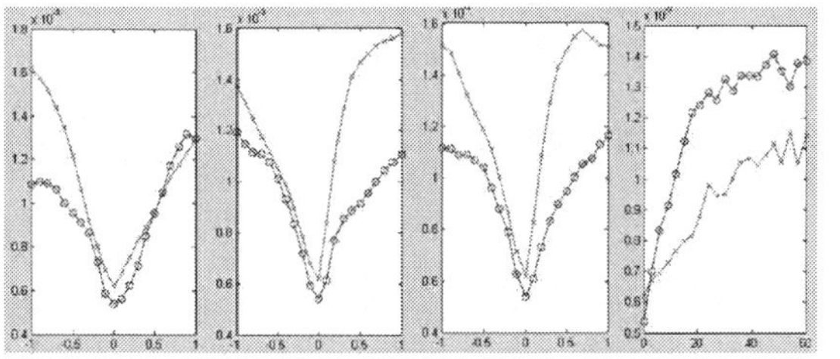

**Fig. 6. Graphs showing the texture 3D reconstruction error for (a) Different yaw angles, (b) Different pitch angles, (c) Combined variation of pitch and yaw angles and (d) Accuracy of point location. The results for the PCA-Based method are shown in blue and results for the Shape Approximation-Based method are shown in red.**

The results show that the Shape Approximation-Based method is less sensitive to reconstructing the geometry of rotated faces than the PCA-Based method. According to the results the shape error for the Shape Approximation-Based method is almost constant for a wide range of rotation angles. In contrast the PCA-Based method achieves the best performance when dealing with frontal faces, but when dealing with rotated faces the performance of the method decreases.

In the case of the intensity error (see figure 6), the PCA-Based method achieves slightly better performance than the Shape Approximation-Based method. In the case of the Shape Approximation-Based the accuracy of texture reconstruction deteriorates for non-frontal faces since for this method the texture of the 3D reconstructed faces is generated through a sampling process. When dealing with rotated faces it is not possible to obtain the overall face texture through sampling. In the case of the PCA-based method, texture is defined through a model-based generative process, which ensures that a reasonable texture estimate can be obtained even in the cases that the face is rotated. However, the results indicate that both methods cannot produce reasonable texture estimates in the presence of significant rotation of the input face.

According to the results of the point location sensitivity test, the performance of both methods deteriorates as the point location error increases. In the case of the shape error, the Shape Approximation methods performs better for point location errors smaller than 30 pixels whereas the PCA-Based method performs better for more extensive point location error (between 30 to 60 pixels).

## VI. Conclusion and Future Work

We have presented our initial work towards the development of a comprehensive framework for evaluating the performance of 3D face reconstruction algorithms. The use of the framework was demonstrated by evaluating the performance of two 3D reconstruction methods – a PCA-Based method and a Shape Approximation-Based method. The evaluation results obtained can lead to the identification of weak points of 3D face reconstruction algorithms that need to be solved in order to facilitate the development of improved algorithms. We believe that our work in this area will lead to the development of the most promising and robust 3D reconstruction approach that can be used as a substitute for expensive 3D imaging equipment, as part of using 3D face processing techniques in real life applications. In the future we plan to build into the initial work in order to deal with the following issues:

*Improved Datasets:* We plan to develop an upgraded 3D face dataset that can be used for extensive tests. For this purpose a database that contains selections of 3D faces captured using different conditions will be compiled.

*Test Scenarios:* We plan to enrich the test procedure with additional test scenarios. For example we plan to include varying illumination tests, where faces to be reconstructed are illuminated with varying illuminations so that the sensitivity of a reconstruction method against different illuminations is tested. Also we plan to evaluate the reconstruction accuracy for face with varying expressions.

*Comparative Performance Evaluation:* Once the framework is completed we plan to run extensive performance evaluation tests in order to assess the performance of other 3D reconstruction algorithms reported in the literature.

## References


[1] Atick, J.J., Griffin, P.A., Redlich, N.A.,. Statistical approach to shape from shading: Reconstruction of 3D face surfaces from single 2d images. Neural Computation. Vol. 8, 1996, pp. 1321-1340.

[2] Blanz, V., Vetter, T., A morphable model for the synthesis of 3D faces. ACM Siggraph. 1999, pp. 187-194.

[3] Blanz, V., Mehl, A., Vetter, T., Seidel, H. A statistical method for robust 3d surface reconstruction from sparse data. Proc. of Int. Symposium on 3D Data Processing, Visualization and Transmission, 2004.

[4] Bowyer K.W, Chang K., and Flynn P., "A survey of approaches and challenges in 3d and multi-modal 3d+2d face recognition," Computer Vision and Image Understanding. vol. 101, no. 1, 2006, pp. 1–15.

[5] Cootes, T.F., Edwards, G.J., Taylor, C.J,. Active Appearance Models. IEEE Transactions of Pattern Analysis and Machine Intelligence, 23, 2001, pp 681-685.

[6] Dryden, L. Mardia, K.V,. Statistical Shape Analysis. John Wiley & Sons, 1998.

[7] Hu, Y., Jiang, D., Yan, S., Zhang, L., Zhang, H,. Automatic 3D reconstruction for face recognition. IEEE Int. Conf. on Automatic Face and Gesture Recognition, 2004.

[8] Lanitis A, Stylianou,, Reconstructing 3D faces in Cultural Heritance Applications. 14th International Conference on Virtual Systems and Multimedia VSMM, (Full Paper). 2008.

[9] Leclercq, P., Liu J., Woodward A., Delmas, P., 2004. Which stereo matching algorithm for accurate 3d face creation?, Int. Workshop on Combinatorial Image Analysis, Lecture Notes in Computer Science, 2004.

[10] Lee, J., Moghaddam, B., Pfister, H., Machiraju, R,. Silhouette-based 3D face shape recovery. Graphics Interface, 2003.

[11] Lee, J., Machiraju, R., Pfister, H., Moghaddam, B.. Estimation of 3D faces and illumination from single photographs using a bilinear illumination model. Proc. of the Eurographics Symposium on Rendering Techniques. 2005, pp. 73-82.

[12] Reiter, M., Donner, R., Langs, G., Bischof, H. Estimation of face depth maps from color textures using canonical Correlation. Computer Vision Winter Workshop, 2006.

[13] Romdhani, S,. Face Image Analysis using a Multiple Features Fitting Strategy. Ph.D. thesis, University of Basel, 2005.

[14] Stylianou G, Lanitis A, 3D Face and Related Applications. Technical Report, Cyprus European University, December 2007.

[15] Stylianou G, Lanitis A, Image Based 3D Face Reconstruction: A Survey. To appear in the International Journal of Image and Graphics, 2008.

[16] Wang, Y., Chua, C. Face recognition from 2d and 3d images using 3d gabor filters. Image and Vision Computing, vol. 23, no. 11, 2005, pp. 1018--1028.

[17] Wen Z. and Huang T.S, 3d face processing, Kluwer Academic Publishers, 2004.

# Multi Dimensional and Flexible Model for Databases

Morteza Sargolzaei Javan
Amirkabir University of Technology ,Tehran Polytechnic, Iran
Farahnaz Mohanna, Sepide Aghajani
Electronic and IT Engineering Departments, University of Sistan & Baluchestan, Zahedan, Iran.
msjavan@aut.ac.ir , F_Mohanna@hamoon.usb.ac.ir , Aghajani.sepide@gmail.com

Abstract

*This article proposes multi dimensional model for databases suitable for both transmissions of data over the networks and data storage in IT/IS systems. In this model there is no need to predict the complete structure of the databases. The structure of these databases can be changeable in any time as needed. In the proposed model, the structure of a record in a database can also be changeable without changing the structure of the other records. Users in this model are able to create any structure that the other databases such as Relational model are not able to do. Furthermore the structure of the tables in the databases can be expanded in all the dimensions. Therefore, the model can minimize the time of data processing and can optimize the capacity of the employed memory in such databases.*

## I. Introduction

Database is a set of data with special definitions and relations and is designed to be managed by an organization. In order to process the data accurately, they must be stored in databases with predefined structures. Today design and modeling of such structures can be carried out using arithmetical models. Relational model is the most common and the easiest one. However, it has some problems and weaknesses, so specialists have made efforts to develop other models using object oriented algorithms. In this article a new model is proposed which is termed flexible and multidimensional databases. This model is designed with different perceptions on the structure of the databases in order to create complex structures very simply. In addition the model is able to process new operations during designing or even running time of the databases.

The remainder of the article is as following: Different aspects of the Relational model, considering its strengths and weaknesses, are explained in section 2. The implementation and the applications of the proposed model are considered in sections 3-6. The most strength of the proposed multi dimensional databases is introduced in section 7. Finally a conclusion of the article is appearing.

## II. Relational Model's Problems

In Relational model, tables are used to store information [1] (Fig.1). As it can be seen in this figure, a table includes personal information with 3 records where in each record, 5 fields exist. Although, the table has a simple structure but some problems maybe arisen as described in the following:

- To store more information [e.g. websites and email addresses] for some of individuals, a new field for each item needs to be created, which may not be used by all individuals (Fig.2). This leaves a number of unused spaces, which is not optimized regarding the capacity of the employed memory for the table.
- On the other hand individuals in above table may have more than one phone number or mobile number, therefore more spaces needed to record additional data.

| P.K | Name | Family | Mobile | Address |
|---|---|---|---|---|
| 1 | Morteza | Javan | 09150000 | Zahedan |
| 2 | Amin | Javan | 09151111 | Zahedan |
| 3 | Ali | Rezaie | 09152222 | Zahedan |

**Fig. 1: A table Based on the Relational Model**

| P.K | Name | Family | Mobile | Address | Site | Email |
|---|---|---|---|---|---|---|
| 1 | Morteza | Javan | 09150000 | Zahedan | www.msjavan.tk | |
| 2 | Amin | Javan | 09151111 | Zahedan | | Javan.it@gmail.com |
| 3 | Ali | Rezaie | 09152222 | Zahedan | | |

**Fig.2: New fields are added to the table in Fig.1 which are not optimized**

Application of conventional storing databases has so far had such the inevitable problems. If during running time new information have to be added to the database, we should return to the designing step again to correct the initial structure of the database with creating a new structure for new information. Suppose Ali has two phone numbers as shown in Fig.3. Therefore a second table must be added including added fields.

T. Sobh (ed.), *Innovations and Advances in Computer Sciences and Engineering*,
DOI 10.1007/978-90-481-3658-2_28, © Springer Science+Business Media B.V. 2010

| P.K | Name | Family | Mobile | Address | Site | Email |
|---|---|---|---|---|---|---|
| 1 | Morteza | Javan | 09150000 | Zahedan | www.msjavan.tk | |
| 2 | Amin | Javan | 09151111 | Zahedan | | Javan.it@gmail.com |
| 3 | Ali | Rezaie | 09152222 | Zahedan | | |

| F.K | Phone |
|---|---|
| 1 | 2510000 |
| 2 | 2520000 |
| 3 | 2410000 |
| 3 | 2511111 |

**Fig.3: Using Foreign key in the Relational model**

In Relational model (Tabular), to create new table for each multi value field, primary and foreign keys needs to be used to relate them [1] (Fig.3). To link two tables in Fig.3 by the Relational model, commands such as 'join' can be used by which the multi values are joined together by applying all the existing keys in the table. Fig.4 represents the resulting table (i.e. by linking two tables in Fig.3).

| P.K | Name | Family | Mobile | Phone | Address | Site | Email |
|---|---|---|---|---|---|---|---|
| 1 | Morteza | Javan | 09150000 | 2510000 | Zahedan | www.msjavan.tk | |
| 2 | Amin | Javan | 09151111 | 2520000 | Zahedan | | Javan.it@gmail.com |
| 3 | Ali | Rezaie | 09152222 | 2410000 | Zahedan | | |
| 3 | Ali | Rezaie | 09152222 | 2511111 | Zahedan | | |

**Fig.4: linking results of using join command for the tables including in Fig.3**

Without understanding the relations between these new entries and previous ones, increasing the number of new entries in a table maybe accompanied by more difficulties. The memory and the cost of such databases may increase rapidly which makes the perception of the data processing more complex (Fig.5).

| P.K | Name | Family | Address | Site | Email |
|---|---|---|---|---|---|
| 1 | Morteza | Javan | Zahedan | www.msjavan.tk | |
| 2 | Amin | Javan | Zahedan | | Javan.it@gmail.com |
| 3 | Ali | Rezaie | Zahedan | | |

| F.K | Phone |
|---|---|
| 1 | 2510000 |
| 2 | 2520000 |
| 3 | 2410000 |
| 3 | 2511111 |

| F.K | Mobile |
|---|---|
| 1 | 09150000 |
| 2 | 09151111 |
| 3 | 09122222 |
| 3 | 09170000 |

**Fig.5: Using foreign keys for multi values fields**

Using link command for linking tables in Fig.5 leads to the table in Fig.6.

| P.K | Name | Family | Mobile | Phone | Address | Site | Email |
|---|---|---|---|---|---|---|---|
| 1 | Morteza | Javan | 09150000 | 2510000 | Zahedan | www.msjavan.tk | |
| 2 | Amin | Javan | 09151111 | 2520000 | Zahedan | | Javan.it@gmail.com |
| 3 | Ali | Rezaie | 09152222 | 2410000 | Zahedan | | |
| 3 | Ali | Rezaie | 09170000 | 2410000 | Zahedan | | |
| 3 | Ali | Rezaie | 09152222 | 2511111 | Zahedan | | |
| 3 | Ali | Rezaie | 09170000 | 2511111 | Zahedan | | |

**Fig.6: Linking of the tables in Fig.5**

## III. Flexible Tables

Considering above discussion, the main problem of Relational model is its limitations in expanding the tables after entering the new entries. This is arisen as the structures of the tables in this model are 2D and non-flexible. We developed the structure of tables flexible and 3D. The proposed model let us to store information with as many as new entries during information processing. The new entries are stored with new structure and without any limitations mentioned in section 2. The detail of the proposed model is as follows. We assume the initial structure of a table in the model to be illustrated in Fig.7.

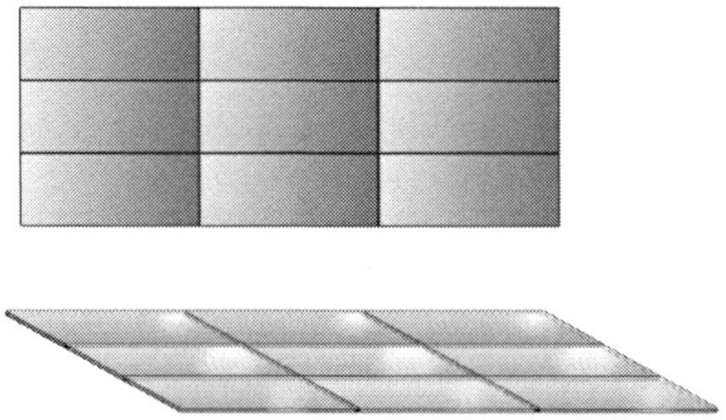

**Fig.7: 3Ds schema of a simple table as an initial table**

In order to add a new field to one of the records in a table, we let the width of the table to be expanded on that field as shown in Fig.8.

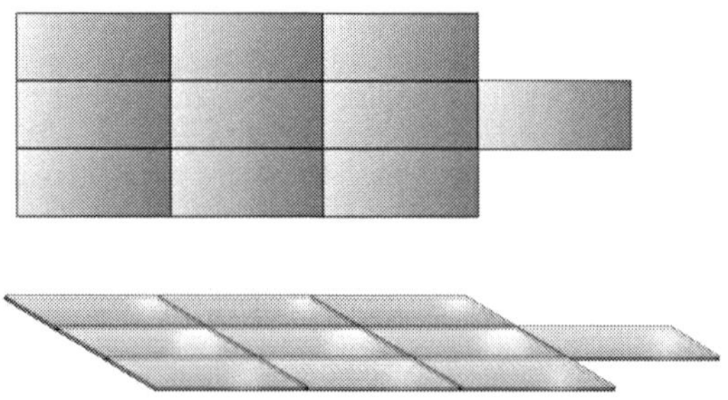

**Fig.8: 3Ds schema of adding a new field to a record in the initial table**

Suppose new value in each field above its current ones need to be added by expanding the height on that field as it is shown in Fig.9.

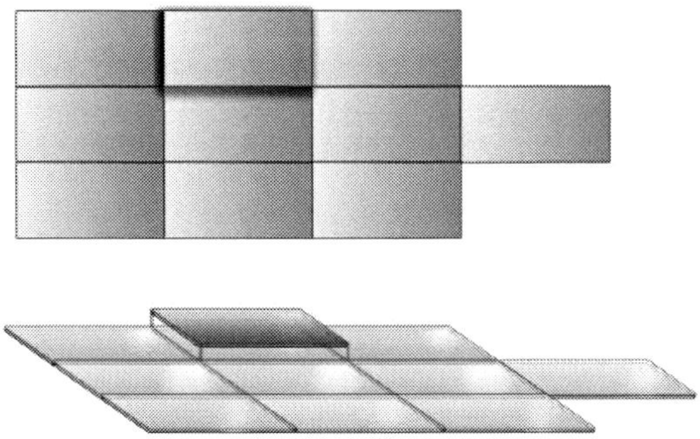

Fig.9: 3Ds schema of adding multi values to a field of a record in the initial table

As it can be seen through Fig.7 to 9, by expanding both the widths and the heights of a table in a database, we can provide flexible and multi dimensional databases. We will discuss that how such a structure for databases will solve all the problems mentioned before.

## IV. Developing the Database with Flexible Tables

By introducing the idea behind the proposed model, we just need to create the initial table including original fields. Then additional fields such as websites and email addresses can then be added to each desired record by applying width expansion. We can also add multi values to each desired field, such as few phone or mobile numbers by creating new cells above the current field and applying the height expansions (Fig.11). The feature of the proposed model for making multi dimensional and flexible databases is as bellows:

- The overhead of redundancy or adding unused spaces has been avoided.
- The overhead of processing 'join' commands has obviated because of the non-cluttered information.
- Because of the non-cluttered information, the analysis and the perception of the structure of databases are simpler.
- For creation of a table in a database, there is no need to predict the whole of the table including fixed number of fields and records.
- For adding new fields during running time of the data processing, there is no need to set new time and to correct the current running time.

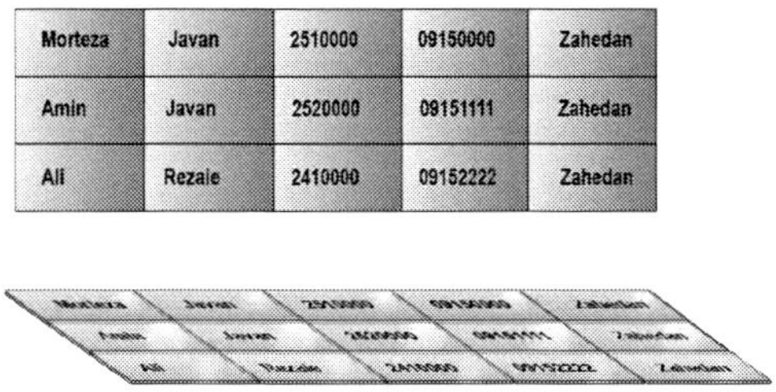


Fig.10: 3Ds schema of a table including multi values base on the Relational model

Fig.11: 3Ds schema of a table including multi values base on the proposed model which is flexible

## V. Implementation of the Proposed Model

In order to store a table with flexible and 3D structure, we have used a structure based on XML programming [4]. Nowadays XML is a markup language such as HTML that is used by many applications and services [5]. This tool can be employed to store information on a database very efficiently [6]. Therefore, in the proposed model, the whole of the desired table are tagged XML technology as following:

```
<PhoneBook>

</PhoneBook>
```

Each record including personal information can then be added to our defined table using the following codes.

```
<PhoneBook>
    <Item>
        <Name> </Name>
        <Family> </Family>
        <Phone> </Phone>
        <Mobile> </Mobile>
        <Address> </Address>
    </Item>
</PhoneBook>
```

The values of each field must also be defined between the start and the end tags.

```
<PhoneBook>
    <Item>
        <Name>Morteza</Name><Family>Javan</Family>
      <Phone>2510000</Phone><Mobile>09150000</Mobile>
        <Address>Zahedan</Address>
    </Item>
    <Item>
        <Name>Amin</Name><Family>Javan</Family>
     <Phone>2520000</Phone><Mobile>09151111</Mobile>
        <Address>Zahedan</Address>
    </Item>
    <Item>
        <Name>Ali</Name><Family>Rezaie</Family>
     <Phone>2410000</Phone><Mobile>09152222</Mobile>
        <Address>Zahedan</Address>
    </Item>
</PhoneBook>
```

So far by this implementation we are able to add new fields to any desired records above. For example we can add new fields of websites and email addresses as follows:

```
<PhoneBook>
    <Item>
        <Name>Morteza</Name><Family>Javan</Family>

    <Phone>2510000</Phone><Mobile>09150000</Mobile>
        <Address>Zahedan</Address>
        <Other>
            <oi name="web">www.msjavan.tk</oi>
        </Other>
    </Item>
    <Item>
        <Name>Amin</Name><Family>Javan</Family>

    <Phone>2520000</Phone><Mobile>09151111</Mobile>
        <Address>Zahedan</Address>
        <Other>
            <oi name="email">javan.it@gmail.com
            </oi>
        </Other>
    </Item>
    ...
</PhoneBook>
```

Similarly, for adding multi values to one of the fields of table shown in Fig.10, we can provide new tags inside each field. So the proposed program will create a multi dimensional and flexible table such as table in Fig.11. This enables us to add new fields and multi values on the databases without changing the running time of the databases.

```
<PhoneBook>
    ...
    <Item>
        <Name>Ali</Name><Family>Rezaie</Family>
        <Phone><vi>2410000</vi>
        <vi>251111</vi></Phone>
        <Mobile><vi>09152222</vi>
        <vi>09170000</vi></Mobile>
        <Address>Zahedan</Address>
    </Item>
</PhoneBook>
```

Note that in proposed model, storing the information according XML, has a lot of benefits including: the independency of the tables structures and data from OS, the independency of the tables structures and data from their applications, and the ability of transferring over the networks.

## VI. The Application of the Proposed Model

Fig.12, 13, and 14 are the snaps from an application based on the flexible tables in our model.

```
<PhoneBook>
    <Item>
        <Name>Morteza</Name>
        <Family>Javan</Family>
        <Phone>2510000</Phone>
        <Mobile>09150000 </Mobile>
        <Address>Zahedan</Address>
        <Other>
            <oi name="web">www.msjavan.tk</oi>
        </Other>
    </Item>
    <Item>
        <Name>Amin</Name>
        <Family>Javan</Family>
        <Phone>2520000 </Phone>
        <Mobile>09151111</Mobile>
        <Address>Zahedan</Address>
        <Other>
            <oi name="email">javan.it@gmail.com</oi>
        </Other>
    </Item>
    <Item>
        <Name>Ali</Name>
        <Family>Rezaie</Family>
        <Phone>
            <vi>2410000</vi>
            <vi>2511111</vi>
        </Phone>
        <Mobile>
            <vi>09152222</vi>
            <vi>09170000</vi>
        </Mobile>
        <Address>Zahedan</Address>
    </Item>
</PhoneBook>
```

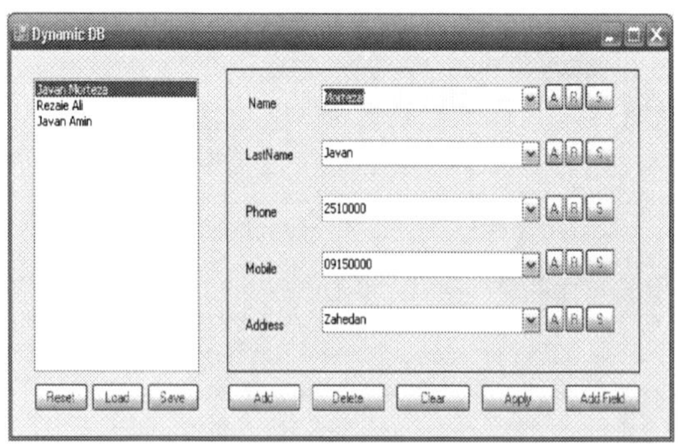


**Fig.12: Developed Interface based on Flexible Tables**

As it is seen in Fig.13, the new fields can easily be added to any record.

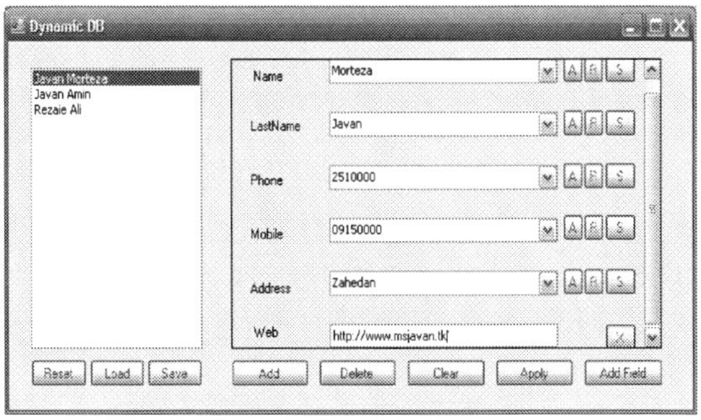


**Fig.13: Adding a new field during running time**

Fig.14 shows the multi values added to any field during running time in the developed interface.

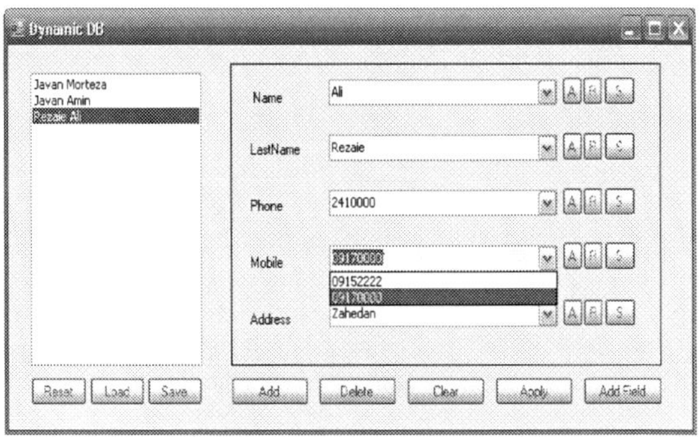


**Fig.14: Adding a new value to a desired field during running time**

## VII. Multi Dimensional Databases

We proposed a method to create flexible tables in a database where the problems arisen by the Relational model was eliminated. However these might be another problem in building the databases. In Fig.9, the new phone numbers added vertically for a person on the table in the database. Suppose each phone belongs to a specific place such as home or office where each phone number may more additional information. In the Relational databases, creation of such this ability is too complex and needs a lot of processing. This problem has already been solved flexible and unlimited dimensions even more than 3Ds. We can easily expand each item in a new dimension without making any complexity in the structure of databases by following. The steps shown below. First each field should be identified by its tags such as:

```
<Phone>2410000</Phone>
```

If the field has multi values, it would be similar to:

```
<Phone>
    <vi>2410000</vi>
    <vi>2511111</vi>
</Phone>
```

If each value needs to be expanded, it should be as follows:

```
<Phone>
    <vi>
        <root>2410000</root>
        <sub name="info">Home Phone</sub>
    </vi>
    <vi>
        <root>2511111</root>
        <sub name="info">Work Phone</sub>
    </vi>
</Phone>
```

More details could be added as follows:

```
<Phone>
    <vi>
        <root>2410000</root>
        <sub name="info">Home Phone</sub>
        <sub name="code">0541</sub>
    </vi>
    <vi>
        <root>2511111</root>
        <sub name="info">Work Phone</sub>
        <sub name="time">
        8-15 every day except fridays</sub>
    </vi>
</Phone>
```

Where each desired item would be expandable using these codes:

```
...
<sub name="info">
  <root>Work Phone</root>
    <sub name="address">
            Zahedan-USB University...
    </sub>
</sub>
...
```

## VIII. Conclusions

In this article, a new approach to make flexible and multi dimensional databases was proposed using XML technology. The model represented new solution to obviate the Relational model's limitations. To obtain multi dimensional databases the model has the ability to create the structures which are impossible to be made by the Relational databases. This model can change the structure of each record, separately, without changing the whole of the table's structure; it also has the ability to change the structure of each field in a table without changing the structure of other fields in this table. Furthermore, the new model can expand a table multi dimensionally in the databases. All these features optimize the memory used by the database as well as the running time.

Clearly that this model is not a replacement for current databases (e.g. Working with huge amount of data), But we could consider it as an extension for current databases infrastructures. Additionally we see that proposed model is useful in a wide range of applications: Data transmission over the networks, developing small scale applications, Prototyping, reporting, analyze and design, online applications, web services, customization and personalization, learning applications and etc. In the future more researches need to be done on the distribution, security, concurrency, and integrity to extend the ability of the model.

## IX. References


[1] Nadira Lammari, Isabelle Comyn-Wattiau and Jacky Akoka, "Extracting generalization hierarchies from relational databases: A reverse engineering approach", Data & Knowledge Engineering, Volume 63, Issue 2, November 2007, From ScienceDirect.com

[2] Andrea Leone and Daoyi Chen, "Implementation of an object oriented data model in an information system", Data & Knowledge Engineering, Volume 22, Issue 12, December 2007, From ScienceDirect.com

[3] Albrecht Schmidt, Florian Waas, Florian Waas and Daniela Florescu. "Why And How To Benchmark XML Databases"

[4] Michael Morrison, Sams, "Sams Teach Yourself XML in 24 Hours", Third Edition, November 14,2005, ISBN: 0-672-32797-X

[5] Microsoft Developer Network's Magazine , "Manipulate XML Data Easily with Integrated Readers and Writers in the .NET Framework"

[6] Elliotte Rusty Harold, "Effective XML: 50 Specific Ways to Improve Your XML;"– Addison Wesley

[7] D. Barbosa, A. Mendelzon,J. Keenleyside,and K. Lyons. ToXgene: a template-based data generator for XML. In Proceedings of the International Workshop on the Web and Databases (WebDB)

[8] Ashraf Aboulnaga, Jeffrey F. Naughton and Chun Zhang, "Generating Synthetic Complex-structured XML Data", Computer Sciences Department, University of Wisconsin - Madison

[9] Microsoft Developer Network –"HOW TO: Implement Common MSXML Tasks in System.xml By Using Visual C# .NET"- English Knowledge Base Articles (PSS ID Number: 330589)

[10] Microsoft Developer Network – "INFO: Roadmap for Programming XML with the DOM-Model Parser in the .NET Framework" - English Knowledge Base Articles (PSS ID Number: 313824)

# Secondary Emotions Deduction from Context

Kuderna-Iulian Bența
Communication Department
Technical University of Cluj-Napoca
Cluj-Napoca, Romania
Iulian.Benta@com.utcluj.ro

Nicoleta Ramona Gibă
Psychology, Personal Development, Social Sciences
Babeş-Bolyai University
Cluj-Napoca, Romania
ramona_giba@yahoo.com

Marcel Cremene
Communication Department
Technical University of Cluj-Napoca
Cluj-Napoca, Romania
Marcel.Cremene@com.utcluj.ro

Ulises Xolocotzin Eligio
Learning Sciences Research Institute
University of Nottingham
Nottingham, United Kingdom
lpxux@nottingham.ac.uk

Anca Rarău
Computer Science Department
Technical University of Cluj-Napoca
Cluj-Napoca, Romania
Anca.Rarau@cs.utcluj.ro

*Abstract*— **Human centred services are increasingly common in the market of mobile devices. However, *affective aware* services are still scarce. In turn, the recognition of secondary emotions in mobility conditions is critical to develop affective aware mobile applications. The emerging field of *Affective Computing* offers a few solutions to this problem. We propose a method to deduce user's secondary emotions based on context and personal profile. In a realistic environment, we defined a set of emotions common to a museum visit. Then we developed a context aware museum guide mobile application. To deduce affective states, we first used a method based on the user profile solely. Enhancement of this method with machine learning substantially improved the recognition of affective states. Implications for future work are discussed.**

***Keywords- affective computing, context awareness, neural networks, basic and secondary emotions.***

## I. Introduction

There is a theoretical distinction between *basic emotions*[1] and *secondary emotions*. In [1], Plutchik proposed that in order to survive, human species developed eight basic (or primary) affective states: acceptance, joy, anger, anticipation, disgust, sorrow, fear and surprise. In this account, combinations of these basic emotions generate 28 secondary emotions, e.g. optimism = anticipation + joy. There is no general agreement in regards of what emotions are basic or secondary. However, the basic-secondary distinction is widely accepted in emotion research.

The field of affective computing has traditionally focused in the recognition of basic affective states (happiness, anger, surprise, fear, sadness, disgust and neutral). In comparison, the recognition of secondary emotions has received less attention. Some advances have been achieved in laboratory settings, e.g. accurate recognition of sadness and amusement from natural expression and personalization over time [2]. But in ever-day life, the contextual information is critical in the definition of secondary emotions e.g. love [3], shame and guilt [4]. As such, the relation between context and secondary emotions is a fairly new interest in the field of affective computing and context awareness mobile technology.

Some researchers claim that the secondary states are exclusively human [5] and culture-dependent [6].

A potential utility of mobile context aware technology is the support of learning in out of school contexts. It allows for control, ownership, learning and continuity between contexts, fun and communication [7]. This utility can be augmented if mobile context aware technology were capable of detecting secondary affective states. It is known that socially intelligent technology can effectively support learning [8] and theoretical studies have shown correlations between affective states and learning outcomes [9]. An affective aware tutoring system that capitalizes on these aspects [10] deals with a set of affective states that are, most of them, secondary (five axis, like anxiety-confidence, frustration-euphoria, with six states each). Moreover, preliminary work [11] focuses on a narrower set of secondary affective states relevant to learning: frustration, confusion and boredom.

[1] The terms emotion and affective state are used interchangeably throughout this paper.

T. Sobh (ed.), *Innovations and Advances in Computer Sciences and Engineering*,
DOI 10.1007/978-90-481-3658-2_29, © Springer Science+Business Media B.V. 2010

This work presents a method to detect secondary emotions from a mobile device in a typical context of informal learning: a museum. We depicted a scenario where the recognition of secondary emotions enhances the adaptability of a context aware mobile device to the user's personal needs:

Jane is a sensitive person. As she walks around the art gallery she likes listening to music assorted to her affective state. A specially designed device is detecting her state (like calm, enthusiasm, interest) but finally she agrees or not with the detected value. The system will request through her smart-phone's mobile connection the song that is most appropriate to her current state, from a long list of affective labelled items stored on a server."

The paper is organized as follows: in the next section we present the related work in secondary affective states detection. Section 3 presents our approach to the relationship between secondary emotions and context. The museum experience we set up as realistic contextual framework is described first. Then we explain the methodology used to define the emotions relevant to this environment and its associated contextual factors. The development of the museum guide test bed is presented in Section 4. In Section 5 we present the procedure and results of two secondary emotion deduction methods. The first is based in a user profile. The second method enhances the first one with a neural network. Future work and conclusions are given in the Section 6 and 7, respectively.

## II. Related Work

Works that define the relationship between secondary and basic emotions are scarce [12][13][14] and/or based on experiments with a very few people (five in [12]). Secondary states are expressed in [12] as a weighted linear combination of the basic states. The weight values are determined by giving values between –100 and 100 for each. In recent work [13], it has been proposed a fuzzy logic based method to build a unified personalized 3D affective model that allows the representation of the secondary states in relation with the basic ones. They claim that knowing the percentage of each basic affective state in the current emotion (detectable with the actual devices) one may classify it as one of the secondary states based on a personal affective model.

The basic emotions have been the main focus for the researchers in affective computing. But there are a few studies on secondary emotion detection, usually related to a learning context. For example, in [14] the authors proposed the use of a leap chair with pressure sensors in an attempt to discover the student engagement level (high and low interest and take a break). They found nine postures that are relevant in detecting engagement and used HMM to train the system. They reported 87.6% recognition with data not included in the training set. In another work, [15] a multiple modalities method (facial, postural and from the current activity (gamming)) is proposed for detecting interest (high and low interest, refreshing, bored, neutral and "other"). An average accuracy of 67.8% is obtained by training a HMM with the values obtained from each separate emotion channel.

Frustration or the need for help is detected from multiple sources using specialised devices like a pressure mouse (frustration), skin conductance sensor (arousal), a posture chair motivation), facial-expression camera (head nod or shake, mouth fidgets, smiles, blink events, and pupil dilations) [16].

## III. The Correlation between Emotions and Context

It is common sense that people uses contextual information to figure out which is somebody's current affective state. There are works showing correlations between environmental factors (context) and emotions [17]. But none, as far as we know, has tried to deduce affective states from context using technology.

### A. *The Contextual Framework: A Museum Experience*

In order to have a realistic and yet easy recordable environment we have chosen first year students in IT&C as a target group to visit an art museum like room organized in the University. Given the situation of exploring art in a gallery, we decided to determine what would be the list of the most important states felt by the visitors, the list of factors that induce those states and the measure each factor would have to induce a certain emotion.

### B. *Data Collection and Analyses*

#### *1) Empirical Observations of the Affective States and Contextual Factors in the Museum Experience*

We observed peoples' affective reactions directly and in video recordings (following facial expressions, movements, body postures and gestures as clues) and we had focus group discussions on feelings when engaging in an art experience in different museums.

Finally we had two outputs: a list of affective states and a list of contextual factors relevant to these affective states. First we defined a list of 22 emotions that appeared the most frequently. We started by using 47 POMS [18] states and then we shrank it to a reasonable list of 22 emotions based on interviewing the ten museum visitors.

The list of emotional states was: Agitated, Amused, Discontented, Curious, Sad, Calm, Aggressive, Depressed, Irritated, Frustrated, Hazy, Alone, Worried, Frightened, Furious, Happy, Indifferent, Relaxed, Interested, Bored, Enthusiastic and Stressed. Secondly we defined a list of 18 binary contextual factors.

#### *2) Predominant Affective States in the Museum Experience*

A questionnaire with the list of 22 emotions plus three blank spaces was designed. We had left three free spaces for other states to be filled in. In correspondence with each state a set of intensity should have been mentioned: very low intensity, low intensity, medium intensity, high intensity, and very high intensity. The tested group consisted in group of 31 persons, 16 of them were females and their average age was 21.7 years. Most of them were students with an IT&C background. All of them were Romanian speakers.

Given the fact that we used a number of persons (31) that is statistically relevant for a well defined target group, we think that the following results can be generalized in respect to the group's characteristics: age (19-25), background (IT&C), gender (balanced), students.

The results indicated that only a few states are felt with more than medium intensity, so we calculated a value for each state taking into account the number of occurrences in the 'intense' and 'very intense cases', with a doubled value for the very intense ones, for example, for the state "Relaxed", 10 persons rated this emotion as 'intense' and 5 persons rates it as 'very intense' and the final rate was 10*1+5*2=20. The most rated states were: Curious, Calm, Happy, Relaxed, Interested and Enthusiastic, as depicted in the Table 1.

TABLE I. THE LIST OF PREDOMINANT STATES IN THE MUSEUM EXPERIENCE FOR IT&C STUDENTS

| State | Rating |
|---|---|
| Relaxed | 36 |
| Calm | 31 |
| Curious | 29 |
| Interested | 28 |
| Enthusiastic | 21 |
| Happy | 12 |

The values for "Curious" and "Interested" are almost the same as the two words meanings are very similar. Among the six states we notice there is just "Happy" that is a basic emotion, the others being secondary. Although we have chosen to include happiness as it had a good rating (12) compared to loneliness (the next item in the rating list), which had only 5.

*3) Contextual Factors Influencing Emotions in the Museum Experience*

We set up a questionnaire to relate the aforementioned set of six affective states (table 1) with the set of 18 binary factors determined from empirical observations as explained above. The tested group was different from the one used to define the list of affective states. They were 31 students, 16 of them females, with an average age of 20.23 years and an IT&C background. The participants were asked to relate each factor to each of the emotions by indicating the extent to what a factor would influence the intensity of a given emotion.

The correlations between factors and context are parameters that change according to target group, kind of museum, person, country but the method of deducing the secondary emotions from context is the same. The data obtained was complex and full explanation is beyond the purpose of this paper. So we focus on two global analyses.

In the first analysis we ranked each factor to notice its participation in inducing intense states. Table 2 shows the results. If positive, the sign in front of the ranking indicates a tendency to induce intense states. If negative, it indicates a tendency to reduce intense states. There are some factors that are ranked near to zero, but they increase the intensity in some states while decreasing the intensity of others.

TABLE II. RANKING FOR FACTORS THAT INDUCE INTENSE EMOTIONS IN AN ART EXPERIENCE.

| No. | Factor | Ranking |
|---|---|---|
| 1 | The novelty of the paintings | +10 |
| 2 | The colors in the paintings | +12 |
| 3 | The shapes in the paintings | +9 |
| 4 | The novelty of the author/authors | -10 |
| 5 | The celebrity of one or more paintings | +60 |
| 6 | The celebrity of one or more authors | +37 |
| 7 | A short time for the visit | -5 |
| 8 | An appropriate duration of time for the visit | +45 |
| 9 | A long time for the visit | +44 |
| 10 | A small number of visitors in the same room | +20 |
| 11 | A medium number of visitors in the same room | +6 |
| 12 | A big number of visitors in the same room | 0 |
| 13 | The presence of familiar persons | +33 |
| 14 | The presence of unknown persons | +1 |
| 15 | A sunny day | +19 |
| 16 | A rainy day | -14 |
| 17 | A cloudy day | -23 |
| 18 | A snowy day | +13 |

In the second analysis, we calculated the probability of each emotion intensity to increase as a function of each listed factor according to the following formula, where $i$ is the index for the intensity level (n=5 levels), $j$ is the index for the factor (m=18 factors) and $k$ is the index for the affective state (p=6 states):

$$\Pi_\kappa = \frac{\sum_{i,j=1}^{n,m} rank_{ij}}{\sum_{i,j,k=1}^{n,m,p} rank_{ijk}}. \quad (1)$$

Table 3 shows the results. There seem to be three emotions that predominate among the 18 factors:

- Interested - average intensity 4 (high intensity) in 8 factors from 18
- Curious - average intensity 4 (high intensity) in 5 factors from 18
- Relaxed - average intensity 3.5 (between medium and high intensity) in 4 factors from 18.

TABLE III. THE PROBABILITY FOR EACH STATE TO OCCUR IN THE ART EXPLORING EXPERIENCE AS A FUNCTION OF THE CONTEXTUAL FACTORS

| State | Probability * 100% |
|---|---|
| Relaxed | 15.596 |
| Calm | 8.869 |
| Curious | 25.075 |
| Interested | 31.804 |
| Enthusiastic | 1.009 |
| Happy | 7.645 |

It is interesting to notice that some states like "Happy" and "Curious" are "negatively" influenced (their intensity is diminished from the reference 'medium intensity' value). In this case the intensity is getting to the value 'low', influenced by the novelty of the author, little time to spend and the rainy weather.

The most important impact on the emotions is noticed in the case of celebrity of the panting and the celebrity of the author/s. The time available has "positive" effects on "Curious", "Interest" and "Calm" inducing high intensity, but has a "negative" effect on "Enthusiasm", "Happy" and "Curious" when it is not enough.

The bad weather[2] (rainy, cloudy) impact negatively the "Relax" and "Calm", rated as "low intensity" by many participants. A big number of factors have decreased the values for the intensity of "Relax" and "Calm". We may notice a positive correlation between "Calm" and "Relax" all over the factors.

"Curious" and "Interest" seem to be induced by the novelty of the paintings, the shapes and the celebrity of the authors and paintings and are rated as "high intensity" by the most participants.

## IV. The Museum Guide Test Bed

The work is based on some previous research done in context aware museum guides [19][20]. Our application has two parts: one is a web based application for profiling the user and the other one is the context aware mobile guide. In the following we explain the functionality of each part.

### A. *The User Profile*

We have designed and implemented a web based application for gathering personal data from the users. It is a simple web application that displays static pages for the questionnaire, dynamically generates pages with questions related to the museum content and it stores the results in a MySQL relational database. The server side is written in PHP.

At first the user has to register and then s/he can have password protected access to the private area. Then the user fills up a questionnaire with 18 questions, one for each 18 contextual factors, allowing the user to indicate the correlation between each factor and the six emotions and for each state, five levels of intensity. Finally another page is generated with questions about: a) the list of known artists in the current gallery; b) the list of painting titles; c) the type of colour preference (warm or cold); d)the preferred shape types (dead or alive nature) where the participant has to make a choice.

### B. *The Context Aware Mobile Guide*

The idea was to have a location aware like museum guide for a test bed in order to be able to monitor the values of the context (number of persons in the room, weather, current painting details etc.) and to predict the current predominant state.

The indoor localisation is still a research issue, even if there are already some good solutions [19][21]. The use of an indoor tracking solution for smart phones was not mandatory for our experiment. However we have had recordings on what was the current painting as the user should indicate the next artwork by pressing direction buttons in order the server to send the appropriate content.

The other context elements monitored were:

- the number of persons located in the same room
- the relations between those persons
- weather information
- the duration

The experimenter would have to set the values for the weather, relations between the persons and the maximum duration of time for the experiment on a web based GUI.

The information about the number of persons simultaneously present in the system was deduced from the number of users currently logged in.

Basically we developed a simple client-server application. On the client side, we had a J2ME application and we used two smart phones (Nokia 6630 and N95) to deploy it. The user entered the name and password and then navigated the content as walking around the room. The pictures where included in the deployment file, but a network connection was needed for the text content to be downloaded locally and displayed and to send the declared affective state back to the server. We used the wireless LAN connection on N95 and the 3G data connection on the Nokia 6630.

On the server side we had a Java application and a very simple communication protocol with the mobile application. For each current painting the mobile application requested to the server a corresponding content. The server accessed the database and sent it. On the other hand a declared affective state was sent to the server and stored in the *Event* table in the database, in addition to current context elements and user id.

## V. Deduction of Secondary Affective States

We proposed two methods for the deduction of the current affective state: the first one is based on the user profile, extracted from the questionnaire (that the user had to fill before using the guide). The second method consists of improving the first one with an expert system. It learns to predict de affective state taking into account the current contextual factors and the users preferences existing in the profile.

### A. *The Profile-based Inference method*

In order to predict the current affective state, we extracted the current contextual factors as a vector F with 18 binary values (for instance given the number of persons is 3 then the value for the $10^{th}$ factor is 1, and for the $11^{th}$ and $12^{th}$ is 0). The users profile extracted from the questionnaire is a three dimensional matrix (factor, affective state, intensity of the affective state) named P. We multiply the F and P matrix and obtain a two dimensional matrix (affective states and intensity). The predicted state would be the one that is the most intense (by adding all the intensity values for one affective state).

[2] The experiments took place in Romania in winter and spring 2008.

*1) Results of the Profile-based inference method*

We recorded 231 events (current contextual factors and the declared state and its intensity) for 8 persons in different test scenarios: single experience, small group (2-5 persons), big group. In some case we simulated time pressure (by reducing the time left for the visit). The weather has been different in the three day when we ran the tests, even if we did not cover all possible situations. The exposed paintings where different as well: some of them were famous because of the content (i.e. “The birth of Venus”) or of the author (i.e. Paul Cezanne), but usually both (i.e. Leonardo da Vinci, “Mona Lisa”). Variations were in terms of content (landscapes, people) or colours (warm and cold), too.

Using the profile extracted from the questionnaire we obtained 40 correct predictions from 231 events, that is 17.316%.

In order to have a reference to compare this result with, we calculated the probability to guess the current affective state: given the fact that the user is in one state (let’s say Happy), the probability of a correct guess is 1/6. The probability for a user to have as current states one of the six states is also 1/6. So the probability of guessing the correct current state, not knowing the current states is 1/6 * 1/6 = 1/32. That is 3.125%.

Comparing the percentage of correct prediction based on the user profile (17.316%) with the percentage of a correct guess (3.125%) we may say it is significant, but not satisfactory.

*B. The Neural Network-based Inference*

We propose to use an expert system in order to increase the prediction rate for the current affective state. We decided to use a neural network, which turned out to be a good choice as you may see in the results section. The first attempt was to use a simple perceptron with 18 factors+30 values extracted from the profile as input and the six affective states as output. We tested the configuration in Matlab. The results were not satisfactory because the big number of inputs. So we designed and tested a multilayer perceptron. Following is a part of the code with the optimum values for the parameters:

```
%% create MLP
Ninp = 48;
Nhid = 8;
inp_domains = repmat([0 9], Ninp, 1);
net = newff(inp_domains, [Nhid Noutp], {‘tansig’ ‘purelin’});
net.trainParam.epochs = 150;
%% train net
net = train(net, inputs, outputs);
```

The number of inputs was 18+30 = 48, the number of hidden layers was 8 (Nhid) and we used *newff* function for the feed forward multilayer perceptron with sigmoid and linear step functions. The number of optimal epochs was 150.

We trained the network with a set of 200 recordings obtained from 8 users. Each recording was a pair of values (input, desired output). As input we had a vector with the actual values for the 18 binary factors and a set of 30 values obtained from the matrix product F and P. The desired output was the declared state (one of the six).

*1) Results of the Neural Network Method*

We have trained the designed neural network, described in Section IV.C.2, with 200 of the 231 recording and we tested it with the remaining 31 values for the percentage of correct prediction.

The results indicate 11 correct predictions out of 31, that is 35.484%. Even if this result is not as good as expected, it is a good improvement compared to the one obtained with the profile based solution.

We also investigated other classification methods. A comparison of the values obtained for the experimental data[3] is represented in Fig. 1.

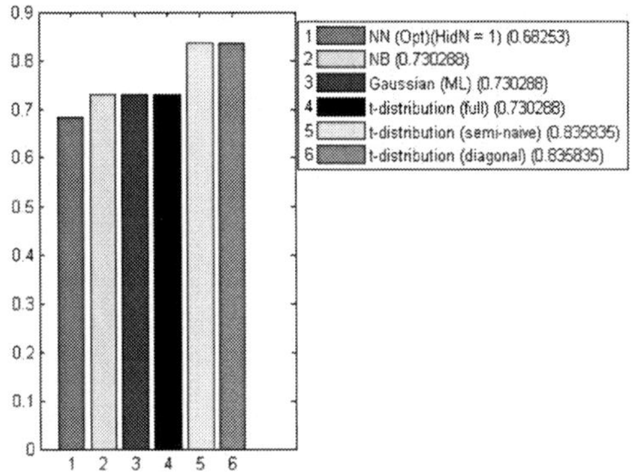


Figure 1. Comparison between different machine learning techniques for the experimental data (the error rate is represented on the y axis)

The input was 200 recordings, 48 input values and 6 classes for the output. The *Naïve Based Classifier* have had 0.730288 classification rate, the *Gaussian Classifier* 0.730288 and for the *t-dist* there were three situations:

a) *full* error rate = *0.730288*
b) *semi-naïve* error rate = *0.835835*
c) *naive* error rate = *0.835835*

The optimal value was obtained by the *Neural Network Classifier* with 8 hidden nodes in one hidden layer. The value obtained is even better then we previously found, that is *0.68253* error rate.

We also performed a SVM[4] test on the data with the following parameters:

```
kernel=‘gaussian’;
kerneloption=1;
C=10000000;
verbose=1;
lambda=1e-7;
nbclass=6;
```

The results indicate a 15.50% of correct classification rate. That can be explained as our inputs are in a large number (48).

---

[3] We used a software tool from http://www.patternrecognition.co.za

[4] We used a software from http://asi.insa-rouen.fr/enseignants/~arakotom/toolbox/index.html

## VI. FUTURE WORK

There are two main directions for future work: one concerns the specification of the secondary emotion recognition model. The other implies the machine learning technique.

In designing the model we plan to extend the user profile with some new parameters like the users' personality type and to use them as input for the neural network.

We think the machine learning could be improved if the user would be able to give feedback regarding the system's prediction correctness in deducing his/her affective state. We plan to develop our application to allow online machine learning based on neural networks and user feedback, for instance by using the GARIC [22] neural fuzzy reinforcement system.

## VII. CONCLUSIONS

In this paper we reported the results of a method for detecting secondary affective states from contextual information and personal profile in relation to a museum experience. On the basis of systematic observation in a realistic environment and self-reports, we defined a set of affective states relevant to the museum experience and defined the contextual factors associated with these states. We used these data to model how the contextual factors contribute to increase or decrease the intensity of people's affective states in a museum. Our results in this regard confirm those obtained in previous research on affect, where it has been found that a person's affective states are sensitive to the characteristics of the context.

Moreover, this approach to model the relation between secondary affective states and context allowed us to deduce the user's current affective state. We proposed two methods: the first one based on user profile solely. The second method improved the first one with a machine learning technique. We obtained 17.3% correct recognition rate with the profile based method and 35.5% using feed-forward multilayer perceptron with one hidden layer as machine learning technique.

We look forward to increase the deduction power of our method by adding new features in the user profile and by using the user's feedback for online machine learning.

## ACKNOWLEDGMENT

We thank our students (A. Fatiol, O.Litan, R.M. Cimpean, E. Ciurea, M. Herculea) for their help in developing the web based and the J2ME application and to all participants in the tests.

This work benefit by the support of the national contract PN2 Idei number 1062 and CNCSIS type A number 1566.

## REFERENCES

[1] R.A. Plutchik, "A General Psychoevolutionary Theory of Emotions. In: Kellerman", R.P.H. (ed.): Emotion: Theory research and experience, Vol. 1, 1980, pp. 3-33.

[2] J. Bailenson, E. Pontikakis, I. Mauss, J. Gross, M. Jabon, C. Huthcerson, Nass and O. John, "Real-time classification of evoked emotions using facial feature tracking and physiological responses", International Journal of Human-Computer Studies, 66(5), 2008 , pp.303-317.

[3] J.B. Nezlek, K. Vansteelandt, I. Van Mechelen, P. and Kupens, "Appraisal-Emotion relationships in daily life", Emotion, 8(1), 2008, pp. 145-150.

[4] E.M.W. Tong, G.D. Bishop, H.C. Enkelmann, P.Y. Why and M.S. Diong, "Emotion and appraisal: A study using ecological momentary assessment", Cognition and Emotion, vol. 21(7), 2007, pp.1361 – 1381.

[5] S. Demoulin, J.-P. Leyens, M.-P. Paladino, R. Rodriguez-Torres, R.-P. Armando and J.F. Dovidio, "Dimensions of "uniquely" and "non-uniquely" human emotions", Cognition and Emotion, 18(1), 2004, pp. 71-96.

[6] R. Harré, "An Outline of the Social Constructionist Viewpoint", Harre, R. (ed.), The Social Construction of Emotions. Blackwell Publishing, Oxford and New York, 1989, pp. 2-13.

[7] K. Issroff, E. Scanlon, and A. Jones, "Affect and mobile technologies: case studies", Proc. The CSCL alpine rendez-vous, workshop 2: Beyond mobile learning, Trinity College, 2007.

[8] N. Wang, L. Johnson, R. Mayer, P. Rizzon, E. Shaw and H. Collins, "The politness effect: Pedagogical agents and learning outcomes", International Journal of Human-Computer Studies, vol. 66(2), 2008, pp.98-112

[9] B. Kort, R. Reilly and R. Picard, "An Affective Model of Interplay between Emotions and Learning: Reengineering Educational Pedagogy-Building a Learning Companion", Proc. Second IEEE International Conference on Advanced Learning Technologies (ICALT'01), 2001.

[10] A. Kapoor, S.Mota and R.W. Picard, "Towards a learning companion that recognizes affect", In Proc. AAAI Fall Symposium, 2001.

[11] S.K. D'Mello, S.D. Craig, B. Gholson, S.Franklin, R. Picard and A. Graesser, "Integrating affect sensors in an intelligent tutoring system", Proc. Affective Interactions: The Computer in the Affective Loop Workshop at 2005 International conference on Intelligent User Interfaces, AMC press, New York, 2005.

[12] T. Yanaru, "An emotion processing system based on fuzzy inference and subjective observations", Proc. 2nd New Zealand Two-Stream International Conference on Artificial Neural Networks and Expert Systems (ANNES '95), Dunedin, New Zealand,1995, pp. 15-21.

[13] K.-I. Benţa, H.-I. Lisei and M. Cremene, "Towards a Unified 3D Affective Model", Proc. Consortium Proceedings of International Conference on Affective Computing and Intelligent Interaction (ACII2007), Lisbon, Portugal, 2007.

[14] S. Mota and R.W. Picard, "Automated posture analysis for detecting learner's interest level", Proc. CVPR Workshop on HCI, 2003.

[15] A. Kapoor, S. Mota and R.W. Picard, "Towards a learning companion that recognizes affect", Proc. AAAI Fall Symposium, 2001.

[16] W. Burleson, "Affective learning companions: strategies for empathetic agents with real-time multimodal affective sensing to foster meta-cognitive and meta-affective approaches to learning, motivation, and perseverance", PhD. thesis, Massachusetts Institute of Technology, 2006

[17] R. Plutchik, The Psychology and Biology of Emotion. Harper Collins New York, 1994.

[18] D.M. McNair, M. Lorr, L.F. Droppleman, "POMS : Profile Of Mood States", Educational and Industrial Testing Service, 1971.

[19] P. Lonsdale, R. Beale and W. Byrne, "Using context awareness to enhance visitor engagement in a gallery space", McEwan, T., Gulliksen, J.andBenyon, D. (eds.), Proc. People and computers xix - the bigger picture. Proceedings of HCI 2005, Springer, London, 2005, pp. 101-112.

[20] O. Stock, M. Zancanaro, P. Busetta, C. Callaway, A. Krüger, M. Kruppa, T. Kuflik, E. Not and C. Rocchi, "Adaptive, intelligent presentation of information for the museum visitor in PEACH", User Modeling and User-Adapted Interaction vol. 17(3), 2007.

[21] http://www.ekahau.com/

[22] H.R. Berenji, P. Khedkar, "Learning and tuning fuzzy logic controllers through reinforcements", Neural Networks, IEEE Transactions on , vol.3, no.5, Sep 1992, pp.724-740.

# EMPOWERING TRADITIONAL MENTORING MATCHING MECHANISM SELECTION USING AGENT-BASED SYSTEM

Ahmad Sofian Shminan, Iskandar Sarkawi, Mohd Kamal Othman & Mardhiah Hayati Rahim
Faculty of Cognitive Sciences and Human Development, Universiti Malaysia Sarawak, 94300 Kota Samarahan, Sarawak.
Tel : 082-581 537, *Email : sasofian@fcs.unimas.my*

## Abstract

*The need for higher technology to assist is clearly evident. Decision Support System is part of an information system that focuses on support and decision making management. For this research, the Intelligent Decision Support System Modelling for Mentoring is the proposed research domain and Faculty of Cognitive Sciences & Human Development, UNIMAS is the research location. The main function of the system is to provide a mechanism to match mentors and mentees based on several criteria of mentors suggested in past studies. The criteria included are the age, marital status, education, gender as well as vocational area of the mentors. The research also applies agent technology in its development of the Decision Support System. The ability of the agent to react freely with the surroundings and taxonomy without user instructions will speed up decision making. RAD methodology using Oracle 9i software is the foundation of implementation. The result from this research shows that this system can be introduced to improve the procedures available and to set the path for a more seamless, to have a better candidate selection and effective mentoring program.*

**Keywords** : Decision Support System, Intelligent Support Mentoring System

## 1. INTRODUCTION

The Higher Education Ministry of Malaysia has continuously showed their efforts in providing a dynamic platform for students to develop and upgrade their multi-disciplinary skills and competencies. This is to ensure that Malaysia produce graduates who are able to work competently; thus, contribute to the development and prosperity of the nation. One of the efforts shown is the establishment of The Committee To Study, Review and Make Recommendations Concerning the Development and Direction of Higher Education in Malaysia on 17$^{th}$ January, 2005 [22,23]. Among the recommendations issued by the committee is the need of a mentor-mentee system to be implemented at higher education institutions:

*"The committee recommends that a mentor-mentee system is to be created to provide opportunities for intellectual and socio-emotional counseling to students in the higher education system. The residential system in teaching and learning should be fully utilised for the attainment of authentic, individual, intellectual and scholarly personalities."*

The committee sees the potential benefits that can be gained through the implementation of mentor-mentee system all parties involve especially the students. The committee really emphasizes the balance growth of students both in academic and non-academic which can be potentially realized through guidance giving session by mentor to mentee. It is also believed that mentor-mentee system is capable in helping the students to be better prepared for the real world [7]. This includes a better preparation to enter working life which demands students to do their very best in helping the nation to become a developed country in 2020.

### 1.1 Problem Statement

The e-mentoring system proposed is intended to take the challenge in realizing the recommendation stated by The Committee To Study, Review and Make Recommendations Concerning the Development and Direction of Higher Education in Malaysia. This is an effort to support and help the nation to become a place where the educational area serves as a great centre in developing the students in all aspects through the guidance received from mentor-mentee relationship.

Universiti Malaysia Sarawak (UNIMAS) is one of the higher education institutions that practices a mentoring system in giving a better services to the student. To be more specific, this system is implemented in the Faculty of Cognitive Sciences and Human Development (FCSHD). In tandem to the programs offered under FCSHD which really concern on the positive growth of students in terms of intellectual, psychological, physiological as well as spiritual, all lecturers of this faculty are required to participate in the mentor-mentees activities. It is a good attempt showed by FSCHD to implement the mentoring approach to assist the students while studying at UNIMAS. Nevertheless, the mentoring system is conducted in a traditional way and does not make use of technological support. Thus, the matching process between mentor and mentee in FCSHD is done using random matching method performed by the coordinator of mentor-mentee program. The usage of random matching process will produce a higher percentage of unsatisfactory mentoring relationship which is deemed by both mentor and mentee [11]. In addition,

T. Sobh (ed.), *Innovations and Advances in Computer Sciences and Engineering*,
DOI 10.1007/978-90-481-3658-2_30, 

other researches also stated that a current practice of random assignment of mentees to mentors is similar to a blind date where it only brings a small chance for the relationship to become successful [3].

It is from this scenario which depicted that there is a need for a certain computerized mentoring management execution mechanism that is more systematic and practical. With the current capability of information technology and networking in every governmental organization, there is no room for excuses to refuse the waves of change which will be the driving force in the administration upheaval. To elaborate further into this problem it is proposed that the DSS, a system capable of generating data to determine decisions, be used. It is the author's opinion that there is a need for further research especially in this country on how DSS along with an integration agent can be used effectively in the field of Human Resource Development.

### 1.2 Objectives

The objectives of this research are:-

i. To identify the methods and analysis processes of matching mentors and mentees process

ii. To identify the categories of agents suitable to be applied as a solution mechanism in the domain of the problem faced.

iii. To propose a DSS design to be integrated with agent technology as a tool to simplify mentors and mentess selection process with these attributes: support decision making, support multi-level management, increase effectiveness, and consist of user-friendly interfaces, and uncomplicated development and modification.

### 1.3 Scope of Research

i. The Mentoring concepts which incorporates abroad scope, thus execution of Intelligent Support Mentoring System (ISMES) will only involve several matching criteria of mentors suggested in past studies, included age, marital status, education and gender.

ii. Faculty of Cognitive Sciences & Human Development, UNIMAS was chosen as the implementation case of this research.

iii. The development of ISMES will involve the development of agents for the purpose of information collection and results generation.

## 2. RESEARCH DOMAIN

Previous studies which has been used as references in this research is divided into two main disciplines, DSS and Agent, and mentoring as a case study in the field of HRD. This literature review will present brief descriptions of all three disciplines. However, the main point of attention in this writing is not to explain every type of two main disciplines. A detailed description on this concept can be obtained by referring to the references provided.

### 2.1 DSS and Agent

In a study stated that a series of research on DSS was already initiated since twenty years ago and it had a strong foundation for further study and development which was pioneered by [1],[4]. DSS provides a support for semi-structured and non-structured decisions, which in turn represents a field of diversity consisting of research in Information System Management, Artificial Intelligence, Organizational research and so forth. Among the early definition of DSS is a system that supports and aids people to make decision at the level of management in semi-structured circumstances. The purpose of DSS is to develop the ability and capability of individuals in decision making but not to replace their human considerations [5].

The development in the field of software and hardware technology has helped the growth of research and development of DSS especially in terms of the computer's capability to reach a conclusion just as humans can. Furthermore, DSS technology has stepped into the realm of artificial intelligence which increases its potentials. A system with intelligent qualities is a system that includes methods and techniques which are able to perform intelligent feats. Research on developing DSS with intelligent features or Intelligent Decision Support System (IDSS) was conducted by many researchers [8],[9].The development of IDSS was done by combining intelligent elements such as expert system, knowledge-based system, ambiguous logic and intelligent agent in proposing response suggestions interactively to the users.

The paradigm shift from DSS to IDSS in fact began when there was lacking research in the development of DSS that is able to facilitate higher level cognitive tasks. These tasks involve human brain activities such as logic reasoning, learning and generating ideas that requires human assessment input [12]. The researcher also added that researches on DSS capability issues in performing high-level cognitive tasks have much been addressed which involved the field of artificial intelligence. Furthermore, intelligent features were also described by [5], saying that the intelligence capability is measured when it includes previous understanding of experiences, quick responses to new situations and applied knowledge of surroundings.

Founded on previous studies. It was found that the need for an ISMES combined with artificial intelligence techniques in the world of human resource development has claimed it's placed. The enhancement in this study is more focussed on three aspects of human resource management and human resouce development, which are training and development, employee recruitment and performance evaluation. The review above has also proven that there is a domination of research in the development of knowledge-based DSS compared to other artificial intelligence techniques. This seemingly supports the opinion expressed by [20] whom argued that the human resource management activities are mostly semi-structured and non-structured which enable the use of a knowledge-based system in aiding decision making.

Nonetheless, other variants of artificial intelligence techniques were also explored and applied as supportive management methods in the practices of HRD and HRM. The research conducted with the application of agent techniques for staff recruitment proved that it has the potential to be developed, even to the extent that it was felt as the most successful research conducted compared to five other studies, [5]. Sadly, overall results indicated that there is a lack of study on agent approach application in the execution of ISMES, which will be an initiative for this researcher to apply in the field of human resource development.

### 2.2 Mentoring Concept

The concept of mentoring is widely used and implemented in various aspects of our lives including the field of education. The approach is believed to be a good way to increase mentees' positive perspectives. Mentoring is a nurturing process where a more skilled or more experienced person act as a role model to teach, sponsor, guide, encourage, counsel and assist a less skilled or less experienced person for the purpose of supporting their professional and personal development [2],[10]. However, to be defined specifically in education field, it brings the meaning that educators utilize the concepts of coaching, counseling and assessment while entrusting responsibility for assisting students in their studies as well as their future career [17],[29].

Relationship between mentor and mentee is a very important element to be considered in ensuring benefits can be gained out of the mentoring practices. A positive bond can be achieved by increasing mutual caring and loyalty between mentor and mentee [27]. They then suggested that these two elements can be increased through decreasing the social distance in mentor-mentee relationship.

There are many benefits that will arise through implementation of mentoring practices. It gives benefits to all parties which include mentor, mentee and institution as well. The benefits the mentor will experience are personal satisfaction as the result of assisting mentee to get through any life difficulties, improve his or her confidence, provides a sense of purpose within himself or herself, improved role satisfaction, have better understanding of trust and mutual support as well as revitalize his or her interest in doing jobs [7],[3]. Mentees also gain benefits from this practices which include having a clear career development, obtain personal supports, increase motivation level, ease loneliness and isolation, better prepared for the real world, receive positive reinforcement, feel challenged to explore new ideas as well as receive helps and positive feedbacks from mentor [7],[11]. Not forgotten, the benefits for institution which includes experience development in management area, increasing mentors' commitment towards the institution, effective management of cost as well as enhancement of communication within the institution [38].

The use of mentoring is helpful in managing desired changes within mentees. and capable to ease transitions and ensuring development if only it is responsive to the strengths, values and needs of both the mentor and mentee [32]. Thus, the pairing process for mentor and mentee must be done effectively and appropriately to ensure that positive results from this relationship will arise optimally. In another situastion, not every educator can be effective mentors as mentoring needs a particular collection of skills, attitudes, knowledge and actions to enable them to become effective mentors [10].

There are several criteria that should be looked into when it comes to matching mentors and mentees. The first criterion is the vocational or work area of the mentor where the research output reveals that most people who want to seek advice will tend to choose someone who has expert knowledge of the relevant vocational field with them [29]. This is to ensure that necessary and sufficient knowledge and experiences can be shared with the mentees through sharing sessions between them. Quality knowledge and experiences relevant to mentee's vocational field is really needed in ensuring that the relationship will produce an effective result [6],[11],[13],[15],[25]. Besides that, age also brings accountability in matching mentors and mentees as an older person posses greater wisdom that will be helpful in easing the guidance-giving sessions with their mentees [14],[15],[25]. Mentees usually tend to select a role-model or a mentor who is older then them and this might be led by the natural behaviour of people in respecting their own parents and teachers [25]. Apart from that, gender selection of mentors by mentees is another aspect that needs to be emphasized in order to produce an effective mentor-mentee relationship [13],[14],[16]. Every mentee has different preference in selecting the gender of the person who will become his or her mentor because of the unique abilities, skills as well as approach in acting as a good role-model to the youngsters [28]. Thus, this criterion should not be abandoned when the process of matching mentors and mentees is conducted.

Educational background of a possible mentor in higher education institution must also be evaluated to ensure that mentees will have assurance and confidence whenever they seek advice from their mentor [14]. Another criterion for matching mentors and mentees is the marital status of the mentors which is proved to be important as people tend to feel comfortable seeking advice from someone who is married [15].

In 1997, past research output recognizes that words have different meanings for different people [39]. Usually, words uttered by mentors are meant to support but in contradiction to this, the words are interpreted as challenge from the mentees' point of view. Mentors need to make their message clear so that mentees will not misunderstood them. Thus, mentors need to be able to determine whether support or challenge is needed by the mentee. Both support and challenge are needed by mentee in order to ensure that the mentor-mentee relationship able to grow. Mentors need to have and acquire skills in giving support and challenge in order to be a good role model to their mentees. The skills for support giving are listening, providing structure, expressing positive expectations, serving as an advocate, sharing experiences and making the relationship special. Besides these, there are other skills for challenge giving which involves setting high standard tasks, constructive hypotheses and engaging in discussions [29],[39].

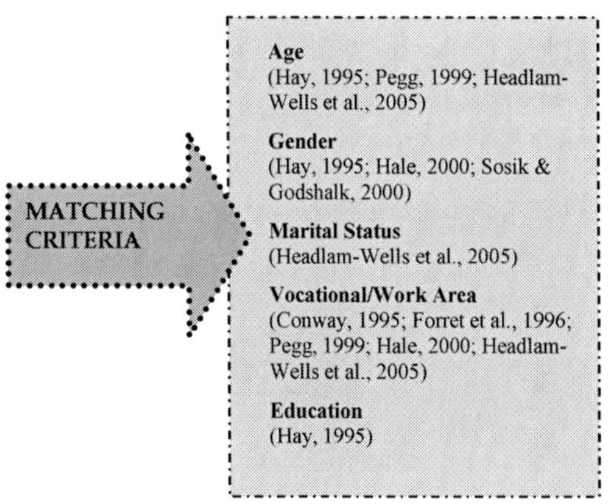


Figure 1.1 Criteria Base Proposition Framework

This section highlighted the important findings from past researchers on the concepts of mentoring, the essential criteria to be looked into in matching mentees and mentors process. The most important element in this chapter is the discussion on the characteristics of mentors where five important criteria have been identified and these criteria will be used as a basis to develop the matching criteria modelling (refer figure 1.1).

## 3. MECHANISM SOLUTION

Design can be defined as a process which uses multiple techniques and principles to define devices, processes or system that represents the real [36]. The system's design was built with the purpose to simplify interpretation process of system requirements as a representation which can simplify code programming activities. Meanwhile, [24] expressed that the purpose of design is to produce a model or entity presentation of the thing to be developed. This prototype system is developed based on a client server computer approach.

The proposal basics of Intelligent Support Mentoring System (ISMES) design was based on the concepts introduced by a few other preceding researchers of this field [27],[35],[36] and [37]. The concepts introduced by these researchers have been elaborated in the reasearch domain review. And so, the proposed design of ISMES consists of three main components, which are the database, model base and user interface. Seeing that ISMES is leaning more towards helping decision making to determine matching between mentors and mentees, the model base components will also have the criteria and conditions required that allows the matching process to take place.

Apart from these components, there is one more vital component which is the decision generation engine. This particular component is the backbone and heart of all three components mentioned above, and it functions to control the overall processes and gives out commands to the data and criteria base, and also forms the interfaces as so the DSS can operate smoothly and achieve its goals. This component will use other components; matching database, criteria base and model base to generate decisions, and then sending this decision to the interface component. So, a diagram of the proposed design of ISMES is shown in Figure 1.2.

The design of this ISMES will involve the integration of agent technology into its execution whereby the agents are developed for the purpose of data extraction and decision generation. The foundation of the DSS and agent design framework is based on the study conducted by [35].

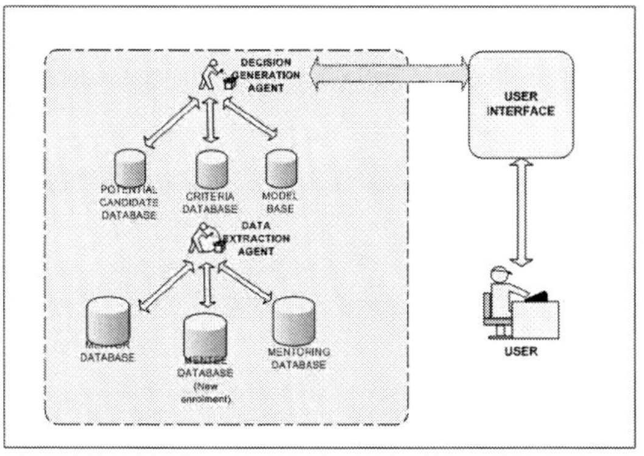


Figure 1.2 Design Proposition of ISMES

## 3.1 Agent Framework

The involvement of a database in a DSS is crucial, and so the process of designing this module must be done thoroughly. Because the functionality of this system is to generate a list of mentors and mentees outputs, a specific database is designed to store the information of candidates which must then be parallel to the 3 main databases.

Data obtained to develop this ISMES are real data which have been used and manipulated for the purpose of generating an output. To make it even more effective, the development of this DSS will include data extraction, where the data extraction agent will be created to extract selected data within the separated database entities.

### *3.1.1 Data Extraction Agent*

Extraction is a program that functions to move data from various sources [35]. The moved data are the data needed by the DSS only and it will be rearranged into a form which the DSS engine can read. The form, state and format of the data sources will meet the measurement in determining the process of extraction. Data extraction process can be done by using many different techniques, whether by modification of programmes, specific programme development or by using artificial intelligence techniques.

For the purpose of data extraction, the approach implemented was an agent-based technique. A successful research on the usage of agent as a method in data collection process in a distributed Database Management System (DBMS) environment was conducted by [8]. One ability of this agent is the ability to learn from its current surroundings. Therefore, the development of agent will require an integration and interaction between the developed model bases within the ISMES. The use of agent as a technique to perform extraction processes and data transformations in a DBMS environment is very significantly reliable and useful [8]. The basic framework presented has combined the design method for agent development and this study will use the approach [8],[34].

### *3.1.2 Decision Generation Agent*

The DSS engine is similar to a human heart, it plays a crucial role in pushing and allowing a DSS to function and achieve the set goals. The transactional medium between the users and the system will be interpreted by the capability of the developed DSS engine. The development of this ISMES decision generation engine will involve developing two categories of decision generation agents, which are the task execution agent and performance score agent. Task execution agent is developed to perform the candidate matching process with the criteria base. While the performance score agent functions to alter the performance score to the format of criteria base range of score. Figure 1.3 shows the architecture of decision generation agent of ISMES. The process of designing the DSS decision generation agent included thorough coding and will only be done once a detailed design is finished. Because a good system is simple and user friendly, the programming of agent must take in consideration the output and interface that are used by the users.

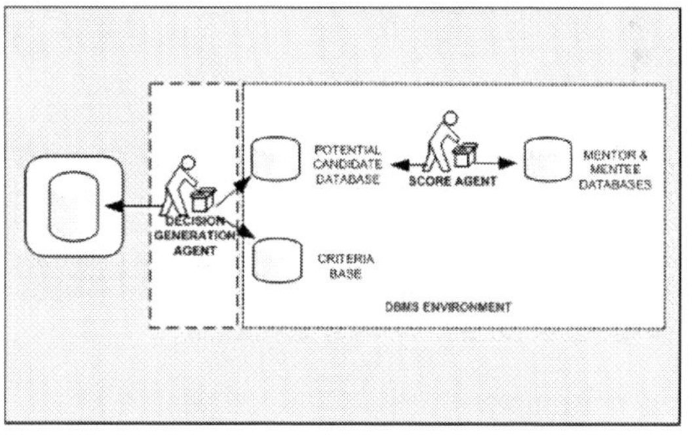


Figure 1.3 Architecture of Decision Generation Agent

# 4. IMPLEMENTATION

This section will pay attention to the implementation of the ISMES prototype based on the case study in FCSHD, UNIMAS. The development of this ISMES is still in its prototype stage and thus, only narrows down on a small scope of users. Even though, the application developed in this study is only in the prototype stage, the data used are real data. From the survey of research conducted, it was found that there were all sorts of software in the market today that is capable and had the features require in developing this intelligent DSS software.

However, the software used in the development of ISMES must be appropriate to the information technological environment of the case study . This is so because there is an importance weighted on the direction of development, such as licensing and existing expertise. The developing software, Oracle Developer 6 and database software, Oracle 9i Microsoft Windows NT version was chosen to develop the prototype.

### 4.1 Agent Implementation Results

#### *4.1.1 Data Extraction*

When record entry or update occurs in the application system's service records, the detection agent in the service record database will automatically go to work, choosing the relevant data and interacting with the sender agent to move the data to the database. Change in data in the record service database will activate the agents to move it automatically into the ISMES database.

#### *4.1.2 Decision Generation*

Every DSS being developed is incomplete without the criteria base. This criteria base is created as the condition for matching process. The condition inputs are based on the criteria set in the model base which uses the criteria (refer figure 1.1). To make the process of data input easier for the users, a code to detect input error is created for every data input boxes. For instance, this input error detection code will pop up during the testing process of components. After going through a few screens, the users will then reach the output screen, this is the interface which will direct the development of ISMES prototype. Because the decision generation is performed by the decision generation agent, the users will only need to select a lecturer staff code to attain a report of the results.

## 5. DISCUSSION

The testing approach implemented for this research is divided into 3 main stages; model efficiency test, components test and user acceptance test. In a computer-based system, apart from the software, there are also other elements like hardware, humans and information which are all involved in a common environment that make the system successful [24]. The summary of results analysis conducted during the testing phase of ISMES prototype indicates that the users agreed on its development because of its simple implementation method which helps Training unit management activities to run smoothly as well as making easy trainee selection. This support from the users is conceived as an encouraging instrument towards the system's implementation in reality. Based on the literature review, the DSS is concluded as a computerized interactive, flexible information system and adjustable developed specially to support the management in finding a solution for their semi-structured and non-structured problems by emphasizing on the support and effectiveness of decision making. Pertaining to this, it is clear that the DSS is indeed capable to act as a mechanism of decision making appropriate in the domain of this research.

Nevertheless, the development of software and hardware technology of today has much helped towards the development and research of this DSS especially with the presence of artificial intelligence technology as a combining element. Research on DSS with the element of artificial intelligence have been conducted by many researchers [5],[18]. The combination of artificial intelligent element with the DSS in generating quality decisions, consequently generating the solution input to the process of decision making. Realizing this matter, the researcher tries to take initiatives to explore the capability of agent technology as an element of artificial intelligence to be integrated into a DSS. Because the DSS itself is capable to solve semi-structured and non-structured problems, just as in this research, it calls for a discussion like this to evaluate the extent of integrating agent with DSS effectiveness compared to the ordinary DSS. The following discussions will include two main issues from the aspects of agent capability as a simplifying tool for data extraction process, and the execution of DSS's decision generation engine.

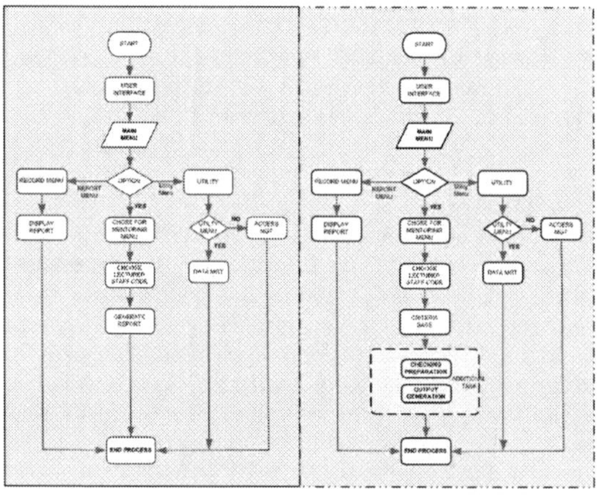

Figure 1.4 Proces Flow Comparison

Based on the main issues discussed, it was found that the integration of agents with DSS gives advantages such as reducing user intervention and time compared to the implementation without integration of agent (refer figure 1.4). It can be concluded here that the exploration of agent's capability for the purpose of data extraction and decision generation engine implementation is an effective approach which can bring about positive implications towards the effectiveness and efficiency of DSS implementation in the practices of human resource management in general. To further support discussion of this research, four comparison features which include selected domain, decision analysis, artificial intelligence technique and form of decision will be the point of referenc. The nature of this discussion has already been elaborated in depths in literature review. Because the research on DSS and agent in the field of training is rare, the following series of discussion is needed to dig out ISMES potentials compared to previous researches [5], [18], [37].

The first argument is from the aspect of decision analysis which shows three other referred studies that did not apply the agent technique. Next, this discussion is continued to other previous studies which used models its implementation. It was found that the research conducted [3] has used a model, differing from other research. The use of this model is rather different in the domain of training and development because the model used was a flight reservation model. In the meantime, in comparing the techniques used, it was found that a knowledge-based system technique was more popular compared to other techniques. While, the form of decision generated by the implementation of researches on DSS, does in fact show a focus for support on the training management processes. Therefore, we can conclude that the implementation of ISMES using a suitable model for the problem domain, and agent-integrated DSS will simplify the process of decision generation.

## 6. CONCLUSION

Certainly the field of information technology does indeed play a significant role in an organization's success. However, changes in the practices of human resource are an issue and challenge in executing the use of software for the purpose of Human Resource Developmentt (HRD). In correspondence to these demands, the development of ISMES is seen as a new dimension which is also a contribution to the world of HRD and HRM research.

## 7. REFERENCES


[1] Ab. Aziz Yusof 2004. *Pengurusan sumber manusia, konsep, isu dan pelaksanaan*. Pearson, Prentice Hall.

[2] Baugh, S. G. and Sullivan, S. E., (2005). Mentoring and career development. Career Development International, Vol. 10 No. 6/7, pp. 425 – 428.

[3] Chao, G.T., Walz, P.M. and Gardner, P.D., (1992). Formal and informal mentorships: a comparison on mentoring functions and contrast with nonmentored counterparts. *Personnel Psychology*, Vol. 45 No. 3, pp. 619-36.

[4] Chestnut, J.A. & Ntuen, C.A 1995. An expert systems for selecting manufacturing workers for training. *Expert systems with applications : An International Journal* 5(2): 130-149.

[5] Cynthia, L. & Webb, H. 2003. Managing training in a technology context. *SIGMIS Conference* 03 *ACM*, hlm 107-115.

[6] Conway, C., (1995). Mentoring in the mainstream. *Management Development Review*, Vol. 8 No. 4, pp. 27 – 29.

[7] Ehrich, L. C. and Hansford, B., (2005). The principalship: how significant is mentoring? *Journal of Educational Administration*, Vol. 44 No. 1, pp. 36 – 52.

[8] Edgar, W., Josef, A., & Wolfgang, E. 2000. Mobile database agents for building data warehouses. *IEEE* 1: 477-481.

[9] Fazlollahi, B., Parikh M.A. & Verma S. 1997. Adaptive decision support systems. http://verma.sfu.edu/profile/adss-97.pdf (4 Julai 2001)

[10] Fletcher, S., (2000). *Mentoring in schools*. Sterling, USA: Stylus Publishing Inc.

[11] Forret, M. L., Turban, D. B., and Dougherty, T. W., (1996). Issues facing organizations when implementing formal mentoring programmes. Gay. B., (1994). What is mentoring? *Education + Training*, Vol. 36 No. 5, pp. 4 – 7.

[12] Gams, M. 2001. A uniform internet-communicative agent, electronic commerce research, 1, *Kluwer Academic Publishers*, hlm. 69-84.

[13] Hale, R., (2000). To match or mis-match? The dynamics of mentoring as a route to personal and organisational learning. *Career Development International*, Vol. 5 No. 4, pp. 223 – 234.

[14] Hay, J., (1995). *Transformational Mentoring*. McGraw-Hill: London.

[15] Headlam-Wells, J., (2004). E-mentoring for aspiring women managers. *Women in Management Review*, Vol. 19, pp. 212 – 218.

[16] Headlam-Wells, J., Gosland, J. and Craig, J., (2005). "There's magic in the web": e-mentoring for women's career development. *Career Development International*, Vol. 10 No. 6/7, pp. 444 – 459.

[17] Johnson, S. K., Geroy, G. D. and Griego, O. V., (1999). The mentoring model theory: dimensions in mentoring protocols. *Career Development International*, Vol. 4 No. 7, pp. 384 – 391.

[18] Kwok, L. F., Lau, W.T. & Kong, S.C. 1996. An intelligent decision support system for teaching duty assignments. *Proceedings on Intelligent Information Systems* hlm. 114-125.

[19] Maimunah Aminuddin. 2002. *Pengurusan sumber manusia*. Kuala Lumpur : Fajar Bakti.

[20] Martinsons, M.G. 1997. Human resource management applications of knowledge-based systems. *International Journal of Information Management* 17(1): 35-53.

[21] Matsatsinis, N.F. & Siskos, Y. 1999. MARKEX : An intelligent decision support system for product development decisions. *European Journal of Operational Research* 113 : 336-354.

[22] Ministry of Higher Education, (2006). *Report: The committee to study, review and make recommendations concerning the development and direction of higher education in Malaysia: Towards Excellence*. Putrajaya, 144.

[23] Ministry of Human Resources of Malaysia Official Website, (2007). <http://jcs.mohr.gov.my/jcs/index.faces#> (2007, August 15).

[24] Power, D.J. 1999. A brief history of decision support systems. http://dssresources.com/history/dsshistory.html (22 Mac 1999).

[25] Pegg, M., (1999). The art of mentoring. *Industrial and Commercial Training,* Vol. 31 No. 4, pp. 136 – 141.

[26] Scandura, T. A., Tejeda, M. J., Werther, W. B. and Lankau, M. J., (1996). Perspectives on Mentoring. *Leadership and Organization Development Journal,* No. 17 No. 3, pp. 50 – 56.

[27] Sprague, R.H. 1993. *A framework for the development of decision support systems : Putting theory into practice*. New Jersey: Prentice-Hall International Editions.

[28] Sosik, J. J. and Godshalk, V. M. (2000). The role of gender in mentoring: applications for diversified and homogeneous mentoring relationships. *Journal of Vocational Behavior,* Vol. 17, pp. 102 - 122.

[29] Stead, R., (1997). Mentoring young learners: does everyone really need a mentor? *Education + Training,* Vol. 39 No. 6, pp. 219 – 224.

[30] Stewart, J. and Knowles, V., (2003). Mentoring in undergraduate business management programmes. *Journal of European Industrial Training*, Vol. 27 No. 2, pp. 147 – 159.

[31] Tannenbaum, S.I., Mathieu, J., Salas, E. & Cannon Bowers, J.A. 1991, Meeting trainees' expectations: the influence of training fulfilment on the development of commitment, selfefficacy and motivation", *Journal of Applied Psychology* 76: 759-69.

[32] Tabbron, A., Macaulay, S., and Cook, S., (1997). Making mentoring work. *Training for Quality*, Vol. 5 No. 1, pp. 6 – 9.

[33] Thayer, P.W & Teachout, M.S 1995. *A climate for transfer model*. Human Resources Directorate: Brooks AFB, Texas.

[34] Traci, J. 1999. A study of autonomous agents in decision support systems. Tesis Dr. Falsafah,.Virginia Polytechnic Institute and State University.

[35] Turban, E., Aronson, J.E. & Liang T.P. 2005. *Decision support systems and intelligent system*. New Jersey: Prentice-Hall International, Inc.

[36] Quinones, M.A. 1997. Contextual influences on training effectiveness. In M.A. Quinones & A. Ehrenstein (Ed.), Training for a rapidly changing workplace. *American Psychological Association*.

[37] Wang, H. 1997. Intelligent agent assisted decision support systems : integration of knowledge discovery, knowledge analysis and group decision support. *Expert Systems with Applications* 12(3): 323-335.

[38] Wilson, J.A. and Elman, N.S., (1990). Organizational benefits of mentoring. *Academy of Management Executive,* Vol. 4 No. 4, pp. 88 – 94.

[39] Woodd, M., (1997). Mentoring in further and Higher Education: Learning from the literature. *Education & Training,* Vol. 39 No. 9, pp. 333 – 343.

# A Method of Evaluating Authentication Mechanisms


Liang Xia[1], Yongjian Yang[2], Yong Wang[3]

1 College of Computer Science and Technology, Jilin University, Changchun 130012, China, xialiang@jlu.edu.cn
2 College of Computer Science and Technology, Jilin University, Changchun 130012, China, yyj@jlu.edu.cn
3 College of Computer Science and Technology, Jilin University, Changchun 130012, China, wangyong@jlu.edu.cn



**Abstract: Based on the analysis and classification of identity factors in application systems, this paper designs a method of quantifying multi factors and an arithmetic of evaluation mechanism. It also adopts the example of application system to evaluate authentication intensity (means the degree of authentication, i.e. strong or weak authentication) and its cost. And the results of the application testify the method is applicable.**



**Key words**: Authentication, Identity Factor, Intensity, Cost, Evaluation


## I. INTRODUCTION

The number of fraud aiming to network is increasing dramatically with the expansion of modern technology and the global superhighways of communication, which results in the loss of billions of dollars worldwide each year[1]. Identity fraud is closely related to authentication.

There are some kinds of identity factors such as username, password and ID card etc. Depending on different identity factors, different authentication methods have different authentication intensity (means the degree of authentication, i.e. strong or weak authentication) and cost.

The evaluation of the authentication intensity needs a specific quantification method. This research tries to construct a quantifiable evaluation method of identity factors based on the literature review of multi factors. The feasibility of this method is then assumed.

## II. AUTHENTICATION MECHANISMS

In this section, we will first review the objects of authentication. Then main factors of authentication and a method of authentication based on main factors are introduced. We then analyze multi factors extended from the main factors.

### A. OBJECTS OF AUTHENTICATION

Usually, Customer's authentication aims to ensure only the real customer can access to the system. Identity fraud may happen possibly once pseudo customers access to the system, which can lead to the loss of the real customers' benefits. Patrick's study, which concentrates on financial fraud done to obtain money from bank accounts and/or to commit credit card, loan, and mortgage fraud[2] can prove this assumption.

The following parts (II.B, II.C and II.D) will introduce the user's identity and how to compose them for further research.

### B. MAIN FACTORS OF AUTHENTICATION

User authentication in computing systems traditionally depends on three factors: something you have, something you are, and something you know[3]. These three kinds of factors are main factors of identity.

### C. A TRADITIONAL METHOD OF AUTHENTICATION

Generally, valid authentication needs at least two kinds of factors. For example, bank machines provide two-factor authentication by requiring a bankcard (something the user has) and a PIN (something the user knows). Business networks may require users to provide a password (something the user knows) and a random number from a security token (something the user has)[4].

### D. MULTI FACTORS

Customer's identity factors can also be extended to some other aspects, such as:


T. Sobh (ed.), *Innovations and Advances in Computer Sciences and Engineering*,
DOI 10.1007/978-90-481-3658-2_31,

- Location-based authentication, such as that employed by credit card companies to ensure a card is not being used in two places at once.
- Time-based authentication, such as only allowing access during normal working hours[4].

According to my assumption, the average trade quantity, which can be found in e-commerce system, may be considered as a multi factor.

In conclusion, the more identity factors are referenced for authentication, the stricter authentication is. Improving authentication methods, including using multi-factor authentication, will be better than simple systems[2].

## III. A QUANTIFICATION METHOD

In this section, we will introduce a method of quantifying the intensity and cost of authentication with the consideration of the analysis of identity factors.

### A. THE CLASSIFICATION OF IDENTITY FACTORS

An ideal method should involve every identity factor, but it is not feasible to do so. Therefore, a set of identity factors will be selected relevant to our study as follows: username, password, ID card, questions-based factor, image-based factor[5,6], fingerprint[7], signature, voice recognition[8], face [9], hand or finger geometry or ear shape [10], unique bio-electric signals, the characteristics of mouse usage patterns [12], key-stroke latencies or dynamics [13, 14], IP of login, time of login, amount of trade, telephone number, email address and on-line security device.

According to the trait of every identity factor, the factors above will be divided into 4 groups.

**Primary Factors**: A set of identity factors (something the user knows) such as username, password, PIN, question-based factor, image-based factor, etc, are seen as the most applicable identity factors for authentication. For example, most commerce systems require username and password to access. And the question-based factor requires customers to answer some private questions such as date of birth, it is also very common for security.

**Senior Factors**: A set of identity factors (something the user is or does) such as fingerprint, DNA sequence, signature, voice recognition, face, hand or finger geometry or ear shape, unique bio-electric signals, etc. Generally, senior factors need extra hardware to support authentication. For example, customers need to input their fingerprint on the special device when they log in system. The authentication intensity is strong using this method because it is hard to theft and copy, at the same time, the cost is very much. It has a disadvantage too: the hardware may be unstable. For instance, biometrics has a perceptible failure rate with respect to permission, with false negatives sometimes being experienced more often than is acceptable [11].

**Dynamic Factors**: This kind of identity factors are defined from some dynamic regularities and habits which are extended from three main identity factors mentioned in II.B. For example, it can be based on the characteristics of mouse usage patterns, key-stroke latencies or dynamics; or signature dynamics, login IP Address , login time , amount of trade etc. The authentication is made based on the records of customer's previous behaviors. It can first design a behavior model according to the memory and summary of customer's special behaviors of access. The system can judge whether the customer's behavior is normal or abnormal according to this model. A warning can arise once an exceptional behavior happens. This method is assumed to have strong authentication intensity and high security because it is difficult to copy the customers' physical records. However, this method can most possibly prolong period of authentication.

**Multi-channel Factors**: A set of identity factors (something the user has) such as telephone/mobile, email, on-line security devices [15], etc. Through these various channels, authentication may be intensified. For example, dynamic one-time password will be sent to customers' email box or mobile phone when they log in the system. This way can solve the problems arose from customers forgetting their own passwords leading to failed login. And it is safer compared to single on-line password system. The fraud can happen only if the deceiver acquire all of the customer's information: username, password, email box, landline and mobile phone etc. But this method is not always available because it strongly relies on extra hardware or special software. Furthermore, it is also possible for the client and server machine to get out of synch, which requires intervention to re-synchronize [16].

To be summarized, all kinds of methods have their advantages and disadvantages, there is not a dream method which can be extremely safe up to now. Ideally various identity factors should be considered in the ongoing development of the authentication.

## B. THE QUANTIFICATION OF IDENTITY FACTORS

The quality of an authentication method is much based on its intensity: the stronger the intensity, the better the method, and vice verse. Therefore, in order to evaluate the method of authentication, we first need to quantify the identity factors. This is a quantitative analysis of the intensity of authentication.

Gilb proposes the three steps for quantification [17]:

- Identification of the quality concerns.
- Assignment of scales of measure to these quality concerns.
- Identification of required quality levels.

According to the three steps above, we will quantify identity factors as follows:

The intensity of Primary Factors (P);

The intensity of Senior Factors (S);

The intensity of Dynamic Factors (D);

The intensity of Multi Channel (H);

The number of Channels (M);

**Quantification of Primary Factors:** Pi depicts the intensity of No.i primary factor. So if No.i primary factor exists, Pi = 1, otherwise, Pi = 0. The sum of all Pi is marked as $\sum Pi$. For example, P1 depicts the intensity of password factor and, P2 depicts the question-based factor, then the intensity sum of two identity factors should be calculated:

$$\sum Pi = P1 + P2 = 1 + 1 = 2$$

This means the result of quantification of authentication method which has two primary identity factors will get 2.

**Quantification of Senior Factors:** Sj depicts the intensity of No.j senior factor. So if No.j senior factor exists, Sj = 2, otherwise, Sj = 0. The sum of all Sj is marked as $\sum Sj$. For example, S1 depicts the intensity of fingerprint factor and, S2 depicts the voice recognition factor, then the intensity sum of two identity factors should be calculated:

$$\sum Sj = S1 + S2 = 2 + 2 = 4$$

This means the result of quantification of authentication method which has two senior identity factors will get 4.

**Quantification of Dynamic Factors:** Dk depicts the intensity of No.k dynamic factor. So if No.k dynamic factor exists, Dk = 2, otherwise, Dk = 0. The sum of all Dk is marked as $\sum DK$. For example, D1 depicts the intensity of IP factor and, D2 depicts the time factor, then the intensity sum of two identity factors should be calculated:

$$\sum DK = D1 + D2 = 2 + 2 = 4$$

This means the result of quantification of authentication method which has two dynamic identity factors will get 4.

**Quantification of Multi Channels:** Mm depicts the No.m channel. So if No.m channel exists, Mm = 1, otherwise, Mm = 0. The sum of all Mm is marked as $\sum Mm$. For example, M1 depicts the channel of primary factor and, M2 depicts the telephone factor channel, then the number of channels should be calculated:

$$\sum Mm = M1 + M2 = 1 + 1 = 2$$

This means the number of channels will get 2.

Also, Hn depicts the intensity of No.n channel. Generally, the information which is send through extra channel is always password, so if No.n channel exists, Hn = 1, otherwise, Hn = 0. The intensity sum of all Hn is marked as $\sum Hn$. For example, H1 depicts the intensity of telephone channel and, H2 depicts the email channel, then the intensity sum of two channels should be calculated:

$$\sum Hn = H1 + H2 = 1 + 1 = 2$$

This means the result of quantification of authentication method which has two channels will get 2.

## C. THE COST OF DIFFERENT QUANTIFICATIOIN METHOD

The costs of authenticating different identity factor (C) can be quantified as follows:

Cn depicts the cost of the No.n identity factor. So if the No.n identity factor exists, Cn = 1, otherwise, Cn = 0. Because accessibility also applies to levels of technical skills and literacy as well as the quality of the user's equipment [18], the cost will be spent more because of the extra hardware. Therefore, I here primarily define Cn as 3 on condition that one kind of factors need extra hardware. The sum of all Cn is marked as $\sum Cn$. For example, C1 depicts the cost of password factor. Because the password factor is implemented by software, so C1 = 1. Also, if customer's fingerprint will be authenticated, the extra cost will be C2, then C2 = 3. Thus the cost of two quantifications should be calculated:

$$\sum Cn = C1 + C2 = 1 + 3 = 4$$

D. QUANTIFICATION ARITHMETIC

Based on the evaluation of authentication intensity (Section III.B) and the cost of authenticating identity factors (Section III.C), arithmetic of intensity of authentication can be designed as follows:

Here, IA depicts intensity of authentication, so

$$IA = (\sum Pi + \sum Sj + \sum Dk + \sum Hn) * \sum Mm \quad (1)$$

The intensity of authentication is related to primary factors, senior factors, dynamic factors, information from multi channels and the number of multi channels. It is linear relationship between intensity of authentication and primary factors, senior factors, dynamic factors and information from multi channels. Also, it is multiple relationship between intensity of authentication and the number of multi channels. Because the extra channel can overtly increase the authentication intensity, the multiple relationship should be reasonable.

And intensity/cost of authentication can be designed as follows:

ICA means Intensity/Cost of Authentication, so

$$ICA = IA/\sum Cn = (\sum Pi + \sum Sj + \sum Dk + \sum Hn) * \sum Mm \;/\; \sum Cn \quad (2)$$

IV. Evaluating Method

In this section, we will evaluate the intensity and cost of authentication depending on 4 websites. Based on the practical application system, the feasibility of this evaluation method will be validated. Considering copyrights, the real name of the 4 websites mentioned in the paper will not be used.

**A - Making Friends Website**

As for Making Friends Website A, two kinds of identity factors are requested: username and password. P1 depicts the authentication intensity of username, P2 depicts the authentication intensity of password. C1 depicts the authentication cost of username, C2 depicts the authentication cost of password. The number of channels is M1 = 1. Therefore, the IA, cost and ICA of this kind of system is:

$$IA = (P1 + P2) \times M1 = (1 + 1) \times 1 = 2$$

$$Cost = C1 + C2 = 1 + 1 = 2$$

$$ICA = 2 / 2 = 1$$

**B - Personal Online Bank**

As for Personal Online Bank B, three kinds of identity factors are requested: username, password and question-based factor. P1 depicts the authentication intensity of username, P2 depicts the authentication intensity of password, P3 depicts the authentication intensity of question-based factor. C1 depicts the authentication cost of username, C2 depicts the authentication cost of password, C3 depicts the authentication cost of question-based factor. The number of channels is M1 = 1. Therefore, the IA, Cost and ICA of this kind of system is:

$$IA = (P1 + P2 + P3) \times M1 = (1 + 1 + 1) \times 1 = 3$$

$$Cost = C1 + C2 + C3 = 1 + 1 + 1 = 3$$

$$ICA = 3 / 3 = 1$$

**C - Enterprise Online Bank**

As for Enterprise Online Bank C, three kinds of identity factors are requested: username, password and online security device. P1 depicts the authentication intensity of username, P2 depicts the authentication intensity of password, H1 depicts the authentication intensity of online security device. The authentication channel is M2. C1 depicts the authentication cost of username factor, C2 depicts the authentication cost of password factor, C3 depicts the authentication cost of online security device. Because online security device needs extra hardware, so C3 = 3. The number of channel is 2, M1 = 1, M2 = 1. Therefore, the IA, cost and ICA of this kind of system is:

IA = (P1 + P2 + H1) × (M1 + M2)= (1 + 1 + 1) × 2 = 6

Cost = C1 + C2 + C3 = 1 + 1 + 3 = 5

ICA = 6 / 5 = 1.2

**D - E-commerce Website**

As for E-commerce Website D, four kinds of identity factors are requested: username, password, IP and telephone. P1 depicts the authentication intensity of username, P2 depicts the authentication intensity of password, D1 depicts the authentication intensity of IP, H1 depicts the authentication intensity of telephone. The telephone authentication channel is M2. C1 depicts the authentication cost of username, C2 depicts the authentication cost of password, C3 depicts the authentication cost of IP, C4 depicts the authentication cost of telephone. Because telephone authentication needs extra hardware, so C4 = 3. The number of channel is 2, M1 = 1, M2 = 1. Therefore, the IA, Cost and ICA of this kind of system is:

IA = (P1 + P2 + D1 + H

The data of four websites above is collected as following table 1:

Table1: Data (IA, Cost and ICA) of four websites

| Website Name | Number of Identity Factors | Number of Channels | Items | Data |
|---|---|---|---|---|
| A | 2 | 1 | I | 2 |
| | | | C | 2 |
| | | | I/C | 1 |
| B | 3 | 1 | I | 3 |
| | | | C | 3 |
| | | | I/C | 1 |
| C | 3 | 2 | I | 6 |
| | | | C | 5 |
| | | | I/C | 1.2 |
| D | 4 | 2 | I | 10 |
| | | | C | 6 |
| | | | I/C | 1.7 |

According to the previous research, we can make the following conclusions:

(1) The authentication intensity will become stronger linearly along with more identity factors;

(2) The cost of authentication will increase linearly along with more identity factors;

(3) The authentication intensity will be multiple along with the number of authentication channels;

(4) The authentication intensity will increase overtly when the senior factor, dynamic factor, dynamic factor and multi channel are compounded, the more combination of layers, the more IA and ICA are;

(5) The stronger authentication intensity, the more cost will be.

These five conclusions indicate the evaluation method designed in this paper is feasible.

## V. CONCLUSION

In summary, we researched and analyzed the identity factors and authentication methods, explored an evaluation methods of authentication mechanism. This method can evaluate the authentication intensity and cost according to quantification values. Therefore, this method can provide quantification references to selecting authentication mechanism. Several examples testify the feasibility of this method.

However, there still exist some factors which should influence the intensity and cost of authentication, for instance, the length of password, the creation method of password and the times of login attempts etc. In general, some systems only allow three attempts before the user is required to renew the authentication key. But Brostoff and Sasse [19] argue that login attempts should be extended to 10 attempts, and further opinion that this will not make the system any more vulnerable to threats from outside the organization. All those factors not being discussed in this paper need many tests and discussion, thus correcting those quantification parameters and arithmetic of this evaluation method to make it more applicable.

## REFERENCES

[1] Bolton, R. J., and Hand, D. J. Statistical Fraud Detection: A Review. *Statistical Science* Vol. 17, 3: 235-255. 2000

[2] Patrick, A. Authentication Technology and Identity Theft. Unpublished manuscript, 01 Jun. 2007 http://www.andrewpatrick.ca/essays/authentication-techn

ology-and-identity-theft/

[3] Brainard, J., Juels, A., Rivest, R., Szydlo, M. and Yung, M. Fourth Factor Authentication: Somebody You Know. *ACM CCS*, 168-78. 2006

[4] Retrieved from http://en.wikipedia.org/wiki/Authentication_factor

[5] Brostoff S, Sasse A. Are passfaces more usable than passwords? A field trial investigation. In: McDonald S, editor. People and computers XIV—usability or else! Proceedings of HCI 2000. Berlin: Springer; 2000. p. 405–24.

[6] Dhamija R, Perrig A. De′ja` vu: a user study using images for authentication. In: Proceedings of USENIX security symposium. Colorado: Denver; p. 45–58. 2000.

[7] L. Hong and A Jain. Integrating faces and fingerprints for personal identification. Lecture Notes in Computer Science, (1351), 1997.

[8] L Bahler, J Porter, and A Higgins. Improved voice identification using a nearest neighbour distance measure. In Proceedings Proc. International Conference of Acoustics, Speech and Signal Processing, pages 321–324, Adelaide., April 19-22 1994.

[9] W. Zhao, R. Chellappa, P. J. Phillips, and A. osenfeld. Face recognition: A literature survey. ACM Comput. Surv., 35(4):399–458, 2003.

[10] M Burger and W Burger. Using ear biometrics for passive identification. In G Papp and R Posch, editors, Proceedings of the IFIP TC11 14th international conference on information security, SEC'98, pages 139–148, Wien, 1998.

[11] Lynne Coventry, Antonella De Angeli, and Graham Johnson. Honest, it's me! Self service verification. In Workshop on Human-Computer Interaction and Security Systems, Fort Lauderdale, Florida, April 2003. ACM.

[12] K Hayashi, E Okamoto, and M Mambo. Proposal of user identification scheme using mouse. In T Okamoto Y Han and S Qing, editors, Proceedings of the 1st International Information and Communications Security Conference, pages 144–148, 1997.

[13] W G de Ru and J H P Eloff. Reinforcing password authentication with typing biometrics. In Proceedings of the IFIP TC11 eleventh international conference on information security, IFIP/SEC'95, pages 562–574, London, UK, 1995. Chapman and Hall.

[14] W G de Ru and J H P Eloff. Enhanced password authentication through fuzzy logic. IEEE Intelligent Systems & their applications, 12(6), Nov/Dec 1997.

[15] S Garfinkel, G Spafford, and A Schwartz. Practical UNIX and Internet Security. O'Reilly, Cambridge, 3rd edition, 2003.

[16] Karen, R. . Quantifying the Quality of Web Authentication Mechanisms A Usability Perspective. 2003, http://www.dcs.gla.ac.uk/~karen/Papers/j.pdf

[17]T Gilb. Advanced requirements specification: Quantifying the qualitative. In PSQT Conference St Paul MN, oct 1999. http://citeseer.ist.psu.edu/332850.html; http://www.pimsl.com/TomGilb/Quantify Quality paper PSQT.pdf.

[18] T Moss. Web accessibility and uk law: Telling it like it is, July 2004. http://www.alistapart.com/articles/accessuk.

[19] S Brostoff and A Sasse. Ten strikes and you're out: Incresing the number of login attempts can improve password usability. In Workshop on Human-Computer Interaction and Security Systems, Fort Lauderdale, Florida, April 2003. ACM.

# ASTRA: An Awareness Connectivity Platform for Designing Pervasive Awareness Applications

**Ioannis Calemis**
Computer Technology Institute
Patras, Greece
calemis@cti.gr

**Achilles Kameas**
Hellenic Open University & Computer Technology Institute
Patras, Greece
kameas@{eap, cti}.gr

**Christos Goumopoulos**
Computer Technology Institute
Patras, Greece
goumop@cti.gr

**Erik Berg**
Telenor Research and Innovation
Trondheim, Norway
erik.berg@telenor.com

*Abstract* - **Awareness systems are a class of computer mediated communication systems that help individuals or groups build and maintain a peripheral awareness of each other. Awareness systems for informal social use are still in their infancy as a technology and as a research area. Such systems promise to address pressing social problems: elderly living alone, families living apart for large parts of the working week, monitoring the well being of an ill relative, etc. The ASTRA platform, which is being developed in the context of the EU research project ASTRA, provides a generalized solution to the development of awareness applications that are based on the concept of pervasive awareness. In this paper, we shall present how smart objects in a person's environment can be used to capture and convey awareness information under this person's control.**

## I. INTRODUCTION

Pervasive awareness systems are computer mediated communication (CMC) systems whose purpose is to help connected individuals or groups to maintain awareness of the activities and the situation of each other. In the domain of group-work where awareness systems were first studied, awareness can be defined as "an understanding of activities of others that provides a context for your own activities" [2]. In a more social context, interpersonal awareness can be considered as an understanding of the activities and status of one's social relations, derived from social interactions and communications with them [5]. Awareness systems promise to address pressing social problems: elderly living alone, families living apart for large parts of the working week, monitoring the well being of an ill relative, etc. [8]

An approach for conceptualization of awareness systems in the current domain research proposes the description of the awareness in reference of the activities that a person is made aware of [13]. Based on this approach, Metaxas and Markopoulos [11] introduced an abstract formal model of awareness systems that incorporates related concepts and supports reasoning regarding social aspects of using awareness systems. Their model draws the basic notions of *focus* and *nimbus* by the work of Rodden [12], who applied them in a spatial model of group interaction, in order to address mutual levels of awareness within a virtual environment. Early works in the domain of informal social communication like the concepts developed by the Presence project [6] or the Casablanca project [7] were created as installations that users could use as they were.

The ASTRA platform, which is being developed in the context of the EU research project ASTRA [1], provides a generalized solution to the development of awareness applications that are based on the concept of pervasive awareness, i.e., where awareness information is automatically generated as a result of using personal and home devices and smart objects, which capture and exchange information about the user semi-autonomously. The ASTRA platform and the assorted end-user tools implement the principles of Theory of Connectedness [17], an extension to the focus - nimbus model. Briefly, focus represents a sub-space within which a person focuses their attention. One is more aware of objects inside one's focus and less aware of objects outside of it. An entity's nimbus is a sub-space within which it makes some aspects of itself available to others. This could be its presence, its identity, its activity or some combination of these.

In this paper, we shall present how smart objects in a person's environment can be used to capture and convey awareness information under this person's control. In the next section, we shall give the basic approach and notions we use in order to represent the problem domain. Then, we shall describe how the solution to the problem is supported by ubiquitous computing technology, give a presentation of the ASTRA awareness platform and provide an example scenario using the proposed technology. Finally, our conclusions are outlined.

## II. BASIC MODELING FRAMEWORK

In order to support the development of awareness applications, we consider that people conduct their activities within an ambient intelligence space using smart objects and that it is possible to access and combine the services offered by these objects. Our approach is based on the following concepts:

- **AmI space:** An AmI space is to a physical space the same as to what an artifact is to an object. To be more precise, an AmI space embeds sensing, actuating, processing and networking infrastructure in a physical (usually closed) space and offers a set of digital services.
- **Artifacts:** An artifact is a tangible object augmented with computation and communication capabilities. Its properties and services are digitally expressed.
- **Services:** These are offered by an artifact or the AmI space. They could be considered as virtual artifacts. Our approach assumes a service-oriented architecture that

T. Sobh (ed.), *Innovations and Advances in Computer Sciences and Engineering*,
DOI 10.1007/978-90-481-3658-2_32, 

enforces a clean service oriented design approach, with a clear distinction between interfaces and implementation [15].

- **Synapse**: A logical connection between services offered in an AmI space. A synapse defines a service composition.
- **Ambient Ecology:** It is the set of artifacts contained in an AmI space and services offered therein; artifacts may be connected through synapses, thus offering more complex services.
- **Spheres:** An activity sphere is deployed over the Ambient Ecology of an AmI space and uses its resources (artifacts, networks, services etc.) to serve a specific goal of its owner. It usually consists of a set of interrelated tasks; the sphere contains models of these tasks and their interaction. The sphere instantiates the task models within the specific context composed by the capabilities and services of the container AmI space and its contained artifacts. In this way, it supports the realization of concrete tasks.

## III. An Example of Ambient Ecologies

Let's consider the following scenario:

Students of the Distance Education University (DEU) usually live in disparate locations all over the country. Each of them has his personal matters, but they all have in common their studies at DEU. Punch and Judy are two students of the "Software Design" Teaching Unit; in the past week they have been collaborating in order to study and submit a common project.

Punch is 33 years old, single, working hard and overcommitted. He likes technology and is keen on using new gadgets he discovers in the shops. Judy is a 27-year old single woman, who lives in a small apartment in Santorini. She is a travel agent, and not so fond of technology. Both have installed in their smart homes an Ambient Ecology to support their study.

Punch's Study sphere consists of the following objects: a Book, a Chair, a DeskLamp and a Desk. All of those objects have been augmented with hardware and software in order to provide their services to the ASTRA system. The Desk can sense light intensity, temperature, weight on it, and proximity of a chair. The Chair can tell whether someone was sitting on it. The DeskLamp can remotely be turned on and off. The Book can tell whether it is open or closed and determine the amount of light that falls on it. Collective artifact operation is accomplished by establishing synapses between the constituent artifacts, in order to realize the following behavior:

**WHEN** this *CHAIR is NEAR the DESK*
AND *ANY BOOK is ON the DESK*,
AND *SOMEONE is sitting on the CHAIR*
AND *The BOOK is OPEN*
**THEN** *TURN the DESKLAMP ON.*

On the contrary, Judy's sphere is rather simple and only uses the services of a Clock, a Lamp and a Picture Frame. Whenever she starts her study, she sets the Clock timer to 90 mins and connects it to the Lamp; after 90 mins, the Clock alarm goes off and forces the Lamp to flash two times, via their connection.

## IV. Awareness System and Applications

The purpose of an awareness application is to convey a person's condition, need or want to a community of users who have subscribed to this application. Usually, an awareness application is developed by a person, who subsequently publishes it to a community, or invites people to subscribe to it.

To the ambient ecology concepts described above we add two basic concepts that originate from modeling of awareness:

- **Focus:** A person's focus is the set of conditions, situations or events that this person is interested in. A person's focus may include another person's nimbus. It is modeled as a set of events that happen in this person's AmI space.
- **Nimbus:** A person's nimbus is the set of conditions, situations or events that this person makes public, i.e. makes them available to become part of some other persons' focus. A person may interact with his nimbus by causing events in his AmI space.

Consider a very simple example. Suppose that Punch every now and then likes to go to the local pub, but he hates going out alone. So he has created a simple awareness application that he calls "out to the pub" and he has invited his friends to join this application.

An awareness application can be considered as a set of conditions and events that convey specific meaning to a defined community of persons. So, regarding the "out to the pub" application Punch has created some rules which signify when he wants to activate this application; for example, he wants to go out when he is not studying or not sleeping, but he does not want to go out when he has company at home. His friends have done the same; of course, each person can create his own rules that activate this application.

So, in order to define an awareness application, a person has to:

- Provide a short textual description of the application and describe its various instances
- Define the conditions that trigger the application and consequently the awareness information to be conveyed – this is his nimbus

- Define the other persons he wants to be aware of this application – they have to include this applications in their focus

So a community is the set of persons that a person allows to have access to his nimbus.

Note that a person may subscribe to an awareness application published by another person. In this case, he has to include this application to his focus.

Based on this framework, we then describe a ubiquitous computing awareness application as an activity sphere, which is instantiated on the ambient ecologies in the different AmI spaces of the various application users [14]. Each instantiation makes use of the different resources in each AmI space and of the artifacts in each ambient ecology and is realized as a set of synapses between the artifacts and the provided services into the AmI space. In order to manipulate their focus, nimbus and awareness applications, people use the artifacts in the AmI space.

## V. ASTRA SYSTEM

In order to support the realization of ubiquitous awareness applications, we have developed a three-tier architecture in which Figure 1:

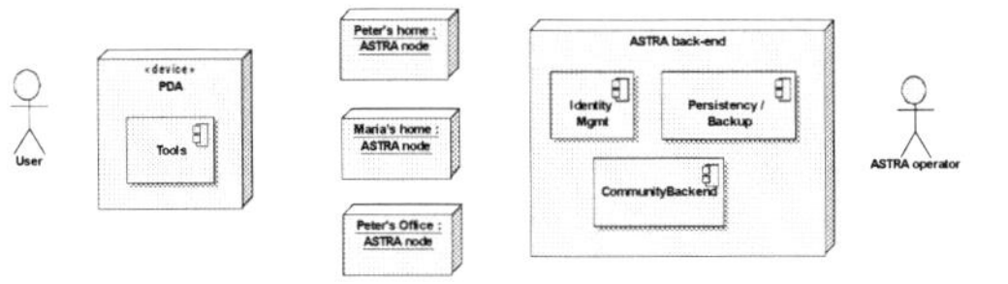


Figure 1. Illustration of ASTRA System

- **End-user tools** implement the presentation layer;
- **Local (user) nodes** are used to support the instantiation of focus and nimbus, and encapsulate the application business logic in terms of rules (application/system logic layer);
- **A centralized remote server** is used to store the awareness application resource descriptions and the community profiles (resource management layer).

Furthermore each local space is comprised of a set of artifacts and services that compose each user's Ambient Ecology.

### *A. Local Node*

A computing system in the person's local space runs the ASTRA node, which provides different end-user services and is responsible for integrating system services, such as context management and service discovery. The platform adopts the Service Oriented Architecture principles (SOA) [15] and makes its resources available as independent services that can be accessed without knowledge of their underlying platform implementation

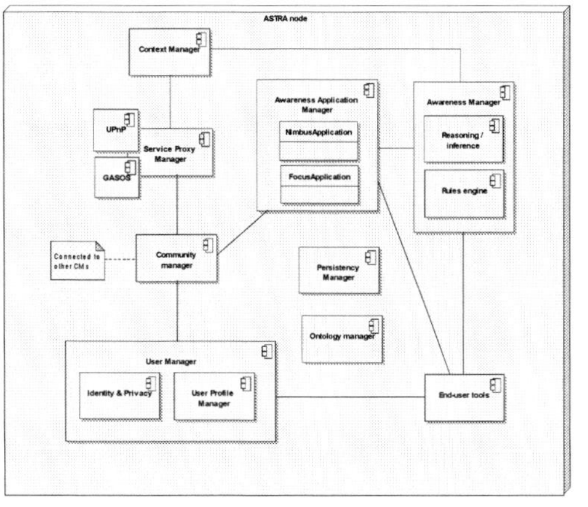


Figure 2. ASTRA Node's Architecture

It has been developed based on the OSGi platform (www.osgi.org) for Service Oriented Computing in Java. OSGi was chosen due to its very elegant and easy way of deploying services. The system is comprised of several different components each one assigned to do a specific task for the whole system.

The functionality provided by the components of the local system is divided in the following categories:

- **Device/Service Manipulation:**

In order to connect the local awareness devices/services to the system we need to identify them, integrate the services to the ASTRA platform and finally analyze and understand their data parameters in order to be able to create applications. This requires, a *Service Proxy Manager* (SPM) which is responsible for discovery of local services. The SPM provides a set of interfaces and implements service proxies for each different communication/middleware technology provided in the local AmI space.

Two service proxies have been implemented in order to connect local artifacts to the system. A GAS-OS service proxy and a UPnP service proxy.

The former service proxy integrates to the ASTRA platform GAS-OS enabled artifacts. GAS-OS enabled artifacts are objects and devices that follow Gadgetware Architectural Style (GAS) [10], a generic architectural style for activity spheres.

GAS adopts the principles of software component technology and service oriented architectures and applies these to the domain of ubiquitous computing, in order to describe the process whereby people configure and use complex collections of interacting artifacts [8]. Each artifact provides its functionality in the form of well-defined services; these are accessible via interfaces. The GAS-OS service proxy tries to identify these services in order to be used by the ASTRA local system.

The later service proxy integrates to the ASTRA platform devices based on the UPnP protocol. Universal Plug and Play (UPnP) is a set of computer network protocols promulgated by the UPnP Forum [16]. The goals of UPnP are to allow devices to connect seamlessly and to simplify the implementation of networks in the home (data sharing, communications, and entertainment) and corporate environments. UPnP achieves this by defining and publishing UPnP device control protocols built upon open, Internet-based communication standards.

All devices and services that are identified by the service proxies are processed by the *Context Manager* whose responsibility is to keep the current state of all artifacts that are identified in a unified format. This knowledge can be accessed in order to connect the local AmI space with awareness applications.

- **Local Awareness Assessment:**

Data gathered by the device/service manipulation components should be combined in order to extract current awareness state of the user. This requires the interaction of the following two components: The *Ontology Manager* (OM) whose responsibility is to provide a common understanding between services and the *Awareness Manager* (AM) which uses the business logic added to the system in terms of rules and infers the current awareness state.

The Ontology Manager's purpose is two-folded: First it tries to provide a common dictionary for identifying similar services. Thus the system may get for example a lamp that provides the service luminosity and another lamp that provides the service light. OM's purpose is to identify the similarity of these to services. Thus if the system for any reason needs to replace a light service the ontology is responsible to provide a list of all matching services for replacement. Secondly Ontology tries to identify higher level meanings considering the data of the specific context. Thus, if the system receives data for a service that describes location that state that the user is at his office, this automatically means that he is also in a specific building in a specific town, in a specific country.

The purpose of the Awareness Manager is to get any changes that affect the data of the system, whether those come from the local devices (local context) or from the Community (change of someone's nimbus) run the rules that the user has specified for his local sphere and take decisions that may affect the current state of the system. Those changes are either transmitted to the Awareness Application Manager (change of local nimbus – let my community know) or to the Context Manager (change of local focus – trigger the appropriate device).

- **Transmission of Awareness Information:**

Any awareness assessment made by the Awareness Manager that affects an application must be transmitted to the involved parties (users or communities). This is done by the Awareness *Application Manager* (AAM) who is responsible for storing and managing the local awareness applications. The AAM supports users in controlling incoming and outgoing awareness information (i.e. their focus and nimbus) by implementing a publish/subscribe model [4]. People who subscribe to the same awareness application are regarded as a virtual community.

An awareness application is a mapping of an awareness state to a service that can be made available to a community, in other words it is a service representation of a specific awareness state. The AAM provisions two types of awareness state applications: nimbus applications and focus applications. Whenever a user wants to share his/her awareness state, she must make a nimbus application. The creation of the application creates the tie with the Awareness Manager, and publishing it will create the link with the Community Manager and the specified community. Any other members of the community can now see that the user has published that awareness state. Every other member can choose to focus on that awareness state, and that is done by creating a focus application.

Apart from the ASTRA system components the local AmI space includes a set of artifacts and services able to perceive user activities (sensing units) or to provide interaction with the system (actuation units) Each of these artifacts and services runs either UPnP [16] or GAS-OS middleware [3]. This enables the discovery of artifact services by the local subsystem and their composition in the context of awareness applications.

An ASTRA user has to define how his focus and nimbus will be realized within an AmI space. For example, Punch has to describe the rules that trigger the awareness application "out to the pub" and also those rules that help the local system deduce his current condition. These rules are defined using the services and states of artifacts in the AmI space. So, Punch could define a rule stating that "when I am not working and it is Sunday evening and I switch my TV set off, then I want to go to the pub". The information necessary for the system to evaluate this rule can be gathered as follows:

- *Not working:* this information describes Punch's current condition; it can be received from a central "point of intelligence" in the house, or deduced as a result of a different set of rules, the description of which lies outside the scope of this paper

- *Sunday morning:* this piece of context refers to time and can be received from a calendar service of the AmI space
- *TV set off:* this other piece of context can be directly retrieved from the artifact TV set

When this rule fires, then an event is sent by the local ASTRA subsystem to the ASTRA server. This event is associated with the "out to the pub" awareness application and is propagated to all of its subscribers by the Awareness Manager.

In a similar manner, Punch can define rules describing how he wants to be notified of events that are caused by the other subscribers in the "out to the pub" application. Examples of such rules are: "when I am in the living room, and the TV set is on display a message on the TV screen, otherwise flash the floor lamp twice", or "when I am in the kitchen, show message in the photo frame", etc.

### *B. Remote Server*

The remote, or back-end, server of the ASTRA platform is responsible for managing identities and providing a backup / persistency solution to components. In addition it also provides the back-end component of the Community Manager, which is responsible for synchronizing community states between the distributed Community Managers of the local ASTRA nodes, as well as providing persistent storage of this state for bootstrapping purposes of the local nodes. This state is comprised of: a) *community members* ***b)*** *shared awareness applications* and c) *shared community space*. The community back-end also supports eventing for changes to any of these parts.

### *C. ASTRA tools*

The ASTRA end user tools use a web interface and connect to the ASTRA platform via a specific API. The tools support user profile management, rule editing, and application management, in a way that semantically relates to the focus / nimbus model, albeit using of a more familiar terminology for end users (Figure 3).

The tools contain the following interlinked modes: *Awareness Connections manager* (where the user can define their Focus or Nimbus), *Pervasive Application manager* (where the user can associate their awareness focus or nimbus to a ubiquitous awareness application), *and User and Communities manager*.

The approach taken in the ASTRA project, when scaled up, has the risk of imposing a heavy semantic load to the user, as she will have to be able to distinguish between various notifications that she will receive and interpret correctly the different events they represent. An obvious solution to this is to use the screens that are available in the ambient ecology (i.e. TV set, mobile, PDA etc) to display semantically rich textual messages, or to use voice synthesis to explain the meaning of the notification.

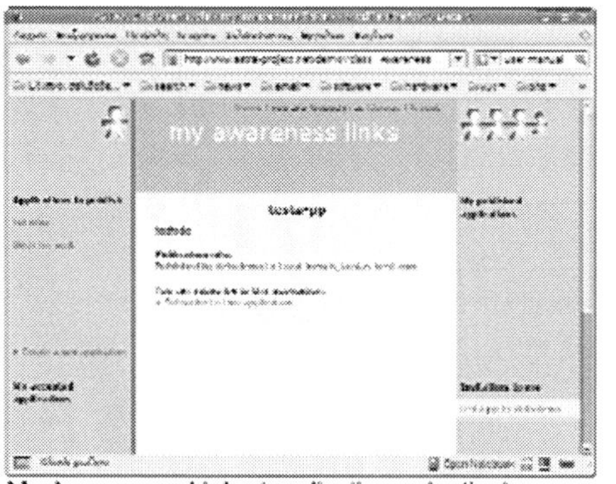

My Awareness Links (application selection)

Figure 3. Sample ASTRA end user tool

Both these approaches, however, are intrusive, in the sense that they will require the person to shift his attention to the event. A more complicated approach that we are looking into in the project is to use ontologies to create semantically rich descriptions of events and services and then use user-defined policies to deal with conflicts of event interpretation. For example, when the notifications for two events from different awareness applications are similar, then more detailed information has to be conveyed, so as the person can distinguish between them.

## VI. AN EXAMPLE AMBIENT AWARENESS APPLICATION

Now we shall develop the previous example, so as to use the ambient ecology to convey awareness information:

. Recently, DEU is offering an awareness service based on the ASTRA platform for a trial period of one academic year to the student of one specific teaching unit. The ASTRA platform enables subscribed users to form communities and to exchange awareness information and applications between the members of a community.

Punch and Judy have taken advantage of the new DEU service. So, Punch created a DEU Study awareness application and Judy subscribed to it. Punch included in his Nimbus the 'Now Reading' state of his sphere and focused his system on Judy's 'Reading' state. On the other hand, Judy included in her Nimbus the state of her eClock and her PictureFrame; her Focus was set on Punch's 'Now Reading' state.

In Punch's side, whenever he turns on his Study sphere, as his eLamp is switched on, his awareness system sets the value of his 'Now Reading state' in his Nimbus. The ASTRA system communicates Punch's Nimbus to Judy. Judy has Focused on Punch's 'Now Reading' state, and has connected it to her PictureFrame; whenever it changes, her eLamp flashes and Punch's picture appears. In parallel, as Punch has set his Focus on Judy's Reading state, whenever she takes a break (as a result of her eClock's timer reaching zero), his eLamp flashes. Figure 4 shows the awareness system configuration described in the example.

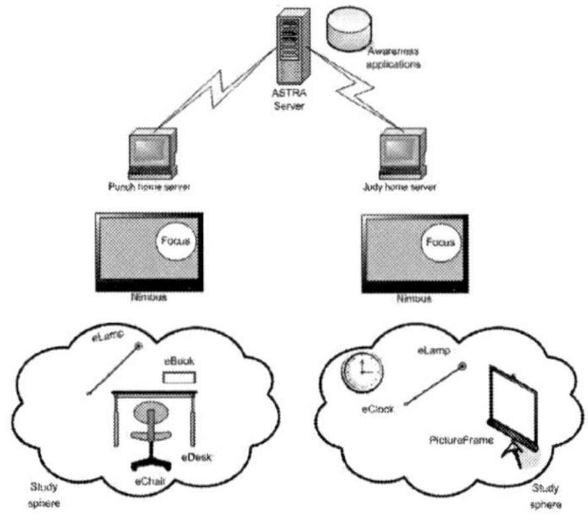


Figure 4. The awareness system of the example

## VII. Conclusions

In this paper, we have presented a service oriented architecture that supports the composition and management of ubiquitous computing awareness applications. The aim of this class of applications is to support communication among people without interfering with their task-at-hand.

A three tier system has been developed to support this service: in the person's local space, interaction among smart objects is achieved with the use of GAS principles; these allow the integration of artifacts running specialized middleware GAS-OS, or other commonly available systems, such as UPnP. In the server side, the specialized ASTRA platform was developed, which offers services for the management of applications and user communities. The architecture presented uses smart objects in the person's space as conveyors of this person's awareness condition or want. The person can configure these objects using special end user tools, which support the discovery of artifact services and their combination in a simple way, as well as the definition of awareness applications using first-order logic rules on these services.

Research on ASTRA project continues in order to evaluate and improve the concepts and tools presented in the paper.

## Acknowledgment

The described work has been partially supported by EU FP6 IST STREP project ASTRA (Awareness Services and Systems – Towards theory and ReAlization), under Grant agreement No. 29266.

## References

[1] ASTRA IST/FET Open project, available in http://www.astra-project.net/

[2] P. Dourish and V. Bellotti, "Awareness and coordination in shared workspaces," In: Proceedings of the 1992 ACM Conference on Computer-Supported Cooperative Work (CSCW '92), ACM, pp. 107-114.

[3] N. Drossos, C. Goumopoulos, and A. Kameas, "A conceptual model and the supporting middleware for composing ubiquitous computing applications". Journal of Ubiquitous Computing and Intelligence, American Scientific Publishers, 1(2), 1-13.

[4] P. Eugster, P. Felber, R. Guerraoui and A. Kermarrec, "The many faces of publish/subscribe". ACM Computing. Surveys, 35, 114–131.

[5] B.A. Farshchian, Presence Technologies for Informal Collaboration, In: G. Riva, F. Davide, W. A IJsselsteijn (Eds.) "Being There: Concepts, Effects and Measurement of User Presence in Synthetic Environments", Ios Press, Amsterdam, The Netherlands, 2003, pp. 209-222

[6] W. Gaver and B. Hooker, "The Presence Project. London", RCA: CRD Publishing

[7] D. Hindus, S.D. Mainwaring, N. Leduc, ,A.E. Hagström, and O. Bayley, , "Casablanca: designing social communication devices for the home". In Proc. CHI 01.

[8] W.A. IJsselsteijn, , J. van Baren and F. van Lanen, "Staying in touch: Social presence and connectedness through synchronous and asynchronous communication media". In Proc. HCII 2003 volume 2.

[9] A. Kameas., S. Bellis, I .Mavrommati, K. Delaney, M. Colley, and A. Pounds-Cornish, "An Architecture that Treats Everyday Objects as Communicating Tangible Components". In Proc. PerCom03.

[10] A. Kameas, I. Mavrommati, I. and P. Markopoulos, "Computing in tangible: using artifacts as components of Ambient Intelligent Environments". In Riva, G., Vatalaro, F., Davide, F. and Alcaniz, M. (eds) Ambient Intelligence, IOS Press, 121-142.

[11] G. Metaxas., and P. Markopoulos., "Aware of what? A formal model of Awareness Systems that extends the focus-nimbus model". In Proc. EIS 2007.

[12] T. Rodden., "Populating the Application: A Model of Awareness for Cooperative Applications". In Proc. CSCW 1996.

[13] K. Schmidt, "The problem with "awareness"": Introductory remarks on "awareness in CSCW". In Proc. CSCW 2002.

[14] I. D. Zaharakis and A. D. Kameas., "Emergent Phenomena in AmI Spaces". The EASST Newsletter, Volume 12 (March 2006 / No. 2006 - 12), pp. 82-96. EASST e.V.

[15] T. Erl, "Service-Oriented Architecture: Concepts, Technology, and Design". Upper Saddle River: Prentice Hall PTR. 2005, ISBN 0-13-185858-0.

[16] UPnP™ Forum, available in http://www.upnp.org/

[17] N. Romero, P. Markopoulos, J. Baren van, B. Ruyter de, I. Wijnand, B. Farshchian, "Connecting the Family with Awareness Systems", Personal and Ubiquitous Computing, Springer-Verlag London, UK, Volume 11, Issue 4, p.1, 2

# Extending OWL-S to nested services: an application to optimum wireless network planning

Alessandra Esposito, Luciano Tarricone, Laura Vallone
Dept. Innovation Eng., University of Salento
Via Monteroni
73100 Lecce – Italy

*Abstract-* **The migration of existing applications to service-oriented paradigm generate services with variegate behaviours and way of interacting with one another. In some cases, available technologies for demarcating services do not provide suited constructs to codify such interactions. As an example, migrating wireless network planning codes towards a service-oriented environment generates a sort of "nested" services, which cannot be codified with available constructs provided by OWL-S. In this paper, we propose an extension of both the OWL-S ontology and the OWL-S API which enables the description of such services. The validity of such an extension has been demonstrated on a real life application in the area of wireless network planning .**

## I. Introduction

Web enabled applications are becoming part of our daily work and life. Nowadays it is possible to teach through the Web, to make business transactions and to experiment complex scientific software.

In such a context, a very appealing perspective has recently emerged. It consists in the (semi-)automatic building of problem-oriented applications, based on the discovery and aggregation of pieces of codes available through the Net. One of the key technologies paving the way to such a revolution, is given by Web Services (WS) [1]. Indeed, WS are potentially discoverable and evocable through the Net, thanks to their capability of describing themselves by means of the so called Web Service Description Language (WSDL, [2]). However, WSDL provides a syntactic description of Web Services but lacks the semantic expressivity needed to represent the requirements and capabilities of Web Services.

To face such a need, a lot of approaches (ontologies, languages, frameworks, …) have been proposed. Such approaches, going under the name of Semantic Web Services (SWS) [3], aim at improving Web Services reuse, discovery and composition capability by enriching them with machine-processable semantics.

Very interesting results have been obtained in several fields [4-7], this increasing the appeal of Web Services and Semantic Web technologies in business and scientific research. As a result, people working in such fields are becoming more and more aware of the utility of integrating such technologies in their applications. In other words, existing stand-alone applications are more and more converted into web-enabled applications, so that they can be located, aggregated and invoked through the Net. However, migrating legacy software towards service-oriented environments, often requires a careful customization process. This is especially true in scientific contexts where traditional compiled languages (such as C or Fortran) have been (and are still) widely used. Moreover, embedding existing codes into Web Services may sometimes result into services showing behaviours whose semantic demarcation may be difficult.

One of the most used standards for demarcating services is OWL-S [8]. The amenability of OWL-S to describe service properties and functioning has been already discussed by the authors in previous works [9-13]. However, in a number of situations, service behaviour cannot be straightforwardly described by OWL-S, this requiring a tailored customization of OWL-S concepts.

In this paper, we discuss the case of a specific complex service whose behaviour is not supported by OWL-S. The example belongs to a real life application, the optimum planning of wireless networks. Starting from this application we propose an OWL-S extension which supplies the OWL-S lack of constructs for describing the needed functional behaviour.

The paper is organized as follows. Section II gives some theoretical background on the optimum planning of wireless networks. Section III describes the service oriented application. Section IV briefly introduces OWL-S. Section V highlights how the OWL-S ontology has been extended. Section V briefly explains how the proposed OWL-S extension can be used to effectively invoke "nested" services. Section VI finally draws some conclusions.

## II. Optimum Wireless Network Planning: Some Background

An optimum wireless network planning (OWNP) system is a software tool able to improve the quality of service and control electromagnetic emissions. In order to accomplish its purpose, an OWNP system must embed some essential features, such as the availability of: a) appropriate models for field prediction and b) efficient optimization tools for the identification of Base Station (BS) locations, power rightsizing, tuning of electrical and mechanical parameters, etc. The former item casts the need of a radiopropagation

T. Sobh (ed.), *Innovations and Advances in Computer Sciences and Engineering*,
DOI 10.1007/978-90-481-3658-2_33, © Springer Science+Business Media B.V. 2010

module (RP), providing a field prediction based on the characteristics of the geographical (topographical) environment and of the EM sources (essentially BS antennas). According to item b), RP capability must be coupled with adequate optimization models and methods (OPT), outputting the optimum choice of the above mentioned variables. In such a way, once the characteristics of the propagation environment are known, the optimum planning system produces the elected locations to install BSs, as well as antenna electrical parameters.

RP models and OPT algorithms design and development require highly skilled competences and know-how.

OPT models generally implement iterative approaches, such as Genetic Algorithms [14] and Tabu Search [15], which have been demonstrated [16-21] to be the most appealing for network planning problems.

The choice of the RP models [22-28], is instead much more critical. Indeed the scientific community has developed a variety of RP models. The simplest, called Free Space Loss (FSL) model, assumes that no obstacles exist between transmitter and receiver. Alternative approaches are often best suited to attack real-life environments. They are usually grouped into empirical, semi-empirical and deterministic models [22-28]. Empirical models (such as COST231OkumuraHata [22-25]) exploit the availability of large collections of data deriving from measurements made in concrete scenarios. Semi-empirical (such as COST231Walfish-Ikegami [25-27]) models combine physical principles with statistical parameters, often providing a good balance between accuracy and complexity. Finally, when dealing with small-sized domains and when very high accuracy is required, deterministic methods (such as Ray-Tracing) can be fruitfully used [29]. In other terms, the choice of the RP model is highly dependent on the geographical environment and computational and accuracy problem requirements.

In such a complex scenario, a cooperative engineering environment enables 1) EM researchers to take advantage from previously developed codes and approaches, thus avoiding to develop them from scratch and 2) personnel not specifically skilled in EM propagation and/or optimization approaches to discover the right service and invoke it.

These requirements are more and more attainable with the SWS technology, according to which RP and OPT modules are distributed over the internet, in the form of Web Services and interact one another, as well as with an end-user having to solve the optimum planning problem.

In order to demonstrate the feasibility of such an environment, we developed a prototype simulating a situation where dispersed OPT and RP services are located and invoked by using semantic tools. As described in the following section, the prototype was implemented on top of a functioning existing OWNP application, this confirming the validity of obtained results.

## III. OWNP SERVICES

As explained in the previous section, the OWNP service oriented application consists essentially of OPT and RP services. OPT services adopt iterative algorithms to optimize the BS parameters. They first calculate a set of candidate solutions, each corresponding to a different network configuration, then evaluate their quality in terms of coverage, field uniformity, compliance with safety standards and other possible relevant parameters. Such quantities depend on the distribution of the field radiated by the BS configuration corresponding to the current solution. Therefore, RP is invoked by the OPT in order to estimate such a field distribution. Depending on the attained cost, a new iteration is started, otherwise the optimization is halted. Fig. 1 proposes a graphical representation of OPT and RP services and of their interaction.

In a real-life situation, the choice of the best suited RP strongly depends on the geographical environment and on computational and accuracy problem requirements. Therefore we considered three RP services corresponding to three RP models, normally adopted in different situations. In detail, we considered the free space loss model , the semi-empirical COST231 Walfish Ikegami model and the COST321 Okumura Hata radiopropagation model.

All the OWNP services were obtained by embedding pre-existing C native codes into services. For this purpose, the open-source Java Native Interface (JNI) [30] was adopted. JNI encapsulates native functions into loadable libraries so that they can be dynamically embedded into java methods. Therefore the code of each OWNP module was embedded into a library. This allowed us to carry out the encapsulation in a straightforward manner and to minimize the adjustments to C native codes.

## IV. OWL-S BASIC CONCEPTS

A number of initiatives [31,32] has been proposed for adding semantic description to WSs. Among them OWL-S (OWL ontology for Web Services) and WSMO [33] are worth to be mentioned. OWL-S is an ontology of services written in OWL. It provides a number of classes and properties useful to describe the properties and capabilities of WSs in unambiguous, computer-intepretable form so to facilitate the automation of a number of tasks, including Web Service discovery, execution, composition and interoperation.

WSMO (Web Service Modelling Ontology) shares with OWL-S the vision that ontologies are essential to enable automatic discovery, composition and interoperation of Web services. Apart from their common vision, however, OWL-S and WSMO follow different methodologies [34-36]. For example, whilst OWL-S explicitly defines a set of ontologies

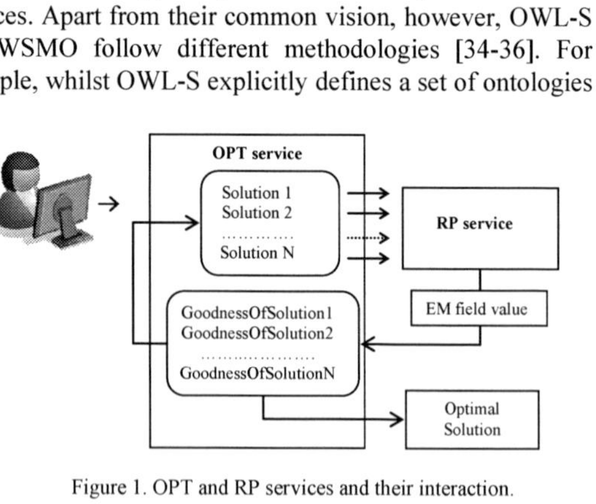


Figure 1. OPT and RP services and their interaction.

to support reasoning about Web services, WSMO defines a conceptual framework within which these ontologies have to be created. Moreover whilst the OWL-S ontologies are written in OWL, WSMO provides a formal language (WSML) for semantically describing relevant aspects of Web Services.

In this work we focus on OWL-S for two main reasons. First of all it is under development since early 2001 and since then it was used by a conspicuous number of research efforts. Moreover OWL-S is developed on top of the standard semantic web technology (RDF, OWL, …) thus being fully integrated in a standardization effort.

OWL-S provides three fundamental types of knowledge about a service: the profile, the model and the grounding (Fig.2). The ServiceProfile class codifies properties, capabilities and functioning of services. It provides the information needed by a client to discover a service. The ServiceModel class codifies the functioning mechanism of a service. It details how a service operates, the conditions under which specific outcomes occur, and, where necessary, the step by step processes leading to these outcomes. Finally, for the purpose of service's invocation, the most relevant part of an OWL-S description is the ServiceGrounding. The ServiceGrounding class provides concrete details on how to access the service, such as communication protocols, message exchange format, address of the service provider.

As explained in Section V, our intervention to OWL-S extends mainly the ServiceModel class. Therefore some more details about this class are provided in the following sub-section.

### A. *OWL-S Service Model*

The ServiceModel class assumes that services are simple entities directly evocable or complex entities that require more than one invocation to be accomplished. Such a distinction is obtained by representing simple services in terms of AtomicProcesses, i.e. of actions that can be carried out in one single step. In other words, an OWL-S atomic process corresponds to a WSDL operation.

On the contrary, complex services are described by means of CompositeProcesses which correspond to actions that require more than one invocation to be executed. Complex services can be composed of atomic or complex processes. This is codified by using typical business-process modelling language constructs such as Sequence, If-then-Else, Iteration and so on, which are codified by the so called OWL-S *control constructs*. A special construct, the so called Perform construct, is used to reference an atomic process inside a composite process. Constructs referring to composite processes are linked through the "components" property to the subprocesses (or control constructs) they are composed of (Fig. 3). For example, the "Choice" control construct calls for the execution of a single control construct from a given bag of control constructs, encoded by the ControlConstructBag class (Fig. 3).

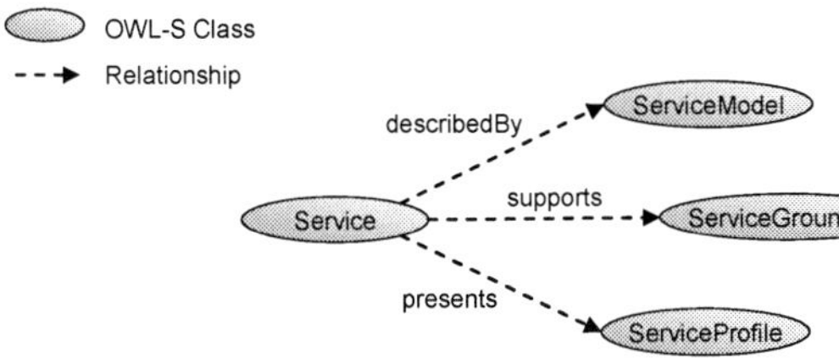


Figure 2 OWL-S fundamental classes.

## V. EXTENDING THE OWL-S ONTOLOGY

As explained in Section III, an OWNP system is mainly made up of RP and OPT services. Their flow of operation must be coded in order to support the automatic invocation of OWNP components. Each RP service exposes a unique operation, i.e. the invocation of the RP native main code. Therefore, they have been represented as atomic processes (Fig. 4). The OPT service is more complex. Although it exposes a unique operation (the invocation of the optimization native code), the optimum planning process is not executable in a single step: it implies the invocation (possibly iterative) of an RP service. Therefore, the OPT process was represented as a composite process (Fig. 4).

As described in the previous section, OWL-S provides a set of control constructs to specify the way in which processes (atomic or composite) are combined.

However, no OWL-S control construct can encode the behaviour of the OPT process. Indeed, although the OPT process involves a call to two operations, each referring to a diverse service (i.e. the OPT service and the RP service), from a client point of view it behaves like an atomic process: the service requester has no visibility into the OPT process execution.

To encode such a case into the OWL-S ontology we introduced a new control construct, named SpawningPerform which extends the Perform class. The SpawningPerform control construct is a special-purpose object which adds to the Perform construct some information about the control construct being spawned (see Fig. 5).

This has been codified by defining the "spawns" property which links the SpawningPerform control construct to the ControlConstruct class. Finally, the following OWL restriction compels individuals of the SpawningPerform class to refer only to atomic processes:

$$\text{SpawningPerform} \equiv \forall\ \text{process:process process:AtomicProcess}$$

Fig. 6 shows how the SpawningPerform construct models

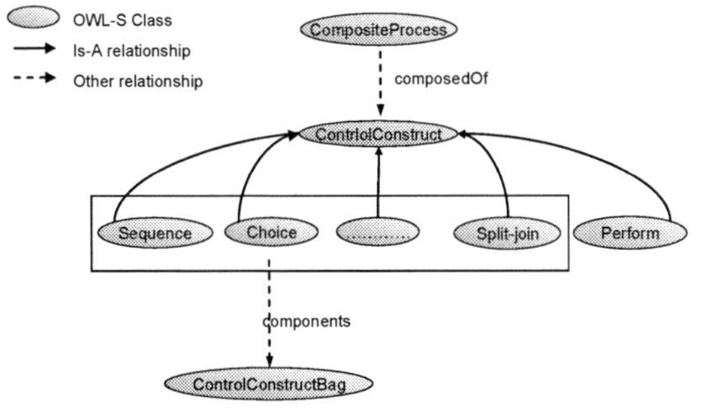


Figure 3 Each CompositeProcess is linked with its control construct via a "composedOf" property. Each control construct, in turn, is associated to a class describing the control constructs it is composed of.

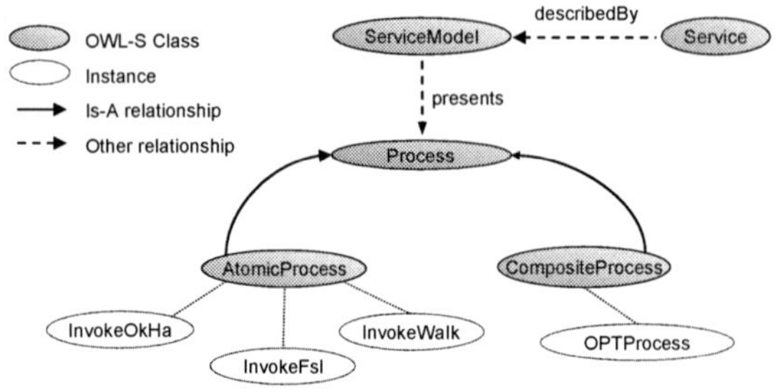


Figure 4 RP services expose a unique operation: they have been represented as atomic processes. The OPT service has been represented as a composite process (named OPTProcess) since it implies the iterative invocation of an RP service.

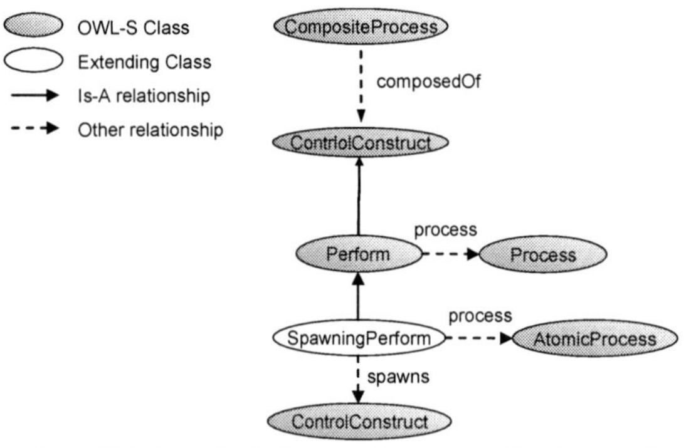


Figure 5 The SpawningPerform control construct codifies the case of an atomic process spawning another control construct.

the optimum planning process. The OPT composite process performs an atomic process (the unique operation exposed by the OPTService, namely "invokeOPT" in Fig. 6) which spawns a nested child process chosen from a bag of processes.

The choice of the RP process to spawn depends on the specific characteristics of the propagation environment and of the EM sources to optimize. Such characteristics have been encoded inside the ontology. More in detail, by taking advantage from the ServiceProfile class, each RP service has been associated to the RP model it implements. Each RP model in turn has been linked to the characteristics of both the propagation environment and the EM sources. By taking advantage from such a description an end user having to simulate a specific environment is guided towards the best suited RP model (and RP service) to use. Such a choice, in the form of a service URI, is passed to the OPT service as input parameter (see following Section).

## VI. Extending the OWL-S API

Once the functional description of the OPTService has been provided, the utilities to interpret it and launch the service must be provided as well. Launching a service described by a composite process means invoking the atomic processes according to the flow specified in its model and stored inside the ontology. For such a purpose, the OWL-S API [37] provides methods to invoke both AtomicProcesses and CompositeProcesses. Therefore, we extended the OWL-S API to support the execution of processes described by a SpawningPerform construct. Such an extension provides the capability of 1) finding the main AtomicProcess pointed by the "process" property; 2) navigating inside the ControlConstruct pointed by the "spawns" property and finding out the process to invoke (chosen among a bag of possible ones); 3) reading the URI of the nested process and setting it as input parameter of the main atomic process and finally invoke it.

## VII. Conclusions

In this paper we discussed the OWL-S amenability to semantically describe the functional behaviour of a particular type of services requiring the evocation of a service from within another one. We proposed an extension to OWL-S, in the form of a new OWL-S construct, to supply the OWL-S lack of constructs for describing the functioning behaviour of such typology of services. An extension to the OWL-S API has been proposed as well. Both extensions were tested and validated on a real life application in the field of wireless network planning.

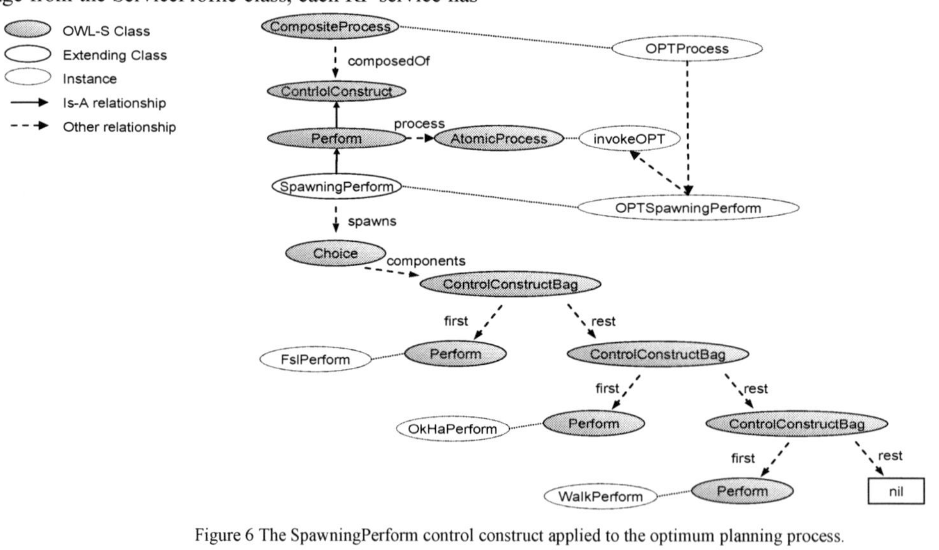


Figure 6 The SpawningPerform control construct applied to the optimum planning process.

## REFERENCES


[1] Cerami, E. 2002, Web Services, O'Reilly & Associates, Inc.

[2] WSDL http://www.w3.org/TR/WSDL

[3] S. McIlraith., T.C. Son and H. Zeng. Semantic Web Services. IEEE Intelligent Systems, Special Issue on the Semantic Web, 16(2):46--53, March/April, 200

[4] E. Motta, J. Domingue, L. Cabral, and M. Gaspari. IRS-II: A framework and Infrastructure for Semantic Web Services. http://www.cs.unibo.it/gaspari/www/iswc03.pdf, 2003.

[5] http://www.win.tue.nl/SW-EL/proceedings.html

[6] E. Hyvönen, Semantic Web Applications in the Public Sector in Finland – Building the Basis for a National Semantic Web Infrastructure, Norwegian Semantic Days, April 26-27, 2006, Stavanger, Norway

[7] A. Léger, L. J.B. Nixon, P. Shvaiko and J. Charlet, Semantic Web Applications: Fields and Business Cases. The Industry Challenges The Research. (2005) Proceedings of the 1st International IFIP/WG12.5 Working Conference on Industrial Applications of Semantic Web (IASW), IFIP vol.188, pp 27-46, 2005.

[8] http://www.daml.org/services/owl-s/

[9] Tarricone, L. and Esposito A. (2004) "Grid Computing for Electromagnetics", Artech House, 2004, pp. 1-290

[10] Esposito A., Tarricone L., Vallone L., Vallone M., Automated grid services composition for CAE of aperture-antenna arrays, Journal of the European Microwave Association (EuMA) - Special Issue on "MMS", ISBN 88-8492-324-7, Vol.4, Issue 1, 2008

[11] Esposito A., Tarricone L. (2006) "Advances in Information Technologies for Electromagnetics", Springer, April 2006, pp.295-326

[12] Esposito A., Tarricone L., Vallone L., Semantic-Driven Grid-Enabled Computer Aided Engineering of Aperture-Antenna Arrays, IEEE Antennas and Propagation Magazine, Vol.48, Issue 2, April 2006

[13] Esposito A., Tarricone L., Vallone. L., New information technologies for the CAE of MW circuits and antennas: experiences with grid services and semantic grids, Mediterranean Microwaves Symposium 2005, September 2005.

[14] D. E. Goldberg, Genetic Algorithm in search, optimization and machine-learning, Addison Wesley, 1992.

[15] F. Glover, M. Laguna, Tabu Search, Kluwer, 1997.

[16] E. Amaldi, A. Capone, F.Malucelli, "Discrete models and algorithms for the capacited location problems arising in UMTS network planning" , Technical Report, D.I.I., Politecnico di Milano.

[17] E. Amaldi, A. Capone, F.Malucelli, "Base station configuration and location problems in UMTS networks", in Proceedings of the 9th International Conference on Telecommunication Systems, Modelling and Analysis 2001, 2001.

[18] J. Zimmermann, R. Höns, H. Mühlenbein, "The Antenna placement problem for mobile radio networks: an evolutionary approach.", in Proceedings of the 8th Conference on Telecommunications Systems, pp 358-366, 2000.

[19] A.M. Vernon, M.A. Beach, J.P. McGeehan, "Planning and Optimization of Third Generation Mobile Networks with Smart Antenna Base Stations", in Proceedings of AP2000, 9-11 April, 2000, Davos.

[20] Michel Vasquez , Jin-Kao Hao, A Heuristic Approach for Antenna Positioning in Cellular Networks, Journal of Heuristics, v.7 n.5, p.443-472, September 2001

[21] L. Brunetta , B. Di Chiara , F. Mori , M. Nonato , R. Sorrentino , M. Strappini , L. Tarricone, "Optimization approaches for wireless network planning", in 2004 URSI EMTS, International Symposium on Electromagnetic Theory, Pisa (Italy) 23-27 May 2004, pp. 182-184

[22] Y. Okumura et a., "Field strength and Its Variability in VHF and UHF Land Mobile Service", in Review of the electrical Communication Laboratory, vol 16 N°9-10, Sept-Oct 1968

[23] M. Hata, "Empirical Formula for Propagation Loss in Land Mobile Radio Services", in IEEE Trans. Veh. Techn., vol. VT-29,N°3 Aug. 1980.

[24] E. Damosso, "Digital Mobile Radio: COST 231 View on the evolution towards 3rd Generation Systems. Bruxelles: Final Report COST 231 Proj.", Eur. Comm., 1998.

[25] http://www.lx.it.pt/cost231

[26] J. Walfisch, H.L. Bertoni, "A Theoretical Model of UHF Propagation in Urban Environments", in IEEE Transaction on Antennas and Propagation, vol 36 n°12 Dec.1988.

[27] F.Igekami, S.Yoshida, T.Takeuchi, M.Umeira, "Propagation factors controlling mean Field Strength on Urban Streets", in IEEE Transaction on Antennas and Propagation, vol AP-26 n°8, Aug. 1984.

[28] S.R. Saunders, "Antennas and Propagation", John Wiley and Sons, LTD 1999

[29] Catedra, M.F., J. Perez, F. Saez de Adana, and O. Gutierrez, "Efficient ray-tracing technique for three dimensional analyses of propagation in mobile communications: application to picocell and microcell scenarios", IEEE Ant. Prop. Mag., Vol. 40, No., 2, April 1998, pp. 15-27

[30] http://java.sun.com/j2se/1.4.2/docs/guide/jni/

[31] http://www.w3.org/Submission/WSMO-related/#owl-s

[32] http://www.wsmo.org/TR/d4/d4.2/v0.1/

[33] http://www.wsmo.org/

[34] http://www.w3.org/Submission/WSMO-related/#owl-s

[35] http://www.wsmo.org/2004/d4/d4.2/v0.1/20040315/

[36] http://www.daml.org/services/owl-s/1.1/related.html

[37] http://www.mindswap.org/2004/owl-s/api/

# A Formal Technique for Reducing Software Testing Time Complexity

Mirza Mahmood Baig
*Department of Computer Science & Information Technology, NED University of Engineering & Technology, Karachi, Pakistan*
mbaig2000@gmail.com

Dr. Ansar Ahmad Khan
*Department of Computer Science & Information Technology, NED University of Engineering & Technology, Karachi, Pakistan*
dransark@yahoo.com

***Abstract-Software testing is a dual purpose process that reveals defects and is used to evaluate quality attributes of the software, such as, reliability, security, usability, and correctness. The critical problem in software testing is time complexity. In this paper we have reduced the time complexity of software testing by using Grover's Search Algorithm [4] for unsorted database.***

*Keywords*: Software Testing; Database Testing; Databases; Grover's Algorithm.

## I. BACKGROUND

Testing is any activity aimed at evaluating an attribute or capability of a program or system and determining that it meets its required results. The difficulty in software testing stems from the complexity of software field. Process and program cannot be tested with a moderate complexity as described in [16]. The purpose of testing can be quality assurance, verification and validation, or reliability estimation and most importantly to reduce time complexity. Testing can be used as a generic metric as well [13, 21].

The aim of the research is to produce high quality software (i.e. free from errors) which are required to identify types of possible tests to determine symptoms of failures of software development process and to develop a new and improved testing strategy to be used during system development phases by utilizing formal concepts.

The proposed research strategy consists of three modules:

- A Formal Notation to describe specifications called "Requirement Specification Language" (RSL);
- A formal test bed (in a matrix form);
- A set of algorithms / tools.

This paper shares the progress and achievement made in the set of algorithm/tools one of such algorithm is called "Grover's algorithm". Using this algorithm we can reduce to a large extent the time complexity which is a major factor (problem) in software testing.

### *A. Conceptual Model of Proposed Testing Strategy*

The following figure summarizes the steps performed in the development of the proposed testing strategy.

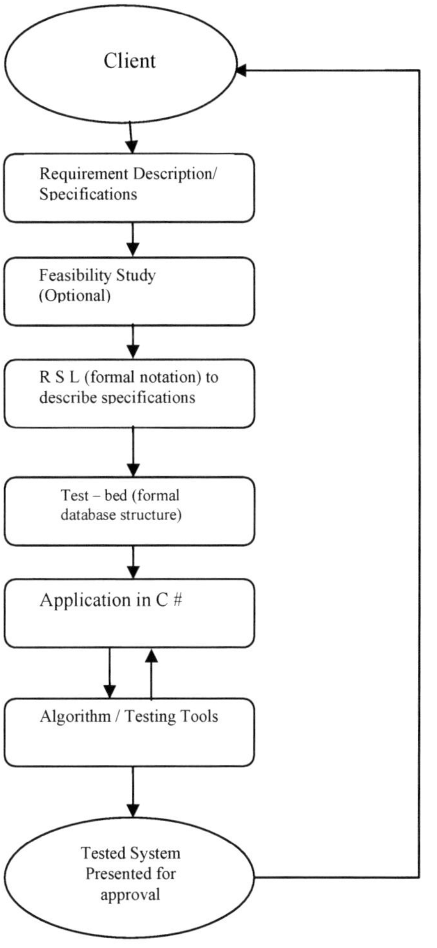


Fig. 1. Proposed Testing Strategy (Flow Diagram)

T. Sobh (ed.), *Innovations and Advances in Computer Sciences and Engineering*,
DOI 10.1007/978-90-481-3658-2_34, 

The Fig.1. Shows an overall Software Testing Strategy which will provide great assistance to system tester. In the first step client gives the requirements to the system tester. The tester first checks the feasibility of the requirement specification (if required). When the requirements specifications in consultation with the client are approved the tester changes the language of the requirements specifications into formal notation called "Requirement Specification Language" (RSL). The syntax of RSL is based on functional analysis which is quite new in the field of software testing. After that the specifications in RSL are entered in pre-designed Test-Bed a formal data structure in the form of Matrix for testing a functional requirement.

As described [4, 14, 17, 27], 50% of the total system development time is spent in testing. To resolve such problem this field required an algorithm or techniques to reduce the testing time and cost. The main problem might be observed of time when database is large and unsorted. As the tester executes test suites as a result the output message may show errors on a particular record since database record are large it is very difficult to reach that record taking lot of time and then required correction inserted. So we studied lot of searching algorithms and finally decide Grover's Algorithm to minimize time and cost because the complexity of the said algorithm is too lesser than other searching algorithms.

## II. INTRODUCTION

Software development is a continuous activity divided into a number of phases, i.e. system definition, analysis, design, coding, implementation, testing and maintenance.

Software testing is the process of executing a program or system with the intent of finding errors as per description in [16]. Software testing process is undertaken to provide correctness, completeness and quality in software. The objective of software testing is to confirm presence of software errors or defects but not their absence. Testing is usually performed for the following purposes:

- To Improve Quality
- For Verification & Validation (V&V)
- For Reliability Estimation

On the basis of available literature as per [1, 3, 4, 6, 10, 11 16, 17, 20, 21, 26, 27] the errors in existing approaches cannot be detected and removed by 100%. As discussed in [3, 7, 16, 17, and 18] so far, only 60% success is achieved. An interesting analogy parallels the difficulty in software testing with the pesticide Paradox: "Every method that you use to prevent or find bugs leaves a residue of subtler bugs against which those methods are ineffectual". Complexity Barrier principle states: "Software complexity (and therefore that of bugs) grows to the limits of our ability to manage that complexity". The complexity and challenge of the software testing process enhances time factor, as per discussions in [4, 7, 14, 27,] 50% of the total system development time is spent in testing, which justifies an improvement in success rate to detect errors and to reduce testing time

Current existing popular system development models, described in [13, 15, 23, 28], to develop software testing consists of a single stage testing process resulting in an increase of time and cost exponentially [9].

As per [7, 22] the typical software testing costs vary from 40% to 85% of the entire system development process. To remedy these shortcomings of cost and time, it is proposed that a formal method for testing be incorporated at each phase in the system development process. As per description in [5, 19, 21, 25] the testing issues defined have shortcoming because of the absence of, formal conceptual model or "a formal testing strategy"

## III. PURPOSE OF ALGORITHM

The purpose of Grover's Algorithm is usually described as searching unsorted databases with N entries in $O(N^{1/2})$ time & using O(LogN) storage space. Searching an unsorted database requires a linear search, which is O(N) in time but this algorithm which takes $O(N^{1/2})$ time i.e. the fastest possible algorithm for searching an unsorted database.[3,4,9,22] It is very useful or more efficient when database (or N) is very large. Like many quantum computers algorithms, Grover's algorithm is probabilistic in the sense that it gives the correct answer with high probability. The probability of failure can be decreased by repeating the algorithm.

As per described in overview more than 50% time is needed in database testing and correction. Trying to reduce this time with the help of said algorithm which is very fast to search unsorted databases. This algorithms has five steps one of the step namely "Walsh Hadamard Transformation Matrix", used in error correcting codes [5, 7, 30]. The Quantum Computer have facility to inverting a function, this algorithm is more well-organized and unambiguous to express it as an "Inverting a function". This means algorithm is based on backtracking from output to input. The inverting a function is related to the searching of a database from output to input because we could come up with a function y=f(x) that produces a particular value of y if x matches a desired entry in a database, and another value of y for other values of x and so on.

The algorithm is based on quantum computing where Quantum Computers are expected in the world market in the near future as per describe in [4, 25]. By application of this algorithm it is expected that the testing time will be considerably reduced. Our existing digital computers are not capable to execute this algorithm because digital computers are based on 0 & 1 but Quantum Computers based on qubits [0 . . . 1]. The qubit is a system which belong to closed interval [0, 1] it takes the values |0>and |1>, Notation like '|>' is called Dirac notation, and combination of intermediate values of 0, 1. The difference between bits and qubits is that qubit can be in a state other than |0> or |1>. It is also to form superposition of two states: |C> = A |0> + B |1> The number

A & B are complex numbers, although for many purposes not much is lost by thinking of them as real numbers as describe in [ 22].

### A. *Application of Grover's Algorithm*

In order to express how the Grover's algorithm contributes in searching the unsorted database, I have automated the steps of the algorithm using C# (C Sharp). The program takes m as input, which is the value of required number of bits, necessary to hold N records and can be given as $\lceil \log_2 N \rceil$. It then automatically generates the $2^m x 2^m$ Hadamard matrix as well as the matrix of inversion about average transformation. Then, it applies the Grover's algorithm as describe in [4] step-by-step and displays the state of the Q-register after each step.

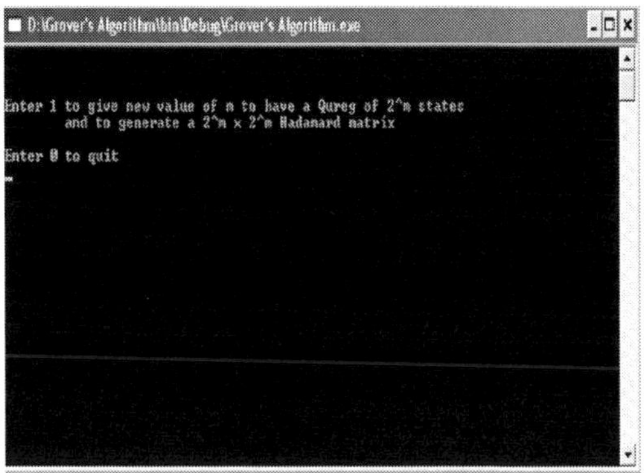


Fig.2. Application prompt to give new value or quit the application

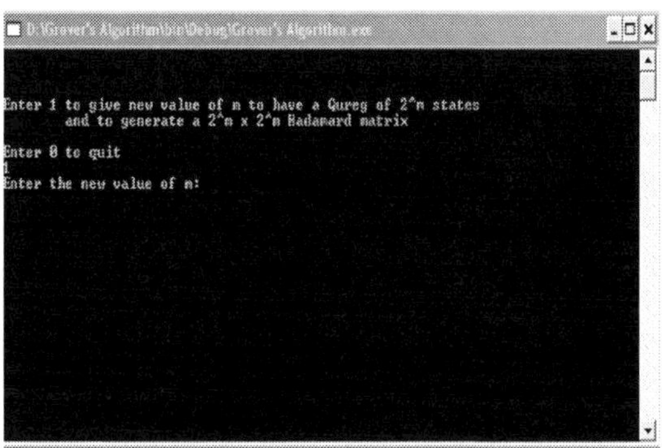


Fig.3. ***Application*** prompt to enter value of m (database items) and to generate the Hadamard matrix of order $2^m$ x $2^m$

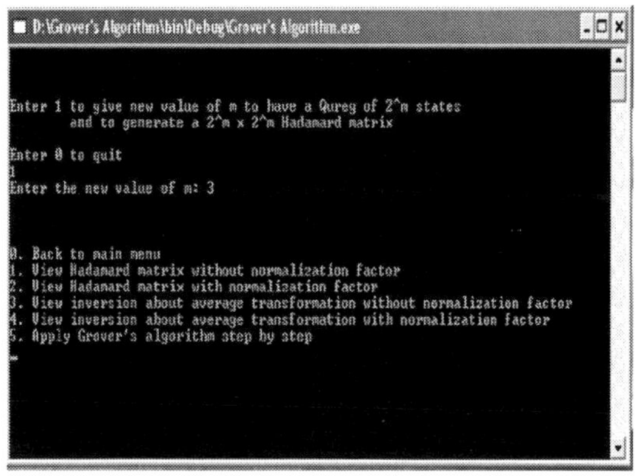


Fig.4. Consider m=3 i.e. the database has N=8 records
Here we have take the small value of 'm' due to understand the each step of the algorithm clearly instead of large value. As we know that the algorithm is most efficient when the data items are very large enough.

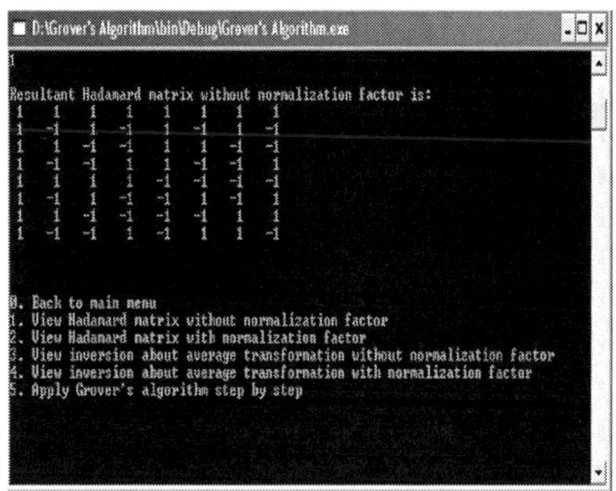


Fig.5. Displays a Hadamard Matrix of order 8x8 without normalization factor. Where, "Hadamard matrix is a square matrix whose entries are either +1 or −1 and whose rows are mutually orthogonal". This means that every two different rows in a Hadamard matrix represent two perpendicular vectors. This matrix is directly used as an error correcting code. The Hadamard matrix can also be used to generate random numbers. The idea of using Hadarmard matrix is quite significant for Software Testing.

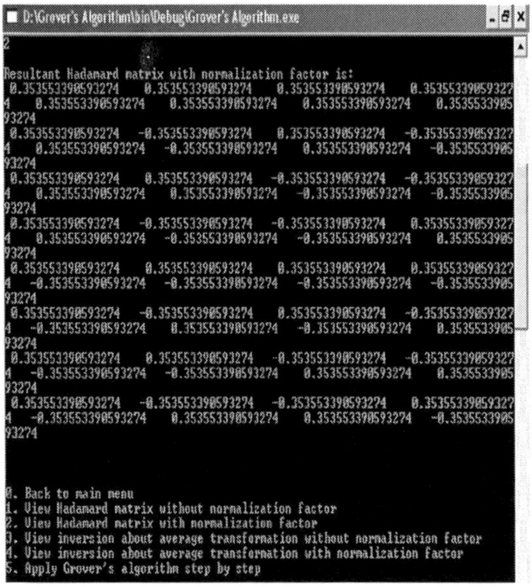


Fig.6. Shows Hadamard matrix after multiplying the normalization factor which is $1/2^{3/2}$

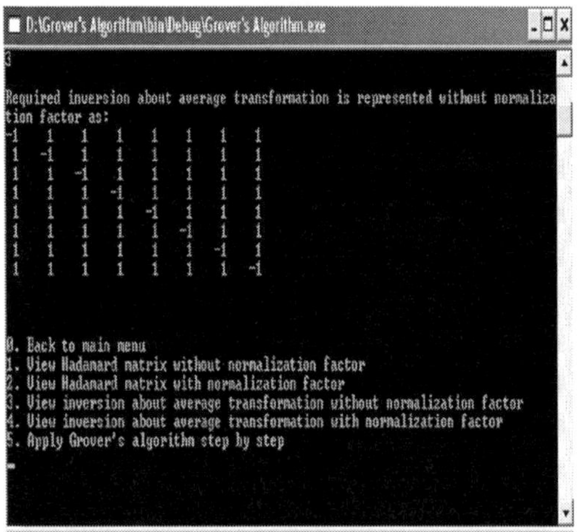


Fig.7. Displays an inversion about average is represented by a matrix without normalization factor. This inversion operator is meant to rotate the phase of a given search record.

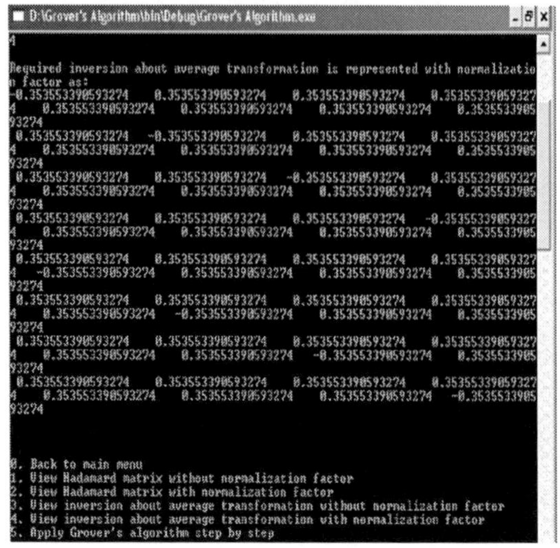


Fig.8. Shows inversion about average with normalization factor

*B. Grover's algorithm is shown step by step*

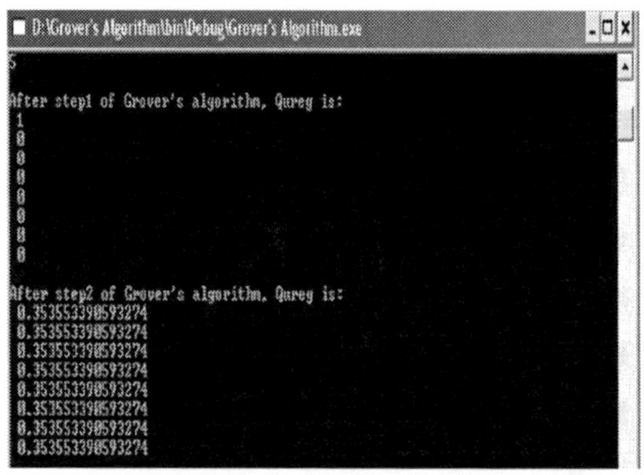


Fig.9. Shows step of Grover's Algorithm

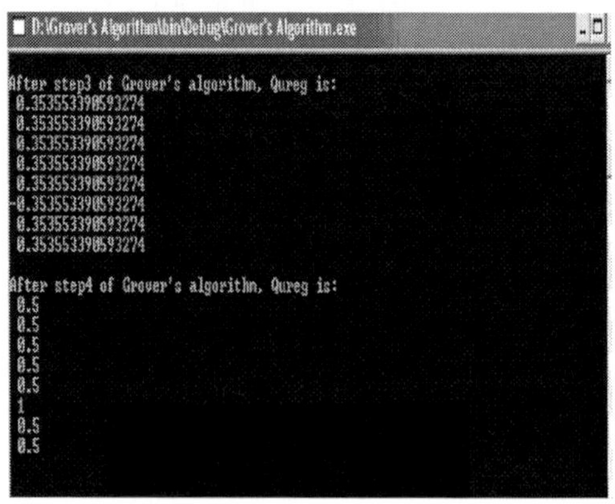

**Fig.10.** Shows the remaining step of Grover's Algorithm. In this figure the row number six with probability one so we will measure the state six which is our required mark state.

The important point to note here is that the complexity statistics of this application program would clearly contradict the desired complexity of Grover's algorithm which is O ($N^{1/2}$). This is because of the fact that the algorithm is dependent on a quantum subroutine that marks a unique state, satisfying the condition C ($S_v$) =1, in unit time, and we know that this is obviously impossible for a classical computer, which we are currently using. This contradiction will be clearly overcome when quantum subroutine is operational.

Since the working procedure of the quantum subroutine is something in a black box for the Grover's algorithm, therefore the demonstration program takes assistance of a random-number generator that marks any record randomly whenever required. However, even this random marking of the record requires a traversing of the database and the worst-case complexity of even this traversal is O (N).

It is, therefore, shown that this program is only for demonstration purposes and is especially useful for large values of m, for achieving the same complexity and optimality as that of the Grover's algorithm, when quantum subroutine becomes operational.

## IV. CONCLUSION

This paper emphasizes on the most important phase of Software Testing, i.e. "Reduction in Testing Time". It is claimed that this time complexity can be achieved by applying a Mathematical algorithm known as, Grover's Algorithm which is O($N^{1/2}$). If quantum subroutine is available (which is expected in near future)? This time complexity will be at least 100 times faster than any other possible algorithm for searching an unsorted data base(If N is very large). For example the number of database items is N=10000 and the error occurred at N=9998. Now this algorithm technique will search the required error item in maximum of 100 iteration. In comparison to this algorithm any other possible algorithm requires a traversing of the data base and worst traversal is of O(N).

## ACKNOWLEDGMENT

We acknowledged the all kind of support by NED University of Engineering & Technology and specially the Vice Chancellor as chairman of Advanced Studies Research Board who always personally supported the research project being undertaken.

## REFERENCES

[1] Cem Kaner, Jack Falk, Hung Quoc Nguyen. *Testing Computer Software*. Wiley. 1999

[2] Cem Kaner, James Bach, Bret Petti Chord, *Lesson Learned in Software Testing*. Wiley, Ist Ed. 2001

[3] C.H. Bennett, E. Bernstein, G. Brassard & U. Vazirani, "Strenghtsand weaknesses of quantum computing" ,to be published in the SIAM Journal on computing 1996.

[4] C. Lavor, L. R. V. Manssur, R. Portugal, "Grover's Algorithm: Quantum Database Search", arXiv: quant-ph/0301079v1, May 25, 2006

[5] D. Deutsch and R. Jozsa, "Rapid solution of problems by quantum computation", Proceedings Royal Society of London, A400, 1992, pp. 73-90

[6] Elisa Ludwig. *Software Testing should reveal errors not avoid them*. Washington information source. 2003

[7] E. Bernstein & U. Vazirani, "Quantum Complexity theory", Proceedings 25th ACM Symposium on theory of computing, 1993, pp. 11-20.

[8] Glenford J. Myers, *The Art of Software Testing* 2nd edition, John Wiley & sons Inc.2004

[9] Grover L.K., "A fast Quantum mechanical algorithm for database search", Proceedings, 28th Annual ACM Symposium on the Theory of Computing, (May 1996) p. 212

[10] Hillar Puskar, *Effective Software Testing, 50 specific ways to improve your Testing*. American Society for Quality, volume 6, 2004

[11] Indian Trends Software Testing. Express co.2005.

[12] Ilene Burnstein, *Practical Software Testing*. Springer, 1st Ed.2003.

[13] Jayesh G Dalal, *Software Testing Fundamentals Methods and Metrics*. American Society for Quality. 2004.

[14] Jeff Tian, *Software Quality Engineering*. Wiley-inter Sc. 2005.

[15] Jackson M, *Software Requirement & Specifications*. Addison-Wesley, 1995.

[16] Jiantao Pan, Software Testing. Carnegie Mellon University. 18-849b Dependable Embedded Systems. 1999.

[17] Lead J, Ronald, *Introduction to Software Engineering*, 2000.

[18] Marie-Claude Gaudel, *Algebraic Specifications and Software Testing: Theory and Application*. University Paris Sud. 1988.

[19] Michal Young, *Software Testing and Analysis: Process, Principles, & techniques*. John Wiley & Sons. 2005.

[20] Mark Last, *Artificial Intelligence Methods in Software Testing*. World Scientific Publishing Company.2004.

[21] Marnie L. Hutcheson, *Software Testing Fundamentals: Methods & Metrics*. Wiley, 2003.

[22] Michael A. Nielsen, Isaac L. Chuang, *Quantum Computation and Quantum Information* Cambridge University Press, 2004.

[23] N. H. PETSCHENIK, Building Awareness of System Testing Issue. Bell Communications Research. Inc. South Plainfield, New Jersey, 1985.

[24] Pressman Roger, *Software Engineering: A Practitioner's approach*. McGraw Hill, 2004.

[25] P. W. Shor, "Algorithm for quantum computation: discrete logarithms and factoring", Proceedings, 35th Annual Symposium on Fundamentals of Computer Science (FOCS), pp. 124-134. 1994.

[26] Ron Patton, *Software Testing*. Sams, 2000.

[27] Rex Black, *Managing the Testing Process:Practica ltools and techniques for Hardware and Software testing*. 2nd edition Rex Black, 2002.

[28] Scott Loveland, Geoffrey Miller, Richard Prewitt Jr., Michael Shannon, *Software Testing Techniques finding the defects that matter*. Charles River Media, 2004.

[29] Summerville Ian, *Software Engineering*. Addison-Wesley. 2002.

[30] S. Georgious, C. Koukouvinos, J. Sebberry, "Hadamard matrices, orthogonal design and construction algorithms", 2002 pp. 133-205 in DESIGN: Further computational and constructive design theory, Kluwer 2003.

# A Multi-Agent Role-Based System for Business Intelligence

Tamer F. Mabrouk
Teaching Assistant of Computer science,
Alexandria High Institute of Engineering & Technology.
*tamer_fm@yahoo.com*
Mohamed M. El-Sherbiny
Prof. of Electronics,
Institute of Graduate Studies and Research, Alexandria University
*mmsherbiny@yahoo.com*
Shawkat K. Guirguis
Prof. of Computer Science and Informatics,
Head, Department of Information Technology,
Institute of Graduate Studies and Research, Alexandria University
*shawkat_g@yahoo.com*
Ayman Y. Shawky
Head, Continuing Education and Training Center,
Alexandria High Institute of Engineering & Technology.
*ayman_shawky@yahoo.com*

***Abstract*-The development of agent-based systems grants developers a way to help investors and organizations in building efficient business systems.**

**In order to obtain relevant, consistent and updated information across a business system, Business Intelligence (BI) philosophy is implemented which is based on maintaining accurate daily information to improve latency from decision making process. In such scenario, Agents are preferred due to their autonomy, mobility and reactivity.**

**This paper addresses the development of a web role-based Multi-Agent System (MAS) for implementing BI taking stock trading as a case study.**

## I. Introduction

During the past several years, organizations have improved their business activities in order to stay competitive on today's market by deploying business management systems.
Organizations tend to make decisions based on a combination of judgment and information from different sources.

Ideally, all related information must be captured and organized together before judgment is established which is a very complex, expensive and time consuming process.

This process can be applied using BI philosophy by translating strategic objectives into measurable indicators which enables the organizations to realize opportunities or threats quickly [1].

The recent applications that implements BI are slightly inflexible, which make users adapt to them instead of conforming to the natural decision-making flow in real life. In addition to that, there is a need for having technical skills and training to implement them. In the end, productivity suffers.

On the other hand, MAS offers an intelligent, user friendly approach for implementing BI, since agents can interact with users, software legacy systems and other agents in order to capture, organize, search, exchange information and even play on behalf of users in decision-making. This leads to the development of an autonomous, mobile and reactive system in a way which is more likely to what happens in real life.

One of the most complex domains in BI is stock trading where millions are investing in buying and selling shares through different stock markets everyday. Stock trading is a dynamic domain that results in a large amount of trading information which must be captured and organized before trading actions are established. Furthermore, this kind of trading requires knowledge, skills and expertise which are expensive and hard to obtain.

In this paper, we discuss the design and implementation of a web role-based Stock Trading Multi-Agent System (STMAS) that implements the BI philosophy in stock trading, taking the Egyptian Exchange (EGX) as a case study.

This paper presents an overview of the system while focusing on agent interactions in the STMAS.

## II. Background

An intelligent agent is an entity that has a certain amount of intelligence and can perform a set of tasks on behalf of the users, due to agent's autonomy, reactivity and mobility [2].

Today's agent-based applications often involve multiple agents in order to maintain complex distributed applications.

Since intelligent agents should be able to play on behalf of the users, it is more likely to depend on roles which help

T. Sobh (ed.), *Innovations and Advances in Computer Sciences and Engineering*,
DOI 10.1007/978-90-481-3658-2_35, © Springer Science+Business Media B.V. 2010

designers and developers in modeling the intelligent agents to perform more tasks than those they are developed for, which is more likely to what happens in the real life, where people learn how to do things and expand their knowledge [3].

Intelligent Agents are designed to interoperate and interact for information gathering and processing tasks. This may include: (a) Locating and accessing information from various on-line sources. (b) Filtering any irrelevant information and resolving inconsistencies in the retrieved information. (c) Integrating information from heterogeneous information sources. (d) Instantaneously adapting to the user's needs. (e) Decision making.

Stock trading is concerned with buying and selling of shares to provide the best possible rate of return for a specified level of risk, or conversely, to achieve a specified rate of return with the lowest possible risk [4].

In real time practice, investors assign a brokerage firm to manage their portfolio. These firms employ a group of brokers and specialists for monitoring, filtering and evaluating the stock market status on a daily bases in order to manage buying and selling shares on the behalf of traders. Unfortunately, this may create a time lag between realizing an investment opportunity and taking advantage of it. In the end, investors suffer.

On the other hand, MAS approach is natural for stock trading as intelligent agents can be very significant due to: (a) Agents don't forget, get tired or make mistakes as human experts sometimes do. (b) Agents can serve all traders all the time while human experts can serve few traders only during his/her working hours. (c) Agents provide permanent and instantaneous documentation of the decision process. (d) Agents can monitor and analyze all the shares in a trading session while human experts can only review a sample.

During the past few years, there have been several agent-based approaches that addressed the issue of stock trading. Most of the agent-based systems have focused on monitoring and extracting the stock market information via the internet to put stock trading strategies as in [5], some researches have focused on the stock management to produce recommendation for a stock buy or sell decision as in [6] and others have partially focused on the stock trading process as in [7]. While implementing these capabilities in a real time trading on behalf of users has not been thoroughly explored.

This complex problem motivates our research in MAS for stock trading. STMAS applies an intuitive trading strategy on behalf of users which has been advised and reviewed by experts in both EGX and several private brokerage firms.

## III. STMAS Architectural Model

STMAS is a novel intelligent distributed architecture emphasis on stock trading. The system was designed and implemented to create a consistent environment where agents can interoperate and interact to capture, filter and organize information resulting in an accurate daily information used for generating buy and/or sell decisions which may be used as a recommendation for the user or in processing a trading action in case the system is trading on behalf of the user.

STMAS borrow the real stock market data from the EGX Website [8]. There are time lags between the system and the real stock data (15 min) due to the data availability on the EGX website. These time lags can be eliminated if proper authorization is granted.

Recently, most of the existing web stock trading systems are either information providers such as Egypt for Information Dissemination (EGID) [9] or simulation systems which aim to educate students or any other interested party in stock trading fundamentals such as STOCK RIDERS simulation game [10].

STMAS was developed and implemented using Java/JSP programming technique on the internet. STMAS Architectural design is shown in Fig. 1.

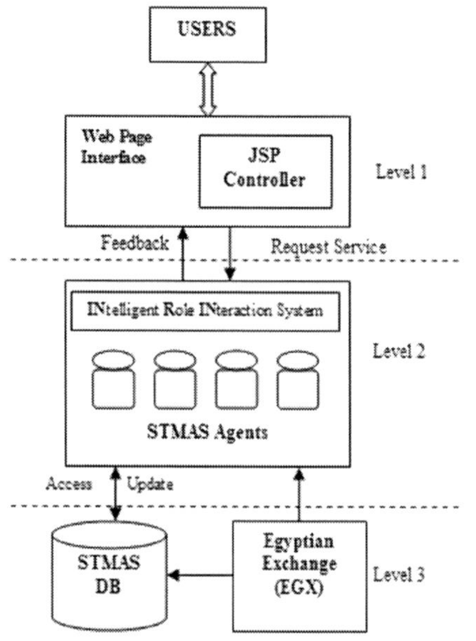


Fig. 1. STMAS architectural design

STMAS is a three-tiered system which consists of:

1. The first level is the JSP controller, which links the users and STMAS agents.
2. The second level consists of the STMAS agents and the **IN**telligent **R**ole **IN**teraction System (INRIN). It is considered as the STMAS kernel.
3. The third level is a data access layer where information about stock market status and system users is stored in STMAS database (STMAS-DB).

## IV. STMAS Major Functions

### A. Stock Information Retrieval

Most of the existing web stock trading systems provide this function, it is not a unique function in our system, but it is the basic requirement for stock trading. STMAS monitors the market status to capture the stock information from the EGX website and organize it to provide the following information. (a) Current market data. (b) Stock quotation. (c) Historical market data in the form of technical charts.

### B. User Portfolio Management

STMAS creates a trading portfolio for each user to provide the following functions. (a) Managing a user's balance in both Egyptian pound and US dollars, where a user can deposit and withdraw from his/her account. This balance is the user's capital for buying shares. (b) Editing the user's shares by monitoring the current status for these shares and calculating the current profit or loss. (c) Editing the user's trading history and showing his/her profit or loss in each trading process.

### C. Technical Analysis Reports

STMAS generates the following reports on daily bases after the market is closed. (a) Resistance and Support levels (Pivot Points). (b) A list of the recommended shares to buy in the next day trading. (c) Risk to Reward ratio. (d) The Highest High and the Lowest Low for n trading day.

### D. Trading Recommendation

STMAS checks the current market status of the ordered shares, and then passes the share's data to the STMAS Rule Engine (STMAS-RE) to recommend whether to buy, sell or hold the share.

### E. Close Price Prediction

The STMAS system can predict whether the following day's closing price would increase or decrease by combining five methods of analyzing stocks. The five methods are On Balance Volume (OBV), Price Momentum Oscillator (PMO), Relative Strength Index (RSI), Stochastic (%K) and Moving Average (MA) [11].

All five methods needed to be in agreement for the algorithm to predict a stock price increase or decrease. If neither of the two predictions were met, then no predictions are made.

### F. Stock Trading

The most important issue in stock trading is choosing the best share to buy and determining the right time to buy or sell the share. In fact, it is a very complex and difficult process, as every trader has his/her own buying and selling strategies.

STMAS provides an intelligent stock trading strategy advised by experts in this domain which depends on a combination of the current market status, technical analysis tools and the system trading recommendations to reduce the trader's work overload.

STMAS trading strategy is an intuitive strategy that adopts assumptions similar to those held by human traders, which can be summarized in to a single rule: "Buy Low, Sell High".

Our trading strategy is based on: (a) Monitoring the current stock status and evolving information needs. (b) Reflecting changing situations and recognizing events. This helps in evolving the trading strategy dynamically in order to gain profits or to stop loss. STMAS can also adapt to other trading strategies based on rules defined by the users themselves.

## V. INRIN: Intelligent Role Interaction System

Roles can be used as a prototype to model real life where the human counterpart can be performed in a way similar to what human may do. This concept was evolved so that it can be applied to agents in order to define a way for performing the common interactions between agents and the working environment to increase agent capabilities, granting adaptabilities and interactions. The concept of Role can be defined as:

*"A Role consists of the initialized/gained knowledge and the capabilities which might be actions and/or reactions that can be acquired and implemented determined by the current state of the agent in the working environment".*

Since, it is not possible for MAS to have a complete platform-independent implementation; our effort was directed to reduce the platform-independent part. In our approach, roles are implemented using Java, a role is composed of:

1) *Role Interface:* Java server page (JSP) is responsible for representing the agent's role interface at runtime. The STMAS' JSPs construction is shown in Fig. 2.

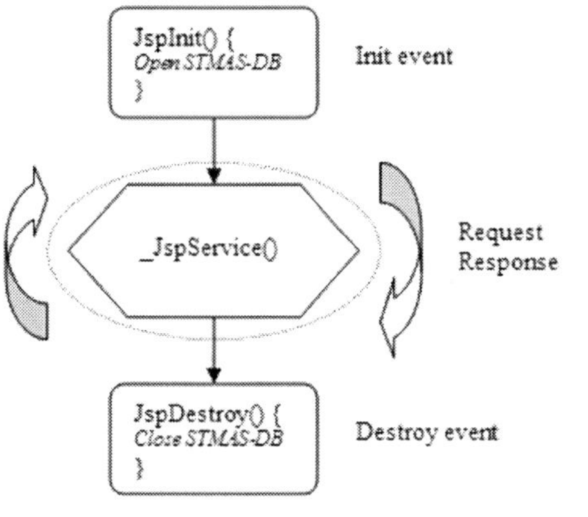


Fig. 2. JSP construction

2) *Role Implementation*: An abstract java class where role features are expressed by static fields and methods. The class that implements a role is a part of a package which represents the role scenario.

This leads to a fact that agent role acquiring may become limited as required role classes must be known at the implementation phase to combine them with the agents. Thus, agents act as separated components.

To overcome this limitation, we have implemented INRIN, where roles are dynamically assigned to agents in runtime. Thus, when a role is required, the INRIN adds both Role-Interface and Role-Implementation at runtime to the agent who is assigned to that role, this is shown in Fig. 3.

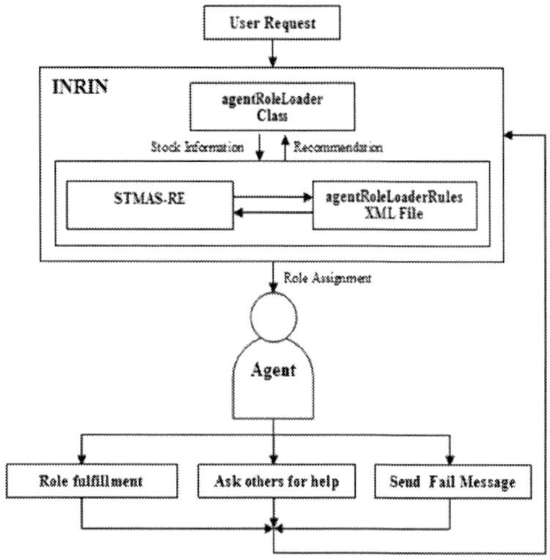


Fig. 3. INRIN system and role assignment

The intelligent component in INRIN is the STMAS-RE, where dynamic role assignment is performed in memory without recompilation using a special class called "agentRoleLoader". The role assignment is shown in the following algorithm:

1. *The user and/or agent request is passed to the STMAS-RE.*
2. *STMAS-RE fetches the corresponding rules for role assignment from an XML file called "agentRoleLoaderRules".*
3. *STMAS-RE fires the fetched rules to translate the user and/or agent request into Role-Interface and Role-Implementation which are assigned to the proper agent to fulfill this role.*

After acquiring a role, agents may successfully perform the role, or may ask other agents for help if needed, or send a fail message back. By this way, agents can take the advantage of local processing in: (a) Reducing the communication latency in roles that acquire high data transfer rate. (b) Making decisions based on the current situation which is more likely to what happens in real life.

## VI. STMAS AGENTS

In our approach, each agent has its own knowledge and capabilities and can communicate only with certain others. Role assignment and agent interactions take place in the INRIN system, where agents are assigned to a role to play or they might communicate with other agents that might partially participate in this role.

During the operational phase, agents have a certain degree of awareness of what others are currently doing, which might be: (a) Fully aware – the agent is fully aware of its currently acquired role. (b) Partially aware – the agent is partially aware of roles that it participates with others. (c) Unaware – the agent is unaware of others' roles that it doesn't participate in it, as each agent has specific roles to play in the system. Currently, the STMAS agents are:

### *A. Trading Information Agent*

The Trading Information Agent (TIAgent) is responsible for monitoring the stock market status through the internet and retrieve the stock information every 15 min from 10:30am to 2:30pm. The retrieved information is stored in the STMAS-DB which is shared among all agents. TIAgent is also responsible for supporting users with stock quotation, current market status and historical market data.

### *B. System Manager Agent*

The System Manager Agent (SMAgent) interacts with the users, receives user requests and passes them to INRIN system where user requests are translated into roles which are assigned to STMAS agents to process. SMAgent is also responsible for user login and registration.

### *C. Role-Interface Generator Agent*

The Role-Interface Agent (RIGAgent) creates Role-Interfaces for agents in order to interact with users to perform their assigned roles.

### *D. Trading Expert Agent*

The Trading Expert Agent (TEAgent) retrieves stock information from the STMAS database to generate the trading recommendations and the technical analysis reports during the trading session. After the market is closed, TEAgent starts to generate the next day trading reports which include: (a) A list of the recommended shares to buy. (b) The close price prediction list which contains whether the shares' closing price would increase or decrease. (c) The fundamental stock analysis reports.

### *E. Broker Agent*

The Broker Agent (BAgent) is responsible for managing the user portfolio and processing the user's buying and/or selling orders based on the system trading strategy if the user chooses to assign BAgent to trade on behalf of him/her or simply to execute the user's trading strategy rules.

During the trading process, BAgent asks the TEAgent for trading recommendations based on the current market status and the fundamental technical analysis reports for shares that the user ordered to buy and /or sell.

## VII. STMAS Agent Interactions

As mentioned in section 3.1, each role consists of initialized and/or gained knowledge and the capabilities which might be actions (scheduled) and/or reactions (delegated by users and/or other agents) during the working environment.

Agent interactions are governed by the INRIN system which is responsible for assigning roles to the STMAS agents. In this section, we will present a sample of agent interaction scenarios.

### *A. Agent Interactions in Information Retrieval Process*

STMAS can provide several forms of information retrieval based on: (a) Scheduled process. (b) User delegation. (c) Interoperation process. These include: Stock information retrieval from the EGX website every 15 min (Scheduled); a given stock quotation of current day (delegated); a given stock history price over a certain period (delegated); a given stock history volume over a certain period (delegated); a given stock close price prediction for the next day trading (delegated); a given stock fundamental analysis data (delegated); a given user's trading history (delegated).

The agent's interaction for getting stock quotation is shown in the following algorithm:

1. *The SMAgent accepts the user request and passes it to the INRIN system.*
2. *The INRIN system accepts the passed request and fires the corresponding rule for stock quote role assignment and assigns the TIAgent to this role.*
3. *The TIAgent sends a request back for the INRIN system for Role-Interface creation which fires the corresponding rule and assigns this role to the RIGAgent.*
4. *The RIGAgent connects to the STMAS-DB using SQL interaction to get a list of all stocks currently traded in the market then generates and loads the Role-Interface for this role.*
5. *The user chooses a stock and sends it back to the TIAgent which connects to the STMAS-DB using SQL interaction to get the required stock quotation.*
6. *Finally, the TIAgent presents the results in a readily form to the user.*

For the other information retrieval processes, another agent interactions scenario exists.

### *B. Agent Interactions in Stock Trading Process*

Agent interactions in stock trading is the most complex interacting process as all agents are interacted together for performing the buying and/or selling orders. The stock trading process is shown in the following algorithm:

1. *The SMAgent accepts the user request and passes it to the INRIN system.*
2. *The INRIN system accepts the passed request and fires the corresponding rule for stock trading role assignment and assigns the BAgent to this role.*
3. *The BAgent sends a request back for the INRIN system for Role-Interface creation which fires the corresponding rule and assigns this role to the RIGAgent.*
4. *The RIGAgent connects to the STMAS-DB using SQL interaction to get a list of all stocks currently traded in the market then generates and loads the Role-Interface for this role.*
5. *The user places a buy order and posts it back to the BAgent which connects to the STMAS-DB using SQL interaction to save this order.*
6. *Step 5 is repeated until the user places all his/her buying orders.*
7. *After the user finishes placing his/her buying orders, For each ordered stock:*
   a. *TEAgent connects to the STMAS-DB using SQL interaction to get the ordered stock current status and the fundamental analysis data previously calculated after the closing session of the previous trading day in order to generate a trading recommendation to the BAgent.*
   b. *Step a is repeated (after the TIAgent gets the current stock information from EGX website every 15 min and saves them in the STMAS-DB) until BAgent processes the user buying order.*
   c. *The BAgent updates the user's balance.*
   d. *The BAgent changes the trading strategy from buying to selling and starts to wait and watches for the TEAgent trading recommendations.*
   e. *Step d is repeated every 15 min until BAgent sells the ordered stock.*
   f. *The BAgent updates the user's balance and trading history.*

## VIII. Implementation and Experimentation

At the time of writing, this research work is still ongoing. The STMAS has been partially implemented using the following experimental platform as shown in Fig. 4.

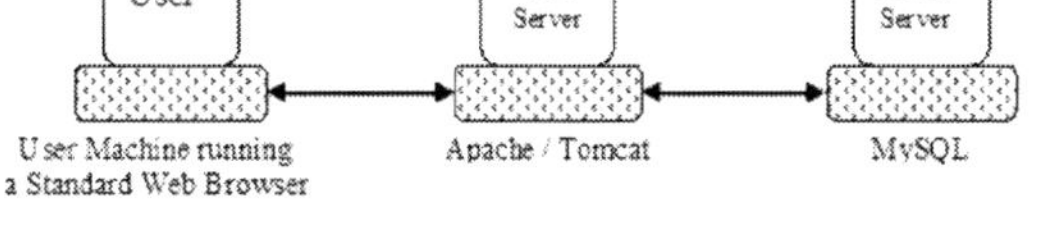


Fig. 4. STMAS experimental platform

STMAS design uses a three-tier architecture built around a web server which is responsible for providing HTTP service and serves both static and dynamic web pages.

Java/JSP programming technique was used in developing and implementing STMAS. Users can interact with the system from the internet through a set of JSP files which are served by Apache/Tomcat web server.

The JSPs, running on the web server, accept users' requests and pass them to the STMAS. According to the user requests, STMAS assigns specific agents to perform the requests on behalf of users.

For the purpose of validating our system prototype, we use real trading data collected from the EGX website on daily bases to test the functions of our system. The data used is the trading data from the date 02/05/2007 till present. A number of experiments have been made, but due to space limitations, an experiment in stock trading is shown.

For the task of stock trading, it is necessary to point out that the STMAS stock trading strategy only aims to help the users in gaining 1% to 3% profit out of their day trading.

Currently, the system simulates the buying and/or selling processes as the permission to process users' orders in the real market is not yet granted.

The chosen experiment was conducted in 08/07/2008 on a stock that will refer to as ABF. It is remarked that the chosen stock was selected from the next day recommended trading list.

The experiment scenario is illustrated in table I. At first, the user placed an order to buy a 1000 share of the stock ABF. The stock opening price was 242.90 L.E. At 11:00 am, the TEAgent recommended the BAgent to buy 500 shares at 243.40 L.E. At 11:30 am, the TEAgent recommended the BAgent to buy the other 500 shares at 246.80 L.E. At 02:00 pm, when the price reached 251.51 L.E., the TEAgent recommended the BAgent to sell all shares. The shares were sold at 245,100 L.E. making a profit of 6,410.00 L.E (2.61 %.)

TABLE I
THE EXPERIMENT SCENARIO

| Time | Current Price | TEAgent Recommendation | BAgent Reaction |
|---|---|---|---|
| 10:30 am | 242.90 | Hold | Wait and Watch |
| 10:45 am | 235.00 | Hold | Wait and Watch |
| 11:00 am | 243.40 | Buy 500 shares | Process Recommendation |
| 11:15 am | 244.43 | Hold | Wait and Watch |
| 11:30 am | 246.80 | Buy 500 shares | Process Recommendation |
| 11:45 am | 246.80 | Hold | Wait and Watch |
| 12:00 pm | 247.65 | Hold | Wait and Watch |
| 12:15 pm | 248.00 | Hold | Wait and Watch |
| 01:00 pm | 247.90 | Hold | Wait and Watch |
| 01:15 pm | 249.15 | Hold | Wait and Watch |
| 01:30 pm | 251.00 | Hold | Wait and Watch |
| 01:45 pm | 255.00 | Hold | Wait and Watch |
| 02:00 pm | 251.51 | Sell all shares | Process Recommendation |

## IX. CONCLUSION

The Stock Trading Multi Agent System (STMAS) is a user-friendly web role-based multi-agent system that has been designed and implemented in the field of business intelligence with emphasis on the stock trading to help investors in buying and selling shares through the Egyptian Exchanges.

The system was developed to serve the Business intelligence philosophy which is based on maintaining an accurate daily information that can be introduced to the users which improves decision making process and realizes opportunities or threats quickly.

Thanks to the agent autonomy, mobility and reactivity, the system offered a practical intelligent solution for applying Business Intelligence philosophy in building intelligent business systems that can extract electronic information from various data sources, linking the useful facts and filtering out irrelevant information and identifying reasonable decisions.

The system is a three-tiered system, the first level is the JSP controller, which links the users and STMAS agents, the second level is the STMAS agents and the INRIN system which is considered as the STMAS kernel and the third level is a data access layer where information about stock market status and system users are stored in STMAS database.

In STMAS, roles are assigned to agents dynamically by the INRIN system which assigns the Role-Interface and Role-Implementation to the proper agent to fulfill the required service.

Currently STMAS consists of five agents to provide a set of services such as new clients' registration, check stock quotation on daily basis, edit stock market historical data, closing price prediction, calculate the fundamental stock analysis data, stock trading, and management of user balance, shares, orders and trading history.

## REFERENCES


[1] Mike Biere, "Business Intelligence for the Enterprise", Prentice Hall PTR, 2003.

[2] N. R. Jennings, M. Wooldridge, eds., "Agent Technology: Foundations, Applications, and Markets", Springer-Verlag, 1998.

[3] G. Cabri, L. Ferrari, L. Leopardi, "Role-based Approaches for Agent Development", Proceedings of the 3rd Conference on Autonomous Agents and Multi Agent Systems (AAMAS), New York, USA, July 2004.

[4] Markowitz, H., "Portfolio selection: efficient diversification of investments", Cambridge, MA: B. Blackwell, second edition, 1991.

[5] Garcia, A., Gollapally, D., Tarau, P. and Simari, G., "Deliberative Stock Market Agents using Jinni and Defeasible Logic Programming", Proc. Of the 14th European Conference on Artificial Intelligence, Workshop on Engineering Societies in the Agents' World (ESAW'00), 2000.

[6] Y. Luo, D.N. Davis, & K. Liu, "A Multi-Agent Decision Support System for Stock Trading" , The IEEE Network Magazine Special Issue on Enterprise Networking and Services, Vol.16, No. 1, 2002.

[7] Raju Tatikunta; Shahram Rahimi; Pranav Shrestha; Johan Bjursel, "TrAgent: A Multi-Agent System for Stock Exchange" , Proceeding of the IEEE/WIC/ACM International Conference on Web Intelligence and Intelligent Agent Technology, 2006.

[8] The Egyptian Exchange (EGX), *http://www.egyptse.com*

[9] Egypt For Information Dissemination (EGID),*http://www.egidegypt.com*

[10] Stock Riders Simulation Game, *http://www.stockriders.com*

[11] Bruce M. Kamich, "How Technical Analysis Works", New York Institute of Finance, 2003.

# LERUS: A User Interface Specification Language

Fernando Alonso, José L. Fuertes, Ángel L. González, Loïc Martínez
Dept. LSIIS
School of Computing. Technical University of Madrid
Madrid, Spain

*Abstract*— **Script or special-purpose languages for describing user interfaces are proving to be very useful. Such languages should be simple and extendible depending on the application domain. In our experience the use of a small language to define user interfaces, together with components and a flexible interface rendering framework reduces the effort it takes to learn to build user interfaces, obviates the need for highly qualified developers, and also very much favours the reuse of earlier development work. This paper describes the LERUS script language and the COREUS framework developed according to these principles and evaluates the user interface development effort saved across a range of projects.**

## I. Introduction

In face of a mounting diversity of devices, platforms and technologies, the design of multi-platform user interfaces is gaining in importance and complexity. Designers are often obliged to create alternative versions of interfaces, which is a costly process. One solution is to build abstract representations of the interfaces to specify device-, technology- and platform-independent interfaces [1]. This way they can focus more on the design model than on the platform-dependent visualization details. Although several abstract interface design languages have been proposed [1], [2], [3], [4], they do not always produce an efficient interface development effort/cost ratio.

User interface programming is a tough task on which a number of aspects, ranging from requirements capture to ergonomics and usability factors, have a bearing. The user interface (UI) is likely to undergo changes during the system's life cycle and users should be able to customize it at execution time. For this reason, the interface programming language should aid interface development and its modification both during development and at execution time. Most user interface design languages are complex; neither are they easily extendible to include new types of interaction objects.

We propose the use of the LERUS (User Relations Specification Language) script language [5] as a basis for programming component- and event-driven UIs. The features of this type of languages are: the interface is described by a script interpreted when the application is loaded; the interface language is defined by the solution components rather than by a grammar; and the scripts are interpreted in a distributed manner through the cooperation of all the solution components. Also we present the COREUS (User Relations Builder) framework [5]. COREUS generates the LERUS-specified UI at run time. It adapts this UI to the platform on which the application is executed. The remainder of the article is organized as follows. Section 2 describes the key UI languages today. Section 3 presents the solution we propose for designing UIs using the LERUS language, together with the COREUS framework. Section 4 presents the Granada's Alhambra as a case study. Section 5 shows the results of applying LERUS-COREUS to several systems, and finally section 6 presents a few concluding remarks.

## II. Related Work

Some of the key user interface (UI) design languages are:

### A. XUL

The XML User interface Language (XUL) [6] is a mark-up language that was designed to describe and create portable user interfaces. The aim was to develop complex multi-platform applications without the need for special tools. XUL is XBL-extendible. XBL (eXtensible Binding Language) is a XML-based mark-up language for implementing reusable components (bindings) that can be bound to elements in other documents. The element with a specified binding (bound element) acquires the new behaviour specified by the binding.

### B. XAL

eXtensible Application Language (XAL) is an open declarative language for building Enterprise Web 2.0 applications. It was designed to work with other leading specifications to enable application development and scalable run time operation. XAL is used to build applications that run on Nexaweb's multi-technology Universal Client Framework (UCF) [7]. With Nexaweb UCF, applications can be deployed on other browsers, in this case using Ajax or Java. XAL defines schema-based APIs to define UI components, to bind UI components to data, to bind business logic to UI events, to modify applications at run time, and to validate user input.

### C. UIML

User Interface Markup Language (UIML) is an XML language for defining UIs. This way the UI can be described in declarative terms. Designers do not exactly specify what the UI is going to look like, but rather what elements it is to show, and how they should behave [2]. The UIML description can be used to generate UIs for different platforms. Different platform capabilities make a complete translation difficult. UIML is used to define the location, the actual interface elements and design of controls, and the actions to take when certain events take place. Interface elements are buttons, lists, menus and other controls that drive a program in a graphical interface. UIML is not easily extendible to accommodate new interaction elements, as it completely defines the grammar.

T. Sobh (ed.), *Innovations and Advances in Computer Sciences and Engineering*,
DOI 10.1007/978-90-481-3658-2_36, © Springer Science+Business Media B.V. 2010

### D. XAML

eXtensible Application Markup Language (XAML) [4] is a declarative XML-based language to describe UIs. XAML is used extensively in .NET Framework 3.0 technologies. In Windows Presentation Foundation, it is used as a UI mark-up language to define UI elements, data binding, eventing, and other features. In Windows Workflow Foundation, workflows can be defined using XAML. Microsoft has built a toolkit as an aid for visually creating interfaces using XAML. The problem with this technology is that it is linked to the .NET platform. Also as the language is fully defined, it is hard to extend to incorporate new components. This is not so much of a handicap (save the dependence on Microsoft's roadmap), as XAML is a Microsoft proprietary technology and Microsoft will incorporate new components.

Clearly, the current trend is to build highly expressive declarative languages to aid interface design. They do make the interface application independent, but lay the flexibility to incorporate new objects into the interface on the line. Additionally, some languages are confined to Web applications, and others are linked to a specific manufacturer.

We bring up the need to define a simple language that is extendible to accommodate new elements. The incorporation of new elements and language extensions should add to, not modify, the existing framework implementation, applying the plug-in or widget principle. The interfaces defined using this language will be processed by a set of components (framework) that will render the interface for a specific platform. The language will not be linked to any specific platform and will be able to be used on any platforms for which there is a framework implementation.

## III. SOLUTION PROPOSAL

The proposed solution is to use the LERUS language to describe the dialogue and specific characteristics of the interface components and apply the COREUS framework to the code generated by LERUS to create and manage the UI at execution time. COREUS has been implemented for Windows, but it can be implemented for any other operating system. The only requirement is for the platform to enable asynchronous message passing and for these messages to be associated with random length character strings. This message passing is used for communication between all the participating elements.

### A. LERUS (User Relations Specification Language)

LERUS [5] is a "light" UI modelling language that can be used to describe the dialogue process, the visual appearance and the space/sound arrangement of prospective interface components. It also specifies the services that the components have to provide. It defines the core of the language and the protocol to which both the interaction components and the service providers should conform for incorporation into COREUS through the appropriate language extensions.

We have defined a flexible and extendible language. Its full definition and interpretation depends on the interface elements. This requires a distributed and collaborative interpretation by all the components participating in the interface. This way any new components can be added to the framework.

#### 1) Characteristics of the language

LERUS is an interpreted language, and the interface can be modified at run time. LERUS is designed to be a rather easy-to-use language that does not require a lot of programming experience and has an uncomplicated syntax. The language is extendible. This implies that interaction objects, together with their interpreters, can be added to the language without having to modify the kernel. To do this, the LERUS interpreter is distributed; the core language is managed by the kernel, and the language extensions of the elements are governed by their own interpreters. The generated interfaces are composed of a collection of independent components that are coordinated by the COREUS kernel. The language is platform independent. This means that it can be used to create and run an interface in any kind of environment. But this does not signify that the specific applications described with LERUS are platform independent.

Thanks to these features, not only does LERUS implement existing technologies, but it will also be able to include descriptions of components that do not yet exist and have unknown information needs.

LERUS has four parts (Figs. 1 and 2): the environment where the dialogue is to be established, the interface elements that take part in the system, the constraints to be satisfied by the interaction elements and service providers, and the reaction of the system (behaviour) to the user or system events.

*Environment* provides the features of the environment that will be used to "filter" any interaction objects or components participating in the user dialogue process. The environment can establish whether the interface is visual, auditory, dual, etc. For each environment, further information, device use preferences, etc., can be provided.

*Interface_Elements* defines the elements in the interaction process. In an interaction process, there will be a series of component types and a set of components for each type. Each component type has some particular information needs for integration in the system. One way of supplying this information is through LERUS. This implies that each new component type integrated into the framework has its own specification language, specified by the respective schema.

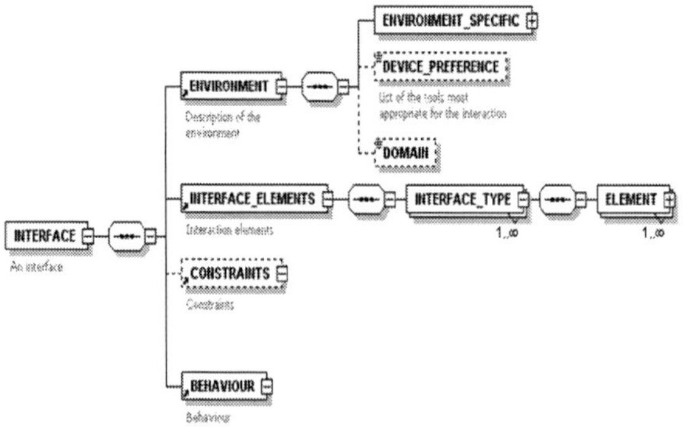


Fig. 1. LERUS structure

```
INTERFACE
 ENVIRONMENT
  TYPE=<Envirotment>
  ENVIRONMENT_PARAMETERS
    <Enviroment_Data>
  END_ENVIRONMENT_PARAMETERS
 END_ENVIRONMENT
 INTERFACE_ELEMENTS
  <Interface_Elements_Specification>
 END_INTERFACE_ELEMENTS
 CONSTRAINTS
   <Constraints_Specification>
 END_CONSTRAINTS
 BEHAVIOUR
  <Behaviour_Specification>
 END_BEHAVIOUR
END_INTERFACE
```

Fig. 2. LERUS outline

*Constraints* specifies constraints on the routine operation of the interaction components (Fig. 3). In some processes, the behaviour of a component depends on the behaviour or state of others. In these cases, a constraint needs to be activated in response to any event causing a change of state in the elements that influence the behaviour of a third party.

*Behaviour* (Fig. 4) describes, first, what conditions any service providers that are party to the interaction have to meet. It specifies what services are required and in what format the information is to be received. Second, it describes how the received events are processed. The processing will be composed of service requests to both registered providers and any interaction components that are part of the interface.

### B. *COREUS (User Relations Builder)*

COREUS is the framework developed by CETTICO [8] to create and adapt user interfaces specified in LERUS at execution time. It is responsible for generating a user interface adapted to the platform on which it runs at execution time. COREUS interprets the LERUS kernel and locates the components and service providers participating in the system. These components and service providers are responsible for defining the LERUS grammar for the system being defined.
Some of the key features of COREUS are:

- The framework components are independent. One feature of the components is that they can provide a number of services. It is easy to add new components or replace a component with an equivalent component that is better for a specific application.
- It uses an interpreted language to represent the interaction process, as suggested in [9].

```
CONSTRAINTS
 TYPE=<TYPE_ID>
 CONSTRAINT=<CONSTRAINT_ID>
  ACTIVATION=<Activation_Event>
  IMPLICATED_ELEMENTS=<Elements_ID>[, Elements_ID>]
  [REPLY=<Event_Id>[, <Event_Id>]
 END_CONSTRAINT
END_TYPE
END_CONSTRAINTS
```

Fig. 3. Constraints

```
BEHAVIOUR
 FUNCTIONS
  WHERE=SERVICE_PROVIDER
    FUNCTION1 [FORMAL_PARAMETER]
    ...
  END_WHERE
  WHERE=SERVICE_PROVIDER
    FUNCTIONN [FORMAL_PARAMETER]
    ...
  END_WHERE
 END_FUNCTIONS
 EVENTS_MANAGEMENT
  WHEN_EVENT=<EVENT_ID>
    PERFORM <ELEM_ID>,<ACTION>,<PARAM_1>...<PARAM_N>
    FUNCTIONI [PARAMETER_LIST]
   ...
  END_WHEN_EVENT
   ...
 END_EVENTS_MANAGEMENT
END_BEHAVIOUR
```

Fig. 4. Behaviour

- COREUS establishes the requirements to be met by a component and service provider for use in the framework. This means that different kinds of components and service providers with special capabilities can be incorporated depending on the state of the art and technological advances.

COREUS can be extended to incorporate new capabilities. The latest version of the framework actually added new functionalities and components to COREUS (i.e. virtual reality components and softbots support). It is now capable of building virtual applications: virtual tourism systems, demonstration systems, teaching applications, etc.

Fig. 5 shows the COREUS architecture and the relationship to the participating modules. The COREUS module generates the user interface: this element implements the LERUS core interpreter, coordinates the different interface components, and maps the description in the script to the rendering in a specific environment at execution time. Along general lines, the steps for generating an interface are:

1. Prepare the work area in which the interaction is to take place.
2. Determine any constraints.
3. Search and select the interaction elements, taking into account the results of steps 1 and 2, the objective or purpose of the interaction element and the services to be provided.
4. Search required services providers to make the system operational.
5. Assemble the interface. This involves locating and assigning the materials that an interaction object should use to generate its *physical representation* within the interface. The materials to be used are determined depending on the component's purpose and the interaction context.

The proposed LERUS-COREUS solution implemented in Windows has been successfully applied to a number of projects as discussed in section V.A. Its use in these projects validated the framework and demonstrated its utility, showing it to be reusable and modular.

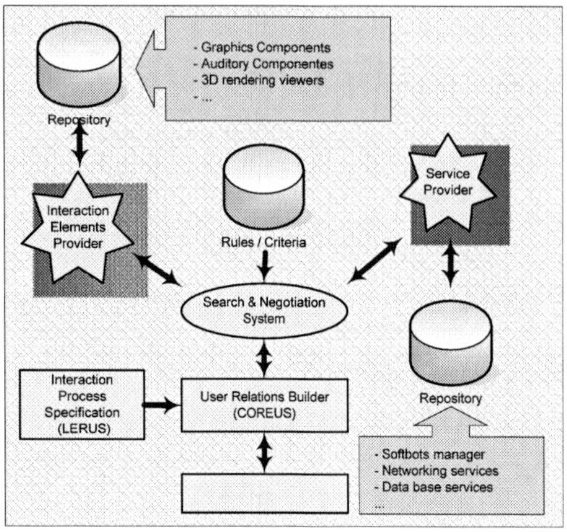


Fig. 5. Diagram of COREUS architecture

## IV. Case Study: Virtual Alhambra

This system was designed to virtually recreate the Alhambra of Granada [10]. It was the first desktop virtual reality project in which the proposed LERUS-COREUS solution was applied. Apart from the traditional interaction components, the project included *softbots* (Fig. 6). Softbots are pseudo intelligent interaction elements. We use them in order to provide two types of interaction with virtual guides: based on the principle of closed questions and geographical location, and based on the use of natural language and ontologies about the place where the visit is taking place. The application was developed for SchlumbergerSema.

The interface was implemented using LERUS scripts, which were supplied to the COREUS implementation for the selected platform. These scripts are the result of modelling the user interface interaction process in the analysis phase. The process of implementation and construction of the interface of "Alhambra Virtual" is described in detail in [10], [11].

Fig. 6. Example of command-driven softbot

Fig. 7 shows fragments of the script that represents the application's interaction process. This is the input for COREUS, which is responsible for materializing and managing the interaction. As the system consists of an application that uses desktop virtual reality technology, the interaction was designed for a 2D environment (monitor). Fig. 7 describes:

- The 2D work space in which the interface is materialized.
- The components that are part of the interaction process (interface_elements). From their definition, the interaction elements range from a button, through an interaction in multi-user virtual worlds, to a softbots-mediated interaction.
- The behaviour of each component and the predefined reaction for the user actions. These reactions can be altered by an adaptive component and constraints specification.

```
INTERFACE
 ENVIRONMENT
  TYPE=2D_ENVIRONMENT
   ENVIRONMENT_PARAMETERS
     TITLE="Alhambra Virtual"
     BACKGROUND_TYPE=IMAGE, alhambra_background
     ...
   END_ENVIRONMENT_PARAMETERS
 END_ENVIRONMENT
INTERFACE_ELEMENTS
 INTERACTION_TYPE ID="SOFT_BOT"
  ELEMENT ID="IDSB_ALHAMBRA_GUIDE"
    WORLD="Alhambra"
    DIALOGUE_TYPE="List of questions"
    USER_CONTROL=YES
    ...
  END_ELEMENT
 END_INTERACTION_TYPE
 INTERACTION_TYPE ID="VR_VIEWER_ELEMENT"
  ELEMENT ID="VIEWER_ID"
   MULTI_USER=YES
   SOFT_BOT=IDSB_ALHAMBRA_GUIDE
   WORLD="Alhambra"
   ...
  END_ELEMENT
 END_INTERACTION_TYPE
 ...
END_INTERFACE_ELEMENTS
BEHAVIOUR
 FUNCTIONS
  WHERE="STANDARD_SERVICES"
   ...
  END_WHERE
  WHERE="Virtual Worlds"
   START_CONVERSATION_WITH_SBOT String
   ...
  END_WHERE
 END_FUNCTIONS
 EVENTS_MANAGEMENT
  WHEN_EVENT="IDS_FLIGHT"
    PERFORM IDS_FLIGHT_MODE, GIVE_STATUS
    PERFORM ID_VIEWER, FLIGHT, PREV_RESULT
  END_WHEN_EVENT
  WHEN_EVENT="IDL_BOTS"
    PERFORM IDL_BOT, GIVE_SELECTED
    PERFORM START_CONVERSATION_WITH_SBOT,PREV_RESULT
  END_WHEN_EVENT
  ...
  END_EVENT_MANAGEMENT
 END_BEHAVIOUR
END_INTERFACE
```

Fig. 7. Interaction process specification

- To specify this behaviour, the domains in which services will be required are described first. In this case, standard and special-purpose services are required to visualize virtual worlds. And the reactions or actions to be taken in response to user actions on any of the interaction components are described after specifying the services.

The construction process is described in Fig. 8. To generate the infrastructure required to establish the interaction process, the first step is to identify the environment where this process is to take place (a 2D representation on a *Windows* platform) (Fig. 6). Next, we need to find out if there is a manager for this environment. This is responsible for locating the required interaction components and for coordinating the behaviours of these components to provide the required interface.

The next step is to locate all the participants and generate the required infrastructure. In this case, we need a set of interaction types for: action selection, language selection, textual information capture, and real-time 3D visualization. For this purpose, all the required interaction components are analysed. This analysis will trigger a series of search processes aimed at locating "candidate" components. The search is done based on the component's interaction capability, the required services, their availability and, obviously, whether the component is able to work in the identified environment. If there are several candidate components for the same interaction type, a selection can be made. The selection will be driven by a series of weighted criteria that will determine the best suited component in each case. The possible selection criteria are: cost, efficiency and physical location.

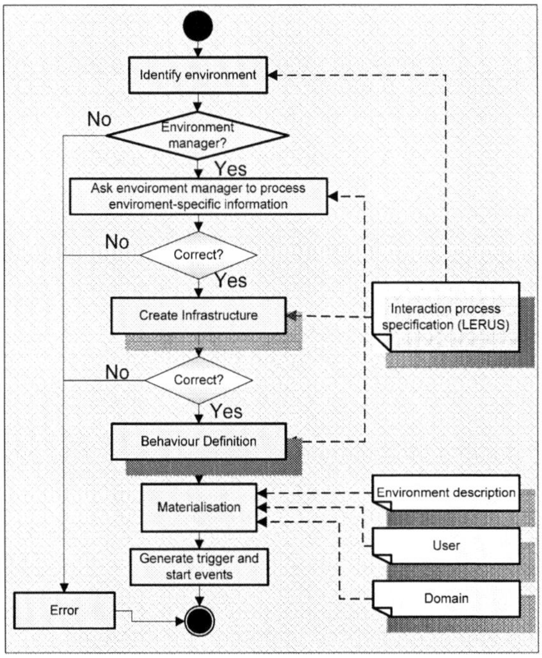


Fig. 8. Construction process

Once the components have been selected, some basic translation rules are applied to transform the interaction types into environment-specific interaction components. Very diverse interaction components need to be located for this sort of interaction process:

- Graphical button for stateless action selection interaction
- Graphical switch for state action selection interaction
- Softbot for language and location selection interaction
- List for state action selection interaction with an unknown number of alternatives
- 3D viewer for rendering the virtual world

Before ending the infrastructure generation phase, the behaviour rules required for correct application operation need to be established. In this case, the rules or constraints are aimed at synchronizing user actions in the virtual world with softbots and components responsible for reporting the user location within the world.

The next phase is to establish what responses need to be provided for the user actions and application changes of state. This will call for services provided by the participant components, general-purpose services and a set of services related to the visualization of multi-user virtual worlds. The procedure for locating possible service providers is similar to how the interaction components were located. For a provider to be selected, it must be able to work in the stipulated environment and with the selected interaction components.

The last phase is to search for the materials required to represent the interface. They include a 3D representation of the hall of the Alhambra and images for concepts such as flying, reproduction, talking to a softbot, conversing with another user, etc. To determine what materials are to be used, knowledge is required of the environment, the user and the domain. This phase outputs what quality the Virtual Alhambra visualization will have and the language to be used by the softbot to dialogue with the user during the virtual guided tour, etc.

## V. RESULTS

There follow some systems built using the proposed solution, and we evaluate the saving in development effort thanks to its application.

### *A. Systems built with COREUS and LERUS*

The proposed LERUS-COREUS solution implemented in the Windows environment has been applied to develop user interfaces for several projects:

- **Tutor** [12]: It is a tutoring system designed to teach hearing-impaired children to communicate. It was developed for the Institute of Migrations and Social Services and was the basis for implementing the first version of COREUS. The teaching process adaptation module exploits the benefits of applying COREUS' model and flexibility to modify and customize the activities depending on the pupil and the pupil's learning curve.
- **DILSE (Spanish Sign Language Dictionary)** [13]: This was one of the first SSL - Spanish dictionaries to enable searching by both SSL (Spanish Sign Language) and Spanish. The sign-based search is founded on a graphical description of the sign. The system was developed for the Spanish National Confederation of the Deaf.

- **Recreation of Mussel Growing** [14]: Its goal is to disseminate how mussels are grown in the Galician estuaries. The project combines the use of virtual reality and education. Note that, thanks to the infrastructures built in earlier projects, this system took less than three months to develop. The system was developed for the Galician Regional Government's Council of Agriculture and Fishery.
- **Sea bass breeding and feeding:** It uses virtual reality to illustrate the operations needed to breed sea bass and gilthead bream for staff training purposes. This system was implemented for the Spanish Ministry of Agriculture and Fishery's Secretary of State of Fishery's Office. To demonstrate LERUS' simplicity, the interface design and application history development were programmed by a holder of an engineering degree in computing and a technician holding a higher vocational training qualification; it was found that despite having different levels of training both were similarly efficient at putting together scenes.

### B. *Development effort evaluation*

The results are supported by customer acceptance of the above systems. Table I shows the summary of user interface development effort (in days) for each system used in the experiments. The effort for generating the materials required for each system is not included.

One key point worth noting is the reusability provided by the proposed solution. This is an aid for systems development. Table I shows that reuse varied from a minimum of 33% to a maximum of 85%. These are really quite substantial figures. There has been a system development effort saving equivalent to the percentage of component reuse.

Thanks to the availability of components developed in earlier systems which is an aid for developing the new systems, the development time trend is downward, although there are occasional peaks due to the need to develop new components.

TABLE I
SUMMARY OF DEVELOPMENT EFFORT

| | Tutor | DILSE I | DILSE II | Alhambra Virtual | Mussel-growing | Sea bass breeding and feeding |
|---|---|---|---|---|---|---|
| Kernel develop. | 140 | 50 | 0 | 20 | 0 | 0 |
| Component develop. | 220 | 90 | 10 | 200 | 30 | 5 |
| General services develop. | 40 | 10 | 0 | 0 | 0 | 0 |
| Domain services develop. | 45 | 30 | 5 | 80 | 55 | 5 |
| Constraints develop. | 20 | 10 | 0 | 3 | 0 | 0 |
| Interface design | 15 | 7 | 2 | 8 | 4 | 2 |
| Model application | 14 | 5 | 1 | 2 | 1.5 | 3 |
| **Total** | 494 | 202 | 18 | 313 | 90.5 | 15 |
| **Total reusable** | 420 | 160 | 10 | 223 | 30 | 5 |
| **Total reusable %** | 85% | 79% | 56% | 71% | 33% | 33% |

## VI. CONCLUSIONS

Recently, a number of special-purpose languages for building user interfaces have been released. These languages aim to help build the user interface —a key element of an application—, make it cheaper to modify, and reduce the interface's dependency on the rendering device. These languages are based on a highly expressive, extensive grammar. Because of their very expressiveness, however, it is hard to learn and use all their capabilities. To use these languages, developers resort to tools that mask their complexity but confine their options to the tools' capabilities. It does not make sense to provide complex languages and tools to account for all possibilities. We take the view that developers will require some facilities or others depending on the application in question. Consequently, we think that it is better to use a simple (small) language, like LERUS, that can be extended depending on the application domain characteristics, in conjunction with the COREUS framework that implements the user interface. The power of the LERUS-COREUS combination is comparable to other tools, whereas its simplicity eases its learning and the development of user interfaces, as demonstrated during the experiments.

## REFERENCES


[1] S. Trewin, G. Zimmermann, and G. Vanderheiden, Abstract representations as a basis for usable user interfaces. Interacting with Computers, 3, pp. 477–506, 2004

[2] UIML: User Interface Markup Language, http://www.uiml.org.

[3] XIML: eXtensible Interface Markup Language; a universal language for user interfaces, http://www.ximl.org.

[4] Microsoft: XAML Syntax Terminology, http://msdn.microsoft.com/en-us/library/ms788723.aspx.

[5] Á. L. González, Modelo para la Generación y Gestión en Tiempo de Ejecución de Procesos de Interacción Hombre-Máquina a Partir de un Lenguaje de Especificación de Relaciones con el Usuario. PhD Thesis dissertation, Technical University of Madrid, http://oa.upm.es/87/, 2003.

[6] Mozilla: XML User Interface Language (XUL) Project, http://www.mozilla.org/projects/xul/.

[7] Nexaweb: XAL – eXtensible Application Language, http://dev.nexaweb.com/home/us.dev/index.html@cid=1784.html.

[8] UPM (Technical University of Madrid): Grupo de Investigación en Tecnología Informática y de las Comunicaciones: CETTICO, http://meminv.upm.es/giweb/GIWEB/Grupo.jsp?idGrupo=19100504.

[9] A. Puerta, and J. Eisenstein, Towards a General Computational Framework for Model-Based Interface Development Systems. In: International Conference on Intelligent User Interfaces (IUI'99), pp. 171–178, ACM Press, 1999.

[10] J. L. Fuertes, Á. L. González, G. Mariscal, and C. Ruiz, Developing Virtual Storytellers for the Virtual Alhambra. In: Cavazza, M., Donikian, S. ICVS 2007. LNCS, vol. 4871. pp.63–74. Springer, Heidelberg, 2007.

[11] J. L. Fuertes, Á. L. González, G. Mariscal, and C. Ruiz, Aplicación de la Realidad Virtual a la Difusión de la Cultura: la Alhambra Virtual. Proc. VI Congreso de Interacción Persona Ordenador (AIPO), INTERACCION'2005, pp. 367–371. Thomson, Granada, 2005.

[12] L. de la Flor, Proyecto Tutor: Componentes Adaptables para la Interacción con el Usuario en Sistemas Inteligentes de Tutoría. Master Thesis, Technical University of Madrid, 2000.

[13] J. L. Fuertes, Á. L. González, G. Mariscal, and C. Ruiz, Bilingual Sign Language Dictionary. In: Miesenberger, K., Klaus, J., Zagler, W., Karshmer, A. ICCHP 2006. LNCS, vol. 4061, pp. 599–606. Springer, Heidelberg, 2006.

[14] El Correo Gallego: Mexi, el Primer Mejillón Virtual. El Correo Gallego, No. 43908, Sección El Tema del Día, 10, 19 September, 2003.

# Mitral Valve Models Reconstructor: a Python based GUI software in a HPC environment for patient-specific FEM structural analysis

A. Arnoldi[a], A. Invernizzi[b], R. Ponzini[b], E.Votta[a], E.G. Caiani[a], A. Redaelli[a]

[a] Bioengineering Department, Politecnico di Milano, Milan, Italy
[b] CILEA, Milan, Italy
arnoldialice@hotmail.com

***Abstract-*** **A new approach in the biomechanical analysis of the mitral valve (MV) focusing on patient-specific modelling has recently been pursued. The aim is to provide a useful tool to be used in clinic for hypotheses testing in pre-operative surgical planning and post-operative follow-up prediction. In particular, the integration of finite element models (FEMs) with 4D echocardiographic advanced images processing seems to be the key turn in patient-specific modelling. The development of this approach is quite slow and hard, due to three main limitations: i) the time needed for FEM preparation; ii) the high computational costs of FEM calculation; iii) the long learning curve needed to complete the analysis without a unified integrated tool which is not currently available.**

**In this context, the purpose of this work is to present a novel Python-based graphic user interface (GUI) software working in a high performance computing (HPC) environment, implemented to overcome the above mentioned limitations. The Mitral Valve Models Reconstructor (MVMR) integrates all the steps needed to simulate the dynamic closure of a MV through a structural model based on human in vivo experimental data. MVMR enables the FEM reconstruction of the MV by means of efficient scientific routines, which ensure a very small time consuming and make the model easily maintainable. Results on a single case study reveal that both FEM building and structural computation are notably reduced with this new approach. The time needed for the FEM implementation is reduced by 1900% with respect to the previous manual procedure, while the time originally needed for the numerical simulation on a single CPU is decreased by 980% through parallel computing using 32 CPUs. Moreover the user-friendly graphic interface provides a great usability also for non-technical personnel like clinicians and bio-researchers, thus removing the need for a long learning curve.**

## I. INTRODUCTION

The mitral valve (MV) is a complex apparatus inserted on the valvular plane through the mitral annulus (MA), which is the support site of two leaflets. The valve is connected to the ventricular myocardium through a net of several branched chordae tendineae that converge into two papillary muscles (PMs). In the last two decades the high prevalence of MV pathologies has induced a growing need for quantitative and patient-specific information on MV biomechanics; these information should be available to cardiac surgeons and useful for an objective assessment of MV diseases and for surgical planning. As compared to traditional engineering in-vitro and animal models, finite element models (FEMs) are an innovative tool able to provide more detailed information on MV biomechanics and, if validated, may be more suitable to predict the characteristics of a given clinical scenario. These features make them a potential turn key in wide range hypothesis testing; thanks to such potential FEMs have already been applied to study the MV normal function [1-3], the biomechanics underlying MV diseases [1] and the effects of surgical corrections [4-7]. However, none of the mentioned studies captures all of the four aspects that drive MV function: morphology, tissues mechanical response, dynamic boundary conditions, and the interaction among MV components and between the MV and the surrounding blood. Nowadays, published models propose idealized geometries and boundary conditions of the MV, in contrast to the complexity and the variability of the MV morphology and dynamics.

Recently, a realistic FEM of a physiological MV has been obtained by the integration of quantitative information from in vivo real time 3D echocardiography (RT3DE), including a detailed mechanical properties description [8].
The model has been successfully tested to simulate the MV closure from end-diastole (ED) to systolic peak (SP), obtaining interesting results. However, the applicability of the procedure presented in [8] would be very limited due to high computational costs of FEM reconstruction and dynamic simulation; moreover the large complexity of manual integration of the modelling procedures restricts to very few people the usability of the new approach.

The present work consists of a possible graphic user interface (GUI) software implementation able to: i) build the FEM of the MV starting from patient-specific RT3DE pre-processed data, ii) perform FEM computation in high performance environment (HPC), iii) analyse FEM results.

## II. DESIGN CONSIDERATIONS

The Mitral Valve Models Reconstructor (MVMR) is an interactive and automated tool built in HPC environment. The user accesses to the software by connecting to a high

T. Sobh (ed.), *Innovations and Advances in Computer Sciences and Engineering*,
DOI 10.1007/978-90-481-3658-2_37, 

performance graphic server for remote rendering and visualization, namely *nodovisual.cilea.it.* A ThinAnyWhere Linux Server (Mercury International Technology) has been installed on the graphical server, ensuring deployment and management of the applications, while ThinAnyWhere Client enables a secure remote access to the server with efficient graphical data compression. Input data are processed in order to reconstruct the FEM of the MV through an interactive interface. Structural analysis computation is then submitted to the computing server *Lagrange*, composed by 208 2-ways nodes, Intel Xeon QuadCore at 3.16 GHz with 16 GB RAM/node, which is ranked at 135$^{th}$ position of Top500, the list of the 500 most powerful computer system in the world.

MVMR has been designed to work in a heterogeneous environment allowing the integration of all procedures (Fig. 1.). The user interacts with the software through a graphic interface where the FEM reconstruction process can be monitored. An e-mail notification service is provided by the visual server when the job will be successfully submitted to the computing server and by PBS 9.1 Pro, a management system for scheduling and managing workload across the cluster, at the beginning and at the end of the computation.

## III. System description

The complete biomechanical analysis of the MV performed by MVMR consists of three main blocks, showed in Fig. 2: Pre-Processing, Processing and Post-Processing.

During the Pre-processing, the input data are loaded once derived from 4D echocardiographic images.

The Processing is composed by FEM reconstruction in terms of geometry and boundary conditions starting from in vivo pre-processed experimental data and by the structural analysis.

Finally, the Post-Processing allows the analysis of FEM results obtained from the numerical computation.

### *A. Pre-Processing*

Transthoracic RT3DE images previously acquired (iE33, Philips, Andover, USA) are pre-processed by a semiautomatic algorithm which tracks frame-by-frame the ED positions of a set of equally spaced points located on the MA. The PMs tips coordinates are also identified at ED frame, while they generally can't be detected in many other RT3DE images. The output phase consists of ASCII files containing the spatial time-dependents coordinates of the characteristics points, used as data input in the FEM reconstruction block. For more details on the procedure see [9].

### *B. Processing: FEM Reconstruction*

The first step of the Processing analysis is the geometrical reconstruction of the MV sub-structures, consisting of the annulus, the papillary muscles, the mitral leaflets and the chordal apparatus, as discussed in the Introduction section.

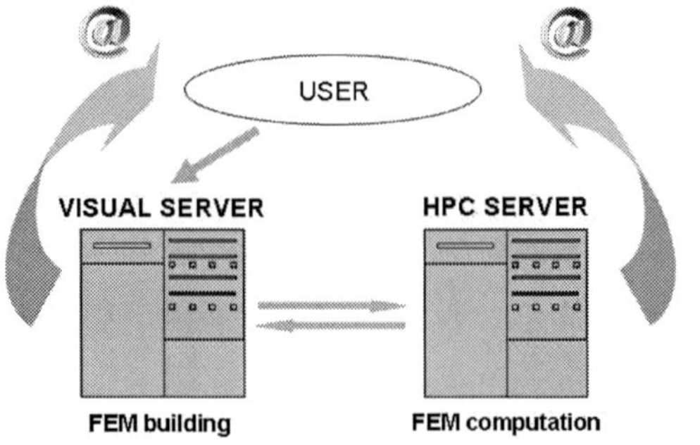


Fig. 1. MVMR architecture.

The software separately computes each component according to the input data and to the realistic hypothesis found in literature.

The annulus profile is reconstructed in all frames by interpolating the points detected on the MA in the Pre-Processing block with sixth order Fourier functions. The continuous profile is partitioned in a set of nodes, later used to define leaflets mesh; nodal displacements from the reference ED frame are then computed and used to define the annular kinematics. During the model building process, MVMR enables the visualization of the mitral annuli in the selected frames, as shown in Fig. 3.

PMs tips are identified as single points, and their position in all frames is estimated through a geometrical criterion verified from in-vivo animals data [10]. PMs displacements from the ED frame are then computed and later imposed as kinematic boundary conditions.

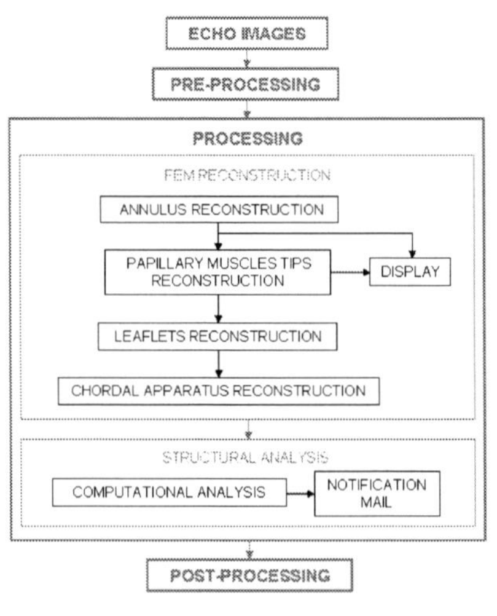


Fig. 2. Processes involved in MV modelling.

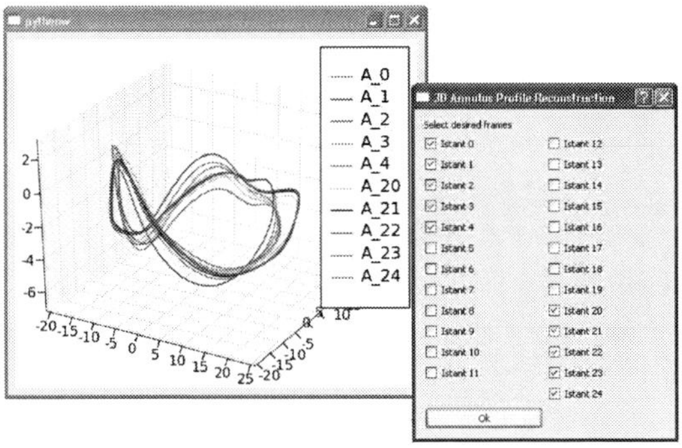


Fig. 3. Graphic display of annulus profiles in the selected frames.

Leaflets geometry at the ED frame is defined from ex-vivo data found in literature and then adapted to annulus size [11]. The user can modify the leaflets extent in different portions on the basis of measured leaflets heights and of proportions between computed leaflets area and the orifice areas, in order to identify the best leaflets configuration for the examined subject (Fig. 4). Anterior and posterior maximum inclinations can also be set according to the RT3DE data. The leaflets surface are meshed with triangular elements by choosing a suitable number of longitudinal nodes. Constant thicknesses of 1.32 mm and 1.26 mm are defined for anterior and posterior elements, in agreement with literature [12].

The chordal apparatus, showed in Fig. 5, is composed by 39 branched chordae, whose positions and insertion on leaflets are referred to ex-vivo data [13], while chordal cross-section areas are set according to literature [14]. The complete system consists of 82 first order branches inserting in leaflets free margin, 50 second order branches (including the strut chordae) inserting in the leaflets belly and 13 third order branches inserting near the annulus.

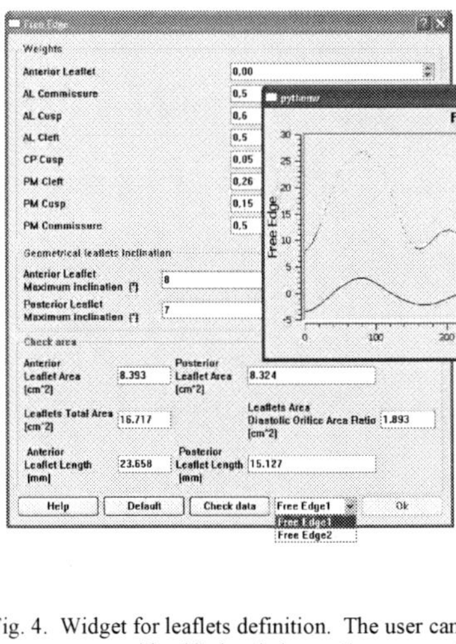


Fig. 4. Widget for leaflets definition. The user can check the configurations identified and choose the more realistic one.

In addition, the MVMR enables the selection of a broken apparatus in order to simulate a pathological condition.

Valvular tissue are described through hyperelastic constitutive models, able to reproduce an elastic and non linear mechanical behaviour.

The blood action on the mitral leaflets was modelled through a physiologic time-dependent pressure curve, applied on the ventricular leaflets surface, with a systolic peak value of 120 mmHg.

The output of this block is a full FEM of the MV, reconstructed according to in-vivo experimental data in terms of geometry and boundary conditions. For all details on the FEM see [8].

## C. *Processing: Structural Analysis*

The reconstructed FEM model is then used to simulate the MV dynamic closure with the commercial solver Abaqus/Explicit 6.7-1 (Dessault Systèmes SIMULIA Corp.). The job submission to the HPC server is carried out through an automated and interactive procedure: for this purpose, MVMR provides a widget, showed in Fig. 6, for the execution of the job. At this level different options such as the specification of an e-mail address for job status notifications, the choice of the number of CPUs for the parallel computation and the definition of the memory requirements for the computational analysis, are available. The job is then submitted to the server and, exploiting the functionality of PBS professional (a workload management system installed on the HPC server machine), the notification of the job execution status is provided.

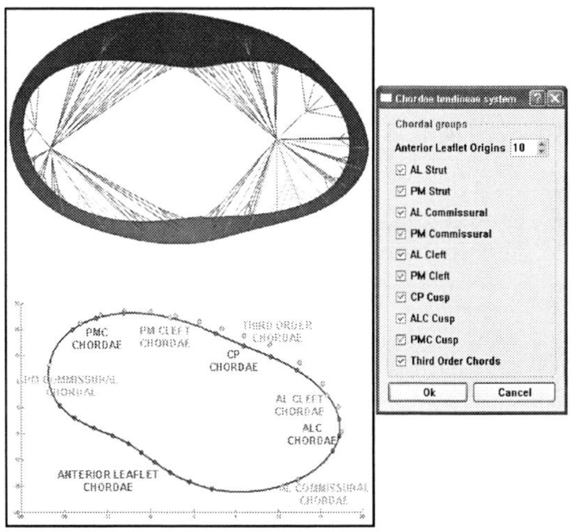


Fig. 5. The complete chordal apparatus (top, on the left); representation of the chordal groups positions (strut chordae are not shown) as regards to the MA (bottom, on the left); widget for the selection of chordal groups included in the FEM (on the right). The user can also specify the number (from 2 to 10) of the chordae tendineae on the anterior leaflet .

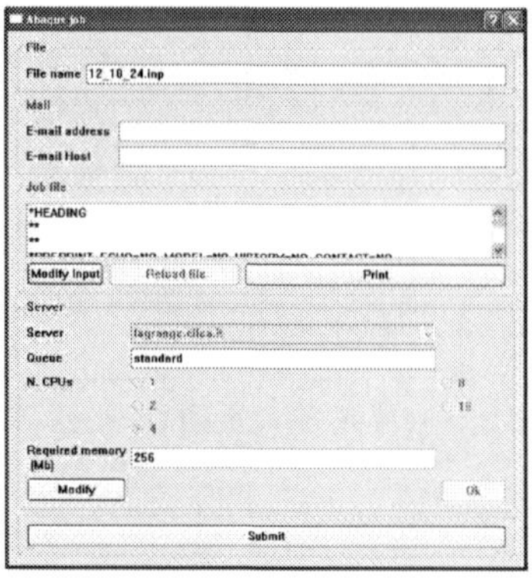

Fig.6. Widget for the job submission to the HPC server.

### D. *Post Processing*

The last main block consists in the visualization of the computational analysis results. By restarting MVMR, the user can download and post-process the produced output of the completed simulation; for this purpose, the output database of the examined job is visualized by Abaqus/Viewer, that will be automatically opened selecting the *Display Results* option, showed in Fig. 7. By means of the Viewer, dynamic and mechanical quantities of interest can be analyzed and the MV closure can be studied.

Simulations on a test case have shown an overall improvement in the model reality compared to standard finite element models found in literature, proving the strength of the implemented method. Quantitative results are comparable to those obtained in computational and experimental models of the MV in terms of stress [1] [5] [6], strain [15] and force acted on the chordae tendineae peak values [16] and of dynamic closure [17]. Contour maps of maximum principal stresses at the systolic peak for the examined subject are reported in Fig. 8.

## IV. PROGRAMMING ENVIRONMENT

MVMR has been developed with the open source high-level programming language Python, version 2.5, supported by the no profit Python Software Foundation. The software is based on a dedicated library, composed by more than 90 functions, that provides the reconstruction of the MV geometry and kinematic boundary conditions from the ASCII input data files and then the FEM building. Due to the need to manipulate complex data structures using efficient numerical routines, scientific Python libraries are also used to optimized time computing and data management. NumPy and SciPy, two of the best-known scientific computation packages using Python programming language, have also been included and used for this purpose. Plotting libraries such as Pylab, allow the display of intermediate results during the model building phase, providing a very interactive and reliable analysis procedure.

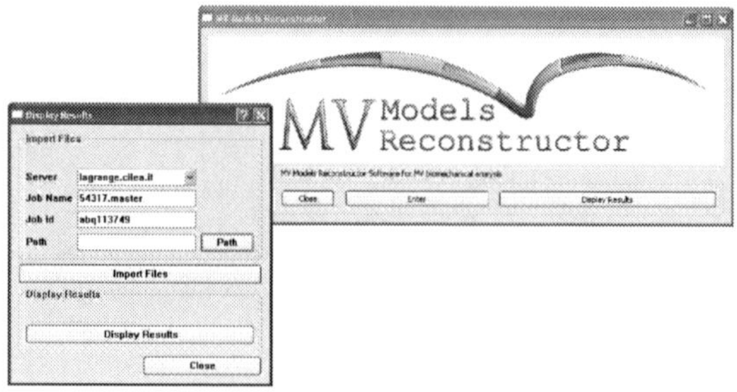


Fig. 7. Widget for the output download and the results visualization.

One of the main advantages of the MVMR is the GUI that enables to easily set, check and modify several biomechanical parameters involved in the MV modelling, and to monitor the overall processing with a great usability. The GUI has been implemented with the cross-platform application framework PyQt4, a high level framework based on Qt4, that is a well-known library for graphical interface building with a binding for the Python language. PyQt4 is organized in several modules; the MVMR is mainly made up with QtCore and QtGui modules that respectively treat non GUI functionality (managing files, directories and streams) and graphical components (widget, buttons, colours, toolbars).

The framework PyQwt has also been included in the software implementation in order to create widgets for scientific and engineering application.

## V. METHODOLOGY STRENGHT AND LIMITATIONS

To the best of the authors' knowledge, there is no published study that develops an automatic GUI for MV models reconstruction integrated in a HPC environment. The main advantage provided by this methodology is the automation of the overall procedure that is itself an integration of very different processes.

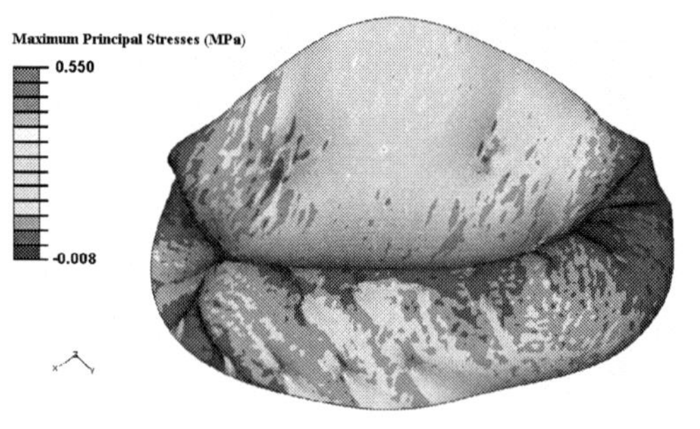


Fig. 8. Maximum principal stresses distribution at the systolic peak on mitral leaflets for an examined subject.

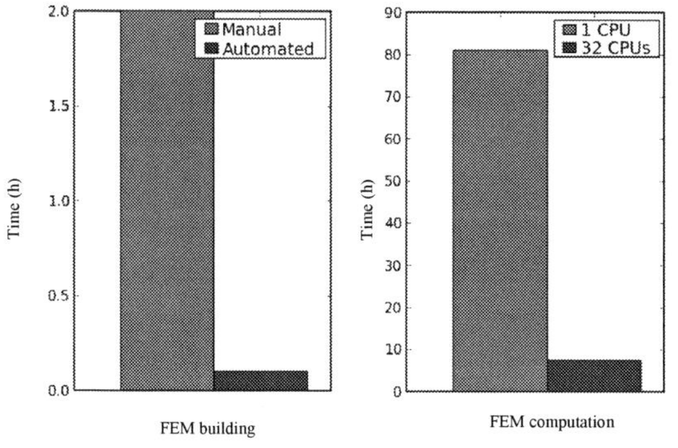


Fig.9. Time consumption for FEM reconstruction (left panel) and for the numerical simulation (right panel).

In Fig. 9. a plot of the processing time per analysis is provided, showing that the improvements are large both in the FEM building and in the FEM calculation procedures. Thanks to the optimization and the integration of all procedures, the MVMR requires a mean time of 6 minutes for the FEM reconstruction, against the 2 hours needed in a manual process, with an improvement of 1900%. The decrease of the time needed for the structural analysis is due to the facilities of the Abaqus/Explicit 6.7-1 solver and to the HPC environment in which the overall MVMR project has been built. A single CPU simulation was completed in about 81 hours, while the analysis time processing with 16 and 32 CPUs in distributed memory mode, was about 13 and 7.5 hours respectively, with an improvement of 980% in the last case.

This project shares the same limitations inherent to all the virtual modelling techniques; however as discussed in the Introduction section, the usability and the reliability of FEM modelling in the biomedical field seems to be very promising and rewarding, when compared to the standard animal and in vitro models.

## VI. Conclusions and Future Plans

The present work shows the benefits provided by the automation and software integration in a heterogeneous HPC environment for mitral valve FEM modelling through a GUI interface. This new tool allows to shrink the time consumption per model reconstruction and computations and potentially permits the access to such powerful technique, also to a non expert user. In this perspective, the enhanced usability and efficiency might trigger a virtuous cycle where the needs, the expectations and the new findings of multidisciplinary users will help the modelling phase itself, by increasing the feasibility of the patient-specific biomechanical analysis of the MV.

## Acknowledgment

This project has been developed at the Consorzio Interuniversitario Lombardo per l'Elaborazione Automatica (CILEA) thanks to a four-month grant provided by the Regione Lombardia (Labor-Lab Project for Technical and Scientific Area).

## References

[1] K.S. Kunzelman et al., "Fluid-structure interaction models of the mitral valve: function in normal and pathological states ", *Philos. Trans. R.. Soc. Lond. B Biol. Sci*, vol. 3622, pp.1393-1406, August 2007.

[2] K.H. Lim, J.H. Yeoh, C.M. Duran, "Three dimensional asymmetrical modelling of the mitral valve: a finite element study with dynamic boundaries", *J. Heart Valve Dis*, vol. 14, pp. 386-392, March 2005.

[3] V. Prot, B. Skallerud, G.A. Holzapfel, "Transversely isotropic membrane shells with application to mitral valve mechanics. Constitutive modelling and finite element implementation", *Int. J. Num. Meth. Eng.*, vol. 71, pp. 987-1008, December 2007.

[4] R.P. Cochran, K.S. Kunzelman, "Effect of papillary muscle position on mitral valve function: relationship to homografts", *Ann Thorac Surg*, vol 66, pp. S155-161, December 1998.

[5] E. Votta et al., "3-D computational analysis of the stress distribution on the leaflets after edge-to-edge repair of mitral regurgitation", *J Heart Valve Dis*, vol. 11, pp. 810-822, November 2002.

[6] F. Dal Pan, G. Donzella, C. Fucci, M. Schreiber, "Structural effects of an innovative surgical technique to repair heart valve defects" *Journal of Biomechanics* , vol. 38, pp. 2460-2471, December 2005.

[7] E. Votta et al., "The geoform disease-specific annuloplasty system: a finite element study", *Ann Thorac Surg* , vol. 84, pp. 92-102, July 2007.

[8] E. Votta et al., "From real time 3-D echocardiography to mitral valve finite element analysis: a novel modelling approach" *Proceedings of the 2008 Computers in Cardiology Conference, Bologna, Italy*, in press.

[9] F. Veronesi et al., "Quantification of mitral apparatus dynamics in functional and ischemic mitral regurgitation using real time 3-dimensional echocardiography", *J. Am. Soc. Echocardiogr.*, vol. 21, pp. 347-354, April 2008.

[10] P. Dagum et al., "Coordinate-Free analysis of mitral valve dynamics in normal and ischemic hearts", *Circulation*, vol. 102[suppl III], pp. III62-III69, November 2000.

[11] K.S. Kunzelman, R.P. Cochran, E.D. Verrier, R.C. Eberhart, "Anatomic basis for mitral valve modelling", *J Heart Valve Dis.*, vol. 3, pp. 491-496, September 1994.

[12] K.S. Kunzelman et al., "Finite element analysis of the mitral valve", *J Heart Valve Dis.*, vol. 2, pp.326-340, May 1993.

[13] J.H.C. Lam, N. Ranganathan, M.D. Silver, "Morphology of the human mitral valve. I. Chordae tendineae : a new classification", *Circulation* vol. 41, pp.449-458, March 1970.

[14] K.S. Kunzelman, M.S. Reimink, R.P. Cochran, "Annular dilation increases stress in the mitral valve and delays coaptation: a finite element computer model", *Cardiovasc Surg.* , vol. 5, pp. 427-434, August 1997.

[15] M.S. Sacks et al., "Surface strains in the anterior leaflet of the functioning mitral valve", *Ann. Biomed. Eng.*, vol. 30, pp. 1281-1290, November 2002.

[16] G.H. Jimenez, D.D. Soerensen, Z. He, J. Rictchie, A.P. Yoganathan, "Mitral valve function and chordal force distribution using a flexible annulus model: an in vitro study", *Ann. Biomed. Eng.*, vol 33., pp. 557-566, May 2005.

[17] T.A. Timek et al., "Ring annuloplasty prevents delayed leaflet coaptation and mitral regurgitation during left ventricular acute ischemia", *J. Thorac. Cardiovasc. Surg.*, vol. 119, pp.774-783, April 2000

# An Intelligible Representation Method For Software Reusable Components

Dr.S.S.V.N.Sharma
Professor
Dept. of Informatics
Kakatiya University,INDIA.
ssvn.sarma@gmail.com

P.Shirisha
Lecturer
Dept. of MCA
KITS,Warangal,INDIA.
rishapakala@yahoo.co.in

## ABSTRACT

Reuse helps the engineers in developing high quality software by reducing the development time,cost and improving productivity. When coming to software industry we are lacking perfect methods for representing the software assets. To reuse components it is necessary to locate the component that can be reused. Locating components, or even realizing they exist, can be quite difficult in a large collection of components. These components need to be suitably classified and stored in a repository to enable efficient retrieval. This paper looks at each of the existing representation and classification techniques and present a representation method for reusable components.

**Key words:** Software reuse, component, classification techniques, reuse libraries

## 1 INTRODUCTION

One of major impediments to realizing software reusability in many organizations is the inability to locate and retrieve existing software components. There often is a large body of software available for use on a new application, but the difficulty in locating the software or even being aware that it exists results in the same or similar components being re-invented over and over again. In order to overcome this impediment, a necessary first step is the ability to organize and catalog collections software components and provide the means for developers to quickly search a collection to identify candidates for potential reuse[2,16].

Software reuse is an important area of software engineering research that promises significant improvements in software productivity and quality[4]. Software reuse is the use of existing software or software knowledge to construct new software[11]. Effective software reuse requires that the users of the system have access to appropriate components. The user must access these components accurately and quickly, and be able to modify them if necessary. Component is a well-defined unit of software that has a published interface and can be used in conjunction with components to form larger units[3]. Reuse deals with the ability to combine separate independent software components to form a larger unit of software. To incorporate reusable components into systems, programmers must be able to find and understand them. Classifying software allows reusers to organize collections of components into structures that they can search easily. Most retrieval methods require some kind of classification of the components. How to classify and which classifications to use must be decided, and all components put into relevant classes. The classification system will become outdated with time and new technology. Thus the classification system must be updated from time to time and some or all of the components will be affected by the change and need to reclassified.

**1.1 Component classification:** The generic term for a passive reusable software item is a component. Components can consist of, but are not restricted to ideas, designs, source code, linkable libraries and testing strategies. The developer needs to specify what components or type of components they require. These components then need to be retrieved from a library, assessed as to their suitability, and modified if required. Once the developer is satisfied that they have retrieved a suitable component, it can then be added to the current project under development. The aim of a good component retrieval system[13] is to be able to located either the exact component required, or the closest match, in the shortest amount of tie, using a suitable query. The retrieved component (s) should then be available for examination and possible selection.

Classification is the process of assigning a class to a part of interest. The classification of components is more complicated than, say, classifying books in a

T. Sobh (ed.), *Innovations and Advances in Computer Sciences and Engineering*,
DOI 10.1007/978-90-481-3658-2_38, 

library. A book library cataloguing system will typically use structured data for its classification system (e.g., the Dewey Decimal number). Current attempts to classify software components fall into the following categories: free text, enumerated, attribute-value, and faceted. The suitability of each of the methods is assessed as to how well they perform against the previously described criteria for a good retrieval system, including how well they manage 'best effort retrieval'.

## 2 EXISTING TECHNIQUES

**2.1 Free text classification:** Free text retrieval performs searches using the text contained within documents. The retrieval system is typically based upon a keyword search[16]. All of the document indexes are searched to try to find an appropriate entry for the required keyword. An obvious flaw with this method is the ambiguous nature of the keywords used. Another disadvantage is that a search my result in many irrelevant components. A typical example of free text retrieval is the grep utility used by the UNIX manual system. This type of classification generates large overheads in the time taken to index the material, and the time taken to make a query. All the relevant text (usually file headers) in each of the documents relating to the components are index, which must then be searched from beginning to end when a query is made. Once approach to reducing the size of indexed data is to use a signature matching technique, however space reduced is 10-15% only.

**2.2 Enumerated classification:** Enumerated classification uses a set of mutually exclusive classes, which are all within a hierarchy of a single dimension[6]. A prime illustration of this is the Dewey Decimal system used to classify books in a library. Each subject area, e.g., Biology, Chemistry etc, has its own classifying code. As a sub code of this is a specialist subject area within the main subject. These codes can again be sub coded by author. This classification method has advantages and disadvantages pivoted around the concepts of a unique classification for each item. The classification scheme will allow a user to find more than one item that is classified within the same section/subsection assuming that if more than one exists. For example, there may be more than one book concerning a given subject, each written by a different author.

This type of classification schemes is one dimensional, and will not allow flexible classification of components into more than one place. As such, enumerated classification by itself does not provide a good classification scheme for reusable software components.

**2.3 Attribute value:** The attribute value classification schemes uses a set of attributes to classify a component[6]. For example, a book has many attributes such as the author, the publisher, it's ISBN number and it's classification code in the Dewey Decimal system. These are only example of the possible attributes. Depending upon who wants information about a book, the attributes could be concerned with the number of pages, the size of the paper used, the type of print face, the publishing date, etc. Clearly, the attributes relating to a book can be:

- Multidimensional. The book can be classified in different places using different attributes
- Bulky. All possible variations of attributes could run into many tens, which may not be known at the time of classification

**2.4 Faceted:** Faceted classification schemes are attracting the most attention within the software reuse community. Like the attribute classification method, various facets classify components, however, there are usually a lot fewer facets than there are potential attributes (at most, 7). Ruben Prieto-Diaz[2,8,12,17] has proposed a faceted scheme that uses six facets.

- The functional facets are: Function, Objects and Medium
- The environmental facets are: System type, Functional area, Setting

Each of the facets has to have values assigned at the time the component is classified. The individual components can then be uniquely identified by a tuple, for example.

<add, arrays, buffer, database manager, billing, book store>

Clearly, it can be sent that each facet is ordered within the system. The facets furthest to the left of the tuple have the highest significance, whilst those to the right have a lower significance to the intended component. When a query is made for a suitable component, the query will consist of a tuple similar to the classification one, although certain fields may be omitted if desired. For example:

<add, arrays, buffer, database manager, *,*>

The most appropriate component can be selected from those returned since the more of the facets from the left that match the original query, the better the match will be.

Frakes and Pole conducted an investigation as to the most favorable of the above classification methods[9]. The investigation found no statistical evidence of any differences between the four different classification schemes, however, the following about each classification method was noted:

- Enumerated classification
  Fastest method, difficult to expand
- Faceted classification
  Easily expandable, most flexible
- Free text classification
  Ambiguous, indexing costs
- Attribute value classification
  Slowest method, no ordering, number of attributes.

## 3 PROPOSED CLASSIFICATION

Whilst it is obvious that some kind of classification and retrieval is required, one problem is how to actually implement this, most other systems follow the same principle: Once a new component has been identified, a librarian is responsible for the classification must be highly proficient with the classification system employed for two reasons. Firstly, the librarian must know how to classify the components according to the schema.

Secondly, a lexicographical consistency is required across the whole of the system. The classification system is separate to the retrieval system, which is for all of the users.

Most established systems tend to rigidly stick with one classification and retrieval scheme, such as free text or faceted. Others tend to use one type of retrieval system with a separate classification system such as the Reusable Software Library, which uses an Enumerated classification scheme with Free Text search.

In this research, we propose a new classification scheme that incorporates the features of existing classification schemes. In this system (Fig. 1) the administrator or librarian sets up the classification scheme. The developers develop and put their components into library. The users who are also developers can retrieve components from the library. Query tracking system can be maintained to improve the classification scheme. The proposed system will provide the following functionality to the users.

- Storing components
- Searching components
- Browsing components

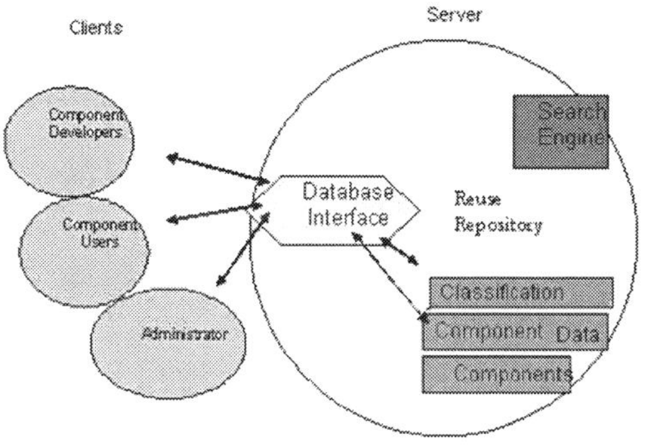


Fig. 1: Proposed system

The librarian task is to establish classification scheme. Each of the four main classification schemes has both advantages and disadvantages. The free text classification scheme does not provide the flexibility required for a classification system, and has too many problems with synonyms and search spaces. The faceted system of classification provides the most flexible method of classifying components. However, it can cause problems when trying to classify very similar components for use within different domains. The enumerated system provides a quick way to drill down into a library, but does not provide the flexibility within the library to classify components for use in more than one way. The attribute value system allows multidimensional classification of the same component, but will not allow any ordering of the different attributes.

Our solution to these problems would be to use an attribute value scheme combined with faceted classification scheme to classify the components details. The attribute value scheme is initially used to narrow down the search space. Among the available components in the repository only components matching with the given attributes are considered for faceted retrieval. Restrictions can be placed upon which hardware, vendor, O.S., type and languages attributes. This will be the restrictive layer of the classification architecture, by reducing the size of the search space. All, some or none of the attributes can be used, depending upon the required specificity or generality required. Then a faceted classification scheme is employed to access the component details within the repository. The proposed facets are similar to those used by Prieto-Diaz, but are not the same.

- Behavior. What the component does.
- Domain. To which application domain the components belong.
- Version. The version of the candidate component

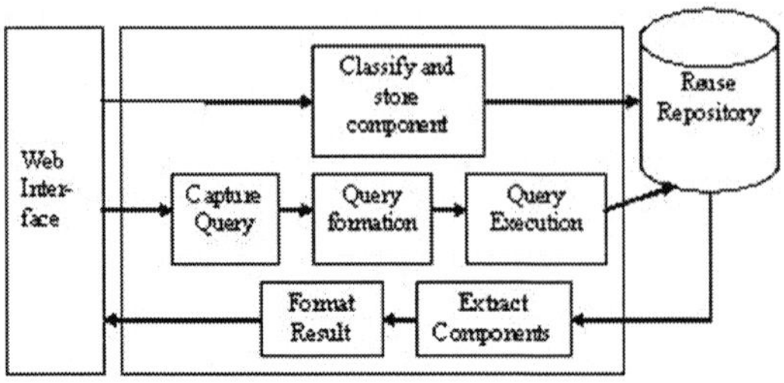


Fig. 2: Architecture of proposed System

## 4 RESULTS

The repository is library of component information, with the components stored within the underlying platforms. Users are provided with interface through which they can upload, retrieve and browse components. User will provide the details of the components to upload components. While retrieving components the user can give his query or he can givedetails so that the system will find the matching components and give the results. Then the user can select his choice from the list of components. The user can select from versions of components also. In addition the descriptions of the components are stored along with components facilitate text searches also.

Figure 3 and 4 shows the experimental prototype screens. The user can browse components by developer, language or platform. In the search screen the user can give his requirement along with fields that are optional

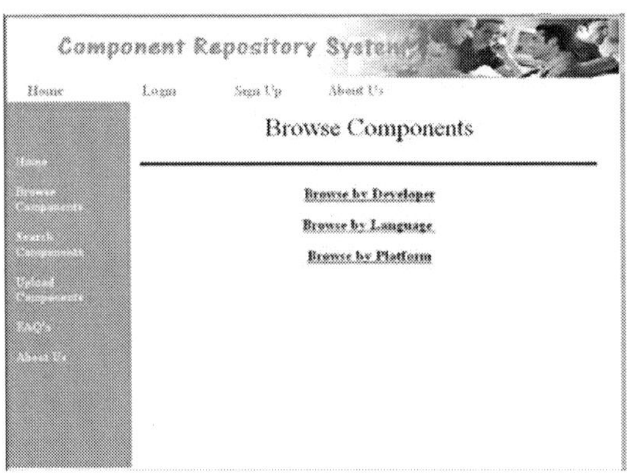


Fig. 3: Screen for browsing components

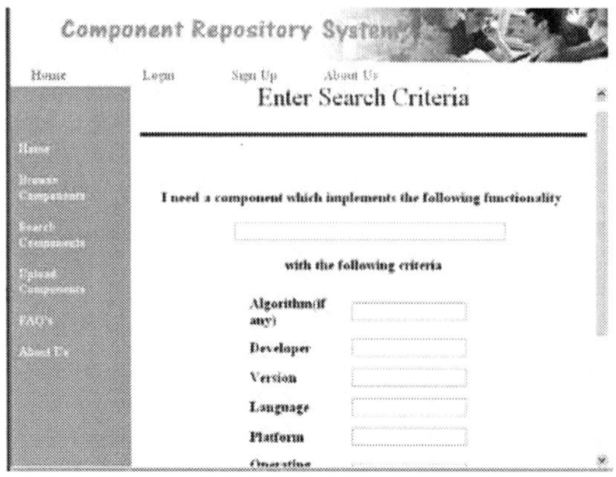


Fig. 4: Screen for searching components

## 5 CONCLUSION

This study presented the methods to classify reusable components. Existing four main methods (free text, attribute value, enumerated and faceted classification) are investigated and presented advantages and disadvantages associated with them. The proposed mechanism takes advantage of the positive sides of each method.The attribute value method is initially used within the classification for specifying the vendor, platform, operating system and development language relating to the component. This allows the search space to be restricted to specific libraries according to the selected attribute values. Additionally, this method will allow the searches to be either as generic or domain specific as required. The functionality of the component is then classified using a faceted scheme.

In addition to the functional facets is a facet for the version of the component. The version of a component is directly linked to its functionality as a whole, i.e. what it does, what it acts upon and what type of medium it operates within. Future work involved with this classification scheme will be to refine the scheme, and formalize it for implementation. A prototyped system for presenting and retrieving software reusable components based on this classification schema is now under implementation.

## REFERENCES


1. Henninger, S., 1997. An evolutionary approach to constructing effective software reuse repositories. ACM Trans. Software Eng. Methodol., 2: 111-150

2. Ruben Prieto-Diaz, 1991. Implementing faceted classification for software reuse. Communication of the ACM,Newyork,USA, 34 (5),pp.88-97.

3. Gerald Kotonya, Ian Sommerville and Steve Hall, 2003. Towards a classification model for component based software engineering Proceedings of the of the 29th Euromicro Conference. IEEE computer society,Washington,DC,USA.pp43.

4. William B. Frakes and Thomas. P. Pole, 1994. An empirical study of representation methods for reusable software components. IEEE Trans. Software Eng., 20: 617-630.

5. Lars Sivert Sorumgard Guttorm Sindre and Frode Stokke, 1995. Experiences from application of a faceted classification scheme. ACM SIGSOFT,Vol20,issue2, April ,1995,Newyork,USA,pp 76-89.ISSN:0163-5948.

6. Jeffrey, S. Poulin and Kathryn P. Yglesias , Nov 1993. Experiences with a faceted classification scheme in a large Reusable Software Library (RSL). In: The Seventh Annual International Computer Software and Applications Conference (COMPSAC'93), pp: 90-99,DOI 10.1109/cmpsac.1993.404220.

7. Ruben Prieto-Diaz, 1991. Implementing faceted classification for software reuseACM Vol 34 issue 5. Newyork ,USA pp88-97.

8. Ruben Prieto-Diaz, 1991. Implementing faceted classification for software reuseACM Vol 34 issue 5. Newyork ,USA pp88-97.

9. Klement, J. Fellner and Klaus Turowski, 2000. Classification framework for business components. Proceedings of the 33rd Hawaii International Conference on System Sciences, Vol 8 ,IEEE Computer society 2000 Washington ,USA ,ISBN 0-7695-0493-0/00 .

10. Vitharana, Fatemeh, Jain, 2003. Knowledge based repository scheme for storing and retrieving business components: A theoretical design and an empirical analysis. IEEE Trans. Software Eng., 29: 649-664.

11. William B. Frakes and Kyo Knag, July 2005. Software reuse research: status and future. IEEE Trans. Software Eng.,Vol 3,issue 7,pp32-36.

12. Prieto-Diaz, R. and P. Freeman, 1987. Classifying software for reuse. IEEE Software, 4: 6-16.

13. Rym Mili, Ali Mili, and Roland T. Mittermeir, 1997. Storing and retrieving software components a refinement based system. IEEE Trans. Software Eng., 23: 445-460.

14. Hafedh Mili, Estelle Ah-Ki, Robert Godin and Hamid Mcheick, 1997. Another nail to the coffin of faceted controlled vocabulary component classification and retrieval. Proceedings of the 1997 Symposium on Software Reusability (SSR'97), May 1997, Boston USA, pp: 89-98.

15. Hafedh Mili, Fatma Mili and Ali Mili, 1995. Reusing software: issues and research directions. IEEE Trans. 1995-Vo.1.ps.gz.

16. Gerald Jones and Ruben Prieto-Diaz, 1998. Building and managing software libraries. Anals of software engineering,5,pp349-414,1998.

17. Prieto-Diaz, Freeman, 1997. Classifying software for reuse. IEEE Software, 4: 6-16.

# Creating Personally Identifiable Honeytokens

Jonathan White
University of Arkansas
jlw09@uark.edu

*Abstract:* **In this paper, a method for creating digital honeytokens consisting of personal information such as names, addresses, telephone numbers and social security numbers is designed and evaluated. We show that these honeytokens can add to the security of digital data by allowing better identification of individuals who are misusing or inappropriately copying personally identifying information. No program currently exists publically to create personally identifying honeytokens; we will show how the design we've proposed will fill this void and be another good tool for the computer security engineer and a vast improvement over previous naïve and manual methods of generating fake identifying information. The main contribution of this work is a new and innovative program to produce PII honeytokens, a description of where they can be deployed, and an analysis that shows they appear realistic.**

## I. INTRODUCTION

The idea of a honeytoken is a new concept to the computer security arena. The honeytoken term was coined by Augusto Paes de Barros in 2003 in a discussion on honeypots. Honeytokens are strongly tied to honeypots, which are generally defined as information system resources whose value lies in unauthorized or illicit use of the resource [1].

Honeytokens are honeypots that aren't physical devices. Instead, they are data objects whose value lies in attracting, and potentially tracing, unauthorized uses of real data [6]. The honeytoken data could be anything that would look attractive or useful to an attacker. Examples would be digital tax returns, lists of passwords, or a database of names, addresses, and social security numbers. The honeytoken objects in and of themselves have no real value; in fact, they are often made to contain no data that is actually valuable or even real. They are often totally synthetic. The importance of the honeytoken is in the fact that they appear worthwhile and useful, without actually being so [2].

### *A. Key properties*

All honeytokens have several common properties no matter what type of digital object they mimic. First and foremost, the honeytoken must appear valuable to an attacker. If the honeytoken doesn't appear worthwhile, it will never be interacted with. Second, the honeytoken must be designed in such a manner so as to not interfere with any real data. The implementers of the honeytokens must be able to tell what is real and what's not, and thus the honeytoken must have unique, identifiable properties.

Furthermore, the honeytokens should be designed so that they appear real to an attacker [9]. If the honeytokens are obviously an imitation, the attacker would have no reason to interact with the honeytoken. Also, if the honeytoken is easily detected, it might also be easily removed by the attacker while still accessing the actual data that the honeytoken was designed to protect.

## II. BACKGROUND

One example that is cited in the literature [3] on how a personally identifiable honeytoken might work involves a hospital's database and a bogus record inserted into that database with the name "John F. Kennedy". Due to governmental regulations in the United States, hospitals are required to enforce patient privacy under severe penalties for infractions. If a hospital finds that someone is accessing patient's personal data without permission, they would like to know this as soon as possible so that they can deal with the threat.

Honeytokens can be used effectively to know when such unauthorized access has occurred. The John F. Kennedy record is loaded into the hospital's database (with the assumption that there is no patient at the hospital with that same name). If it is detected that someone has attempted to access this record, this access is a violation of patient privacy, as John F. Kennedy never was a patient and no valid user would have a reason to view that data. The hospital needs to set up a monitoring system to catch any access to the honeytokens, and once revealed, investigate why someone was looking at the records. Stored procedures and other layers of the database management system would need to be in place to ensure the honeytokens don't interfere with valid uses. In this example, access to the honeytoken file implies that the database is not being accessed in a valid, authorized method.

These honeytokens need to be produced in a large enough quantity so that they can be distributed throughout areas where malicious users will encounter them. For example, when an individual attempts to locate patient data using a search engine, honeytoken results could be returned with each search in order to track users who are attempting to access data illicitly. While users can make honeytokens manually, an automatic program that makes provably realistic honeytokens would be a great benefit as the amount of honeytokens in use increases.

T. Sobh (ed.), *Innovations and Advances in Computer Sciences and Engineering*,
DOI 10.1007/978-90-481-3658-2_39, 

While no program existed to make these honeytokens, examples were present in the literature ([2], [3], and [4]) that advocated their use. Also, individuals have been known to manually create fake personal data, but this is an error prone, nonscientific method that is difficult to do in large quantities[10]. There was a demonstrated use and need for a program like we've made. This program will encourage more companies to use honeytokens in their databases to add another layer of protection to people's privacy.

## III. PROGRAM CREATION

Having identified the need and potential uses, we began the process of creating a program that could construct honeytokens that consisted of personally identifying honeytokens to fill this need. The goal of the program design was to make digital honeytokens that looked enticing to an attacker, realistic enough to fool an attacker into thinking they were genuine and also to incorporate uniquely identifiable elements so as to maintain distinguishability between the honeytoken and the real data.

The steps in the design included identifying the data attributes the honeytoken should posses, gathering information about the potential values for each of these attributes (including the relative distribution of each attribute in the sample population), defining what relationships needed to exist between fields in order for the honeytokens to look realistic, and, finally, coding a program that uses this information to generate the honeytokens.

### *A. Identification of valuable data*

After looking at what data is often stolen and used maliciously by phishers and other intruders in identity theft ([3], [8]), we identified names, addresses, telephone numbers, social security numbers, dates of birth, genders, income levels, ethnicities, credit card numbers, and email addresses as particularly important data attributes that every real person would more than likely possess. We were able to locate suitable statistics on many of these domains. Other attributes, such as medical histories, purchasing preferences, and entertainment preferences, while important to many groups, require domain specific knowledge that is difficult to collect and quantify, and we weren't able to aggregate any data in these areas.

While ethnicities, credit card numbers and email addresses are identified as valuable attributes that personally identifying honeytokens should possess, we have not implemented these fields at this time. Future versions of the program will realize these attributes once more accurate information is located on their distributions in the population. A listing of the fields implemented, including the total unique values for each, is in table 1.

TABLE 1: ATTRIBUTES

| *Attribute* | *Unique Values* |
|---|---|
| | |
| Last Name | 88,795 |
| First Name | 5,496 |
| Gender | 2 |
| Street Address | 2,999,700 |
| Zip Code | 29,472 |
| City | 16,627 |
| State | 50 |
| Approx. Latitude | 16,627 |
| Approx. Longitude | 16,627 |
| Birth Date | 36,500 |
| Telephone Number | ~865,000,000 |
| Social Security Number | ~700,000,000 |
| Occupation | 713 |
| Hourly Salary | 100,000 |
| Yearly Salary | 100,000 |

### *B. Gathering relevant data*

Each of the identified attributes has distinctive properties; a telephone number is very different than a social security number, even though they are superficially comparable. Nonetheless, commonalities were identified that needed to be maintained in order for the design goals to be met. They include the potential values for each field, the relative frequency that each of these values appears in the population, and the dependencies between the fields in each individual honeytoken.

Each attribute only has a limited set of potential values. First names, social security numbers and zip codes are very specific; giving an incorrect or invalid value would indicate that the honeytoken isn't genuine. Misspelled, made up, or transposed names are also easily detected, especially if a computer is used to detect counterfeits. These errors should be avoided, and data must be gathered on the potential values for each field.

All of the attributes have a unique distribution. Several last names, such as Smith and Johnson, are much more common than others in the United States and several occupations, such food service and retail sales, are much more common than others. When honeytokens are used, they should have a realistic distribution of elements [2]. If the honeytokens possess rare names or occupations a high percentage of the time, this is an indication that they are fake. Also, if the honeytokens are to be used in a database that is from a particular state or city, certain addresses will be much more common than others. For example, if the database consists of a list of California voters, a certain percentage should come from Los Angeles, which is a heavily populated city. If the honeytokens all come from a small town or a particular geographic area, they can be detected and more easily removed.

Some attributes have dependencies on other fields in the honeytoken that must be maintained in order for realism to be achieved. For example, the area code and office prefix of telephone numbers depend on the particular city and state the person lives in. The yearly and hourly salaries depend on the type of occupation and where that person lives. So, dishwashers in California make more than dishwashers in Kansas, but dishwashers in California should not make more than lawyers in Kansas or the honeytokens will look fabricated. Again, these relationships must be maintained or the honeytoken will look fabricated. See table 2 for a list of some of the most common values for a few of the honeytoken attributes with the frequency that they appear in the population.

TABLE 2: ATTRIBUTE DISTRIBUTIONS

| *Attribute* | *Most common* | *% of Pop.* |
|---|---|---|
| Last Name | Smith | 1.006 |
| | Johnson | .810 |
| | Williams | .699 |
| | | |
| First Name | James | 1.659 |
| | John | 1.636 |
| | Mary | 1.314 |
| | | |
| Birth Month | August | 8.987 |
| | July | 8.721 |
| | October | 8.546 |
| | | |
| Occupation | Food Service | 7.894 |
| | Retail Sales | 3.054 |
| | Cashier | 2.647 |

### *C. Social security numbers*

The effort in obtaining information about the potential values, frequencies, and dependencies of each field was very demanding. The U.S. Census Bureau and other government agencies were a great help in this regard. The social security attribute was the hardest attribute to construct; an in depth description of this important field follows.

A social security number is typically given at birth in America. This nine-digit number serves to uniquely identify the individual. Social security numbers are used for bookkeeping purposes by various governmental divisions within America, and they are one of the most misused and sought after pieces of information by phishers and other criminals.

An American social security number consists of three parts [16]. The first three digits are determined by what state the person applied for the social security number in (typically the state they were born in). The middle two digits range from 01 to 99, but they are not given out consecutively. The middle two digits depend on how many people have ever applied for a social security number in that state. For example, if 100,000 people have applied for social security numbers in Alaska, they will have much smaller middle two digits when compared to a state that has had several million people apply for social security numbers. The last four digits serve to make the entire social security number unique. According to the Social Security Bureau, the last four digits are given out at random.

A data file that had all the first three potential digits which are available for each state in America was found. Since 1971, the first three digits that a person gets are randomly selected from all the possible first three digits that are available for the state. In all, there are around 750 possible first three digits currently in use. So, when the honeytoken generator program manufactures an SSN, the first process that is performed is to select an appropriate first three digits at random from all the available first three digits for the state that the person was born in, which is assumed to be already known.

A data file from the Social Security Bureau's website that gave the order in which the middle two digits are given was used to make the honeytokens more realistic. The middle two digits are not given out consecutively, as was mentioned above. They are given out starting out at 01, then 03, 05, 07, 09 are given, then all the evens greater than 10 are given, then all the evens less than 10 are given, and finally all the odds from 11 to 99 are given. The honeytoken generator program produces social security numbers that conform to this requirement.

The middle two digits are a function of how many people have been born in each state. People who were born when the Social Security Bureau was first formed would tend to have the lowest available middle two digits, and people born fairly recently would tend to have the highest available middle two digits. A data file from the Social Security Bureau that listed of the highest middle two digits that had been given out as of the year 2005 for each of the available first three digits was employed. Also, information about how the age distribution of the population of America was located, and this information was able to be used to give very realistic middle two digits, given what year a person was born. It was assumed that the age distribution for each state was close to the national average.

When the honeytoken generator program produces the middle two digits of an SSN, the program takes in as parameters the first three digits of the SSN and the age of the person. Based on the highest middle two digits that have currently been given out for those particular first three digits, the program calculates what middle two digits a person with that age would likely have had given the age distribution of people in the United States. For example, if the highest middle two digits for SSN 123 was 99 (the highest possible given), a person born in 2005 would receive the middle two digits 99. A person born in 1905 would probably receive the middle two digits 01 and a person born in 1950 would receive middle two digits of 60. This happens because the population distribution of the U.S. is skewed. If however, the highest middle two digits for SSN 123 was 09 (only 5 are actually available, 01, 03, 05, 07, and 09), then a person born in 1905 would still receive 01, a person born in 1950 would receive 05 or 07, and a person born in 2005 would receive 09.

The last four digits of the SSN are what enable the SSN to be unique for each individual. Each of the possible first three – middle two – last four combinations starts at a different random number, which is generated dynamically when the program begins. The program is guaranteed to never generate the same social security number for different individuals. A mathematical analysis of the Social Security attribute is expounded upon in the following section.

*D. SSN properties*

$SSN \equiv F_3 \cap M_2 \cap L_4$, where

$$001 \le F_3 \le 750;$$

$$01 \le M_2 \le 99;$$

$$0001 \le L_4 \le 9999 \text{ and}$$

$$F_3, M_2, L_4 \in Z.$$

The probabilities for $F_3$, $M_2$, and $L_4$ are as follows:
Given $S_k$ and $B_n$ where $S_k$ = state of birth and $B_n$ = year of birth, then

$$P\{L_4\} = \frac{1}{N_1} \text{ where } N_1 = 9998.$$

$$P\{F_3\} = \frac{P\{S_k\}}{N_c}; N_c = \sum \text{ Potential } F_3 \text{ for state k}$$

$$P\{S_k\} = \frac{N_k}{N_T} \text{ and } N_k = \sum \text{ People in state k and}$$

$$N_T = \sum \text{ People in all states}.$$

$$P\{M_2\} = P\{S_k\} * P\{B_n\} \text{ where}$$

$$P\{B_n\} = \frac{C_n}{W_n}; C_n = \sum \text{ Children born in year n, and}$$

$$W_n = \sum \text{ Women of childbearing age in year n.}$$

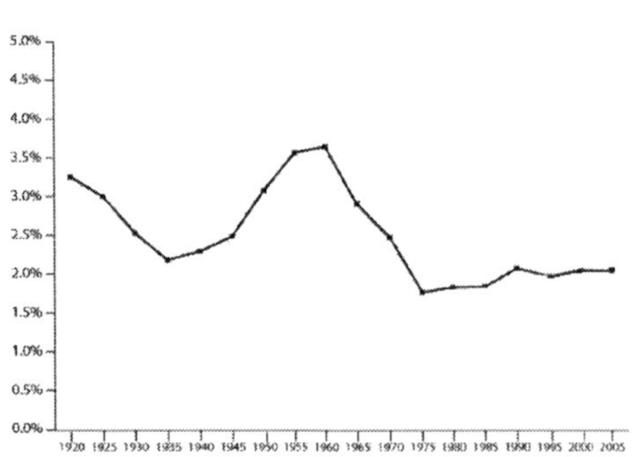


Fig. 1: *$P\{B_n\}$*, the relative population growth, in the US from 1920 to 2005. Notice the 'baby boomer' years from 1950 to 1965.

*E. Telephone numbers and other attributes*

Various phone companies throughout the United States give out telephone numbers. Each phone number is unique to a telephone line, so by calling a number you can be assured that you are reaching a particular place or person. The North American Numbering Plan Administration (NANPA), which is a part of the U.S. government, is in charge of making sure that no phone company gives out the same number to two different people. A telephone number consists of an area code, a three digits prefix, and a four digit house code. Data from NANPA that had every area code and prefix for most cities in America was used to generate realistic phone numbers. The area code is determined by the zip code that the person lives in. Each zip code has only one area code, which models real life. This makes it very easy to generate the area code for an American.

The prefix of a telephone number depends on the city that the particular person lives in. There are many available prefixes per city, sometimes even over one hundred in the case of a large city like Dallas or Chicago. Given the city, the program selects an available prefix. The program treats every prefix as being equally likely in the population.

The last four digits are almost exactly the same as the last four digits of the social security number. The last four digits for a given area code – prefix pair serve to make the telephone number unique, and they are given out at random. Like SSNs, the program will not give out the same telephone number to more than one person.

The other attributes, such as names and addresses, were made in a similar manner. Data was collected on the distributions, and this knowledge was used to make the honeytoken production. Information about these fields is summarized in table 2.

## IV. PROGRAM IMPLEMENTATION

In all, over 600,000 individual facts were gathered that were used in the resulting honeytoken program. Each time the program is invoked, the current time stamp is used to seed the program. If the user ever should wish to generate the same honeytokens, they can use this time stamp as a key. Otherwise, the program generates different honeytokens each time it is executed.

The program then interacts with the user to specify how many honeytokens are desired, what the file should be named, and whether the honeytokens are fixed width or character delimited. The user then chooses which of the 16 attributes (first name, address, social security number, etc…) should be present and what the geography of the honeytokens should be. This last property bears some explanation.

At times, the honeytokens must be put into data that has very specific geographic properties. For example, if the honeytokens are to be used in a state university, the user may want to specify that the honeytokens should only come from that state. The user can choose the entire United States, particular states or particular city/state combinations that the honeytokens should come from. The program still maintains the population distribution of that region when it generates the honeytokens. For example, if the user requests that the honeytokens should only come the state of Texas, a majority of the honeytokens will be from Houston, Dallas, and Austin, because these are the largest cities in the state. This models real life.

## V. PROGRAM EVALUATION

We then began an evaluation of the personally identifying honeytokens that the program produced. We wanted to show that the created honeytokens appeared realistic to a human. Since the term 'realistic' is vague, we chose to define our success or failure with a mathematical metric. Realism would be achieved if the honeytokens were detected by the humans with a distribution no greater than 10% from the baseline benchmark mean. These benchmarks, and how they were chosen, will be described in the sections that follow.

In order to test the realism of the honeytokens, we selected at random real names, addresses, and phone numbers from an online telephone book. We then used the honeytoken program to generate several thousand honeytokens that had similar attributes, namely that the honeytokens all had addresses and phone numbers from the same states that the real names and addresses came from. Future testing will need to be done to test the realism of the social security number and occupation attributes.

We then contacted one hundred human evaluators who agreed to be testers. We performed two basic types of tests; in the first, the evaluators weren't told how many individuals were honeytokens. Using this methodology, we presented them with several lists with 20 names, addresses, and phone numbers. The evaluators were told to circle the names and addresses that they thought were forgeries. One of the lists that they received contained only real people, though they weren't told this. The value that they selected as fake was used as a measurement benchmark. Then the testers were given several more lists, each containing one, five, or ten honeytokens out of a list of 20 people. They were also presented a list that was completely honeytokens. If the evaluators weren't able to distinguish the real individuals from the honeytokens, the number of people that they circled in each of these tests would be close to the benchmark value that had been established. They received different lists each time so that they wouldn't have any knowledge about the people on the lists.

In the second type of test, the testers were told that a certain number of the individuals that they were looking at were forgeries. The evaluators were asked to select the individuals that they thought weren't real, given three lists that contained one, five, and ten honeytokens respectively. For this type of test, the benchmark would be the percentage of honeytokens that would be found if the human just selected at random.

### *A. Results*

The results of the tests were very encouraging. In the first test, when the evaluators were given a list of 20 real names, on average they thought that 3.67 of the 20 were made up. When presented with 3 lists that did contain honeytokens, the number that they thought were honeytokens was very close, well within ten percent of what was selected from the mean. In the cases of 1, 5, and 10 inserted honeytokens in a list of twenty names, the volunteers thought that 3.46, 3.08, and 3.55 of them were fake, respectively. This was incredibly close to the benchmark.

On the completely fake list of 20 honeytokens, the humans selected 3.96 of them on average as fake. This was the largest amount detected as honeytokens, but it was still near to the benchmark value, differing by slightly less than 8%. This result leads us to believe that the testers detected some subtle differences in the lists when presented with data that was completely fabricated, and we've identified some of these reasons in the analysis section that follows why this might be the case. The results of this test are summarized in Figure 2.

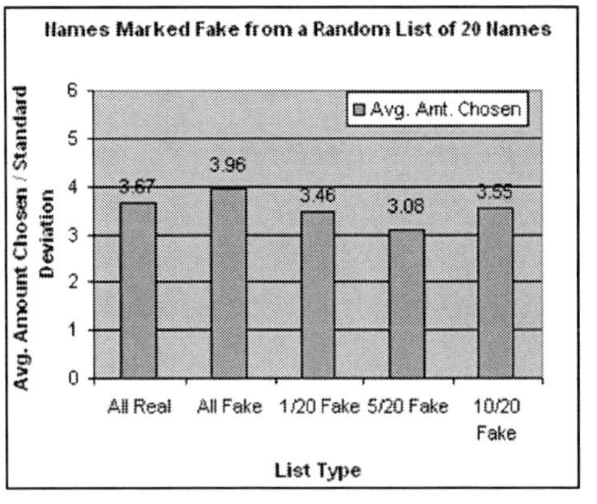


Fig. 2: Results of First Tests

The results of the second test also showed that the evaluators had a difficult time selecting the honeytokens, even when told that a certain amount were fake. Of the testers that took the test which had 1 in 20 fake records, only one tester successfully identified the honeytoken. On the other tests, the evaluators appeared to be selecting names at random, and the amount that they selected correctly was very close to what would have been expected had they been selecting randomly. The results of this test are summarized in Figure 3.

In all, over 5000 data points were analyzed by the evaluators in over 250 separate tests. While more sophisticated methods of testing will be used in the future, it was important for the honeytokens to pass this first type of test because honeytokens are designed to appear realistic to humans, first and foremost. If the honeytokens had appeared fake to untrained humans, they certainly would not have appeared valuable to a more sophisticated attacker.

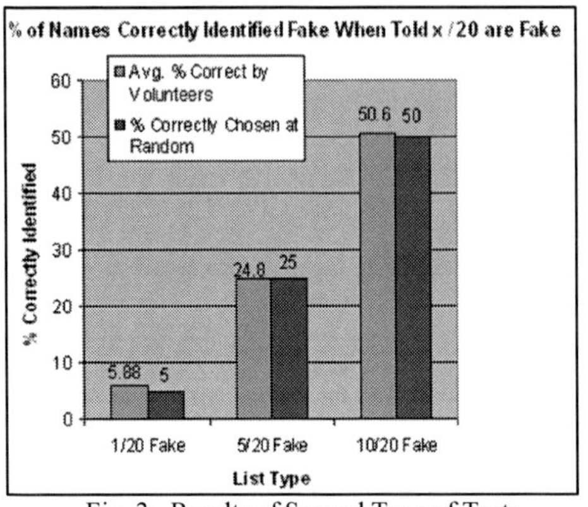


Fig. 3: Results of Second Type of Tests

### *B. Analysis & Needed Improvements*

After the evaluators took the tests, they generally commented on how difficult it was to select the honeytokens. The testers made statements about how all the data looked the same, even though some were fake. The results support that the conclusion that the evaluators had very little insight into what was real and what was a honeytoken.

### *C. Future Work*

Future work is needed in the personally identifying honeytoken arena. We would like to test the social security number and occupation attributes for realism. We would also like to add the other attributes that were identified in section 3.1. In addition, we would like to make the honeytoken program available to the public, so that others can use these types of honeytokens.

More work also needs to be done to ensure that the honeytokens do in fact add to the security of personally identifiable datasets. We would like to deploy our honeytokens into an actual database and document how the DBMS can be configured to remove the honeytokens for valid uses, and how an IDS can be set up to track and notify the security team if the honeytokens are ever detected.

## VI. CONCLUSION

In conclusion, our initial experiments have shown that it is possible to make realistic honeytokens consisting of personally identifiable information. We have documented how we created a program using public information in order to make these realistic honeytokens.

We have identified several important properties that all honeytokens should posses, and we have proven that the honeytokens that our program manufactures conform to these requirements. Specifically, we have shown that the honeytokens that are produced appear realistic, that they are easy to produce, and that they contain unique properties to identify them from the real data.

More work needs to be done with personally identifying honeytokens, as the area has several potential benefits to the security of data, and these benefits need to be explored in further detail. Insider threats to computer systems are very real [5] and they take a large toll on an organization's resources [7]. Honeytokens are a good tool that can be used to provide better accountability and security for an organization's personally identifying data.

## References

[1] I. Mokube, and M. Adams "Honeypots: concepts, approaches, and challenges," *ACM-SE 45: Proceedings of the 45th annual southeast regional conference*, New York, NY, USA, 2007, pp. 321 – 326.

[2] L. Spitzner, "Honeytokens: The Other Honeypots," http://www.securityfocus.com/infocus/1713 (current Nov. 18, 2005).

[3] C. McRae, and R. Vaughn, "Phighting the Phisher: Using Web Bugs and Honeytokens to Investigate the Source of Phishing Attacks," *40th Annual Hawaii International Conference on System Sciences (HICSS'07)*, 2007, pp. 270c.

[4] N. Thompson, "New Economy, the 'Honeytoken' an Innocuous Tag in a File Can Signal an Intrusion in a Company's Database," *NY Times*, April 2003.

[5] R. Chinchani, A. Iyer, S. Ngo, S. Upadhayaya, "Towards a theory of insider threat assessment," *DSN 2005: Proceedings of the 2005 International Conference of the Dependable Systems and Networks,* Yokohama, Japan, June 28 – July 1, 2005.

[6] G. Dhillon, "Violation of safeguards by trusted personnel and understanding related information security concerns," *Computers and Security,* 2001, Vol. 20, pp. 165 – 172.

[7] D. Straub, R. Welke, "Coping with systems risk: security planning models for management decision making," *MIS Quarterly*, 1998, Vol. 22, issue 4, pp. 441.

[8] J. Yuill, M. Zappe, D. Denning, F. Feer. "Honeyfiles: deceptive files for intrusion detection," *Information Assurance Workshop, 2004. Proceedings from the Fifth Annual IEEE SMC* , vol., no., pp. 116-122.

[9] C. Valli, "Honeypot technologies and their applicability as an internal countermeasure", *International Journal of Information and Computer Security*, Inderscience Publishers, Geneva, Switzerland, 2007, pp. 430 – 436.

[10] C. Clifton and D. Marks., "Security and privacy implications of data mining", *ACM SIGMOD Workshop on Research Issues on Data Mining and Knowledge Discovery,* May 1996, pp. 15-19.

# The role of user experience on FOSS acceptance

M. Dolores Gallego
University of Pablo de Olavide
Ctra. Utrera, km.1
Seville, 41013 Spain

Salvador Bueno
University of Pablo de Olavide
Ctra. Utrera, km.1
Seville, 41013 Spain

***Abstract* - Free and open source software (FOSS) movement essentially arises like answer to the evolution occurred in the market from the software, characterized by the closing of the source code. Furthermore, some FOSS characteristics, such as (1) the advance of this movement and (2) the attractiveness that contributes the voluntary and cooperative work, have increased the interest of the users towards free software. Traditionally, research in FOSS has focused on identifying individual personal motives for participating in the development of a FOSS project, analyzing specific FOSS solutions, or the FOSS movement itself. Nevertheless, the advantages of the FOSS for users and the effect of the demographic dimensions on user acceptance for FOSS have been two research topics with little attention. Specifically, this paper's aim is to focus on the influence of the user experience with FOSS the FOSS acceptance. Based on the literature, user experience is an essential demographic dimension for explaining the Information Systems acceptance. With this purpose, the authors have developed a research model based on the Technological Acceptance Model (TAM).**

## I. Introduction

From a professional, academic, business and political standpoint, few topics are as current as the development and implementation of free and open source software (FOSS). The changes introduced by FOSS in the software industry have been surprising, and represent a radical change of perspective in developmental business models and software distribution. This change has turned FOSS into one of the most debated topics among software users and analysts [1].

In recent years, FOSS use has rapidly grown among organizations and users, thanks to the advantages that it offers when compared to proprietary software [2]. As a consequence of its evolution, a great amount of research has been done on FOSS. Traditionally, this research has focused on, either the identification of the personal motives of the people who participate in the development of an FOSS project [3] [4] [5] [6] [7], the analysis of specific solutions that are developed by the FOSS movement [8] [9] [10] [11] [12], or on the FOSS movement, itself [13] [14] [15] [16] [1] [17] [18] [19] [20] [21].

However, the profile of the FOSS user or the influence of the demographic dimensions the acceptance towards this software solution has received very little attention. For this reason, our purpose is to analyze the effect of the user experience on the acceptance for FOSS. User experience is one of the essential demographic dimensions for explaining the Information Systems acceptance. For this development, we have considered the Technology Acceptance Model (TAM) [22] which provides the theoretical and methodological framework capable of explaining the acceptance for FOSS. With this objective in mind, we have carried out a study on users of the Linux operating system. We consider that the TAM application for users of the Linux Operative System serves as a representative sample of potential FOSS users. Thus, we understand that the conclusions reached from our specific study will allow to uncover the factors that influence user acceptance of any technology based on FOSS.

## II. TAM Methodology and Hypothesis.

The TAM model developed by [22] has been widely applied with the purpose of understanding the conduct and motivational factors that influence Information Systems and Technologies (IS/IT) adoption and use. Just as [23] indicate, only ten years after the model publication, the Social Science Citation Index listed more than four-hundred articles that had cited both articles which introduced TAM methodology, [22] and [24]. Since then, the model has become well established as a powerful and efficient tool for predicting user acceptance.

The TAM model is an adaptation of the Theory of Reasoned Action (TRA) proposed by [25] to explain and predict the behavior of organizational members in specific situations. TAM adapts the TRA model to provide evidence for the general factors that influence IS/IT acceptance in order to help determine user behavior towards a specific IS/IT. This powerful model allows for a contrast in behavior on the part of the user and is based on four fundamental variables or constructs which are: perceived usefulness (PU), perceived ease of use (PEU), intention to use (IU) and usage behavior (UB).

Independent of the internal constructs of the TAM model, FOSS users consider that the acceptance for FOSS is influenced by some external variables. Therefore, our goal is to identify the external constructs that influence the intention of use a FOSS solution. With this objective in mind, we have taken as a reference the study elaborated previously by the authors [26]. Based on this work, we consider suitable to include four external constructs to the TAM model. These variables are: system capability (SC), software flexibility (SF), software quality (SQ) and social influence (SI).

T. Sobh (ed.), *Innovations and Advances in Computer Sciences and Engineering*,
DOI 10.1007/978-90-481-3658-2_40, © Springer Science+Business Media B.V. 2010

In a same way, based on the TAM model proposed by [26] about FOSS usage behavior, we formulate an acceptance model that was validated in the cited work with a sample with 347 Linux users. This model explains the user behavior for FOSS solutions in a 39.1% (Fig. 1). We asked to the users of the sample the years of experience with FOSS for completing our study.

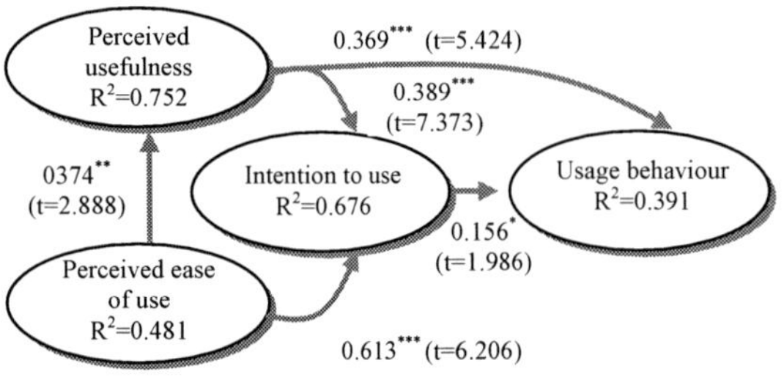


Fig. 1. Reduced research model. Source: [26]

The experience with FOSS already had been including in other similar studies. The experience with FOSS allows to users the assimilation of information about this software solution and to increase the knowledge on its advantages. In this sense, the user experience influences the user perceptions for FOSS. Some researchers, such as [27] [28] or [29], suggest that the user experience with a technology improves the perceptions and attitudes towards the use. Based on this point of view, we formulate the first working hypotheses of this research. These hypotheses are stated in the following way:

Hypothesis 1 (H1): The user experience with FOSS has a positive effect on perceived ease of use for FOSS

Hypothesis 2 (H2): The user experience with FOSS has a positive effect on perceived usefulness for FOSS.

Hypothesis 3 (H3): The user experience with FOSS has a positive effect on intention to use for FOSS

Hypothesis 4 (H4): The user experience with FOSS will have a positive effect on usage behavior for FOSS

## III. Sample Information and Preliminary Analysis

We have selected Linux FOSS users for the sample of our study. We understand that these users are a representation of all FOSS users, and the results obtained could extrapolate any solution based on FOSS. In this sense, we turned to The Linux Counter website, where we were able to access the contact information of the users that were registered in the website. The information provided to us happened to be organized by geographic areas. We selected the European countries which tended to have a higher number of registered users. Within each area, the selection made was completely at random. A total of 1,736 study invitations were sent out by electronic mail to registered Linux users in eleven different European countries. In the end, we received 363 survey responses. Of those, 347 were complete and valid for our study. This number represents a response rate of twenty percent.

In order to measure each one the variables included in the TAM model developed for our study, we carried out a review of the literature that allowed to identify items for each one of the constructs. The survey and the selection of the sample used for the study already were validated [26].

In order to cover the main objective of this study, the users of the sample indicated the years of experience with FOSS. The feedback obtained allows realizing a descriptive analysis about the user experience with FOSS and classifying the sample in three categories of users (see Table I). Besides, we have could can observe that the average years of experience with FOSS is very high, (7.43 years). This data shows the loyalty of FOSS users towards this type of technology.

TABLE I
DESCRIPTIVE ANALYSIS

| Mean | S.D. | Categories | Number of participants |
|---|---|---|---|
| | | Less than 5 years | 106 |
| 7,43 | 3,33 | Between 6 - 9 years | 131 |
| | | More than 10 years | 110 |

## IV. Analysis and Findings

In order to agree upon all the hypotheses collected in the research model, a causal analysis have been developed. The research models were tested by structural equation modeling (SEM) using Lisrel 8.51 with maximum-likelihood estimation. The parameters that represent the regression coefficients among the constructs are indicated with the symbol $\gamma$ if the relationship it represents is between an independent construct and a dependent one. If the relationship established is between two dependent constructs, it is indicated with the symbol $\beta$.

$$PEU = \gamma_1 \, SC + \gamma_2 \, SI + SF \, \gamma_3 + \varepsilon_1$$

$$PU = \gamma_4 \, SQ + SC \, \gamma_5 + SI \, \gamma_6 + \beta_1 \, PEU + \varepsilon_2$$

$$IU = \beta_2 \, PEU + \beta_3 \, PU + \varepsilon_3$$

$$UB = \beta_4 \, IU + \beta_5 \, PU + \varepsilon_4$$

For the Lisrel, we were able to prove how the model very adequately explains the variance in the perceived ease of use ($R^2$=0,481), perceived usefulness ($R^2$=0,752), intention to use ($R^2$=0,676) and usage behavior ($R^2$=0,391) [26]. Based on these findings, we have developed a causal analysis to each group of user experience with FOSS. After, we will compare the findings. The comparison analysis will allow to obtain significant discussions about the influence of user experience the FOSS acceptance. We also use Lisrel 8.51 with maximum-likelihood estimation.

## V. Research Model Test

The user experience toward a certain type of technology has generally a positive effect on the perceptions and attitudes towards the use. In our particular research, we want to test the positive influence of the user experience with FOSS on perceived ease of use (H1), perceived usefulness (H2), intention to use (H3) and usage behavior (H4) for FOSS. Finally, in order to test satisfactorily all the hypotheses, we have divided the sample in two sub-samples (see Table II). This decision is adopted to obtain sufficiently wide sub-samples for reaching estimations of the regression equations with the software Lisrel [30].

The hypotheses were tested for p <0.05*, p <0.01**, p <0.001*** (based on t(499); t(0.05;499) = 1.967007242; t(0.01; 499) = 2.590452926; t(0.001;499) = 3.319543035).

TABLE II
MODEL HYPOTHESES CONTRAST

| Interval of year | Relationship | γ or β | t-student | R2 | R2 |
|---|---|---|---|---|---|
| Up to seven years of user´s experience with FOSS | SC→ PEU | 0.394 | 4.435 | 0.446 | 0.481 |
| | SI→ PEU | 0.0298 | 0.539 | | |
| | SF→ PEU | 0.118 | 2.034 | | |
| | PEU→ PU | 0.440 | 3.164 | 0.820 | 0.752 |
| | SQ→ PU | 0.775 | 4.756 | | |
| | SC→ PU | 0.0904 | 0.593 | | |
| | SI→ PU | 0.00596 | 0.0852 | | |
| | PEU → UB | 0.677 | 4.883 | 0.675 | 0.676 |
| | PU → UB | 0.412 | 4.772 | | |
| | PU → IU | 0.355 | 3.458 | 0.463 | 0.391 |
| | UB→ IU | 0.279 | 2.613 | | |
| More than seven years of user´s experience with FOSS | SC→ PEU | 0.565 | 4.041 | 0.676 | 0.481 |
| | SI→ PEU | 0.0274 | 0.604 | | |
| | SF→ PEU | 0.114 | 3.697 | | |
| | PEU→ PU | 0.384 | 1.124 | 0.777 | 0.752 |
| | SQ→ PU | 0.714 | 3.283 | | |
| | SC→ PU | 0.788 | 1.571 | | |
| | SI→ PU | 0.0500 | 0.544 | | |
| | PEU → UB | 0.737 | 4.198 | 0.685 | 0.676 |
| | PU → UB | 0.350 | 5.396 | | |
| | PU → IU | 0.316 | 3.960 | 0.319 | 0.391 |
| | AU→ IU | 0.0176 | 0.160 | | |

These findings show the high significance of the relationship posed in hypothesis H1, between perceived ease of use and user experience. Specifically, the sub-group formed by users with experience with more than seven years has an explained variance of 0.676. The group of users with less than seven years of FOSS experience has an explained variance of 0.446. In both case the level of significance is very high. According to these findings, hypothesis H1 is tested significantly.

Regarding to the perception of usefulness, we can observe how the users with less experience perceive more usefulness (0.820) than the users with major experience (0.777). These results don't verify the positive relationship defined in the hypothesis H2. For that, we can´t accept the hypothesis H2.

Finally, with respect to the hypotheses H3 and H4, which define the positive effect between the FOSS experience with the usage behavior and the intention to use an FOSS solution, we can´t accept them. In these sense, the hypotheses H3 and H4 have not been tested significantly.

Nevertheless, based on these findings, we can't state that the user experience with FOSS has a negative effect on perception of usefulness, usage behavior or intention to use a FOSS solution. Only we can confirm that these relationships aren't positive or don't exist a significant difference.

## VI. DISCUSSIONS

Even though research on FOSS has proliferated in recent years, the acceptance of this type of technological solution on behalf of the users had not been tackled. Thus, the main objective that we pose with this research is to analyze the influence of the user experience with FOSS the acceptance towards this type of technology. With this aim, we have formulated a Technology Acceptance Model based on a previous study of the authors. Besides, we have identified relevant discussions about the influence of user experience the FOSS acceptance.

First, we have observed the particularity of FOSS. Specifically, the positive relationship between user experience and perceived ease of use for FOSS (H1) has been tested significantly. Nevertheless, there aren't significant differences about the perception of usefulness, intention to use or user behavior between users with different level of experience (hypotheses H2, H3 and H4).

Second, with these findings we cannot affirm that exist negative relationships with respect the perception of usefulness, intention to use or user behavior with the level of experience. For that, these findings show the particularity of the FOSS movement.

Thirds, based on the findings, we can affirm that the FOSS governmental organizations and developers must favorer the flow of FOSS information for fomenting the use of these solutions. With this in mind, we think that the efforts of organizations for increasing the number of FOSS users must be orientated to users who only apply proprietary solutions.

## REFERENCES

[1] A. Fuggetta, "Open source software—an evaluation". Journal of Systems and Software Vol. 66 (1), pp. 77–90, 2003

[2] C. Ruffin, and C. Ebert, "Using open source software in product development: a primer", IEEE Software Vol. 21 (1), pp. 82-86, 2004.

[3] A. Bonaccorsi and C. Rossi, "Comparing motivations of individual programmers and firms to take part in the Open Source movement. From community to business", http://opensource.mit.edu/papers/bnaccorsirossimotivationlong.pdf, 2003 [25-03-2006].

[4] G. Hertel, S. Niedner, and S. Herrmann, "Motivation of software developers in Open Source projects: an Internet-based survey of contributors to the Linux kernel", Research Policy Vol. 32 (7), pp. 1159-1177, 2003.

[5] Y. Ye and K. Kishida, "Toward an Understanding of the Motivation of Open Source Software Developers", International Conference on Software Engineering (ICSE2003). Portland - Oregon (EE.UU.), 2003.

[6] A. Hars and S. Ou, "Working for Free? Motivations for Participating in Open-Source Projects" International Journal of Electronic Commerce Vol. 6 (3), pp. 25-40, 2002.

[7] R. Ryan, and E. Deci, "Intrinsic and Extrinsic Motivations: Classic Definitions and New Directions", Contemporary Educational Psychology Vol. 25 (1), pp. 54-67, 2000.

[8] M. Federman, "The Penguinist Discourse: A critical application of open source software project

management to organization development", Organization Development Journal Vol. 24 (2), pp. 89-100, 2006.

[9] M. Fink, "The business and economics of Linux and Open Source". Ed. Prentience Hall PTR, Upper Saddle River - New Jersey (EE.UU.), 2003.

[10] N. Franke and E. Von Hippel, "Satisfying heterogeneous user needs via innovation toolkits: the case of Apache security software", Research Policy Vol. 32 (7), pp. 1199-1215, 2003.

[11] M. Mustonen, "Copyleft–the economics of Linux and other open source software", Information Economics and Policy Vol. 15 (1), pp. 99-121, 2003.

[12] G. Carbone and D. Stoddard, "Open source enterprise solutions", Ed. Wiley, Nueva York (EE.UU.), 2001.

[13] X. Shen, "Developing Country Perspectives on Software: Intellectual Property and Open Source - A Case Study of Microsoft and Linux in China", International Journal of IT Standards & Standardization Research Vol. 3 (1), pp. 21-43, 2005.

[14] R. Van Wendel and T. Egyedi, "Handling variety: the tension between adaptability and interoperability of open source software". Computer Standards & Interfaces Vol. 28 (1), pp. 109-121, 2005.

[15] B. Dwan, "Open source vs closed", Network Security Vol. 5, pp. 11-13, 2004.

[16] A. Bonaccorsi and C. Rossi, "Why Open Source software can succeed". Research Policy Vol. 32 (7), pp. 1243-1258, 2003.

[17] S. Krishnamurthy, "A managerial overview of open source software", Business Horizons, Vol. 46 (5), pp. 47-56. 2003

[18] K. Lakhani and E. Von Hippel, "How open source software works: "free" user-to-user assistance", Research Policy Vol. 32 (6), pp.923-943, 2003.

[19] J. West, "How open is open enough?: Melding proprietary and open source platform strategies", Research Policy Vol. 32 (7), pp. 1259-1285, 2003.

[20] J.P. Johnson, "Open Source Software: Private Provision of a Public Good", Journal of Economics & Management Strategy Vol. 11 (4), pp. 637-662, 2002.

[21] W. Scacchi, "Understanding the Requirements for Developing Open Source Software Systems", IEE Proceedings--Software Vol. 149 (1), pp. 24-39, 2002.

[22] F.D. Davis, "Perceived usefulness, perceived ease of use, and user acceptance of Information Technology". MIS Quarterly Vol. 13 (3), pp. 319-340, 1989.

[23] V. Venkatesh and F.D. Davis, "A theoretical extension of the technology acceptance model: four longitudinal field studies". Management Science Vol. 46 (2), pp. 186-204, 2000.

[24] F.D. Davis, R.P. Bagozzi and P.R. Warshaw, "User acceptance of computer technology: a comparison of two theoretical models", Management Science Vol. 35 (8), pp. 982–1003, 1989.

[25] M. Fishbein and I. Ajzen, "Belief, Attitude, Intention, and Behavior: An Introduction to Theory and Research". Addison-Wesley. New York, 1985.

[26] M.D. Gallego, P. Luna, and S. Bueno, "Designing a forecasting analysis to understand the diffusion of open source software in the year 2010", Technological Forecasting and Social Change Vol. 75 (5), pp. 672-686, 2008.

[27] M. Igbaria and A. Chakrabarti, "Computer anxiety and attitudes towards microcomputer use", Behavior and Information Technology Vol. 9 (3), pp. 229-241, 1990.

[28] M. Igbaria and S.A. Nachman, "Correlates of user satisfaction with end user computing: an exploratory study", Information and Management Vol. 19 (2), pp. 73-82, 1990.

[29] R. Agarwal and J. Prasad, "Are Individual Differences Germane to the Acceptance of New Information Technologies?", Decision Sciences Vol. 30 (2), pp. 361-391, 1999.

[30] A. Boomsma, "The robustness of Lisrel against small sample sizes in factor analysis models", In K.G. Ed. Joreskog y H. Wold, Systems underindirect observation. Amsterdam, Holanda, 1982.

# Using Spectral Fractal Dimension in Image Classification

J. Berke
Dennis Gabor Applied University
H-1115, Budapest, Etele street. 68. HUNGARY

*Abstract*-**There were great expectations in the 1980s in connection with the practical applications of mathematical processes which were built mainly upon Fractal Dimension (FD) mathematical basis. Significant results were achieved in the 1990s in practical applications in the fields of information technology, certain image processing areas, data compression, and computer classification. In the present publication the so far well known algorithms calculating fractal dimension in a simple way will be introduced (CISSE SCSS 2005), [6] as well as the new mathematical concept named by the author 'Spectral Fractal Dimension - SFD'. Thus it will be proven that the SFD metrics can directly be applied to classify digital images as an independent parameter. Independent classification methods will be established based on SFD (SSFD – Supervised classification based on Spectral Fractal Dimension, USFD - Unsupervised classification based on Spectral Fractal Dimension). Using mathematical methods, estimation will be given to a maximum real (finite geometric resolution) SFD value measurable on digital images, thus proving the connection between FD and SFD as well as their practical dependence.**

## I. INTRODUCTION

In the IT-aimed research-developments of present days there are more and more processes that derive from fractals, programs containing fractal based algorithms as well as their practical results. Our topic is the introduction of ways of application of fractal dimension, together with the spectral fractal dimension, the possibilities of the new mathematical concept, and the introduction of its practical applications in image processing and remote sensing.

## II. THE FRACTAL DIMENSION

Fractal dimension is a mathematical concept which belongs to fractional dimensions. Among the first mathematical descriptions of self similar formations can be found von Koch's descriptions of snowflake curves (around 1904) [20]. With the help of fractal dimension it can be defined how irregular a fractal curve is. In general, lines are one dimensioned, surfaces are two dimensioned and bodies are three dimensioned. Let us take a very irregular curve however which wanders to and from on a surface (e.g. a sheet of paper) or in the three dimension space. In practice [1-3], [9-24] we know several curves like this: the roots of plants, the branches of trees, the branching network of blood vessels in the human body, the lymphatic system, a network of roads etc. Thus, irregularity can also be considered as the extension of the concept of dimension. The dimension of an irregular curve is between 1 and 2, that of an irregular surface is between 2 and 3. The dimension of a fractal curve is a number that characterises how the distance grows between two given points of the curve while increasing resolution. That is, while the topological dimension of lines and surfaces is always 1 or 2, fractal dimension can also be in between. Real life curves and surfaces are not real fractals, they derive from processes that can form configurations only in a given measure. Thus dimension can change together with resolution. This change can help us characterize the processes that created them.

The definition of a fractal, according to Mandelbrot is as follows: A fractal is by definition a set for which the Hausdorff-Besicovitch dimension strictly exceeds the topological dimension [20].

The theoretical determination of the fractal dimension [1]: Let $(X, d)$ be a complete metric space. Let $A \in H(X)$. Let $N(\varepsilon)$ denote the minimum number of balls of radius $\varepsilon$ needed to cover $A$. If

$$FD = \lim_{\varepsilon \to 0} \left\{ Sup \left\{ \frac{LnN(\bar{\varepsilon})}{Ln(1/\bar{\varepsilon})} : \bar{\varepsilon} \in (0, \varepsilon) \right\} \right\} \tag{1}$$

exists, then $FD$ is called the fractal dimension of $A$.

The general measurable definition of fractal dimension (FD) is as follows:

$$FD = \frac{\log \frac{L_2}{L_1}}{\log \frac{S_1}{S_2}} \tag{2}$$

where $L_1$ and $L_2$ are the measured length on the curve, $S_1$ and $S_2$ are the size of the used scales (that is, resolution).

## III. SPECTRAL FRACTAL DIMENSION

Nearly all of the practical methods measure structure, the definition and process described above gives (enough) information on the (fractal) characteristics of colours, or shades of colours. Fig. 1 bellow gives an example.

T. Sobh (ed.), *Innovations and Advances in Computer Sciences and Engineering*,
DOI 10.1007/978-90-481-3658-2_41, 

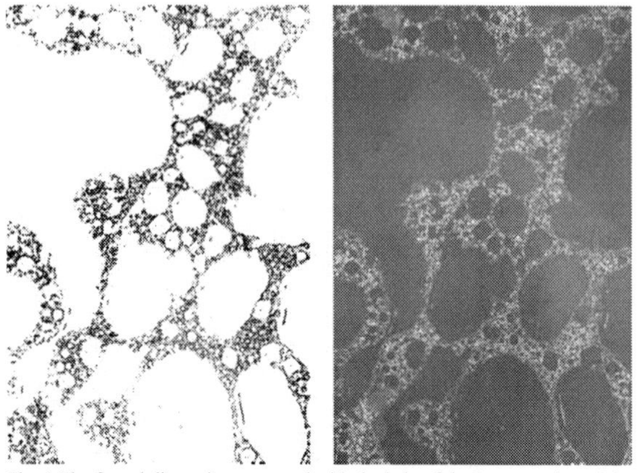

Fig. 1 The fractal dimension measured with the help of the Box counting of the two images is the same (FD=1.99), although the two images are different in shades of colour. The measured SFD of the two images show an unambiguous difference (SFD left side image=1.21, SFD right side image=2.49)

Measuring with the help of the Box counting the image on the right and the one on the left have the same fractal dimension (FD=1.99) although the one on the left is a black and white (8 bit) image whereas the other one on the right is a 24-bit coloured image containing shades as well - the original images can be found at http://www.digkep.hu/sfd/index.htm/, [24] -, so there is obviously a significant difference in the information they contain. How could the difference between the two images be proven using measurement on the digital images?

Let spectral fractal dimension (SFD) be [6], [7]:

$$SFD = \frac{\log \frac{L_{S2}}{L_{S1}}}{\log \frac{S_{S1}}{S_{S2}}} \quad (3)$$

where $L_{S1}$ and $L_{S2}$ are measured spectral length on N-dimension colour space, $S_{S1}$ and $S_{S2}$ are spectral metrics (spectral resolution of the image).

In practice, N={1, 3, 4, 6, 8, 10, 12, 32, 60, 79, 126, 224, 242, 254, 488, 498, ... }, see Table 1.

TABLE 1

NUMBER OF LAYERS OR BANDS IN PRACTICAL IMAGES

| Type of images | Number of layers or bands in practice /N/ |
|---|---|
| Black and white or greyscale image | 1 |
| RGB, YCC, HSB, IHS colour space image | 3 |
| Traditional colour printer CMYK space image | 4 |
| Landsat TM, Landsat ETM satellite images | 6 |
| Professional photo printers space image | 6, 8, 10, 12 |
| DAIS7915 VIS_NIR or DAIS7915 SWIP-2 airborne sensors | 32 |
| COIS VNIR satellite sensor | 60 |
| DAIS 7915 all airborne sensor | 79 |
| HyMap airborne sensor | 126 |
| AVIRIS airborne sensor | 224 |
| Hyperion satellite sensor | 242 |
| AISA Hawk airborne sensor | 254 |
| AISA Eagle airborne sensor | 488 |
| AISA Dual airborne sensor | 498 |

In practice the measure of spectral resolution can be equalled with the information theory concept of $\{S_i=1, \ldots, S_i=16$, where $i=1$ or $i=2\}$ bits.

Typical spectral resolution: (4)

- Threshold image -1 bit
- Greyscale image - 2-16 bits
- Colour image - 8-16 bits/bands

On this basis, spectral computing is as follows:

1. Identify which colour space the digital image is
2. Establish spectral histogram in the above space
3. Half the image as spectral axis
4. Examine valuable pixels in the given N-dimension space part (N-dimension spectral box)
5. Save the number of the spectral boxes that contain valuable pixels
6. Repeat steps 3-5 until one (the shortest) spectral side is only one (bit).

In order to compute dimension (more than two image layers or bands and equal to spectral resolution), the definition of spectral fractal dimension can be applied to the measured data like a function (number of valuable spectral boxes in proportion to the whole number of boxes), computing with simple mathematical average as follows [7]:

$$SFD_{measured} = \frac{n \times \sum_{j=1}^{S-1} \frac{\log(BM_j)}{\log(BT_j)}}{S-1} \quad (5)$$

where

- n – number of image layers or bands
- S - spectral resolution of the layer, in bits – see Eq. 4
- $BM_j$ - number of spectral boxes containing valuable pixels in case of j-bits
- $BT_j$ - total number of possible spectral boxes in case of j-bits

The number of possible spectral boxes ($BT_j$) in case of j-bits as follows:

$$BT_j = (2^S)^n \quad (6)$$

With Eqs. (5) and (6) the general measurable definition of spectral fractal dimension is as follows, if the spectral resolution is equal to all bands (SFD $_{\text{Equal Spectral Resolution}}$ – $SFD_{ESR}$), [7]:

$$SFD_{ESR} = \frac{n \times \sum_{j=1}^{S-1} \frac{\log(BM_j)}{\log((2^S)^n)}}{S-1} \quad (7)$$

If the spectral resolution is different in bands/layers, the general measurable definition of spectral fractal dimension (SFD $_{\text{Different Spectral Resolution}}$ – $SFD_{DSR}$) is as follows [7]:

$$SFD_{DSR} = \frac{n \times \sum_{j=1}^{(\min(S_i))-1} \frac{\log(BM_j)}{\log(2^{(\sum_{k=1}^{n} S_k)})}}{(\min(S_i))-1} \quad (8)$$

where,

- $S_i$ - spectral resolution of the layer i, in bits

During computing:

1. Establish the logarithm of the ratio of BM/BT to each spectral halving
2. Multiply the gained values with *n* (number of image layers or bands)
3. Find the mathematical average of the previously gained values

A computer program that measures SFD parameter has been developed in order to apply the algorithm above. The measuring program built on this method has been developed in C++ environments. The SFD results measured by the program are invariant for identical scale pixels with different geometric positions in case the number of certain scales is the same and shade of colour is constant.

Successful practical application of SFD at present [4-12], [16-18], [24]:

- Measurement of spectral characteristics of multispectral and hyperspectral satellite images
- Measurement of spectral characteristics of multispectral and hyperspectral airborne images
- Psychovisual examination of image compressing methods
- Temporal examination of damage of plant parts
- Classification of natural objects in multispectral and hyperspectral satellite and/or airborne images
- Virtual Reality based 3D terrain simulation

## IV. SFD AS METRICS

Beneath it will be proven that spectral fractal dimension-SFD generally defined above (3) and the $SFD_{ESR}$ és $SFD_{DSR}$ introduced for different spectral resolutions are metrics, that is, it satisfies the following conditions:

1. non-negative definite, that is

$$\rho(P_1, P_2) \geq 0$$
$$\rho(P_1, P_2) = 0 \quad \text{if } P_1 = P_2$$

2. symmetric, that is

$$\rho(P_1, P_2) = \rho(P_2, P_1)$$

3. satisfies triangle inequality, that is

$$\rho(P_1, P_3) \leq \rho(P_1, P_2) + \rho(P_2, P_3)$$

*Statement 1 - SFD according to (3) is non-negative definite*

Let

$$\rho_{SFD} := SFD(A+P) - SFD(A) \tag{3.1}$$

where A is an obtional subset of N dimension image plane, whereas P is an optional point of N dimension image plane.

Then, if $P \in A$ that is, the value or intensity of P (N dimension) equals a value or intensity of a point in set A, then

$$\rho_{SFD}(A, P) = 0 \text{, as } \rho_{SFD}(A) = \rho_{SFD}(A+P).$$

$\rho(P_1, P_2) \geq 0$ condition is also satisfied, as SFD defined by (3) monotonously grows with the same S and n.

*Statement 2 - SFD according to (3) is symmetric*

This condition in the present case means that

$$\rho_{SFD}(A, P) = \rho_{SFD}(P, A) \tag{3.2}$$

is satisfied.

Let then

$$\rho_{SFD} := SFD(A+P) - SFD(A)$$

If $P \in A$ then

$$SFD(A) = SFD(A+P) \text{ that is } \rho_{SFD} = 0$$

If $P \notin A$, but $P \in K$, where K is the set of points of the image plane

$$SFD(A+P) > SFD(A) \text{ that is}$$

$$\rho_{SFD} = SFD(A+P) - SFD(A) > 0\,!$$

*Statement 3 - SFD according to (3) satisfies triangle inequality*

The above statement in the present case:

$$\rho_{SFD}(A, P_2) \leq \rho_{SFD}(A, P_1) + \rho_{SFD}(A, P_2) \tag{3.3}$$

Let according to 3.1

$$\rho_{SFD} := SFD(A+P) - SFD(A)$$

Then,

$$\rho_{SFD}(A, P_2) \leq [SFD(A+P_1) - SFD(A)] + [SFD(A, P_2) - SFD(A)]$$

that is

$$SFD(A, P_2) - SFD(A) \leq [SFD(A+P_1)] + [SFD(A, P_2)] - 2SFD(A)$$

Simplified:

$$0 \leq SFD(A+P_1) - SFD(A)$$

which does satisfy, as SFD defined by (3) monotonously grows with the same S and n.

Another condition to be satisfied by metrics is regularity. This means that the points of a discreet image plane are to be evenly dense. This condition, in the case of digital images, is usually fulfilled, or can be considered so.

Based on the above calculations we have accepted that the

correspondence (3) as well as (7) and (8) give metrics, thus can directly be used for classifying digital images.

## V. Estimation of SFD for Digital Images with Finite Spatial Resolution

Bellow, practical correspondence will be given for SFD measurements applicable for finite spatial resolution images (images supplied by CCD and CMOS sensors are all such). In the case of correspondence (3), (7) or (8), it can be directly identified that

$$0 \leq SFD \leq n$$

that is, the value of SFD can be between 0 and the number of channels/layers used in the measurement.

For further estimation, let us make use of the fact that the number of pixels in a digital image is known, let it be K,

$$K = X * Y$$

where

K – is the number of pixels of the image

X - is the width of the image in pixels

Y - is the length of the image in pixels

If

$$K \geq BT_j$$

that

$$SFD_{\max} = n \tag{9}$$

but if

$$K < BT_j$$

then the maximum of different spectral pixels is the number of pixels, then

$$SFD_{ESR-MAX} = \frac{n \times \left( \sum_{j=Z}^{S-1} \frac{\log(BM_j)}{\log((2^S)^n)} + (Z-1) \right)}{S-1} \tag{10}$$

where it is true that

$$1 \leq Z \leq S-1$$

and $Z$ is selected so that in the case of $Z-1$, $K \geq BT_j$ be true.

Let us see what correspondence (10) means in some practical cases (Table 2.):

TABLE 2

SPATIAL AND SPECTRAL RESOLUTION IN PRACTICAL IMAGES

| | Human eye 1. | Human eye 2. | Phase One P65+ | Hasselblad H3DII-50 | EOS 1Ds MIII |
|---|---|---|---|---|---|
| x | | | 8984 | 8176 | 5616 |
| y | | | 6732 | 6132 | 3744 |
| S | 21 bit | 21 bit | 16 bit | 16 bit | 14 bit |
| K | 120000000 | 2000000 | 60480288 | 50135232 | 21026304 |
| n | 3 | 3 | 3 | 3 | 3 |
| SFD | 2,3282 | 2,0486 | 2,5710 | 2,5597 | 2,6416 |
| SSRR | **48,8927** | **43,0207** | **41,1352** | **40,9558** | **36,9827** |

Based on (10), SFD depends on S, n and K, thus its value is different depending on CCD/CMOS sensor type. In the case of 2-dimension sensors n=3, thus SFD will be dependent on only 2 parameters (S, K). With fixed pixel images (K=21MP), with increase in the value of S, that of SFD will decrease (Table 3.).

TABLE 3

THE CHANGE OF max(SFD) WITH FIXED K (K=21mp)

| S /bit/ | 16 | 14 | 12 | 10 | 8 |
|---|---|---|---|---|---|
| max (SFD) | 2,5078 | 2,6416 | 2,7812 | 2,9135 | 3,0000 |

Thus, SFD in itself is not a characteristic parameter of a sensor. Only K or only S are also not characteristic of the sensor independently. If correspondence (10) is multiplied by (S-1), the following value will be gained:

$$SSRR_{CCD-CMOS} = n \times \left( \sum_{j=Z}^{S-1} \frac{\log(BM_j)}{\log((2^S)^n)} + (Z-1) \right) \tag{11}$$

The value of this quantity monotonously grows, if any two of S, K, or n are fixed and the third value grows. (11) includes all three characteristic parameters, so it can by itself be a characteristic value of any digital image sensors.

## VI. Conclusion

When examining digital images where scales can be of great importance (image compressing, psychovisual examinations, printing, chromatic examinations, etc.) SSRR is suggested to be taken among the so far usual (eg. sign/noise, intensity, size, resolution) types of parameters (eg. compression, general characterization of images). Useful information on structure as well as shades can be obtained applying the SSRR parameter.

Several basic image data (aerial and space photographs) consisting of more than three bands are being used in practice. There are hardly any accepted parameters to characterize them together. I think SSRR can perfectly be used to characterize (multi-, hyper spectral) images that consist of more than three bands. On the basis of present and previous measurements it can be stated that SFD are significant parameter in the classification of digital images as well (SSFD – Supervised classification based on Spectral Fractal Dimension, USFD - Unsupervised classification based on Spectral Fractal Dimension). SFD can be an important and digitally easily measurable parameter of natural processes and spatial structures [7], [16], [24].

The applied method has proven that with certain generalization of the Box method fractal dimension based measurements – choosing appropriate measures- give practically applicable results in case of optional number of dimension.

It is therefore suggested that, in the case of digital image sensing devices, (S-1)*max(SFD) value, as Spatial and Spectral Resolution Range - SSRR, which includes all three important parameters (number of sensor pixels, spectral resolution, number of channels), be launched.

In the case of multispectral or hyperspectral images, the SFD value can be identified by bands/layers. Drawing a graph from these values, unique curve(s) can be obtained of the original images or the objects on them. These curves can be used independently as well, e.g. in the case of plants and minerals directories can be set up, similarly to spectrum directories based on reflectance. An advantage is that it gives information directly on the image recorded by the detector, as well as that analysing the curves (Fig. 2), information on image noises caused by the atmosphere can be obtained in the case of remote sensing aerial and space devices [17].

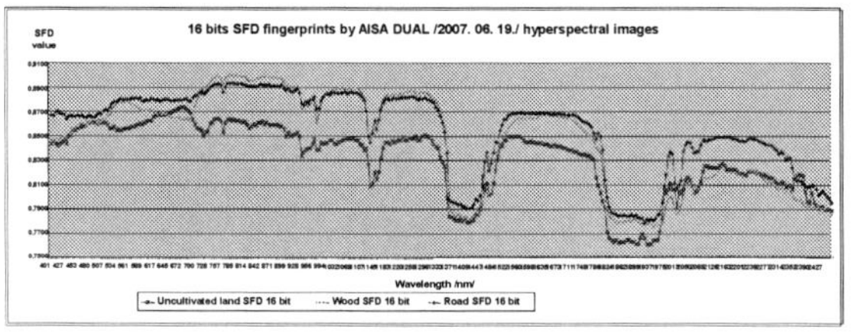


Fig. 2 SFD wavelength-based 16-bit spectral curves "fingerprints" of uncultivated land, wood and road based on AISA Dual aerial data

## REFERENCES


[1] Barnsley, M. F., Fractals everywhere, Academic Press, 1998.

[2] Barnsley, M. F. and Hurd, L. P., Fractal image compression, AK Peters, Ltd., Wellesley, Massachusetts, 1993.

[3] Batty, M. and Longley, P. Fractal cities, Academic Press, 1994.

[4] Berke, J., Fractal dimension on image processing, *4th KEPAF Conference on Image Analysis and Pattern Recognition*, Vol.4, 2004, pp.20.

[5] Berke, J., The Structure of dimensions: A revolution of dimensions (classical and fractal) in education and science, 5th International Conference for History of Science in Science Education, July 12 – 16, 2004.

[6] Berke, J., Measuring of Spectral Fractal Dimension, Advances in Systems, Computing Sciences and Software Engineering, Springer pp. 397-402., ISBN 10 1-4020-5262-6, 2006.

[7] Berke, J., Measuring of Spectral Fractal Dimension, Journal of New Mathematics and Natural Computation, ISSN: 1793-0057, 3/3: 409-418, 2007.

[8] Berke, J. and Busznyák, J., Psychovisual Comparison of Image Compressing Methods for Multifunctional Development under Laboratory Circumstances, WSEAS Transactions on Communications, Vol.3, 2004, pp.161-166.

[9] Berke, J., Spectral fractal dimension, Proceedings of the 7th WSEAS Telecommunications and Informatics (TELE-INFO '05), Prague, 2005, pp.23-26, ISBN 960 8457 11 4.

[10] Berke, J., Wolf, I. and Polgar, Zs., Development of an image processing method for the evaluation of detached leaf tests, Eucablight Annual General Meeting, 24-28 October, 2004.

[11] Berke, J. and Kozma-Bognár V., Fernerkundung und Feldmessungen im Gebiet des Kis-Balaton I., Moorschutz im Wald / Renaturierung von Braunmoosmooren, Lübben, 2008.

[12] Berke, J., Polgár, Zs., Horváth, Z. and Nagy, T., Developing on Exact Quality and Classification System for Plant Improvement, Journal of Universal Computer Science, Vol.XII/9, 2006, pp. 1154-1164.

[13] Burrough, P.A., Fractal dimensions of landscapes and other environmental data, Nature, Vol.294, 1981, pp. 240-242.

[14] Buttenfield, B., Treatment of the cartographic line, Cartographica, Vol. 22, 1985, pp.1-26.

[15] Encarnacao, J. L., Peitgen, H.-O., Sakas, G. and Englert, G. eds. Fractal geometry and computer graphics, Springer-Verlag, Berlin Heidelberg 1992.

[16] Kozma-Bognár, V., Hegedűs, G., and Berke, J., Fractal texture based image classification on hyperspectral data, AVA 3 International Conference on Agricultural Economics, Rural Development and Informatics, Debrecen, 20-21 March, 2007.

[17] Kozma-Bognár, V. and Berke, J., New Applied Techniques in Evaluation of Hyperspectral Data, Georgikon for Agriculture, a multidisciplinary journal in agricultural sciences, Vol. 12./2., 2008, preprint.

[18] Kozma-Bognár, V., Hermann, P., Bencze, K., Berke, J. and Busznyák, J. 2008. Possibilities of an Interactive Report on Terrain Measurement. Journal of Applied Multimedia. No. 2/III./2008. pp. 33-43., ISSN 1789-6967. http://www.jampaper.eu/Jampaper_ENG/Issue_files/JAM080202e.pdf.

[19] Lovejoy, S., Area-perimeter relation for rain and cloud areas, Science, Vol.216, 1982, pp.185-187.

[20] Mandelbrot, B. B., The fractal geometry of nature, W.H. Freeman and Company, New York, 1983.

[21] Peitgen, H-O. and Saupe, D. eds. The Science of fractal images, Springer-Verlag, New York, 1988.

[22] Schowengerdt, R. A. 2007. Remote Sensing Models and Methods for Image Processing. Elsevier. ISBN 13: 978-0-12-369407-2.

[23] Turner, M. T., Blackledge, J. M. and Andrews, P. R., Fractal Geometry in Digital Imaging, Academic Press, 1998.

[24] Authors Internet site of parameter SFD - http://www.digkep.hu/sfd/index.htm.

# A Novel Method to Compute English Verbs' Metaphor Making Potential in SUMO

Zili Chen, Jonathan J Webster, Ian C Chow

*Department of Chinese, Translation & Linguistics, City University of Hong Kong, 81 Tat Chee Avenue, Kowloon Tong, Kowloon, Hong Kong*

**Abstract** - A general practice in research of metaphor has been to investigate metaphor's application and computation based on its behavior and function in different contexts. This paper investigates the postulate of verb's built-in property of Metaphor Making Potential (MMP), thus being an initiatory context-free experiment with metaphor. A case study of a selected group of English core verbs has been conducted. A new algorithm is proposed to operationalize the assessment of English verb's MMP in the framework of WordNet and SUMO. A hypothesis is set up to testify the validity of verb's MMP.

## I. INTRODUCTION

Metaphorical computation continues to pose a significant challenge to NLP and machine translation. Up to now, tentative achievements have been attained in the machine understanding of metaphors generally through rule-based and statistical-based approaches. Among them, knowledge representation based methods are predominant [1]. These methods mainly employ as their working mechanism knowledge representation based ontologies, such as The Suggested Upper Merged Ontology (SUMO), which has been exploited by Ahrens, Huang, et al in their doing metaphorical computation [2, 3].

SUMO, an effort of the IEEE Standard Upper Ontology Working Group with the support of Teknowledge, contains terms chosen to cover all general domain concepts needed to represent world knowledge. Related to this, the metaphor-making potential in language relies on crossing domain attributes. Ahrens & Huang's research with SUMO and metaphor has focused on specific domain metaphors [2, 3], thus failing to make full use of SUMO's overall domain coverage.

Since verb maintains the core for language processing, as believed by some linguists and philosophers, and previous work on metaphor computation was focusing on noun metaphors, or verb's collocations (either pre- or post- verb), now my question is, would it be possible to look into the verb itself for its metaphorical property? With this query in mind, SUMO comes into sight since it has a well developed hierarchy of domain concepts.

This paper conducts an in-depth case study of a selected group of English core verbs in the framework of WordNet and SUMO. In seeking ways to operationalize the assessment of English verbs' property of MMP, an algorithm is proposed based on the WordNet lexical representation and SUMO ontology. A pilot experiment is carried out, manually with a small sample size of 20 most frequent English verbs obtained from BNC, TIME Magazine, CCAE (previously ANC) and Brown Corpus. A hypothesis based on Lakoff view of metaphor as a result of "our constant interaction with our physical and cultural environments" [4, 5] is also set up to testify the validity of verb's metaphor-making potential, i.e. higher frequency verbs possess bigger MMP. As a study both theory and application-oriented, this paper also shows that an ontology-based approach is more objective than an

T. Sobh (ed.), *Innovations and Advances in Computer Sciences and Engineering*,
DOI 10.1007/978-90-481-3658-2_42, © Springer Science+Business Media B.V. 2010

intuition-based approach in generating insights into verbs' metaphorical property.

## II. METAPHOR, CONCEPTUAL METAPHOR AND METAPHORICAL COMPUTATION

Metaphor study has gone through three major stages from Aristotle's Comparison and Substitute View, through Richard and Black's Interaction View to finally the current Conceptual View. Meanwhile, Chinese linguists have for the most part limited their investigation of metaphor to its rhetorical and psychological properties.

In 1979, G. Lakoff and M. Johnson set out to develop a new theory called Conceptual Metaphor (CM), in which they identified metaphors as central, not peripheral, to the study of language, understanding, and the meaningfulness of everyday experience. They argue that human thought processes and conceptual system are metaphorically defined and structured; and "the essence of metaphor is understanding and experiencing one kind of thing in terms of another." Differing from the objectivist's view of inherent property, CM's conceptual system is the product of how we interact with our physical and cultural environments. Furthering the definition of a concept and changing its range of applicability is possible because metaphor-driven categorization and recategorization render the open-endedness of concept. Thus we should expect the most efficient way to investigate those Interactional Properties and their underlying internal cross-domain alignment of prototypes is to examine how they are projected by the category oriented SUMO hierarchy.

Recent research in metaphorical computation has been focused primarily on the English Language. The approaches mainly fall into two categories: rule-based approaches and statistical-based approaches. The former stems from conventional theories of metaphor in linguistics, philosophy and psychology, including specifically metaphor semantics, possible worlds semantics and logic, knowledge representation. And the latter dwells on corpus linguistics and employs statistical based techniques. Those papers are all limited to the study of metaphor's behavior and function in different contexts.

## III. ENGLISH VERB'S MMP CALCULATION

### *A. Research Justification and Design*

In line with the above consideration, we plan to carry out an in-depth investigation into a selected group of English core verb's self-contained metaphorical traits through mapping their senses in WordNet to SUMO's domain-aligned hierarchy.

Lakoff argues that verbs, as well as words of other classes, develop their new metaphorical meanings and usages from their root meanings through interaction with their surroundings [4, 5]. But illustration and validation of this phenomenon depends on linguists' introspection and inference. Investigating this phenomenon using SUMO's hierarchy will provide a de facto computable ground for understanding verbs' self-contained metaphorical nature. Moreover, the centrality of verbs for language progression and processing has often been emphasized. Also in the field of first language acquisition and second language acquisition, a similar conclusion was arrived that the verb system is "the central feature of the target language" that influences the acquisition of increasingly more complex structures [6].

SUMO has more than 1000 terms, 4000 axioms and 750 rules. A verb in SUMO hierarchy has different senses located in different levels of concepts under Entity. Verbs differ from each other in that each verb's senses' depth to the root differs from that of other verbs [7, 8, 9]. Calculation of these differences resembles computation of words' semantic distance, semantic similarity and semantic relatedness. There are currently dozens of calculators to measure words' semantic distance/similarity/relatedness, most of which rest on WordNet. Representative measures are Lin and Wu & Palmer etc. They assign different weights on words' width, depth, information content, etc., thus output different effects [10, 11]. All those measures calculate the semantic distance by computing the shortest edges between two words. My tentative measure varies from the above mainly in that the objects to be computed are not words themselves, but their senses.

### B. Research Methodology

#### Identification of the selected list of English core verbs

A simple method shown to be very useful to delimit a group of core verbs is frequency ranking (e.g. the normal practice is the 10, 20, 50, or 100 most frequent verbs) within a particular word class; frequency ranking of general purpose corpus will be considered for trimming the list of core verbs. Specifically, the British National Corpus (BNC), TIME Magazine, Corpus of Contemporary American English (previously named the BYU American National Corpus) created by Mark Davies at Brigham Young University, and the book "Frequency Analysis of English Usage" based on the earlier Brown Corpus by W. Nelson Francis are consulted for English verbs' general purpose frequency ranking. We filtered and finalized a list of 20 most frequent verbs for our pilot study.

#### Mapping Verb's WordNet Senses to SUMO Concepts

Adam Peace et al have already mapped a word's WordNet senses to its SUMO corresponding concepts [12]. The above 20 verbs' corresponding concepts and their distribution were manually recorded.

#### Algorithmic Consideration

The Depth of a verb's WordNet sense is defined as the minimum edge count of paths from the root *Entity* in SUMO hierarchy to this said sense. We define the depth of sense *i* as $DP(S_i)$ and normalize a verb's MMP as

$$MMP(Verb) = \sum_{i=1}^{m} \frac{DP(S_i)}{Max_{DP(S)}};$$

$$DP(S_i) = Min(Edges(Path_j) \mid 1 \leq j < TotalPaths)$$

where $m$ is the number of senses mapped to SUMO hierarchy concepts, $S_i$ is the i-th sense in $m$, $DP(S_i)$ is the depth of $S_i$ in SUMO, $Max_{DP(S)}$ is maximum depth of SUMO's hierarchy, $Edges(Path_j)$ is edge count of $Path_j$ of $S_i$ and $TotalPaths$ is the total number of paths of $S_i$.

## IV. RESULTS AND DISCUSSION

Before the experiment, what has been anticipated is that the higher frequent verbs would possess the more metaphorical potential, which is based on the belief that a more utilized verb is involved in more interactions, thus tends to incur more metaphorical usages [4, 5]. Result of this preliminary study shows that the hypothesis is generally true as shown by the trend line in FIG. 1.

Mann-Kendall method [13] is used to further test whether verbs' MMP has a downward trend in correlation with verbs' frequency ranking. Kendall test is a nonparametric test rule and insensitive to extreme value and thus fits the feature of the experimental data (MMP(Verb1), .., MMP(Verb20)) as a sample of independent and non-normally distributed random variables. Its null hypothesis $H_0$ is that there is no trend in the top 20 verbs' Metaphor Making Potential *MMP(Verb)*. The Kendall test rejected the $H_0$ by showing that there is a significant downward trend at the 0.05 level for the top 20 verb's MMP.

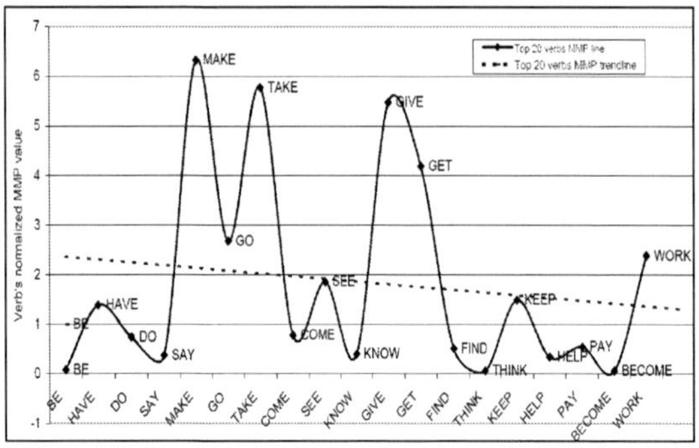


**FIG.1 Top 20 most frequent verbs MMP distribution**

Moreover, we also observed some interesting phenomenon. Verbs like *give, take, make, get* which are positioned in the middle based on frequency rank at the top in terms of their MMP value; while verbs *be, do, say* which are ranked at the top based on frequency now at the bottom in terms of their MMP ranking. Further investigation reveals that those verbs ranking higher in terms of metaphorical potential fall into the verb categories of Possession and Production; while those ranking lower in metaphorical potential (with the exception of *say*) all fall into the verb category of General Dynamic [14]. This finding suggests that verbs' MMP trait is closely linked to verbs' functional categories.

The small size of the samples analyzed however precludes the possibility of hastily drawing any generalizations. Instead, we anticipate that such should be possible after conducting a future study into verbs' metaphorical traits based on a large sample size analyzed using SUMO.

## V. Summary

This study is both theory- and application-oriented. A new method is proposed to study a word's intrinsic metaphorical property and SUMO as an ontology benchmark is validated. We have as well observed that higher frequency verbs generally possess greater metaphor making potential; while the verb's MMP on the other hand is also strongly influenced by its functional category.

One of the future tasks is to expand the sample size of core English verbs to produce a stronger validation; another is to apply this method to other classes of words to generate the contour of a word's MMP trait.

## REFERENCE

[1] Zhou Changle, Yang Yun, Huang Xiaoxi. Computational mechanisms for metaphor in languages: a survey. Journal of Computer Science and Technology, vol.22, no.2, pp.308-319. 2007.

[2] Ahrens K, Huang C R, Chung S F. Conceptual metaphors: Ontology-based representation and corpora driven mapping principles. In Proc. ACL Workshop on Lexicon and Figurative Language, Sapporo, Japan. pp.35-41. 2003

[3] Ahrens k. When love is not digested: Underlying reasons for source to target domain pairing in the contemporary theory of metaphor. In Proc. 1st Cognitive Linguistics Conference, Taipei, pp.273-302. 2002

[4] Lakoff G. and Johnson M. Metaphors we live by. Chicago: The University of Chicago Press. 1980

[5] Lakoff G. The Contemporary Theory of Metaphor. Metaphor and Thought, 2nd Edition. Cambridge: Cambridge University Press, Ortony A (ed.), pp.202--251. 1993

[6] Viberg, A.: Crosslinguistic perspectives on lexical organization and lexical progression. In Hyltenstam, K. & Viberg, A. (eds.), Progression & Regression in Language: Sociocultural, Neuropsychological, & Linguistic Perspectives. Cambridge University Press, 340-385. 1993

[7] Niles, I., and Pease, A. Towards a Standard Upper Ontology. In Proceedings of the 2nd International Conference on Formal Ontology in Information Systems (FOIS-2001), Chris Welty and Barry Smith, eds, Ogunquit, Maine, October 17-19, 2001.

[8] Pease, A., Niles, I., and Li, J. The Suggested Upper Merged Ontology: A Large Ontology for the Semantic Web and its Applications. In Working Notes of the AAAI-2002

Workshop on Ontologies and the Semantic Web, Edmonton, Canada, July 28-August 1, 2002.

[9] Ian C Chow, Jonathan J Webster, Mapping FrameNet and SUMO with WordNet Verb: Statistical Distribution of Lexical-Ontological Realization, micai, pp. 262-268, Fifth Mexican International Conference on Artificial Intelligence (MICAI'06), 2006.

[10] D. Lin. An information-theoretic definition of similarity. In Proceedings of the International Conference on Machine Learning, Madison, August. 1998

[11] Z. Wu and M. Palmer. Verb semantics and lexical selection. In 32nd Annual Meeting of the Association for Computational Linguistics, Las Cruces, New Mexico, pages 133–138. 1994

[12] http://sigma.ontologyportal.org:4010/sigma/KBs.jsp

[13] Kendall, M.G.. Rank Correlation Methods. Hafner, 145pp. 1962

[14] Levin Beth. English Verb Class and Alternations: A Preliminary Investigation. Chicago: University of Chicago Press. 1993

# A Numerical Construction Algorithm of Nash and Stackelberg Solution for Two-person Non-zero Sum Linear Positional Differential Games*

Anatolii F. Kleimenov
*Inst. of Math. and Mech., Ural Branch of RAS, 16, S.Kovalevskaja street, Ekaterinburg, 620219, Russia (e-mail: kleimenov@imm.uran.ru)*

Sergei I. Osipov
*Ural State University, 51, Lenin Ave., Ekaterinburg, 620083, Russia (e-mail: sergei.osipov@usu.ru)*

Dmitry R. Kuvshinov
*Ural State University, 51, Lenin Ave., Ekaterinburg, 620083, Russia (e-mail: evetro@2-u.ru)*

**Abstract.**
**The report proposes a numerical method of Stackelberg and Nash solutions construction in a class of differential games. It is based upon results of the positional antagonistic differential games theory developed by N. N. Krasovskii and his scientific school. The method transforms a non-antagonistic game into so-called non-standard optimal control problem. Numerical solutions for Stackelberg games are constructed by an algorithm developed by S. Osipov. For Nash solution construction we build auxiliary bimatrix games sequence. Both algorithms make use of known antagonistic game value computation procedures and are ultimately based upon computational geometry algorithms including convex hull construction, union, intersection, and Minkowski sum of flat polygons. Results of numerical experiment for a material point motion in plane are presented. The point is moved by force formed by two players. Each player has his personal target point. Among the obtained results, there is a Nash solution such, that along the corresponding trajectory the position of the game is non-antagonistic at first, and then becomes globally antagonistic starting from some moment of time.**

## I. INTRODUCTION

Various approaches for computation of solutions in differential games exist, see e.g. [1,2,3,4]. Many of them suggest numeric methods. For example, such algorithms proposed for antagonistic games are discussed in papers [5,6], as well as in other studies of the same and other authors. Comparing to this, there are distinctly less research concerning non-antagonistic games. This paper describes an algorithm for Nash equilibrium solutions and Stackelberg solutions in a two-person linear differential game with geometrical constraints for players' controls and terminal cost functionals of players. The algorithm and the program for Nash solutions was mainly implemented by D. Kuvshinov.

The paper is organized as follows. Section II contains a problem statement. Section III describes a common method for Nash and Stackelberg solutions construction based on reduction of the original problem to non-standard problems of (optimal) control. Section IV presents a description of two algorithms. The first one builds a Nash solution with the help of an auxiliary bimatrix game. The second one solves the Stackelberg problem by approximating admissible trajectories via repetitive intersections of stable bridges and local attainability sets. A description of the program implementation and precision parameters used by it is given in Section V. Results of numerical experiment for material point motion in plane are presented in Section VI. The point is moved by force formed by two players. Each player has his personal target point. Among the obtained results, there is a Nash solution such, that along the corresponding trajectory the position of the game is non-antagonistic at first, and then becomes globally antagonistic starting from some moment of time. Finally, Section VII proposes possible research perspectives.

## II. PROBLEM STATEMENT

Let the dynamics of a two-person non-antagonistic positional differential game be described by

$$\mathbf{x}'(t) = \mathbf{A}(t)\mathbf{x}(t) + \mathbf{B}(t)\mathbf{u}(t) + \mathbf{C}(t)\mathbf{v}(t), \qquad (1)$$
$$\mathbf{x}(t_0) = \mathbf{x}_0,\ t \in [t_0, \theta],$$

where $\mathbf{x} \in R^n$ is a phase vector. Matrices $\mathbf{A}(t)$, $\mathbf{B}(t)$ and $\mathbf{C}(t)$ are continuous and have dimensions $n\times n$, $n\times k$ and $n\times l$, respectively. Controls $\mathbf{u} \in P \subset R^k$ and $\mathbf{v} \in Q \subset R^l$ are handled by Player 1 and Player 2, respectively. Sets $P$ and $Q$ are assumed to be convex polyhedrons. The final time $\theta$ is fixed.

The goal of Player 1 is to maximize cost functional $\sigma_1(\mathbf{x}(\theta))$, while Player 2 must maximize cost functional $\sigma_2(\mathbf{x}(\theta))$. Functions $\sigma_1 : R^n \to R$ and $\sigma_2 : R^n \to R$ are continuous and concave.

* The report is partially supported by Russian Foundation for Basic Research, grant 06–01–00436.

T. Sobh (ed.), *Innovations and Advances in Computer Sciences and Engineering*,
DOI 10.1007/978-90-481-3658-2_43, © Springer Science+Business Media B.V. 2010

It is assumed that both players know value $\mathbf{x}(t)$ at the current moment of time $t \in [t_0, \theta)$. Then formalization of players' strategies in the game could be based upon the formalization and results of the positional antagonistic differential games theory from [2,7]. According to this formalization (see also [4]) we consider strategies from the class of pure positional strategies as pairs of functions. Strategy of Player 1 $U$ is a pair $\{\mathbf{u}(t, \mathbf{x}, \varepsilon), \beta_1(\varepsilon)\}$, where $\mathbf{u}(\cdot, \cdot, \cdot)$ is an arbitrary function that maps a position $(t, \mathbf{x})$ and a positive precision parameter $\varepsilon$ to an element of $P$. The function $\beta_1 : (0, \infty) \rightarrow (0, \infty)$ is a continuous monotone one and satisfies the condition $\beta_1(\varepsilon) \rightarrow 0$ when $\varepsilon \rightarrow 0$. The function $\beta_1(\cdot)$ has the following sense. For a fixed $\varepsilon$ the value $\beta_1(\varepsilon)$ is the upper bound for step of a subdivision of the interval $[t_0, \theta]$, which Player 1 uses for forming step-by-step motions. Strategy $V \div \{\mathbf{v}(t, \mathbf{x}, \varepsilon), \beta_2(\varepsilon)\}$ of Player 2 is defined analogously. Thus a pair of strategies $(U, V)$ generates a motion $\mathbf{x}[t; t_0, \mathbf{x}_0, U, V]$. The set of motions $X(t_0, \mathbf{x}_0, U, V)$ is non-empty (see [7]).

*A. Nash Solutions*

A pair of strategies $(U^N, V^N)$ is called a *Nash equilibrium solution* (*N-solution*) if for any motion $\mathbf{x}^*[\cdot] \in X(t_0, \mathbf{x}_0, U^N, V^N)$, any $\tau \in [t_0, \theta]$, and any strategies $U$ and $V$ the following inequalities are held

$$\max \sigma_1(\mathbf{x}[\theta; \tau, \mathbf{x}^*[\tau], U, V^N]) \leq$$

$$\min \sigma_1(\mathbf{x}[\theta; \tau, \mathbf{x}^*[\tau], U^N, V^N]),$$

$$\max \sigma_2(\mathbf{x}[\theta; \tau, \mathbf{x}^*[\tau], U^N, V]) \leq$$

$$\min \sigma_2(\mathbf{x}[\theta; \tau, \mathbf{x}^*[\tau], U^N, V^N]). \quad (2)$$

The operations of min and max in (2) are taken in the sets of corresponding motions. Trajectories of (1) generated by an N-solution are called *N-trajectories*.

*B. Stackelberg Solutions*

We denote by $S1$ a game under the following assumptions. Player 1 chooses his strategy $U$ before the game and informs Player 2 about this choice. Player 2 knowing the announced strategy $U$ chooses a *rational strategy* $V$ in order to maximize cost functional $\sigma_2$. Player 1 is a leader here and Player 2 is a follower. Such a game is called a *hierarchical* one. The task is to find strategy $U^{S1}$ of Player 1, which paired with a rational Player 2 strategy $V^{S1}$, provides the maximum of the cost functional $\sigma_1$. Then a pair of strategies $(U^{S1}, V^{S1})$ is called *Stackelberg solution* (*S1-solution*) in hierarchical game with first player as a leader.

A game, where Player 2 acts as a leader and Player 1 is a follower, is denoted by $S2$. It is analogous to $S1$ game. The corresponding Stackelberg solution (S2-solution) is a pair $(U^{S2}, V^{S2})$ in hierarchical game with Player 2 as a leader.

A trajectory generated by S*i*-solution is called S*i*-trajectory.

## III. Non-Standard Problems Of Control

At first we will consider antagonistic positional differential games $\Gamma_1$ and $\Gamma_2$. The dynamics of both games is described by (1). In game $\Gamma_i$, Player $i$ maximizes his cost functional $\sigma_i(\mathbf{x}(\theta))$, and Player 3-$i$ counter-acts to this goal. It is known from [2,7], that both games $\Gamma_1$ and $\Gamma_2$ have universal saddle points

$$\{\mathbf{u}^{(i)}(t, \mathbf{x}, \varepsilon), \mathbf{v}^{(i)}(t, \mathbf{x}, \varepsilon)\}, \quad i=1,2, \quad (3)$$

and continuous value functions

$$\gamma_1(t, \mathbf{x}), \ \gamma_2(t, \mathbf{x}). \quad (4)$$

It was shown in [4] that finding N- and S*i*-solutions could be reduced to finding solutions of following non-standard problems of control.

*Problem 1.* Find a pair of measurable functions $\mathbf{u}(t)$ and $\mathbf{v}(t)$, $t \in [t_0, \theta]$, guaranteeing the fulfillment of conditions

$$\gamma_i(t_*, \mathbf{x}(t_*)) \leq \gamma_i(t^*, \mathbf{x}(t^*)), \ i = 1,2, \quad (5)$$

where $t_* \in [t_0, \theta)$ and $t^* \in (t_*, \theta]$.

*Problem 2.i* ($i$ = 1,2). Find a pair of measurable functions $\mathbf{u}(t)$, $\mathbf{v}(t)$, $t \in [t_0, \theta]$, providing a maximum of cost functional $\sigma_i(\mathbf{x}(\theta))$ holding the following condition

$$\gamma_{3-i}(t, \mathbf{x}(t)) \leq \gamma_{3-i}(\theta, \mathbf{x}(\theta)) = \sigma_{3-i}(\mathbf{x}(\theta)), \quad (6)$$

for $t \in [t_0, \theta]$.

Let piecewise continuous functions $\mathbf{u}^*(t)$, $\mathbf{v}^*(t)$, $t \in [t_0, \theta]$, generate a trajectory $\mathbf{x}^*(t)$ of (1). Consider strategies of Player 1 and Player 2 $U^o \div \{\mathbf{u}^o(t, \mathbf{x}, \varepsilon), \beta^o{}_1(\varepsilon)\}$ and $V^o \div \{\mathbf{v}^o(t, \mathbf{x}, \varepsilon), \beta^o{}_2(\varepsilon)\}$, where

$$\mathbf{u}^o(t, \mathbf{x}, \varepsilon) = \begin{cases} \mathbf{u}^*(t) & \text{if } \|\mathbf{x} - \mathbf{x}^*(t)\| < \varepsilon\varphi(t), \\ \mathbf{u}^{(2)}(t, \mathbf{x}, \varepsilon) & \text{if } \|\mathbf{x} - \mathbf{x}^*(t)\| \geq \varepsilon\varphi(t), \end{cases} \quad (7)$$

$$\mathbf{v}^o(t, \mathbf{x}, \varepsilon) = \begin{cases} \mathbf{v}^*(t) & \text{if } \|\mathbf{x} - \mathbf{x}^*(t)\| < \varepsilon\varphi(t), \\ \mathbf{v}^{(1)}(t, \mathbf{x}, \varepsilon) & \text{if } \|\mathbf{x} - \mathbf{x}^*(t)\| \geq \varepsilon\varphi(t), \end{cases} \quad (8)$$

for all $t \in [t_0, \theta]$. Functions $\beta_i(\cdot)$ and a positive increasing function $\varphi(\cdot)$ are chosen so that the inequality

$$\| \mathbf{x}[t; t_0, \mathbf{x}_0, U^o, \varepsilon, \Delta_1, V^o, \varepsilon, \Delta_2] - \mathbf{x}^*(t) \| < \varepsilon\varphi(t) \quad (9)$$

holds for $\varepsilon > 0$, $\delta(\Delta_i) \leq \beta_i(\varepsilon)$, $t \in [t_0, \theta]$.

Usually in literature strategies $\mathbf{u}^{(2)}(t, \mathbf{x}, \varepsilon)$, $\mathbf{v}^{(1)}(t, \mathbf{x}, \varepsilon)$ are called *punishment strategies*. The following theorem is true.

*Theorem.* Let controls $\mathbf{u}^*(\cdot)$ and $\mathbf{v}^*(\cdot)$ be a solution of Problem 1 (or Problem 2.*i*). Then the pair of strategies $(U^o, V^o)$ (7,8,9) is an N-solution (or an S*i*-solution).

Generally it is very difficult to find the whole set of solutions for non-standard problems described above. Therefore the report presents an algorithm for constructing only some N-solution. The algorithm is essentially constructing an N-trajectory.

*Remark.* In [8] a classification of positions $(t, \mathbf{x})$ in non-antagonistic positional differential game is proposed. It introduces three types of positions: non-antagonistic, locally antagonistic and globally antagonistic. Situations when position type changes along N-trajectories are of special interest.

## IV. ALGORITHMS

Algorithms proposed in the paper for finding S- and N-solutions are based upon various computational geometry algorithms. Unfortunately, at the present time these procedures are only two-dimensional, so it is additionally assumed, that cost functionals of players $\sigma_i$ depend on two selected components solely. Because of the fact, all the problems considered could be reduced to problems in plane.

### *A. Algorithm for N-solution*

This algorithm is based on the procedure (see [8]), which uses the principle of non-decrease of player payoffs, maximal shifts in the direction best for one player and then another player, and Nash equilibrium in auxiliary bimatrix games made up for each step of a subdivision of the time interval.

The procedure implies that Player $i$ is interested in increasing the function $\gamma_i(t, \mathbf{x})$ along the solution trajectory as in (5).

Let a position $(t, \mathbf{x})$ be given. We fix $h > 0$ and put $\tau(t, h) = \min\{t + h, \theta\}$. Points of maximum for functions $\gamma_1(t, \mathbf{x})$ and $\gamma_2(t, \mathbf{x})$ in the $h$-neighborhood of the position are denoted by $w^1(\tau(t, h))$ and $w^2(\tau(t, h))$, respectively.

Consider vectors

$$\mathbf{s}^1(t, \mathbf{x}, h) = w^1(\tau(t, h)) - \mathbf{x},$$
$$\mathbf{s}^2(t, \mathbf{x}, h) = w^2(\tau(t, h)) - \mathbf{x}.$$

We find vectors $\mathbf{u}_{10}(t, \mathbf{x}, h)$, $\mathbf{v}_{10}(t, \mathbf{x}, h)$, $\mathbf{u}_{20}(t, \mathbf{x}, h)$, and $\mathbf{v}_{20}(t, \mathbf{x}, h)$ from conditions

$$\max_{\mathbf{u}\in P, \mathbf{v}\in Q} \mathbf{s}^{1\mathrm{T}} [\mathbf{B}(t)\mathbf{u} + \mathbf{C}(t)\mathbf{v}] = \mathbf{s}^{1\mathrm{T}} [\mathbf{B}(t)\mathbf{u}_{10} + \mathbf{C}(t)\mathbf{v}_{10}],$$
$$\max_{\mathbf{u}\in P, \mathbf{v}\in Q} \mathbf{s}^{2\mathrm{T}} [\mathbf{B}(t)\mathbf{u} + \mathbf{C}(t)\mathbf{v}] = \mathbf{s}^{2\mathrm{T}} [\mathbf{B}(t)\mathbf{u}_{20} + \mathbf{C}(t)\mathbf{v}_{20}],$$
$$\gamma_1(t, \mathbf{x}(t)) \le \gamma_1(\tau(t, h), \mathbf{x}[\tau(t, h); t, \mathbf{x}(t), \mathbf{u}_{20}, \mathbf{v}_{20}]),$$
$$\gamma_2(t, \mathbf{x}(t)) \le \gamma_2(\tau(t, h), \mathbf{x}[\tau(t, h); t, \mathbf{x}(t), \mathbf{u}_{10}, \mathbf{v}_{10}]). \quad (10)$$

Inequalities in (10) come from the condition of non-decrease of functions $\gamma_i(\cdot, \cdot)$ along a motion (5), and maxima are searched only for vectors that hold these inequalities.

Now we construct an auxiliary bimatrix 2×2 game $(A, B)$, where the first player has two strategies: "to choose $\mathbf{u}_{10}$" and "to choose $\mathbf{u}_{20}$". Similarly, the second player has two strategies: "to choose $\mathbf{v}_{10}$" and "to choose $\mathbf{v}_{20}$". Payoff matrices of the players are defined as follows:

$$A = (a_{ij})_{2\times 2}, \; B = (b_{ij})_{2\times 2}$$
$$a_{ij} = \gamma_1(\tau(t, h), \mathbf{x} + (\tau(t, h) - t)(\mathbf{A}(t)\mathbf{x} + \mathbf{B}(t)\mathbf{u}_{i0} + \mathbf{C}(t)\mathbf{v}_{j0})),$$
$$b_{ij} = \gamma_2(\tau(t, h), \mathbf{x} + (\tau(t, h) - t)(\mathbf{A}(t)\mathbf{x} + \mathbf{B}(t)\mathbf{u}_{i0} + \mathbf{C}(t)\mathbf{v}_{j0})),$$
$$i, j = 1,2.$$

Bimatrix game $(A, B)$ has at least one Nash equilibrium in pure strategies. It is possible to take a Nash equilibrium as both players' controls for semi-interval $(t, \tau(t, h)]$. Such an algorithm of players' controls construction generates a trajectory, which is an N-trajectory. When $(A, B)$ has two equilibria (11 and 22), then one is chosen after another in turn (they are interleaved).

We take a solution of $(A, B)$ and try to fit both players' controls to maximize shift along the motion direction of this equilibrium, while holding inequalities in (10). It is called *BM-procedure* here. To provide Nash equilibrium in the game, controls generated by BM-procedure (giving *BM-trajectory*) must be paired with punishment strategies. In this case, each player watches for the trajectory and applies a punishment strategy when the other evades following BM-trajectory. BM-procedure usually gives better results for both players than the original $(A, B)$ game based algorithm, but players must agree to follow BM-trajectory before the game will start.

### *B. Algorithm for S1-solution*

The general idea behind the algorithm is to search $\max_\alpha \sigma_1(\mathbf{x}_\alpha[\theta])$, where $\mathbf{x}_\alpha[\theta]$ are node points of a grid constructed for a *set of admissible trajectories final states* $D_1$. This $\mathbf{x}_\alpha[\theta]$ serves an endpoint for S1-trajectory, which is then built back in time (controls $\mathbf{u}(t)$, $\mathbf{v}(t)$ may be found simultaneously). A $D_1$ approximation is constructed by a procedure that builds sequences of attainability sets (step-by-step in time) repeatedly throwing out the positions, which do not satisfy (6). The procedure is described in brief below. More details (some designations differ) about it could be found in [9].

A special notion from the theory of antagonistic positional differential games called *maximal stable bridge in pursuit-evasion game* is used. The aim of the follower in this game is to drive the phase vector to Lebesgue's set for the level function (for a chosen constant $c$) of his cost functional. (In order to find positions satisfying the inequality $\gamma_2(t, \mathbf{x}) \le c$.) Note, that any position in the bridge holds the inequality, but positions outside the bridge do not.

$W^c_{1,t}$ designates an approximation of a bridge (in the pursuit-evasion game) section for the time moment $t$ = const. The following discrete scheme is used for building admissible trajectories pipe $G^c_1$ section approximations $G^c_{1,tk}$ (here $k$ runs through some time interval subdivision, $k = 0$ corresponds to $t_0$ moment and $k = N$ corresponds to $\theta$):

$$G^c_{1,tk+1} = [\, G^c_{1,tk} + \delta_k(\mathbf{B}(t_k)P + \mathbf{C}(t_k)Q) \,] \setminus W^c_{1,tk+1}, \quad (11)$$

where $G^c_{1,t0} = \{\mathbf{x}_0\}$, $\delta_k = t_{k+1} - t_k$.

Operation $A + B = \{a + b \mid a \in A, b \in B\}$ denotes Minkowski sum of sets $A$ and $B$.

We iterate through a sequence of $c$ values to make up $D_1$ of corresponding sequence of $D^c_1 = G^c_{1,\theta} \cap W^c_{1,\theta}$ using (11) as follows:

1. We have some initial step value $\delta c_0 > 0$ constrained by $\delta c_{\min} \le \delta c$.
2. Let $D_1$ = empty set, $c = c^{\max} = \max_{\mathbf{x}\in R^n} \sigma_2(\mathbf{x})$, $\delta c = \delta c_0$.
3. Build a pipe $G^c{}_1$ and a set $D^c{}_1$ as in (11).
4. Supplement $D_1 := D_1 \cup \{(\mathbf{x}, c) \mid \mathbf{x} \in D^c{}_1\}$.
5. If $\delta c \ge \delta c_{\min}$ then we choose the next $c$ value:
   - if $\mathbf{x}_0 \in W^c{}_{1,t0}$ then 1) return to the previous value $c := c + \delta c$, 2) decrease step $\delta c$;
   - take next value $c := c - \delta c$;
   - repeat from item **3**.
6. Quit.

$D_2$ set for $S2$ game may be built in the same way.

The procedure presented is called *D-procedure* here.

One example of S-trajectories numerical computation results was presented in [10]. A program used for value function calculation is based on results of [5,6].

## V. PROGRAM IMPLEMENTATION

The current implementation was written in C++ and builds upon a C++ wrapper library, which uses polygon tessellation facilities from *OpenGL Utility Library* (GLU) plus our own Minkowski sum implementation. Examples of the next section were obtained with GLU being used for polygons processing. Advantages of GLU include straightforward standard API in C, which is simple to use in almost any programming environment. Many implementations exist (usually bundled with operational systems), both proprietary and open source.

Despite positive results achieved by the current implementation, it is considered to be a temporary solution. Another library (which is an open source project) was tested, as a possible future base for our algorithms. It was *Computational Geometry Algorithms Library* (CGAL), which goal "is to provide easy access to efficient and reliable geometric algorithms in the form of a C++ library" (see http://www.cgal.org). While GLU is a convenient external component (usually represented by a DLL), CGAL provides a complex framework to expand upon and is not bounded by hardware-supported double precision arithmetics.

In the case of S-solutions, OpenMP 2.0 was adopted (discrete scheme (11) is run for different $c$ values in parallel). Tests on a machine with Intel Core 2 Duo processor demonstrated twofold run-time improvement for two-threaded computation runs against one-threaded runs. We plan to test cluster computation in near future. High computation power may help to make a close study of D-procedure behaviour with different precision parameters and under different arithmetics.

The following paragraph describes precision parameters used by the implementation.

Each step of (11) tends to grow polygons' sizes almost geometrically, so it is necessary to throw out some vertices after each step or at least some steps. Currently the simplest method is used: we just set a minimal edge length ($\text{min}_{\text{edge}}$) and a minimal sine of angle between two edges absolute value ($\text{min}_{\text{sin}}$) and throw out vertices, which don't fit these parameters. Due to limitations of machine real arithmetics (11) may also give small artefact polygons, which are swept out by setting minimal polygon area ($\text{min}_{\text{area}}$). Another important parameter is tolerance for $D^c{}_1 = G^c{}_{1,\theta} \cap W^c{}_{1,\theta}$ operation ($\text{tol}_{\text{arc}}$). In other words, it defines how far from $W^c{}_{1,\theta}$ resulting vertices may lie. Experience showed, that robustness of D-procedure mainly depends on correlation between $\text{min}_{\text{edge}}$ and $\text{tol}_{\text{arc}}$ parameters. One should not await good results if $\text{tol}_{\text{arc}}$ is too low for a selected $\text{min}_{\text{edge}}$ value.

BM-procedure implementation has only one precision parameter: payoff tolerance ($\text{tol}_{\text{payoff}}$), which defines a bound for accumulated violation value of (5), i.e. $-\sum\min\{\gamma_i(t_{k+1}, \mathbf{x}_{k+1}) - \gamma_i(t_k, \mathbf{x}_k), 0\}$ must be not greater than $\text{tol}_{\text{payoff}}$ ($i$ = 1,2) for computed BM-trajectory. BM-procedure is quite robust and even $\text{tol}_{\text{payoff}} = 0$ may fit well in many cases. However, under a condition of limited-precision arithmetics small $\text{tol}_{\text{payoff}}$ non-zero values proved to be useful.

## VI. AN EXAMPLE

The following vector equation

$$\xi'' = \mathbf{u} + \mathbf{v},\ \xi(t_0) = \xi_0,\ \xi'(t_0) = \xi'_0,$$

$$\xi,\ \mathbf{u},\ \mathbf{v} \in R^2,\ \|\mathbf{u}\| \le \mu,\ \|\mathbf{v}\| \le \nu,\ \mu > \nu, \tag{12}$$

describes motion of a material point of unit mass on the plane ($\xi_1$, $\xi_2$) under action of a force $\mathbf{F} = \mathbf{u} + \mathbf{v}$. Player 1 (Player 2) who governs the control $\mathbf{u}$ ($\mathbf{v}$) needs to lead the material point as close as possible to the given target point $a^{(1)}$ ($a^{(2)}$) at the moment of time $\theta$. Then players' cost functionals are

$$\sigma_i(\xi(\theta)) = -\|\xi(\theta) - a^{(i)}\|, \tag{13}$$

$$\xi = (\xi_1, \xi_2),\ a^{(i)} = (a^{(i)}{}_1, a^{(i)}{}_2),\ i=1,2.$$

By taking $y_1 = \xi_1$, $y_2 = \xi_1$, $y_3 = \xi'_2$, $y_4 = \xi'_2$ and making the following change of variables $x_1 = y_1 + (\theta - t)y_3$, $x_2 = y_2 + (\theta - t)y_4$, $x_3 = y_3$, $x_4 = y_4$ we get a system, which first and second equations are

$$x'_1 = (\theta - t)(u_1 + v_1),$$
$$x'_2 = (\theta - t)(u_2 + v_2). \tag{14}$$

Further, (13) can be rewritten as

$$\sigma_i(\mathbf{x}(\theta)) = -\|\mathbf{x}(\theta) - a^{(i)}\|,\ \mathbf{x} = (x_1, x_2),\ i = 1,2. \tag{15}$$

Since the cost functional (15) depends on variables $x_1$ and $x_2$ only, and the right-hand side of (14) does not depend on other variables, one can conclude, that it is sufficient to consider only reduced system (14) with cost functionals (15).

Then initial conditions for (14) are given by

$$x_i(t_0) = x_{0i} = \xi_{0i} + (\theta - t_0)\xi_{0i},\ i = 1,2.$$

It can easily be shown that value functions in antagonistic differential games $\Gamma_1$ and $\Gamma_2$ are given by formulae

$$\gamma_1(t, \mathbf{x}) = -\|\mathbf{x} - a^{(1)}\| - (\theta - t)^2(\mu - \nu) / 2,$$

$$\gamma_2(t, \mathbf{x}) = \min\{-||\mathbf{x} - a^{(2)}|| + (\theta - t)^2(\mu - \nu) / 2, 0\},$$

and universal optimal strategies (3) are given by

$$\mathbf{u}^{(i)}(t, \mathbf{x}, \varepsilon) = (-1)^i \mu(\mathbf{x} - a^{(i)}) / ||\mathbf{x} - a^{(i)}||,$$
$$\mathbf{v}^{(i)}(t, \mathbf{x}, \varepsilon) = -(-1)^i \nu(\mathbf{x} - a^{(i)}) / ||\mathbf{x} - a^{(i)}||.$$

Let the following conditions be given: $t_0 = 0$, $\xi_0 = (0.2, 0.5)$, $\xi'_0 = (-0.3, -1.3)$, $\mu = 1.4$, $\nu = 0.6$, $a^{(1)} = (2.5, 4.5)$, $a^{(2)} = (4.5, -3.5)$. Two variants of target points were considered:

(V1) $\theta = 2$ and hence $\mathbf{x}_0 = (-0.4, -2.1)$;

(V2) $\theta = 2.5$ and $\mathbf{x}_0 = (-0.55, -2.75)$.

Time step for N-trajectory was 0.0005. $D_i$ and S-trajectories were built with time step 0.005. Other precision parameters are the following: $\min_{edge} = 0.0003$, $\min_{sin} = 10^{-6}$, $\min_{area} = 10^{-3}$, $tol_{arc} = 10^{-3}$, $tol_{payoff} = 10^{-8}$, $\delta c_0 = 0.05$, $\delta c_{min} = 0.0005$. V1 computation took 22 minutes, V2 – 39 minutes on AMD Athlon 64 2GHz processor (double precision machine arithmetics in AMD64 mode).

Fig.1 and Fig.2 show computed trajectories for variant V2. Fig.1 depicts S1-, S2-trajectories and N-trajectory generated by BM-procedure in $(\xi_1, \xi_2)$ plane. Symbols $S^1$ and $S^2$ on figures denote endpoints of S1- and S2-solution trajectories, respectively, while $N$ denotes the N-solution trajectory endpoint. On Fig.1 symbol "×" is used to show a position $(\xi_1(t^*), \xi_2(t^*))$, which corresponds to time moment $t^* = 1.775$. Starting from this moment, position of the game, calculated along the N-trajectory, changes its type from non-antagonistic to globally antagonistic. The fact of position type change is well-illustrated on Fig.2, where the N-trajectory is shown in the plane $(x_1, x_2)$. Since $t^*$ the trajectory of the reduced system lies on the line segment that connects target points $a^{(1)}$ and $a^{(2)}$.

Fig.3 shows computed trajectories for variant V1: S1, S2 and N-trajectory in $(\xi_1, \xi_2)$ plane, as well as N-trajectory of the reduced system in $(x_1, x_2)$ plane. Position of the game is non-antagonistic along the trajectory here.

On all figures a border of relevant $D = (D_1 \cap D_2) \setminus R$ set is depicted, where $R$ is a half-plane not containing point $\xi_0$ and bounded by a line drawn through points $a^{(1)}$ and $a^{(2)}$. Set $D$ contains all N-trajectories endpoints, but in general case, there may be also points, which are not endpoints of any N-trajectories.

## VII. CONCLUSION

Study in the field of non-antagonistic positional differential games has a great number of directions. We enumerate three promising development directions of the research presented in the paper.

Firstly, it seems to be fairly easy to transparently generalize N- and S-solution algorithms to make them work with non-linear systems like

$$\mathbf{x}'(t) = \mathbf{F}_1(t, \mathbf{x}(t), \mathbf{u}(t)) + \mathbf{F}_2(t, \mathbf{x}(t), \mathbf{v}(t)).$$

Secondly, software development may lead to a powerful framework for simplifying solutions computation in a class of differential games. The last program implementation uses techniques of generic programming, which is common for modern C++ software like CGAL. For example, the early experience allows to suggest that incorporating CGAL facilities can give literally new dimension to our algorithms: polygons could be changed to polyhedrons without deep changes in generic procedures constructing the solutions. On the other hand, it should provide a flexible and convenient interface for future modernizations.

Finally, an algorithm approximating all N-trajectories endpoints set (or Pareto unimprovable part of its bound) is planned. If this could be done, one might choose there an endpoint, which is optimal in some sense. Then it is possible to build an N-solution leading to the endpoint chosen using back propagation similar to that used here for S-trajectories building.

## REFERENCES


[1] T. Basar and G.J. Olsder, *Dynamic Noncooperative Game Theory.* NY: Acad. Press, 1982.

[2] N.N. Krasovskii and A.I. Subbotin, *Game-Theoretical Control Problems.* NY, Berlin: Springer-Verlag, 1988.

[3] A.N. Krasovskii and N.N. Krasovskii, *Control under Lack of Information.* Berlin: Birkhäuser, 1995.

[4] A.F. Kleimenov, *Positional Differential Nonantagonistic Games.* Ekaterinburg: Nauka, Urals Branch, 1993 (In Russian).

[5] E.A. Isakova, G.V. Logunova and V.S. Patsko, "Stable bridges construction in linear differential game with fixed final time," in *Algorithms and programs for linear differential games solutions*, A.I. Subbotin and V.S. Patsko, Eds. Sverdlovsk: Ural Sci. Center of Acad. Sci. of USSR, 1984, pp. 127 – 158 (In Russian).

[6] V.A. Vahrushev, A.M. Tarasiev and V.N. Ushakov, "An algorithm of union and intersection of sets in plane," in *Control with guaranteed result*, A.I. Subbotin and V.N. Ushakov, Eds. Sverdlovsk: Ural Sci. Center of Acad. Sci. of USSR, 1987, pp 28—36 (In Russian).

[7] N.N. Krasovskii, *Control of a Dynamical System.* Moscow: Nauka, 1985 (In Russian).

[8] A.F. Kleimenov, *Solutions in a nonantagonistic positional differential game*, J. Appl. Maths Mechs, Vol. 61, No. 5, 1997, pp. 717—723.

[9] A.F. Kleimenov and S.I. Osipov, "Computation of Stackelberg trajectories in a class of two-person linear differential games with terminal players' payoffs and polygonal constraining for controls," IFAC Workshop on Control Applications of Optimization, Preprints, Oxford: Elsevier Science Ltd., 2003, pp. 201—205.

[10] A.F. Kleimenov, S.I. Osipov, A.S. Cherepov and D.R. Kuvshinov, *A Numerical Solution for a hierarchical differential game of two persons*, Proc. of Ural State Univ., No. 46, 2006, pp. 69 – 78 (In Russian).

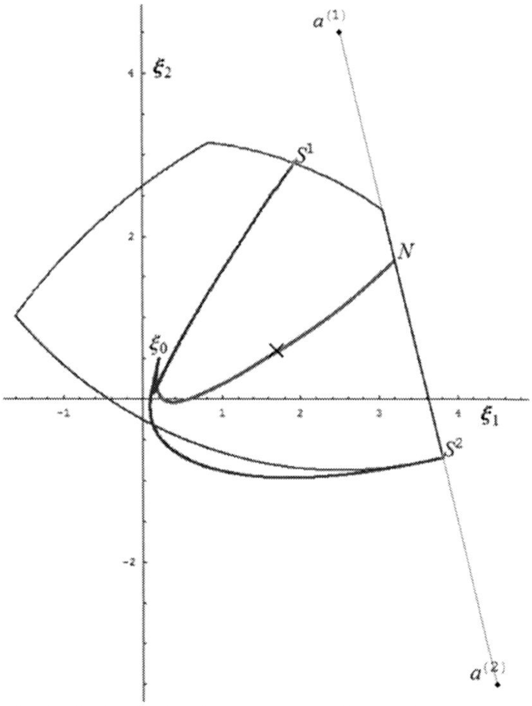


Fig. 1 V2: Optimal trajectories: N- and S-solutions

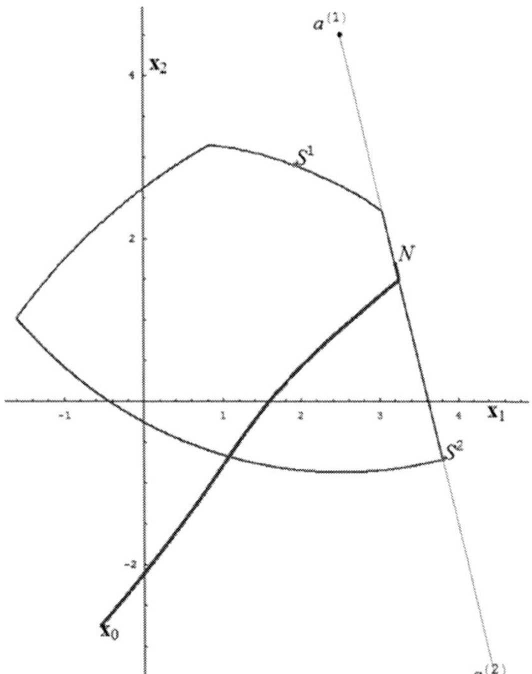


Fig. 2 V2: N-trajectory for the reduced system

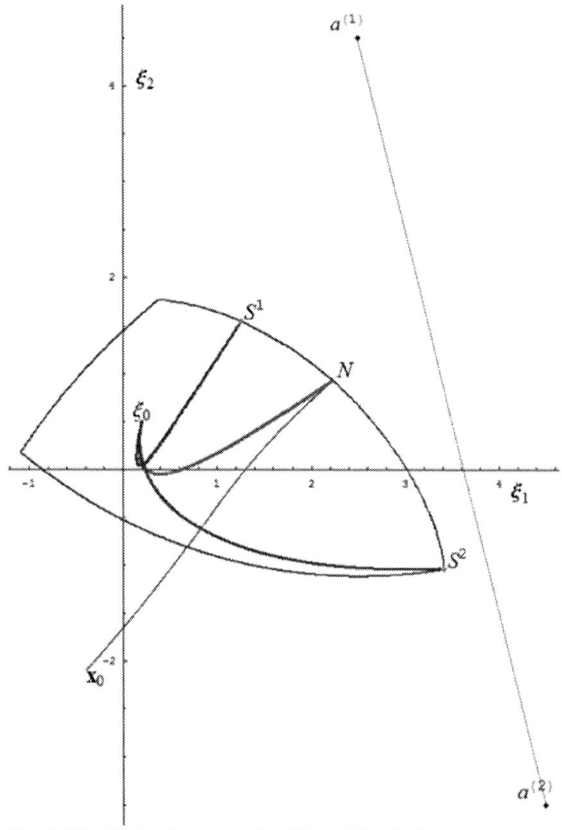


Fig. 3 V1: Optimal trajectories: N- and S-solutions

# Computer Simulation of Differential Digital Holography

Krešimir Nenadić, Marina Pešut, Franjo Jović, Ninoslav Slavek
Faculty of Electrical Engineering in Osijek
Kneza Trpimira 2B
Osijek, 31000, Croatia

*Abstract* - **This paper presents a complete procedure of the differential digital holography process. The method is based on digital hologram construction and reconstruction. Holograms and reconstructs are calculated by applying Fourier Transformation on bitmap images. The elaborated algorithm allows amplification of small differences between two images of the same size. Advantages and disadvantages of the described process are discussed.**

## I. Introduction

Digital holography has many aspects of applications [1-3]. The specific area of application that is of interest to us is the possible amplification of small differences between two image patterns. We have named this method differential digital holography (DDH). This method relies on calculation of holograms and reconstructs in order to amplify small image differences [4]. Both holograms and image reconstructs are obtained by computer simulation processes. Holograms obtained by computer simulation are called computer-generated holograms (CGH). Computer simulations consist of intensive calculations of pixel luminance intensities on hologram or reconstruct planes. Due to computational power of processors that are available today, these intensive calculations do not present any problem.

There is a possibility to perform all calculations using graphic processors which have more scalable architecture and greater computational power [5]. This arises from the greater number of transistors a Graphical processing unit (GPU) contains. They have more pipelines and more specialized circuits to perform intensive mathematical calculations [6].

The other possibility to speed up calculations is to use digital signal processor (DSP) and to program it to perform specialized mathematical calculations necessary for this method.

All calculations for this experiment were performed on personal computer with following specifications:

- CPU: Intel Pentium IV 3.4 GHz,
- RAM: 2 GB DDR2.

Images used in experiments were of the same size 300×300 pixels. Calculation time depends on number of pixels with black colour. Greater number of black pixels means shorter calculation time. Black pixels in image represent obstacle in light path and light source will not pass through black pixels. Gray coloured pixels represent obstacle that source light can pass through with certain loss in intensity depending on halftone of gray colour. White pixels do not present any kind of obstacle and light source can pass through them without any losses.

Section II describes differential digital holography method, i.e. hologram construction, difference hologram calculation and image reconstruction.

Experiment and results are given in section III.

Conclusion and guidelines for future research are given in section IV.

## II. Differential Digital Holography

Differential digital holography method consists of three steps:

1. a) constructing a hologram from the first image,
   b) constructing a hologram from the second image,
2. calculating the difference hologram,
3. reconstructing image from difference hologram.

All three steps were performed as computer simulation. Image format for storing images, holograms and image reconstructs in computer memory are gray scale 8-bit bitmap. Advantage of chosen bitmap form is that every pixel is represented by one characteristic value – colour or, in our case, intensity.

According to [3] there are three domains:

1. image (object) domain,
2. hologram domain,
3. reconstruct (image) domain.

Calculation process when hologram is obtained from image values is called hologram construction. This is a process of transition from image domain to hologram domain. Reverse process, when reconstruct or image is calculated from hologram, is called reconstruction process. Image (object) domain and image reconstruct domain are in fact equal domains, i.e. they represent image while hologram domain represents transformed values.

### A. *Digital holography – computer-generated hologram*

Digital holography method relies on feature of the hologram process to project every single pixel of the image to a hologram. Digital hologram is constructed by superposition of spherical waves. Intensity, or amplitude, values of the hologram in (x, y) position of the pixel is given by:

$$I_{H(x,y)} = \sum_{i,j=0}^{N-1} I_{O(i,j)} \sin\left(2\pi \frac{d}{\lambda}\right), \quad (1)$$

T. Sobh (ed.), *Innovations and Advances in Computer Sciences and Engineering*,
DOI 10.1007/978-90-481-3658-2_44, 

where $I_H(x, y)$ is pixel intensity of hologram at coordinates (x, y), $I_O(i, j)$ is pixel intensity of image at coordinates (i, j), d is distance between image and hologram and $\lambda$ is light source wavelength used to illuminate image. Distance d is given by following expression:

$$d = \sqrt{z^2 + (x-i)^2 + (y-j)^2} \quad (2)$$

where z is distance between parallel image and hologram plane, (x, y) are coordinates of pixel in hologram plane and (i, j) are coordinates of pixel in image plane.

*B. Difference hologram*

Difference hologram is calculated by subtracting respective pixel values (intensities) of two holograms. Both holograms have to be of the same size because subtraction is performed at pixel level. The chosen bitmap form alleviates calculation. Following equation shows how to obtain pixel values of difference hologram:

$$I_{dH(x,y)} = \left| I_{H1(x,y)} - I_{H2(x,y)} \right| \quad (3)$$

Difference hologram pixel values cannot have negative values. Bitmap picture values for 8-bit gray scale bitmap are given in range from 0 to 255. Value 0 represents black or no luminance, value 255 represents white or maximum luminance and intermediate values represent gray halftone luminance. That is why equation (3) has to have absolute value expression.

*C. Image reconstruction*

Image reconstruction process is also performed by computer simulation according to equation (1) while image intensity pixel values and hologram pixel intensity values exchange positions. Equation (1) then becomes:

$$I_{R(x,y)} = \sum_{i,j=0}^{N-1} I_{H(i,j)} \sin\left(2\pi \frac{d}{\lambda}\right) \quad (4)$$

where $I_R(x, y)$ is pixel intensity of reconstructed image at coordinates (x, y). The expressions $I_O$ (image) and $I_R$ (reconstruct) are from the same domain.

*D. Simulation process*

Figure 1 shows complete computer simulation process for differential digital holography. Process of hologram construction and image reconstruction is more processor demanding then calculating difference hologram. Computer algorithms for differential digital holography simulation are listed in appendix.

Program in listing 1 in appendix shows algorithm for hologram construction. Similar algorithm is applied for image reconstruction. Algorithm is partially optimized because it tests image pixel values and for pixel values containing 0 does not perform transformation. Program in listing 2 in appendix shows algorithm for calculating difference hologram. Some program languages have built-in functions to access pixel values at specified coordinates in bitmap image. These values are number of row and column in bitmap image.

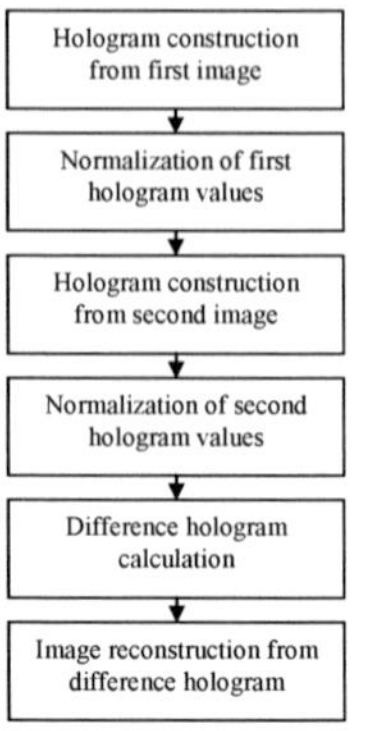


Fig. 1. Computer simulation process flow diagram

Hologram intensity pixel values are normalized immediately after construction process. This operation is performed because pixel intensity values in images, holograms and image reconstructs have to be in the range from 0 to 255 for 8-bit gray scale bitmaps. Process of pixel data normalization is performed immediately after hologram construction and after image reconstruction.

## III. EXPERIMENT

Images, holograms and image reconstructs in this experiment were all 300×300 pixels. Holograms can be of different size than images or reconstructs but all holograms have to be of the same size because difference hologram calculation is performed on paired pixels of two holograms.

For this experiment purpose we have created bitmap image 300×300 pixels containing logo of Faculty of Electrical Engineering in Osijek shown in figure 2a. One image contains only logo mentioned before and all other images will also contain some other drawings like dots or lines. Figures 2b and 2c shows images with logo and another pattern.

Holograms are constructed according to (1) from all three images from figure 2 and they are shown in figure 3.

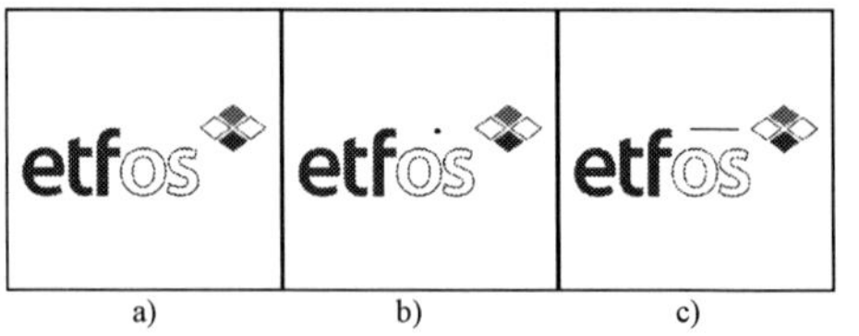


a) b) c)

Fig. 2. Original images:
a) logo; b) logo and dot; c) logo and line

Images in figure 3 represent corresponding holograms from figure 2. All three holograms have dominant gray colour. This effect can be explained due to hologram feature to project (map) given pixel of original image to all pixels in hologram.

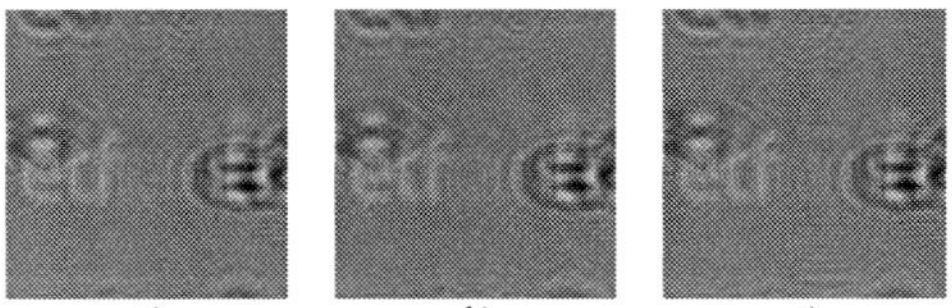

a) b) c)

Fig. 3. Holograms of images from figure 3
a) logo; b) logo and dot; c) logo and line

Mapping intensity depends on several factors:
- pixel intensity in original image,
- light source wavelength - $\lambda$,
- distance z between image and hologram plane.

Figure 4 shows image reconstructs obtained from holograms in figure 3.

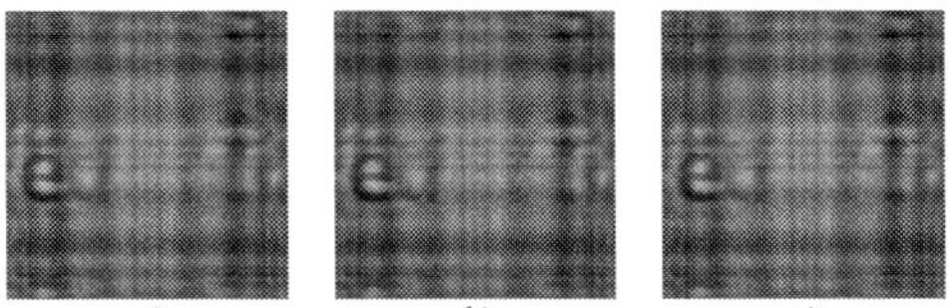

a) b) c)

Fig. 4. Image reconstructs from holograms in figure 3
a) logo; b) logo and dot; c) logo and line

Difference holograms are calculated according to (3) and they are shown in figure 5.

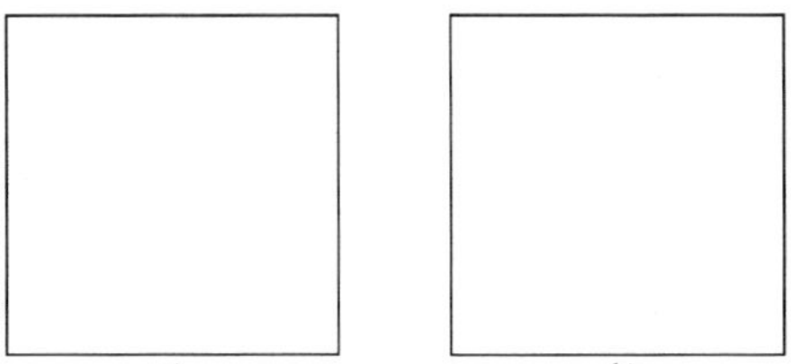

a) b)

Fig. 5. Difference holograms
a) 3a-3b; b) 3a-3c

Figure 5a represents difference hologram calculated from holograms in figure 3a and 3b while figure 5b represents difference hologram calculated from holograms in figure 3a and 3c. Both of difference holograms just appear to be white. Their pixel intensities have values lower than 255 which is value for maximum illumination in bitmap image. Confirmation of this statement can be obtained by reconstructing images from difference holograms. Figure 6 shows image reconstructs calculated from difference holograms in figures 5a and 5b.

Images in figure 6a and 6b confirm our hypothesis that differences between two picture patterns are amplified. Another phenomenon can be perceived. Characteristic curves are formed around place where the pattern difference is situated in image reconstructs. Curves look like water waves around pattern difference. This confirms another hypothesis.

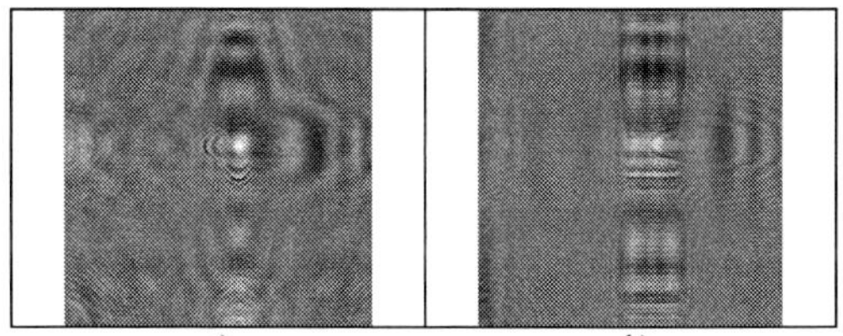

a) b)

Fig. 6. Image reconstructs from difference holograms in figure 5

Hologram feature to map (project) given pixel from image plane on every pixel on hologram plane is used to amplify pattern difference. Pixel values from pattern difference are projected to every pixel in hologram. That resulted in extending pattern difference through all pixels both in hologram and reconstruct.

## IV. Conclusion

We have demonstrated in this paper that computer simulation of differential digital holography process can detect and amplify small differences between two pictures with similar, but not identical, patterns. The main advantage of differential digital holography process is pattern difference amplification. Calculation process during computer simulation is processor demanding.

Differential digital holography algorithm shown in appendix 1 can be rearranged and applied on parallel computer with multiple processors. In that way we could have much shorter time of calculation depending on number of processors in parallel system.

Another way to speed up calculation is to perform all calculations on graphical processing unit on graphic cards in personal computers. This arrangement requires good knowledge in Direct X or OpenGL programming.

Differential digital holography method can find application in different kind of manufacturing processes [3]. There is some effort in applying this method in ceramic tile manufacturing process for pattern defect detection [1,2]. Efforts are also maid to circumvent computer processor in all intensive calculations of hologram and image reconstructs.

Further investigations and experiments will focus to determine influence of shifts and rotations of picture patterns in differential digital holography method.

## References


[1] K. Nenadić, T. Keser, and F. Jović, "Small Surface Defects Detection on Ceramic Tiles Using Digital Holography" *MIPRO 2007 Proceedings*, vol. III, CTS & CIS, pp. 33, Opatija, Croatia, 2007.

[2] K. Nenadić, I. Novak, J. Job, F. Jović and Ž. Jagnjić, "A Possibility of Applying Differential Digital Holography in Manufacturing Process", *Proceedings ELMAR 2006*, pp. 103, Zadar, Croatia, 2006.

[3] F. Jović, J. Job, M. Pešut and M. Protrka, "Holographic Coding of Process Actions and Interactions", *Technical Gazette*, vol. 14 (2007), pp. 3-9

[4] F. Jović, J. Job and Z. Radoš, "Detecting Small Changes in Process Images with Digital Holography", *18th Meeting on Mathematical Modeling of Materials Processing with Lasers*, Igls / Innsbruck, Austria, 2005.

[5] N. Masuda, T. Ito, T. Tanaka, A. Shiraki, and T. Sugie, "Computer generated holography using a graphics processing unit," *Opt. Express* 14, 603-608 (2006)

[6] L. Ahrenberg, P. Benzie, M. Magnor, and J. Watson, "Computer generated holography using parallel commodity graphics hardware," *Opt. Express* 14, 7636-7641 (2006)

## APENDIX

```
/* Listing 1
  Hologram construction algorithm
  Image size: n x n
  Distance between image and hologram: z
  Image intensities values:   A[i,j]
  Hologram intensities values: H[x,y]
*/
for(i=0; i<n; i++)
  for(j=0; j<n; j++)
    if (A[i,j]!=0)
      for(x=0; x<n; x++)
        for(y=0; y<n; y++)
        {
          d = sqrt(z*z+(x-i)*(x-i)+(y-j)*(y-j));
          H[x,y]=H[x,y]+A[i,j]*sin(2*Pi*d/λ);
        }

/* Listing 2
  Difference hologram calculation algorithm
  bmpSUB1 – first hologram bitmap
  bmpSUB2 – second hologram bitmap */
for(i=0; i<n; i++)
{
  for(j=0; j<n; j++)
  {
    if(bmpSUB1.GetPixel(i,j).B > bmpSUB2.GetPixel(i,j).B)
      x = 255 - (bmpSUB1.GetPixel(i,j).B - bmpSUB2.GetPixel(i,j).B);
    else
      if (bmpSUB2.GetPixel(i,j).B > bmpSUB1.GetPixel(i,j).B)
        x = 255 - (bmpSUB2.GetPixel(i,j).B - bmpSUB1.GetPixel(i,j).B);
      else
        x=255;
    bmpSUB3.SetPixel(i,j,Color.FromArgb(x,x,x));
  }
}
```

# Evaluation of Case Based Reasoning for Clinical Decision Support Systems applied to Acute Meningitis Diagnose

Cecilia Maurente
Facultad de Ingeniería y Tecnologías
Universidad Católica del Uruguay
Montevideo, Uruguay
cmaurente@ucu.edu.uy

Ernesto Ocampo Edye
Facultad de Ingeniería y Tecnologías
Universidad Católica del Uruguay,
Montevideo, Uruguay
eocampo@ucu.edu.uy

Silvia Herrera Delgado, MD
Centro de Referencia Nacional de VIH-SIDA,
Hospital Pereira-Rossell
Montevideo, Uruguay
Silvia@qualisyss.com

Daniel Rodriguez García
Departamento de Ciencias de la Computación
Universidad de Alcalá de Henares
Alcalá de Henares, España
daniel.rodriguez@uah.es

*Abstract*

**This work presents a research about the applicability of Case Based Reasoning to Clinical Decision Support Systems (CDSS), particularly applied to the diagnosis of the disease known as Acute Bacterial Meningitis.**

**In the last few years, the amount of information available to the medical doctor, who usually finds himself in the situation of making a diagnosis of one or more diseases, has dramatically increased. However, the specialist's ability to understand, synthesize and take advantage of such information in the always-little time during the medical act remains to be developed.**

**Many contributions have been made by the computer sciences, especially those by Artificial intelligence, in order to solve these problems. This work focuses on the diagnose of the Acute Bacterial Meningitis, and carries out a comparative assessment of the quality of a Clinical Decision Support System made through Case Based Reasoning, in contrast to an already existing CDSS applied to the same task, but developed using a technique called Bayesian expert system.**

*Keywords: Intelligent Systems, Expert Systems, Case Based Reasoning, Clinical Decision Support Systems, Clinical diagnose, Artificial Intelligence..*

## I. Introduction

During the clinical usual practice of evaluating a patient and making a clinical diagnosis, the problem of the analysis of signs and symptoms in the patient and the usage of available reference information (related to similar cases, their respective analysis and diagnosis) commonly shows up. In virtue of this and taking into account the reference information (that includes mainly previous experience), the clinical doctor develops and tests a series of hypothesis, eventually reaching a diagnosis or a group of differential diagnoses. Based on these, and generally also on protocols as well as standardized or commonly accepted guidelines, the doctor designs and indicates an appropriate treatment, or else orders ulterior examinations that might pose a threat to the patient's health, and can also be of a considerably higher cost.

The amount of information related to similar cases and the recommended diagnosis and procedure for each of them as well as its complexity has increased drastically. Although this represents a great help to the doctor when it comes to making a clinical assessment and a diagnosis, it requires the doctor's availability of attention and concentration on the information in order to be able to synthesize, analyze and utilize it. Apart from that, it mainly requires a fair deal of time, which is not commonly at the disposal of doctors during the clinical

T. Sobh (ed.), *Innovations and Advances in Computer Sciences and Engineering*,
DOI 10.1007/978-90-481-3658-2_45, 

assessment. The available time in a visit to the doctor has not changed significantly in the last few years, if any, it has been reduced. These restrictions can be summed up in two problems: limited rationale and time.

Computer sciences have been applied, during the last 40 years or more, in different ways in order to extend the rationale and help use the available time to take advantage of the information more effectively. For this reason, multiple Artificial Intelligence techniques have been put into practice: pattern analysis, neuronal network, expert systems, Bayesian networks among others.

One of the most recent techniques, which presents several interesting characteristics, is the one known as Case Based Reasoning and it is based on the so common associative paradigm among experts: similar solutions correspond to similar problems.

A recurrent problem of the Clinical Decision Support Systems is that its main approach is computer science-based, applied to a specific working area, in this case the clinical diagnosis, which is performed by the expert, in this particular case, the doctor. This computer science-based approach used in the majority of cases is completely different from the way in which these experts carry out their daily job and, even more from how they develop their reasoning and inference or association processes. As a consequence, many of these systems turn out to be virtually futile due to the difficulty of operation that the expert is required to deal with.

The aim of this research is to demonstrate that the CBR technique is appropriate for the development of CDSSs, used in an independent way or combined with other AI techniques besides proving that its usage allows the development of more accepted and usable CDSSs in this field, in contrast to an already existing reference system, based on Bayesian inference.

## II. Clinical Decision Support Systems

### A. Main concepts, historical examples of CDSSs

A Clinical Decision Support System is, according to [2], "an algorithm based on computer science that helps the clinical doctor in one or more steps during the process of diagnose". These are systems that provide information, to help and advise doctors in the process of making a decision of diagnosis. They suggest a range of diagnoses for the expert to adapt that information using his knowledge and experience to the definitive diagnosis.

When using these systems, the interaction between the expert and them is paramount as the system cannot work by itself. It needs to be fed with enough, clear and precise information. The differentiated specific diagnosis is the result of an elaboration made by the doctor who combines his own knowledge and experience with the information provided by the CDSS.

Signs and Symptoms shown by the patient are received as input, and using the knowledge incorporated in the system, the experience and reasoning of the expert, it is elaborated, as output, a list of possible diagnoses, eventually considered according to their certainty.

Two of the first Decision Support Systems that appeared in the marketplace were MYCIN[3] and PROSPECTOR[4]. MYCIN is a system developed at Stanford University, based on rules, designed to diagnose and recommend a treatment for blood infections. The knowledge is represented as a group of IF-THEN rules which have associated to them a certainty factor. PROSPECTOR, (applied to geology instead of medicine) is a system that allows the assessment of places according to diverse criteria: presence of beds and deposits, assessment of geological resources and the selection of a drilling spot. It uses the Bayes theorem as main mechanism to assess the probability of the occurrence of a certain event.

### B. Application context – Acute Bacterial Meningitis diagnosis

This investigation is developed using as application case the Acute Bacterial Meningitis (ABM) diagnosis in pediatric patients. Based on de assessment of the signs and symptoms related to this disease, the doctor must develop the corresponding diagnosis, distinguishing between the different possible differential diagnoses.

### C. The disease: Acute Bacterial Meningitis

This disease has a high rate of morbidity in pediatric patients and also produces important sequels. It can be seen either in an isolated way or in an epidemic one, and it is of utmost importance to make both an early diagnosis and an immediate treatment.

### D. Signs and Symptoms.

As explained in detail in [5], there are, at least 24 signs and symptoms that can be found in a patient with ABM, in an independent or combined form. Such signs and symptoms have different levels of significance in the composition of the clinical presentation that leads to the diagnosis of the disease. In [5] can also be found the combinations of these signs and symptoms according to the way they are assessed by the corresponding doctor, for infants and over two-years-old patients.

### E. Differential Diagnoses

The ABM diagnosis is complicated as other diseases present a combination of similar signs and symptoms, what we define as "differential diagnoses". Among these alternative diseases are found: Acute Viral Meningitis, Tuberculous Meningitis, Encephalitis, Brain Abscess, Meningism, Proximity Meningeal Reaction, Meningeal Hemorrhage, Brain tumor. The doctor has to clearly identify the existence of ABM among all these.

### F. The reference system: Acute Bacterial Meningitis diagnosis Expert System based on a Bayesian inference engine ABMDES

The Acute Bacterial Meningitis Diagnose Expert System (ABMDES) used as a reference in comparison to the performance of the proposed Case Based Reasoning system (Acute Bacterial Meningitis Case Based Diagnostic System – AMBCBDS), has been described in [5]. It uses a Bayesian inference engine to suggest the differential diagnoses, each with its corresponding certainty level.

In order to work, the ABMDES uses a database of real cases from pediatric patients that have visited the doctor. Using this data, the probability of the existence of each of the differential diseases (*"PI"*) has been specified. Afterwards, in each disease, for each sign or symptom, two levels of probability have been identified: The probability that the patient suffers from the disease as he has actually shown the symptoms ("PS") and the probability that the patient shows the symptom without having the disease ("PN"). Through a simple user interface, the doctor inputs in the application the group of symptoms that he finds in the patient. From this data, the system, applying the Bayes theorem repeatedly in its inference engine, calculates the accumulated probabilities of the existence of the different possible diseases.

## III. Case Based Reasoning

### A. *Introduction – general concepts*

Case based reasoning (CBR) is a methodology utilized for the solution of problems and learning within the AI area. Its invention dates back to the late 1970s [6]; certain results could be tracked down from Psychology, where it is demonstrated that on several occasions, human beings solve their problems based on their past experiences, rather than on a profound knowledge of the topic in question. For instance, doctors look for groups of known symptoms, engineers take many of their ideas from previously successful solutions, and programmers reutilize abstract schemes they already know [7;8]. The fundamental concept on which this methodology is based is... *"similar solutions correspond to similar situations or problems"*.

A Case Based Reasoning System (CBRS) consists in, from a base of experiential knowledge (previous cases rightfully identified with their corresponding solutions), analyze the existing correlation with the new suggested problem and, in virtue of the correspondences, adapt and propose the nearest solution. Instead of using an explicit model of the problem for the inference process, it simply utilizes the experience captured in the same way the expert usually inputs and processes it. Another characteristic that differentiate these systems from other approaches of expert systems is the increasing learning, that is given in an automatic and almost transparent way due to the fact that the retained cases are stored as new cases [8;9].

When a new problem appears, the CBRS looks for a previously occurred problem whose description is the most similar taking into consideration the presented characteristics. The solution to that problem is used as a basis to generate the solution to the new problem.

### B. *Fundamental principles.*

The CBRS can be defined as a cyclic process named ***"the four Rs"*** [10]: **Recover** the most similar cases, **Reutilize** the cases that might solve the problem, **Revise** the proposed solution if necessary, **Retain** the new solution as part of a new case.

*What is a case? "A case is a contextualized piece of knowledge representing an experience".* It contains the previous lesson and the context in which that lesson can be applied [10]. It can also be defined as "a complete description of the problem, with its respective solution and also an assessment of the solution's efficiency" [11].

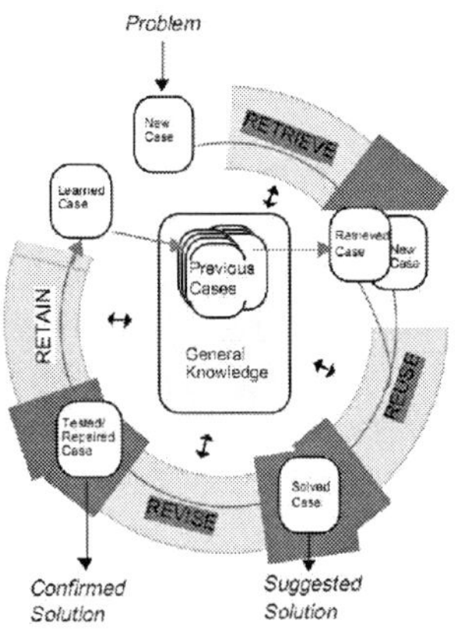


**Figure 1 The "four Rs" CBR Cycle, taken from [1]**

*How can a case be stored?* Case storage is a very important aspect that has a direct impact on the design of the CBRS. Some aspects are to be taken into account when creating a case base: the structure and representation of the cases, the memory model used to organize the case base and the selection of indices used to identify each case [7]. The case base is to be organized in manageable structures that support efficient searches and recovery methods. For this purpose, a wide range of possibilities can be used: text files, relational databases, xml files, etc. and, in order to access them rapidly, indices and manifold algorithms. Cases may represent different sorts of knowledge that can be stored in different representation formats, such as objects, semantic webs, tables, etc.

*How can a case be recovered?* This process could be divided into three tasks: Identifying the characteristics or the indices that describe the new problem; locating the relevant cases and choosing the best candidate, or candidates, among the most relevant cases. Two of the most currently used techniques are: recovery of the closest neighbor, and inductive recovery [2;12].

*What does the adaptation consist in?* Usually, when a case is recovered, an analysis is carried out to determine the similarity with the presented problem. The adaptation consists in identifying the differences between the recovered case and the current case and afterwards, applying mechanisms (formulas, rules or others) to those differences as to obtain the final solution.

Generally, there are two types of adaptation: structural adaptation, which consists in applying rules and formulas directly to the stored solution, and the derived adaptation, which consists in reutilizing the rules and formulas that generated the recovered solution in order to generate the new solution[10].

*What does the revision of a case consist in?* After the case has been adapted, it is convenient to verify that the differences with the new one were taken into account. If the obtained solution to the new problem is not the correct one, it is feasible to repair it and in this way, learn from mistakes. Basically two steps are taken: the solution is assessed and its applicability to the real case is determined, and the case to be stored is repaired.

*What does the retention consist in?* This process consists in incorporating what is useful from the new solution to the knowledge. This involves: Selecting the case information to be retained, in which way to retain it, and how to integrate it to the structure of the memory.

### C. Some existent applications of CBRSs

The CBRS have been applied in multiple contexts. Some reference examples of CBRS applied to CDSS can be:

- CASEY [13] It is a diagnosis system for heart conditions.
- PROTOS [14] It learns to classify auditory disorders based on the description of the symptoms, analyses result and clinical history.
- PAKAR [15] It is a system that identifies the possible causes of construction pitfalls and suggests corrective measures.

## IV. Research Hypothesis

*"A CBRS applied as a help to the clinic diagnosis of the disease known as Acute Bacterial Meningitis is more effective, precise, flexible and intelligent than the ABMDES"*

To prove this claim, a CDSS has been developed using CBR, we'll call it Acute Bacterial Meningitis Case Based Diagnose System - ABMCBDS. Both the new one and the reference – ABMDES – systems, have been fed with data taken from a database of the cases of patients (real ones), and the result of the execution of both programs has been classified and processed.

## V. Case Based Reasoning System implementation for the ABM diagnosis.

### A. Proposed System

A CBRS was developed applied to the diagnosis of the Acute Bacterial Meningitis of children under the age of twelve months (henceforth ABMCBDS). Previous to its construction, a signs-and-symptoms subgroup was selected. This subgroup is representative of the total of signs and symptoms considered for the ABMDES. This has been done, among other things, to simplify the construction process. The signs and symptoms therefore selected, based on the opinion of an expert in this field, are either highly significant for the choice of a diagnosis among all other available diagnoses, or relatively ambiguous, found in the majority of differential diagnoses, with different importance levels.

The following table indicates these signs and symptoms, and it also displays the specificity level (Very specific –VS, Specific –S, Not specific –NS) of the symptom of the disease (indicated by the clinical doctor) as well as the weight to be taken into account in order to carry out similarity calculations.

**Table 1 Case signs and symptoms**

| Sign or symptom | Specificity | Weight |
|---|---|---|
| Convulsions | S | 0.65 |
| Consciousness decrease | VS | 0.9 |
| Fever | NS | 0.3 |
| Bulging fontanelles | VS | 0.9 |
| Irritability | VS | 0.9 |
| Facial Palsy | NS | 0.3 |
| Meningeal signs (Neck and body stiffness) | VS | 0.9 |
| Purpuric signs in the skin | S | 0.55 |
| Somnolence | VS | 0.9 |
| Vomits | NS | 0.3 |

It is important to note that at this stage, the system does not consider the strength of the symptoms, but just their existence or absence.

The group of differential diagnoses to be taken into consideration is then selected. These are:

a. *Acute Bacterial Meningitis.*
b. *Acute Viral Meningitis.*
c. *Tuberculous Meningitis.*
d. *Encephalitis.*
e. *Brain Abscess*
f. *Meningism.*
g. *Proximity Meningeal Reaction.*
h. *Meningeal Hemorrhage.*
i. *Brain Tumor.*

The ABMCBDS is a cyclic process consisting of several phases: recovery of the most similar cases, reutilization of such cases, a revision of the proposed solution and, the retention of the new solution. The system was developed using the JColibri Framework[16;17].

### B. Knowledge representation

In CBR systems, the case is typically comprised of three components: a problem, a solution to it and sometimes an assessment of the solution's properties. In ABMCBDS, each case represents the situation of a medical visit: the "problem" consists of the description of the signs and symptoms shown by the patient (the "clinical feature"); the "solution" represents the diagnosis given by the doctor in that particular situation; and the "assessment" indicates how accurate a diagnosis is the one given (that is to say, if the system has proposed the diagnosis the expert was expecting, and not a differential diagnosis).

The clinical feature is represented as a "*compound attribute"[17;18]*, which is composed of *"single attributes"* and other components. Each single attribute has a name, a type of data (which permit their comparative assessment; in these case, all data are Boolean type), and the weight (whose incidence affects the similarity calculations). Compound attributes only have a name and a certain weight.

### C. *Case recovery – similarity.*

During the ABMCBDS execution, a new visit is registered by the expert (the doctor) as a new case. This new case has only the "problem" part, the clinical feature. The ABMCBDS then proceeds to the recovery of the most similar case/s. In order to do this, the clinical feature of the visit is compared to all other clinical features that compose the Knowledge Base, calculating the similarity to each of them [19]. This similarity calculation is accomplished using a global similarity function for all compound attributes, and local similarity functions for single attributes. The similarity function used for single attributes is that of equality. The global function of a compound attribute is calculated as the weighed summation of the local functions.

Given two cases or situations T and S, the similarity between both is:

$$\text{Similarity}(T, S) = \left(\sum_{i=1}^{n} f(T_i, S_i) * w_i\right) / n$$

In which:

- n is the number of signs and symptoms of each case,
- i is an individual sign or symptom from 1 to n, being n the total amount of symptoms that can exist (a fixed value in the current application) so the symptom referred by this index is always the same in every case (e.g. "fever")
- f is the local or global similarity function for the attribute I (single or compound attribute) in T and S
- W is the weight of the sign or symptom

Comparisons are done between cases on the bases of existence or absence of symptoms.

Additionally, a similarity threshold is defined to delimit the quantity of cases returned by the system, and it was defined as the 85% of the highest similarity value.

### D. *Revision*

Once the most similar cases are recovered, the recovered solutions are to be revised. In this stage the expert doctor gives her decision as regards the differential diagnoses, also providing the information about the accuracy assessment of the ABMCBDS. The doctor will then indicate whether the proposed solution is the correct one (*"success"*) or not (*"failure"*), and, if it is not, also the diagnosis she deems correct. The system's learning is based not only in success but also in failures and mistakes.

### E. *Retention and learning*

Once the case is revised, the diagnosis and its corresponding assessment are obtained, and it is ready to be incorporated to the knowledge base. The solution – inferred and proposed diagnosis – to the new presented case could be either from previous success (the solution is correct) or from failure (the solution is not correct).

For the demonstrative implementation of ABMCBDS, simple text files have been used as data structures to store the cases. These are stored sequentially, in separate locations, and the information of each one is registered as a commas-separated text. The main advantage of these kinds of structures is that it is easy to implement and to understand. A further advantage is that adding cases is rather simple and fast, and its insertion order is 1. However, it is clearly not the adequate representation for a large-scale production system as the case recovery turns out to be rather slow when the number of cases is high (the order is N) and it lacks indexation mechanisms.

## VI. SIMULATIONS AND OBTAINED RESULTS.

In order to proceed to the comparison of both ABMCBDS and ABMDES, a case base is constructed. This case base is built with the aid of the expert doctor, using the defined signs, symptoms and differential diagnoses. For the development of the case base for the ABMCBDS Montecarlo's method is applied, utilizing it for the simulation of the disease, sings and symptoms probabilities from [5].

Once statistically calculated the appropriate size of the sample, the next step is to compose and extract 51 cases, which were inputted as entry in both systems. For each inputted case, the result (proposed diagnosis) was registered by both systems, and was compared to the expert's own diagnosis, to determine whether the result coincided with the one the expert was expecting or not. With these data, the level of *"precision"* or *"accuracy"* of each system was determined.

As an experiment to compare the ability of both systems to capture experience or knowledge from the expert (another of the studied dimensions), seventeen cases were chosen at random from the case base. The average time that the expert needed (calculated size of the sample) to carry out the visits in ABMCBDS (which implicitly incorporates knowledge and learning) and to build the production rules for the knowledge representation in ABMDES, was taken.

The third test was with reference to the tolerance (Figure 2). The aim of this test was to analyze the impact that the degradation of the entry information would have on each system. With the expert doctor, we were able to analyze and define those symptoms that are usually more difficult to detect, or those whose correct interpretation largely depends on the experience of the doctor.

The "system's effectiveness" is referred to the amount of right answers, that is to say, the answers that verify what the expert had said. The obtained results verify that ABMCBDS is at least 20% more efficient than ABMDES.

The accuracy is inversely proportional to the amount of the system's failures. It was verified that ABMCBDS was more accurate than ABMDES.

The intelligence is defined as the speed at which the system learns. Based on the accomplished experiment, it was verified that the average time that the system required to learn was at least 40% less in ABMCBDS than in ABMDES.

The flexibility of a system is defined as the tolerance it presents towards the lack of specificity of a case. This can occur due to the different abilities of doctor to detect signs and symptoms, or to the doctor's experience. The ABMDES is more sensitive to the absence of a symptom; there is more degradation when there is lack of precision (Figure 2).

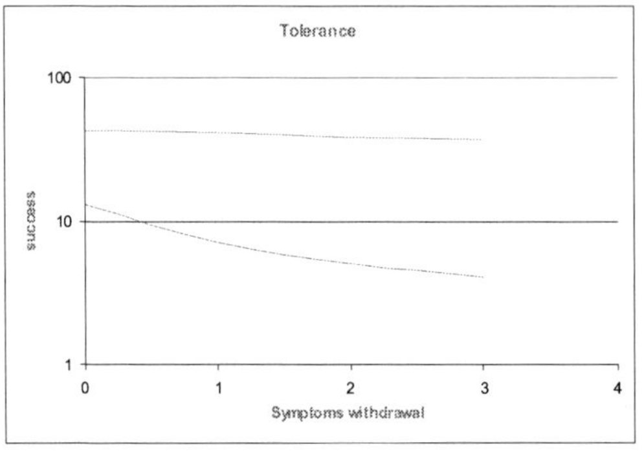


**Figure 2 Tolerance to lack of symptoms**

## VII. Conclusions

The experiments carried out with ABMCBDS and their verification by doctors with great experience in diagnosing the diseases in question, allow us to conclude that this Artificial Intelligence approach applied to the construction of Clinical Decision Support Systems results interesting indeed, given its effectiveness, its learning abilities, and its capacity for capturing the expert's experience.

As stated previously in the Introduction, these kinds of DCSSs are not intended to substitute the expert action, but to help her to analyze and synthesize the huge amounts of experience information that is currently available when dealing with diagnosis situations

In comparison to the already existing reference system ABMDES, built based on a Bayesian inference engine for the diagnosis of the same diseases, the experiments allow us to state:

- ABMCBDS is at least 20% more effective than ABMDES.
- ABMCBDS is at least 20% more accurate than ABMDES.
- ABMCBDS is at least 40% more intelligent than ABMDES.
- ABMCBDS is at least 20% more flexible than ABMDES.

Finally, the ABMCBDS has shown less degradation at the lack of precision of some signs or symptoms that may be difficult to assess, depending on the expert's level of experience.

## References

[1] Aamodt, A. and Plaza, E., "Case-Based Reasoning: Foundational Issues, Methodological Variations, and System Approaches," *IOS Press*, vol. Vol. 7: 1 pp. 39-59, 2004.

[2] Berner, E. S. Clinical Decision Support Systems Theory and Practice. 1998. Springer.

[3] John McCarthy. Some Expert System Need Common Sense. 1984. Standford University.

[4] Hart, P. E., R.O.Duda, and M.T.Einaudi, "Prospector-- A Computer-Based Consultation System for Mineral Exploration," *Ματηεμαticχαλ Γεολογψ*, vol. 10 pp. 589-610, 1978.

[5] Ocampo, E., Herrera, S., Machado, F., and Ruibal, A. Diseño y construcción de sistema experto de diagnóstico de meningitis aguda supurada, basado en máquina de inferencia bayesiana. 2003. CISIC. 1-10-2003.

[6] Kolodner, J. Case - Based Reasoning. 1993. Morgan Kaufmann.

[7] Pal, S. P. and Shiu, S. C. K., *Foundations of Soft Case-Based Reasoning* New Jersey: Wiley Interscience, 2004.

[8] David B.Leake, "CBR in context: The Present and Future," AAAI Press / MIT Press, 1996, pp. 1-35.

[9] San Miguel Carrillo, J. A. Introducción al Razonamiento Basado en Casos. 2007. Práctica de Inteligencia Artificial II. Universidad de Valladolid.

[10] Watson, I., *Applying Case-Based Reasoning: Techniques for Enterprise Systems* California: Morgan Kaufmann , 1997.

[11] Otavio Alvares, L. Raciocínio Baseado em Casos. 2006. Informática UFRGS.

[12] Díaz Agudo, M. B. Una aproximación ontológica al desarrollo de sistemas de razonamiento basado en casos. 6-40. 2002.

[13] Phyllis Koton. Using experience in learning and problem solving. 1989. Massachusetts Institute of Technology, Laboratory of Computer Science.

[14] Porter, B. W. and Bareiss, E. R. PROTOS: An experiment in knowledge acquisition for heuristic classification tasks. In Proceedings of the First International Meeting on Advnces in Learning (IMAL), Les Arcs. 1986.

[15] Watson, I. D. and Abdullah, S. Developing Case-Based Reasoning Systems: A Case Study in Diagnosing Building Defects. IEE Colloquium on Case-Based Reasoning: Prospects for Applications, Digest. 1994. In, Proc.

[16] Recio García, J. A., Diaz Agudo, B., and González Calero, P. JColibrí 2 Tutorial. 2008. Universidad Complutense de Madrid.

[17] Recio, J. A., Anto, o Sánchez, Díaz, gudo, B., Gonz, ez-Cal, and o, A. jCOLIBRI 1.0 in a nutshell. A software tool for designing CBR systems*. 2005. Spain, Universidad Complutense de Madrid.

[18] Sánchez Ruiz Granados, A. A. jColibrí: estado actual y posibles mejoras. 2004. Universidad Complutense de Madrid, España.

[19] Boagaerts, S. and Leake, D. Facilitating CBR for Incompletely-Described Cases: Distance Metrics for Partial Problem Descriptions. 2004. Indian University.

# Information Systems via Epistemic States

Alexei Y. Muravitsky
Louisiana Scholars' College
Northwestern State University
Natchitoches, LA, U.S.A.
*E-mail*: alexeim@nsula.edu

*Abstract* – **We develop a multi-valued logic approach to data base design, that are tolerant to inconsistent information. Also, we discuss possible knowledge transformers in one type of such data bases.**

## 1 PRELIMENARIES

In 1975, Nuel Belnap outlined a strategy of how a four-valued logic can be used in representing knowledge in an artificial agent (the computer) with the possibility of reporting to it contradictory information. (Cf. [3], reprinted in [2], § 81.) One can think of it as a data base tolerant to inconsistency. Usually, data bases contain information in the form of statements or in such a form, when statements can be retrieved. Belnap's idea was different. Instead of saving in the data base pieces of information in the form of statements like *If a bird is a penguin then that bird does not fly* the computer assigns truth values to the statements, which arrive at the computer as information reports and stores in the data base the truth values of their components which have not been already present, as well as corrects those which are already in the memory. Accordingly, the two main questions arise: 1) What should the computer do, if several assignments valuate a statement as true? and 2) How should the computer correct currently stored assignments, if they are effected by a new message?

As the reader will see soon, working with truth values is convenient when contradictory messages enter the computer. For example, it might well be that two agents report to the computer that a statement $p$ is true and not true. Instead of storing two reports in its memory the computer changes the current truth value of $p$ accordingly. The goal of the present paper is to propose a machinery of how the computer "should" act to work with assignments of atomic components instead of whole statements. First, we address the question how to assign the truth values to the components of a compound statement $A$ if an agent reports to the computer that $A$ is true (or $A$ is false). Of course, some reports may contradict one another. This issue will also be discussed.

We postpone the discussion of whether the order, in which the reports enter the data base, matters. Of course, it depends on the character of the operation which makes changes, and this question will be considered later on. Now we have to notice that two truth values, in the scope of which the classical logic operates, is certainly not enough. To accommodate two contradictory reports we need al least three truth values. One more (at least) truth value can be envisaged when we take into account Closed-World Assumption. According to this assumption, what is not contained in the data base is regarded false. We will rather be assuming that what is not in the data base is unknown.

Thus we arrive at Belnap's four truth values: $\boldsymbol{t}$ (*true*), f (*false*), $\top$ (*overknown*) and $\bot$ (*unknown*). Bringing new truth values into consideration we have to define the value of a compound statement which is the formal counterpart of a report the computer receives from outside.

At this point, it is clear that we confine ourselves in the scope of propositional logic. For the sake of simplicity we ground our formal language on the infinite set of (propositional) letters, $p$, $q$, $r$ (with or without indices), called also *atomic formulas*, and (logical) connectives $\wedge$ (*conjunction*), $\vee$ (*disjunction*) and $\neg$ (*negation*), which stand for '...*and*...', '*either*...*or*...' and '*it is not the case that*...', respectively. Sometimes we will need to focus on the letters occurring in a statement $A$. Given $A$, we denote the set of those letters by $E(A)$.

Now we have to learn how a valuation, which is saved in the computer, is going to change when the data base receives the report that a compound statement $A$ is true (or false). The classical logic teaches us that a truth value of a compound statement depends on the values of its components and that dependence is governed by the classical logic truth tables. If we accept these two principles – and we do – we have to define the connectives as algebraic operations on the set $\{\boldsymbol{t}, \boldsymbol{f}, \top, \bot\}$.

We are about to spell out Belnap's proposal. However, before introducing it, we want to emphasize that some conditioning in doing such a definition is unavoidable. There are different ways to arrive at the classical truth tables and all of them are relative and, in fact, have been the subject of criticism.

Belnap suggested to keep the classical logic definitions for the values

T. Sobh (ed.), *Innovations and Advances in Computer Sciences and Engineering*,
DOI 10.1007/978-90-481-3658-2_46, © Springer Science+Business Media B.V. 2010

$\{t, f\}$ and then extend these definitions to

$\{t, f, \top, \bot\}$ in such a way that all connectives understood as operations on the last set be *Scott-continuous*.[1]

Dana Scott introduced at first in [13] his definition of continuity for functions defined between *complete partially ordered sets* (*cpo*'s for short). (See definition also in [5] or in [6] or in [4]. Thus in order to define the connectives as Scott-continuous operations we have to arrange first the set

$\{t, f, \top, \bot\}$ as a cpo. Belnap did it in a somewhat remarkable way. First, he arranged the values in the

set $\{t, f, \top, \bot\}$ as depicted in Figure 1, motivating this arrangement by approximation feeling regarding the four truth values.

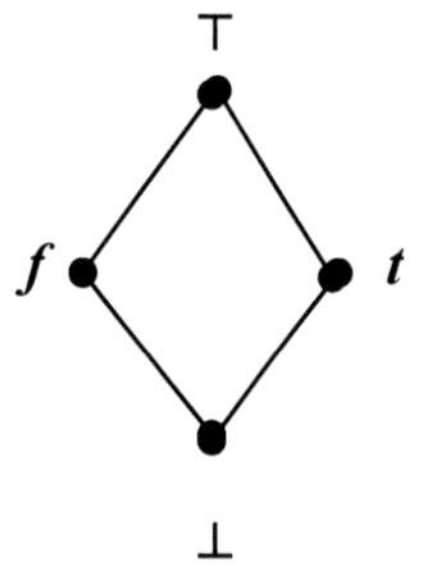


Figure 1: LATTICE **A4**

We write $x \sqsubseteq y$ and say that $x$ *approximates* $y$, when $x$ is at a lower level than $y$ or $x=y$, for any two truth values $x$ and $y$. Belnap calls this lattice *approximation lattice* **A4**.

Then, implementing Scott Thesis, Belnap showed that there is no other way to extend the classical definitions to the new set in order to have all connectives Scott-continuous as defining them as in the tables below.

| $x \wedge y$ | $t$ | $f$ | $\top$ | $\bot$ |
|---|---|---|---|---|
| $t$ | $t$ | $f$ | $\top$ | $\bot$ |
| $f$ | $f$ | $f$ | $f$ | $f$ |
| $\top$ | $\top$ | $f$ | $\top$ | $f$ |
| $\bot$ | $\bot$ | $f$ | $f$ | $\bot$ |

[1] In [Bel 1975] (see also [ABD 1992], § 81) Belnap calls his proposal *Scott Thesis*.

| $x \vee y$ | $t$ | $f$ | $\top$ | $\bot$ |
|---|---|---|---|---|
| $t$ | $t$ | $f$ | $\top$ | $\bot$ |
| $f$ | $f$ | $f$ | $f$ | $f$ |
| $\top$ | $\top$ | $\top$ | $\top$ | $f$ |
| $\bot$ | $\bot$ | $f$ | $f$ | $\bot$ |

| $x$ | $\neg x$ |
|---|---|
| $t$ | $f$ |
| $f$ | $t$ |
| $\top$ | $\top$ |
| $\bot$ | $\bot$ |

We introduce another partial order on the set $\{t, f, \top, \bot\}$, defining $x \leq y$ if and only if $x \wedge y = x$, which is equivalent $x \vee y = y$.

Thus, with respect to $\leq$, we get another lattice (Figure 2), which Belnap calls *logical lattice* **L4**.

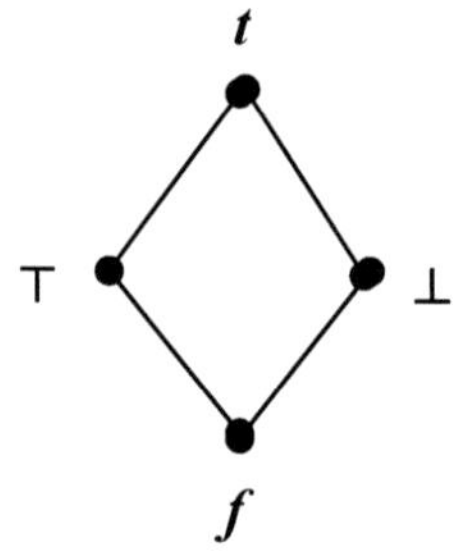


Figure 2: LATTICE **L4**

Two orders on the set of Belnap's truth values play different roles in organizing the inner structure of the units of the data base and their interconnection in it. We will discuss first this inner structure and give the definition of an inner unit of the data base.

Suppose the computer receives the report that the statement $p \vee q$ is true. If the arrived massage is only about the truth value of $p \vee q$, the computer has to decide of how to valuate the components. If it does not have any additional information, it could add to the data base four assignments, which valuate $p \vee q$ as true. This is where we need **L4**. With respect to **L4**, these assignments are: $s(p)=t$ and $s(q)=f$; $t(p)=f$ and $t(q)=t$; $v(p)=\bot$ and $v(q)=\top$; $w(p)=\top$ and $w(q)=\bot$. Since the computer does not know, which is the case, it has to store all these assignments. Thus a data base unit can be thought as a collection of assignments. Such a collection we call an *epistemic state* (of the data base). However, in practice, we never deal with infinite assignments.

We call an assignment *finite* if only finitely many letters are assigned with the values from {***t***, ***f***, ⊤} and all others are assigned with ⊥. The set of letters to which an assignment s assigns values unequal to ⊥ is denoted by *E*(s). But even a finite collection of finite assignments may contain more information than we need. Given such a collection ε and a statement *A*, Belnap, [3], gave his breakthrough definition of *the value of A at the epistemic state* ε as follows:

$$\varepsilon(A)=\sqcap\{v(A) \mid v\in\varepsilon\},$$

where ⊓ is the operation of the greatest lower bound of a set in lattice **A4**.

On the set of all epistemic states (which will be used merely as an auxiliary tool), we define an order relation, doing it in two steps. At first we define the partial order on the set of all assignments as follows:

s≤t if and only if for any letter *p*, s(*p*)⊑t(*p*) in **A4**.
Then, we define

$\varepsilon_1 \preccurlyeq \varepsilon_2$ if and only if for any s∈$\varepsilon_1$, there is t∈$\varepsilon_2$ such that s≤t.

The last ordering is a preorder, because the anisymmetric property may not be fulfilled. Anyway, a finite epistemic state containing only finite assignments, if it is not empty, has the minimal assignments with respect to ≼. Let us denote the set of minimal assignments of ε by **m**(ε) and the procedure of formation of **m**(ε) we call *minimization of* ε.

According to the observation made in [8], for any finite epistemic state ε, containing only finite assignments, and any statement *A*, ε(*A*)=**m**(ε)(*A*). Also, **m**(**m**(ε))=**m**(ε).

We call a finite epistemic state with finite assignments *minimal* if **m**(ε)=ε and regard all minimal states as possible data base's units. Actually, our data base stores all its current information in one minimal epistemic state. Thus all emphasis is made on how the computer moves from one minimal state to another.

We can consider (at least) three types data base transformers. The first will change a current state to another one when the computer receives the message that a statement *A* is true, the second one makes changes when the message reports that *A* is false. And the third type of transformers makes corrections as follows. Given statement *A* and *B*, when *s* in the current state valuates *A* true the transformer makes such updates of s to t so that t valuates *B* true. Moreover, the computer should not loose the information it has had before the transformation. In other words, it changes s to t by increasing the values of the former with respect to the order ⊑ in **A4**. For these types of transformers, we introduce the following notation [*A*:***t***](ε), [*A*:***f***]( ε) and [*A*→*B*](ε), respectively. Other transformers can also be considered, but we limit ourselves with these three.

When Belnap was defining his truth tables above, he was guided by Scott Thesis. Now, thinking of defining data base transformers (above and, possibly, others) we have to ground our definitions on a similar principle. Even if we simply add a new information to an old one, we believe that this addition without any change works well enough. However, if we know nothing about *A*, that is, ε(*A*)=⊥, then the message that *A* is true (or it is false) should change ε.

Referring above to Belnap's Scott Thesis, we did not spell it out in full. Belnap's idea in its full form can be expressed as follows: *There exist only Scott-continuous functions*. This principle presupposes that the domain of such a function should be arranged as a cpo. In the case of Belnap's four values, we have **A4**. What should we take for the date base state transformers mentioned above? And why Scott-continuous functions are better than any others? First we discuss the first question.

Some work in this direction has been done. To explain it we begin with the partial order ≤ defined on the set of all assignments. We denote this partially ordered set by **AS**. The following facts are known.

**Proposition 1** ([8]) **AS** *is a complete lattice with operations* $\wedge\{s_i \mid i\in I\}(p)=\sqcap\{s_i(p) \mid i\in I\}$ *and* $\vee\{s_i \mid i\in I\}(p)= \sqcup\{s_i(p) \mid i\in I\}$, *where* ⊓ *and* ⊔ *greatest lower and least upper bounds in* **A4**, *respectively.*

The role of **AS** will not be seen for a while. Now we turn to minimal states, the set of which along with relation ≼ is denoted by **AFE**.

**Proposition 2** ([8]+ [10]) **AFE** *is distributive lattice with operations meet and join defined as* $\varepsilon_1\wedge\varepsilon_2$=**m**($\varepsilon_1\cup\varepsilon_2$) *and* $\varepsilon_1\vee\varepsilon_2$=**m**({s∨t | s∈$\varepsilon_1$ and t∈$\varepsilon_2$}), *respectively.*

When we have defined **AFE**, we can observe, at least in principle, all possible states of the data base. Unfortunately, **AFE**, taken by itself, cannot serve as a domain for Scott continuous functions because it is not a cpo. (Cf. [8].) Thus we have to embed **AFE** into a broader environment. But we should do it with caution, because the functions $[A{:}\boldsymbol{t}]$, $[A{:}\boldsymbol{f}]$ and $[A \rightarrow B]$ are supposed to be Scott-continuous in the broader environment and must be closed in **AFE**.

The following solution was suggested in [8]. First we define the following equivalence on the set of all epistemic states:

$\varepsilon_1 \asymp \varepsilon_2$ if and only if for any statement $A$, $\varepsilon_1(A) = \varepsilon_2(A)$.

The class $|\varepsilon| = \{\varepsilon' \mid \varepsilon' \asymp \varepsilon\}$ is arranged by the following relation:

$$|\varepsilon_1| \ll |\varepsilon_2| \text{ if and only if any statement } A, \; \varepsilon_1(A) \sqsubseteq \varepsilon_2(A) \text{ in } \mathbf{A4}.$$

We call this structure the *generalized epistemic states* and denote it by **AGE**.

**Proposition 3** (([8]+ [10]) **AGE** *is a cpo. Moreover,* **AGE** *is a co-atomic Heyting algebra and* **AFE** *is its sublattice* (*up to isomorphism*).

Now we turn to operations. We will explain how $[A{:}\boldsymbol{t}]$ can be defined and refer the interested reader to [8] and [10], where he can find additional information, including the definitions of the two other operations.

As to $[A{:}\boldsymbol{t}]$, we first define operation $\mathbf{T}(A) = \{s \in \mathbf{AS} \mid \mathrm{t} \sqsubseteq s(A), E(s) \subseteq E(A)\}$ which is an epistemic state. Now we define

$$[A{:}\boldsymbol{t}](|\varepsilon|) = |\varepsilon| \vee |\mathbf{T}(\mathrm{A})|,$$

where $\vee$ is the join operation in **AGE** with respect to the order $\ll$.

**Proposition 4** (([8]+ [10]) *Operation* $[A{:}\boldsymbol{t}]$ *is Scott-continuous and closed on* **AFE**. *Moreover, being restricted to* **AFE**, $[A{:}\boldsymbol{t}](\varepsilon) = \varepsilon \vee \mathbf{m}(\mathbf{T}(A))$, *where* $\vee$ *is the join operation in* **AFE** *with respect to the order* $\preccurlyeq$.

At this point we can ask the question: What have we achieved? We think of the data base as of a pebble which moves from one point to another within **AFE**. The operations that change the current position of the pebble are Scott-continuous. And it is the time to ask, why Scott-continuous operations are better than any others. The property which makes Scott-continuous operations so attractive is that they act very smoothly. Suppose a minimal state $\varepsilon$, i.e. $\varepsilon \in \mathbf{AFE}$, is the join of its approximations, i.e. $\varepsilon = \bigvee\{\varepsilon_i \mid i \in \mathrm{I}\}$. Then, for example, $[A{:}\boldsymbol{t}](\varepsilon) = \bigvee\{[A{:}\boldsymbol{t}](\varepsilon_i) \mid i \in \mathrm{I}\}$. Since **AGE** is a cpo, the last equality holds even for an infinite set of approximations from **AFE**, providing that they form a directed set. In case we have infinitely many approximations, $\varepsilon$ may no longer belong to **AFE**, but, then it belongs to **AGE**. The last remark is based on the following

**Proposition 5** ([8]) **AFE** *is* (*up to isomorphism*) *an effective basis of* **AGE**.

That is, not only the join of any directed set of approximations from **AFE** lies in **AGE**, but also every element in **AGE** can be represented as the join of some elements from **AFE**.

At this point of our exposition, we want to raise two questions. The first: Is there a convenient way to describe **AFE** and operation $[A{:}\boldsymbol{t}]$ in it? The second question may seem too abstract: What if the Belnap's set of truth values does not meet our goal. Can this approach be applied to other sets of truth values. We address these questions in Sections 2 and 3, respectively. We will illustrate the approach presented in Section 3 on the example of Kleene's 3-valued strong logic.

## 2 IFORMATION SYSTEMS

In [14] Dana Scott introduce the notion of. an *information system* is a triple $(\boldsymbol{D}, \boldsymbol{Con}, \vdash)$, where $\boldsymbol{D}$ is a nonempty set of tokens, $\boldsymbol{Con}$ is the finite nonempty subsets of $\boldsymbol{D}$ and $\vdash$ (*entailment*) is a binary relation on $\boldsymbol{Con} \times \boldsymbol{D}$. The following axioms define the information system:

1) For any $d \in \boldsymbol{D}$, $\{d\} \in \boldsymbol{Con}$;
2) $u \in \boldsymbol{Con}$ and $v \subseteq u \Rightarrow v \in \boldsymbol{Con}$;
3) $d \in \boldsymbol{D}$ and $u \in \boldsymbol{Con}$ and $u \vdash d \Rightarrow u \cup \{d\} \in \boldsymbol{Con}$;
4) $d \in u$ and $u \in \boldsymbol{Con} \Rightarrow u \vdash d$;
5) For all $x \in v$, $\mathrm{u} \vdash x$, and $v \vdash d \Rightarrow \mathrm{u} \vdash \mathrm{d}$.

Scott proposed to use information systems to define *domains* (in the sense of [6] or [5] or [4]). Since **AGE** is a domain, in virtue of Proposition 5, we can try to define **AGE**, as well as **AFE**, through an information system.

Let $\boldsymbol{D}$ be $\boldsymbol{Fm}$, that is, the set of formal statements, $\boldsymbol{Con}$ be all finite nonempty sets of statements and $\vdash$ be defined as follows:

$u \vdash A$ if and only if the conditional $\wedge u \rightarrow A$ is derivable in $E_{fde}$, *the first degree entailment* from [1].

**Proposition 6** ([11]) *The triple* (***D***, ***Con***, $\vdash$) *defined above is an information system.*

If we define for all nonempty $x \subseteq \boldsymbol{D}$,

$$x^* = \{A \mid x \vdash A\},$$

then $(\{x^* \mid x \subseteq \boldsymbol{D}\}, \subseteq)$ is the domain associated with the information system (***D***, ***Con***, $\vdash$).

**Proposition 7** ([11]) *Partially ordered sets* $(\{x^* \mid x \in \boldsymbol{Con}\}, \subseteq)$ *and* **AFE** *are isomorphic. Partially ordered sets* $(\{x^* \mid x \subseteq \boldsymbol{D}\}, \subseteq)$ *and* **AGE** *are isomorphic.*

Thus the sophisticated relation $\preccurlyeq$ on **AFE** can be reduced to entailment $\vdash$ and set inclusion.

**Proposition 8** (([11])) *Operation* $[A{:}\boldsymbol{t}]$ *can be defined on* $(\{x^* \mid x \in \boldsymbol{Con}\}, \subseteq)$ *as* $[A{:}\boldsymbol{t}](x^*) = (x \cup \{A\})^*$.

## 3 More Truth Values Structures and Generalization of the Main Constructions

In this section we will address the second issue mentioned in Section 2. Namely, we propose a construction which covers **AGE**, along with **AFE**, as well as many other similar constructions based on other sets of truth values. As an example of application of this new construction, we take Kleene's 3-valued (strong) logic.

To describe the new construction in abstract terms we start with the arrangement of possible (always finite number) truth values.

An *epistemic structure* $\mathcal{T}$ = ( ***E***; $\wedge$, $\vee$, $\neg$, ***f***, ***t***, «) is an algebraic system with two binary operations, one unary operation, two constants and one binary relation, respectively, satisfying the following conditions:

1) $(E, \sqsubseteq)$ is a finite meet-semilattice with respect to $\sqsubseteq$, such that any nonempty directed subset of $E$ has a least upper bound;
2) Operations $\wedge$, $\vee$ and $\neg$ are monotone with respect to $\sqsubseteq$;
3) $\neg f = t$ and $\neg t = f$.

Belnap's **A4-L4** structure is an example of an epistemic structure. Another example of it is a well-known Kleene's 3-valued logic. Indeed, Let the values in $\{\boldsymbol{t}, \boldsymbol{f}, \bot\}$ be arranged by $\sqsubseteq$ as depicted in Figure 3.

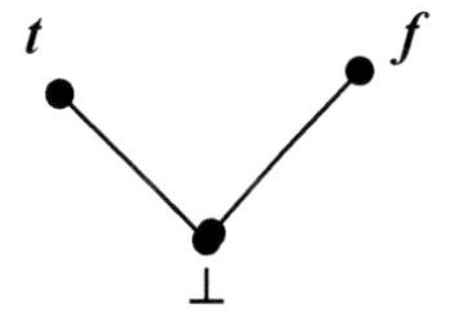

Figure 3: Lattice **A3**

Now according to the tables:we define operations $\wedge$, $\vee$ and $\neg$ as follows:

| $x \wedge y$ | ***t*** | ***f*** | $\bot$ |
|---|---|---|---|
| ***T*** | ***t*** | ***f*** | $\bot$ |
| ***F*** | ***f*** | ***f*** | ***f*** |
| $\bot$ | $\bot$ | ***f*** | $\bot$ |

| $x \vee y$ | ***t*** | ***f*** | $\bot$ |
|---|---|---|---|
| ***T*** | ***t*** | ***t*** | ***t*** |
| ***F*** | ***t*** | ***f*** | ***f*** |
| $\bot$ | ***t*** | ***f*** | $\bot$ |

| $x$ | $\neg x$ |
|---|---|
| ***t*** | ***f*** |
| ***f*** | ***t*** |
| $\bot$ | $\bot$ |

As to $\wedge$ and $\vee$, their definitions by the tables above correspond the meet and join operations on the lattice in Figure 4.

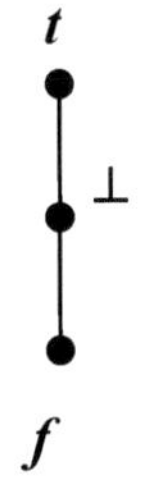

Figure 4: Lattice **L3**

The reader must have noticed conversion to **A4** to **L4**, as well as **A3** to **L3**. One should turn the approximation lattice clockwise at 90°. This is the way of how one can get bilattices, used in Artificial Intelligence. Indeed, **A4** is a bilattices, the simplest one, but **A3** is not, because the bilattices must have a top element. In general, every finite bilattices is an epistemic structure, but not vice versa.

Our generalization is grounded on the observation that **AGE** is an effectively represented domain (in the sense of [6], [7] or [5]) with the basis **AFE**. From now on we understand the set of all assignments **AS** (Section 1) with respect to any fixed epistemic structure.

**Theorem 1** *Lattice* **AS** *(with the relation* $\leq$ *on it) is an effectively represented domain, where the set of compact elements coincides with the set of finite assignments.*

Now let **AS*** be the powerdomain of the domain **AS**. (See definition of powerdomain, e.g., in [7].) It is well known that the compact element of a powerdomain form its effective basis.

Our next goal is illustrate how powerful for our purposes is the powerdomain construction. Of course, this abstract view from a height can be useful if we want to observe the entire space of all possible stages the data base may take, for example, if we want to see in which direction it can develop. On the other hand, we have to learn how to come from this abstract view to a concrete one determined by its initial point, that is, by its set of truth values.

In what follows, we test the abstract approach with our cases above – with the pairs **A4-L4** and **A3-L3**. For our purposes, we denote the set of all assignments associated with **A4** by $\mathbf{AS_4}$ and the set of those associated with **A3** by $\mathbf{AS_3}$.

**Theorem 2** **AGE** *is isomorphic to* $\mathbf{AS_4}$* *and* **AFE** *is isomorphic to the sublattice of the compact elements of* $\mathbf{AS_4}$*.

Next we define a domain via an information system, as we did it in Section 2. However, now we use **A3** as a starting point. The logic system which plays the key role here is the following extension of the first degree entailment:

$$E_3 = E_{fde} + A \wedge \neg A \rightarrow B \vee \neg B.$$

The definition of entailment $\vdash_3$ is made up relative to $E_3$:

$u \vdash_3 A$ if and only if the sequent is derivable in $E_3$.

Similar to Proposition 6, we prove the

**Theorem 3** *The triple* $(\boldsymbol{D}, \boldsymbol{Con}, \vdash_3)$ *is an information system.*

Now let $x^* = \{A \mid x \vdash_3 A\}$.
Similar to Proposition 7, we obtain the

**Theorem 4** *Partially ordered sets* $(\{x^* \mid x \in \boldsymbol{Con}\}, \subseteq)$ *and* $\mathbf{AS_3}$* *are isomorphic.*

## REFERENCES


[1] A.R.Anderson and N.D. Belnap, **Entailment: the Logic of Relevance and A Necessity**, Vol. 1, Princeton Univ. Press, 1975.

[2] A.R. Anderson, N.D. Belnap, and J.M. Dunn. **Entailment: the Logic of Relevance and Necessity**, Vol. 2, Princeton Univ. Press, 1992.

[3] N.D. Belnap. *A Useful Four Valued Logic*. In: J.M. Dunn and G. Epstein, editors, Modern Uses of Multiple-Valued Logic, **Proceedings of International Symposium on Multiple-Valued Logics**, 5th, Indiana University, 9-37, D. Readel Publ. Co., 1975.

[4] B.A. Davey and H.A. Priestley, **Introduction to Lattices Order**, Cambridge Univ. Press, 2nd edition, 2002.

[5] G. Gierz, K.H. Hofmann, K. Keimel,, Lawson, M. Mislove, and D.S. Scott, **Continuous Lattices and Domains**, Cambridge University Press, 2003.

[6] G. Gierz, K.H. Hofmann, K. Keimel, J.D. Lawson, M. Mislove, and D.S. Scott, **Compendium of Continuous Lattices**, Springer-Verlag, 1980.

[7] C.A. Gunter and D.S. Scott. *Semantic Domains*. In: J.van Leeuwen, editor, **Handbook of Theoretical Computer Science**, Vol. B: "Formal Models and Semantics," pages 635-674, Elsevier, 1990.

[8] Y.M. Kaluzhny and A.Y. Muravitsky. *A Knowledge Representation Based on the Belnap's Four-Valued Logic*, **Journal of Applied Non-Classical Logics**, Vol. 3, no. 2, 1993, 189-203.

[9] S.C. Kleene. **Introduction to Metamathematics**, P. Noordhof of N.V., Groningen, 1952.

[10] A.Y. Muravitsky, *A Framework for Knowledge-Based Systems*, **Journal of Applied Non-Classical Logics**, Vol. 6, no. 3, 1996, 263-286.

[11] A.Y. Muravitsky. *Knowledge Representation as Domain*, **Journal of Applied Non-Classical Logics**, Vol. 7, no. 3, 1997, 343-364.

[12] A.Y. Muravitsky. *Logic of Information Knowledge: a New Paradigm*. In: R. Asatiani, K. Balogh, G. Chikoidze, P. Dekker, and D. de Jongh, editors, **Proceedings of the Fifth Tbilisi Symposium on Language, Logic and Computation**, ILL – University of Amsterdam, 2004, 121-128.

[13] D.S. Scott, *Continuous Lattices*, **Lecture Notes in Mathematics**, Vol. 274, 1972, 97-136.

[14] D.S. Scott, *Domains for Denotational Semantics*, **Lecture Notes in Computer Science**, no. 140, 1982, 577-613.

# A Practical Application of Performance Models to Predict the Productivity of Projects

Carla Ilane Moreira Bezerra[1,2], Ciro Carneiro Coelho[2], Carlo Giovano S. Pires[2], Adriano Bessa Albuquerque[1]

[1] University of Fortaleza (UNIFOR) – Masters Degree in Applied Computer Sciences (ACS)
Washington Soares Avenue, 1321 - Bl J Sl 30 - 60.811-341 - Fortaleza – Ce - Brazil

[2] Atlantic Institute, Chico Lemos Street, 946, 60.822-780 - Fortaleza – CE, Brazil

*Abstract-* **In the traditional project management discipline, performance tracking is based on analysis and comparisons between the performance in a given moment of the project and the planned performance. The quantitative project management proposed in CMMI model allows predicting the current and future performance based on performance models. The Six Sigma methodology supports these models through statistical tools that are suitable for the quantitative management implementation. This paper presents a case in the definition of performance models based in CMMI and Six Sigma and their application in productivity prediction on the projects of a software organization.**

## I. Introduction

The continuous search for products and services more perfects make some organizations to research and to implement many techniques, tools and standards able to increase the quality of their products. Inside this context, where the organizations aim to reduce the costs and rework, increase the clients' satisfaction and productivity and improve the products quality, the Capability Maturity Model Integration (CMMI) [1] and Six Sigma [2] may be seen as good options to help the companies.

Many organizations are adopting the Six Sigma as a strategy to reach both low levels of maturity and high levels from CMMI [3].

The Six Sigma and CMMI have compatibles objectives and the Six Sigma may be executed on the macro and micro levels of the organization, interacting with elementary graphical tools and advanced statistical tools [4].

The process area "Organizational Process Performance" is of the maturity level 4 of CMMI and belongs to the Process Management area. Its purpose is to establish and maintain a quantitative understanding of the performance of the organization's set of standard processes in support of quality and process-performance objectives, and to provide the process performance data, baselines, and models to quantitatively manage the organization's projects [5].

Process performance models are used to estimate or predict the value of a process performance measure from the values of other process and product measurements. The process performance models typically use process and product measurements collected throughout the life of the project to estimate progress toward achieving objectives that cannot be measured until later in the project's life [1].

On this context, this paper presents a definition and a practical application of performance models in a software organization to foresee the projects productivity and to be able to compare the results of the models with the traditional methods used on the initial estimation of productivity.

On the Section II is presented the level 4 of CMMI. On the Section III is presented contents related to Six Sigma and the DMAIC methodology. On the Section IV, is presented the definition of the organization's performance models. On the Section V, is presented a practical application, on the organization's projects, of the models defined, comparing the obtained results with anterior data. Finally, on the Section VI, is concluded this paper and present the lessons learned of this experience.

T. Sobh (ed.), *Innovations and Advances in Computer Sciences and Engineering*,
DOI 10.1007/978-90-481-3658-2_47, © Springer Science+Business Media B.V. 2010

## II. CMMI Level 4

The CMMI [6] is a maturity model of software created by the Software Engineering Institute (SEI). It guides the organizations on the implementation of continuous improvements on their software development process.

This model has 5 maturity levels. The level 4 has the objective to establish a process quantitatively managed, in other words, a controlled process through the use of statistical tools or quantitative techniques [1]. This purpose is reached executing the following process area: Organizational Process Performance (OPP) and Quantitative Project Management (QPM).

The purpose of the OPP is to create a quantitative understanding of the organizational process using data, baselines and performance models to the relevant processes of the organization. This understanding results from the quantitative control established by the process area QPM, which defines the subprocesses that should be quantitatively controlled, establishes and monitors the quality and process performance objectives, selects the measurement and analysis to the selected subprocesses, uses techniques to understand the subprocesses variation and registers the data obtained from the quantitative control [1].

So the objective of the performance models is to permit the predictability of future performance of the processes thorough others attributes of the process and products. These models describe relationships between attributes of the process and work products [7]. Performance models are used, mainly, on the estimations that are useful to the planning and monitoring of the projects [8].

## III. The Six Sigma and the DMAIC methodology

Nowadays, one of the important methodologies utilized by the companies on the world is the Six Sigma. It is a method, which the objective is support the reduction or elimination of the errors, defects or fails in a process. This methodology also aims to reduce the instability of the process and may be applied on many economic activities sectors [9].

The Six Sigma joins a rigorous statistical approach to a set of tools that are used to characterize the sources of variations to demonstrate how this knowledge may control and improve the results of the process [10].

Due to the Six Sigma supports statistical control process, helping the organizations to apply practices of management more advanced and to improve their processes performance, it began to be very relevant to reach higher maturity levels. The implementation of DMAIC (Define, Measure, Analyze, Improve and Control) projects collaborates so much to this success.

The DMAIC methodology encompasses: define, measure, analyze, improve and control. On the phase "Define" is necessary to identify the problem and the existent opportunities to solve it according to the clients' requirements. On the phase "Measure", we should verify the actual situation through quantitative measurements of the performance to the subsequent decisions base on facts. On the phase "Analyze", we should determine the causes of the obtained performance and analyze the existent opportunities. After this analysis is possible to perceive points to improve the performance, that may be implemented on the phase "Improve". On the phase "Control" we should guarantee the desired improvement, through the control of the implemented process performance [12].

## IV. Definition of performance models

On this section, we present the definition of performance models based on the practices of the CMMI process área: "Organizational Process Performance". We use the DMAIC methodology to construct them and they were applied in projects of the Atlantic Institute, a medium-sized enterprise, assessed on CMMI level 3, which the actual goal is reach the maturity level 5.

We construct the productivity and defects density models through a DMAIC project, which had the objective to improve the general productivity and reduce the defect density of the organization's projects systemic tests.

The DMAIC steps help to construct the processes performance models, through the use of statistical tools. On the construction of the model, the factor to be addressed and controlled on the process or subprocess is named "Y" and the factors that will influence the "Y", that may be other measures, are named "X".

Firstly were defined stable and reliable baselines to the test and project management processes. The baselines establishment permitted begins the models construction, on the phase "Analyze" of the DMAIC project, using the statistical method Multiple Linear Regression. This method associates a dependent variable to many independent variables, demonstrating to be efficient to well behave variables or when two or more independent variables do not present any correlation.

On the Table I are defined the measures of performance of each model, showing the Ys and Xs.

Table I
MODELS VARIABLES AND MEASURES.

| Models | Variables | Measures |
|---|---|---|
| Model to Defect Density | Y | Defect Density in Systemic Tests (DDST) |
| | X | Percentage of Defects in Technical Revisions (PDTR) |
| | | Unit Test Coverage (UTC) |
| Model to Productivity | Y | General Projects Productivity (GPP) |
| | X | Defect Density in Systemic Tests (DDST) |
| | | Level of the Requirements Unstableness (LRU) |
| | | Level of Continuous Integration Utilization (LCIU) |
| | | Level of Experience (LEX) |
| | | Development Environment (DENV) |

The construction of the models used reliable data from the measures presented on Table 1 and others measures that could be important to them.

On the Minitab tool we performed multiple regressions until find out consistent models that had determination coefficients (R-Sq(adj)) higher than 60%. Many statistical works affirm that on these cases the model is more reliable and the factors of model's equation have more influence.

After a lot of tests using the measures data, the models were considered valid. The model to defect density more adequate has R-Sq(adj) = 72.2% and equation:

$$DDST = 1.8955 - 0.5087*PDTR - 1.6020*UTC$$

In case of the model to productivity, the higher level of explication has R-Sq(adj) = 69.8% and include the model to defect density, with the following equation:

$$GPP = 32.087 - 3.637*DDST + 11.71*LRU - 9.451*LCIU - 0.8187*LEX*DENV$$

The Fig. 1 shows the factors of the integrated prediction model, where de model to defect density integrates the model to productivity.

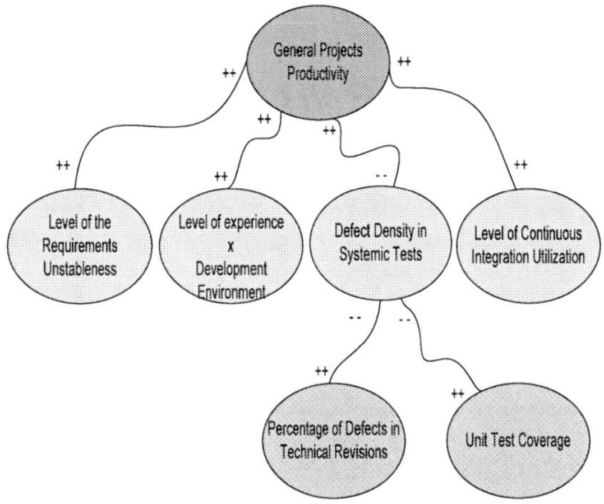


Fig. 1. Factors of the Integrated Prediction Model

## IV. PERFORMANCE MODELS APPLICATION

Initially, we chose five projects of the organization to test the constructed model, aiming to verify the model's efficacy in relation to the predictability of the projects final productivity and to compare the obtained results to the percentage of the difference between the planned productivity on the beginning of the projects and the final productivity.

The performance models equations were calibrated using real values of the projects to obtain estimations more true. On the model to defect density were inserted value related to the measures: Percentage of Defects in Technical Revisions (PDTR) and Unit Test Coverage (UTC).

On the model to productivity were inserted values related to the measures: Defect Density in Systemic Tests (DDST), Level of the Requirements Unstableness (LRU), Level of Continuous Integration Utilization (LCIU), Level of Experience (LEX) and Development Environment (DENV).

Some parameters were established, like: average, goal, superior limits and inferior limits for each baseline, as may be seen on Table II. These parameters support the projects to define the estimated values of the variables Xs to calibrate the model. The result of the model will be a more real prediction of the projects productivity and defect density. The average was calculated through the average of the projects that encompasses each baseline.

The goal and the limits (superior and inferior) were obtained through the use of simulations, to guarantee a more aggressive goal to the organization.

Table II
PARAMETERS OF THE BASELINES

| Performance Baselines | Goal | Average | Superior Limit | Inferior Limit |
|---|---|---|---|---|
| Defect Density in Systemic Tests (DDST) | 0.26 | 0.37 | 0.45 | 0.00 |
| Percentage of Defects in Technical Revisions (PDTR) | 0.70 | 0.51 | 0.90 | 0.52 |
| Unit Test Coverage (UTC) | 0.80 | 0.74 | 0.90 | 0.74 |
| General Projects Productivity (GPP) | 17.36 | 19.91 | 24.00 | 9.00 |
| Level of the Requirements Unstableness (LRU) | 0.05 | 0.07 | 0.15 | 0.00 |
| Level of Continuous Integration Utilization (LCIU) | 1.00 | 1.00 | - | - |
| Level of Experience (LEX) | 3.00 | 2.00 | - | - |
| Development Environment (DENV) | 2.00 | 2.00 | - | - |

The approximation of the productivity calculated by the model in relation to the final productivity of the project was measured in percentage and was compared with the difference (percent) between the planned productivity on the beginning of the project and the final productivity. Table III presents this result.

The comparison of the Productivity Deviation of the Model and the Productivity Deviation of the Conventional Technique shows that the smaller percentages are obtained from the utilization of the performance model to productivity. So the productivity estimation using models is significantly more precise than using traditional techniques, for example, utilize the organization's historical average. Whereas the middle deviation of the conventional techniques was 30.88%, the deviation of the model to productivity was 9%.

Table III
DEVIATION OF THE PRODUCTIVITY USING THE MODEL X DEVIATION OF PRODUCTIVITY USING CONVENTIONAL TECHNIQUES

| Projects | Deviation of Productivity using the Model | Deviation the Productivity using Conventional Techniques |
|---|---|---|
| Project 1 | 7.44% | 13.79% |
| Project 2 | 7.89% | 28.76% |
| Project 3 | 7.15% | 29.96% |
| Project 4 | 11.26% | 57.58% |
| Project 5 | 11.32% | 24.33% |

We can conclude that the application of models is the better way to obtain estimations more precise and therefore improve the clients' satisfaction, reduce the variability of timeline, cost and productivity on software organizations.

## VI. CONCLUSION

The organizations look for, day after day, better practices related to the development and management of software projects, aiming to increase their competitiveness through, mainly, of the improvement of their level of timeline, cost and quality predictability [8].

The traditional project management uses only analysis comparing measures collected in some period with the planned on the beginning of the projects. However, it is impossible to foresee values of measures on the future. Due to this fact, the mature organizations use processes performance models to improve their estimations.

One of the biggest difficulties that organizations meet to reach the high levels of maturity is to define the performance models to be used on the projects quantitative management.

This work presented the definition of performance models in a software organization and evaluated them in real software projects, observing that the results of models application are better than the results of the estimation using conventional methods.

As this experience was very important and rich to the organization, we present bellow some difficulties met during the process of define and apply the performance models and the main lessons learned. In this way we want to help others organizations to define better performance models and apply them faster:

- Difficulty to stabilize the organization's data base to reach reliable baselines. The organization, aiming to verify the integrity of the collected data, should begin, early, to review the projects measures using criteria;
- A small quantity of points may difficult the construction of performance models, because we do not get a model with an acceptable level of explication. Choosing measures with greater granulation on the projects and projects with many iterations may contribute to increase the points on the baselines of the organizational data;
- Difficulty to find correlations between variables that might influence on the choice of the factor to improve the construction of the model. When choose the factors (Xs) that may influence the problem to be addressed (Y), is important that the organization have a previous knowledge of the correlated measures. For example, we may say that the experience of the team will always influence, directly, the productivity factor.
- Another great difficult found is the few experience of the project managers with the contents and activities involved on the projects quantitative management. Besides, the few understanding of the benefits that performance models can add to their management, like the possibility to foresee future risks and to anticipate actions to mitigate these possible risks. The managers should be formally trained on quantitative management applied to projects management and to be informally trained by a consultancy and/or the software process group during the beginning of the quantitative management on the project.

## REFERENCES


[1] CMU/SEI, "Capability Maturity Model Integration, version 1.2. CMMI for Software Engineering (CMMI-SW/IPPD, v1.2) Staged Representation", Software Engineering Institute (2006).

[2] Tayntor, C. B., *Six Sigma Software Development,* Flórida, Auerbach, (2003).

[3] Siviy, J.; Penn, M. L.; Harper, E., "Relationships Between CMMI and Six Sigma". Available in: http://www.onesixsigma.com/node/3795. Acessed on 2008/03/15.

[4] Dennis, M., *The Chaos Study*, The Standish Group International, (1994).

[5] Ahern, D. M.; Clouse, A.; Turner, R.. *CMMI Distilled: A Practical Introduction to Integrated Process Improvement,* 2nd edition: Addison Wesley (2003).

[6] Chrissis, M. B.; Konrad, M.; Shrum, S., *CMMI: Guidelines for Process Integration and Product Improvement*, 2nd edition, Boston, Addison Wesley (2006).

[7] Kulpa, M.K., Johnson,K.A., *Interpreting the CMMI – A process improvement approach*, CRC Press LLC (2003).

[8] SOFTEX, " MPS.BR – Guide of Implementation – Part 6: Level B – version 1.0" . Available in: " www.softex.br".

[9] Smith, B.; Adams, E., "LeanSigma: advanced quality", In: Proceedings of the 54th Annual Quality Congress of the American Society for Quality, Indianapolis, Indiana (2000).

[10] Watson, G. H., *Cycles of learning: observations of Jack Welck*, ASQ Publication (2001).

# An Approach to Evaluate and Improve the Organizational Processes Assets: the First Experience of Use

Adriano Bessa Albuquerque[1], Ana Regina Rocha[2]

[1]University of Fortaleza (UNIFOR)
Washington Soares Avenue, 1321 - Bl J Sl 30 - 60.811-341 - Fortaleza – Ce – Brazil

[2]COPPE/UFRJ – Federal University of Rio de Janeiro
POBOX 68511 – ZIP21945-970 – Rio de Janeiro, Brazil

*Abstract-* **Software development organizations must improve their products' quality, increase their productivity, reduce the projects' costs and increase the projects' predictability aiming to maintain their competitiveness in the market. So, it is essential to invest on software process and support approaches to continually improve the processes on this volatile market. This paper presents an approach to evaluate and improve the processes assets of software development organizations, based on internationally well-known standards and process models. This approach is supported by automated tools from the TABA Workstation and is part of a wider improvement strategy constituted of three layers (organizational layer, process execution layer and external entity layer). Moreover, this paper presents the first experience of use of the approach.**

## I. INTRODUCTION

Nowadays, the world's software industry increases because the software becomes part of many products and activities. In the United States, between 1995 and 1999, the tax of investments in software was four times higher then in the period of 1980-85 [1].

The growth of the Brazilian industry of software is also representative. According to the last publication of "Quality and Productivity on Brazilian Software Sector" [2], the participation on market of the software products and computing technical services in Brazil rose from 42% to 51% during the period of 1991/99, considering all the computing sector.

However, the software organizations need improve the quality of their products, increase the productivity, reduce the costs and increase the predictability of the projects to continue in this promising market, where the changes of the clients needs are constants and the increase of the competitiveness. This scenario demands that the processes of the organizations improve continually.

In face of this context and of the knowledge that the quality of the software products is influenced by the quality of the software processes used to develop them [3], the industry of software and the academy are investing more and more in researches related to software process. Besides, as the market is volatile and its level of exigency increases day by day, the processes should stay all the time in a state of continuous improvement [4] [5].

The Indian software organizations were globally recognized because they invested, mainly, on the quality of software processes and on training. In 2004, 275 Indian companies had received certifications of quality or had been assessed on CMMI and more than 80 companies were near to be certificated or assessed on a maturity model [6].

Nevertheless, the improvement of software processes comprehend complex issues, being fundamental support them in an efficient and organized way.

This paper presents an approach to evaluate and improve the processes assets in the organizational layer, which is part of the strategy in layers to define, evaluate and improve software processes, implemented on TABA Workstation. Moreover, it presents the results of a first experience of use.

Following this introduction, Section II presents contents about software process improvement. Section III presents the strategy in layers to define, evaluate and improve processes. Section IV presents

T. Sobh (ed.), *Innovations and Advances in Computer Sciences and Engineering*,
DOI 10.1007/978-90-481-3658-2_48, 

the proposed approach and the results of the first experience. Section V finally concludes the paper.

## II. SOFTWARE PROCESSES IMPROVEMENT

In this work, we will adopt the definition of ISO/IEC 12207 [7] to software process: a set of interrelated activities which transforms inputs into outputs to product software.

However, the organizations need to consider, rigorously, some aspects to obtain the expected results with the definition, implementation and improvement of software processes

We can emphasize the following aspect: The process and the responsibilities to execute the activities should be clearly defined. Besides, it is fundamental to represent the process with a very good level of usability, because it helps to institutionalize the process on the organization.

Adequate methods and techniques to support the activities is also an aspect very important to consider. They may facilitate the execution of the activities and increase the performance of the process and the project's productivity.

Also, a clear definition of the inputs and outputs (work products) of the activities and the support of automated tools were identified as important aspects.

In the face of the complexity involved in software processes, that are influenced by technical, cultural and environmental factors, some standards, models and other approaches to guide the organizations to define and improve processes were created or evolved [4] [5] [7] [8], [9], [10], because some aspects must be addressed rigorously.

Furthermore, the success of an Improvement Program depends of some factors that must be considered adequately. For example: (i) provide sufficient resources [11]; (ii) analyze, frequently, the return on investment [12]; (iii) exist a long stated period compromise with the investment on process improvement [13] (iv) customize the Improvement Program to the organizations characteristics [11]; (v) adjust the improvements objectives to the business strategies objectives [14]; (vi) not consider only technical factors [11]; (vii) invest on the human resources qualification [15]; (viii) obtain the engagement of all collaborators [11]; (ix) take advantage of the organization's existents knowledge [11]; (x) provide support of knowledge management approaches [11].

## III. STRATEGY IN LAYERS TO DEFINE, EVALUATE AND IMPROVE SOFTWARE PROCESSES

In 2006, a research group of COPPE defined the Strategy in layers to define, evaluate an improve software processes. Nowadays (Fig. 1), the strategy comprises three layers: external entity layer, organizational layer and processes execution layer.

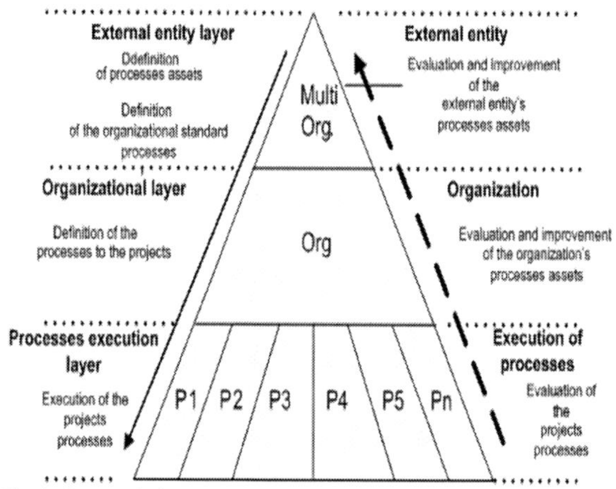


Fig. 1. Strategy in layers to define, evaluate and improve software processes

According to Fig. 1, the interaction of the three layers is collaborative and cyclical. The external entity has a set of processes assets which is used to define the organization standard processes. This set of assets needs to be always improved.

The organization defines the processes that will be executed on the projects, adapting the standard processes. After they have been adapted, they are executed.

The improvement of the processes also comprises the three layers. However, it is important to say that most of the data analyzed on the organizational layer come from TABA Workstation's tools, like: Avalpro [16], Acknowledge [17], Metrics [18] and AdaptPro [19]. TABA Workstation is an enterprise-oriented Process-centered Software Engineering Environment (PSEE) created to support the definition, deployment and software process improvement [20].

In relation of process improvement, the layers interact in the way presented bellow:

**(i) processes execution layer and organizational layer:** a set of data from the executed processes is analyzed on the organizational layer. They may be collected from the following sources:

- **processes adequacy evaluation:** encompasses the defined process evaluation, which always occurs at the end of an activity execution (supported by Avalpro [18]) and the organizational processes evaluations, executed when is necessary. Three types of adequacies are considered: "training adequacy, "tools adequacy" and "templates adequacy");

- **processes adherence evaluation:** periodically performed during the processes execution. It tries to identify when the activities are not been executed like had been defined. Always when non-conformities are identified, an action plan should be defined;

- **work products adherence evaluation:** every time a work product is created or modified, the evaluation is executed;

- ***post-mortem* analysis:** evaluation that is performed at the end of the projects. Its objective is to identify strengths, weakness and lessons learned of the project (supported by Avalpro [16]);

- **processes monitoring indicators:** these indicators are produced from the monitoring measures collected during the processes execution (supported by Metrics [18]);

- **lessons learned:** more important processes lessons learned (supported by Acknowledge [17]);

- **guidelines:** guide the execution of the processes activities. They are almost always updated and sometimes are incorporated to the processes (supported by Acknowledge [17]);

- **processes changes rationales:** if some modifications are done when the standard process is been adapted to the defined process, the project manager must justify it (supported by AdaptPro [19]);

- **processes changes demands:** anyone who has any kind of involvement with the processes may demand modifications. The demands are described on a specific template.

The results of official MPS.BR [4] and SCAMPI [5] assessments may contribute with the analysis too.

**(ii) organizational layer and external entity layer:** the processes problems identified on the organizational layer, their root-causes and improvements to be implemented, besides the results of MPS.BR or CMMI assessments are sent to the external entity. These data, from many organizations, are analyzed and improvements are identified to the external entity's assets.

Fig. 2 presents the interface of the strategy's layer

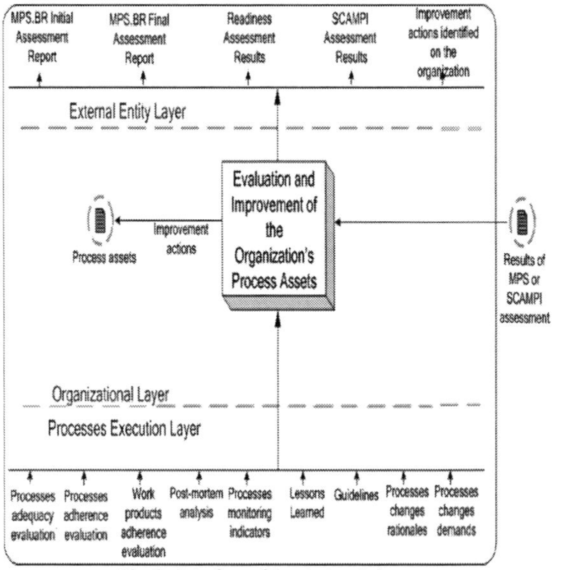


Fig. 2. Interface of the strategy's layers

## IV. APPROACH TO EVALUATE AND IMPROVE THE PROCESSES ASSETS ON THE ORGANIZATIONAL LAYER AND THE EXPERIENCE OF USE

The approach on the organizational layer comprises the following phases: (1) Identify improvements objectives; (2) Analyze data; (3) Identify improvements; (4) Analyze and prioritize improvements; (5) Implement improvements; (6) Define preventives actions; and (7) Incorporate lessons learned.

Bellow, the phases are mentioned, besides the approach's experience of use in a medium size company from Rio de Janeiro between March and August of 2007 is described. The objective of the experience was to find out evidences of viability and inadequacy aspects of the approach.

**Phase 1 - Identify improvements objectives:** the purpose of this phase is to identify the improvements objectives to the organization's processes. These objectives may be reach higher levels on a maturity model (vertical improvement) or make changes on the processes to improve the productivity, the adequacy to the organization or the performance (horizontal improvement), or both of them.

In this phase, the results of a meeting with the organization's directors were defined hierarchically the following objectives: (i) reach the level F of the MPS.BR maturity model; (ii) improve the processes adequacy and (iii) enlarge the use of the processes in all the organization.

**Phase 2 - Analyze data:** the purpose of this phase is to identify the problems that must be solved, because are making difficult the organization achieve their improvements objectives. As the problems are organizational, it is necessary to analyze data from more than only one project.

In a meeting, one of the organization's directors and the coordinator of the Improvement Program defined the following business objectives: (i) reduce the level of rework, that is an historical problem and (ii) reduce the projects costs to increase the organization's profits.

At the same meeting, were defined the following product quality objectives to the organization: (i) Improve the usability, because the company develops websites; and (ii) Increase the products level of reliability, reducing the quantity of defects.

From the objectives described above, the Improvement Program coordinator defined and prioritized the following processes, which the data would be analyzed on the actual improvement cycle: Project Management, Requirements Management and Measurement.

Then, three types of problem were established to be considered: (i) adequacy: related to the level of adequacy of the processes or activities ("Training Inadequacy", "Tools Support Inadequacy" and "Templates Inadequacy"); (ii) usability: related to the difficulty to understand the description of the activities e (iii) relevance: related to the execution of not necessary activities.

Indeed, three projects and the following sources of evidence were selected: (i) processes adequacy evaluation; (ii) processes adherence evaluation; (iii) post-mortem analysis; (iv) processes monitoring indicators; (v) lessons learned; (vi) processes changes rationales; and (vii) processes changes demands.

At first, we analyzed the obtained results in the processes adequacy evaluation, using the Matrix to Analysis of Problems, created according with the qualitative analysis approach named "Content Analysis". The data were registered on TABA Workstation.

Then, the following problems were selected to be considered in the actual improvement cycle: (i) Project Management: Tools Support Inadequacy and Templates Inadequacy; (ii) Requirements Management: Training Inadequacy; and (iii) Measurement: Tools Support Inadequacy.

Finally, the problems were presented to the company's directors, aiming the commitment of them.

**Phase 3 - Identify improvements:** the purpose of this phase is to identify the improvements to be implemented on the organization processes assets. For this, we decide to use a collaborative approach. The participants were the employees that somehow interacted with the processes.

To facilitate achieve the objectives we used pre-defined cause-effect diagram to the problems that would be considered on the meetings. There, the root-causes of the problems, lessons learned and the following improvements opportunities were identified and selected to be considered in the actual cycle:

**Project Management:** (i) promote trainings on MS-Project and (ii) define action plans only to high relevant facts.

**Requirements Management:** (i) define an activity to obtain the approval of use cases from the developers, test analyst and quality assurance analyst, before they are implemented and (ii) promote trainings in requirements management also to the developers.

**Measurement:** implement others ways on the company to increase the culture of collect data.

**Phase 4 - Analyze e prioritize improvements:** the purpose of this phase is to analyze, prioritize and select the improvements to be implemented. Initially, to deepen the analysis, we applied a SWOT Analysis in all of the improvements to know factors that may facilitate or difficult the implementation of them.

After the SWOT Analysis, the coordinator of the Improvement Program prioritized the improvements using the Matrix to Prioritize Improvements. Although this matrix comprises nine criterions, we decide to use only five: (i) urgency to implement the improvement; (ii) impact of the improvement on the process performance; (iii) satisfaction of those who use the improved process; (iv) resources (financial, human and technology) necessaries to the improvement implementation; and (v) simplicity of the operations to implement the improvement. Each criterion was evaluated using a scale comprised of the following values 3, 5 and 7.

The improvement that obtained the higher level of priority was that related with the use cases (Requirements Management). The others obtained the same rank. Finally, the Improvement Program coordinator decided to implement all the improvements.

**Phase 5 - Implement improvements:** the purpose of this phase is to implement and institutionalize the improvements selected on the anterior phase, including the planning, execution and evaluation of pilot projects.

As this phase had already been evaluated by COPPE in experiences of processes implementation (consultancy) [21], it was not evaluated, but a report containing the results was elaborated to be sent to the external entity.

**Phase 6 - Define preventive actions:** the purpose of this phase is to define actions that may prevent the organization against imminent problems. For this, we decide to know the relations of the causes of the problems.

We chose to use the Matrix to Find-out Relationships, adapted from the Matrix to Discovery, suggested by Bacon and presented in [22]. Besides, we used circles to represent the strengths of the influences among the causes, similarly in the Matrix of Distances, illustrated in [23]. It aimed to facilitate the identification of influence zones.

Finally, we defined for most strength relations, the possible effects and preventive actions.

**Phase 7 - Incorporate lessons learned:** the purpose of this phase is to register the lessons learned captured during the execution of the phases, aiming to be reused in future situations. At this first experience, many lessons learned were captured and registered.

### *A. Synthesis of the improvements opportunities to the approach*

After the execution of each phase, we evaluated the results of the experience and identified a set of improvements opportunities to the proposed approach (Table I).

Table I
SYNTHESIS OF THE IMPROVEMENTS OPPORTUNITIES

| Phase | Improvements Opportunities |
|---|---|
| **Phase 2 – Analyze data** | • Know the perception of the collaborators related to the organization's product quality.<br>• Institutionalize a Client Forum, to discuss about the organization's product quality and identify problems of relationship. |
| | • Implement a tool on the TABA Workstation to consolidate automatically the results of the adequacy evaluations. |
| **Phase 3 – Identify improvements** | • Establish a guideline mentioning the possibility of a improvement begin to be considered on the actual improvement cycle, if it is relevant to the company, although it was not related to one of the processes selected to the cycle. |
| **Phase 4 – Analyze and prioritize improvements** | • Provide more than one approach to analyze and prioritize the improvements.<br>• Define different weights to the criterions of the Matrix to Prioritize Improvements. |
| **Phase 6 – Define preventives actions** | • Use risks that occurred on projects and the reports of the processes adherence evaluation to identify imminent problems.<br>• Include activities to define, execute and manage the preventive action plans. |

## V. CONCLUSION AND FUTURE WORKS

The results of the first experience demonstrated that the approach was effective, supporting the organization to identify and prioritize improvements. The phases guide the execution of the approach adequately, providing knowledge that helps the collaborators. Besides, the methods used also support the socialization of knowledge in the company.

However, a defined process more detailed was necessary to guide better the execution of the activities.

After these experiences, we may explore the following further works: (i) Define a more detailed process to evaluate and improve the processes assets in the organizational layer (with the improvements incorporated); (ii) Define and perform a formal case study to validate the defined process; and (iii) Develop a tool on the TABA Workstation to support the process.

## REFERENCES

[1] ARAUJO, E. E.. R., MEIRA, S. R. L., "Competitive Insertion of Brazil on the International Market of Software", available at http://www.softex.br/portal/_publicacoes/publicacao.asp?id=806.

[2] MCT/SEPIN, *Quality and Productivity on the Brazilian Software Sector – 2001*, Brasília (2002).

[3] ISO/IEC, "ISO/IEC 25000: Software engineering – Software product Quality Requirements and Evaluation (SQuaRE) – Guide to SQuaRE" (2005).

[4] SOFTEX, "Brazilian Software Process Improvement – General Guide version 1.2", available at http://www.softex.br/mpsbr.

[5] CMU/SEI, "CMMI for Development version 1.2.", CMU/SEI-2006-TR-008 (2006).

[6] NASSCOM, Quality Summit, available at http://www.nasscom.org/eventdetails.asp?id=148.

[7] ISO/IEC, "ISO/IEC PDAM 12207: Information Technology – Amendment 2 to ISO/IEC 12207" (2004).

[8] ISO/IEC, "ISO/IEC 15504-4: Information Technology – Process Assessment, Part 4: Guidance on use for Process Improvement and Process Capability Determination" (2004).

[9] KOMI-SIRVIÖ, S., *Development and Evaluation of Software Process Improvement Methods*, Espoo 2004, VTT Publications 535 (2004).

[10] MARTINS, V. M., DA SILVA, A. R., "ProPAM: SPI based on Process and Project Alignment", In: Proceedings of the IRMA International Conference, Vancouver, May (2007).

[11] DYBA, T., "Factors of Software Process Improvement Success in Small Organizations: An Empirical Study in the Scandinavian Context", In: Proceedings of the ESEC/FSE'03, 2003, pp. 148-157, Helsinki.

[12] ERDOGMUS, H. et al., "Return on Investment", IEEE Software, May/June, 2000, pp. 18-22.

[13] VARKOI, T., "Management of Continuous Software Process Improvement", In: Proceedings of the International Engineering Management Conference (IEMC '02), 2002, pp. 334-337, Cambridge, August.

[14] HEFNER, R., TAUSER, J., "Things They Never Taught You in CMM School", In: Proceedings of the 26th Annual NASA Goddard Software Engineering Workshop, 2001, pp. 27-29, November.

[15] CATER-STEEL, A. P., "Low-rigour, Rapid Software Process Assessments for Small Software Development Firms", In: Proceedings of the 15o. Australian Software Engineering Conference (ASWEC'04), 2004, pp. 368-377, Melbourne, April.

[16] ANDRADE, J. M. S., *Evaluation of Software Process in ADSOrg*, Dissertation of M.Sc., COPPE/UFRJ, Rio de Janeiro, RJ, Brazil (2005).

[17] MONTONI, M., *Knowledge Acquisition: an application on the software process*, Dissertation of M.Sc., COPPE/UFRJ, Rio de Janeiro, RJ, Brazil (2003).

[18] SCHNAIDER, L. et al., "MedPlan: an approach to measurement and analysis on projects of software development", In: Proceedings of the SBQS 2004, 2004, Brasília.

[19] BERGER, P., *Instantiation of Software Process on Configured Environment in the TABA Workstation*, Dissertation of M.Sc., COPPE/UFRJ, Rio de Janeiro, RJ, Brazil (2003).

[20] MONTONI, M. et al., "Enterprise-Oriented Software Development Environments to Support Software Products and Processes Quality Improvement", In: Proceedings of the 6th International Conference on Product Focused Software Process Improvement (PROFES 2005), 2005, pp. 370-379, Oslo, June.

[21] FERREIRA, A. I. F. et al., "ISO 9001:2000, MPS.BR Level F e CMMI Level 3: a software process improvement strategy on company BL", In: Proceedings of SBQS 2006, 2006, pp. 375-382, Vila Velha.

[22] MOLES, A., *The Scientific Creation*, São Paulo, Perspectiva Press (1971).

[23] MOLES, A., The. Sciences of the Imprecise, Rio de Janeiro, Civilização Brasileira Press (1995).

# A Multiple Criteria Approach to Analysis and Resolution of the Causes of Problems on Software Organizations

Francisca Márcia G. S. Gonçalves[1,2], Adriano Bessa Albuquerque[1], Daniele de Oliveira Farias[2], Plácido Rogério Pinheiro[1]

[1]University of Fortaleza (UNIFOR) – Masters Degree in Applied Computer Sciences (ACS)
Av. Washington Soares, 1321 - Bl J Sl 30 - 60.811-341 - Fortaleza – Ce – Brazil
[2]Serviço Federal de Processamento de Dados (SERPRO)
Av. Pontes Vieira, 832 – São João do Tauape - 60.130-240 - Fortaleza – Ce – Brazil

*Abstract-* **The methodologies used in analysis and resolution of problems vary widely according to the considered situations. Due to the fact they are generic, the implementation of the existent approaches is difficult when is necessary adapt it to a specific context, for example, in software organizations. Besides, when the problems are complex, this process becomes more critical. So, the use of multiple criteria approaches to support decision-making, specially, the causal analysis process is extremely useful. In face of this context, this paper proposes an approach based on multiple criteria to support the analysis and resolution of causes of problems, in software organizations. The purpose is to understand the relations between the causes and the results, to find out which causes are more relevant and, consequently, improve the results.**

## I. Introduction

The quality of the software organizations products and services determines the success in the actual competitive environment. In face of the transformations occurred on the last decades, because of the globalization and the continuous innovations, the actions predictability and the repetitiveness of the problems that already occurred is not possible anymore.

The technical complexity of the services of this kind of organizations demands, besides specialized people, techniques and processes to the resolution and analysis of the problems causes. The problems solutions using, only the experience, is no feasible. Problems more complex need deepest analysis using methodologies and techniques to obtain better results.

Aiming to execute their activities, the decision makers are sometimes in conflict, because the decision may encompass big risks.

In most of the times, a decision maker incorporates his intrinsic values, becoming a very personal and subjective decision. Probably, the decision will be vague and imprecise, almost always excluding relevant variables.

So, the use of approaches of multiple criteria decision analysis to support the causal analysis of these problems becomes extremely useful, because this methodology provides a better adaptation to practical decision context, permitting evaluate, integrally and using criteria, a lot of data, interactions and goals

This kind of analysis is important because may consider the alternatives of the actions according of many point of views.

In face of this situation, this paper presents an approach, based on multiple criteria decision analysis, to support the analysis and resolution of the causes of the problems in software organizations, aiming to understand the relations of the causes and the results, to find out and eliminate the more relevant causes and, consequently, improve the results.

Section II presents contents related to analysis and resolution of problems causes. Section III describes the multiple criteria decision analysis. Section IV presents the proposed approach to analyze and resolve the causes of the problems in software organizations.

T. Sobh (ed.), *Innovations and Advances in Computer Sciences and Engineering*,
DOI 10.1007/978-90-481-3658-2_49, 

Section V presents the initial results of an experience of use. And section VI concludes and presents further works.

## II. Analysis and Resolution of Problems Causes

According to HOSTANI [1], a problem is the difference between the actual situation and the ideal one.

In the context of process, problem may be seen as a lack of conformity to the defined standards or a management situation that needs to be improved [2].

However, some problems are considered complex, because they need a big effort to structure them, in other words, are those where many people have to be involved on the decision and have a lot of subjectivity, with many factors to be thought about [3].

When a problem is identified, the analysis to find out its causes begins. KEPNER and TREGOE [4] said that in each stage of the analysis of the problem, the informations began to appear while the process goes in direction of what is wrong, being identified the problem, then, the possible causes and finally, the more probable cause. After this, corrective actions related to the problem should be defined.

## III. Multiple Criteria Decision Analysis

The organizations have many complex decision problems, mainly, management problems. The majority of the real situations are characterized by the existence of many objectives or "desires" to be achieved.

When the decision is intrinsically related to multiples point of views (criteria) and the choice of any alternative depends of the analysis of different point of views or "desires", it is a multiple criteria problem.

So, as we know that decision-making is a hard task and that, normally, the decision should satisfy the conflict of objectives and criteria, do not accept the subjectivity may be a difficulty to the problem solution.

If it is truth, that searching to the objectivity is an important worry, it is crucial do not forget that decision-making is before anything, a human being activity, sustained by the notion of value, so, the subjectivity is omnipresent [5].

In this way, the multiple criteria approaches are very important, mainly on the cases where exists conflicts between the decision makers, or when the perception of the problem by all the involved actors is still not totally consolidated [6].

The multiple criteria approaches try to reduce the subjectivity on decision-making, in face of multiple criteria using mathematical calculus. Their objective is to help the decision makers to analyze the informations and look for a better strategy of management.

According to ENSSLIN et al. [6], all of the existent multiple criteria approaches have two basic purposes: support the process of choice, ordination or classification of potential actions and incorporate multiple aspects on this process.

The methodology used on this work was MCDA - Multiple Criteria Decision Aid, which was created aiming to prioritize the human factors on analysis of complex problems.

We used the software Hiview to implement this methodology. This software permits the decision makers, define, analyze, evaluate and justify their preferences related to the alternatives, based on their attributes. It helps to structure a problem, in a simple way, specifying the criteria to be used on the choice of the alternatives and attributing importance levels (weights) to them.

The alternatives are evaluated comparatively to each criterion and a value of preference is attributed to each one of them and to each criteria. Besides, it permits change the judgments and compare graphically the obtained answers, permitting the actors involved on the decision, to rethink and, if necessary, to ratify the recommended decision.

The Hiview, when used on a process to support a decision-making, permits the evaluation of models obtained from multiple criteria methodologies, which use an additive aggregation function, like the MACBETH methodology, used on this work.

## IV. Multiple Criteria Approach to Analyze and Resolve Causes of Problems in Software Organizations

The proposed approach is structured using the PDCA Cycle, defined by Deming. The motivation was to help the teams to solve problems using a well known methodology. Each phase of the PDCA cycle encompasses steps and sub-steps which describe what the individuals have to do. The proposed phases that require a major effort are: Plan (P) and Do (D), because they are more critical and complex.

### A. Phases of the approach

**Plan:** the objective of this phase is define a plan to solve the considered problem (Table I).

Table I
STEPS OF PHASE "PLAN"

| Step | Description |
|---|---|

| | |
|---|---|
| **P1 – Describe the problem** | Comprises collect useful and necessary informations about the problem to help the next steps, such as: historical; moment when it was identified, frequency of occurrence, impact, restrictions, sources and categories. |
| **P2 – Make the team** | Comprises select the members of the team responsible to analyze and solve the causes of the problem. For this, the following role should be defined: (i) Owner; (ii) Facilitators; (iii) decision makers; and (iv) Participants. |
| **P3 – Define the goals** | Comprises define the goals to be reached with the problem resolution. |

**Do:** on this phase (Table II) are conducted activities, according with the planned on the anterior phase. The multiple criteria approach is used on this phase (*Sub-step D1.3* to *Sub-step D1.5*).

Table II
STEPS OF PHASE "DO"

| Step | Description |
|---|---|
| **D1 – Define the causes of the problem** | **Sub-step D1.1: Investigate the potential causes of the problem:** comprises investigate what are the potential causes of the problem, performing brainstorming sessions with the involved people, using for this, cause and effect diagrams. |
| | **Sub-step D1.2: Consolidate the results:** comprises list all the identified potential causes and group them, to eliminate the less probable causes. Paretto graphics also should be constructed to improve the visibility of the obtained results. |
| | **Sub-step D1.3: Identify the elementary causes:** comprises construct the point of view tree. |
| | **Sub-step D1.4: Evaluate the causes:** comprises compare the many causes of the problems, to choose the more conscious and adequacy. It is necessary prioritize the causes for each category (criteria) and attribute values to each pair. |
| | **Sub-step D1.5 – Analyze the results:** comprises analyze to support the evaluation of the results. Some kinds of analysis should be made, like: sensibility analysis, comparative maps and others graphical analysis, available on the tool (Hiview), if necessary. |

| | |
|---|---|
| **D2 – Analyze the possible solutions related with the causes of the problems** | Comprises make a brainstorming session to capture the solutions that are related with the causes of the problem. |
| **D3 – Define and execute an Action Plan** | Comprises define a plan with the necessary actions to deal with the problem and execute these actions. |

**Check:** on this phase is performed the management of the action plan. Besides, from the analysis of the data collected on the anterior phase, the actual results are compared with the established goals (Table III).

Table III
STEP OF PHASE "CONTROL"

| Step | Description |
|---|---|
| **C1 – Manage the action plan and evaluate the results** | Comprises monitor the results, aiming of evaluate if the actions are obtaining the desired effects and, after the execution of the actions, analyze if the problem was solved and if the goal was reached. The results are monitored during a long time to observe if there is any kind of recurrence. |

**Act:** on this phase, the obtained results are communicated and the lessons learned are disseminated and becomes an organizational asset (Table IV).

Table IV
STEPS OF PHASE "ACT"

| Phase | Description |
|---|---|
| **A1 – Communicate the results** | Comprises communicate the obtained results and the lessons learned to all collaborators of the organization. These informations may be useful to deal with similar problems in others projects, helping to be proactive. The problems which were not solved and the goals not reached with their respective analysis also should be communicated. |
| **A2 – Incorporate lessons learned** | Comprises analyze the identified lessons learned and evaluate the needs of changes of the organizational processes, guidelines and policies. The obtained gains should be shared to all of the organization. |

## V. Experience of Use

The proposed approach was applied in a Brazilian federal public organization. The company provides information technologies and communications services to the public sector and is considered as one of the biggest on this sector, in the Latin America.

The company has a process architecture that guides the organization on the development of Information Technology solutions. It was constructed using some important references and models, like the RUP (Rational Unified Process) and CMMI. The structure of the architecture comprises macro-activities. One of them is named Decision Analysis and Resolution, which has the purpose to support the decision-making.

However, this macro-activity was defined to deal with any kind of decision and it is neither only related to solve problems nor consider the level of complexity of the problems.

*B. A general description of the experience*

On the public sector is usual the occurrence of complex problems, because, normally, there are: conflict of interests, concentration of power, aspects related with low motivation, centralized decisions and incentive policies to the employees. In this situation, a decision-making is a hard task that requires knowledge, confidence and coherence.

In this way, is necessary the utilization of an approach to support the managers on the analysis of the causes of complex problems, becoming more coherent and realistic.

Bellow, we present, briefly, the results of the proposed approach application.

- **Plan.**

**STEP P1 – Describe the problem:** aiming to control the productive effort of the collaborators, the company monitor the worked hours using the software SGI. Each collaborator should register the hours effectively worked according to the rules defined in a document named "Systematic of appropriation". Appropriation is a term used by the organization that means the registration of the productive hours (really worked) of the collaborators. The indicator "% of appropriation in projects by team" was established to monitor the projects and is collected since January of 2007. Then, were created others two indicators to help the control of the appropriation: (i) "% of appropriation in services by team", to inform how much of the effort was related to services and (ii) "% of inoperativeness by team" to inform how much of the effort was not appropriated. This proposed approach will be used to support the following problem: "High percent of inoperativeness on Team 1". This problem has been identified since January of 2008. The indicator was inside the limit control only in February and from this month the problem presents a tendency of growth. The proposed approach was used on the solution of these problems, which were classified on the following categories: people, process, organization and technology.

**STEP P2 – Make the team:** we analyzed the roles that influenced directly or indirectly on the results, or might contribute with informations, suggestions and knowledge. All the collaborators which performed the following roles were members of the team: Senior Manager, Chief of Division, Coordinator of the Process Group, Coordinator of Measurement and Analysis, Measurement Analyst, Coordinator of Quality Assurance, Project Leader and Developers.

**STEP P3 – Define the goals:** the details of the indicator's goal are presented on Table V. It was defined by the facilitator and the decision makers.

Table V
DETAILS OF THE GOAL

| **Goal: % of inoperativeness on Team 1 inside the control limits (10% a 20%)** | |
|---|---|
| **Sub-goal** | **Time Limit** |
| **SG1** – Reduce 15% of the inoperativeness. So, the result of the indicator in October should be at most, 28%. | Oct/08 |
| **SG2** – Reduce 8% of the inoperativeness, in such a way that the result of the indicator in November will be inside the control limits (at most 20%). This percentage may be adjusted according to the results obtained in September. | Nov/08 |

- **Do.**

**STEP D1 – Define the causes of the problem**

**Sub-step D1.1 – Investigate the potential causes of the problem:** the facilitator performed an individual brainstorming session with 15 collaborators involved directly on the problem, using a cause and effect diagram divided on the four categories defined on Step P1. When the facilitator handed the cause and effect diagram model out to each member of the selected team, he explained the objective of the brainstorming and taught how they should utilize the diagram. On this sub-step, the facilitator observed that the main difficult of the team was to classify some causes, since they may be associated to more than one category. So, we guide them to classify analyzing which category was more strongly related to the cause. After all the participants had identified the causes of the problem, we listed those that were inserted on the diagrams: (i) 34 causes related to people, (ii) 16 related to processes, (iii) 6 related to organization, and (iv) 22 related to technology.

**Sub-step D1.2 - Consolidate the results:** after listing all the identified causes, the facilitator groups the repeated or similar causes to consolidate the results. Firstly, the groups were numbered to maintain the traceability of the group and to know how many causes were identified. Then, the groups were revised by the collaborators that would go to decide and, finally the results were consolidated.

**Sub-step D1.3 - Identify the elementary causes:** we constructed a cause and effect diagram to support the selection of the elementary causes. On this diagram, we highlighted the causes considered that had more impact on the problem (Fig. 1).

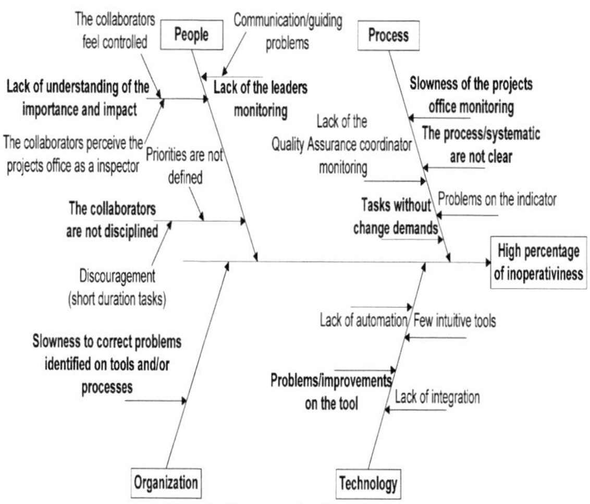


Fig. 1. Cause and effect diagram

**Sub-step D1.4- Evaluate the causes:** on this sub-step, the facilitator had a meeting with the decision makers, aiming to construct the matrix of value judgment and obtain the scales of local preference to each of the fundamental point of view (categories). They followed the MACBETH methodology and the procedures of questions [6]. During the construction of these matrixes, occurred some cases of cardinal inconsistence, solved through discussions, changing, consequently, some of the judgments. Fig. 2 presents an example of a matrix.

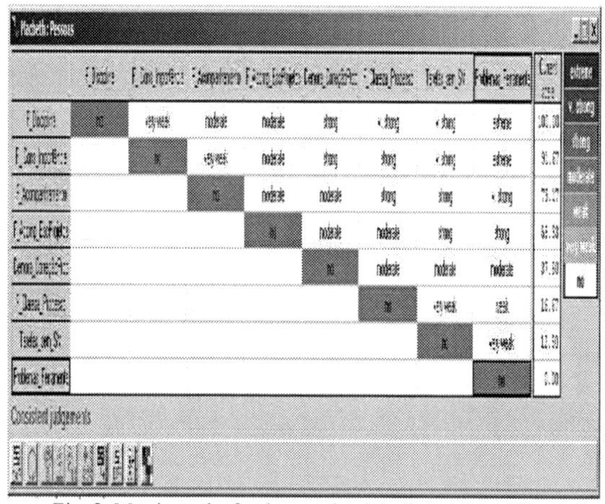

Fig. 2. Matrix to the fundamental point of view from "People"

After the construction of the matrixes to each fundamental point of view, we constructed the matrix of value judgment to the problem (inoperativeness) (Fig. 3). This matrix has the purpose to identify the relevance of each criterion, resulting in a scale of weights (column Current Scale).

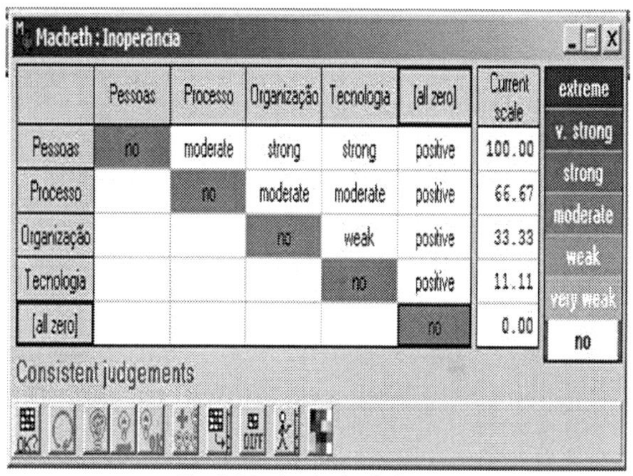


Fig. 3. Matrix of value judgment

**Sub-step D1.5: Analyze the results**: the decision makers analyzed the results to verify the need to modify the values obtained in some option. For this, we used the sensibility analysis of the Hiview tool, generating a graphic to each criterion. It permits analyze both the variations of the total value attributed to the actions and the consequences of these variations. The Fig. 4 presents an example.

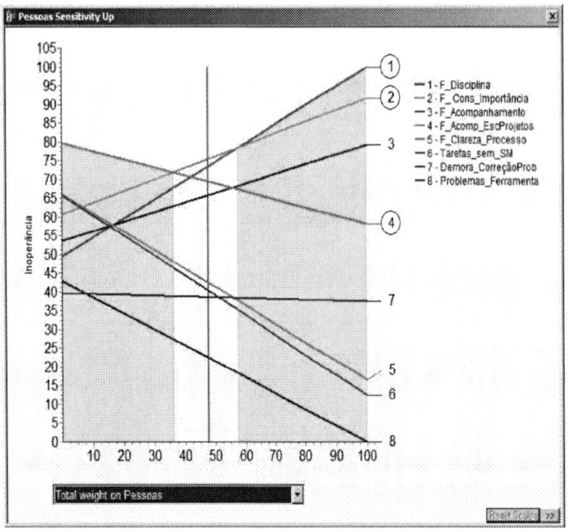


Fig. 4. Graphic of sensibility analysis

The axis X represents the index of criteria relevance. The vertical line parallel to the axis Y (value 47.6 on the axis X) indicates the level of relevance of this criterion to the organization. The axis Y represents the index of relevance of the actions.

**STEP D2 - Analyze the possible solutions related with the causes of the problems:** after the analysis of the results, the facilitator and the decision makers performed a brainstorming session to list the possible solutions to the problem. On this moment, 68 suggestions of all involved (individual brainstorming) were analyzed.

**STEP D3 – Define and execute an action plan:** On this step an action plan was defined with the following informations: (i) What to do; (ii) Why to do; (iii) Who will do; (iv) Where to do; (v) How to do; and (vi) When to do.

The others phases have not been executed completely yet, because of the time, but in some weeks we will execute them and analyze the results.

## VI. CONCLUSION AND FURTHER WORKS

Analyzing the execution of the phases "Do" and "Execute", the approach can be considered adequate. However some difficulties were met: (i) in the sub-step **D 1.1 - Investigate the potential causes of the problem**, sometimes the facilitator did not get to do the individual brainstorming session, because the participants could not participate. Nevertheless, this was better than have only one brainstorming with all the team, since with the impartiality of the facilitator, the participants felt more comfortable and presented informations that de decision makers still did not know. Besides, the informations captured during the construction of the diagrams were very useful to the understanding of the perceptions and difficulties of each participant.

In relation to the sub-step **D1.2 - Consolidate the results**, the facilitator met difficulties to group some causes, because they were related to the same group. However, firstly consolidate the results to be then reviewed by the decision makers was advantageous because the execution of this step became more agile.

The others steps/sub-steps were executed without any kind of relevant difficulties.

The Hiview tool proved to be adequate, eliminating mathematic calculus, optimizing time, permitting richer analysis with rapid construction of graphics and supporting a powerful sensibility analysis.

Finally, the support of the owner was essential both to provide resources and to stimulate the commitment of the involved collaborators.

On the future we will execute the others phases of the approach and analyze them. Besides, we intend to execute the same approach on other project with a different problem to compare the results and try to have some important insights.

## REFERENCES

[1] HOSOTANI, K., *The QC Problem Solving Approach Solving Workplace Problems the Japanese Way*. 3A Corporation, Tokyo, Japan (1992).

[2] SMITH, G.F., *Quality Problem Solving*. Milwaukee: ASQ Quality Press (1998).

[3] CHURCHILL, J., *Complexity and Strategic Decision-Making*. London: Sage Publications (1990).

[4] KEPNER, C. H., TREGOE, B. B., *The rational administrator – a systematic approach to resolve problems and to making-decision*. São Paulo: Atlas (1981).

[5] BANA AND COSTA, C. A.; CHAGAS, M. P., "A career choice problem: an example of how to use MACBETH to build a quantitative value model based on qualitative value judgments", European Journal of Operational Research, vol. 153, no. 2, 2004, pp. 323-331.

[6] ENSSLIN, L.; ENSSLIN, S. R. ; DUTRA, A., "MCDA: A Constructivist Approach to the Management of Human Resources at a Governmental Agency". International Transactions in Operational Research, United States, v. 7, 2000, pp. 79-100.

# A Secure Software Development Supported by Knowledge Management

Francisco José Barreto Nunes, Adriano Bessa Albuquerque

University of Fortaleza (UNIFOR) – Masters Degree in Applied Computer Sciences (ACS)
Washington Soares Avenue, 1321 - Bl J Sl 30 - 60.811-341 - Fortaleza – Ce – Brazil

*Abstract*- **Organizations that want increase their profits from reliable and secure software product need to invest in software security approaches. However, secure software is not easily achieved and the actual scenario is that investments in software development process improvement do not assure software that resist from attacks or do not present security vulnerabilities. The PSSS may help obtaining secure software as it proposes security activities to be integrated into software development life cycles. This paper resumes the application of the PSSS and proposes the support of a knowledge management environment based, specially, on security inspections of the artifacts generated during the processes execution. This will improve how the security aspects are being considered during the development of secure software, will make more effective the use of PSSS in producing secure software and will help to establish the security as a important aspect on the organizational culture.**

## I. Introduction

The growing need for software products to support business processes has motivated considerable research in the improvement of software development processes. In this sense, information security and security engineering increase their importance to become part of the business processes and the systems supporting these processes, in order to protect corporate assets and information. According to CERT [1], software security defects are the main concerns that security professionals deal with.

The principle: "Security design and implementation" as a characteristic of information systems was stated by [2].

Information security is related to many models and standards, like SSE-CMM [3], ISO/IEC 15408 [4], ISO/IEC 27002 [5], and OCTAVE (The Operationally Critical Threat, Asset, and Vulnerability Evaluation) [6]. This paper proposes the Process to Support Software Security (PSSS) based initially on the activities derived from these models and standards. In addition, this paper describes briefly the results of the application of this process in a software development project, and finally, proposes an approach based on security inspections and knowledge management to support the PSSS.

Following this introduction, Section II describes briefly the security models and standards, Section III describes the activities of the PSSS, Section IV analyzes the results of an experience of use of PSSS; Section V describes contents related to knowledge management and the proposed approach to support the PSSS execution, based on security inspections and knowledge management. Section VI presents the conclusions of this paper.

## II. Information Security Models and Standards

### *A. SSE-CMM*

SSE-CMM is a process reference model that describes security features of processes at different levels of maturity. The scope encompasses:

• The system security engineering activities for a secure product or a trusted system addressing the complete lifecycle;

• Requirements for product developers, secure systems developers and integrators, organizations that provide computer security services and computer security engineering;

• Applies to all types and sizes of security engineering organizations.

A previous version of SSE-CMM was adapted and became ISO/IEC 21827 [7].

### *B. ISO/IEC 15408*

ISO/IEC 15408 presents a set of criteria to evaluate the security of products. This standard claims that a

T. Sobh (ed.), *Innovations and Advances in Computer Sciences and Engineering*,
DOI 10.1007/978-90-481-3658-2_50, 

development process should include security in the development environment and security in the developed application.

It presents a set of requirements that should be satisfied to make software more secure: security functional requirements and security assurance requirements.

*C. ISO/IEC 27002*

ISO/IEC 27002 aims to preserve information confidentiality, integrity and availability in such type of business scenario.

*D. OCTAVE*

According to ALBERTS et *al.* [6], OCTAVE is a comprehensive, systematic, context driven, and self-directed approach, that requires a small, interdisciplinary analysis team of business and information technology personnel from the organization to lead its evaluation process.

## III. Process to Support Software Security (PSSS)

The PSSS was designed to follow the iterative and incremental life cycle approach. Its activities may be adapted to function effectively within the organizational development process. It is an important aspect to have each activity as integrated as possible into the life cycle phases and one approach to reach this integration is to apply each activity in parallel with the software development life cycle phases.

The PSSS considers 37 activities grouped in a set of 11 subprocesses (Fig. 1).

The subprocesses are described bellow.

- **Plan security:** The Security Engineer identifies the security objectives of the software to be developed and prepares the project security plan. The activities include: Developing security plan; Planning processing environments; Planning security incidents management.
- **Asses security vulnerability:** The Security Engineer identifies and describes, for each iteration, the system security vulnerabilities related to the environment where the system would operate. In order to identify effectively the security vulnerabilities, the Security Engineer and the Software Engineer should organize interviews with executives, managers, and operational staff. The activities include: Identifying security vulnerabilities; Analyzing identified security vulnerabilities.

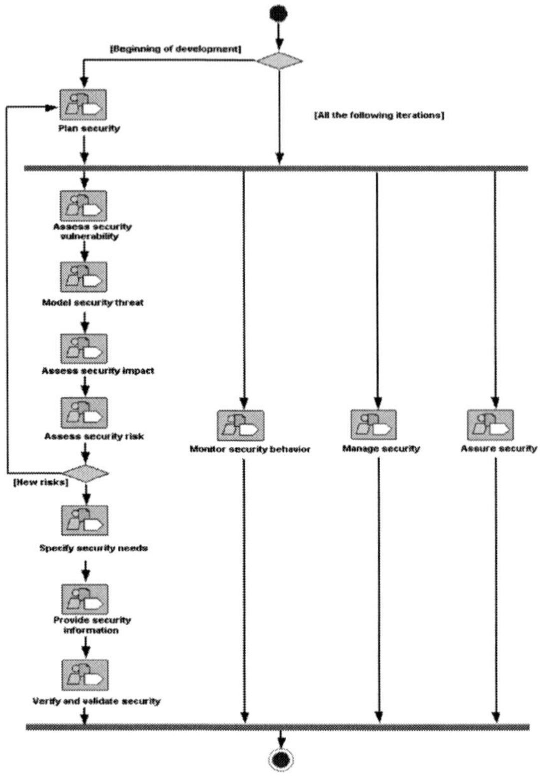


Fig.1. Process to support software security

- **Model security threat:** The Security Engineer is responsible to execute security threat identification and implement abuse cases and attack trees. Then, he selects an adequate approach to classify these threats and organizes interviews with executives, managers, and operational staff to develop strategies to reduce the impact of these threats. The activities include: Identifying security threats; Classifying security threats; Developing strategies to reduce security threats.
- **Assess security impact:** The Security Engineer, Software Engineer, and users (customer) are responsible to prioritize the critical activities influenced by the software. Security Engineer and Software Engineer review software's security artifacts. Based on the previous information and information about security vulnerabilities and threats, the Security Engineer and users identify security impacts from unwanted incidents and detailed information about these impacts. The activities include: Treating critical activities for security; Reviewing system security artifacts; Identifying and describing security impacts.

- **Assess security risk:** The Security Engineer and Software Engineer, with the help of the user, identify security exposure by evaluating and prioritizing security risks. The activities include: Identifying security exposure; Assessing security exposure risk; Prioritizing security risks.
- **Specify security needs:** The Security Engineer, with the help of the user and of the customer, is responsible for understanding customer's security needs and for developing a high-level security view of the software. This high-level view helps to define the security requirements. Finally, the Software Engineer mediates between customers and Security Engineer to obtain agreement about the security requirements. The activities include: Understanding customer security needs; Capturing a high-level security view of the system; Defining security requirements; Obtaining agreement about security requirements.
- **Provide security information:** The Security Engineer acts as a security mentor to the project team identifying necessary information and making it available. For example, architects could need information about security in Web services. In this case, the Security Engineer interacts with Software Architects to give the information. Security information would be considered kind of information which has impact on, is necessary to support, or helps members of a software security project. The activities include: Understanding information security needs; Identifying security constraints and considerations; Providing security alternatives; Providing support requirements.
- **Verify and validate security:** The Security Engineer, with the help of the Software Engineer, defines security verification and validation approach that involves plan elaboration, scope, depth, and tests. Then, the Security Engineer, with the help of the quality assurance team, performs the security verification and validation, reviews and communicates the results. The Security Auditor assesses the security verification and validation to check whether the activities are being performed correctly. The activities include: Defining security verification and validation approach; Performing security verification; Performing security validation; Reviewing and communicating security verification and validation results.
- **Manage security:** The Security Engineer deals with and controls additional security services and system components. The Security Engineer identifies training needs and educational programs about information security and about the process to support software security. Finally, he manages the implementation of security controls in the software being developed. The Project Manager helps the Security Engineer with the management of all the activities of the PSSS. The activities include: Managing security services and components; Managing security training and education programs; Managing the implementation of security controls.
- **Monitor security behavior:** The Security Engineer, with the help of the Software Engineer, is responsible for the following activities: Analysis of events with security impact; Identification and preparation of the incidents response; Monitoring changes in environments, in security vulnerabilities, threats, and risks, and in their characteristics; and Reviewing software security behavior to identify necessary changes. The Security Auditor assesses the activities described below to identify irregularities and problems. Besides that, he is responsible for: Reassessment of the changes in environments and in security vulnerabilities, threats, and risks; and Performing security audits. The activities include: Analyzing events with security impact; Identifying and preparing the answer to relevant security incidents; Monitoring changes in security threats, vulnerabilities, impacts, risks, and environment; Reviewing system security condition to identify necessary changes; Performing security audit.
- **Assure security:** The stakeholders should receive information to assure that their expectations are satisfied in relation to the effective application of the PSSS and the security of the software being developed. Besides that, the Security Engineer and the Software Engineer, with the help of the customer, should develop a strategy to guarantee the maintenance of security assurance. The Security Auditor executes an impact analysis based on security changes to assure that no change compromises software security. Finally, the Software Engineer and the

Security Auditor control the security assurance evidences that confirm this maintenance. The activities include: Defining security assurance strategy; Performing security change impact analysis; Controlling security assurance evidences.

## IV. APPLICATION OF THE PSSS

The PSSS was applied in a software development project of an access control and audit web-based system that, among other functions, defines and controls user access and registers the actions of these users.

Firstly, each activity was described and explained to the principal stakeholders, then, evaluated to verify if they could be really applied to the project and, afterwards, a simple version of a security plan was prepared.

Next, the security team with the help of the Software Engineer identified and analyzed security vulnerabilities and identified security threats and needs.

With these security needs and with the help of abuse cases and attack trees, it was possible to identify a set of security requirements: (i) Prevent the creation of unsafe passwords; (ii) Prevent wrong system access and (iii) Prevent the change of registered audit information.

The main problems identified were:

- Lack of knowledge to implement in its entirety the activities related to threat modeling;
- Insufficient time for the teams to get used to the PSSS and its activities;
- Need for additional resources to implement effectively the PSSS.

The main advantages gained by applying and following the PSSS were:

- Software security was considered during the system development which included the definition of security activities and artifacts;
- Identification and definition of security requirements;
- Limited project resources were effectively applied based on security assessments and major negative security impacts.

## V. THE KNOWLEDGE MANAGEMENT APPROACH PROPOSED TO SUPPORT THE PSSS

One of the improvements to be implemented in the PSSS is the execution of security inspections using checklists. These inspections are to be applied in artifacts produced during software processes (including tests activities) and the checklists should be suitable to the kind of artifact. This initiative may diminish the amount of security non conformities and vulnerabilities and help to consolidate the value of security on the organization's culture.

Fig.2 presents an example where the security inspections may be performed.

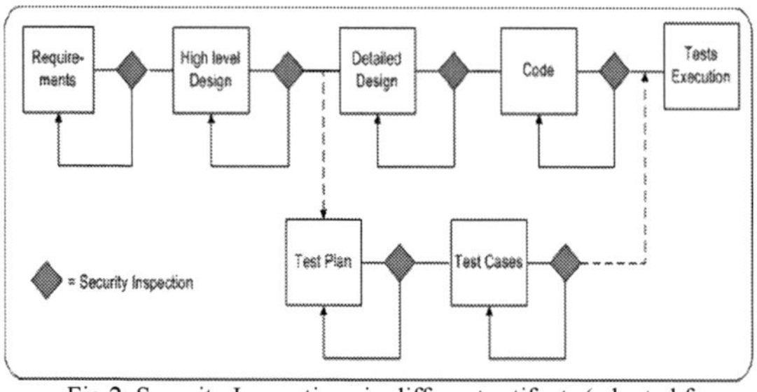


Fig.2. Security Inspections in different artifacts (adapted from [8])

According to GILB and GRAHAM [9], inspections increase the productivity between 30% and 50% and reduce the time of development between 10% and 30%.

BOEHM and BASILI [10] said that software inspections capture nearly 60% of the artifact's defects.

Applying organized inspections may produce others benefits [8]: **(i) Learning** - experienced inspectors may detect defect patterns and define best practices to the organization; **(ii) Integration of the inspections and the defect prevention** - know where and when defects occur may help to establish contingence plans; **(iii) Intelligible products:** the authors, knowing that their artifacts will be inspected, will produce artifacts easier to understand; **(iv) Defective data help to improve the software process defined to the project:** analyzing the root-causes of found defects is possible to modify the process aiming their occurrence again.

However, to support these inspection is important the development of a knowledge management environment, because the act of improve a process is directly related to the act of learn.

Improve practices of software is create new knowledges or improve ancient knowledges and institutionalize on the software organization [11].

According to [12] knowledge is information combined with experience, context interpretation and reflection. It is a valorous kind of information which is ready to be used in decisions and actions.

A Knowledge Management Environment should be composed of an organizational memory, with a lot of knowledge bases, activities and tools to support the activities. Bellow, we will describe each one of the proposed activities.

In relation to the activity "Knowledge identification", it should be identified the knowledge related to security that should be part of the knowledge base. Considering the purpose of PSSS and the use of security inspections, the non conformities and lessons learned identified during the security inspections are the most important knowledge item that should be part of the knowledge base. Besides, the security problems identified during the projects also may be another item.

The activity "Knowledge acquisition" should be responsible to define the moment and the procedures to collect the knowledge and also to collect them.

However, it is important define the kind of knowledge item that will be stored on the organizational memory. They may be: guidelines, lessons learned, non-conformities, problems solutions, modification's rationale etc. [13, 14].

Besides, it is also very relevant to define the organization's sources of knowledge, like: interview, measurements, *etc*. On this approach, the security inspector will collect the knowledges during the security inspections and on the analysis post-mortem.

The "Knowledge construction" will be responsible to characterize, connect and package the acquired knowledge before being inserted into the knowledge base. All the knowledge captured on the inspections or analysis post-mortem will be used to define security baselines or guidelines (other kind of knowledge).

ALTHOFF et al. [14] considered relevant only one specialist characterize the knowledge, because it may improve the recover of knowledges acquired in similar projects. They also defined some kinds of relations that may be used to connect knowledges: related to, respond to, justify, question, generalize, suggest, pros and cons.

The knowledge dissemination refers to the presentation of the knowledge to a person, always when necessary and the spread of the knowledges, which were recently stored on the organizational memory. Intelligent agents may support the knowledge dissemination, however some requirements are fundamental to obtain good results, like: characterize formally the knowledge, personalize the presentation, link the knowledges to the processes activities and use an adequate media to speed and amplify the dissemination.

PROBST et al. [13] said that the best solution to the knowledge dissemination is the hybrid, joining the technology and the face-to-face contact, because may increase the appearance of new knowledges.

On this approach, initially, the "Knowledge dissemination", will be reactive. The user will search in the base, always when some knowledge is necessary, considering similar projects. One of the techniques that may be used to support the reuse of knowledge is Case-based Reasoning approach [15].

In the context of PSSS, the reuse of knowledge will depend if the projects will be categorized using characteristics which allow the identification of the level of similarity among these projects (type of paradigm, life cycle, *etc.*). This will guide the effective reuse of experiences, like: the identification of preventive actions during the security inspections and the identification of security methods suitable for specifics cases.

The knowledge utilization refers to the use of knowledges by the users. HENNINGER [16] said that the main problem on knowledge management is neither capture nor store knowledges, but use them to support the execution of current activities.

In relation to the "Knowledge utilization", the knowledges will be used not only in others inspections. For example, the knowledge captured in one inspection or analysis post-mortem may be used in a different moments, like on the risks definition during the planning of the project and on the definition of the software's requirements.

The knowledge maintenance is executed based on the users' evaluation. Analyzing the results, the knowledge manager may exclude obsolete or never used knowledge.

On this approach, the "Knowledge maintenance" will be performed, periodically, to assess the knowledge that is still important and useful. When a knowledge is identified as obsolete bys the auditor, this knowledge will be removed. He will also, periodically, try to find out the perception of the others security auditors in relation of the utility of the stored knowledges.

Fig.3 shows the representation of the integration between knowledge management and PSSS.

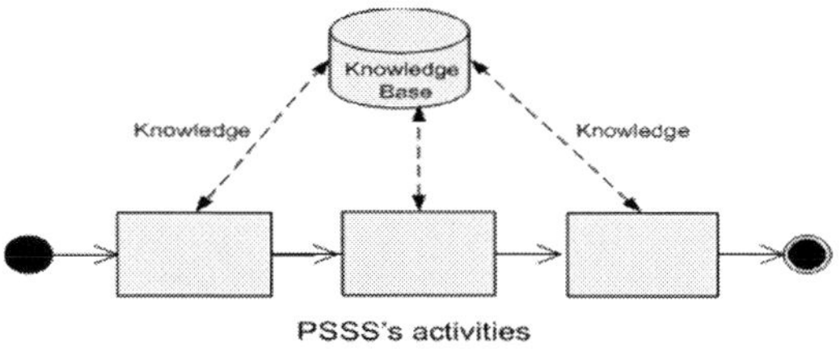


Fig.3. PSSS supported by Knowledge Management

## VI. CONCLUSION AND FUTURE WORKS

The PSSS could be seen as an important tool to improve the effectiveness of software security projects.

The development and organization of the PSSS tried to help in the production of more secure software as it protects the confidentiality, integrity, and availability of processed and stored information.

This research aims, initially, to define an approach to support the PSSS execution, based on security inspections and knowledge management.

The analysis of the security problems identified in many projects may help identifying security problems pattern of behavior organizational level). These informations may also produce baselines and guidelines to suggest best solutions related with security's methods and approaches into a software development process.

Another advantage of this approach is that on iterative development, knowledge acquired in previous iterations may be reused in the next ones.

In addition, whether the knowledge base for an organization is well organized and maintained, it may be used as infrastructure to implement a learning organization focused on software security.

On the future, a tool will be developed to support the execution of the PSSS activities and a knowledge management environment will be developed to be integrated to this tool, aiming to support the PSSS.

Finally, we will develop a security checklist to each important artifact generated on the development process and define the questions based both on the problems already identified on the organization and on the results of experimental studies related to security on software development found on the literature.

## REFERENCES

[1] CERT, "Coordination Center Statistics". available at: www.cert.org/stats/cert_stats.html.

[2] OECD, "Organisation for Economic Co-operation and Development. Guidelines for the Security of Information Systems and Networks: Towards a Culture of Security, page 13, Principle 7, Security design and implementation". available at: www.oecd.org.Center Statistics.

[3] SSE-CMM, "System Security Engineering – Capability Maturity Model, Version 3". available at: www.sse-cmm.org.ordination.

[4] ISO/IEC 15408-1, "Information technology – Security techniques – Evaluation criteria for IT security" (2005).

[5] ISO/IEC 27002, Information technology – Security technical - Code of practice for information security management (2005).

[6] ALBERTS, C. et al., "OCTAVE - The Operationally Critical Threat, Asset, and Vulnerability Evaluation", Carnegie Mellon – Software Engineering Institute. available at: www.cert.org/octave.

[7] ISO/IEC 21827, "Information technology - Systems Security Engineering - Capability Maturity Model" (2002).

[8] KALINOWSKY, M., SPINOLA, R., "Introduction to Software Inspections", Software Engineering Magazine, special edition, 2007, pp. 68-75.

[9] GILB, T., GRAHAM, D., *Software Inspection*, Addison-Wesley (1993).

[10] BOEHM, B. W., BASILI, V.R., "Software Defect Reduction Top 10 List.", IEEE Computer 34 (1), 2001, pp. 135-137.

[11] ARENT, J., NøBJERG, J., PEDERSEN, M. H., "Creating Organizational Knowledge in Software Process Improvement", In. Proceedings of the 2nd International Workshop on Learning Software Organizations (LSO 2000), pp. 81-92. (2000).

[12] MARKULA, M., "Knowledge Management in Software Engineering Projects", In: Proceedings of the 11th International Conference on Software Engineering and Knowledge Engineering, Kaiserslautern, Germany, June (1999).

[13] PROBST, G., RAUB, S., ROMBARDT, K., *Managing Knowledge - Building Blocks for Success*, John Wiley & Sons Ltd.(2000).

[14] ALTHOFF, K.-D et al., "Managing Software Engineering Experience for Comprehensive Reuse", In: Proceedings of the Eleventh Conference on Software Engineering and Knowledge Engineering, Kaiserslautern, Germany, June 1999, Knowledge Systems Institute, Skokie, Illinois, USA (1999).

[15] ALTHOFF, K-D., BIRK, A., VON WANGENHEIM, C. G., TAUTZ, C., *Case-Based Reasoning for Experimental Software Engineering*, IESE-Report No. 063.97/E, Fraunhofer IESE (1999).

[16] HENNINGER, S., 2001, Keynote Address: Organizational Learning in Dynamic Domains, In: Proceedings of the Learning Software Organization, 2001, pp. 8-16.

# Mobile Application for Healthcare System - Location Based

Sarin Kizhakkepurayil
Dept. of Electrical Engineering and Computer Science, Texas A&M University – Kingsville
Kingsville, TX-78363, USA
sarinkp@hotmail.com

Joon-Yeoul Oh
Dept. of Computer Information Systems, Texas A&M University – Kingsville
Kingsville, TX-78363, USA
kfjo000@tamuk.edu

Young Lee
Dept. of Electrical Engineering and Computer Science, Texas A&M University – Kingsville
Kingsville, TX-78363, USA
young.lee@tamuk.edu

*Abstract* - **Advancements in Information Technology are frequently adapted in the field of health-care services for better performance. Rapid evolution in positioning technologies lead the growth of mobile services to value-added, localized, personalized and time critical services offering connectivity at anytime, in anywhere. This paper proposes a Location Based Mobile Healthcare System Architecture and Mobile User Application integrated with latest technology in wireless communication; Personalization and customization of such services based on the Location Based Services and the profiles of mobile users. The proposed architecture and application allow the health-care providers and patients to connect across the globe. The movement of services provided to the patient from desktop platform to mobile technology can bring significant impact on health care services.**

## I. INTRODUCTION

Information Technology (IT) has been playing a key role in Telemedicine as providing efficient healthcare services, such as medical robots, automated pharmacy systems, bar coding, etc. [1][2]. Even though the services are fully provided for the doctors, the patients has no or limited services. The most up to dated technologies should provide the services in more flexible manner, such as allowing the patients to access the services for their needs. Mobile communication with Global Positioning System (GPS) and Geocoding, which is the process of retrieving latitude and longitude of a location, can be a solution for satisfying the patients' needs [3] [4][5].

In the last decade, mobile communication and GPS have enjoyed unprecedented growth all over the world. The location technology, such as GPS, is a crucial tool for providing the right service, at the right time, in the right location to the mobile customers. The ability to pinpoint the location of an individual has an obvious and vital value in the context of emergency services [6]. Delivering personalized services to a mobile user based on their location are called Location Based services (LBS) [4]. The LBS can provide nearby doctors from the current location, map directions, find nearby health centers and locate health events tailored to the particular area. All these services can be provided based on the mobile user profiles and preferences which make the meaningful services available to the user.

Developing a tailor-made application on the user device for the healthcare service is a challenge because of the tremendous growth in the number of networks, applications and mobile devices [3]. The application must run with personalized and customized different user profiles, different context of use and different devices, which are the basic factors of usability of service [4].

This paper proposes a Location Based Mobile Healthcare System Architecture and Mobile User Application for the healthcare systems. With the application, the registered patients can view medications, lab reports, schedule appointment, send message to doctor and other location based services, such as finding healthcare centers and getting information about the local health events. An important fact of this application is to allow service providers to reach their customers with most relevant and accurate services at anytime and anywhere.

The location based mobile healthcare architecture and application help both the physicians and patients to connect to each other and give patients a choice of location based services along with other basic healthcare services. This application also helps to reduce medication errors. Prescription to a patient can be made with his profile where allergic and age factors can be checked. The users also can manage their account through web interface.

T. Sobh (ed.), *Innovations and Advances in Computer Sciences and Engineering*,
DOI 10.1007/978-90-481-3658-2_51, © Springer Science+Business Media B.V. 2010

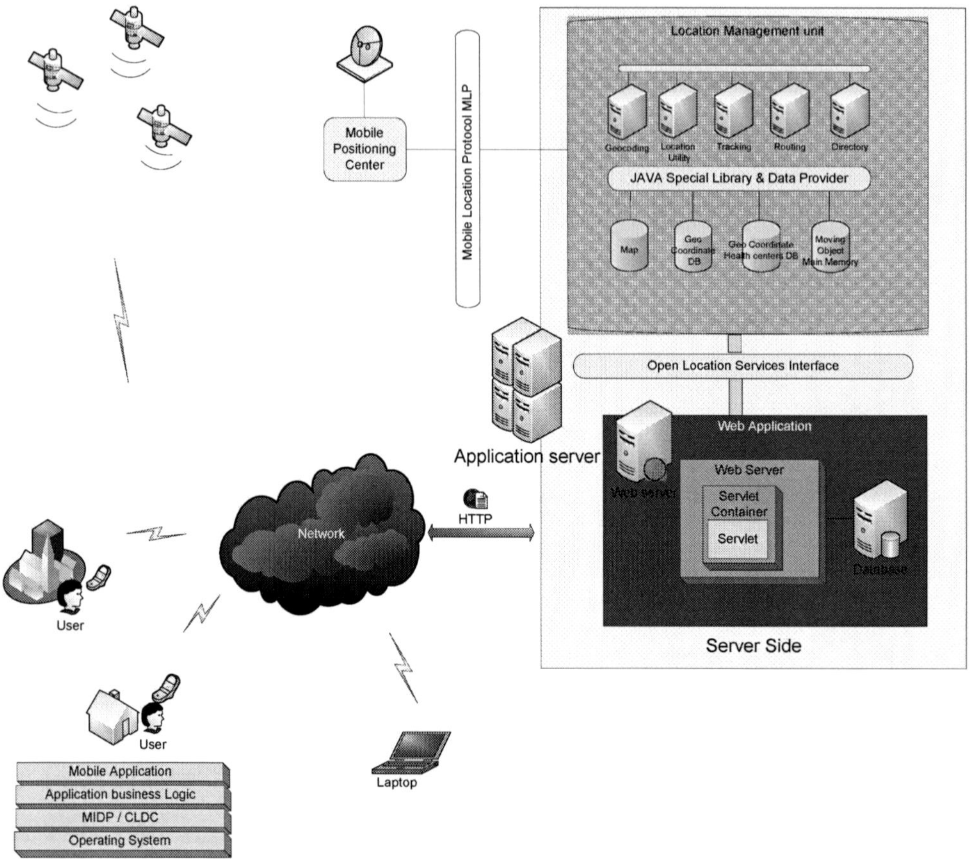


Figure 1: Location based mobile healthcare system architecture

The architecture of location based mobile healthcare services is described with a diagram in the next section. The background of mobile user application and its snapshots are provided. This paper also discusses the result and future study.

## II. LOCATION BASED MOBILE HEALTHCARE SYSTEM ARCHITECTURE

Figure 1 illustrates the detailed architecture of the system. The server side of the application consists of a web application and a composite application in an application server, which manages patient database transaction, location based health care services, and the services that are irrespective to the location of user.

The system uses an Application Server which has LBS libraries and can provide third party solutions for location based services. A good example of such server can be Oracle9iAS wireless from Oracle Corporation [7]. The location management unit contains the servers for geo coding, location utility, tracking, routing and directory. The wireless package from Oracle includes methods to invoke server calls which can resolve the geographical coordinates of the requesting mobile. The server is configured with database which can map geo coordinates with users, doctors, health centers and other various entities in the system. When users request wireless service, the following occurs:

1) The wireless device connects to the gateway, passing the URL of the requested service.

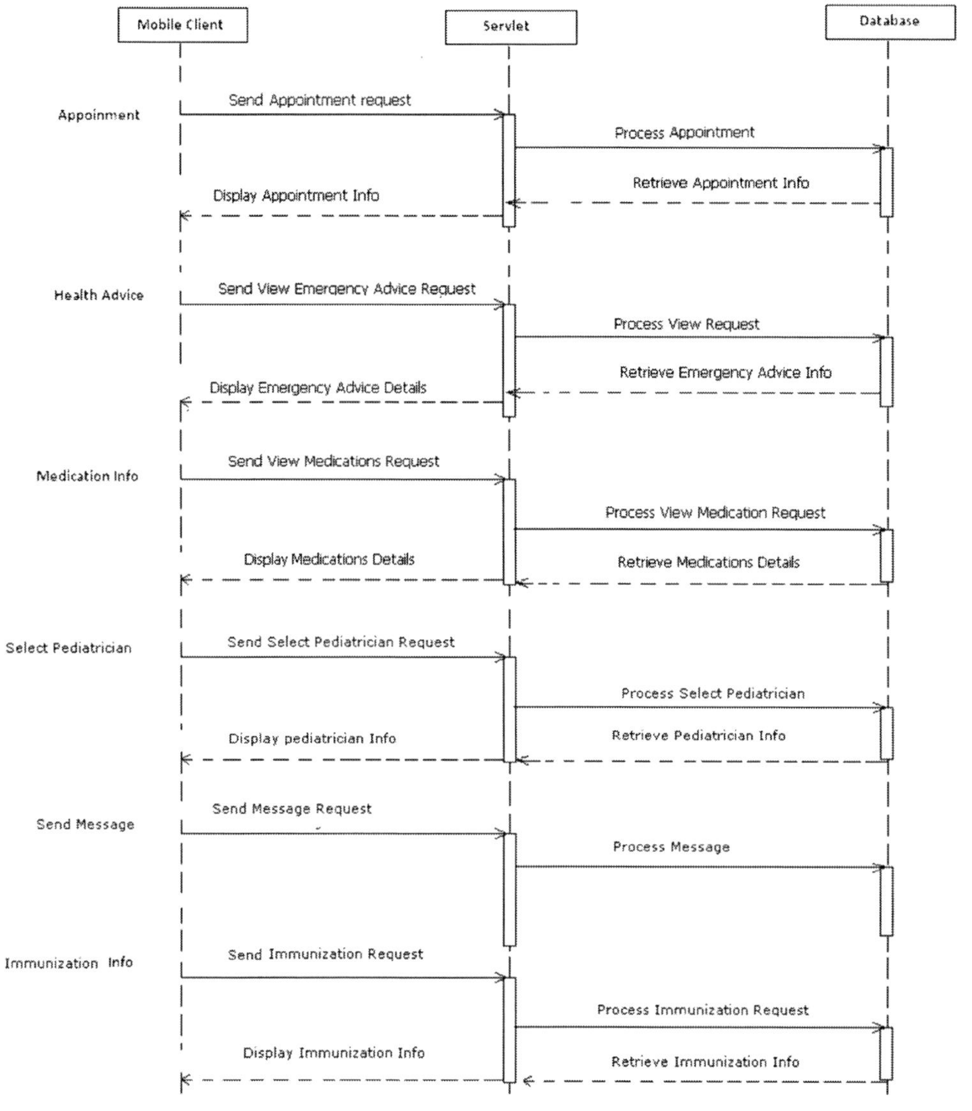


Figure 2: Sequence of User request and server responses

2) The gateway collects the device profiles, such as subscriber ID, device ID, user agent profile from the wireless network.
3) The gateway submits a URL request to Application Server, passing the device profiles.
4) Application Server normalizes the request and forwards the request to the target URL.
5) Application Server transforms the response from the target URL to the requesting device markup language.
6) The gateway sends the response from Application Server to the device.

The target URL specifies the deployed web server inside the application Server. Users can also access web interface of the application to manage their accounts. Web application is deployed in the application server and JSP (Java Server Pages) is used for generating user requests. Back end of the system, the database is implemented with MySQL and simple query language is used for performing database transactions.

However, there are some limitations to be considered on the delivered information and the wireless devices, such as the small screen and low bandwidth. To overcome these limitations, the user profiles can be resolved on server side and the mapping service is done on a database before providing services to a user device. Since there are different capabilities on the users' devices, such as high or low performance devices, the provided service is customized on the server to fit the user's device. The server also can customize the information based on the current user preferences.

The following steps are used for testing the architecture and the user application. The first step of this project was to build a primitive form of the system with minimal functionalities. Secondly, the mobile application was built for pediatric services and then the domain was extended to include services for all the age group.

## III. CASE STUDY

### *A. Mobile User Application*

A primitive form of the application was built to perform pediatric services. The client side application was developed using J2ME (Java 2 Micro Edition) technology, which is a collection of technologies and specifications, used to implement applications for mobile devices. It is a secure, portable, robust, standard application platform used to design, develop, deploy, test, and debug mobile applications [8]. The major components used for this project were frame, textbox and menu.

Connected Limited Device Configuration (CLDC) is a fundamental part of the architecture of the J2ME. The J2ME technology is delivered in API bundles that are called configurations, profiles, and optional packages. A J2ME application environment includes both a configuration like CLDC and a profile like the Mobile Information Device Profile (MIDP) [9]. CLDC configuration provides the most basic set of libraries and virtual-machine features that must be present in each implementation of the J2ME environment and MIDP is a set of standard APIs that support a narrower category of devices within the framework of a chosen configuration. When CLDC coupled with a profile such as the Mobile Information Device Profile (MIDP), it provides a solid Java platform for developing applications to run on devices with limited memory, processing power, and graphical capabilities [10].

The HTTP user request is framed by concatenating all the parameters and a suitable header, which can be later parsed on server side, is added. Figure 3 and 4 show the snapshots of application run on a simulator. Once the users login with username and password successfully authenticate with server, then the service page is displayed. As seen on Figure 4, if one service is selected among all the services listed on the page and then it invokes the required function. Figure 4 shows the stages of scheduling for an appointment with the physician.

### *B. Result*

The mobile application built for pediatric services includes functionalities for scheduling appointment, retrieving immunization records, selecting pediatrician, prescription details, health advice and messaging physician. Application was installed in a java based mobile and functional testing was done on each of these services. The application successfully communicated with the web server and all the user queries were processed. The updated database was checked with administrator privilege. The snapshots of the test results are illustrated in Figure 3 and 4.

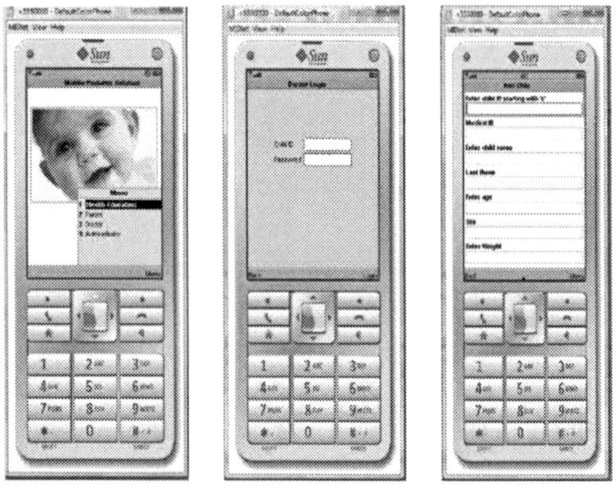

Figure 3: Application, Login Page, and Add Child Form

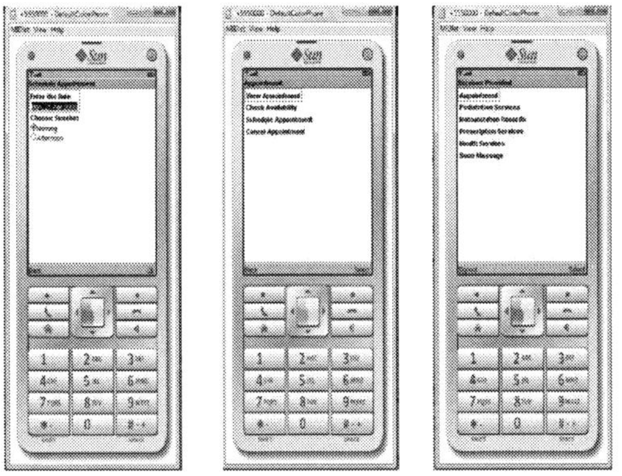

Figure 4: Parent Services, Appointment Services, and Schedule Appointment

## IV. CONCLUSION

In this study, we have presented the mobile user application and the architecture for a mobile healthcare service that can provide location based service to the user. The primary benefit of this tool is its ability to connect both the physicians and patients and give patients a choice of location based services along with other basic healthcare services. We demonstrated the ability of the proposed architecture through a case study to investigate whether the mobile application for pediatric services can be used for scheduling appointment, retrieving immunization records, selecting pediatrician, prescription details, health advice and messaging physician. The test results showed that the mobile application developed for pediatric services had eased the users to access the services and get timely response from a service provider. Although the model is at its initial stage, fully functional software has great potential to serve patients needs and it can give new ways for healthcare service providers to showcase their services and manage their resources.

## V. DISCUSSION AND FUTURE STUDY

In this paper, the JSP (Java Server Pages) was used on the application server. However, EJB (Enterprise JavaBeans) can be used for bring out more reusability and organized approach than the JSP. For the security issues, the knowledge of a person's location can be used by an adversary as the vital medical data is transported over open network, so a suitable data encryption algorithm should be used.

## ACKNOWLEDGMENT

The authors would like to express sincere thanks to Mr. Haridas Puthiyapurayil, Senior Software Engineer, Abbott Diagnostics, San Jose, CA; Minto Xavier, Reswin R Nath, Sheena George, Govt.RIT, India for their great work on developing location based application for health care services .

## REFERENCES

[1] Rainu Kaushal, Kenneth N. Barker, David W. Bates, "How Can Information Technology Improve Patient Safety and Reduce Medication Errors in Children's Health Care?", *article Arch Pediatr Adolesc Med*, vol. 155, pp. 1002-1006, September 2001.

[2] Sapal Tachakra, X.H. Wang, Robert S.H. Istepanian, Y.H. Song, "Mobile e-Health: The Unwired Evolution of Telemedicine", *Telemedicine Journal and e-Health,* vol. 9, pp. 247-250, November 2003.

[3] Chiew-Lian Yau and Wan-Young Chung – "IEEE 802.15.4 Wireless mobile application for healthcare system", 2007 International Conference on Convergence Information Technology, *IEEE 2007,* pp. 1433-1435.

[4] Heechang Shin, Vijayalakshmi Atluri, Jaideep Vaidya - "A profile anonymization model for privacy in a personalized location based service environment", Ninth international conference on Data Management, *IEEE 2008,* pp. 73-74.

[5] S.H.Chew, P.A.Chong, E.Gunawan, K.W.Goh, Y.Kim and C.B.Soh, "A hybrid mobile based patient location tracking system for personal healthcare applications", EMBS Annual International conference , New York city, USA, Aug 30- Sept 3, 2006, *IEEE 2006,* pp. 5188-5190.

[6] Nokia Mobile Phones, "Mobile Location Services", *Nokia Technical white paper*, Finland, pp. 5-9, August 2001, retrieved on 11/20/2008 from http://www.nokia.com/A4140028

[7] Prabuddha Biswas, Mike Horhammer, Song Han and Chuk Murray, "Leveraging location based services for mobile applications", *Oracle technical white paper,* pp. 1-16, June 2001.

[8] "J2ME Building blocks for mobile devices", Technical white paper, *Sun Microsystems*, CA, pp. 1-16, May 19, 2000.

[9] Mobile Information Device Profile (MIDP), JSR 37, JSR 118, retrieved on 11/20/2008 from http://java.sun.com/products/midp/

[10] Connected Limited Device Configuration (CLDC), JSR 30, JSR 139 Overview, retrieved on 11/20/2008 from http://java.sun.com/products/cldc/overview.html

# A General Framework for Testing Web-Based Applications

Saeid Abrishami, Mohsen Kahani
Computer Engineering Department, Ferdowsi University of Mashhad
s-abrishami@um.ac.ir, kahani@um.ac.ir

***Abstract-*** *Software testing is a difficult task for web based applications due to their special features like multi-tier structure and emergence of new technologies (e.g. Ajax). In recent years, automatic testing of web based applications has been emerged as a promising technique to tackle the difficulties of testing these types of applications and several frameworks have been proposed for this purpose. But the most important problem of these frameworks is the lack of generality for different types of tests and programming environments. In this paper, we proposed a general framework for automatic testing of web based applications, that covers all aspects of different types of testing in an arbitrary web based application.*

## I. INTRODUCTION

Frequent use of the internet for crucial tasks, creates serious concern for the quality of web-based software systems. Web based system tends to change rapidly, due to emergence of new technologies (like Ajax) and the demands of users. In such a highly variable environment, manual testing of softwares is a hard and time consuming task, since automated software testing is an inevitable choice for testing the web-based software.

Software testing methods are traditionally divided into black box testing and white box testing. Black box testing treats the software as a black-box without any understanding of internal behavior. It aims to test the functionality according to the requirements. Thus, the tester inputs data and only sees the output from the test object. White box testing, however, is when the tester has access to the internal data structures, code, and algorithms. White box testing methods include creating tests to satisfy some code coverage criteria, and also can be used to evaluate the completeness of a test suite that was created with black box testing methods. In recent years the term grey box testing has come into common usage. This involves having access to internal data structures and algorithms for purposes of designing the test cases, but testing at the user, or black-box level.

Several techniques have been proposed for the testing of web-based applications as both, research proposals and commercial tools. It is roughly possible to categorize such techniques into three groups[1]: (1)functional testing techniques, supporting requirement-base testing; (2) structural techniques, supporting some form of white box testing based upon the analysis and instrumentation of source code; and (3) model-based techniques, which exploit a navigation model of the application.

Different frameworks aim to construct an infrastructure for automatic testing of web-based applications. Sampth et al. [2] proposed a framework that uses user session logs to generate test cases and then a replay tool sends these generated test cases to the server and collects the results. These results send to a test oracle that compares them to the expected results. This framework also uses a Coverage Analysis Tool measure the adequacy of the test suit, using statement and method coverage testing.

Zhu [3] proposed a framework for testing of web services. In this framework, each web service should be accompanied by a testing service. In addition to these testing services, testing tool vendors and companies have independent testing services to perform various kinds of test tasks like to generate test cases, to measure test adequacy, to extract various types of diagrams from source code and so on. The trusted independent test services can call the testing services belong to a web service, to access the internal information (such as source code).

Chu et al [4] presented a testing framework called FAST (Framework for Automating Statistics-based Testing) based on a method called statistical testing. Statistical testing techniques involve exercising a piece of software by supplying it with test data that are randomly drawn according to a single, unconditional probability distribution on the software's input domain. This distribution represents the best estimate of the operational frequency for the use for each input.

One of the most important parts of this framework is Automated Test Data Generator that is responsible for generating test cases. The approach adopted in this paper is specification of the input domain of a software by means of a SIAD (symbolic input attributed decomposition) tree which is a syntactic structure representing the input domain of a piece of software in a form that facilitates construction of random test data for producing random output for quality inspection.

In a similar manner, the specification of each product unit (output) is addressed by the SOAD (symbolic output attributed decomposition) tree. A SOAD tree can be used as a tool for describing the expected result which satisfies the user's requirement and as a basis for analyzing the product unit automatically, without a test oracle. The Quality Analysis module analyzes the product units and finds the "Defective Outputs". At the end, the defect rate has been computed using a binomial distribution.

MDWATP (Model Driven Web Application Testing Program) is a framework proposed by Li et al. [5] for testing web application based on model driven testing. It uses 4 principal models based on UML2.0 standard. The first model is System Under Test (SUT) View that presents the model of the system being tested. So the navigation is a basic characteristic of web applications, they used a navigation model proposed by

T. Sobh (ed.), *Innovations and Advances in Computer Sciences and Engineering*,
DOI 10.1007/978-90-481-3658-2_52, © Springer Science+Business Media B.V. 2010

Lucca et al. and also by Ricca and Tunella, named Web Application Navigation Model (WANM).

WANM depicts the navigation relations among the client pages, and hyperlinks and forms in each client page. Based-on the SUT View, test cases are automatically or semi-automatically generated. The generated test cases are described in the Test Case View, that is a model extends UML sequence diagram. Each test case is a sequence of client pages to be accessed. After that, the process and environment of test execution are modeled in the Test Execution View. In the test execution view, two kinds of models are defined: the test deployment model that extends the deployment diagram of UML2, and the test control model that extends the UML activity diagram.

After the execution engine automatically executes test cases based-on models described in the Test Execution View, Finally test results are represented in the Test Result View. Test result model is defined to save and present test results. And besides being shown in reports, test results would be associated with test case models.

But current frameworks have two major problems: (1) they concentrate on special aspects of testing and there is no general framework that contains all elements and types of testing for web-based applications, and (2) application developers don't like to share their internal information, models and source codes with others (including external application testers) and this makes white box testing so difficult or even impossible. In this paper we have proposed a framework that solves these two problems. This framework has been explained in section 2.

## II. The General Framework

In this section we review our proposed framework for software testing. The general framework and its components have been shown in figure 1.

**STOL language:** As the components of the framework has to exchange information with each other (and with the external world), we require a common language for this purpose. We define STOL (Software Testing Ontology Language), an XML-based language with the required terms and relations (ontology) for software testing. It has been used for explaining program models, test case input ranges, etc.

**Testers:** various testers included in the framework for different purposes. Before we examine these testers in detail, note that these testers belong to different strategies: Black Box, White Box and Grey Box testing. In the case of Black Box testing, the tester directly communicates with the application itself. But in the other two cases, additional information needed that must be provided by the programmer or directly extracted from the source code. In this situation, the tester communicates with the wrapper of application to obtain the required information (e.g. the application model). The wrapper and its usages will be explained in the following. Here is a non-exhaustive list of testers:

- Functional Tester: This is a Black Box tester that checks if the application behaves as expected. This tester simply applies test cases to the application directly, and passes the generated results to the oracle to compare them with the expected behavior.

- Code Coverage Tester: this is a White Box tester that determines the adequacy of test cases by assessing the level of coverage of the structure they reach. Two common forms of code coverage are *function coverage*, which reports on functions executed and *statement coverage*, which reports on the number of lines executed to complete the test. Besides standard coverage measures, other coverage criteria can be defined for web-based applications that we discuss them in Model-Based tester. As this tester requires an instrumented version of application, it must communicate with the wrapper interface, asking for enabling the instrumented code (see Wrapper below). In some programming languages like java, the instrumentation can be done directly on the object code, hence there is no need to the source code (or the wrapper in our case).

- Model-Based Tester: a model describes some aspects of the system under test. The model is usually an abstract presentation of the application's designed behavior. Models have two usages in testing of web applications: automatic generation of test cases (see Test Case Generator below), and structural testing of application. In the latter, a high level representation of application, like the navigation model is used. Navigation model describes a web application using its composing pages (static and dynamic) and navigation links. Using this model, some coverage criteria such as page coverage and hyperlink coverage can be tested, in order to determine the adequacy of test cases. But how the tester acquires the model? The models can be built manually by the application programmer, or created automatically from the application (source) by running a special program. In each case, this is the responsibility of the wrapper to provide this model to the tester (see Wrapper below) via an appropriate interface, using STOL language.

- Stress Tester: this is a Black Box testing which checks for the stress the applications can withstand. The idea is to create an environment more demanding than the application would experience under normal work loads, and to see how application responds to this extreme load.

- Load Tester: this is a Black Box Testing in which the application is tested against heavy loads or inputs such as testing of web sites in order to find out at what point the application fails or at what point its performance degrades. Load testing operates at a predefined load level, usually the highest load that the system can accept while still functioning properly.

- Security Tester: security testing is carried out in order to find out how well the system can protect itself from unauthorized access, hacking, cracking, any code damage etc. In the case of web applications, the application must be tested under familiar attacks like SQL injection, session hijacking, XSS attacks and so on.

**Test Case Generator:** this component generates the inputs to the desired testers. In the case of a web application, test cases consist of URLs, name-value pairs (input data) and user actions (like clicking on a button).

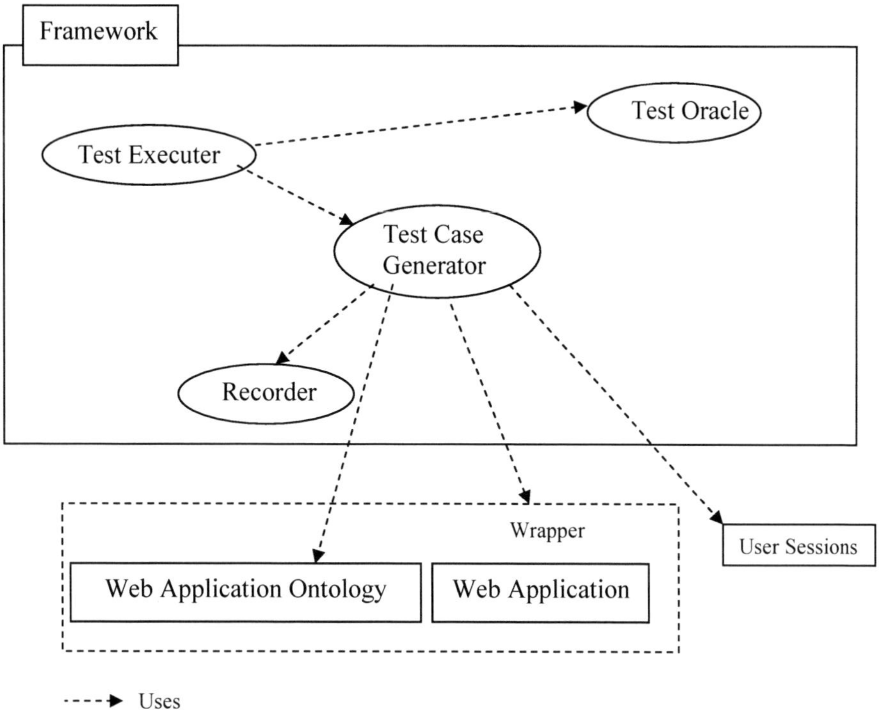


Figure 1. Architecture of the proposed framework

This component is a critical part for automating the test process, since we have anticipated four independent modules for this component in the framework.

1. The first module provides record and playback features that allow testers to record interactively user actions and replay it back any number of times (using Executer component), and comparing actual results to those expected (using Oracle component).

2. The second module exploits the user session logs of a web server as a means of generating test cases in a large scale. Each user request logged in the web server usually consists of IP address, time stamp, the requested URL, cookies, name-value pairs in GET/POST requests and the referrer URL. By following the subsequent requests in a predetermined time interval from a particular IP address, a complete user session can be extracted and used as a new test case.

3. The third module uses different models and specifications of the application, like UML models, navigation model, Z-specifications and etc. to generate the test cases. The models or specifications can be obtained from the proper interface of Wrapper, using STOL.

4. And finally, the fourth module uses ontology-based test case generation approach, in which the ontology of the application domain is utilized to generate test data. For instance, in a web forum application, a light-weight ontology can be developed which describes the users, and then, different parts of the application, e.g. form input fields, can be annotated with concepts of this ontology. In such case, test-cases for those parts can be generated automatically based on the ontology.

**Test Oracle:** generates the desired outputs for the test cases and compares the actual results of tests, to the desired ones to determine the degree of success. In the case of web applications, the desired outputs mainly consist of HTTP responses returned from the web server. Generation of the desired outputs depends on the method selected for test case generation.

**Executer:** this component sends the test cases to the web server and receives the responses, and finally sends them to the Oracle to be evaluated.

**Wrapper:** An important problem in white-box or gray-box testing of an application is the lack of knowledge about internal structure of that application. Most programmers don't share their valuable source codes with other, and this is a problem for tests that require the source code. Because we want to present a framework to provide testing service for external application, we need to tackle with this problem. Our solution is simple:

external applications should provide required wrappers to provide enough information for the framework to perform the test. Each time a tester requires a service that is beyond the functional services of the application (i.e. requires the source code), calls the appropriate API of the wrapper and receives that service. Some of theses services are:

- Activate the instrumented mode: some testers likes Code Coverage Tester require an instrumented version of application that includes instrumentation codes measure the coverage obtained during test. These testers have to activate the instrumented mode, before starting the test.
- Getting various application models: testers like Model-Based Tester require application models (e.g. navigation model) and they can obtain these models by calling this service with the name of their desired model. This service has an API that returns the list of models represented by this service.

To produce the wrapper, an application is distributed for each web programming platform (e.g. PHP, Java ...). Developers can produce the wrapper by running this application on their source code, and add the resulting codes to their programs.

## III. Conclusion and Future Works

In this paper a general framework for automatic testing of web-based application has been proposed. This framework contains different testing techniques, including black-box and white-box testing, which make it a comprehensive testing framework. Also we proposed a Wrapper for each application that provides the required internal information to the testers. This Wrapper solves the problem of lack of trust between application developers and the external testers. Furthermore, we define a language and ontology called STOL (Software Testing Ontology Language) for communication between different parts of the framework.

For the future, we plan to (1) Define the details of different components of the framework and connections between them then, (2) implement and test the performance of the framework using current open source tools and programs written by team members.

## Acknowledgment

This work has been supported by a grant by Iran's Telecommunication Research Center (ITRC), which is hereby acknowledged. The authors also appreciate the support of WTLab at FUM.

## References

[1] Filippo Ricca, Paolo Tonella, "Web Testing: a Roadmap for the Empirical Research," *wse*, pp. 63-70, Seventh IEEE International Symposium on Web Site Evolution, 2005

[2] Sreedevi Sampath, Valentin Mihaylov, Amie Souter, Lori Pollock, "Composing a Framework to Automate Testing of Operational Web-Based Software," *icsm*, pp. 104-113, 20th IEEE International Conference on Software Maintenance (ICSM'04), 2004

[3] Hong Zhu, "A Framework for Service-Oriented Testing of Web Services," *compsac*, pp. 145-150, 30th Annual International Computer Software and Applications Conference (COMPSAC'06), 2006

[4] P. Dhavachelvan, G.V. Uma, *"Multi-agent-based integrated framework for intra-class Testing of object-oriented software"*, Applied Soft Computing 5, pp. 205–222, 2005.

[5] Nuo Li, Qin-qin Ma, Ji Wu, Mao-zhong Jin, Chao Liu, "A Framework of Model-Driven Web Application Testing", Proceedings of the 30th Annual International Computer Software and Applications Conference (COMPSAC'06), 2006

# Integrated Reverse Engineering Process Model

Ghulam Rasool, Ilka Philippow
Software Systems/Process Informatics Group
Technical University of Ilmenau Germany
Ghulam.rasool|ilka.philippow@tu-ilmenau.de

**Abstract-The continual process of technology, new business requirements and stakeholder's needs escort to frequent migration from legacy systems to more powerful, dedicated, secure and reliable computing systems. The reverse engineering techniques and tools are used to extract the structure of existing legacy systems starting form the implementations and going back to design, architecture and requirements. Software engineers are using different methodologies, recovery techniques and tools according to nature, complexity and size of software. The existing recovery techniques and tools are not integrated with each other. This paper introduces novel software reverse engineering process model integrated with different recovery approaches, tools and traceability links to extract different artifacts from legacy systems.**

**Keywords**: Reverse engineering, architecture recovery, program understanding, reverse engineering process.

## I. INTRODUCTION

Due to long life cycles, frequent maintenance and technology evolutions, the software systems that were developed 10-15 years ago becomes candidate for re-engineering. As an alternative of developing completely new systems from scratch, it's better to reuse the existing system components and libraries in the new system development which definitely can save the cost as well as time. History shows that documents normally generated during the development of systems are inconsistent with the source code due to frequent maintenance in the existing systems. Source code is frequently the only complete and reliable software artifact available for reverse engineering legacy systems. According to IBM survey report of different legacy applications, 250 billion lines of source code are maintained in 2000[1]. It is also reported in another study that old languages are still not dead and 80% of IT systems are running on legacy platforms [2]. So the maintenance cost of software is increasing with respect to time and the use of reverse engineering is getting more and more attention in the field of legacy and embedded applications. Mostly existing tools are supporting the object oriented languages but they lack support for procedural languages which really requires their restructuring, refactoring and reengineering as mentioned above.

Reverse engineering can extract design information from source code at higher level of abstraction. Ideally the abstraction level should be as high as possible. "The completeness of a reverse engineering process refers to the level of detail that is provided at an abstraction level. In most cases, the completeness decreases as the abstraction process increases [3].

There is no standard and well-defined software reverse engineering process model widely accepted and used by the reverse engineering community like the standard process models used for forward engineering. In forward engineering, software goes through several phases of system development including from requirement specifications to design and implementation. Different process models are generally used by the developers to manage complexity, time constraints and communicate information to many stakeholders. The reverse engineers define their own process model according to their requirements, nature of source code and available sources of information. The used process models are not integrated with the recovery techniques and tools. The process of examining the existing system can be facilitated with effective integration of various recovery techniques and available tools. The new tools can be developed according to requirements of the process.

The reverse engineering process is dependent on ones cognitive abilities, preferences, familiarity with application domain and on set of support facilities provided by reverse engineering techniques. The process must help to generate different views, diagrams, graphs, metrics and text descriptions that represent the system at higher level of abstraction. So we introduce a novel and well-defined reverse engineering process model which will be integrated with different techniques and tools.

## II. RELATED WORK

The software engineering process used for recovering different artifacts from legacy systems is defined according to nature of software and requirement of maintainer. The success of recovery task will entirely depend on the well defined reverse engineering process model. The state of source code and other documents should also be kept in front before defining the process model. The complex code without comments and other sources is not a good candidate for reverse engineering. The high level reverse engineering process consists of extraction, abstraction and presentation phases. Asif [4] used REAM for recovering different artifacts from the legacy code. The author has defined various models to extract information from different sources but models are not integrated with the recovery techniques and tools.

Abbattista et. al [5] applied their reverse engineering process model to understand the static and dynamic aspects of the source code. The author has major focus on just the ease of maintenance of the software. The major activities are manually performed in the described model without integration of tools.

Murphy has defined the software reflection model [6] which indicates where the source code model and conceptual

T. Sobh (ed.), *Innovations and Advances in Computer Sciences and Engineering*,
DOI 10.1007/978-90-481-3658-2_53, © Springer Science+Business Media B.V. 2010

model differ. The major focus is to note converges and divergences in the extracted source code and conceptual model. The presented model is similar to our reflection model but limited only to source code and conceptual model. Our reverse engineering process model will be used to review the results of conceptual model, source code model, design model, functional model and feature models for refining our pattern specifications iteratively to extract different artifacts from the legacy systems.

Kosche[12] has pointed out that more than 24 recovery techniques are used by the reverse engineers for extracting artifacts from legacy applications. Some of recovery techniques may be integrated with our process model according to need of maintainer.

Mader et al [7] used traceability link management approach which is integrated with software development methods and tools. The author has defined different rules for maintaining traceability links from requirements to design and implementation. The traceability links are maintained only for forward engineering and no support for reverse engineering process. Current approaches are establishing traceability links between code and documents using information retrieval techniques. Our vision is to integrate the rule based traceability links with reverse engineering process model and tools. The activities of reverse engineering process model will be integrated with traceability links. The missing capabilities of rational rose can be recovered by maintaining links between the various artifacts of different models.

## III. THE SOFTWARE REVERSE ENGINEERING LIFE CYCLE MODEL

A key success to reverse engineering is a well-defined reverse engineering process model that could be used for extracting different artifacts from legacy applications. A common downfall of reverse engineering projects is due to poor or non existent planning data and not a standard well defined process model. Our major focus is to integrate the process model with the recovery techniques, tools and traceability links. The process model may vary according to the need and requirements of developers and certain step can be customized. The abstract view of process model is shown in fig1.

We defined reverse engineering process model which consists of following sub- models.

- Domain Model.
- Mapping Model.
- Source code Model.
- Design Model.
- Functional Model.
- Feature Model.
- Reflection Model.

### A. DOMAIN MODEL

A domain model is the sketch of the key concepts within the domain which is used to clarify the concepts and convey ideas. A domain model is the abstraction behind what the system does. The domain knowledge helps understanding and overcoming the complexity of the recovered information about the systems. The knowledge about the commonalities and variability within a system helps the maintainer to establish necessary abstractions. The domain model can be build by the integration of following recovery approaches.

Domain based recovery approaches [8].
Concept analysis approaches [9].
Program comprehension based approaches [10].
Feature modelling approaches [11].
Metrics based approaches [12].

The objectives and possible source of information for above recovery approaches and domain model may be obtained from following:

- Identification of goals.
- Artifacts collection.
- Existing Documents.
- Expert Knowledge.
- Maintenance reports.
- Source code.
- End-users
- Visual Modelling.

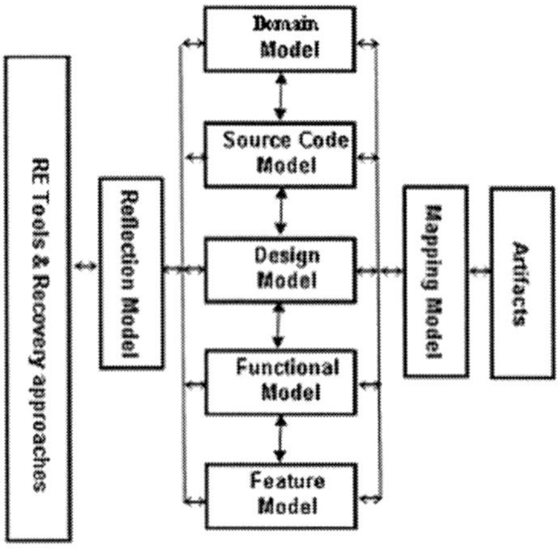


Fig 1 RE process model.

It is imperative to use above aids before starting the reverse engineering process. Reverse engineering is a time consuming activity and it should be clear at the start how far to go as a trade off against the time and cost.

For example the figure 2 shows the domain model of ZBRENT[13] which is a single root finder for polynomials with the combination of open and bracket method.

### B. MAPPING MODEL

The Mapping Model bridges between all the models and help to consolidate the other models. It is used to map the components and entities identified during high level model with the entities available in the source code at various level of abstraction. Here user can define different pattern specifications and map these patterns with the source code entities. The matched results are compared with the information already available in the domain model.

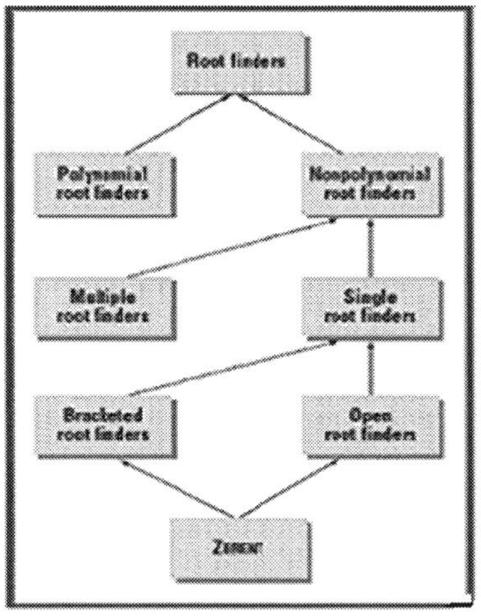


Fig2 ZBRENT domain model[13]

. The patterns may be refined to extract the low level details from the source code. The user iteratively maps the different entities of source code like files, classes, functions and variables to determine the relationship between entities in the source code and entities available in the domain model. The user can map different classes with the packages and may compare the result with the information available in the domain model. The naming conventions in the system artifacts can be used to map patterns using regular expressions. The technique used by Murphy et. al [14] can be used for mapping purpose. The following example shows the mapping of different source code entities with the entities of conceptual model.

```
(package\s*(\w)+)|(JClM)
C:\stefen\EinkaufszentrumProjekt        *.java
\sAwk\s
C:\stefen\EinkaufszentrumProjekt        *.*
\sGui\s
C:\stefen\EinkaufszentrumProjekt        *.class
```

### C. SOURCE CODE MODEL

The source code model may be integrated with program comprehension based [10] and pattern based approaches [15]. The success of program comprehension and pattern extraction depends on the correct use of domain knowledge. The source code model is developed iteratively with the help of domain model and mapping model. Source code model gives detailed information about the structure of the source code. The model associates the entities with directories and files in which code is organised. The use of reverse engineering tools is helpful to build the source code model. The design and architecture related information is extracted in the source code model by using mapping technique. The pattern specification language is used iteratively to extract information used for source code model. The following abstract pattern is used to extract all the classes from the package of Java GUI application.

a:
```
(package\s*(\w)+)|(JAllclasses)
```

b:
```
(JClasModifiers)?\s*((class)((extends)\s*(\w)
+)?\s*((implements)\s*(\w)+)?(\s*(,)\s*(\w+)
```

c:
```
((class)\s*(\w)+\s*\{)
```

The definition of JAllclasses pattern in the above pattern specification is abstracted and shown in pattern b. Similarly the definition of class pattern in pattern b is abstracted and shown in pattern c. The abstracted pattern definitions are helpful for defining new pattern specifications. The artifacts extracted from different procedural languages to build the source code model are presented in [16].

### D. DESIGN MODEL

The design model may be integrated with design pattern recovery [17] and architecture pattern recovery approaches [18]. The information extracted from domain, mapping and source code models is used to build the design model. The existence of architecture and design patterns in source code gives valuable information about the design decisions. The class, component and package diagrams of UML can be used to convey information about structure of the software. It's very difficult to extract all the design information only from source code if source code is not well organised which totally depends on the discipline of programmers. The existing pattern mining tools like DP++[19] and SPOOL[20] etc can be used to extract different design patterns from the legacy code. These design patterns can provide the important decisions made during the system development.

### E. FUNCTIONAL MODEL

The information extracted form domain model, source code model and design model is further used to build the functional model. The use case description can be extracted directly from the source code using different mapping techniques. The use cases do not provide information of non-functional requirements but help to comprehend it. We can define mapping between entities in the source code model (Packages, classes, methods) and entities in the conceptual model (concepts, hypothesis). The comments can be extracted from the source codes which are helpful to improve and verify the understanding about the functionality. Use-case recovery can be automated with proper tools support. The ontology based techniques may also be used to match the patterns of source code with documents to extract the use cases.

### F. FEATURE MODEL

The feature model is integrated with feature mapping methodology FODA [21].The feature model is build from the all the above mentioned models by realising different requirements of users as extracted during different models. Feature model is the major source of communication between the stakeholders and the developers of the systems. All the functional and non-functional requirements are mapped to different features for building feature model.

### G. REFLECTION MODEL

The Reflection model takes input from other models and indicates where the source code model, domain model, design model, functional and feature model differs and make a note of the convergences and divergences. The hypotheses made during different models are refined iteratively by comparing results of different activities. The existing tools may also be helpful for analysis and refinement of pattern specifications to build iteratively the reflection model and the other models.

The detailed activities of all the models are shown in Fig3. The defined process model is customizable and the user can define different models according to his requirements and maintenance task. The user can skip the certain activities according to the available resources for the legacy system. The major focus is on the integration of various recovery techniques and tools with the activities of process model to support in different legacy applications

## IV. TRACEABILITY SUPPORT IN MODELS

Traceability between models in a system can be defined as "The capability of finding, and following specific changes to a model, or between models throughout all phases of development" [22]. Traceability gives the information about how and why the system has reached its current state. The developers can reuse the knowledge of their experienced colleagues with the help of traceability. Mader et al [7] has defined traceability link model which is integrated with the UP. There is no existing model that can support the traceability in the reverse engineering. The lack of backward traceability is the major requirement for software design and architecture restructuring. So the above defined reverse engineering process will be integrated with traceability links to support the connection between the different artifacts from implementation to the requirements. These links will be helpful for refactoring and understanding the architecture of the existing system

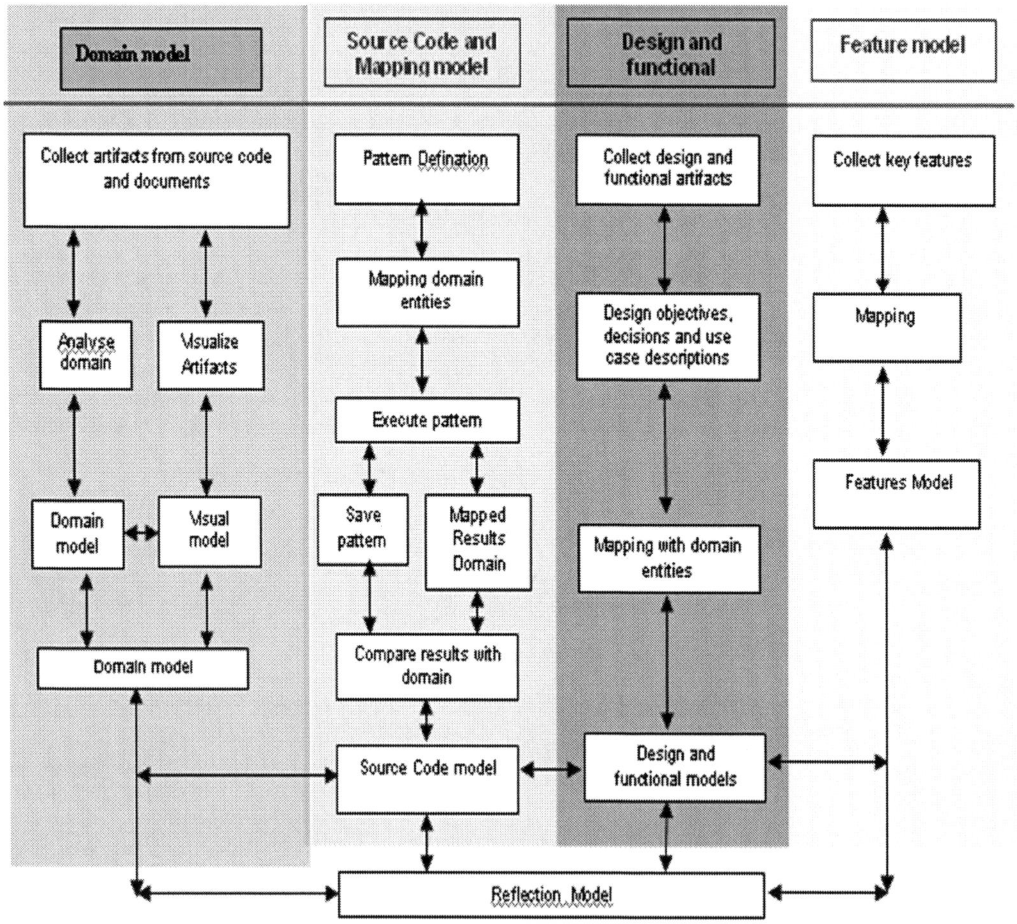


Fig 3 Detailed overview of activities and artifacts in IREPM

## V. CONCLUSIONS

We have anticipated the possibilities of integrating different recovery techniques, tools and backward traceability links with the reverse engineering process model. Moreover, the proposed model is still not applied completely in any real application. However the partial results obtained from different activities of some models are very encouraging. We have extracted different artifacts from legacy applications using domain, mapping and source code models very successfully. The extracted artefacts are presented in [16]. The efficiency and effectiveness of model will be tested for different large systems. We hope that user may extend, modify and tailor our model to his specific needs. Missing activities should be added, those not needed may be ignored and others may be changed.

## ACKNOWLEDGMENT

I would like to thanks to Prof. Dr. Metthias Riebisch for his precious comments and suggestions for this research.

## REFERENCES

[1] Erlikh, L., "Leveraging legacy system dollars for E-business". *(IEEE) IT Pro*, May/June 2000, pp.17-23.

[2] IBM Executive Brief, Legacy Transformation: Finding New Business Value in Older Applications, 2005. http://www-306.ibm.com/software/info/features/bestofboth/.

[3] Pressman, Roger S., Software Engineering: A Practitioner's Approach, McGraw Hill, 1997.

[4] Nadim Asif, "Software Revere Engineering", ISBN 969-9062-00-2 Pakistan, 2006.

[5] Fabio Abbattista, Gregorio M.G. Fatone, Filippo Lanubile, Giuseppe Visaggio, "Analyzing the Application of a Reverse Engineering Process to a Real Situation", In Proc. Of Third Workshop Program Comprehension, Washington D.C., pp. 62-71, Nov. 1994.

[6] Murphy G, Notkin D, " A Re-engineering with Reflection Model", A Case Study", *Computer 30*, 8, 1997, pp. 29-36.

[7] Patrick Mäder, Ilka Philippow, Matthias Riebisch, "A Traceability Link Model for the Unified Process" International Association for Computer & Information Science, 2007. IEEE Computer Society, 2007, pp. 700-705.

[8] J.M. DeBaud, B. Moopen, S. Rugaber, "Domain Analysis and Reverse Engineering," http://www.cc.gatech.edu / reverse/papers.html [accessed on March, 2007.

[9] Hongji Yang, *"Software evolution with UML and XML"*, Idea Group Publishing, 2005, page 58, 2004.

[10] A.Von MayrHauser ,A.M. Vans, "Program Comprehension during Software Maintenance and Evolution", IEEE Computer, Vol. 28. pp 44-55, August 1999.

[11] Ilian Pashov, "Feature Based Methodology for supporting architecture refactoring and maintenance of long-life software systems", PhD thesis , TU Ilmenau 2004.

[12] Rainer Koschke, Gerardo Canfora, Jörg Czeranski, "Revisiting the ΔIC approach to component recovery", Science of Computer Programming, Volume 60, Issue 2 (ISSN:0167-6423 ), Pages: 171 - 188 , April 2006.

[13] Rugaber S, Stirewalt K "Model-Dreiven Reverse Enginering", IEEE Software vol 21, issue 4,July-August 2004,pp.45-53.

[14] Murphy G, Notkin D " Leightweight lexical Source Model Extraction" ACM transactions on Software Engineering and methodology Vol 5, 1996.

[15] R.Keller, R.Schauer, S.Robitaille, P.Page, " Pattern Based Reverse-Engineering of Design Components". In *Proc. 21st International Conference on Software Engineering*, 1999.

[16] Ghulam Rasool, Nadim Asif, " Software Artifacts Recovery Using Abstract Regular Expressions", In Proc of 11th IEEE Multitopic Conference , 28-30 December 2007 , Comsats Institute of IT Lahore Campus.

[17] Ilka Philippow, Detlef Streitferdt, Matthias Riebisch, Sebastian Naumann, "An approach for reverse engineering of design patterns", Accepted: 29 January 2004/Published online: 29 April 2004 – Springer-Verlag 2004.

[18]. Martin Pinzger, Harald Gall, "Pattern-Supported Architecture Recovery", In Proceedings of 10th International Workshop on Program Comprehension (IWPC '02)", 27-28 June,2002 Paris, pp. 53-61.

[19] Bansiya J (1998) Automatic Design-Pattern Identification.,Dr. Dobb's Journal. Available online at: http://www.ddj.com

[20] Keller RK, Schauer R, Robitaille S, "Pattern based reverse engineering of design components", In Proceeding of the 21st International Conference On Software Engineering. IEEE Computer Society Press, pp 226–235.

[21] Pashov, I., Riebisch, M.," Using Feature Modeling for Program Comprehension and Software Architecture Recovery". In Proc of 11th Annual IEEE International Conference and Workshop on the Engineering of Computer Based Systems (ECBS2004)", May 24-27, 2004, Brno, Czech Republic, pp. 406-117

[22] http:// www.iasahome.org/c/portal/layout?p_l_id=PUB.1.335.

# Achieving Consistency and Reusability in Presentation Layer Design using Formal Methods and Design Patterns

Faheem Sohail, Farooq Zubairi, Nabeel Sabir and Nazir Ahmad Zafar

Faculty of Information Technology, University of Central Punjab
Plot No. 31-A, Main Gulberg, Lahore, PAKISTAN
Tele: +92-42-5755314-7; Fax: +92-42-5710881
Emails: {faheem.sohail, farooq.zubairi, nabeel.bloch, dr.zafar}@ucp.edu.pk

***Abstract***—Enterprise applications are frequently designed and developed using a layered approach where the entire application is divided into layers. Each layer in such architecture performs a specific and well defined function. The first and foremost of these layers is the presentation layer. Object oriented design patterns exist for this layer that provide reusable solutions to the commonly occurring design issues. These patterns are frequently used to build the presentation layer application frameworks which are in turn used to develop enterprise applications. Formal methods are techniques which are used to construct mathematically verifiable models of computerized systems. We have applied formal methods to presentation layer patterns which has resulted a verifiable recipes for solving presentation layer design problems. Our formal models are described using VDM++ specification language, and analyzed and validated using the VDM++ Toolbox.

***Keywords***— Enterprise applications, Reusability, Formal methods, Design patterns, VDM++

## I. Introduction

The discipline of software engineering was born out of necessity. During the 1960's, the capabilities of computer hardware and end user expectation grew tremendously. Newer approaches that were more scalable and better defined were required for building larger and more complex software systems in a quicker and cost effective way. These approaches form the software engineering discipline. Software architecture design is a vital component of software engineering and deals with designing software specifications of software systems. A good software design must be simple, consistent, reusable and repeatable.

Design patterns, which are the basis of modern software architecture design, play a vital role in achieving the above mentioned desired characteristics. The patterns discussed in this paper can be classified as "enterprise application design patterns" due to their role in the development of modern object oriented enterprise applications. In this paper, we propose modeling enterprise patterns using formal methods. Formal methods are mathematical techniques based on the principles of set theory and predicate logic which are used to build verifiable models of computerized software systems. It is expected that the combination of patterns and formal methods will result formal reusable software building blocks. We believe that when these components will be used for modeling really complex systems, it must retain the formal essence of its basic constructs.

The fusion of formal methods with software architecture and design patterns is not a new concept. This can be seen from the works on formalizing software architectures that show that an Architecture Description Language based on a formal abstract model of system behavior can provide a practical means of describing and analyzing software architectures and architectural styles [1]. Alencar et al., have introduced a formal approach to model design patterns at many different architectural levels including pipes-and-filters design pattern, layered system and others [2]. Extensive work has been done to formally model behavioral design patterns [3], [4] and object oriented frameworks [5]. New formal languages have been proposed specifically for describing and reasoning about object oriented software architectures, designs, and patterns. In [6], an example of such languages is discussed. Our work, however, is on patterns used in building enterprise applications, particularly, the presentation layer, which is different from what has been done so for.

Formal model of "intercepting filter" and the "front controller" design pattern are described in this paper. The intercepting filter pattern is similar to the "pipes and filters" architectural style in which the application can have a set of pre and post processors that act on any incoming request. The front controller pattern, however, is a delegation pattern that delegates processing responsibility based on the nature of the request. Rest of the paper is structured as given below. In sections II, significance of software design patterns is argued. In section III, the use of design patterns in enterprise applications is discussed. An identification of relationship between formal methods and design patterns is done in

T. Sobh (ed.), *Innovations and Advances in Computer Sciences and Engineering*,
DOI 10.1007/978-90-481-3658-2_54, 

section IV. In section V, formalization of the design patterns is described using VDM++. Finally, conclusion and future work are discussed in section VI.

## II. Significance Of Software Design Patterns

The use of patterns in any field indicates progress of human understanding [7]. In computer science, rather more specifically in software engineering, patterns are a key element of software architecture because patterns can be used to derive architectures, such as a mathematical theorem can be proved from a set of axioms [8]. Software design patterns are a mechanism for expressing recurrent design structures [4] and can be repeatedly used to address similar design problems in similar architectures. Design patterns have been extracted and repetitively applied in diverse sub domains such as software architecture [9], general object oriented designs [10], web applications development, enterprise application development [11], enterprise application integration [12], software implementation [13], software development for portable devices [14], embedded systems [15] and many more.

An object oriented design pattern is "a reusable and abstract solution to a commonly occurring design problem". This solution is defined in terms of classes, their instances, the relationships between the instances and the delegation of responsibility amongst them [4]. The most important term in this definition is "design problem". A design problem relates to the structural, creational, behavioral or other aspects of a software system as opposed to the algorithmic or computational aspects. We can classify design patterns on the basis of the problem to be solved. For example, structural patterns define ways to compose objects to obtain new functionality of a system. Similarly, creational patterns address issues related to object construction. The behavioral patterns are specifically concerned with the communication between the objects of a system [16].

Some other properties of the software design patterns can be listed as: (i) are abstract, (ii) cannot be coded but have their implementations, (iii) are well documented and their use improves the readability of the code which in turn increases maintainability, (iv) promote reusability, and (v) increase compliance to the object oriented principles.

## III. Design patterns in Enterprise Applications

Enterprise software applications provide internal services to an enterprise in some focused business domain such as human resource or accounting, or could also provide services to customers of the enterprise in question. Being a class of computer software, all enterprise applications share a certain set of properties which are discussed below.

Modern enterprise applications are distributed, multi-tiered and modular in nature and need to have high reliability. All enterprise applications have a data source which in most cases is a relational database management system (RDBMS) but can also have other sources such as XML, Excel etc. Multiple data sources are common and access to it will be required by hundred of concurrent users. All these aspects of enterprise applications make designing and maintaining an enterprise application a challenge. While developing enterprise applications, the focus is on a good structural design which is easily maintainable.

Enterprise applications are frequently divided into layers each of which performs a specific and well defined functionality. The outermost layer of an enterprise application is the presentation layer which serves the purpose of an interface between the application and the outside world. Next is the service or the business layer, which performs the core functionality of the system. The data access layer hides the data source from the service layer allowing us to change the data source without impacting the rest of the application. Fowler [11] describes how this layering mechanism increases modularity, decreases unnecessary coupling between components, increases cohesion and promotes reusability, resulting in a maintainable and well structured system. The patterns that belong to the enterprise suite are always limited to a layer of the application. In this paper, we present the formal model of two most useful presentation layer design patterns, the "Intercepting Filter" and the "Front Controller". These patterns are a part of the enterprise pattern catalog [17] published by Sun Microsystems.

## IV. Formal Methods and Design Patterns

Formal methods constitute a branch of software engineering that incorporates the use of mathematics, specifically, discrete mathematics, for software development. Bowen describes the importance of formal methods in the following way. "Just because a system has passed unit and system testing, it does not follow that the system will be bug free. That is where formal methods offer considerable advantageous over more traditional methods when developing systems where high integrity is required. Formal methods allow us to propose properties of the system and to demonstrate that they hold. They allow us to examine the system behavior and convince ourselves that all possibilities have been anticipated [18]."

Due to increasing influence of computers in all aspects of our lives, and the fact that modern computer hardware allows complex computer systems to be developed [19], the impact of failure of such systems has increased. Critical systems can be divided into three broad categories. Safety critical systems are those whose failure results in human injury or loss of life. Business critical systems impact the financial standing of an enterprise. Finally, mission critical systems are whose failure impairs the goal of a given mission [20].

"If formal modeling is applied at the level of style, analyses can be extended to any system that conforms to the style [1]". This statement best captures the motivation for formalizing design patterns. In software development, the "level of style" is the design of the system which is expressed in terms of a multitude of patterns. When these patterns are formalized and then implemented, their implementations inherently contain the original analyses that were made during the formalization of the pattern. This "trickling down" of formal detail from an abstract design pattern to a concrete implementation is the primary impetus for formalizing patterns.

Our focus is on enterprise applications patterns which lie squarely in the domain of business critical systems. Most enterprise applications handle large amounts of data on which complex processes and rules operate on [11]. Failure on part of the application to provide reliable and accurate services to the stakeholders may result in irreversible damage to the financial status or the well being of the enterprise. Hence, it is logical to apply formalism to the most basic design construct of such applications, for example, the enterprise application design patterns. Uniformity and consistency of modeling approaches is desired in software development. Consider an enterprise application which we model using a combination of formal techniques and informal techniques. When implemented, our application will surely contain a large number of open source and commercial components and frameworks resulting in a mix of formalism and in-formalism in our code base. We believe that such a mix is undesirable because a faulty component used in our system may cause our critical system to collapse. Therefore, the best case scenario would be for framework developers to use formal design patterns to model their wares resulting in an all encompassing formal model. Another way to put across this idea is through the well known English idiom, "a chain is only as strong as its weakest link".

## V. Formalizing Design Patterns

The structure of the pattern is a sequence of relationships between classes which are defined in terms of a UML sequence diagram. Any implementation of a pattern can only be verified for compliance with the original pattern if the implementation's structure is the same as of the pattern. Our formalization methodology works on the concept discussed above. The VDM++, which is an object oriented formal language is used to model our base object orientation model and patterns.

### *A. Formal Specification of Object Orientation*

Our efforts to formalize design patterns would be flawed and incomplete if we do not formalize object orientation first. This approach has been adopted by several others who have formalized patterns [4], [3]. We have intentionally kept the scope of our object oriented formal model limited as opposed to more comprehensive approaches pursued by other researchers. Only those concepts of object orientation that are core to our work have been formalized currently. This model however will be expanded as part of our future work in this direction.

***Method Signature***

```
class MethodSignature
types
parameter :: token
instance variables
accessModifier : Visibility;
name : Global'mSignature;
parameters : set of parameter;
returnType : token;
end MethodSignature
```

***Interface:*** An interface can be considered as a set of method signatures. We ensure as part of the pre condition that the same signature is not repeated in a given interface.

```
class Interface
instance variables
public signatures : set of signature;
public name : Global'interfaceName;
operations
addMethodSignature(sig : Signature)
ext rd signatures
pre sig not in set signatures
post signature = signature union {sig};
end Interface
```

***Class Definition:*** An object oriented class can be considered as a set of interfaces which the class implements and a set of signatures that are part of the class definition.

```
class ClassDef
instance variables
name : seq of char;
interfaces : set of Interface;
signatures : set of Signature;

operations
containSignature()b : bool
ext rd signatures : set of Signature
pre sig <> null and len signatures>0
post true;

end ClassDef
```

***Relation:*** A relationship class describes the type of relationship that could exist between two given classes. Patterns discussed in this paper can be modeled by defining association relationships between the participating classes. These relationships are dictated by the pattern structure.

```
class Relation
types
relations = Association | Aggregation |
Composition;
instance variables
relationship : (ClassDef-> ClassDef)-> relations
end Relation
```

***Pattern:*** An object oriented design pattern can be considered as a sequence of called methods or operations. The specification described below has an operation by the name of *isValidPattern*. It checks if the sequence of relationships of a particular pattern is valid or not.

```
class Pattern
types
```

```
pType = InterceptingFilter | FrontController;
instance variables
relations : seq of Relation;

operations
addRelation(rel:Relation)
ext wr relations : seq of Relation
pre rel <> null and rel not in set relations
post relations = relations ^ [rel];

isValidPattern(relation : Relation`relations,
classA : ClassDef, classB : ClassDef)flag:bool
ext rd relations : seq of Relation
pre elems relations > 0
post forall r in set relations &
rel(classA,classB,relation) in set
r(i).relationship;
end Pattern
```

### *B. Formalization of Presentation Layer Patterns*

#### 1) Intercepting Filter Pattern

The intercepting filter pattern is the first pattern of the presentation layer. The primary inspiration of it is to have multiple preprocessors or "filters" each of which performs a specific and well defined task. In essence this design pattern is the preprocessor variant of the "pipes and filters" architectural style [21] useful in applications that receive input from external world in the form of HTTP requests. The same pattern is also used frequently for post processing.

The best way to describe the interceptor filter pattern is via the airport analogy. For example, a passenger has to board a flight but before he is able to travel, he has to complete a number of predefined and sequential activities each of which is unrelated and independent of each other. Transforming UML sequence diagram to mathematical relations results in the following:

((Class ↔ Class) ↔ Relation Type)

1 ((Client, Filter Manager), Association)

2 ((Filter Manager, Filter Chain), Association)

3 ((Filter Chain, Filter), Association)

4 ((Filter Chain, Target), Association)

***Intercepting Filter Pattern:*** The intercepting filter pattern is a subclass of pattern. The *isValidPattern* method of the parent class is used to validate the pattern.

```
class InterceptingFilter is subclass of Pattern
operations
isValidPattern(relation : Relation`relations,
classA : ClassDef, classB : ClassDef)flag:bool
pre true
post Pattern` isValidPattern(relation, classA,
classB);
end InterceptingFilter
```

**`Filter Chain:`** The filter chain class is a component class of the pattern and is a subclass of the *ClassDef* class. The filter chain is composed of a sequence of filters. Each filter is called in order and is validated in the *processFilter* operation as defined below.

```
class FilterChain is subclass of ClassDef
instance variables
public filters : seq of Filter;
operations
containSignature()b : bool
pre true
post ClassDef`containSignature();

addFilter(filter : Filter)
ext rd filters : seq of Filter
pre filter not in set filters
post filters = filters ^ [filter];

processFilter(request : token,response : token)
ext rd filters : seq of Filter
pre elems filters > 0
post let filter = hd filterChain and filterChain
= tl filterChain in filter.containSignature(sig)
and filter.doFilter(request,response,filterChain);
end FilterChain
```

**`Filter:`** The filter class is another component of the intercepting filter design pattern. Every filter called has its own structure validated from *isWellDefinedClass* operation.

```
class Filter is subclass of ClassDef
instance variables
name:seq of char;
interfaces : set of Interface;
operations
containSignature()b : bool
pre true
post ClassDef`containSignature()
execute()done:bool
pre let fClass = mk_ ClassDef(name,interfaces) in
isWellDefinedClass(fClass) and containSignature()
post TRUE;

doFilter(servletRequest:token,servletResponse:token
,filterChain: FilterChain)
pre let fClass = mk_ ClassDef(name,interfaces) in
isWellDefinedClass(fClass) and containSignature()
post TRUE;
end Filter
```

**`Filter Manager:`** The filter manager class contains the filter chain. The process filter operation is used to start the execution process.

```
class FilterManager is subclass of ClassDef
instance variables
filterChain : FilterChain;
operations
containSignature()b : bool
pre true
post ClassDef`containSignature();
processFilter(filter : Filter,request : token ,
response:token)
pre filterChain <> null
post filtersChain.processFilter(request, response)
and let target = mk_ Target() in
target.execute(request ,response);
end FilterManager
```

**`Target:`** The target class is the final one called after all filters are successfully called and executed.

```
class Target
operations
execute(servletRequest:token,servletResponse:token)
pre true
post true
end Target
```

### 2) Front Controller Pattern

Once, a requestor's request has passed through the intercepting filter, the request is handled by the front controller pattern that deals with delegation of computational responsibility based on the nature of the problem to be computed.

For example, a bank serves as an analogy for the front controller design pattern. Based on the type of request, the receptionist guides the customer to the relevant desk where his request is fulfilled. Transforming UML sequence diagram to mathematical relations results in the following:

((Class ↔ Class) ↔ Relation Type)
1 ((Client, Controller), Association)
2 ((Controller, Dispatcher), Association)
3 ((Controller,View), Association)
4 ((Controller, Helper), Association)

**`Client:`** Like the previous pattern, the client class is checked if it's well defined. The *doGet* operation is the entry point of execution for a client request.

```
class ClientPage is subclass of ClassDef
instance variables
controller : Controller;
operations
containSignature()b : bool
pre true
post ClassDef`containSignature();
public doGet(HttpServletRequest : token,
HttpServletResponse : token)
pre controller <> null and
let fClass = mk_ClassDef(controller.name,
controller.interfaces) in
isWellDefinedClass(fClass) and
controller.containSignature()
post controller.processResponse(HttpServletRequest,
HttpServletResponse);
end ClientPage
```

**`Controller`**: The controller class, which is a major component of the front controller pattern, is checked for class validity. Similarly, the other classes, i.e., dispatcher, helper and command have simple checks for structure validity.

```
class Controller is subclass of HttpServlet
instance variables
page : View;
name:seq of char;
interfaces : set of Interface;
operations
public init(ServletConfig : token)
pre true
post HttpServlet'init(ServletConfig);
public processResponse(HttpServletRequest : token,
HttpServletResponse : token)
pre true
post let helper = mk_ RequestHelper()
in let command = helper.getCommand() in
command.containSignature() and
page = command.execute(HttpServletRequest,
HttpServletResponse) and
page.containSignature()and
dispatch(HttpServletRequest,
HttpServletResponse,page);
public dispatch(HttpServletRequest : token,
HttpServletResponse : token)
pre true
post true;
end Controller
```

**`Dispatcher`**

```
class RequestDispatcher is subclass of ClassDef
operations
containSignature()b : bool
pre true
post ClassDef`containSignature();
public forward(HttpServletRequest : token,
HttpServletResponse : token)
pre HttpServletRequest<>null
post true;
end RequestDispatcher
```

**`Helper`**

```
class RequestHelper is subclass of ClassDef
instance variables
httpServletRequest : token;
command : Command;
operations
containSignature()b : bool
pre true
post ClassDef`containSignature();
public getCommand()c:Command
pre true
post c = command;
end RequestHelper
```

**`Command`**

```
class Command is subclass of ClassDef
instance variables
page : View;
operations
containSignature()b : bool
pre true
post ClassDef`containSignature();
public execute(HttpServletRequest : token,
HttpServletResponse : token)mypage: View
pre HttpServletRequest<>null and
HttpServletResponse<>null
post mypage = page;
end Command
```

## VI. Results and Conclusion

In this paper, we have presented a formal specification of two enterprise presentation layer design patterns. The objective of this research is proposing approaches to formalize systems at an abstract design level rather than giving its implementation detail at the early stages of system's development. As formal methods are powerful at an abstract level, that is why, their use will increase reusability of the formal specification of the design for any of its implementations in addition to increasing confidence ensuring its correctness. We observed that the formal nature of this design will trickle down to any implementation of the design of a system. After formal description of design patterns, we believe that these reusable components will result in consistent and uniform architectures that use formalism across the entire application instead of just formal custom code supported by informal reusable components.

We proposed a formal methodology rather a detailed approach followed in [9], [3]. We first provided a formal specification of the basic object oriented constructs and then reused it while constructing formal models of the patterns. Some similar work can be found in [22], [23], [24] which was taken as starting point for this resaerch.

The relative ease by which we were able to formalize our patterns using VDM++ can be attributed to the fact that an object oriented formal specification language was used to model object oriented constructs and concepts. A lot of concepts, that we would have had to model ourselves in a non-object oriented language such as Z, came inherently while using VDM++. The formal description of the next layer design pattern of the enterprise suite is under progress and will appear soon in our future work.

## VII. References


[1] Allen, R. J., "A Formal Approach to Software Architecture", CMU Technical Report CMU-CS-97-144, Vol. 6, 1997.

[2] Alencar, P.S.C., Cowan, D.D. and Lucena, C.J.P., "A Formal Approach to Architectural Design Patterns", FME96: Industrial Benefit and Advances in Formal Methods, 1996.

[3] Reynoso, L. and Moore, R., "Gof Behavioral Patterns: A Formal Specification", United Nations University, International Institute for Software Technology, 2000.

[4] Cechich, A. and Moore, R., "A Formal Specification of GoF Design Patterns", International Institute for Software Technology, The United Nations University, Vol. 1, 1999.

[5] Crnkovic, I., et al., "Object Oriented Design Frameworks: Formal specification and some Implementation Issues", Proc. Of 4th IEEE Int'l Baltic Workshop, 2000.

[6] Eden, A.H., "Formal Specification of Object-Oriented Design", International Conference on Multidisciplinary Design in Engineering CSME-MDE, Montreal , 2001.

[7] Coad, P., "Object Oriented Patterns", Communications of the ACM, Vol. 5 (4), p102-103, 1994.

[8] Beck, K., and Johnson, R., "Patterns Generate Architectures", ECOOP'94, Springer-Verlag, 1994.

[9] Shaw, M., "Some Patterns for Software Architecture", Vlissides, Coplien & Kerth (eds.), Pattern Languages of Program Design, Vol. 2, Addison-Wesley, pp. 255-269, 1996.

[10] Gamma, E., Helm, R. Johnson, R and Vlissides, J., "Design Patterns: Elements of Reusable Object Oriented Software", Addison Wesley Professional Computing Series, 1995.

[11] Fowler, M., "Patterns of Enterprise Application Architecture" Addison-Wesley, 2003.

[12] Hohpe, G. Woolf, B., and Brown, K., "Enterprise Integration Patterns: Designing, Building, and Deploying Messaging Solutions", Addison-Wesley, 2004.

[13] Beck, K., "Implementation Patterns", Addison-Wesley, 2008.

[14] Narsoo, J., and Nawaz M., "Identification of Design Patterns for Mobile Services with J2ME", The Journal of Issues in Informing Science and Information Technology, Vol. 5, 2008.

[15] Gay, D., et al., "Software Design Patterns for TinyOS", ACM Trans, Embedd Computing, Vol. 6 (4), 2007.

[16] Bowen, J.P., and Stavridou, V., "Safety-Critical Systems, Formal Methods and Standards", IEE/BCS Software Engineering Journal, Vol 5, 1993.

[17] Alur, D., Crupi, J., and Malks, D., "Core J2EE Patterns: Best Practices and Design Strategies", Prentice Hall/Sun Microsystems Press, 2003.

[18] Bowen, P., and Hinchey, M., "Ten Commandments of Formal Methods", IEEE Computer Society, Vol. 10, 1995.

[19] Bowen, J., Stavridou, V. P., "Safety-Critical Systems, Formal Methods and Standards", IEE/BCS Software Engineering Journal, Vol. 1(2) 1993.

[20] Charatan, Q., and Kans, A., "Formal Software Development from VDM to Java", Palgrave Macmillan, 2003.

[21] Garlan, D., and Shaw, M., "An Introduction to Software Architecture", CMU Software Engineering Institute Technical Report, CMU/SEI-94-TR-21, ESC-TR-94-21, Vol. 6, 1994.

[22] Reza, H. and Grant, E., "A Formal Approach to Software Architecture of Agent-base Systems", Int'l Conference on Information Technology, Vol. 1, pp. 591-595, 2004.

[23] Dong, J., et al., "Composing Pattern-based Components and Verifying Correctness", Journal of Systems and Software, Vol. 80 (11), pp. 1755-1769, 2007.

[24] Taibi, T., "Formal Specification of Design Patterns Relationships", Proc. of 2nd Int'l Conference on Advances in Computer Science and Technology, pp. 310–315, 2006.

# First Level Text Prediction using Data Mining and Letter Matching in IEEE 802.11 Mobile Devices

B. Issac
Swinburne University of Technology (Sarawak Campus),
Jalan Simpang Tiga, 93576 Kuching, Sarawak, MALAYSIA.
Email: bissac@swinburne.edu.my

***Abstract*-When a user uses an 802.11 wireless mobile device like a Personal Digital Assistant (PDA) to type in a message, typing every word fully can be quite tedious. This paper investigates into a first level text prediction scheme through data mining and letter matching which could help the user with writing messages using predictive approach, to be used for short message service text or for email text. The performance results of simulation show that the results are quite encouraging and the accuracy of first level text prediction is above 80% for a huge sample, where a given word is mapped to a subset of possible next words, in fixed and flexible scenarios. Though the prediction is tried with English language, it can be applied to other similar languages.**

## I. Introduction

There has been a lot of research going on in text prediction that could help the users with predicted text as they write in any mobile device. In our paper we want to discuss on ways by which text prediction can be done as the user writes text in a message to be sent by an 802.11 enabled wireless device. The 802.11 mobile devices would contact a central server where all the past history of the user's typed text is stored, along the all the possible dictionary words as 'neutral words'. This happens in regular and frequent intervals (of the order of ms) through the access point of association. Based on this past data repository for a user, the future words for that user can be predicted. The research assumes that the user is regularly moving around in an area like his office.

The paper is organized as follows. Section 2 is on related works and technologies, section 3 is the text prediction method using data mining and letter matching, section 4 is the Java simulation performed and section 5 is further improvements and section 6 is the conclusion.

## II. Related Work and Technologies

Wang, Liu and Zhong in their paper presents preliminary investigations concerning the use of Simple Recurrent Network (SRN) in Chinese word prediction. They explore the architecture introduced by J. L. Elman for predicting successive elements of a sequence. This model is based on a multi-layer architecture and contains special units, called context units which provide the short-term memory (STM) in the system. Based on this model, they constructed a modular SRNs to predict Chinese word at two levels. The first level network predicts the major category of the next word, and then the next possible word is predicted at the second level network. Also, the specific encoding schemes were described in the paper [1]. Even-Zohar in his paper discusses on a classifier. The goal of a classifier is to accurately predict the value of a class given its context. Often the number of classes 'competing' for each prediction is large. Therefore, there is a need to 'focus the attention' on a smaller subset of these. He investigates the contribution of a 'focus of attention' mechanism using enablers to the performance of the word predictor. He then describe a large scale experimental study in which the approach presented is shown to yield significant improvements in word prediction tasks [2]. Nakamura and Shikano present a study of English word category prediction based on neural networks. Using traditional statistical approaches, it is difficult lo make an N-gram word prediction model to construct an accurate word recognition system because of the increased demand for sample data and parameters to memorize probabilities. To solve this problem, NETgrams, which are neural networks for N-gram word category prediction in text, are proposed. NETgrams can easily be expanded from Bigram to N-gram networks without exponentially increasing the number of free parameters. Training results show that the NETgrams are comparable to the statistical model and compress information. Results of analyzing the hidden layer (micro features) show that the word categories are classified into some linguistically significant groups [3]. Yoshiaki Ajioka and Yuichiro Anzai presents the prediction of next alphabets and words of four sentences by adaptive junction. They think that the feedback-type neural networks are effective methods to realize sequential processing. They have studied Adaptive Junction which is one of feedback-type neural networks recognizing spatio-temporal patterns. This paper demonstrates that Adaptive Junction networks performing the chain reaction with 1-degree feature patterns can behave prediction of next alphabets or words of four sentences, "A MAN CRIES", "A BABY CRIES", "A DOG BARKS" and "A CAT MEWS". These results indicate that the chain reaction must play an important role for such cognitive behavior as prediction [4]. Fu and Luke in their paper focus on integrated prosodic word prediction for Chinese TTS. To avoid the problem of inconsistency between lexical words and prosodic words in Chinese, lexical word segmentation and prosodic word prediction are taken as one process instead of two independent tasks. Furthermore, two

T. Sobh (ed.), *Innovations and Advances in Computer Sciences and Engineering*,
DOI 10.1007/978-90-481-3658-2_55, © Springer Science+Business Media B.V. 2010

word-based approaches are proposed to drive this integrated prosodic word prediction: The first one follows the notion of lexicalized hidden Markov models, and the second one is borrowed from unknown word identification for Chinese. The results of their primary experiment show these integrated approaches are effective [5]. Dong, Tao and Xus discusses on prosodic word prediction using lexical information. As a basic prosodic unit, the prosodic word influences the naturalness and the intelligibility greatly. Although the relation between the lexicon word and the prosodic word has been widely discussed, the prosodic word prediction still can not reach a high precision and recall. In this paper, the study shows the lexical features are more efficient in prosodic word prediction. Based on careful analysis on the mapping relationship and the difference between the lexicon words and the prosodic words, this paper proposes two methods to predict prosodic words from the lexicon words sequence. The first is a statistic probability model, which efficiently combines the local POS and word length information. Experiments show that by choosing appropriate threshold this statistic model can reach a high precision and high recall ratio. Another is an SVM-based method [6].

## III. Text Prediction Using Data Mining And Letter Matching

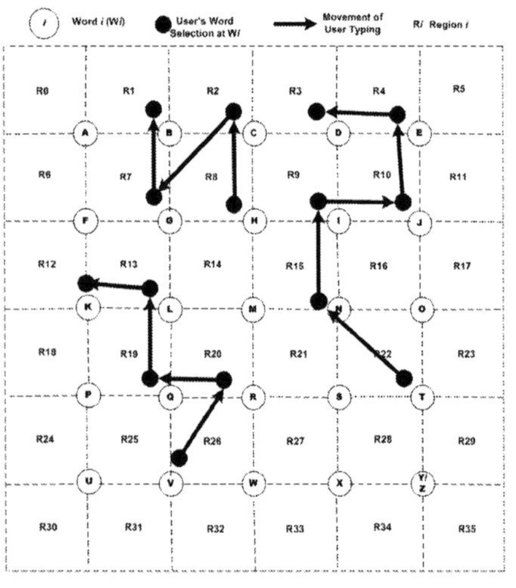


Figure 1. A broad and generic view of the user typing movement from one word to the other, placed in a matrix format (for fixed scenario). Each node represents the first letter of the word.

Our focus in the paper is only the first level word prediction. If the word typed doesn't indicate any match, a neutral dictionary word would be suggested based on word pattern. The data mining approach consists of three phases: user text pattern mining, generation of text pattern rules using the mined user patterns, and the word prediction [7]. The next word prediction of users is predicted based on the text/word pattern rules in the last phase. For example, if the word to be predicted is in region x (which can be known once the user presses the first letter of to be predicted word), the only possible predictions are say, {2, 6, 7}. Using the extracted text/word patterns from the database of the lines stored in the server, the next word prediction is done based on the frequency rank of the word transitions.

### *A. User Word Pattern Mining*

Once the server knows the initial word that the user typed, it locates all the immediate neighbours of that word. Consider for example a path: 19(22) → 13(15) → 8(9) → 9(10) → 4(4) → 3(3), where values in parenthesis represent the regions where prediction is made. The starting word is using a word beginning with alphabet or letter 19 (or t) and is in region 22. The prediction moves to region 15 then and gets to the word with letter 13 (or n). The movement of prediction is decided initially by the highest frequency of the neighbouring words. When the user types the first letter, based on that letter the region can be confirmed. Once in a region we assume that the user will only change to other word options within that region. This concept could be extended to cases where within a region (or with a node), the prediction can happen to all the possible *n* neigbouring words starting with that first letter.

Once the prediction moves from one region (current region) to the other region (next region), we need to apply the formula: *Current word {all neigbhour words}* ∩ *Next Region {all words}* = *{S}*, where S is the set of probable next words. To find the predicted next word, as per the rule given above, take the intersection of neighbour word set of word with starting letter 19 (or t) and neighbour word set of region 15. This is given as: word starting with letter 19 {neigbours} ∩ R15 = {13, 14, 18, 23, 24} ∩ {7, 8, 12, 13} ={13}. Thus the predicted next word is starting with letter 13 (or n), which is correct as the user prediction. Predicted word transition = 19→13 (or t→n).

From word starting with letter 13 (or n) in region 15, the prediction moves to region 9 and would have a transition to word starting with letter 8 (or i). Word starting with letter 13 {neigbours} ∩ R9 ={7, 8, 12, 9, 14, 17, 18, 19}∩ {2, 3, 7, 8} = {7, 8}. Since we have 2 entries 7 and 8, we need to do pattern mining. From word starting with letter 13, the transition can be to word starting with letter 7 (which is h) or 8 (which is i). So 13→7 and 13→8 patterns are searched in the sorted database and their frequency count is noted as in table 1. The path 13→8 is selected because of higher frequency count. Word starting with letter 8 (or i) is the next predicted word. Predicted path = 19→13→8 (or t→n→i). This procedure is repeated for other word nodes.

Consider a path 22(26)→21(25)→15(18)→16(19). Following our earlier calculation, word starting with letter 22{neigbours} ∩ R25 ={16, 17, 18, 21, 23} ∩ {15, 16, 20, 21} ={20, 21}. Here the prediction transition moves from word starting with letter 22 (or w) in region 26 to word starting with letter 21 (or v) in region 25, as 22→21 has higher frequency count (of 272) than 22→16 (where the count was 162), like we discussed before. Predicted word transition is correct and thus is: 22→21 (or w→v).

TABLE I. TEXT PATTERN MINING FOR 2 NEIGBOURING WORDS

| No. | Path | Frequency |
|---|---|---|
| 1 | 13→7 | 238 |
| 2 | 13→8 | 367 |

Then the prediction moves to region 18. So, the word starting with letter 21{neigbours} ∩ R28 ={15, 16, 17, 20, 22} ∩ {10, 15} ={15}. So the prediction moves to word starting with letter 15 (or p) in region 18. Prediction is correct and thus is: 22→21→15. From there the word prediction moves to region 19. Following our earlier calculation again, word starting with letter 15{neigbours} ∩ R19 ={10, 11, 16, 20, 21} ∩ {10, 11, 15, 16} ={10, 11, 16}.

Note there are there possible prediction transitions here, which are 15→10 (or p→k), 15→11 (or p→l) and 15→16 (or p→q). In such a situation, we would like to do the data mining, slightly differently as in table 2. The final score for 15→10 = 707 + x1*0.5 + x2*0.5, for 15→11 = 40 + x3*0.5 + x4*0.5 and for 15→16 = 54 + x5*0.5 + x6*0.5. Here, we are taking a corruption factor of 0.5 that has to be multiplied with the 'indirect transition counts'. When we consider the transition 15→10 (or p→k), transitions like 15→11→10 (or p→l→k) and 15→16→10 (or p→q→k) are termed as indirect transitions, as they have the source and destination words correct. Total score = score of 2 word transition + corruption factor * {sum of scores of 3 word transition}. In this case, we found our prediction not correct (to be from 15→10 (or p→k), instead of the correct path 15→16 (or p→q)). But the concept will definitely work for other cases as the word path database gets populated.

TABLE II. TEXT PATTERN MINING FOR THREE NEIGHBOURING WORDS

| No. | Path | Frequency |
|---|---|---|
| 1 | 15→10 | 707 |
| 2 | 15→11→10 | *x1* |
| 3 | 15→16→10 | *x2* |
| 4 | 15→11 | 40 |
| 5 | 15→10→11 | *x3* |
| 6 | 15→16→11 | *x4* |
| 7 | 15→16 | 54 |
| 8 | 15→10→16 | *x5* |
| 9 | 15→11→16 | *x6* |

### B. Generation of Word Prediction Rules

An example of word prediction rule generation from our previous examples is given in table 3. Word prediction rule can thus be generated as in table 3, after the mining is done for specific word transitions. These prediction rules differ depending on the direction of prediction movement. Since we incorporate the concept of region, once the prediction enters a region, it gets associated to one of the neighbouring words, depending on the direction of next node or word movement. That narrows down the probability of error as the movement can happen to either 1, 2 or 3 words.

TABLE III. WORD PREDICTION RULE GENERATION

| No. | Region Movement | Predicted Word Transition |
|---|---|---|
| 1 | Word with 13:R15→R9 | Word with 8 (correct) |
| 2 | Word with 22:R26→R25 | Word with 21 (correct) |
| 3 | Word with 15:R18→R19 | Word with 10 (incorrect) |

### C. Prediction of Word

Once the word prediction rules are generated, the predicted next word is the one whose path or transition has the highest frequency. In the example that we are discussing, based on table 3, the next predicted word would be word starting with letter 8 (or i), as per the first rule in table 3.

## IV. Simulation Performed

Simulation program was written in Java and the simulation has been done with a matrix formation of word possibilities, to check the accuracy of the first-level text prediction. It was done to check the accuracy for the first level prediction of each word with the first letter as any of the 26 English alphabets. The simulation is run to create 10,000 lines of words and this can be stored in a central server, as history of user behaviour. This data set is used for data mining and to create the word/text prediction patterns. Later, a test set of 10 to 20 random text lines were created (with a maximum of six to eight words and a minimum of three words) and tested for prediction accuracy. Two scenarios are considered here.

### A. Fixed Scenario

The assumptions made in this scenario are as follows:

1. The words are arranged as nodes in a matrix form (with 5 rows and 5 columns), with possible transition from one word to the next word, placed as neighboring nodes.
2. Fixed words are stored in the node.
3. A word will have only a fixed number of maximum eight neighbors.
4. Transition is only permitted to a neigbouring node.
5. When the user keys in the first letter, the region can be confirmed and if he changes the first letter, he can only change it to a subset of words within that region.

Refer to table 4 that show 20 random samples. The numbers 0-25 are mapped to a-z respectively. Accuracy is shown for the first level prediction, where by the text prediction reaches a node, at the highest level for that alphabet. The predicted line contains, for example entries like 10(12)[10, 1] (equivalent to k(12)[k,1]) in row 2 of table 1. The 10 (or k) is the predicted next word and 12 within the normal parenthesis is the region. Inside the square brackets, 10 (or k) is the actual/original text and 1 is the frequency rank of the predicted word. For prediction success the frequency rank would be 1.

TABLE IV. PREDICTION TEXT TABLE USING DATA MINING AND LETTER MATCHING FOR 10 RANDOM LINES.

| Predicted Line (Predicted Text No [Original Text No, Rank]) | Prediction Accuracy (%) |
|---|---|
| 10(12)[10]→5(6)[5, 1] | 100 |
| 16(19)[16]→10(12)[10, 1]→5(6)[5, 1] | 100 |
| 5(6)[5]→0(0)[0, 1] | 100 |
| 15(18)[15]→10(12)[10, 1] | 100 |
| 17(20)[17]→11(13)[11, 1]→5(6)[5, 1]→0(0)[0, 1] | 100 |
| 21(25)[21]→15(18)[15, 1]→10(12)[10, 1] | 100 |
| 6(7)[6]→1(1)[1, 1]→0(0)[0, 1] | 100 |
| 18(21)[18]→13(15)[13, 1]→8(9)[8, 1]→3(3)[3, 1]→2(8)[7, 2] | 75 |
| 2(2)[2]→1(1)[1, 1]→0(0)[0, 1]→5(6)[5, 1] | 100 |
| 3(3)[3]→2(2)[2, 1]→1(7)[6, 2]→5(6)[5, 1]→0(0)[0, 1] | 75 |

The average accuracy of DM with LM prediction was found to be 95% for 10 paths as in table 1. But the average text prediction accuracy for 10,000 lines of words checked with a maximum of six words in a line was found to be 81.95%.

We also wanted to compare our approach with two other baseline schemes. The first prediction method is called the Transition Matrix (TM) prediction scheme. In this method, a word to word transition matrix is formed by considering the previous word to word transitions of users. The predictions are based on this transition matrix by selecting the $x$ most likely words as the predicted words. We used TM for performance comparison because it makes predictions based on the previous word transitions of the user [8]. Assuming x=1 (as the previous scheme also used x=1), the average accuracy of TM was found to be 73.33% for 10 paths as in table 2. But the average TM prediction accuracy for 10,000 lines of words checked with a maximum of six words in a line was found to be 52.49%. Refer to table 5.

TABLE V. PREDICTION TEXT TABLE USING TRANSITION PREDICTION FOR 10 RANDOM LINES.

| Predicted Line (Predicted Text No [Original Text No, Rank]) | Prediction Accuracy (%) |
|---|---|
| 10[10]→5[5, 1] | 100 |
| 16[16]→15[10, 3]→5[5, 1] | 50 |
| 5[5]→0[0, 1] | 100 |
| 15[15]→10[10, 1] | 100 |
| 17[17]→16[11, 3]→10[5, 3]→0[0, 1] | 33.33 |
| 21[21]→20[15, 3]→10[10, 1] | 50 |
| 6[6]→5[1, 2]→0[0, 1] | 50 |
| 18[18]→13[13, 1]→8[8, 1]→3[3, 1]→2[7, 4] | 75 |
| 2[2]→1[1, 1]→0[0, 1]→5[5, 1] | 100 |
| 3[3]→2[2, 1]→1[6, 4]→5[5, 1]→0[0, 1] | 75 |

The second prediction method is the Ignorant Prediction (IP) scheme [9]. This approach disregards the information available from movement history. To predict the next word for a user, this method assigns equal transition probabilities to the eight neighboring words. It means that prediction is performed by randomly selecting $m$ words. We have taken $m$ to be the maximum no. of neighbouring words. The value in the parenthesis in the text prediction shows the corrected text number. The average accuracy of this scheme was found to be 10.83% for 10 paths as expected, and was quite inconsistent. But the overall average IP prediction accuracy for 10,000 lines of words checked with a maximum of six words in a line was found to be 19.3%. Refer to table 6 that show a sample of 10 lines.

TABLE VI. PREDICTION TEXT TABLE USING IGNORANT PREDICTION FOR 10 RANDOM LINES.

| Predicted Line (Predicted Text No [Original Text No] | Prediction Accuracy (%) |
|---|---|
| 10[10]→11[5] | 0 |
| 16[16]→21[10]→11[11] | 50 |
| 5[5]→11[0] | 0 |
| 15[15]→21[10] | 0 |
| 17[17]→22[11]→6[5]→11[0] | 0 |
| 21[21]→22[15]→21[10] | 0 |
| 6[6]→11[1]→6[0] | 0 |
| 18[18]→12[13]→17[8]→9[3]→9[7] | 0 |
| 2[2]→3[1]→0[0]→1[5] | 33.33 |
| 3[3]→7[2]→6[6]→1[5]→11[0] | 25 |

A comparison line graph in figure 2 shows the prediction accuracy for the three different schemes, for the predicted lines discussed before. It is very clear that our proposal on text prediction using data mining (DM) and letter matching (LM) which averages 93.5% prediction accuracy is better compared to TM (53.17% accuracy) and IP (19.67% accuracy) schemes for a random sample of 20 lines.

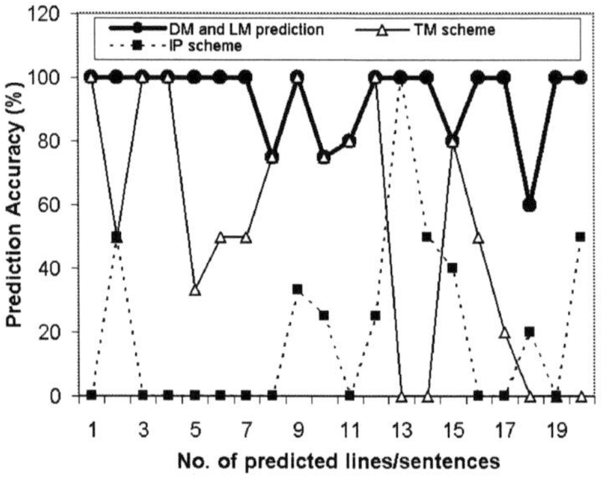


Figure 2. Under fixed scenario, the prediction accuracy graph for the random sample of 20 lines for text prediction through data mining and letter matching, through TM and through IP schemes.

### *B. Flexible Scenario*

The assumptions made in this scenario are as follows:

1. The words are NOT arranged as nodes in a matrix form. The prediction of a next word starting with any 26 English alphabets could move to any other word starting with 26 alphabets or its subset (like high frequency transition words).
2. A word may have all possible word neighbours (starting with any 26 alphabets) or a set of high ranking eight neighbors.
3. Transition is permitted to any neigbouring node or the high frequency transition words.
4. When the user keys in the first letter, the possible next word can be confirmed and if he changes the first letter, he can change it to different word with any letter.

The simulation was thus done for a more realistic case where, transition from one word could happen to 26 possible words, representing all letters in English. For 1000 random lines tested, the prediction accuracy of DM with LM prediction was the highest (87.5%), followed by low values for TM (7.69%) and IP (3.95%) schemes. For DM with LM prediction, we allowed a subset of possible eight high frequency transition words from a given word. This was decided based on the frequency of word 1 to word 2 transitions. If the initial prediction is wrong, then the next word is displayed, based on the initial first letter typed. For TM scheme only the highest ranking word transition was considered and for IP scheme a random transition was made out of 26 possibilities. That explains why TM and IP performance looks very poor as shown in figure 3.

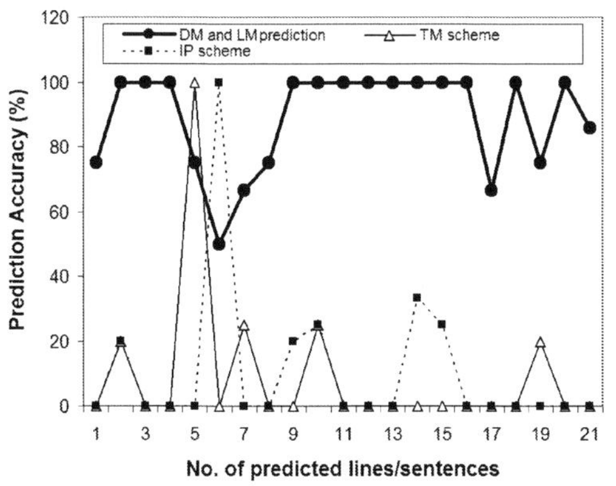


Figure 3. Under flexible scenario, the prediction accuracy graph for the random sample of 20 lines for a 'more realistic' text prediction through data mining and letter matching, through TM and through IP schemes.

## V. FURTHER IMPROVMENTS

We can further improve our prediction to a much more realistic scenario as follows. A word node can have maximum $n$ neighbors (with high frequency transition). A neighbor is a word collection having the same initial alphabetical pattern. Once the user types in a letter, the prediction can be more focused into a node with word collection that matches the user's typed letter in the word being written. After the first word is typed, the word with the highest frequency rank (of word 1 → word 2) transition is displayed. He may choose that word or types in a new letter. In that case, the highest frequency word starting with that letter is displayed. The process would happen recursively for the word collection within a node, where $n$ neighbors (with high frequency transition) would be formed conceptually as in figure 4.

## VI. CONCLUSION

This paper investigates into text prediction in mobile devices to help users to type words in messages that are sent through 802.11 mobile devices, like a PDA. Since typing all the letters and words in a mobile device can be quite tedious, some form of text prediction can help which uses the user's past typing behaviour and selection of words. The proposed first-level text prediction through data mining and letter matching is checked through simulation for fixed and flexible scenarios and the results were quite encouraging, when compared to other baseline schemes. Though the prediction is done with English language, it can be easily extended to any similar language. The next step in this research would be to look at the recursive aspect of prediction within each node.

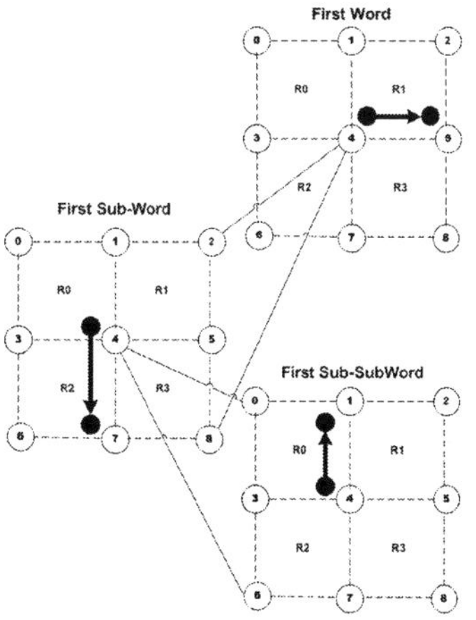


Figure 4. Once the user types a word, conceptually around the central word (here node 4), eight words with the highest frequency is chosen virtually, based on the word 1 → word 2 transitions. Each node can recursively probe into a similar pattern to find the exact match.

## REFERENCES

[1] M. Wang, W. Liu, and Y. Zhong, Simple recurrent network for Chinese word prediction, In proceedings of International Joint Conference on Neural Networks, 1993, pp.263-266.

[2] Y. Even-Zohar, Using focus of attention in classification. In proceedings of XX International Conference of the Chilean Computer Science Society, 2000, pp. 109-116.

[3] M. Nakamura and M. Shikano, A study of English word category prediction based on neutral networks. In proceedings of International Conference on Acoustics, Speech, and Signal Processing, 1989, pp.731-734.

[4] Y. Ajioka and Y. Anzai, Prediction of next alphabets and words of four sentences by adaptive junction. In proceedings of International Joint Conference on Neural Networks, 1991, pp.8-14.

[5] G. Fu and K. K. Luke, Integrated approaches to prosodic word prediction for Chinese TTS. In proceedings of International Conference on Natural Language Processing and Knowledge Engineering, 2003, pp.413-418.

[6] H. Dong, J. Tao, and B. Xu, Prosodic word prediction using the lexical information. In proceedings of IEEE International Conference on Natural Language Processing and Knowledge Engineering, 2005, pp.189-193.

[7] G. Yavas, D. Katsaros, O. Ulusoy and Y. Manolopoulos, A Data Mining Approach for Location Prediction in Mobile Environments. Data and Knowledge Engineering, 2005, pp. 121–146, (2005).

[8] Rajagopal, S., Srinivasan, R.B., Narayan, R.B., Petit, X. B. C: GPS-based Predictive Resource Allocation in Cellular Networks. In proceedings of IEEE International Conference on Networks, 2002, pp. 229–234.

[9] A. Bhattacharya and S. K. Das, LeZi-Update: An Information-Theoretic Approach to Track Mobile Users in PCS Networks, ACM Wireless Networks, 2002, pp. 121-135.

# Visualization of Large Software Projects by using Advanced Techniques

Juan Garcia, Roberto Theron, Francisco Garcia
University of Salamanca, Spain
{ganajuan, theron, fgarcia}@usal.es

*Abstract*- **Large Software Projects need to be represented in a conceptual way for humans to understand. Nevertheless, most of the time the diagrams involved are far from clear and understandable. The use of advanced visualization techniques is useful for large projects and representations in order to clarify the abstraction process. This paper proposes the use of well known visualization techniques to enrich UML models in the structural and hierarchical representation of these types of projects. A first prototype has been developed to test the proposal and the results are described below.**

**Keywords**: Software Visualization, Treemap, Table Lens

## I. INTRODUCTION

Modeling is an activity that has been carried out over the years in software development. Defining a model makes it easier to break up a complex application or a huge system into a simple one. Many proposals have been made during years, the most important being Rumbaugh, Booch and Jacobson methodologies [1], [2]. A standardized way of modeling was needed, and OMG (Object Management Group) [3] proposed the Unified Modeling Language (UML) which is now the most widely used language of modeling. The UML 2 Specification [4], [5] defines six diagram types that represent static application structure. Class diagrams being the most mainstay of object oriented analysis and design. UML 2 class diagrams show the classes of the system, their interrelationships (including inheritance, aggregation and association), and the operations and attributes of the classes. Class diagrams are used for a wide variety of purposes, including both conceptual / domain modeling and detailed design modeling [6]. For the conceptual model classes are depicted as boxes with three sections, the top one indicates the name of the class, the middle lists the attributes and the third lists the methods.

A class diagram is mainly composed of Classes, Responsibilities, Associations and Inheritance relationships. In a large project, the number of classes, responsibilities, associations and relationships can be huge, which means that the class diagram can be hard to explore.

Nowadays, open source software is becoming more popular to software developers and researchers, due to the advantages it offer to the developer's community. Open Source repositories like *sourceforge* [7] are intended to make accessible a wide variety of software projects to everyone interested in them. These software projects are commonly developed without cost, and the source code is made available for public collaboration.

In order for this code to be modified, the structure, interrelationships, class diagrams, etc need to be known. This can be done by providing the UML diagrams involved in the analysis and design phases. However, it is not common to provide this information, so a reverse engineering process is commonly used to get back these diagrams. There are a lot of UML tools that can be used to reach this goal. Reference [8] gives an overview of some UML tool implementations, and discusses the different layout algorithms implemented by them. These implementations of CASE tools, do not provide standardized layout algorithms drawing UML diagrams. This means that each tool produces different diagrams as a result of applying its own algorithm. Furthermore, the complexity of implicit large software projects makes more the abstraction process difficult. Another problem appears when there are a lot of inheritance and association relations that make the class diagram difficult to explore.

This work aims to propose a new way to visualize large class diagrams by using some advanced visualization techniques without losing the UML-like perspective, to simplify understanding of static software

T. Sobh (ed.), *Innovations and Advances in Computer Sciences and Engineering*,
DOI 10.1007/978-90-481-3658-2_56, © Springer Science+Business Media B.V. 2010

structures. These techniques are combined into a visualization which the user can interact with and manipulate to find information or to explore the diagram.

The next section summarizes some related work. Section 3 summarizes some important aspects of the visualization techniques and explains advantages of using the techniques. Section 4 analyzes a case study and finally section 5 discusses conclusions and future work.

## II. RELATED WORK

There are some works about combining visualization techniques and UML. Brian Berenbach [9] proposed some techniques and heuristics to analyze large UML models, without focusing on diagrams.

Reference [10] focuses on the use of Virtual Reality for software visualization in order to comprehend very large software systems, to support the development, maintenance and reverse engineering phases of software engineering, by using a highly interactive, collaborative and immersive visualization environment. This is an interesting project, but the use of 3D models is time-processor and memory resources expensive when dealing with very large software projects. Furthermore, the use of Virtual Reality requires an environment that is not supported by the CASE tools, and cannot be easily added as a plugin for example.

Reference [11] proposes a framework to emphasize the tasks of understanding and analysis during development and maintenance of large-scale software systems. These tasks include development activities, maintenance, software process management and marketing. Based on these specific tasks, the user needs to obtain different levels or types of understanding of the software.

Ref. [12] proposes an architecture based on two concepts: datasets and operations. As in classical SciViz dataflow, operations can read and write parts of datasets. This proposal is based on the use of clustered graphs applied on Reverse Engineering Process. This is a traditional graph representation which has the well known problems of crowded graphs which become unreadable.

## III. USING VISUALIZATION TECHNIQUES

### USING TREEMAPS TO ORGANIZE INTO A HIERARCHY

Treemap is a space filling visualization technique widely used for large hierarchical data sets. It was first proposed by Shneiderman during the 1990's. It graphically represents hierarchical information via a two dimensional rectangular map, providing compact visual representations of complex data spaces through both area and color. A treemap works by dividing the display area into a nested sequence of rectangles whose areas correspond to an attribute of the data set. This effectively combines aspects of a Venn diagram and a pie chart and can represent both hierarchical structure and each element's quantitative information simultaneously, thus utilizing 100 percent of the designated screen area. They have been applied to a wide variety of domains ranging from financial analysis [13], [14] gene ontology [15] and sports reporting [16]. Due to the natural squared shape of treemap nodes, it can be easily used to represent classes as the UML definition requires.

The treemap is used to represent the hierarchical structure of the classes in the project, avoiding the classical use of lines to represent hierarchical relations, which makes the diagram less crowded with lines and clearer to understand.

### FOCUS + CONTEXT AND SEMANTIC ZOOM FOR DETAILED INFORMATION

Information Visualization often deals with data of which users have no mental image. Visualization imposes a graphical structure –a mapping from data to screen space- on the data that users have to learn. It is therefore necessary to change this mapping as little as possible; but often there is not enough space on the screen to display all information with enough detail. Focus + Context (F + C) methods make it possible to show more detailed or targeted information, and at the same time gives users a sense of where in the data the zoomed-in, more detailed, or pointed out information is [17]. This technique can perform Semantic Zoom, which means that the focused item is shown with specific information that is not visible when the item is not focused. Multi-focus is also used in Table Lens to select more than one class. This is useful for selecting all related classes to a selected.

### USING TABLE LENS FOR AN OVERVIEW OF INTERNAL COMPONENTS

Reference [21] proposes Table Lens Focus+Context technique. This technique is based on a distorted Table that is able to manage a huge number of data, more than the simple spreadsheets can, without losing framing context. In the Table Lens, a Degree of Interest function maps from a cell address to an interest level, and each of the two dimensions has an independent function.

## IV. CASE STUDY

Our prototype was proven by using as an input a Java™ jar file containing a huge amount of class files. These files were loaded into memory using Java™ reflection. Diverse files were used, including *lucene-1.4.3.jar* the apache lucene project v1.4.3, which contains 211 classes and can be freely downloaded. Other used file was *servlet.jar* the servlet api jar file, with 55 files. Two of our own projects were used too, the largest with 291 classes, including classes from *iText.jar* file which is used to create pdf files from a java class[1]. Treemaps can be used to represent some thousands of classes as well. Java reflection was used to deal with jar files in order to load the classes.

### REPRESENTING THE HIERARCHY

The project's hierarchy is represented by using the treemap technique. Treemap was fed with all the loaded classes in order to visualize them. This is done by taking all the classes starting with the superclass Object and recursively subclasses are nested into their superclass. Then for each subclass the process is repeated resulting in some nested treemaps. The treemap can be colored by considering different options as the package, number of methods, attributes or none. Fig. 1 shows the result of the hierarchical representation of a Java™ jar file with almost 300 classes. The process to build the treemap starts with the superclass Object in the highest level which is represented as a treemap containing only this class. Then, the second level treemap is built by taking the direct subclasses of Object. This process is repeated for each subclass in the jar file, resulting in nested treemaps according to the hierarchy. Then the inheritance relations are implicitly represented by self-contained elements. Each class is represented as the typical UML class diagram shape.

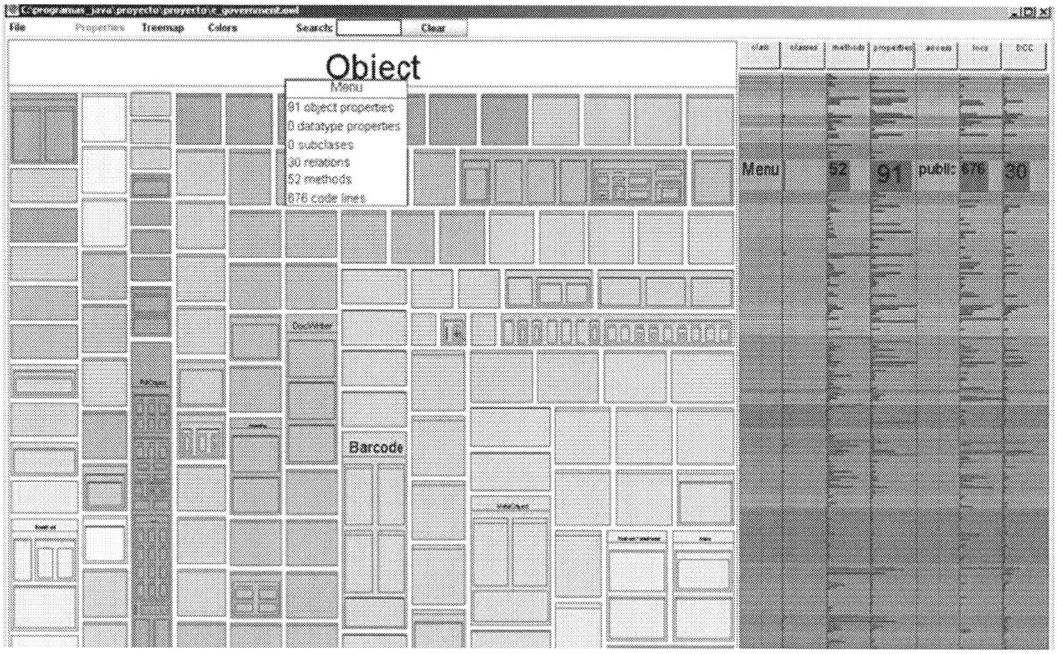


**Fig. 1 a general view of treemap hierarchy and table lens**

### REPRESENTING RELATIONSHIPS

Inheritance relationships are represented in the treemap in a direct way by nested rectangles representing subclasses. Other types of relationships are represented by using Bézier Curves. [18] [19] [20] explain the use of this type of curves to represent graph edges. Basically having a source and destiny points, instead of using a simple line to connect them, two control points are calculated. These control points define the curve amplitude, and the line is colored by using a range of two colors indicating the source and destiny. The source is colored in red and goes through the edge, the color varies in tone to blue. This helps to

---

[1] *It can be freely downloaded at http://www.lowagie.com/iText/*

avoid the typical arrowhead that makes the diagram overcrowded with lines and harder to understand. Fig. 2 shows the treemap but focuses on the relations of one class. The same class is selected in the Table Lens and all related classes are red coloured in their header, in order to make them more visible. Not related classes are blured in order to related ones being highlighted.

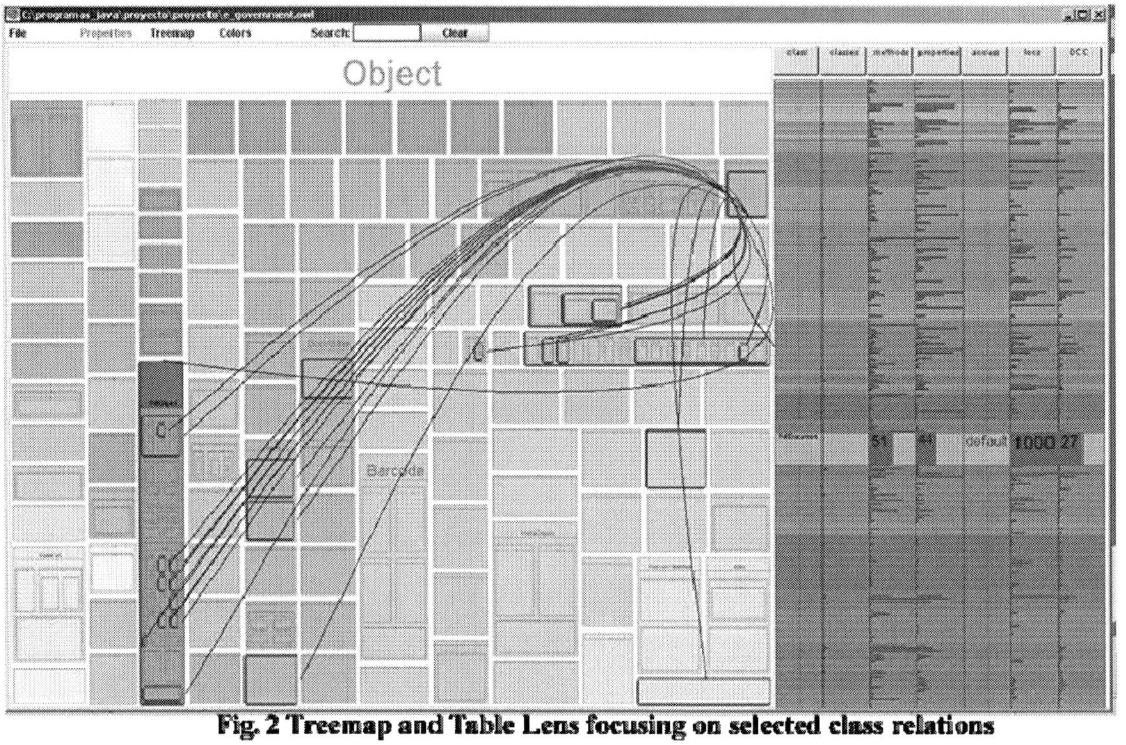

Fig. 2 Treemap and Table Lens focusing on selected class relations

AN OVERVIEW OF INTERNAL COMPONENTS

Treemap is useful to visualize the whole structure, but we do not know nothing about the internal components classes until we select a class. We can be interested in an overview of these components without being so detailed to select a class, or a general idea about the ontology population. Table lens can help us to have this overview. Table Lens technique has been motivated by the particular nature of tables. The most salient feature of a table is the regularity of its content: information along rows or columns is interrelated, and can be interpreted on some reading as a coherent set, e.g. members of a group or attributes of an object. The use of Table Lens makes the larger tables more understandable. The visualization uses Focus+Context technique that works effectively on tabular information because it permits the display of crucial label information and multiple distal focal areas. This technique is used to represent general aspects of the classes, like the number of declared methods, properties, internal classes, access modifier. Fig. 1, Fig. 2 and Fig. 3 show the view of the table lens on a selected class. The Table Lens displays information about all classes. It starts by using the same color as in the treemap which is related to a specific value, then to the number of declared classes, the number of properties and methods are represented, as well as the access modifier, total of line codes and finally the value of DCC (Direct Class Coupling). Table Lens can be ascending or descending ordered by any column. This is useful to detect those classes that have the most or the less value for any parameter such as attributes, methods or lines of code if source code is available.

Some desing patterns can be discovered by doing an analysis. For instance, Fig. 2 shows that selected class is closely related with polymorphism of classes. It seems that heterogeneity of classes is common due to there are relations from selected class to some other classes and many of their own subclasses.

Table lens is descending-ordered by DCC (Direct Class Coupling) value. Direct Class Coupling refers to the measurement of the interdependencies between classes of the design. A high coupling value has a negative influence on almost every quality attribute. The most negative effect it has on reusability, flexibility, understandability and extendibility. The higher the value the more negative its effect. DCC is depicted on Fig. 3 with curves in treemap.

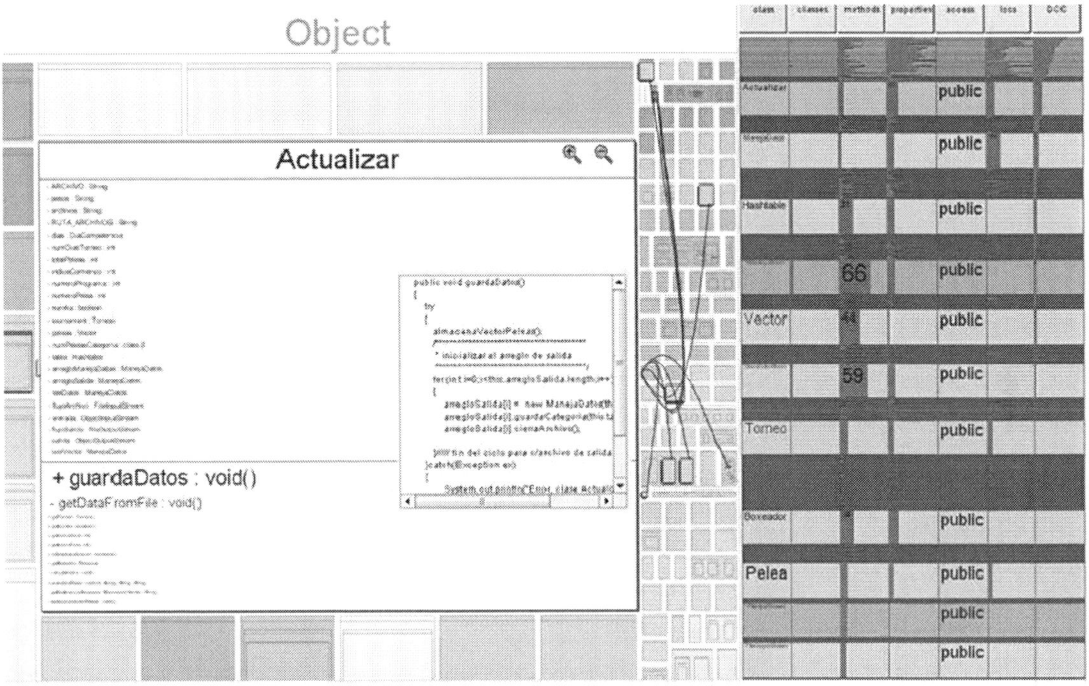


**Fig. 3 Semantic Zoom on a selected class and source code method**

## SEMANTIC ZOOM OVER A SELECTED CLASS

Semantic zoom is done by distorting the treemap class dimensions. It focuses on the selected class by resizing all the other ones, and giving more space to selected class. The focused class can be explored in its content, methods, properties and source code . The fish eye technique is used to visualize the properties and methods information when needed. Fig. 3 shows a semantic zoom on a selected class like a UML classical view. The attributes and methods are displayed and the source code of each method is shown in a separated window. The same class is focused on the Table Lens panel and all related classes are focused as well, using a multi-focus technique. Selected class relations are shown as curves and classes are "came to the front" while other classes are blured.

## CONCLUSIONS AND FUTURE WORK

This proposal offers new possibilities to explore large-scale software projects that have not been used yet. Most of the time open source software diagrams are not provided and a reverse engineering process is needed. The visualization techniques used have been applied in other fields with excellent results. Currently it is being adapted to be used in the Software Engineering field. 300 classes is a good parameter to test, projects with some thousands of classes are going to be tested as well.

This first prototype aims to visualize in a comprehensible way, the whole hierarchy of this kind of projects, allowing searching and focusing on certain classes.

The use of the visualization treemap style makes the abstraction of an unknown hierarchy clearer for the user. By combining different techniques, the user is able to explore the whole structure or part without losing the context. UML-like class views help the user as it is the most widely used representation, and the possibility to explore the source code of methods is useful too. Including DCC metric is a first step to evolve as an analysis tool.

There is a lot of work to be done, starting by representing interesting software developing aspects like cohesion and coupling metrics. Javadoc comments and annotations if present in the source code can be shown to help the user to familiarizes themselves with it. More analysis can be done by using metrics, more software metrics would be desirable to be included.

## ACKNOWLEDGEMENTS

This work was supported by the "Ministerio de Educación y Ciencia (projects TSI2005-00960 and CSD 2007-00067)".

## REFERENCES

[1] G. Booch, "Object-Oriented Analysis and Design with Applications", *Addison-Wesley, 2004.*

[2] G. Booch, J. Rumbaugh and I. Jacobson, "Unified Modeling Language User Guide The", *Addison-Wesley, 1998.*

[3] Object Management Group (OMG), http://www.omg.org.

[4] OMG, "OMG Getting started with UML", http://www.omg.org.

[5] OMG, "Unified Modeling Language (UML) version 2.1.2", http://www.omg.org.

[6] S. Ambler, "The Object Primer: Agile Model-Driven Development with UML 2.0", *Cambridge University Press, 2004.*

[7] Open Source Software repository, http://sourceforge.net.

[8] H.Eichelberger and J. W. Gudenberg, "UML Class Diagrams – state of the Art in Layout Techniques", Wurzburg University, 2003.

[9] B. Berenbach, "The evaluation of large, complex UML Analysis and Design models", *Proceedings of the 26th. International Conference on Software Engineering (ICSE04)", 2004 .*

[10] J. I. Maletic, A. Marcus, and M.L. Collard, "Visualizing Software in an Immersive Virtual Reality Environment".

[11] J. I. Maletic, A. Marcus, and M.L. Collard, "A task oriented view of software visualization", *Proceedings of the First International Workshop on Visualizing Software for Understanding and Analysis (VISSOFT.02)", 2002 .*

[12] A.Telea, A. Maccari and C. Riva, "An open visualization toolkit for reverse architecting", Proceedings of the 10th. International Workshop on Program Comprehension (IWPC02), 2002.

[13] W. A. Jungmeister, "Adapting treemaps to stock portfolio visualization", *1992.*

[14] M. Wattenberg, "Map of the market", http://www.smartmoney.com/marketmap, 1998.

[15] K. Babaria, "Using treemaps to visualize gene ontologies", *University of Maryland, 2001.*

[16] Jin and Banks, "Tennis Viewer: a browser for competition trees", *IEEE Computer Graphics and Applications,* 17(4), 1997, pp. 63-65.

[17] R. Kosara, S. Miksch and H. Hauser, "Focus+Context taken literally", *IEEE Computer Graphics and Applications, 2002.*

[18] J. D. Fekete, D. Wang, N. Dang, A. Aris and C. Plaisant, "Overlaying Graph Links on Treemaps", *Information Visualization 2003 Symposium Poster Compendium, IEEE (2003), pp. 82-83.*

[19] N. Wong, S. Carpendale and S. Greenberg, "EdgeLens: An Interactive Method for Managing Edge Congestion in Graphs", *IEEE*, 2003, pp. 51-58.

[20] D. Holten, "Hierarchical Edge Bundles: Visualization of Adjacency Relations in Hierarchical Data", *IEEE Transactions on visualization and computer graphics*, Vol. 12, NO. 5, September-October 2006.

[21] R. Rao and S. K. Card, "The Table Lens: Merging Graphical and Symbolic Representations in an Interactive Focus+Context Visualization for Tabular Information", *Proceedings of the ACM SIGCHI Conference on Human Factors in Computing Systems*, Boston MA. ,April 1994.

# A multi level priority clustering GA based approach for solving heterogeneous Vehicle Routing Problem (PCGVRP)

M.Mehdi S.Haghighi , M.Hadi Zahedi and Mehdi Ghazizadeh

haghighi@ieee.org, zahedi@ieee.org and Mo_gh@stu-mail.um.ac.ir

Computer Department, Faculty of Engineering, Ferdowsi University of Mashhad, Iran

## Abstract

This research presents a two phase heuristic - evolutionary combined algorithmic approach to solve multiple-depot routing problem with heterogeneous vehicles. It has been derived from embedding a heuristic-based two level clustering algorithm within a MDVRP optimization framework. In logistic applications, customers have priority based on some logistic point of view. The priority levels of customers, affect distribution strategy specially in clustering level. In this research we have developed an integrated VRP model using heuristic clustering method and a genetic algorithm, GA, of which operators and initial population are improved. In the first phase of the algorithm, a high level heuristic clustering is performed to cluster customers serviced by a special depot. Next, a low level clustering is done for each depot to find clusters serviced by a single vehicle. Likewise other optimization approaches, the new formulation can efficiently solve case studies involving at most 25 nodes to optimality. To overcome this limitation, a preprocessing stage which clusters nodes together is initially performed to yield a more compact cluster-based problem formulation. In this way, a hierarchical hybrid procedure involving one heuristic and one evolutionary phase was developed.

Keywords: Vehicle Routing, Clustering, Evolutionary Algorithm, Genetic Algorithm.

## 1. Introduction

The classical vehicle routing problem (VRP) involves routing a fleet of vehicles, each visiting a set of nodes such that every node is visited exactly once and by exactly one vehicle, with the objective of minimizing the total distance traveled by all vehicles. It has been investigated exhaustively and has been proved to be np-hard [1, 8]. The classical VRP has variants depending on tasks performed and on some constraints. The problem is to be addressed under the conditions that all vehicles must return to the node of origin (depot) and that every node will be visited exactly once [1]. We seek to develop a unified heuristic that addresses the MDVRP, solves it within reasonable limits of accuracy and computational time and is also applicable to the classical VRP and VRP with back-hauls (VRPB). When embedded in the planner's software, it can be a valuable tool towards providing service of high quality with low cost[7]. Such a problem exists in distribution of foods in urban areas. In such problems, single or multiple depots may exist. To solve such problems, we introduce a multi level clustering method. Cluster may be described as connected region of a multi-dimensional space containing a relatively high density of points, separated from other such regions by a region containing a relatively low density of points. This definition of group is an excellent reason to use cluster analysis in the resolution of VRPs. Recognizing groups of customers can be a good start to obtain good VRP solutions [2, 7]. In this solution, a high level clustering is done by finding customer

T. Sobh (ed.), *Innovations and Advances in Computer Sciences and Engineering*,
DOI 10.1007/978-90-481-3658-2_57, © Springer Science+Business Media B.V. 2010

clusters assigned to a depot. A low level clustering will find clusters of customers each assigned to a single vehicle based on their demands and priority. In logistic applications, customers have priority based on some logistic point of view. The priority levels of customers, affect distribution strategy specially in clustering stage. Next, an evolutionary algorithm, finds optimum route for each customer cluster assigned to a single vehicle.

The remainder of the paper is organized as follows. Section 2 describes the problem based on mathematical descriptions. According to the descriptions, some basic and important constraints are defined in this section. Part 3 and 4, introduce new hierarchical approach to group customers based on some heuristic criteria. The hierarchical approach, finds hierarchically, clusters of customers to be serviced. in part 5, an evolutionary algorithm based on population is described to find minimum route in each cluster. Finally, section 6 compares the results, obtained by the algorithm executed on known benchmark problems, with best results obtained by other problems on same data.

## 2. Formal Definition

Let us consider a routing network, represented by the directed graph G{I, P, A}, connecting customer nodes $I = \{i_1, i_2, \ldots, i_n\}$ each one with coordinates $I_{ix}$ , $I_{iy}$ and priority Ti and depot nodes $P = \{p_1, p_2, \ldots, p_l\}$ with coordinates $P_{ix}$ , $P_{iy}$ through a set of directed edges $A = \{(i, j), i, j \in (I \cup P)\}$. The edge $(i, j) \in A$ is supposed to be the lowest cost route connecting node i to node j. At each customer location $i \in I$, a fixed load $w_i$ is to be delivered . Matrix W denotes demands or loads each customer should receive. A fleet of heterogeneous vehicles $V = \{v_1, v_2, \ldots, v_m\}$ with different cargo-capacities $(q_{v,p})$ housed in multiple depots $p \in P$ are available to accomplish the required delivery tasks. Each vehicle v must leave from the assigned depot $p \in P$, deliver the full load to several supply points and then return to the same terminal p. Then, the route for vehicle v is a tour of nodes $r = (p, \ldots, i, (i + 1), \ldots, p)$ connected by directed edges belonging to A that starts and ends at depot p assigned to vehicle v. Assigned to each depot $p_i$ , there are some clusters of customer nodes $C = \{ c_{i,j,k} \}$ where $C_{i,j}$ denotes customer node k in cluster j serviced in the depot i by some vehicle. $R_i$ is maximum range of customers can be serviced by depot i. $S_i$ is maximum distance of customers in cluster I from center point of the cluster t.

## 3. Hierarchical approach

There is no doubt that the multiple-depot heterogeneous fleet VRPTW is very difficult to solve through a pure optimization approach [2]. In fact, even simpler vehicle routing problems are among the most difficult class of combinatorial optimization problems. Current fastest algorithms can discover the optimal solution for single-depot homogeneous fleet VRPTW problem instances. However, heuristics usually lack robustness and their performance is problem dependent. Instead, optimization algorithms offer the best promise for robustness [2]. Given the enormous complexity of large VRPTW problems, however, it does not seem realistic to apply pure optimization methods. Instead, we can focus on hybrid solution algorithms that can be as robust as the optimization methods and capable of discovering good solutions for large problems. In this work, it has been developed a hierarchical hybrid solution approach that integrates a heuristic clustering procedure into an optimization framework based on evolutionary algorithms [2]. Clusters of nodes are first formed, then these clusters are attached to vehicles and sequenced on the related tours and finally the routing for each individual tour in terms of the original nodes is separately optimized based on an evolutionary algorithm. In this way, a three phase VRP hierarchical hybrid approach has been defined. Since the problem is used in logistic areas where customers have different priorities, before clustering takes place, the priorities should be considered so that customers with higher priorities are clustered before lower priority customers, therefore serviced by vehicles available sooner. Vehicles are also sorted by capacity and availability, so higher priority customers are clustered first and serviced by available vehicles sooner.

## 4. Clustering algorithm

Clustering algorithm is done in two steps: high level and low level. In high level clustering, customers serviced by a depot are clustered. Then in low level clustering, customers serviced by a single vehicle are clustered. Before clustering begins, customers and vehicles lists are sorted based on some criteria. First, customers are sorted based on decreasing order of priorities and demands. In logistic situations, priority has very important effect in algorithm. Next, high level clustering is done for each depot P. Another criterion in high level clustering is Euclidian distance between depot and each customer. In this level, a radius R is defined so that no customer in the cluster has radius more than R. The constraints defined in this level are as follows:

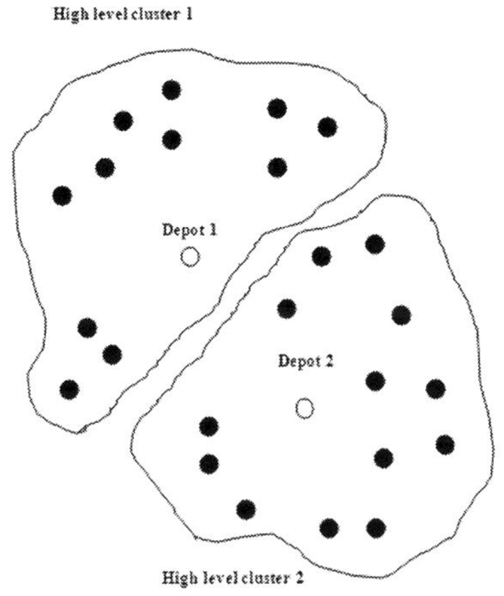


Figure 1: High level clustering

In low level clustering, for each of the clusters found in high level clustering process, vehicles are selected based on the criteria mentioned earlier, next, customers which can be serviced by a single vehicle are clustered . To do this, one high priority customer in the high level cluster is selected as center point. Next, neighbors of the center point are found in the region determined by radius R'. The step continues until total customer demands reaches to the capacity of the selected vehicle. The constraints defined in this level are as shown in equation 3, 4 and 5.

$$\sum_{j=1}^{p}\sum_{k=1}^{m} C_{i,j,k} <= \sum_{k=1}^{m} q_k \quad (3)$$

$$\sum_{i \in C_{j}} W_i <= q_k \quad (4)$$

$$\sqrt{(q_{jx} - I_{x})^2 + (q_{jy} - I_{y})^2} <= S_i \quad (5)$$

$$\sqrt{(I_{tx} - P_{tx})^2 + (I_{ty} - P_{ty})^2} <= R_i \quad (1)$$

$$\sum_{t \in CP_i} W_t <= \sum_{f \in V_p} q_f \quad (2)$$

This process is repeated until no other customer remains without being serviced or no other vehicle is available (First Algorithm).

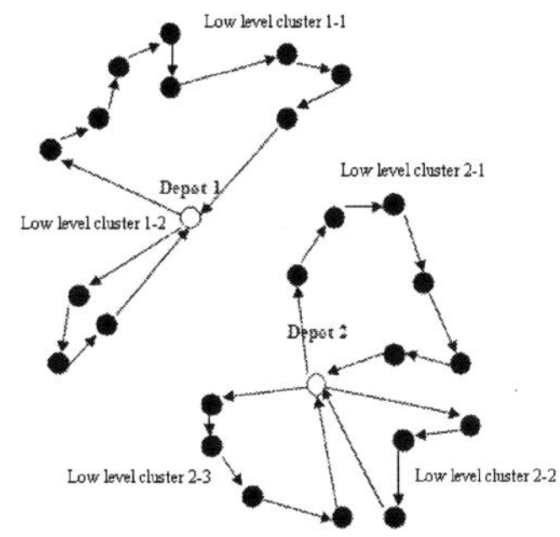


Figure 2: Low level clustering

First Algorithm: Clustering.

*1- Sort list of customers based on their priority and demand T, W , then sort list of vehicles based on their availability and capacity q then sort list of depots P based on their priority.*

*2- Select a depot $P_i$ to form high level cluster $CP_i$ .*

*If no such $P_i$ found, terminate the algorithm.*

*3- Select a customer I from list of customers which satisfies conditions of equation 1 and 2:*

*If no such customer found*

*Terminate high level cluster*

*Creation for $P_i$ , continue from 4 .*

*Else Add I to cluster $CP_i$, continue from 3.*

*4- Select a customer as center point in high level cluster $CP_j$ with highest priority to form cluster $C_j$ for $P_i$ .*

*If no such customer point is found, continue from 2.*

*5- Select a vehicle $V_j$ from list of vehicles for high level cluster j based on their availability and priority.*

*If no vehicle found ,*

*terminate cluster creation for depot $P_i$,*

*continue from 2 .*

*Else continue from 6.*

*6- select a customer $I_k$ from list of customers to satisfy the condition in equation 4 :*

*If no customer found to satisfy the condition,*

*terminate cluster creation for $C_j$ continue from 4 .*

*Else continue from 7.*

*7- add customer $I_k$ to cluster $C_j$ continue from 6.*

## 5. Route finding (Evolutionary step)

By creation of clusters in high and low levels, it is time to find optimum (minimum) route in each low level cluster, so that the vehicle assigned to each cluster can visit customers with lowest cost. Finding such routes needs execution of a population based evolutionary algorithm in each cluster in a way that every individual in the population, shows a route in the cluster[3-6]. Basic assumptions and definitions of the evolutionary algorithm will be described.

Route optimization is done based on distance between the nodes showing customer demands. Each individual is a list of node numbers showing a route in the cluster. Each population is formed as a group of such individuals. Fitness function computes fitness of each individual or route in a cluster from equation 6.

$$Fitness\ (inlen) = \frac{inlen - minlen}{maxlen - minlen} \quad (6)$$

Where, inlen is length of the individual, maxlen is maximum route length of individuals and minlen is minimum route length of individuals.

1 2 3 4 5 6 7 8 9 10 11 12 13

4 2 3 1 5 6 7 8 9 10 11 12 13

4 2 3 7 5 6 1 8 9 10 11 12 13

4 2 3 7 5 6 10 8 9 1 11 12 13

4 2 3 7 5 6 10 8 9 13 11 12 1

Figure 3: Single parent two point crossover

Crossover operator selected for the algorithm is special type of two point crossover. In the algorithm, according to crossover probability, the individuals for crossover selected. Next, randomly select one crossover point and the second point is selected by some step ahead. Then, these two nodes are swapped to generate a new individual replaced by its parent. The same process is done for other individuals selected according to crossover probability. Another operator is elitism. In this algorithm, with a normal distribution, low fitness individuals are replaced by highest fitness one. After enough iteration, the individual with highest fitness will be returned as the best rout in the cluster.

3 5 7 6 9 11 23 8 13 2 4

3 5 8 6 9 11 23 7 13 2 4

Figure 4: Mutation

Mutation takes place for the individuals with low fitness. If fitness of an individual is lower than some threshold randomly chosen, the individual is replaced by the one with highest fitness. Then mutation takes place on the replaced one by selecting two points randomly, and then swaps them.

Second Algorithm: Evolutionary step.

*1 - Form distance matrix D from customer cluster C.*

*2 - Make population pop by randomly generate paths in the cluster $C_i$ as individuals.*

*3 - Compute length of each random path (individual) $L_i$.*

*4- Repeat instructions,5 thru 10, n times.*

*5 - Find fitness of each individual $F_i$.*

*6 - Find individuals with minimum (MinF) and Maximum (MaxF) fitness.*

*7 - Randomly choose individuals with low fitness.*

*8 - Replace selected individuals by the one with maxF fitness.*

*9 - Mutate replaced individuals by swapping two randomly selected nodes of the individual.*

*10 – Make new Childs by crossover then replace them by their parents.*

## 6. conclusion

The algorithm has been tested on different data sets based on Augerat benchmarks (see table 1). As mentioned earlier, the PCGVRP uses nodes based on their priorities. Priority can affect the results even the results obtained by the same algorithm. In some cases, it may cause the results to be longer than the case with no priority. In table 1, the best results obtained from other algorithms shown in right most columns where nodes have no priority. But, the results obtained by PCGVRP are based on priority of nodes. Since the main goal of the algorithm in logistic areas is to obtain minimum length route, based on the priority defined for the nodes, only the total route length of all vehicles is measured for comparison. Comparison of the results with priority and without priority shows that in some cases, priority may cause routes to be longer, but in logistic situations, it is very important to service higher priority nodes before lower priority nodes despite more route length.

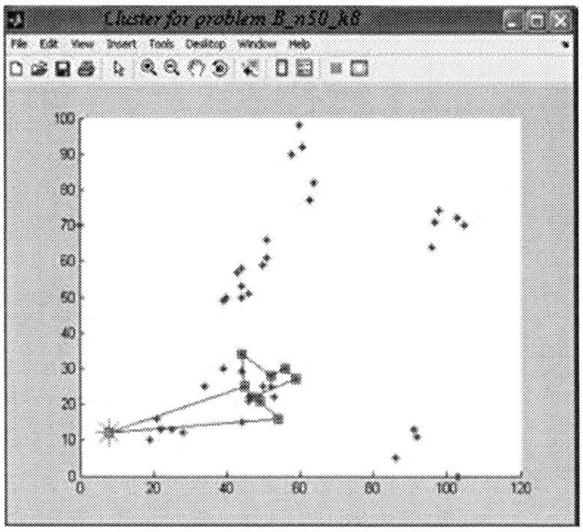


Figure 5: a Cluster for problem B_n50_k8 .

| Problem | # of Vehicles | Vehicle Capacity | # of Nodes | PCHLVRP Results (with priority) | Best Results by previous works (without priority) |
|---|---|---|---|---|---|
| A_n32_k5 | 5 | 100 | 31 | 818 | 784 |
| A_n36_k5 | 5 | 100 | 35 | 808 | 799 |
| A_n44_k5 | 5 | 100 | 43 | 1021 | 937 |
| A_n48_k7 | 7 | 100 | 47 | 940 | 1073 |
| A_n53_k7 | 7 | 100 | 52 | 1170 | 1010 |
| A_n60_k9 | 9 | 100 | 59 | 1551 | 1408 |
| A_n64_k9 | 9 | 100 | 63 | 1488 | 1402 |
| A_n69_k9 | 9 | 100 | 68 | 1862 | 1168 |
| A_n80_k10 | 10 | 100 | 79 | 1834 | 1764 |
| A_n60_k9 | 9 | 100 | 59 | 1179 | 1408 |
| A_n80_k10 | 10 | 100 | 79 | 1722 | 1764 |
| B_n31_k5 | 5 | 100 | 30 | 465 | 672 |
| B_n35_k5 | 5 | 100 | 34 | 596 | 955 |
| B_n41_k6 | 6 | 100 | 40 | 704 | 829 |
| B_n45_k6 | 6 | 100 | 44 | 648 | 678 |
| B_n50_k8 | 8 | 100 | 49 | 828 | 1313 |
| B_n51_k7 | 7 | 100 | 50 | 838 | 1032 |
| B_n63_k10 | 10 | 100 | 62 | 1208 | 1537 |
| B_n68_k9 | 9 | 100 | 67 | 1090 | 1304 |
| B_n78_k10 | 10 | 100 | 77 | 1218 | 1266 |

Table 1: Results based on Augerat et al benchmarks

# References


[1]. Jean-Charles Créput and Abderrafiaâ Koukam .The memetic self-organizing map approach to the vehicle routing problem . Springer-Verlag Soft Comput ,2008.

[2]. Rodolfo Dondo and Jaime Cerda. A cluster-based optimization approach for the multi-depot heterogeneous fleet vehicle routing problem with time windows. European Journal of Operational Research 176 , 2007.

[3]. S. Salhi and R. J. Petch. A GA Based Heuristic for the Vehicle Routing Problem with Multiple Trips. Springer Science, Business Media , 2007.

[4]. Chung-Ho Wang a, Jiu-Zhang Lu . A hybrid genetic algorithm that optimizes capacitated vehicle routing problems. Elsevier, Expert Systems with Applications ,2008.

[5]. Christian Prins. A simple and effective evolutionary algorithm for the vehicle routing problem. Elsevier, Computers & Operations Research 31,2004.

[6]. István Borgulya. An algorithm for the capacitated vehicle routing problem with route balancing. Springer-Verlag, 2008.

[7]. K. Ganesh, T.T. Narendran. CLOVES: A cluster-and-search heuristic to solve the vehicle routing problem with delivery and pick-up. European Journal of Operational Research 178,2006.

[8]. Caroline Prodhon, Solving the capacitated location routing problem,Springer Verlag, 2001.

# BWN - A Software Platform for Developing Bengali WordNet

Farhana Faruqe Mumit Khan
Center for Research on Bangla Language Processing,
Department of Computer Science and Engineering,
BRAC University, 66 Mohakhali, Dhaka, Bangladesh
E-mail: farhanadiba@gmail.com, mumit@bracu.ac.bd

**Abstract-Advanced Natural Language Processing (NLP) applications are increasingly dependent on the availability of linguistic resources, ranging from digital lexica to rich tagged and annotated corpora. While these resources are readily available for digitally advanced languages such as English, these have yet to be developed for widely spoken but digitally immature languages such as Bengali. WordNet is a linguistic resource that can be used in, and for, a variety of applications from a digital dictionary to an automatic machine translator. To create a WordNet for a new language however is a significant challenge, not the least of which is the availability of the lexical data, followed by the software framework to build and manage the data. In this paper, we present BWN, a software framework to build and maintain a Bengali WordNet. We discuss in detail the design and implementation of BWN, concluding with a discussion of how it may be used in future to develop WordNets for other languages as well.**

## I. INTRODUCTION

One measure of the "digital maturity" of a natural language is the richness and diversity of linguistic resources available for that language – from the simple digital dictionaries to the complex annotated corpora – needed for advanced natural language processing applications such as automatic machine translation. Bengali, despite being very widely spoken [8], is only just beginning to see the development of these linguistic resources. One such important resource is WordNet, a lexical semantic database for a language [1]. The basic building block of WordNet is a synonym set or *Synset*, a word sense with which one or more synonymous words or phrases are associated. Each synset in WordNet is linked with other synsets through the lexical relations synonymy and anotnymy, and the semantic relations hypernymy, hyponymy, meronymy, troponymy, etc. The applications of WordNet range from creating digital lexica to performing word-sense disambiguation in automatic machine translation. The synonym set {পাখি, গগনগতি, খেচর, চিড়িয়া, নভৌকা, পংখি, পখী, পক্ষধর, পক্ষালু, পক্ষী, পতগ, পত্রী, বিহগ, বিহঙ্গ, বিহঙ্গ} and {পাখি, জমির একক বিশেষ, ৩০ কানি ভূমি, ২৬/৩৩/৩৫ শতাংশ, অঞ্চল একক} for example can serve as an unambiguous differentiator of the two meanings of "পাখি". Such synsets are the basic entities in WordNet; each sense of a word is mapped to a separate synset in the WordNet, and synset nodes are linked by semantic relationships, such as hypernymy. Building a WordNet for a language faces two primary challenges – creating the lexical data, and the software framework to store and manage that data. The primary focus of this paper is on the design and implementation of BWN, which is a framework to enable building and using Bengali WordNet.

The design of Bengali WordNet closely follows that of the English WordNet [2]. The software design that we detail in this paper allows the linguists to import lexical data in a "batch" mode, and then allows for visual querying and editing of all the relations in the data. The basic design to support data import and then subsequent queries is relatively simple; however, support for incrementally building the WordNet and for editing the data using a visual interface are two key features of BWN, and these complicate the design in a significant way.

We start by looking at the current approaches for building a new WordNet and discuss our methodology, and then discuss the design and implementation of BWN. We then conclude with a look at what the future holds for BWN.

## II. RELATED WORK AND METHODOLOGY

There are two common approaches for building a WordNet for a target language: (i) a top-down approach, using an existing WordNet in a source language to seed the linguistic data for the target language WordNet, and (ii) a bottom-up approach, where the linguists create the WordNet synsets from scratch without depending on an existing one. The first approach has been tried for a number of WordNets [3][4][5][6][7]. Using a source WordNet as the base, Barbu et al., Chakrabartu et al., and Farreres et al. generate target WordNet by mapping the synsets of the two languages. For the sysnsets to be mappable however, concepts in the source WordNet must exist in the target WordNet being built. Additionally, a significant amount of language resource is required for building a WordNet in this method. For example, a set of synsets strictly aligned with the source WordNet must exist before the new WordNet can be built. This is a significant drawback of building a WordNet from an existing one.

Given that there is a well-defined and well-designed English WordNet, one would be tempted to use that to map the synsets and build a Bengali WordNet of a reasonable quality. However, for that to be successful, there must first be significant level of linguistic similarity between the two languages, which is not

T. Sobh (ed.), *Innovations and Advances in Computer Sciences and Engineering*,
DOI 10.1007/978-90-481-3658-2_58, © Springer Science+Business Media B.V. 2010

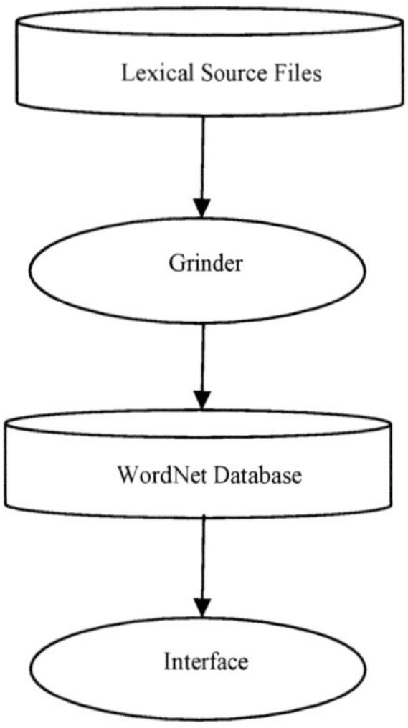


Fig. 1. Block diagram of WordNet system

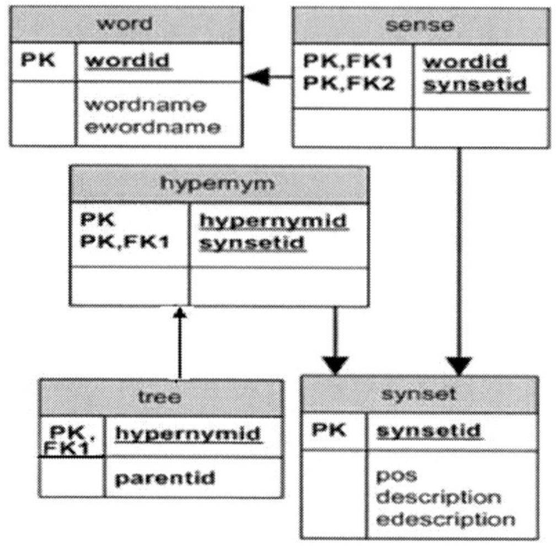


Fig. 2. Block diagram of the WordNet database

the case. In addition, Bengali word senses need to be clearly identified in an English-Bengali-English dictionary, which is also not available. Even if there were a rich tagged corpus, a WordNet can be created semi-automatically. Again, we do not have such a resource available.

There have been many other recent attempts at building a WordNet quickly, such as creating lexical networks by using the web or some well-structured corpora such as Wikipedia, or the BNC corpus. All of these require linguistic resources not yet available for Bengali, leaving us with the bottom-up approach as the most practical one.

Considering the challenges with the first approach, a simpler approach is by using the bottom-up approach, in which we build a WordNet by starting with the words in the target language and not by using an existing WordNet. For BWN, we started by translating the ontology, and chose words using a frequency list from a newspaper corpus. These synsets are compiled in lexical source files, which are then injected into the WordNet database using a "grinder", and the resulting system can then be used through a set of interfaces. We discuss the details of this in the next section.

## III. Design and Implementation

Generally, a WordNet software system is comprised of four parts as shown Fig. 1: lexical source files, grinder, WordNet Database and the interface to WNDB to build, use and edit the WordNet. This is the same structure that we follow in BWN as well.

### A. Lexical Source Files

These files contain the synsets that are manually compiled by the lexicographers, and are used to eventually populate the WordNet database. In a WordNet, nouns, verbs, adjectives and adverbs are organized into synsets, which are further arranged into a set of lexical source files by syntactic category. This is where the all the linguistic information is kept, typically hand-crafted by the linguists. The schema used for nouns in the lexical source file is shown below:

*Word name□Word name(english) □ description□*
*Pos□||description(english)||*
*Hypernyms:*
*Synonyms:*

And a sample "noun" record is shown below.

কাজ। work□কিছু করা বা তৈরির লক্ষে সরাসরি কার্যক্রম। বিশেষ্য।
||work -- (activity directed toward making or doing something)||
hypernyms:| কার্যক্রম | কৃতকর্ম | ঘটিত বিষয় | মনস্তাত্বিক-বিষয় | বিমূর্তন | বিমূর্ত-সত্তা | সত্তা |
synonyms: কর্ম, কর্মকাণ্ড, কাজ, কাজকাম, কাম, কার্য

### B. Grinder

The grinder is used to convert these lexical source files in a form that can be injected into the WordNet Database (WNDB). Basically, it parses and processes the text from the lexical source files into records, and then stores each record in the WNDB.

### C. WordNet Database (WNDB)

WNDB is the heart of WordNet for any language. For BWN, the basic design is similar to "Wordnet SQL Builder" [2], shown in Fig. 2. However, as we shall soon see, there are significant differences under the hood, primarily to support incremental building of the database, and editing of the synsets directly via the user interface. One of the design goals is to ensure that WNDB is extensible to new lexical relations between synsets. In addition, in the word table, we store the English word that can be used to link to other WordNets such as the EuroWordNet in the future. In the sense table, both the word and the synset are mapped together. In the synset table, we generate an ID for a synset but do not create the synset itself. We regenerate the synset at run-time from the sense and word tables, which serves a very important role in the case of an edit or update operation.

*D. WNDB Interface*

There are essentially three different interactions with WNDB, the underlying WordNet database. The first is to create the initial database using the lexical source files, and then to incrementally update the database, which is a feature that significantly contributes to the database schema complexity; the second is to use the database to query the data; and, the third is to edit the lexical data, which is the other reason behind the database schema complexity. In this rest of this section, we look at each of these interactions in detail.

TABLE 1
Word table after data entry

| wordid | wordname | ewordname |
|---|---|---|
| 1 | কাজ | Work |
| 2 | কর্ম | Work |
| 3 | কাজকাম | Work |

TABLE 2
Synset table after data entry

| synsetid | description | edescription | pos |
|---|---|---|---|
| 1 | কিছু করা বা তৈরির লক্ষে সরাসরি কার্যক্রম | work -- (activity directed toward making or doing something) | বিশেষ্য |

TABLE 3
Sense table after data entry

| wordid | synsetid |
|---|---|
| 1 | 1 |
| 2 | 1 |
| 3 | 1 |

*1) Update WNDB*: The Grinder takes each record from the lexical source file, then splits the text according to the database field and then stores it into the database. The process starts with reading each record from a lexical source file. To illustrate the process, let us take the following sample record in a lexical source file:

" কাজ□work□কিছু করা বা তৈরির লক্ষে সরাসরি কার্যক্রম□বিশেষ্য□
||work -- (activity directed toward making or doing something)||
hypernyms:| কার্যক্রম | কৃতকর্ম | ঘটিত বিষয় | মনস্তাত্ত্বিক-বিষয় | বিমূর্তন | বিমূর্ত-সত্তা
| সত্তা | synonyms:কর্ম, কাজকাম"

After splitting the text, the grinder updates the word table with the value of *wordid* (auto incremented integer), *wordname*, and *ewordname*. Each synonym word is also entered into the word table, (see Table 1).

The Grinder then updates the synset table with *synsetid* (auto incremented integer), *description*, *edescription*, and *pos* (see Table 2).

The Grinder then updates the sense table with those *wordids* and the particular *synsetid* (see Table 3).

To update the hypernym table (Table 4), we need the *synsetid* of that particular record and its corresponding *hypernymid*; because each synset, with the exception of "entity/ সত্তা", may have one or more hypernyms. For that, we have to match each hypernym with the *wordname* field's value in the word table and then take the *wordid*; with this *wordid*, we have to find out the *synsetid* (because the *hypernymid* is nothing but a *synsetid*) from the sense table. Here we assume that all of these hypernym words already exist in the word table.

TABLE 4
Hypernym table after data entry

| Synsetid | Hypernymid |
|---|---|
| 1 | 2 |
| 1 | 3 |
| 1 | 4 |
| 1 | 5 |
| 1 | 6 |
| 1 | 7 |
| 1 | 8 |

The tree table (Table 5) keeps track of the parent of each hypernym word, because hypernymy relates each child to its parent. Then the Grinder updates the *tree* table with *hypernymid* and *parentid* (which is also a *synsetid*). Since "entity/সত্তা" does not have a parent, its *parentid* is given a value of 0 (zero) to indicate that.

TABLE 5
Tree table after data entry

| hypernymid | parentid |
|---|---|
| 2 | 3 |
| 3 | 4 |
| 4 | 5 |
| 5 | 6 |
| 6 | 7 |
| 7 | 8 |
| 8 | 0 |

At this point, a specific complication may arise because of BWN's support of incremental update – there may be some hypernym words that do not currently exist in the WordNet Database. As we have noted earlier, this is one of the key features of BWN, and one that contributes significantly to the complexity of the design. We still have to enter these words into the database because the hypernym and tree tables' values are fully dependent upon the *synsetid*. However, the currently entered record is only partially complete, which is why we have to mark it as such. We do that by marking it with a special tag, "hypernym", to be updated later when its corresponding entity record is encountered in the lexical data. When this record eventually comes as an entity, we update the record tagged as a "hypernym" with its complete value. In fact, we have to consider all the synonym words, and not just the entity word, because the previously entered hypernym word may exist as part of a synset. Let us illustrate this with the following example record:

"কালবিন্দু□ measure, quantity, amount তা□ক্ষণিক সময়□ বিশেষ্য।||point, point in time -- (an instant of time)||
hypernyms: |বিমূর্তন | বিমূর্ত-সত্তা | সত্তা | synonyms: কালবিন্দু"

After entering the data word table (see Table 6), synset table (see Table 7), and the sense table (see Table 8) as discussed earlier, the data looks like the following:

TABLE 6
Word table after data entry

| wordid | wordname | Ewordname |
|---|---|---|
| 12 | কালবিন্দু | measure, quantity, amount |

TABLE 7
Synset table after data entry

| synsetid | description | Edescription | Pos |
|---|---|---|---|
| 9 | তা□ক্ষণিক সময় | measure, quantity, amount | বিশেষ্য |

TABLE 8
Sense table after data entry

| wordid | Synsetid |
|---|---|
| 12 | 9 |

TABLE 9
Word table after data entry

| wordid | wordname | ewordname |
|---|---|---|
| 13 | বিমূর্ত-সত্তা | hypernym |

TABLE 10
Synset table after data entry

| synsetid | description | Edescription | Pos |
|---|---|---|---|
| 10 | বিমূর্ত-সত্তা | hypernym | বিশেষ্য |

TABLE 11
Sense table after data entry

| wordid | Synsetid |
|---|---|
| 13 | 10 |

TABLE 12
Hypernym table after data entry

| synsetid | hypernymid |
|---|---|
| 9 | 6 |
| 9 | 10 |
| 9 | 8 |

TABLE 13
Tree table after data entry

| hypernymid | parentid |
|---|---|
| 6 | 10 |
| 10 | 8 |
| 8 | 0 |

TABLE 14
Word table after updated data

| wordid | wordname | ewordname |
|---|---|---|
| 13 | বিমূর্ত-সত্তা | Abstract Entity |

TABLE 15
Synset table after updated data

| synsetid | description | edescription | pos |
|---|---|---|---|
| 10 | শুধুমাত্র বিমূর্ত (দৈহিক রূপহীন) অস্তিত্ব আছে এমন সত্তা | abstract entity -- (an entity that exists only abstractly) | বিশেষ্য |

Now suppose that one of the hypernym words "বিমূর্ত-সত্তা" does not yet exist in the database; in this case, we have to enter this word into the word table (see Table 9) and the synset table (see Table 10), generating the *wordid* and the *synsetid*; then, we have to enter it in the sense table (see Table 11) with the generated *wordid* and *synsetid*.

Then we insert the value into the hypernym table (see Table 12) and the tree table (see Table 13) as discussed earlier.

Now, later one, when this "বিমূর্ত-সত্তা" hypernym word shows up as an entity, we have to update the word and the synset tables with new value, while the sense table remains the same.

For example, the following records add the hypernym word as an entity:

অমূর্ত-সত্তা □ Abstract Entity □শুধুমাত্র বিমূর্ত (দৈহিক রূপহীন) অস্তিত্ব আছে এমন সত্তা□বিশেষ্য□||Abstract Entity -- (an entity that exists only abstractly)||hypernyms:সত্তা | synonyms: অরূপ-সত্তা, নিরূপ-সত্তা, বিমূর্ত-সত্তা, সত্তা□

Here "বিমূর্ত-সত্তা" comes as part of a synonym, and not as an entity name. Now we have to update the word table (see Table 14) and the synset table (see Table 15) with the new value.

The rest of the entry – the entity name, the synonym, and the hypernym will be entered in the same manner as discussed earlier.

*2) Using WNDB:* The second interface to the WNDB is for querying the data in WNDB, as shown in Figures 3 and 4. A typical scenario is the following:

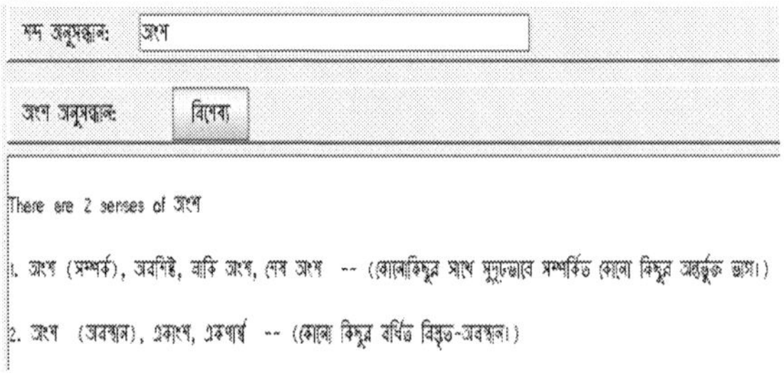


Fig. 3 Result of a search option

TABLE 16
Word table

| wordid | wordname | ewordname |
|---|---|---|
| **20** | অংশ (সম্পর্ক) | part, portion |
| **25** | অংশ (অবস্থান) | region, part |

TABLE 17
Sense table

| wordid | synsetid |
|---|---|
| 20 | **17** |
| 25 | **19** |

TABLE 18
Sense table

| wordid | synsetid |
|---|---|
| 20 | 17 |
| 21 | 17 |
| 22 | 17 |

TABLE 19
Word table

| wordid | Wordname | ewordname |
|---|---|---|
| 20 | অংশ (সম্পর্ক) | part, portion |
| 21 | অবশিষ্ট | part, portion |
| 22 | বাকি অংশ | part, portion |

TABLE 20
Synset table

| synsetid | description | edescriptio n | pos |
|---|---|---|---|
| 17 | কোনোকিছুর সাথে সুদৃঢ়ভাবে সম্পর্কিত কোনো কিছুর অন্তর্ভুক্ত ভাগ | …… | বিশেষ্য |
| 19 | কোনো কিছুর বর্ধিত বিস্তৃত-অবস্থান | …… | বিশেষ্য |

TABLE 21
Hypernym table

| synsetid | hypernymid |
|---|---|
| 17 | 8 |
| 17 | 9 |
| 17 | 10 |

TABLE 22
Tree table

| hypernymid | parentid |
|---|---|
| 8 | 9 |
| 9 | 10 |
| 10 | 0 |

1. User enters the query text into query field as shown in the following figure.

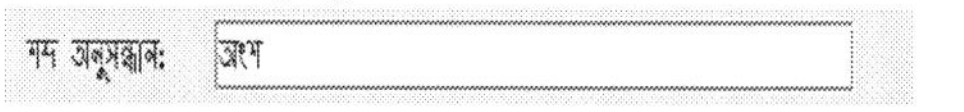


2. The WNDB search engine first finds the sense (or senses) of that given word from word table (see Table 16), then maps the *wordid* to the *synsetid* from the sense table (see Table 17), and then returns those *synsetids*.
   In this example, "অংশ" has two senses (each word represents a single value, as mentioned earlier).
   So, the returned *synsetids* are 17 and 19.
3. For each of the resulting *synsetids*, we have to find all the *wordids* from the sense table. To create a synonym set, we have to find all the *wordnames* from the word table after matching the *wordids* for a particular synsetid. Tables 18 and 19 show these procedures. Here, we consider only one *synsetid*, 17. For *synsetid* 17, the synonyms are {অংশ (সম্পর্ক), অবশিষ্ট, বাকি}.
4. Then, we find the description for each *synsetid* from the synset table (see Table 20) with those *synsetids*. Then the search result is shown in Fig 3.
5. To view a noun's hypernymy relation, as shown in Fig 4. The application execute steps 2-4 for each sense, and then, within each sense, it performs the following steps:
   a. It finds the *hypernymids* from the hypernym table (see Table 21) for the specific *synsetid.*
   b. The application also has to track each of the hypernym's parent from the tree table (see Table 22) to track the child-parent relation.
6. After performing steps 5 (a) and (b), it shows the hypernym from child to parent order.

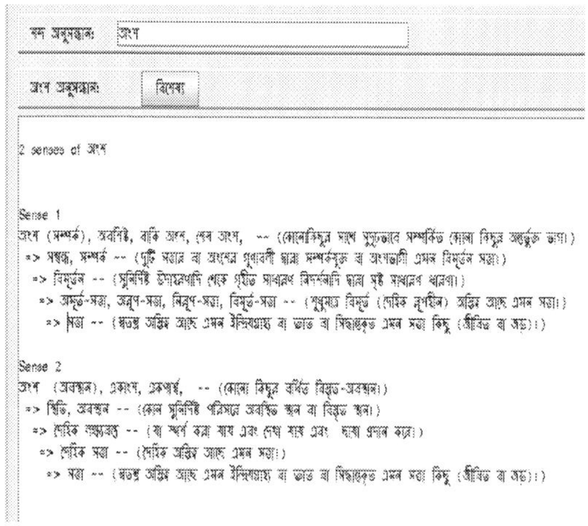


Fig. 4. Hypernym relation of a noun

*3) Editing WNDB:*

BWN supports editing any existing record through a user interface shown in Fig 5. It also supports a limited version of delete operation, because an unrestricted deleted may destroy the underlying tree. If the user wants to delete a record, there are three cases to consider:

- If the record has synonym, then we can delete it (updates only the word table);
- If the record is used as a hypernym entry then we cannot delete it without risking relational integrity;
- If the record is not used as a hypernym entry, then we can delete that record, which affects all tables except the tree table.

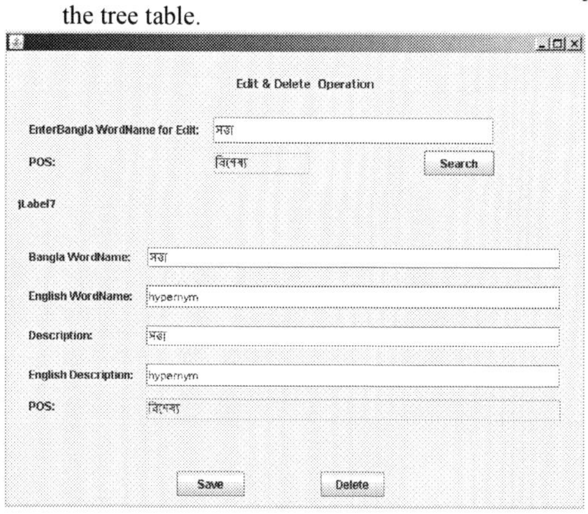


Fig. 5. Edit interface

## IV. Conclusion and Future Work

We present the design and implementation of BWN, a software framework for developing a Bengali WordNet. BWN at the basic level supports building the WordNet database from lexical source files using a grinder, and then supports querying the data using an interface; in addition, it has two key features not found in other designs support for incremental building of the WordNet database, and for editing the WordNet data using an interface. These two key features significantly contribute to the complexity of the design and implementation of BWN. BWN makes no assumption about the underlying language, so it should be extendable to other languages as well. Future work will focus on two fronts – improving the interface to the underlying WordNet database such as creating Webservice and .NET bindings, and to link to non-Bengali WordNets such as the Hindi and Euro WordNets.

## V. Acknowledgment

This work has been supported in part by the PAN Localization Project (www.PANL10n.net) grant from the International Development Research Center, Ottawa, Canada. The Center for Research on Bangla Language Processing (CRBLP) is supported in part by IDRC and Microsoft Corporation.

## VI. References

[1] Fellbaum C. "WordNet: An Electronic Lexical Database", MIT press. 1998.

[2] "wordnet sql builder", http://wnsqlbuilder.sourceforge.net/schema.html, last accessed: 16 Oct 2008

[3] Manish Sinha, Mahesh Reddy and Pushpak Bhattacharyya, "An Approach towards Construction and Application of Multilingual Indo-WordNet", 3rd Global Wordnet Conference ( GWC 06), Jeju Island, Korea, January, 2006.

[4] Piek Vossen, "EuroWordNet: A Multilingual Database with Lexical Semantic Networks", Computational Linguistics, Volume 25, Number 4, September 1999.

[5] Farreres, Xavier, German Rigau and Horacio Rodriguez. "Using WordNet for building WordNets." In: Proceedings of the COLING/ACL Workshop on Usage of WordNet in Natural Language Processing Systems, Montreal, 1998.

[6] Barbu, Eduard and Verginica Barbu Mititelu, "Automatic Building of Wordnets" In: Proceedings of the International Conference Recent Advances in Natural Language Processing, Borovets, Bulgaria, pp. 329-332, 21-23 September 2005.

[7] Debasri Chakrabarti, Gajanan Rane and Pushpak Bhattacharyya, Creation of English and Hindi Verb Hierarchies and their Application to English Hindi MT, International Conference on Global Wordnet (GWC 04), Brno, Czeck Republic, January, 2004.

[8] Wikipedia contributors, Bengali language, Wikipedia, The Free Encyclopedia; 2008 Oct 27, 17:06 UTC [cited 2008 Oct 28]. Available: http://en.wikipedia.org/w/index.php?title=Bengali_language&oldid=248011954

# Robust Learning Algorithm for Networks of Neuro-Fuzzy Units


Yevgeniy Bodyanskiy[1], *Senior Member, IEEE*, Sergiy Popov[1], *Senior Member, IEEE*, and Mykola Titov[2]

[1]Control Systems Research Laboratory
Kharkiv National University of Radio Electronics, Ukraine
bodya@kture.kharkov.ua, serge.popov@gmx.net

[2]Khartep LLC
Kharkiv, Ukraine
office@khartep.com.ua



***Abstract*—A new learning algorithm based on a robust criterion is proposed that allows effective handling of outliers. The obtained results are confirmed by experimental comparison in the task of short-term electric load forecasting.**


## I. Introduction

A Neuro-Fuzzy Unit (NFU) introduced by Ye. Bodyanskiy and S. Popov [1] is a further development of a Neo-Fuzzy Neuron (NFN) by T. Yamakawa et al. [2]. It is a nonlinear computational structure (Fig. 1) consisting of nonlinear synapses (Fig. 2) followed by a summation unit and a nonlinear activation function that forms its output.

The NFU's output is calculated in the following manner

$$y = \psi\left(\sum_{i=1}^{n}\sum_{j=1}^{h} w_{ji}\mu_{ji}(x_i)\right) = \psi\left(\sum_{i=1}^{n} f_i(x_i)\right) = \psi(u), \qquad (1)$$

where $\psi(\cdot)$ is a nonlinear activation function, e.g. hyperbolic tangent or sigmoid, $x_i$ are inputs, $\mu_{ji}$ – grades of membership, $w_{ji}$ – synaptic weights, $h$ – number of fuzzy intervals, $n$ – number of inputs, $y$ – output.

Grades of membership depend on the distance between the input $x_i$ and centers $c_{ji}$:

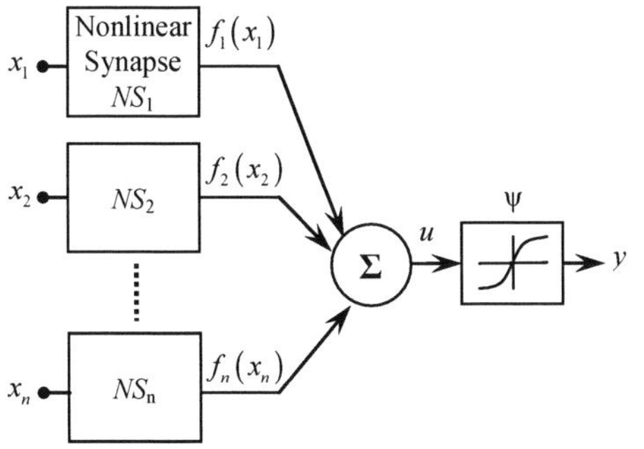


Fig. 1. Neuro-Fuzzy Unit (NFU)

$$\mu_{ji}(x_i) = \begin{cases} \dfrac{x_i - c_{j-1,i}}{c_{ji} - c_{j-1,i}}, x_i \in \left[c_{j-1,i}, c_{ji}\right]; \\ \dfrac{c_{j+1,i} - x_i}{c_{j+1,i} - c_{ji}}, x_i \in \left[c_{ji}, c_{j+1,i}\right]; \\ 0 \text{ otherwise}, \end{cases} \qquad (2)$$

$$\sum_{j=1}^{h}\mu_{ji}(x_i) = 1, \forall i. \qquad (3)$$

Given the current active fuzzy interval $p$, the output of the nonlinear synapse can be expressed in this way:

$$f_i(x_i) = \sum_{j=1}^{h} w_{ji}\mu_{ji}(x_i) = w_{pi}\mu_{pi}(x_i) + w_{p+1,i}\mu_{p+1,i}(x_i) =$$
$$= \frac{c_{p+1,i} - x_i}{c_{p+1,i} - c_{pi}} w_{pi} + \frac{x_i - c_{pi}}{c_{p+1,i} - c_{pi}} w_{p+1,i} = a_i x_i + b_i, \qquad (4)$$

where $a_i = \dfrac{w_{p+1,i} - w_{pi}}{c_{p+1,i} - c_{pi}}, b_i = \dfrac{c_{p+1,i}w_{pi} - c_{pi}w_{p+1,i}}{c_{p+1,i} - c_{pi}}$.

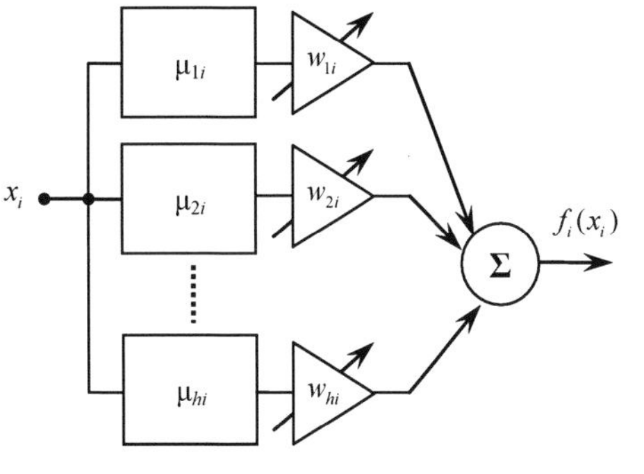


Fig. 2. Nonlinear synapse


T. Sobh (ed.), *Innovations and Advances in Computer Sciences and Engineering*,
DOI 10.1007/978-90-481-3658-2_59, © Springer Science+Business Media B.V. 2010

Introduction of additional (compared to NFN) nonlinearity at the neuron's output provides automatic limiting of unit's output range that is very important for building multilayer networks.

Weight adjustment of NFU is performed with respect to the following quadratic criterion

$$E(k)=\frac{1}{2}(d(k)-y(k))^2=\frac{1}{2}e^2(k)=$$
$$=\frac{1}{2}(d(k)-\psi(u(k)))^2=$$
$$=\frac{1}{2}\left(d(k)-\psi\left(\sum_{i=1}^{n}\sum_{j=1}^{h}w_{ji}\mu_{ji}(x_i(k))\right)\right)^2=$$
$$=\frac{1}{2}\left(d(k)-\psi\left(\sum_{i=1}^{n}w_i^T\mu_i(x_i(k))\right)\right)^2=$$
$$=\frac{1}{2}\left(d(k)-\psi\left(w^T\mu(x(k))\right)\right)^2, \qquad (5)$$

where $k$ is a discrete time, $d(k)$ – reference signal, $e(k)$ – learning error, $w_i=(w_{1i},w_{2i},\ldots,w_{hi})^T$,
$\mu_i(x_i(k))=(\mu_{1i}(x_i(k)),\mu_{2i}(x_i(k)),\ldots,\mu_{hi}(x_i(k)))^T$,
$w=(w_1^T,w_2^T,\ldots,w_h^T)^T$,
$\mu(x(k))=(\mu_1^T(x_1(k)),\mu_2^T(x_2(k)),\ldots,\mu_n^T(x_n(k)))^T$.

To minimize (5), a gradient descent learning algorithm can be applied

$$w(k+1)=w(k)-\eta(k)\nabla_w E(k)=$$
$$=w(k)+\eta(k)e(k)\frac{\partial e(k)}{\partial u(k)}\nabla_w u(k)=$$
$$=w(k)+\eta(k)e(k)\psi'(u(k))\mu(x(k)), \qquad (6)$$

where $\eta(k)$ is a learning rate.

## II. Robust Learning Algorithm

The use of NFU is very attractive for robust signal processing applications. This is because its nonlinear properties can be easily adjusted via parameters of membership functions in nonlinear synapses. In this way, large outliers can be removed and influence of less extreme inputs can be dampened appropriately. But these measures affect only forward signal pathway. On the other hand, backward signal pathway employed during network training also requires attention, if outliers are present in the training data.

Learning algorithms based on quadratic criteria like (5) are highly sensitive to deviations of the data distribution from Gaussian law. In case of different kinds of anomalous observations, gross errors, disturbances with heavy tail distributions, learning algorithms based on quadratic criteria become less efficient.

In such situations robust estimation methods [3] are more appropriate. Some of them are already used for neural network learning [4-7]. Among different goal functions used in robust estimation, the following criterion [8] is widely used

$$E^R(k)=\gamma^2\ln\left(\cosh\frac{e(k)}{\gamma}\right), \qquad (7)$$

where $\gamma$ is a scalar parameter, which is usually chosen empirically, that defines sensitivity to anomalous errors. Fig. 3 shows superimposed plots of criteria (5) and (7) with $\gamma=1$.

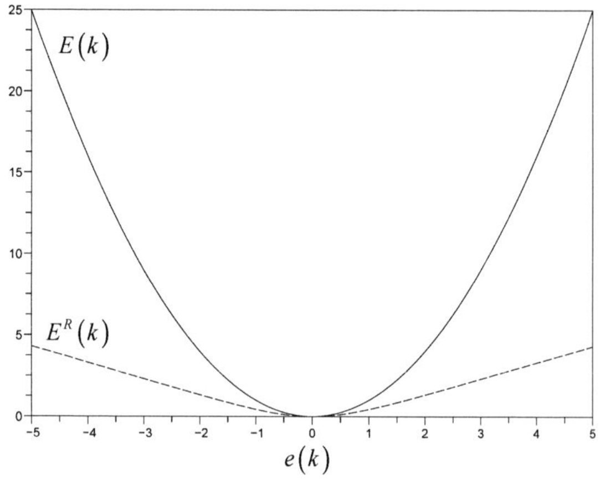


Fig. 3. Plots of criteria (5) – solid line and (7) – dotted line, $\gamma=1$

As far as

$$\frac{\partial E^R(k)}{\partial e}=\gamma\tanh\frac{e(k)}{\gamma}, \qquad (8)$$

NFU learning algorithm (6) based on robust criterion (7) can be written as

$$w(k+1)=w(k)-\eta(k)\nabla_w E^R(k)=$$
$$=w(k)+\eta(k)\gamma\tanh\frac{e(k)}{\gamma}\psi'(u(k))\mu(k)=$$
$$=w(k)+\eta(k)\delta^R(k)\mu(k), \qquad (9)$$

where $\delta^R(k)=\gamma\tanh\frac{e(k)}{\gamma}\psi'(u(k))$.

If $\psi(u(k))=\tanh(u(k))$, then $\psi'(u(k))=1-y^2(k)$ and $\delta^R(k)=\gamma\tanh\frac{e(k)}{\gamma}(1-y^2(k))$.

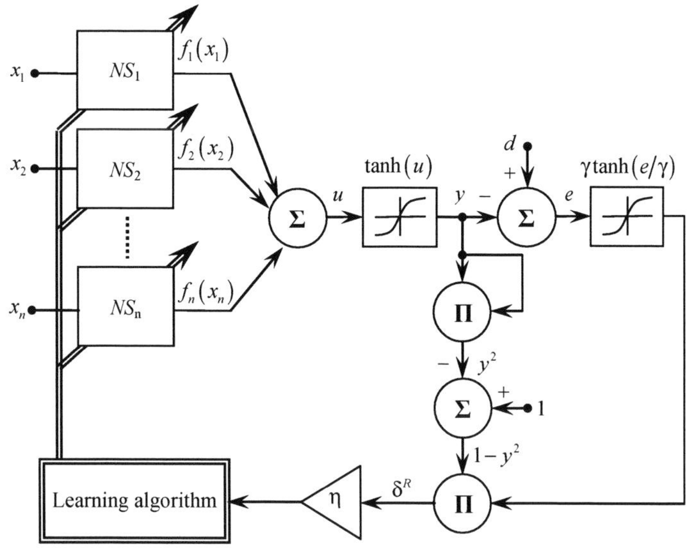


Fig. 4. Block diagram of NFU learning based on robust criterion

It can be seen from (5), (7), and (8) that

$$\lim_{e(k)\to\infty} \frac{\partial E(k)}{\partial e} = \lim_{e(k)\to\infty} e(k) = \infty \tag{10}$$

and

$$\lim_{e(k)\to\infty} \frac{\partial E^R(k)}{\partial e} = \lim_{e(k)\to\infty} \gamma \tanh \frac{e(k)}{\gamma} = \gamma, \tag{11}$$

i.e. anomalous errors are dampened by hyperbolic tangent squashing function.

Fig. 4 shows a block diagram of a single NFU learning based on robust criterion (7). Extension to networks of NFUs is straightforward. An interesting feature of this algorithm is that both forward (processing) and backward (learning) signal pathways are subject to nonlinearities of the same type – hyperbolic tangent. This may, in particular, simplify hardware implementation of the whole system (neural network and its learning circuits), because only one type of nonlinear circuit will be required.

## III. Experimental Results

We validate our theoretical results by conducting the following experimental study based on the short-term electric load forecasting problem. Given a data set of 26280 values describing three full years (2005-2007) of electricity consumption in Lugansk region (Ukraine), we compare performance of two approaches to its prediction: based on criterion (5) and based on robust criterion (7). The data set is split into training and test sets, containing 17520 (years 2005-2006) and 8760 (year 2007) values respectively. For the sake of simplicity we perform only one hour ahead forecasting, for which a simple model consisting only of a single NFU is sufficient. Model inputs $x_1, x_2, \ldots, x_n$ correspond to delayed observations of the analyzed time series.

The plot of the data set (Fig. 5) clearly shows the presence of outliers caused by measurement errors, peak loads and other factors. Due to their random nature, outliers are unpredictable and cause large prediction errors. If these errors are directly used to drive the learning process (as in criterion (5)), this will distort the model parameters and hence deteriorate the prediction quality on normal data that is a very undesirable property. Application of criterion (7) alleviates this problem by dampening large errors and thus preventing large changes of the parameters during learning phase.

The above is confirmed by comparison of prediction errors given in Table 1. Mean Absolute Percentage Errors (MAPE) are reported, which is a standard error measure in short-term electric load forecasting problems. Application of the proposed robust learning algorithm lowered the prediction errors by about 0.2% both on training and test sets that is quite tangible as it translates to quantities of about 0.5 MW of load for the given energy system.

TABLE 1
Prediction Errors (MAPE)

| | **Training set** | **Test set** |
|---|---|---|
| **Quadratic learning** | 2.29667% | 2.34525% |
| **Robust learning** | 2.10321% | 2.17349% |

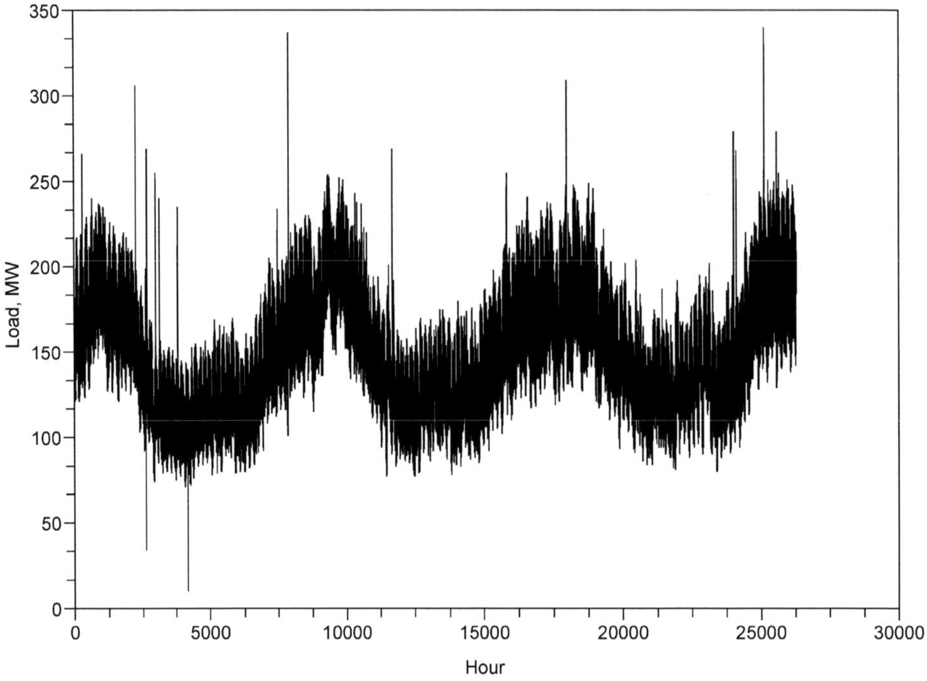


Fig. 5. Electricity consumption in Lugansk region

## IV. Discussion

We developed a new robust learning algorithm that is applicable to single NFUs and networks of NFUs. Thus both forward (processing) and backward (learning) signal pathways have robust properties. Inputs are processed by membership functions in nonlinear synapses whose properties may be easily adjusted by their parameters to limit gross inputs and dampen less extreme values. The learning signal is also dampened by hyperbolic tangent function, hence alleviating the influence of outliers present in the training data on the network weights and improving generalization capabilities of the network. The use of hyperbolic tangent functions both in forward and backward signal pathways may be beneficial, in particular, for hardware implementation of the neural network and its learning circuits because only one type of nonlinear circuit will be required.

The proposed theoretical results are confirmed by experimental study based on the short-term electric load forecasting problem. Application of the robust learning algorithm substantially lowered the prediction errors both on training and test sets.

## References

[1] Y. Bodyanskiy and S. Popov, "Neuro-fuzzy unit for real-time signal processing," in *Proc. IEEE East-West Design & Test Workshop (EWDTW'06)* Sochi, Russia, 2006, pp. 403-406.

[2] T. Yamakawa, E. Uchino, T. Miki, and H. Kusanagi, "A neo-fuzzy neuron and its applications to system identification and prediction of the system behavior," in *Proc. 2nd Int. Conf. Fuzzy Logic and Neural Networks* Iizuka, Japan, 1992, pp. 477-483.

[3] W. J. J. Rey, *Robust Statistical Methods.* Berlin-Heidelberg-New York: Springer, 1978.

[4] A. Cichocki and R. Unbehauen, *Neural Networks for Optimization and Signal Processing.* Stuttgart: Teubner, 1993.

[5] C.-C. Chuang, S.-F. Su, and C.-C. Hsiao, "The Annealing Robust Backpropagation (ARBP) Learning Algorithm," *IEEE Trans. Neural Networks,* vol. 11, pp. 1067-1077, September 2000.

[6] D. S. Chen and R. C. Jain, "A Robust Back Propagation Learning Algorithm for Function Approximation," *IEEE Trans. Neural Networks,* vol. 5, pp. 467-479, May 1994.

[7] J. T. Connor, "A Robust Neural Network Filter for Electricity Demand Prediction," *Journal of Forecasting,* vol. 15, pp. 437-458, 1996.

[8] P. W. Holland and R. E. Welsch, "Robust regression using iteratively reweighted least squares," *Communication Statistics – Theory and Methods,* vol. A6, pp. 813–827, 1977.

# An Experience of Use of an Approach to Construct Measurement Repositories in Software Organizations

Solange Alcântara Araújo, Adriano Bessa Albuquerque, Arnaldo Dias Belchior, Nabor Chagas Mendonça

University of Fortaleza (UNIFOR) – Masters Degree in Applied Computer Sciences (ACS)
Washington Soares Avenue, 1321 - Bl J Sl 30 - 60.811-341 - Fortaleza – Ce – Brazil

*Abstract*- **It is important that estimates related to cost, size, effort, time and quality of the projects and/or products of a software organization be feasible, and that organization's commitments towards those estimates may be properly monitored. Implementing a measurement and analysis program in a software organization helps to support these activities. However, the institutionalization of this program requires construct a measurement repository to store measurement data.**

**This paper presents the requirements for the construction of an organizational measurement repository based on the CMMI and MPS.BR maturity models, and proposes an approach for the construction of this repository satisfying those requirements. As an example of use, an organizational measurement repository based on the approach has been defined to a software organization.**

## I. Introduction

Nowadays, the complexity of software increases from day to day. To deal with software development into this scenario, the managers are adopting new tools, processes and methods, to support the decision analysis. The measurement is very important to the management activities in software engineering, because it helps to assure that the developed products satisfies the objectives established on the project [1].

According to GOLDENSON et al. [2] the professionals do not understand clearly how to use the measurement in a better way and the organization's guidelines are also not easy to have a real comprehension.

Measurement and analysis activities are considered in many relevant approaches, standards and maturity models, like: GQM [3], GDSM [4], PSM [5], ISO/IEC 15939 [6], MR MPS.BR [7] and CMMI [8]. The GQM (Goal-Question-Metric) and the GDSM (Goal-Driven Software Measurement) guide the identification of metrics. The PSM (Practical Software & System Measurement) and ISO/IEC 15939 establish a process of software measurement. The CMMI and MR- MPS.BR (Reference Model for Brazilian Software Process Improvement) have a process related with the measurement and analysis discipline.

This paper presents the requirements to a measurement repository to be used at levels 2 and 3 of CMMI and an approach to construct measurement repository in software organizations. Besides, it presents an experience of use of this approach.

Section II presents contents, methods and maturity models related to measurement programs. Section III describes the requirements to a measurement repository to levels 2 and 3 of CMMI and the proposed approach to construct it. Section IV describes an experience of use of the approach. Section V presents the conclusions and further works.

## II. Measurement and Analysis

The measurement captures information about attributes and entities [9]. The entities are objects we observe on the world and attributes are properties or characteristics, like: age, weight, and so on.

T. Sobh (ed.), *Innovations and Advances in Computer Sciences and Engineering*,
DOI 10.1007/978-90-481-3658-2_60, © Springer Science+Business Media B.V. 2010

Metrics and indicators are used to conduct the measurements. Metric maps the attributes of the entities from the real world to formal entities [9]. An indicator may be used like a metric or a combination of them [10]. In this work we will use measure instead of metric.

In software engineering, the measurement is fundamental to understand, monitor, control, foresee and improve the development process software [1].

The purpose of Measurement and Analysis (MA) is to develop and sustain a measurement capability that is used to support management information needs [8]. Table I presents their goals and practices.

Table I
SPECIFIC GOALS AND PRACTICES

| Specific goals | Specific practices |
|---|---|
| SG1 Align Measurement and Analysis Activities | SP 1.1 Establish Measurement Objectives |
| | SP 1.2 Specify Measures |
| | SP 1.3 Specify Data Collection and Storage Procedures |
| | SP 1.4 Specify Analysis Procedures |
| SG2 Provide Measurement Results Measurement and Analysis Activities | SP 2.1 Collect Measurement Data |
| | SP 2.2 Analyze Measurement Data |
| | SP 2.3 Store Data and Results |
| | SP 2.4 Communicate Results |

An organization at the level 2 of maturity must measure the work products to know their quality and the processes to monitor their performance.

As in level 2, the processes may be implemented differently in each project, so each project may define particular activities and artifacts related to measurement and analysis. Besides, the projects may choose to store data and results in a specific repository [8].

At level 3, each project has a defined process which is an adaptation of the organizational standard process. The organization must satisfy all the objectives of level 2 and establish and sustain a set of organizational process assets, for example, the organizational measurement repository.

This repository is used to store and communicate the organizational measures of the processes and work products [1].

## III. MEASUREMENT REPOSITORY TO LEVELS 2 AND 3 OF CMMI

This section identifies the requirements of a measurement repository to levels 2 and 3 of CMMI and proposes a possible structure and content to meet these requirements.

### *A. Requirements and measurement repository to level 2 of CMMI*

After the analysis of the CMMI process area "Measurement and Analysis", a set of requirements was established to level 2 (Table II). Each requirement is identified by the number "2" and is identified the practice or objective and process area that it try to satisfy. Besides, is defined his obligation or not.

Table II
REQUIREMENTS TO LEVEL 2

| Requirements | CMMI |
|---|---|
| 2.1 Store data related to the estimation of the parameters of the project planning. Ex.: cost, effort, size and period of conclusion.<br>**Obligation:** Yes | **PA project planning**<br>SP 1.2 – Establish estimates of work product and task attributes.<br>SP 1.4 – Determine estimates of effort and cost. |
| 2.2 Store data from the monitoring and controlling the performance of project progress: | **PA measurement and analysis**<br>SP 2.3 – Store data and results. |
| (i) Monitoring the actual values of planning parameters Ex.: progress, size, cost, effort and time (planned versus actual);<br>**Obligation:** Yes | **PA project monitoring and control**<br>SP 1.1 – Monitor project planning parameters |
| (ii) Monitoring the commitments of the project;<br>**Obligation:** No, if the monitoring was done analyzing the planned and done commitments. | **PA project monitoring and control**<br>SP 1.2 – Monitor commitments |
| (iii) Monitoring the risks of the project;<br>**Obligation:** No, if the monitoring was done analyzing the risks. | **PA project monitoring and control**<br>SP 1.3 – Monitor project risks |
| (iv) Monitoring the management of the project's data;<br>**Obligation:** No, if the monitoring was done revising the activities of data management. | **PA project monitoring and control**<br>SP 1.4 – Monitor data management |
| (v) Monitoring the involvement of relevant stakeholders.<br>**Obligation:** No, if the monitoring was done analyzing the involvement of the stakeholders. | **PA project monitoring and control**<br>SP 1.5 – Monitor stakeholder involvement |
| 2.3 Store data related to the monitoring and control of processes of level 2.<br>**Obligation:** No | **GP 2.8 – project monitoring and control** |

| | |
|---|---|
| 2.4 Store information to permit the understanding and utilization of data. Ex.: measurement plan, metrics specification, data collection and storage procedures, analysis procedures, analysis results storage specification.<br>**Obligation:** Yes | **PA Measurement and analysis**<br>SP 1.1 – Establish measurement objectives.<br>SP 1.2 – Specify measures.<br>SP 1.3 – Specify data collection and storage procedures.<br>SP 1.4 – Specify analysis procedures<br>SP 2.3 – Store data and results.<br>SP 2.4 – Communicate results. |

Although, at level 2 is not necessary the existence of an organizational measurement repository, the data and results must be stored in some repository.

At level 2, the measurement repository may be a part of the project's repository. However, sometimes it is not possible to store them on the project's repository, because some of the measures and items may be commons to many projects or have data from the organization.

One of the requirements of the measurement repository defined to level 2 (Table 3.1) demands the storage of the necessaries informations to facilitate the understanding and utilization of the stored data. So, beyond the storage of the data, the measurements specifications and the analysis results, it is also important to store the measurement plan, which may have complementary informations to the specification. The measurement repository at level 2 encompasses:

- **Measurement plan:** similar to others plans, the measurement plan may be apart of the project plan or be a section of it. At level 2, is necessary that all projects plan their specific measurement activities.
- **Measurement specification:** among the measures stored on the measurement repository at level 2, should have measures to address the requirements defined on Table 2. The measures must be defined from the indicators, which are defined from the measurement goal, which is based on the organization's goals or necessity of informations.

- **Measures and results of analysis:** this approach uses the "Template to indicators" to register the measures and the consolidated results of indicators analysis.

*B. Requirements and measurement repository to level 3 of CMMI*

At level 3, is necessary an organizational measurement repository. On this work, the repository created at level 2 will be the basis for the repository of the level 3. The organizations may implement one of the following solutions: (i) the organizational measurement repository is formed by others repositories created at level 2. In this situation, each project's repository could have only their information; and (ii) all the measures information is stored in only one organizational measurement repository

Table III presents the requirements that were inserted to the requirements defined at level 2 (Table II).

Table III
REQUIREMENTS TO LEVEL 3

| Requirements | CMMI |
|---|---|
| 3.1. Store historical data to support the estimation of the projects parameters.<br>**Obligation:** Yes | **PA integrated project management**<br>• SP 1.2 – Use organizational process assets for planning project activities. |
| 3.2. Store organizational measures and related information on the organizational measurement repository.<br>**Obligation:** Yes | **PA organizational process definition**<br>• SP 1.4 – Establish the organization's measurement repository; and<br>• GP 3.2 – Collect improvement information. |

When the requirement 3.2 of the repository at level 3 is satisfied, the measurement repository begins to have a set of common measures of the projects and the organization.

The utilization of historical data is only possible if the repository has informations that permit the project managers find similarities and differences between the projects. So, besides the templates defined at level 2, we created a new one, to register the projects characteristics (identification, project status *etc.*)

At level 3 exists two types of measurement plan: the organizational measurement plan and the project's measurement plan. On this approach, were inserted information related to risk management, a need of level 3, and the list of projects that must use the organizational measurement plan.

## IV. AN EXPERIENCE OF USE

We used the approach on a software organization that had been recently certified in ISO 9001:2000, there was 3 years that had been assessed successfully at level 2 of SW-CMM and was defining processes to be assessed at level 3 of SW-CMMI.

We utilized the following methodology to evaluate the feasibility of the measurement repository construction: firstly, we constructed a measurement repository that satisfied the needs of level 2 and 3, then we analyzed the improvement suggestions of the organization's collaborators and finally, we identified improvement opportunities from an official CMM assessment.

The repository was constructed by the organization's measurement group. One of the authors of this paper was responsible to coordinate this group, which decided to implement an organizational repository and auxiliary repositories.

The organizational measurement repository was stored at MC2 tool [11], a collaboration and knowledge environment, which provides mechanisms to exchange and store experiences.

During the implantation of the repository, data and information of twelve projects were inserted on it.

The measures were defined based on the organization's Strategic Plan, the recurrent technical and management problems and the improvement plan.

Table IV presents the defined organization's goals and measurement goals.

Table IV
GOALS AND MEASUREMENT GOALS

| Goals | Measurement goals |
|---|---|
| 1 – Satisfy the development period of the project. | 1 – Evaluate the productivity.<br>2 – Monitor the time of project's activities.<br>3 – Monitor the project's risks.<br>4 - Evaluate the planning. |
| 2 - Satisfy the project's cost. | 5 – Monitor the project's cost.<br>4 – Evaluate the planning.<br>6 – Analyses historical data. |
| 3 – Monitor the utilization of the processes on the organization. | 7 – Monitor processes that manage the projects.<br>8 – Monitor organizational processes.<br>9 – Monitor engineering processes. |

The measurement group decided do not implement the measurements to the requirements related to monitoring the data management and involvement of stakeholders, because monitor the comparison of the planned and the accomplished was considered sufficient.

For the definition of the measures from the goals and information needs we used the GDSM approach [4], which was very useful and important.

After the measures have been identified, they were registered using the "Template for specification of measures", improved by the measurement group. Table V shows an example of a measure's specification.

Table V
EXAMPLE OF A MEASURE'S SPECIFICATION

| **General informations** | |
|---|---|
| **Name of the measure** | Percent of requirements per status. |
| **Objective or needs of informations** | Satisfy the requirements and needs of the client. |
| **Measurement goal** | Evaluate how the requirements and needs of the projects were satisfied. |
| **Definition of the measure** | Measure the percent of use cases in each one of the status: proposed, approved, detailed, valid, implemented, tested and ratified. |
| **Type of measure** | Base. |
| **Formula of calculus** | Percent of requirements (status) = (total of requirements per status) / (∑total of requirements per status) *100. |
| **Name of the associated indicator** | Percent of requirements per status. |
| **Unity of measure** | Percent. |
| **Does it compose the organizational measurement repository?** | No. |
| **Informations to collect and storage measures** | |
| **Data sources** | Spreadsheet – Traceability matrix, Table of requirements per status. |
| **Procedures to collect and storage measures** | **How to collect:** the measure is stored on the spreadsheet.<br>**Where to store:** the indicator is stored on the spreadsheet. |
| **Period and moment of the collection** | Monthly (January– December). |
| **Responsible** | Project's coordinator |
| **Procedure to verify** | Verify if the percent of the total of requirements per status is equal to 100%. |
| **Informations to analyze the measures** | |
| **Is an indicator** | Yes. |
| **Parameters** | Status with the higher percent of requirements. |
| **Procedures to analyze** | **How to analyze:** analyze the situation of the requirements and compare with the planned, justifying the divergences and changes suggestions.<br>**Corrective actions:** if necessary, create an action plan, searching solutions to avoid delay on the project.<br>**Where to store**: the results of the analysis are stored on the project's performance report. |

| | |
|---|---|
| **Period and moment to analyze** | After the data collection. |
| **Responsible** | Project's coordinator |
| **Reports** | Project's Performance Report (including pizza chart of the percent of use cases in each status). |
| **Informations to communication** | |
| **Procedures** | Presentation of the project's performance report on weekly meetings to technical monitoring, on monthly meetings to monitoring with the supervisors and on the monthly meetings to monitoring with the company's directors. |
| **Period and moment to communicate** | Monthly, during the monitoring meetings. |
| **Responsible** | Project's coordinator. In the meetings with the company's directors, the responsible is the project's manager. |
| **Destination** | Project's team, in the weekly meetings to technical monitoring and managers, on monthly meetings to monitoring with the supervisors and on the monthly meetings to monitoring with the company's directors. |

Table VI presents an example of a measure's specification filled to the measure "percent of requirements per status". Although the model to specify measures was modified, we did not adapt the proposed model to present the indicators.

Each project of the organization, when it began, was registered on the organizational measurement repository, but it was updated periodically. Beyond the characteristics defined on the approach, the indicators and lessons learned were registered too. New characteristics were also inserted: client name, project manager and coordinator name, planning measures *etc.*

A measurement plan to the organization was defined and also a measurement plan to the projects (a section of the project plan, named "Indicators"). On this, the project manager may define others indicators, beyond those defined on the organization's measurement plan, if necessary, during the project planning.

Table VI
EXAMPLE OF AND INDICATOR

<table>
<tr><td>Name of the indicator</td><td>Percent of requirements per status</td></tr>
<tr><td>Period of collection</td><td>2008/03/01</td></tr>
<tr><td>Collected data</td><td><table>
<tr><th>Status</th><th>% of Use Cases in the status</th></tr>
<tr><td>Proposed</td><td>0,00%</td></tr>
<tr><td>Approved</td><td>50,00%</td></tr>
<tr><td>Detailed</td><td>23,00%</td></tr>
<tr><td>Valid</td><td>17,00%</td></tr>
<tr><td>Designed</td><td>10,00%</td></tr>
<tr><td>Implemented</td><td>0,00%</td></tr>
<tr><td>Tested</td><td>0,00%</td></tr>
<tr><td>Ratified</td><td>0,00%</td></tr>
</table></td></tr>
<tr><td>Graphical representation</td><td>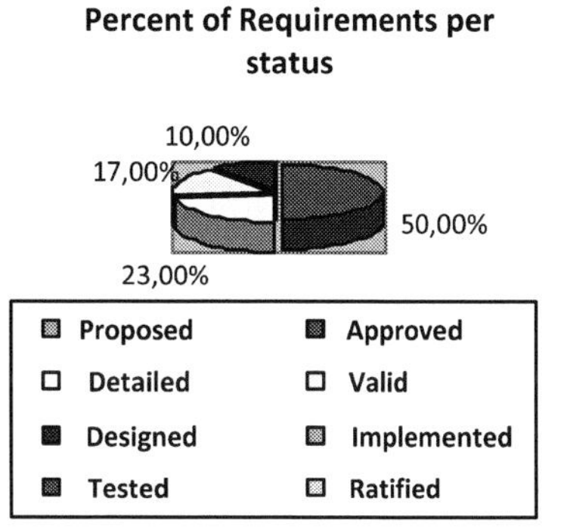
</td></tr>
<tr><td>Complementary Informations to the analysis</td><td>Requirements status:<ul>
<li>proposed – functional and non functional requirements and the use case model were elaborated;</li>
<li>approved – functional and non functional requirements and the use case model were approved;</li>
<li>detailed – the use case specification was elaborated;</li>
<li>valid – the use case specification was approved;</li>
<li>designed – the analysis and design of the use cases were elaborated;</li>
<li>implemented – the codification and unit tests of the use case were concluded;</li>
<li>tested – the use case was validated by the systemic test; and</li>
<li>ratified – the use case was validated by the acceptance test.</li></ul></td></tr>
<tr><td><b>Results of the analysis</b></td><td><<b>Analysis:</b> analyze which situation the major of the requirements are and compare with the planned, justifying the divergences and changes suggestions.><br><br><<b>Corrective actions:</b> if necessary, create an action plan, searching solutions to avoid delay on the project..></td></tr>
</table>

After we have specified the measures, we observed that the periodicity of the collections was very distinct, being very hard to collect, monitor and control the indicators. So, we decided to review the periodicity, categorize the indicators, and create a performance report for each of the indicators categories.

### A. Discussion

After the construction of the repository using the proposed approach, the organization was submitted to a class A assessment and obtained the maturity level 3 of CMMI-SW version 1.1. The Final Findings did not point out any weakness related to the measures defined, neither to the organizational measurement repository. This result showed that the constructed repository achieved its purpose, validating the proposed approach.

Then, some collaborators listed the following benefits of the measurement repository. (i) permitted the comparison of data from different projects, (ii) improved the data interpretation, (iii) permitted the access control of the measures; (iv) permitted the communication of data collected to relevant stakeholders; (v) improved the integrity and accuracy of the stored measures; (vi) supported the planning and estimation from the historical base; (vii) supported projects monitoring and control; (viii) reduced the number of estimations errors.

The great challenge we faced was the restrict quantity of commercial tools to support the measurement repository and that could be integrated with other tools. However, the MC2 facilitated the configuration of the repository and satisfied all the requirements defined.

The measurement group established the following lessons learned related to the implementation of the measurement repository: (i) all the projects characteristics must be defined from the measures; (ii) the measures must be defined with the participation of those who will collect and analysis the measures; (iii) the characterization of the indicators facilitates the collection, monitoring, control and analysis of the indicators; (iv) the reviews of measures, before their storage on the repository, are very important to guarantee the data integrity; (v) the improvement suggestion increased when the measurement and analysis activities began to be executed; and (vi) the support of the company's directors is indispensable to the configuration and utilization of the organizational measurement repository.

Nowadays, the repository has data from 80 projects. All the measures initially defined are still used, being refined constantly. Some new measures were defined, but those defined on this work continue satisfying the objectives and needs of information established before.

## V. Conclusion and Further Works

On this work, we identify a set of requirements to a measurement repository adherent to levels 2 and 3 of CMMI and proposed an approach to construct measurement repository in software organizations. The experience of use of this approach permitted us to know what was adequate and to identify improvements.

A second application of the approach is beginning in a public organization, which will be assessed on CMMI level 2 and we also intend to apply this approach on organizations that use different tools from MC2. Besides, it is very important evaluate the repositories and the approach to know the level of adherence to the needs of level 4 and level 5 of CMMI.

## References

[1] FLORAC, W. A., CARLETON, A. E., *Measuring the Software Process: Statistical Process Control for Software Process Improvement*, Addison-Wesley (2000).

[2] GOLDENSON, D. R., JARZOMBEK, J., ROUT, T., "Measurement and Analysis in Capability Maturity Model Integration Models and Software Process Improvement", Crosstalk: The Journal of Defense Software Engineering, 2003.

[3] SOLINGEN, R. V, BERGHOUT, E., *The Goal/Question/Metric Method: A Practical Guide for Quality Improvement of Software Development*, McGraw Hill (1999).

[4] BORGES, E. P., *A Model to Measure Software Development Process*, UFMG (2003).

[5] PSMSC, "Practical Software and Systems Measurement: A Foundation for Objective Project Management", available at: http://www.psmsc.com.

[6] ISO/IEC 15939, "Software Engineering – Software Measurement Process", ISO/IEC JTC1 SC7 (2002).

[7] MR-MPS, "MPS.BR General Guide, version 1.2", available at www.softex.br.

[8] CMU/SEI, "Capability Maturity Model Integration, CMMI for Development, Version 1.2", CMU/SEI-2006-TR-008.ESC-TR-2006-008, Pittsburgh, Software Engineering Institute – Carnegie Mellon University (2006).

[9] FENTON, N. E.; PFLEEGER, S., *Software Metrics: a Rigorous and Practical Approach*, PWS Publishing Company (1997).

[10] RAGLAND, B., "Measure, Metric or Indicator: What's the Difference?", Crosstalk, vol. 8, nº 3, 1995.

[11] SECREL, "MC2 – strategic knowledge", available at http://www.mc2.com.br.

# A Visualization-based Intelligent Decision Support System Conceptual Model

Dr. Hawaf Abdalhakim
Faculty of computers and Information,
Helwan University, Egypt
Dr_hawaf@yahoo.com

Mohamed Abdelfattah
Faculty of computers and Information,
El Minea University, Egypt
drmohabdo@yahoo.com

***Abstract:*** **the development of Intelligent Decision Support Systems IDSS is till requiring much effort to cope with both: the complexity of today decisions and daily flood of risky decisions. However evolving a covenant intelligent components and visualization aspects in IDSS are big challenges to be developed; but it would provide a muzzy decision taker an insight, preference, and much capability during a decision choice. This paper opt the advanced information visualization schemes for both decision's fact finding and decision taking processes. It proposes a conceptual model for IDSS that integrates DSS and dynamic information visualization within enterprise functionality. And it has introduced a three-module IDSS conceptual model that assembles a model base subsystem, fact finding subsystem, and dynamic visualization subsystem as a practitioner solution. The paper will focus on information visualization as an emerging computing relevant to be incorporated in order to model hidden facts in data and to expose such patterns in a visual manner. Finally this work formulates a viability of implementing such architecture and presents the conclusions.**

**Keywords: IDSS, Interactive information visualization IIV, IDSS visual interfaces. A muzzy decision taker, IV-IDSS**

## *I. INTRODUCTION*

Information systems (IS) applications that support decision-making processes and problem-solving activities have evolved over the past decades. In the 1970s, these applications were simple and based on 3rd generation languages, and spreadsheet technology. During the 1980s, decision-support systems incorporated a combination of optimization models, which originated in the operations research and management science.

DSS include many activities in order to create solution's alternatives: e.g. analysis, deduction, projection, comparison, simulation, optimization etc. [1]. In performing these essential activities, DSS utilize many types of the quantitative models. They may be linear programming, integer programming, network models, goal programming, and simulation, statistical models etc. Such models should be implemented in DSS model base subsystems via model management facilities.

DSS have recently emerged multiple criteria decision making (MCDM) model embedded DSS and knowledge-based [2].Many technological and organizational developments have exerted impact on this evolution. The Web has enabled inter-organizational DSS. Intelligent Decision Support Systems (IDSS) [2] are computer-mediated system that can assist managerial decision making by presenting information and interpretations for various alternatives. IDSS are usually regarded as one of the major implementations of Business Intelligence (BI).

While the human brain can process a limited amount of information, it instantly can recognize hundreds of different visual objects; hence visualization is an important process by which numerical data can be converted to meaningful images. This conversion can be computationally done. Information visualization (IV) became an IS area that has received an increased amount of attentions in developing techniques for exploring databases, attempting to extract a relevant hidden relationships among variables or among causes and effects [3]. The emerging results from IV community can be an important contribution to IDSS community if it can provide novel techniques enable IDSS to utilize a wide range of the available information in databases. Unfortunately information visualization is still facing several serious challenges.

In this paper several information visualization techniques are examined. We have noticed that both structured data and semi-structured data are considered for relevantly integrating IDSS with IV, and to provide more power knowledge source for decision making.

Many research efforts have focused on how to transform the business process data; whereas decision makers face the challenge of understanding such underlying data and finding out important relationships from which they can draw conclusions. Thus IV interface should support and improve the entire problem-solving phases, not only data transformation.

Information overloads burden, with an increasing pressure to perform tasks more quickly and in better manner create a muzzy decision taker, hence interactive visualized information is being worth during decision making activities subject to other limitations in time and cost. Hence, we address 2D Interactive Information Visualization IIV for this concern.

This research tackles these limitations and trying to overcome such shortcomings, and architecting a developmental model for IV-IDSS. In this paper, we present a novel decision support system, called IV-IDSS, which mainly originates from focusing on user-specific needs. Therefore, the first aspect of the system is the provision of visualizations and interactivity in a well understandable way for non-expert users.

T. Sobh (ed.), *Innovations and Advances in Computer Sciences and Engineering*,
DOI 10.1007/978-90-481-3658-2_61, 

The goal of our paper is to contribute in the next generation of IDSS which would be featured by incorporating: the incorporated-database with information visualization functionalities, supported decision making via means of intelligent information visualization IIV, and providing options for its explorative analysis. The focusing on visual interactive graph rather than upon looking at a hard graph generated from the answered queries is an advantage. Such featured IDSS would be proper to adaptability and applicability for a wide range of problems, and can be customized for many business sectors such as higher education, medical care, and banking decisions, as a workable version.

The rest of this paper is organized as follows. Section 2 discusses IDSS issues in a related work. Section 3 motivations for information visualization. Section 4 briefly describes the proposed system. The implementation and architecture of the system are introduced in section 5. Section 6 concludes the paper and projects a future work.

## *2. A DISCUSSION FOR **I**DSS ISSUES **T**HROUGHOUT **A** **R**EVIEWED RELATED WORK*

Most research efforts reported in the last decade tried to fill the gap of developing IDSS by recommending and exploiting intelligent tool. They addressed a significant new class of decision making discipline.

Basically, DSS concept has extremely broad definitions; its definition was depending on authors' point of view. For instance use, user, and goal view; interface characteristics view; time horizon and system objective view; type of problem structure view, system functions view; systems components view; and development process view were given by Gorry and Scott-Moton 1971, Little 1970, Alter 1980, Moore and Chang 1980, Bonzek at el 1989, and Keen 1980 respectively [4, 46] respectively. From these definitions, one may define DSS as a computer based information system that can support users during a decision life cycle for a semi structured problem.

IDSS definition has been float up as 'a computer-based information system that provides knowledge using analytical decision models, and providing access to data and knowledge bases to support effective decision making in complex problem domains' [13], which was agreed with basic concept of an IDSS as an integration of classical decision support including information access and analytical decision models with knowledge-based subsystem including reasoning and inference and go along with the concept that IDSS may use models built by an interactive process, supports all phases of decision-making, and may include a knowledge component [16].

Dong-Ling Xu *et al.*, [25] introduced an Intelligent Decision System (IDS) as a general-purpose multi-criteria decision analysis tool based on a new methodology called the Evidential Reasoning (ER), Hence, one main constituent of IDSS is knowledge. Accordingly another synonym for DSS was knowledge-based systems. Knowledge based systems (KBS) embody the knowledge of experts, it help to in manipulate expertise to solve problems at an expert level of performance [7]. KBS refers to use knowledge for reasoning, i.e. KBS was the transient state to IDSS. While the letter" I "in IDSS transient state was standing for Integrated, Interactive or Intelligent concepts; the intelligent concept is this research intention. Another progress in the IDSS direction was based on the argument that, there are many different types of IDSS, so some researchers suggested a replacing of the model base management subsystem with an Expert Systems (ES), or by other intelligent decision making functionality in order to enhance the model base management system (MBMS). Others suggested improving user interfaces via artificial intelligence AI e.g. using natural language processing or similar techniques[6, 7]. Also recommendations to IDSS were to support a wider range of decision issues such uncertainty cases, making recommended alternatives, handling complicated domains, and assessing the impact of the recommended alternatives [6]. Other enhancements were proposed by Marakas [8] to improve explanations, justifications, and formalization in knowledge organization. Arguments by McGregor [9] who uses an Agent-based DSS attached to the model base instead of replacing it.

Another main constituent of IDSS is visualization, the computer supported visualization enables humans to perceive, use, and communicate abstract data and amplify cognition [10]. Data visualization is categorized as either structural or functional. Structural categories focus on the form of the graphic material rather than its content [11, 36]; in contrast, functional taxonomies focus on use and purpose.

Business Intelligence (BI) trend helps in gathering, management, and analysis of large amounts of data [28, 29], and knowledge discovery in database (KDD). It attempts to extract relevant and interesting hidden relationships that can be existed among variables or between causes and effects [31]. An ideal visualization would provide a set of views [17, 19, 20, 30] using many of knowledge discovery KD approaches [20]. But the KD searching and extraction is a difficult and exhaustive process [24] and Data Mining (DM) remains poorly integrated with the decision support [5, 32] Visual data mining will be a step forward [27].

The implications upon this research according to the surveyed sample are:

- Information management is still facing problems in how to discover, and use facts, i.e. detect important rules among data, particularly when the data set is complex and uncertain. The rule-mining algorithms can generate a large number of rules, but they cannot be digested by human as similar as to any voluminous data set.
- Interactive Information visualization IIV in IDSS takes less attention, since much of the existing IDSS researches have a

limited work to show how to help decision makers understand and use the discovered rules, since for practical reasons, understanding over a set of rules before trusting them and using their mining outcomes is an important issue.

Doing much effort for those IIV problems is a challenge. In this concern "the large resulting rule set", and its associated ''hard to understand'' problem, the Object Oriented OO rule pruning and interesting rules selection approaches may help to deal with these problems. But a consequence problem is a 'rule behavior' problem. Whereby data are changed over time and the mined rules in the past may be invalid in the future, such consequence makes each rule has to have a behavior history over time.

- Ready-made Software like statistical packages and spreadsheets have static nature since the data is a read once give a hard displays. The user cannot easily interact with the visualization. Interactive visualizations IV should allow users to get free hand to amend a visualized displays as similar as displayed data.

However, this reviewed sample is short and we have discussed it from different point of views, it addresses decision takers' issues for much Intelligence DSS, and much visualized IDSS interface due to the human cognitive limitation. The utilization of IV as a discoverer of a new category of hidden information would increase the efficiency of decision making life cycle. In the next sections we would Identify, address, and conceptualize the developed work to empower the IV constituent in IDSS life cycle, either a decision-taking step would be semi or completely decided by a human, or completely decided by an intelligent agent with less human interfere.

### *3. MOTIVATIONS **W**ORK **T**OWARDS **I**NFORMATION VISUALIZATION*

The time line for visualization shows that: static 2D drawings of information were stared thousands yeas ago, 3D Graphics started with Ivan Sutherland were in 1962, scientific visualization early examples were in 1981, and Information Visualization early examples in 1990 Major efforts for IV are started in 1992

Fisher [33] explained the importance of bringing the human in the problem-solving loop to enhance the solution procedure. A visual representation of the problem enables a human expert to identify 'patterns' and come closer to the best possible solution [35].

Brady et al. [34] described an interactive optimization system can solve a facilities location problem. Their algorithm drew circular regions on the screen, indicating prospective facility location sites, with the user having to identify a point in the intersection of the regions. Cullen et al. [42] described another interactive optimization system for solving the vehicle routing problem. In this system the users could specify new routes as the starting points. Pirkul et al. [39] described a human-machine cooperative approach for solving the capacitated P-median problem. Eick and Wells [40] gave a comprehensive survey of the core features of interactive graphics and investigated how familiar plots like histograms, quantile diagrams and scatterplots could be made interactive.

Another area of visualization is animation of algorithms. The main idea behind algorithm animation is, as the algorithm executes, a picture of the current state of the algorithm is updated when 'interesting' events occur. Lustig [38] used a three-dimensional polytope of either a three variable linear programming problem or a projection of a large problem in three dimensions. Each pivot of the algorithm was illustrated as a movement from one vertex to another along the polytope. Gay [36] approached an algorithm-animation to interior point method and showed how the algorithm iteratively distorted the polytope. Gleick [37] visualized the behavior of Newton's method for finding the roots of nonlinear functions.

These examples illustrated the power of IV in providing insights into the behavior of well-established algorithms. Stasko [41] showed that the wide use of techniques can have some useful applications. Jones [35] has introduced an important technique in the solution of large-scale optimization problems.

Although the evolution in information visualization is presenting many technological future challenges, but perhaps the greatest challenge will in retaining an intimate involvement and understanding of end users. This challenge is a great issue since IV enable decision makers to fast analyze large quantities of information (hundreds of entities), aiding quick understanding of data distributions and rapid detection of patterns.

Finally, Information visualization will be a powerful trend to communicate and navigate through outcomes from intelligent visualization processes.

Hence we have featured that there are three important roles for information visualization in the IDSS:

1- Mission identification i.e. discovering a problem or finding out an opportunity and at which data domain

2- Offering decision making backend services.

3- Interactively managing visual outcomes on iterative modes.

### *4. APROPOSAL **IDSS M**ODEL*

In this section a conceptual model for IV-IDSS is identified as Visualization–Based IDSS system. It is described in a three-level abstraction, as given in section 4.1, 4.2, and 4.3 respectively. Section 4.1 presents the first level of abstraction, where Figure 1 shows the main panel that can capture the necessary information used during decision life cycle, as long as keeping the decision criteria to manage the progress onto decision environmental states.

Section 4.2 presents building blocks architecture and the functionality of the proposed IV-IDSS services. Section 3.3 presents IV-IDSS implementation concerns

#### *4.1 IV-IDSS Main Panel*

IV-IDSS Main Panel is given in Figure 1. It consists of four sub windows the first is to guide whether the decision be for recognizing a problem, or searching for an opportunity, it specifies the transparent decision's objects e.g. data entity type, which one be domain which one be in co domain, their argument and standards etc The second is for acquiring a knowledge about the hidden relationships (facts) in question which can guide the search mechanisms, it is in turn has two levels, basic level is mandatory options and the second is optionally advanced options. The third is to choose which decision modeling schemes according to end user's interest; it provides simple level schemes and advanced level schemes. The fourth is to specify when and how a decision is alerted. The system's graphical user interface has been designed with the objective to encourage both strictly guided as well as freewheeling user interaction modes.

In IV-IDSS, recognizing problem phase supports the identification and formulation of a problem to be solved, since a problem finding is the key to effective decision-making, while IV-IDSS problem solving phase is a process of using information, knowledge and intuition to solve a problem previously identified, both phases are using exploring activities followed a detailed specific analysis. To understand complexity, statistical and data mining reasoning are powered.

Guiding panel's interactive mode is backed by a very simple tools box.

*4.2 IV-IDSS Services building blocks architecture and its functionality*

The IDSS have to offer maximize human interaction, formalization of organizing knowledge, optimization and Projection explanations and justifications for specific recommendation. Such IDSS would also keep track on the effect of actions upon environmental states and changes in states caused by different decisions. A proposed IV-IDSS architecture has the backing of four basic components as shown in figure 2 and briefly described as follow:

*4.2.1 Data management component*

IV-IDSS' database management mainly contains a DBMS-based Module, it is managed by a DBMS, It can provide data retrieval, and updating. An IDSS database is a collection of current and historical data sets from a number of applications. These data are usually extracts of operational databases, so using the IDSS does not interfere with online critical operational systems, but it can use its own data warehouse.

*4.2.2 Model base component*

This module includes statistical, management science models, these quantitative models offer the system's analytical and projection capabilities to support reading and understanding states. Optimization models, such as linear programming and dynamic programming, may be adopted to determine the optimal resource allocation to maximize or minimize a certain objective concern.

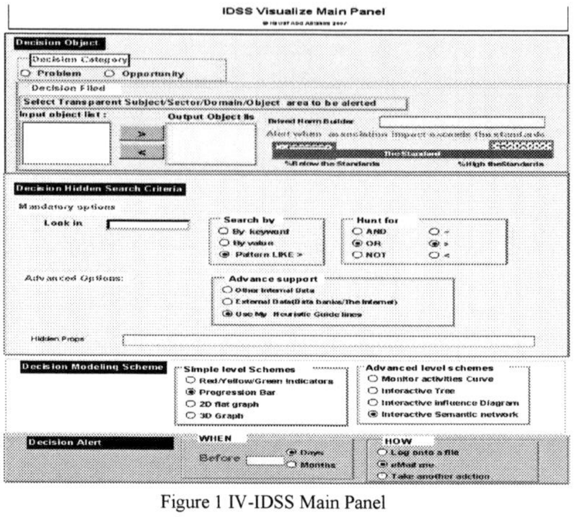


Figure 1 IV-IDSS Main Panel

*4.2.3 Inference engine component*

This IV-IDSS' module consists of pattern searching and reasoning subsystems, it is a fact finding module; It invokes knowledge base subsystem (KBS). KBS in turn contains facts, definitions, heuristics and computational procedures applicable to a certain problem domain. The inference engine accesses both the database and the knowledge base during the reasoning processes. It would exploit any up-to-date interpretation technology to decide how to apply the rules to infer a new state or a conclusion.

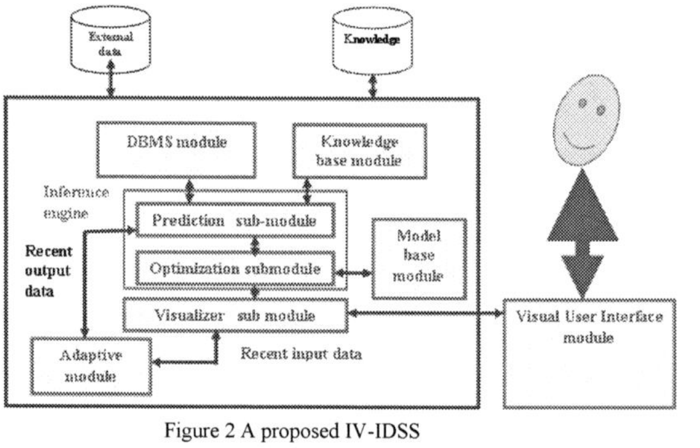


Figure 2 A proposed IV-IDSS

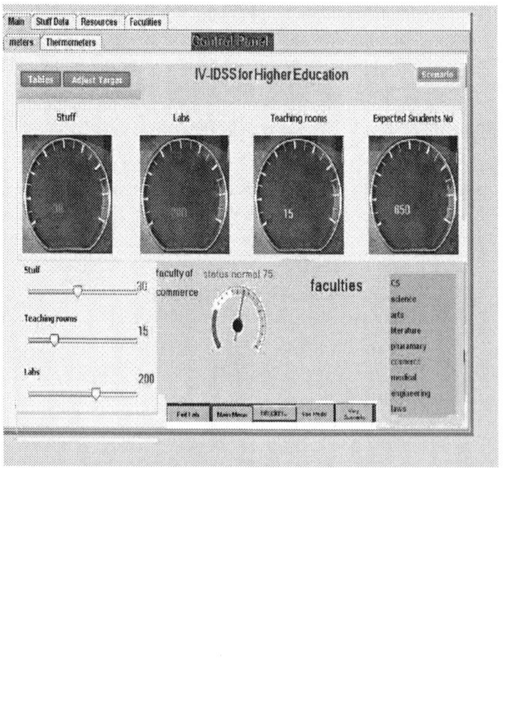

*4.2.4 Visualizer component*

IV-IDSS' Visualizer is a back end server used by IV-IDSS' main panel interface. Its callable elements enable users to get acquainted with the computational concept, evaluate various scenarios and can iterate toward a desired output. The user is guided through the sequence of preparation, fine-tuning, simulation, and analysis steps, with context-specific assistance throughout a free hand interaction.

## *5. IMPLEMENTATION IDENTIFICATION*

Many suggestion options can be introduced for IV-IDSS implementation; it is finally enterprise-dependent option. The authors are concerning with experimental within a research bases and not dominate to a commercial domain. Hence the system has been implemented as a database-enabled application with the multilayered client-server architecture as given in figure 3, in orders to fulfill the requirement of the IDSS technical features, it should support multi-user access.

This three-tier architecture option is the common assumption in which we have: 1) A database server holds the operational data (either internal for reformatted from externals resources), knowledge of user heuristics. A web server uses number of web servlets to handle decision making life cycle. It finally can provide a generic Decision-base i.e. scratch pad database used iteratively for further during trails decision cycle as along as transient states enhancements. 2) The client side is supported by Java applets code that provides the intelligent GUI to be applied. The GUI has to present the results in IV analyzable form, relevant to a muzzy-end-user. Such visualized information graphs e.g. counters, curves, sliders, and red-yellow-green spotlights (or its vagueness) should be accessible by a muzzy-end-user.

The used colouring would add a semantic to the visualized information. As Figure 4 shows, systems can be fitted with visualizations of control panels, which provide simple-hyper toolbox of 2D knobs, dials, levers, and switches that show states of the situation.

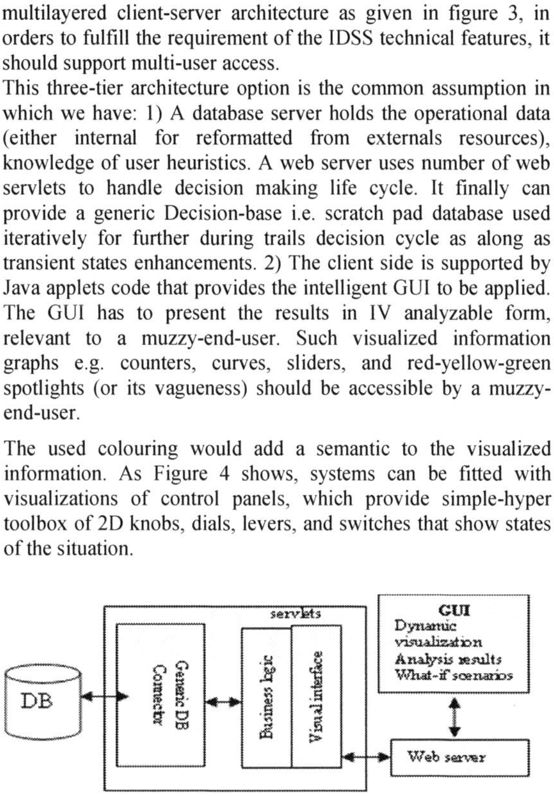


Figure 3 System

## *6. CONCLUSION*

In this paper we have conceptualized a framework for a visualization-based IDSS, which may cope with the progression in both: users' cognitive needs and the modern visualization computing power. It is referred herein as IV-IDSS. We have based IV-IDSS architecture upon: 1) the reviewed work given in this paper context, 2) authors background [44], and 3) the acquired needs from focus Groups. We conclude that the big-challenges, in IDSS field, are Information visualization IV limitations in offering predicts for both: what the affected decision-state will be, and what reactions upon different decisions scenarios are. In this study we also have emphasized on building learning-curve knowledge of decision makers as a part of the developed IV-IDSS. We consider a continuous beak time of decision makers by stating inference mechanisms to associate option-base visualization toolbox in order to enhance IV-IDSS interactivity. IV-IDSS model exploits new capabilities in object oriented programming to make 2D graph accessible. To answer the future IDSS issues, we recommend that IDSS must conduct: 1) incorporated knowledge base mechanisms. 2) intelligent visualiz

Figure 4 Screen Shots of the IV-IDSS System

what-if, and why-not analysis, 4) providing solution- state and solutions-path explanations which would increase information perception.

## REFERENCES

1] Sprague, R.H., Jr and Carlson, E.D. "Building effective decision support system"s, Englewood Cliffs, *NJ: Prentice Hall. (A classic DSS textbook with data-dialogue-model framework,* 1982.

[2] Eom, S.B. "Decision support systems implementation research: review of the current state and future directions", *Proceedings of the Ninth International Conference on Information Systems Development* 2000.

3] K. Zhao, B. Liu," Visual analysis of the behaviour of discovered rules", in: *ACM SIGKDD International Conference on Knowledge Discovery and Data Mining (KDD 2001),*San Francisco, USA, 2001.

[4] Efraim Tuban, Jay E. Aronson, Decision support systems and intelligent systems, 6th *international edition, , P96-99, Prentice Hall,* 2001.

[5] Guangzhou , Intelligent decision support system based on data mining: foreign Trading Case Study; *IEEE International Conference on Control and Automation, CHINA - May 30 to June* 1, 2007 .

[6]. Efraim Turban, E., and Aronson, J. E., "Decision support systems and Intelligent Systems." *6th ed. Prentice Hall* ,2006.

[7]. Sauter, V.," decision support systems: An applied managerial approach", *John Wiley & Sons, Inc* ,1997.

[8]. Marakas, G.M., "Decision support systems in the 21st century". *PrenticeHall* ,1999.

[9]. McGregor, C., and Kumaran, S." An Agent-Based System for trading partner management in B2B e-Commerce", *12th International Workshop on Research Issues in Data Engineering: Engineering e-Commerce/ e-Business Systems (RIDE™02). IEEE*, 2002.

[10] McCormick BH, DeFanti TA, Brown MD. "Visualization in scientific computing—a Synopsis". *IEEE Comput Graph Applic* 7(7):61–70,*1987.*

[11] Bertin J. Semiologie graphiques, 2nd ed. Paris, France,Gauthier- Villars; Berg WJ." Semiology of graphics", *Madison, WI: University of Wisconsin Press*, 1983.

[12] Dorian Pyle, "Business modeling and data Mining", *Morgan Kaufmann Publishers*, 2003.

[13] Klein M, Methlie LB. "Expert systems: a decision support approach with applications in management and finance". *Wokingham, England:Addison-Wesley*,1995.
[14] Rauch-Hindin WB. 'A guide to commercial artificial intelligence." *Englewood Cliffs, NJ: Prentice Hall;* 1988.
[15] Madjid Tavana, "Intelligent flight support system (IFSS): a real-time intelligent decision support system for future manned spaceflight operations at Mission Control Center", *Advances in Engineering Software journal ,2004.*
[16] Efraim Turban., Jay, A. "Decision Support Systems and Intelligent Systems", *fifth ed. Simon and Schuster Company, Upper Saddle River, NJ.*1995
[17] Fayyad, U., et al. "The KDD process for extracting useful knowledge", *volumes of data Communications of the ACM, 39(11). Mena, J. Decision support and data warehouse systems. Singapore: McGraw- Hill International Editions*, 2000.
[18] A. Inselberg," Data mining, visualization of high dimensional data", *ACM SIGKDD International Conference on Knowledge Discovery and Data Mining (KDD 2001), Proceedings of the Workshop on Visual Data Mining, San Francisco, USA, pp. 65–81*, 2001.
[19] U.M. Fayyad, G.G. Grinstein, "Introduction, in: Information Visualization in Data Mining and Knowledge Discovery", *Morgan Kaufmann, Los Altos, CA*, pp. 1–17, 2001.
[20] Tamio Shimizu, Marly Monteiro de Carvalho , "Strategic Alignment Process and Decision Support Systems: Theory and Case studies', IRM *Press Idea Group Inc.*, 2006.
[21] S. Makridakis, S.C. Wheelwright, and R.J. Hyndman, 'Forecasting, Methods, and Applications", *John Wiley & Sons*, 1998.
[22] I. Kopanakis, B. Theodoulidis,' Visual Data Mining and Modeling Techniques', *ACM SIGKDD International Conference On Knowledge Discovery and Data Mining (KDD 2001), Proceedings of the Workshop on Visual Data Mining, San Francisco, USA*, pp. 114–128,2001.
[23] Tom Soukup Ian Davidson "Visual Data Mining: Techniques and Tools for Data Visualization and Mining" *Wiley Publishing, Inc*, 2002.
[24] Ahlberg, C. and Schneiderman, B., Visual Information Seeking: Tight Coupling of Dynamic Query Filters with Starfield Displays, *Proc. ACM SIGCHI, p. 313-317*, 1994.
[25] Dong-Ling Xu, Grace McCarthy, Jian-Bo Yang, Intelligent decision system and its application in business innovation self assessment *,Decision Support Systems journal*, 2006.
[26] Shneiderman, B., Dynamic Queries for Visual Information Seeking, *IEEE Software., 11, p. 70-77*, 1994.
[27] Ahlberg, C. and Wistrand, E., IVEE: Information Visualization and Exploration Environment, *Proc. IEEE Info. Vis., p. 66-73*, 1995.
[28] S. Kudyba, R. Hoptroff, Data Mining and Business Intelligence: A Guide to Productivity. *Idea Group Inc*, 2001.
[29] D J Power, Decision Support Systems: Concepts and Resources for Managers. *Quorum/Greenwood*, 2002.
[30] U Fayyad, G Piatetsky-Shapiro, P Smyth, "From Data Mining to Knowledge Discovery in Databases," *AI Magazine, vol.17, no.3, pp. 37-54*, 1996.
[31] S Delisle, "Towards a Better Integration of Data Mining and Decision Support via Computational Intelligence, *" Proceedings of 16th International Workshop on Database and Expert Systems Applications*, pp. 720-724, 2005.
[32] LIU Qiong-xin, LIU Yu-shu, ZHENG Jim-jun , Multi-Agent Based IDSS Architecture and Negotiation Mechanism *IEEE*, 2003.
[33] M.L. Fisher, Interactive optimization, Annals of Operation Research 5 (1986) 541– 546.
[34] S.D. Brady, R.E. Rosenthal, D. Young, Interactive graphical minimax location of multiple facilities with general constraints, AIIE Transactions 15 (3) (1984) 242–254.
[35] C.V. Jones, Visualization and optimization, ORSA Journal on Computing 6 (3) (1994) 221– 250.
[36] Intel, Array visualizer, www.intel.com/sfoftware/products, 2008
[37] J. Gleick, Chaos: Making a New Science, Viking, New York, 1988.
[38] I. Lustig, Applications of interactive computer graphics to linear programming, Proceedings of the Conference on Impact of Recent Computer Advances in Operations Research, 1989, pp. 183– 189.
[39] H. Pirkul, E. Rolland, R. Gupta, VisOpt: a visual interactive optimization tool for p-median problems, Decision Support Systems 26 (3) (1999) 209– 233.
[40] S.G. Eick, G.J. Wills, High interaction graphics, European Journal of Operational Research 81 (1995) 445– 459.
[41] J. Stasko, A practical animation language for software development, Proceedings of the IEEE International Conference on Computer Languages, IEEE Computer Society Press, Los Alamos, CA, 1990, pp. 1– 10.
[42] F.H. Cullen, J.J. Jarvis, H.D. Ratliff, Set partitioning based heuristics for interactive routing, Networks 11 (1981).
[43] Brath R. and M. Petters, Visualization spreadsheets, DM Direct, Jan, 2006
[44] A. Hakim Hawaf, Does Decision Maker's Preferences Influence upon DSS Success (Sample Study of Egypt's Governorates), scientific bulletin of Statistical Studies and research, Cairo University, 2002
[45] Efraim Turban, Jay E. Aronson, Decision support Systems and Intelligent Systems, 8th *international edition, , P88, Prentice Hall*, 2007.

# Analysis of selected component technologies efficiency for parallel and distributed seismic wave field modeling

Kowal A., Piórkowski A., Danek T., Pięta A.
Department of Geoinfomatics and Applied Computer Science,
AGH University of Science and Technology, Cracow, Poland.
*{pioro, tdanek}@agh.edu.pl*

***Abstract*—In this article efficiency of using component-based software for seismic wave field modeling is presented. The most common component solutions like: .NET, Java and Mono were analyzed for various operating systems and hardware platform combinations. Obtained results clearly indicate that the component approach is able to give satisfactory results in this kind of applications, but global solution efficiency can strongly depends on operating system and hardware. The most important conclusion of this work is that for this kind of computations and at this stage of component technology development there are almost no differences between commercial and free platforms.**

***Index Terms*—wave field modeling, component platforms**

## I. Introduction

Seismic wave field modeling is very useful on various stages of seismological investigations or seismic exploration. It can be used during planning, processing and interpretation parts of seismic surveys [1]. In seismology modeling plays an important role in earthquake analysis and velocity inversion [2]. Unfortunately modeling process is very time consuming, even for simplified equations. One of the best methods to overcome this disadvantage is parallelization of computations. Luckily elastic or acoustic wave equations and their finite difference method (FDM) solutions are very good examples of computational problems which are easy to parallelize by domain decomposition. The other problem is life cycle of created program codes which is usually very short. New solutions are frequently created and new software and hardware solutions are introduced. Fortunately, mostly because of high code reuse ratio and hardware independence, using of component technologies can make progress faster and easier in this area of applied computer science.

## II. Component technologies

Component-based software engineering is a new way of software production that gives portability, security and independence of hardware. Another feature - code reuse - enables cooperation for coders. There are numerous solutions for this domain, although the most advanced and popular are Sun Java [3] and MS .NET [4]. Although MS .NET is a commercial platform, there is an alternative, open source solution, called Mono [5]. It enables to run .NET applications in various operating systems.

The main advantage of component environments is a managed code. Managed code is executed under the management of virtual machine (Java Virtual Machine for Sun Java and .NET Framework for MS .NET), so code is portable and hardware independent. A intermediate, precompiled managed code is called as *bytecode* for Sun Java and CLI for MS .NET. The process of translating intermediate code to native code of processor seems to be time-consuming, but modern compilers and virtual machines can optimize executing effectively.

The communication between components is very important theme. It enables to communicate a pair or group of components transparently in a network or Internet. Each of environments has own communication solution - Sun Java has RMI [6], MS .NET - .NET Remoting. These technologies can be used in creating a cluster for seismic modeling.

## III. Test environment and applications

### A. Hardware

The test computations were performed in two different environments.

The first environment was a single computer (notebook), which has following parameters:

- single processor (Intel Pentium M 1,86 GHz),
- 1 GB RAM.

The second environment was a cluster of 30 PC computers, which have following parameters:

- processor with Hyper-Threading Technology (Intel Pentium 4 2,8 GHz),
- 1 GB RAM,
- Gigabit Ethernet network adapter.

All nodes in cluster were connected by Gigabit Ethernet switch.

T. Sobh (ed.), *Innovations and Advances in Computer Sciences and Engineering*,
DOI 10.1007/978-90-481-3658-2_62, 

### B. Operating Systems

We have tested efficiency component seismic computing in following operating systems: Linux (Fedora Core 3, kernel: 2.6.12-1.1381 [smp]), MS Windows 2000 (SP4), MS Windows XP (SP2) and MS Windows Vista.

### C. Component environments

The code of seismic modeling was written in Java language for Sun Java SE virtual machine (version 1.6.0). The part in C# language was written for .NET Framework, version 2.0 and Mono, version 1.2.6.

New version of Mono – Mono 2.0 – has been newly released on Oct 06 2008 – some of issues for previous version had been found and reported during implementation of this project.

## IV. SEISMIC MODELING

### A. Basics of seismic modeling

We decided to test our component solutions for the case of acoustic wave equation which can be written as:

$$\frac{\partial^2 p}{\partial t^2} - c^2\left(\frac{\partial^2 p}{\partial x^2} + \frac{\partial^2 p}{\partial z^2}\right) = f(x,z,t) \qquad (1)$$

where p(x, z) is pressure, c(x, z) is velocity of acoustic wave, t is time and f denotes function which describes pressure change in a source point.

We used classic Alford finite different solution [7]. In this scheme can obtain pressure results for each point i,j of computation grid in time k+1 through neighboring points in time k and k-1 as follows:

$$p_{i,j}^{k+1} = 2(1-2\gamma^2)p_{i,j}^{k} - p_{i,j}^{k-1} + \gamma^2(p_{i+1,j}^{k} + p_{i-1,j}^{k} + p_{i,j+1}^{k} + p_{i,j-1}^{k}) \qquad (2)$$

where $\gamma = c\Delta t/\Delta h$, $\Delta t$ is time sampling interval, $\Delta h$ is distance between grid points in x and z directions. The stability criterion for above scheme is: $\gamma = 1/2^{-2}$. To avoid reflections from model borders in we used typical absorbent boundaries condition by Reynolds [8].

### B. Serial case

In this test we analyzed simple two dimensional model with two layers. It was 500 meters long and 500 meters deep. Depth to the layers border was 250 meters. Velocity was 1000 m/s in upper layer and 2000 m/s in lower. Acoustic wave source was localized at 250 meter in horizontal direction and 20 meter in vertical. Distance between computational grid points was 1 meter which gave 250 000 FDM points in total.

The main goal of this test was evaluation of the efficiency of analyzed runtime environments for seismic wave field modeling. We used serial C# and Java version of computational code. Results for various operating systems (MS Windows Vista, MS Windows XP, MS Windows 2000 and Linux Fedora Core 3), programming platform and hardware are presented in table I and figure 1.

TABLE I.
TIME OF SEISMIC WAVE FILED MODELING COMPUTATIONS FOR SERIAL CASE.

| | Avg. times in E1 [ms] | | | Avg. times in E2 [ms] | |
|---|---|---|---|---|---|
| | MS Vista | MS XP | FC 3 | MS 2000 | FC 3 |
| .NET | 80,07 | 75,291 | - | 97,707 | - |
| Mono | 66,675 | 64,154 | 64,514 | 134,304 | 79,545 |
| Java | 55,926 | 54,05 | 54,742 | 116,019 | 81,459 |

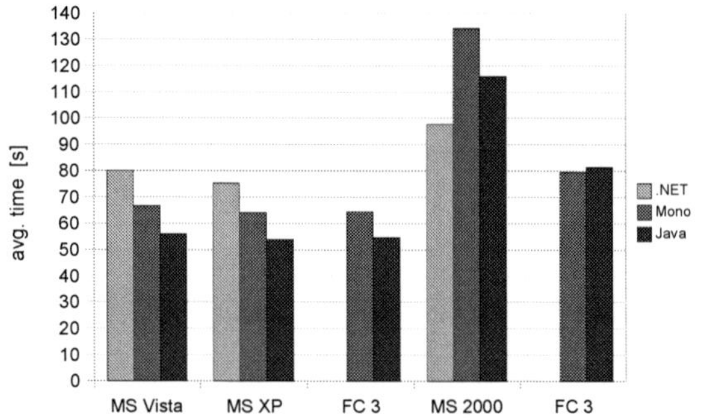


Fig 1. Graphical comparison of times of computations for various operating systems (MS Windows Vista, MS Windows XP, MS Windows 2000 and Linux Fedora Core 3), programming platform and hardware.

### C. Parallel and distributed case

In this experiment we used exactly the same setup like in serial case. We decided to use domain decomposition which is the most natural way of parallelization of this kind of partial differential equations. Global computational domain was divided into subdomains (Fig 2.), which were later assigned to P (up to eight) independent processors for concurrent equation solving.

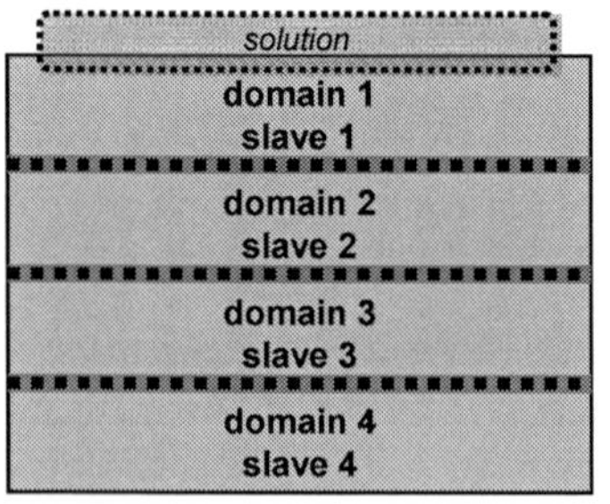


Fig 2. The domain decomposition schema.

It usually makes computational process much faster and limits the risk of local RAM exceeding which can ruin code efficiency. In this test run we used master-slave network scheme (Fig. 3.). One of the machines hosted master process, other hosted only slave processes. Master was responsible for creation of numerical model representation, domain division and data sending whereas slave processes performed computation and were exchanging domain border nodes after every time step.

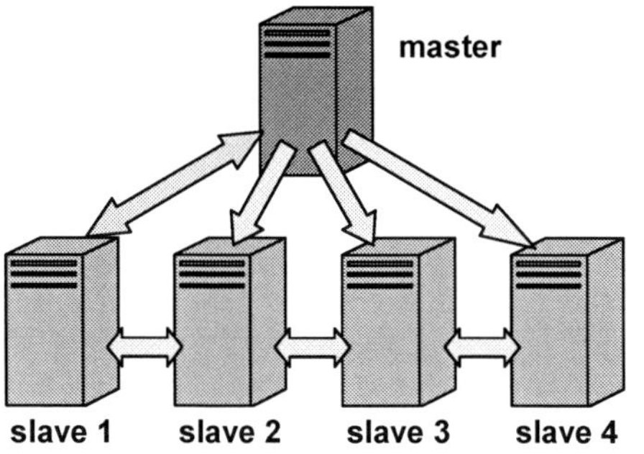


Fig 3. The environment for distributed seismic wave field modeling.

All computations were performed for float type elements. Master slave communications occurred only at the beginning of computations when subdomains were sent and at the end when results were received. The most important and time consuming were intra slave communications which had to be done about 4 thousand times during one program run. Of course such a dense communication can strongly limit usage of CPUs especially in case of smaller models. For presented model slaves with one neighbor had to send and receive 2004 bytes of border data, whereas slaves with two neighbors had to send and receive two times more. All test runs were done in described earlier hardware environment number two. Fragments of test codes are presented in table II. Times of computations are shown in tables IIIa, IIIb and in figures 4,5,6,7.

TABLE IIIA
MINIMAL TIME S OF COMPUTATIONS IN DISTRIBUTED EXPERIMENT FOR VARIOUS SYSTEMS AND COMPONENT PLATFORMS

a)

| No of proc | Minimal time of processing [s] | | | | |
|---|---|---|---|---|---|
| | .NET | Mono | Java | Mono | Java |
| | Win 2000 | Win 2000 | Win2000 | FC 3 | FC 3 |
| 1 | 96,42 | 134,88 | 107,41 | 76,94 | 77,54 |
| 2 | 52,78 | 71,88 | 57,71 | 42,53 | 47,53 |
| 3 | 38,61 | 51,22 | 43,11 | 32,71 | 35,84 |
| 4 | 31,28 | 42,09 | 35,13 | 27,52 | 28,37 |
| 5 | 27,22 | 36,41 | 30,33 | 25,06 | 25,16 |
| 6 | 24,44 | 33,09 | 26,58 | 23,12 | 22,93 |
| 7 | 23,44 | 32,28 | 24,3 | 24,08 | 21,52 |
| 8 | 22,42 | 30,72 | 22,41 | 22,04 | 20,88 |

TABLE II
FRAGMENTS OF JAVA AND C# CODES USED IN SERIAL AND DISTRIBUTED TESTS

```
if (prec == 0)//2th order
{
 for (i = 2; i <= nx - 2; i++)
  for (j = 2; j <= nz - 2; j++)
    pp[i][j] = (float)(2.0 * p[i][j] - pm[i][j]
     + ((dtr * dtr) / (ds *   ds)) * V[i][j]
     * V[i][j] * (p[i + 1][j] + p[i - 1][j]
     + p[i][j + 1] + p[i][j - 1] - 4.0 * p[i][j])
     + s[i][j]);
}
else //4th order
{
 for (i = 2; i <= nx - 2; i++)
  for (j = 2; j <= nz - 2; j++)
    pp[i][j] = (float)((2.0 -
     5.0 * Math.pow((V[i][j] * dtr) / ds, 2))
     * p[i][j] - pm[i][j] + (4.0 / 3.0)
     * Math.pow((V[i][j] * dtr) / ds, 2)
     * (p[i + 1][j] + p[i - 1][j] + p[i][j + 1]
     + p[i]  [j - 1]) - (1.0 / 12.0)
     * Math.pow((V[i][j] * dtr) / ds, 2)
     * (p[i + 2][j] + p[i - 2][j] + p[i][j + 2]
     + p[i][j - 2]) +   s[i][j]);
}
```

```
if (prec == 0)//2th order
{
 for (i = 2; i <= nx - 2; i++)
  for (j = 2; j <= nz - 2; j++)
    pp[i][j] = (float)(2.0 * p[i][j] - pm[i][j]
     + ((dtr * dtr) / (ds *   ds)) * V[i][j]
     * V[i][j] * (p[i + 1][j] + p[i - 1][j]
     + p[i][j + 1] + p[i][j - 1]
     - 4.0 * p[i][j]) + s[i][j]);
}
else //4th order
{
 for (i = 2; i <= nx - 2; i++)
  for (j = 2; j <= nz - 2; j++)
    pp[i][j] = (float)((2.0 - 5.0 *
     Math.Pow((V[i][j] * dtr) / ds, 2))
     * p[i][j] - pm[i][j] + (4.0 / 3.0)
     * Math.Pow((V[i][j] *  dtr) / ds, 2)
     * (p[i + 1][j] + p[i - 1][j] + p[i][j + 1]
      + p[i][j - 1]) - (1.0 / 12.0) *
     Math.Pow((V[i][j] *   dtr) / ds, 2)
     * (p[i + 2][j] + p[i - 2][j] + p[i][j + 2]
     + p[i][j - 2]) + s[i][j]);
}
```

TABLE IIIB
AVERAGE TIME S OF COMPUTATIONS IN DISTRIBUTED EXPERIMENT FOR VARIOUS SYSTEMS AND COMPONENT PLATFORMS

b)

| No of proc. | Average time of processing [s] | | | | |
|---|---|---|---|---|---|
| | .NET | Mono | Java | Mono | Java |
| | Win 2000 | Win 2000 | Win2000 | FC 3 | FC 3 |
| 1 | 96,6 | 135,06 | 107,72 | 77,16 | 77,74 |
| 2 | 52,99 | 72,57 | 57,92 | 43,13 | 52,29 |
| 3 | 38,86 | 52,22 | 43,41 | 34,7 | 37,13 |
| 4 | 31,71 | 42,71 | 35,39 | 28,29 | 29,08 |
| 5 | 27,89 | 36,78 | 30,61 | 25,61 | 26,78 |
| 6 | 24,95 | 33,66 | 27,22 | 23,43 | 23,66 |
| 7 | 25,62 | 33,94 | 25,09 | 24,49 | 22,18 |
| 8 | 24,3 | 31,21 | 23,35 | 22,51 | 22,43 |

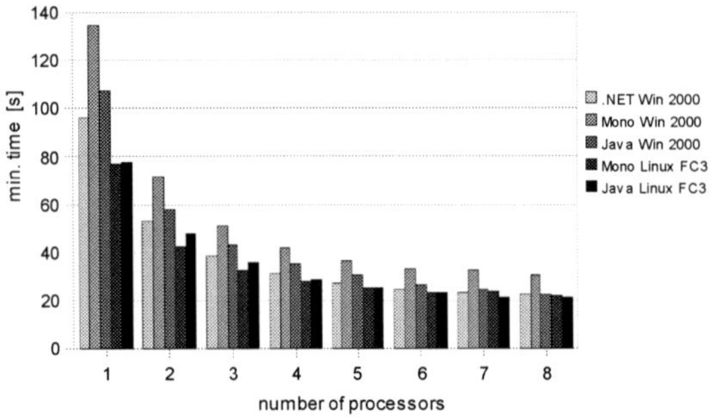


Fig. 4 Relation between minimal times of seismic wave filed modeling and number of processors for various operating systems and programming platforms.

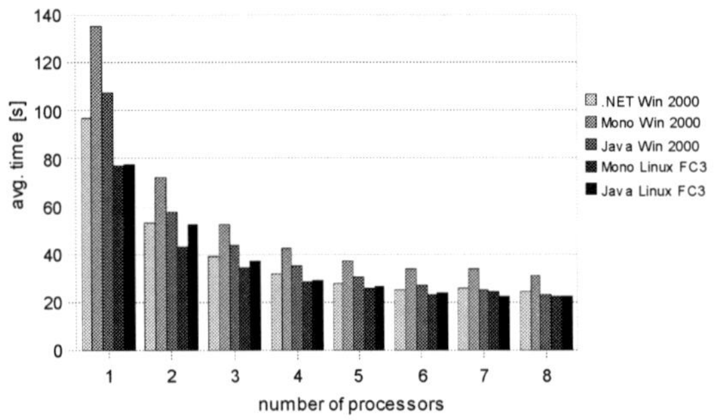


Fig. 5 Relation between average times of seismic wave filed modeling and number of processors for various operating systems and programming platforms.

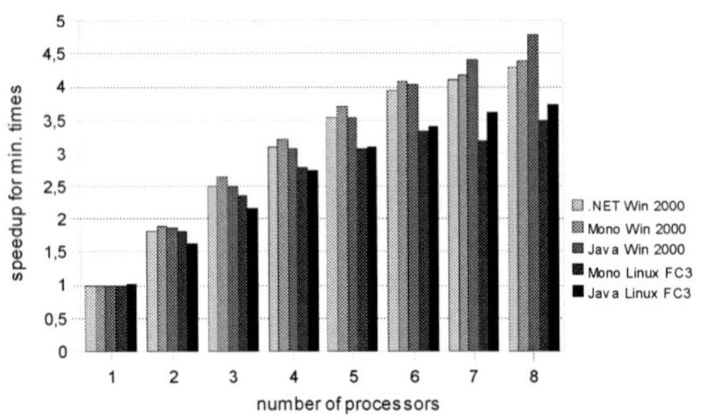


Fig. 6. Relation between minimal speedups of computation and number of processors for various operating systems and programming platforms.

## V. Discussion

For the serial case the shortest time was reached in Mono under control of Linux Fedora Core 3, in MS Windows 2000 the fastest computation was in .NET. The times of processing in distributed environments prove valid decomposi-

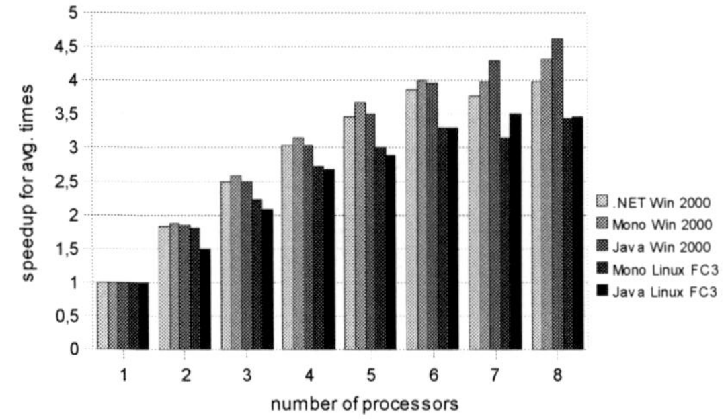


Fig. 7. Relation between average speedups of computation and number of processors for various operating systems and programming platforms.

tion of algorithm - the average and minimal times of computations were smaller for higher number of processors. The shortest time - 20,88s - was reached in Sun Java under control of Linux Fedora Core 3 for 8 processors. For MS Windows 2000 the shortest time was reached in Sun Java for 8 processors too. The highest average speedup was 4,61 in Java under control of ms Windows 2000, under Linux Fedora Core 3 in Sun Java too - 3,47. For conventional C codes usual speedup for 8 processors is about 6 to 6.5.

## VI. Summary

Scientific codes written in the conventional way often suffers from low portability, limited code reuse and short life cycle. In this paper, we show that various free and commercial component technology can be effectively use to eliminate these problems. Obtained speedup results are very promising. Of course typical C codes are still faster but the difference is not as big as it used to be few years ago.

## Acknowledgment

This work was financed by the AGH - University of Science and Technology, Faculty of Geology, Geophysics and Environmental Protection as a part of statutory project number 11.11.140.561.

## References

[1] Pietsh K., Marzec P., Kobylarski M., Danek T., Lesniak A., Tatarata A., Gruszczyk E. 2007. Identification of seismic anomalies caused by gas saturation on the basis of theoretical P and PS wavefield in the Carpathian Foredeep, SE Poland Acta Geophysica 55 (2).
[2] Debski W , Danek T , Pieta A , Lesniak A 2008. Waveform Inversion through the Monte Carlo Sampling. ESC 31st General Assembly Crete 2008.
[3] Reilly D., Reilly M.: Java Network Programming and Distributed Computing. Addison Wesley, 2002.
[4] MacDonald M.: Microsoft .NET Distributed Applications: Integrating XML Web Services and .NET Remoting. Microsoft Press, 2003
[5] Schonig H.J, Geschwinde E.: Mono Kick Start. Sams, 2003.
[6] Pitt E., McNiff K.: java.rmi: The Remote Method Invocation Guide. Addison Wesley, 2001.
[7] Alford R.M., Kelly K.R., Boore D.M. 1974. Accuracy of finite-difference modeling of the acoustic wave equation. Geophysics, 39(6).
[8] Reynolds A. C., 1978, Boundary Conditions for the Numerical Solution of Wave Propagation Problems, Geophysics, 43.

# Modified Locally Linear Embedding based on Neighborhood Radius

Yaohui Bai

School of Electronics, Jiangxi University of Finance and Economics, Nanchang, 330013, China

E-Mail: byhnpu@163.com

*Abstract*-**As a nonlinear dimensionality reduction technology, locally linear embedding is a kind of very competitive approach with good representational capacity for a broader range of manifolds and high computational efficiency. However, LLE and its variants determine the neighborhood for all points with the same neighborhood size, without considering the unevenly distribution or sparsity of data manifold. This paper presents a new performance index-ratio of neighborhood radius to predict the unevenly distribution or sparsity of data manifold, and a new approach that dynamically determines the neighborhood numbers based on the ratio of neighborhood radius, instead of adopting a fixed number of nearest neighbors per data point. This approach has clear geometry intuition as well as the better performance, compared with LLE algorithm. The conducted experiments on benchmark data sets validate the proposed approach.**

## I. INTRODUCTION

Most real data lies on a low dimensional manifold embedded in a high dimensional space. The task of manifold learning is to recover the meaningful low-dimensional structures hidden in high dimensional data. Classical techniques, such as Principal Component Analysis (PCA)[1] and Multidimensional Scaling (MDS)[2], are efficient to find the true structure of the data when dealing with a linear manifold. These methods are theoretically simple and easy to implement. However, an assumption has been taken in these methods: the data lies in a linear or almost linear sub space of the high-dimensional space and the embedding can be obtained using these linear methods. Obviously, this assumption is too restrictive as many real data cannot satisfy the linear assumption.

Recently, a new class of nonlinear embedding techniques has been designed to discover the structure of high-dimensional data that lies in a nonlinear high-dimensional manifold and find their embedding in a low dimensional Euclidean space. Some of the more prominent examples of nonlinear manifold learning algorithms are ISOMAP[3], Locally Linear Embedding (LLE)[4] and Laplacian Eigenmaps [5]. Among these, LLE is an effective nonlinear algorithm. Compared with others, the notable advantages of LLE are: i) only two parameters should be set;ii) it can reach global minimization without local minima; iii) it can well preserve the local geometry of high dimensional data in the embedded space.

The primary idea of the LLE algorithm is to preserve the relationships between neighbors in manifold data and map the high dimensional data in low dimensional Euclidean space. Assuming that each data point lies on a locally linear patch of a manifold, LLE characterizes local geometry by representing each data as an approximate affine mixture of its neighbor points. The local geometry of each patch can be described by the reconstruction weighs with which a data point is reconstructed from its nearest k neighbors.

The LLE algorithm was demonstrated on a number of artificial and real world data sets. It has many variants, such as Hessian LLE[6], incremental LLE[7], supervised LLE[8,9], etc. However, LLE and its variants are based on the assumption that the whole data manifold is evenly distributed. These approaches determine the neighborhood for all points with the same neighborhood size, without considering the uneven distribution or sparsity of data manifold. To solve this problem, this paper proposes a new performance index-ratio of neighborhood radius to predict the uneven distribution or sparsity of data manifold. Based on the ratio, this paper also proposes an improved LLE algorithm that can dynamically selects the neighborhood numbers for each point on manifold according to the structure of the manifold.

T. Sobh (ed.), *Innovations and Advances in Computer Sciences and Engineering*,
DOI 10.1007/978-90-481-3658-2_63, © Springer Science+Business Media B.V. 2010

The remainder of the paper is organized as follows. Section 2 proposes an improved LLE algorithm based on the definition of ratio of neighborhood radius. In section 3 experimental results on synthetic data sets are presented. Finally, section 4 gives the conclusion.

## II. DYNAMICAL SELECTION OF NEIGHBORHOOD NUMBERS

LLE largely depends on the neighborhood structure. It always assumes that the whole data manifold is evenly distributed, and determines the neighborhood for all points with the same neighborhood size. However, the assumption that the real world data is evenly distributed is not always true, and always unevenly distributed or sparse, as shown in Figure 1. This makes the selection of same neighborhood size for all points unsuitable. In Figure 1, the distribution of data in high dimension space is unevenly and sparse. The data can be divided into two groups, and there are some outliers between the two groups data. For this data, when applying the original LLE algorithm, the property of uneven distribution of data is not considered and it will select same neighborhood numbers for all points. Such as, LLE selects 6 neighbor points for the point $x_1$, $x_2$ and $x_3$, respectively. In fact, $x_3$ is an outlier, and should be deleted in the procedure of reconstruction. For this shortcoming of LLE, we propose a new method to dynamically select the neighborhood numbers, based on the definition of neighborhood radius and the ratio of neighborhood radius.

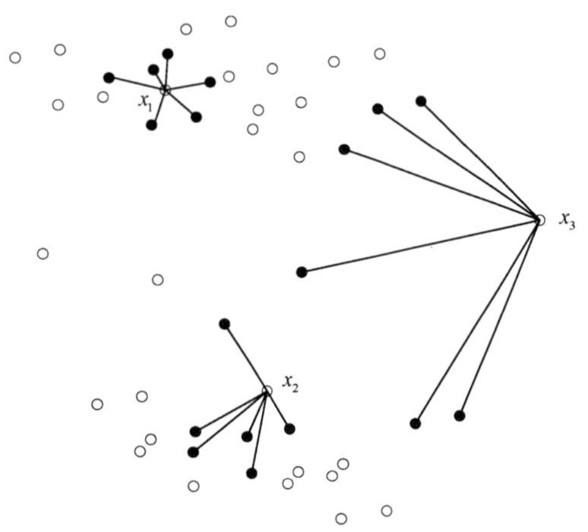


Figure 1: Neighbors selection of LLE algorithm. The neighborhood numbers of all the points is 6.

Definition 1: the neighborhood radius of point $X_i$ is defined as follows:

$$radius(X_i) = \frac{1}{K}\sum_{j=1}^{k} disp(X_i, X_{ij})$$

Where K is the number of neighbors, $X_{ij}$ is the $K$ nearest neighbors, and $disp(X_i, X_{ij})$ is the Euclidean distance between $X_i$ and $X_{ij}$.

Definition 2: the ratio of neighborhood radius is defined as follows:

$$ratio(X_i) = \frac{|radius(X_i) - radius(X_j)|}{radius(X_i)} \qquad i \neq j$$

The two definitions can be used as a performance index to predict the non-uniformity or sparsity of dataset, and identify the outliers. Under the ideal condition, the ratio of neighborhood radius for even distribution dataset should be equal to zero. With the ratio value increasing, the non-uniformity of dataset became serious. If the ratio value is greater than a threshold, we can think the point is an outlier. Based on this idea, these two definitions can be used to dynamically select the neighborhood numbers according to the data distribution. The method is described as follows:

**Step 1**. Find the data set of k nearest neighbors for each point $X_i$ by Euclidean distances;

**Step 2**. Compute the neighborhood radius and the ratio of neighborhood radius for each point $X_i$, if the ratio is greater than a constant $C$, it is that some point in the neighborhood of $X_i$ is too far away from $X_i$, and it may be an outlier point, so, delete it from neighborhood of $X_i$.

**Step 3**. Repeat step 2. If the neighborhood numbers is less than a constant $\alpha$, then this point is an outlier itself, delete this point from the original data.

Using this method, the neighborhood numbers is not fixed, and can be dynamically selected according to the ratio of neighborhood radius. For the three points in Figure 1, set their neighborhood radius as $r_1$, $r_2$ and $r_3$, respectively. After dynamically adjusting, the neighborhood numbers is shown in Figure 2. At this point, the neighborhood numbers of $x_1$ is still six, the neighborhood numbers of $x_2$ changes to 4, and point $x_3$ just has only one nearest neighbor. So, we can think $x_3$ as an outlier, and delete it from the original data.

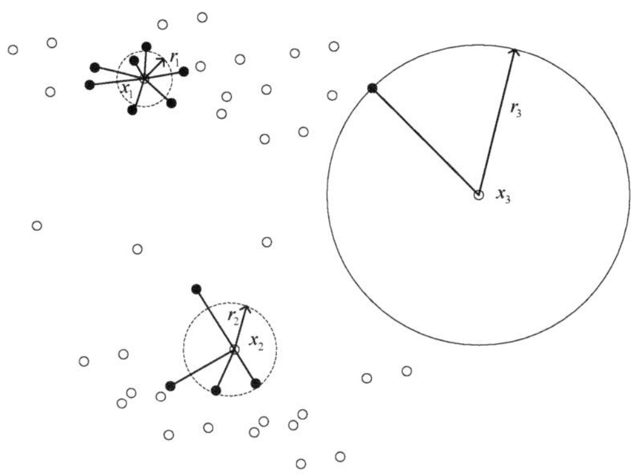


Figure 2: Neighbors selection results based on neighborhood radius. The neighborhood numbers of point $x_1$ is 6, point $x_2$ is 4. however, point $x_3$ has only one neighbor, and can be seen as an outlier.

According to the idea above, a new algorithm R-LLE is proposed to improve the LLE.

Algorithm R-LLE($X, K, d, C, \alpha$)

**Input- parameters:-** $X = \{X_i \in R^D\}$ is the high dimensional data with $N$ points, $K$ is the number of neighbors, $d$ is embedding dimensionality, $C$ and $\alpha$ is the constant above.

**Output:** it gives $N$ points $Y \in R^d$ ,- -

**Step-1.-**Find k nearest neighbors for each point $X_i$ by Euclidean distance, compute the neighborhood radius and the ratio of neighborhood radius so as to dynamically adjust the number of neighbors and determine the outliers.

**Step-2.-**Compute constrained weights $W_{ij}$ that best linearly reconstruct $X_i$ from its neighbors. Reconstruction errors are measured by the cost function:

$$\varepsilon(W) = \sum_i \left\| X_i - \sum_j W_{ij} X_j \right\|^2$$

Subject to constraints: $W_{ij} = 0$ , if $X_i$ and $X_j$ are not neighbors and $\sum_j W_{ij} = 1$

**Step-3.-**Compute embedded coordinates $Y_i$ for each $X_i$. Projections are found by minimizing the embedding cost function for the fixed weights:

$$\varphi(Y) = \sum_i \left\| Y_i - \sum_j W_{ij} Y_{ij} \right\|^2$$

Under the constraints $\sum_i Y_i = 0$ and $\frac{1}{N} \sum_i Y_i^T Y_i = I$.

## III. EXPERIMENTAL RESULTS

In order to demonstrate the performance of our proposed R-LLE algorithm, several experiments are shown here. All the experiments are running under Matlab. It involves the following four algorithms: ISOMAP, LLE, Hessian LLE and R-LLE. These algorithms have tested in two datasets: the "swiss roll" and "swiss roll with hole". In all the experiments, the neighborhood size K is set to be 12 for LLE, HLLE and R-LLE, respectively, and set to be 7 for ISOMAP. For R-LLE, the constant C and $\alpha$ is set to be 10 and 3, respectively.

We test the performance of all four algorithms on a random sample of 1000 points in three dimensions of the two dataset. The results, as seen in Figure 3 and Figure 4, show the different performance of four algorithms. From the comparison of these two figures, it is obviously that HLLE embeds the result almost perfectly into two-dimensional space, ISOMAP is unstable, causes a strong dilation of the missing sampling region and warps the rest of the embedding, and however, R-LLE algorithm improves the performance of LLE significantly, and shows a certain degree of stability.

Furthermore, we test the time performance of all four algorithms. Data sets with different numbers of samples, that is, [500; 1000; 1500; 2000; 2500], were generated from "swiss roll" data. The results of time performance are shown in Table 1. From the Table, we can see that R-LLE has a good time performance , nearly as the same as the time performance of LLE. With numbers of samples increasing, it's running time increase slowly. But ISOMAP is sensible for the size of data, and may be unsuitable for large-scale data.

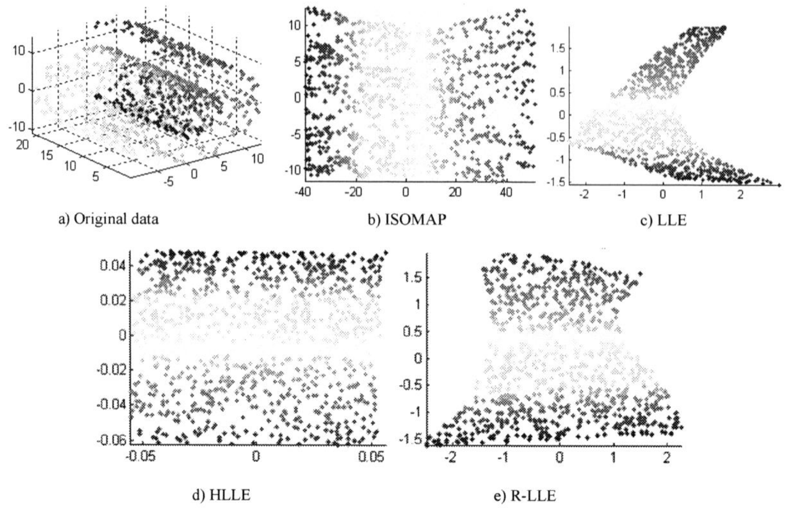


Figure 3 Embedding results on sampled Swiss roll dataset

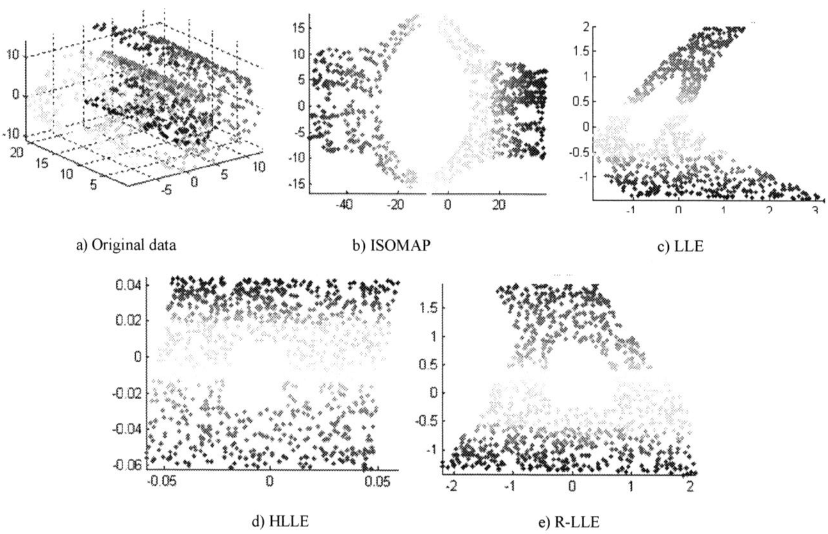


Figure 4 Embedding results on sampled Swiss roll with hole dataset

TABLE 1 AVERAGE RUNNING TIME OF THE FOUR APPROACHES(S)

| Data set size | 500 | 1000 | 1500 | 2000 | 2500 |
|---|---|---|---|---|---|
| *ISOMAP* | *11.7272* | *107.5583* | *380.7617* | *910.0671* | *1813.1953* |
| *LLE* | *0.2811* | *2.0996* | *4.5169* | *9.224* | *15.634* |
| *HLLE* | *2.6622* | *16.7907* | *55.3249* | *139.4282* | *227.4115* |
| *R-LLE* | *0.5541* | *2.2041* | *4.6485* | *12.2434* | *18.0287* |

## IV. CONCLUSIONS

This paper presents a new locally linear embedding approach to deal with unevenly distributed manifolds. It can be dynamically select the neighborhood numbers based on the ratio of neighborhood radius. Experimental results show that the proposed method improve the performance of LLE algorithm and can achieve better embedding result.

## ACKNOWLEDGMENT

This work is supported by Humanities and Social Sciences Planning Project of Chinese Ministry of Education (Project No. 07JA630090).

## REFERENCES

[1] I.T. Jolloffe. Principal Component Analysis. New York: Speinger-Verlag, 1989.

[2] T. Cox and M. Cox. Multidimensional Scaling. Chapman and Hall, 1994.

[3] J. Tenenbaum, V. de Silva, and J. Langford. A Global Geometric Framework for Nonlinear Dimensionality Reduction. Science, vol. 290, pp. 2319-2323, Dec. 2000.

[4] S. Roweis and L. Saul. Nonlinear Dimensionality Reduction by Locally Linear Embedding. Science, vol. 290, pp. 2323-2326, Dec. 2000.

[5] M. Belkin and P. Niyogi. Laplacian Eigenmaps for Dimensionality Reduction and Data Representation. Neural Computation, vol. 15, no. 6, pp. 1373-1396, 2003.

[6] D. Donoho and C. Grimes. Hessian Eigenmaps: New Locally Linear Embedding Techniques for High-Dimensional Data. Proc. Nat'l Academy of Sciences, vol. 100, no. 10, pp. 5591-5596, 2003.

[7] Kouropteva, Olga, Okun, Oleg; Pietikainen, Matti. Incremental locally linear embedding. Pattern Recognition,38, 1764-1767, 2005.

[8] de Ridder D, Kouropteva O, Okun O, et al. Supervised locally linear embedding. Lecture Notes in Artificial Intelligence, pp. 333-341,2003.

[9] Xin Geng , Zhan Dechuan , Zhou Zhihua. Supervised nonlinear dimensionality reduction for visualization and classification. IEEE Trans on SMC B, 35 (6) : 1098 – 1107, 2005.

# A Digital Forensics Primer

Gavin W. Manes*, *Member, IEEE*, Elizabeth Downing, Lance Watson
Avansic 401 S. Boston #1701 Tulsa, Oklahoma 74103
Gavin.Manes@Avansic.com

***Abstract***— **Digital forensics experts are increasingly called upon for business, legal, and criminal investigations that include evidence extracted from digital devices. The use of digital evidence in the justice system has become a cornerstone for both civil and criminal cases in this country. The proper collection and analysis of digital information is critical to the usefulness of that evidence, and it is crucial to possess knowledge of both basic and advanced forensics analyses to consider during the investigative process.**

## I. INTRODUCTION

THE proliferation of digital devices in the modern world has created significant hurdles for their collection and preservation for forensics purposes. Digital forensics experts are increasingly called upon for business, legal, and criminal investigations that include evidence extracted from digital devices. The growing numbers of electronic devices that integrate digital data storage components have exacerbated the issue of forensically sound investigations. Devices such as cell phones, digital cameras, and digital music players, along with laptops and desktop computers store information using a variety of file systems, media technologies, and data formats. The sheer variety of these storage possibilities presents major hurdles for those wishing to collect and preserve sensitive information. When digital forensics experts gather a "snapshot" of information, they must do so detailed and methodical manner. In fact, they employ many of the same procedures as forensics scientists, since any or all evidence collected can be used in discovery, depositions, or trial.

Digital forensics collection and analysis has existed since the advent of computers. However, the collection and investigation of these devices has evolved from the simple analysis of diskettes to complex treatments of large amounts of data that may exist on computer hard drives. The situations in which digital devices must be analyzed have also broadened, from tracking down pedophiles and hackers to catching fraudulent employees, competitors with proprietary intellectual property, and even terrorist organizations that use technology to perpetrate their crimes. The use of digital evidence in the justice system has become a cornerstone for both civil and criminal cases in this country.

This paper presents a general overview of the devices that can be investigated by forensics professionals, the collection process, and a number of analyses that can be performed on digital information. It also includes a summary of the current laws relating to digital evidence.

## II. DEVICES TO INVESTIGATE

Although computers are the most commonly investigated devices, forensics experts can examine a large array of digital devices. These include:

TABLE 1
LIST OF DIGITAL DEVICES

| | |
|---|---|
| Removable Media | Floppy Disks<br>CD/DVD<br>Media Cards |
| Portable Electronic Devices | MP3 players<br>GPS systems<br>Cell Phones<br>PDAs |
| Digital Cameras | Photo/Video Timestamps<br>Photo/Video Metadata |
| Faxes and Printers | Fax History<br>Print Logs |
| Communications Devices | PBX<br>Voicemail Systems |
| Automobiles | Event Data Recorders<br>Navigation Systems |

The amount of digital evidence available during any particular case is increasing as the majority of modern communication takes place digitally. E-mails, instant messages, text messages, and even telephone conversations over the Internet pass through a variety of electronics devices. Any of these can be forensically investigated to determine user activity or inactivity. Moreover, most users do not even realize the detailed information that is being digitally captured.

## III. COLLECTION

Digital forensics professionals are responsible for the collection, analysis, and investigation of electronic evidence from digital devices [1]. The proper duplication of a hard drive is critical to the authenticity of the information gathered from

T. Sobh (ed.), *Innovations and Advances in Computer Sciences and Engineering*,
DOI 10.1007/978-90-481-3658-2_64, © Springer Science+Business Media B.V. 2010

that drive, making the physical act of collection extremely important. Standard forensic collection methods involve capturing the entire media in a bit-by-bit manner, and cryptographically sealing to the evidence to prevent tampering. It is also crucial that the evidence be captured without modification: simply copying the information via "drag-and-drop" is an unacceptable method of duplication [2].

Proper handling of evidence is the first step in any digital forensics proceeding. Forensics professionals begin the entire process of collection by filling out a chain of custody when they arrive on site. This includes the name of personnel involved, time and place of collection, materials, names and serial numbers of all equipment and programs used. The chain of custody initiates the process of tracking that particular piece of evidence, which will continue throughout the entire time it is in the possession of the forensics professional.

A log is kept of all evidence seized at a particular location, and a report is written that describes each action taken by the collection specialist. This is very important in order to document any technical problems encountered during collection, as well as the particular course of that collection. This information can be referred to in the case of any problems or questions should the case be taken to court.

Any collection specialist or forensics examiner can be called to testify as a fact witness in court, making their collection and examination records all the more important. Additionally, they may need to provide credentials and the appropriate licensing for the state in which the investigation was performed. Their past experience is often scrutinized, and they must be able to answer cross examination questions about a particular case with confidence.

## IV. Analyses to Consider

Investigation is the next step in the forensics process. The examination of evidence collected from a computer involves searching through hundreds of thousands of documents, including system files and other operational programs. Additionally, there are a number of specific analyses that help ensure the information being produced represents a complete picture of user activity. Similarly, metadata and data wiping are substantial issues for both the forensics professional and the court. Analysis of the Windows registry, event logs, and link files can provide valuable information about use of a digital device.

Forensic investigations typically use keyword/pattern based searches to find relevant information. Due to the volume of information on a typical computer hard drive, general search terms can result in a very large return: words such as "legal", "company", or "stock", for instance, would likely return a majority of irrelevant hits. The use of specific phrases can significantly narrow down the search returns. Additionally, if the company or individual uses a particular naming convention for Microsoft Office or other documents, the specific file path can be more readily identified. Identifying a date range and the particular program used to create a document in question can also help reduce search hits.

The request for all emails on a computer is relatively common, but can prove expensive and very time consuming to search. It is far more efficient to request specific results, such as emails during a certain time period, or sent to a particular individual. Note that if a specific phrase or unusual word can be searched for, the process can return more relevant hits.

One of the capabilities of a digital forensics examination is the inclusion of deleted documents for any searches performed. Due to the bit-by-bit nature of the forensics copy made at the beginning of the process and the forensics software used to investigate that copy, deleted data can be searched [3]. When information is deleted from a digital device, the file itself is not erased but rather the computer's reference point for the file. It is much like removing a card from the library's card catalogue, but not taking the book off the shelf. The computer then marks the space containing the file as available. If a large amount of data is added to the drive, there is a chance that the "deleted" file can be overwritten. However, the use of very large hard drives in most current computers makes the chance of overwriting very unlikely, and fragments of documents can often be found even a year after normal computer use.

### *A. Advanced Forensics Methods*

An analysis of the Windows registry can prove useful in tracking the access of removable media such as thumb drives, CD or DVD drives, or floppy disks. It can also show if any external devices were plugged into a computer. The ability to track information transfer in this manner can prove useful in cases involving the theft of proprietary information. Information about instant messaging (IM) can also be found in the Window registry, which can be very difficult to find in any other location due to the configuration of IM programs. Additionally, the install and uninstall dates of programs can be found, which can help in cases where fraud or subterfuge is suspected or when the dates of program use are key. Windows registry analysis also has the ability to show the most recently viewed and used files. If digital information is preserved promptly and properly after an incident, this type of analysis can yield important information about computer use. These pertinent files and analyses can be easily performed by a reputable digital forensics professional.

Event logs capture information such as time, date, and content of CD burning events, file access times, or shortcut information. Both applications and the operating system of a computer can use an event log to note program errors such as failure to start or close properly or a breach of security. The presence of this type of information can be used to correlate data from other programs or files present on the computer.

Event logs can be found through data carving during the examination phase of the forensics process.

Link file analysis shows which files were accessed but that do not appear anywhere else on the computer. These files typically exist on computers in a network environment, where many files are shared. Link files list files that were accessed from a desktop computer that exist on remote devices or that no longer exist on the computer. Link file analysis typically contains the created, accessed, and modified times for the actual files that were linked to, rather than the created, accessed, and modified times for the link file itself.

*B. Data Wiping Programs*

An increasing number of forensic investigations are turning up evidence of the use of data wiping programs. Individuals wishing to hide important data or those that use these tools as a part of their regular computer maintenance have a variety of options for data wiping operations. Some software is available for free online, and other off-the-shelf programs are available in a wide range of prices. However, many of these programs do an incomplete job and leave traces of their presence: some even leave log files containing the dates and times they were executed.

The Department of Defense has issued a standard for data wiping that mandates writing over the data seven times, after which it is considered unrecoverable [4]. Although deleted information is never technically gone, it becomes more economically infeasible to recover it as the number of wipes increases. Damaged or overwritten hard drives can be investigated by microscope to visually read 1's and 0's on the platters, but this can cost up to hundreds of thousands of dollars and is therefore rarely employed.

Users can also employ a variety of data hiding techniques for data or programs. The specific techniques employed depend on the party they are trying to hide from, including other users, administrators, authorities, or forensic investigators. Savvy computer users can attempt to change filenames or extensions, or even to the file signature. Seasoned forensics examiners can often uncover attempts at subterfuge, particularly if the data is properly collected and examined shortly after the incident.

*C. Metadata*

Digital documents in their native formats contain information that their paper counterparts do not including how, when, and by whom a document was created, modified and transmitted. This history of a document is called metadata and can offer valuable clues regarding user interaction with documents and files. Metadata can also be used to authenticate a document. Most users do not know the extent of the information being captured, which can include hundreds of items. Metadata is not visible on a printed version of an electronic document.

There are a number of different types of metadata, including email, document, application-specific, and operating system. The metadata captured for emails is vast, including hundreds of items such as the sender, recipient, carbon copy and blind carbon copy fields, the date and time an email was sent and received, the filenames of attachments, whether the email had been forwarded, etc. As the most common form of communication in the business world, emails are invaluable documents in business investigations or legal proceedings. Metadata is often the deciding factor in the "he said, she said" type of arguments so often employed in fraud or propriety information theft cases.

Document metadata varies based on the specific program settings and operations, but typically the date and time of creation, modification, and last access to the document are all captured, as well as the document size, file location, and some editing information. Microsoft Word captures a great deal more metadata, including keywords, track changes editing information, comments, statistics about saving, number of times the document has been revised, how long the document has been open for editing, the number of characters, words, lines, and pages, the last date and time it was printed, template information if used, owner of the program (typically the company), and more. Word also has the unique feature for the user to set custom metadata fields for the program to capture. Other applications have metadata specific to that particular program's function, such as spreadsheets recording calculations or equations. Email headers can provide far more information than the fields obvious to the user, including the route taken by the mail from sender to recipient and file attachment information.

*D. Metadata In Court*

One of the issues addressed by the newly changed Federal Rules includes the production of documents in their native format as they are stored on the device, which may mean including metadata [5]. Although it is a common practice in modern E-Discovery, the process of converting all documents from a digital device to TIF may not survive under the rule changes. Metadata can be extremely useful in the process of document review during litigation, as the creators of documents or the dates of creation can be used as sorting criteria. Indeed, metadata is a case where paper documents provide far less useful information than digital documents. The inclusion of metadata can also be a concern regarding privilege, and whether this information should be captured is a topic of discussion during the discovery planning conference.

There can be difficulties with the accuracy of metadata, as this information is easily changed if strict preservation protocols are not observed. Any instance where a document is opened can alter the last metadata information captured. This type of information can sometimes be forged, placing importance on the appropriate authentication of key data. The capture of metadata can also dramatically increase the size and scope of discovery requests, making it necessary to carefully choose what is relevant to a particular case.

The presence of metadata can unfortunately be unnecessarily revealing to those in the practice of business or law, as a variety of metadata-analyzing software is available. This is particularly crucial given the only elementary steps many take in redacting digital information: several PDF redaction programs have well-published vulnerabilities. The National Security Agency's "Redacting with Confidence" document helps to sanitize metadata for documents converted from Word to PDF [6]: Adobe Acrobat version 8 also has a redaction feature. Metadatarisk.org provides a great deal of information regarding metadata's advantages, disadvantages, and potential uses [7].

## V. LAWS RELATING TO DIGITAL EVIDENCE

Recent changes to the Federal Rules of Civil Procedure have legitimized the use of digital evidence in court. In particular, the new rules indicate that digital documents are given the same weight and status as paper documents in terms of production [5]. Forensics professionals should have a working knowledge of these rules in order to ensure that the evidence they collect and analyze will be admissible in court. These rule changes underscore the fundamental shift of modern litigation towards the inclusion of electronic information in the legal process. Although the implications of these rule changes have begun to coalesce, the demand will only increase for properly performed data collection and digital forensics investigations. As with any other forensic evidence, the admission of digital forensics information into cases will continue to be governed by the Daubert rules [8].

Forensics can be useful to legal professionals and businesses wishing to collect and preserve digital information for use in court cases or internal investigations. Employment law cases and other civil issues are the most common situations seen by civil forensics professionals. Indeed, some estimates indicated that 85% of all crimes contain an element of digital evidence [9].

## VI. CONCLUSION

The field of digital forensics has recently enjoyed explosive growth, and the techniques for collection, investigation, and storage of digital information from investigations are rapidly evolving.

The treatment of digital documents in the court system is evolving as rapidly as the technology used to create them. Although the changes to the Federal Rules are an important first step in the acceptance of the nature of modern communications, the specifics of their use have yet to be fully ironed out in practice. Digital documents have the capability to reveal a great deal of information, but must be carefully and properly preserved and authenticated and thoroughly examined in order to take advantage of these characteristics.

## REFERENCES

[1] Arkfeld, M. R. (2005), Electronic Discovery and Evidence, Law Partner Publishing, LLC, Phoenix, AZ.

[2] National Institute of Standards and Technology (2004), 'Digital Data Acquisition Tool Specification,' nist.gov, 4 October 2004.

[3] National Institute of Standards and Technology (2005), 'Computer Forensics Tool Testing (CFTT),' cftt.nist.gov, 4 October 2004.

[4] DoD Directive 5220.00, "National Industrial Security Program Operating Manual", February 2nd, 2006. http://www.dtic.mil/whs/directives/corres/html/522022m.htm

[5] Federal Rules of Civil Procedure, December 2006.

[6] National Security Agency (2006), 'Redacting with Confidence: How to Safely Publish Sanitized Reports Converted From Word to PDF,' nsa.gov, 13 December 2005.

[7] "Security Best Practices". Retrieved 18 September 2008 from Metadata Risk website: http://www.metadatarisk.org

[8] ***Daubert v. Merrell Dow Pharmaceuticals (92-102), 509 U.S. 579 (1993)***

[9] American Bar Association. "ABA Digital Evidence Project Survey on Electronic Discovery Trends and Proposed Amendments to the Federal Rules of Civil Procedure, Preliminary Report." February 2005.

# Semantic Enrichment: The First Phase of Relational Database Migration

Abdelsalam Maatuk,M.Akhtar Ali, Nick Rossiter
School of Computing, Engineering & Information Sciences, Northumbria University,Newcastle upon Tyne, UK.
Email:{abdelsalam.maatuk;akhtar.ali;nick.rossiter}@unn.ac.uk

***Abstract*—Semanticenrichment is a process of analyzing and examining a database to capture its structure and definitions at a higher level of meaning. This is done by enhancing a representation of an existing database's structure in order to make hidden semantics explicit. In contrast to other approaches, we present an approach that takes an existing relational database as input, obtains a copy of its meta data and enriches it with as much semantics as possible, and constructs an enhanced Relational Schema Representation (RSR). Based on RSR, a Canonical Data Model (CDM) is generated,which captures essential characteristics of targetdatabases(i.e., object-based and XML) suitablefor migration.We have developed an algorithm for generatingCDM from an enhanced relational representation of an input relational database.Aprototype has been implementedandexperimentalresultsare reported.**

***Keywords* — Semanticenrichment, Relationaldatabases, Object-baseddatabases,XMLSchema**

## I. Introduction

Object-Oriented DataBases (OODBs), Object-Relational DataBases (ORDBs) and eXtensible Markup Language (XML) have become mainstream because they offer more functionality andflexibilitythan traditional Relational DataBases (RDBs). The advantages provided by these relatively newer technologies and the dominance of RDBs and their weaknesses in handling complex data have motivated a growing trendformigratingRDBs into OODBs, ORDBs and XML instead ofdesigning them from scratch [3,5,11]. This can be accomplished through reverse engineering an RDB into a conceptual schema that is enriched with its semantics and constraints. The result can be converted into another database according to a target platform. However, the question is:which of the new databases is most appropriate to use? So there is a need for an integrated method that deals withdatabasemigration fromRDB to OODB/ORDB/XML in order to provide an opportunity for exploration, experimentation and comparison among the alternative databases. The method should assist in evaluating and choosing the most appropriate target database to adopt for non-relational applications to be developed according to required functionality, performance and suitability. Such techniquescould help increasetheacceptanceof the newer approachesamong enterprises and practitioners. However, the difficulty facing this method is that it is targetingthreedatabasesthatare conceptually different. Due to theheterogeneity among the three target data models, a canonical model is needed tobridge the semantic gap among them. We believe that it isnecessary todevelop aCanonical Data Model (CDM) to facilitateourapproachfor migrating RDBs into object-based/XML databases[15]. TheCDM should be able to preserve and enhance RDB's integrity constraints anddata semantics tofit in with target databases'characteristics. Consequently, additional domain semantics need to be investigated,e.g.,relationclassificationsandrelationship identification.

Our aim in this paper is to present an approach in whichnecessary (explicit) semantics(e.g., relation and attribute names, keys, etc.) about a given RDB could be inferred,leading to the construction ofan enriched structure called Relational Schema Representation (RSR). The RSR constructs are then classified to produce a CDM, which is enhancedby additional(implicit)data semantics (e.g., classes and attributes classification, and relationship names, types, cardinalities, inverse relationships). More specifically, our aim is to construct an RSR from the extracted logical relational schema as part of a process called Semantic Enrichment (SE).SEresults in generatinga furtherenriched structure (i.e., CDM), which isusedformigratingRDBsintotargetdatabases. Canonical modelsdesigned for database integration should have semanticsat leastequal to any ofthelocal schemas to be integrated [19]. Similarly, our CDM is designed to upgrade the semantics level of RDB and to play the role of an intermediate stage formigratingRDBsto OODB/ORDB/XML acting on both levels:schematranslation anddataconversion.Its constructs are classified to facilitate the migration into complex target objectsavoiding theflat one-to-oneandcomplicated nesting conversions.Through the CDM, well-structured target databases can be obtained without proliferation of references and redundancy.However, its richness may not be fully exploited due to the relatively limitedexpressiveness of the input RDB. Consequently, someobjectconceptsprovided by target database, e.g., behavioural aspects, get less attention inourCDM.

The rest of the paper is organised as follows. SectionIIprovides an introduction to the SE process. SectionIIIdescribes how to construct an RSR based on meta data extracted from an existing RDB. SectionIVpresents the CDM definition and how to generate it from an RSR and an existing RDB. We evaluate our method in SectionV. The related workis presented in SectionVI, and SectionVIIprovides conclusions.

## II. Overview of Semantic Enrichment

Semantic Enrichment (SE)is a process of analyzingdatabasesto understandtheirstructure and meaning, andtomake hidden semantics explicit.Conflicts

T. Sobh (ed.), *Innovations and Advances in Computer Sciences and Engineering*,
DOI 10.1007/978-90-481-3658-2_65, © Springer Science+Business Media B.V. 2010

in naming have tobe resolved, and attributes and interrelationships amongstdata have to be deduced.In our approach, the SEprocess involves the extraction of data semantics of an RDB to be represented in RSR followed by conversion into a much enriched CDM. This facilitatesthemigration into new target databases without referringrepeatedlyto the existing RDB. The main benefit from usingRSR andCDM together is that an RDB is read and enriched once while the results can be used many times to serve different purposes (e.g.,schema translation, data conversion). Fig.1 shows the schematic view of the SE process.The process starts by extracting the basicmetadatainformation about an existingRDB inorder to construct RSR, which is designed in such a waytoease key matching foritsconstructs classification. To get the best results,itispreferredthattheSE processisapplied to the schema in 3rd Normal Form (3NF). A relation that is not in 3NF may have redundant data, update anomalies or no clear semantics of whether it represents one real world entityor arelationship type. The nextmajorstepisto identify the CDM constructs based on classification of RSR constructs, including relationships and cardinalities, which are performed through data access. Lastly, the CDM structure is generated.

## III. EXTRACTINGRSR FROM AN RDB

In this section, wedefineRSR, as a representation of an RDB's metadata, tobe used as a source of information for CDM generation.Basic information needed to proceed with the SEprocessincludesrelation names and attributeproperties(i.e.,attribute names, data types,length, default values, and whether the attribute isnullable).

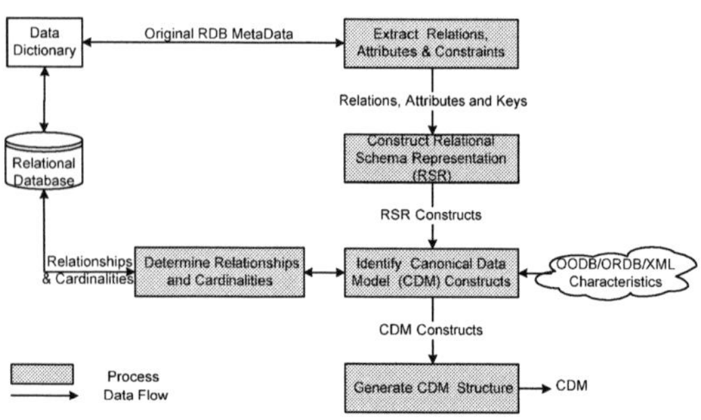


Fig. 1.Schematicview of the Semantic Enrichmentprocess

Moreover,themostimportant information needed is about keys including Unique Keys (UKs).We assume that datadependencies are represented byPrimary Keys(PKs) and ForeignKeys(FKs) as for each FK value thereisan existing,matched PK value, which can be considered as a value reference. The inverse of an FK is called an Exported Key(EK). EKs play an important role as regards toobject-baseddatabases, which support bi-directional relationships. Theuserhelpmight be necessary to provide any missingsemantics.

**Definition 1:**An RDB schema is represented in ourapproach as a set of elements$RSR$,where:

$RSR := \{R \mid R := \langle rn, Arsr, PK, FK, EK, UK \rangle\}$,

- $rn$denotes the name of relation$R$,
- $Arsr$ denotesaset of attributes of$R$ : $Arsr := \{a \mid a := \langle an, t, l, n, d \rangle\}$, where $an$ is an attribute name, $t$ its type, $l$ data length, $n$ nullable or not('y'|'n') and $d$ a default value if given.The function$A(R)$returns a set containing only$an$component ofattributes.
- $PK$ denotes R'sPK: $PK := \{\alpha \mid \alpha := \langle pa, s \rangle\}$,where$\alpha$represents one PK attribute,$pa$is an attribute name and$s$is a sequencenumber in the case of a composite key;however,$s$isassigned 1 in the case of a single valuedkey.The function $P(R)$ returnsasetcontaining only$pa$componentofthePK of$R$.
- $FK$ denotesaset ofFK(s) of $R$: $FK := \{\beta \mid \beta := \langle er, \{\langle fa, s \rangle\}\rangle\}$,where$\beta$represents one FK(whether it is single or composite),$er$is the name ofan exporting (i.e., referenced) relation that containsthe referenced PK, $fa$ is anFKattribute name,and $s$ is a sequence number.Thefunction $F(R)$ returnsaset containing only $fa$ componentofallFKs of $R$ .The function $fk(\beta)$ returns a setcontaining$fa$componentofone FKof$R$.
- $EK$ is a set ofEK(s) of$R$: $EK := \{\gamma \mid \gamma := \langle ir, \{\langle ea, s \rangle\}\rangle\}$, where $\gamma$ represents one EK, $ir$ is the name of an importing (i.e., referencing) relationthat contains the exported attribute name $ea$ (i.e.,FKattribute).The function $E(R)$ returnsaset containing only$ea$componentofall EKs of $R$ .The function $ek(\gamma)$ returns a set containing$ea$componentofone EK of$R$.
- $UK$ is a set ofUK(s)of$R$: $UK := \{\delta \mid \delta := \{\langle ua, s \rangle\}\}$, where$\delta$represents one UK, $ua$ is an attributename and$s$is a sequence number.
- The function*getRSRrelation*( $r$ ) returns the $RSR$ relation that corresponds to the relation name$r$. To get an element of a composite construct, we use "." notation, e.g.,$R.rn$.

The main purpose behind constructing an RSR is to read essential metadata into memory outside the database's secondary storage.An efficient RSR construction overcomes the complications that occur during matching keys in order to classify relations, e.g., strong or weak, attributes, e.g., non-key attribute (NK) and relationships, e.g., M:N, inheritance, etc. Each relation $R$ is constructed withitssemanticsas one element, which is easily identifiableanduponwhich set theoretic operations can be applied formatching keys. Each of $R$ 's elements describes a specificpart of $R$ (e.g., $Arsr$ describes $R$ 'sattributes). Animportant advantage of RSR is that it identifies the set$EK$ therefore adding more semantics to an RDB's metadata.The$EK$holds keys that are exported from$R$tootherrelations.

Consider the databaseshown in Fig.2.PKs areunderlinedandFKsare marked by "*". Table I gives the RSRconstructedfromthedatabaseshowing only`Emp`, `Salaried_emp`, `Dept`and`Works_on`relations.

**Emp**(<u>eno</u>,ename, bdate, address, spreno*,dno*)
**Salaried_emp**(<u>eno</u>*, salary)
**Hourly_emp**(<u>eno</u>*, pay_scale)
**Proj**(<u>pnum</u>, pname, plocation, dnum*)
**Dept**(<u>dno</u>, dname, mgr*, startd)
**Dept_locations**(<u>dno*, location</u>)
**Works_on**(<u>eno*,pno*</u>)**Kids**(<u>eno*, name</u>, sex)

Fig. 2.Sample input RDB schema

TABLE I
RESULT OF RSR CONSTRUCTION

| *r n* | *A rsr* | | | | | *PK* | | *FK* | | | *EK* | | | *UK* | |
|---|---|---|---|---|---|---|---|---|---|---|---|---|---|---|---|
| | *an* | *t* | *l* | *n* | *d* | *pa* | *s* | *er* | *fa* | *s* | *ir* | *ea* | *s* | *ua* | *s* |
| Emp | eno | int | 25 | n | | eno | 1 | Emp | spreno | 1 | Salaried_emp | eno | 1 | | |
| | ename | char | 40 | n | | | | Dept | dno | 1 | Hourly_emp | eno | 1 | | |
| | bdate | date | | y | | | | | | | Works_on | eno | 1 | | |
| | address | char | 40 | y | | | | | | | Dept | mgr | 1 | | |
| | spreno | int | 25 | y | | | | | | | Kids | eno | 1 | | |
| | dno | int | | n | | | | | | | Emp | spreno | 1 | | |
| Salaried_emp | eno | int | 25 | n | | eno | 1 | Emp | eno | 1 | | | | | |
| | salary | int | | y | | | | | | | | | | | |
| Dept | dno | int | | n | | dno | 1 | Emp | mgr | 1 | Emp | dno | 1 | mgr | 1 |
| | dname | char | 40 | n | | | | | | | Proj | dnum | 1 | | |
| | mgr | int | 25 | n | | | | | | | Dept_locations | dno | 1 | | |
| | startd | date | | y | | | | | | | | | | | |
| Works_on | eno | int | 25 | n | | eno | 1 | Emp | eno | 1 | | | | | |
| | pno | int | | n | | pno | 2 | Proj | pno | 1 | | | | | |

## IV. GENERATIONOFCDM FROM RSR

This section presents theCDMdefinitionand its algorithm. We concentrate on how to identify CDM constructs using information provided by RSRand how to generate relationships and cardinalities amongclasses using data instances. RSR constructs areclassified, enriched and translated into CDM. CDMspecificationsare based on the similarities among object-based and XML datamodels. Similarities produce naturalcorrespondences that can be exploited tobridge the semantic gap among diverse data models. This provides abasis for theCDM that can be usedas an intermediate representation to convert an RDB intomore than one target database. The definition of CDM isprovided in SubsectionA. The algorithm to generateCDM from RSR and an RDB is presentedin SubsectionB.

### A. Definition of CDM

CDM has three concepts: class, attribute andrelationship. Attributes define class structure, whereas relationships define a set of relationship types. The model is enriched by semantics from an RDB such as PKs, FKs, attributes length, etc. Besides, the modelhas taken into consideration features that are provided byobject-based and XMLdatabasessuch as association, aggregation and inheritance. However, the CDM is independent of an RDB, from which it has taken semantics as well as any target databases to which it might be converted. Real world entities, multi-valued and composite attributes, and relationshiprelations areall represented as classes in CDM. Object-baseddatabasesencapsulate static(i.e., properties)and dynamic aspects(i.e., methods)of objects. However, dynamic aspects will get less attention in CDM compared to static aspects becausean RDB does not support methods attached to relations. Static aspects involve a definition of class, and its attributes and relationships. CDM classes are connected through relationships.

**Definition 2:** TheCDM is defined as a set of classes, $CDM{:}=\{C|C{:}=\langle cn, cls, abs, Acdm, Rel, UK\rangle\}$, where each class$C$has a name$cn$, a classification$cls$and whether it is abstract or not$abs$. Each $C$ has a set of attributes $Acdm$, a set of relationships $Rel$ and a set of UKs$UK$.

- Classification ( $cls$ ): A class $C$ is classified intoninedifferent kinds of classes(according to relationship participations), which facilitate its translation intotarget schemas:
    1. RegularStrong Class (RST): main class,
    2. Secondary Strong Class (SST): super-class,
    3. Subclass (SUB): subclass,
    4. Secondary Subclass (SSC):inherited subclass,
    5. Regular Relationship Class (RRC): M:Nrelationship class without attributes,
    6. Secondary Relationship Class (SRC): referenced RRC, M:Nrelationship with attributes,orn-ary relationships,where n>2,
    7. Multi-valued Attribute Class (MAC): classrepresents multi-valued attributes,
    8. Composite Attribute Class (CAC): classrepresents composite attributes, and
    9. Regular Component Class (RCC): componentclass inarelationshiprather than its wholeclass.
- Abstraction ( $abs$ ): A superclass is abstract (i.e.,$s$:=`true`) when all its objects are members of its subtypes. Instances of an abstract type cannot appearin database extension but are subsumed into instancesof its subtypes. A class is not abstract (i.e., $s$ :=`false`) when all (or some of) its corresponding RDBtable rows are not members of other subtables.
- Attributes ($Acdm$): A class$C$has a set of attributes ofprimitive data type, $Acdm{:}=\{a|a:=\langle an, t, tag, l, n, d\rangle\}$ ,where eachattribute $a$ has a name $an$ , data type $t$ and a$tag$,whichclassifies attributes into a non-key 'NK', 'PK', 'FK'or both PK and FK attribute 'PF'. Each $a$ can havea length $l$ and may have a default value$d$whereas$n$indicates that$a$is nullable or not.
- Relationships( $Rel$ ): Each class $C$ has a set of relationships $Rel{:}=\{rel|rel{:}=\langle RelType, dirC, dirAs, c, invAs\rangle\}$. Eachrelationship $rel$is defined in$C$with another class$C'$, through a set of attributes $dirAs$ using relationship type $RelType$ and cardinalities$c$. $RelType$can have the following values:

'*associated with*' for association, '*aggregates*' for aggregation, and '*inherits*' or '*inherited by*' forinheritance. CDM does not support multipleinheritances, as target databasestandardsdo not allowa concrete subtype to have more than one concrete super-type; hence, a subclass inherits only from one superclass. The $dirC$ is the name of the related class $C'$ participating in the relationship, and $dirAs$ is a set containing the attribute names representing the relationship in$C'$,whereas the inverse relationship attribute names in$C$are contained in the set$invAs$.Cardinality$c$is defined by*min..max*notation to indicatetheminimumandmaximumoccurrence of$C'$object(s) within$C$objects.

*B. Algorithm for Generation of CDM*

This subsection presents the**GenerateCDM**algorithm shown in Fig.3. Given an RSR and RDB dataas input, thealgorithm goesthroughamainloop to classify RSR constructs and generates their equivalents in CDM (ref point 1). Using key matching the algorithmclassifieseachrelation $R$ of the inputRSR,itsattributesandrelationships.Abstraction of each class in CDMis checkedusing the*checkAbstraction*function.The set of unique keys $R.UK$ remains unchanged.Eachrelation $R$ is classifiedandmapped intocorrespondingclass $C$ (ref point 2).We assume that relation kinds/relationships participation are represented in RDB by meansof PKs/FKs matching,i.e., keys composition of each other. Other representations may lead to different target constructs.For instance,$R$is a mainrelationif its PK is not fully or partially composite of anyFKs,i.e., $P(R) \cap F(R) = \emptyset$; $R$ is a subclass, ifits PK is entirely composite of the primary key of a superclass relation(i.e.,we assume one relation for each superclass and one for each subclass in inheritance representations).Similarly $R$ is a weakrelation if its PK isapartialcomposite oftheprimary keyof a strong relation.Severalfunctionsare used to facilitatethe CDM constructclassifications, e.g.,$DFK(R)$functionreturns the numberofdisjointFKs,if $R$ is a relationship relation, and*DangKs*$(R)$returns the numberofdanglingkey attributes in the case that $R$ is a weak entity relation. Using the*classifyAttributes*function, attributes of$R$are identified and mapped (with other properties, e.g., data type) into attributesof$C$(ref point 3).Attributesare classifiedinto non-key attribute, PK attributes or FK attributesusing $tag$ .Using $PK, FK$ and $EK$ setsof $R$, all relationships are identified, classified and their cardinalities determined, and then mappedinto CDMas association, inheritance or aggregation. $FK$ set(ref point 4)showsrelationships (i.e.,'*associated with*','*inherits*') whichareestablished from other relations side,when $R$ contains FKs, whereas $EK$ set(ref point 5),helpsto identify relationships (i.e.,'*associated with*','*aggregates*'and'*inherited by*') when $R$ isadominant(referenced)relation. Cardinality$c$of eachrelationshipis determined by querying datain a completedatabase. The function*deterCard*determines$c$when $R$ contains FKs,andthe*deterInverCard*functionreturns theinverse$c$when $R$ is referenced by other relations.After generating CDM, we translate it intoobject-based/XML

```
algorithmGenerateCDM(rsr:RSR)returnCDM
cdm:CDM :=∅
1.foreachrelationR∈rsrdo
r: relation name :=R.rn
A': set[nonKeyattribute] :=A(R) − (P(R) ∪ F(R))
2.if(P(R) ∩ F(R) = ∅)thencls:= RST// classify classes
else if(DFK(R)>1)then
if(DFK(R)= 2andA' = ∅andE(R) = ∅)then
cls:= RRCelsecls:= SRC
else if(P(R) ⊆ F(R))thencls:= SUB
else if(F(R) − P(R) = ∅andE(R) = ∅)then
if(DangKs(R) = 1andA' = ∅)thencls:=MAC
elsecls:=CAC
elsecls:= RCC
end if
3.Acdm:=classifyAttributes(R.Arsr)
4.foreachforeign keyβ ∈ R.FKdo
// identify and classify relationships
Re:=getRSRrelation(β.er)
re: relation name :=Re.rn
if(fk(β) ⊈ P(R)orDFK(R) ≥ 2)then
RelType:= 'associated with'
else if(P(R) ⊆ fk(β))thenRelType:= 'inherits'
end if
c:=deterCard(r, fk(β))//determine cardinality
Rel: = Rel ∪ {⟨RelType, re,P(Re), c, fk(β)⟩}
end for//define one relationship
5.foreachexported keyγ ∈ R.EKdo
// identify and classify inverse relationships
Ri:=getRSRrelation(γ.ir)
ri: relation name :=Ri.rn
if(ek(γ) ⊈ P(Ri)orDFK(Ri) ≥ 2)then
RelType:= 'associated with'
elseif(ek(γ) ⊆ P(Ri))then
if(ek(γ) ≠ P(Ri))thenRelType:= 'aggregates'
elseRelType:= 'inherited by'
if(cls= RST)thencls:= SST
elsecls:= SSC
end if
end if
c:=deterInverCard(r, ri, ek(γ)) // inverse cardinality
        Rel: = Rel ∪ {⟨RelType, ri, ek(γ), c, P(R)⟩}
end for
abs:=checkAbstraction(R)
        cdm: = cdm ∪ {⟨r, cls, abs, Acdm, Rel, R.UK⟩}
endfor//addoneclasstocdm
returncdm
end algorithm
```

Fig. 3. The**GenerateCDM**Algorithm

schemas, details of which can be found in our technical report [16].

Consider theRSRshown inTable I to the input to the**GenerateCDM**algorithm. Fig.4showsthe resulting CDM for only**`Emp`**and**`Dept`**classes. TheCDM's class**`Emp`**has attributes:*ename, eno, bdate,address, spreno*and*dno*. Other properties (e.g., attributes'types, tags, default values) are not showndue to lackofspace. The class**`Emp`**is'*associated with*'classes:**`Dept`**,**`Works_on`**and with itself. Moreover, it'*aggregates*'**`Kids`**class and'*inherited by*'**`Salaried_emp`**and**`Hourly_emp`**classes.

Relationships with cardinalitiesare definedinCDMclasses as: $RelType\{invAs \leftrightarrow dirC(dirAs)c$ ( $\leftrightarrow$ indicates bidirectional associationand$\leftarrow$indicates aggregation).

```
Emp[Acdm:= {ename, eno, bdate, address,
spreno, dno}, Rel:= {associated with{
dno↔Dept(dno)1..1, eno↔Dept(mgr)0..1,
spreno↔Emp(eno)1..1, eno↔Emp(spreno)0..*,
eno↔Works_on(eno)1..*},
aggregates{eno←Kids(eno)0..*},
inherited by{Salaried_emp,Hourly_emp}}]

Dept[Acdm:= {dname, dno, mgr, startd},
Rel:= {associated with{mgr↔Emp(eno)1..1,
dno↔Emp(dno)1..*, dno↔Proj(dnum)1..*},
aggregates{dno←Dept_locations(dno)1..*}}]
```

Fig. 4.Sample generated CDM schema

## V. Experimental study

To demonstratethe effectiveness and validity ofthe CDM, a prototype has been developedusing Java 1.5,realizingthe algorithmpresented in this paper.Wesetuptwo experiments to evaluate our approach by examining the differencesbetweensource RDB and target databasesgenerated bythe prototype.Theexperimentswererun on a PC with Pentium IV 3.2 GHz CPU and 1024 MB RAM operatingunder Windows XP Professional.We measureddatabase equivalences, includingsemantics preservation,loss of dataandredundancy, and integrity constraints. Full detailsaboutthe experimentscan befound in [14,15].

In the firstexperiment, wetest schema information preservation by comparing the target schemas resultingfromour prototype and those generated from other manual-based mapping techniques.A schema is correct if all concepts of underlying model are used correctly with respect to syntax andsemantics [10]. In general, the resultsfrom thedatabase engineering process could be validated against the results that are obtainedmanually by a knowledgeable person[7].Sowe claimthat the CDM generated from an RDB is correct when target schemas generated basedonit are equivalent to the schemas mapped from the same RDB by otherapproaches. TheCDM is then validated as a representation of an existing RDB. The secondexperiment wasa query-based experiment basedon the BUCKYbenchmark [4]. We have translated the benchmarkqueries into equivalent versions in OODB and XML and run them ontheirnative systems,observingany differencesin resultsregarding data content and integrity constraintsequivalence.

After evaluating the results,our approachis shown to be feasible, efficient and correct.Given that all approachesthathave beencomparedto our approach, in the first experiment,are manual techniques, which give the user an opportunity to use all features of target models to result in well-designed physical schemas, we found that our approach, which is fully-automatic has the ability to generate a more accurate and intuitive target schemas. The CDM, which preservesan enhanced structure of an existing RDB, is translatable into any of the three target schemasand the queries return identical results.Therefore, target databases are generated without loss or redundancy of data. Moreover, many semantics can be converted fromRDB into the targets,e.g., association, aggregation and inheritance with integrity constraints enforced on the target database. Some update operations are applied on the databases to show that integrityconstraints in the RDB are preserved in the target database. However, we cannot cover automatically referential integrity on REFs that are in nested tables in ORDB because Oracle does not have a mechanism to do so; this integrity could be preserved once the schema is generated, e.g., using triggers.In addition, the keys of XML elements may notbevalid for other element(s),which would substitute themin instance document. This is becauseXpath 2.0 is not schema-aware.

## VI. Related Work

Inferring a conceptual schema from a logical RDB schema has been extensively studied by many researchers in the context ofdatabase reverse engineering [5,2,3,6,18],semantic enrichment[19,12] andschema translation[9,20,11]. Such conversions are usually specified by rules, which describe how to derive RDBs constructs (e.g., relations, attributes, keys), classify them, and identify relationships among them. Semantic information is extracted by an in-depthanalysis of relations in an RDBschema together with their data dependencies into a conceptual schema model such as Entity-Relationship Model (ERM), UML, object oriented and XML data models. Data and query statements are also used in some work to extract somesemantics. However, most of the work has been focused onschema translation ratherthandata conversionwith an aim to generate one target data model based on its conceptual schema or other representations as an intermediate stage for enrichment. In addition, the existing work does not provide a complete solution for more than one target database,for either schema or data conversion.A classification on database migration techniques can be found in our work [13].

An approach that focuses on deriving an Extended ERM (EERM) from an existing RDB is presented in [6]. The process recovers domain semantics throughclassification of relations, attributes and key-based inclusion dependencies using the schema. However, expert involvement is required to distinguish betweensimilar EERM constructs. The approach discussed in [3] extracts a conceptualschema by analysing equi-join statements. The approach uses a join conditionand a distinct keyword for elimination of attributes during key location.Ref [2]developed algorithms that utilise data to derive all possible candidate keysfor identifying foreign keys of eachgiven relation in a legacy RDB. Thisinformation is then used to construct what is termed as RID graph, which includes all possible relationships among RDB relations.Ref [11]introducesa method in which data semantics are extracted from an RDB into anEERM, which is then mapped into a conceptual XML Schema Definition language (XSD) graph that captures relationships and constraints among entitiesin an EERM. Finally, the XML logical schema is extracted from the XSD.A model, called BLOOM, is developed, which acts like a CDM for federateddatabase management systems [1]. Its goal is to upgrade the semantic level oflocalschemas of different databases and facilitate their integration. A method,which improves an RDB schema semantically and translates it into a BLOOMschema,is described in [5].Ref [18]proposesa procedure for mapping an RDB schema into an Object-Modelling Technique (OMT) schema [18].Ref [9]developsa method for converting an RDB schema into a model, calledORA-SS[8], which is then translatedinto XML Schema. However, they adoptan exceptionally deepclusteringtechnique, which is prone to error.

Although current conceptual models, e.g., ERM or UMLmaybe usedas a CDM indatabase migration, we argue that they do not satisfy the characteristics andconstructs of more than one target data model and do not support data representation. Some important semantics (e.g., inheritance, aggregation) have notbeen considered in some work, mainly due their lack of support either in sourceor target models, e.g., ERM and DTD lack support for inheritance. UML shouldbe extended by adding new stereotypes or other constructs to specify ORDBand XML models peculiarities [17,20] and it is still weak and not suitable tohandle the hierarchical structure of the XML data model [11]. Several dependentmodels were developed at specific applications, and they are inappropriate tobe applied to generate three different data models; e.g.,BLOOM wasdefined for different schemas to be integrated in federated systems and ORA-SShas been designed to support semi-structured data models.

In contrast, the CDM described in this paper can be seen as an independent model, which embraces object oriented concepts with rich semantics thatcater forobject-relationaland XML data models. It preserves a variety of classification forclasses, attributes and relationships, which enableusto represent the target complex structuresin an abstract way.Classes aredistinguishedasabstractclassesand concrete classes.Relationships are defined intheCDM in a waythatfacilitatesextractingand loading of dataduringdata conversionincluding defining andlinking objects using user-definedobjectidentifiers. Moreover, it provides non-OODBkey concepts (i.e., FKs, null and UKs)and explicitly specifieswhetherattributesare optionalor required using null values.Because of thesecharacteristics, our CDM can facilitate the migration of an existing RDB intoOODB/ORDB/XMLduringbothschematranslationand dataconversion phases.

## VII. Conclusion and Future Works

This paper presents an approach tothesemantic enrichment,in which necessary data semantics abouta given RDB are inferred andenhanced to producean RSR. The RSR constructs arethen classified to generate a CDM, which provides adescription of the existing RDB's implicit and explicitsemantics. The generated CDM is asound source ofsemantics and is a well organised data model, whichforms the starting point for the remaining phases ofdatabase migration. In addition to considering mostimportantcharacteristics of target models, the CDM preserves all data semantics that can possibly be extracted from an RDB, e.g., integrity constraints, associations,aggregations andinheritance. Moreover, the CDMrepresents a key mediator forconverting an existing RDBdata into target databases. It facilitates reallocationofattributes in an RDB to theappropriate values in a targetdatabase. A prototype has been implemented based on thealgorithm proposed in this paper for generating OODB,ORDBandXML schemas.Our approach has been evaluated by comparing the prototype's outputs with the results ofexistingmethods. We found that the resultswere comparable.Therefore, we conclude that the source and target databaseswereequivalent. Moreover, the results obtained demonstrate thatour approach, conceptually and practically, is feasible,efficientand correct.Our future research focus is on data specific manipulation (e.g., update/query)translations and further prototyping to simplify relationship names that are automatically generated.

## References

[1] A.Abelló,M.Oliva,M. E.Rodríguezand F.Saltor,*The Syntax of BLOOM99 Schemas*, TR LSI-99-34-R Dept LSI,1999.

[2] R.Alhajj,"Extracting theextendedentity-relationshipmodel from alegacyrelationaldatabase,"*Inf. Syst.*vol.28,pp.597-618, 2003.

[3] M.Andersson, "Extracting anentityrelationshipschema from arelationaldatabase throughreverseengineering," in*13thint.conf. on the ERapproach*,pp.403-419, 1994.

[4] M. Carey,et al.,"TheBUCKYobject-relational benchmark,"*SIGMOD Rec*, vol. 26,pp. 135-146, 1997.

[5] M.Castellanos,F.Saltorand M.García-Solaco, "Semanticallyenrichingrelationaldatabases into anobjectorientedsemanticmodel," in*DEXA*,pp.125-134, 1994.

[6] R.Chiang,T.Barronand V. C.Storey,"Reverseengineering ofrelationaldatabases: Extraction of an EERmodel from a relationaldatabase,"*Data Knowl. Eng*, vol.12, pp.107-142, 1994.

[7] R. Chiang, T. Barron and V. C. Storey, "A framework for the design and evaluation of reverse engineering methods for relational databases,"*Data Knowl. Eng.*, vol. 21, pp. 57-77, 1996.

[8] G.Dobbie,X.Wu,T.Lingand M.Lee,*ORA-SS: Object-Relationship-Attribute Model for Semistructured Data*,TR 21/00 National University of Singapore,2001.

[9] W.Du,M.Li,L.Tokand W.Ling, "XML structures forrelational data,"*WISE*, vol.1, pp.151-160, 2001.

[10] C. Fahrner and G. Vossen, "A survey of database design transformations based on the Entity-Relationship model,"*Data Knowl.Eng.*,vol. 15,pp. 213-250,1995.

[11] J.Fongand S. K.Cheung, "Translatingrelationalschema into XML schema definition withdata semanticpreservation and XSDgraph,"*Inf. & Soft. Tech.*, vol. 47, pp.437-462,2005.

[12] U.Hohensteinand V.Plesser, "Semanticenrichment: A firststep to providedatabase interoperability,"*WorkshopFderierte Datenbanken*,pp.3-17, 1996.

[13] A. Maatuk,M.A. Ali, and N. Rossiter,"Relational database migration: A perspective," in*DEXA*, vol. 5181, pp. 676-683,2008.

[14] A. Maatuk,M.A. Ali, and N. Rossiter,*Relational database migration:AnEvaluation*,Technical report,http://computing.unn.ac.uk/staff/cgma2/papers/Migper.pdf, 2008.

[15] A. Maatuk,M.A. Ali, and N. Rossiter, "An integrated approach to relational database migration," in*IC-ICT2008*, 6pp, in press,http://computing.unn.ac.uk/staff/cgma2/papers/icict08.pdf, 2008.

[16] A. Maatuk, M. A. Ali, and N. Rossiter,*Aframeworkfor relational database migration*,Technical report,http://computing.unn.ac.uk/staff/cgma2/papers/RDBM.pdf,2008.

[17] E.Marcos,B.Velaand J. M.Cavero, "Extending UML forobject-relationaldatabase design,"in*4th int.conf. on theunifiedmodelinglanguage*, vol.2185, pp.225-239, 2001.

[18] W. J.Premerlaniand M. R.Blaha, "Anapproach for reverseengineering ofrelationaldatabases,"*Commun. ACM*, vol.37, pp.42-49, 1994.

[19] F.Saltor, M.Castellanosand M.García-Solaco "Suitability of datamodels ascanonicalmodels for federateddatabases,"*SIGMOD*,vol.20,pp.44-48, 1991.

[20] B.Vela and E. Marcos, "Extending UML to represent XML schemas,"*CAiSE*short paper proceedings, 2003.

# The Impact of the Prototype Selection on a Multicriteria Decision Aid Classification Algorithm

Amaury Brasil, Plácido R. Pinheiro, André L. V. Coelho
University of Fortaleza (UNIFOR) – Master of Applied Computing
Av. Washington Soares, 1321 - Bl J Sl 30 - 60.811-341 - Fortaleza – Brasil
{abrasil}@gmail.com, {placido, acoelho}@unifor.br

*Abstract* - **This paper presents an experimental analysis conducted over a specific Multicriteria Decision Aid (MCDA) classification technique proposed earlier by Goletsis et al. Different from other studies on MCDA classifiers, which put more emphasis on the calibration of some control parameters related to the expert's preference modeling process, this work investigates the impact that the prototype selection task exerts on the classification performance exhibited by the MCDA model under analysis. We understand that this sort of empirical assessment is interesting as it reveals how robust/sensitive a MCDA classifier could be to the choice of the alternatives (samples) that serve as class representatives for the problem in consideration. In fact, the experiments we have realized so far, involving different datasets from the UCI repository, reveal that the proper choice of the prototypes can be a rather determinant issue to leverage the classifier's performance.**

## I. Introduction

A classification problem refers to the assignment of a group of alternatives to a set of predefined classes, also known as categories. During the last decades these problems have been tackled using a high variety of statistical and machine learning techniques. Recently, the area of Multicriteria Decision Aid (MCDA) [1] has also brought about new methodologies and techniques to solve these problems. The main difference between the MCDA classification methods and others coming from related disciplines (for instance, artificial, neural networks, Bayesian models, rule-based models, decision trees, etc.) [2] lies in the way that the MCDA methods incorporate the decision maker's preferences into the categorization process.

According to Doumpos and Zopounidis [3], in MCDA the classification problems can be divided into two distinct groups: ordinal sorting problems and the nominal sorting problems. The first one is used when the problem has its classes (groups) defined in an ordinal way. A good example of this type of problem is the bankruptcy risk evaluation problem [4]. In this problem, there is an ordinal definition of the groups, since it is obvious that for a decision maker that the healthy firms are in better situation than the bankrupt ones. The second type of problem refers to the assignment of alternatives into classes that do not present a preferential order.

When the problem to be solved is an ordinal sorting problem, the MCDA methods usually introduce a fictitious alternative (sample), called a reference profile, in order to delimit the boundary between two consecutive groups. Conversely, in a nominal sorting problem, the boundaries between classes cannot be properly defined beforehand as, usually, the knowledge about lower and upper boundary samples is not readily available. To overcome such difficulty, Belacel [5] has cast a new interpretation to the role of the reference profiles (also known as prototypes) in his PROAFTN method: That of a good representative sample for a specific class.

As pointed out by Zopounidis and Doumpos [6], the great majority of works conducted on the MCDA classification theme has focused on the development of novel MCDA classification methods, not giving much emphasis on characterizing and comparing their distinctive problems. Likewise, the authors also advocate that future research on this field should consider a more deep investigation into some important practical issues, such as the analysis of the interdependencies of the control parameters of the algorithms, the statistical validation of the generated models, the analysis of performance over large data sets, and the establishment of links between MCDA classifier models and those coming from related disciplines, such as Pattern Recognition, Machine Learning, and Data Mining [2].

In this context, this work investigates the impact that the prototype selection task exerts on the classification performance exhibited by a MCDA model while coping with a nominal sorting problem. We understand that this sort of empirical assessment is pertinent as it reveals how robust/sensitive a given MCDA classifier could be to the choice of the alternatives (samples) that serve as class representatives for the problem in consideration. In fact, the experiments we have realized so far, involving the MCDA classification model proposed earlier by Goletsis et al [7] and different datasets taken from the UCI repository [8], reveal that the choice of the prototypes can be a key issue to be properly dealt with in order to leverage the classifier's performance.

The rest of the paper is organized as follows. The next section presents an overview of some related work on MCDA classification. The third section outlines the main conceptual ingredients of the MCDA classification algorithm under consideration. The next section provides details of some of the experiments we have conducted so far, discussing the main results achieved. Finally, the last section concludes the paper and brings remarks on future work.

## II. Related Work

ELECTRE TRI [9] is a member of the family of ELECTRE [10] methods. The ELECTRE methods are based on the outranking relation techniques, and the ELECTRE TRI, specifically, was designed to solve classifications problems. The objective of ELECTRE TRI is to assign a discrete set of alternatives into groups that are defined in an ordinal way. This method introduced the concept of reference profile as a fictious alternative that is a boundary between two consecutive groups.

The N-TOMIC method [11], was developed for addressing classification problems when the groups are defined in an ordinal way. The method pre-specifies nine different groups or classes to which the alternatives should be assigned. The groups indicate the

T. Sobh (ed.), *Innovations and Advances in Computer Sciences and Engineering*,
DOI 10.1007/978-90-481-3658-2_66, © Springer Science+Business Media B.V. 2010

aspects related to the performance of the alternatives (high performance, low performance, inadequate performance, etc.) in relation to two reference profiles. The nine classes are basically settled into three different categories: good alternatives, uncertain alternatives, and bad alternatives. The concept of reference profile used in the N-TOMIC outranking relation model assumes a different meaning from what was originally described in ELECTRE-TRI. Instead of a boundary between the classes, in N-TOMIC, a reference profile denotes either a "good" or "bad" alternative.

Different from ELECTRE-TRI and N-TOMIC approaches, PROAFTN [5] presents itself as an MCDA method suitable to cope specifically with nominal sorting problems. PROAFTN defines a fuzzy indifference relation that measures the strength of the affirmation "alternative $a$ is indifferent to prototype $p$". To determine this, some computations based on the ELECTRE-TRI are realized. This method has adapted the concept of reference profile, to the prototype one. Instead of representing the upper and the lower bounds of a boundary alternative, the prototype can be considered a good representative of a specific group.

In the Machine Learning field, a particular classification algorithm known as $k$-NN [12] is similar to some MCDA classification algorithms. This algorithm calculates the Euclidian distance that can be weighted, between an alternative to be classified and each training neighborhood alternative. The new alternative will be assigned to the most frequent class among the $k$ neighbors. A problem that have been widely investigated that it is concerned to the $k$-NN, is the difficulty that it has to deal with large datasets due to the computational costs involved. To solve that issue, some research was developed to apply instance selection to reduce the number of alternatives of a dataset. For the $k$-NN, methods such as: CNN [13], ENN [14], VSM [15], and Multedit [16] have been successfully applied over the instance selection problem.

## III. GOLETSIS' MCDA CLASSIFICATION MODEL

The methods presented in the last section have been successfully applied to real world problems. The major difficulty in applying these methods, however, is that, in order to produce models that comply with the decision maker's expectations, a set of control parameters, such as threshold variables, weights, coefficients, etc., needs to be properly set in advance, which turns out to be a hard task to be dealt with. Some authors, like Belacel [5] and Jacquet-Lagrèze and J. Siskos [17] have already provided some alternatives to counter this sort of drawback, although their solutions seem to be rather specific to the contexts that were investigated and yet no general recipes are available to be deployed in all methods.

The MCDA classification method that we have chosen for investigation was that proposed by Goletsis et al [7]. Like PROAFTN [5], this method makes use of prototypes to serve as references against which the new alternatives are compared (matched) with. One distinctive aspect of this scheme with respect to other MCDA-based ones is that it presents less control parameters to be adjusted (only some thresholds and criteria weights). In what follows, we provide further details of Goletsis' algorithm.

Analytically, the model can be defined as in the following way: Let $A$ be the finite set of alternatives, $F$ the set of $n$ features (in the nominal sorting problem it is also known as criteria), with $n$>=1, $w$ the weight of a specific criterion, $\sum_j w_j = 1$, $C$ is the set of categories of a problem where $C = \{C^1, C^2, C^3, \ldots, C^k\}$ and $K > 1$, and $B^h = \{b_p^h \mid 1, \ldots, L^h$ and $h = 1, \ldots, K\}$ the set of prototypes of the category $C^h$, where $b_p^h$ represents the $p$ prototypes of the category $C^h$ and $L^h$ the number of the prototypes of this category. Each alternative in $A$ and $B$ is characterized by a feature vector $\overline{g}$ containing its feature values for all $n$ criteria in the $F$ set. Each alternative is compared with each prototype $b_p^h$ under each criterion $j$.

As described by Goletsis et al. [7], during this comparison the first thing to be computed is the Similarity Index ($SI_j(a, b_p^h)$). This index is calculated for each criterion, and its objective is to model the criteria into a five zone similarity index. In order to compute this index, two thresholds must be specified.

The first threshold that needs to be specified is the similarity threshold, $q_j$, that represents the maximum allowed criterion difference $|g_j(a) - g_j(b_p^h)|$ between the alternatives and the prototypes. Using this, the alternatives can be judged similar under a specific criterion.

The second threshold used by the ($SI_j(a, b_p^h)$) computation is the dissimilarity threshold, $p_j$, representing the minimum allowed criterion difference between an alternative $a$ and prototype $b_p^h$. This threshold needs to be defined in order to considerate the criteria totally dissimilar.

The similarity index ($SI_j(a, b_p^h)$) is computed as described below:

$$SI_j(a, b_p^h) = 1, \quad \text{if} \, |g_j(\text{a}) - g_j(b_p^h)| <= q_j \tag{1}$$

$$SI_j(a, b_p^h) = \left(\frac{|g_j(\text{a}) - g_j(b_p^h)| - p_j}{q_j - p_j}\right), \quad \text{if} \, q_j < |g_j(\text{a}) - g_j(b_p^h)| < p_j \tag{2}$$

$$SI_j(a, b_p^h) = 0, \text{ if } |g_j(\text{a}) - g_j(b_p^h)| \geq 0 \tag{3}$$

After the computation of the similarity index, the next step is to compute the concordance index (CI). This index indicates the overall similarity concordance of an alternative $a$ with a prototype $b_p^h$. This index is computed as follows:

$$CI(a, b_p^h) = \sum_j w_j \, SIj\,(a, b_p^h) \tag{4}$$

Each alternative will have its CI computed for all prototypes of each class. After that, the next step is the computation of the membership degree (MD) of an alternative $a$ to a category $h$. The membership degree applies the best CI of $a$ to all prototypes of $h$. The MD is computed as follows:

$$MD(a, C^h) = max\ \{\ CI(a, b_1^h), ..., CI(a, b_{Lh}^h)\ \} \quad (5)$$

Finally, the last step is the assignment of the alternative $a$ to a category $C(a)$ with the maximum MD calculated to all the groups of prototypes. The formula is presented below.

$$C(a) = arg\ max_h\ d(a, C^h) \quad (6)$$

Goletsis' method was first applied in the ischemic beat classification problem [7]. Aiming to overcome the parameters adjustment by the expert, Goletsis et al. [7] incorporated to its method a genetic algorithm. Besides the emphasis that the MCDA methods put in the estimation of some preferential parameters values, this paper had it focus on the experimental analysis of the prototypes selection.

## IV. Experiments and Results

### A. Experiments

The objective of the experiments is to demonstrate, for different datasets, the model's sensitivity to the prototypes selection as their impact over the classifier's performance. The experiments contemplated 5 different datasets (Winsconsin Breast Cancer, Iris, Zoo, Balance Scale and Lungs Cancer) that are available at the UCI repository [8]. In order to achieve these objectives 10 different prototypes groups were randomly defined for each dataset.

For the experiments, 30 different parameters values were arbitrated, so that the classifier's sensitivity to the prototypes could be demonstrated for more than a single set of thresholds. In this context, an application's execution is composed by 30 classifier's iterations that maintain the prototypes groups and varies only the parameters values between them.

After the generation of 10 prototypes groups, the respective prototypes' alternatives were removed from the original dataset, so that they would not serve as references against itself during the classification process. The number of prototypes chosen was 10% of the original dataset. From that number, the prototypes were separated into their classes respecting their original distribution in the dataset.

### B. Results

This section presents the results generated for the experiments described. To resume the results produced by an application's execution over a dataset, the mean and the standard deviation of the 30 classifier's iteration were computed.

The figure 1 shows the 10 application's execution with their means and standard deviations for the Winsconsin Breast Cancer dataset. This is a real dataset and it consists of a total of 699 instances (alternatives) and 10 features (criteria) including its class value. The purpose of the Goletsis' method is to classify those alternatives into two different classes (benign or malign).

| Winsconsin Breast Cancer | | | | | |
|---|---|---|---|---|---|
| | 1st | 2nd | 3rd | 4th | 5th |
| Mean | 66.32 % | 71.65 % | 63.48 % | 69.74 % | 69.12 % |
| Std. Deviation | 27.5 | 21.31 | 24.96 | 27.5 | 27.86 |
| | 6th | 7th | 8th | 9th | 10th |
| Mean | 66.28 % | 69.49 % | 67.68 % | 57.62 % | 70.09 % |
| Std. Deviation | 26.21 | 27.4 | 27.01 | 28.64 | 30.05 |
| | | | | | |
| Global Mean | 66.38 % | | | | |

Fig. 1. Winsconsin Breast Cancer Dataset

For this dataset, the results reveal that the second execution's mean correctly classified 71.65% of the alternatives. On the other hand, the ninth execution's mean assigned only 57.62% of the alternatives correctly, which represents a difference of 14.03% between those executions.

Figure 2 presents the results of 10 application's executions over the Iris dataset. This dataset consists of 50 samples from each of three species of Iris flowers (setosa, virginica and versicolor). For each sample, four attributes were measured to determine from which specie the sample belongs to.

| Iris | | | | | |
|---|---|---|---|---|---|
| | 1st | 2nd | 3rd | 4th | 5th |
| Mean | 77.53 % | 67.84 % | 68.6 % | 72.6 % | 76.93 % |
| Std. Deviation | 6.54 | 4.16 | 6.92 | 6.09 | 4.53 |
| | 6th | 7th | 8th | 9th | 10th |
| Mean | 81.64 % | 74.97 % | 79.06 % | 67% | 82.6 % |
| Std. Deviation | 7.45 | 7.51 | 5.88 | 4.37 | 5.09 |
| | | | | | |
| Global Mean | 74.88 % | | | | |

Fig. 2. Iris Dataset.

The results demonstrate that, for this dataset, the impact over the prototypes selection is very significative. The difference between the best and the worst groups of prototypes in the classifier's performance is of 15.6%.

The third experiments included the Zoo dataset that consists of 17 attributes, including its class value. This is an artificial dataset that has its alternatives assigned into 7 different classes.

Figure 3 demonstrates the importance of selecting the prototypes for the classifier's performance. For this dataset, the results reveals that half of the groups of prototypes selected had their classifier's performance below 24%, while the other half had their performance over 31%. Another aspect that can be found from this experiment is the significative difference of 19.63% between two executions.

| Zoo | | | | | |
|---|---|---|---|---|---|
| | 1st | 2nd | 3rd | 4th | 5th |
| Mean | 23.33 % | 31.32 % | 22.93 % | 18.18 % | 23.21 % |
| Std. Deviation | 26.9 | 24.84 | 26.78 | 25.4 | 27.1 |
| | 6th | 7th | 8th | 9th | 10th |
| Mean | 36.7 % | 17.39 % | 37.02 % | 31.45 % | 35.9 % |
| Std. Deviation | 23.56 | 24.99 | 23.04 | 25.65 | 22.74 |
| | | | | | |
| Global Mean | 27.73 % | | | | |

Fig. 3. Zoo Dataset.

The fourth analysis included the use of the Balance Scale Tip dataset. This dataset contains 625 alternatives with 5 attributes including its class values (tip to the right, tip to the left or balanced).

For this dataset, the prototypes selection had its impact minimized in relation to the others. As it can be seen in figure 4, most part of the executions present the classifier's performance close to the global mean, but there is still a big gap between the 7th execution and the global mean with a difference of 24.6%.

| Balance Scale Tip | | | | | |
|---|---|---|---|---|---|
| | 1st | 2nd | 3rd | 4th | 5th |
| Mean | 40.41 % | 44.33 % | 43.01 % | 39.46 % | 42.86 % |
| Std. Deviation | 18.08 | 18.84 | 19.17 | 16.79 | 19.85 |
| | 6th | 7th | 8th | 9th | 10th |
| Mean | 36.7 % | 17.39 % | 37.02 % | 31.45 % | 35.9 % |
| Std. Deviation | 18.75 | 19.22 | 18.05 | 17 | 18.17 |
| | | | | | |
| Global Mean | 41.99 % | | | | |

Fig. 4. Balance Scale Tip Dataset.

The last experiments were realized over the Lungs Cancer dataset. This dataset is composed by 32 instances with 57 different attributes that can be assigned into 3 different classes. Figure 5 also demonstrates the influence that the selection of the prototypes can exerts in the final result. For this dataset, the greater difference between two groups of prototypes was of 11.88%.

Another important factor to be analyzed is the difference between the standard deviations of each application's execution. For some prototypes the thresholds and weights values can produce a high variation for the classifier's performance, while for others the executions do not present a high oscillation. For the eighth and tenth executions this difference is over than 12%.

| Lungs Cancer | | | | | |
|---|---|---|---|---|---|
| | 1st | 2nd | 3rd | 4th | 5th |
| Mean | 27.81 % | 31.35 % | 26.56 % | 23.75 % | 32.08 % |
| Std. Deviation | 10.55 | 13.42 | 12.49 | 15.17 | 10.31 |
| | 6th | 7th | 8th | 9th | 10th |
| Mean | 34.27 % | 25% | 25.63 % | 33.23 % | 35.63 % |
| Std. Deviation | 9.59 | 10.64 | 19.29 | 9.69 | 7.09 |
| | | | | | |
| Global Mean | 29.53 % | | | | |

Fig. 5. Lungs Cancer Dataset.

## VII. CONCLUSIONS AND FUTURE WORK

Different from most literature studies, that put their emphasis only in the parameters optimization, this paper reveals that the prototypes selection is another element of the MCDA classifiers that can directly influence its performance. To demonstrate that, the Goletsis' classifier was implemented, so that some experiments could be realized for different datasets.

For each dataset different groups of prototypes were defined, so that when they are submitted to the same thresholds and weights values an analysis over their performance can be produced. The results demonstrated that the prototypes can exerts a determinant impact over the classifier's performance depending of its selection.

To solve that problem, we will employ two distinct ways of selecting these elements. The first way will be based on an indirect technique, which needs a decision analyst to input its preferential information, known as ELECTRE IV [18]. This method will be responsible for ranking the alternatives according to each class, so that the best ranked alternatives will serve as prototypes. The second approach that will be applied to solve the prototype selection task is a direct or automated technique. This approach is based on a custom Genetic Algorithm [19] that will automatically select the best prototypes for a given problem. Besides that, some aspects of the implemented MCDA algorithm will be modified with the purpose of increase the classifier's performance. Finally, other MCDA algorithms will be implemented to offer new analysis over the MCDA classifier's aspects.

## ACKNOWLEDGMENT

The authors are thankful to FUNCAP (Fundação Cearense de Apoio ao Desenvolvimento Científico e Tecnológico) for the support they have received for this project.

## REFERENCES

[1] B. Roy, *Multicriteria Methodology for Decision Aiding*, Kluwer Academic Publishers, 1996.

[2] I. H. Witten and E. Frank, *Data Mining: Practical Machine Learning Tools and Techniques*, 2nd ed, Morgan Kaufmann, 2005

[3] M. Doumpos and C. Zopounidis, *Multicriteria Decision Aid Classification Methods*, Springer, 2002.

[4] R. Slowinski and C. Zopounidis, Application of the rough set approach to evaluation of bankruptcy risk, *International Journal of Intelligent Systems in Accounting*, vol. 4, pp. 27-41, 1995.

[5] N. Belacel, Multicriteria assignment method PROAFTN: Methodology and medical applications, *European Journal of Operational Research*, vol. 125, pp. 175-183, 2000.

[6] C. Zopounidis and M. Doumpos, Multicriteria classification and sorting methods: A literature review, *European Journal of Operational Research*, vol. 138(2), pp. 229-246, 2002.

[7] Y. Goletsis, C. Papaloukas, D. I. Fotiadis, A. Likas, and L. K. Michalis, Automated ischemic beat classification using genetic algorithms and multicriteria decision analysis, *IEEE Transactions on Biomedical Engineering*, vol. 51(10), pp. 1717-1725, 2004.

[8] C. L. Blake and C. J. Merz, UCI Repository of machine learning databases, 1998, available at: http://www.ics.uci.edu/~mlearn/MLRepository.html.

[9] W. Yu, ELECTRE TRI: Aspects methodologiques et manuel d'utilisation, Technical report, Universite de Paris-Dauphine, 1992.

[10] B. Roy, Classement et choix en présence de points de vue multiples: La méthode ELECTRE, *Revue Francaise d'Informatique et de Recherche Operationnelle*, vol. 8(2), pp. 57-75, 1968.

[11] M. Massaglia and A. Ostanello, N-TOMIC: A decision support for multicriteria segmentation problems, *Lecture Notes in Economics and Mathematics Systems*, vol. 356(2), pp. 167-174, 1991.

[12] B. V. Dasarathy, Nearest Neighbor (NN) Norms: NN Pattern Classification Techniques, *IEEE Computer Society Press*, 1990.

[13] P. E. Hart, The condensed nearest neighbor rule, *IEEE Transactions on Information Theory*, vol. 14, pp. 515-516, 1968.

[14] D. L. Wilson, Asymptotic properties of nearest neighbor rules using edited data, *IEEE Transactions Systems, Man and Cybernetics*, vol. SMC(2), pp. 408-421, 1972.

[15] D. G. Lowe, Similarity metric learning for a variable-kernel classifier, *Neural Computing*, vol. 7(1), pp. 72-85, 1995.

[16] P. A. Devijver and J. Kittler, *Pattern Recognition: A Statistical Approach*, Prentice-Hall, 1982.

[17] E. Jacquet-Lagréze and J. Siskos, Preference disaggregation: Twenty years of MCDA experience, *European Journal of Operational Research*, vol. 130(2), pp. 233-245, 2001O. Larichev, Psychological validation of decision methods. Journal of Applied Systems Analysis, 130:233–245, 2001.

[18] B. Roy and B. Hugonard, Ranking of suburban line extension projects on the Paris metro system by a multicriteria method., *Transportation Research*, vol. 16, pp. 301-312, 1982.

[19] Z. Michalewicz, *Genetic Algorithms + Data Structures = Evolution Programs*, Springer, 1996.

# Information Handling in Security Solution Decisions

Md. Abdul Based
Department of Telematics, NTNU, Norway
based@item.ntnu.no

***Abstract-* A security solution (SS) is a mechanism, process or procedure to handle security problems for an organization. The decision makers are responsible to choose and implement the SS for their organizations. For the selection of a decision, handling of information plays a very important role. The decision makers collect information both in explicit and implicit form, then take their decision based on trusting or distrusting that collected information. The way of collecting information and the way of using it are not well structured for them. Sometimes they do know how to collect information, but do not collect and analyze information in a structural way while making their security solution decisions (SSDs). Very often they collect information as knowledge, experience, and recommendation in both forms (explicit and implicit). This paper focuses on SSDs and in particular, how information is gathered and used in such decision processes. This paper also works on trust, how trust can reflect the status of a certain piece of information based on knowledge, experience, and recommendation. This paper conducts a survey to investigate how the decision makers (experienced and inexperienced participants in the survey) use empirical data (explicit information) and their knowledge and experience (implicit information) to deal with SSDs. The survey further studies the effect of implicit information in the answers provided by the experienced participates and observes that the variation in the answers provided by the experienced participants is larger than the answers provided by the inexperienced participants.**

## I. Introduction

A particular piece of information might be the key of a certain SS. The principal issues in the SS processes are: how the decision makers choose a SS, how they handle information in the processes, and how they use a particular piece of information. The decision makers should also consider other characteristics of information system such as functionality, cost, and time. This paper investigates how information (explicit and implicit) is used by different types of decision makers to take a decision in such context. Fig. 1 shows the two dimensions of information (explicit and implicit). The explicit information is well documented and completely expressed, can be used from a repository. However, for the implicit information there is no such repository. Implicit respectively explicit information is sometimes referred to as tacit respectively non-tacit information. That is, tacit information refers to the information and/or knowledge that individuals hold that is not outwardly expressed and non-tacit information is the complete expressed information.

Information handling (in this paper) is defined as how the decision makers gather and use information in SSs. Decision makers take some independent information as input, and then if they trust the information they take one decision, and if they do not trust the information then they take another decision. Decision makers should also consider the variations of the decisions; misuse or problem of the decision, solutions, and cost/effect vs. estimated cost/effect. In many cases, decision makers get implicit information form several information sources (observable sources or subjective sources [1]). They deal with this implicit information to take their security decisions explicitly. Hence, information handling is a challenging job from security point of view, because, using implicit information to take explicit security decision is not an easy task. This requires knowledge, recommendations, experience and expertise also. Section II of this paper describes knowledge and trust in SSDs, section III presents a SSD process, and section IV presents the survey conducted on information handling in SSDs. This paper concludes with section V, where conclusions and future works are discussed.

## II. Knowledge and Trust in Security Solution Decisions

Knowledge, experience, and recommendations can be regarded as information in SSD. In this case, experience [2] can be defined as the result of earlier activities that has proof and acceptance. Recommendations [2] can be human opinions or Intrusion Detection System (IDS) log-files or other files like firewall log-files or honeypots. Knowledge is also the experience of the stakeholder or stored prior experience from similar SSD [2]. Tetsuo Sawaragi, Liu Yuan, and Yajie Tian developed a framework [3] to convert tacit knowledge into explicit information and applied it into a heuristic algorithm to improve the flexibility of general algorithm to solve complex problems. In that framework, they have also shown how tacit knowledge from experts' can be reused later by adding tacit knowledge incrementally in their framework. This is important in the sense that this information can be reusable when needed. Simon and Janet [4] described a procedure for expressing knowledge, theorizing from it, identifying data suitable for testing theories, and the value to a business of the outcome based on 'theorise-inquire' technique.

The decision makers depend on trust to take their SS decisions finally. There is no explicit technique available to specify trust, measure trust, or reason about trust. Setting a

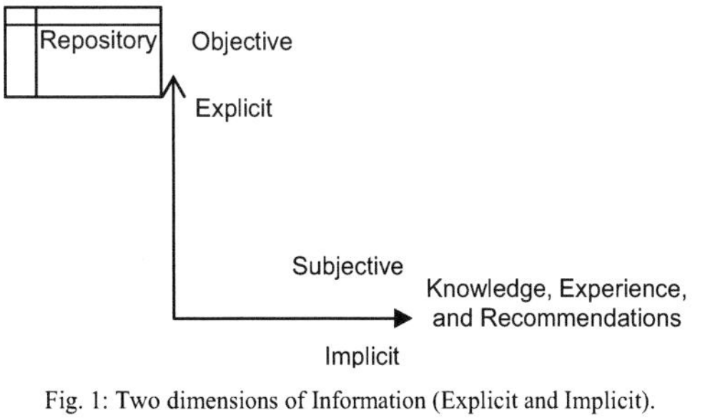


Fig. 1: Two dimensions of Information (Explicit and Implicit).

T. Sobh (ed.), *Innovations and Advances in Computer Sciences and Engineering*,
DOI 10.1007/978-90-481-3658-2_67, © Springer Science+Business Media B.V. 2010

particular trust level to a particular piece of information depends on the sources of the information. If a particular piece of information is collected from the sources those are under direct control of the central administrator or central manager, that information can be considered as trusted. Sometimes a piece of information from friendly or other sources is also considered as trusted. These friendly sources are not under direct control of central administrator or manager. These sources may have their own control, so trusting these sources raises questions about trust. Sometimes it becomes very difficult to set a level for a particular piece of information. That is, defining completely trusted or distrusted information are rather complex. This introduces the concept of uncertainty in trust for a particular piece of information. Therefore, while dealing with trust, uncertainty could also be under consideration. Setting or defining trust for a particular piece of information or for a system is not sufficient. The trust should be evaluated. Indrajit Ray and Sudip Chakraborty [2] described how to evaluate trust value in terms of a vector of numbers. Knowledge, recommendations, and experience play an important role in trust. These three parameters may change the level of trust. These parameters can be evaluated by using the vector model [2] and based on the vector model trust can be managed [5] properly.

## III. Security Solution Decision Process

This section presents SSD as a process in a structural way (shown in Fig. 2). This process starts with collection of the requirements for the security purposes. This implies what is the present situation, what are the problems that are needed to deal with, and what are the requirements. After getting this information, decision makers (network manager or network administrator) should do a manual processing. Manual processing describes and analyzes the security requirements and then finds out the components or elements that are related to the security requirements at that period. These requirements are used as the raw data in Fig. 2. For example, the network manager or network administrator should find out how many servers they are using with details of those servers, how many computers/workstations are connected to the network, and what are the protection strategies they have currently. Protection strategies means Intrusions Detection Systems, Firewalls, mail filtering mechanisms, encryption of database for privacy, antivirus, security of data, and awareness.

After manual processing, decision makers should perform risk assessment and management activities. Risk assessment and management activities are the prerequisite of the security decision activities. Decision makers can follow the guidelines and recommendations from the existing standards and frameworks [6-19] to perform risk assessment and management activities. The choice of selection of frameworks and standards depends on the current requirements and situations for the organizations. The results from risk assessment and management phase will be the input for the decision makers in the decision process as indicated in Fig. 2. This is not the only input to consider, decision makers should consider some other important factors as well before taking the final decisions. To get a better result from the decision, decision makers should consider both type of information (implicit and explicit). This information can be collected as recommendation, experience, and knowledge as described in section II.

After collecting results from risk assessment and management, results are integrated with implicit and explicit information; then decision makers should manage trust properly. Trust plays an important role to accept or reject a particular piece of information. After performing the above processes, decision makers can make a security decision. If the decision that is taken by the decision makers fits well with the requirements by the organization it will be accepted. Then it should be updated considering future scenarios. If the decision does not fit well then there may be two cases. If the relatively experienced or expert decision makers with considering both explicit and implicit form of information make the decision, the decision might require slight changes. Therefore, the decision makers should repeat the decision process only as shown in Fig. 2. The other case may happen for the inexperienced decision makers not considering both explicit and implicit information. In that case, decision makers may repeat the whole process.

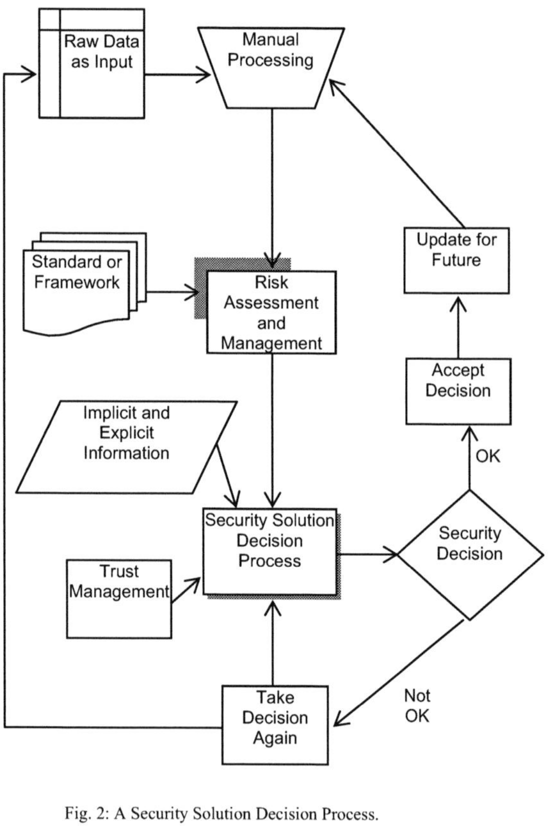


Fig. 2: A Security Solution Decision Process.

## IV. A SURVEY ON INFORMATION HANDLING IN SECURITY SOLUTION DECISIONS

The data collection method in the survey was a combination of questionnaire and interview. At the beginning of the survey, a scenario (with empirical data [21]) was presented to the participants (interviewees). The participants read the scenario and answered the questionnaires accordingly. In the first questionnaire, they rated themselves based on their experience level related to the given scenario. This information helps to divide the participants into two groups based on their experience level. Those participants who have no experience in the related field, their experience level is low, and they are in one group. The other participants who have working experience in the related field, their experience level is medium or high (depending on their year of experience), and they are in another group. Those who worked within the area for less than five years are considered experienced with medium experience level and those who worked for at least 5 years are considered experienced with high experience level. It is important to notice here that the empirical data provided to the participants during the survey is considered as explicit information. The knowledge and experience used by the participants in the survey is considered as implicit information.

This survey observes that those participants who had experience level low; they filled the questionnaires based on the empirical data. That is, these participants answered the questionnaires based on the explicit data, since almost all of them were with no practical experience in that field. On the other hand, the participants with experience level medium or high, they just had a look or just ignored the empirical data. That is, they were focusing on their previous experiences (implicit information) to fill the questionnaires rather than the empirical data (explicit information). To analyze the data, t-distribution was chosen, since the number of participants (24 participants) in the survey was small [22]. By using t-distribution, we can limit the participants' choice of probability distribution as this distribution is defined by minimum, most probable and maximum values [21]. According to t-distribution for smaller samples, there is no significant difference in the variance of the number of virus or worms provided by the participants with experience level low, medium or high. Out of 15 t-tests, there was significant difference in two cases (maximum number of anticipated virus or worm attacks per month and per year for questions 2 and 3 respectively) [shown in TABLE I]. The t-values for these two tests are 2.3226 and 2.387, and the effect sizes are 0.0010 and 3.847 respectively. One reason for this might be the participants adjusted their minimum and most probable value according to the empirical data and their prior experience, while they tended to give a high maximum value. This is called uncertainty about the question [21] and the level of uncertainty was high in this case because their level of experience was mainly low and medium. Moreover, there was a variation of the number of years of experience of the participants. Out of the 8 participants with experience level medium and high, 6 of them were with experience less than 5 years, and only 2 of them were with 5-10 years of experience.

Fig. 3, Fig. 4, and Fig. 5 show that the variance in the anticipated number of virus or worm attacks for the participants with medium or high experience level is higher than that of the participants with low experience level, though, according to t-distribution the difference is not significant. One of the reasons for this high variance in the number of anticipated virus or worm attacks provided by the experienced participants (with experience level medium or high) might be the overconfidence of the participants. Overconfidence found in many other researches [23-33] of economics and psychology. The possibilities and reasons for which the decision makers might be overconfident to make their decisions are discussed in the following.

In research community, overconfidence is so prevalent; it would be surprising if decisions makers were not overconfident. In decision making, the experienced decision makers might be overconfident by expressing themselves as experts with great confidence or they believe that they are brighter and more educated. Another reason might be the nature of the judgment task. The judgment task in decision making is very challenging and overconfidence tends to increase with the difficulty of this judgment task. Furthermore, learning from experience is difficult than one might think. People tend to overweight evidence that supports their position and underweight evidence that undermines it. This tendency is called confirmation bias [33]. Decision makers might tend to ignore disconfirming, they did not realize that they might be wrong in the past. Another reason why it is hard to learn from experience is that we exaggerate the predictability of past events. This tendency is called hindsight bias [33]. There is evidence that both of these biases are stronger when predictions are vague and outcome feedback is ambiguous [34 and 35]. The experienced participants worked in some companies or institutions. They suffered from institutional constraints under which they were operated. They do not appear to face effective social sanctions to encourage them to minimize the ambiguity and vagueness of their predictions. In addition, they do not suffer noticeable penalties for expressing extreme confidence. This supports them to be overconfident since they worked under social and institutional constraints that invite overconfidence [33]. Another reason for which the decision makers might be overconfident is selection bias. This survey selected the participants who were PhD students or Masters' students doing their thesis in a university. The selection of the participants for this survey was convenience sampling [36]. This survey selected the nearest and most convenient persons as subjects in the data collection methods. Other participants like experienced personnel from network section or experienced participants from industry could be more useful.

TABLE I
T-VALUES FROM 15 T-TESTS

| Question No. | Minimum | Most Probable | Maximum |
|---|---|---|---|
| Q1 | 1.5396 | 1.3723 | 2.1162 |
| Q2 | 1.9501 | 1.5356 | **2.3226** |
| Q3 | 0.627 | 1.3127 | **2.387** |
| Q4 | 0.4781 | 1.414 | 0.5020 |
| Q5 | 0.0000 | 0.3682 | 0.7072 |

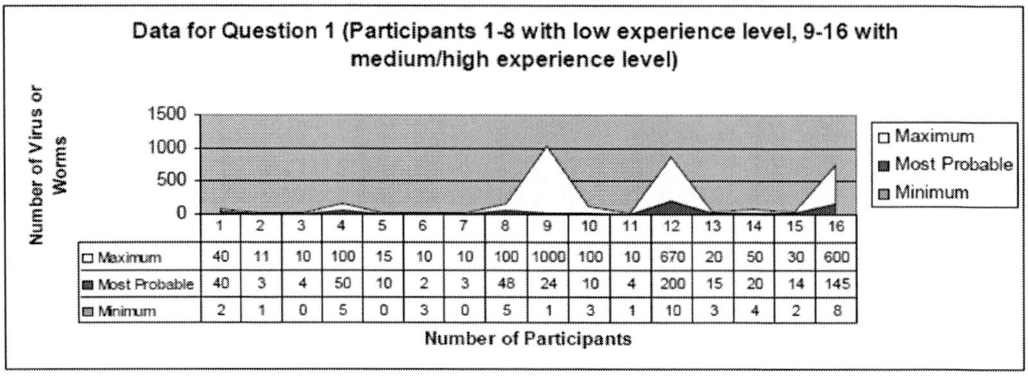


| | 1 | 2 | 3 | 4 | 5 | 6 | 7 | 8 | 9 | 10 | 11 | 12 | 13 | 14 | 15 | 16 |
|---|---|---|---|---|---|---|---|---|---|---|---|---|---|---|---|---|
| Maximum | 40 | 11 | 10 | 100 | 15 | 10 | 10 | 100 | 1000 | 100 | 10 | 670 | 20 | 50 | 30 | 600 |
| Most Probable | 40 | 3 | 4 | 50 | 10 | 2 | 3 | 48 | 24 | 10 | 4 | 200 | 15 | 20 | 14 | 145 |
| Minimum | 2 | 1 | 0 | 5 | 0 | 3 | 0 | 5 | 1 | 3 | 1 | 10 | 3 | 4 | 2 | 8 |

Fig. 3: The variance in the number of anticipated virus or worm attacks per day is lower for the participants with low experience level (1-8) than that of the participants with medium or high experience level (9-16).

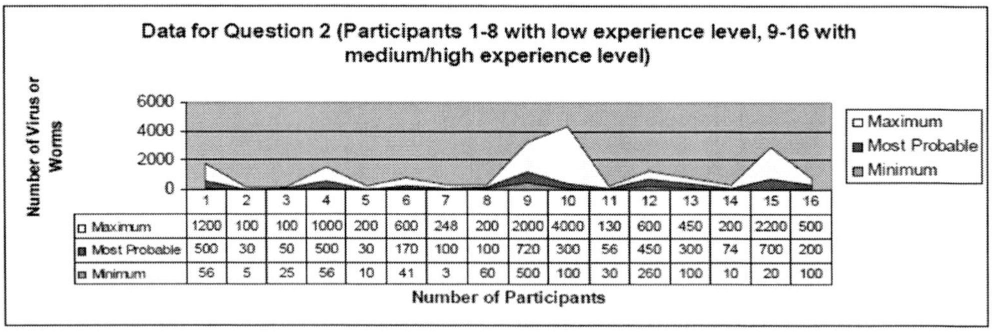


| | 1 | 2 | 3 | 4 | 5 | 6 | 7 | 8 | 9 | 10 | 11 | 12 | 13 | 14 | 15 | 16 |
|---|---|---|---|---|---|---|---|---|---|---|---|---|---|---|---|---|
| Maximum | 1200 | 100 | 100 | 1000 | 200 | 600 | 248 | 200 | 2000 | 4000 | 130 | 600 | 450 | 200 | 2200 | 500 |
| Most Probable | 500 | 30 | 50 | 500 | 30 | 170 | 100 | 100 | 720 | 300 | 56 | 450 | 300 | 74 | 700 | 200 |
| Minimum | 56 | 5 | 25 | 56 | 10 | 41 | 3 | 60 | 500 | 100 | 30 | 260 | 100 | 10 | 20 | 100 |

Fig. 4: The variance in the number of anticipated virus or worm attacks per month is lower for the participants with low experience level (1-8) than that of the participants with medium or high experience level (9-16).

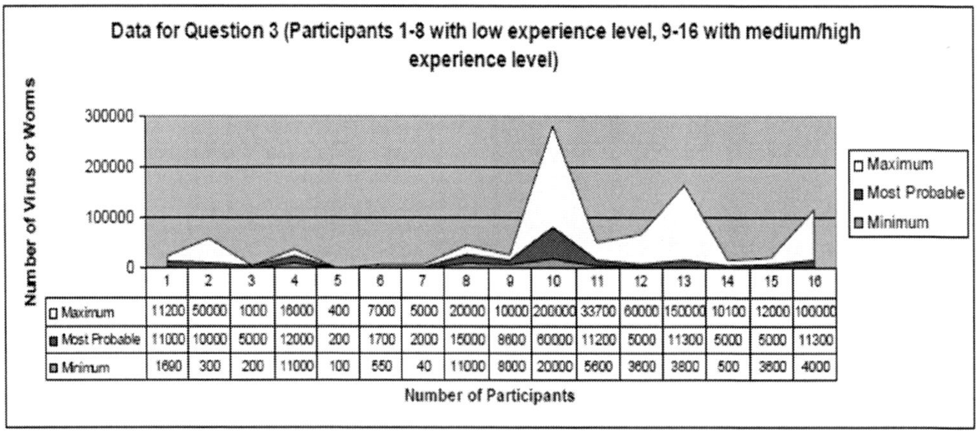


| | 1 | 2 | 3 | 4 | 5 | 6 | 7 | 8 | 9 | 10 | 11 | 12 | 13 | 14 | 15 | 16 |
|---|---|---|---|---|---|---|---|---|---|---|---|---|---|---|---|---|
| Maximum | 11200 | 50000 | 1000 | 16000 | 400 | 7000 | 5000 | 20000 | 10000 | 200000 | 33700 | 60000 | 150000 | 10100 | 12000 | 100000 |
| Most Probable | 11000 | 10000 | 5000 | 12000 | 200 | 1700 | 2000 | 15000 | 8600 | 60000 | 11200 | 5000 | 11300 | 5000 | 5000 | 11300 |
| Minimum | 1690 | 300 | 200 | 11000 | 100 | 550 | 40 | 11000 | 8000 | 20000 | 5600 | 3600 | 3800 | 500 | 3600 | 4000 |

Fig. 5: The variance in the number of anticipated virus or worm attacks per year is lower for the participants with low experience level (1-8) than that of the participants with medium or high experience level (9-16).

## V. Conclusions and Future Work

The results from the survey are inconclusive concerning the impact of overconfidence in terms of experience. However, the experienced decision makers' (with medium or high experience level) can be considered overconfident simply because they are human and subject to the biases as described earlier. If the decisions of experienced decision makers are indeed better than the decisions of their less experienced decision makers, then the more positive self-evaluations will not be considered as overconfidence. It is unambiguous that experienced decision makers can make better security decisions. Altogether, the data provide some support to the hypothesis that the variance in the answers whether the number of anticipated virus or worm attacks provided by the participants with low experience level is lower than that of the participants with experience level medium or high. However, for a final answer on this hypothesis, data from the experts of industry is required. It is common that overconfidence decreases if there is more information. Since there was not a lot of information in the empirical data, that could be one reason of overconfidence.

In conclusion, the decision makers may gather information as knowledge, experience, and recommendation from directly observable sources or from subjective sources to make security decisions. That is, they gather both explicit and implicit information. The survey observed that the inexperienced decision makers make their decisions based on the explicit information they have with them. They simply trust the explicit information. The experienced decision makers focus on the implicit information rather than the explicit information since they have knowledge and experience in the similar activities. They do not want to trust the explicit information without proper reasoning; they implement their own knowledge and experience. The survey also observes that the variation in the decisions for experienced decision makers was high due to overconfidence. So, as a decision maker one should take into account the effect of the overconfidence while making any SSD. If the experienced decision makers do not aware of the effect of overconfidence on their knowledge and experience, this might produce ambiguous results in their SSDs.

## Acknowledgment

This work is carried out as part of the VRIEND (Value-based security Risk Mitigation in Enterprise Networks that are Decentralized) project funded by Sentinels, a joint initiative of the Dutch Ministry of Economic Affairs, the Netherlands Organization for Scientific Research Governing Board (NOW-AB) and the Technology Foundation STW, and supported by Philips Electronics, AkzoNobel, Corus, DSM and Hoffmann Strategic Risk Management. I would like to thank Prof. Louise Yngström, Information and Communication System Security, Department of Computer and System Sciences, Royal Institute of Technology (KTH), Stockholm, Sweden and Dr. Siv Hilde Houmb, Information Systems Group, University of Twente, Enschede, Netherlands for their support and fruitful discussions.

## References

[1] Siv Hilde Houmb, "*Decision Support for Choice of Security Solution: The Aspect-Oriented Risk Driven Development (AORDD) Framework*", Trondheim, November 2007.

[2] Indrajit Ray and Sudip Chakraborti, "*A Vector Model for Developing trustworthy Systems*", Colorado State University, Fort Collins, CO 80523, USA.

[3] Tetsuo Sawaragi, Liu Yuan, and Yajie Tian, "*Human-Machine Collaborative Knowledge Creation: Capturing Tacit Knowledge by Observing Experts' Demonstration of Load Allocation*", in CSM/KSS'05, Knowledge Creation and Integration for Solving Complex Problems, 29-31 August, 2005.

[4] Simone Stumpf, and Janet McDonnell, "*Data, information and Knowledge Quality in Retail Security Decision Making*", 3rd International Conference on Knowledge Management (IKNOW'03), Graz, 2-4 July, 2003.

[5] Indrajit Ray, Sudip Chakraborti, and Indrakshi Ray, "*VTrust: A Trust Management System Based on a Vector Model of Trust*", Colorado State University, Computer Science Department, Fort Collins, CO 80523, USA.

[6]. ISO 15408:2006 Common Criteria for Information Technology Security Evaluation, Version 3.1, Revision 1, CCMB-2006-09-001, CCMB-2006-09-002 and CCMB-2006-09-003, September 2006.

[7] Department of Defense. DoD 5200.28-STD: Trusted Computer System Evaluation Criteria, August 15, 1985.

[8] Department of Trade and Industry. The National Technical Authority for Information Assurance, June 2003. http://www.itsec.gov.uk/.

[9] Government of Canada. The Canadian Trusted Computer Product Evaluation Criteria, January 1993.

[10] Australian/New Zealand Standards, AS/NZS 4360:2004 Risk Management, 2004.

[11] ISO/IEC 17799:2000 Information Technology – Code of Practise for Information Security Management. http://www.iso.ch, 2000.

[12] ISO/IEC 13335: '*'Information Technology Guidelines for the Management of IT Security'*'. http://www.iso.ch/, 2001.

[13] Applying COSO's ERM- Integrated Framework, http://www.coso.org, Access Date: September 29, 2007.

[14] IT Governance Institute, "*COBIT 4.0 Exert*", http://www.cobit.org/, http://www.isaca.org/, Access Date: September 21, 2007.

[15] Information Technology Infrastructure Library, http://www.itil-officialssite.com/, Access Date: September 20, 2007.

[16] Christopher Alberts, Audrey Dorofee, "*Managing Information Security Risks: The OCTAVE Approach*", Addison- Wesley Publication, ISBN-10: 0-321-11886-3.

[17] B. Barber and J. Davey, "*The use of the CCTA Risk Analysis and Management Methodology, CRAMM in Health information Systems*". In K. Lun, P. Degoulet, T. Piemme, and O. Rienhoff, editors, In Proceedings of MEDINFO'92, North Holland, 1992, pp. 1589-1593.

[18] Australian/New Zealand Standard for Information Security Management, AS/NZS 4444:1999.

[19] Australian/New Zealand Standard for Risk Management, AS/NZS 4360:1999.

[20] R. Kazman, M. Klein, and P. Clements, ATAM: "*Method for architecture evaluation*", Technical report CMU/SEI-2000-TR-004, CMU/SEI'', http://www.sei.cmu.edu/pub/documents/00.reports/pdf/00tr004.pdf, 2000.

[21] S. H. Houmb, O.-A. Johnsen, and T. Stålhane, "*Combining Disparate InformationSources when Quantifying Security Risks*", 1st Symposium on Risk Management and Cyber-Informatics (RMCI.04), Orlando, FL, pages 128.131. International Institute of Informatics and Systemics, July 2004.

[22] Murray R. Spigel, Donna Difranco, "*Schaum's Electronic Tutor Statistics*", Second Edition.

[23] Lukas menkhoff, Ulrich Schnidt, and torsten Brozynski, "*The Impact of Experience on Risk Taking, Overconfidence, and Herding of Fund Managers: Complementary Survey Evidence*", University of Hannover, ISSN 0949-9962, April 2005.

[24] Lichtenstein, S., B. Fischhoff, and L.D. Phillips, "*Calibration of Probabilities: The State of the Art to 1980*", in D. Kahneman, P. Slovic, and A. Tversky (ed.), Judgment under Uncertainty: Heuristics and Biases, Cambridge University Press, 306-334, 1982.

[25] Langer, E.J. and J. Roth, "*The Illusion of Control*", Journal of Personality and Social Psychology, 32, 311-328, 1975.

[26] Locke, P.R., and S.C. Mann, "*House Money and Overconfidence on the Trading Floor*", Working Paper, George Washington University, December 2001.

[27] Christoffersen, S. and S. Sarkissian, "*Location Overconfidence*", Working Paper, McGill University, November 2003.

[28] Heath, C. and A. Tversky, "*Preference and Belief: Ambiguity and Competence in Choice under Uncertainty*", Journal of Risk and Uncertainty, 4, 5-28, 1991.

[29] Frascara, J., "*Cognition, Emotion and Other Inescapable Dimensions of Human Experience*", Visible Language, 33, 74-87, ISSN-0022-224, 1999.

[30] Maciejovsky, B. and E. kirchler, "*Simultaneous Over- and underconfidence: Evidence from Experimental Asset Markets*", Journal of Risk and Uncertainty, Springer, Volume 25, Pages 65-85, July 2002.

[31] Glaser, M., T. Langer, and M. Weber, "*On the Trend Recognition and Forecasting Ability of Professional Traders*", University of Mannheim, CEPR Discussion Paper DP 3904, May 2003.

[32] Glaser, M., T. Langer, and M. Weber, "*Overconfidence of Professionals and Lay Men: Individual Differences Within and Between Tasks*", Working Paper Series, University of Mannheim, April 26, 2005.

[33] Erik Anger, "*Economics as Experts: Overconfidence in theory and practice*", Journal of Economic Methodology, Volume 13, Pages 1-24, March 2006.

[34] Robin, Mathew, "*Psychology and Economics*", Journal of Economic Literature 36: Volume 36, Pages 11-46, March 1998.

[35] Fischhoff, Baruch, "*Learning from Experience: Coping with hindsight bias and ambiguity*", in J. Scott Armstrong (ed.), Principles of Forecasting: A handbook for researchers and practitioners, Boston: Kluwer, pp. 543-54, 2001.

[36] Claes Wohlin, Per Runeson, Martin Host, magnus C. Ohlsson, Bjorn Regnell, Anders Wesslen, "*Experimentation in Software Engineering: An Introduction*", Kluwer Academic Publishers, Boston/Dordrecht/London.

# Identifying Connected Classes for Software Reuse and Maintenance

Young Lee
*Department of Electrical Engineering & Computer Science*
*Texas A&M University-Kingsville, Kingsville, TX*
young.lee@tamuk.edu

Jeong Yang
*Department of Electrical Engineering & Computer Science*
*Texas A&M University-Kingsville, Kingsville, TX*
kujy2000@tamuk.edu

Kai H. Chang
*Department of Computer Science & Software Engineering*
*Auburn University, Auburn, AL*
kchang@eng.auburn.edu

**ABSTRACT-Automated identification of reusable and maintainable software components based on dependency is explored. We develop an automated tool, *JamTool* [2], and describe how this tool can guide a programmer through measuring dependency of a program for software reuse and maintenance. We demonstrate *JamTool*'s ability through a case study to investigate whether this tool can be used to identify the connected classes for software reuse and maintenance.**

## I. INTRODUCTION

There are valuable open source programs written by other programmers in software companies and they are open to the public on the Internet, but it is desirable to have a tool that retrieves reusable and maintainable software components.

In this paper, we describe the tabular representation as a display technique for reusable units and maintainable units. This technique is implemented and integrated in *JamTool* (Java Measurement Tool) [2, 3]. With this automated measurement tool, a user can learn about the architectural information specifically for reuse and maintenance. By using *JamTool* with tabular representation, the user can find reusable and maintainable codes.

This paper is organized as follows. In Section 2 we present an automated metric tool. Section 3 presents a case study to capture the difference between two consecutive versions, and review the results of the case study. We conclude, in Section 4, the research contributions of this work.

## II. JAMTOOL

The intended application domain for *JamTool* is small-to-middle sized software developed in Java. The acceptance of Java as the programming language of choice for industrial and academic software development is clearly evident. The overall system architecture of the *JamTool* is shown in Figure 1, in which solid arrows indicate information flow. The key components of the architecture are: 1) User Interface, 2) Java Code analyzer, 3) Internal Measurement Tree, 4) Measurement Data Generator, and 5) Measurement Table Generator.

Each key component works as a subsystem of overall system. The Java Code Analyzer syntactically analyzes source code and builds an Internal Measurement Tree (IMT) which is a low level representation of classes, attributes, methods, and relationships of the source code. Then the Measurement Data Generator takes the IMT as an input, collects the measurement data, and generates the size, complexity, coupling and cohesion metrics of classes in the original source code. Those measurement results as well as the other metrics are displayed in a tabular

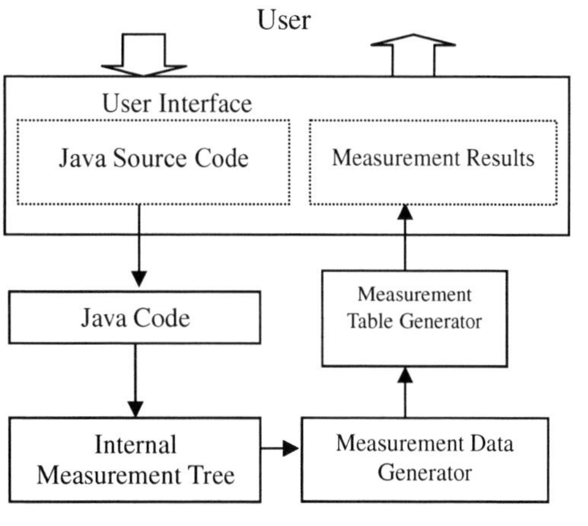


Fig. 1. Architecture of JamTool

T. Sobh (ed.), *Innovations and Advances in Computer Sciences and Engineering*,
DOI 10.1007/978-90-481-3658-2_68, 

representation through the Measurement Table Generator subsystem. With this interface of tabular form, software developers can easily analyze the characteristics of their own program.

In this paper, we focus on the Measurement Result Tables generated by Measurement Table Generator. Because of the results in the empirical study [4], we choose two versions of open source software, *JFreeChart* [1], compare the coupling metrics in the tables for both versions, and examine how we can identify reusable and maintainable components from the target software.

Some important tables are reusable unit, maintainable unit, and connected unit tables. In a reusable unit table, each class in the first column depends on classes in other columns since the class uses the others, and in a maintainable unit table, each class in the first column is used by classes in other columns. Thus the classes in the same row make a special reusable unit and maintainable unit. In a connected unit table, we identify coupled classes in coupling metrics and connected attributes and methods in cohesion metrics. In this way of representation, we could easily recognize which classes need more/less effort when they are needed for reuse, modify, update or fix. This could definitely help programmer in developing and maintaining a program. We discuss the tables in detail in the following case study.

## III. CASE STUDY

This chapter presents a case study to investigate if the reusable and maintainable components can be used to capture the difference between two consecutive versions on the evolution of *JFreeChart* [1]. According to the empirical study reported in the previous paper [4], there was a big change of coupling metric values from 0.9.3 to 0.9.4. We choose those two versions to compare with metrics and the measurement result tables obtained by *JamTool.*

Reusable unit is a collection of a target class and its related classes we should reuse together. Identifying reusable components means that each class has its own reusable unit with other classes which the class depends on. The identification of a reusable unit of classes requires an understanding of the relation of classes in a software system. A maintainable unit contains a target class and its related classes we should test together. Reusable unit and maintainable unit are necessary to understand software structure and more importantly, to serve as a source of information for reuse and maintenance.

### *A. Reusable Unit Table*

Reusable unit table is to present how much a class depends on other classes. In Figure 2 (a), the first column, A, displays all classes in the selected project. A class in column A uses the classes in columns to its right. The classes in the same row make a special reusable unit.

For instant, in the second and third rows, we see that class *AbstractRenderer* depends on a class *ChartRe…*, and class *AbstractTitle* depends on four classes *AbstractT…, Spacer, TitleCh…, and TitleCh….* This dependency means that, for example, if programmer wants to use a certain class (*AbstractTitle*), then he/she must use the other classes in the reusable unit (*AbstractT…, Spacer, TitleCh…, and TitleCh…*) since they are used by the certain class (*AbstractTitle*). Therefore, if a class depends on too many other classes, it is obvious that such a class is difficult to be reused.

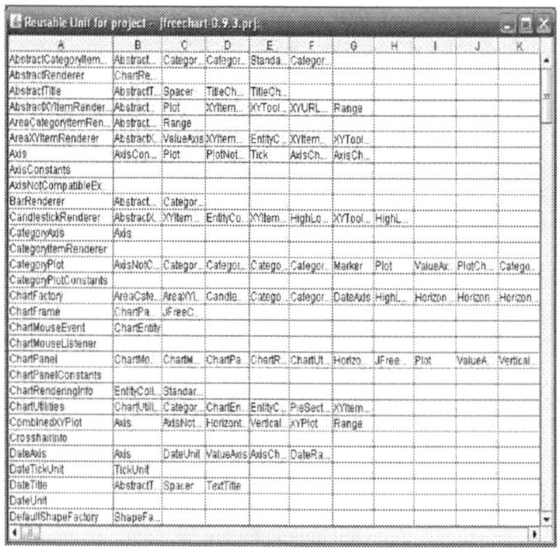


(a) Reusable unit in version 0.9.3

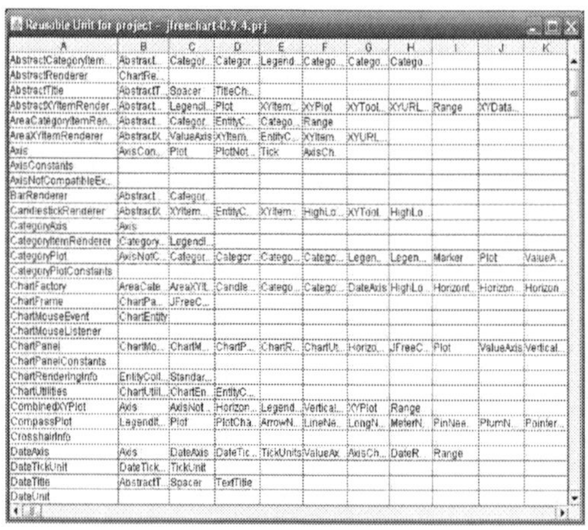


(b) Reusable unit in version 0.9.4

Fig. 2: Reusable unit

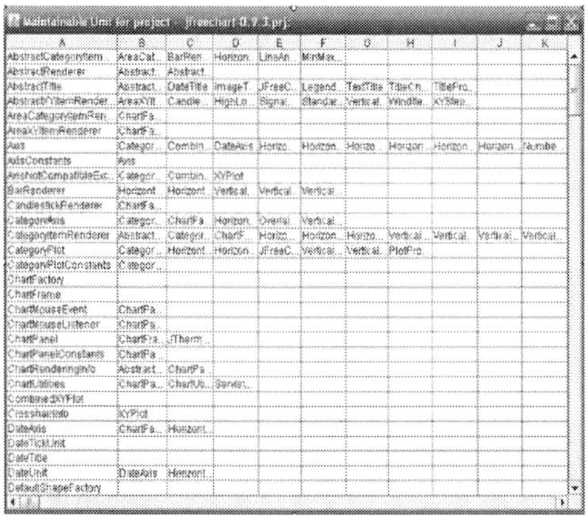

(a) Maintainable unit in version 0.9.3

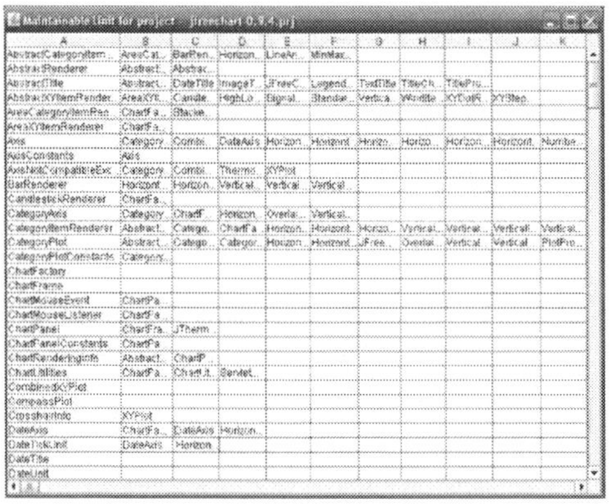

(a) Maintainable unit in version 0.9.4

Fig. 3. Maintainable unit

Figure 2 shows the reusable units in versions 0.9.3 and 0.9.4. From these unit tables, progression of the reusable units is captured. Some classes like *CategoryPlot, Chartfactory*, and *ChartPanel* have too many classes in their reusable units in both versions and some have changed. For example, class *AbstractCategoryItemRender* depends on five classes in version 0.9.3, but seven classes in version 0.9.4.

## B. *Maintainable Unit Table*

Maintainable unit is to present how many classes depend on a specific class. All classes in the selected project are displayed in the first column, A, and each class in that column is used by the classes in other columns, thus the classes in the same row are identified as a maintainable unit. For instant in Figure 3 (a), two classes *Abstract*…and *Abstract*… in the second row use class *AbstractRenderer*, thus if you want to modify or update *AbstractRenderer*, you must test tow classes, *Abstract*… and *Abstract*… as well. Therefore if there are too many classes in a maintainable unit, it is very hard to maintain that specific class.

Figure 3 shows maintainable units in two versions. From these maintainable units, we can capture the progression how classes depend on a particular class. Some classes like *Axis, CategoryItemRenderer*, and *CategoryPlot* are used by too many classes in both versions and some are not. Class *DateTickUnit* has no classes that depend on it in version 0.9.3, but 2 classes (*DateAxis. HorizonDateAxis*) depend on it in version 0.9.4.

## C. *Connected Unit*

We define a connected unit as the classes that are coupled together. In a connected unit table, all classes directly and indirectly coupled together are displayed in the same column, thus a set of classes in the same column is a connected unit. A connected unit is likely to be of interest to the user in finding software units that can be reused. We build a connected unit by identifying coupled classes in the coupling metrics and the connected attributes and methods in the cohesion metrics. A user should consider reusing the connected classes together in a new application. In that sense, the connected classes are a reusable unit.

For instance in Figure 4 (a), only three classes, *StandardToolTipsCollection, ToolTip*, and *ToolTipsCollection*, in column D are coupled to each other. Class *DatasetChangeListener* in column E could be a dead code because there is no relation to other classes in the project. By observing connected units, we may also discover connection patterns. For example, if a project is composed of an application program and libraries, an investigation of the connected unit will tell how the application program uses a library function. In that sense, this type of connection pattern is a use pattern.

Figure 4 shows part of connected units of *JFreeChart* in two versions. From these connected units, we found that version 0.9.3 establishes a main connected unit which has 224 classes out of a total of 257 classes as shown in column A in Figure 4 (a), and a minor connected unit with 3 classes in column D of Figure 4 (a). Three classes (*StandardToolTipsCollection, ToolTip*, and *ToolTipsCollection*) belong to the same package named "com.jrefinery.chart.tooltips". There are also 11 independent classes, e.g., *DatasetChangeListener* in column E, which have no relation to other classes.

We also found that version 0.9.4 has a main connected unit with 254 classes out of a total of 275 as shown column A in Figure 4 (b), and a minor connected unit with 3 classes in column K of Figure 10 (b).

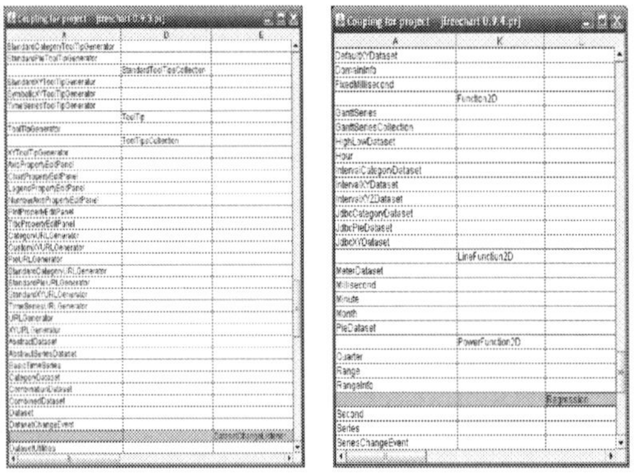

(a) Connected unit in 0.9.3 (b) Connected unit in 0.9.4

Fig. 4. Connected unit

These three classes (*Function2D, LineFunction2D, PowerFunction2D*) belong to the same package named "com.jrefinery.data ". There are 18 independent classes which have no relation to other classes in Figure 4 (b). The independent classes for both versions are listed in Table 1.

Table I.
Independent classes in two versions

| 0.9.3 (11 classes) | 0.9.4 (18 classes) |
|---|---|
| JFreeChartInfo,<br>PlotException,<br>DatasetChangeListener,<br>Values,<br>XisSymbolic,YisSymbolic,<br>DataPackageResources,<br>DataPackageResources_de,<br>DataPackageResources-es,<br>DataPackageResources_fr,<br>DataPackageResources_pl | DataUnit, JFreeChartInfo,<br>PlotException,<br>ChartChangeListener,<br>LegendChangeListener,<br>lotChangeListener,<br>TitleChangeListener,<br>JFreeChartResource,<br>DatasetChangeListener,<br>Regression,<br>Values, XisSymbolic,<br>YisSymbolic,<br>DataPackageResources,<br>DataPackageResources_de,<br>DataPackageResources-es,<br>DataPackageResources_fr,<br>DataPackageResources_pl |

*D. Summary*

The goal of this case study is to compare and analyze two versions of *JFreeChart* at class level. Specifically, it aims to answer the following questions:

- How can the differences between them be compared and detected in terms of reusable and maintainable units?
- How can the huge information of source code be filtered and compared in the context of software reuse and maintenance?

In this case study, we analyzed the differences between the metrics of two versions using *JamTool* and found overall trend of metrics of *JFreeChart* in versions 0.9.3 and 0.9.4. From the comparison and analysis of two versions of *JFreeChart*, we summarize the following findings:

- By comparing reusable units and maintainable units in version 0.9.3 and version 0.9.4, we found newly added or removed classes to the reusable unit and maintainable unit.
- By analyzing connected unit, we found that most classes are directly or indirectly related to each other and they form one main connected unit. But we also found minor connected units with 3 classes, and 11 and 18 independent classes which have no relations to other classes in versions 0.9.3 and 0.9.4, respectively.

## IV. CONCLUSION

In this study, we have presented an automated measurement tool, *JamTool*, for software reuse and maintenance. The primary benefit of this tool is its ability to automatically capture the dependency among the classes and give informative feedback on software reuse and maintenance by reusable/maintainable units.

We believe that various tabular techniques provide structural information for software reuse and maintenance (e.g., reusable unit and maintainable unit). By inspecting these tables, software developers are able to detect reusable software components from libraries and existing open sources by using library documents or inspecting the source code. To reuse software components from exiting application source code, a user should learn the source code before using it. Measuring relationship of the software components is useful to overview the software and to locate the reusable software components. From the measurement results, the user may decide whether or not he/she should reuse the software.

By browsing reusable units and maintainable units, a developer can learn how to reuse certain software entity and how to locate problematic parts. The application of this easy-to-use tool significantly improves a developer's ability to identify and analyze quality characteristics of an object-oriented software system.

Based on the findings, we conclude that the Measurement Result Tables produced by *JamTool* can be used in the following tasks:

- To locate reusable units that should be reused together.
- To locate maintainable units that should be consider for modification.
- To locate connected units that should be consider to be packaged together.

In the future, we consider an empirical test to extract reusable and maintainable units from the professional library programs (e.g., Java Foundation Class) and to determine whether these features can be used to classify the reusable software.

## REFERENCES

[1] http://www.jfree.org/jfreechart/

[2] Lee, Young, Chang, Kai H., and Umphress, D., Hendrix, Dean, and Cross, James H., "Automated Tool for Software Quality Measurement", The 13th international conference on software engineering and knowledge engineering, 2001

[3] Lee, Young, Chang, Kai H., and Umphress, D., "An Automated Measurement Environment: Retrieving Reusable Software Components Based in Tabular Representation", The 40th ACM Southeast Conference, 2002

[4] Lee, Young, Yang, Jeong, and Chang, Kai H. "Quality Measurement in Open Source Software Evolution", The Seventh International Conference on Quality Software, 2007, Portland, Oregon

[5] Lee, Young, "Automated Source Code Measurement Environment For Software Quality", Doctorial Dissertation, Auburn University, 2007

[6] Lee, Young, and Yang, Jeong, "Visualization of Software Evolution, International Conference on Software Engineering Research and Practice, 2008

[7] Briand, L., Daly, J., and Wust, J., "A Unified Framework for Coupling Measurement in object-oriented Systems," IEEE Trans. on software eng., vol. 25, no. 1, 1999

[8] McCabe, T. J. "A Complexity Measure," IEEE Trans. on Software Eng., SE-2(4), pp 308-320, Dec. 19

# Usability Design Recommendations: A First Advance

Marianella Aveledo
Facultad de Informática
Universidad Politécnica de Madrid
Campus de Montegancedo 28660
Madrid, Spain
maveledo@usb.ve

Agustín De la Rosa
Facultad de Informática
Universidad Politécnica de Madrid
Campus de Montegancedo 28660
Madrid, Spain
adelarosa@almumnos.fi.upm.es

Ana M. Moreno
Facultad de Informática
Universidad Politécnica de Madrid
Campus de Montegancedo 28660
Madrid, Spain
ammoreno@fi.upm.es

**Abstract- This paper presents some guidelines to help software designers to design architectures that support particular usability features. We focus on features with high impact on software functionality and therefore on software design, like Feedback or Undo/Cancel. We have adopted an empirical approach to propose some design recommendations to be used to build the above usability features into specific architectures.**

## 1 Introduction

### *1.1. Purpose*

Usability is a software quality attribute listed in most classifications [1][2]. Usability is everything that plays a part in assuring that specific users can use a product to their satisfaction in order to effectively and efficiently achieve specific goals in a specific context of use [3]. Accordingly, usability goals tend to cover a wide range of system features related to criteria like satisfaction and learning. In this context, it is not surprising that usability is gaining more and more recognition as a critical factor for software system acceptance [4].

Over the last twenty years, usability in the software development field has mainly been related to how to present information to the user. Recently, though, it has been shown that usability has implications beyond the user interface and affects the entire system [5][6]. This conjures up the idea that usability should be dealt with in the early stages of the development process like any other quality attribute. Along these lines, *Juristo*, *Moreno* and *Segura* [7] suggest adding the usability features that have the biggest impact on the functionality of the software under construction to the functional requirements. This they term functional usability features. Based on these results, this paper suggests some design recommendations for building some of the usability features most commonly identified by HCI authors [8][9][10][11][12][13] into a software system.

We took an empirical approach to this problem. We built functional usability features into several software systems with a specific architecture and abstracted the design recommendations resulting from adding the above mechanisms. This paper is a first attempt at defining design guidelines for building usability elements into a software system. It needs to be validated on other types of applications. Even so, this early input is potentially of foundational interest, as it supplements the information provided by HCI authors about the visible part of usability features and, as discussed below, the information in the software engineering literature about how to elicit usability recommendations.

### *1.2. Background*

From a software engineering perspective, this article is underpinned by the work of *Juristo*, *Moreno* and *Segura* [7]. In that paper they proposed guidelines for eliciting and specifying usability requirements for what they termed *functional usability features*. These guidelines, called "usability elicitation and specification guidelines", match with the functional usability features of Feedback, Undo/Cancel, User input errors prevention, Wizard, User profile, Help, and Command aggregation.

Additionally, the HCI knowledge used to put together the above patterns is grounded in [14], also defining the responsibilities of the different stakeholders involved in the process.

### *1.3. Usability mechanisms under consideration*

HCI literature mentions a great amount of usability features that can be built into a software system. However, authors appear to agree to some extent on the importance of two families of features, namely Feedback [8][9][10][11][15] and Undo/Cancel [10][11][12][13][16]. Note that software users generally have no trouble in appreciating the effect of these usability elements. Additionally, according to [7], the incorporation of these features tends to have a big impact on design.

Each of these families is further divided into subfamilies or usability mechanisms, as discussed in [7]. For example, Feedback can be divided into System Status Feedback (to report system status changes to users), Progress Feedback (to inform users about the evolution of time-consuming tasks), Interaction Feedback (to let users know that the

T. Sobh (ed.), *Innovations and Advances in Computer Sciences and Engineering*,
DOI 10.1007/978-90-481-3658-2_69, 

system has heard the request) or Warning (to warn users about actions with serious consequences). In view of the breadth of this field and taking into account space constraints, we will focus here on providing some design guidelines and recommendations for the most representative usability mechanisms of each family, specifically System Status Feedback and Progress Feedback, from the Feedback family and Global Undo and Cancel, from the Undo/Cancel family.

## 2 DESIGN RECOMMENDATIONS

As already mentioned, in this paper, we are going to suggest some recommendations on how to design software functionalities needed to incorporate particular usability features.

For that aim, we will examine the functionality under study and then we will describe *generic structures* that we have inferred from several case studies. Then we will instantiate such generic recommendations in a particular case study showing what changes need to be made to the design to incorporate the different usability mechanisms.

Those design recommendations can be used generically in similar projects or at least referenced as an outline that can be tailored, with only minor changes, to architectures with similar features.

In view of their flexibility and applicability under different circumstances, these structures have been called "usability design guidelines" or "usability design recommendations". They are a supplement for the usability requirements elicitation patterns examined in [7].

### 2.1. Software Functionality Used as Example

The *Easy Theatre* system is a Web-based e-commerce application offering users a number of features for administering and managing information on tickets for theatre performances and simulating ticket booking and purchasing transactions.

The *purchase theatre tickets* option is precisely the functionality that will be taken as an example and to which the different usability mechanisms will be added. Generally, this transaction involves users selecting the show they would like to see and paying for their tickets by credit card over the Internet.

### 2.2. Global Undo

Global undo is a usability mechanism offering users some guarantees of being able to restore the system status after they have done the wrong thing. This mechanism basically establishes that the system should register, store and order the operations executed by a user during a work session and undo them in reverse order [10].

One of the most widespread and effective techniques for implementing the above functionality in a software system is to design the architecture according to the dictates of the design pattern known as *Command* [17]. This applies especially to highly interactive systems. By encapsulating the communications between user and system in *command objects* it is possible to store the user actions in time order within a stack and then execute them in reverse order [18].

As Fig. 1 shows, the design recommendation prescribes building a structure containing a set of classes that implement the `Command` interface. Through its `execute()` method each specific class will know which object knows how to execute the exact command action. The `Client` class creates a command every time it needs to execute an order. It associates this command with the `receiver` through the `Invoker`. The instances of the `Invoker` class store the different commands.

As the invoker is capable of storing and managing the command objects in a particular order, they will be able to used later (to create the *undo* effect). Fig. 2 shows the interactions between classes.

When a system needs to be able to undo operations, it will be good idea to follow this recommendation to design its structure. To get a clearer idea, compare the diagrams in Fig. 1 and Fig. 2 with Fig. 3 and Fig. 4 below, where the design recommendation is applied to a case study. These figures represent an instantiate of the previous design recommendations in the *Easy Theatre* application. Before using the system to buy tickets, users (*customers*) have to choose which show they want to see and how many seats they require. To do this, they create a reservation containing this information. To get a better understanding of the relationships between concepts, look at the unshaded part of the diagram shown in Fig. 3 dealing with seat booking.

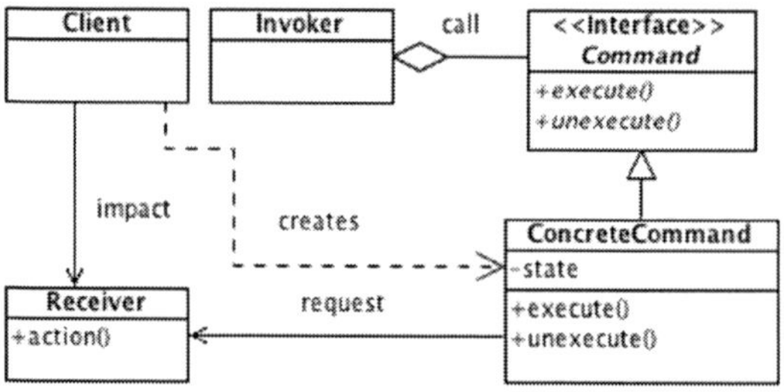


**Fig. 1.** *Command* pattern structure.

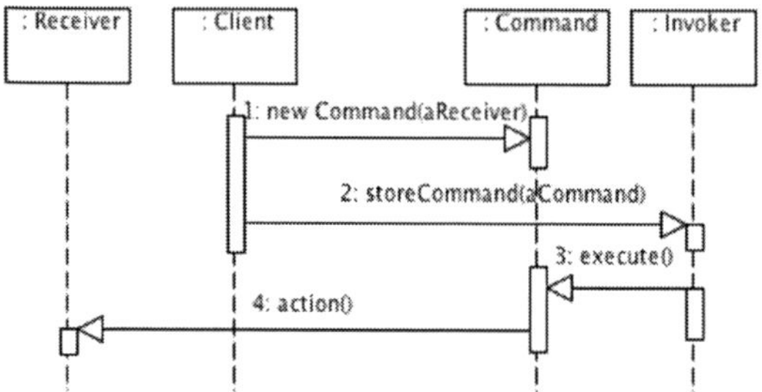


**Fig. 2.** Generic undo. Objects interacting through the *Command* pattern to perform an operation. The `unexecute()` method is used for *undo*.

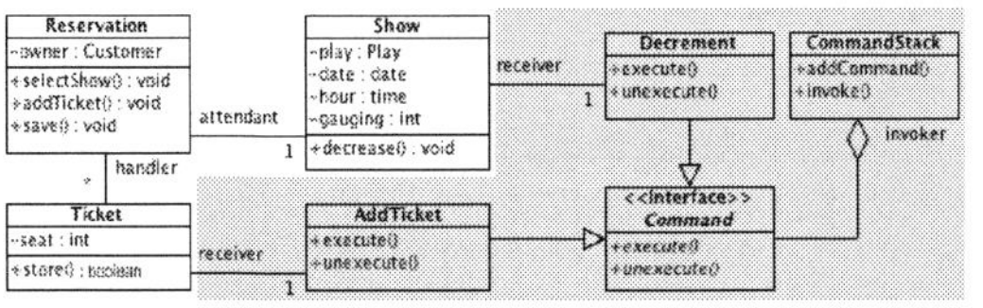


**Fig. 3.** Class diagram for seat booking. The unshaded area is the original structure without Global Undo. The shaded part shows the classes added to implement the Global Undo usability mechanism.

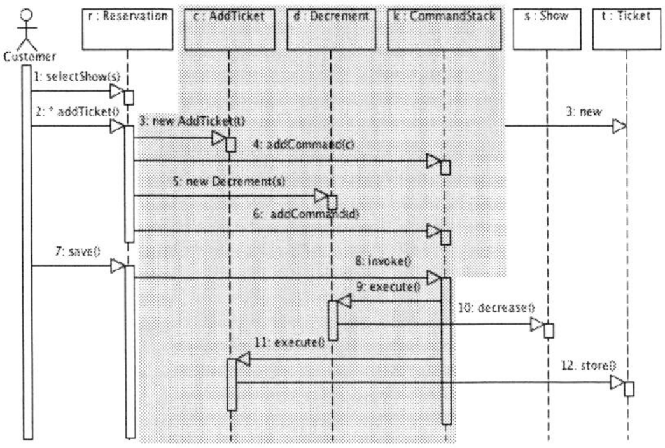


**Fig. 4.** Ticket booking sequence diagram without usability (unshaded area). The shaded area includes the interactions that support the usability mechanism.

The unshaded part of Fig. 4 shows how users apply `selectShow()` to specify the selected performance for the booking. The `Reservation` class reacts to this message by creating as many `Ticket` instances as specified. When users have finished, they use `save()` to confirm that they want to confirm the reservation. At this point this number is subtracted from the vacant seats for the show, and the reservation and the tickets are stored permanently.

Before confirming the reservation, users —who have added a number of tickets their reservations— may decide that they do not want the last, the last two or, in the worst case, any of the tickets. Therefore, some sort of provision should be made for cancelling the changes without having to delete the whole reservation. This is the goal of the *Global Undo* mechanism: enable users to undo the last action taken in an environment where an error or sudden change of mind could affect a long sequence of steps.

Looking at Fig. 3 again, the shaded area shows the changes required to incorporate this usability mechanism. The *command* pattern participant classes have been included: the `CommandStack` materializes as a stack storing different command types. The shaded part in Fig. 4 shows that the objects implementing the `Command` interface use the `execute()` method to carry out tasks on their receivers. To undo actions, they must implement and use the `unexecute()` method.

### *2.3. System Status Feedback*

The *System Status Feedback* mechanism suggests that it is important to keep users informed about what the system is doing about their interactions. This way, users will know at any time when their actions are satisfactorily performed and no error conditions will go unnoticed. This feedback is usually presented as visual confirmation or warning messages, either in the work space or through dialog boxes or status bars.

To implement System Status Feedback, consider the diagram shown in Fig. 5. First, identify the classes whose instances interact to carry out a particular task about which the user is to receive *feedback*. Then, adapt these classes to work with a class containing the system status (ranging from a warning message to a more complex object).

The instances of classes A and B each do a job in the sequence diagram in Fig. 5. `ClassB` generates system status information (of type `SystemStatus`), which it returns to `ClassA` for processing (usually to send information to the user).

As seen before in Fig. 4, when usability is omitted, the `Reservation` instance could interpret the value returned by the `store()` operation and show an operation-dependent message to users.

But, in this case, there is no way of knowing what the exact error was. For example, if a fault occurred during a connection to the database, `store()` would return the *'false'* status, which is the same result as it would return due to a server memory fault.

One possible solution to the above problem would be to modify the specification of the `Ticket` class to return a datum with a greater range of possible values (e.g., a character) and interpret it later as code. However, this gives `Reservation` the added responsibility of decoding, interpreting and possibly reacting in different ways to each of the detected values.

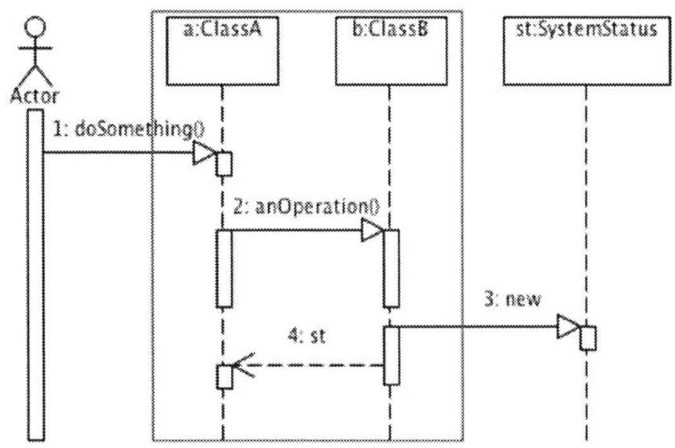


**Fig. 5.** Generic sequence diagram including a system status class to cooperate with existing functionality.

Fig. 6 shows a possibly better way of improving the above solution by encapsulating the system status messages in objects that can be generated within the interaction sequence.

Following this design recommendation, the objects involved in generating each status set its actual characteristics. In this case, this encapsulation materializes through the `SystemStatus` class, which has two attributes. One is a numerical value (`code`) that can be used to identify the status type and the other is a text (`message`) that will eventually be displayed to the user.

The `store()` operation –which previously returned a *boolean* value– now returns an instance of the status to the interacting classes (look at Fig. 7). These classes can store and then use this status instance, for example, by getting a text message (`toString()`) and showing it to the user display, a dialog box, status bar, etc.

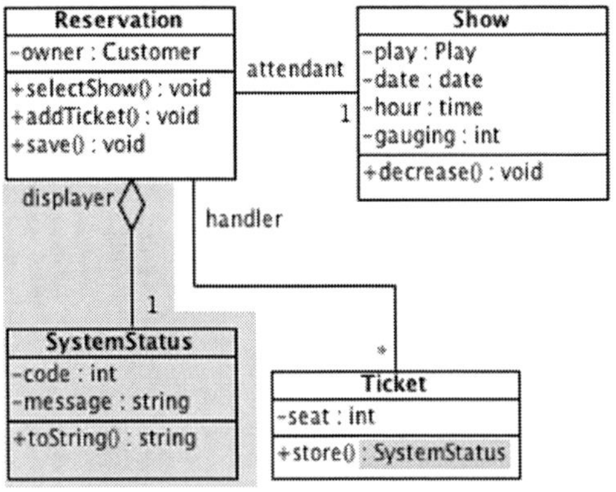


**Fig. 6.** *System Status Feedback* implemented to create reservations. Modification of the class diagram to include the shaded `SystemStatus` class and change the `store()` operation return value.

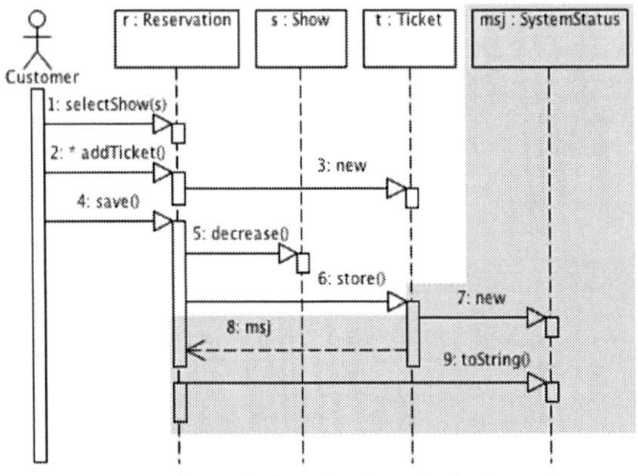


**Fig. 7.** Sequence diagram implementing the mechanism *System Status Feedback* in reservations.

### *2.4. Progress Feedback*

The *Progress Feedback* usability mechanism specification monitors complex or costly operations that users tend to have to wait quite a long time for [8].

One possible design guideline is related to the addition of the *Observer* design pattern elements [19].

This way, the objects of the classes responsible for carrying out the operations requiring progress indicators notify the object that the operation is in progress and, for each notification, the progress bar itself finds out and reports to the user whether progress starts, stops or advances, as shown in Fig. 8.

To comply with this design recommendation, we will, in this case, have to identify the objects of the architecture that are to interact with the progress bar. Then we will have to adapt these objects to register the progress bar as an *Observer*. This way, every time there is a change, the progress bar will be able to find out what the new status is and update accordingly, as the interactions in Fig. 9 show.

Let's see now how the *Progress Feedback* has been included in our case study by adding a progress bar to the *Easy Theatre* tickets purchase transaction. Looking at the class diagram in Fig. 10, the components in the shaded area are the elements added to implement the usability mechanism and are equivalent to a structure based on the *Observer* design pattern. To buy a ticket via the Internet, system users must have previously created a reservation as described earlier. They then have to enter their credit card details (`setCardData()`) and authorize the system to debit their card (`pay()`). Transparently, the system will communicate with another external system mediating with banking institutions to authorize credit card payments (`validate()`). As it belongs to a distributed environment, this communication is usually an operationally slow process and users will have to wait a while. Users should not be able to take other actions within the same context until they have received the decision on whether or not the means of payment has been accepted.

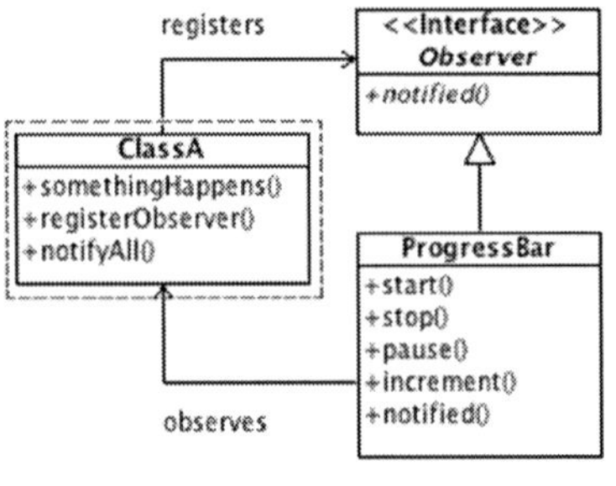


**Fig. 8.** Design of part of the structure according to the Observer pattern. The dashed line represents the class to be "observed".

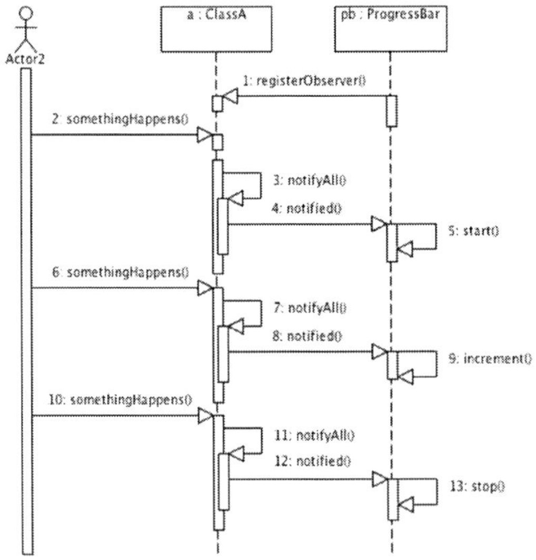


**Fig. 9.** Generic progress bar sequence diagram, illustrating how the progress bar starts, increases and stops.

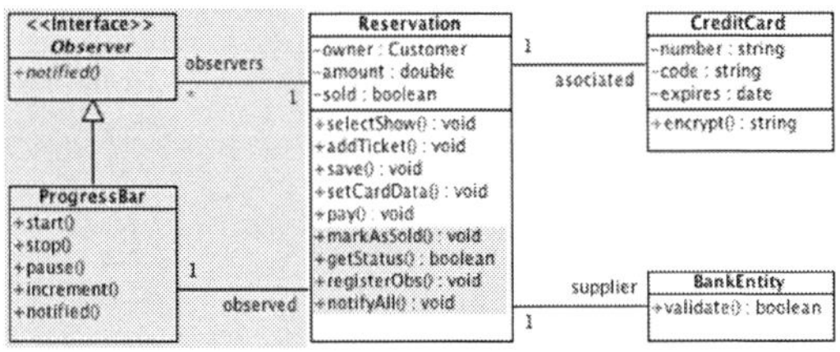


**Fig. 10.** Implementation of a progress bar following the *Observer* pattern. Ticket purchase transaction components: credit card (`CreditCard`) and symbolic credit authorizer external system (`BankEntity`).

To add the *Progress Feedback* mechanism, `Reservation` creates a new instance of `ProgressBar` at the start of the operation, which it registers as its *observer* (`registerObs()`). When payment starts, the progress bar receives a notice before each important change in the transaction. In response, the progress bar uses `getStatus()` to find out what the reservation status is. If the reservation has not been marked as sold, it means that communication is starting and the progress bar starts (`start()`); otherwise, the transaction has finished and the progress bar will have to stop (`stop()`) before any further action can be taken.

### 2.5. Cancel

In any software system, complex, high latency operations tend to modify the system status as they are carried out.

Apart from enabling users to invert the effect of the operations (*undo*) and monitor progress (*progress feedback*), the system should be able to stop progress under different circumstances (e.g. when, half way through a sequence, users are no longer confident about its outcome or when the waiting time is longer than originally expected).

The *Cancel* usability mechanism describes this feature. It states that some way should be provided of suspending any operation behaving like an atomic transaction, assuring that, in doing so, the changes to the system status are no different than if the transaction had never taken place.

This usability mechanism can be construed as being built on the implementation of other features. This way, *Cancel* will be very often implemented at the same time as adding a progress bar, a command-based operational mechanism (like the one implemented above for the *undo* mechanism) or even both.

The general recommendation for designing a system providing the cancel feature is to design this feature at the same time as the *Undo* and *Progress Feedback* mechanisms [16]. To do this, the following points should be considered:

1. Each *specific command* of the *Undo* mechanism will have to be modified to store the system status, if any, as will the `execute()` and `unexecute()` methods to assure that the status is synchronized (See Fig. 11).
2. On the behaviour front, some way of introducing and recognizing the cancellation message should be set up through interactions among observers and observed, as shown in Fig. 12. In order to do this, the progress bar will have to provide a method for alerting sequence participants to the cancellation of the task (`doom()`). The objects notifying the progress bar should then notice that the sequence has been suspended and execute *undo*, restoring all the statuses modified up to that point.

Going back to our example, cancellation would be added to a transaction from *Easy Theatre* in the procedure for deleting unsold reservations that have expired. The user gives the order and the system performs a "soft delete" of each obsolete reservation located through a search.

In this process, one reservation is deleted at each step (change of system status) and the removal or "physical delete" is confirmed at the end, when the marked reservations are entirely removed (this is equivalent to the *COMMIT* order in a database manager).

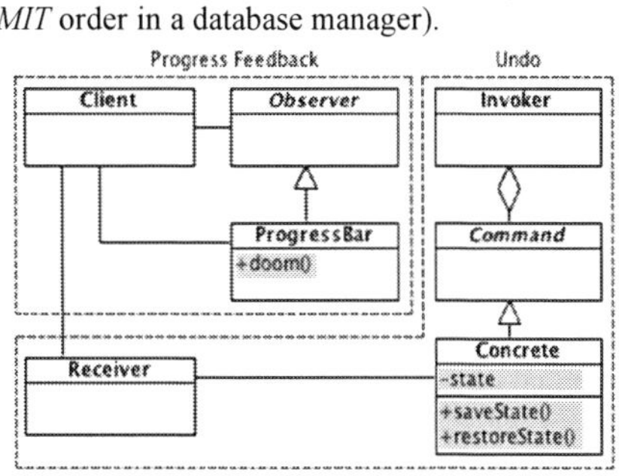


**Fig. 11.** Previously known Progress Feedback and Global Undo generic construction blocks, together for implementing Cancel usability mechanism. Shadowed area shows modifications; the rest was omitted.

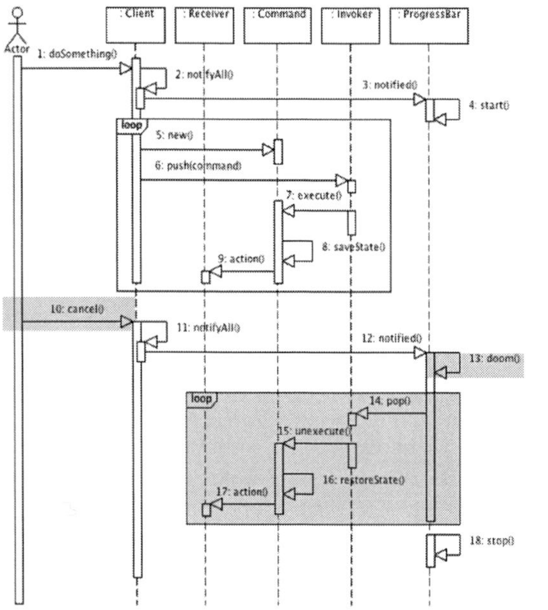


**Fig. 12.** Interactions between objects in a transaction within a system with *cancellability*. The shaded area illustrates the changes added by this usability mechanism.

If the user were to cancel the operation, the `doom()` progress bar method would start up, interacting to execute the commands in reverse order and gradually returning the system to its original status.

## 3 Conclusions

Desirable usability features should be taken into account at all times during the software system development process, both at the start, to be able to effectively elicit the requirements, during design to ease construction and reduce the need for rework, and even at the end to validate and test systems.

Adding different features has an impact on architecture design. The dimension of the impacts depends on the kind of mechanism to be implemented, when during the life cycle they are added and the problem type. But while it is true that all this requires an additional effort, it is no less so that it is more beneficial to weigh up these points as early on as possible.

In this paper we have established some recommendations aimed at guiding software developers through the design of architectures implementing usability features. Different design recommendations can be provided for the same mechanism; we have just presented ones that have been proved in a few systems. Their definition is not yet very detailed, and they are only valid for systems with characteristics similar to the example shown here. However, thanks to their simplicity, they are easily adaptable to other applications with just a few modifications.

A future goal is to conduct an impact study to complement the definition of the usability mechanisms and their design recommendations to somehow be able to more or less formally quantify the impact of their incorporation on systems development.

## References

[1] B. Boehm, J.R. Brown, H Kaspar , M. Lipow, G.J: Macleod and M.J. Merritt, "Characteristics of Software Quality", North Holland, 1978

[2] ISO "ISO 9126-1 Software Engineering-Product Quality-Part1: Quality Model", 2000.

[3] "ISO-Std 9241-11: Ergonomic Requirements for office Work with Visual Display Terminals. Part11: Guidance on Usability," International Organization for Standardization ISO,1998.

[4] L.M. Cysneros, V.M. Wemeck, A. Kushniruk, "Reusable Knowledge for Satisfying Usability Requirements," Proceedings of the 2005 13th IEEE International Conference on Requirements Engineering.

[5] N. Juristo, A.M. Moreno, M. Sanchez-Segura, "Analysing the impact of usability on Software Design," Journal of Systems and Software, Vol. 80, Issue 9, 2007, pp.1506-1516.

[6] L. Bass, B. John "Linking Usability to software Architecture Patterns Through General Scenarios", The Journal of Systems and Software.Vol 66(3) 2003,pp.187-197.

[7] N. Juristo, A.M. Moreno, M. Sanchez-Segura, "Guidelines for eliciting usability Functionalities," IEEE Transactions on Software Engineering, Vol. 33, N°. 11, Nov. 2007, pp. 744-758.

[8] D. Hix, H.R. Hartson, "Developing Use Interfaces: Ensuring Usability Through Product and Process," J. Wiley & Sons, 1993.

[9] B. Shneiderman, "Designing user interface: Strategies for Effective Human-Computer Interaction," third edition, Addison Wesley, Melon, Park, CA, 1998.

[10] J. Tidwell, "Designing Interfaces," O´Reilly Media Inc. 2005.

[11] M. Welie, "Amsterdam Collection of Patterns in user interface Design," *http://www.welie.com*

[12] S.A. Laasko, "User interface Designing Patterns," *http://www.cs.helsinki.fi/u/salaakso/patterns/index_tree..html*

[13] Brighton "Usability Pattern Collection," *http://www.cmis.brighton.ac.uk/research/patterns/*

[14] M. Aveledo , A.M. Moreno, "Responsibilities in the Usability Requirements Elicitation Process," in the Proceedings of the 12th World Multi-Conference on Systemics,Cybernetics and Informatics, Jun. 2008, Vol. I, pp.232-237.

[15] L. Constantine, L. Lockwood, "Software for Use. A Practical Guide to the Models and Methods of Usage-Centered Design," Addison-Wesley, 1999.

[16] J. Nielsen, "Usability Engineering" Morgan Kufmann,1993.

[17] H. Chen, "Design & Implementation, The Command Pattern," *http://www.cs.mcgill.ca/~hv/classes/CS400/01.hchen/doc/*

[18] J. Boutelle, R. Sinha, "Avoiding an extreme makeover," *http://www.oracle.com/technology/pub/articles/masterj2ee/j2ee_wk5.html*

[19] H. Chen, "Design & Implementation, The Observer Pattern," *http://www.cs.mcgill.ca/~hv/classes/CS400/01.hchen/doc/*

# THE STATUS QUO OF 3G APPLICATION IN CHINA

Naipeng DING†, Cui WANG†, Guangming JIA†

†School ofManagement,Shanghai university

99 Shangda Rd., Baoshan District

Shanghai, 200444,China,

npding@staff.shu.edu.cn;hywc1234@163.com;mingguangjia@yahoo.com.cn

***Abstract*-In this paper, webriefly introduce the three mainstream 3G technology:WCDMA,CDMA2000,TD-SCDMA,an ddiscussthe 3G strategy in businesses. With regard to the future development of 3G , wemainly describe 3G industry value chain and the various components of the value of the distribution. Finally,based on 3Gm-commerce platform, we presentsomeadvantagesof e-business in3G environment.**

Ininformationage,telecommunications is an important area of professional growth and tendsto bean indispensable tool in people's living and working environment. Customersdonotonlydemand forvoice communications business,browsing www,and sending&receiving e-mails any more.Instead, theyneedto getnecessaryinformationto support their decisionsat any time, anywherein aleisureway. Accordingly,3G, asa newand prevalenttechnologyapplication, is being considered in Chinese businesses at present.

## 1.FEATURESAND APPLICATIONSOF 3G

3G is the abbreviation of the 3rd Generation Telecommunication.It refers to the combination of wireless communications and many multimedia communications on theInternet, showing a new generation of mobile communication systemwithhigh speed and broadband data transmission.Compared to 1Gand2G which takeanalog signaltransmission technologyand digital signaltransmissionas their core technology,3G has its own unique features such as high-speed broadbandexperience, various application services, personalizedproducts andcustomizedservices,and seamless access to the Internet. Itspotentialgoalsincludedesign of consistencyacrossthe world,compatibility with a variety of fixed network services,high quality of service; smaller size of theterminal,global roaming capability,anda wide range of multimedia functions and operations of the terminal.Welistthe followingsasa comparisonamong1G, 2G,and 3G.

TableOne: differencesamong1G, 2G and 3G

| Difference | 1G | 2G | 3G |
|---|---|---|---|
| **Speed of data transferring** | No | 9.6kb/s | Indoor:2Mbps<br>Outdoor:384kbps<br>Travelling:144kbps |
| **Network system** | Analogsignal | GSM<br>TDMA | WCDMA<br>CDMA2000<br>TD-SCDMA |
| **Main business model** | Voice | Voice | Voice<br>Data |
| **Technology** | Simulation technology and FDMA | Circuit Switching | Packet Switching |

For 3G, therewerethree mainstream technical standardsin China.WCDMA, CDMA2000 and

T. Sobh (ed.), *Innovations and Advances in Computer Sciences and Engineering*,
DOI 10.1007/978-90-481-3658-2_70, © Springer Science+Business Media B.V. 2010

TD-SCDMAbecame the mainstream standards of3Gmobiletelecoms.

*i)WCDMATechnology*

WCDMA technologywas known as anevolution from GSM technology to 3G.It adaptedto a wide range of transmissionrates,providedflexiblebusinessrunning,andallocateddifferent resources. GSM, a global mobiletelecomtech,exploreda wide space for WCDMAapplication.Experts fromChina Academy of Telecommunication Research(CATR) expectedWCDMA users be more than CDMA 1X users by 2009in China. China Unicom tested WCDMA based networking in the past months in several provinces in China.

*ii)CDMA2000Technology*

CDMA2000 technologywasupgradedinto 3G from CDMAOne structure.Compared toWCDMA standard architecture, CDMA2000putscore network, wireless network into relatively modules, eachof whichevolves independently onitsownside, avoiding running conflict with other modules.China Telecom focused on CDMA2000 tech deployment.

*iii)TD-SCDMATechnology*

TD-SCDMA standard developed by Datang Telecom Group. It wasthe first complete communication technology standard in China. The standard integrates now the world's leading technology such as intelligent wireless technology, synchronous CDMA and software radio.TD formed a comparatively complete industrial chainto integratesystems, terminals, chips ,software ,and so on.China Mobile applied TD-SCDMA tech in its network environment although DT-LTE, a new standard, was in its progress.

## 2. 3G BUSINESS DEVELOPMENT STRATEGY ANALYSIS

In differentoccasion,3Gwasendowed withdifferent meanings.The most popular "3G business" concept referedto the business based on3Gnetwork. Therewere types of conversation, cross-category, flow, background, voice, information, content, location, entertainment, etc.To achieve sustained andhealthy development of 3G,companiesshouldpromoteitsbusiness applicationscustomers expect. If therewasno support for rich business applications, huge investment in 3G networks wouldbe difficult to obtain good returns, or even causedthe investment risk..

For the development of China's 3G business, we shouldbebased on our own national conditions and take different promotion strategy at different markets. In the past two years, under the conditions that 3G licenseswerenot yet issued, the two major domestic mobiletelecomoperatorswere committed to actively speeding up the 2G and 2.5G mobile value-added business development, paving the way for 3G markets.

### *A. VoiceCommunications can beUsed as anImportantEntryPoint for 3GMarket..*

Although the 3G data servicewas the key point, the voice business wouldcontinue to play an important role that couldn't be underestimated. First of all voice communicationwas the most basic needs and wouldnot be changed with technology updating.Secondly,3G network operating equipmentwas typical of economies of scale, 3G applications should start with voice business which made a huge towel of market share.Itnot only enhancedthe operator's network equipment utilization, fully utilizing operators' fixed assets, but also seizedmarket share towels.Through the promotion of 3G business, the customer's attentionwas enhanced.The new brand for operatorwas established.Userswere developed.

### *B.During theTransitionalPeriod ofNetwork, theDevelopment ofDual-modeTerminal isVeryCritical for 3GBusinessDevelopment.*

3G technology standard versionswere evolving constantly, due toconfront potentialproblems with the new standards in areas such as roaming and compatibility.The major operatorswere still undertaking a comprehensive assessment of existing technologies in order to carry out a reasonable scale of the networkconstruction. In this case, the user must have dual-mode terminals to ensure the continuity of network services.

*C.Characteristics of Mobile Communications Business DevelopmentwereQuiteObvious,That isPersonalization andLocalization.*

Customs in the world are quite different.As a result, customers in different countries demand a greater diversity of mobile communications businesses. China's 3G operatorsshouldwork in accordance withdomesticconditions to find out a suitable way for the development of China's 3G business users.

*D.SubdivideCustomers,Businesses andIndividuals toProvide theUsers'CustomizedBusiness.*

For the time being, China Mobilewas showing anexample in subdividing userswiththree major brands - GSM, M-Zone as well as Shenzhou Xing. Targeted at different customers, China Mobile's hugenumber ofcustomerswere divided into three groups through customer brands. At present, the classification for China Mobile customers' demandwas on the framework level.The problems facing the future 3G operators especially the 3G mobile data business promotion,wouldbe quite severe, by then customerswouldrequire more personalized business. As a result, on the business mode level, China's major operators should pay more attention to the user's personality and diversified needs, and continue to develop new services based on the original brand of value-added services, and enrich the connotation of brand ,in order to meet our customers' differentiated, personalized services.

*E.FlexiblePricingStrategy.*

Due to the pricing strategies of the operators in the marketing strategy occupies quite an important position, as a result, to develop a comprehensive and reasonable price system is very important. In general, pricing program should follow the following principles:

Depending on the user's spending habits, a variety of business should be properly tied to form a product package or product brand to adapt to different user groups[1], and a reasonable price should be made largely depending on the different brand users' economic situation.

Clear, simple and user-friendly business package and contents togetherwas a simple pricing program.By doing this, potential customers wouldnot be scared away by the complicated charges. Specific charges also brought a certain convenience to calculate the cost of communication.

In favor of operator's long-term development. During making a proper price of new 3G business, wetook short-term business benefits into consideration, as well as long-term sustainability of corporate earnings. For 3G business hadthe economies of scale, so prices reflectedthe value of the services.Pricing levels and methods should be able to encourage the use of multi-user enterprise services.

*F.EncourageConsumers toExperience the newMobileValue-addedServices toFoster the use ofConsumerHabits[5].*

## 3. 3G INDUSTRY VALUE CHAIN AND THE DISTRIBUTION OF BENEFITS ANALYSIS

*A.3GIndustryValueChainAnalysis*

The development of 3G depends on the maturity of technology and formation of competition pattern. Technically, 3G has been comparatively mature; next we are going to discuss competition--3G value-chain issues. At present, China's 3G value chain has taken shape, as graph one displays. Throughout the value chain, operators delivered the products provided by content providers to customers though portals and the so called 'bridge': access service providers. At present, operators stand at the core of the entire value chain, and become the leading force in the whole industrial chain, however, with the further technical development; the status of content provider will be more and moreimportant. It is worth noting that end-users pay the bill for many providers and operators, their needs in this value chain, plays the role of the compass.

*i)ContentProviders*

Content service providers mainly include content producers and content integrators, they interact and cooperate together to provide users with various forms of content. Content producers are those businesses that produce a large number of specific mobile content. The production of its content contains both simple text and complex images, according to the characteristics of contents, the right form of content and the right way of expression are chosen, which is a very complex process. Content integrator refers to enterprises which transform a variety of mobile service content into a form needed by customers. Characteristics of mobile content service determine that the general electronic content can not be sold directly as mobile commerce products; it needs a certain processing in order to become mobile products and services sent to the user.

*ii)Portaland Access Service Providers*

Portal and access service providers provide an exchanging "bridge" for content providers and operators, it offers content providers interface accessible to wireless network, to ensure their smooth entry into the wireless network transmission systems and reach for users finally.

*iii)Operators*

Wireless network operators work between content service providers and users, providing them with information transmission highway, to ensure that they can successfully carry out the exchange of information.

*iv)Infrastructureand Platform Providers*

Infrastructure and platform providers mainly refer to infrastructure manufacturers and application developers. Infrastructure manufacturers are the economic entities which provide a variety of wireless network equipmentfor wireless network operators, and application developers are the economic entities which provide a variety of necessary wireless communication network applications for wireless network operators.

*v)Paymentand Security Service Providers*

Pay and security service providers include mainly security guarantees, pay-payment of support services. As the mobility of the mobile terminal equipment which shows a high degree of personalization is very strong, and for its ability to store the user's private information, so naturally this kind of equipment is more easily to become tool of payment. However, the security of wireless data transmission network is a lot worse, by comparison, pay for security and other service providers can ensure the information security, the proper transferring as well as carrying value-added services smoothly.

*vi)Terminal Manufacturers*

Terminal manufacturers are entities who are committed to provide users with mobile terminals as well as applications on terminal equipment and good service interface. We all know that the emergence of 3G bring new challenges for terminal manufacturers, in order to meet users' demand, terminal manufacturers must explore new technologies and research., or else they will be difficult to survive[6].

*B.Analysisofthe Distributionof 3GInterestsinthe Industry Value Chain*

With the development of market as well as the of technology, 3G industry value chain will surly be more complicated, each link of the entire value chain is crucial, each component in the value chain restrain and influence mutually, only by combining each other ,a win-win situation will be created. If a organic combination is wanted in the whole value chain, we must do a good job in the key point, that is distribution of value.

At present, network operatorswereat the core of the value distribution.However, if too much effortin the industry value chainswasputin the exclusive pursuit of monopoly profits, industrialupstream and downstream value chain wouldn't be able to form an effective balance. As a result, scientific and reasonable share of the value wouldpromote the integration and win-win situation of the industry chain.As for the industrial chain business models, several modelswere shown asfollows:

*i)Distribution ofBenefit*

For each segment in the value chain, joint response to market risk can increase the resistance to the risk of the entire chain, as a result, for the key link in the value chain, cooperation agreements should be signed, and revenue generated by users should be shared in accordance with a certain percentage, which is a good choice in the distribution of benefits.

*ii)Fees for theChannelServices*

Opennesstocontent service providers.content service providers carry out billing.Network operators offer platform for content service providers who will pay fees for the channeluse.

*iii)Proceeds with Operators*

Operators provide billing services for the content service providers and carry out the share of proceeds with them.

## 4. 3G AND MOBILE COMMERCE

M-commerceperformeda variety of activities through the network, the use of mobile terminals. M-commerce requiredthe realization of a large number of mobile terminals. According to the survey offormerChina's Ministry of Information Industry, China's mobile phone users continuedto maintaina rapid growth, the followingwere the number of mobile phone users in China from October 2002 to October 2007(Table Two).

Due tothelarge population,theproportionof China's mobilephone userswas low, but difference in the regionswas obvious.The eastern,central and westernwere decreased.Wesawfrom theFigureOnethat compared to the

**TableTwo: China's total mobile phonesubscribers (Unit:10 thousand)**

| Month and yearforChina's total mobile phone users | Total number |
|---|---|
| OCT., 2002 | 19583.3 |
| OCT., 2003 | 25693.8 |
| OCT., 2004 | 32503.4 |
| OCT., 2005 | 38304.2 |
| OCT., 2006 | 44902.1 |
| OCT., 2007 | 53144.7 |

Source: Ministry of Information Industry, People's Republic of China

same period last year, the eastern, central and westernnew mobile phone subscribers increased, but the growth rate in central and western mobile phone userswas faster thanthatin the eastern.Thismeans that the eastern part hada very high proportionof mobile phone users and tendedto be saturated.As a result,development of the mobile business in the easternareabecame an inevitable trend.

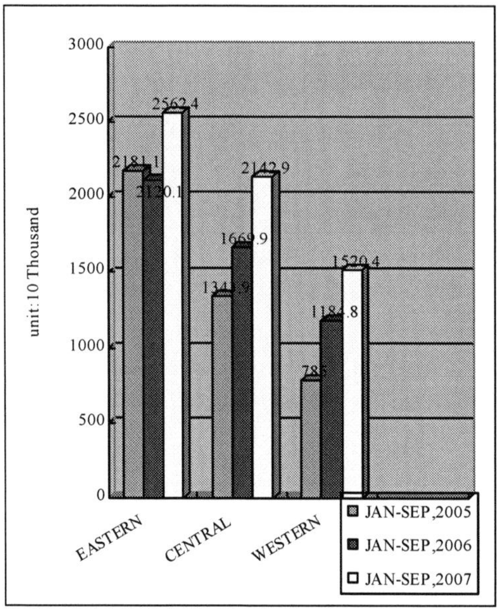


**FigureOne:Comparison amongEastern,Central andWesternAreasnew mobile phone subscribers**

Source: Ministry of Information Industry, People's Republic of China

Therewasno doubt that end-users can quickly browse the Web through the high-speed of 3G which will facilitate the users who want to carry out business activities; the mobility of 3G also brings a lot of convenience to users; coupled with its security features on the identity, it eliminates a lot of concern for us users. As a result, it can be said that 3G builds a better service platform for the development of mobile business.

## 5. CONCLUSION

Under the conditions that the global mobile telecom industry entered the era of 3G, we should combine the research results, implementation and practical experience of leaders in various 3G areas, and grasp comprehensively of the mobile telecom industry chain on a rational-analysis basis.Firstly, after regroupingof China's telecom companies, telecom industry was developing toward effective market, efficient mechanism, and impartialsupervision.Secondly, 3G technology-enabled businesses combine networks, products, and services streamlessly. Based on 3G,LTEand4Gwere near future goals. Thirdly, supervision systemand policy surveillance means came out of3G running environment and became more urgent.Relevant issues include market entry, interconnection, service fee auditing, added value and foreign capital and so on. Fourthly, new telecom regulations and laws confronteddomesticsupervisionorganizations.Furthermore, we should stand at a strategic height to create a platform for the exchange of high-quality among the parties who pay attention to and take participation in the development of 3G in China,imposing an important role in promoting the development of China's 3G industry chain, business models and investment opportunities. China's 3G will be glory under such a background.

## ACKNOWLEDGMENT

This work partially was backed by research project*Value Proposition of Service Role onM-commerce ValueChain*undercontract A10-0104-07-404.

## REFERENCES

[1]. L.Luo,etal.*TheThirdGenerationMobileCommunicationTechnologyAndBusiness.*1st ed., 2005. Posts & Telecom Press,pp3-11,pp.162-164

[2]. T.H.Tian."3G is coming: How should we deal with?"*Technology of Telecom Network..* 2005(1)

[3]. B.M.Guo."Research significance of abroad 3G business for China's 3G development".*ModernTelecommunicationsTechnology.* 2007(2)

[4]. Z.Li."3G development situation and prospects".*Network and Information.* 2007(5)

[5]. T.Zhang, et al."Rational pricing model in 3G era".*ModernTelecommunicationsTechnology.* 2007(2)

[6]. Y.F.Yuanet al.*M-Commerce.*1st ed., 2006(6), Tsinghua University Press, pp.109-119

# A new approach for critical resources allocation

**Facundo E. Cancelo, Pablo Cababie, Daniela López De Luise, Gabriel Barrera**
ITLab, Universidad de Palermo, Ciudad Autónoma de Buenos Aires, C1188AAB, Argentina
aigroup@palermo.edu

**Abstract - This paper presents a solution based on Artificial Intelligence using Multi-objective Genetic Algorithms to optimize the allocation of teachers and classrooms. The implementation was created in order to optimize the process in both cases, allowing them to compete so as to establish a balance and arrive at a feasible solution quickly and efficiently.**

## I. Introduction

This work is based on research that is being conducted by the ITLab of Universidad de Palermo [17]. The aim is to improve the actual system of allocation of teachers and classrooms within this institution at the beginning of each academic year. For that purpose, the automatization of this task is proposed in order to determine the best distribution of classrooms that will, in turn, optimize the use of spaces and the allocation of teachers. At present, this task is done manually: the existing resources are distributed at the discretion of the person in charge. However, due to the amount of data and the variables involved in the process the task is costly and complex and results are not always satisfactory: sometimes classrooms are too big or too small to hold a certain number of students. Besides, the teacher's schedules may not be considered, so they may be asked to give a lesson regardless of their availability. This process of allocation that seems to be easily solved by considering the number of students per class, the capacity and the number of classrooms available has been the source of various researches.

It is important to mention the fact that the different existing variables such as classrooms, availability of teachers and courses they may teach, among others, have posed a problem for which there is no solution so far, partly because there has almost been no research registered in the area.

The rest of this paper is organized in different sections: Section II presents an analysis of the products and of the existing alternatives in the market; Section III explains the structure and overall architecture of Gdarim in connection with its implementation, codification and design. Finally, Section IV presents the conclusion and future work.

## II. BACKGROUND

There is still no product that may be used to determine an appropriate distribution of classrooms, according to the teacher's availability and the characteristics of the building. A number of versions of software was developed. All of these versions aimed at assisting employees in the manual allocation of classrooms by providing a visual aid, but no artificial intelligence was involved and, as a consequence, the distribution process was not optimal [2], [4], [5]. There exist products that use mathematical methods, such as simplex [8]. Recently, a number of records and documents that showed the beginning of research work to implement artificial intelligence was found [9].

Most of the existing solutions only assist people by allowing them to have a visual display of spaces, but they are not softwares that rely on artificial intelligence.

- Visual Classroom Scheduler [2]: Application developed in Java language to distribute classrooms in the spaces available.
- Classroom Scheduler [5]: It allows people to have a fairly clear view of the distribution of classrooms. It is an Open Source software that provides the source code, so it can be modified and adapted according to the specific needs of any institution.
- phpScheduleIt [4]: It is also an Open Source software. It is not useful to allocate resources but it can select new classrooms should the need for a particular room arise, as in the case of special events or laboratories. In the face of an outline and a timetable, this software tracks changes in order to minimize the impact. It was developed in PHP language (PHP Hypertext Pre-processor).
- Softaula [3]: This software has four license types. Some of its features are students' registration and assignment of free classrooms to new students through automatic processes. This product integrates many aspects such as academic management, statistics, marketing and accounting. Still, it does not improve the distribution of classrooms and the allocation of teachers.

Other organizations have attempted to improve their manual allocation processes by integrating different softwares in their computer systems. Such is the case of Universidad nacional del Comahue (Argentina) and the municipality of Pereira (Colombia):

- SIU Guaraní [6]: Software of Universidad nacional del Comahue. This system handles and processes all the information regarding students from the moment of their enrollment to their graduation. This information is complemented by other processes, such as management of classrooms, dates of examinations and jurors' nominations. As regards exams and courses, students can register online. Any other information concerning the students' enrollment in different courses, courses of studies, exams, registration of equivalent subjects and exams taken can also be searched on the Internet.
- Matrícula [1]: This software was developed by the municipality of Pereira in Colombia, so that the educational institutions may have the academic information of their students in an integrated system.

T. Sobh (ed.), *Innovations and Advances in Computer Sciences and Engineering*,
DOI 10.1007/978-90-481-3658-2_71, © Springer Science+Business Media B.V. 2010

But there is neither an optimal use of spaces nor artificial intelligence applied to manage resources efficiently.

The implementation of these systems helps to lighten the academic paperwork so as to avoid bureaucracy. This in turn allows the authorities of those institutions to rely on tools that assist them in the control of the academic management. While all these alternatives are a significant improvement in the process of allocation of classrooms, none of these solve the problem completely

Initially for the development of this framework were used files accessed and parsed. They were created as the information they contained needed to be stored for subsequent executions. At this time, the way the data persists is starting to change to a database model. This information is the result of continuous surveys that were done in the university to detect parameterization problems and implementing new restrictions. The files and forms were initially loaded with fictitious information for a specified school and one particular day. Once the system is properly stabilized, the files will be loaded with all the information of classrooms, teachers and resources of the whole university for every day of the week.

These adjustments and improvements are supervised and supported by those who are taking care of the task manually at the moment.

## III. GDARIM STRUCTURE

After a deep analysis of surveyed data, the trouble can be reduced to an allocation of resources problem; with a large number of parameters that are closely intertwined and cannot be processed by the human mind altogether. Therefore, this paper proposes the use of the Epsilon MOEA algorithm [11] in order to optimize all objectives simultaneously, since this algorithm makes it possible to work with the number of parameters mentioned. This new technology also allows a future addition of new goals to be optimized. The language chosen for this development is Java because of its portability, its efficiency and its characteristics as an Open Source.

### *A. Modules*

The prototype contains three modules:

#### *A.1 Application Programming Interface*

An API (Application Programming Interface) information module Fig. 2 has the domain values of the problem. The API information module manages the configuration of the MOEA algorithm such as the necessary restrictions, the tests and the specific weightings for each goal within a problem.

#### *A.2 Genetic Algorithm motor module*

A GA motor module Fig. 3: it implements the core of the genetic algorithm and the intelligence of the solution. It includes the administration of populations, the process of genetic operators and the joint assessment of the goals to be optimized.

This is the algorithm in high level language

1. Stabilization phase:
   a. The first population is codified based on the domain values of the problem for each assignment of all the classrooms in one giving day.
   b. Evaluation of the index of restrictions that an individual may face.
   c. While the index of restrictions keeps high:
      i. The assignments are mutated for each individual.
      ii. New evaluation after the mutated operation.
      iii. Recalculate the index of restrictions.
   d. Step to the next phase.

2. Processing stage:
   a. While the level of fitness remains low:
      i. Two individuals are crossed over according to the AG-defined operations.
      ii. The assignments are mutated for each individual.
      iii. New evaluation after the AG-defined operations.
      iv. Create de new generation with a sample of the result, depending to the fitness functions.
      v. Establish the degree of dominance between individuals according to the fitness. Those who dominate other and are no-dominated, represents the elite population.
   b. Shows the elite population as the possible solution to the problem.

#### *A.3 API allocation module*

An API allocation module records any possible allocation of spaces as well as the solution generated by the algorithm.

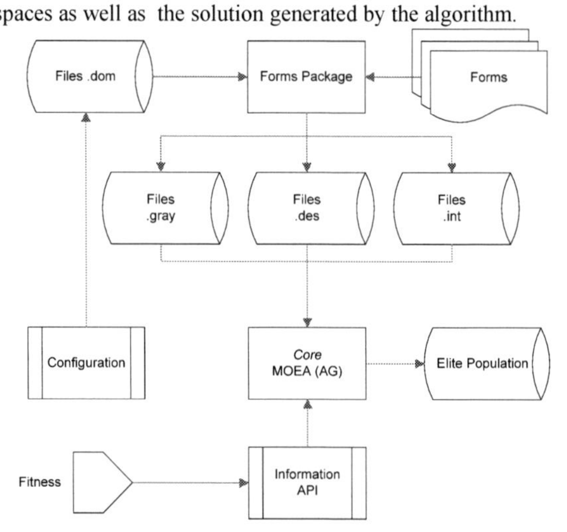


Fig. 1. Gdarim general architecture.

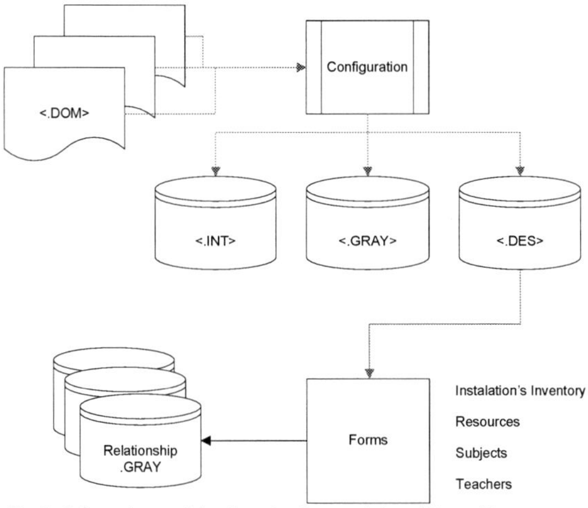


Fig. 2. Information module where the characteristics of the problem are shown

*B. Goals*

At first, it is important to determine the kind of problem that should be tacked in order to optimize the use of resources: classrooms and teachers. The first goal is to use as less classrooms as possible without interfering with the number of existing courses to be attended by students. The second goal focuses on the teachers' availability in order to assign courses in an atomic and consecutive row, if possible. The optimization of the first goal could clash with that of the second. To avoid this, it is necessary to configure the weighting of each goal so that they both compete in terms of their importance and a balance is finally achieved.

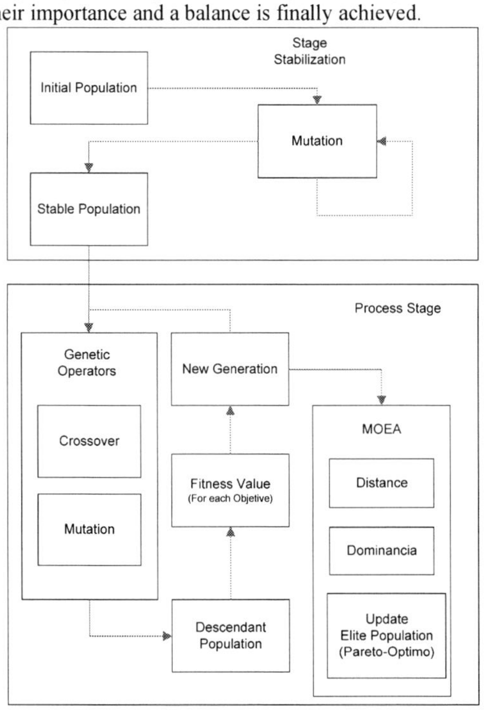


Fig. 3. GA motor module, core of the algorithm genetic

*C. Restrictions*

In order to narrow down solutions, a series of restrictions should be defined, including for example, the fact that a classroom which is not in use or that does not have the required resources cannot be assigned. In the case of teachers, a same person cannot be assigned two different courses at the same time, or a course that is beyond his or her specialization.

*D.Loading process*

The information is entered through several forms. One example is the teachers' form (Fig. 4), which reflects different information such as the employees' basic data, courses to be given and the teachers' availability. In the facilities form (Fig. 5), all the information connected to classrooms is loaded: existing classrooms, the university branches to which they belong, their seating capacity, the kind of classrooms they are and the resources they have. There is also a form for new courses (Fig. 6), which shows the course quota, the probable enrollment quota (number of students enrolled), the preferred university branches and a list of resources required for the classes.

A number of parameters will be established to define a representation for the allocation of one classroom (Table 1).

The information is stored in files that are used by the application to identify all the possible values that can receive each of the parameters that make up the gene. The composition of this gene specifies the parameters that are evaluated to represent the problem (Table 1). They are the domain values of the problem.

Table 1: Example of a gene for a classroom and a teacher

| GENE |
| --- |
| Classroom number |
| Course |
| Teacher |
| Shift |
| Branch |
| Etc. |

*E Process*

Initially, the files are populated with the domain values of the problem, which means a discretization of all the possibilities for each parameter included within the solution. Then, an initial population is randomly created in order to achieve one or more optimal results. Conceptually, the key to solve the problem is in one or more individuals that constitute composition assignments of all the classrooms at all possible times for a given day. However, the problem of creating random solutions lies in the fact that there is a high likelihood that these solutions may face one or more restrictions. If such is the case, the population is invalidated by what is termed "premature mortality" since by definition a restriction overrides the individual as a possible solution.

*E.1 Penalized restriction*

A possible solution to avoid overridden individuals is to implement penalized restrictions: theoretically, a restriction will initially behave like a penalty but it will not invalidate the individual. The drawback of this method is that it only avoids the initial population's invalidation but does not allow for improvement.

*E.2 Weighted Mutation*

Another way of solving the problem is by increasing the rate of mutation, which is directly proportional to the index of restrictions that an individual may face. The increase will make it possible to explore within the space search and gradually achieve a decline in the restrictions. As restrictions are fulfilled to a lesser extent, and due to the exploratory task, the level of mutation decreases until it reaches a stable value, which is configured as the number of restrictions by individual.

*F. Stabilization phase of initial population*

At this stage, a stable population is required, according to the non-fulfillment of excessive restrictions of the problem. There is no need for a fitness evaluation or for an individuals' crossing, since it is necessary to find as much heterogeneous solutions as possible. This way, the system focuses on the search of a starting point to determine the optimum, with the selection and mutation of individuals to decrease the rate of restrictions. Both the weighting of the mutation and the restrictions decrease too. The result is that restrictions reach an expected value.

It should be pointed out that this step may take several iterations or generations. However, the processing cost is not critical for this specific problem. It is also essential to have a valid and diversified population to begin with the search of an optimal solution.

*G. Processing stage*

The stabilized initial population is codified in genomes that are reproduced, mutated (genetic operator), crossed over (genetic operator), replaced and evaluated according to AG-defined operations [7]. Also, it changes according to an applied a series of fitness functions that determine the alternative that may best solve the problem (scoring individuals). The project has configuration files in text format that can be accessed manually to see the information already loaded. They are located in a folder called "CFGS". There are files with extension ".dom" containing domain values of the problem (starting point of the project). These files are updated and populated thanks to the entries in the forms. At the same time, interrelations between different parameters are established. Then, a converting process codifies these files and generates a Gray codification [16] in the "gray/AG" folder and a decimal codification in the "ints" folder.

The Gdarim problem is made up of twenty binary variables (Gen) and a number of restrictions. Most of these restrictions are taken from the forms. The rest of them are inferred from defined specific problems, such as overlapping of courses or teachers allocated in different branches in consecutive classes. The algorithm of the engine has a data administrator that runs the logical access. Since this manager was implemented with a Singleton Pattern [15], there is no alternative access to such information.

The MOEA algorithm optimizes the use of classrooms and teachers. As both objectives can be conflicting [11], [12], [13] a fitness function (1) is applied for classrooms and another (2) for teachers.

$$f(x)=\sum(AV-AN) \quad (1)$$

AV represents the non-allocated classrooms and AN, the empty ones, either because they require maintenance or because of any other reason.

$$f(y)=\sum(SHD-SMax)+\sum(HsD_t-HsD_u) \quad (2)$$

SHD is the separation time of the teacher. SMax is the maximum separation between contiguous schedules. HSD(t) represents the teacher's availability. HsD(u) is the hourly load of the teacher.

Advantages and disadvantages identified in both functions [10] are tested. For this research, optimality means to minimize the fitness function of both objectives, under the following criteria:

- The lowest the rate of unused classrooms is, the worse fitness score for classrooms is obtained.
- The more availability a teacher may have, the worse fitness score for teachers is obtained.

In order to narrow down the search space, the algorithm works with one or several functions that shape the restrictions of the problem.

For each population, and individual fitness score is estimated and each of their objectives is considered. Then, a sample of individuals is extracted. This sample is based on a probability obtained by a configuration file and it is processed by genetic operators like crossover and mutation depending on another probability given in the configuration.

For the crossover operation, two individuals are chosen. A cutoff point is established in order to determine which parameters may be taken from the first individual and which from the second, so as to compose two descendants. The crossover allows for a deep search of solutions.

In the process of mutation, only an individual is considered: one of its parameters is mutated, transforming its value into some other that may be valid within the domain of the problem. Besides, the mutation process eases the way in the search for a variety of solutions instead of narrowing down the possibilities with a small number of answers. As a result of this process, there is a second population, descended from the original one and which includes individuals created by a crossover process on the one hand, and others that are the product of mutations, on the other hand.

Thus, the initial and the descended population have to be tested in order to establish the degree of dominance between them. The fitness score helps determine which population may dominate which. The dominant one may be part of the elite population. Therefore, this elite population represents individuals with the best fitness score and non-dominated solutions, that is to say, they are part of the Pareto-optimal [13]. Then, the best n candidates are selected from among the remaining individuals (n represents the size of the population). These n candidates will shape a new generation that will start a new cycle. Within each cycle, the best candidates, which try to enter the elite population, are processed, either as new members or as replacements for one or more previous candidates.

Therefore, the cycle is repeated until a defined number of times in the configuration or an acceptable solution are reached. There may be more than one resulting solution to the problem and the user will decide which one to apply.

## IV. CONCLUSIONS AND FUTURE WORK

The next stage of the investigation will be focused primarily on the general configuration of the system to deliver results in a more efficient way. This will require the definition of weighted values to penalize every restriction in particular. In turn, this will set in motion a relativization process among restrictions. It will be also necessary to change the weighting of the goals by means of a static setting that will oppose a dynamic process weighting at a runtime, that is to say, the weight or relevance of each goal within a defined problem.

In addition, distance will be considered a system parameter that will be used to evaluate and discard any possible solutions. This parameter might be defined and configured according to the user's needs. For this reason, an adjustment of the setup of the existing distances among different solutions will be required in order to establish a certain level of comparison and membership to the environment or hypercube homogeneity. Finally, tests will be conducted with real data to evaluate the results and improve the existing weightings and configurations in order to enhance the system response and its effectiveness.

## V. REFERENCES


[1] Sistema Matricula, Pereira Educa, Colombia, http://www.pereiraeduca.gov.co/modules.php?name=Matricula&file=matricula_valoraciones

[2] Visual Classroom Scheduler, http://www.vss.com.au/index.asp

[3] Softaula, http://www.softaula.com/es/prod/prod_comparativa.asp

[4] PHPScheduleIt, software opensource, http://sourceforge.net/projects/phpscheduleit/

[5] Classroom Scheduler, software Open Source, http://sourceforge.net/projects/cr-scheduler/

[6] "SIU Guaraní", Universidad Nacional de Comahue, Tecnologías de la Información, http://www.uncoma.edu.ar/

[7] "Introducción a la computación evolutiva", Anselmo Perez Serrada, 1996

[8] "Modelos de despacho eléctrico económico – ambiental", Pablo Pizarro. Universidad de Mendoza, 2006.

[9] "Modelado de la Distribución de Espacios Físicos mediante Algoritmos Evolutivos", C.A. Delrieux, IX WICC, 2007.

[10] "Optimización Multiobjetivos del Proceso de Torneado", Ing. R. Quiza Sardiñas, Matanzas, 2004.

[11] "Twenty Years of Evolutionary Multi-Objetive Optimization: A Historical View of the Field", Carlos A. Coello Coello. Mexico, D.F., Nov. 11, 2005.

[12] "Visualization and Data Mining of Pareto Solutions Using Self-Organizing Map", S. Obayashi et al., 980-8577. Japan.

[13] "Introducción a la Optimización Evolutiva Multiobjetivo", Carlos A. Coello Coello, Sept. 2002, http://neo.lcc.uma.es/

[14] Anselmo Perez Serrada, "Una Introducción a la Computación Evolutiva", 1996, p. 17.

[15] Martin Fowler, "Analysis Patterns: Reusable Object Models".

[16] Gray code definition, http://en.wikipedia.org/wiki/Gray_code

[17] "Asignación de Aulas y Distribución Optima de Espacios". Cababie P, Cancelo F. López De Luise.

# Numerical-Analytic Model of Multi-Class, Multi-Server Queue with Nonpreemptive Priorities

Mindaugas Snipas, Eimutis Valakevicius
Department of Mathematical Research in Systems
Kaunas University of Technology
Kaunas, LT - 51368, Lithuania
**eimval@ktu.lt**
m.snipas@stud.ktu.lt

***Abstract* - We consider a multi-class, multi-server queuing system with preemptive priorities. We distinguish three groups of priority classes that consist of multiple customer types, each having their own arrival and service rate. We assume Poisson arrival processes and exponentially distributed service times. The performance of the system is described in event language. The created software automatically constructs and solves system of equilibrium equations to find steady state probabilities. We suggest a numerical-analytic method to estimate the probabilities. Based on these probabilities, we can compute a wide range of relevant performance characteristics, such as average number of customers of a certain type in the system and expected postponement time for each customer class.**

## I. Introduction

Multi-class, multi-server priority queuing systems can arise in various practical applications, for example, telecommunication and computer systems, production, logistics etc. There is quite some literature on multi-server priority queuing systems, see e.g. both for preemptive priorities [and for nonpreemptive priorities [1-6]. Various queuing systems are solved using analytical approach, especially matrix-geometric or matrix-analytical methods. However, since analytical approach is effective for queuing systems with infinite waiting space (so called insensitive systems), usually it is more difficult to find exact solutions for systems with limited waiting space. In this paper, we apply the numerical-analytic approach [7], which enables to model queuing systems with various limitations on waiting space.

To construct a model we need to describe the performance of the considering queuing system in the event language. It allows automating some stages of the model. The created software in C++ generates the set of possible states, the matrix of transitions among states, constructs and solves equilibrium equations and finally computes relevant performance measures according given formulas.

## II. Conceptual Model of Queuing System

Consider a multi-class, multi-server queuing system shared by $N$ customer classes, numbered $1,\dots,N$. A number of class indicates the priority rank (class 1 has highest priority and class $N$ has the lowest priority). Priority rule is nonpreemptive. Class $i$ customers arrives in the system according to a Poisson process with rate $\lambda^i$. The service times of class $i$ customers are exponentially distributed with mean $1/\mu^i$. Within each class, service discipline of customers is a First Come First Served (FCFS*)*. Queue length of each class customers can have various limitations.

## III. Numerical Model of the System

Consider a queuing system with 3 priority classes – high, medium and low, and 2 servers, which we call first and second. In addition, we consider limitation $L$ ($L \in N$) for summary waiting space of each customer class. The system's performance is described in the event language, using methods from [1].

The set of system states is the following:

$N = \{(n_1, n_2, n_3, n_4, n_5)\}, n_1 + n_2 + n_3 : L; n_4 : 3; n_5 : 3,$

$n_1, n_2, n_3, n_4, n_5 \in Z^+,$

where

$n_1$ – the number of high priority customers in the queue;

$n_2$ – the number of medium priority customers in the queue;

$n_3$ – the number of low priority customers in the queue;

$n_4$ – indicates the state of the first server (0 – if server is empty; 1 – if high priority customer is being served; 2 - if medium priority customer is being served; 3 – if low priority customer is being served);

$n_5$ – indicates the state of the second server (0 – if server is empty; 1 – if high priority customer is being served; 2 - if medium priority customer is being served; 3 – if low priority customer is being served);

The following events can occur in the system:

$E = \{e_1, e_2, e_3, e_4, e_5, e_6, e_7, e_8, e_9\},$

where

$e_1$ – a high priority customer arrived to the system with intensity $\lambda^h$;

T. Sobh (ed.), *Innovations and Advances in Computer Sciences and Engineering*,
DOI 10.1007/978-90-481-3658-2_72, © Springer Science+Business Media B.V. 2010

$e_2$ – a medium priority customer arrived to the system with intensity $\lambda^m$;

$e_3$ – a low priority customer arrived to the system with intensity $\lambda^l$;

$e_4$ –a high priority customer was served in the first server with intensity $\mu^h$;

$e_5$ –a medium priority customer was served in the first server with intensity $\mu^m$;

$e_6$ – a low priority customer was served in the first server with intensity $\mu^l$;

$e_7$ – a high priority customer was served in the second server with intensity $\mu^h$;

$e_8$ – a medium priority customer was served in the second server with intensity $\mu^m$;

$e_9$ – a low priority customer was served in the second server with intensity $\mu^l$;

The set of transition rates between the states is the following:

$INTENS = \{(\lambda^h, \lambda^m, \lambda^l, \mu^h, \mu^m, \mu^l)\}$

The description of system behavior using event language is presented below.

$e_k$: $k = 1,2,3$
**if** ( $n_1 - n_2 + n_3 < L$ **and** $n_4 = 0$ )
  **then** $n_4 \leftarrow k$;
**end if**
**else if** ( $n_1 + n_2 + n_3 < L$ **and** $n_5 = 0$ )
  **then** $n_5 \leftarrow k$;
**end if**
**else if** $n_1 + n_2 + n_3 < L$
  **then** $n_k \leftarrow n_k + 1$
**end if**
**Return** Intens $\leftarrow \lambda^k$

$e_{3+k}$: $k = 1,2,3$
**if** $n_4 = k$
  **then**
    **if** $n_1 > 0$ **then** $n_1 \leftarrow n_1 - 1$; $n_4 \leftarrow 1$;
    **else if** $n_2 > 0$ **then** $n_2 \leftarrow n_2 - 1$; $n_4 \leftarrow 2$; **end if**
    **else if** $n_3 > 0$ **then** $n_3 \leftarrow n_3 - 1$; $n_4 \leftarrow 3$; **end if**
    **else** $n_4 \leftarrow 0$
    **end if**
**end if**
**Return** Intens $\leftarrow \mu^k$

$e_{6+k}$: $k = 1,2,3$
**if** $n_5 = k$
  **then**
    **if** $n_1 > 0$ **then** $n_1 \leftarrow n_1 - 1$; $n_5 \leftarrow 1$;
    **else if** $n_2 > 0$ **then** $n_2 \leftarrow n_2 - 1$; $n_5 \leftarrow 2$; **end if**
    **else if** $n_3 > 0$ **then** $n_3 \leftarrow n_3 - 1$; $n_5 \leftarrow 3$; **end if**
    **else** $n_4 \leftarrow 0$
    **end if**
**end if**
**Return** Intens $\leftarrow \mu^k$

The created software generates the possible set of states, constructs the transition matrix among them, creates and solves Kolmogorov-Chapman system of linear equations and estimates steady state probabilities $\pi(n_1, n_2, n_3, n_4, n_5)$. Various characteristics for current system can be calculated using the estimated probabilities $\pi(n_1, n_2, n_3, n_4, n_5)$. For example, average queue length of each type customers:

$$E(L) = \sum_{n_1} \sum_{n_2} \sum_{n_3} \sum_{n_4} \sum_{n_5} n_i \cdot \pi(n_1, n_2, n_3, n_4, n_5)$$

## IV. Modeling Results

3 class and 2 servers queuing system with limitation on the total number of customers in the queues was modeled. The numerical-analytic approach, depending on complexity of considered system, can require a large amount of calculations and computer resources. For example, if the set of states is described as

$N = \{(n_1, n_2, n_3, n_4, n_5)\}, n_1 + n_2 + n_3 : L; n_4 : 3; n_5 : 3,$

it is easy to prove that the total number of states equal to

$$|N| = \frac{(L-1)(L-2)(L-3)}{3!} \cdot 4 \cdot 4 = \frac{8}{3}(L+1)(L+2)(L+3) \cdot$$

For example, if $L = 10$ then system has $4576^2$ states! We chose limitation on summary waiting space $L = 7$. We estimated average queue length for each customer class ( $E(L^h)$, $E(L^m)$ and $E(L^l)$ for high, medium and low priority queues respectively ) and general loss probability $P(Loss)$ (i.e., probability, that customer of any class will not be served ).

Numerical-analytic modeling results were compared with simulation results using ARENA simulation software. During each simulation session total amount of more than 5 500 000 customer arrivals were generated. We used PC with AMD Athlon 64 X2 dual core processor 4000+ 2.10 GHz, 896 MB of RAM physical address extension. Intensity rates, modeling results and calculation times are in Table 1.

## V. Summary

Numerical-analytical method enables to model various queuing systems with various limitations on waiting space. The obtained results showed that it requires less calculation time and gives higher accuracy than standard simulation software to model certain queuing systems.

## References

[1] A. Sleptchenko, A. Harten, M. Heijden "An exact solution for the state probabilities of the multi-class, multi-server queue with preemptive priorities, Queuing systems", *Queueing Systems: Theory and Applications* vol. 50, pp. 81-107, 2005.

[2] H. R. Gail, S. L. Hantler., B. A. Taylor "On preemptive Markovian queue with multiple servers and two priority classes" *Mathematics of Operations research*, vol. 17(2), 365-391, 1992.

[3] A. Sleptchenko,ˊ I.J.B.F. Adan, G.J. van Houtum. "Joint queue length distribution of multi-class, single-server queues with preemptive priorities" http://alexandria.tue.nl/repository/books/581962.pdf

[4] M. Harchol-Balter, T. Osogami, A. Scheller-Wolf, A. Wierman Multi-server queueing systems with multiple priority classes *Queueing Systems: Theory and Applications* vol. 51, pp. 331-360, 2005.

[5] E. P. C. Kao, S. D. Wilson "Analysis of nonpreemptive priority queues with multiple servers and two priority classes" *European Journal of Operational Research*, 118, 181-193, 1999.

[6] D. Wagner "Waiting time of a finite-capacity multi-server model with nonpreemptive priorities" *European Journal of Operation Research*, Vol. 102, 227-241, 1997.

[7] Valakevicius, E., Pranevicius, H. An algorithm for creating Markovian Models of Complex Systems, In Proceedings of the 12th World Multi-Conference on Systemics, Cybernetics and Informatics, Orlando, USA, June-July 2008, pp. 258-262.

TABLE I
MODELING RESULTS

| Parameters | | | Method | $E(L^h)$ | $E(L^m)$ | $E(L^l)$ | $P(Loss)$ | Time, s. |
|---|---|---|---|---|---|---|---|---|
| $\lambda^h = 1$ | $\lambda^m = 0.8$ | $\lambda^l = 0.5$ | Numerical | 0.0721 | 0.0925 | 0.0801 | 0.0011 | 76 |
| $\mu^h = 2$ | $\mu^m = 3$ | $\mu^l = 4$ | Simulation | 0.0721 | 0.0924 | 0.0797 | 0.0011 | 311 |
| $\lambda^h = 2$ | $\lambda^m = 1.6$ | $\lambda^l = 1$ | Numerical | 0.4111 | 0.7778 | 1.0708 | 0.0767 | 76 |
| $\mu^h = 2$ | $\mu^m = 3$ | $\mu^l = 4$ | Simulation | 0.4099 | 0.7786 | 1.0683 | 0.0764 | 312 |
| $\lambda^h = 4$ | $\lambda^m = 3.2$ | $\lambda^l = 2$ | Numerical | 0.6154 | 1.2690 | 3.7636 | 0.4434 | 77 |
| $\mu^h = 2$ | $\mu^m = 3$ | $\mu^l = 4$ | Simulation | 0.6153 | 1.2690 | 3.7668 | 0.4432 | 312 |

# Project Prioritization as a Key Element in IT Strategic Demand Management

Igor Aguilar Alonso
School of Computer Science,
Technical University of Madrid,
Madrid, Spain
iaguilar@zipi.fi.upm.es

José Carrillo Verdún
School of Computer Science,
Technical University of Madrid,
Madrid, Spain
jcarrillo@fi.upm.es

Edmundo Tovar Caro
School of Computer Science,
Technical University of Madrid,
Madrid, Spain
etovar@fi.upm.es

**Abstract. This paper describes demand management and its life cycle, taking into account the IT priorities model and project prioritization as a key demand management element.**

**One of a company's main objectives is to get the maximum benefit from its businesses in the shortest possible time. To do this, it is necessary to properly manage customer demand and satisfy customer needs. Therefore, it has to properly prioritize projects in accordance with criteria established by the company or through the IT portfolio management office, giving priority to projects that will most benefit the company if run.**

**To achieve these goals, liaison between senior business executives and the chief information officer responsible for managing IT should be good.**

## I. INTRODUCTION

The information technology (IT) strategy varies from one organization to another and must be applied to all IT departments. IT departments are managed by the chief information officer (CIO).

The best CIOs use a set of sophisticated techniques to develop strategies to ensure that the basic IT operations work well at the lowest possible cost, giving added value to the enterprise. Released from basic operating problems, they are free to focus on maintaining effective working relationships and create thorough business knowledge, recruit the best talent for their teams, and maintain a clear vision of the industry and the evolution of technology to create added value for the enterprise. This ensures that the CIO is able to identify, promote and implement significant business change initiatives.

The CIOs, the chief executive officer (CEO) and the board of directors are responsible for conducting strategic planning for the enterprise.

The project prioritization is one of the key business stages, and is directly related to the stages of the demand process life cycle. The CIO will work actively with other senior executive members to develop the enterprise's project prioritization, thus ensuring that demand management is efficient.

According to the demand classification [1] —(1) strategic demand, (2) tactical demand and (3) operational demand—, strategic demand is the most complex to manage. This is why the CIO and the board of directors should have thorough business knowledge to implement and carry out very detailed planning.

This paper presents project prioritization as a key element in IT demand management.

Section (1) is an introduction setting out the importance of IT strategic demand management for achieving strategic business planning objectives. Section (2) defines strategic demand and the demand process life cycle. Section (3) presents the IT priorities model and alignment of IT with business objectives. Section (4) presents the key components for project prioritization and assessment. Section (5) presents the project prioritization, followed by the conclusions.

## II. STRATEGIC DEMAND

Strategic demand is managed through the project portfolio. This portfolio manages ideas spawning new businesses or project innovation. Strategic demand "represents the most significant opportunity to increase business value". [2]. Project portfolio management (PPM) is an event-based process that can evaluate, prioritize and monitor projects.

Pressure to maximize profit can come from internal factors, i.e. business partners, board of directors or other business units, or from external factors, such as regulatory frameworks, market competition, etc. For this reason, there needs to be a very close link between strategic planning and project processes from the very start.

Strategic demand management depends on (1) strategic planning units, (2) strategic process planning, (3) resource location, (4) budget location, (5) project selection and implementation and (6) project "post-mortem" metrics to improve good practice in the organization. By implementing these best practices, we can manage demand at a management process level, such as:

- Clearly identify strategic objectives: strategy development is highly collaborative, involving both IT and business executives [3]
- Use an event-based decision-making process: projects are evaluated, selected, prioritized, funded and reviewed based on their potential risk-adjusted value in the context of the organization's strategic objectives [4]

T. Sobh (ed.), *Innovations and Advances in Computer Sciences and Engineering*,
DOI 10.1007/978-90-481-3658-2_73, © Springer Science+Business Media B.V. 2010

- Take a life cycle approach to investment and profit realization: portfolio investments are managed through their entire economic life cycle to deliver the optimal value through implementation, adoption, and eventual retirement [5].

### A. *Demand Process Life Cycle*

The demand management is a cyclical process that begins with strategic planning and goes through a number of different stages, ending with value management, as shown in Figure 1.

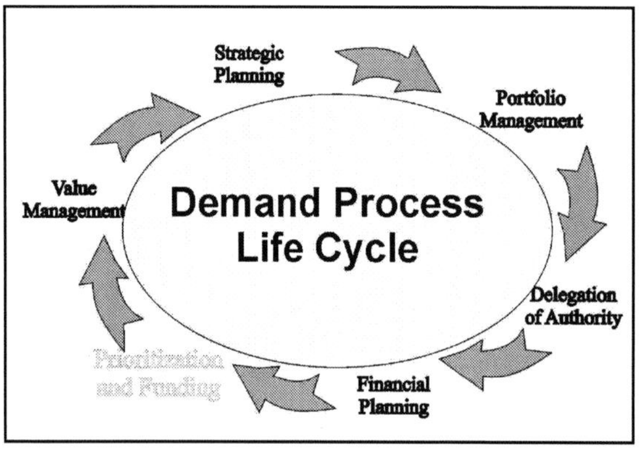


Fig. 1. Demand process life cycle.

This process takes place within the disciplined process of IT governance and is more often applied when the enterprise implements best management process practices and the company has reached an advanced level of maturity.

The demand management life cycle [6] has the following stages: (1) strategic planning, (2) portfolio management, (3) delegation of authority, (4) financial planning, (5) prioritization and funding, and (6) value management.
Now that we know what the demand process life cycle is, we need to find out how mature the enterprise is with a view to carrying out these projects. A mature enterprise manages demand efficiently by delivering quality products on schedule and within budget, using the planned financial resources.

## III. IT PRIORITIES MODEL

The alignment of IT goals with an enterprise's business is one of the priorities of IT and is located at level IV within the IT priorities. Fig. 2 shows an organization's priority levels for IT use, according to a hierarchy that is comparable to Maslow's human needs [7].

Levels I, II and III have an internal target, based on operations, infrastructure and applications. Levels IV and V are geared to IT governance and have an external target. Business-IT alignment, strategic planning, project prioritization and business decision-making by the board of directors fall at level IV.

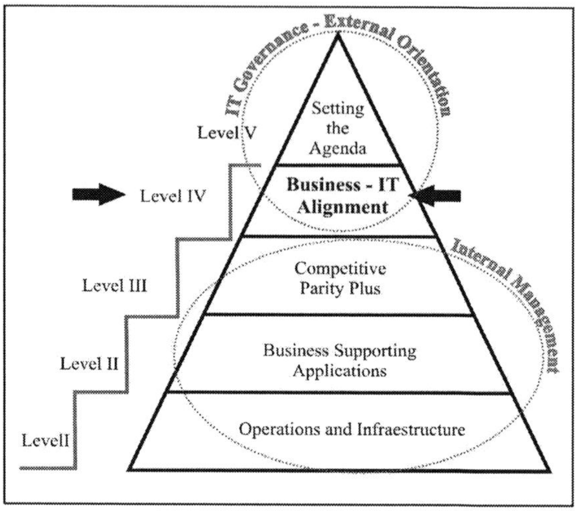


Fig. 2. IT Priority levels

It is necessary to take into account IT demand management as well as the IT priorities model to achieve business objectives. Both play a very important role in IT governance.

### A. *Business – IT Objectives Alignment*

The alignment of corporate strategic plans with business unit strategic plans is a key requirement for business executives. At this level, all decisions are made in keeping with the business priorities. This is reflected at all levels through staffing, budget, projects and applications architecture.

The business priorities should provide a standpoint for answering the following questions:

- What hardware should be purchased?
- What staff should be hired?
- What are the staffing levels for different tasks and activities?
- What vendors are our partners?
- How is capital allocated and how are budgets drafted?
- What projects are going to be run?
- What projects can be carried out simultaneously?

After analyzing each of these questions, it is necessary to prioritize projects and implement the projects that more beneficial for the enterprise.

## IV. KEY COMPONENTS FOR PROJECT PRIORITIZATION

There are four key components [8] to be taken into account to complete a preliminary assessment of the projects and continue with other work, as shown in Fig. 3. Each of the key components is described in more detail in the following. The projects are then assessed, which would be equivalent to a preliminary prioritization. The prioritization process incorporates the best thinking of the IT and business leaders.

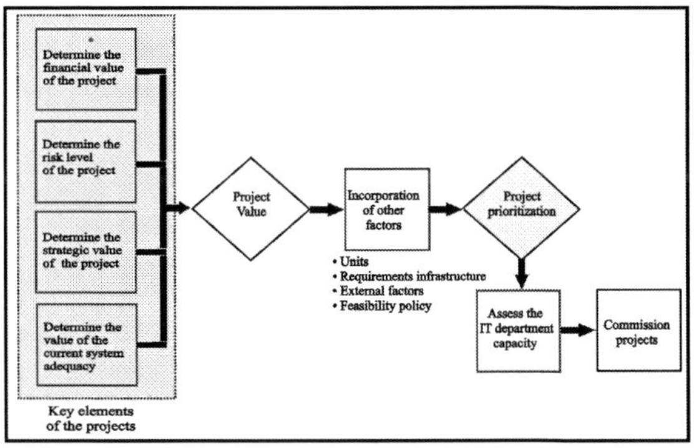


Fig. 3. Project prioritization process.

## A. *Project Financial Assessment*

The financial value is measured as the flow of benefits that a project delivers or can offer compared with the cost of generating those profits. Fig. 4 shows a framework for assessing the value of the project based on its associated financial value.

Currently, there are many methods to calculate the financial value of a given project, including return on investment (ROI), investment value, project amortization period, project net present value (NPV) and internal rate of return (IRR). All the individual methods for measuring project return use the same inputs to understand the value of the project costs and benefits. Project costs usually come in two categories: time-mediated project costs and change-mediated project costs (e.g., changes in the underlying operating model). Similarly, profits have deferred and immediate components. Usually, most costs come under the time category, whereas benefits tend to come in the shape of savings or income.

| FINANCIAL VALUE | | Feature | Value |
|---|---|---|---|
| HIGH | Level 4 | • Project has direct revenue benefits<br>• Payback is 0 - 2 years<br>• Benefits significantly outweigh costs.<br>• Project value and solid assumptions are high.<br>• Project has direct or indirect revenue improvement, or cost reduction benefits. | |
| | Level 3 | • Project has direct revenue benefits.<br>• Payback is 2 - 5 years.<br>• Benefits outweigh the costs.<br>• Project value and most assumptions are reasonable.<br>• Projects has direct or indirect revenue improvement or cost Reduction benefits. | |
| | Level 2 | • Project generates no direct revenue.<br>• Cost reduction potencial .<br>• Project may provide information valued for improved control of business.<br>• Project offers productivity improvements.<br>• Payback is difficult to measure. | |
| LOW | Level 1 | • Project generates no revenue.<br>• Limited cost reductions.<br>• Project generates. | |

Fig. 4. Project financial value.

To measure project value a simple measure of costs and benefits is preferable, and the key cost and benefit parameters should be clearly documented.

We can summarize the financial value of a project in mathematical terms, represented by the net present value (NPV) of a product in the equation (1).

$$NPV = (DP + CP) - (DC + CC) \quad (1)$$

where DP are deferred profits; CP are current profits, DC are deferred costs and CC are current costs.

Companies engaged in projects to develop new technological products employ the following financial methods to calculate the financial value of a project. One is the expected commercial value (ECV) of a project [9]. This method seeks to maximize the value of a business portfolio, subject to certain budget constraints, introducing the notion of risk and probability, as shown in equation (2).

$$ECV = [(PV*PCS\text{-}C)*PTS\text{-}D] \quad (2)$$

Also used is the productivity index (PI) [10]. This is similar to the method in equation (2), which shares many strengths and weaknesses. This method is applied to maximize the financial value of a portfolio for certain resource constraints.

$$PI = ECV*PTS/R\&D] \quad (3)$$

where PV is the present value, PTS is the probability of technical success, PCS is the probability of commercial success, D are the development costs, C are the commercialization costs (C) (launch), PV is the present value of future project earnings (at today's discounted prices) and R&D are dollar resources.

Also note that the estimated financial value should be based on cash flow, whereas the accounting rules include investment in hardware and software in the project.

The chief financial officer (CFO) and BU managers generally need to know how the project will affect the budget for the current fiscal year. They can then prepare for changes that are being developed as part of the project. In some cases, as long as the project costs can be capitalized, the impact on annual budget flow should be minimal. The CFO needs to know the project cash layout schedule to effectively manage the corporate reserves, ensuring that cash flow is available when required for resources outside the project (hardware and software).

## B. *Project Risk Valuation*

The next critical step in the process of prioritizing projects is the valuation of the impact of risk on company priorities. The risk in delivering the total project has a great impact on organizations' ability to realize the benefits of the project. Some IT project risk categories include project complexity, user training and technology to be used. Fig. 5 shows the risk factors to be assessed when determining the project value.

| RISK LEVEL | | Project complexity | User preparation | Technology | Value |
|---|---|---|---|---|---|
| LOW | Level 4 | •Understandable and well documented requirements.<br>•Well defined scope. | • Resources are available and capable.<br>• Ownership and accountability are clear. | •Application built on stable platform in which IT team has significant experience.<br>• Limited impact on operations. | |
| | Level 3 | •Good understanding of the requirements.<br>•Limited potential impact of different regions or customers. | •Resources are available.<br>•Ownership and responsability are shared.<br>• Key issues tracked with impact understood. | • Enough time to test changes to technical environments and train appropriate staff.<br>• Release strategy easily understood and controlled. | |
| | Level 2 | • Support applications for new business practice not well understood.<br>• Limited experience with business functions. | • Ownership is unclear or limited.<br>• Accountability is shared across broad areas.<br>• Isues not tracked - no Understanding of impact. | • Significant restructuring of technical environments with limited time for testing or training.<br>• Release strategy results in two or more versions of production code. | |
| HIGH | Level 1 | •Increases block functionality is not clearly defined.<br>•No clear understanding of the requirements for IT.<br>•Many outstanding issues. | •Key resources spread across many projects.<br>• No ownership or project or benefits.<br>•Limited accountability.<br>•Many outstanding issues. | •Aplication one of first delivered on new technical platform.<br>• Spans multiple technical platforms.<br>• Technology has unproven performance. | |

Fig. 5. Project risk valuation.

Depending on the specific project type under examination, different factors may be appropriate for assessing project risks.

It is very important to take risk assessment into account, covering internal IT department as well as external risks that could harm the business.

Because risk is an element that will always be present in the business, the board and IT strategy steering committee must adopt the measures necessary to manage risk by establishing appropriate policies to detect threats, vulnerabilities, etc., and react quickly and promptly mitigate and prevent harm to the business.

## C. *Strategic Valuation of a Project*

The strategic value of a project is defined as the impact that a project is to have on external entities, particularly customers and suppliers. The project's strategic value gives a better idea of what new capabilities the projects has to offer the company to improve its competence and for it to work more effectively with project customers and suppliers.

The strategic value should be estimated according to the BU user inputs as well as the viewpoints of the senior management. Fig. 6 shows a scorecard used to assess the strategic value of a business project.

To develop correct strategies, the board of directors should take into account new products being developed, new markets, the economy, technology to be used and take into account the new potential business risks, such as increased competition and other factors.

The development of appropriate strategies will enable the business – IT objectives alignment, and thus create added value for the business. Senior executives must have the right tools, like reporting, dashboards, scorecards, etc., to make this strategic valuation, thereby assuring that decision-making meets the business needs..

The results of the analysis should rank the project according to its strategic value in relation to the most and least competitive projects.

| STRATEGIC VALUE | | Feature | Value |
|---|---|---|---|
| HIGH | Level 4 | • Establish industry leader to driving the sustainable increase in market share.<br>• Helps strengthen long-term relationship with customers or suppliers.<br>• Creates sustainable, hard-to-emulate competitive position.<br>• Competitive parity is difficult to match. | |
| | Level 3 | • Significant improvement in customer and/or supplier relations.<br>• Creates tempory competitive advantage or positioning.<br>• The competitive parity will lag by one to two years. | |
| | Level 2 | • Helps to keep up with the market or provide a competitive response.<br>• Sustains marketplace credibility.<br>•Provides better information.<br>- Sales increase.<br>- Better customer service support.<br>- Cost reduction. | |
| LOW | Level 1 | • Little or no strategic impact.<br>• No external impact on customers or suppliers.<br>• Competitors will easily copy capabilities. | |

Fig. 6. Strategic project value.

## D. *Valuation of Current Systems Adequacy*

The adequacy of current systems is also a critical factor to take into account when determining the project value. Fig. 7 shows a system / process scorecard for several levels of adequacy. This card can be used to judge the current system's level of competence coverage.

Where there are already systems in place, they can provide appropriate competence for the organization, minimizing, or at least reducing, the benefits of increased investment and effort. A trend has been observed within IT teams to seek incremental improvements on existing systems rather than undertaking new expensive (and potentially high-risk or effort-intensive) initiatives.

| SYSTEM ADEQUACY | | Systems | Process | Value |
|---|---|---|---|---|
| HIGH | Level 4 | • System handling current function is adequate for next 1-2 years. | • Process handling current function is adequate for next 1-2 years. | |
| | Level 3 | • System handling current function is adequate for next 7-12 months. | • Process handling current function is adequate for next 7-12 months. | |
| | Level 2 | • System handling current function is adequate for next 0 - 6 months. | • Process handling current function is adequate for next 0 - 6 months. | |
| LOW | Level 1 | • No system currently handling the function proposed by the project.<br>• Current system handling function is unstable or not delivering necessary functionality. | • No process currently handling the function proposed by the project.<br>• Current process handling function is unstable or not delivering necessary functionality. | |

Fig. 7. Current systems adequacy.

The adequacy of current systems is therefore a good metric for understanding the overall priorities of a project. Overall competence can be divided into two categories for measurement: systems technology and system support processes and procedures.

## V. PRIORITIZING PROJECTS

The project prioritization is the process by which the IT department determines, jointly with the CIO, the board and IT strategy steering committee, which projects will generate the highest value for the company and how many can be run simultaneously, considering the organization's IT capacity.

The project prioritization process is a key element for effective IT demand management, enabling the IT department to succeed and overcome many difficulties.

Project prioritization generates the following benefits for the IT department and provides added value to businesses.

- The IT department will always implement the highest value projects and invest adequate resources to develop the project.
- The IT department's capacity analysis ensures that the permitted number of simultaneous projects will be manageable for the IT department and the organization's ability to absorb change.
- Project prioritization ensures that project goals and scope are clearly defined; assigning responsibilities for the success of the project.

After assessing the four key project elements —financial value, risk, strategic value and systems adequacy— the projects are prioritized overall based on the preliminary valuations. The approved projects can be arranged in a 2 x 2 matrix.

The first assessment is based on a comparison between the strategic and financial value of a project. Fig. 8 shows an assorted project classification. The comparisons between these values indicate the high value projects, i.e. which projects will most benefit the organization.

Projects with relatively low financial and strategic values are in the bottom left cell. These projects probably should not be considered in the list of high priority projects and should possibly be reworked.

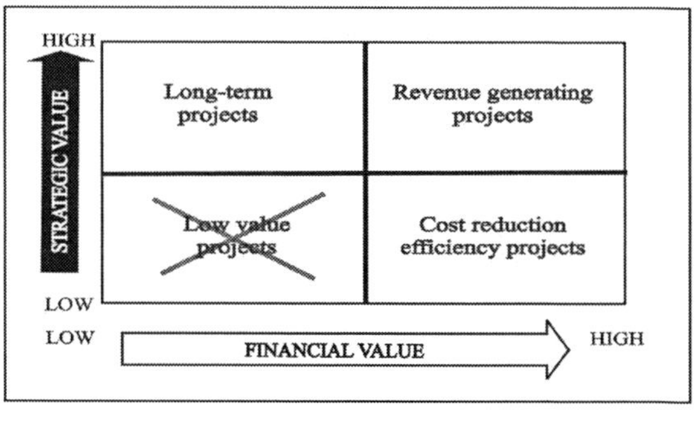


Fig. 8. Financial versus strategic valuation.

The second project evaluation, shown in Fig. 9, takes place after the low-value projects have been rejected.

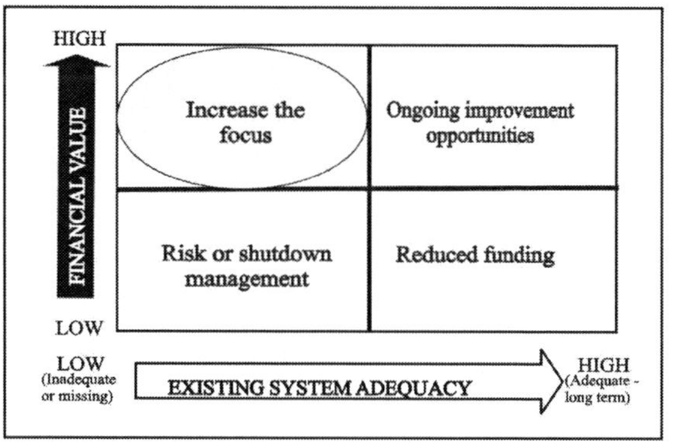


Fig. 9. Evaluation of systems adequacy versus financial value.

The assessment will be based on the financial value and the adequacy of current systems, Each cell in Fig. 9 suggests a series of actions for the project classed in that cell.

Projects that are assessed and are located in the top left cell are high-value projects and are considered to be high priority projects that have inadequate systems support capabilities. They should be further focalized by the organization.

The third evaluation should be of future projects that need to increase the focus and be classified based on "executability" (cost, speed, risk) versus financial value.

Fig. 10 shows this analysis. The purpose of this classification is to ensure that easy-to-develop high value projects are ranked highest. This ensures that the company will reap the benefits of projects at the earliest possible time. Projects that are classed in the top right cell are considered "treasures". These projects are easy to develop and guarantee a high financial value. This is where the important projects are classed.

These are projects to which the IT department should give top priority. Many projects that are classed in this phase of the analysis are probably major projects, as shown in Fig. 10 of the dashboard, in the top right cell.

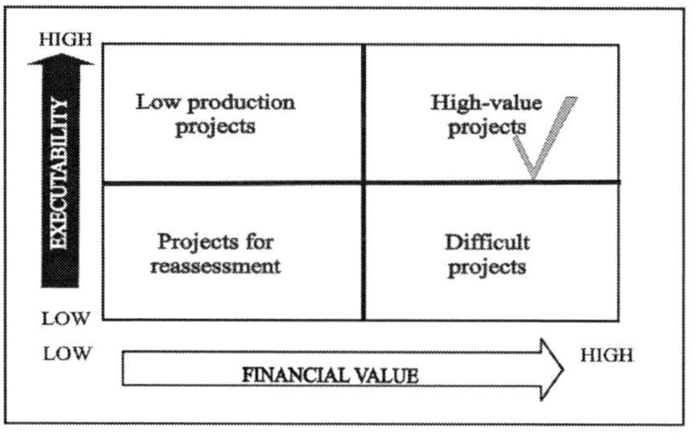


Fig. 10. Workability versus financial value.

This simple prioritization is useful for the final ranking of projects.

In each case, the result of the prioritization must pass a "rigorous" control. If the analysis produces bias or non-essential results, and a careful verification of the cases yields no change in the outcome, common sense should always prevail over the priorities generated by the methodology.

Note that at present there are many project prioritization models and models used by technological development projects that prioritize projects according the following criteria [11]:

- Strategic alignment
- Strategic leverage
- Technical success probability
- Technical feasibility
- Commercial success probability
- Product/competitive advantage
- Market appeal
- Reward
- Synergies (leveraging core competencies)
- Risk vs. return

After each project has been evaluated, the results of each variable should be recorded in a form used to prioritize projects.

The inventory of approved projects must be updated with the results of each element. The total score for each project can be calculated and sorted from highest to lowest priority. The priorities listing and ranking are added to periodically and reassessed as new projects are added or when projects have been completed.

After project prioritization, the results must be delivered to the board of directors for final approval.

The CIO must take into account the capacity of the IT department, if it has the necessary means to implement the project, taking into account human resources, appropriate technology (hardware and software), and the necessary infrastructure. If the IT department does not have all these resources, it will have to come up with an alternative solution to remedy these deficiencies, e.g. sign up infrastructures or specialized staff.

The CFO must see to it that the company has enough financial resources to implement the project or otherwise provide adequate alternatives to prevent the project from being held up after it has started.

After the prioritized projects have been approved by the board of directors, the CEO will give his approval. These projects will then be ready for implementation, and project managers will be given the responsibility to execute them

## VI. CONCLUSIONS

The conclusions from this work are as follows:

- Project prioritization as part of the demand management life cycle is regarded as one of the keys to an enterprise's business success.
- It is very important for the strategic planning committee to meet regularly to evaluate company performance and be able to make the necessary adjustments to strategies for ongoing projects.
- It is very hard to speak of IT successes and failures if projects were not prioritized, aligning the objectives of the projects portfolio with business objectives. If this condition is not met, an organization is unlikely to be able to find out whether the projects it develops maximize the profit shareholders expect from IT investments.
- At present, the challenge is to build a business model for which the CIO, the senior executives and the business line managers are responsible.
- The IT Portfolio Management Office plays an important role within the enterprise and is responsible for the prioritizing projects, deciding which will most benefit the enterprise, which projects will be implemented first, tracking ongoing projects, making necessary corrections to projects, constantly managing the projects awaiting implementation and new projects that are added to the portfolio.
- It is very important for project prioritization to take into account the four key project elements: financial value, the risks of the project, the project strategy and the adequacy of current systems. First the value of each element should be determined. Then the project should be assessed as a whole. Other project-dependent factors can then be added to arrive at the final prioritization of the projects.

## REFERENCES


[1]. S. Cray, "How it must shape and manage demand", Forrester, Cambridge, June 15, 2006.

[2]. B. Andrew, "Defining the MOOSE in the it room", Forrester, Cambridge, October 18, 2005.

[3]. S. Cray, "Relationship managers: Focal points for innovation", Forrester, Cambridge, June 15, 2006.

[4]. V. Margo, "What successful organizations know about project management", Forrester, Cambridge, May 26, 2006.

[5]. G. Chip, V. Margo, "Moving up the portfolio management ladder", Forrester, Cambridge, March 22, 2004.

[6]. C. Susan, "Maturing your demand management practices", Valuedance Counting on Technology, Publisher Compuware, January 2007.

[7]. I. Aguilar, J. Carrillo, E. Tovar, "The importance of IT strategic demand management in achieving the objectives of the strategic business planning", 2008 International Conference on Computer Science and Software Engineering (CSSE 2008), in press.

[8]. John, B., Jon P., Nicholas G. C., "The executive's guide to information technology", Second Edition, Publisher: John Wiley & Sons Pub, March 23, 2007, pp 564-567.

[9]. G. Cooper, R., Scott, Edgett, J.: Stage-Gate Inc.: Portfolio Management for New Products, Picking the Winners. 2007

[10]. G. Cooper, R., Scott, Edgett, J.: Stage-Gate Inc.: Portfolio Management for New Products, Fundamental for New Products Success. 2007

[11]. G. Cooper.: Stage-Gate Inc.: Managing Technology Development Projects. 2007- IEEE Engineering management review, Vol.35, N° 1, First quarter 2007

# Greylisting method analysis in real SMTP server environment – Case-study

Tomas Sochor

Department of Informatics and Computers, University of Ostrava, Czech Republic

*Abstract* - **Greylisting is a method for protection against unsolicited electronic mail messages (SPAM) based on the combination of white lists and temporary deferring of incoming messages from new sources. The idea is quite simple and the method proved to be surprisingly efficient but there are some issues to be concerned with. One of the most important issues is possible efficiency decay through a long-term period. Therefore the investigation of the efficiency of the method throughout longer period was done Also various other factors of the greylisting method application were studied and are discussed.**

## I. INTRODUCTION

Electronic mail services based on SMTP protocol that is the most popular electronic mail transmission protocol in the world have been suffering from unsolicited messages (usually called SPAM) for a long time and the amount of unsolicited messages is growing rapidly (see [1]). Even worse, not only total number of SPAM messages is growing, also the portion of unsolicited messages in the total amount of all messages sent seems to be growing. Therefore most electronic mail service providers look for tools how to eliminate those unsolicited messages. Such tools are usually called anti-SPAM tools or methods.

There are various approaches to eliminate SPAM resulting in lot of methods. Some of them are based on the distinguishing of the contents of the message (like words often used in SPAM messages), the message form (HTML code) or combination of both (HTML message with white letters on white background) or similar approaches (see e.g. [2], [3] or [4]). Only minority of SPAM elimination methods is based on handling the message being delivered. One member of this minority is greylisting.

As previous works showed (e.g. [5]) there are many problems associated with measuring the efficiency of anti-SPAM methods used. A case-study describing implementation of greylisting method (in combination with other methods) is discussed in this article.

The subject of case study is the SMTP server of the University of Ostrava. Greylisting (sometimes also spelled "graylisting") is used as one of methods of elimination of SPAM getting into users' mailboxes at the University of Ostrava since fall 2006. The method proved to be successful in lowering the amount of SPAM in mailboxes dramatically because while before greylisting implementation almost 70 % of all incoming e-mail messages were identified as SPAM by our mail filtering, after greylisting implementation the ratio dropped to approx. 10 %. Such dramatic decrease looks very well but due to the nature of greylisting method as discussed later the stability of results is an important issue. Therefore the main goal of the study was to investigate longer-term results.

The greylisting method at the protected SMTP server operates as follows:

1. The protected SMTP server receives the request to deliver an e-mail message into one of his inboxes from certain sending SMTP server (hereinafter called "SMTP client" or "client").
2. Instead of immediate delivery the protected server refuses the delivery of the message temporarily (usually with 450 error code and the text where the reason is announced as "greylisted") sent to the SMTP client.
3. The sending SMTP client should repeat the attempt to send the message again after certain period. The length of this period depends on the sending server. The minimum period for accepting the repeated delivery is set at the protected server and it is usually recommended to be 5 minutes (see [6]). The protected server usually advertises this period in its temporary error message sent to the SMTP server that dispatched the message being subject to temporary refusal. If such repeated attempt occurs in the allowed period the identification of the sender and the recipient in the form of a new triplet (see below) is stored in in the automatic white list (as explained below).

It should be noted that the procedure described above is applied only to messages from unknown sources (i.e. sources not yet listed in the automatic white list). The protected SMTP server keeps and maintains so-called automatic where list where "legal" senders of e-mail messages are listed. For senders from the list the procedure above is not applied and their messages are delivered without delay.

The identification of the sources of messages is usually performed using so-called triplet consisting of:

- sender IP address taken from the IP packet header when TCP connection is established,
- sender e-mail address transmitted by the sender during an initial SMTP dialog, and
- e-mail address of the recipient also taken from initial SMTP dialog.

There are also some variants of greylisting using different composition or evaluation of the triplet as follows:

- **Full** method as described above,
- **Weak** greylisting variant taking only certain part of the IP address into account (usually 24 bits). Both variants (full and weak) above will be referred hereinafter as "**basic**" method.

T. Sobh (ed.), *Innovations and Advances in Computer Sciences and Engineering*,
DOI 10.1007/978-90-481-3658-2_74, 

- **Simple** greylisting based only on the sender IP address (e-mail addresses are ignored),
- **Reverse** algorithm that replaces the IP address of the sending SMTP server with their domain name (with optional stricter treatment of sending servers without reverse DNS record thus more suspicious; such combination of weak greylisting method to SMTP sending servers with a valid DNS record and full greylisting method applied to clients without valid DNS record or with dynamically assigned IP address is called **smart**),
- **sqlgrey** algorithm that allows delivery of all messages from the same source (IP address and the sender's e-mail address) after successful delivery of one message to any recipient,
- and a range of selective variants that rely on cooperation with other (usually quite sophisticated) systems of evaluation of potential SPAM sources (SMTP clients that send messages) with the subsequent application of greylisting only to suspicious sources.

The above variants were investigated in the study described here in limited scope as described later. New variants of greylisting or new approaches are being developed continuously as illustrated for example in [5].

The idea of greylisting as a method for protecting against receiving SPAM messages is based on the assumption that SPAM sources are very simple pieces of software optimized for fast automatic operation so that most of them is not able to cope with the fact that the receiving server refuses the delivery from the the SPAMming SMTP server as an unknown source for certain period of time. When regular SMTP server is sending the message and it gets such an answer the server is able to repeat the request after certain time. Most SPAM producing SMTP servers do not do so but one can assume that the SPAMming SMTP servers are going to be more sophisticated so that the efficiency of the method could go down. Therefore probably the most important question associated with greylisting is whether the efficiency of the method is stable throughout long-term periods or not.

## II. OBJECTIVES

Based on the fact above the following issues to examine were formulated:

- real value of delay of greylisted messages (it means the delay from the first refused attempt to deliver the message and the attempt that was accepted),
- the amount of SPAM detected and refused (the ratio between the number of refused messages and the total number of messages), and
- long-term tendency of the amount of detected and refused SPAM-suspicious connections.

The implementation of greylisting at the main SMTP server of the University of Ostrava resulted in some questions to arise that should be answered. Therefore the aim of the study was to investigate how the implementation of greylisting could affect the quality of e-mail service for users. Also the efficiency of the method was be studied for the reasons mentioned above.

### *A. Greylisting parameters*

The behavior of the greylisting on the specific SMTP server from the point of view of an ordinary user and the efficiency of the greylisting method depend on the following parameters:

- Length of initial delay – this is the period starting at an unsuccessful attempt to deliver the message that is temporarily refused.
- New triplet lifetime – this parameter determines the maximum period for which the incoming server waits for repeated attempt for delivery from sender of the message.
- Verified triplet lifetime – this parameter determines the period for which the verified triplet is stored in the automatic white list. The period countdown is restarted at every successful delivery of a message from the specific source to the same recipient.

### *B. Parameter values influence*

Each of the parameters listed above influences the results achieved using greylisting. The detailed discussion of the influence is as follows:

**1. Length of initial delay:** As mentioned above, the value of the delay is usually recommended to be 5 minutes or more. Despite the fact that original recommendation were about 1 hour (see [6]), nowadays for practical reasons the minimum recommended value of 5 minutes is not exceeded in most implementations of greylisting unless the longer delay is really necessary because the longer the minimum delay is set the higher likelihood of user complaints due to delayed delivery of some incoming messages. In some cases it is however recommended to set the value to higher values. One of possible reasons could be the fact that greylisting is implemented as a first (most external) layer of multilayer anti-SPAM system. If one of the following layers uses is based on IP address blacklisting (blocking) the longer initial delay increases the likeliness that the IP address of SPAM source passing through the greylisting is at the blacklist at the time of accepted delivery.

**2. New triplet lifetime:** The value of the new triplet lifetime specifies the upper limit of the period in which repeated attempt to deliver the message will be successful. Its value must be obviously significantly longer than the length of initial delay but this is condition is still too general to be a practical guide for setting it. The results described and discussed later showed that the value should not be too low. One of older recommendation was 4 hours (e.g. [6]) but it seems too short period from the point of view of results obtained in this study because non-neglectable portion of messages could stay undelivered if the period is too short. On the other hand the value should not be too long because the amount of memory for storing all new triplets could raise significantly.

**3. Verified triplet lifetime:** The value of the verified triplet lifetime is the maximum period for which the verified lifetime stays in the automatic whitelist. The period depends significantly on the messages delivered to the protected SMTP server. Usually the period slightly longer than 31 days is recommended to avoid greylisting of periodically sent e-mails (the period of sending such e-mails is usually not longer than one month , i.e. max. 31 days, but in certain circumstances the period could be even several days longer (for detail see e.g. [6]).

### *C. Greylisting efficiency evaluation*

When one tries to classify the accuracy of greylisting method or other common measures (see [7]), a problem consisting in the fact that from the incoming SMTP server's point of view there are no incorrectly classified messages should be resolved. This is due to the fact that for any message incoming to the SMTP server that is greylisted (i.e. the message delivery is refused temporarily) there are two possible situations:

- either the sender repeats the delivery after a while (longer than minimum period set for repeated delivery at the incoming SMTP server) and the message is received into the recipient's mailbox, or
- it fails to do so (either the repeated attempt does not occur at all or it occurs too soon or too late).

In case of failure such a message never enters into the addressed SMTP server anyway. If it occurs, the failure should be indicated at the sender's mailbox. The sender could then attempt to resend the message because he learns about the failure. Nevertheless for the purpose of efficiency examination it is useless because such cases of manual resending could not be detected in reliable association with the message as sent originally.

Therefore the decision was made to measure other quantities that probably could characterize the behavior of the greylisted server from the point of view of service for users. From the above it implies that certain delay could result from greylisting in the delivery of legal messages. The length of the delay depends both on the value of minimum acceptable delay as adjusted on the protected SMT server and on the behavior of the sending SMTP server. Our observations showed that there is a great variability in time when sending servers repeat the attempt to deliver the refused messages. Sometimes the time does not exceed several minutes but in some cases even tens of hours delay occurred as shown later.

## III. METHODS

The study was performed on real data from the main SMTP server of the University of Ostrava. Performing similar study with test set of data was also considered as an alternative but the fact that the method is based on continual learning of sources of received messages made it practically impossible. For such experiment sufficiently large sets of sending SMTP servers would be necessary and they could be difficult to perform. The simulation study of this problem is being prepared however.

The use of real data implies certain limitations. The first one could be in the protection of sensitive personal data. The risk of unauthorized use of the collected data from greylisting was eliminated primarily by organizational tools but in fact only data of low sensitivity were the object of the study.

The other limitation consists in the fact that in real server there is no control of computing environment where the messages come from. Therefore significant disturbances in the measured data were observed. They were smoothed later with standard mathematical tools.

The above discussed parameters of the greylisting method were used at the SMTP server of the University of Ostrava as follows:

- length of initial delay – 5 minutes (excluding experiments focused to investigate the influence of the value of this parameter to the behavior of the method),
- new triplet lifetime – 24 or 48 hours (24 hours for short-term measurements, 48 hour for long-term),
- verified triplet lifetime – 35 days.

The basic method of investigation was the long term automatic logging of all SMTP server activities associated with greylisting and subsequent processing of log files. The length of the studied period was 14 months.

The studied server operated the greylisting implemented in postfix SMTP server with postgrey script in version 1.24. The script is written in PERL language as a policy server (i.e. after processing the postgrey script passes the decision whether the message should be accepted or refused to the postfix process).

The resulting log files were loaded into a MySQL database using own PHP scripts. Further processing was performed in a spreadsheet.

Also some measurements were made focused to find the optimum length of initial delay and to the comparison of various method of greylisting as described above.

## IV. RESULTS

### *A. Short-term measurements*

The first part of results is represented by a comparison of various methods (variants) of greylisting. Certain of the most important variants were selected from the list in the chapter I, namely full, sqlgrey and reverse methods.

TABLE I
BASIC COMPARISON OF VARIOUS GREYLISTING MODIFICATIONS

| Method | full | sqlgrey | reverse |
|---|---|---|---|
| **Unique triplets** | 34 099 | 34 207 | 31 948 |
| % of triplets with at least 1 delivered message | 7.21% | 5.85% | 8.09% |
| % of triplets with more than 1 delivered message | 1.44% | 1.91% | 1.23% |
| | | | |
| **No. of connections** | 37 081 | 36 965 | 34 631 |
| - allowed deliveries | 12.20% | 11.60% | 12.65% |
| - attempted in the init. delay | 0.46% | 0.30% | 0.49% |
| - refused connections | 87.33% | 88.10% | 86.86% |
| | | | |
| **Accepted deliveries** | | | |
| - based on static WL | 70.23% | 62.07% | 71.53% |
| - based on AWL | 15.14% | 29.95% | 13.22% |
| - delayed deliveries | 14.63% | 7.98% | 15.25% |
| **Delays (sec.)** | | | |
| - minimum | 300 | 301 | 301 |
| - maximum | 83 215 | 82 451 | 81 823 |
| - median | 1 750 | 1 681 | 1 703 |

TABLE 2
BASIC COMPARISON OF VARIOUS GREYLISTING MODIFICATIONS – CONTD.

| Method | full | sqlgrey | reverse |
|---|---|---|---|
| **Delay distribution (with initial delay subtracted)** | | | |
| + 1 min | 3.02% | 3.22% | 2.99% |
| + 5 min | 6.50% | 9.06% | 11.83% |
| + 15 min | 22.21% | 19.59% | 16.47% |
| + 1 hr | 44.86% | 41.81% | 46.71% |
| + 4 hrs | 12.39% | 21.05% | 11.98% |
| + 12 hrs | 6.65% | 3.80% | 7.49% |
| + 24 hrs | 4.38% | 1.46% | 2.54% |

All these methods were tested for the period of one week at the backup SMTP server with the parameters as specified above. Obtained results are summarized in the Tables 1 and 2.

As it can be seen from the table 1, the conditions for all three tests were comparable. Number of unique triplets and number of connections are comparable despite the fact that in third case (reverse algorithm) there is approx. 5% decrease. This difference can be neglected because only relative efficiency values are listed as results.

As one can see the efficiency of all three algorithms was approx. 87%. This relatively high efficiency confirms older data (e.g. [6]). There is just one significant difference in values in the table making the difference between methods, namely significantly lower value of rate of messages delivered with delay (less than 8% in case of sqlgrey and almost 15% in case of other methods).

The difference is due to the fact that sqlgrey method handles the automatic white list (AWL in the table) in a different way than the full and reverse methods. The results showed that sqlgrey was significantly more efficient while the full and reverse algorithms are roughly comparable (no significant difference was observed). On the other hand it should be noted that the period of data collection was quite short (one week) and that results for individual methods could be influenced by the slight differences in composition of the sets of incoming messages that could not be avoided due to real server test environment. Therefore additional measurements should be done to confirm whether there are some significant differences or not.

Fig. 1. Rate of delivered and delayed messages

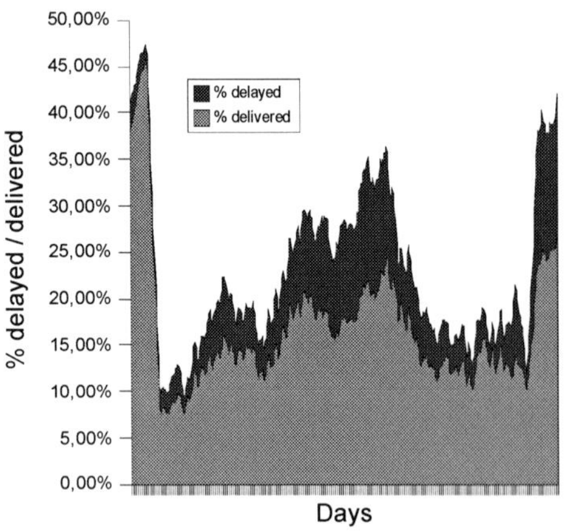


TABLE 3
MONTHLY NUMBERS AND AVERAGES OF CONNECTIONS AND MESSAGESS

| Month | No. of connections | avg. | Delivered messages | avg. percentage | Delayed messages | avg. percentage |
|---|---|---|---|---|---|---|
| 1 | 1 240 833 | 40 027 | 454 768 | 36,7% | 27 208 | 2,2% |
| 2 | 1 093 736 | 39 062 | 91 218 | 8,3% | 25 868 | 2,4% |
| 3 | 771 592 | 24 890 | 98 703 | 12,8% | 33 945 | 4,4% |
| 4 | 729 078 | 24 303 | 104 653 | 14,4% | 34 343 | 4,7% |
| 5 | 771 302 | 24 881 | 103 457 | 13,4% | 35 175 | 4,6% |
| 6 | 515 392 | 17 180 | 97 385 | 18,9% | 41 299 | 8,0% |
| 7 | 631 719 | 20 378 | 107 484 | 17,0% | 60 157 | 9,5% |
| 8 | 626 097 | 20 197 | 130 847 | 20,9% | 74 133 | 11,8% |
| 9 | 568 302 | 18 943 | 100 479 | 17,7% | 40 133 | 7,1% |
| 10 | 904 397 | 29 174 | 116 763 | 12,9% | 38 572 | 4,3% |
| 11 | 921 050 | 30 702 | 109 147 | 11,9% | 29 049 | 3,2% |
| 12 | 893 700 | 28 829 | 113 760 | 12,7% | 34 726 | 3,9% |
| 13 | 946 280 | 30 525 | 167 605 | 17,7% | 81 487 | 8,6% |
| 14 | 571 508 | 28 575 | 142 361 | 24,9% | 80 664 | 14,1% |

## *B. Long-term results*

The most important part of results of our study consists in data from long-term measurement of greylisting parameters.

For these measurements the main SMTP server was used where weak variant of greylisting is used (first 3 bytes of the IP address is taken into account) and combination with fixed white-list is used so that the IP of the SMPT sender is stored into the white-list after 5 successful deliveries.

The results were evaluated from two points of view as mentioned above. First view was the rate of delivered messages from all attempts as well as the ration of delayed messages. The other view was to distribution of the delay in the set of messages delivered with delay.

The ratio of delivered messages demonstrated very high variations on daily basis being between 3 and almost 50%. Similar variation was in messages delivered with delay – the daily average varied among less than 1% and more than 26% of all attempts.

The summary of results is shown in the Fig. 1 where the graph of dependence of rate of delivered and delayed messages throughout the period of measurement illustrates the measured results. From the graph as well as Table 3 where summary results are listed it can be seen that despite significant variations the ratio of delivered messages (upper line in the graph) in the total number of connections is stable at high values. It means that for the period of logging the greylisting kept being highly efficient tool for avoiding SPAM delivery.

The distribution of delay differs significantly on day-to-day basis which is likely due to different sets of messages sent to the SMTP servers in different days, weeks or seasons. Therefore only summary results and the evolution in time scale are shown. The distribution of delays is shown at the Fig. 1.

TABLE 4
SUMMARY DISTRIBUTION OF DELAY FOR DELAYED MESSAGES

| Delay | 5 – 10 min | 10 – 20 min | 20 – 30 min | 30 – 60 min | 1 – 2 h | 2 – 6 h | 6 – 12 h | > 12 h |
|---|---|---|---|---|---|---|---|---|
| No. of messages | 173969 | 364019 | 35929 | 27959 | 11826 | 10954 | 3976 | 7388 |
| % rate | 27% | 57% | 6% | 4% | 2% | 2% | 1% | 1% |

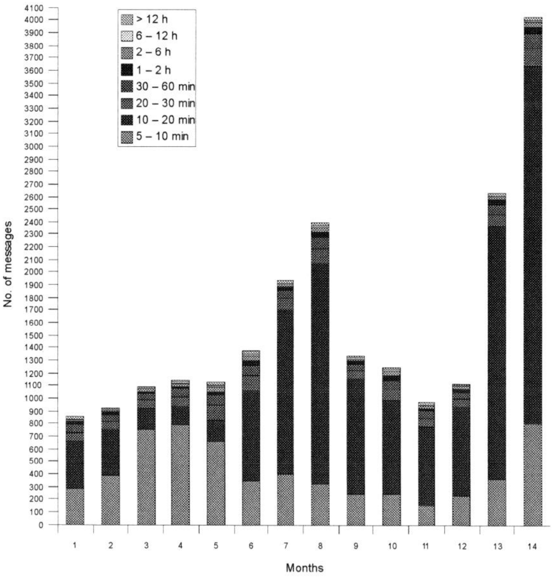


Fig. 2. Delay distribution averaged on monthly basis

As shown in the Fig. 2 as well as illustrated in the summary table in Table 4 more than 97 % of delayed messages were delivered in period up to 6 hours from the first attempt to deliver the message. On the other hand it is easily seen that even after 12 hours after the first attempt there is significant amount of messages still not delivered (from those that were finally delivered). Therefore it cannot be recommended that the new triplet lifetime that determines the end of period for repeated attempt to deliver the message is shorter than 24 hours (like in [6] where even the value of 4 hours is recommended).

## V. CONCLUSIONS

The aim of the study was to show how the greylisting method behaves in long term period. The results can be summarized as follows:

- Despite possible pessimistic expectations greylisting does not demonstrate any measurable loss of efficiency.
- There are some open questions regarding greylisting implementation especially in connection with the optimum values of main greylisting parameters – too aggressive values do not seem to be good choice.
- Due to the fact that real environment measurement are subject to great variations due to the unstable nature of e-mail messages sent it is highly recommended that subsequent work are done in simulated environment.

As mentioned above the obtained results and their interpretation is limited by the fact that in real environment results are subject to variations that depend on the real set of incoming messages. This is the reason why great variation can be observed in the results presented here too. Greylisting method as very efficient front-end (to be combined with other subsequent method for detailed analysis of suspicious messages) anti-SPAM method deserves greater attention from the point of view of its investigation with special focus to optimum parameter adjustment, proper variant selection and long-term efficiency stability.

## REFERENCES

[1] Associated Press, "Study: spam costs businesses $13 billion." January 2003. [online] <http://www.cnn.com /2003/TECH/biztech/01/03/spam.costs.ap/index.html> [2008-10-15]

[2] Chih-Ping Wei, Hsueh-Ching Chen, Tsang-Hsiang Cheng, "Effective spam filtering: A single-class learning and ensemble approach." *Decision Support Systems*, Vol. 45, Issue 3, June 2008, pp. 491-503

[3] J.R. Méndez, D. Glez-Peña, F. Fdez-Riverola, F. Díaz, J.M. Corchado, "Managing irrelevant knowledge in CBR models for unsolicited e-mail classification." *Expert Systems with Applications*, Vol. 36, Issue 2, Part 1, March 2009, pp. 1601-1614

[4] G. Sakkis, I. Androutsopoulos, G. Paliouras, V. Karkaletsis, C.D. Spyropoulos, P. Stamatopoulos, A memory-based Approach to anti-spam filtering for mailing lists. Information Retrieval, 6, 2003, pp. 49-73

[5] J. Carpinter and R. Hunt, "Tightening the net: A review of current and next generation spam filtering tools" *Computers & Security* 25 (2006), pp. 566 – 578

[6] E. Harris, The next step in the spam control war: greylisting. 2003. [online] <http://projects.puremagic.com/ greylisting/whitepaper.html> [2008-10-15]

[7] M. R. Islam, W. Zhou, M. Guo and Y. Xiang, "An innovative analyser for multi-classifier e-mail classification based on grey list analysis." *Journal of Network and Computer Applications*. 2008, In press.

# Using Formal Methods in Component Based Software Development

Sajad Shirali-Shahreza
Sharif University of Technology
shirali@ce.sharif.edu

Mohammad Shirali-Shahreza
Sharif University of Technology
shirali@cs.sharif.edu

***Abstract-*** **Reusing the programs which have already been developed can shorten the production time and reduce costs and expenses. One of the important issues in software reuse is finding a program or a program component in a program library which has been already developed and using it in the new program. In view of precision and automation that formal methods can provide, we can use formal methods in retrieval appropriate components from the software libraries. In this paper, some of the works done in the field of retrieval of the components from the libraries by the help of formal methods have been surveyed and reviewed.**

***Keyword-*** **Component-Based Software Development, Component Retrieval, Formal Methods, Reuse, Software Engineering**

## I. Introduction

Due to expansion of the companies and developing great number of programs, the engineers have found out that there are some components in the programs which are developed several times while have few differences in various programs. Therefore, it seems logical and reasonable to reuse some components of the programs which have been previously developed. This matter has been put forth and named REUSE.

The preliminary methods of reuse emphasized reuse of small components such as functions and then the reuse of abstract data type was put forth.

The next stage in reuse was to use larger components of the programs. In this method the program is broken into some components. Each component which has been previously produced and exists in the library is extracted from the library and used; otherwise the component is produced and not only it is used in the present program but also it is placed in the library for probable usage in the future programs.

Among the advantages of this method, we can refer to the reduction of the period of time required for the development and presentation of the product to the market (Time-To-Market), reduction of production costs and improvement reliability of the components.

But reuse of the components has its own difficulties. The most important one among these difficulties is to find a component in the component library for reuse. The goal of this paper is to survey this problem and how to solve it using formal methods.

In the next section some of the existing problems in the reuse of the components, especially the problems which can be solved by formal methods, will be enlisted and described and some of the solutions presented for handling them will be reviewed. For each problem, we are providing a number of references which are describing the problem in more details and can be used for further study.

In section 3 the methods which have tried to solve the present difficulties in the reuse of the components by formal methods will be reviewed, especially the subject related to finding appropriate component in the library. In the final section a conclusion and a summary of the results is presented.

## II. Software Reuse Problems

One of the most important problems in reuse is the procedure of finding components and using them in new programs. For reusing the produced components in future, at first we must describe a component in such a way that we can search among them and then create a library containing the produced components. Ultimately we should design and implement a method suitable for finding an appropriate component in the library which meets the requirements of new program. In this section, we will describe five problems and provide a number of references which are dealing with the problem in more detail.

### *A. Search among the Components Library*

One of the key problems in software reuse is searching in the library to find desired components. Even if we have a library of software components that can be used in new products, but cannot find a component in the library that meets our requirements, the software library has no value.

In this field we can imitate the human behavior. In other words, we analyze how program-developers find the concerned component manually and use the results in designing automatic methods. Such study is done in [1].

In paper [2] some of the systems presented for management of the components libraries are reviewed and their advantages and disadvantages are also compared.

### *B. Definition of a component*

Another important task in software reuse is defining components. We must define the task that each component does to be able to search among components for finding a component.

T. Sobh (ed.), *Innovations and Advances in Computer Sciences and Engineering*,
DOI 10.1007/978-90-481-3658-2_75, © Springer Science+Business Media B.V. 2010

One of simplest methods is using Natural Language, but due to ambiguity existing in natural language, the search for finding a suitable component turns out to be a sophisticated and complicated process.

In paper [3] a method is presented which is employed for extraction of a specification from natural language description and then search among components. In this method, descriptive logic is used for specifying components and case-based reasoning methods are used for searching.

In paper [4], the idea of using the anthologies is employed to describe and retrieve the components.

The different idea presented in [5] is specifying each component in various levels of abstraction in the library. Then it can retrieve components in the response to queries described in different levels of abstraction.

### C. *Expansion of Libraries*

Another problem in software reuse is managing the size of library and be able to work with large software libraries, especially distributed libraries.

Search for a component in a distributed library containing lots of components will be a difficult task. In [6] an effort is made to solve this problem through presentation of a method named REBOOT.

Another problem concerning the distributed libraries is the matter relating comparison and finding similar components in various libraries. In paper [7] a method is presented for finding similar components in various libraries based on the name or specification of components.

Perhaps some companies want to use each others' libraries in accordance with a mutual agreement. Here the procedure of using the library belonging to the other company will be a sophisticated and intricate problem which is studied in [8].

### D. *Development of Components as Time Passes*

Software components are usually evolved and matured during software production life. As a result, the definition of the components must be updated during time. This is complicated task.

In paper [9] a method based on meta-component is presented which can handles the problems relating development of components and various versions of a component and also the matter concerning the distributiveness of a library.

### E. *Categorization of Components*

Categorizing components in the library can ease the search process for finding components. Usually the categorization of components is carried out manually. But categorization will be very troublesome and problematic in case of large libraries. As a result, automatic methods are required.

A sample of the works done in this field is to employ the neural networks for categorization of components [10]. Another similar work is described in [11]. In [12] a method is proposed using neural networks for finding standards (criteria) for separation of components.

## III. Formal Methods Applications

One of the active persons in the field of formal method usage for software reuse is Dr. David Hemer. In the work he done for using formal methods in retrieval of components from libraries as his PhD project and reported in his PhD dissertation [13], he created two principal systems. The first system handles the duty of changing components and adaptation of it to the needed requirements. The second system has been created for search in the library for finding the components suitable for the query.

In paper [14] a general method is presented for automatic adaptation of components and modules. Adaptation is required to change a component available in the library so that it can be used in new program. In paper [15] a method is proposed for assessing the compliance of the user's query with specification of components which have been created from state-based components. In paper [16], the methods of finding a suitable component for a query among the components created by object-oriented methods are reviewed and then a new method is presented.

In paper [17] an effort is made to use the present methods for compliance such as the first-order logic used for specifications and developed in recent years, for adaptation of a component in the library for usage in new product.

In paper [18] a method has been presented for specification and retrieval of components in the library. This method uses some concepts of formal methods, but in general it is a non-formal method which is using XML for saving the specification of a component. In [19] another method is described for formal stating of components and their retrieval using LSL (Larch Shared Language), which is specifying the preconditions and post conditions of components operation in addition to their operation.

One of the comprehensive works done in the field of formal methods for retrieval and compliance of the component with the respective and concerned use is the SPARTACAS Framework [20]. This system is based on the specification of components and has been used in design of embedded systems and signal processing systems practically. In this framework, the specifications are described using the Rosetta modeling language [21].

As we said earlier, one of the most important usages of formal methods is to use them for specification and finding components in the library.

In paper [22] a method is proposed for specification and retrieval of components from the libraries using Object-Z language.

In paper [23] an effort is made to combine formal methods with current methods in order to enjoy the advantages of the existing methods along with the precision provided by the formal methods.

In many cases there is not a precise reply to a query for search [21] and the goal is to find components which can meet the concerned requirements. In paper [24] the writer suggests to use the fuzzy theory along with formal methods for search in libraries in order that the permission for lack of definiteness can be given and items other than full matches.

An important matter in choosing a component for using it in the program is the degree of matching of component with that part in which the component must be used. In [25] a formal method has been presented for measuring the degree of matching of component with the concerned use of this component. In paper [26], four quantitative measurements have been presented in order to specify how much change each of the components need in order that they can be used in the new program.

In paper [27] a method is presented for reviewing consistency of the components with each other in a program created by changing and combining a number of components. In paper [28], the authors have also explained the procedure of combining this system with component-retrieval systems and new programs creating systems. Moreover, we can use formal methods in embedded systems or even in hardware systems which previously-built components are used and need features such as synchronous operation [29].

Along the creation of formal methods for specification and retrieval of components from the libraries, a problem remains unsolved and that is the specification of components manually. In paper [30] a method is presented to automatically create specification for components which have been already created and there is no specification for them and we cannot create manual specification for them. Methods like this word can enable us to use previously built components which do not have specification in software library for reuse. Besides retrieval stage, we can also use formal methods in reverse engineering like methods proposed in [31] and [32].

## IV. Conclusion

Reusing the previously built components in developing new programs is highly demanded, because it can shorten the product development time and reduces the costs. But software reuse is not an easy task and we are facing a number of problems in software reuse. In this paper, we try to present the main problems in software reuse, their solutions and current research directions. The aim of this paper is to introduce formal method usage to software engineers who are using component based software development and also provide the current active fields of formal methods for persons who want to do research in this field.

In section 2, we define the major problems in software reuse and some references for more study on them. Then we are reviewing different works which are done on formal method usage in component based software development in section 3. Section 3 provides a brief overview of researches done in this field.

In previous sections, we review the advantages of formal methods. On the other hand, they are a number of problems in using formal methods. We will mention some of these problems here, which is showing the area for further research on formal methods.

A principal problem of formal methods is the difficulty of specification a components or requirements in a formal manner for most of engineers. This problem limits the usage of formal methods. Moreover, using formal methods is usually expensive, especially costs required for training employees, which limits their usage to cases which they are more important parameters than costs such as cases in which the safety of the system is very important. For example formal methods are used in medical or aviation systems because the security and safety are very important in these applications.

But it seems that in comparison to the problems facing the non-formal methods, such as uncertainty about the authenticity of the performance of the system or incapability to meet the requirements, the formal methods will be widely used in future.

In view of the achieved advances in the field of automation of process of retrieval of components from the library and their matching with requirements and in a more general viewpoint due to automation of program-development process from its specification, it does not seems that the fully-automatic systems of program development from the specifications will reach the commercial use in few years.

In general, regarding to the works done recent years in the field of formal methods, especially after 1990, perhaps the usage of formal methods will be expanded in various fields of software engineering such as software reuse and component based software development. Moreover, formal methods can help us to solve some of the problems facing the software industry today, such as dissatisfaction of customers for incapability to meet the requirements and high costs of support.

## References


[1] S. Takada , Y. Otsuka, K. Nakakoji, and K. Torii, "Strategies for seeking reusable components in Smalltalk," *Proceedings of the Fifth International Conference on Software Reuse,* Victoria, BC, Canada, June 1998, pp. 66-74.

[2] S. Atkinson, "Cognitive deficiencies in software library design," *Proceedings of the Asia Pacific Software Engineering Conference and International Computer Science Conference (APSEC '97 and ICSC '97),* December 1997, pp. 354-363.

[3] M.R. Girardi and B. Ibrahim, "A software reuse system based on natural language specifications," *Proceedings of Fifth International Conference on Computing and Information (ICCI '93),* Sudbury, Ontario, Canada, May 1993, pp. 507-511.

[4] R. Meling, E.J., Montgomery, P. Sudha Ponnusamy, E.B. Wong, and D. Mehandjiska, "Storing and retrieving software components: a component description manager," *Proceedings of the 2000 Australian Software Engineering Conference,* Canberra, Australia, April 2000, pp. 107-117.

[5] J.L.B. Justo, "A repository to support requirement specifications reuse," *Proceedings of the Information Systems Conference of New Zealand,* Palmerston North, New Zealand, October 1996, pp. 53-62.

[6] G. Sindre, E.A. Karlsson, and T. Staalhane, "A method for software reuse through large component libraries," *Proceedings of Fifth International Conference on Computing and Information (ICCI '93)*, Sudbury, Ontario, Canada, May 1993, pp. 464-468.

[7] A. Michail and D. Notkin, "Assessing software libraries by browsing similar classes, functions and relationships," *Proceedings of the 1999 International Conference on Software Engineering*, Los Angeles, CA USA, May 1999, pp. 463-472.

[8] N.K. Agarwal, D.C.C. Poo., and T.K. Yong, "Component-based Development of MILLS: A Case Study in the development of an Inter-library Loan Software System," *Proceedings of the 13th Asia Pacific Software Engineering Conference (APSEC 2006)*, Bangalore, India, December 2006, pp. 37-44

[9] G.E. Da Silveira and S.L. Meira, "A metacomponent model to support the extensibility and evolvability of networked applications," *Proceedings of 34th International Conference on Technology of Object-Oriented Languages and Systems (TOOLS 34)*, Santa Barbara, CA, USA, July-August 2000, pp. 185-194.

[10] D. Merkl, A.M. Tjoa, and G. Kappel, "Learning the semantic similarity of reusable software components," *Proceedings of Third International Conference on Software Reuse: Advances in Software Reusability*, Rio de Janeiro, Argentina, November 1994, pp. 33-41.

[11] H. Ye and B.W.N. Lo, "Towards a self-structuring software library," *IEE Proceedings-Software*, April 2001, vol. 148, pp. 45-55.

[12] C. Clifton and L. Wen-Syan, "Classifying software components using design characteristics," *Proceedings of the 10th Knowledge-Based Software Engineering Conference*, Boston, MA, USA, November 1995, pp. 139-146.

[13] D. Hemer, *A Unified Approach to Adapting and Retrieving Formally Specified Components for Reuse*, Ph.D. Dissertation, University of Queensland, Australia, March 2000.

[14] D. Hemer, and P. Lindsay, "Specification-based retrieval strategies for module reuse," *Proceedings of the 2001 Australian Software Engineering Conference*, Canberra, Australia, August 2001, pp. 235-243.

[15] D. Hemer, "Specification matching of state-based modular components," *Proceedings of the Tenth Asia-Pacific Software Engineering Conference*, 2003, pp. 446-455.

[16] F. Feiks and D. Hemer, "Specification matching of object-oriented components," *Proceedings of the First International Conference on Software Engineering and Formal Methods*, September 2003, pp. 182-190.

[17] D. Hemer, "Specification-based retrieval strategies for component architectures," *Proceedings of the 2005 Australian Software Engineering Conference*, March-April 2005, pp. 233-242.

[18] U. Praphamontripong and G. Hu, "XML-based software component retrieval with partial and reference matching," *Proceedings of the 2004 IEEE International Conference on Information Reuse and Integration (IRI 2004)*, Nov. 2004, pp. 127-132.

[19] J. Penix and P. Alexander, "Using formal specifications for component retrieval and reuse," *Proceedings of the Thirty-First Hawaii International Conference on System Sciences*, Kohala Coast, HI, USA, Jan. 1998, vol.3, pp. 356-365.

[20] B. Morel, and P. Alexander, "Automating component adaptation for reuse," *Proceedings of the 18th IEEE International Conference on Automated Software Engineering*, October 2003, pp. 142-151.

[21] B. Morel and P. Alexander, "SPARTACAS: automating component reuse and adaptation," *IEEE Transactions on Software Engineering*, September 2004, vol. 30, no. 9, pp. 587-600.

[22] S. Atkinson, "Modelling formal integrated component retrieval," *Proceedings of the Fifth International Conference on Software Reuse*, Victoria, BC, Canada, June 1998, pp. 337-346.

[23] B. Fischer, "Specification-based browsing of software component libraries," Proceedings of the 13th IEEE International Conference on Automated Software Engineering, Honolulu, HI, USA, October 1998, pp. 74-83.

[24] S.A. Ehikioya, "A formal model for the reuse of software specifications," *Proceedings of the 1999 IEEE Canadian Conference on Electrical and Computer Engineering*, Edmonton, AB, USA, May 1999, vol.1, pp. 283-288.

[25] W.C. Chu, "The integration and adaptation of reusable components through semantic interface analysis," *Proceedings of the Eighteenth Annual International Computer Software and Applications Conference. (COMPSAC 94)*, Taipei, Taiwan, Nov. 1994, pp. 252-257.

[26] L.L. Jilani, J. Desharnais, M. Frappier, R. Mili, and A. Mili, "Retrieving software components that minimize adaptation effort," *Proceedings of the 12th IEEE International Conference Automated Software Engineering*, Incline Village, NV, USA, November 1997, pp. 255-262.

[27] W.C. Chu, and H. Yang, "A formal method to software integration in reuse," *Proceedings of 20th International Computer Software and Applications Conference (COMPSAC '96)*, August 1996, pp. 343-348.

[28] C.T. Chang, W.C. Chu, C.S. Liu, and H. Yang, "A formal approach to software components classification and retrieval," *Proceedings of the Twenty-First Annual International Computer Software and Applications Conference (COMPSAC '97)*, Washington, DC, USA, August 1997, pp. 264-269.

[29] P.S. Roop, A. Sowmya, and S. Ramesh, "A formal approach to component based development of synchronous programs," *Proceedings of the Asia and South Pacific Design Automation Conference 2001 (ASP-DAC 2001)*, Yokohama, Japan, January-February. 2001, pp. 421-424.

[30] W.C. Chu, and H. Yang, "Component reuse through reverse engineering and semantic interface analysis," *Proceedings of the Nineteenth Annual International Computer Software and Applications Conference (COMPSAC 95)*, Dallas, TX, USA, August 1995, pp. 290-296.

[31] G.C. Gannod and B.H.C. Cheng, "A specification matching based approach to reverse engineering," *Proceedings of the 1999 International Conference on Software Engineering*, Los Angeles, CA USA, May 1999, pp. 389-398.

[32] G.C. Gannod, Y. Chen, and B.H.C. Cheng, "An automated approach for supporting software reuse via reverse engineering," *Proceedings of the 13th IEEE International Conference on Automated Software Engineering*, Honolulu, HI, USA, October 1998, pp. 94-103.

# Costs and Benefits in Knowledge Management in Czech Enterprises

P. Maresova, M. Hedvicakova
Department of Economics
Faculty of Informatics and Management, University of Hradec Kralove
Rokitanskeho 62, 500 03 Hradec Kralove, Czech Republic

***Abstrakt-*** **New approaches in management are based on knowledge. Knowledge becomes the most significant form of capital of enterprises. Companies need to appraise costs and benefits when introducing knowledge management.**

**The aim of this paper is to present one of the possible solution of the given problem in the Czech Republic.**

## I. INTRODUCTION

Nowadays entrepreneurial environment is characteristic by being increasingly competitive, which is also reinforced by the globalization and the related expansion of the free market. *Knowledge* becomes the most significant form of capital of enterprises, traditional resources of capital (money, land, technology) depend on the knowledge capital and this dependency has been increasingly deepened. As the need of knowledge grows, the significance of knowledge management grows as well, since it is the tool to manage knowledge in organizations. *Knowledge management* is the tool that can, to a substantial extent, decide about the near future of Czech enterprises. Organizations that are strongly dependent on management of information and knowledge concerning company processes include especially [1]:

- companies with large research and development (e.g. industrial ICT companies with large engineering (production and engineering construction companies),
- companies that depend on exact procedures, expert knowledge and their documentation (pharmaceutic, chemical and medical companies),
- consultation companies, auditorial and advisory companies,
- software companies and system integrators,
- insurance companies and bank institutions

Knowledge management in the form of purposeful, systematic and system activities is a relatively new concept. The question then is, in what way can it be introduced into the company. This task can be solved by means of using and applying the existing methods of knowledge management introduction. However, we can encounter various problems here. As an example: although these methods exist today, it is still difficult to find a complex methodology, which would more or less cover all possible perspectives of the introduction process [2]. Another problems might be [3]:

- unpreparedness of the technical infrastructure to keep, share and spread knowledge,
- difficulty of appointing the person responsible for knowledge management,
- oversaturation by unnecessary knowledge,
- unwillingness of employees to share knowledge,
- disagreement about benefits of knowledge management among the organization's management.

Last but not least, it is difficult to appraise costs and benefits when introducing knowledge management. It is well known that companies prefer to realize projects, where they can calculate indicators such as investment return and the like.

This article outlines one of the possible proposals of solving the given problem in the Czech Republic.

## II. THE SITUATION OF THE KNOWLEDGE MANAGEMENT IN ENTERPRISES IN THE CZECH REPUBLIC

In 2002, a research of the situation of the knowledge management in Czech enterprises was conducted by the company Per Partes Consulting. The research included 149 respondents across all industries based on the field classification according to the CZECH TOP 100. The questionnaire was focused on finding out, to what extent Czech enterprises are involved in knowledge management and if they want to deal with it in future, or possibly what problems they have to face and in what way they plan to solve them.[13]. The results of the research proved that knowledge management has been coming into the awareness of Czech managers as a tool of managing a substantial part of intellectual capital. As it is further apparent from the answers obtained in the research, a great majority of companies deal with or plan to deal with knowledge management in the near future. Managers of Czech enterprises know that the competitiveness reinforced by the entry of CZR into the European Union compels them more strongly to engage in the use and expansion of intellectual capital. The Czech Republic will become a country „expensive" in the European style and enterprises will not be able to take advantage of cheap labour forces, especially with regard to employees with university education.[13]

Czech enterprises also have been fighting, as it was found out from the results of the pilot research, with a series of problems in the KM area. Certain problems appear in the area of procuring new knowledge.

T. Sobh (ed.), *Innovations and Advances in Computer Sciences and Engineering*,
DOI 10.1007/978-90-481-3658-2_76, 

The most important barrier for application of knowledge management is unwillingness of employees to share knowledge, as it was stated by almost 62% of respondents of the research. More than 40% respondents state that one of the risks is also oversaturation by unnecessary knowledge, that is: they see a parallel of the information explosion in the area of knowledge. What almost 35% of respondents see as a barrier is the disagreement about benefits of knowledge management among members of the company's management. Another significant problem is the difficulty to calculate costs and benefits of the realization of such a „project" as is the introduction of knowledge management. The solution of this problem will enable an easier calculation of costs and a subsequent enumeration of benefits, which could otherwise be - and have remained to be so far - frequently underestimated.

Even researches from the academic field in the Czech Republic have shown this focus on knowledge management. It is attested by many publications from academic workers and also by projects in progress or by past projects. The subject of knowledge management is commonly taught at universities, not only in magister but also in doctoral study programs.

Currently a research is taking place at the University of Hradec Králové in cooperation with the company Per Partes Consulting, s.r.o. and it deals with the very topic: „ Measurement of costs and benefits in knowledge management". The objective of this research is to create a proposal for defining costs and benefits in knowledge management on the basis of an extensive literary exploration of both local and foreign literature, questionnaire survey and personal consultations in companies. Currently, one of the possible methods has been chosen that will be very probably modified with respect to specific needs of Czech enterprises.

## III. Possibilities of Measurement in Knowledge Management

Desire for measurement of knowledge has its roots in the conception of social sciences, in which models for subsequent management of social processes are created, or for their prognostication at least. This way arguments are obtained to implement interventions into spontaneous social processes. In connection with research of management, we talk about two basic categories of measurement indicators, which are hard and soft metrics.

Hard metrics, that is objectively measurable indicators, observe development of company objectives, activities and processes. Their basic characteristics are easy measurability, they are available without additional costs and they can be mostly converted to financial formulations. Soft metrics serve for measurement and evaluation of the level of individual processes or functional areas of the company, namely in the way of an audit, that is by means of expert evaluations, questionnaire surveys or interviews with competent employees. They are drawn up in accord with the purpose of their use, for instance for the purpose of being useful for evaluation of the extent of fulfilment of particular objectives in a given area.

Metrics, which you can most often encounter in the area of knowledge management, are stated in the Table No.1. This list served as a basis for the selection of metrics appropriate for the environment of Czech enterprises. This selection is realized in the framework of the research at the University of Hradec Králové in cooperation with the company Per Partes Consulitng, s.r.o. The research is in the phase of processing information and it is possible that after the planned consultations with companies and after finding out their opinion on the selected proposal of formulation of costs and benefits, the method will be modified.

TABLE I
AN OVERVIEW OF POSSIBLE METHODS OF MEASUREMENT OF KNOWLEDGE MANAGEMENT [13]

| | |
|---|---|
| **Scandia Navigator** | [Truneček, 2004], [Mládková, 2005], [Bontis, 2000], [Dalkir, 2005], [Bennet, 2001], [Kankanhalli, 2004], [Boyett, 2001], [Maier, 2007], [BEI Consulting, 2003] |
| **Value Chain Scorecard** | [Truneček, 2004], [Mládková, 2005] |
| **Total Value Creation** (TVC) | [Truneček, 2004], [Mládková, 2005] |
| **Accounting for Future** (AFF) | [Truneček, 2004], [Mládková, 2005] |
| **Intangible Assets Monitor** (IAM) | [Truneček, 2004], [Mládková, 2005], [Bontis, 2000] [Bennet, 2001], [Kankanhalli, 2004], [Boyett, 2001], [Maier, 2007], [Boughzala, 2004], [BEI Consulting, 2003] |
| **Tobin's Ratio Indicator** | [Truneček, 2004], [Mládková, 2005], [Maier 2007], [Grossman 2005] |
| **VAIP – Value Added Intellectual Potential** | [Truneček, 2004], [Mládková, 2005] |
| **Knowledge Intensity** | [Truneček, 2004], [Mládková, 2005] |
| **Rannier's Subtotal** | [Truneček, 2004], [Mládková, 2005] |
| **Strategic Stakeholder system of PM** according to Atkinson and Waterhouse | [Šiska, 2005] |
| **Measuring based on TQM** | [Šiska, 2005] |
| **Strategic scorecard according to CIMA** – support for | [Šiska ,2005] |

| supervisory bodies of enterprises | |
|---|---|
| **IC index** | [Bontis, 2000], [Bennet, 2001], [Kankanhalli, 2004], [Maier, 2007], [Boughzala, 2004], [BEI Consulting, 2003], [Grossman, 2005] |
| **Technology brooker** | [Bontis 2000] |
| **Balanced score card** (BSC) | [Dalkir, 2005], [Bennet, 2001], [Truneček, 2004], [Mládková, 2005], [Kankanhalli, 2004], [Boyett, 2001], [Maier, 2007], [Bergeron, 2003], [Boughzala, 2004], [BEI Consulting, 2003], [Graef, 1997] |
| **The House of Quality Method** | [Dalkir, 2005], [Kankanhalli, 2004] |
| **Benchmarking** | [Dalkir, 2005], [Kankanhalli, 2004], [Kahn, 2004], [Bergeron, 2003], [APAQ, 2003], [Graef, 1997] |
| **ROI** | [Dalkir, 2005],[Boyett, 2001], [Bergeron, 2003], [Tobin, 2004], [BEI Consulting, 2003], [APAQ, 2003], [Graef, 1997] |
| **Time Value** | [Beregeron, 2003] |
| **Incremental Value** | [Bergeron, 2003] |
| **Knowledge management Assessment Tool** (KMAT) | [Boughzala, 2004], [Eirma, 1999], [Grossman, 2005], [Dalkir, 2005] |
| **Knowledge Maturity Model** | [Boughzala, 2004], [Ermine, 1999] |
| **Dynamic Value of Intangible Capital** | [Boughzala, 2004], [Bonfour, 2000] |
| **Cost-Benefit Analysis (CBA)** | [BEI Consulting, 2003], [Phusavat ,2007], **[Diraby, 2005],** [Völkel, 2008], [Sassone, 1988], [Foltýnová, 2007], [SCFM, 2008], [Layard, 1994], [Ochrana, 2001], [Sieber, 2004] |

## IV. A PROPOSAL OF FORMULATION OF COSTS AND BENEFITS OF KNOWLEDGE MANAGEMENT FOR THE CZECH REPUBLIC

What seems to be one of the appropriate methods for Czech conditions is the Cost-Benefit Analysis (CBA). CBA is a type of ratio approach in decision-making processes. In this analysis, all benefits, advantages, the positives of one side of the equation or of an imaginary scale are summed up and all costs, disadvantages and negatives on the other side are summed as well. Outlined impacts of the action are subsequently aggregated, converted to cash flows and included into the calculation of decisive indicators, on the basis of which one can decide, whether the project's final outcome is beneficial for the company. In the case of comparing two or more investments, the calculated indicators make it possible to set their order or to determine the preference of one project over another [14].

In connection with this method, it is necessary to characterize basic terms, which the method works with.

*Effects proceeding from the investment* – all impacts on observed subjects that are brought about by the investment action. They can exist in a financial or non-financial form (or possibly intangible). From the point of view of a certain subject, they can have a positive nature (benefits), negative nature (costs) or neutral nature (they will not influence the subject in any ways).

*Costs*– all negative impact on the observed object(s) or on a group of them. These are negative effects following from the investment.

*Benefits*– all positive impacts on the observed object(s) or on a group of them. These are positive effects following from the investment.

*Beneficient* – any subject or a group of them (including the investor or resp. the applicant), who are influenced by both positive and negative effects following from the investment.

In the framework of this method, certain steps are characterized, which could be used in the Cost-Benefit Analysis. The sequence of individual steps is not entirely rigid, as well as their limitation, nevertheless, these phases of CBA processing are sequenced in a logical order and their random shifting could complicate the way to achieve valid results. The recommended procedure for CBA processing can be summarized into the following 11 steps [14]:

- defining the essence of the project,
- delineating the structure of beneficients,
- describing differences between an investment and zero variant,
- determining and possible quantification of all relevant costs and benefits for all phases of the project,
- singling out complementary „inestimable" costs and benefits and their description in words,
- converting „estimable" costs and benefits into cash flows,
- determination of the discount rate,
- calculation of criteria indicators,
- sensitivity analysis,

- evaluation of the project on the basis of the calculated criteria indicators, inestimable effects and the sensitivity analysis (indicators e.g.: FRR - Financial Rate of Return, ERR - Economic Rate of Return, NPV – net present value, IRR - Internal Rate of Return and Cost-Benefit Ratio,
- decision about acceptability and financing of the investment.

The starting point of the economical analysis are cash flows used in the financial analysis.

When determining indicators of economical performance, it is necessary to apply certain modifications, one of them for example is the modification of prices. Apart from distortion caused by taxes or by external elements, prices can also be deviated from the balance of an market able to compete (that is of an effective market) by other factors: systems of monopolies, business obstacles, regulation of work, incomplete information etc. In all such cases the observed market (financial) prices are misleading, therefore it is necessary to substitute them with accounting (shadow) prices, which reflect costs of entry opportunities and willingness of consumers to pay for outputs. Accounting prices are calculated by using equalizing coefficients for financial prices. If it is hardly realizable to determine shadow prices, models of so-called substitute markets are created, which enable us to infer shadow prices [6]. Other detailed information about procedures of this analysis can be found for example in [5], [6], [7]. In CBA costs and benefits are constructed during the whole life of a project, or respectively of an investment action being realized. The following general rules are valid for acceptance of the project. [5]:

$$\sum_{t=0}^{T} \frac{B_t - C_t}{(1+r)^t} > 0 \qquad (1)$$

Symbols in the equation (1) are:

t ... given time period,

T ... final time horizon, when the project ends its economic viability,

$B_t$ ... benefit in the period t,

$C_t$ ... costs in the period t,

r ... discount rate.

From the above-mentioned relation it follows that the project is economically beneficial, when the discounted value of benefits surpasses the discounted costs. If we regard financial evaluation of costs and benefits, the resulting effect of the formulated project is qualified by this relation:

$$E = \frac{B}{C} \qquad (2)$$

Symbols in the equation (2) are:

E ... resulting effect,

B ... benefit from the project during the whole period of its viability,

C ... costs of implementation of the project and costs during the whole period of the viability of the project.

CBA is very often used for evaluation of public finance projects and public infrastructure. In most cases of public finances, the evaluated project has a character of a public commodity, for the use of which the user does not pay directly and the investor and the future operator do not expect an indirect benefit, for example such as better services, satisfaction or better life conditions of population etc. It is not always simple to evaluate the anticipated benefit in money. The anticipated advantage is often converted to some measurable value, for example by adding up remuneration costs in new job positions, decree of the value added tax, local fees and the decree of future local taxes etc.

For an illustration of an example from the entrepreneurial environment, imagine a company that is considering a purchase of a software. A purchase of a program equipment brings not only direct costs on its acquisition and increase in work productivity, but also many other aspects, which the very Cost-Benefit Analysis can evaluate. The negatives („cista") of the project may for example be the price of the software, costs on consultants, installation and training of users. The positives („benefits") of the project may be an improved company process leading to savings in production costs, ameliorating of the decision-making process and an increase in moral of employees for the reason of a better satisfaction from working with a new thing. It is obvious that the mentioned positives of the project are hardly evaluated in money. A frequent problem of CBA is that the costs are tangible and financially appraisable. On the other hand, positives are often intangible and it is difficult to measure them, namely for instance when they concern intellectual values. In such cases, when it is very difficult to achieve an evaluation of an expert, evaluating scales are often used, which will assign some value both to the negatives and to the positives. In the final result, the summation of the values on the side of the negatives is compared with the summation of the values on the side of the positives [8].

The method of the Cost-Benefit Analysis (CBA) is not an entirely new method, CBA follows from the main current of the economical theory (neoclassic economy), first attempts to introduce it in evaluation of projects appeared already in 1930s in USA [9]. Especially in foreign literature, you can find its application in projects with different focus - not focusing only on the public sector. This area is also knowledge management, or its parts. For example Max Völkel applies the Cost-Benefit Analysis in the area of creation and sharing of knowledge in an organization [11]. Peter G. Sassone used this method to introduce an information system [10], T.E. El-Diraby with colleagues applied knowledge management in the area of infrastructure and subsequently his benefits were evaluated by the method of CBA [15]. The company BEI Consulting has ranked this method among one of the possibilities of evaluation of investment return in projects of knowledge management [16]. K.Phusavat used this analysis for e-learning in the area of knowledge management [12].

## V. Conclusion

High-quality and adequately managed knowledge in an organization is necessary nowadays for its effective operation, they become a significant competitive advantage and they are therefore also an effective tool of the competitive fight. Knowledge have thus become a main source of wealth, at the same time they are a source of quite a big inequality among people.

Knowledge management provides a certain set of instructions about how to effectively work with knowledge. As it follows from the research in past years [17], Czech enterprises are interested in knowledge management, but they encounter many obstacles that finally discourage them from its consistent implementation. One of these obstacles may be problems of evaluation of costs and benefits, which is crucial in deciding about any project. It is a difficult task just in the area of introduction of knowledge management, especially for the reason that many benefits are hardly financially appraisable. This can convince many companies that introduction of knowledge management will not bring a sufficient benefit.

One of the possible proposals of a formulation of costs and benefits in the area of knowledge management in the Czech Republic is the method of Cost-Benefit Analysis. This method is used exactly in the projects, where the anticipated benefit is not always easily appraisable in money. In the Czech Republic it has been applied in projects of public administration so far, but it was many times applied abroad in the very area of knowledge management.

## References


[1] P Hujnak, Plenty of data, leck of information and all most no knowledge, (*Hodně dat, málo informací a skoro žádné znalosti)*,1999, [on-line], Available from http://petr.hujnak.cz.

[2] V. Bures, Knowledge management and process of its implementation (*Znalostní management a proces jeho zavádění – Průvodce pro praxi), Grada Publishing,* 2007, s.216, ISBN 8024719788

[3] P Hujnak, *Knowledge in enterprise*, (Znalosti v akci), System On Line, [on-line], Available from http://www.systemonline.cz/site/trendy/hujnak2.htm, 2000

[4] P. Sieber, *Cost – Benefit Analysis* (Analýza nákladů a přínosů metodická příručka), Ministerstvo pro místní rozvoj, 2004, [on-line], Available from http://www.strukturalni-fondy.cz/uploads/old/ 1083945131cba_1.4.pdf

[5] F. Ochrana, Evaluation of the governances project, (*Hodnocení veřejných zakázek a veřejných projektů)*, ASPI Publishing, s.r.o., Praha, 2001, s. 220, ISBN 80-85963-96-5

[6] European Commission, *Guide to cost-benefit analysis of investment projects,* 2002, [on-line], Available from http://europa.eu.int/comm/ regional_policy/sources/docgener/guides/cost/guide02_en.pdf

[7] Layard, R., Glaister, S. „*Cost – Benefit Analysis"* Cambridge, Cambridge University Press, 1994, ISBN: 0-521-46674-1

[8] SCFM, *Cost-Benefit Analysis,* 2008, [on-line], Available from http://www.finance-management.cz/080vypisPojmu.php?X=Cost Benefit+Analysis+CBA&IdPojPass=57

[9] H. Foltynova, *Cost – Benefit Analysis* (Analýza nákladů a přínosů a možnosti jejího využití pro aplikaci na cyklistickou infrastrukturu, konference národní strategie rozvoje cyklistické dopravy ČR), 2007, [on-line], http://www.cyklostrategie.cz/download/tema3-11.pdf

[10] P. G. Sassane, *Cost benefit analysis of information systems a survey of methodologies*, Association for Computing Machinery, 1988, [on-line], Available from http://personal.stevens.edu/~dgonzal3/mis620/ articles/CostBenefitAnalysis.pdf

[11] M. Völkel, A. Abecker, *Cost-Benefit Analysis For The Design Of Personal Knowledge Management Systems*, International Conference on Enterprise Information Systems, 2008, [on-line], Available from http://mavenrepo.fzi.de/xam.de/2008/2008-06-16-ICEIS-PAPER-costmodel-voelkel.pdf

[12] K. Phusavat, Benefit and cost analysis of e-learning for knowledge management: *the Royal Thai Government, Int. J. Knowledge Management Studies, Vol. 1, Nos. 3/4, 2007*, [on-line] http://learning.ncsa.uiuc.edu/papers/AHRD2002_wentling-park.pdf

[13] P. Maresova: *Costs and Benefits of the implementing knowledge in Czech Republic,* in press.

[14] P. Sieber, *Cost–Benefit Analysis* (Analýza nákladů a přínosů metodická příručka), Ministerstvo pro místní rozvoj, 2004, [on-line], Available from http://www.strukturalni-fondy.cz/uploads/old/ 1083945131cba_1.4.pdf

[15] El-Diraby, B. Abdulhai, and K.C. Pramod. *The application of knowledge management to support the sustainable analysis of urban transportation infrastructure,* University of Toronto, Toronto, ON M5S 1A4, Canada, 2005, [on-line], Available from http://pubs.nrc-cnrc.gc.ca/rp/rppdf/l04-115.pdf

[16] BEI CONSULTING, 1. *Estimating return on investment (ROI). For. Knowledge management (KM) initiatives:.* An Information Technology (IT) Perspective, 2003, [on-line] Available from http://www.bei-consulting.com/papera.pdf

[17] P. Hujnak, *Knowledge in enterprise* (Znalosti v akci – přínosy managementu znalostí pro řízení podniků), Systémová integrace 2003, Praha, 2002

# Ontology-based representation of activity spheres in ubiquitous computing spaces

Lambrini Seremeti, Achilles Kameas

*Hellenic Open University*
*23 Sahtouri Str*
*26222 Patras Greece*
*&*
*DAISy group, Computer Technology Institute*
*N. Kazantzaki Str., University of Patras campus*
*26500 Patras Greece*

*kameas@{eap, cti}.gr*

**Abstract-An Ambient Intelligence (AmI) space embeds sensing, actuating, processing and networking infrastructure in a physical (usually closed) space and offers a set of services in its digital space. In order to model the way everyday activities are carried out within an AmI space, we introduce the notion of "activity sphere". In this paper we show how the use of ontologies has helped us realize an activity sphere and we discuss how ontological components (ontologies and alignments) can be reused to support the realization of overlapping activity spheres.**

## I. INTRODUCTION

Ubiquitous computing environments involve a variety of objects, augmented with sensors, actuators, processors, memories and wireless communications. These augmented objects are called "artifacts" [7] and they tend to overcharge humans with complex interaction. In addition to objects, spaces also undergo a change, towards becoming augmented "Ambient Intelligence" (AmI) spaces [6]. An AmI space embeds sensing, actuating, processing and networking infrastructure in a physical (usually closed) space and offers a set of services in the digital AmI space. Ambient Intelligence pushes forward a vision, where technology is integrated into everyday objects, with the intent of making users' interaction with their surrounding environment simpler and more intuitive.

In order to design flexible and "smart" applications, it is useful to take advantage of the ontologies of these various artifacts available in an AmI space. It is expected that artifact providers will develop different ontologies adapted to their products, or will extend some standard ontologies.

Moreover, since applications evolve in ever changing environments in which artifacts can fail and new ones can appear, there is no way to freeze, once and for all, the ontologies that are relevant and available at a particular moment.

Therefore, in order to properly operate in an AmI space, applications have to be expressed in terms of generic features, that are matched against the actual environment. This matching process can take advantage of ontology matching, since similar artifacts are likely to be used by similar applications.

Thus, reconciling various ontologies and storing the results obtained from previous interactions, should help these applications in sharing and reusing the established alignments, as well as, the predefined ontologies.

## II. AMBIENT INTELLIGENCE SPHERES

In order to model the way everyday activities are carried out within an AmI space, we introduce the notion of "Ambient Intelligence sphere" or "activity sphere". An activity sphere is intentionally created by an actor (human or agent), in order to support a specific activity. The sphere is deployed over an AmI space and uses its resources (augmented objects, networks, services). An activity usually consists of a set of interrelated tasks; the sphere contains models of these tasks and of their interaction. The sphere can also form and use a model of its context of deployment (the AmI space), in the sense that it discovers the services offered by the infrastructure and the contained objects. It instantiates the task models, within the specific context composed by the capabilities and services of the container AmI space and its contained objects. In this way, it supports the realization of concrete tasks in the form of ubiquitous computing applications.

The configuration of a sphere could be realized in two ways, explicit and tacit. In the former mode, people configure spheres by explicitly composing artifact affordances, based on the visualized descriptions of the artifact properties, capabilities and services. To operate in this mode, people must form explicit task models and translate them into augmented object affordances; then they must somehow select or indicate the artifacts that bear these affordances. The independence between an object and a service is maintained, although there do not exist clear guidelines regarding the degree of visibility (of system properties and seams), that a sphere should offer to people.

T. Sobh (ed.), *Innovations and Advances in Computer Sciences and Engineering*,
DOI 10.1007/978-90-481-3658-2_77, © Springer Science+Business Media B.V. 2010

The tacit mode operates completely transparently to the user and is based on the system observing the user's interactions and actions within the sphere. In an ideal AmI space, people will still use the objects in their environment, in order to carry out their tasks. Because objects and spaces are augmented, they can monitor user actions and record, store and process information about them. Then, they can deduce user goals or habits and pro-actively support people's activities within the sphere (i.e. by making the required services available, by optimizing the use of resources, etc). The sphere can learn user preferences and adapt to them, as it can adapt to the configuration of any new AmI space that the user enters. To achieve this, the encoding of task- and context-related metadata is required, as well as the encoding of the adaptation policies, which will be used by the task realization mechanisms. In fact, the role of these metadata is crucial, as the new ubiquitous computing paradigm moves beyond the data-centric object oriented approach, to a more human-centric task based one.

In the following, we shall propose an ontology-based description of an activity sphere, and then the reuse of ontologies and alignments for the realization of another activity sphere which has common components with the previous one.

## III. Realizing An Activity Sphere Using Ontologies

An ontology is usually defined as "a formal, explicit specification of a shared conceptualization" [4]. A "conceptualization" refers to an abstract model of some phenomenon in the world, which identifies the relevant concepts of that phenomenon. "Explicit" means that the type of concepts used and the constraints on their use, are explicitly defined. "Formal" refers to the fact that the ontology should be machine readable. "Shared" reflects the notion that an ontology captures consensual knowledge, that is, it is not private of some individual, but accepted by a group. Thus, an ontology is a structure of knowledge, used as a means of knowledge sharing within a community of heterogeneous entities. It is a set of interconnected entities.

Each activity sphere is a set of heterogeneous artifacts described by independently developed ontologies. In order to achieve efficient communication between the artifacts, a technique for ontology matching is required. This will make the ontologies of the interacting artifacts semantically interoperable.

Ontology matching is the process of finding relationships, or correspondences between entities of different ontologies. Its output, is a set of correspondences between two ontologies, that is, relations holding, or supposed to hold, between entities of different ontologies, according to a particular algorithm, or individual. This set of correspondences, is called an alignment. According to [3], the ontology alignment process is described as: given two ontologies, each describing a set of discrete entities (which can be classes, properties, rules, predicates, or even formulas), find the correspondences, e.g. equivalences or subsumptions, holding between these entities.

These alignments can be further used, in order to result, through the ontology merging process, in the top level sphere ontology which realizes an activity sphere. When performing ontology merging, a new ontology is created, which is the union of the source ontologies. The new ontology, in general, replaces the original source ontologies. The challenge in ontology merging, is to ensure that all correspondences and differences between the ontologies are reflected in the merged ontology.

In this section, a simple real ubiquitous computing application will be described. As is the usual case, in order to carry everyday activities, people will look for services or objects they can use, select the most appropriate ones and combine the respective services into functioning connections, which they can manually adapt, in order to optimize the collective functionality. After describing the application, we describe the ontologies of each artifact and then we apply ontology alignment to develop the top level sphere ontology, through a merging process.

Suppose that we are using two objects, a lamp and a chair, to support the "Meeting Awareness" task. According to this, when the chair is occupied, the lamp is turned on. This simple sphere can be extended by adding more chairs; when more than one chairs are occupied, then the system deduces that a meeting is taking place.

The eChair and eLamp ontologies are shown in Fig. 1. An eLamp is an Object which has PhysicalProperties such as

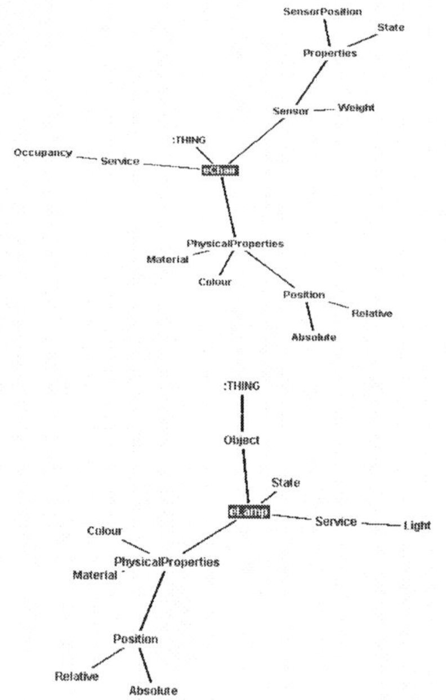


Fig. 1. Two artifact ontologies (representing an eLamp and an eChair)

Colour, Material and its Position, defined, either by using absolute, or relative to another object co-ordinates. It also has a State, which describes if the eLamp is On or Off, while it provides Light as its Service. An eChair is another Object, which similarly has some PhysicalProperties, such as its Material, Colour and Position. It also has a Sensor, which has Properties, such as its State and SensorPosition and it measures Weight. An eChair provides Occupancy as a Service. Similar ontologies are embedded in the other artifacts.

We apply the ontology alignment algorithms that are embedded in the PROMPT Suite [9], a plugin of Protégé (http://protege.stanford.edu), to deduce the higher level sphere ontology through the alignment/mapping (Fig. 2) and the merging process. iPROMPT and AnchorPROMPT, are two of the tools in the suite, that support semi-automatic ontology merging. The first one is an interactive ontology merging tool, that guides the user through the merging process, presenting him suggestions for next steps and identifying inconsistencies and potential problems, while the second tool uses a graph structure of the ontologies, to find any correlation between concepts and to provide additional information for iPROMPT. This ontology can then be used to realize the "Meeting Awareness" in different contexts. Fig. 2 shows the alignments/mappings found between the two ontologies. Whereas alignments merely identify the relations between ontologies, mappings focus on the representation and the execution of the relations for a certain task. Each alignment/mapping provides us with suggestions in order to accept or reject them, according to our application. If we accept all or some of the recommended alignments, according to our application, a new merged ontology will result, that contains the information from both our original ontologies and which describes the sphere ontology (Fig. 3).

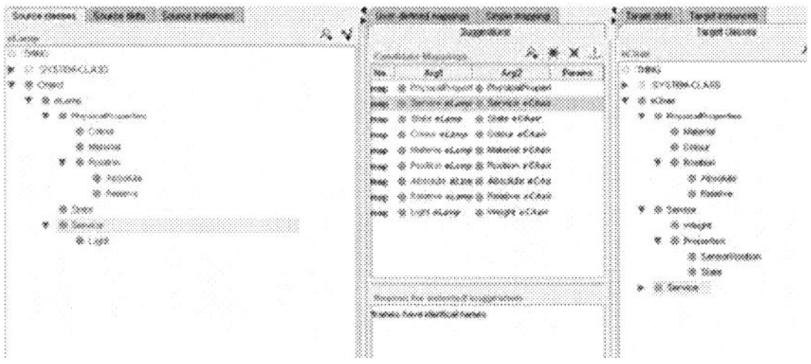

Fig. 2. Alignments/mappings between the two ontologies

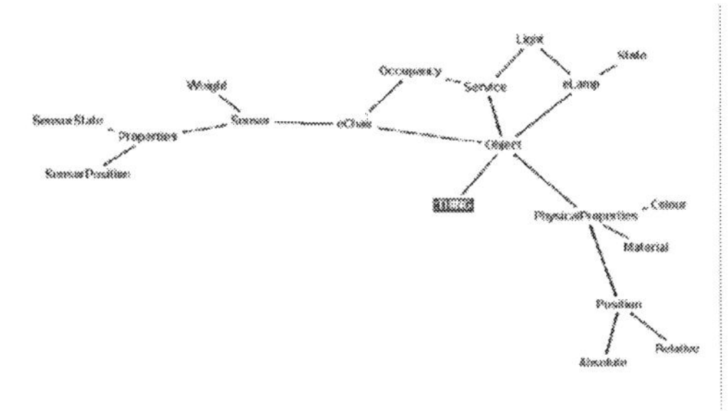


Fig. 3. The new merged ontology that describes the sphere

## IV. Overlapping Activity Spheres

Each activity sphere has a dual hypostasis: one concerning how this sphere conceives itself and the other one concerning how it is conceived by the other activity spheres. In that sense, each activity sphere has an introvert "self" that encompasses the short-term context information and an extrovert "self", in order to communicate with another activity sphere.

Its introvert self is a set of heterogeneous ontologies, which describe the independently developed and autonomous components that participate in this sphere and the connections between these ontologies, that is, the provided alignments. Thus, we consider the introvert self of a sphere, as a "repository" of ontological resources, that is ontologies and alignments.

Its extrovert self is its temporal view, as a whole, that another activity sphere can "see". It is a merged ontology, which contains all the information, provided by the connected/aligned ontologies, that is, all the information provided by the contained ontological components in a sphere. Thus, we can reuse the ontological components that are contained in one sphere, in order to realize another sphere, if the two spheres are referred to the same activity from different points of view, or to overlapping activities.

Many distinct or overlapping activity spheres can coexist in the same AmI space, or similar activity spheres can be realized in different AmI spaces. For this reason, it is more convenient to reuse the explicit information (contained ontologies and alignments) that describes an activity sphere, into another activity sphere.

## V. Realizing Overlapping Activity Spheres By Reusing Ontological Components

In this section, two simple real ubiquitous computing applications will be described. The previously described "Meeting Awareness" activity sphere and the "Studying" activity sphere. In the "Studying" activity sphere, suppose that we are using four objects, a lamp, a chair, a book and a desk, to support this activity sphere. According to this, when the chair is occupied and it is near the desk and the book is open on the desk, the lamp is turned on. These two activity spheres are considered as overlapping, because of the participation of the same or similar artifacts, into them. Thus, we can reuse knowledge from the "Meeting Awareness" activity sphere, in order to build the "Studying" activity sphere. This knowledge is embedded into ontological components, such as artifact ontologies and their alignments.

The top level sphere ontology is shown in Fig. 4. It describes the interconnected entities of an eObject. Here, an eBook, an eDesk, an eLamp and an eChair are eObjects. They have PhysicalProperties, such as ObjectPosition, Material, Colour, NumberOfPages. The ObjectPosition can be characterized as RelativeObjectPosition or Absolute-ObjectPosition. The NumberOfPages is referred only to an

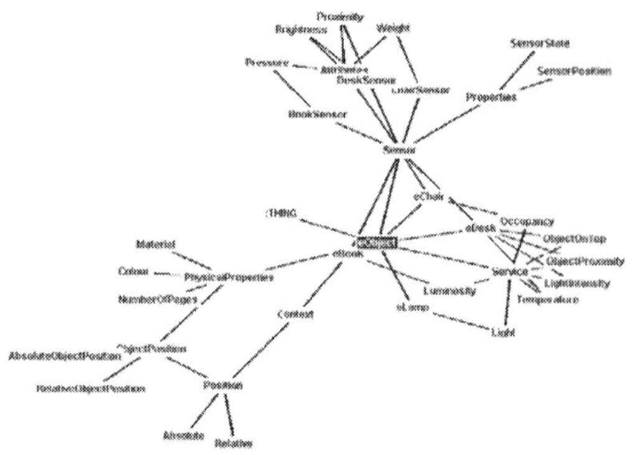


Fig. 4. The top level sphere ontology, which describes the studying activity

eBook, whereas Material, Colour and ObjectPosition classes are referred to the other eObjects. The Context of the eBook is referred to its Position and can be Relative, for instance OnDesk or Absolute and it may be in an Open or Closed State. The eBook, the eDesk and the eChair have sensors, such as ChairSensor, DeskSensor and BookSensor. Each Sensor has Properties, such as SensorState and SensorPosition, and it measures Attributes, such as Pressure, Brightness, Proximity and Weight. Weight is measured by the ChairSensor, Proximity and Brightness are measured by the DeskSensor and Pressure is measured by the BookSensor. Each eObject provides Services, such as Luminosity, Light, LightIntensity, Temperature, ObjectOnTop, ObjectProximity and Occupancy. Occupancy is the Service that is provided by the eChair. Light is the Service that is provided by the eLamp and the other services are provided by the eDesk. This ontology can be used to realize the studying activity in different contexts. It can be the result of pair-wise merging processes between the artifact ontologies that are shown in Fig. 5, after finding the appropriate alignments between these ontologies.

Another less time consuming way for realizing this activity sphere, is to reuse the pre-existing artifact ontologies that describe the eLamp and the eChair objects (Fig. 1) and the alignment found between these two ontologies (Fig. 2), from the previously described activity sphere of the "Meeting Awareness" and connect this explicit information with the information provided by the eBook and eDesk ontologies.

## VI. Related Work

Ontologies are usually used in ubiquitous computing in two ways:

1) As resources of structured information describing knowledge of the ubiquitous computing field. There do not exist proposals of ontologies that describe the whole knowledge of a ubiquitous computing domain, but there are some attempts that conceptualize parts of this domain. These attempts allow developers to define vocabularies for their individual applications. Some prominent examples of these conceptualizations are OWL-S [8], which is an OWL-based ontology for describing services and FOAF (Friend-Of-A-Friend) [1], which describes the personal profile information and social relationship among groups of peers. These ontologies are usually integrated with other ontologies, widely accepted, with the goal of knowledge re-use.

2) As software artifacts used in the development of ubiquitous computing applications, or during application execution. In this case, ontologies are parts of the system software architecture. In these ontology based systems, the software architecture is characterized by the use of one or more ontologies, as central elements of the system. GAIA [10], CoCA [2], CoBrA [5] are some prominent examples of ubiquitous computing systems that use ontology models. All these ontology-based systems have built static heavyweight domain ontologies to represent the primitives of ubiquitous computing. These ontologies are used to represent, manipulate, program and reason with context data and they aim to solve particular ubiquitous computing problems, such as context representation, service modeling and composition, people description, policy management, location modeling, etc. However, in the ubiquitous computing domain, it is difficult for applications to share context information, because this demands to dynamically implement new applications with the characterization of context previously made, because this characterization of context depends on its use.

Our aim was to introduce an ontology-based notion, general enough for being used by different ubiquitous computing applications, specific enough for encompassing only the necessary information for an application and flexible enough for supporting the dynamic nature of ubiquitous computing applications. To this end, we introduced the concept of activity sphere and we employed ontology merging and alignment mechanisms to capture and store, in an ontology repository, information needed by task-based activity spheres, in order to represent their environment, goals, states, events and available resources.

## VII. Conclusion

Each activity sphere has embedded ontological components, such as ontologies and alignments. Ontologies are used for the semantic description of the physical features, the provided services and the embedded sensors of an artifact and alignments describe semantically its context, that is, the way the artifacts that participate in a sphere are connected. Ontologies and ontology alignments are the foundations of this world of activity spheres, in order to deal with semantic interoperability, dynamic nature, context awareness and adaptive services that are some of the issues that are involved in this "world". These ontological components can be reused for realizing other spheres that describe overlapping activities, in the sense that they contain some artifacts that are the same.

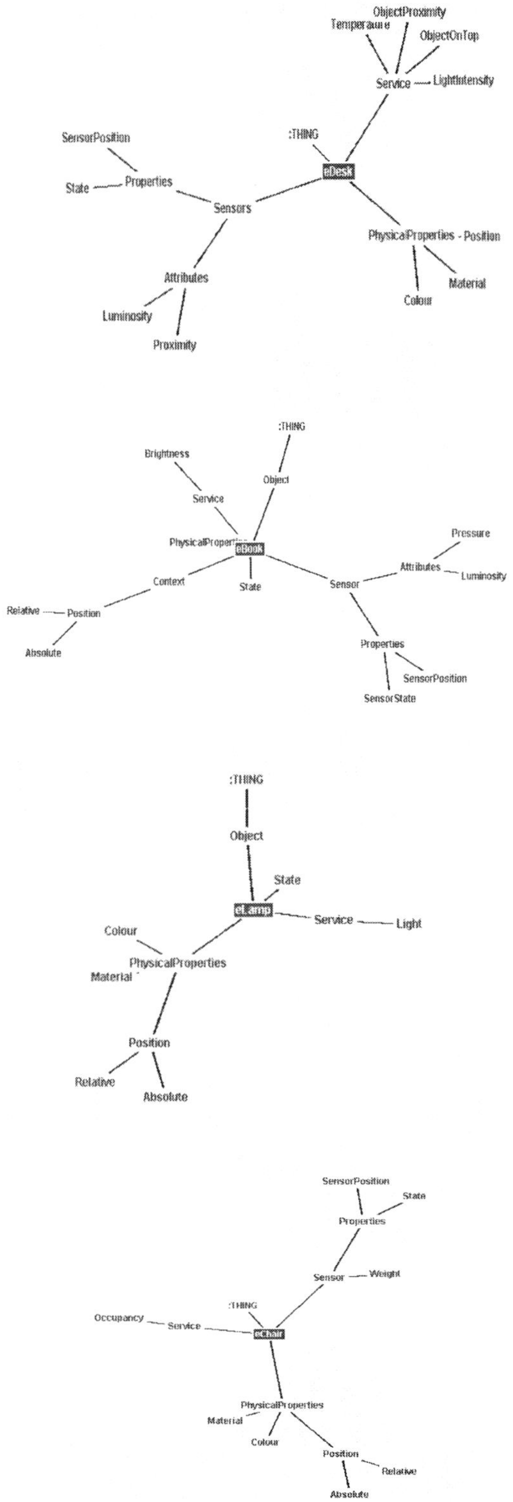


Fig. 5. The artifact ontologies (eChair, eBook, eLamp and eDesk) that can be used for building the merged ontology of Fig. 4

## ACKNOWLEDGEMENTS

Part of the research described in this paper was conducted in the context of ATRACO project (ICT-216837). The authors would like to thank all the partners in ATRACO for their valuable input and support.

## REFERENCES

[1] Dumbill E., "Finding friends with XML and RDF: FOAF", Available at http://www-106.ibm.com/developerworks/xml/library/x-foaf..html, 2002.

[2] Ejigu D., Scuturici M., and Brunie L., "CoCA: A collaborative Context-Aware Service Platform for Pervasive Computing", 4th IEEE International Conference on Information Technology: New Generations, ITNG'07, pp. 297-302, Las Vegas, USA, 2007.

[3] Euzenat J., and Schvaiko P., *Ontology Matching*, Springer-Verlag, Berlin Heidelberg, 2007.

[4] Gruber T. "A translation approach to portable ontologies", Knowledge Acquisition, 5(2), pp. 199-220, 1993.

[5] Kagal L., Korolev V., Chen H., Joshi A. and Finin T., "Centaurus: a framework for intelligent services in a mobile environment", in Proc. 21st International Conference on Distributed Computer Systems, Washington DC: IEEE Computer Society, pp. 195-201, 2001.

[6] Kameas A., Bellis S., Mavrommati I., Delaney K., Colley M., and Pounds-Cornish A., "An Architecture that Treats Everyday Objects as Communicating Tangible Components", in Proc. 1st IEEE International Conference on Pervasive Computing and Communications (PerCom03), Fort Worth, USA, 2003.

[7] Kameas A., Mavrommati I. and Markopoulos P., "Computing in tangible: using artifacts as components of Ambient Intelligent Environments", in Riva G., Vatalaro F., Norman D.A., (1998), The Invisible Computer, MIT Press, 2004.

[8] Martin D., Burstein M., Hobbs J., Lassila O., McDermott D., McIlraith S., Narayanan S., Paolucci M., Parsia B., Payene T., Sirin E., Srinivasan N., and Sycara K., "OWL-S: Semantic Markup for Web Services", W3C Member Submission 22, November 2004.

[9] Noy N. and Musen M., "PROMPT: Algorithm and tool for automated ontology merging and alignment", in Seventeenth National Conference on Artificial Intelligence (AAAI-2000), Austin, TX, 2000.

[10] Roman M., Hess C.K., Cerqueira R., Ranganathan A., Campbell R. H., and Nahrstedt K., "GAIA: A Middleware Infrastructure to Enable Active Spaces", IEEE Pervasive Computing, vol. 1, No. 4, pp. 74-83, 2002.

# Image Decomposition on the basis of an Inverse Pyramid with 3-layer Neural Networks


Valeriy Cherkashyn
Student Member, IEEE,
CARTEL, Université de Sherbrooke
Sherbrooke, QC, J1K 2R1, Canada
valeriy.cherkashyn@usherbrooke.ca

Roumen Kountchev
Dept. of RCVT
Technical University-Sofia
Sofia 1000, Kl. Ohridski 8, Bulgaria
rkountch@tu-sofia.bg

Dong-Chen He
CARTEL
Université de Sherbrooke
Sherbrooke, QC, J1K 2R1, Canada
dong-chen.he@usherbrooke.ca



***Abstract.*** **The contemporary information technologies and Internet impose high requirements on the image compression efficiency. Great number of methods for information redundancy reduction had already been developed, which are based on the image processing in the spatial or spectrum domain. Other methods for image compression use some kinds of neural networks. In spite of their potentialities, the methods from the last group do not offer high compression efficiency.**

**New adaptive method for Image Decomposition on the basis of an Inverse Pyramid with Neural Networks is presented in this paper. The processed image is divided in blocks and then each is compressed in the space of the hidden layers of 3-layer BPNNs, which build the so-called Inverse Difference Pyramid.**

**The results obtained for a group of similar images were also compared with the results obtained by JPEG2000 compression method.**



***Key words:*** *Image Pyramidal Decomposition, Image Representation, Image Compression, Neural Networks.*


## I. INTRODUCTION

The demands towards the efficiency of the methods for image representation in compressed form are getting higher together with their wide application and this is the basis for further elaboration and development.

The classic methods [1-6] could be classified in the following main groups:

- deterministic and statistical orthogonal linear transforms (DFT, DCT, WHT and KLT, SVD, PCA correspondingly);
- discrete wavelet transforms (DWT, Embedded Zerotree Wavelet, Multi Resolution Seamless Image Database –MrSid, Enhanced Compression Wavelets – ECW, etc.;
- transforms based on prediction (LP, Adaptive LP, etc.); vector quantization (VQ, Adaptive VQ, Multistage Predictive VQ, etc.);
- fractal transforms (IFS, Quadtree Partitioned IFS, etc.)
- and decompositions, based on various pyramids, such as GP/LP, RLP, RSP/RDP, IDP, etc.

The analysis of these methods shows that for image compression are usually used:

- deterministic orthogonal transforms;
- linear prediction with fixed coefficients [1,4,6].

For example, in the standard JPEG is used discrete cosine transform (DCT) and linear prediction (LP), and the standard JPEG2000 [3] is based on the discrete wavelet transform (DWT). The transformed image data are compressed with entropy coding [6], implemented as a combination of various methods for lossless coding (RLC, AC, LZW, etc.).

The transforms DCT, DWT and LP are deterministic and the values of the coefficients of their matrices do not depend on the processed image content. For this reason, the compression efficiency is low when the correlation between the image content and the corresponding transform functions is low.

The statistical transforms [1,4] are more efficient than the deterministic ones, but they have higher computational complexity. Similar disadvantage have the fractal methods [2,6], which together with this are not enough efficient when images with unclear texture regions are processed.

The famous pyramidal image decompositions are usually implemented with digital filters with fixed decimation and interpolation coefficients [2,4] or use some kind of transform, such as for example the Multiple DCT [5], i.e. these methods are not well conformed to the image content.

A group of methods for image representation, based on the use of artificial neural networks (NN) [7-14] had recently been developed. Unlike the classic methods, this approach is distinguished by higher compression ratios, because together with the coding, NN training is performed. The results already obtained show that these methods can not successfully compete the still image compression standards, JPEG and JPEG2000 [3].

For example, the Adaptive Vector Quantization (AVQ), based on SOM NN [8, 13], requires the use of code books of too many vectors, needed to ensure high quality of the restored image and this results in lower compression.

In this paper is offered new adaptive method for inverse pyramidal decomposition of digital images with 3-layer BPNNs. The results obtained with the method modeling show significant visual quality enhancement for group of similar static images in comparison with results for the single images from the same group.


T. Sobh (ed.), *Innovations and Advances in Computer Sciences and Engineering*,
DOI 10.1007/978-90-481-3658-2_78,

One of the first encouraging results for a similar method of research has been received in [16].

The paper is arranged as follows: in Section II is described the method for adaptive pyramidal image representation, in Section III is given the algorithm simulation; in Section IV are given some experimental results, and Section V is the Conclusion.

## II. METHOD FOR ADAPTIVE PYRAMIDAL IMAGE REPRESENTATION

### *2.1. Pyramidal Decomposition Selection*

The basic advantage of the pyramidal decomposition in comparison with the other methods for image compression is the ability to perform "progressive" transfer (or storage) of the approximating image obtained for every consecutive decomposition layer. In result, the image could be first restored with high compression ratio and relatively low quality and permits quality improvement on request.

The classic approach for progressive image transfer is based on the Laplasian pyramid (LP) [4] combined with the Gaussian (GP).

In this paper is offered new approach for pyramidal image representation, based on the so-called *Adaptive Inverse Difference Pyramid* (AIDP). Unlike the non-adaptive Inverse Difference Pyramid (IDP) [5] it is built in the non-linear transformed image space using a group of NNs.

The AIDP is calculated starting the calculation from the pyramid top, placed down, and continues iteratively with the next pyramid layers.

Representation thus constructed, has some important advantages when compared to the LP: easier implementation of the progressive image transfer and compression enhancement in the space of the hidden layers of the corresponding NN.

### *2.2. Description of the Inverse Difference Pyramid*

Mathematically the digital image is usually represented as a matrix of size $H \times V$, whose elements $b(x,y)$ correspond to the image pixels; $x$ and $y$ define the pixel position as a matrix row and column and the pixel brightness is $b$.

The halftone image is then defined as:

$$[B(x,y)] = \begin{bmatrix} b(0,0) & b(0,1) & \cdots & b(0,H-1) \\ b(1,0) & b(1,1) & \cdots & b(1,H-1) \\ \vdots & \vdots & & \vdots \\ b(V-1,0) & b(V-1,1) & \cdots & b(V-1,H-1) \end{bmatrix} \quad (1)$$

In order to make the calculation of the pyramidal image decomposition easier, the matrix is divided into $K$ blocks (sub-images) of size $m \times m$ ($m=2^n$) and on each is then built a multi-layer IDP.

The number $p$ of the IDP layers for every block is in the range $0 \leq p \leq n-1$. The case $p = n-1$ corresponds to complete pyramidal decomposition comprising maximum number of layers, for which the image is restored without errors (all decomposition components are used).

The IDP top (layer $p=0$) for a block of size $2^n \times 2^n$ contains coefficients, from which after inverse transform is obtained its worse (coarse) approximation.

The next IDP layer for the same block (the layer $p=1$) is defined from the difference between the block matrix and the approximation, divided into 4 sub-matrices of size $2^{n-1} \times 2^{n-1}$ in advance.

The highest IDP layer (layer $p=n-1$) is based on the information from the pixels in all the $4^{n-1}$ difference sub-matrices of size $2 \times 2$, obtained in result of the $(n-1)$-time division of the initial matrix into sub-matrices.

In correspondence with the described principle, the matrix $[B_{k_0}]$ of one image block could be represented as a decomposition of $(n+1)$ components:

$$[B_{k_0}] = [\tilde{B}_{k_0}] + \sum_{p=1}^{n-1} [\tilde{E}_{k_{p-1}}] + [E_{k_n}] \quad (2)$$

for $k_p = 1,2,..,4^p K$ and $p = 0,1,...,n-1$.

Here $k_p$ is the number of the sub-matrices of size $m_p \times m_p$ ($m_p = 2^{n-p}$) in the IDP layer $p$; the matrices $[\tilde{B}_{k_0}]$ and $[\tilde{E}_{k_{p-1}}]$ are the corresponding approximations of $[B_{k_0}]$ and $[E_{k_{p-1}}]$; $[E_{k_n}]$ is the matrix, which represents the decomposition error in correspondence with Eq. (2), for the case, when only the first $n$ components are used.

The matrix $[E_{k_{p-1}}]$ of the difference sub-block $k_{p-1}$ in the IDP layer $p$ is defined as:

$$[E_{k_{p-1}}] = [E_{k_{p-2}}] - [\tilde{E}_{k_{p-2}}], \quad (3)$$

for $p = 2,3,...,n-1$. In this case $p = 1$:

$$[E_{k_0}] = [B_{k_0}] - [\tilde{B}_{k_0}] \quad (4)$$

The matrix $[E_{k_{p-1}}]$ of the difference sub-block in the layer p is divided into $4^p K$ sub-matrices $[E_{k_p}]$ and for each is then calculated the corresponding approximating matrix $[\tilde{E}_{k_p}]$. The submatrices $[\tilde{E}_{k_p}]$ for $k_p = 1,2,...,4^p K$ define the next decomposition component $(p+1)$, represented by Eq. (2). For this is necessary to calculate the new difference matrix and then to perform the same operations again following the already presented order.

*2.3. Image representation with AIDP-BPNN*

The new method for image representation is based on the IDP decomposition, in which the direct and inverse transforms in all layers are performed using 3-layer neural networks with error back propagation (BPNN) [7].

The general BPNN structure in AIDP was chosen to be a 3-layer one of the kind $m^2 \times n \times m^2$, shown in Fig. 1. The input layer is of $m^2$ elements, which correspond to the input vector components; the hidden layer is of $n$ elements for $n<m^2$, and the output layer is of $m^2$ elements, which correspond to the output vector components.

The input $m^2$-dimensional vector is obtained in result of the transformation of the elements $m_{ij}$ of each image block of size $m \times m$ into one-dimensional massif of length $m^2$ using the "meander" scan, shown in Fig. 2.

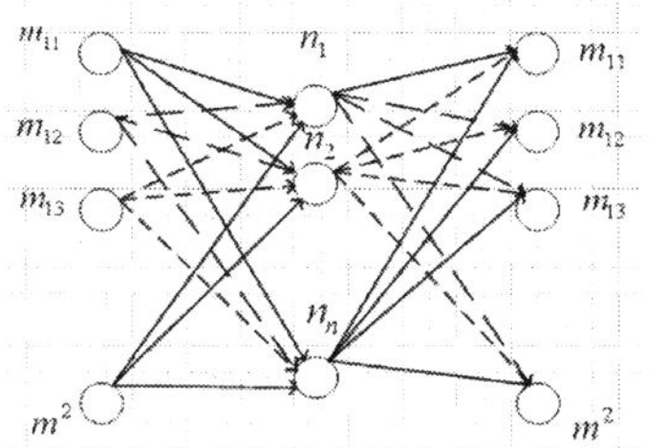

Fig. 1. Three-layer BPNN with $n_n < m^2$ neurons in the hidden layer and $m^2$ neurons in the input and in the output layer

In order to obtain better compression the processed image is represented by the sequence of $m^2$-dimensional vectors $\vec{X}_1, \vec{X}_2, ..., \vec{X}_K$, which are then transformed into the $n$-dimensional vectors $\vec{h}_1, \vec{h}_2, ..., \vec{h}_K$ correspondingly.

The components of the vectors $\vec{h}_k$ for $k=1,2,..K$ represent the neurons in the hidden layer of the trained 3-layer BPNN with $m^2 \times n \times m^2$ structure. In the output NN layer the vector $\vec{h}_k$ is transformed back into the $m^2$-dimensional output vector $\vec{Y}_k$, which approximates the corresponding input vector $\vec{X}_k$.

The approximation error depends on the training algorithm and on the participating BPNN parameters. The training vectors $\vec{X}_1, \vec{X}_2, ..., \vec{X}_K$ at the BPNN input for the AIDP layer $p=0$ correspond to the image blocks.

For the training was chosen the algorithm of Levenberg-Marquardt (LM) [7,8], which ensures rapidity in cases, when high accuracy is not required and as a result is suitable for the presented approach.

One more reason is that the data necessary for the training has significant volume and information redundancy, but this does not make worse the training with the LM algorithm and influences only the time needed (i.e. it becomes longer).

The parameters of the 3-layer BPNN define the relations between the inputs and the neurons in the hidden layer, and between the neurons from the hidden and the output layer. These relations are described using weight matrices and vectors, which contain threshold coefficients, and with functions for non-linear vector transform.

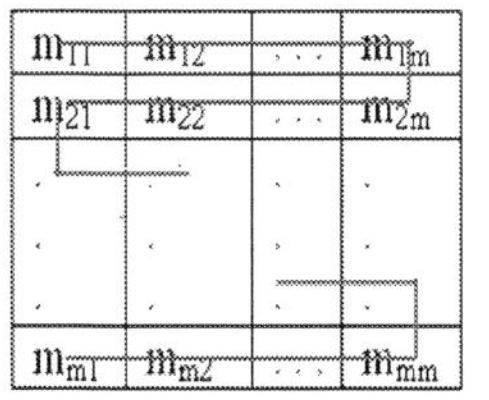

Fig. 2. The "meander" scan applied on mage block of size $m \times m$

The relation between the input $m^2$-dimensional vector $\vec{X}_k$ and the corresponding $n$-dimensional vector $\vec{h}_k$ in the hidden BPNN layer for the AIDP layer $p=0$ is:

$$\vec{h}_k = f([W]_1 \vec{X}_k + \vec{b}_1) \text{ for } k=1,2,..K, \quad (5)$$

where $[W]_1$ is the matrix of the weight coefficients of size $m^2 \times n$, which is used for the linear transform of the input vector $\vec{X}_k$; $\vec{b}_1$ is the $n$-dimensional vector of the threshold coefficients in the hidden layer, and $f(x)$ is a linear activating sigmoid function, defined by the relation:

$$f(x) = 1/(1+e^{-x}). \quad (6)$$

In result the network performance becomes partially non-linear and this influence is stronger when $x$ is outside the range [-1.5, +1.5].

The relation between the n-dimensional vector $\vec{h}_k$ of the hidden layer and the $m^2$-dimensional BPNN vector $\vec{Y}_k$ from the AIDP layer $p=0$, which approximates $\vec{X}_k$, is defined in accordance with Eq. (5) as follows:

$$\vec{Y}_k = f([W]_2 \vec{h}_k + \vec{b}_2) \text{ for } k=1,2,..K, \quad (7)$$

where $[W]_2$ is a matrix of size $n \times m^2$ representing the weight coefficients used for the linear transform in the hidden layer of the vector $\vec{h}_k$, and $\vec{b}_2$ is the $m^2$-dimensional vector of the threshold coefficients for the output layer.

Unlike the pixels in the halftone images, whose brightness is in the range $[0,255]$, the components of the input and output BPNN vectors are normalized in the range $x_i(k), y_i(k) \in [0,1]$ for $i=1,2..,m^2$.

The components of the vector which represents the neurons in the hidden layer $h_j(k) \in [0,1]$ for $j=1,2..,n$ are placed in the same range, because they are defined by the activating function $f(x) \in [0,1]$. The normalization is necessary, because it enhances the BPNN efficiency [8].

The image representation with AIDP-BPNN is performed in two consecutive stages:

1) BPNN training;

2) Coding of the obtained output data.

For the BPNN training in the AIDP layer $p=0$, the vectors $\vec{X}_k$ are used as input and reference ones, with which are compared the corresponding output vectors. The comparison result is used to correct the weight and the threshold coefficients so that to obtain minimum MSE. The training is repeated until the MSE value for the output vectors becomes lower than predefined threshold.

For the training of the 3-layer BPNN in the next ($p>0$) AIDP layers are used the vectors obtained after the dividing of the difference block $[E_{k_{p-1}}]$ (or sub-block) into $4^pK$ sub-blocks and their transformation into corresponding vectors.

The BPNN training for each layer $p>0$ is performed in the way already described for the layer $p=0$.

In the second stage the vectors in the hidden BPNN layers for all AIDP layers are coded losslessly with entropy coding. The coding is based on two methods:

- Run-Length Coding (RLC) and
- Variable length Huffman coding [6].

The block diagram of the pyramid decomposition for one block of size $m \times m$ with 3-layer BPNN for the layers $p=0,1,2$ and entropy coding/decoding is shown in Fig. 3.

When the BPNN training is finished, for each layer $p$ are defined the corresponding output weight matrix $[W]_p$ and the threshold vector $[b]_p$.

The entropy coder (EC) compresses the data transferred to the decoder for the layer $p$, i.e.:

- The vector of the threshold coefficients for the neurons in the output NN layer (common for all blocks in the layer $p$);
- The matrix of the weight coefficients of the relations between the neurons in the hidden layer towards the output BPNN layer (common for all blocks in the layer $p$);
- The vector of the neurons in the hidden BPNN layer, personal for each block in the layer $p$.

In the decoder is performed the entropy decoding (ED) of the compressed data. After that the BPNN in the layer $p$ is initialized setting the values of the threshold coefficients for the neurons in the output layer and of the weight coefficients for the neurons, connecting the hidden and the output layers.

At the end of the decoding the vector of the neurons in the hidden BPNN layer for each block is transformed into corresponding output vector. The obtained output vectors are used for the restoration of the processed image.

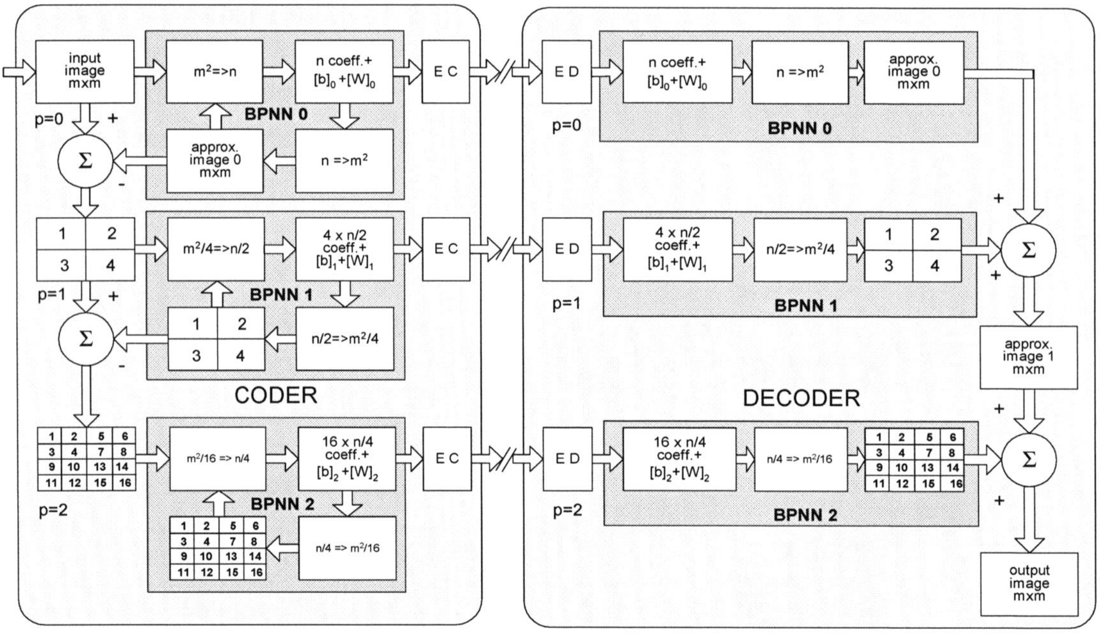


Fig. 3. Block diagram of the 3-layer inverse pyramidal image decomposition with 3-layer BPNN. Here $[b]_p$ – the vector of the threshold coefficients in the output layer for $p=0,1,2$; $[W]_p$ – the matrix of the weight coefficients between the hidden and the output BPNN layer for $p=0,1,2$.

## III. SIMULATION OF THE AIDP-BPNN ALGORITHM

For the simulation of the AIDP-BPNN algorithm is necessary to perform the following operations:

- Transformation of the input data into a sequence of vectors;
- Selection of the BPNN structure;
- BPNN creation and initialization of its parameters;
- BPNN training using the input vectors so that to obtain the required output vectors;
- Testing of the AIDP-BPNN algorithm with various test images and evaluation of their quality after restoration (objective and subjective).

The AIDP-BPNN algorithm consists of the following four basic steps [15]:

1. preliminary preparation of entrance data;
2. neural network training;
3. coding;
4. decoding.

For example, the compressed data size for the same layer (*p=0*) of the test image *Grayscale_foret010032.bmp* is 1510 B (the result is given in Table 1). Taking into account the size of the original image, is calculated the compression ratio *CR*=52.21.

The Peak Signal to Noise Ratio for the first test image *Grayscale_foret010032.bmp* for *p=0* (Table 1) is *PSNR*=23.45 dB. In the same table are given the compression ratios obtained with AIDP-BPNN for other 18 test images of same size (224×352). It is easy to see that for the mean compression ratio *CR* =52.13 is obtained *PSNR*>22.52 dB, i.e. the visual quality of the restored test images is suitable for various applications. Better image quality is obtained when the next pyramid layers are added.

In Table 2 are given the results for the group of 18 test images "forest" after AIDP-BPNN compression, implemented with MATLAB. The results obtained (Table 2) show that AIDP-BPNN method performance for group of similar test images surpasses that for fixed images (Table 1).

Fig. 4. Original test images "forest".

Fig. 5. The restored test image "forest" by AIDP-BPNN metohod.

## IV. EXPERIMENTAL RESULTS

The experiments with the AIDP-BPNN algorithm were performed with test images of size 224×352, 8 *bpp* (i.e. 78 848B), and original images are presented in fig.4.

In the AIDP layer p=0 the image is divided into K blocks of size 8×8 pixels, (K=1232). At the BPNN input for the layer *p=0* is passed the training matrix of the input vectors of size 64×1232=78 848. In the hidden BPNN layer the size of each input vector is reduced from 64 to 8.

The restoration of the output vector in the decoder is performed using these 8 components, together with the vector of the threshold values and the matrix of the weight coefficients in the BPNN output layer. For the layer *p=0* the size of the data obtained is 83 456 B, i.e. - larger than that of the original image (78 848 B). As it was already pointed out, the data has high correlation and is efficiently compressed with entropy coding.

Table 1. Results obtained for 18 test images "forest" after AIDP-BPNN compression (*Cr*), 78 848 Bytes for each original image

| File name | *CR* | *PSNR [dB]* | *RMSE* | Bit-rate/ pixel (*bpp*) | Compressed file (Byte) |
|---|---|---|---|---|---|
| *1* | 52.21 | 23.45 | 17.36 | 0.1532 | 1510 |
| *2* | 52.42 | 23.05 | 17.94 | 0.1526 | 1504 |
| *3* | 51.53 | 19.10 | 28.27 | 0.1552 | 1530 |
| *4* | 50.54 | 17.14 | 35.45 | 0.1583 | 1560 |
| *5* | 52.35 | 22.71 | 18.68 | 0.1528 | 1506 |
| *6* | 52.35 | 19.72 | 26.32 | 0.1528 | 1506 |
| *7* | 53.06 | 22.28 | 19.61 | 0.1508 | 1486 |
| *8* | 52.49 | 23.40 | 17.25 | 0.1524 | 1502 |
| *9* | 52.85 | 31.63 | 06.69 | 0.1514 | 1492 |
| *10* | 51.80 | 21.55 | 21.33 | 0.1544 | 1522 |
| *11* | 52.21 | 21.92 | 20.45 | 0.1532 | 1510 |
| *12* | 52.21 | 22.28 | 19.62 | 0.1532 | 1510 |
| *13* | 52.35 | 19.87 | 25.88 | 0.1528 | 1506 |
| *14* | 51.73 | 19.50 | 27.00 | 0.1546 | 1524 |
| *15* | 52.43 | 22.31 | 19.55 | 0.1526 | 1504 |
| *16* | 52.08 | 23.67 | 16.71 | 0.1536 | 1514 |
| *17* | 52.49 | 28.07 | 10.07 | 0.1524 | 1502 |
| *18* | 51.20 | 23.63 | 16.80 | 0.1563 | 1540 |
| The average result for all 18 images: | | | | | |
| | 52.13 | 22.52 | 20.28 | 0.1535 | 1513 |

Thus, the results received show higher degree of compression CR=63.21 for a group of images in comparison with average compression ratio CR=52.13 obtained for each image separately.

The NN architecture used for the experiments comprises for the zero level 64 neurons in the input layer, 8 neurons in the hidden layer, and 64 neurons in the output layer. The chosen ratio for the input vectors was correspondingly: 80% for *Training*; 10 % for *Validation* and 10% for *Testing*.

The group of static images presented on fig. 6 creates an image of size 224x352x18, that is 1 419 264 B.

Table 2. Results obtained for a group of 18 test images "forest" after AIDP-BPNN compression (1 419 264 B for this group; Fig.6)

| | *Image Size* | *CR* | *PSNR [dB]* | *RMSE* | Compressed file |
|---|---|---|---|---|---|
| *original* | 6336 x 224 | - | - | - | - |
| *AIDP-BPNN* | 6336 x 224 | 63.21 | 21.35 | 17.36 | 22452 |
| *JPEG2000* | 6336 x 224 | 63.00 | 21.02 | 22.66 | 22528 |

Fig. 6. Comparaison with a) original group of similar images; b) approximate group of similar images by BPNN; c) approximate group of similar images by LuraWave SmartCompress 3.0 – JPEG2000

## V. CONCLUSION

In this paper is presented one new approach for still image adaptive pyramid decomposition based on the AIDP-BPNN algorithm. The algorithm modeling was used to evaluate its performance for the group of 18 test images "forest". The results obtained show that for the tested group of images it ensures higher compression ratio than for the case when they were processed individually.

The AIDP-BPNN method is asymmetric (the coder is more complicated than the decoder) and this determines it mostly in application areas which do not require real time processing i.e. applications, for which the training time is not crucial.

The hardware implementation of the method is beyond the scope of this work. The experiments for the AIDP-BPNN algorithm were performed with sub-blocks of size 8x8 pixels. The computational complexity of the method depends on the training method selected.

The new method offers wide opportunities for application areas in the digital image processing, such as the progressive transfer via Internet, saving and searching in large image databases, remote sciences, etc.

The results obtained in this work confirm the teoretical presumption, that images of high resolution (satellite images, multispectral images) possess high degree of data redundancy. This allows the efficient use of the AIDP-BPNN method to compression of high resolution images in remote sensing applications.

## ACKNOWLEDGMENT

This paper was supported by the CARTEL - Centre d'Application et de Recherche en Teleldetection, Universite de Sherbrooke, Canada, *http://www.usherbrooke.ca/cartel/ and* the National Fund for Scientific Research of Bulgarian Ministry of Education and Science, Contr. VU-I 305.

## REFERENCES

[1] K. Rao, P. Yip, Ed. *The Transform and Data Compression Handbook.* CRC Press LLC, 2001.

[2] M. Barni, Ed., *Document and Image Compression.* CRC Press Taylor and Francis Group, 2006.

[3] T. Acharya, P. Tsai, "JPEG2000 Standard for Image Compression", John Wiley and Sons, 2005.

[4] R. Gonzales, R. Woods, *Digital Image Processing*, Prentice Hall, 2002.

[5] R. Kountchev, V. Haese-Coat, J. Ronsin. "Inverse Pyramidal Decomposition with Multiple DCT", *Signal Processing. Image Communication,* Elsevier 17, pp. 201-218, 2002.

[6] D. Salomon, *Data Compression*, Springer, 2004.

[7] St. Perry, H. Wong, L. Guan. *Adaptive Image Processing: A Computational Intelligence Perspective*, CRC Press LLC, 2002.

[8] Y. H. Hu, J.N. Hwang (Eds.), *Handbook of Neural Network Signal Processing*, CRC Press LLC, 2002.

[9] R. Dony, S. Haykin, *Neural Network Approaches to Image Compression*, Proceedings of the IEEE, vol. 23, No. 2, pp. 289-303, 1995.

[10] A. Namphol, et al., "Image Compression with a Hierarchical Neural Network", *IEEE Trans. on Aerospace and Electronic Systems*, vol. 32, No. 1, pp. 327-337, 1996.

[11] S. Kulkarni, B. Verma, M. Blumenstein, "Image Compression Using a Direct Solution Method Based Neural Network", *10-th Australian Joint Conference on AI, Perth, Australia*, pp. 114-119, 1997.

[12] J. Jiang. "Image Compressing with Neural Networks - A survey", *Signal Processing: Image Communication, Elsevier*, vol. 14, No. 9, pp. 737-760, 1999.

[13] N. Kouda, et al., "Image Compression by Layered Quantum Neural Networks", *Neural Processing Letters*, *16*, pp. 67-80, 2002.

[14] *International Journal on Graphics, Vision and Image Processing* (GVIP), Special Issue on Image Compression, *www.icgst.com*, 2007.

[15] V. Cherkashyn, R. Kountchev, D.-C. He, R. Kountcheva. "Adaptive Image Pyramidal Representation", *IEEE Symposium on Signal Processing and Information Technology, ISSPIT*, Dec. 2008 (In press).

[16] V. Cherkashyn, N. Hikal, R. Kountchev, Y. Biletskiy. "Image compression based on the invers difference pyramid with BPNN", *Proc. of IEEE International Symposium on Industrial Electronics (ISIE'06)*, pp. 645-648, July 2006.

# Testing Grammars For Top-Down Parsers

A.M. Paracha and F. Franek
paracham@mcmaster.ca,franek@mcmaster.ca
Dept. of Computing and Software
McMaster University, Hamilton, Ontario

ABSTRACT
**According to the software engineering perspective, grammars can be viewed as *"Specifications for defining languages or compilers"*. They form the basics of languages and several important kinds of software rely on grammars; e.g. compilers and parsers, debuggers, code processing tools, software modification tools, and software analysis tools. Testing a grammar to make sure that it is correct and defines the language for which it is developed is also very important.**

**We implemented Purdom's algorithm to produce test data automatically for testing the MACS [1] grammar (LL(1) grammar) and the parser. Two different implementations of Purdom's algorithm were carried out in this project, one in Java and the other in C++; test strings and other analysis data automatically generated by these implementations are given in the paper.**

## I. INTRODUCTION

During the past several years, complexity of compilers has grown much and so is the importance of testing them. [10], [12],[13] Compiler essentially is a software tool and hence its testing should fulfill all the software testing criteria. Testing is the process of finding errors in an application by executing it. It is one of the essential and most time consuming phases of software development. Hence a lot of effort is directed to fully automate this process, making it more reliable, repeatable, less time consuming, less boring and less expensive.

A compiler is a computer program that accepts a source program as input and produces either an object code, if the input is correct, or error messages, if it contains errors. Compilers are tested to establish some degree of correctness. The basic testing relies on ***Test Cases***. A test case for a compiler should have [2]:

1. A test case description.
2. A source program for which the output of the compiler under observation is verified.
3. An excepted output.

In this project, we were testing MACS compiler whose top-down parser is based on an LL (1) grammar, so our test data are program fragments correct with respect to the MACS grammar used. Test cases should cover all possible valid and invalid input conditions. One of the major problems in generating test cases is the completeness of coverage and the potentially unfeasible size of the test data.

When generating test data for compilers they should cover all the syntax and semantic rules of the language and generate errors in all possible contexts. If upon executing a test case, the output matches the excepted one (including the error messages generated), then the compiler passed the test. On the other hand, if the generated output and/or errors if applicable do not match, the compiler has errors and should be corrected. [14]

The rest of the paper is organized as follows. Section 2 gives details about grammars and how to test them. Section 3 presents issues regarding parsers and their testing, and section 4 presents Purdom's algorithm in detail. In section 5 we give details about our implementation and the results generated. Section 6 concludes the paper and gives remarks on future research directions.

## II. TESTING GRAMMARS

Compilation is the process of transformations of the input program written in a source language to a program in the target language. Traditionally (since the advent of Algol 60 programming language), the source language is specified by means of a formal grammar. A grammar is the main input for the test case generation process. There are a wide variety of grammars, however two play an important role in programming language design and the compilation process, namely context free grammars (used to define the control constructs of the language), and regular grammars (used to define lexical terms).

A grammar defines both a language and provides a basis for deriving elements of that language, grammar is considered both as a program and a specification. [3]

A ***Context Free Grammar*** (CFG) is a set of (possibly) recursive rewriting rules (also called productions) used to generate strings of alphabet symbols in various patterns. There are four major components in a grammar **<*N*, *T*, *s*, *P*>**, where ***N*** is a finite set of so-called non-terminal symbols (non-terminals); ***T*** is a set of so-called terminal symbols (terminals or tokens), which does not intersect with ***N***; ***s*** is an

[1] MACS is a name for a programming language used in the forthcoming book of Franek on compilers. MACS is an acronym for McMaster Computer Science

T. Sobh (ed.), *Innovations and Advances in Computer Sciences and Engineering*,
DOI 10.1007/978-90-481-3658-2_79, © Springer Science+Business Media B.V. 2010

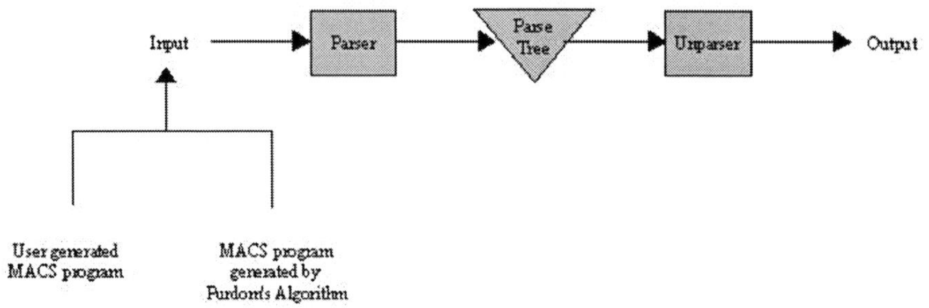


Fig. 3.1 Validation Testing

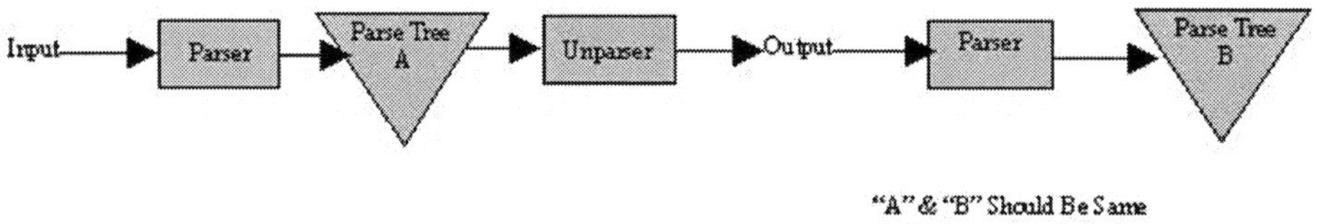


Fig. 3.2 Testing the Parser

initial symbol from *N* (starting non-terminal), and ***P*** is a finite subset of the set ***N*** × **(*N* U *T*)***. Pairs from the set ***P*** are called ***grammar rules*** or ***productions***. A rule (***n***, ***m***) is usually written as ***n*** → ***m***. The set of all sequences derivable from the starting symbol *s* of the grammar G and containing only terminal symbols is called the ***language generated*** by the grammar G and is denoted as L(G). Such sequences are called ***language phrases*** or ***language sentences***. A context-free grammar only defines the syntax of a language; other aspects such as the semantics cannot be covered by it. Every symbol of the grammar defines some finite set of attributes, each production corresponds to a set of rules for evaluating these attributes. [7],[9]

To derive a program (sentence) in a language using a grammar, we begin with the start symbol ***s*** and apply the appropriate productions to non-terminals, as rewriting rules until we are left with a sequence consisting entirely of terminals. This process generates a tree whose root is labeled by the start symbol, whose internal nodes are labeled by the non-terminals and whose leaves are labeled by the terminals. The children of any node in the tree correspond precisely to those symbols on the right-hand-side of the corresponding production rule. Such a tree is known as a parse tree, the process by which it is produced is known as parsing. [7], [8], [9]

Testing a grammar for errors is very difficult. Grammars should be tested to verify that they define the language for which they were written, they should be tested for completeness, that is every non-terminal must have some derivation to be converted into a string of terminals, and that every rule is needed. Detection of errors in a grammar at an early stage is very important as the construction of compiler depends on it.

For compiler design, the grammar used is supposed to be unambiguous. An ambiguous grammar is the one that has more then one parse tree for some sentence. Ambiguity in grammar is to be removed because ambiguous grammars are problematic for parsing and can lead to unintended consequences for compilation, in essence detracting from the intended definition of the language. We have to check the grammar for ambiguity and if needed, to disambiguate it at an early stage.

There are no practical means to check the dynamic semantics of a language defined by a context-free grammar. This problem is addressed by program validation and verification. Here we are strictly concerned with the syntax analysis only.

For generating program test cases for a parser based on context free grammar, the grammar is in fact an ideal tool. The rules of the grammar if combined in a random or regular way can be used to generate sentences (or fragments of sentences) for that language. If a language is intended to be a computer programming language, then the sentences represent programs or their fragments written in the language.

## III. TESTING PARSER

Parsing, or syntax analysis, is one of the activities performed by the front end of a compiler and hence finds use in many software engineering tools such, as automatic generation of documentation, coding tools such as class browsers, metrication tools, and tools for checking code styles. Automatic re-engineering and maintenance tools, as well as tools to support refactoring and reverse-engineering also typically require a parser at the front end.

Parsing is the process whereby a given program is checked against the grammar rules to determine (at least) whether or not it is syntactically correct. [4]

Software testing activities are organized on how the test data is generated. Based on this criterion, there are two basic types of testing:

***Black Box Testing:*** The test cases are developed without considering the internal structure of the software under investigation; only the desired behavior is considered and tested for.

***White Box Testing:*** The test cases are prepared with the full knowledge of the inner structure of the software to be tested. One of the goals of the white box testing is to ensure that every control flow path in the software is exercised and tested. Applying this approach to testing parsers leads to the requirement that every production rule of the grammar is used at least once. [8] ,[14]

Test data for a parser is a program that uses all the production rules of the underlying grammar. Generating such programs manually is difficult and error prone (and seldom complete), so we need to have a method for automatic generation of test data for the parser.

As mentioned above, the grammar used for parsing and for generating the test data must be unambiguous to guarantee reasonable results.

While testing the parser, the input of the parser and the output of the unparser (see Figure 3.1) should both agree. Of course they can differ in indentation and other white space as these are generally ignored by compilers, redundant parentheses in expressions, and many other aspects. In other words, the input and the output must be similar, though not necessarily the same. If the input and the output are compared by humans, these differences can be easily determined and checked. However, for an automated comparison we need a more elaborate setting to determine if the input and the output are similar, see Figure 3.2

If the input and the output are not properly similar, we might have one of the following possibilities:

- The parser is incorrect.
- The unparser is incorrect.
- Both are incorrect.

## IV. PURDOM'S ALGORITHM

Purdom in 1972 [1],[15] proposed a method for testing compiler by automatically generating test programs on the basis of the grammar with the main objective of using each language rule at least once. According to him, a set of sentences using all the language rules has a good choice of exercising most of the compiler code or tables. As all the programming languages are context sensitive, this method only confirms the syntactical aspect and there is no guarantee that these programs will execute correctly.

Also Purdom's algorithm focus on verifying the compiler correctness not interested in checking the efficiency, performance and other aspects. Purdom's algorithm generates sentences that are correct with respect to the context-free grammar of the language, but may be inconsistent with respect of the contextual constraints such as variable declarations and use of identifiers. It only verifies the syntax analyzer of the compiler.

We used Purdom's algorithm to generate test cases for the parser of MACS compiler validation. It produces a set of sentences from the given context-free grammar of MACS, which are then used as test input for the parser.

***How Purdom's Algorithm Work***

Given a set of terminals, a set of non-terminals, a starting symbol, and a set of productions, it generates a shortest program in the language so that every rule has been used at least once. The emphasis is on speed and performance. The productions are used as rewrite rules for generating sequences of grammar symbols. The initial sequence consists of the start symbol. Repeatedly, a non-terminal in the sequence is replaced by the right-hand-side symbols of a rule that has the non-terminal as its left-hand-side. The process terminates when all symbols in the sequence are terminals. Since a MACS program is not a sequence of terminals, but a sequence of lexemes, the sequence of terminals is then translated (in a somehow arbitrary way) to a sequence of lexemes, i.e. a MACS program.

The output of Purdom's algorithm can be used only to check the syntax, but it generates short test suites rapidly and efficiently. One of the goals of Purdom's algorithm is to keep the length of the sentences as short as possible.

## V. IMPLEMENTATION OF PURDOM'S ALGORITHM AND RESULTS

We have used Purdom's algorithm to test the parser for MACS grammar. It is an LL (1) grammar and thus can be used for top-down predictive parsing. Specification of the grammar will be available upon request. The grammar has 77 terminals, 90 non-terminals, and 301 productions.

In our implementations, intermediate data structures needed to generate sentences are:

- Production number used to rewrite the symbol resulting in the shortest terminal string (SLEN).

- Production number used to introduce the symbol into the derivation of shortest string (DLEN)

In the implementations, the lengths of strings are calculated as:
For Terminals: Length=1 (We opted for this compromise, since the actual length of corresponding lexemes cannot often be determined. For instance, for the terminal COMMA we know that the length will always be 1 as there is only one lexeme for it ','. On the other hand, the lexemes for ID can be of any length and so we would not know how to assign the lengths to the terminal ID)

For Non-terminals: Length=No. of terminals+ No. of steps for the derivation

We have implemented the algorithm in both C++ and Java. Both implementations are very closely related to the implementation and reformulation of Purdom's algorithm given by Malloy and Power in [5] and [6]. However, we ran into several problems with their implementation of phase III and so we had to choose our own approach to rectify it. In phase I and II, Purdom's algorithm generates several intermediate arrays, which are used to hold the intermediate results. Our implementation consists of three phases, each producing the following results:

### *A. Phase I (Shortest String Length)*

In the first phase of Purdom's algorithm, we take the set of terminals, non-terminals, language rules and the start symbol "S", and for each symbol it calculates the following pieces of information which will be used later in our sentence generation procedure.
Input:

| | |
|---|---|
| terms.asc | set of terminals |
| nonterms.asc | set of non-terminals |
| grammar.asc | production rules |

- SLEN: An array containing entries for all the symbols in the grammar (terminals & non terminals). At the start it is initialized as:
  Non-terminals: set the value to infinity
  Terminals: set the value to 1 (will remain unchanged.)
  We start rewriting non-terminals using the production rule, which gives us the shortest length. At the end of the phase we get the shortest length of terminal string for each symbol.
  ***Length of string=No. of steps + No. of characters in the string***
- RLEN: An array containing entries for each rule, it gives the length of the shortest terminal string which we get using that rule. Again, the length is the sum of the steps taken in the derivation and the number of terminals in the resulting string.
- SHORT: For each non-terminal we maintain an array containing the production number, which gives us the shortest terminal string.

We can check the grammar by the end of phase I. If any entry of SLEN is infinity or if SHORT contains –1, it is an indication that the grammar is ambiguous; there are some productions that are never used for deriving strings or some non-terminals have no rewritten rules (the grammar is incomplete). It is one of the unique methods to detect these errors in a context free grammar.

### *B. Phase II (Shortest Derivation Length)*

Second phase uses the SLEN and RLEN computed in the previous phase and produces DLAN and PRE, to be used by the final phase.
Input: SLEN and RLEN

- DLEN: For each non-terminal, it gives the length of the shortest terminal string, used in its derivation.
- PREV: Contains the rule number use to introduce a non-terminal in the shortest terminal string derivation.

We calculate these two arrays for all non-terminals except the starting symbol. For the starting symbol the PREV should be –1, as it cannot be introduced by any rule and DLEN should be the same as SLEN.

At the end of this phase, DLEN should not be infinity (which in the programs is represented by the maximum possible integer value MAX_INIT) for any non-terminal and PREV should not be equal to –1 except for the Starting symbol. If this happens the grammar is erroneous.

### *C. Phase III (Generate Sentence)*

In the third phase, the sentences for the given language are generated. First we push the start symbol on the stack and as long as the stack contains some elements, we keep on popping the top-most element and rewrite it using a production rule.
Input: SHORT and PREV

### *C1. Choose A Rule*

The goal is to use each production rule at least once. In these implementations rules are selected on the basis of the values in PREV and SHORT, whenever a rule with a low value of PREV or SHORT is found then we replace the existing one with it giving the minimum length sentences.

For a non-terminal A at the LHS, if a rule **A → α** exists, which has not yet used then we choose it. If more than one rule exists, then we choose the one with the lowest value of PREV and SHORT.

Else if a derivation **A** ==> **α** ==> $\gamma_1 \mathbf{B} \gamma_2$ exists such that B is a non- terminal not on the stack and a rule B → β exists which has not been used, then use A → α production which will then be rewritten using any of the α rules. [11]

For each non-terminal on the stack, we maintain the following arrays:

- ONST: Contains the occurrences of non-terminals on the stack .At the end it should be zero, their should not be any unused symbol on the stack.
- ONCE: For each non-terminal, an array is maintained such that it contains any one of the following values:
    1. READY: The production number previously in ONCE has been used and the next time this non-terminal will be rewritten using a different production.
    2. UNSURE: The value of ONCE calculated in the last loop is not sure.
    3. For some non-terminals it is the production number used to introduce that symbol in shortest string derivation and for some non-terminals this is not true.
    4. FINISHED: The non-terminal is rewritten using all possible productions and can't be used in any other way.
    5. INTEGER: Containing the production number used to rewrite the symbol in some useful derivation.

    In our implementation we have used an integer array for ONCE where the following values are represented as:

    | | |
    |---|---|
    | Ready | -1 |
    | Unsure | -2 |
    | Finished | -3 |

    From 0 to onwards are the rules numbers.
- MARK: For each rule it contains either true or false but at end of this phase all the entries should be true, which shows that each and every rule is used at least once, the main requirement of Purdom's algorithm.
- STACK: The stack should be empty at the end of this phase and all the symbols should be used in making sentences.

| SLEN | Non-terminals | Infinity |
|---|---|---|
| | Terminals | 1 |
| RLEN | Rule Number | Infinity |
| SHORT | Non-terminal | -1 |

Table 5.1 Initialization of Phase-I

| DLEN | Infinity |
|---|---|
| PREV | -1 |

Table 5.2 Initialization of Phase-II

| ONST | Zero |
|---|---|
| ONCE | READY |
| MARK | False |
| STACK | Push all the terminals and non terminals on to the stack |

Table 5.3 Initialization of Phase-II

| Purdom's generated Sentences | MACS Test Cases | Syntactically Correct | Semantically Correct |
|---|---|---|---|
| VOID CLASSNAME_DOT ID_LP CLASSNAME ID COMMA CONST CLASSNAME ID RP SEMICOL | void A.a( A a1, const A b); | X | X |
| VOID CLASSNAME_DOT ID_LP BOOL ID RP SEMICOL | void A.a(bool b); | X | X |
| PUBLIC VOID CLASSNAME_DOT ID COMMA CLASSNAME_DOT ID SEMICOL | public void A.a,A.b; | | |
| CLASS CLASSNAME SEMICOL | class A; | X | X |
| CLASS ID LB RB | class a{} | X | X |
| CLASS ID EXTENDS CLASSNAME SEMICOL | class a extends A; | X | X |

| | | | |
|---|---|---|---|
| CLASSNAME_DOT CLASSNAME_LP RP SEMICOL | A.B( ); | X | X |
| MAIN_LP CONST STRING LS RS ID RP LB RB | main( const string [] a ){ } | X | X |
| PUBLIC SHARED VOID CLASSNAME_DOT ID SEMICOL | public shared void A.a; | X | |
| PUBLIC CONST VOID CLASSNAME_DOT ID SEMICOL | public const void A.a; | X | |
| PRIVATE VOID CLASSNAME_DOT ID SEMICOL | private void A.a; | X | |
| PRIVATE SHARED VOID CLASSNAME_DOT ID SEMICOL | private shared void A.a; | X | |
| PRIVATE CONST VOID CLASSNAME_DOT ID SEMICOL | private const void A.d; | X | |
| SHARED VOID CLASSNAME_DOT ID SEMICOL | shared void A.a; | X | |
| SHARED PUBLIC VOID CLASSNAME_DOT ID SEMICOL | shared public void A.a; | X | |
| SHARED PRIVATE VOID CLASSNAME_DOT ID SEMICOL | shared private void A.a; | X | |
| SHARED CONST VOID CLASSNAME_DOT ID SEMICOL | shared const void A.a; | X | |
| CONST VOID CLASSNAME_DOT ID SEMICOL | const void A.a; | X | |
| CONST PUBLIC VOID CLASSNAME_DOT ID SEMICOL | const public void A.a; | X | |
| CONST PRIVATE VOID CLASSNAME_DOT ID SEMICOL | const private void A.a; | X | |
| CONST SHARED VOID CLASSNAME_DOT ID SEMICOL | const shared void A.a; | X | |
| MAIN_LP LB RB | main({ }) | X | |

Table 5.4 Sentences Generated By the Algorithm

VI. SUMMARY AND FUTURE RESEARCH

Although parser is one of the most important subset in compiler design, not much attention is given to parser design and especially to parser testing. There is a need to do more work in this area.

Purdom's algorithm is one of its kinds in generating test cases for parser; it is a complete method for testing small grammars. The test cases generated can be extended manually to incorporate some semantic aspects of the language for the complete validation of underlying compiler. However, more work is required to cover the language syntax and semantics in a systematic and formal way.

We implemented Purdom's algorithm using two different languages Java and C++ and have achieved the same set of sentences. The sentences generated are mostly syntactically and semantically correct with a few exceptions which are semantically incorrect. We have extended the test cases we get from the algorithm and combined the language semantics to validate both the static and dynamic aspects of the MACS grammar, providing the user with the guarantee that the MACS compiler is error free up to our best knowledge.

Our future research includes testing the most advanced features of MACS compilers to guarantee that it deals properly with more complex semantics features of the language such as data definitions parts of the program.

REFERENCES

[1] P. Purdom, "A Sentence Generator For Testing Parsers", BIT, vol 12:366-375, April 1972.
[2] A.S. Boujarwah and K. Saleh, "Compiler Test case generation methods: a survey and assessment", Information and software technology vol 39(9):617-625, May 1997.
[3] B. A. Malloy and J. F. Power, "Metric–Based Analysis of Context Free Grammars", Proceedings 8th International Workshop on Program Comprehension, IEEE Computer Society: Los Alamitos, CA, 171-178, 2000.
[4] B. A. Malloy and J. T. Waldron, "Applying Software Engineering Techniques to Parser Design: The Development of a C# Parser", ACM International Conference Proceeding Series; Vol. 30:75-8, 2002.
[5] B. A. Malloy and J. F. Power, "An interpretation of Purdom's algorithm for automatic generation of test cases", In 1st Annual International Conference on Computer and Information Science, Orlando, Florida, USA, October 3-5 2001.
[6] B. A. Malloy and J. F. Power, "A Top-down Presentation of Purdom's Sentence Generation Algorithm", National University of Ireland, Maynooth, 2005.
[7] A. V. Aho, R. Sethi, and J. D. Ullman, "Compilers: Principles, Techniques and Tools", Addison-Wesley, 1986.
[8] J. Riehl, "Grammar Based Unit Testing for Parsers", Master's Thesis, University of Chicago, Dept. of Computer Science.
[9] S. Aamir, "Verification and Validation Aspects of Compiler Construction", Master's thesis, McMaster University, Dept. of Computing and Software, April 2007.
[10] A. Boujarwah and K. Saleh, "Compiler test suite: evaluation and use in an automated environment", Information and Software Technology vol 36 (10): 607-614, 1994.
[11] A. Celentano, S. Regizzi, P.D. Vigna, and C. Ghezzi, "Compiler testing using a sentence generator", Software- Practice and Experience, vol 10:897-913, June 1980.
[12] K.V. Hanford, "Automatic generation of test cases", IBM System Journal. 242-258,1970.
[13] C.J. Burgess, and M. Saidi, "The Automatic Generation of Test Cases for Optimizing Fortran Compilers", Information Software Technology, vol 38:111-119., 1996
[14] J. B. Goodenough, "The Ada Compiler Validation Capacity", 1981.
[15] F. Bazzichi and I. Spadafora, "An automatic generator for compiler testing", IEEE Transactions on Software Engineering, SE-8(4):343-353, July 1982

# Accessing Web Based Multimedia Contents for the Visually Challenged: Combined Tree Structure and XML Metadata

Victoria Christy Sathya Rajasekar, Young Lee, Barbara Schreur
Department of Electrical Engineering and Computer Science,
Texas A&M University Kingsville, USA
Email: {ksvr012, young.lee, barbara.schreur}@tamuk.edu

***Abstract*-We present an Integrated Multimedia Interface for the VIsually Challenged (IMIVIC), a multimedia data visualization technique for a web browser that combines a tree structure and XML metadata. We discuss two different approaches to provide an efficient multimedia interface for the blind people, which turned out to be important in the construction of IMIVIC. The first approach is the implementation of web contents with a tree structure, which integrates many different web pages into a single page. The second approach is the creation of external metadata to describe dynamically changing multimedia contents. We demonstrate our approach in an on-line training website and show how this combined approach helps visually challenged users access multimedia data on the web, with appropriate test.**

## I. Introduction

Technology has reached a peak to a "No web, No world" situation. Now-a-days websites have become very attractive with rich multimedia content that changes dynamically without user interaction. Are these websites accessible by people with visual disabilities? The answer is unfortunately, No. The major difficulty in designing websites for them stems from, the nature of textual and pictorial information that has to be presented.

One of the major problems for the visually challenged, in accessing the web page is Navigation. Blind users are facing difficulties while moving from one page to another page, due to web pages that are arranged linearly in a website. Considerable advancements have been made so far to enable navigation in a simple and user-friendly way. We have implemented a tree structure in the website to avoid this problem.

Another problem is to access the web based dynamically changing multimedia contents. Generally, screen readers like JAWS, Webanywhere, and IBM's AI Browser, are used by the blind people to browse the internet. The screen reader finds it difficult to recognize rapid changes in the multimedia contents and thus the blind and visually impaired will not receive all available information [1].

Two proposed techniques have been integrated in this research to overcome these problems and to provide an efficient interface for the visually challenged. The first technique is implementing a tree structure and the second is creating external metadata for dynamically changing multimedia contents.

In tree structure, there is a single integrated web page in which the structure of the entire web pages in a website is displayed in a hierarchy [2]. Since the hierarchy of the website is already known, the user can directly open a web page and move to different levels. In the second approach, a metadata scheme for multimedia content description is provided. The multimedia contents include flash video, swish file, or any image/movie file which changes dynamically in a web page [1].

This paper is organized as follows: in section II, past researches on creating accessible websites are given; while in section III the two proposed techniques are described. The tree structure is explained with a sample in subsection A, and the metadata description using XML is given in subsection C. The evaluation of the IMIVIC is given in section IV with appropriate screen shots. Various tests held on this web application are shown in section V, with different tasks and its results and discussions. The conclusion and future works are given in section VI followed by the references.

T. Sobh (ed.), *Innovations and Advances in Computer Sciences and Engineering*,
DOI 10.1007/978-90-481-3658-2_80, 

## II. Related Works

Some of the advancements for the visually disabled people are Screen readers, Braille, Note takers, Optical Character Recognition, etc. The screen reader produces voice output for text displayed on the computer screen as well as keystrokes entered on the keyboard. Thus the text given in the website can be converted into speech and the user can access it. Certain image descriptions are given as alternate text so that the screen reader converts it into speech [7]. In case of multimedia contents, the screen reader instructs the user to navigate to the audio/video content. But once the audio starts playing it becomes difficult for the users to listen to the screen reader. Research has been done to sidestep such problems by providing required shortcut keys [2].

'Window-Eyes' is screen reader software, which reads out the contents shown in the computer, and helps the blinds use the programs. Some programs like, Eudora, Ease reader, are specially designed for the visually impaired for emailing and for reading e-books. IBM has developed AI browser to allow users to access the Internet.

The major problem discussed earlier is navigating through the web. This problem has been analyzed and required measures are suggested. Some of them are, using JAWS with IE to directly jump from the current position to the next hyperlink element, to the next interactive control, generating alternative views of the webpage, and creating list of hyperlinks in the page. But all these are not systematic and well integrated. So an efficient way of navigation is to provide a single integrated webpage view which can be dynamically and interactively varied based on the page structure [2]. Some websites provide accessibility through 'Text only' option, which eliminates all images and graphics and gives only the links and textual contents of the web page. This gives easy access to the blind users.

## III. Approach

The Integrated Multimedia Interface for the Visually Challenged is an integrated research to provide an efficient interface for the visually disabled. The two approaches integrated in this research, are:

(i) Implementation of tree structure in a website to enable simple navigation;

(ii) Creation of external metadata to describe the dynamically changing multimedia content.

These two approaches are discussed in detail in following sections.

### *A. Tree Structure*

A website can be completely accessed by a visually disabled user only if the necessary contents of the pages are easy to navigate. The user should be able to browse the websites without any difficulties. There is no benefit in designing a website that is not easily accessible by the user.

Websites contain several web pages in a linearized form. Each and every link in a web page leads to a completely new web page opening in the browser, which makes navigation complicated. The user is unaware of the structure of the page and finds it difficult to move around the website. This is the major problem for the blind users, which can be overcome by using tree structure in web pages.

A tree structure in a website integrates all linear web pages into a single page. This has been implemented in this research by using Microsoft .net framework with Java script. Physically there is only one web page that is a single integrated page, consisting of two frames [11]. One contains the tree structure of the web pages and the other frame displays the web page that is being selected from the tree. Since the hierarchy of the website is known before, the user can directly open a web page and move to different levels. The users can create their own view of the web page by expanding or collapsing the nodes of the tree structure. Expanding nodes result in displaying the page in frame 2 and collapsing the nodes hides the pages.

### *B. Sample Tree Structure*

The single integrated page consists of two frames, where the first frame has the actual tree structure and the second frame displays the pages as per the nodes selected in the tree structure. If a node is selected in the first frame, that particular page will open in the second frame, as shown in Fig. 1.

In this structure, a sample IMIVIC Site is shown. The 'IMIVIC SITE' is the parent node. It has two child nodes that are further subdivided. The 'Entertainment' node is divided into nodes Shopping.pdf and Downloadable Games.zip. The 'Online Training' node consists of Outlook videos, Sample Tests and a Help File. The IMIVIC logo.jpeg, and Acknowledgement.doc are under the parent node.

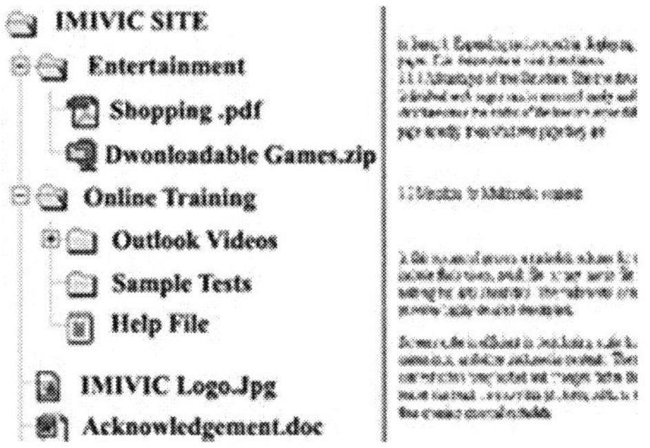


Fig. 1. Sample Tree Structure [11]

The nodes are expandable and collapsible as shown here. The node 'Entertainment' is expanded completely, but the node 'Online Training' is not. It has other pages into the 'Outlook Videos', which are collapsed. Thus the users can show or hide the pages and create their own view, and can directly access the page which is an efficient navigation.

The tree structure is used in websites due to the following advantages. (i) Individual web pages can be accessed easily and quickly, (ii) Users can create their own view of the tree structure since the nodes of the tree are expandable and collapsible, (iii) Users can return to the home page directly from whatever page they are.

### *C. Metadata for Multimedia contents*

In this research a metadata scheme for multimedia content description is provided. The multimedia contents include flash video, swish file, and any image/movie file that changes dynamically in a web page. Metadata is data about data. The multimedia content description files which are the external metadata are created using XML which provides highly detailed description.

Screen reader is efficient in deciphering static html pages but not the dynamic contents like flash animations, and other multimedia contents. These cannot be rendered as speech because the multimedia content is too complicated and changes rapidly in the web page and has no alternate text that the screen reader can read [12]. To solve this problem, XML is used to describe the dynamically changing web content, thus creating external metadata.

### *D. Using XML for Metadata*

XML is the most common metalanguage for web to carry data and is broadly used to create description of dynamically changing multimedia contents. XML allows users to create their own tags which are very flexible and much more customizable. XML tags can be made more descriptive, thus it is possible for the web designer to meaningfully describe the data type in any set of tags. Using XML, both ALT and LONGDESC can be included in an image. Similarly the metadata about the dynamically changing multimedia content points to an alternative description which is elaborate. Thus the metadata includes title, short tile, subject, keywords, description, language, comments, scene description, creation/modification time, creator, etc [10]. More and more information can be stuffed into this metadata using XML. Screen reader recognizes this description and reads out to the users, thus making the multimedia contents accessible.

## IV. EVALUATION OF IMIVIC

The two techniques explained in the previous sections are implemented as IMIVIC Online Training application. This provides online training on Microsoft Office Outlook 2007 for the blind and visually impaired.

### *A. Tree Structure of IMIVIC*

The webpage consists of two frames, one at the left side of the page, contains the tree structure and the other contains the actual web pages integrated together. This second frame displays the webpage as per the link selected in the tree structure. The screenshot of the webpage that illustrates the tree structure is shown in Fig. 2.

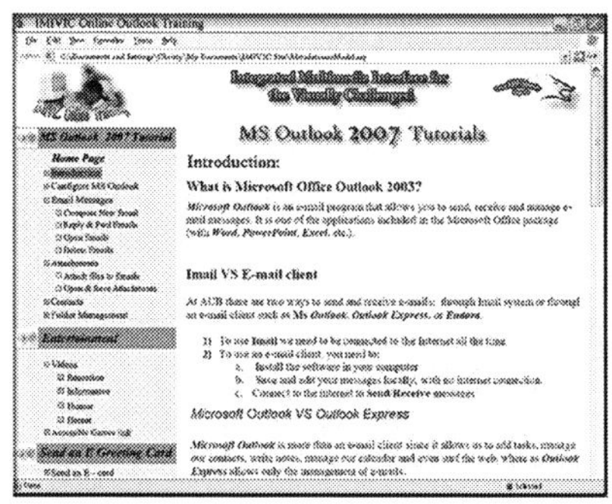

Fig. 2. Screenshot illustrates Tree Structure

In the given tree structure, the parent node is 'MS Outlook 2007 Tutorials' which contains the other child nodes 'Introduction', 'Configure MS Outlook', 'Email Messages', 'Attachments', and so on. The node 'Email Messages' is further divided into 'Compose New Email', 'Reply & Fwd Emails', Open Emails, and 'Delete Emails'. And the node 'Attachments' consists of, 'Attach Files to Emails', and 'Open & Save Attachments' nodes. Likewise, all the nodes are arranged in a hierarchy.

In Fig. 2, the link 'Introduction' is selected by the user and the corresponding webpage is opened in the other frame. Then the screen reader reads the contents of the web page. The 'TAB' key is used to move from one link to another, and the 'H' key is used to jump between headings. The homepage link can be used by simply pressing 'CTRL+H' keys.

Apart from outlook training, the web application also contains some links for entertainment. The Entertainment section contains different video collections categorized by the type of video. The screenshot above clearly shows the structure of node 'Entertainment' which contains 'Videos and Accessible Games Links'. The nodes, Education, Informative, and Humor come under the node Video which is expanded. Shooting and Puzzles are links under the node Accessible Games Links which is collapsed and hence they are not visible. Thus by expanding and collapsing nodes of the web page, users can create their own view.

### B. Video Description

Fig. 3 illustrates the concept of external metadata, which describes the dynamically changing multimedia contents.

Fig. 3. Screenshot illustrates Video Description

It describes contents like a flash video, swish file, mpeg video, and gif file, in detail and the description is displayed in the web page so that the screen reader will read it for the user.

In this web application, 'Entertainment' section contains different video collections categorized and one of the videos is shown here with the description in Fig. 3. The link 'Informative' is selected from the 'Entertainment' section and the video is opened in the second frame. Appropriate keyboard shortcuts have been provided for each and every audio control.

The description of this video is included in the webpage by using XML tags. And the actual description of the video is shown at the bottom of the webpage in the form of text. This text which describes the video is then read by the screen reader. This is called text video since the video is described in the form of text. The description of the video is shown.

## V. TESTING

To test our web application, we have created a testing environment of 50 people who are blinds and visually impaired. We divided them into two equal groups and allowed 25 people to access a test site which does not have the tree structure and another 25 people to access our IMIVIC Online Outlook Training site. We assigned three similar tasks for both the groups which are explained below in detail.

### A. Task I

The first task is to go to the node 'Accessible Games Link' and open a game either from nodes 'Puzzles' or 'Shooting'. These links contain the list of websites that contains accessible games. The users can access any of these games and then should go to the page 'Compose New Email'.

IMIVIC users took less than 30 seconds to go to the game links within the tree structure. Most of the users found Puzzles to be interesting and since the links they selected are under the tree structure they could open them in the same webpage. After the game is done, the users simply hit the keyboard shortcuts to open the 'Compose New Email' link. 21 out of 25 people successfully navigated through the website and completed the given task in less than one minute.

On the other hand, users who used the test site without the tree structure found it difficult to go to the games links as the web pages are opened in separate pages. Since the users do not know the complete map of the website, some did not know in which page they are. Users have to come back to the home page and go to the page 'Compose New Email'. 19 out of 25 users could navigate through the webpage in one and a half minutes.

### B. Task 2

The second task is to learn to compose a new email from the given tutorials and send an email from MS Office Outlook 2007. After the users come to the page,

'Compose New Email', following the tutorials the users should create a new email message and send it.

Users found the step-by-step processes interactive and learned the keyboard shortcuts to send an email. The users participated in this experimental test are both kinds of users who are well experienced in using the computer and internet and also people who are not familiar with the internet. 20 out of 25 from the first group and 19 out of 25 from the second group completed the given task successfully. Almost same number of people from both the groups completed this task in the given time, since the same tutorial page is used in the test site and the IMIVIC site.

### *C. Task 3*

The third task is to open a video from the category 'Informative' under the node 'Videos' and give the summary of the video they watched including the feedback on how accessible these videos are and how could they manage the audio controls.

Users could access the Informative Video directly with the appropriate shortcuts and enjoyed the video with complete access. We mean complete access because users could understand the complete video even if there is no audio provided for some scenes. The external metadata provided describes all attributes of the video. Users control the audio so that the audio from the screen reader and the audio from the video would not overlap each other. This needed some practice and since this was their first time to use our website, all users could not follow the shortcuts to control the audio. 20 out of 50 people could not figure out the controls and they missed the synchronization at first. But they found it easy after they got well acquainted.

### *D. Results and Discussions*

Thus the IMIVIC Online Training has been tested successfully by the visually challenged for accessing the multimedia contents of the webpage and navigating efficiently. Task 1 in this experimental test proves that, the tree structure is very efficient compared to the linear webpage as the pages from different levels can also be accessed directly. The IMIVIC users opened the web pages within the single integrated page directly without going back and forth the website. The users had a clear idea about the whole structure of the website. Whereas the test site users took considerable amount of time to figure out how to navigate through the site.

The online training for Outlook 2007 trains the users with appropriate keyboard shortcuts to configure the email account, compose, send, and reply emails, maintain address book and managing attachments, etc. The second task proves that the tutorials have been very useful as most of the users from both the groups completed their task in a given time.

The external metadata allows the users to understand the dynamically changing multimedia contents of the webpage which was considered almost impossible for the blind users. Since the description provided is in text, there is no pain to embed them into the video like audio descriptions. In the third task, the users were able to understand the video well comparing to the other websites like 'www.youtube.com', as the detailed descriptions are provided using the external metadata. Even though the required audio controls are given to avoid the synchronization problems, some users were not able to avoid it. When the screen reader reads the video description to the users, the audio that comes along with the video was overlapping with it. The users need little practice to get used to the audio controls so that they can browse the multimedia contents efficiently. The experienced users could use the audio controls well and get all possible information from the website without any loss.

The website has been tested under different browsers, like JAWS, Webanywhere and IBM's AI browser to check whether the users are able to navigate efficiently. The website has also been tested under different operating systems like PC and MAC.

The options 'Text only' and 'Font resize' make the website more accessible. All the images are provided with ALT text and also LONG-DESC is given wherever it is necessary. The users tested these entire criterions and found the website efficient.

## VI. Conclusion And Future Works

Thus the tree structure is implemented in the website for online tutorials and the multimedia contents are rendered using external metadata techniques. From the tests done and results discussed, we can conclude the following:

(i) Tree structure is faster than the linear web pages and avoids the complexity of navigation.

(ii) External metadata for the dynamically changing multimedia contents of the web pages provides easy understanding.

The tree structure is efficient comparing to the linear web pages. Since all the web pages are integrated in a single page, web pages from any different levels can be accessed directly. The major advantage of this research is that the multimedia content description in text is more efficient than the audio description and is cost effective. Thus this approach can be used in any web application to provide the users with a better browsing environment.

The approach implemented in this research can be improved in many ways. One such improvement is enabling voice recognition in the browser, so that the browser can recognize the keyboard shortcuts and invoice of the user and act as per user request. Thus the user can simply say the link he/she would like to go without worrying about the hierarchy. For example, to go to the page, 'Compose New Email', the user does not have to say 'Email Messages' and then say 'Compose New Email'. Instead, the user can directly say 'Compose New Email'. This way the difficulty of using TAB/H keys to move from one link to another can be avoided. With proper training of the voice recognizer, the browsing environment can be interactive and user friendly.

## REFERENCES

[1] Hisashi Miyashita, Daisuke Sato, Making Multimedia Content Accessible for Screen Reader Users, IBM Research, Tokyo Research Laboratory, 16th international World Wide Web conference, 2007.

[2] Esmond Walshe, Barry McMullin, Research Institute for Networks and Communications Engineering (RINCE), Accessing Web Based Documents Through a Tree Structural Interface, In Proceedings of ICCHP 2004.

[3] Yahoo's Accessibility Improvement, http://www.petitiononline.com/yabvipma/, accessed on 03/06/2008

[4] Geoff Freed, The Web Access Project - Access to Multimedia on the Web, National Center for Accessible Media, 1997

[5] http://www.pcworld.com/article/id,116934-page,1/article.html, accessed on 4/09/2008

[6] Second Life for the Visually Impaired, http://blindsecondlife.blogspot.com/2007/10/ rnib-web-access -centre.html, accessed on 02/06/2008

[7] Kristy Williamson, Steve Wright, Don Schauder, and Amanda Bow, The Internet for the Blind Visually Impaired, JCMC 7 (1), October 2001.

[8] Patrick Roth, Lori Petruccil, AB – Web: Active audio browser for visually impaired and blind users, Department of Computer Science, University of Geneva, ICAD'98.

[9] Kozo Kitamura, Cheiko Asakawa, Speech Pointer, A Non-visual User Interface Using Speech Recognition and Synthesis, IEEE International Conference on Systems, Man, and Cybernetics, 1999

[10] http://ncam.wgbh.org/salt/guidelines/sec4.html, accessed on 09/22/2009

[11] http://www.w3.org/Amaya/, accessed on 09/25/2009

[12] http://www2002.org/CDROM/alternate/334/, accessed on 09/25/2009

# M-Business and Organizational Behavior

## Applying Behavioral Theory Ideas to Mobile Interface Design

Olaf Thiele
University of Mannheim
Germany
thiele@uni-mannheim.de

***Abstract*—Just ten years ago programming interfaces were built according to technological possibilities. Search results had to be displayed as textual lists, due to a lack of graphical support. Later, when searching for a music album, search results could be enhanced by showing the cover to simplify recognition. Today, mobile devices are mainly used in private contexts (e.g. on the IPhone), corporate use, like Blackberry devices, is just emerging. Whether the same principles apply for corporate as well as private us is questionable, therefore insight from behavioral theory might aid in deriving corporate mobile interfaces by combining these insights and existing approaches from private use. Special attention is given to mobile information visualization.**

**Behavioral Theory has been developed and evaluated for over sixty years. Its main ideas evolve around bounded rationality of the people, decision making processes and conflicting interests. As these are main themes in the offline business world, members of an organization cannot abandon them using mobile devices in the online world. Therefore, taking organizational considerations into account, mobile interface design might be improved.**

***Keywords-Mobile Human-Computer Interaction, M-Business, Behavioral Theory, Information Visualization.***

## I. INTRODUCTION

Cell phone use has been increasing steadily over the last years. While US households spent three times as much money on land lines than cell phones in 2001, 2007 has been the first year that more money is spent on cell phones than on land lines [1]. Typically, early private technical adopters are eager to be first using technology, while organizational users are more hesitant and follow later. A recent example is the IPhone. Originally, the device was meant solely for personal use. Just recently, Apple started to offer business services for the IPhone platform in addition to the existing personal use. The new 2.0 firmware offers Microsoft Exchange synchronization to allow for push e-mail similar to Blackberry devices [2].

As more and more business applications need to be programmed or ported to mobile devices it remains unclear whether consumer interface guidelines are well suited for the task. While it is clear that many of these design guidelines apply for both consumer as well as business users, this paper aims at showing differences using insights from organizational science.

Differences are due to the surrounding environment of the user. While consumers or other private users are usually intrinsically motivated to use mobile devices (e.g. a kid really wants to buy the newest computer game), business users are embedded in their organizational context (e.g. stocks have to be replaced considering price, shipment, organizational factors, ...). If a private buyer is not sure whether to buy a certain good he might want, a business customer might have to buy nevertheless, because of her boss or pressure by co-workers. Conflicting views within the organization might further hinder the process.

## II. BEHAVIORAL THEORY

Behavioral theory is part of the organizational science research field. Organizational science is not one single coherent theory, but consists of many theories about how organizations act as they do. The research field might be compared to the metaphor of blindfolded scientists exploring an elephant. While one holds a foot he might call it thick and round, another scientist touching an ear could find it to be soft and flat. Both are right. Due to the complex nature of organizations, this approach offers insights on many aspects of organizational behavior.

Behavioral theory can be distinguished from both classical organization theory as well as classical decision theory. Classical organization theory (e.g. Taylorism) focused on the human within the organization as a simple machine. Taylor, for example, believed in stopping times for certain tasks to find an optimum. Then workers suited for this work had to complete the tasks within the allotted time frame. Taylor is often named together with Fordism, due to his insights on mass production around the 1900s [3]. Behavioral theory, in contrast, sees humans and their environment as being much more complex, especially in dealing with modern information technology. Classical decision theory defines humans as "homi oeconomici". All aspects of human behavior can be modeled through variables. This theory is popular among economic researchers, but in behavioral theory humans are also seen as being too multifaceted to find simple parameters to model them. Thus, insights from behavioral theory are not simple formulas on how to design interfaces, but rather guidelines that should be kept in mind. We introduce the main concepts in the next two sections.

### *A. Bounded Rationality*

In contrast to classical economic decision theories, the behavioral theory argues that individuals never have the full objective overview of the situation when making decisions in

T. Sobh (ed.), *Innovations and Advances in Computer Sciences and Engineering*,
DOI 10.1007/978-90-481-3658-2_81, © Springer Science+Business Media B.V. 2010

an organization. Some simple cases might exist where only one alternative looks like the right one, but in many business decisions information about the underlying processes is diffuse. Herbert Simon, one of the founders of the theory, therefore states that we should not believe that "human beings are always or generally rational" [4, p.61]. This does not mean that they do not wish to act rationally, but the mere abundance of information and its processing are limited by our cognitive capacity. Some more factors add to the bounded rationality: incomplete knowledge of the situation, difficulty to value future events and limited consideration of decision alternatives. Therefore, decisions are made on the basis of a limited, simplified model. March and Simon "speak of rationality only relative to some specified frame of reference" [5, p.139].

Individuals apply some tricks in order to cope with bounded rationality. They abandon their search for optimal solutions in favor of satisfying ones. "An example is the difference between searching a haystack to find the sharpest needle in it and searching the haystack to find a needle sharp enough to sew with it" [5, p.140]. Further tricks include simplifying the definition of the situation, applying simple decision rules or habitual behavior for routine decisions.

Especially in software development, many programmers know their software inside out; their users in contrast do not. The users might refrain to these simplification techniques, due to the mass of options given by the software. The behavioral theory terms this simplification technique as "satisficing" [5, p.169].

### B. *Problemistic Search*

Main parts of the theory deal with how individuals interact with other organizational members to reach their goals. This can be through conflicts, coalitions or simply be defining standard operating procedures. A good overview is given in the book "A Behavioral Theory of the Firm" [6, Chap.6]. In this context the notion of problemistic search is most relevant. "By problemistic search we mean search that is stimulated by a problem (usually a rather specific one) and is directed toward finding a solution to that problem" [6, p.123]. The authors distinguish problemistic search from mere curiosity and search for understanding and therefore describe a typical business situation. While the authors had search processes for traditional offline organizational problems in mind, it can be assumed that the theory is applied in the online world as well. Three assumptions are made concerning the search process:

- **Search is motivated.** Search within a company is always driven by a problem. Typically, a problem is resolved by finding a satisfying solution. According to the theory this does not have to be an optimal one.
- **Search is simple-minded.** Organizational members usually search for solutions "near" to the problem. The notion of "near" can be transferred into two simple rules. Either one searches in the neighborhood of the problem or in the neighborhood of the current alternatives.
- **Search is biased.** Solutions are sought from a personal point of view. The point of view reflects past training and experience as well as communication within the organization.

These two concepts, bounded rationality and problemistic search, are now applied to mobile interface design.

## III. Designing User Interfaces

When designing mobile interfaces for business use, the programmer should keep in mind that corporate users typically want to solve a task. As this is done more and more in mobile contexts, bounded rationality and problemistic search offer the following insights.

### A. *Mobile Interfaces*

One of the main problems in mobile application development is the limitation in screen size and processing power of the devices used. When taking bounded rationality as a starting point, applications should not try to give optimal solutions but instead offer a satisfying solution to the problem. The most prominent example for the problem on how to exchange information on the move is short text messages (SMS) sent through cell phones. Even though SMS is usually associated with teenagers, today organizational members are using them as well. Germany's chancellor Merkel "typically communicates with her ministers by text messages from her cell phone" [7].

Other applications were not quite as successful. Accessing e-mails on mobile devices, for example, has been possible for almost ten years. But the introduction of push e-mails together with a well known keyboard led to the breakthrough of the Blackberry. Having a regular keyboard in a miniaturized form is definitely not the optimal solution but satisfies the users' needs.

Generalized guidelines are those found in most human computer interface textbooks. Human interface guidelines give specialized information for mobile devices (e.g. Microsoft's Smartphone Guidelines [8]). Taken to the organizational context, corporate mobile software applications should not try to give an optimal solution, but a "satisficing" one.

Problemistic search applied to mobility might involve searching with a traditional search engine on a mobile device. As many business users are accustomed to use a classical search engine, they transfer this behavior to the mobile device [9]. At the same time search engines offer simplified versions of their web pages. Google, for example, displays no side ads on mobile devices as they would hardly be noticed. They are placed on top of the search results page. Another example for a good implementation is Amazon's "other users bought"-feature. When looking for a book other books in the proximity are presented. Transferred to context adaptation in mobile scenarios or pervasive computing, these research fields are able to implement the ideas of problemistic search. As search is always motivated by a problem, the context data gathered by the device and the accompanying software allow for guessing what the motivation might be.

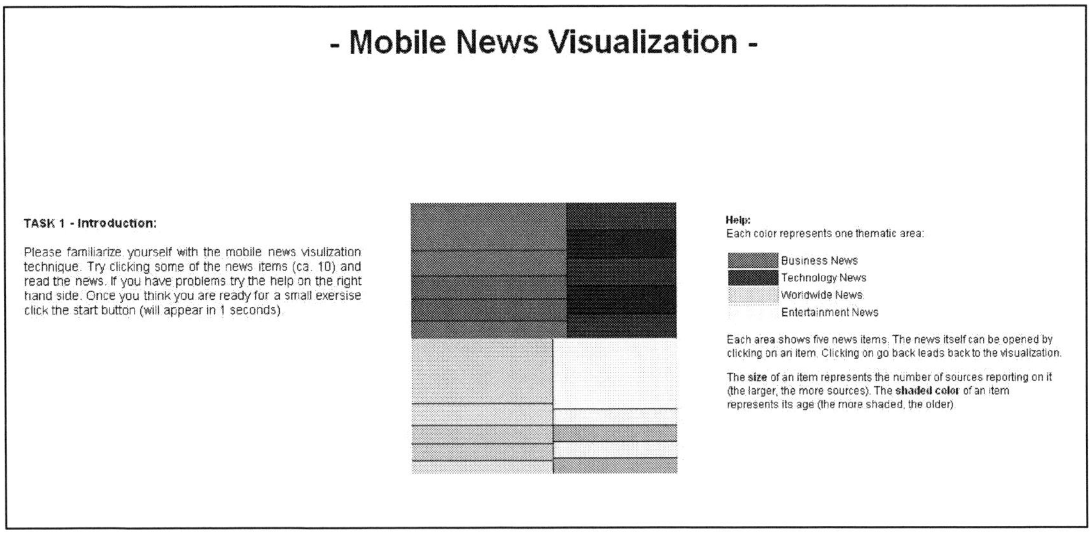


Figure 1. Screenshot Mobile News Visualization

For example, if a business consultant visits customer A, problems will typically evolve around that customer. As search is simple-minded, the context-aware software might be able to offer alternatives that are "near" the current alternative.

Finally, as search is biased, the software might have information stored on whether the user understands technical documentation or whether rather abstract information should be returned.

## IV. Information Visualization

While many success stories can be told for designing mobile interfaces in general, the subarea of mobile information visualizations has not yet produced largely successful applications. This is similar to desktop information visualizations: many visualizations have been proposed, most of them have been evaluated positively, but only few desktop visualizations are used regularly. Due to these facts, this application area might learn from behavioral theory. Typical assumptions made prior to developing a visualization include that people want complex visualizations of the underlying data. Showing a thousand nodes first and zooming in later is quite common. Furthermore, it is assumed that users are willing to learn and are eager to accept a complex visualization technique, because they realize the future potential.

According to behavioral theory the interfaces should instead address bounded rationality. Users are not comfortable with a complex visualization and will immediately try to simplify it. Thus, the original intention to show all information at once is almost never used. In a past experiment we tried to address this problem by giving participants of our study simple information visualizations. Figure 1 shows a visualization of the news given at news.google.com. The four colored areas represent the thematic areas business, technology, worldwide and entertainment. The larger a box, the more sources are reporting this article. The brighter the color, the more recent an article is. In our evaluation people generally liked the interface, but were not totally satisfied [10].

One possible conclusion could be that we did not incorporate the ideas of problemistic search. When participants were searching for certain news, we faked their motivation. They were solving problems to earn money in the experiment. When offering results, we did not show only news "near" the problem. Finally, we did not take their bias into account. Past training or experience played no role in the study.

One of the few popular visualization techniques for mobile devices is a game. Tap Tap Revolution for the iPhone works through touching stars that are moving towards the player on three separate cords according to the song played [11]. First, it is simple to understand within seconds. Second, it works with songs stored on the iPhone and therefore draws upon past experiences and the environment of the player. Software applications analogous to such games would enrich the design of business software in future applications.

## V. Discussion

We simply highlighted two aspects of the behavioral theory. Interesting observations like garbage can decision making were left out. We recommend an excellent paper on decision making within universities [12]. Furthermore, we plan to evaluate our findings through an eye-tracking experiment.

## VI. References

[1] D. Sarkar, Cellphone spending surpasses land lines. USA Today, 18.12.2007.

[2] Apple, Apple Announces iPhone 2.0 Software Beta. http://www.apple .com/pr/library/ 2008/03/06iphone.html, 03.06.2008.

[3] A. Kieser and M. Ebers, Organisationstheorien. Kohlhammer, 2006.

[4] H. A. Simon, Administrative Behavior. The Free Press, 1947.

[5] J. G. March and H. A. Simon, Organizations. John Wiley and Sons Inc., 1958.

[6] R. M. Cyert and J. G. March, A Behavioral Theory of the Firm. Prentice-Hall, 1963.

[7] TimeMagazine, "Angela merkel: http://www.time.com/time/magazine/article/0,9171,1187212,00.html," 2006.

[8] Microsoft, "Smartphone user interface guidelines: http://msdn.microsoft.com/en-us/library/ ms854546.aspx," 2005.

[9] M. Kamvar and S. Baluja, "Deciphering trends in mobile search," Computer, vol. 40, no. 8, pp. 58–62, 2007.

[10] O. Thiele and D. Thoma, "Evaluation of information visualizations vs. language proficiency," in Second IASTED International Conference on Human-Computer Interaction, Chamonix, France, March 14-16, 2007, 2007.

[11] TTR, "Tap tap revolution: http://cre.ations.net/creation/taptaprevolution," 2008.

[12] M. D. Cohen, J. G. March, and J. P. Olsen, "A garbage can model of organizational choice," Administrative Science Quarterly, vol. 17, no. 1, pp. 1–25, 1972. [Online]. Available: http://www.jstor.org/stable/2392088

# A Policy-based Framework for QoS Management in Service Oriented Environments

Elarbi Badidi[1], M. Adel Serhani[1], Larbi Esmahi[2]
[1]College of Information Technology, United Arab Emirates University, PO.Box. 17551, Al-Ain, United Arab Emirates, Email: {ebadidi,serhanim}@uaeu.ac.ae
[2]School for Computing & Information Systems, Athabasca University,1 University Drive, Athabasca, Alberta, T9S 3A3, Canada, Email: larbie@athabascau.ca

*Abstract-* **The successful integration of the Service Oriented Architecture (SOA) in large distributed environments greatly depends on their support of quality of service (QoS) management. The aim of QoS management is to guarantee diverse QoS levels to users issuing requests from a variety of platforms and underlying networks. In this paper, we present our policy-based framework for QoS management in SOA environment with both traditional and mobile users. The framework is based on a QoS broker that is in charge of mediating between service requesters and service providers, and carrying out various QoS management operations. It is also in charge of handling appropriately mobile users that are using various handheld devices to request services.**

## I. Introduction

With the advent of service oriented computing and the prevalent deployment of business applications on the web to reach a large base of customers, many organizations have moved their business online. Service oriented computing together with Web technologies intend to facilitate business collaboration and application integration on a global scale. Web Services are the current most promising technology based on the idea of service oriented computing. They provide the basis for the development and execution of business processes distributed over the network and available via standard interfaces and protocols.

Another development in the area of service oriented computing is the unprecedented rise in the number of mobile workers using a variety of devices including laptops and handheld devices, such as PDAs and SmartPhones, to consume online services. Modern mobile devices are often fully equipped with broad capabilities. Most of these devices support several wireless communication technologies including Wi-Fi, Bluetooth, GPRS, and EDGE. They also come with advanced multimedia capabilities including streaming, and the ability to play several audio and video formats. These devices offer now browsing capabilities that go beyond the simple WAP protocol, to support HTML-based Web sites.

Due to this rapid growth, quality of service (QoS) is becoming a key feature in Web services competition. End-to-end quality management of Web services is a critical matter that is highly related to the changing and dynamic environment of Web services. Most of the research works on QoS support in Web services have focused on the enumeration of QoS requirements and mechanisms for QoS management. Also, some research efforts have been conducted to address the issues of QoS manageability, and security in Service Oriented Architectures (SOA) in general. The success of QoS support for Web services depends largely on the scalability of the proposed mechanisms as the number of clients and servers is expected to be very high in a SOA.

To cope with this concern, we propose a policy-based framework for QoS management of Web services that is capable of supporting at runtime the QoS management phases (QoS specification, QoS negotiation, QoS monitoring, and guarantee). This framework provides support for mobile users using various handheld devices. Its policy-based management allows dealing with issues of authentication, authorization, QoS policy, user profile management, and mobile devices profile management.

## II. Objectives

As a result of the growing interest in QoS support in Web services, the limited number of QoS management architectures in SOA environments, and the increasing use of mobile devices to consume online services, our aim in this work is to develop a framework that meets the following requirements:

1. Provide a QoS model for the specification of the quality of services in Web services. The model should allow providers to specify the classes of services they can deliver to their users.
2. Allow description, publication, and discovery of Web services based on their QoS attributes.
3. Provide support for QoS management operations such as QoS-based service selection, QoS negotiation, and QoS monitoring.
4. Provide support for mobile users, with QoS requirements, using various handheld devices.
5. Allow user and device profile management.
6. Enable monitoring the provision of QoS agreed between provider and users.
7. Allow the implementation of various policies such as authorization policies, policies for monitoring services and quality of service, and policies for QoS-enabled Web service selection.

To meet the above requirements, we chose a broker-based architecture for our framework, an architecture that has been successfully used in many distributed systems. Our interest in

T. Sobh (ed.), *Innovations and Advances in Computer Sciences and Engineering*,
DOI 10.1007/978-90-481-3658-2_82, © Springer Science+Business Media B.V. 2010

using brokers is motivated by the fact that they have been used for a while in SOA to mediate between services providers, service consumers, and partners. They have also been extensively used in multimedia systems and in mobile computing systems to deal mainly with the issue of QoS management [1][2][3].

## III. Related Work

QoS brokers have been proposed in many research works in QoS management for Web services. The authors in [4] have designed and implemented a Web Service QoS broker system, which monitors QoS, in terms of availability, performance, and reliability, of Web Services. In [5], the authors propose a broker based framework for the composition of QoS-aware Web services with QoS-constraints. The broker's components implement dynamic service composition, service selection and adaptation. The authors in [6] describe a Web service architecture providing QoS management using QoS brokers. The interaction with the UDDI is performed through the brokers. Brokers publish in the UDDI QoS information obtained from providers and help customers select services according to their functional and QoS requirements. In [7], a broker based architecture has been proposed to handle Web services selection with QoS-constraints.

All these works share some common goals with our work. However, most of them do provide support for only some of the functionalities and concerns we raised in the objectives section. None of these works has considered the issue of handling mobile users that have different requirements as they use various devices with limited capabilities. Also, none of these works has considered a policy-based management approach to use policies at several levels: authentication, User profile management, QoS specification, QoS monitoring, and service policies in general.

In the following section we present background information on recent research initiatives regarding the description of mobile device capabilities, as well as background information on policy-based management.

## IV. Background

### A. Web Services on Mobile Devices

Mobile services access is still suffering today from interoperability and usability problems. This is to some extent attributable to the small physical size of the screens of mobile devices. It is also partly due to the incompatibility of many mobile devices with not only computer operating systems, but also the format of much of the information that is delivered to mobile devices.

The W3C Mobile Web Initiative (MWI) is a new initiative established by the W3C to develop best practices and technologies for creating mobile-friendly content and applications. The goal of the initiative is to enable the access to the Web from mobile devices and to make it more reliable and accessible. This typically requires the adaptation of Web content based on the device capabilities. The W3C has published guidelines (Best Practices, W3C mobileOK checker service) for mobile content [8]. The MWI Device Description Working Group is actively tackling the problem of device diversity by setting up a *repository of device descriptions* [9]. Authors of Web content may use the repository to adapt their content to best suit the requesting device.

The OMA (Open Mobile Alliance) specification defines the User Agent Profile (UAProf) to describe capability and preference information of wireless mobile devices [10]. This information is to be used mainly by content providers to generate content in an suitable format for the specific device. It is based on the generic framework W3C CC/PP (Composite Capabilities/ Preference Profiles)[11]. CC/PP defines a schema for the description of a device's profile, which is composed of components that describe characteristics of hardware, software, network, and so on. A CC/PP profile can be used to adapt Web contents to a specific device.

A UAProf file describes the capabilities of a mobile handset, including Vendor, Model, Screen size, Multimedia Capabilities, Character Set support, and more. It consists of seven components, which are *Hardware Platform*, *Software Platform*, *Network Characteristics*, *Browser UA*, *WAP Characteristics*, *Push Characteristics*, and *MMsCharacteristics*. User agent profiles are stored in a server called the *Profile Repository*. Very often a profile repository is maintained by a mobile device manufacturer. For example, the user agent profiles describing the capabilities of Nokia cell phones are stored in a profile repository maintained by Nokia. The URL that points to the user agent profile of a mobile device can be found in the headers of requests sent by the mobile device. Very often the URL is located in the x-wap-profile header or the Profile header. For example, the value of the x-wap-profile header generated by a Nokia N80 cell phone is:

"http://nds.nokia.com/uaprof/NN80-1r100.xml"

### B. Policy-based Management

Policy-based management had been mainly used for network management. A policy is a set of conditions imposed on a subject that permits or prohibits actions. These conditions may apply to the subject, the target, or the state of the system. A policy may be the trigger that allows actions to be performed when conditions are applicable.

The IETF Policy Framework Working Group has developed a policy management architecture that is considered the best approach for policy management on the Internet [12]. Figure 1 shows the framework components, which are: *Policy management service*, *Dedicated policy repository*, *PDP (policy decision point)*, *PEP (policy enforcement point)*, and *LPDP (local policy decision point)*.

Policy-based management may play a crucial role in the management of Web services in general and QoS in particular. The context of Web services is however different from the network context as the intrinsic components are different from one context to the other. In Web services environments, only a Policy Manager, which plays the role of the Policy Decision Point, and a Policy Repository in which policies are to be stored, will be required to handle policies.

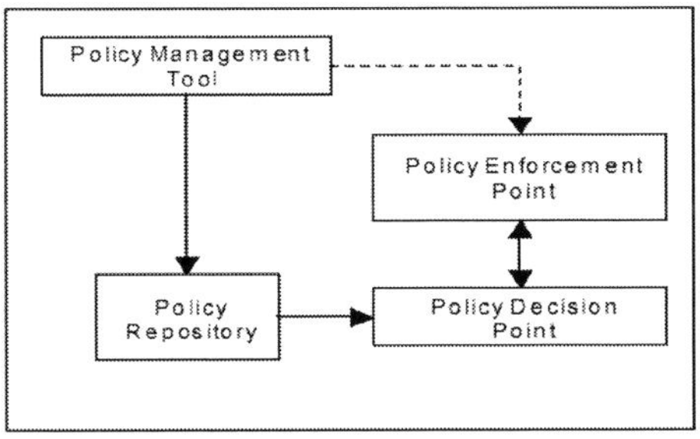


Fig. 1. The Policy Architecture [13]

Utilization of policies in Web services environments has been recognized since the specification of the first standards for Web services. The WS-Policy specification was proposed by IBM, BEA, SAP, Microsoft, and others, to simply define a framework and mechanisms for constructing policy expressions that describe the requirements for accessing a Web service [14].

In this policy-based model for Web services, individual requirements or capabilities of a policy subject are declared using XML policy assertion elements. Policy assertions are the building blocks of policies. Each assertion describes an atomic aspect of a service's requirements. A policy expression can be comprised of one or more policy assertions assembled in Policy alternatives using logical policy operators. This expression can also be associated with a Web service resource, such as a service or endpoint, using WSDL or other mechanisms defined in WS-PolicyAttachment [15].

## V. QoS Management Framework

To address the issues related to handling requests from mobile devices and policy-based management in addition to the already mentioned requirements, our proposed framework is founded on a QoS broker service. The aim of the broker is to play the interface between service providers and requesters. Providers specify the interfaces of their QoS-enabled Web services in WSDL and publish them through an extended UDDI registry [16][6]. Clients may then search the UDDI registry for Web services that are able to deliver their required services with QoS requirements. Given that Web services providers and clients do not have normally the capabilities to negotiate, manage, and monitor QoS, they delegate management tasks, such as Web services selection and QoS negotiation, to the QoS broker.

Figure 2 depicts the main components of our framework. The broker is composed of several components, which cooperate in order to deliver personalized services to both normal users and mobile users with various devices. These components are: *Admission Manager*, *Request Dispatcher*, *QoS Negotiator*, *QoS Monitoring Manager*, *Profile Manager*, and *Policy Manager*. They are under the control of the *Coordinator* component and they allow carrying out several management operations: Admission control, QoS negotiation, QoS-based service selection, QoS monitoring, User profile management, and policies management. These management operations are described in the subsequent sub-sections.

The backend databases maintain information about services' policies, user profiles and preferences, and dynamic QoS information.

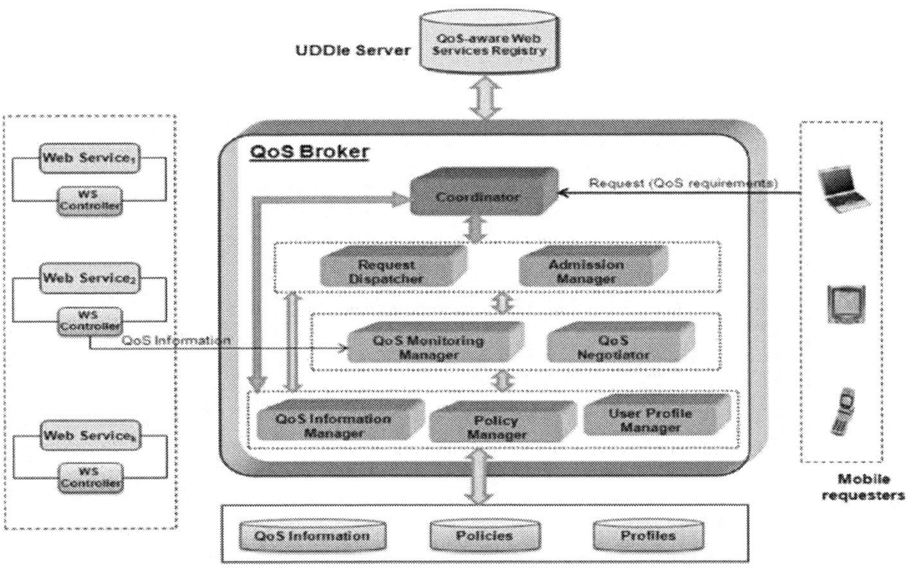


Fig. 2. Framework components

### A. *QoS Specification*

Specification of QoS that providers can deliver may be performed by incorporating QoS parameters, such as response time, throughput and cost, into the WSDL service description. This is the QoS model supported by the UDDIe [16]. Recent works [6] [17] have proposed extensions to the Web services Policy Framework (WS-Policy) to represent QoS policies of Web services. WS_Policy does not specify how policies are discovered or attached to a Web service. The WS_PolicyAttachment specification [15] defines such mechanisms, especially for associating policy with WSDL artifacts and UDDI elements. Figure 3 depicts an example of a QoS policy defined with a policy assertion representing the response time supported by a Web service.

In our framework, QoS policies are also stored in the policies repository as well as the other policies concerning authentication, authorization, user profile and preferences management, and mobile devices profile management.

### B. *Admission Control*

The Admission Manager classifies incoming requests and verifies the provisioned classes of QoS. It is responsible of determining whether the received requests are allowed to use the requested services. This means that Web services access is denied to requests from users who did not negotiate the level of QoS with the selected Web services providers.

### C. *QoS-based Service Selection*

The Request Dispatcher is in charge of implementing different policies for the selection of the web service that will deliver the user required service with the required QoS. These policies can range from simple policies, as random and round robin, to complex ones taking account of the current state of servers in terms of availability, load, and level of quality of service they can deliver. The Request Dispatcher has to perform the match-making of the required QoS with the QoS stored either in the UDDIe or in the policies repository. A queuing model for QoS-based service selection is described in our previous work [18].

### D. *QoS Negotiation*

The negotiation process is carried out by the QoS Negotiator in order to reach an agreement, concerning the QoS to be delivered to the user. First, the user notifies the broker about its required service and its required class of QoS. Based upon available QoS information, the Request Dispatcher component selects a suitable server capable of satisfying the required QoS. Then, this server is approached to determine whether it will be able to satisfy the required class of QoS given its current conditions. Next, a contract is signed by the client and the provider. The contract specifies the service to be provided by the provider to the client, the guaranteed QoS, the cost of the service, and the actions to be taken when there is a violation of the agreed QoS.

If the selected server is not able to deliver the required QoS, the broker selects another server and reiterates the negotiation process.

```
<wsp :Policy...
      xmlns :wsp= "schemas.xmlsoap.org/ws/2004/09/policy"
      xmlns:qosp="garnize:8080/schema/qospolicy"
   <wsp:ExactlyOne>
     <wsp:All>
       <qosp:ResponseTime
             xmlns:qosp="garnize:8080/schema/qospolicy"
             operation="get"
             specification="uddi:uddi.org:qos:attribute:
             responsetime">45</qosp:ResponseTime>...
     </wsp:All>
   </wsp:ExactlyOne>
</wsp:Policy>
```

Fig. 3. Example of a QoS policy [6]

If no server is available to satisfy the required QoS, the client is informed so that it may reduce its QoS expectations or wait until the conditions of the system may allow obtaining the required level of service. Figure 4 shows the interactions of the broker components with the user and the WS provider. These interactions are conducted via the Coordinator component.

### E. *QoS Monitoring*

QoS monitoring is very crucial to assure the compliance of delivered QoS with the contracted QoS and to take appropriate actions when violation of the contract is detected. QoS monitoring in our framework is carried out by the QoS Monitoring Manager and by the QoS Monitor of the Web Service Controller (Figure 5).

The QoS Monitoring Manager continually observes the level of QoS level rendered to clients. QoS parameters, such as response time and availability, that need to be observed are specified in the contract. Observation is achieved through periodic measurement of these QoS parameters at some observation points at the server side and at the client side. QoS violation is detected when the measured value of a QoS parameter does not meet the requirements of the agreed one.

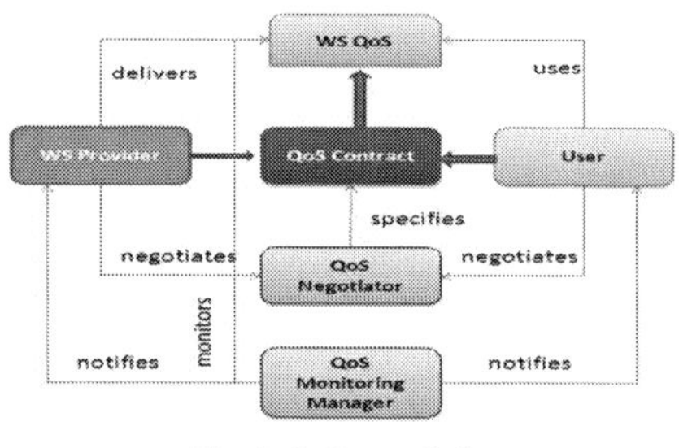


Fig. 4. QoS negotiation

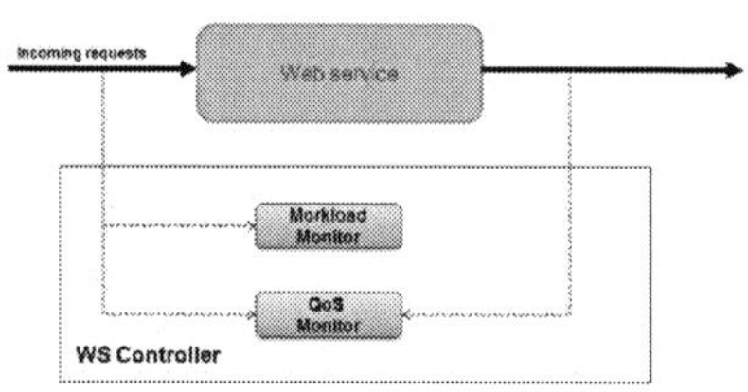


Fig. 5. Web service controller

In this case, both the client and the provider are notified about the violation.

Details about QoS violations are stored in the broker QoS Information Base for statistical usage, and they include: type of violation, name of violated QoS parameters, time of violation, and cause of violation. They may also be used to perform eventual dynamic adaptation of QoS.

### *F. Profile Management*

The Profile Manager is responsible for managing users' profiles, which include their preferences, in terms of personalized services, current location, and required QoS. It is also responsible for negotiating the user profile and authorized services, as well as the devices profiles, that are described using UAProf and CC/PP. This is described in the scenario section.

### *G. Policies Management*

The Policy Manager is responsible for maintaining authorization policies, and polices for monitoring services and quality of service. It receives access control requests, processes them against a set of policies that define how the services are to be used by its mobile users, and returns access control responses.

Services are allocated by the Policy Manager based on the time of the day, the mobile user authorization privileges', availability of services, and any other factors that the administrator may specify when composing the policy. The policy Manager can allow or deny access to services. It controls the extent to which a user (eventually a mobile user) can use the service (e.g, access to limited functions or access to the whole functions of the service). A Policy Manager is very crucial for protecting the mobile user privacy.

## VI. SCENARIO OF MOBILE USER – BROKER INTERACTIONS

Figure 6 depicts a scenario of the interactions between a mobile user and the QoS broker components. The QoS broker has the ability to initiate and control interactions with Web services, within its domain of control, and with service requesters depending on its current needs. This is done by exchanging information messages using the SOAP protocol. Reliable exchange of messages may be implemented using WS-ReliableMessaging specification [19].

1. The mobile user submits a request, including the CC/PP profile of the mobile device in use.
2. After processing the user's authentication on an authorized device, the Coordinator requests the user profile from the User Profile Manager. It also requests the selection of a target Web service from the Request Dispatcher.
3. The coordinator requests policies of the selected Web service from the Policy Manager.
4. If the user profile is available locally in the profile repository, for example because the user had previously used some services within the domain of control of the QoS broker, the Coordinator may decide whether the mobile user request can be processed. This decision is based on the mobile user profile and the requested Web service policies.
5. If the requested profile is not available locally, the mobile user is requested to provide more information, such as service preferences and required levels of QoS, so that a new profile will be created for the mobile user.
6. If the requested service by the mobile user can be delivered to the user, the Coordinator requests from the QoS Negotiator to determine the final profile by negotiating the level of QoS to be delivered.
7. After getting the final profile of the user, the Coordinator forwards the user request to the selected Web service in order to be serviced.

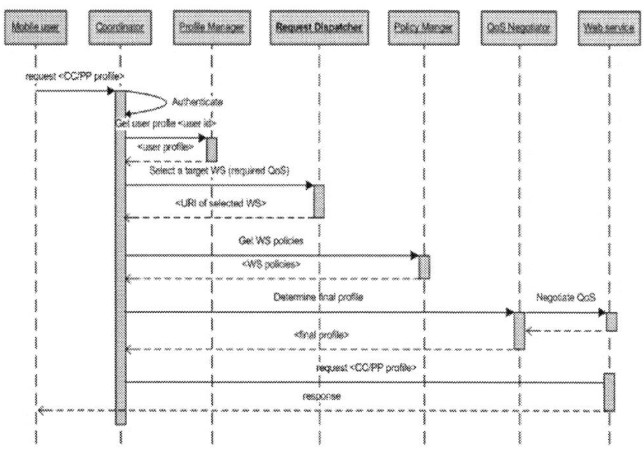


Fig. 6 - Scenario of interactions

## VII. Conclusion

In this paper, we have presented a new framework for the management of the quality of service in SOA environment, which is based on a QoS broker to mediate between Web services providers and users. The framework is capable of handling mobile users using various handheld devices. Policies are a crucial part of the framework. They are used at different levels: Authorization, QoS specification, QoS service monitoring, and description of service policies.

A prototype of our proposed framework is under development. The implementation platform includes: NetBeans, the UDDIe registry [16] with MySQL, and Apache Neethi, which provides a general framework for developers to use WS Policy. Some components of the QoS broker such as the QoS Monitoring Manager and the Request Dispatcher with basic selection policies have been partially implemented in our previous work. QoS monitoring is achieved using observers that are called SOAP handlers. As a future work, we intend to extend our proposed architecture with security policies related to SOA using standards such as WS_Security and WS_Policy.

## References

[1] G. Stattenberger and T. Braun, "QoS provisioning for mobile IP users," in H. Afifi and D. Zeghlache, editors, Conference on Applications and Services in Wireless Networks, ASW 2001, Paris, July 2001.

[2] V. Marques et al., "An architecture supporting end-to-end QoS with user mobility for systems beyond 3rd generation," http://newton.ee.auth.gr/summit2002/papers/SessionW9/2602138.pdf

[3] D. Chalmers and M. Sloman, "A survey of quality of service in mobile computing environments," *IEEE Communications Surveys*, vol. 2, no. 2, 1999.

[4] G. Yeom and D. Min, "Design and implementation of Web services QoS broker," in Proceeding of The International Conference on Next Generation Web Services Practices (NWeSP 2005), 2005, pp. 459- 461.

[5] Y. Tao and K.J. Lin, "A broker-based framework for QoS-aware Web service composition," in Proceedings of The 2005 IEEE International Conference on e-Technology, e-Commerce and e-Service, 2005 (EEE'05), pp. 22-29.

[6] D. Zuquim Guimares Garcia and M.B. Felgar de Toledo, "A Web service architecture providing QoS management," in Proceeding of The Fourth Latin American Web Congress (LA-WEB'06), 2006, pp. 189-198.

[7] M.A. Serhani, R. Dssouli, A. Hafid, and H.A. Sahraoui, "A QoS broker based architecture for efficient Web services selection," in Proceeding of the International Conference on Web Services (ICWS 2005), pp. 113-120.

[8] W3C, "W3C mobileOK Checker [Beta]," http://validator.w3.org/mobile/

[9] W3C, "MWI device description working group," http://www.w3.org/2005/MWI/DDWG/

[10] Open Mobile Alliance, "WAG UAProf, version 20-Oct-2001," http://www.openmobilealliance.org/tech/affiliates/wap/wap-248-uaprof-20011020-a.pdf

[11] W3C, "Composite Capability/Preference Profiles (CC/PP): Structure and vocabularies 2.0," W3C Working Draft 30 April 2007. http://www.w3.org/TR/2007/WD-CCPP-struct-vocab2-20070430/

[12] IETF Network Working Group, "Policy framework architecture," http://tools.ietf.org/html/draft-ietf-policy-arch-00

[13] D.C. Verma, S. Calo, and K. Amiri, "Policy-based management of content distribution networks," *IEEE Network Magazine* (2002), vol. 16, pp. 34-39.

[14] S. Bajaj, et al., "Web services policy 1.5 – framework," W3C Candidate Recommendation 28 February 2007. http://www.w3.org/TR/2007/CR-ws-policy-20070228/

[15] W3C, "Web services policy attachment," http://www.w3.org/Submission/WS-PolicyAttachment".

[16] A. ShaikhAli, O.F. Rana, R. Al-Ali, and D.W. Walker, "UDDIe: an extended rRegistry for Web services," in Proceedings of The IEEE Symposium on Applications and the Internet Workshops, Jan 2003, pp. 85 - 89.

[17] S. Chaari, Y. Badr, and F. Biennier, "Enhancing Web service selection by QoS-based ontology and WS-policy," in Proceedings of The 2008 ACM Symposium on Applied Computing (SAC 2008), Fortaleza, Ceara, Brazil, 2008, pp. 2426-2431.

[18] E. Badidi, L. Esmahi, and M.A. Serhani, "A queuing model for service selection of multi-classes QoS-aware Web services," in Proceedings of The 3rd IEEE European Conference on Web Services (ECOWS'05), Sweden, November 2005, pp. 204-212.

[19] OASIS, "Web services reliable messaging (WS-1 ReliableMessaging) Version 1.1," OASIS Standard, 14 June 2007, http://docs.oasis-open.org/ws-rx/wsrm/200702/wsrm-1.1-spec-os-01.pdf

# An Attacks Ontology for computer and networks attack

F.Abdoli, N.Meibody, R.Bazoubandi
{fateme.abdoli,neda.meibody,reihaneh.bazoubandi} @gmail.com

***Abstract-*** **In this paper we introduced our designed attacks ontology. The proposed ontology is in the domain of Denial of Service attack. We studied and peruse great number of network connection that caused a Denial of Service attack, specially those connections in the KDD cup99 dataset. We used Protégé software for building this ontology. For checking the consistency and accuracy of the designed ontology, we use Racer software and also to test the ontology we use KDD cup99 test dataset. Finally we use Jena framework and SPARQL query language to inference and deduction across the attacks ontology.**

**Keyword:** Ontology, Denial of service attack.

## I. INTRODUCTION

Since intrusion detection was introduced in the mid-1980s, intrusion detection system has developed for almost twenty years to enhance computer security. High false negative and false positive prevent using intrusion detection system practically. In these years the computer scientists attempt to solve this problem and reduce these false rates. They use so many methods and techniques to improve these systems, such as: Data mining [7], State Transition diagrams [6], Clustering [5], Classification [4] and Neuro-Fuzzy methods [8] …. And try to reduce false rate and increase their reliability.

Victor Raskin et al[3] inaugurate new field in Information Security, they discussed about using "Ontology" in Information Security and its advantages. They believed that ontology is an extremely promising new paradigm in Computer security field. They say by using ontology we have a strong classification tools for unlimited events.

The remainder of this paper is to organize as follows: Section 2 presents related work in this domain. Section 3 presents our proposed ontology. Section 4 is about our experimental test. Section 5 is about future work, and finally we have conclusion in the end section.

## II. Related work

Ontology-based intrusion detection techniques: Denker et al. [10][11] and Raskin et al. [3] developed security ontology for DAML+OIL [13] in order to control access and data integrity of Web resources respectively. In applying ontologies to the problem of intrusion detection, the power and utility of the ontology is not realized by the simple representation of the attributes of the attack. Instead, the power and utility of the ontology is realized by the fact that we can express the relationships between collected data and use those relationships to deduce that the particular data represents an attack of a particular type.

Jeffrey Undercoffer and et al. [1][2] produced an ontology specifying a model of computer attack. Their ontology is based upon an analysis of over 4,000 classes of computer intrusions and their corresponding attack strategies and it is categorized according to system component targeted, means of attack, and consequence of attack and location of attacker. They argue that any taxonomic characteristics used to define a computer attack be limited in scope to those features that are observable and measurable at the target of the attack. They present their model as a target-centric ontology.

Moreover, specifying an ontological representation decouples the data model defining an intrusion from the logic of the intrusion detection system. The decoupling of the data model from the Intrusion Detection System logic enables non-homogeneous Intrusion Detection System's to share data without a prior agreement as to the semantics of the data. So designing and developing an ontology in this area must be so useful for the computer and network security.

## III. Proposed ontology

There is several ways for building ontology for a special domain. For example, we can reuse old ontology, which is available in that domain by rebuilding them. Another way for building ontology is using available taxonomy in that domain and builds related ontology based on that

T. Sobh (ed.), *Innovations and Advances in Computer Sciences and Engineering*,
DOI 10.1007/978-90-481-3658-2_83, 

taxonomy[10]. There is few ontologies in the domain computer and network attacks and they are neither comprehensive nor good for our purpose. For this reason we chose the second way for building ontology. For this intention we use the taxonomy which is introduced by Hansman et al. [9]. They proposed taxonomy consists of four dimensions which provide a holistic taxonomy in order to deal with inherent problems in the computer and network attack field. The first dimension covers the attack vector and the main behavior of the attack; in this dimension attacks can be categorized in the following group: Viruses, Worms, Buffer overflow, Denial of service attacks, Network attacks, Password attacks and Trojans etc. The second dimension allows for classification of the attack targets, it says the target of an attack, for example, hardware or software. Vulnerabilities are classified in the third dimension and payloads in the fourth.

For finding attacks scenario ant their behavior and effect in the target we studied more than 244000 records of the network connections status, which lead to Denial of Service attacks. Most of the records are about Smurf and Neptune attacks. By these studies we found attacks relationship and modified the properties of the classes of ontology.

For designing our proposed ontology we use Protégé software that is free and open source [14].

Figure1. High Level Illustration of the Proposed Ontology

Figure 1 presents a high level graphic illustration of our proposed ontology. Designed ontology has one main class: attack class. This class contains all kinds of computer attacks and has so many subclasses and branches.

For the simplicity and reducing the complexity and the amount of work of our designed ontology, we only expand it in the domain of Denial of Service attacks. Figure 2 illustrate the Denial of Service class of our ontology.

For checking the consistency and accuracy of the designed ontology, we use Racer software, which is the most popular software for checking the consistency of designed ontology and also to test the ontology we use KDD cup99 test dataset. Finally we use Jena framework and SPARQL query language to inference and deduction across the attacks ontology for more assurance of the accuracy of that ontology. We discussed about the experimental result in the next section.

## IV. Experimental comparisons among related algorithms on KDD 99

Our experimental dataset was the KDD Cup 1999 Data[15], which contained a wide variety of intrusions simulated in a military network environment. The dataset is about 4 gigabytes of compressed tcpdump data of 7 weeks of network traffic. The simulated attacks fell in one of the following four categories: (1) Denial of Service, (2) R2L — unauthorized access from a remote machine, (3) U2R — unauthorized access to local Root privileges by a local unprivileged user and (4) Probing — surveillance and other probing for vulnerabilities. Our designed ontology is in the domain of Denial of Service attacks, for this reason, we focus on these type of attacks; For example, a SYN flood, smurf, teardrop, ping–of–death, etc. In summary KDD cup99 dataset contain some attack types and 41 features for each of them.

We try to classify this dataset by the designed ontology; Experimental results are presented in Table1

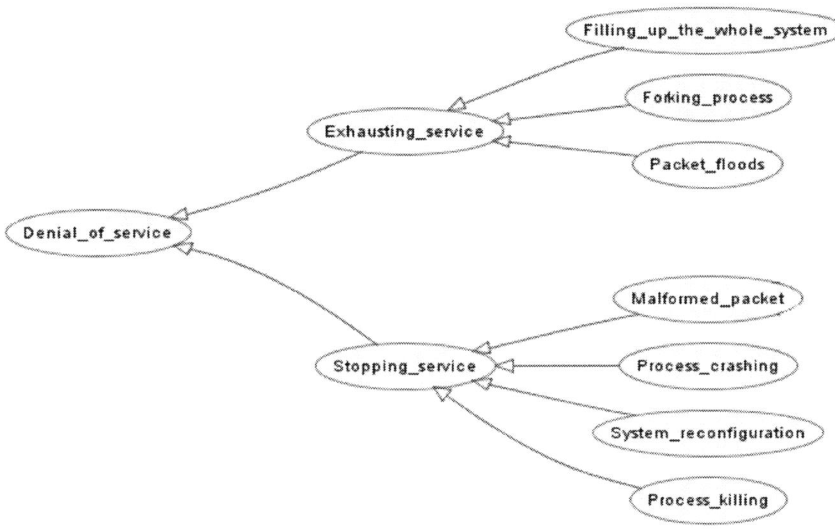


Figure2. Illustration of the Denial of Service class

| Detection ratio | Number of patterns | class |
|---|---|---|
| 97.5% | 100563 | Normal (not DoS) |
| 99.9% | 391458 | DoS |

Table1. The classification results for the proposed attacks ontology

For more assurance we use this ontology in the classification field of network base Intrusion Detection System and then calculating the false ratio. This Intrusion Detection System use Jena framework and SPARQL query language to inference and deduction across the attacks ontology. With the getting results we depict ROC [1] diagram for the proposed system. By the ROC diagram we can found the effect of the system's parameters changes on the system's evaluations measures and trace the results for depicting ROC diagram. Figure 3 show the depicted ROC diagram.

[1] Receiver Operating Characteristic

## V. Future work

Our future work will focus on the improvement of proposed attack ontology in intrusion detection domain; it means that we want to improve it to contain all kinds of attack not just DoS attacks.

## VI. Conclusion

This paper introduced a novel ontology for computer and network attacks and intrusions. The proposed ontology is in the domain of Denial of Service attack. We used Protégé software for building ontology. For checking the consistency and accuracy of the designed ontology, we use Racer software and also to test the ontology we use KDD cup99 test dataset. Finally we use Jena framework and SPARQL query language to inference and deduction across the attacks ontology. Experimental results show the accuracy of the proposed ontology.

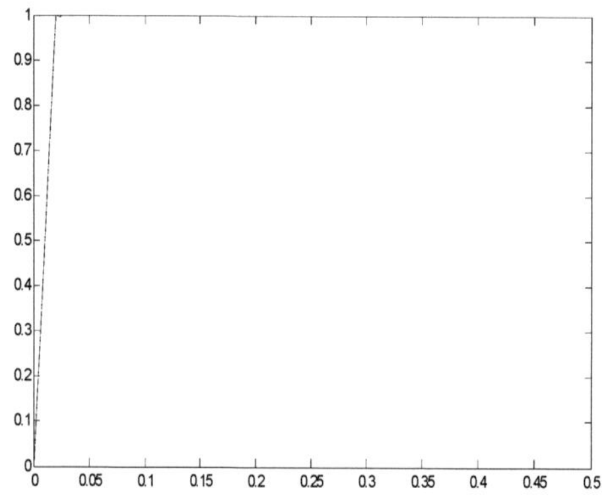


Figure3. Illustration of ROC diagram

## REFERENCES

[1] Undercoffer. J, Joshi. A, Pinkston. J, Modeling Computer Attacks: An Ontology for Intrusion Detection, Springer, pp. 113–135, 2003.

[2] J. Undercoffer, A. Joshi,, T. Finin, and John Pinkston, "A target centric ontology for intrusion detection: using DAML+OIL to classify intrusive behaviors", Knowledge Engineering Review, Cambridge University Press, pp. 23-29, January, 2004.

[3] V. Raskin, C. Helpenmann, K. Triezenberg, and S. Nirenburg, "Ontology in information security: a useful theoretical foundation and methodological tool", New Security Paradigms Workshop, ACM Press, pp. 53-59, Cloudcroft, NM, 2001.

[4] Gomez J., Dasgupta D., "Evolving Fuzzy Classifiers for Intrusion Detection", Proceeding Of 2002 IEEE Workshop on Information Assurance, United States Military Academy, West Point NY, June 2001.

[5] Guan Y., Ghorbani A. And Belacel N., "Y-means: A Clustering Method for Intrusion Detection", Proceedings of Canadian Conference on Electrical and Computer Engineering. Montreal, Quebec, Canada. May 4-7, 2003.

[6] Ilgun K., Kemmerer R.A., and Porras P.A., "State Transition Analysis: A Rule-Based Intrusion Detection Approach," IEEE Transaction on Software Engineering, Vol 2, No 3, 21(3), March 1995.

[7] Lee W., Stolfo S.J., Mok K., "A data mining framework for building intrusion detection models", Proceedings of IEEE Symposium on Security and Privacy, pp 120 – 132, 1999.

[8] Mohajerani M., Morini A., Kianie M. "NFIDS: A Neuro-Fuzzy Intrusion Detection System", IEEE 2003.

[9] Simon H, Ray , A taxonomy of network and computer attacks, Elsevier, Computers & Security (2005) 24, 31e43

[10] G.Denker, L. Kagal, T. Finin, M. Paolucci, K. Sycara, " Security for DAML Web Services: Annotation and Matchmaking," The Semantic Web (ISWC 2003), LNCS 2870, Springer, pp. 335- 350, 2003.

[11] N. Tuck, T. herwood,B. Calder, G.Varghese, " Deterministic Memory- Efficient String matching algorithms for Intrusion Detection," Twenty- third Annual Joint Conference of the IEEE Computer and Communications Societies ( INFOCOM 2004), pp. 2628- 2639, 2004.

[12] Deborah L. McGuinness, Ontology Come of Age, spinning the semantic web,2003.

[13] DAML+ OIL. Available at: http:// www. daml.org/2000/12/daml+oil.daml

[14] http://protege.stanford.edu

[15] kdd.ics.uci.edu/databases/kddcup99/kddcup99.html

# Information Quality and Accessibility

Owen Foley* and Markus Helfert**

* Galway Mayo Institute of technology/Computing, Galway, Ireland. Email: owen.foley@gmit.ie

** Dublin City University/Computing, Dublin, Ireland. Email: markus.helfert@computing.dcu.ie

*Abstract*—**This research examines the relationship between information quality dimensions across information quality frameworks. An examination of the literature reveals that several information quality frameworks have been developed in an attempt to measure the phenomenon of information quality. These frameworks consist of information quality dimensions. Current research has placed much emphasis on dimensions such as accuracy, completeness and consistency. However little if any research has been conducted with respect to the consistency of dimension measures across frameworks? The literature also points out that research into conceptual dimensions is limited. This research endeavours to address these shortfalls by examining the accessibility dimension. The research is conducted within the context of information quality frameworks and assessment methodologies. Over the last number of years, access methods to information systems have also evolved. This has resulted in a diverse number of architectures accessing multiple information systems. Much research has concluded that accessibility is an influence on information quality. An experimental research methodology is employed to tackle the research questions. The research to date has examined different information systems' access methods and their affect upon information quality dimensions. The affect upon other dimensions that make up the information quality framework is measured. The findings to date indicate that the timeliness dimension is most affected. The restriction of access to information systems via web services is also significant.**

*Keywords*—**Information Quality; Accessibility; Experiment**

## I. Introduction

The increased interest in Information Quality (IQ) stems directly from a critical reliance by enterprises at all levels on information systems for completion of their core business activities. The access to Information Systems has also increased with widespread deployment of Web Servers. Many of these businesses have mission critical information systems, yet have no measure of the quality of the information that they derive their business and profit from. In stark commercial terms Colin White [1] president of Intelligence Business Strategies estimates that poor quality customer data cost U.S. business $611 billion a year and that 50% of the companies examined by the research have no plans for improving data quality. Many researchers have examined IQ, as a result.

However the measurement of the accessibility dimension has to date been by and large a subjective measure of the users' perceptions. This it is argued does not provide an adequate or objective enough measure and has the potential to skew results unfavourably. In our research we aim to address the lack of research in this area and focus on the development of an objective measure that has the potential to improve IQ. The perceptions of users, novice or expert will always have the potential to vary, however the objective measure of the accessibility dimension taken with the subjective view can it is argued provide a more accurate measure. This paper proposes a framework for these objective measures, thus contributing to improve IQ.

This paper is structured as follows. An initial examination of the concept of Data Quality is presented followed by identification of the key dimensions from previous research. The concept is then set in context with respect to these dimensions. A discussion of the value and necessity of metrics for these dimensions is followed by an outline of the challenge with respect to the measurement in particular of the accessibility dimension. A proposed framework is put forward along with the presentation of interim results followed by conclusions and the research challenges that are presented.

## II. Information Quality – Definition

The importance of data like many commodities is only ever realised when there is scarcity or complete lack of it. Quality data is a vital commodity in a modern knowledge driven economy. The concept of IQ has received a significant amount of attention from non Information Systems professional since the terrorist attacks on The World Trade Centre in 2001. The lack of quality data or 'bad' data was a contributing factor in failing to prevent the attacks [3]. The non accessibility to vital information with respect to engineering details was a contributing factor. The lack of joined up communication systems to share the available information among many of the emergency services came in for criticism. The aim of IQ research is to minimise the impact of inadequate information.

The concept of IQ must be examined in context. Previous database research concentrated on intrinsic values such as accuracy [4] and completeness [5] and did not consider the data in context. The independent examination of data argues Strong et al [6] does not offer a true reflection of quality. The concept of fitness for use as described by Tayi et al [7] is a definition that implies that the concept of IQ is relative to the use of the data with respect to a particular task. There are many dimensions that make up IQ [2] as outlined in Table 1 below.

T. Sobh (ed.), *Innovations and Advances in Computer Sciences and Engineering*,
DOI 10.1007/978-90-481-3658-2_84, 

TABLE I. INFORMATION QUALITY DIMENSIONS

| Dimension |
|---|
| Accessibility |
| Appropriate Amount of Data |
| Believability |
| Completeness |
| Concise Representation |
| Consistent Representation |
| Ease of Manipulation |
| Free-of-Error |
| Interpretability |
| Objectivity |
| Relevancy |
| Reputation |
| Security |
| Timeliness |
| 'Understandability' |
| Value-Added |

## III. THE ACCESSIBILITY DIMENSION

There is no simple definition of accessibility. Users and IS professionals have [8] indicated to researchers that they consider it fundamental to the concept of IQ. In its simplest form Loshin [9] has described it in terms of ease of access to the information in the system and access to the complete amount of information required. He further asserts that information must be in a format that allows for satisfactory presentation. Accessibility argues Lee et al [10] is simply the 'ease of attainability of the data'. The research completed by Wang et al [2] have indicated that it is a separate IQ category with the characteristics being described as 'Accessibility' and 'Access Security'. The definition has been further added to by Batini et al [11] where it is described as 'the ability to access the data from his/her own culture, physical status / functions and technologies available'. This definition is followed by guidelines with respect to web site design and interface layout that maximises ease of access.

These definitions while offering an understanding of the concept does not provide a comprehensive definition that allow for accurate measurement as a dimension. The above definitions are reliant on interested and informed users in order to obtain a true reflection of the accessibility as a dimension of IQ. The research presently being conducted is examining these characteristics with a view to implementing objective measures and overcoming the problem of subjective measurement.

TABLE II. VIEWS OF INFORMATION QUALITY

| Academics | Practitioners |
|---|---|
| Accessibility | Reliability of delivery |
| Ease of Operations | Security |
| Security | Privacy |
| System Availability | Accessibility |
| Transaction Availability | 'Obtainability' |
| Privileges | Flexibility |
| 'Usableness' | Robustness |
| 'Quantitativeness' | |
| Assistance | |
| Convenience of Access | |
| Ease of Use | |
| 'Locatability' | |

## IV. INFORMATION QUALITY MEASUREMENTS

This research has examined the dimensions as outlined by Wang et al [2] with a view to identifying corresponding measures or metrics. The following dimensions summarized in Table 3 are a sample of the metrics that have been identified in the literature. The purpose of the table is to indicate measurements that are currently in use with respect to IQ and identify areas for more detailed analysis with respect to the accessibility dimension. The objective measures of dimensions such as accuracy and consistency could possibly provide a basis for building an objective measure with respect to the accessibility dimension, in a web environment.

TABLE III. IQ DIMENSIONS AND ASSOCIATED METRICS

| Dimension | Metric |
|---|---|
| Appropriate Amount of Data | Measure of User Requests for data change. |
| Believability | Measure of Consumer Expectation |
| Completeness | Sampling, Tracking |
| Concise Representation | Ratio, Min or Max, Weighted Average |
| Consistent Representation | Column Analysis Domain Analysis, Statistical Analysis |
| Ease of Manipulation | Ratio, Min or Max, Weighted Average |
| Free-of-Error | Ratio, Min or Max, Weighted Average |
| Interpretability | Gap Analysis, Cronbach's Alpha. 1 to 10 scale of user satisfaction |
| Objectivity | Meets users expectations |
| Relevancy | Functional Score Card |
| Reputation | Meets users expectations |
| Security | Protection of privacy Existence of encryption, Storage Policy, Security Constraints |
| Timeliness | Multiple Channel Measurement |
| 'Understandability' | Rating of effectiveness |
| Value-Added | Gap Analysis, Cronbach's Alpha. |

## V. Towards Defining Suitable Metrics

The (IEEE) [20] define a metric as a 'quantitative measure of the degree to which a system component or process possesses a given attribute'. Metrics thus require the following attributes a measurable property, a relationship between that property and what we wish to know and a consistent, formal, validated expression of that relationship.

Good metrics therefore should possess the following characteristics [20]

Persuasive, consistent/objective, consistent in use of units/dimensions, programming language independent and gives useful feedback

## VI. A Proposed Metric for Accessibility

The involvement of the user in the process of IQ is central to a successful implementation. This argues Strong et al [6] must be the case as data consumers have more choices and control over their computing environment and the data they use. The access methods to data has altered radically since the framework outlined by Wang et al [2] was formulated. Access to data is now possible via a myriad devices and platforms. The accessibility dimension therefore needs to be examined in this context. The users subjective opinion of the dimension can it is argued vary considerably depending on the architecture and platform some of which the user may have little if any knowledge. This research poses the following questions specifically with respect to the accessibility dimension. How can accessibility be measured in an objective manner? How can platform or architecture be taken into account? Is it possible to use TIQM techniques to ensure on going improvement?

Table 6 outlines a framework for objective metrics that could be implemented in a web environment. These initial metrics are chosen as they conform to the requirements of a good metric previously outlined. They also take into account the modern dynamic environment that information systems find themselves. The previous research conducted by Wang [2] and subsequent research based upon the 16 dimensions which they identified as central to IQ have not taken architecture into account, they have queried the user independent of technology or architecture.

TABLE IV. Objective Measures Accessibility

| **Proposed Metric** |
|---|
| Meta data analysis at back end, web server and client. |
| Failed Logins |
| Incorrect Username / Password |
| Failed Client calls to web server |
| Failed Client calls from web server |
| Failed connections to Database |
| Failed Network Authentication |

The users were surveyed prior to controlling any of the above meta-data. They were surveyed as per an accessibility subset of the AIMQ developed by [12] as outlined in table 4 with the results being noted. The meta-data item were restricted individually and the users were then required to assess the level of data quality based on the same subset of the AIMQ methodology.

## VII. Experiement

Data can be collected in a number of ways in order to answer research questions. It can be gathered by direct observation or reported by the individual. Fisher et al [8] indicate that systematically collecting data to measure and analyze the variation of one or more processes forms the foundation of statistical process control. In the case of an experiment a variable is manipulated and the corresponding effect on the other variables is noted [7]. Fisher et al [8] also point out that a statistical experiment is a planned activity where variables that have the potential to affect response variables are under the control of the researcher. According to Bernard [4] a statistical experiment has five distinct stages as outlined below.

- Formulate the hypothesis.
- Intervention of control group at random.
- Measure the response variables.
- Introduce Intervention.
- Measure the response variables again.

In order to ascertain the impact of accessibility dimension as an IQ dimension, we examine four IQ dimensions across three architectures and two IS domains.

*IQ Dimensions:* As IQ is a multidimensional concept the impact on individual dimensions is examined in the experiment. For our research, we selected four dimensions that are common across IQ frameworks free-of-error, completeness, consistency and timeliness. In order to measure IQ, a subset of the questions from the AIMQ [16] methodology are employed. The specific survey questions with respect to free-of-error, completeness, consistency and timeliness were used.

*Architectures:* Web, Client Server, Work Station

*Domains:* The two IS domains are a library system and a student exam result system. The major areas of functionality of both systems were employed during the experiment. Three different access methods were used namely workstation, client server and web. These are used on day to day operation of both systems. All users were also day to day operators of the systems.

The experiment sets different levels of accessibility and measures the corresponding effects on the four dimensions as the example outlined in table 4 below. Three levels of accessibility are manipulated in the experiment basic, intermediate and advanced. Basic accessibility has no restrictions set while the advanced level is stringent. Table 4 is an example of a table used to summarize survey data gathered for a particular IS domain where an advanced level of accessibility is set.

As this is a pilot study the number of participants was limited. There were twenty seven participants for the library system and eighteen for the student exam result system. The results recorded are the average scores for the twenty seven participants of the Library IS and eighteen participants of the student exam system IS. The experiment was conducted over a two day period in March 2008.

## VIII. RESEARCH ANALYSIS

Research Question 1: *How does accessibility impact on other dimensions in an IQ framework?*

The results of the experiments to date indicate that accessibility levels have an impact on IQ dimensions in the initial framework examined. The key findings of the initial experiment indicate that as accessibility levels are manipulated the IQ dimension predominantly affected is that of timeliness. As the level of accessibility became more advanced the users survey results with respect to the timeliness dimension were less and less satisfactory. Fisher et al [8] point out that if data is not accessible then quality will decrease because information can not be accessed in a timely fashion. This research indicates that accessibility as a dimension of IQ can have different levels and the more access is restricted the greater the dissatisfaction with the timeliness dimension. It is not merely two states of accessible and inaccessible.

A closer examination of the timeliness dimension is warranted. There is an increase in satisfaction in the survey results with respect to timeliness as the level of accessibility is lessened. At a high level of accessibility the satisfaction with timeliness is 46% for web access, 48% for client server and 56% for work-station. This is an average satisfaction of 50% with the timeliness dimension. This average increases to 61.6% for intermediate and 85.6% for a basic level of accessibility. The results for the student exam system IS domain display a similar pattern with an average of 55% satisfaction with the timeliness dimension when there is an advanced level of accessibility where as at a basic level of accessibility the satisfaction was at 83%. The other dimensions surveyed; free-of-error, completeness and consistency did not radically change across IS domain.

Research Question 2: *Do current IQ frameworks provide valid and reliable measures?*

The research to date has indicated that a number of frameworks do not even consider accessibility as a dimension. The initial experiment on Wang and Strongs [25] framework indicates that accessibility does impact on IQ. The omission of the dimension from a framework it is suggested fails to give a complete view of IQ. The application of the research model to subsequent frameworks will clarify further this finding.

Research Question 3: *Is the impact of accessibility consistent across IQ frameworks?*

This question will be fully analysed when subsequent frameworks are examined by the research model.

Research Question 4 *What impact do different access methods (architectures) have upon IQ dimensions?*

The initial results of the experiment indicate that information systems architecture affects its IQ. Specifically users' satisfaction with IQ dimensions when web architecture was employed. It compared less favourably with client server or workstation architectures. The application of the research model to subsequent frameworks will clarify further this finding.

## IX. LIMITATIONS

Although the research revealed interesting results, the current research is a pilot study concentrating only on a subset of dimensions. An initial assessment of IQ is done by means of a questionnaire. In addition the number of participants was limited for this phase of the research. The initial experiments are limited to the impact of four dimensions. Currently, the results of the pilot study are descriptive and the data is analyzed qualitatively. The questionnaire for assessment is from the consumer's perspective only.

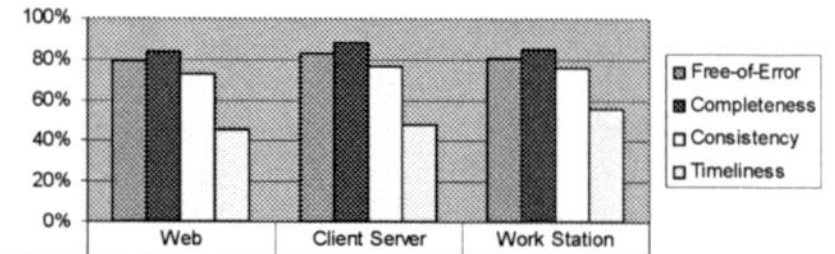


| | Web | Client Server | Work Station |
|---|---|---|---|
| Free-of-Error | 79% | 83% | 81% |
| Completeness | 84% | 88% | 85% |
| Consistency | 73% | 77% | 76% |
| Timeliness | 46% | 48% | 56% |

Figure 1. Library IS Domain Access Level Advanced

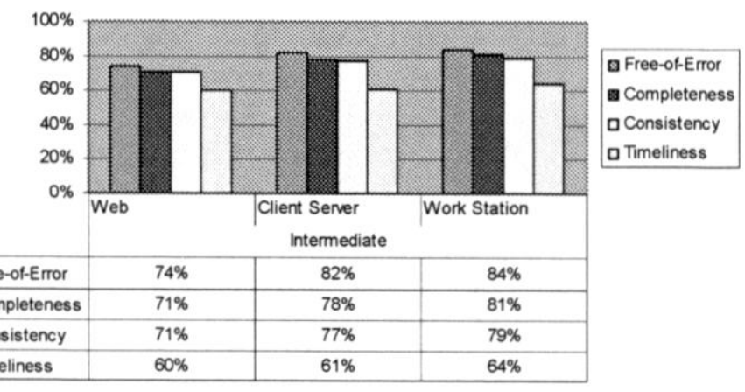


| | Web | Client Server | Work Station |
|---|---|---|---|
| | | Intermediate | |
| Free-of-Error | 74% | 82% | 84% |
| Completeness | 71% | 78% | 81% |
| Consistency | 71% | 77% | 79% |
| Timeliness | 60% | 61% | 64% |

Figure 2. Library Domain Access Level Intermediate

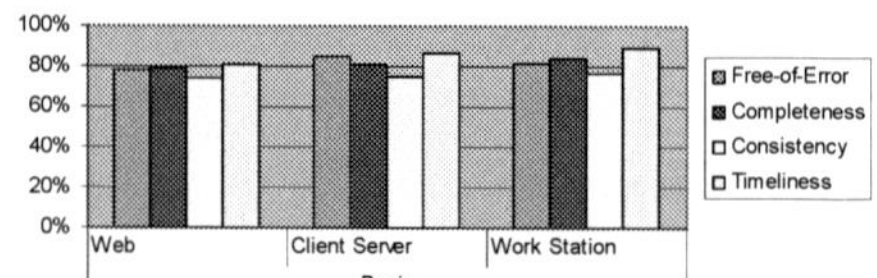


| | Web | Client Server | Work Station |
|---|---|---|---|
| | | Basic | |
| Free-of-Error | 78% | 85% | 82% |
| Completeness | 79% | 81% | 84% |
| Consistency | 74% | 75% | 77% |
| Timeliness | 81% | 87% | 89% |

Figure 3. Library Domain Access Level Basic

This research argues that these figures indicate a strong relationship between IQ and the objective attributes of the accessibility dimension as examined in this and previous research. It is possible to measure these attributes independently of the user. Each of the control elements can be measured via transcript logs at database and operating system level.

The necessity for exclusive reliance on the user to measure the accessibility dimension of IQ it is argued should now not be a requirement. The occurrence of failed logins, failed database connections and web server access will it is argued reduce the level of IQ.

## X. Future Research and Conclusions

With this research we contribute to analysis of accessibility as an IQ dimension. Although frequently mentioned, foremost research lacks in explaining the impact accessibility has on IQ and its relation to other IQ dimensions. We conducted an experiment in which accessibility is examined in the context of information system use. To allow for this examination, we proposed a model that considers accessibility and its impact on information quality. The model is validated with respect to information quality frameworks and assessment methodologies. The findings of the experiment indicate that the user's satisfaction with timeliness as an IQ dimension is related to the level of accessibility. Interestingly, the other dimensions examined free-of-error, consistency and completeness did not indicate significant change through out the experiment. The employment of web architecture was also a factor.

In further research, we intend to address the limitation identified above. Indeed, we plan to examine all dimensions in a number of frameworks to ascertain if dimensions other than timeliness are affected by accessibility levels. Furthermore, we aim to further examine the literature with respect to the timeliness dimension along with its relationship as it pertains to the accessibility dimension. It is intended to use the proposed research model to examine other IQ frameworks, dimensions and assessment methodologies along with inferential statistic techniques with a view to finding out if the affect on timeliness are consistent across IQ frameworks.

This research suggests that the ever increasing emphasis that business, IS and IT professionals place on restricted access will need careful consideration. The very real concern of correct access to IS needs to consider many factors and this research it is argued demonstrates the necessity for IQ to be considered when these policies are being drawn up and put in place. The solving of an access problem may indeed lead to an unforeseen IQ problem.

## References


[1] Redman, T.C., *Data Quality The Field Guide*. 2001: Digital Press.

[2] Wang, R.Y. and D.M. Strong, *Beyond Accuracy: What Data Quality Means to Data Consumers*. Journal of Management Information Systems, 1996. **12**(4): p. 5-34.

[3] Fisher, C., et al., *Introduction to Information Quality*. 3rd ed. 2006, Boston: MIT.

[4] Olson, J.E., *Data Quality - The Accuracy Dimension*. 2003: Morgan Kaufmann.

[5] Yang, L.W., et al., *Process Embedded Data Integrity*. Journal of Database Management, 2004. **15**(1): p. 87-103.

[6] Strong, D.M., L.W. Yang, and R.Y. Yang, *Data Quality in Context*. Communications of the ACM, 1997. **40**(5): p. 103-110.

[7] Tayi, K.G. and D.P. Balou, *Examining Data Quality*. Communications of the ACM, 1998. **41**(2): p. 54-57.

[8] Pipino, L.L., L.W. Yang, and R.Y. Wang, *Data Quality Assessment*. Communications of the ACM, 2002. **45**(4): p. 211-218.

[9] Loshin, *Enterprise Knowledge Management - The Data Quality Approach*. 2001: Morgan Kaufmann.

[10] Lee, Y.W., et al., *Journey to Data Quality*. 2006: MIT.

[11] Batini, C. and M. Scannapieco, *Data Quality Concepts, Methodologies and Techniques*. 2006: Springer - Verlag.

[12] Lee, Y.W., et al., *AIMQ: a methodology for information quality assessment*. Information and Management, 2002. **40**: p. 133-146.

[13] Cha-Jan Chang, J. and W.R. King, *Measuring thePerformance of Information Systems: A Functional Scorecard*. Journal of Management Information Systems, 2005. **22**(1): p. 85-115.

[14] Kahn, B.K., D.M. Strong, and R.Y. Wang, *Information Quality Benchmarks: Product and Service Performance*. Communications of the ACM, 2002. **45**(4): p. 184-192.

[15] Pradhan, S., *Beleiveability as an Information Quality Dimension*, in *MIT Information Quality Conference 2005*. 2005, MIT, Boston.

[16] Pierce, E.M., *Assessing Data Quality with Control Matrices*. Communications of the ACM, 2004. **47**(2): p. 82-86.

[17] Shankaranarayan, G., Z. Mostapha, and R.Y. Wang, *Managing Data Quality in Dynamic Decision Environments:* Journal of Database Management 2003. **14**(4): p. 14-32.

[18] Cappiello, C., C. Francalanci, and B. Pernici, *Data Quality Assessment from the User's Perspective*, in *IQIS*. 2004, ACM: Paris.

[19] Mens, T. S. Demeyer. *Future Trends in Software Evolution Metrics*. in *4th International Workshop on the principles of Software Engineering* 2001. ACM Press.

[20] Sommerville, I., *Software Engineering*. 6th ed. 2001: Addison-Wesley.

[21] Codd, E.F., *A Relational Model of Data for Large Shared Data Banks*. Communications of the ACM, 1970. **13**(6): p. 377-387

# Engineering Autonomous Trust-Management Requirements for Software Agents: Requirements and Concepts

Sven Kaffille and Guido Wirtz
Distributed and Mobile Systems Group
University of Bamberg
Feldkirchenstraße 21, 96052 Bamberg, Germany
{sven.kaffille|guido.wirtz}@uni-bamberg.de

*Abstract*—**Due to the need for open and distributed computer systems the demand for trust and autonomous trust management by software agents has become an important topic. This led to the development of different frameworks to support trust management by autonomous software agents. These frameworks enable software agents to autonomously form trust, make trust decisions, and update trust, but are often limited to certain application domains. Currently there has been little effort (e. g. [1], [2]) to integrate trust modeling into phases and models of Agent Oriented Software Engineering (AOSE). So one step to further fit AOSE methodologies to develop open agent societies is to integrate a systematic approach to develop trust models. This paper describes a systematic approach to identify requirements on software agents in order to autonomously manage trust. For this purpose the extendable agent-oriented development process of O-MaSE (Organization-based Multi-Agent System Engineering) and its modeling concepts are extended.**

## I. Introduction

In the past decade integration of trust in open and distributed computer systems such as peer-to-peer and multi-agent systems (MAS) has been extensively researched. The demand for trust concepts in computer science and especially in MAS is generated by the need for technologically and organizationally open MAS which for example are used to form virtual organizations ([3], [4]) or open marketplaces where autonomous artificial agents make deals on behalf of their human users. Agents in an open MAS cannot fulfill their tasks alone and depend on other agents that may potentially have harmful intentions. In order to be autonomous and honor autonomy of interaction partners agents must be enabled to make decisions about how and when to interact with whom [5]. For this purpose they require a notion of trust.

It is claimed that Agent Oriented Software Engineering (AOSE) is especially suited to deal with distributed, open, and situated software systems which become more and more important today. These systems bring together different stakeholders and their (potentially selfish) goals and business needs. The most important and distinctive feature of agents is their autonomy. On the one hand an autonomous agent must be able to make autonomous decisions and on the other hand autonomy of other agents and dependence on other agents confine this autonomy. While many frameworks (e. g. [6], [7], [8]) to support autonomous trust formation and decisions by software agents were developed in the last decade, there has been little effort (e. g. [1], [2]) to integrate trust modeling into AOSE phases and models. So one step to further fit AOSE methodologies to develop open MAS is to integrate a systematic approach to develop trust models.

This paper examines requirements on trust modeling in AOSE and proposes new concepts to be integrated in agent modeling languages. For this purpose the next section examines what we understand as trust. The subsequent section examines properties of open agent societies and their relation to trust and AOSE. Afterwards - illustrated by an example- we introduce new concepts we consider useful to model requirements on agents that autonomously manage trust relationships. The paper concludes with shortly demarcating our approach from related work and identifying future work.

## II. Trust

Trust is constituted by a) the beliefs a truster has about the willingness and capability of another agent (trustee) to possess properties on which the truster depends in a certain situation and b) the decision to rely on the trustee in this situation ([9], [10]). There is an inherent risk that a trustee does not possess the expected properties or it does not want to provide them to the truster, so that the truster may sustain a loss when it interacts with the trustee. Agents have relationships to other agents either through direct or indirect interactions. We call an agent relationship in which trust decisions have to be made trust relationships. A trust decision requires a truster to form beliefs about willingness and capability of a trustee. In order to form these beliefs a truster needs information about the trustee and the situation. Information about a trustee is used to estimate the trustee's trustworthiness. This information characterizes the trustee and can be collected from different sources e. g. experience in past similar interactions, recommendations, or reputation of trustee. Depending on several factors (e. g. number of interactions, reliability of recommender) the trustworthiness may be more or less reliable from the subjective view of the truster. Therefore, a truster must be confident with estimated trustworthiness in a trust relationship

T. Sobh (ed.), *Innovations and Advances in Computer Sciences and Engineering*,
DOI 10.1007/978-90-481-3658-2_85, 

which may be in an unstable or stable state. A relationship is stable if the truster - according to its information - can estimate trustworthiness based on its subjective assessments and regard this as reliable. This does not mean that estimated trustworthiness matches actual trustworthiness of a trustee. If trustworthiness estimation is really accurate arises during or after an interaction with a trustee (this is the inherent problem of trust).

As can be seen trust is context-dependent (type of interaction with trustee, institutional and environmental factors dependent) ([11], [10]) and trust in an agent in a certain context cannot necessarily be transfered to another context. Which information is relevant in a context and how it can be obtained in order to estimate trustworthiness depends on the type of trust required. Trust is concerned with dependability (availability, reliability, safety) and security (integrity, confidentiality) attributes of assets belonging to the truster which can be negatively affected by the trustee. Therefore, a truster needs information about which attributes are affected, why and how a trustee may affect them, and how the truster can evaluate the outcome of an interaction.

### A. Types of trust

Four different types of trust can be distinguished ([11], [10]).

- *Provision/Delegation trust*: The truster needs to use a service/delegates a task (a task can either be a goal, an action or both[9]) to the trustee and requires the trustee not to fail (completely or partially) in providing the service or performing the task. The truster is interested in availability and reliability of provided services or performed tasks.
- *Access trust*: The truster owns (or is responsible for) a resource that is accessed by the trustee and the resource may be misused or damaged by the trustee. The truster is for example interested in the availability, integrity, and - in case of data and their value - confidentiality of properties of the resource.
- *Identity trust*: The truster believes that the identity of the trustee is as claimed. As identity trust is the foundation for the other types of trust to be applied to trustees, we only further examine access and provision trust in this paper, as usually trustees must be identifiable to develop personal (provision and access) trust.
- *Context trust*: Is not directly related to the trustee, but to the environment and institutional context of the truster, as e. g. mechanisms that prevent the trustee from performing harmful actions, norms associated with sanctions or insurances that cover losses for the truster. Therefore, context trust is complementary to personal trust in the three trust types described above.

Personal trust in the trustee (provision, access, or identity) may not be sufficient or necessary for a truster to interact with the trustee. Instead a truster may rely on the environment or agent society as e. g. for identity trust with help of certificates as is done in centralized or distributed public-key infrastructures. Access trust may be put into mechanisms that protect the resources of the truster and prevent the trustee from unauthorized actions as e. g. preserving confidentiality of data by encryption. If provision trust is required, trust can partially be put into the context, as well. As in both cases an autonomous trustee cannot be forced to act as desired it may be necessary that norms or insurances are in place that deal with situations in which the trustee is not able or willing to behave as expected.

### B. Dynamics of Trust

Development of trust in a trust relationship can be divided into three phases. Theses phases are what we understand as trust management and may completely or partially be performed by an autonomous software agent depending on characteristics of the trust relationship.

- Trust formation [12]: The truster tries to establish a trust relationship with a trustee. For this purpose it must be able to reliably estimate the trustee's trustworthiness. This may be done with help of reputation, recommendations, or personal experience with the trustee. If these factors are not sufficient the relationship cannot be regarded as stable.
- Trust decision: The truster decides to trust (or not) the trustee in the current situation based on the trustee's trustworthiness and information about the current situation available to it. The truster may decide to interact with the trustee, even if personal trust is low. The outcome of the interaction will provide more information about the trustee (after or while interacting), which may strengthen or weaken the relationship and make trustworthiness estimation more reliable from the view of the trustee.
- Trust update/evolution [12]: The truster evaluates the outcome of the interaction and updates the information about the trustee's trustworthiness and the state of the trust relationship. Therefore, this phase is closely coupled with and required for trust relationship formation, as it provides means to create information about personal experience.

Trust formation and trust update/evolution are concerned with the beliefs of the truster about the trustee, while trust decision uses these beliefs and beliefs about the context in order to decide if the truster should interact with an interaction partner (see definition above). In order to form and update these beliefs the agent must be able to collect and interpret information about the trustee and the outcome of interactions with the trustee.

### C. Example Case Study

As a simple example we consider an agent society that facilitates File-Service-Provider agents to sell storage space over the Internet to other agents (File-Service-Users) that need to be able to autonomously and flexibly expand or reduce the storage available to them according to the needs of their users. The File-Service-User agents can provide payment information (e. g. credit card information) to the File-Service-Provider agents in order to pay for the desired service. The agent society additionally consists of agents that provide yellow page services and infrastructure to identify interaction partners. These agents constitute part of the infrastructure of the agent society and are owned by a third party trusted by all users whose agents participate in the agent society.

Trust in File-Service-Providers cannot be established off-line, as owners of File-Service-Users do not know owners of File-Service-Providers and cannot establish a trust relationship with them. Therefore, a trust relationship between File-Service-Providers and File-Service-Users has to be managed on-line and consists of provision and access trust. The File-Service-Providers provide storage service to the File-Service-Users, which requires them to store files owned by the File-Service-Users. The file service has to be available to the File-Service-Users, when they need their files (provision trust). While the files are stored by the File-Service-Providers they may access the files. Integrity of files should not be destroyed and contents of files should be kept confidential.

## III. Trust in Open Agent Societies and AOSE

In recent years it has become common to view MAS as agent societies, which are designed with help of explicitly defined social concepts, such as roles and norms ([13], [14]). An agent society is an MAS in which agents occupy at least one explicitly defined social role and behavior of agents is governed by a set of explicitly defined norms. The conception of open MAS as agent societies has brought the distinction between physically possible behavior and socially expected (obliged), acceptable (permitted) or not-acceptable (forbidden) behavior of agents into AOSE. Not every physically possible behavior may be socially expected or acceptable. During development and operation of an open agent society it has to be considered that members of an agent society may violate norms.

Openness of agent societies can be characterized from two perspectives: technical and organizational. Technical openness is concerned with open (possibly standardized) interfaces for agents and MAS, and freedom of architecture. The question of organizational openness is a question of organizational control. Organizational openness can be characterized with help of the owners (stakeholders) of an agent society and its constituting agents.

In [15], [16] Davidsson identifies categories of agent societies as closed, semi-closed, semi-open, and open agent societies. In a closed society no new members can join it at runtime. In semi-closed societies an interface to the society is provided by the society owner, through which agents of agent owners different from the society owner can interact with agents in the society. Semi-open agent societies are similar to the previous type, but an agent joining the society can directly interact with the agents already in the society. Before it can join, it has to register with a society gatekeeper which decides about the membership. In open agent societies there is not necessarily a society owner, as it is difficult to control of which members the society consists. Furthermore, it is difficult to define a concise set of roles for society members, as new interaction patterns may emerge at runtime and only generic roles can be defined [16]. Even new norms, roles, and communication protocols can emerge. Therefore in our work we consider only semi-closed and semi-open agent societies and refer to them as open agent societies in the following. This ensures that there exists the possibility to establish authentic identities (at least pseudonyms) for agents participating in the agent society.

Designers of open agent societies have to take into account the possibility of norm violations, as they cannot control behavior of agents participating in the society. Therefore, open agent societies require mechanisms to establish a kind of social order. In [17] it is argued that social control is required to achieve social order. Social control is "*any action of an agent aimed at (intended or destined to) enforcing the conformity of the behavior of another agent to some social norm*" [17].

Castelfranchi criticizes that computer scientists try to increase or establish social order in MAS or open agent societies by technical prevention (e. g. security measures), total control, or imposing norms[1] on the agents ("*formal-norm illusion*"), which is not possible and self-defeating [17]. One possible mechanism of social control is trust. In order to create dependable open agent societies designers have to take into account trust relationships from early phases of agent society development to implementation of societies.

State of the art in AOSE methodologies (e.g. [18], [19]) is to model a MAS as an organization that consists of groups, in which agents are organized and play roles to achieve the goals of the MAS. The central idea of organizational modeling is the abstraction from concrete agents with help of the roles agents play in a group in order to facilitate open systems. The examination of AOSE methodologies by [20] and [21] yields the following result: Most methodologies are concerned with analysis and design of MAS with a fixed number of agents, a static structure, fixed interactions and relationships between agents. Therefore these methodologies lead to implemented agent systems that can only deal with dynamic and unpredictable environments in a limited way. One shortcoming is that it is not considered that agents may deviate from the behavior prescribed by the design of the system. Current approaches lack concepts to model trust relationships, which can be used as foundation to deal with dynamic relationships between agents.

More recent approaches such as [13], [22] try to improve on that by modeling MAS as open agent societies or electronic institutions. They explicitly incorporate norms and sanctions, as they acknowledge that norms may be violated at runtime of an MAS. But they have no standard or a limited understanding of software engineering, as they only focus on the concepts and not on the process in which these concepts are incorporated. In [13] it is stated: "*One way to inspire trust in the parties is by incorporating the regulations (**norms**) in the society architecture that indicate the type of behavior expected from*

[1] i. e. norms are explicitly specified and the agents are forced to comply to the norms without a possibility of violation, so that they are not norms, but constraints.

*role enactors"*. Norms can help to establish context trust (institutional trust), but it is desirable to have only required norms. In order to know which norms are required the trust relationships required among agents have to be identified, so that it can be decided which assets in the agent society must be protected by norms. Furthermore, security mechanisms can be identified in order to protect as many assets as possible in trust relationships (environmental trust). So the amount of risk in a trust relationship that exist due to malicious intent or inability of the trustee and can be minimized. Additionally, trade-offs between addressing risks by trust management, norms, or security mechanisms can be made depending on their cost and complexity.

One suitable method to develop organization-based agent societies is O-MaSE ([23], [19]) which has currently been made customizable in order to select appropriate models and steps to develop organization-based MAS [24]. For this purpose the O-MaSE process has been described in terms of the OPEN Process Framework [25]. This facilitates extension of O-MaSE with additional steps and makes the O-MaSE process more concise and understandable. The process currently covers two phases: Initiation and Construction. This paper deals with the Initiation phase which is divided into the activities of Requirement Engineering and Analysis. The Requirements Engineering phase translates system requirements into goals. The purpose of the Analysis phase is to identify and define the roles of agents in the organization (or society) and their responsibilities. In the full O-MaSE process the initiation phase consists of the following tasks:

- Model Goals: Creation of a goal model that defines goals which the agent organization (or society) should fulfill.
- Model Domain: Definition of a domain model that captures domain entities, their attributes, and their relationships with help of a class diagram.
- Goal Refinement: Refinement of goals with temporal relationships (precedence and trigger. See [24] for the detailed O-MaSE meta model) and the domain entities they concern.
- Model Roles: In this task goals are assigned to roles that members of the organization being modeled play in order to fulfill them.

We propose to add two additional steps into O-MaSE. The first is to analyze where trust relationships are required that (are at least partly) managed by autonomous software agents and what constitutes these trust relationships. This task yields trust management requirements on agents playing certain roles and requirements on the agent environment and society to provide security mechanisms (such as digital signatures, encryption etc.) and which assets have to be protected by norms with help of associated sanctions or rewards. The second (which is not covered in this paper) "Norm Analysis" (as described in [13]) is to analyze the requirements on the normative system, which can be derived from the identified trust relationships. These two steps will make O-MaSE more suitable to model open agent societies.

The following questions have to be answered in the Trust Analysis Task:

1) Where are autonomously managed trust relationships required?
2) What type of trust is required (one trust relationship may consist of many trust types)?
3) Which dependability and security attributes of the resources involved have to be considered to be at risk?
4) Why/How are these assets at risk through malicious goals or inabilities of a trustee?
5) What can be done about a particular risk? Answers may be: Prevent the goals of the trustee to be realizable e.g. by security mechanisms (Environmental Trust). Align the incentives of the trustee with those of the truster with help of norms and associated sanctions/rewards (Institutional Trust).
6) For risks that are addressed by trust: Who (user, agent, trusted third party) is responsible to establish trust relationships/make trust decisions/update trust relationships? This requires also to answer the question: Who has access to which information and who must be able to interpret results of interaction?

Note, that answers to these questions introduce new goals to be achieved by the stakeholder of the truster, the agent, other agents (respectively their roles), or new roles in the agent society. Consider for example our case study, where a File-Service-User requires to establish a trust relationship with a File-Service-Provider or must select a trustworthy File-Service-Provider, it can be necessary to introduce the goal for the trustee to retrieve reputation information or recommendations. Furthermore it may be imaginable to introduce a certification authority that certifies File-Service-Providers as trustworthy. Furthermore, it may not be possible for a truster to update trustworthiness, as it may not be able to interpret the outcome of an interaction. If we extend our example so that Service-Providers can for example sell physical products to other agents, which are then shipped to the owner of the agent. In such a trust relationship the truster (a software agent) is hardly able to evaluate the performance of the trustee as it cannot inspect and evaluate the condition of the product. Therefore the owner of a buying agent (truster) has to evaluate it and provide this information to his agent in order to facilitate subsequent autonomous trust decisions. The introduction of new goals and roles requires repeated performance of the tasks initially carried out in the requirements phase, which is no problem, as O-MaSE tasks have been defined to be used in an iterative fashion.

## IV. Concepts to Model Trust Requirements

The meta model to support our extension of O-MaSE is shown in figure 1. In order to model goals and their relations concepts from O-MaSE and KAOS [26] have been incorporated into our meta-model. Only the relevant parts of the O-MaSE meta model have been included and the goal model has been adapted to fit goals as they are defined in the KAOS approach.

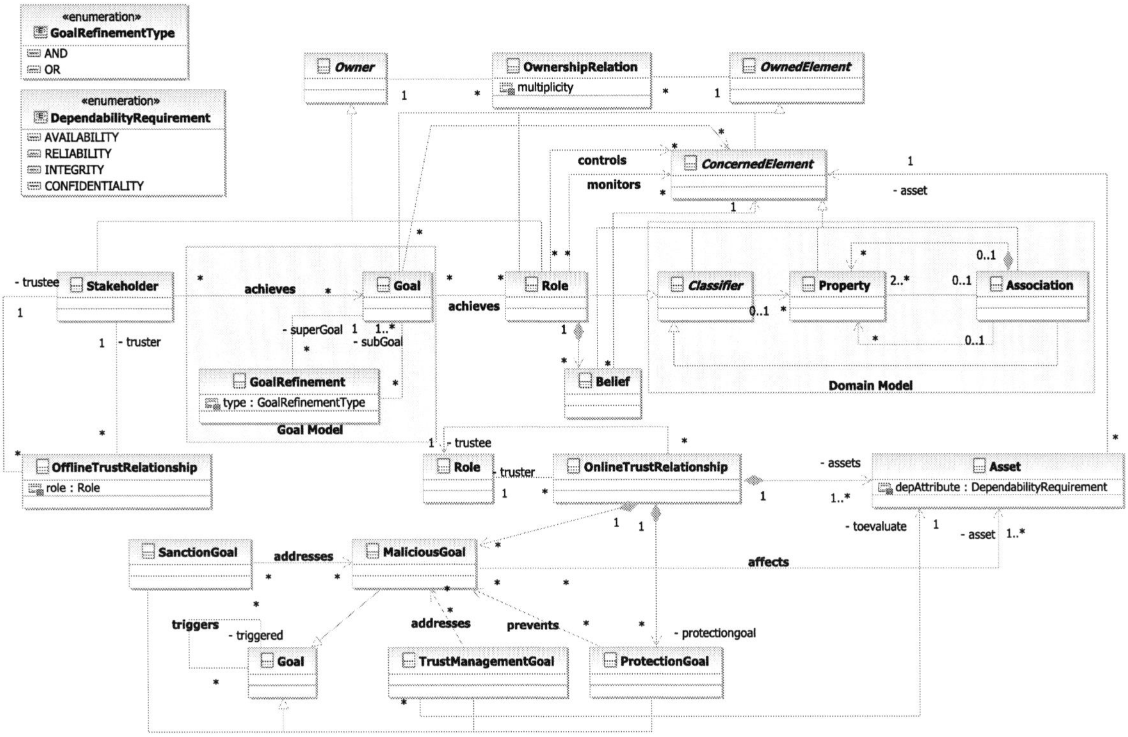


Fig. 1. Concepts to model trust requirements

As explained in section II an important question regarding trust is who owns which goals and which resources and who achieves these goals or accesses the resources. The classes `Owner` and `OwnedElement` in figure 1 are introduced in order to model ownership between stakeholders and roles (respectively the agents playing those roles), roles and goals, resources such as data and services. The latter are modeled in O-MaSE with help of UML class diagrams. Only important concepts from UML regarding objects are depicted in figure 1. Different roles may require to own, control, or monitor different parts of instances of a `Classifier`. For this purpose the abstract Class `ConcernedElement` has been introduced. If a role owns such an element it can control it. Furthermore, control implies that the role is able to monitor instances of the referenced `ConcernedElement`.

A simplified domain model of our file service example is shown in figure 2. The File-Service-User is for example able to control parts of the association between the file service and files. As an additional constraint this should only be the case for the files it owns.

In order to answer question 1 from the previous section it is necessary to know what kind of relationship exists between owners of File-Service-User and File-Service-Provider agents. As they do not know each other there is no relationship. Another example is the owner of the agents providing the yellow pages. Between this owner

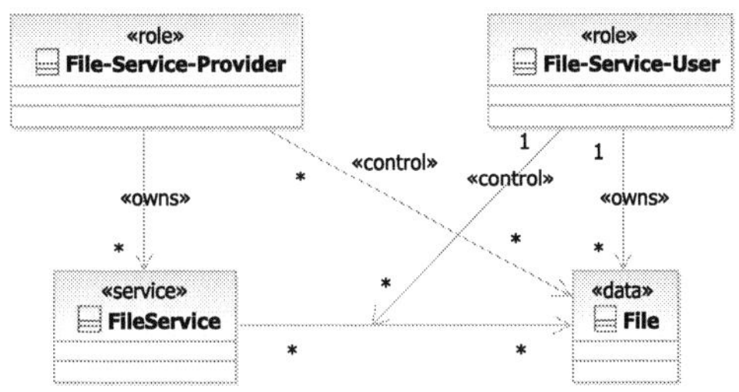


Fig. 2. Example domain model

and the owner of File-Service-Users and File-Service-Providers a trust relationship is established off-line. This can be modeled with help of the `Stakeholder` and `OfflineTrustRelationship` by setting the role attribute of the latter to the role of the yellow pages service. The role referenced in `OfflineTrustRelationship` must be owned by the `Stakeholder` at the trustee end of the association.

For the relationship between File-Service-User (truster) and File-Service-Provider (trustee) an `OnlineTrustRelationship` is required, which has to be managed by an agent playing the File-Service-User attribute. In this relationship provision trust and access trust is required, which answers question 2 of the previous section. The truster is interested in access trust, as it owns files that can be controlled by the trustee by "navigating" along the owner relation and the association between files, when the truster decides to store files at the trustee's file service. Furthermore, provision trust is required as the truster owns the goals to store and retrieve files from the file service, which are performed by the File-Service-Provider. The answer to question 3 can be given with help of the trust types. Regarding access trust the truster is concerned with integrity and confidentiality of its files. Additionally, the availability of a file service is important for the truster, which can be modeled with help of the concept `Asset` associated with an `OnlineTrustRelationship`. Once the assets are

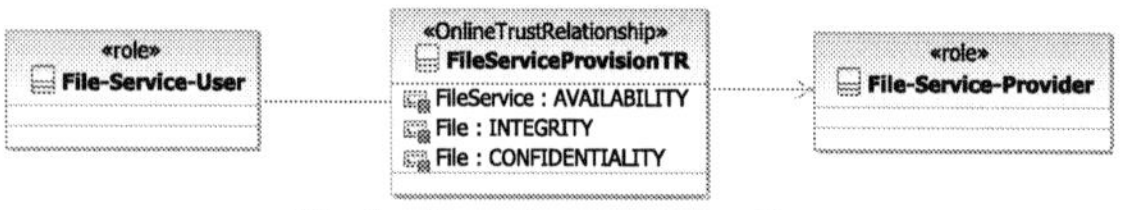


Fig. 3. Example trust relationship.

identified question number 4 of the previous section can be answered with help of the proceeding describe in [26]. For this purpose first a `MaliciousGoal` to make the service unavailable to the truster is introduced. The role that owns and possibly achieves this goal is the trustee involved in the trust relationship. A reason why the trustee wants to achieve this goal may be higher level goals, as e.g. collecting money without providing the desired service or it wants to personally harm the truster.

Question 5 can be answered in different ways, which are successful in making malicious goal impossible to different extends. It may not be possible to prevent the realization of the goal. In this case either trust or norms with associated sanctions have to be employed. The malicious goal to make the service unavailable may be addressed directly by a `TrustManagementGoal` to evaluate availability. This goal can be achieved autonomously by a truster, as it can measure availability. As this is only possible after an interaction with a service provider has been initiated, further goals may be introduced, as e. g. collecting reputation or recommendations. Furthermore, to complement autonomous trust management, it may be required to remove incentives for higher level goals by introducing norms with associated sanctions. In this case a `SanctionGoal` can be introduced that is modeled with help of the `trigger` association between goals (from O-MaSE). In the case of the integrity requirement on a file a similar approach may be pursued. For this purpose file integrity must be made observable by a truster e. g. by means of signatures. In the case of confidentiality of files `ProtectionGoal` can be introduced that requires a truster to encrypt all files. A protection goal ensures that an affected asset is protected. Question 6 from the previous section is answered with help of the classes (as e. g. a file) involved in evaluating trust. If a truster is able to monitor and interpret all relevant properties there is no need for its owner to participate in trust management. When a `TrustManagementGoal` is assigned to the owner of a role a suitable interface between agents playing the role and their owners has to be provided.

`SanctionGoals` introduced in Trust Analysis require to extend parts of requirements models created in earlier tasks, as e.g. roles to perform sanctions have to be defined. Norms that describe expected behavior and sanctions have to be introduced in the Norm Analysis task. Sanctions should require sufficient evidence to be applied, as otherwise a truster may create an unfounded sanction that harms the trustee. Furthermore, a truster in a trust relationship must trust agents playing sanctioning roles to effectively sanction misbehaving trustees. This introduces new trust relationships into a trust model. The same applies for `TrustManagementGoals` that involve reputation or recommendations, which may require to extend existing or introduce new roles and create trust relationships with them regarding recommendation provision for other trust relationships.

## V. RELATED WORK

To our knowledge this is the second proposal to integrate trust modeling in the early phases of AOSE. The first is the Secure-Tropos approach presented in [1], [27] which is an extension of Tropos [28]. This approach can be seen as complementary to our approach as it mainly targets the early requirements phase and relations between actors in the application domain and these actors and the system to be developed. In a first step dependencies among actors in an application domain are analyzed. Dependencies can exist because of goals, tasks, and resources (called *service* in the Tropos methodology). Each dependency entails a trust relationship. Secure-Tropos helps to identify which services have to be considered in these trust relationships and depending on the type of service and who owns the service which kind of trust relationship is required. Secure-Tropos trust distinguishes in permission trust and execution trust. The former corresponds to our notion of access trust and the latter to provision trust.

When trust relationships among actors have been identified the system to be developed is introduced as a new actor and trust relationships between actors and the system are identified. The system is represented by one or more actors that have dependencies to each other and the actors from its environment identified in the previous step. Trust relationships among these new actors can then be identified.

But what is required to manage trust relationships autonomously by software agents is not addressed in Secure-Tropos i. e. the requirements on what has to be done to manage trust by single actors (the agents) and their stakeholders. Furthermore, there is no notion of ownership between actors in the environment of the system and the actors (agents) within the system as in our approach. An integration of [1] with our approach seems to be desirable and should be examined further. The early requirements phase explicitly addresses the needs of actors in the environment of the system which is not addressed in the O-MaSE initiation phase that corresponds roughly with the late requirements phase.

## VI. Conclusion and Future Work

This paper introduced new concepts that can be used to characterize trust in open agent societies in the early activities of the O-MaSE method. Unfortunately, a standardized AOSE methodology is still not available, but we hope that the concepts we developed are not only limited to the O-MaSE method, as many AOSE methods use similar concepts.

Due to space limitations and as we focused on introduction of concepts, we have only informally presented a small case study as a first proof of concept. The applicability of our approach to a more complex application has to be examined. Furthermore, automatic reasoning support about trust requirements similar to that in [1] will be provided, in order to automatically evaluate the consistency of our trust models.

A very important step is to examine how trust models have to evolve in the design phase of AOSE (e. g. integration with [2]) in order to be later implemented in software agents with help of suitable trust and reputation frameworks like [6], [7] and [8]. These models will be integrated into the O-MaSE Framework, as well. In this context it is also necessary to extend our model, in order to describe how different attributes of assets affect trust decisions. For instance, in our example case study files may have an attribute describing their confidentiality. Depending on this attribute it may not be necessary to encrypt files. Furthermore, it may be very expensive to check integrity of each file each time a File-Service-User retrieves one of its files. This behavior may be adjusted according to the trust in the File-Service-Provider in order to really benefit from trust management. Another important research direction is to examine simulation of trust models during the design phase in order to evaluate their validity and appropriateness to enable software agents to make trust decisions at runtime.

## References


[1] P. Giorgini, F. Massacci, J. Mylopoulos, A. Siena, and N. Zannone, "ST-Tool: A CASE tool for modeling and analyzing trust requirements," in *Proceedings of the Third International Conference on Trust Management (iTrust 2005)*, ser. LNCS, vol. 3477. Springer Verlag, 2005, pp. 415–419.

[2] S. Kaffille and G. Wirtz, "Modeling the static aspects of trust for open mas," in *Proceedings of the International Conference on Intelligent Agents, Web Technologies and Internet Commerce*, 2006.

[3] T. J. Norman, A. Preece, S. Chalmers, N. R. Jennings, M. Luck, V. D. Dang, T. D. Nguyen, V. Deora, J. Shao, A. Gray, and N. Fiddian, "Conoise: Agent-based formation of virtual organisations," in *Proceedings of 23rd SGAI International Conference on Innovative Techniques and Applications of AI*, 2003.

[4] A. Grünert, S. Kaffille, and G. Wirtz, "A proposal for a decentralized multi-agent architecture for virtual enterprises," in *The Nineteenth International Conference on Software Engineering and Knowledge Engineering (SEKE 2007)*, 2007.

[5] S. D. Ramchurn, D. Huynh, and N. R. Jennings, "Trust in multi-agent systems," *Knowl. Eng. Rev.*, vol. 19, no. 1, pp. 1–25, 2004.

[6] C. Bryce, P. Couderc, J.-M. Seigneur, and V. Cahill, "Implementation of the secure trust engine," in *iTrust*, 2005, pp. 397–401.

[7] J. Sabater, M. Paolucci, and R. Conte, "Repage: REPutation and ImAGE among limited autonomous partners," *Journal of Artificial Societies and Social Simulation*, vol. 9, 2006.

[8] S. König, S. Kaffille, and G. Wirtz, "Implementing ReGreT in a decentralized multi-agent environment," in *Fifth German Conference on Multiagent System Technologies (MATES 2007)*, 2007.

[9] C. Castelfranchi and R. Falcone, "Principles of trust for MAS: Cognitive anatomy, social importance, and quantification," in *ICMAS '98: Proceedings of the 3rd International Conference on Multi Agent Systems*. Washington, DC, USA: IEEE Computer Society, 1998, p. 72.

[10] A. Jøsang, R. Ismail, and C. Boyd, "A survey of trust and reputation systems for online service provision," *Decision Support Systems*, vol. 43, no. 2, pp. 618–644, 2007.

[11] T. W. A. Grandison, "Trust management for internet applications," Ph.D. dissertation, Imperial College of Science, Technology and Medicine, London, 2003.

[12] C. M. Jonker and J. Treur, "Formal analysis of models for the dynamics of trust based on experiences," in *Proceedings of the 9th European Workshop on Modelling Autonomous Agents in a Multi-Agent World: Multi-Agent System Engineering (MAAMAW-99)*, F. J. Garijo and M. Boman, Eds., vol. 1647. Berlin: Springer-Verlag: Heidelberg, Germany, 30– 2 1999, pp. 221–231.

[13] V. Dignum, "A model for organizational interaction: based on agents, founded in logic," Ph.D. dissertation, Universiteit Utrecht, 2003.

[14] J. Vazquez-Salceda, *The Role of Norms and Electronic Institutions in Multi-Agent Systems*, ser. Whitestein Series in Software Agent Technologies and Autonomic Computing. Birkhäuser, 2004, no. XVIII.

[15] P. Davidsson, "Categories of artificial societies," in *ESAW '01: Proceedings of the Second International Workshop on Engineering Societies in the Agents World II*. London, UK: Springer-Verlag, 2001, pp. 1–9.

[16] P. Davidsson and S. Johansson, "On the potential of norm-governed behavior in different categories of artificial societies," *Computational and Mathematical Organization Theory*, vol. 12, no. 2, pp. 169–180, 2006.

[17] C. Castelfranchi, "Engineering social order," in *ESAW*, 2000, pp. 1–18.

[18] R. Cervenka, I. T. Cansky, M. Calisti, and D. Greenwood, "AML: Agent Modeling Language Toward Industry-Grade Agent-Based Modeling," in *Agent-Oriented Software Engineering V: 5th International Workshop, AOSE 2004*, ser. Lecture Notes in Computer Science, J. Odell, P. Giorgini, and J. P. Müller, Eds. Springer, 2004, vol. 3382 / 2005, pp. 31–46.

[19] S. A. DeLoach, "Engineering organization-based multiagent systems." in *SELMAS*, 2005, pp. 109–125.

[20] I. Partsakoulakis and G. Vouros, "Importance and Properties of Roles in MAS Organization: A review of methodologies and systems," in *Proceedings of the workshop on MAS Problem Spaces and Their Implications to Achieving Globally Coherent Behavior*, 2002.

[21] G. Cabri, L. Ferrari, and L. Leonardi, "Agent Role-based Collaboration and Coordination: a Survey About Existing Approaches," in *2004 IEEE International Conference on Systems, Man, and Cybernetics*, 2004.

[22] J. Vazquez-Salceda, V. Dignum, and F. Dignum, "Organizing multiagent systems," *Autonomous Agents and Multi-Agent Systems*, vol. 11, no. 3, pp. 307–360, November 2005.

[23] S. A. DeLoach, "Modeling organizational rules in the multi-agent systems engineering methodology," in *AI '02: Proceedings of the 15th Conference of the Canadian Society for Computational Studies of Intelligence on Advances in Artificial Intelligence*. London, UK: Springer-Verlag, 2002, pp. 1–15.

[24] J. C. Garca-Ojeda, S. A. DeLoach, Robby, W. H. Oyenan, and J. Valenzuela, "O-mase: A customizable approach to developing multiagent development processes." in *AOSE*, ser. Lecture Notes in Computer Science, M. Luck and L. Padgham, Eds., vol. 4951. Springer, 2007, pp. 1–15.

[25] D. G. Firesmith and B. Henderson-Sellers, *The OPEN Process Framework*. Addison-Wesley, 2002.

[26] A. van Lamsweerde, "Elaborating security requirements by construction of intentional anti-models," in *ICSE '04: Proceedings of the 26th International Conference on Software Engineering*. Washington, DC, USA: IEEE Computer Society, 2004, pp. 148–157.

[27] P. Giorgini, F. Massacci, J. Mylopoulos, and N. Zannone, "Requirements engineering for trust management: Model, methodology, and reasoning," vol. 5, no. 4, pp. 257–274, 2006.

[28] P. Bresciani, P. Giorgini, F. Giunchiglia, J. Mylopoulos, and A. Perini, "Tropos: An agent-oriented software development methodology," *Journal of Autonomous Agents and Multi-Agent Systems*, vol. 8, no. 3, pp. 203–236, May 2004.

# Evaluation of Semantic Actions in Predictive Non-Recursive Parsing

José L. Fuertes, Aurora Pérez
Dept. LSIIS
School of Computing. Technical University of Madrid
Madrid, Spain

*Abstract*— **To implement a syntax-directed translator, compiler designers always have the option of building a compiler that first performs a syntax analysis and then transverses the parse tree to execute the semantic actions in order. Yet it is much more efficient to perform both processes simultaneously. This avoids having to first explicitly build and afterwards transverse the parse tree, which is a time- and resource-consuming and complex process. This paper introduces an algorithm for executing semantic actions (for semantic analysis and intermediate code generation) during predictive non-recursive LL(1) parsing. The proposed method is a simple, efficient and effective method for executing this type of parser and the corresponding semantic actions jointly with the aid of no more than an auxiliary stack.**

## I. Introduction

A parser uses a context-free grammar (*G*) to check if the input string syntax is correct. Its goal is to build the syntax tree for the analyzed input string. To do this, it applies the grammar rules. The set of valid strings generated by this grammar is the language (*L(G)*) recognized by the parser.

An LL(1) parser is built from an LL(1) grammar. The symbols of the grammar are input into a stack. The non-terminal symbol on the top of the stack and the current symbol of the input string determine the next grammar rule to be applied at any time.

A syntax-directed translator is built by defining attributes for the grammar symbols and semantic actions to compute the value of each attribute depending on others. This translator performs syntax analysis, semantic analysis, and code (intermediate or object) generation tasks.

Semantic action execution [1] can be easily integrated into several different parser types. But if you have designed a compiler with a predictive non-recursive LL(1) parser, you will find that attributes for grammar symbols that have been removed from the LL(1) parser stack are required to execute most of the semantic actions [2].

One possible solution is to build the parser tree and then transverse this tree at the same time as the semantic actions are performed. The attribute values are annotated in the tree nodes. Evidently, there is a clear efficiency problem with this solution. It also consumes an unnecessarily large quantity of resources (memory to store the whole tree, plus the node attributes, execution time…), not to mention the extra work on implementation. For this reason, a good approach is to evaluate the semantic actions at the same time as syntax analysis is performed [3] [4].

Semantic actions can be evaluated during LL parsing by extending the parser stack. The extended parser stack holds action-records for execution and data items (synthesize-records) containing the synthesized attributes for non-terminals. The inherited attributes of a non-terminal *A* are placed in the stack record that represents that non-terminal. On the other hand, the synthesized attributes for a non-terminal *A* are placed in a separate synthesize-record right underneath the record for *A* in the stack [5].

In this article, we introduce an algorithm for a top-down translator that provides a simpler, more efficient and effective method for executing an LL(1) parser and the corresponding semantic actions jointly with the aid of no more than an auxiliary stack.

The remainder of the article is organized as follows. Section II reviews the notions of top-down translators. Section III describes how the proposed top-down translator works, and section IV introduces an algorithm to implement this translator. Section V shows an example of how this method works. Finally, section VI outlines some conclusions.

## II. Related Work

This section reviews the concepts of top-down parsers and translation schemes that can be used to cover semantic and code generation aspects.

### A. Top-Down Parser

A parser applies context-free grammar rules [6] to try to find the syntax tree of the input string. A top-down parser builds this tree from the root to the leaves. At the end of the analysis, the tree root contains the grammar's start symbol and the leaves enclose the analyzed string, provided this is correct.

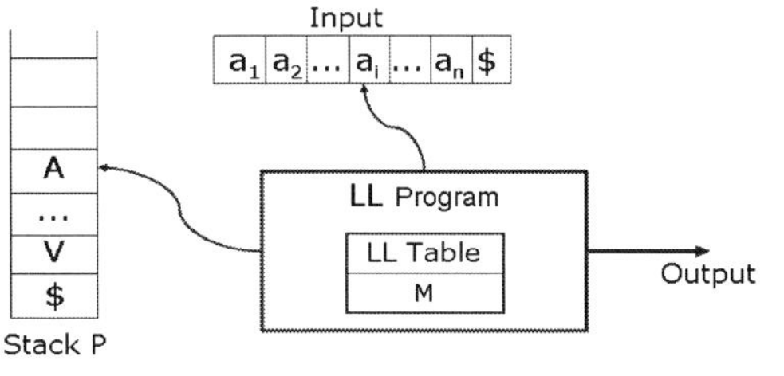


Fig. 1. Overview of an LL parser

T. Sobh (ed.), *Innovations and Advances in Computer Sciences and Engineering*,
DOI 10.1007/978-90-481-3658-2_86, © Springer Science+Business Media B.V. 2010

| *M* | ... | $a_i$ | ... | $ |
|---|---|---|---|---|
| ... | ... | ... | ... | ... |
| A | ... | A→XYZ | ... | ... |
| ... | ... | ... | ... | ... |

Fig. 2. LL(1) parsing table.

Additionally, a compiler parser always has to produce the same parser tree for each input string. In the case of an LL(k) parser, the mechanism used to assure that there is only one rule applicable at any time is an LL(k) grammar. This grammar finds out which rule has to be applied by looking ahead at most *k* symbols in the input string. The simplest grammar of this type is LL(1). LL(1) finds out which rule to apply by looking no further than the first symbol in the input string.

An LL parser (Fig. 1) uses a stack (*P*) of grammar symbols and a table (*M*). The table (*M*) stores information on which rule to use to expand a non-terminal symbol on the top of the stack (*A*) depending on the current input symbol ($a_i$).

As initially configured the stack contains a symbol to indicate the bottom of stack (*$*) with the grammar's start symbol on top.

The rows of the LL(1) parsing table (Fig. 2) contain the non-terminal symbols. The table columns include the terminal symbols (set of valid input symbols) plus the end-of-string symbol (*$*). The table cells can contain a grammar production or be empty. If the parser accesses an empty cell, there is a syntax error in the input string. By definition, an LL can evidently never have more than one rule per cell.

If there is a non-terminal symbol, *A*, on top of the stack, the parser inspects the current input symbol, $a_i$, and looks up the matching table cell, *M [A,* $a_i$*]*, as shown in Fig. 1. This cell contains the rule to be applied (see Fig. 2). Then the non-terminal symbol *A* that was on top of stack *P* is removed and replaced by the right side of the applied rule. The symbol that was in the left-most position on the right side of the production (in this case *X*) is now on top of the stack (see Fig. 3). This is the next symbol to be expanded.

If there is a terminal symbol on top of the stack, it must match the current input symbol. If they are equal, the parser takes out the symbol on top of the stack and moves ahead to the next symbol in the input string. Otherwise, the parser discovers that the syntax of the input string is incorrect.

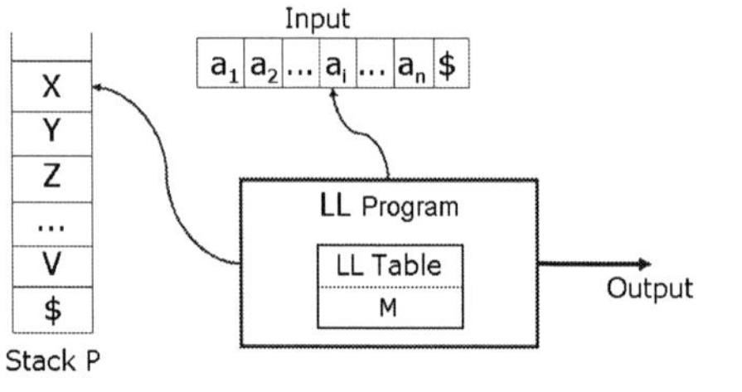


Fig. 3. Configuration of the parser after expanding rule *A → X Y Z.*

## B. *Translation Schemes*

A context-free grammar accounts for the syntax of the language that it generates but cannot cover aspects of the semantics of this language. For example, let rule (1) be:

S → id := E (1)

Rule (1) reproduces a language's assignation sentence syntax perfectly. But it is no use for checking whether the expression and identifier types are compatible or, conversely, the programmer is trying to assign an invalid value to that identifier.

A translation scheme is a context-free grammar in which attributes are associated with the grammar symbols and semantic actions are inserted within the right sides of productions [1]. These semantic actions are enclosed between brackets *{ }*. The attributes in each production are computed from the values of the attributes of grammar symbols involved in that production [7].

So, a translation scheme can include semantic information by defining:

- as many attributes as semantic aspects need to be stated for each symbol
- semantic actions that compute attribute values.

For rule (1), for example, the type attribute would be used for both the identifier (*id*) and the expression (*E*), and it would need to check that *id.type* is equal to or compatible with *E.type.*

There are two kinds of attributes: synthesized and inherited [8]. An attribute is synthesized if its value in a tree node depends exclusively on the attribute values of the child nodes. In any other case, it is an inherited attribute. In rule (2), for example, *A.s* is synthesized and *Y.i* is inherited.

A → X {Y.i:= g (A.i, X.s)} Y Z {A.s:= f (X.s, Y.i)} (2)

An L-attributed translation scheme assures that an action never refers to an attribute that has not yet been computed. An L-attributed translation scheme uses a subset of attributes [9] formed by:

- all the synthesized attributes
- inherited attributes for which the value of an attribute in a node is computed as a function of the inherited attributes of the parent and/or attributes of the sibling nodes that are further to the left than the node.

Whereas the *Y.i* and *A.s* attributes in rule (2) above meet this requirement, the attribute *X.i* would not if the rule included the semantic action *X.i:= h (A.s, Z.i).*

# III. PROPOSED TOP-DOWN TRANSLATOR

In this section we introduce the design of the proposed top-down translator that can output the translation (the intermediate or object code in the case of a compiler) at the same time as it does predictive non-recursive parsing. This saves having to explicitly build the annotated parse tree and then transverse it to evaluate the semantic actions (perhaps also having to build the dependency graph [10] to establish the evaluation order).

We use an L-attributed translation scheme as a notation for specifying the design of the proposed translator. To simplify translator design, we consider the following criteria:

*Criterion 1.* A semantic action computing an inherited symbol attribute will be placed straight in front of that symbol.

*Criterion 2.* An action computing a synthesized attribute of a symbol will be placed at the end of the right side of the production for that symbol.

For example, (3) would be a valid rule:

$$X \rightarrow Y_1\ Y_2\ \{Y_3.i := f(X.i,\ Y_1.s)\}\ Y_3\ Y_4\ Y_5\ \{X.s := g(Y_3.i,\ Y_4.s)\} \qquad (3)$$

To generate the proposed top-down translator the LL(1) parser is modified as follows. First, stack *P* is modified to contain not only grammar symbols but also semantic actions. Second, a new stack (*Aux*) is added. This stack will temporally store the symbols removed from stack *P*. Both stacks are extended to store the attribute values (semantic information).

Let us now look at how the attribute values will be positioned in each stack. To do this, suppose that we have a generic production $X \rightarrow \alpha$. This production contains semantic actions before and after each grammar symbol, where $\alpha \equiv \{1\}\ Y_1\ \{2\}\ Y_2 ...\ \{k\}\ Y_k\ \{k+1\}$.

Fig. 4 shows the parser stack *P* and the auxiliary stack *Aux*, both augmented to store the symbol attributes. For simplicity's sake, suppose that each grammar symbol has at most one attribute. If it had more, each position in the extended stacks would be a register with one field per attribute.

Suppose that these stacks are configured as shown in Fig. 4, with semantic action *{i}* at the top of stack *P*. This means, as we will see from the algorithm presented in section 4, that this semantic action should be executed. There is a pointer to the top of each stack. After executing the semantic action *{i}*, there will be another pointer to the new top (*ntop*) of stack *P*.

Because of the above-mentioned Criterion 1, the semantic action *{i}* uses an inherited attribute of *X* and/or any attribute of any symbol $Y_j$ ($1 \leq j < i$) on the right side of the production to compute the inherited attribute of $Y_i$. If $i = k + 1$, the action *{i}* computes the synthesized attribute of *X*, because of Criterion 2. The following then applies.

- *Case 1.* The semantic action *{i}* computes the inherited attribute of $Y_i$.
  The symbol $Y_i$ will be in stack *P*, right underneath action *{i}*, which is being executed. Thus, $Y_i$ will be the new top (*ntop*) of stack *P* at the end of this execution. The reference to an inherited attribute of $Y_i$ can be viewed as an access to stack *P* and, specifically, position *P [ntop]*.
- *Case 2.* The semantic action *{i}* contains a reference to an attribute of *X*.

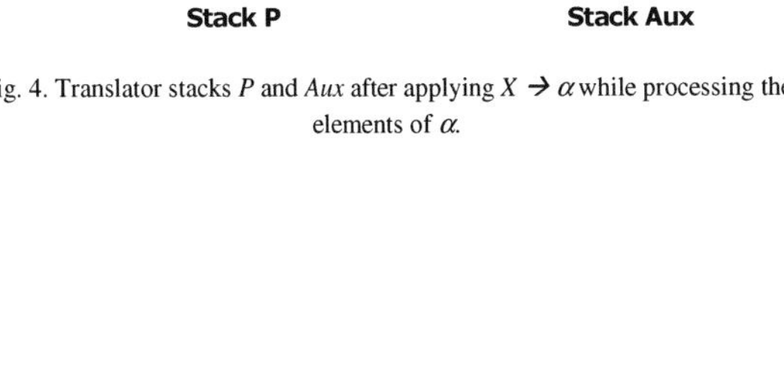


Fig. 4. Translator stacks *P* and *Aux* after applying $X \rightarrow \alpha$ while processing the elements of $\alpha$.

By definition of the L-attributed translation scheme, this will always be a reference to an inherited attribute of *X*. Only if $i = k + 1$ will there be a reference to a synthesized attribute of *X*. As *X* will have already been removed from stack *P* when the rule $X \rightarrow \alpha$ was applied, the symbol *X* will have been entered in stack *Aux*. All the grammar symbols $Y_1$, $Y_2$... $Y_{i-1}$ (preceding the semantic action *{i}*) will be on top of *X*. These symbols will have been removed from *P* and inserted into *Aux*. Then any reference in *{i}* to an attribute of *X* can be viewed as an access to stack *Aux*, specifically, position *Aux [top – i + 1]*.

- *Case 3.* The semantic action *{i}* contains a reference to an attribute of some symbol of $\alpha$.
  By definition of the L-attributed translation scheme, this attribute will necessarily belong to a symbol positioned to the left of action *{i}*, i.e. to one of the symbols $Y_1$, $Y_2$... $Y_{i-1}$. These symbols will have already been moved from stack *P* to stack *Aux*. Then any reference to an attribute of any of these symbols of $\alpha$ can be viewed as an access to stack *Aux* taking into account that $Y_{i-1}$ will be on *Aux [top]*, $Y_{i-2}$ will be on *Aux [top – 1]*... $Y_1$ will be on *Aux [top – i + 2]*.

The translator is implemented by programming the semantic actions and inserting them into the parser code. These semantic actions were written in terms of grammar symbols and attributes in the translation scheme. They now have to be rewritten in terms of accesses to the exact stack positions containing the values of the symbol attributes referenced in each semantic action.

The two criteria are designed merely to simplify the translator design. But the method would also work provided that the semantic action computing an inherited attribute of a symbol is located before, but not necessarily straight in front of, that symbol (Criterion 1). It would also be operational if the semantic action computing a synthesized attribute of, the symbol located on the left side of the production (*X.s*) is not at the right end of the production (Criterion 2) but depends exclusively on symbol attributes to its left. Therefore, we could also use rule (4) instead of rule (3). In the first semantic action, attribute $Y_3.i$ will be positioned in the middle of stack *P*, specifically *P [ntop – 1]* in this case. The second semantic action is also a valid action because *X.s* does not depend on $Y_5$. The referenced attributes will be in stack *Aux* ($Y_3.i$ at *Aux [top – 1]*, $Y_4.s$ at *Aux [top]* and *X.s* at *Aux [top – 4]*.

$$X \rightarrow Y_1\ \{Y_3.i := f\ (X.i,\ Y_1.s)\}\ Y_2\ Y_3\ Y_4\ \{X.s := g\ (Y_3.i,\ Y_4.s)\}\ Y_5 \qquad (4)$$

## IV. Top-Down Translator Algorithm

Having established the principles of the proposed top-down translator, we are now ready to present the algorithm. This algorithm is an extended version of the table-driven predictive non-recursive parsing algorithm that appears in [1].

The algorithm uses a table *M'*. This table is obtained from the LL parser table *M* by substituting the rules of the grammar *G* for the translation scheme rules (which include the modified semantic actions for including stack accesses instead of attributes). Then the proposed top-down translator algorithm is described as follows:

*Input.* An input string $\omega$, a parsing table $M$ for grammar $G$ and a translation scheme for this grammar.

*Output.* If $\omega$ is in $L(G)$, the result of executing the translation scheme (translation to intermediate or object code); otherwise, an error indication.

*Method.* The process for producing the translation is:

1. Each reference in a semantic action to an attribute is changed to a reference to a position in the stack ($P$ or *Aux*) containing the value of this attribute. Then the translation scheme is extended by adding a new semantic action at the end of each production. This action pops as many elements from the stack *Aux* as grammar symbols there are in the right side of the production. Finally, this modified translation scheme is incorporated into table $M$, leading to table $M'$.
2. Initially, the configuration of the translator is:
   - $\$S$ is in stack $P$, with $S$ (the start symbol of $G$) on top,
   - the stack *Aux* is empty, and
   - $\omega\$$ is in the input, with *ip* pointing to its first symbol.
3. **Repeat**
   **Let** $X$ be the symbol on top of stack $P$
   **Let** $a$ be the input symbol that *ip* points to
   **If** $X$ is a terminal **Then**
     **If** $X = a$ **Then**
       **Pop** $X$ and its attributes out of stack $P$
       **Push** $X$ and its attributes onto stack *Aux*
       **Advance** *ip*
     **Else** Syntax-Error ()
     **If** $X$ is a non-terminal **Then**
       **If** $M[X, a] = X \rightarrow \{1\}\ Y_1\ \{2\}\ Y_2 \ldots \{k\}\ Y_k\ \{k+1\}$
         **Then**
         **Pop** $X$ and its attributes out of stack $P$
         **Push** $X$ and its attributes onto stack *Aux*
         **Push** $\{k+1\}, Y_k, \{k\} \ldots Y_2, \{2\}, Y_1, \{1\}$ onto stack $P$, with $\{1\}$ on top
       **Else** Syntax-Error ()
     **If** $X$ is a semantic action $\{i\}$ **Then**
       **Execute** $\{i\}$
       **Pop** $\{i\}$ out of stack $P$
   **Until** $X = \$$ and $Aux = S$

## V. EXAMPLE

To illustrate how the method works let us use a fragment of a C/C++ grammar (Fig. 5) designed to declare local variables. Fig. 6 shows the translation scheme for this grammar.

Based on the translation scheme, apply step 1 of the algorithm described in section 4 to build the modified version of the translation scheme (see Fig. 7). In Fig. 7, references to the inherited attributes *L.type* in rule (*1*) and rule (*5*) and *R.type* in rule (*4*) have been changed to references to new top (*ntop*) of stack $P$. References to other attributes have been replaced with references to stack *Aux*. New semantic actions (calls to the *Pop* function) are included to remove symbols that are no longer needed from stack *Aux*. The number of symbols to be removed is the number of grammar symbols on the right side of the production. This number is passed to the *Pop* function.

Table 1 shows table $M'$ for the modified translation scheme that includes references to stack positions instead of attributes. We have numbered the semantic actions with the production number and, if necessary, with a second digit showing the order of the semantic action inside the production. For instance, action *{1.1}* represents *{P [ntop]:= Aux [top]}*, the first action of production *(1)*, whereas action *{1.2}* represents *{Pop (3)}*, the second action of production *(1)*.

To illustrate how the translator works, consider the input: '*float x, y;*'. This string is tokenized by the scanner as the input string $\omega \equiv$ *float id, id;*.

```
(1)  D→ T L ;
(2)  T→ int
(3)  T→ float
(4)  L→ id R
(5)  R→ , L
(6)  R→ λ
```

Fig. 5. Grammar for C/C++ variables declaration.

```
(1) D→ T {L.type:= T.type}
        L ;
(2) T→ int {T.type:= integer}
(3) T→ float {T.type:= float}
(4) L→ id {insertTypeST (id.ptr, L.type);
           R.type:= L. type}
        R
(5) R→ , {L.type:= R.type}
        L
(6) R→ λ {}
```

Fig. 6. Translation scheme for C/C++ variables declaration.

```
(1) D→ T {P[ntop]:= Aux[top]}
        L ; {Pop (3)}
(2) T→ int {Aux[top-1]:= integer; Pop (1)}
(3) T→ float {Aux[top-1]:= float; Pop (1)}
(4) L→ id {insertTypeST (Aux[top], Aux[top-1]);
           P[ntop]:= Aux[top-1]}
        R {Pop (2)}
(5) R→ , {P[ntop]:= Aux[top-1]}
        L {Pop (2)}
(6) R→ λ {}
```

Fig. 7. Modified translation scheme for C/C++ variables declaration including references to stack positions.

TABLE 1
TABLE M' FOR THE MODIFIED TRANSLATION SCHEME ILLUSTRATED IN FIG. 7.

| *M'* | id | int | float | ; | , | $ |
|---|---|---|---|---|---|---|
| **D** | | D → T {1.1} L ; {1.2} | D → T {1.1} L ; {1.2} | | | |
| **T** | | T → int {2} | T → float {3} | | | |
| **L** | L → id {4.1} R {4.2} | | | | | |
| **R** | | | | R → λ | R → , {5.1} L {5.2} | |

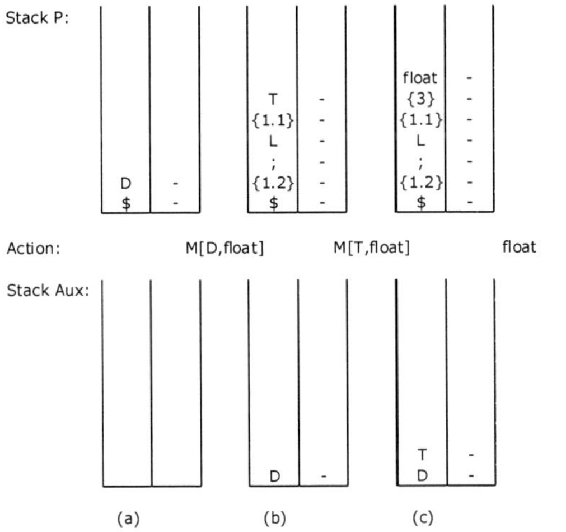

Fig. 8. First stack configurations (the input string is '*float x, y;*').

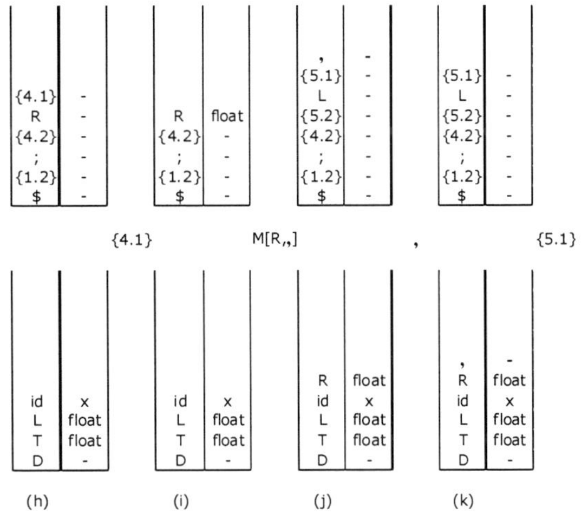

Fig. 10. Stack configurations 8 to 11 (', *y;*' is in the input).

The stacks are initialized (Fig. 8(a)) with *$* and the grammar start symbol (*D*). Figs. 8 to 13 illustrate the different configurations of the extended stacks *P* and *Aux* (stack *P* is positioned above stack *Aux* throughout, and the action taken is stated between the two stacks).

As *D* (a non-terminal) is on top of stack *P* and *float* is the first symbol of $\omega$, check *M [D, float]*. This gives the production $D \rightarrow T\,\{1.1\}\,L\,;\,\{1.2\}$. Therefore, move *D* from stack *P* to stack *Aux* and push the right side of the production onto stack *P* (Fig. 8(b)).

As shown in Fig. 8(c) we find that there is a terminal *float* on top of stack *P*. As this matches the current input symbol, transfer it from stack *P* to stack *Aux* and move *ip* ahead to point to the next input symbol (*id*).

The next element on top of stack *P* is the semantic action *{3}* to be executed. This action is *{Aux [top – 1]:= float; Pop (1)}*. First, insert *float* as the attribute value of symbol *T* (the second symbol from the top of stack *Aux*). Then execute the *Pop* function, which removes one element (*float*) from the top of stack *Aux* (Fig. 9(e)).

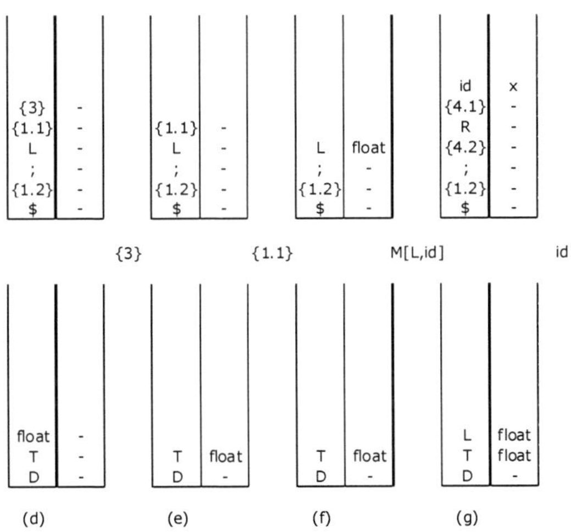

Fig. 9. Stack configurations 4 to 7 ('*x, y;*' is in the input).

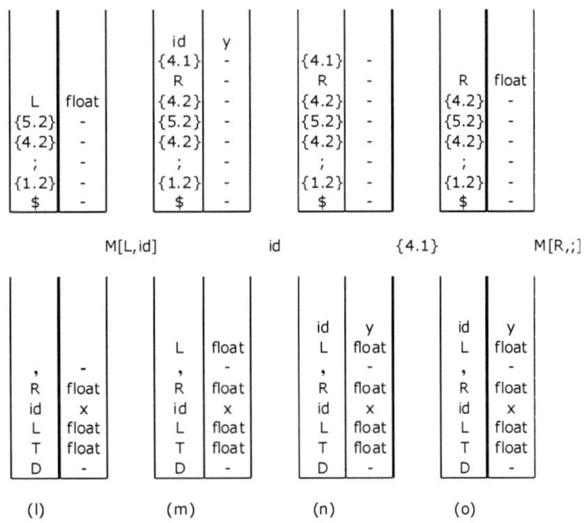

Fig. 11. Stack configurations 12 to 15 ('*y ;*' is in the input).

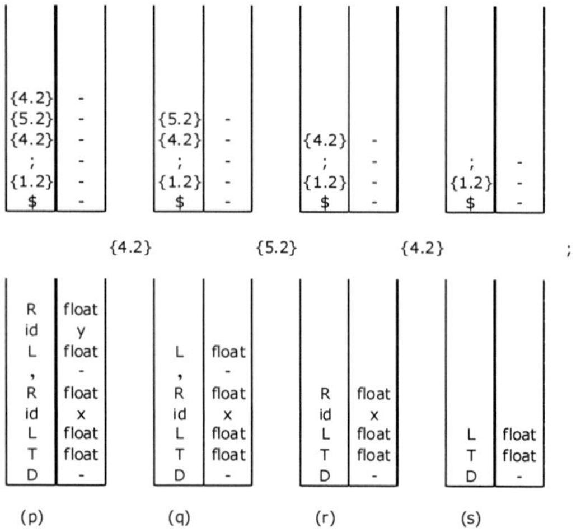


Fig. 12. Stack configurations 16 to 19 (';' is in the input).

As shown in Fig. 9(g), the *id* token has an attribute gathered directly from the scanner. This attribute is handled as a reference to the symbol table entry for this identifier. We represent the attribute using just the name of the identifier in the source code ($x$). Later the semantic action *{4.1}* is on top of stack $P$ (Fig. 10(h)). Its execution copies *float* from stack *Aux* to stack $P$ (*{P [ntop]:= Aux [top – 1]}*) as the attribute value of symbol $R$. The analysis continues as shown in Figs. 10 to 12.

In addition, two actions executed in this example (specifically, semantic action *{4.1}* executed in Fig. 10(h) and Fig. 11(n)) will have included type information about the $x$ and $y$ float identifiers in the compiler's symbol table.

Fig. 13(t) shows the execution of action *{1.2}* that removes three symbols from the stack *Aux*. Fig. 13(u) represents the algorithm exit condition.

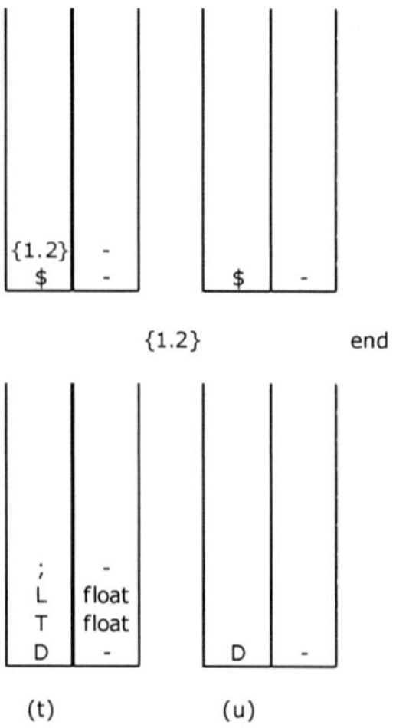


Fig. 13. Final stack configurations having read the whole input string.

## VI. Conclusions

We have introduced a simple method and an algorithm to manage the evaluation of semantic actions in predictive non-recursive top-down LL(1) parsing. The main advantage of this method is that it can evaluate semantic actions at the same time as it parses. The main goal of this simultaneous execution is to save compiler resources (including both execution time and memory), since a compiler of this kind no longer needs to explicitly build a complete tree parser.

This method has been taught in a compilers course at the Technical University of Madrid's School of Computing for the last 6 years. As part of this course, students are expected to build a compiler for a subset of a programming language. About one third of students used this method, with very encouraging results. The method has proved to be easy to implement and understand.

The two criteria imposed in section 3 are merely to simplify the translator design. But the method is general and can be applied to any L-attributed translation scheme for an LL(1) grammar. Additionally, the tests run by students from our School on their compilers have shown that it is an efficient and simple way to perform the task of top-down syntax-directed translation.

## References


[1] A.V. Aho, R. Sethi, J.D. Ullman, *Compilers. Principles, Techniques and Tools*, Addison-Wesley, 1985.

[2] R. Akker, B. Melichar, J. Tarhio, "Attribute Evaluation and Parsing", *Lecture Notes in Computer Science*, 545, Attribute Grammars, Applications and Systems, 1991, pp. 187-214.

[3] T. Noll, H. Vogler, "Top-down Parsing with Simultaneous Evaluation of Noncircular Attribute Grammars", *Fundamenta Informaticae*, 20(4), 1994, pp. 285-332.

[4] K. Müller, "Attribute-Directed Top-Down Parsing", *Lecture Notes in Computer Science*, 641, Proc. 4th International Conference on Compiler Construction, 1992, pp. 37-43.

[5] A.V. Aho, M.S. Lam, R. Sethi, J.D. Ullman, *Compilers. Principles, Techniques and Tools*, 2nd ed., Addison-Wesley, 2007.

[6] N. Chomsky, "Three models for the description of language", *IRE Transactions on Information Theory*, 2, 1956, pp. 113-124.

[7] T. Tokuda, Y. Watanabe, *An attribute evaluation of context-free languages*, Technical Report TR93-0036, Tokyo Institute of Technology, Graduate School of Information Science and Engineering, 1993.

[8] D. Grune, H.E. Bal, C.J.H. Jacobs, K.G. Lagendoen, *Modern Compiler Design*, John Wiley & Sons, 2000.

[9] O.G. Kakde, *Algorithms for Compiler Design*, Laxmi Publications, 2002.

[10] S.S. Muchnick, *Advanced Compiler Design & Implementation*, Morgan Kaufmann Publishers, 1997.

# Pair Hidden Markov Model for Named Entity Matching

Peter Nabende, Jörg Tiedemann, John Nerbonne
Department of Computational Linguistics,
Center for Language and Cognition Groningen,
University of Groningen, Netherlands
{p.nabende, j.tiedemann, j.nerbonne}@rug.nl

*Abstract* – **This paper introduces a pair-Hidden Markov Model (pair-HMM) for the task of evaluating the similarity between bilingual named entities. The pair-HMM is adapted from Mackay and Kondrak [1] who used it on the task of cognate identification and was later adapted by Wieling et al. [5] for Dutch dialect comparison. When using the pair-HMM for evaluating named entities, we do not consider the phonetic representation step as is the case with most named-entity similarity measurement systems. We instead consider the original orthographic representation of the input data and introduce into the pair-HMM representation for diacritics or accents to accommodate for pronunciation variations in the input data. We have first adapted the pair-HMM on measuring the similarity between named entities from languages (French and English) that use the same writing system (the Roman alphabet) and languages (English and Russian) that use a different writing system. The results are encouraging as we propose to extend the pair-HMM to more application oriented named-entity recognition and generation tasks.**

*Keywords*: ***Named entity, Similarity Measurement, Hidden Markov Model, pair-Hidden Markov Model***

## I. INTRODUCTION

The existence of a gigantic number of Named Entities (NEs)[1] across languages necessitates proper handling of names in cross-lingual Natural Language Processing (NLP) tasks. As an example of a problem associated with NEs; In Machine Translation (MT), many systems will rely on the existence of a bilingual lexicon or a dictionary to process cross-language queries [2]. It is well known that bilingual lexicons comprise a very tiny percentage of NEs; a Machine Translation system would therefore fail or perform poorly if an unseen NE is encountered in a query. The task of measuring the similarity between cross-lingual NEs finds its main application in the identification or extraction of term translations ([4], [2]) to help deal with unseen terms. Other NLP tasks that can benefit from NE similarity measurement are Cross Language Information Retrieval (CLIR), Question Answering (QA), and Bilingual lexicon construction. Recent work on finding NE translations and measuring similarity between NEs is divided into two approaches: those that consider phonetic information and those that do not. Lam et al. [2] argue that many NE translations involve both semantic and phonetic information at the same time. Lam et al. [2] therefore developed an NE matching model that offers a unified framework for considering semantic and phonetic clues simultaneously within two given entities in different languages. In their work [2], they formulate their problem as a bipartite weighted graph matching problem. Similarity measurement has also been done for transliterations. Hsu et al. [6] measure the similarity between two transliterations by comparing the physical sounds through a Character Sound Comparison (CSC) method. The CSC method is divided into two parts: a training stage where a speech sound similarity database is constructed including two similarity matrices; and a recognition stage where transliterations are switched to a phonetic notation and the similarity matrices are applied to calculate the similarity of different transliterations. Pouliquen et al. [7] compute the similarity between pairs of names by taking the average of three similarity measures. Their similarity measures are based on letter n-gram similarity. They calculate the cosine of the letter n-gram frequency lists for both names, separately for bi-grams and for tri-grams; the third measure being the cosine of bigrams based on strings without vowels. Pouliquen et al. [7] do not use phonetic transliterations of names as they consider them to be less useful than orthographically based approaches. Because of various limiting factors, however, Pouliquen et al. [7] obtain results that are less satisfactory than for language-specific Named Entity Recognition systems. The precision obtained from their results was nevertheless higher.

We propose a pair-HMM that takes in as input NE pairs in their original orthographic representation for evaluation. When a Romanization system or phonetic representation step is required, the pair-HMM can still be used for evaluating the Romanized or phonetically represented input [5].

In the second section we introduce the pair-HMM, in the third section we describe how the pair-HMM trained and test results, we conclude in the fourth section with pointers to future work.

[1] There are three different types of Named Entities[3]: entity names, temporal expressions, and number expressions; in this paper we only consider entity names (that is person, location, and organization names)

T. Sobh (ed.), *Innovations and Advances in Computer Sciences and Engineering*,
DOI 10.1007/978-90-481-3658-2_87, 

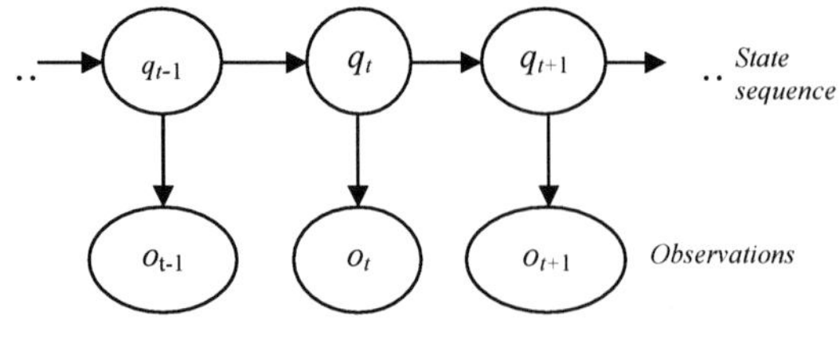


Fig. 1: An instantiation of a HMM

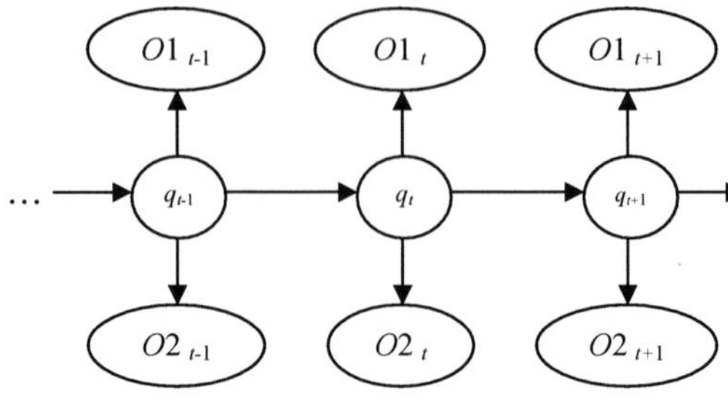

Fig. 2: An instantiation of a pair-HMM

## II. pair-Hidden Markov Model

Pair Hidden Markov Models belong to a family of models called Hidden Markov Models (HMMs) which are in turn derived from Markov models. Markov models originated from the work of Andrei Markov in 1907 when he began the study of a chance process where the outcome of a given experiment can affect the outcome of the next experiment [8]. A markov model comprises a set of states, each state corresponding to an observable (physical) event, and arcs. Rabiner [9] described how the concept of a Markov model can be extended to include the case where the observation is a probabilistic function of the state. In other words, the resulting model (HMM), is a doubly embedded stochastic process with an underlying stochastic process that is not observable. Figure 1 illustrates an instantiation of a HMM. HMMs became powerful statistical tools for modeling generative sequences that can be characterized by an underlying Markov process that generates an observable sequence. The key difference between a HMM and Markov model is that we can not exactly state what state sequence produced the observations and thus the state sequence is "hidden". It is, however, possible to calculate the probability that the model produced the sequence, as well as which state sequence was most likely to have produced the observations. When using HMMs, two assumptions are made. The first, following the first-order Markov Assumption is formally specified as [10]:

$$\forall t \leq n: P(q_t \mid q_{t-1},...,q_1) = P(q_t \mid q_{t-1}) \quad (1)$$

where $q_t$ represents the state at time $t$. The second is called the Independence assumption and states that the output observation at time $t$ ($o_t$) is dependent only on the current state; it is independent of previous observations and states:

$$P(o_t \mid o_1^{t-1}, q_1^t) = P(o_t \mid q_t) \quad (2)$$

Generally, an HMM $\mu$ is specified by a tuple $\mu = (A, E, \Pi)$, where $A$ represents the transition probabilities specified in a transition matrix, $E$ specifies emission probabilities and $\Pi$ specifies initial probabilities associated with starting in a given state.

Durbin et al. [11] made changes to a Finite State Automata (FSA) and converted it into an HMM which Durbin et al. [11] called the pair-HMM. Observations in a pair-HMM are formed by a couple of sequences (the ones to be aligned) and the model assumes that the hidden (that is the non-observed alignment sequence $\{q_t\}_t$) is a Markov chain that determines the probability distribution of the observations [12]. The main difference between the pair-HMMs and the standard HMMs lies in the observation of a pair of sequences (Figure 2) or a pairwise alignment instead of a single sequence. The model we adapt in this work is a modification of the pair-HMM adapted by Mackay and Kondrak [1] to the task of computing word similarity. Mackay and Kondrak's [1] pair-HMM was successfully adapted to the task of comparing different Dutch dialects by Wieling et al. [5]. Mackay and Kondrak's pair-HMM is shown in figure 3. The pair-HMM (Figure 3) has three states that represent the basic edit operations: substitution (represented by state "M"), insertion ("Y"), and deletion ("X"). In this pair-HMM, probabilities both for emissions of symbols from the states, and for transitions between states are specified. "M", the match state has emission probability distribution $p_{x_i y_j}$ for emitting an aligned pair of symbols $x_i$:$y_j$. States X and Y have distributions $q_{x_i}$ and $q_{y_j}$ for emitting symbols $x_i$ and $y_j$ respectively against a

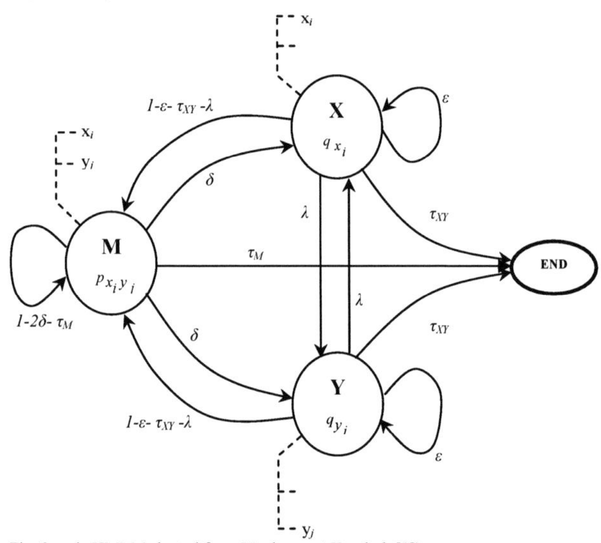


Fig. 3: pair-HMM (adapted from Mackay and Kondrak [1])

gap. The model has five transition parameters: $\delta$, $\varepsilon$, $\lambda$, $\tau_M$, and $\tau_{XY}$ as shown in Figure 3.

Two main modifications were made by Mackay and Kondrak [1] to the model developed by Durbin et al. [11]: first, a pair of transitions between states "X" and "Y" whose parameter is denoted by $\lambda$ in figure 3 was added and the probability for the transition to the end state $\tau$ was split into two separate values: $\tau_M$ for the substitution state and $\tau_{XY}$ for the gap states X and Y.

Although we have maintained some of the assumptions used in previous work on the pair-HMM, there are more modifications that are worth considering. Firstly Mackay and Kondrak maintain the assumption used by Durbin et al. (1998) concerning the parameter δ, that is, the probability of transiting from the substitution state to the deletion or insertion state is the same. Similarly, the probability of staying in the gap state X or gap state Y is the same. These assumptions are reasonable for the case where the alphabets of both the languages are identical (Mackay, 2004). If the language alphabets are not identical, then the insertion and deletion states should be distinct, with various emission probabilities from the gap state X and gap state Y. Also, the transition parameter from the substitution state to the gap state X should be different from the transition parameter from the substitution state to the gap state Y. In the same vein, the transition parameter of staying in the gap state X should be different from the transition parameter of staying in the gap state Y, and the transition parameter between the gap states should be different. As a result, the pair-HMM that should be suitable for the similarity task in our work should have the parameters as illustrated in the figure 4.

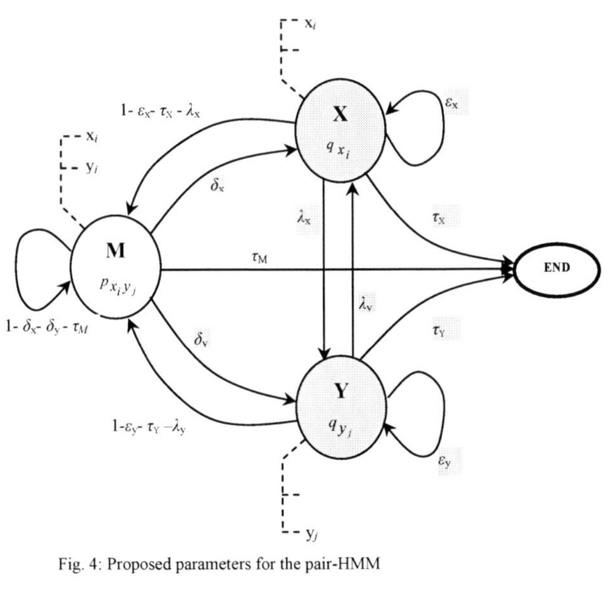


Fig. 4: Proposed parameters for the pair-HMM

We have only included a modification to the pair-HMM concerning the emissions in the gap states in such a way that it considers separate alphabets used in the two states. For example, an emission in the state X will be that for only a symbol from one language alphabet against a gap. Likewise, an emission in the state Y will be that for only a symbol from the other language alphabet against a gap.

## III. SIMILARITY MEASUREMENT USING pair-HMM

Basically, the pair-HMM works by computing a score for input comprising two words from two different languages. The scores can then be used in evaluating the accuracy of the model or in a particular similarity measurement application. The scores are computed using the initial, emission and transition probabilities which are determined through a training procedure.

### A. *pair-HMM Training*

To train a pair-HMM, we need a collection of pairs of names that are considered to be relevant or standard translations of each other. Different methods can be used in estimating the parameters of the pair-HMM. Arribas-Gil et al. [12] review different parameter estimation approaches for pair-HMMs including: Numerical Maximization approaches, Expectation Maximization (EM) algorithm and its variants (Stochastic EM, Stochastic Approximation EM). In their work [12], Maximum Likelihood estimators are proved to be more efficient and to produce better estimations for pair-HMMs when compared with Bayesian estimators on a simulation of estimating evolutionary parameters. The Baum-Welch algorithm (BW) had already been implemented for training the pair-HMM that we adapted and it is used for training the pair-HMM in our work. The BW algorithm falls under the EM class of algorithms that all work by guessing initial parameter values, then estimating the likelihood of the data under the current probabilities. These likelihoods can then be used to re-estimate the parameters, iteratively until a local maximum is reached. A formal description of the BW algorithm for HMMs obtained from Hoberman and Durand [17] is shown in the following:

Algorithm: Baum-Welch

Input: A set of observed sequences, $O^1, O^2, \ldots$

Initialization: Choose arbitrary initial values for model parameters, $\mu(\pi, p_{ij}, e_i(.))$

score = $\sum_d P(O^d \mid \mu')$

Repeat

{

$\mu = \mu'$ , $S = S'$

For each sequence, $O^d$,

{

/* calculate "probable paths" $Q^d = q_1^d, q_2^d, \ldots$ */

calculate $\alpha_i(t)$ for $O^d$ using the Forward algorithm calculate $\beta_i(t)$ for $O^d$ using the Backward algorithm

calculate the contribution of $O^d$ to A = $[p_{ij}]$

calculate the contribution of $O^d$ to E

}

$$p_{ij} = \frac{A_{ij}}{\sum_l A_{il}} \quad (3)$$

$$e_i(v) = \frac{E_i(v)}{\sum_\tau E_i(\tau)} \quad (4)$$

score = $\sum_d P(O^d \mid p_{ij}, e_i()).$

}

Until (the change in score is less than some predefined threshold.)

To determine the transition and emission estimation formulas for the pair-HMM, we sum over each pair position, and over all possible sequences. If we let $h$ represent the index of the pair we are using and forward and backward variables are represented by $f$ and $b$ respectively; we obtain the following expressions for the transition and emission estimations respectively in the substitution state [10]:

$$A_{kl} = \sum_h \frac{1}{P(O \mid \mu)} \sum_i \sum_j f^h_{(i,j)}(k) p_{kl} e_l(x^h_{i+1}, y^h_{j+1}) b^h_{(i+1,j+1)}(l)$$

$$E_k(O^{xy}) = \sum_h \frac{1}{P(O \mid \mu)} \sum_{i \mid x^h_i \in O^{xy}} \sum_{j \mid y^h_j \in O^{xy}} f^h_k(i,j) b^h_k(i,j)$$

The equations for the insertion and deletion state will have slightly different forms [10]. For example, insertions only need to match $y_j$ with the emission $O^{xy}$, since $y_j$ is emitted against a gap. In the estimation for the transition probability, when we end in an insertion or deletion state, we only change the index for one of the pairs and we use the emission probability for a symbol from one word against a gap.

In training the pair-HMM, the training input comprised names for geographical places from two different languages obtained from the GeoNames[2] data dump. We extracted 850 English-French pairs of names and 5902 English-Russian pairs of names. For the English-French dataset, we used 600 pairs of names for training. From the English-Russian dataset, we used 4500 pairs of names. The remaining pairs of names where reserved for evaluating the accuracy of the algorithms used in the pair-HMM. For the English-Russian dataset, we made a modification to the pair-HMM used in previous work so that two separated alphabets belonging to the two languages can be specified. In this paper, we denote $V_1 = \{v_{1_i}\}$ for $i = 1, ..., m$ as the alphabet of symbols in one writing system and $V_2 = \{v_{2_j}\}$ for $j = 1, ..., n$ as the alphabet of symbols in the other writing system. We automatically generated each alphabet used in the pair-HMM from the available dataset for a particular language, for example, the Russian alphabet was generated from the collection of Russian names only. For a given language, we expect to have additional symbols included in the alphabet including diacritics that can be used in stressing pronunciations. For the English-Russian data set, we generated 76 symbols for the English language alphabet while 61 symbols were generated for the Russian language alphabet. For the English-French data set we had 57 symbols for each language.

Following the execution of the Baum-Welch algorithm implemented for the pair-HMM, 282 iterations were needed for the algorithm to converge on the English-French input while it took 861 iterations on the English-Russian input.

[2] http://download.geonames.org/export/dump

### B. *Evaluation of the Viterbi and Forward Algorithms*

Two algorithms have been implemented in the pair-HMM system to estimate similarity scores for two given strings. We did not follow an application-oriented approach for evaluation, instead we only evaluate the performance of the two algorithms implemented in the pair-HMM. We used two measures: Average Reciprocal Rank (ARR) and Cross Entropy (CE).

Figure 5 illustrates the design of the approach leading to the calculation of ARR. In Figure 5, the similarity measurement model comprising the pair-HMM will give a similarity score to the pair of names in the input. The scores can then be sorted in decreasing order and the position at which the actual relevant transliteration for the target name appears in the sorted scores is taken as the rank. A higher rank increases the accuracy of the model. The expressions for Average Rank (AR) and ARR used in this paper follow from Voorhess and Tice [13]:

$$AR = \frac{1}{N} \sum_{i=1}^{N} R(i) \quad (5)$$

$$ARR = \frac{1}{N} \sum_{i=1}^{N} \frac{1}{R(i)} \quad (6)$$

where $N$ is the number of testing data (pairs of NEs), and $R(i)$ is the rank of the correct answer (NE pair) in the set of answers associated with the $i^{th}$ testing data.

For the English-French dataset, we tested only 2 algorithms, that is the log versions of the Viterbi and Forward algorithm. For

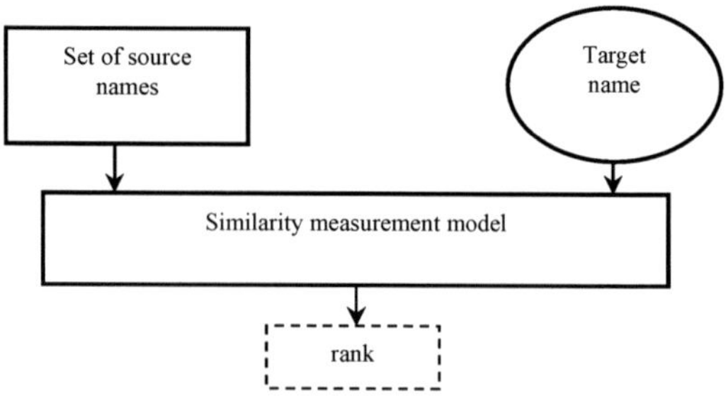


Fig. 5: Ranking approach

the English-Russian data set we tested 4 algorithms: the Viterbi and Forward algorithms in their basic form and the log versions for the two algorithms. The log versions of the algorithms are used to deal with any errors that could arise due to assigning very low probabilities to input string pairs. The ARR results are shown in Table 1 and Table 2 for English-French and English-Russian test data respectively. The ARR results for the logarithmic (log) versions for both the Viterbi and Forward algorithms for the English-French and English-Russian test data show that the Viterbi algorithm consistently performs slightly better than the Forward algorithm. However, for the English-Russian data set, the ARR results show that the basic Forward algorithm performs slightly better than the basic Viterbi algorithm[3].

We also evaluated the algorithms used in the pair-HMM based on Cross Entropy. Cross Entropy involves measuring the model entropy for both the Viterbi and Forward algorithm. "CE is used to compare the effectiveness of different training regimes (estimates of probabilities on the same test data). CE is useful when we do know the actual probability distribution $p$ that generated some data. CE enables us to sue some model of $p$ denoted as $m$ as an approximation to $p$" [14]. For the pair-HMM the CE is specified by the following equation:

$$H(p,m) = \lim_{le\to\infty} - \sum_{(x\in V_1, y\in V_2)} \frac{1}{le} p(x_1 : y_1, \ldots, x_{le} : y_{le}) \log m(x_1 : y_1, \ldots, x_{le} : y_{le})$$

where $x_i : y_i$ for $i$ = 1,…,$le$ represents a pair of output symbols in a pair of strings that are being evaluated with the symbols $\{x_i\}$ from an alphabet $V_1$ and symbols $\{y_i\}$ from an alphabet $V_2$. $le$ represents the length of a given observation sequence for the situation that restricts the length of all observation sequences to the same length.

TABLE 1
ARR RESULTS FOR ENGLISH-FRENCH TEST DATA

| Algorithm | ARR ($N$ = 164) |
|---|---|
| Viterbi log | 0.8099 |
| Forward log | 0.8081 |

TABLE 2
ARR RESULTS FOR ENGLISH-RUSSIAN TEST DATA

| Algorithm | ARR ($N$ = 966) |
|---|---|
| Viterbi | 0.8536 |
| Forward | 0.8545 |
| Viterbi log | 0.8359 |
| Forward log | 0.8355 |

[3] We expected to get similar results when the log computation is introduced and when it is not used. However, that was not the case, despite conducting a check on the software for the log versions of the algorithms. This may be attributed to errors resulting from the computer system when doing log computations. It would therefore be more reliable to consider the values obtained from the algorithms in their basic form and not when a log computation is used.

TABLE 3
CE RESULTS FOR ENGLISH-RUSSIAN TEST DATA

| Algorithm | CE ($n$ = 1000) |
|---|---|
| Viterbi log | 32.2946 |
| Forward log | 32.2009 |

For the pair-HMM, we can approximate an expression for the per name-pair CE for the model $m$ to the expression (7):

$$H(p,m) = \lim_{le\to\infty} - \frac{1}{n} \sum_{(x\in V_1, y\in V_2)} \frac{1}{le} \log m(x_1 : y_1, \ldots, x_{le} : y_{le}) \quad (7)$$

where $n$ in equation (7) is the total number of paired observation sequences considered for calculating the cross entropy of the pair-HMM for a given observation set.

Between two models $m_1$ and $m_2$, the more accurate model will be the one with the lower cross entropy. The Cross Entropy values for two algorithms used (Forward and Viterbi) in the pair-HMM are shown in the Table 3.

The Cross Entropy results show that the Forward algorithm is slightly more accurate than the Viterbi algorithm. This should be the case since the Forward algorithm considers all possible alignments and in so doing considers more information than the Viterbi algorithm that looks for only one best alignment.

## IV. CONCLUSION

Despite the assumptions made concerning the pair-HMM used in this paper, experimental results show encouraging accuracy values. The ability to train the pair-HMM, and to use the pair-HMM for measuring similarity on strings that use completely different alphabets also shows that it is feasible to use the pair-HMM in generating transliterations between two languages that use completely different writing systems.

We propose to extend our research on the pair-HMM in the following ways: Firstly, it is desirable to incorporate more modifications in the pair-HMM for the case of languages that use different writing systems and determine whether improvements will be obtained with regard to the accuracy of the model on the name matching task. Modification of the pair-HMM in this case can involve adding more parameters based on the requirements of the task and / or extending the structure of the model to incorporate additional information. Secondly, although the accuracy values are encouraging, it is important to evaluate the pair-HMM against other models for the name matching task. A different direction would involve considering different Dynamic Bayesian Network (DBN) structures such as those used by [16] on the task of cognate identification. Since DBNs are considered as generalizations of HMMs, it would be interesting to see whether there are any improvements that can be achieved by considering different DBN structures. Finally, discriminative methods have been shown to achieve exceptional performance on separate cognate identification tasks [18] as compared to traditional orthographic measures like Longest Common Subsequence Ratio and Dice's coefficient. It should be interesting to evaluate the performance of the DBN techniques against discriminative methods.

## References


[1] W. Mackay and G. Kondrak, "Computing Word Similarity and Identifying Cognates with Pair Hidden Markov Models," *Proceedings of the Ninth Conference on Computational Natural Language Learning (CoNLL)*, pp. 40-47, Ann Arbor, Michigan, 2005.

[2] W. Lam, S-K. Chan and R. Huang, "Named Entity Translation Matching and Learning: With Application for Mining Unseen Translations," *ACM Transactions on Information Systems*, vol. 25, issue 1, article 2, 2007.

[3] N. Chinchor, "MUC-7 Named Entity Task Definition," *Proceedings of the 7th Message Understanding Conference (MUC-7)*, 2007.

[4] C-J. Lee, J.S. Chang and J-S.R. Juang, "Alignment of Bilingual Named Entities in Parallel Corpora Using Statistical Models and Multiple Knowledge Sources," *ACM Transactions on Asian Language Information Processing*, vol. 5, issue 2, 2006, pp. 121-145.

[5] M. Wieling, T. Leinonen and J. Nerbonne, "Inducing Sound Segment Differences using Pair Hidden Markov Models. In J. Nerbonne, M. Ellison and G. Kondrak (eds.), *Computing and Historical Phonology: 9th Meeting of ACL Special Interest Group for Computational Morphology and Phonology Workshop*, Prague, pp. 48-56, 2007.

[6] C-C. Hsu., C-H. Chen, T-T. Shih and C-K. Chen, "Measuring Similarity between Transliterations against Noise and Data," *ACM Transactions on Asian Language Information Processing*, vol. 6, issue 2, article 5, 2005.

[7] C.M. Grinstead and J.L. Snell, *Introduction to Probability*, 2nd Edition, AMS, 1997.

[8] B. Poliquen, R. Steinberger, C. Ignat, I. Temnikova, A. Widiger, W. Zaghouani and J. Žižka, "Multilingual Person Name Recognition and Transliteration. *Revue CORELA, Cognition, Represéntation, Language*, 2005.

[9] L.R. Rabiner, "A Tutorial on Hidden Markov Models and Selected Applications in Speech Recognition," *Proceedings of the IEEE*, vol. 77, issue 2, pp. 257-286, 1989.

[10] W. Mackay, *Word Similarity using Pair Hidden Markov Models*, Masters Thesis, University of Alberta, 2004.

[11] R. Durbin, S.R. Eddy, A. Krogh and G. Mitchison, *Biological Sequence Analysis: Probabilistic Models of Protein and Nucleic Acids*. Cambridge University Press, 1998.

[12] A. Arribas-Gil, E. Gassiat and C. Matias, "Parameter Estimation in Pair-hidden Markov Models," *Scandinavian Journal of Statistics*, vol. 33, issue 4, pp. 651-671, 2006.

[13] E.M. Voorhees and D.M. Tice. The TREC-8 Question Answering Track Report. *In English Text Retrieval Conference (TREC-8)*, 2000.

[14] D. Jurafsky and H. Martin. *Speech and Language Processing: An Introduction to Natural Language Processing, Computational Linguistics, and Speech Recognition*, 2nd Edition, Pearson Edn Inc., Prentice Hall, 2009.

[15] C-J. Lee, J.S. Chang and J-S.R. Juang. A Statistical Approach to Chinese-to-English Back Transliteration. *In Proceedings of the 17th Pacific Asia Conference*, 2003.

[16] G. Kondrak and T. Sherif. Evaluation of Several Phonetic Algorithms on the Task of Cognate Identification. In *Proceedings of the Workshop on Linguistic Distances*, pages 43-50, Association for Computational Linguistics, Sydney, 2006.

[17] D. Durand and R. Hoberman. HMM Lecture Notes, Carnegie Mellon School of Computer Science. Retrieved fromhttp://www.cs.cmu.edu/~durand/03711/Lectures/hmm3-05.pdf on14th Oct. 2008.

[18] S. Bergsma and G. Kondrak, "Alignment-Based Discriminative String Similarity". In *Proceedings of the 45th Annual Meeting of the Association of Computational Linguistics*, pp. 656-663, Czech Republic, June 2007.

# Escaping Death – Geometrical Recommendations for High Value Targets

Zeeshan-ul-hassan Usmani
Department of Computer Science
Florida Institute of Technology
zusmani@fit.edu

Eyosias Yoseph Imana
Department of Electrical Eng.
Florida Institute of Technology
eyoseph2007@fit.edu

Daniel Kirk
Department of Aerospace and Mech. Eng.
Florida Institute of Technology
dkirk@fit.edu

**Abstract** – *Improvised Explosive Devices (IED) and Suicide Bombing have become the headline news for every other day. It is a growing global phenomenon that is happening from Iraq to Afghanistan, Syria to United Kingdom and Spain to Pakistan. An average US soldier in Iraq receives 6 to 23 blast wave shocks over his/her tenure in Iraq. Pakistan has witnessed 120 suicide bombing attacks in last six years. While various attempts have been made to assess the impact of blast overpressure on buildings and animals, little has been done on crowd formation, crowd density and underlying geometry to mitigate the effects. This paper is set to make geometrical recommendations to reduce the casualties in case of such an incident for high value targets, like mosques and army facilities in the frontline countries fighting the global war on terrorism. A virtual simulation tool has been developed which is capable of assessing the impact of crowd formation patterns and their densities on the magnitude of injury and number of casualties during a suicide bombing attack. Results indicated that the worst crowd formation is street (Zig-Zag) where 30% crowd can be dead and 45% can be injured, given typical explosive carrying capacity of a single suicide bomber. Row wise crowd formations was found to be the best for reducing the effectiveness of an attack with 18% crowd in lethal zone and 38% in injury zones. For a typical suicide bombing attack, we can reduce the number of fatalities by 12%, and the number of injuries by 7%. Simulation results were compared and validated by the real-life incidents and found to be in good agreement. Line-of-sight with the attacker, rushing towards the exit, and stampede were found to be the most lethal choices both during and after the attack. These findings, although preliminary, may have implications for emergency response and counter terrorism.*

## I. INTRODUCTION

Suicide bombing is an operational method in which the very act of the attack is dependent upon the death of the perpetrator [12]. A suicide attack can be defined as a politically motivated and a violent-intended action, with prior intent, by one or more individuals who choose to take their own life while causing maximum damage to the chosen target. Suicide bombing has become one of the most lethal, unforeseeable and favorite modus operandi of terrorist organizations. Though only 3% of all terrorist attacks around the world can be classified as suicide bombing attacks, these account for 48% of the casualties [12]. The average number of deaths per incident for suicide bombing attacks is 13 over the period of 1980 to 2001 (excluding 9/11). This number is far above the average of less than one death per incident across all types of terrorism attacks over the same time period [7]. In Israel, the average number of deaths per incident is 31.4 over the period of November 2000 to November 2003 [8]. The average number of deaths in Pakistan is 14.2 over the period of 2006 and 2007. Suicide bombers, unlike any other device or means of destruction, can think and therefore can detonate the charge at optimal location with perfect timings to cause maximum carnage and destruction. Suicide bombers are adaptive and can quickly change targets if forced by security risk or the availability of better targets. Suicide attacks are relatively inexpensive to fund and technologically primitive, as Improvised Explosive Devices (IEDs) can be readily constructed. Suicide bombing works most of the time and requires no escape plan [4].

Past research has focused on developing psychological profiles of suicide bombers, understanding the economical logic behind the attacks [6,7,8], explaining the strategic and political gains of these attacks, their role in destabilizing countries [2, 5], and the role of bystanders in reducing the casualties of suicide bombing attacks [8, 10]. The specifics of the actual crowd formation and orientation of the bomber with respect to the crowd has not been examined. The presented simulation examines variables such as the number and arrangement of people within a crowd for typical layouts, the number of suicide bombers, and the nature of the explosion including equivalent weight of TNT and the duration of the resulting blast wave pulse. There are two models of the simulation: one uses the basic model of blast wave overpressure discussed in Section 2, while, the other also incorporates the effects of partial and full blockers as discussed in Section 3.

The goals of the analysis are to determine optimal crowd formations to reduce the deaths and/or injuries of individuals in the crowd, to determine what architectural and geometric changes can reduce the number of casualties and injuries, and what is the correlation between variant crowd densities and formations with the weight and pulse duration of the explosives? The main objective of our research is to explore and identify crowd formation precautions that when followed will minimize the number of deaths and injuries during a suicide bombing attack.

## II. EXPLOSIVE MODEL

In order to model the effects of a suicide bomber on a given crowd formation, it is essential to properly model the deleterious properties of the blast waves themselves. A conventional bomb generates a blast wave that spreads out

T. Sobh (ed.), *Innovations and Advances in Computer Sciences and Engineering*,
DOI 10.1007/978-90-481-3658-2_88, © Springer Science+Business Media B.V. 2010

spherically from the origin of the explosion. The strength of the blast wave decreases exponentially with distance [13]. Although the physics of blast waves are complex and nonlinear, a wave may be broadly characterized by its peak overpressure (pressure above atmospheric) and the duration of the positive phase of the blast event, as shown in Figure 1.

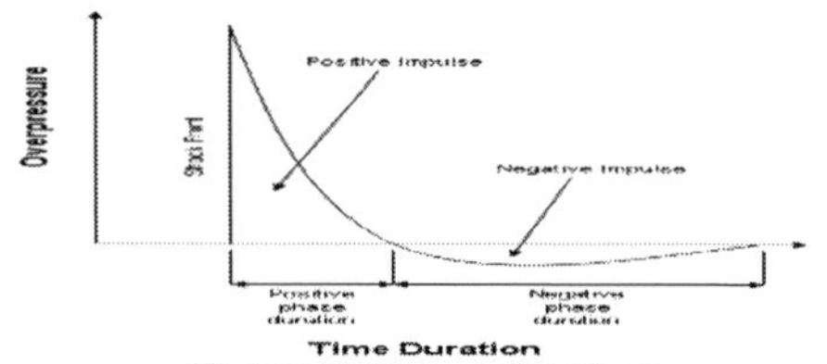

Figure 1: Ideal Pressure-Time Curve

Experimental and theoretical means have been used to obtain important parameters associated with blast waves. A theoretical analysis for peak overpressure utilizes the same mathematical approach as for a planar shock wave, but includes the effects of spherical divergence and the transient nature of the blast event [3, 9]. As an example, values for the peak overpressure generated in a standard atmosphere for the blast wave generated from a one pound spherical charge of TNT are shown in Figure 2.

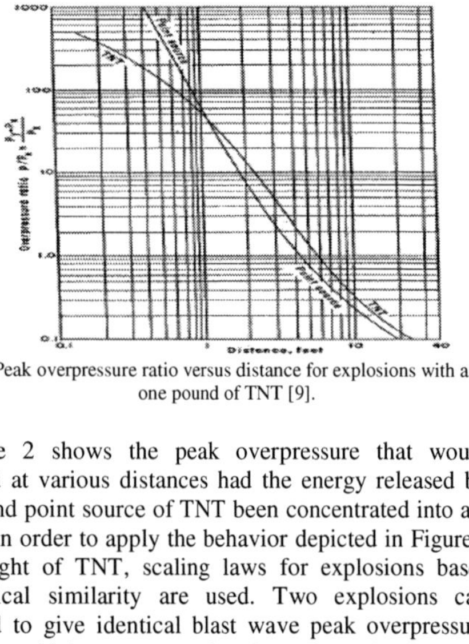

Figure 2: Peak overpressure ratio versus distance for explosions with a yield of one pound of TNT [9].

Figure 2 shows the peak overpressure that would be expected at various distances had the energy released by the one pound point source of TNT been concentrated into a point source. In order to apply the behavior depicted in Figure 2 for any weight of TNT, scaling laws for explosions based on geometrical similarity are used. Two explosions can be expected to give identical blast wave peak overpressures at distances which are proportional to the cube root of the respective energy release [9]:

$$\frac{d}{d_0} = \left( \frac{W}{W_0} \right)^{\frac{1}{3}} \tag{1}$$

The energy release factor is contained in a ratio $(W/W_0)^{1/3}$, where W is the energy release, or amount of TNT, in the explosion to be described, and $W_0$ is that of a reference amount of TNT, such as the one pound explosion shown in Figure 2. By using this scaling law, the distance at which a given peak overpressure is produced by a reference explosion may be scaled up or down to provide a corresponding distance for other explosions. All simulations considered in this study use the one pound TNT curve shown in Figure 2. However different explosives can also be considered by modifying the overpressure versus distance history or by utilizing data specific to the explosive composition.

Impulse is also an important aspect of the damage-causing ability of the blast, and may become a controlling factor for short duration, small yield explosives. The significant portion of the impulse is associated with the positive phase. The decay of blast overpressure does not follow a typical logarithmic decay relation, because the overpressure drops to zero in finite time (Figure 1). A quasi-exponential form, in terms of a decay parameter, α, and of a time, t, which is measured from the instant the shock front arrives, the pressure can be given as:

$$p = p_0 \left( 1 - \frac{t}{t_d} \right) e^{\frac{-\alpha t}{t_d}} \tag{2}$$

Where p is the instantaneous overpressure at time t, $p_0$ the maximum or peak overpressure observed when t is zero, and, $t_d$, the time duration. The decay parameter is also a measure of intensity of the shock system. Equation (2) may also be used in the simulation if the decay parameter, α, is specified, for example to determine the evolution of the positive phase duration as a function of distance from the explosive center.

In order to tie together the influence of peak overpressure and duration to injury and fatality probability, a series of data curves were utilized. Figure 3 shows the fatality curves predicted for 70-kg man applicable to free-steam situations where the long axis of the body is perpendicular to the direction of propagation of the shocked blast wave.

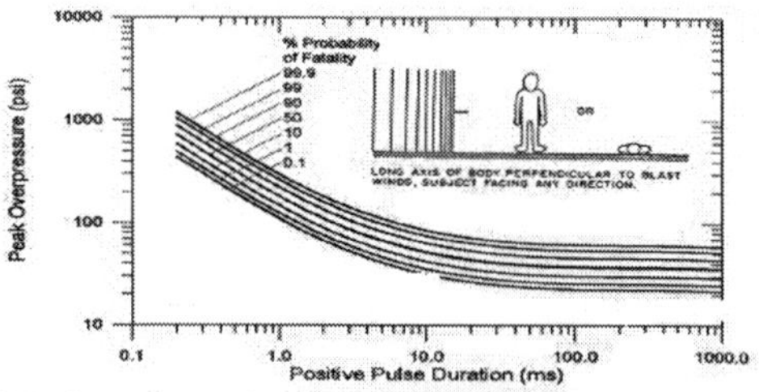

Figure 3: Fatality curves as a function of blast wave peak overpressure and positive pulse duration [3].

The inputs to the simulation for the suicide bomber are the equivalent amount of TNT that the bomber is carrying and the initial duration (or α if equation (2) is used) of the blast wave at the bomber's location. Specifying the amount of TNT, using the scaling law of equation (1), and the overpressure versus distance curve of Figure 2, then allows for the calculation of the peak overpressure at any distance away from the bomber. Using this peak overpressure and the increasing duration given by the scaled baseline data set a new duration of the blast wave can be calculated at any distance away from the bomber. Using these two pieces of information and injury or fatality probability curves, such as Figure 3, an estimate of the injury or fatality levels at any location of the bomber can be calculated for various crowd formations.

## III. Blockers and Zones

Blockage or human shields present in the crowd can play an important role in the event of suicide bombing attack. A person in front of you, and thus providing a blockage in the line-of-sight between you and the suicide bomber, can actually save your life by taking most of the shrapnel coming from the explosion or by consuming the part of the blast wave overpressure PSI. Spatial distribution of individuals in the crowd can significantly alter the casualty toll, and different crowd formations can yield different outcome with the same amount and kind of explosives.

This section presents the models to find the exact number of full and partial blockers between each person and the point of explosion. Persons on the line of sight between a given target and the blast point are termed as full blockers. The partial blockers are those who are not on the line of sight but their body width can cover some part of the body of the person from the blast projectiles. Imagine a short person of 4 feet standing in front of a tall 6 feet 10 inches person, or a person standing next to you, these persons, while, not covering you completely can provide partial blockage. To the best of our knowledge, this study is the first instance of introducing partial blockers in the blast wave simulation. Figure 4 presents the blockage model. Each person in the area is modeled by a line segment, where mid-point is the position of the person and the length is the specified width of the person.

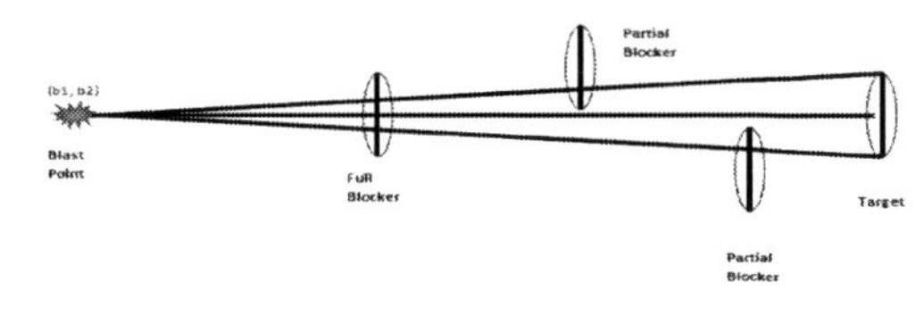


Figure 4. Full and Partial Blockers

Each line in the model is represented by the coordinates of the two end points. The line between the mid-point of the target and the blast point is called the line-of-sight. Each target is represented by a line segment termed as body-width-line. The triangle whose base is the body-width-line of the target and opposite vertex is the blast point and is termed as blast triangle.

The line segment between the blast point ($b_1$, $b_2$) and the center of the target ($t_{01}$, $t_{02}$) is constructed and its slope is calculated. Assuming all the people are going to face towards the blast point at the moment of blast, the body-width-line of the target will be perpendicular to the line of sight. The slope of this line is minus of the slope of the line of sight. Using simple coordinate geometry, one can easily determine the end points of the body-width-line of the target $((t_{11},t_{12}):(t_{21},t_{22}))$ having the mid-point of the line ( $(t_{01},t_{02})$ ), the body width and the slope of the line. Finding the end points of the body-width-line of the target, we can easily construct the two other sides of the blast triangle. All other people's body-width-line is assumed to have the same slope as the slope of the body-width- line of the target. Taking this slope and the position coordinate and the width, it is trivial to determine the end points of the body-width-line of each person.

It is also worth noting that all infinite slopes are approximated by $\pm 1\times 10^6$ in the code. To determine the blockage we have to see if the body-width-line (representing a person) is intersecting with either the line-of-sight or the sides of the blast triangle. If a body-width-line is intersecting the line of sight, the person represented by this line is taken as full blocker. Else, if it is intersecting with either of the sides of the blast triangle, the person will be considered as a partial blocker. Otherwise the person has no interaction with the target. Figure 5 shows full, partial and no blockers.

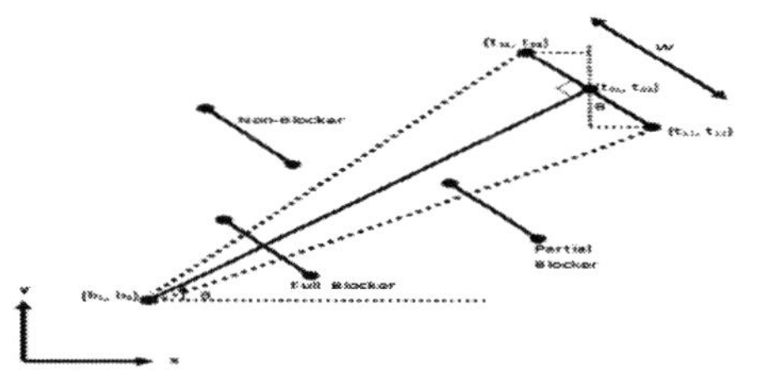


Figure 5. Full, Partial and No Blockers

Injuries that occur as a result of explosions can be grouped into several broad categories, as illustrated in Figure 6.

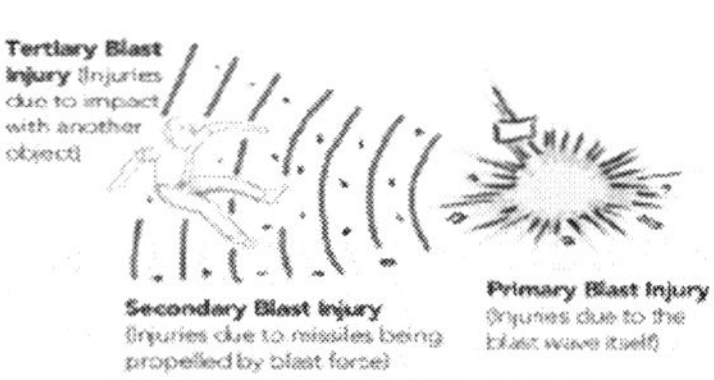


Figure 6: Categories of Blast Injuries

We have divided our results in six zones, three for lethality and three for injuries. Lethal zone one refers to 99% probability of death, lethal zone two represent 50% probability of death and the zone three counts the persons with 1% probability of death. Similarly, injuries are divided into three zones. Injury zone one represents the persons who are getting 60 or more overpressure PSI, zone two refers to more than 40 and less than 60 overpressure PSI and zone three for more than 20 and less than 40 overpressure PSI. In general, 60 PSI means severe injury like missing body parts, amputation, brain or heart rupture or Abbreviated Injury Score (AIS) 3. 40 PSI usually refers to the rupture of air-filled organs like lungs and kidney or AIS 2, and 20 PSI is usually responsible for minor bruises and ear-drum rupture or AIS 1. Persons below the range of 20 PSI are generally unharmed.

Self-explanatory Figure 7 provides the details of the respective impacts of the full and partial blockers on the lethal and injury zones. For example, a person within the 50% lethality zone blocked by a full blocker will be unharmed or the same person blocked by a partial blocker will be downgraded to the lethal zone 3 (1% probability of death).

| Lethal Zones | No Blocker | Full Blocker | Partial Blocker |
|---|---|---|---|
| 99% | Dead 99% | Dead 50% | Dead 99% |
| 50% | Dead 50% | Unharmed | Dead 1% |
| 1% | Dead 1% | Unharmed | Unharmed |
| **Injury Zones** | | | |
| 60 PSI | Injured 60 PSI | Injured 20 PSI | Injured 40 PSI |
| 40 PSI | Injured 40 PSI | Unharmed | Injured 20 PSI |
| 20 PSI | Injured 20 PSI | Unharmed | Unharmed |

Figure 7. Full and Partial Blockers Impact

## IV. SIMULATION

The simulation is programmed in Visual C++. We choose the Visual C++ programming language due to its extensive library of graphics and geometry functions (to generate the Cartesian grid with agents) and exceptional coverage of code integration with other third party tools like MatLab® (to code the blast overpressure and explosion models). In the simulation, it is assumed that the crowd is uniformly distributed throughout the area. The explosive range is determined by its weight. Specific simulation inputs are the number of individuals in the vicinity, walking speed of the attacker, time associated with the trigger, crowd formation, pulse duration and the total weight of TNT that is detonated. Additionally the arrival time of the explosive pressure front to travel from the bomber to any given location may also be calculated within the simulation.

We have considered mostly "open space" scenarios to serve as the basis for our crowd formation types (e.g., mosques, streets, concerts etc.). Type of injury caused by overpressure depends on whether overpressure occurs outdoor in open air or within buildings and whether they cause collapse of a building or other structure.

There are nine different settings a user can choose from the simulation main screen to estimate the outcome of an attack for a particular crowd formation. There are formations for Conference, Market, Street, Bus, Concert, Hotel, Shopping Mall, Mosque and University Campus. These nine settings were derived from the findings of Mark Harrison, where the majority of the suicide bombing attacks from November 2000 to November 2003 in Israel, occurred on the Streets, Cafeterias, Buses or other open spaces [8]. Users can also define number of participants (victims), number of attackers (suicide bombers), bomb strength (TNT weight in pounds), and bomb-timer (if any). Figure 8 shows the selection menu for crowd formation styles, and Figure 9 shows the display after the blast is simulated.

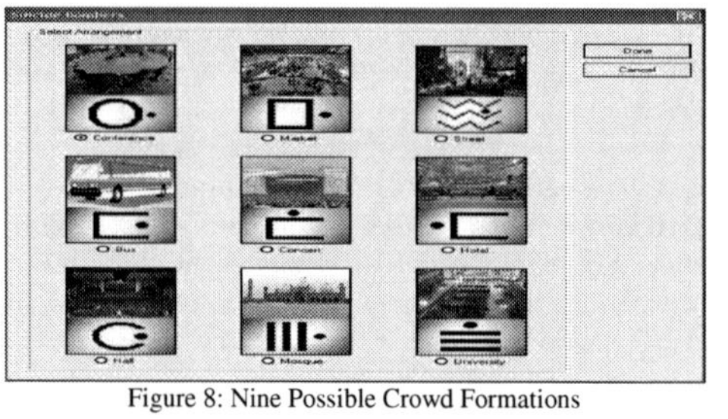

Figure 8: Nine Possible Crowd Formations

The simulation takes care of the beam and line-of-sight adjustments in cases of uneven surfaces (e.g., concert stage, mosque or shopping mall etc.). We have not considered physical objects (like wall, tree, furniture etc.) as obstacles or means to harm people at this point of time. The suicide bomber is a pedestrian in all cases and the explosion does not originate from a moving vehicle. The reason for choosing a suicide bomber location in almost all cases (except in Street scenario) on the entrance or exit gate was based upon the recent attacks in Iraq, Israel and Pakistan where suicide bombers detonated their bombs at the gates of mosques and restaurants.

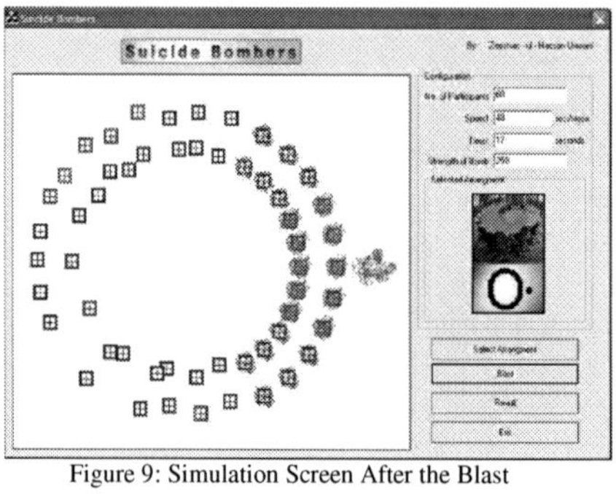


Figure 9: Simulation Screen After the Blast

The display depicts the casualties by red colored icons, those with injuries in light red colored icons, and those who remain unharmed in the attack are shown in blue colored icons. Thus, there are three states of victims after the blast: dead, injured and unharmed (but in panic and contributing in stampede).

There are two models of the simulation: Level 0 and Level 1. Level 0 (L0) is the basic simulation of blast wave without blockage (full or partial), while the Level 1 (L1) model is with the full and partial blockage. We have divided the models to examine the exact impact of blockers in the end results.

## V. RESULTS AND VALIDATION

We run the simulation for average case loss scenario for both models (L0 and L1). The weight of the explosives being used in the simulation for following results ranges from 1 lb to 30 lbs. The numbers of participants were from 20 to 100 and the pulse duration used for each set of simulation was from 0.5 milliseconds to 2 milliseconds. We have also run the simulation for sensitivity analysis for bigger crowds like with 500 to 1000 participants. The overall impact of blast on number of participants get stabilized as the participants increases. For example, the average of participants in the lethal zone were 10 with 20 total participants (50%) and 358 for total participants of 1000 (35.8%). These findings are parallel with Moshe Kress findings in [10]. The simulation was performed for all nine crowd formations with same number of participants and same weight of explosives. Following are the average results of 200 simulation runs for each crowd formation with different explosive mass, pulse duration and the number of participants.

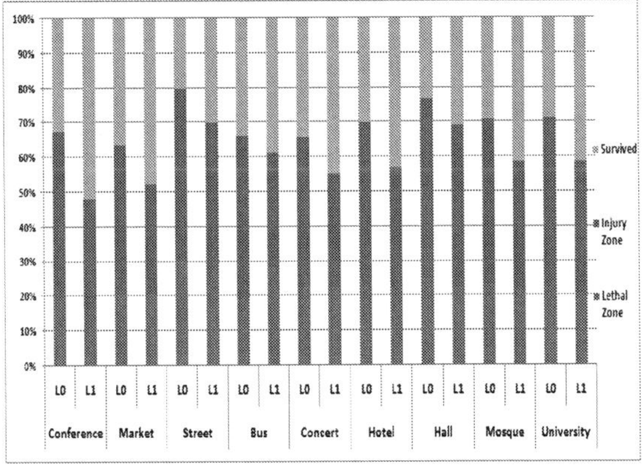


Figure 10. Casualties and Crowd Formations

Figure 10 summarizes the findings of percentage of persons in lethal zone and injury zone with given crowd formations. Each pair of formation in Figure 10 represents L0 and L1 model. It is clear to see that L1 model with blockers have less number of dead and injured people. We have also performed the simulation runs with 40 and 50 lbs of explosives (though it is uncommon to see the pedestrian suicide bombing attack of that magnitude). The relationship between the increase in the percentage of casualties and injuries with the amount of explosive is observed to be piecewise linear. This relationship is logical since augmenting the explosive material will increase the overpressure pounds per square inch (psi) in the vicinity.

The average deadliest crowd formation for casualties is found to be the Street (Zig-Zag) scenario, where 30% of the participants were in lethal zone and 45% in injury zones. Row wise crowd formations like Market were found to be the best for reducing the effectiveness of an attack, with on average 18% crowd in lethal zone and 38% in injury zones. Thus by only changing the way crowd forms, we can reduce the deaths by 12% and injuries by 7% on average. This is really useful where we have the control to form the crowd, like on airports by placing them in queues, banks, cafeteria, and stadium or in presidential debates or political rallies. One of the reasons of that dramatic change in the casualties is that in row wise formations, there are fewer people in the direct line-of-sight with the bomber and more people also provide the human shield to others by blocking the blast wave.

To validate our results and to see how close they are with the real-life incidents, we have compiled the database of every single suicide bombing attack in Pakistan from January 1995 to October 2008. Figure 11 shows the comparison of the average number of persons killed and injured in all of the simulation runs against the suicide bombing attacks in Pakistan. The real-life averages come from mostly open-space scenarios with a single pedestrian suicide bomber. For the sake of consistency, we have excluded the suicide bombing attacks in close environments like bus or with multiple suicide bombers or ones carried out with the help of automobile.

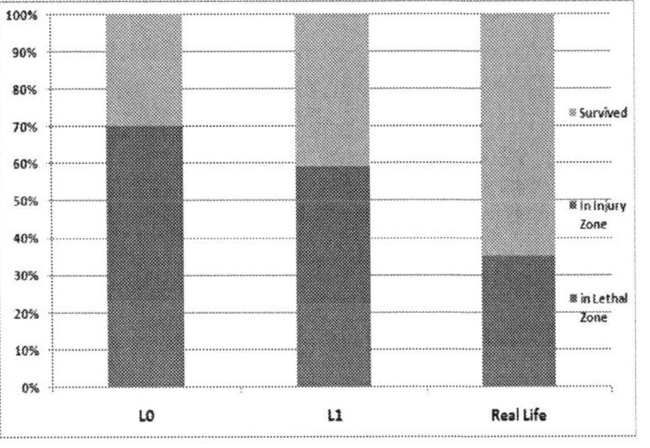


Figure 11. Models Comparison

Clearly, the Level 1 model with blockers is more close to real-life results than Level 0 with no blockers. The average injury per fatality ratio in real-life incidents is 2.18, that is, for every dead person there are 2.18 injured people. The number is pretty much consistent in the history of modern world, where we have 2.6 injuries per fatality ratio in Vietnam War, 2.8 in Korean War, 1.8 in World War I, and 1.6 in World War II. Simulation models on the other hand had produced 1.9 Injuries per fatality in L0 model and 1.6 for L1 model. And there is couple of reasons for that. First the current simulation does not include non-human objects that can provide shields to human beings from blast overpressure like cars, buildings, plants etc. Second, the current simulation only accounts for TNT explosive, while in the real-life there are quite a few mixtures of explosives being used. For example, RDX and TNT mixture in the recent suicide bombing attack in Pakistan that claims the life of former Prime Minister Benazir Bhutto and the mixture of Ammonium Nitrate and RDX in Oklahoma City bombings. Third, simulation is not giving the exact number of dead and injured people; instead it is giving the number of people in lethal and injury zones based on their probabilities of death and injury. For example, a person in lethal zone 3 with 1% chances of being dead is most likely to be injured and not dead, similarly a person in Injury zone 3 with 20 PSI can be unharmed. There are demographical, environmental and physical characteristics as well, that play an important roll in the overall toll. For example, an infant next to fire cracker can die while a muscular six and half feet person with 250 lbs of weight can survive 1 pound of TNT explosion. We expect more realistic results with the 3D simulation and incorporation of non-human shields, refraction waves and physical characteristics. However, this simulation can provide a good upper bound estimate of the number of dead and injured for emergency preparedness, triage of patients and required medical and ambulances facilities for such an event.

The results are in good agreement for the death count but little bit off for injury counts. Beside the aforementioned reasons, one of the reasons for this difference can be totally political, where governments are tend to show the manipualted figures to minimize the aftereffects (like riots, revenge etc) by victim's supporters or a huge cryout in the home state. For example, 4,000 soldiers have been died in Iraq so far, since the invasion of the country by US forces in October 2003. Media only concentrate on the dead, while, little has known about more than 250,000 injured soldiers [14]. An injured soldier costs atlesat three times more than the dead soldier economically to the country. Government has to pay disability and social security allownces, it is a loss of one worker from the labor force, thus a loss of one statical value of life, and the injured also need a caretaker, therefore another loss of one statistical value of life. Given the current geo-political conditions of the world and US on going war in Iraq and Agahnistan, it is more necessary than ever to examine and employed the technologies to reduce the rate of the injured and the dead. Another reason of the gap in the number of injured might be the level of injuries – a victim who has the minor injury and was able to walk away may not have been included in the actual count of injuires in the reallife events.

Announcing the threat of suicide bombing in the crowd can only make the condition and the causality toll much worse. People will panic and thus increase the possibility of more victims in the line-of-sight with the suicide bomber than before. People will also try to rush towards the exit gates (thus coming closer to suicide bomber in majority of cases) and there will be high chances of a stampede.

## VI. CONCLUSIONS AND FUTURE WORK

There are a number of lessons we can learn from the analysis of this suicide bombing simulation with given crowd formation styles. For example, we can reduce the number of fatalities by 12% and number of injuries by 7% by switching the crowd formation from Zig-Zag to Row-Wise formation styles. To avoid the stampede in possible crowd formation, we could arrange more exit points than normally available to such crowds. Suggestions can also be made for architectural design changes in the buildings to reduce the count, for example by placing entrance and exit gates by *X* feet away from the main venue, victims can be reduced by *Y*% (the values depends on environment, crowd information and the weight of explosive). The results can also help to plan for post-disaster management, for example, how many ambulance and doctors we will need if something like this happen to given crowd or how to direct the crowd to behave or run towards particular exit by announcing it through loudspeakers. In the light of these findings, the crowd can be manipulated in real-life by imposing formation guidelines like queues at the airport or by placing the chairs in particular order that will block the line-of-sight with any perspective attacker with the rest of the crowd.

The simulation and findings are limited in that it only incorporates the primary injuries. Future plans are to add secondary effects (e.g., injuries by fire, debris, etc.) to better approximate the real world environment and provide more valid comparisons with the data of suicide bombing attack aftermaths [11]. We will also add the flexibility to create the user defined crowd formations with variable number of entrances and exits in the future. This paper provides an interesting direction for future research to take in investigating the catastrophic event of the suicide bomber attack in hopes of making the world a safer place.

## REFERENCES


1. Air Force, "U.S. Air Force Installation Force Protection Guide", 2004
2. Azam, Jean-Paul, "Suicide-bombing as inter- generational investment", Public Choice. Springer. 122, 177-198, 2005.
3. Cooper, Paul W., "Explosive Engineering", Wiley-VCH, 1996
4. FEMA, "Explosive Blast Manual", Section 4.1. Federal Emergency Management Association Press, 2004.
5. Ganor, Boaz, "The Rationality of the Islamic Radical Suicide Attack Phenomenon", Countering Suicide Terrorism, Institute of Counter Terrorism, 2000.
6. Gupta, Dipak K. & Mundra, Kusum, "Suicide Bombing as a Strategic Weapon: An Empirical Investigation of Hamas and Islamic Jihad. Terrorism and Political Violence", 17, 573-598, 2005.
7. Harrison, Mark, "The Immediate Effects of Suicide Attacks", University of Warwick, 2004.
8. Harrison, Mark, "Bombers and Bystanders in Suicide Attacks in Isreal 2000 to 2003", Studies in Conflict and Terrorism. 29, 187-206, 2006.
9. Kinney, Gilbert & Graham, Kenneth, "Explosive Shocks in Air", 2nd Sub edition. Springer, 1985.
10. Kress, Moshe, "The Effect of Crowd Density on the Expected Number of Casualties in a Suicide Attack", Wiley Periodicals, 2004.
11. Lester, David, Yang, Bijou & Lindsay, Mark. "Suicide Bombers: Are Psychological Profiles Possible?", Studies in Conflict and Terrorism. 27, 283-295, 2004.
12. Pape, Robert A., "Dying to Win: The Strategic Logic of Suicide Terrorism", Random House, 2005
13. Irwin, R. J., Lerner, M. R., Bealer, J. F., Mantor, P. C., Brackett, D. J., and Tuggle, D. W., "Shock after blast wave injury is caused by a vagally mediated reflex", Journal of Trauma, Volume 47, p. 105-110, 1999.

Stiglitz, Joseph and Bilmes, Linda, "Three Trillion Dollar War, Norton and Company, July 2008

# Sentiment Mining Using Ensemble Classification Models

Matthew Whitehead and Larry Yaeger
Indiana University
School of Informatics
901 E. 10th St.
Bloomington, IN 47408
{mewhiteh, larryy}@indiana.edu

**Abstract**

*We live in the information age, where the amount of data readily available already overwhelms our capacity to analyze and absorb it without help from our machines. In particular, there is a wealth of text written in natural language available online that would become much more useful to us were we able to effectively aggregate and process it automatically. In this paper, we consider the problem of automatically classifying human sentiment from natural language written text. In this sentiment mining domain, we compare the accuracy of ensemble models, which take advantage of groups of learners to yield greater performance. We show that these ensemble machine learning models can significantly improve sentiment classification for free-form text.*

## I. INTRODUCTION

The amount of information being generated and stored on computers is growing rapidly [1], [2]. The explosion of the popularity of the internet and world wide web has all but eliminated the barrier for entry into publishing. For virtually no cost, just about anyone can record their thoughts on blogs, message boards, and personal web pages. More recently, many so-called Web 2.0 business models rely solely upon users to generate all of their meaningful content. Many of these websites do little more than make content publishing, organization, and access as simple as possible. The result is a massive increase in the amount of widely available human-generated, natural language data.

This wealth of data has the potential to significantly alter the way that we access, process, and use information. Increasingly, the challenge has become one of trying to make sense of the information available and organize it in such a way that it can used to maximum benefit.

Much of the difficulty is that so much of the useful data being generated by users online is generated in a communication medium that is easiest for humans to create and process, namely natural language. Because of this, much of the challenge lies in developing computer software that can process written natural language, aggregate it, organize it, and present it back to humans in meaningful and, often, more succinct ways.

A subset of this very difficult problem of natural language processing that is of particular interest is determining the sentiment of a piece of written text. The world wide web has numerous sites dedicated to collecting user opinions about a wide variety of subjects. There are sites for political opinions, movie reviews, restaurant reviews, music reviews, and more. A system that could automatically determine user opinions from a number of sources and then present a report showing the results in aggregate would be highly useful.

Consider the problem of trying to find the definitive performance of Bach's cello suites. Without a useful sentiment mining tool that aggregates user reviews, a user must read through numerous opinions of cello CDs to reach a final conclusion about which one to buy. This is time-consuming and inefficient. If the user could instead glance at a chart of cello CDs that shows how positive or negative the reviews were for that CD, then the search process would be much simplified. The system could present opinion results for thousands of reviews in just seconds.

In order for a tool like this to be useful, it must maintain a high level of classification accuracy. If reviews used in the system are not properly identified as being either positive or

T. Sobh (ed.), *Innovations and Advances in Computer Sciences and Engineering*,
DOI 10.1007/978-90-481-3658-2_89, 

negative, then the system quickly becomes more trouble than it is worth.

Machine learning algorithms are a natural fit for solving this sort of problem and building a useful classification system. Using a bag of words representation for user reviews allows the problem to be cast as a binary classification problem. In this case, the two categories are *positive* and *negative* and they reflect the sentiment of the accompanying user review. Of course reviews need not be entirely positive or negative, so we could also classify reviews using real numbers to show the degree to which each review is positive or negative. We chose to consider the simpler binary classification setup because we sampled real-world data and found that, at least for some bodies of data, very few reviews were truly mixed. For the datasets used in this study, only about 10% of sampled reviews had neither a strong positive nor a strong negative opinion.

In the recent past there has been considerable interest in the use of ensemble techniques in machine learning. Ensemble techniques, such as bagging [3] and boosting [4], use groups of learners to outperform a single learner.

In this paper, we show how a variety of ensemble methods perform in the sentiment mining domain. We believe that the resulting classification accuracies are high enough to be usable in practical settings.

## II. RELATED WORK

### *Sentiment Mining*

The area of sentiment mining (also called sentiment extraction, opinion mining, opinion extraction, sentiment analysis, etc.) has seen a large increase in academic interest in the last few years. Researchers in the areas of natural language processing, data mining, machine learning, and others have tested a variety of methods of automating the sentiment analysis process.

Pang et al. [5] researched sentiment mining using a binary unigram representation of patterns. In this representation, training patterns are represented by the presence/absence of words instead of by the count of total word occurrences. They tested a variety of algorithms for classification and found that a support vector machine had the highest accuracy of 82.9% using a movie reviews dataset. In later work, Pang and Lee [6] report improvement by adding a preprocessing filter to remove objective sentences which allowed the classifier to focus only on subjective sentences, raising the accuracy to 86.4%.

Whitelaw et al. [7] proposed improving sentiment mining pattern representations by using appraisal groups. They define appraisal groups as "coherent groups of words that express together a particular attitude, such as 'extremely boring', or 'not really very good'." By combining a standard bag-of-words approach with appraisal groups they report a 90.2% classification accuracy.

Snyder and Barzilay [8] describe an algorithm that breaks up reviews into multiple aspects and then provides different numerical scores for each aspect. This would be helpful for mixed reviews that explicitly describe those aspects which are good or bad. For example, a movie reviewer may like a movie's acting and special effects, but find its plot poorly conceived.

### *Ensemble Learning*

Ensemble learning techniques have been shown to increase machine learning accuracy by combining arrays of specialized learners. These specialized learners are trained as separate classifiers using various subsets of the training data and then combined to form a network of learners that has a higher accuracy than any single component.

Ensemble techniques increase classification accuracy with the trade-off of increasing computation time. Training a large number of learners can be time-consuming, especially when the dimensionality of the training data is high. Ensemble approaches are best suited to domains where computational complexity is relatively unimportant or where the highest possible classification accuracy is desired.

A number of different approaches to building ensemble learners have been proposed. There are numerous ways to build ensemble systems and there are several decisions to be made which affect the performance of the final model [9]:

- How are subsets of the training data chosen for each individual learner? Training subsets can be chosen by random selection, by examining which training patterns are difficult to classify and focusing on those, or by other means.
- How are classifications made by the different individual learners combined to form the final prediction? They can be combined by averaging, majority vote, weighted majority vote, etc.
- What types of learners are used to form the ensemble? Do all the learners use the same basic learning mechanism or are there differences? If the learners use the same learning algorithm, then do they use the same initialization parameters?

One of the first ensemble machine learning techniques was bootstrap aggregating, also called bagging [3]. Bagging requires the construction of a number of classifiers using randomly chosen subsamples of the training data. By choosing different random subsamples for each classifier, some input patterns are repeated and some are not chosen at all. This allows each particular classifier the ability to focus on its training subset and the resulting ensemble is less likely to suffer from being stuck in a local optimum.

[10] describes another ensemble technique called the random subspace method. Instead of choosing subsets of training patterns like in bagging, the algorithm randomly chooses subsets of training features.

Another popular ensemble technique is boosting [4], which also has many variants [11], [12]. Boosting is an iterative process where each successive classifier's training subset is chosen based on the performance of the previously trained classifier. If the previous classifier had difficulty properly classifying a particular training pattern, then that pattern is more likely to be chosen to be included in the current classifier's training set. This allows the system to build learners which focus on those difficult training patterns. This method forces each learner to act as a specialist for classifying its particular region of the data space.

## III. EXPERIMENTAL SETUP

In order to gauge the performance of ensemble techniques in the domain of sentiment mining, we set up classification accuracy tests to compare ensembles against standard, single machine learning models. If ensemble techniques are useful in this domain, then we would expect a higher level of classification accuracy. If classification accuracy does not increase, or only increases a negligible amount, then the added complexity and computational overhead of using an ensemble of classifiers would outweigh the benefit.

### *Datasets*

We chose to perform our tests on several datasets that we collected and one used by Snyder and Barzilay [8]. Their dataset consists of restaurant reviews written by many different users along with labels denoting whether each review had positive or negative sentiment. The full dataset used by Snyder and Barzilay has 3488 reviews, but has a preponderance of positive reviews, so for this binary classification task we chose to use a subset of 1476 reviews containing exactly half positive reviews and half negative reviews. The full dataset also has separate ratings for the various aspects of the restaurant being reviewed (e.g. different ratings for service and food quality), but we only used each restaurant's overall rating. We also tested with four other datasets that we collected from Amazon.com, lawyerratingz.com, and tvratingz.com. These datasets are available online (http://www.cs.indiana.edu/~mewhiteh/html/opinion_mining.html).

| Review Text | Sentiment |
|---|---|
| This is a truly, truly great restaurant... | Positive |
| I found the CD boring to listen to. | Negative |
| This laptop is cheap, but slow. | Negative |

*Table 1: Example raw training patterns*

Since the datasets consist of natural language reviews (see Table 1), they needed to be converted to numeric vector representations before any machine learning algorithm would be able to use them to learn. We chose to use a unigram (bag-of-words) conversion. Using this kind of conversion, each machine learning input vector has one input value per word in the lexicon. We created the lexicon by including every unique word seen in any review and then removing stopwords and words that occurred only once.

To reduce the large size of the lexicons, we used the odds ratio method to select those words which were likely to help in classification. For each term in the full lexicon, the following odds ratio value was calculated:

$$\text{odds ratio} = \frac{p(1-q)}{q(1-p)}$$

where $p$ was the fraction of positive reviews containing the target word and $q$ was the fraction of negative reviews containing the target word. We then selected the words which had the highest absolute value odds ratios, a good measure of effect size for binary classification, under the assumption that those words would be most useful for distinguishing positive reviews from negative reviews. This allowed us to produce smaller lexicons that were only about 10% the size of the originals.

Each value in an input vector is the count of occurrences of a particular word in a certain review mapped to fit the range [-1.0, 1.0]. The maximum count of a word in any single review over all reviews maps to the value 1.0 and the minimum count to the value -1.0. Note that many of the values in an input vector are -1.0 since there are typically many words in each lexicon that do not occur in a particular review. These vectors of word occurrence counts coupled with converted class labels form the final training dataset.

*Procedure*

First, we ran a single support vector machine (SVM) learning algorithm on the full training datasets. To model the support vector machines, we used the libsvm library [13]. The results from these tests act as a baseline to compare with the results from our ensemble tests.

Next, we wrote our own implementations of several popular ensemble algorithms: bagging, boosting, random subspace, and bagging random subspaces. Each of these ensemble methods used SVMs for their base models.

For the bagging ensemble models, we trained groups of 50 individual models each with different training subsets. Individual model classifications from all models in the ensemble were combined by averaging to produce the final result for each test pattern.

We implemented the AdaBoost boosting algorithm [4] since it has been shown to have a high level of performance for many different problems. We set the number of iterations, T, to be 50, but also used the boosting variation of ending the process early given a high enough error from the last trained model.

For all the tests, 50 models were used for both the random subspace ensembles and the bagging random subspace ensembles. In the subspace tests, each model was given a randomly chosen subset of pattern features. In the bagging random subspace ensemble, each model was given a subset of training patterns and each pattern had a subset of training features.

| Algorithm | camera | laptop | lawyer | restaurant | tv |
|---|---|---|---|---|---|
| Single Model | 83.40 | 88.94 | 83.11 | 83.65 | 82.53 |
| Bagging | 84.69 | 89.11 | 83.45 | 85.41 | 82.98 |
| Boosting | 82.23 | 86.66 | 85.04 | 81.49 | 81.74 |
| Subspace | 85.80 | 90.00 | 84.56 | 85.95 | 84.26 |
| Bagging Subspace | 86.80 | 90.00 | 84.36 | 86.62 | 84.08 |

*Table 2: Classification accuracies (% correctly classified)*

To compare classification accuracy, we used K-fold cross validation with 10 folds to provide statistically stable accuracy estimates. This was done by first splitting up the dataset into 10 randomly chosen subsets. Then 10 trials were performed with each trial being a test on a single subset using the other 9 subsets as training data. The results of all 10 trials were then averaged to produce the final classification accuracies as reported.

## IV. RESULTS

Table 2 shows the accuracy of the various setups. In general, the ensemble methods performed well and often outperformed the single model SVM.

The bagging ensemble was consistently equal to or better than the single SVM model across all the datasets. This is an encouraging result since this means a bagging ensemble can be used with the reasonable assumption that it will not hurt performance on sentiment mining datasets. It did not seem to have problems overfitting and losing generalization capabilities.

The random subspace and bagging random subspaces ensembles did even better than the bagging ensembles, often achieving the best overall accuracies of any of the methods. These techniques that use subsets of available features seem to be quite effective on the tested sparse datasets containing a large number of available features.

The boosting ensembles did not perform as well as the other ensemble methods with the exception of its performance on the lawyer dataset. The sparsity and noisy structure of the datasets appears to have caused the boosting models to overfit and degrade generalization performance.

When deciding upon which model to use for a real-world application, one should consider the accuracy requirements and time constraints. If time and computational resources are not an issue or the highest possible classification accuracy is desired, then the bagging subspace model seems to be the best choice.

## V. CONCLUSION

We have shown that ensemble learning techniques can increase classification accuracy in the domain of sentiment mining, though the choice of ensemble method affects accuracy. The random subspace and bagging subspaces models typically had the highest classification accuracies. If one wants to create a model with the highest possible classification accuracy for sentiment mining, then ensemble methods should be considered.

The only drawback of these methods is that they increase the computational time required for training and classification. For all the ensemble methods, a group of many different learners must be trained as opposed to a single learner that is used to make all classifications. This makes the training time of ensemble techniques roughly $S{\cdot}t$, where $S$ is the size of the ensemble and $t$ is the time it takes to train a single learner. This time increase may be prohibitive in cases where $t$ is already very large, so a computation time vs. accuracy decision would have to be made. However, the bulk of the computational expense is a one-time occurrence, applying only to the learning phase, and ensemble learners can be sufficiently fast at classification that the up-front expense is well justified for the increased accuracy in actual use.

In the future we intend to empirically test other ensemble models, including ways to increase computational efficiency while still maintaining the benefits of using a combined group of classifiers. It would also be interesting to investigate different ways of representing the user reviews (e.g. using other n-gram models, or building a lexicon that consists entirely of polarity words) so that the various machine learning algorithms would have less noise with which to contend.

## REFERENCES


[1] Lyman, P., and Varian, H. 2000. How much information. http://www.sims.berkeley.edu/how-much-info.

[2] Lyman, P., and Varian, H. 2003. How much information 2003. http://www.sims.berkeley.edu/how-much-info-2003.

[3] Breiman, L. 1996. Bagging predictors. *Machine Learning* 24(2):123–140.

[4] Schapire, R. E. 2002. The boosting approach to machine learning: An overview.

[5] Pang, B., and Lee, L. 2004. A sentimental education: Sentiment analysis using subjectivity summarization based on minimum cuts. In *ACL*, 271–278.

[6] Pang, B.; Lee, L.; and Vaithyanathan, S. 2002. Thumbs up? sentiment classification using machine learning techniques. *CoRR* cs.CL/0205070.

[7] Whitelaw, C.; Garg, N.; and Argamon, S. 2005. Using appraisal groups for sentiment analysis. In Herzog, O.; Schek, H.-J.; Fuhr, N.; Chowdhury, A.; and Teiken, W., eds., *Proceedings of the 2005 ACM CIKM International Conference on Information and Knowledge Management, Bremen, Germany, October 31 - November 5, 2005*, 625–631. ACM.

[8] Snyder, B., and Barzilay, R. 2007. Multiple aspect ranking using the good grief algorithm. In *Proceedings of NAACL HLT*, 300–307.

[9] Polikar, R. 2006. Ensemble based systems in decision making. *IEEE Circuits and Systems Magazine* Third Quarter:21–45.

[10] Ho, Tin Kam. 1998. The random subspace method for constructing decision forests. IEEE Transactions on Pattern Analysis and Machine Intelligence, 20(8): 832-844.

[11] Demiriz, A., and Bennett, K. P. 2001. Linear programming boosting via column generation.

[12] Domingo, and Watanabe. 2000. Madaboost: A modification of adaboost. In *COLT: Proceedings of the Workshop on Computational Learning Theory, Morgan Kaufmann Publishers*.

[13] Chang, C. C., and Lin, C. J. 2001. *LIBSVM: a library for support vector machines*. http://www.csie.ntu.edu.tw/~cjlin/libsvm.

# Parallelization of Finite Element Navier-Stokes codes using MUMPS Solver

Mandhapati P. Raju
Research Associate, Case Western Reserve University, Cleveland, OH, 44106, USA
E-mail: raju192@gmail.com

*Abstract*- **The study deals with the parallelization of 2D and 3D finite element based Navier-Stokes codes using direct solvers. Development of sparse direct solvers using multifrontal solvers has significantly reduced the computational time of direct solution methods. Although limited by its stringent memory requirements, multifrontal solvers can be computationally efficient. First the performance of MUltifrontal Massively Parallel Solver (MUMPS) is evaluated for both 2D and 3D codes in terms of memory requirements and CPU times. The scalability of both Newton and modified Newton algorithms is tested.**

## I. Introduction

Discretization of Navier-Stokes equations involves a large set of non-linear equations, requiring high computing power in terms of speed and memory. The resulting set of weak form (algebraic) equations in such problems may be solved either using a direct solver or an iterative solver. The direct solvers are known for their generality and robustness. The direct solution methods generally involve the use of frontal algorithms [1] in finite element applications. The advent of multifrontal solvers [2] has greatly increased the efficiency of direct solvers for sparse systems. They make full use of the high computer architecture by invoking level 3 Basic Linear Algebra Subprograms (BLAS) library. Thus the memory requirement is greatly reduced and the computing speed greatly enhanced. Multifrontal solvers have been successfully used both in the context of finite volume problems [3-5] and finite element problems [6]. The disadvantage of using direct solvers is that the memory size increases much more rapidly than the problem size itself [6]. To circumvent this problem, out-of-core multifrontal solvers [7] have been developed which has the capability of storing the factors on the disk during factorization. Another viable alternative is to use direct solvers in a distributed computing environment

The system of non-linear equations obtained from the discretization of Navier-Stokes equations is usually solved using a Newton or a Picard algorithm. Newton algorithms are known for their quadratic convergence behavior. When the initial guess is close to the final solution, Newton achieves quadratic convergence. In this paper, only the Newton algorithm is used. In using direct solvers, factorization of the left hand side matrix is the most time consuming step. To avoid factorization during every iteration, a modified Newton is used in which the factorization is done only during the first iteration. The left side matrix evaluated for the first iteration is retained and is not changed during the subsequent iterations. Only the right hand side matrix is updated during each iteration step. The right hand side vector is appropriately modified to give the final converged solution. Since the factorization is done only during the first iteration, the subsequent iterations are extremely cheap. It usually requires more number of iterations to obtain the overall convergence. So there is a trade off between the computational time per iteration and the number of iterations to obtain the final convergence. Although the convergence rate is lower compared to the Newton iteration, the savings in computational time per iteration is so high that it can more than compensate the decrease in the convergence rate.

MUMPS [8-10] and SUPERLU [11] are amongst the fastest parallel general sparse direct solvers that are available under public domain software. A detailed description of the various features and algorithms employed in these packages can be found in [12]. MUMPS is found to be much faster compared to SUPERLU, although its scalability is low compared to that of SUPERLU. In this paper, parallelization is achieved using a MUltifrontal Massively Parallel Solver (MUMPS) on a distributed environment using MPI. The linear system of equations is evaluated on different processors corresponding to the local grid assigned to the processor. The right hand side vector is assembled on the host processor and is input to the MUMPS solver. On exit from the MUMPS solver, the solution is assembled centrally on the host processor. This solution is then broadcast to all the processors. In the context of modified Newton algorithm, the LU factors evaluated during the first iteration are reused and the solution of the linear system with the new right hand side vector is solved. The performance of the solver in terms of scalability and memory issues for both two-dimensional and three-dimensional problems are discussed in detail.

## II. Mathematical Formulation

The governing equations for laminar flow through a two-dimensional rectangular duct are presented below in the non-dimensional form.

$$\frac{\partial u}{\partial x}+\frac{\partial v}{\partial y}=0, \tag{1}$$

$$\frac{\partial}{\partial x}\left(u^2\right)+\frac{\partial}{\partial y}(uv)=-\frac{\partial p}{\partial x}+\frac{\partial}{\partial x}\left(\frac{2}{\mathrm{Re}}\frac{\partial u}{\partial x}\right)+\frac{\partial}{\partial y}\left(\frac{1}{\mathrm{Re}}\left(\frac{\partial u}{\partial y}+\frac{\partial v}{\partial x}\right)\right), \tag{2}$$

T. Sobh (ed.), *Innovations and Advances in Computer Sciences and Engineering*,
DOI 10.1007/978-90-481-3658-2_90, 

and

$$\frac{\partial}{\partial x}(uv)+\frac{\partial}{\partial y}(v^2)=-\frac{\partial p}{\partial y}+\frac{\partial}{\partial x}\left(\frac{1}{\mathrm{Re}}\left(\frac{\partial u}{\partial y}+\frac{\partial v}{\partial x}\right)\right)+\frac{\partial}{\partial y}\left(\frac{2}{\mathrm{Re}}\frac{\partial v}{\partial y}\right), \quad (3)$$

where $u, v$ are the $x$ and $y$ components of velocity, $p$ is the pressure. The bulk flow Reynolds number, $\mathrm{Re}=\frac{\rho U_o L}{\mu}$, $U_o$ being the inlet velocity, $\rho$ the density, $L$ the channel length, , and $\mu$ is the dynamic viscosity. Velocities are non-dimensionalized with respect to $U_o$, pressure with respect to $\rho U_o^2$.

The boundary conditions are prescribed as follows:

(1) Along the channel inlet:

$$u=1;\ v=0. \quad (4)$$

(2) Along the channel exit:

$$p=0;\ \frac{\partial u}{\partial x}=0;\ \frac{\partial v}{\partial x}=0. \quad (5)$$

(3) Along the walls:

$$u=0;\ \ v=0. \quad (6)$$

The governing equations for laminar flow through a three-dimensional rectangular duct are presented below in the non-dimensional form. In three-dimensional calculations, instead of the primitive $u,v,w,p$ formulation, penalty approach is used to reduce the memory requirements.

$$\frac{\partial u}{\partial x}+\frac{\partial v}{\partial y}+\frac{\partial w}{\partial z}=0, \quad (7)$$

$$\frac{\partial}{\partial x}(u^2)+\frac{\partial}{\partial y}(uv)+\frac{\partial}{\partial z}(uw)=\lambda\frac{\partial}{\partial x}\left(\frac{\partial u}{\partial x}+\frac{\partial v}{\partial y}+\frac{\partial w}{\partial z}\right)+\frac{\partial}{\partial x}\left(\frac{2}{\mathrm{Re}}\frac{\partial u}{\partial x}\right) \quad (8)$$
$$+\frac{\partial}{\partial y}\left(\frac{1}{\mathrm{Re}}\left(\frac{\partial u}{\partial y}+\frac{\partial v}{\partial x}\right)\right)+\frac{\partial}{\partial z}\left(\frac{1}{\mathrm{Re}}\left(\frac{\partial u}{\partial z}+\frac{\partial w}{\partial x}\right)\right),$$

$$\frac{\partial}{\partial x}(uv)+\frac{\partial}{\partial y}(v^2)+\frac{\partial}{\partial z}(vw)=\lambda\frac{\partial}{\partial y}\left(\frac{\partial u}{\partial x}+\frac{\partial v}{\partial y}+\frac{\partial w}{\partial z}\right)+\frac{\partial}{\partial x}\left(\frac{1}{\mathrm{Re}}\left(\frac{\partial u}{\partial y}+\frac{\partial v}{\partial x}\right)\right) \quad (9)$$
$$+\frac{\partial}{\partial y}\left(\frac{2}{\mathrm{Re}}\frac{\partial v}{\partial y}\right)+\frac{\partial}{\partial z}\left(\frac{1}{\mathrm{Re}}\left(\frac{\partial v}{\partial z}+\frac{\partial w}{\partial y}\right)\right),$$

and

$$\frac{\partial}{\partial x}(uw)+\frac{\partial}{\partial y}(vw)+\frac{\partial}{\partial z}(w^2)=\lambda\frac{\partial}{\partial z}\left(\frac{\partial u}{\partial x}+\frac{\partial v}{\partial y}+\frac{\partial w}{\partial z}\right)+\frac{\partial}{\partial x}\left(\frac{1}{\mathrm{Re}}\left(\frac{\partial u}{\partial z}+\frac{\partial w}{\partial x}\right)\right) \quad (10)$$
$$+\frac{\partial}{\partial y}\left(\frac{1}{\mathrm{Re}}\left(\frac{\partial v}{\partial z}+\frac{\partial w}{\partial y}\right)\right)+\frac{\partial}{\partial z}\left(\frac{2}{\mathrm{Re}}\frac{\partial w}{\partial z}\right).$$

where $u, v, w$ are the $x$, $y$ and $z$ components of velocity,. The bulk flow Reynolds number, $\mathrm{Re}=\frac{\rho U_o L}{\mu}$, $U_o$ being the inlet velocity, $\rho$ the density, $L$ the channel length, $\mu$ is the dynamic viscosity and $\lambda$ is the penalty parameter. Velocities are non-dimensionalized with respect to $U_o$.

The boundary conditions are prescribed as follows:

(1) Along the channel inlet:

$$u=1;\ v=0; w=0. \quad (11)$$

(2) Along the channel exit :

$$\frac{\partial u}{\partial x}=0;\ \frac{\partial v}{\partial x}=0;\ \frac{\partial w}{\partial x}=0. \quad (12)$$

(3) Along the walls:

$$u=0;\ \ v=0;\ \ w=0. \quad (13)$$

## III. MUMPS SOLVER

The most time consuming part is the solution of the set of linear equations. To solve these linear systems, we use a robust parallel solver MUMPS. It employs multifrontal techniques to solve the set of linear equations on parallel computers on a distributed environment using MPI. It relies on Level II and Level III optimized BLAS routines. It requires SCALAPACK and PBLACS routines. In this paper, the vendor optimized INTEL Math Kernal Library is used. The Software is written in Fortran 90 and has a C interface available.

The solution of the set of linear equations takes place in 3 essential steps

(i) Analysis step: MUMPS offers various built in ordering algorithms and interface to external packages such as PORD [13] and METIS [14]. The Matrix is analyzed to determine an ordering and a mapping of the multifrontal computational graph is constructed. This symbolic information is passed on from the host to all the other processors.

(ii) Factorization: Based on the symbolic information, the algorithm tries to construct several dense sub matrices that can be processed in parallel. The numerical factorization is carried out during this step.

(iii) Solution step: Using the right hand side vector, the solution vector is computed using the distributed factors.

All these steps can be called separately or as a combination of each other. This can be exploited to save some computational effort during the solution of subsequent iterations in the solution of a set of nonlinear equations. For example if the structure of the matrix does not change during every iteration, the analysis step can be skipped after evaluating once. Similarly, if the left hand matrix does not change, both the analysis and the factorization steps can be skipped.

## IV. PARALLEL IMPLEMENTATION

The MUMPS solver is implemented using the MPI library, which makes the code very portable and usable on both, shared and distributed memory parallel computers. The parallelization is done internally in the code. The calling program should also be in a parallel environment to call the code. In the present formulation, each element is assigned to particular processor such the elements are equally (or almost equally) distributed amongst all the processors. The computation of the matrix coefficients and the right hand side vector are done in parallel corresponding to the set of local elements. Evaluation of the Jacobian matrix and the right hand side vector in a parallel environment is crucial for problems, which consume lot of time for the evaluation of matrix coefficients.

During the progress of overall iterations, the different set of linear equations obtained during every iteration is solved by successive calls to the MUMPS. For the modified Newton's algorithm, the left hand matrix remains the same (numerically). So both the analysis and the factorization steps are skipped during the subsequent iterations. Since the factorization is most costly step, it leads to a significant amount of savings in time for the subsequent iterations. The performance of Newton and Modified Newton's method is tested.

While implementing the finite element code in a parallel environment with the MUMPS code, the element matrix entries are calculated locally on each of the processors. Although the facility for element matrix input is available, only the option of centralized element entry is available in the current versions of MUMPS solver. To facilitate distributed matrix input (necessary for improving the parallel efficiency), the local element entries are converted into sparse matrix triplet entries in coordinate format and are input in a distributed fashion (using ICNTL(5) = 0 and ICNTL(18) = 3). There will be lot of duplicate entries due to contribution of all the neighboring elements at a given grid point. MUMPS solver automatically sums up all the duplicate entries. Different ordering can be chosen by using different values for ICNTL(7). The different ordering options that are available within MUMPS solver are (i) Approximate minimum degree (AMD), (ii) Approximate minimum fill (AMF), (iii) PORD, (iv) METIS, (v) Approximate Minimum degree with automatic quasi-dense row detection (QAMD).

## IV. Performance Of Mumps

Preliminary experiments have been done to study the effect of different ordering routines on the performance of MUMPS solver both in terms of memory and computational time. Table 1 shows the performance of different ordering routines for two-dimensional codes. Table 1 shows the comparison of different ordering routines for a 300x300 mesh using 12 processors. Results indicate the both PORD and METIS perform well in minimizing the computational time requirements but METIS performs well in terms of memory requirements. Based on this observation, METIS ordering is used for all subsequent calculations. Table 2 shows the performance of Newton's and modified Newton's method using MUMPS solver for a two-dimensional channel flow problem. Results indicate that modified Newton's method performs better than Newton's method. This is due to the fact that in modified Newton's method, factorization is done only during the first iteration. During the subsequent iterations, factorization is skipped and only the solution step is performed. This decreases the convergence rate, thereby increasing the number of iterations to obtain convergence. However, the solution step being computationally inexpensive compared to the factorization step, the overall solution time is less compared to the Newton's step. However it should be noted that the memory requirement for the Newton and modified Newton's method are the same. Table 2 shows that the memory requirement does not vary linearly with the number of processors. It behaves erratically. Table 2 also shows that the MUMPS solver does not scale so well. Using 20 processors, the computational time is approximately halved compared to the computational time using 2 processors. It does not scale much beyond 6 processors. The scalability of MUMPS solver for two-dimensional problems is observed to be poor.

| Ordering | CPU time/ iteration (sec) | Memory (MB) | |
|---|---|---|---|
| | | avg | max |
| PORD | 16.6 | 1525 | 1640 |
| METIS | 5.1 | 1560 | 1820 |
| AMD | 5.5 | 1586 | 1923 |
| AMF | 6.7 | 1592 | 2000 |
| QAMD | 5.3 | 1603 | 2056 |

Table 1: Comparison of the performance of different orderings in MUMPS solver for a 300x300 mesh on 12 processors.

| # of processors | Time to solve (Seconds) | | Memory Requirements (MB) | | Ordering |
|---|---|---|---|---|---|
| | Newton | Modified Newton | max memory on one processor | Total memory | |
| 2 | 35.7 | 27.4 | 412 | 806 | Metis |
| 4 | 23 | 18.4 | 259 | 977 | Metis |
| 6 | 21 | 16.8 | 227 | 1082 | Metis |
| 8 | 20 | 15.3 | 178 | 1209 | Metis |
| 10 | 19.2 | 14.8 | 159 | 1394 | Metis |
| 12 | 18.6 | 14.6 | 157 | 1700 | Metis |
| 14 | 19.1 | 15 | 171 | 2039 | Metis |
| 16 | 18.4 | 13 | 156 | 2352 | Metis |
| 18 | 18.3 | 13 | 145 | 2534 | Metis |
| 20 | 18 | 12 | 141 | 2675 | Metis |

Table 2: Comparison of the performance of Newton and Modified Newton's methods using MUMPS solver for a 200x200 mesh.

Table 3 shows the performance of MUMPS solver using different ordering routines for a three dimensional channel flow problem. The table shows that METIS ordering is a better choice both in terms of computational speed and memory requirement. Hence METIS is used for all the subsequent computations of three dimensional problems. For a 100x20x20 mesh, memory was not sufficient to run on 2 processors. The table shows that the scalability of MUMPS solver for three-dimensional problems is better than that for the two-dimensional problems. When the number of processors increased from 4 to 20, the computational time of Newton's method has reduced to a factor of 3.6 approximately, while that of modified Newton's method has reduced to a factor of 3 approximately. The maximum memory requirement for a single processor has reduced to a factor of 4. The use of MUMPS solver for three dimensional problems seems to be promising. However, memory requirement is a serious limitation for solving three dimensional problems using direct solvers. Out-of-Core MUMPS solver is currently under development by the CERFACS group. This will potentially increase the applicability of MUMPS solver to large three dimensional problems.

| Ordering | CPU time/iteration (sec) | Memory (MB) | |
|---|---|---|---|
| | | Avg | max |
| PORD | 41 | 1385 | 1612 |
| METIS | 38 | 1286 | 1400 |
| AMD | 105 | 2296 | 2496 |
| AMF | 93 | 1246 | 1425 |
| QAMD | 102 | 2296 | 2593 |

Table 3: Comparison of the performance of different orderings in MUMPS solver for a 100x20x20 mesh on 12 processors.

| # of processors | Time to solve (Seconds) | | Memory requirements (GB) | | Ordering |
|---|---|---|---|---|---|
| | Newton | Modified Newton | max memory on one processor | Total memory | |
| 4 | 184 | 89 | 2 | 7.3 | metis |
| 6 | 147 | 77.2 | 1.6 | 8.6 | metis |
| 8 | 112 | 61.6 | 1 | 7.5 | metis |
| 10 | 91 | 53.4 | 1.5 | 13.9 | metis |
| 12 | 99 | 42.5 | 1.4 | 15.4 | metis |
| 14 | 68 | 37.2 | 1.45 | 17 | metis |
| 16 | 58 | 37 | 0.7 | 9.6 | metis |
| 18 | 52 | 34.8 | 0.63 | 10.2 | metis |
| 20 | 50 | 31.1 | 0.56 | 9.9 | metis |

Table4: Comparison of the performance of Newton and Modified Newton's methods using MUMPS solver for a 100x20x20 mesh.

## V. CONCLUSIONS

Finite element based Navier-Stokes codes are parallelized using MUMPS solver. Both Newton and modified Newton's algorithms are used. It is observed that modified Newton's method leads to savings in computational time compared to the Newton's algorithm. It is also observed that METIS ordering enhances the performance of MUMPS solver both for two-dimensional and three-dimensional problems. MUMPS solver does not scale well for two-dimensional problems but it scales better for three dimensional problems.

## REFERENCES


[1] Irons, B.M. (1970) 'A frontal solution scheme for finite element analysis,' Numer. Meth. Engg, Vol. 2, pp. 5-32.

[2] Davis, T. A. and Duff, I. S. (1997) 'A combined unifrontal/multifrontal method for unsymmetric sparse matrices', ACM Trans. Math. Softw. Vol 25, Issue 1, pp. 1–19.

[3] Raju, M. P. and T'ien, J. S. (2008) 'Development of Direct Multifrontal Solvers for Combustion Problems", Numerical Heat Transfer, Part B, Vol. 53, pp. 1-17.

[4] Raju, M. P. and T'ien, J. S. (2008) "Modelling of Candle Wick Burning with a Self-trimmed Wick", Comb. Theory Modell., Vol. 12, Issue 2, pp. 367-388.

[5] Raju, M. P. and T'ien, J. S. (2008) "Two-phase flow inside an externally heated axisymmetric porous wick", Vol. 11, Issue 8, pp. 701-718.

[6] P. K. Gupta and K. V. Pagalthivarthi, Application of Multifrontal and GMRES Solvers for Multisize Particulate Flow in Rotating Channels, Prog. Comput Fluid Dynam., vol. 7, pp. 323–336, 2007.

[7] Scott, J. A. "Numerical Analysis Group Progress Report", RAL-TR-2008-001.

[8] P. R. Amestoy, I. S. Duff and J.-Y. L'Excellent, Multifrontal parallel distributed symmetric and unsymmetric solvers, in Comput. Methods in Appl. Mech. Eng., 184, 501-520 (2000).

[9] P. R. Amestoy, I. S. Duff, J. Koster and J.-Y. L'Excellent, A fully asynchronous multifrontal solver using distributed dynamic scheduling, SIAM Journal of Matrix Analysis and Applications, Vol 23, No 1, pp 15-41 (2001).

[10] P. R. Amestoy and A. Guermouche and J.-Y. L'Excellent and S. Pralet, Hybrid scheduling for the parallel solution of linear systems. Parallel Computing Vol 32 (2), pp 136-156 (2006).

[11] Xiaoye S. Li and James W. D. (2003) " A Scalable Distributed-Memory Sparse Direct Solver for Unsymmetric Linear Systems", ACM Trans. Mathematical Software, Vol. 29, Issue 2, pp. 110-140.

[12] Gupta, A, "Recent advances in direct methods for solving unsymmetric sparse systems of linear equations", ACM transaction in Mathematical Software, 28(3), 301-324, 2002.

[13] Schulze J. (2001) "Towards a tighter coupling of bottom-up and top-down sparse Matrix ordering methods", BIT Numerical Mathematics, Vol. 41 Issue 4, pp. 800-841.

[14] Karypis, G. and Kumar, V., 1999, "A Fast and High Quality Multilevel Scheme for Partitioning Irregular Graphs," SIAM J Scientific Computing, Vol. 20, pp. 359-392.

# Shapely Functions and Data Structure Preserving Computations

Thomas Nitsche

*Abstract*—**In parallel and distributes systems we have to consider aspects of communication, synchronization and data movement besides the actual algorithmic solution. In order to enable the clear separation of these aspects we have to separate the structure or shape of data structures from the data elements themselves. While the latter can be used to describe algorithmic aspects, the former enables us to derive the communication.**

**The notion of shape and data elements and how they can be separated are defined for arbitrary container types, i.e. parameterized, nested combinations of algebraic data types and arrays. This programming oriented separation of concerns serves as the basis for formally defining the notion of *shapely functions*. Such functions preserve the data structure, i.e. shape, and thus allow data parallel computations. We distinguish weakly and strongly shapely functions. We have used this as the semantic basis for handling overlapping data distributions not only over arrays as in commonly used approaches but over arbitrary (container) types.**

## I. Introduction

Programming parallel and distributed systems requires programmers to consider the distribution of the work respectively the data onto the different processors, as well as synchronization and data exchange, i.e. communication, between them [1]. For this reason the extraction of the data elements from a given data structure, their distribution and communication has to be handled in a parallel program. This can be done explicitly by the programmer in a low-level way as in MPI [2], by the system or combinations thereof.

The decision which data will be communicated and when depends largely on the characteristics of the parallel computer such as its network topology, bandwidth, etc. To achieve maximum efficiency, many low-level machine- and algorithm-specific details have to be considered. The resulting parallel program is highly problem- and machine-specific, which makes it error-prone if programmed by explicit message-passing [3] and difficult to port to another parallel machine or to reuse the code for another program. Thus, parallel programs should not be written for a specific parallel machine but rather parameterized using certain architectural parameters. The number of available processors is one such architectural parameter; others are its memory, the network parameters, etc.

In order to achieve a higher level of programming effectiveness and coordination of parallel activities, we can use algorithmic skeletons [4], [5], [6], [7] as a kind of collective parallel operation instead of directly using low-level (point-to-point) communication operations. Well known examples for such generic operations include *map*, which applies a function to all data elements, and *reduce*, which combines all data elements to a "sum" using some binary operator:

$$\mathrm{map}(f)([a_1, \ldots, a_N]) = [f(a_1), \ldots, f(a_N)] \quad (1)$$

$$\mathrm{reduce}(\oplus)([a_1, \ldots, a_N]) = a_1 \oplus \ldots \oplus a_N \quad (2)$$

Another operation is the *filter* function that removes all elements from a list that do not satisfy a given predicate:

$$\mathrm{filter}(p)(\langle a_1, \ldots, a_N\rangle) = \langle a_i \mid p(a_i) = \mathrm{true}\rangle \quad (3)$$

Operations on locally available data elements correspond to purely local operations operating merely on the data elements themselves, while accesses to elements residing on other processors imply communication requirements, where the structure of the data type determines the communication structure. If we thus separated the data elements from the data structure (i.e. shape) itself, we could describe the parallel operations independently from the actual data types and use the shape for deriving the communication [8].

The remainder of this paper is organized as follows. Section II describes the concept of shape in an informal way. In order to make this paper self-contained, we summarize its formal definition and basic properties in Section III. On this basis we introduce the notion of shapely functions in Section IV. Finally, Section IV discusses related work and concludes.

## II. Shape and Content of Data Structures

Every data structure has two aspects: its shape and its content (or data). The content, for example, is the set of elements in the fields of a matrix, in the nodes of a tree, and so on, while the shape corresponds to the index range of a matrix or the node structure of a tree. Separating shape from content allows us to handle the computational operations on the data elements independently of their mapping and ordering aspects.

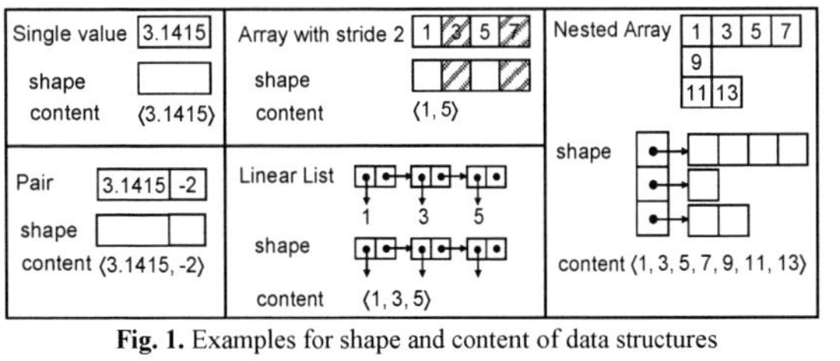


**Fig. 1.** Examples for shape and content of data structures

Thomas Nitsche is a research scientist at the Research Institute for Communication, Information Processing and Ergonomics (FGAN/FKIE), Neuenahrer Straße 20, 53343 Wachtberg, Germany; e-mail: nitsche@fgan.de.

T. Sobh (ed.), *Innovations and Advances in Computer Sciences and Engineering*,
DOI 10.1007/978-90-481-3658-2_91, © Springer Science+Business Media B.V. 2010

Let us look at some examples (cf. Fig. 1). The simplest one is a single data element, e.g. a real value like $\pi$. Its shape is a placeholder for a real value, while its content corresponds to the value of the element. Similarly, the shape of a tuple, e.g. a pair *&(3.1415, -2)* consisting of a real (*3.1.415*) and an integer (*-2*) value, corresponds to the storage place for the values. Its content is the list of the data values. Note that the values have different types (*real* and *int*, respectively), i.e. we must encode them by the the sum of the data types. Thus, formally, the content $\langle in_1(3.1415), in_2(-2)\rangle$ is of type *list[real + int]*.

In these simple examples, the shape corresponds directly to the type of the data. The same holds for homogeneous arrays. MPI allows strides in arrays, i.e. not all array elements are used but only those in a regular interval. For the array with stride 2 in Fig. 1, only the elements with even index are within the shape. Thus the content only consists of the values 1 and 5.

For nested, inhomogeneous arrays, the shape is merely the nested array structure. The list of data elements can be obtained by flattening [9], i.e. by putting all data elements into a single list. For algebraic data types, we have to consider the structure, i.e. the graph of their cells, as the shape. In the case of linear lists, this is merely the sequence of cons cells.

### A. Shape Theory

Shape theory [10], [11] formally describes the separation of shape and data and how data is stored within data structures. Shapely types can be formally described as a pullback in the sense of category theory.

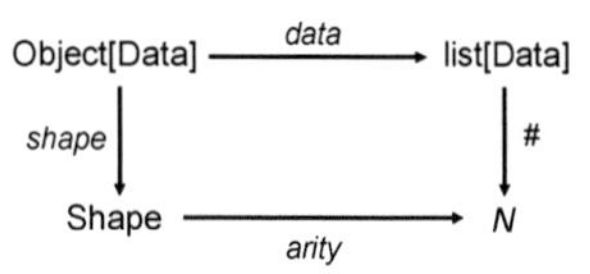


**Fig. 2.** Separation of shape and data

The semantics of Fig. 2 are as follows. Given two morphisms *arity: Shape* → *N* and *#: list[Data]* → *N* (where # computes the length of a list), then there exists an – up to isomorphisms – uniquely determined object *Object[Data]* with morphisms *shape: Object[Data]* → *Shape* and *data: Object[Data]* → *list[Data]* such that the above diagram commutes, i.e. it satisfies $arity \circ shape = \# \circ data$. This means that the data structure *Object[Data]* can be constructed from its shape and the list of its data elements.

### B. Representing Shapes

In the language FISh (Functional + Imperative = Shape) [12] the shape of a data structure is merely its size, i.e. the number of data elements, together with the shape (i.e. size) of the elements. This allows the description of nested homogeneous vectors, all subvectors being required to have the same shape and hence the same length. The shape information thus describes the memory consumption of a certain element.

This is similar to the language SAC (Single Assignment C) [13]. It offers multidimensional, homogeneous arrays and – as in APL – dimension-invariant array operations [14]. Arrays in SAC consist of two parts: a shape vector holding the array sizes in the different dimensions, and a separate data vector containing the data values. Consider, for example, a two-dimensional 2×3 matrix *A* with data elements 1,...,6. It is defined as *A = reshape([2,3], [1,2,3,4,5,6])*. Then, *shape(A)* yields the shape vector *[2, 3]* with the sizes within each dimension. This shape vector can be used to generate other shapes and hence new arrays. Note, however, that only homogeneous vectors are allowed in SAC, i.e. all subvectors of a nested matrix must be of the same size.

In Nesl [15] and Nepal [16], [17], the subvectors of a nested array may have different lengths, thus allowing us to model also sparse matrices. Internally, the nested vectors are subject to the flattening transformation and are represented by a shape vector containing the sizes of the different subvectors and a data vector [9]. For example, the nested vector *[[1,2,3], [], [4], [5,6]]* is represented by the size vector *[3,0,1,2]* and the data vector *[1,2,3,4,5,6]*.

ZPL [18] offers the concepts of regions that are abstract index sets with no associated data and grids as abstractions of processor sets [19]. Since ZPL is an array processing language, regions are based on arrays and indexes.

### C. Shapes of Container Types

We do not want to deal only with arrays or lists but also with arbitrary algebraic data types, so the size or the index vector is not sufficient in our case. Instead, we have to pay attention to how elements are stored within a data structure. Formalizing our notion of shapes, we allow arbitrary container types, i.e. arbitrary (nested) combinations of parameterized algebraic data types and arrays.

**Definition 1 (Container Type).** The set of *container types* *C(V,B)* over base types *B* and type variables *V* is defined as

| | | |
|---|---|---|
| C(V,B) → | B | Base type |
| \| | V | Type variable |
| \| | C(V,B) × ... × C(V,B) | Product |
| \| | C(V,B) + ... + C(V,B) | Sum ∇ |

A container type $t[V] \in C(V,B)$ is an algebraic data type parameterized with the set of type variables *V*. $\pi_j$: $t_1 \times ... \times t_k \rightarrow t_j$ denotes the projection of the *j*-th component, while $in_j$: $t_j \rightarrow t_1 + ... + t_k$ is the usual injection function. The final object, i.e. the one-elementary set $1 = \{*\}$, corresponds to the empty product, while the initial object ∅ corresponds to the empty sum. Arrays are just special cases of product *types* $C(V,B)^n$, so we can omit them in the definition of shape and data functions. Base types and arguments for the type parameter variables can be arbitrary types including function types. *C(∅,B)* is the usual set of (unparameterized) algebraic data types together with arrays. Abbot et al. [20] give a more general, categorical definition of containers. In the case of locally Cartesian closed categories, this is equivalent to the notion of shapely types [10].

The idea behind the concept of container types is that the

values of such a parameterized type represent the data structure, while the type variables serve as holes or placeholders for the data elements stored within the data structure.

Examples for container types include pairs, linear lists, and (non-empty) search trees that store data elements of type $\alpha$ according to a key element of type *int*:

$$\text{pair}[\alpha,\beta] = \alpha \times \beta \quad \in C(\{\alpha,\beta\}, \varnothing) \qquad (5)$$

$$\text{list}[\alpha] = 1 + \alpha \times \text{list}[\alpha] \quad \in C(\{\alpha\}, \varnothing) \qquad (6)$$

$$\text{tree}[\alpha] = \alpha + \text{tree}[\alpha] \times \text{int} \times \text{tree}[\alpha] \in C(\{\alpha\}, \{\text{int}\}) \qquad (7)$$

Thus the structure of a container type *C(V,B)* represents the shape, while the parameters of the type variables *V* contain the data elements. The elements of the base types contain additional information about the data structure, i.e. its shape. This holds, e.g., for the search tree, where the key elements are part of the shape.

## III. Formalizing Shape and Data

### A. Definition of Shape

We define a function ***shape*** that removes the data elements from a given data structure and leaves the structure – with the data replaced by * – alone. We call the resulting structure ***shape***. Its size, i.e. the number of * elements (as placeholder for date elements) is the ***arity***, while the (omitted) data elements can be extracted using the function ***data***. To generate a data structure from a shape and a list of data elements, we use the function ***re_cover***. Their definitions are shown in Fig. 3.

Let $b_i \in B_j \in B$, $p_i \in P_j \in P$, $V_j \in V$, $x, x_i \in t[P]$, $s, s_i \in t[1]$, $ds, d_i \in list[P_1+\ldots+P_m]$:
Then shape: $t[P] \rightarrow t[1]$, data: $t[P] \rightarrow list[P_1+\ldots+P_m]$ are defined as follows:

| | | |
|---|---|---|
| $shape_{Bj}(b_i)$ | $= b_i$ | (base type) |
| $shape_{Vj}(p_i)$ | $= *$ | (data value) |
| $shape_{T1\times\ldots\times Tk}((x_1,\ldots, x_k))$ | $= (shape_{T1(x1)},\ldots, shape_{Tk(xk)})$ | (product) |
| $shape_{T1+\ldots+Tk}(inj(x))$ | $= in_j(shape_{Tj}(x))$ | (sum) |

| | | | |
|---|---|---|---|
| $data(b_i)$ | $= \langle\rangle$ | $arity(b_i)$ | $= 0$ |
| $data(p_i)$ | $= \langle in_j(p_i)\rangle$ | $arity(*)$ | $= 1$ |
| $data((x_1,\ldots,x_k))$ | $= data(x_1)$ -\|\|-…-\|\|-$data(x_k)$ | $arity((s_1,\ldots,s_k))$ | $=\Sigma_{i=1}^{k} arity(s_i)$ |
| $data(in_j(x))$ | $= data(x)$ | $arity(in_j(s))$ | $= arity(s)$ |

$re\_cover(b_i, \langle\rangle) = b_i$
$re\_cover(*, in_j(p_i)) = p_i$
$re\_cover((s_1,\ldots,s_k), \langle d_1,\ldots,d_l\rangle) = (y_1,\ldots,y_k)$
where $y_i = re\_cover(s_i, \langle d_{\Sigma_{j=1}^{i-1} arity(s_j)+1},\ldots, d_{\Sigma_{j=1}^{i} arity(s_j)}\rangle)$
$re\_cover(in_i(s), ds) = in_i(re\_cover(s, ds))$

**Fig. 3.** Functions shape, data, arity and re_cover

For example, the shape of a pair $d = \&(3.1415, -2) \in$ *pair[real, int]* is $shape(d) = (*,*) \in pair[1,1]$. Its list of data elements is $data(d)=\langle in_1(3.1415), in_2(-2)\rangle \in list[real+ int]$, where $in_j$ encodes the corresponding data type, i.e. *real* or *int*.

Analogously, the shape of the two-element list of natural numbers $\langle 1,2\rangle$ and that of the list of lists $\langle\langle 1,2\rangle, \langle 3,4,5,6\rangle\rangle$ is merely a list $\langle *,*\rangle$ with two * elements. It is obtained by removing the data elements, i.e. the natural numbers or the sublists, and replacing them by *. The arity (size) of the shape, i.e. the number of (removed) data elements, is equivalent to the length 2 of the list. The list of data values is, in our example, the original list itself, so the function *data* is the identity here.

### B. Consistency Conditions

The application *re_cover(s, ds)* of the partial function *re_cover* is only defined if it is provided with enough data from *ds*, which requires $\#(ds) \geq arity(s)$. Furthermore, the data elements of *ds* must have the proper type to rebuild a correct data structure, which can be expresssed as *shape(ds) = data(s)*. We call this property **consistency** of shape and data.

**Definition 2 (Consistency).** Let $ds \in list[P_1+\ldots+P_m]$ be a list of data elements, $s \in t[1]$ be a shape and $d \in t[P]$ a data structure.

1) *ds* is called **consistent** with *s* iff *shape(ds) = data(s)*.
2) *ds* is called consistent with *d* iff *ds* is consistent with the shape of *d*, i.e. *shape(ds) = data(shape(d))*. ∇

Consistent data lists can be used to reconstruct a correctly typed data structure from a shape structure (Def. 2-(1)), or they can be used to replace all data elements in a data structure, yielding a correctly typed new data structure (Def. 2-(2)). If a function on data lists operates in such a way that it keeps this type information, we call this function **type-preserving** or, in the case of arbitrary data types, **shape-preserving**.

**Definition 3.** $f\colon t[P] \rightarrow t[P']$ is called **shape-preserving** iff $shape = shape \circ f$. ∇

### C. Properties of Shape and Data

The function *data* is defined by an in-order traversal of the data structure. Similarly, we can describe the functions defined in Fig. 3 as special cases of a general *traversal* operation on the data structure or its shape, as shown in [21]. It is thus sufficient to define a traversal on a corresponding data type. This can then be used to derive implementations for extracting the data elements and the shape as well its reverse *re_cover* operation. We use this for handling overlapping data distributions with arbitrary container types, i.e. nested algebraic data types (see Fig. 4 or [22] for details).

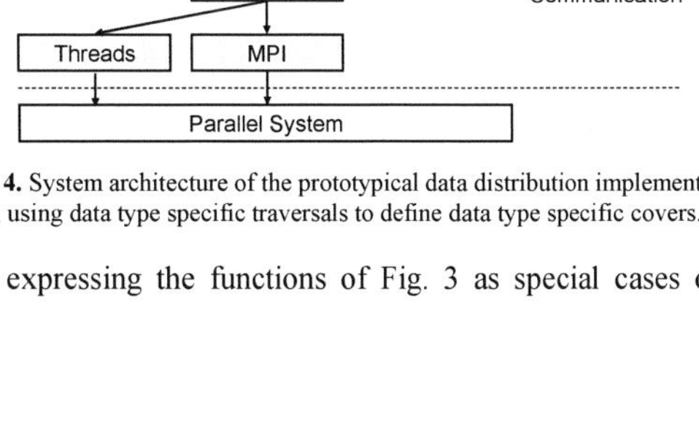


**Fig. 4.** System architecture of the prototypical data distribution implementation, using data type specific traversals to define data type specific covers.

By expressing the functions of Fig. 3 as special cases of a

general traversal operation, we can also show that *re_cover* is indeed the inverse function of *shape* and *data*:

**Theorem 4 (Inverse of shape and data).**

1) re_cover ° (shape × data) = $id_{t[P]}$
2) (⟨shape, data⟩ ° re_cover)(s, ds) = (s, ds), if ds consistent with s. □

*Proof.* Follows directly from the corresponding description of these functions as a traversal and the traverse-fusion law. □

Since the size of a shape corresponds to the number of *data* elements extracted from the data structure, the diagram in Fig. 5 commutes.

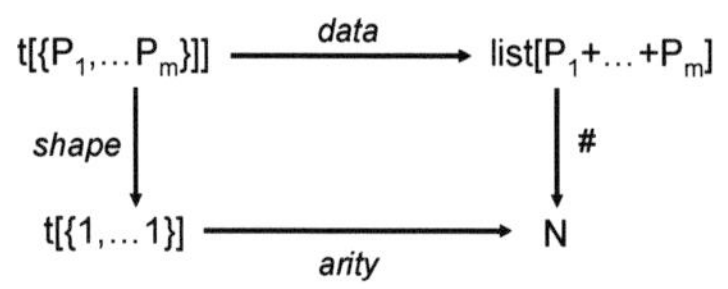


**Fig. 5.** Relationship between shape and data elements for container types

Moreover, not only the number of data elements extracted from a data structure matches the size (or arity) of its shape, but even the types of the elements are such that we can reconstruct the data structure from the elements (and their shape). The data elements selected from a data structure are therefore obviously consistent with the data structure itself because they are of the corresponding number and types.

**Proposition 5 (Consistency of Shape and Data).**

1) *data(d)* is consistent with *shape(d)*, i.e. data ° shape = shape ° data.
2) arity ° shape = # ° data □

*Proof. By structural induction (see [21]).* □

## IV. Shapely Functions

Now that we have formalized the notion of shape and data, we can use this to address the question of shapeliness. This ensures that a data structure is only computed in a somehow local manner. In particular, it means that changes in local values do not influence remote values unless explicitly by a foreign reference and that the data distribution is not affected.

### A. Weakly Shapely Functions

A function is (weakly) *shapely* if the shape of the result only depends on the shape of its input but not on the data values within the structure [10]. An example of a (weakly) shapely function is *map*. It updates every element in a certain data structure (e.g. list, tree or array) leaving the data structure, i.e. the shape, unchanged. *Filter*, on the other hand, removes from a list all elements that fail to satisfy a certain predicate. It is non-shapely because the size of the resulting list depends on the values of the data elements, so the shape, i.e. the list structure, of the result is data-dependent.

**Definition 6 (Weakly Shapely Function).** A function *f: t[P]→t'[P']* is called **weakly shapely** iff there exists a function $f_s$*: t[1]→t'[1]* on shapes such that $f_s$ *°shape = shape °f.* ∇

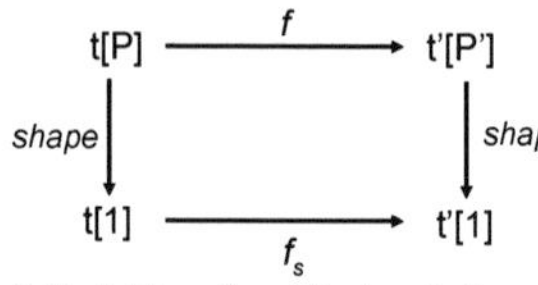


**Fig. 6.** Definition of weakly shapely functions

A function *f: t[P]→t'[P']*is thus (weakly) shapely if it operates on the shape *shape(d)* of the data structure *d∊t[P]* without taking into account values of the data elements. The function $f_s$ can be regarded as the part of *f* operating on the shape, i.e. $f_s$ can be derived from *f* by ignoring (or forgetting) the data part (cf. Fig. 6).

The function $f_s$ is thus defined independently of the actual data values, so even if we re-order, i.a. permute, them within the data structure, the condition of Def. 6 still holds. However, the reordering can only be done in a type-preserving manner, i.e. we are only allowed to exchange data elements $d_i$ and $d_j$ if they have the same type. In this case, we can even replace the data elements with any values with the correct type.

**Theorem 7 (Weakly Shapely Function Condition – Data Independence of Shape Calculations).** The following statements are equivalent:

1) *f* is weakly shapely.
2) There exits a function $f_s$ such that for any type-preserving permutation function *permute: list[P₁+...+P_m] →list[P₁+...+P_m]* it holds
   $f_s$° shape = shape° f° re_cover° (shape×(permute° data))
3) There exists a function $f_s$ such that for all data structures *d∊t[P]* and lists of data elements *ds∊ list[P₁+...+P_m]* consistent with *d*, it holds:
   ($f_s$° shape)(d) = (shape° f° re_cover)(shape(d),ds) □

Condition (2) is a special case of condition (3).

*Proof.* **(1) ⇒ (3)** Let $f_s$ such that $f_s$ *°shape = shape °f.* Further, let *d∊t[P]* and *ds∊ list[P₁+...+P_m]* consistent with *d*. Then *re_cover(shape(d), ds)* is defined and it holds

($f_s$°shape)(d) = ($f_s$°$\pi_1$)(shape(d), ds) (def. $\pi_1$)
= ($f_s$°$\pi_1$°$id_{t[1]\times list[P1+...Pm]}$)(shape(d), ds) (def. id)
= ($f_s$°$\pi_1$°(shape×data)°re_cover)(shape(d),ds) (Theorem 4-(2))
= ($f_s$°shape°re_cover)(shape(d),ds) (def. $\pi_1$)
= (shape°$f_s$°re_cover)(shape(d),ds) (precondition)

**(3) ⇒ (2)** Let $f_s$ such that ∀*d'*, ∀*ds'* consistent with *d'*, it holds *($f_s$° shape)(d') = (shape ° f ° re_cover)(shape(d'),ds')*. Further, let *permute* be type-preserving, i.e. *shape°permute= shape*, and *d∊t[P]*. Then, for *ds:=permute(data(d))* it holds

shape(ds) = (shape(permute)(data(d))
= (shape°permute)(data(d)
= shape(data(d)) = data(shape(d)).

Thus *ds* is consistent with *d*. Condition (2) follows now directly:

($f_s$° shape)(d') = (shape°f°re_cover)(shape(d),ds) (precond.)

= (shape°f°re_cover)(shape(d), permute(data(d))) (def. ds)
= (shape°f°re_cover°(shape×(permute°data))(d).

**(2) ⇒ (1)** Let $permute:=id_{list[P1+...+Pm]}$. It is obviously a type-preserving permutation function, and it satisfies

$f_s$° shape = shape°f°re_cover°(shape×(permute°data)) (precond.)
= shape°f°re_cover°(shape×data) (permute=id)
= shape°f°$id_{t[P]}$ (Th. 4-(1))
= shape ° f □

*B. Strongly Shapely Functions*

Weakly shapely functions correspond to the notion of shapely functions used by Jay et al. [24]. They ensure that the shape of the result is data-independent. However, if we also pay attention to the order of elements within the data structure because this may affect the mapping of subobjects to the available processors, we need a stronger notion of shapeliness. In this case we require that the position of a certain data element $d_i$ within the data structure $d$ – which is, according to our definition of the *re_cover* function, equivalent to the index $i$ within the list of data elements *data(d)* of $d$ – may not depend on the data values themselves.

*Example*. Consider, for example, the selection of the *smallest* element of a list. Unless the list is already sorted, we must check every list element in order to get the result. The index of the smallest element must be calculated from the list values. The shape of the result is always a single data value and thus independent of the data values. We can thus define a shape function $f_s$ which is a constant function and thus obviously satisfies the condition of Def. 6, this implies weak shapeliness here. But since the resulting value depends on the values of *all* data elements in the list, the selection can *not* be regarded as a strongly shapely operation.

Another situation is the selection of the *last* element of a list. Here the shape function $f_s$ is constant as before which implies weak shapeliness. This operation is even strongly shapely because the index of the selected element is independent of the values of the list, depending only on its length, i.e. the shape.

The function *last* can thus be computed on each processor in parallel, independently of the other processors. We only have to send the value of the last element if required by another processor. The owner of the *smallest* element, on the other hand, must be computed dynamically each time by accessing all data elements. Thus the communication structure for access to the *last* element can be derived statically or at least once at the beginning of the program execution, which is not possible for the *smallest* element.
*(End of Example)*

For this reason, we also introduce the notion of *strong shapeliness*. It implies both the shape's independence of the data elements and that for the computation of data elements only the shape and the old data values are used, but not other data elements possibly stored at another processor. A function is thus ***strongly shapely*** if

1) the resulting shape depends only on the input shape and not on the input data values (i.e. the function is weakly shapely), and
2) the order and the values of the data elements within the resulting shape are not dependent on arbitrary data but only on the input shape and their own local data values.

Due to space limitations and in order to simplify the presentation, we omit the formal definition of strongly shapely functions here. It can be found in [23]. We only note the following proposition:

**Proposition 8 (Special Cases).**
1) $f$ is strongly shapely ⇒ $f$ is weakly shapely.
2) $f$ is shape preserving ⇒ $f$ is weakly shapely. □

*Proof.* (1) follows directly from the definition of strong shapeliness. (2) follows from immediately by using $f_s = id$. □

*C. Examples of Shapely and Non-Shapely Functions*

The shape functions defined in Fig. 3 are strongly shapely [23]. By defining $f_s$=*id* we see that the *map*-skeleton is weakly shapely. For the proof of its strong shapeliness we refer to [23]. It thus preserves the data structure of its result and allows an easy parallelization. The *reduce*-skeleton is only weakly shapely with constant shape, so $f_s=\lambda x.*$ proves its shapeliness. However, the *filter* function is not even weakly shapely, because it is changing the shape of its result in a data dependent manner.

*D. Composition of Shapely Functions*

To efficiently analyze functions to be parallelized, it is important that shapeliness is preserved by function composition. This ensures that we can check the shapeliness of the functions independently from each other. Their composition is than shapely is well, if the corresponding base functions are shapely.

**Proposition 9 (Function Composition Preserves Shapeliness).** Let $f$: $t_1[P_1] \rightarrow t_2[P_2]$ and $g$: $t_2[P_2] \rightarrow t_3[P_3]$ be weakly (strongly) shapely functions.

Then $g$ °$f$ is weakly (strongly) shapely as well. □

*Proof.* Let $f$: $t_1[P_1] \rightarrow t_2[P_2]$ and $g$: $t_2[P_2] \rightarrow t_3[P_3]$ be weakly shapely functions. According to Def. 6, there exist functions $f_s$: $t_1[1] \rightarrow t_2[1]$ and $g_s$: $t_2[1] \rightarrow t_3[1]$, such that $f_s$ °*shape* = *shape* °$f$ and $g_s$ °*shape* = *shape* °$g$.

Then the function $g$°$f$: $t_1[P_1] \rightarrow t_3[P_3]$ satisfies $g_s$ °$f_s$ °*shape* = $g_s$ ° *shape* ° $f$ = *shape* ° $g$ ° $f$, and hence proves the weak shapeliness of $g$°$f$.

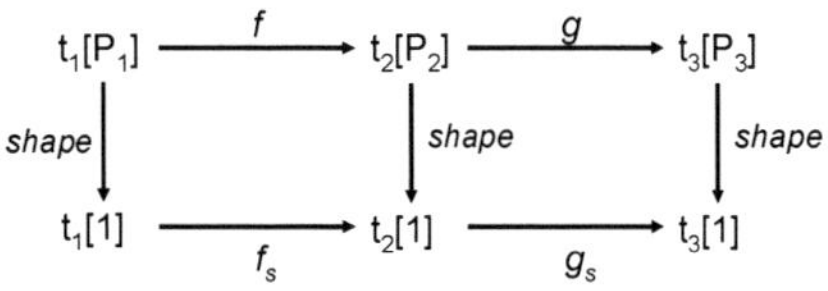


**Fig. 7.** Composition of weakly shapely functions

The case of composing strongly shapely functions is demonstrated analogously by defining a proper shape transformation function from *g* and *f*.

□

## V. Conclusion

Examples for exploiting shape in parallel programming are diverse. They include, among others, the data field model for indexed collections of data [25], PEI [26], or the FISh language [12]. The shape of data structures, or more specifically the sizes and indices of distributed arrays and matrices, are also handled in the (internal implementation of) the skeletons of the Muesli library [27].

Beckmann and Kelly adopt an interesting approach in optimizing regular matrix computations. They optimize the parallel execution of library calls like that of the Basic Linear Algebra Subroutines (BLAS) at runtime. Their basic idea is to suspend the actual execution of array operations, and optimize the combination of (hopefully) all library calls of the program. Since the arguments of library calls are known at runtime, they know the exact sizes of each matrix and how the output of one library call is used in later calls. This allows them to distribute the data in a way that might be non-optimal for one specific library call, but achieves better overall performance because additional data redistributions can be avoided [28], [29].

However, common approaches only work for one specific data type: the languages HPF, SAC [13], ZPL [18] or FISH [11], [12] operate on arrays or matrices, while NESL [15] and Nepal [16] use (nested) vectors. Other data types have to be encoded explicitly (by the user) [30].

We formalized the notion of shape and data elements and how they can be separated as well as used to reconstruct a valid data structure again. The approach works for arbitrary container types, i.e. parameterized algebraic data types, arrays and arbitrary nested combinations thereof.

In this paper we formally defined the notion of shapely functions, i.e. functions that change their shape only in a data-independent manner [10]. We introduced the notion of weak and strong shapeliness. While weak shapeliness ensures that the shape of the result is data-independent, strong shapeliness also ensure that the order and the values of data elements only depend on the shape and local data values.

This allows the data elements and the shape to be processed independently from each other. We can especially distribute the data onto different processors, with the shape encoding the communication structure between them. We have used this as the basis to handle overlapping data distributions [23]. Since for shapely functions the shape can be calculated without having to know data from other processors, this allows pre-calculated communication schedules as well as security checks in distributed environments [31]. Since shapeliness is preserved by function composition, we can efficiently analyze functions to be parallelized independent from each other [32].

## References


[1] T. Nitsche, "Coordinating computation with communication", in *Proc. COORDINATION 2006*, vol. 4038 of LNCS, Springer, Jun. 2006, pp. 212–227.

[2] M. Snir, S. W. Otto, S. Huss-Ledermann, D. W. Walker, and J. Dongarra, *MPI – The Complete Reference*, MIT Press, 1998.

[3] S. Gorlatch, "Send-receive considered harmful", *ACM Transactions on Programming Languages and Systems*, vol. 26, no. 1, pp. 47–56, 2004.

[4] M. I. Cole, *Algorithmic Skeletons: Structured Management of Parallel Computation*, MIT Press, 1989.

[5] M. I. Cole, "Bringing skeletons out of the closet: a pragmatic manifesto for skeletal parallel programming", *Parallel Computing*, vol. 30, no. 3, pp.389–406, 2004.

[6] J. Darlington, Y. Guo, H.W. To, and J. Yang, "Functional skeletons for parallel coordination", in *Proc. Euro-Par'95*, vol. 966 of LNCS, 1995.

[7] F. A. Rabhi and S. Gorlatch, Ed., *Patterns and Skeletons for Parallel and Distributed Computing*, Springer, 2003.

[8] T. Nitsche, "Deriving and scheduling communication operations for generic skeleton implementations", *Parallel Processing Letters*, vol. 15, no. 3, pp. 337–352, Sep. 2005.

[9] G. Keller and M. Simons, "A calculational approach to nested data parallelism in functional languages", in *Proc. Asian Computing Science Conf.*, vol. 1179 of LNCS, Springer, Dec. 1996, pp. 234–243.

[10] C.B. Jay, "A semantics for shape", *Science of Computer Programming*, vol. 25, no. 2-3, pp. 251–283, 1995.

[11] C. B. Jay, "Costing parallel programs as a function of shapes", *Science of Computer Programming*, vol. 37, no. 1–3, pp. 207–224, May 2000.

[12] C. B. Jay and P. A. Steckler, "The functional imperative: shape!", in *Proc. ESOP'98*, vol. 1381 of LNCS, Springer, 1998, pp. 139–53.

[13] C. Grelck and S.-B. Scholz, "SAC – a functional array language for efficient multi-threaded execution", *International Journal of Parallel Programming*, vol. 34, no. 4, pp. 383–427, Aug. 2006.

[14] C. Grelck and S.-B. Scholz, "SAC – from high-level programming with arrays to efficient parallel execution", *Parallel Processing Letters*, vol. 13, no. 3, pp. 401–412, 2003.

[15] G. E. Blelloch, "Nesl: A nested data-parallel language (3.1)", Carnegie Mellon Univ., Tech. Rep. CMU-CS-95-170, Sept. 1995.

[16] M. M. T. Chakravarty, G. Keller, R. Leshchinskiy, and W. Pfannenstiel, "Nepal - nested data parallelism in Haskell", in *Proc. Euro-Par'01 Parallel Processing*, vol. 2150 of LNCS, Springer, 2001, pp. 524–534.

[17] R. Leshchinskiy, "Higher-order nested data parallelism: semantics and implementation", PhD thesis, Technical Univ. of Berlin, Dec. 2005.

[18] B. L. Chamberlain, S.-E. Choi, S. J. Deitz, and L. Snyder, "The high-level parallel language ZPL improves productivity and performance", in *Proc. IEEE International Workshop on Productivity and Performance in High-End Computing*, 2004.

[19] S. J. Deitz, B. L. Chamberlain, and L. Snyder, "Abstractions for dnamic data distribution", in *Proc. IEEE Workshop on High-Level Parallel Programming Models and Supportive Environments*, 2004.

[20] M. Abbott, T. Altenkirch, and N. Ghani, "Categories of containers", in *Proc. FOSSACS'03*, vol. 2620 of LNCS, Springer, 2003, pp. 23–38.

[21] T. Nitsche, "Separation of Shape and Data", in *Innovations and Advanced Techniques in Computer and Information Sciences and Engineering*, pp. 455–461, Springer, 2007.

[22] T. Nitsche, "Skeleton implementations based on generic data distributions", in *Constructive Methods for Parallel Programming*, vol. 10 of *Advances in Computation: Theory and Practice*, Nova Science, 2002.

[23] T. Nitsche, "Data distribution and communication management for parallel systems", PhD thesis, Dept. of Comp. Sci. and Electr. Eng., Technical Univ. of Berlin, Dec. 2005.

[24] C.B. Jay and J.R.B. Cockett, "Shapely types and shape polymorphism", in *Proc. ESOP'94*, vol. 788 of LNCS, pp. 302–316, Springer, 1994.

[25] J. Holmerin and B. Lisper, "Development of parallel algorithms in Data Field Haskell", *Proc. Euro-Par'00*, vol. 1900 of LNCS, Springer, 2000.

[26] E. Violard, "Typechecking of PEI expressions", in *Proc. Euro-Par'97 Parallel Processing*, vol. 1300 of LNCS, Springer, 1997, pp. 521–529.

[27] H. Kuchen, "A skeleton library", in *Proc. Euro-Par'02 Parallel Processing*, vol. 2400 of LNCS, Springer, Aug. 2002, pp. 620–629.

[28] O. Beckmann, "Interprocedural optimisation of regular parallel computations at runtime", PhD thesis, Imperial College, London, Jan. 2001.

[29] P. H. J. Kelly and O. Beckmann, "Generative and adaptive methods in performance programming", *Parallel Processing Letters*, vol. 15, no. 3, pp. 239–256, Sep. 2005.

[30] G. Keller and M. M. T. Chakravarty, "Flattening trees", in *Proc. Euro-Par'98 Parallel Processing*, vol. 1470 of LNCS, 1998, pp. 709-719.

[31] T. Nitsche, "Secure communication in distributed environments – Ensuring consistency of communication structures by shapeliness analysis", presented at the Spring School on Security, Marseille, April 2005.

[32] T. Nitsche, "Shapeliness analysis of functional programs with algebraic data types", *Science of Computer Programming*, pp. 225–252, 2000.

# IraqComm and FlexTrans: A Speech Translation System and Flexible Framework

Michael W. Frandsen, Susanne Z. Riehemann, and Kristin Precoda
SRI International, 333 Ravenswood Ave, Menlo Park, CA 94025
michael.frandsen@sri.com, susanne.riehemann@sri.com, precoda@speech.sri.com

***Abstract*-SRI International's IraqComm system performs bidirectional speech-to-speech machine translation between English and Iraqi Arabic in the domains of force protection, municipal and medical services, and training. The system was developed primarily under DARPA's TRANSTAC Program and includes: speech recognition components using SRI's Dynaspeak engine; MT components using SRI's Gemini and SRInterp; and speech synthesis from Cepstral, LLC. The communication between these components is coordinated by SRI's Flexible Translation (FlexTrans) Framework, which has an intuitive easy-to-use graphical user interface and an eyes-free hands-free mode, and is highly configurable and adaptable to user needs. It runs on a variety of standard portable hardware platforms and was designed to make it as easy as possible to build systems for other languages, as shown by the rapid development of an analogous system in English/Malay.**

## I. Overview

The IraqComm system translates conversations between speakers of English and Iraqi Arabic. The speech recognition components are speaker-independent and noise-robust. The system has a vocabulary of tens of thousands of English and Iraqi Arabic words taken from the domains of force protection, municipal services, basic medical services, and personnel recruiting and training. Performance on other topics is related to the degree of similarity with the training domains.

SRI's Flexible Translation (FlexTrans) Framework is highly adaptable to the user's needs. It has an intuitive graphical interface, an eyes-free/hands-free mode, and many practical features. It was designed to facilitate the Spoken Language Communication and Translation System for Tactical Use (TRANSTAC) program's goal of rapidly developing and fielding robust, reliable systems in new languages.

The system has been running on ruggedized laptops in Iraq since early 2006. It currently runs on standard Windows computers and can be adapted to various specialized hardware platforms (see Section 5).

The purpose of this paper is to describe the architecture of a speech-to-speech translation framework that was designed to be flexible, configurable, and adaptable both to user needs and to the characteristics of different languages. We do give some details about the speech recognition and translation components of a particular version of the IraqComm system, and provide an example and some discussion of translation quality. But the main focus is on system architecture, user interface, system configurability, adaptability to different hardware and other languages, and lessons learned in the course of system development.

## II. System Architecture

In this paper we will refer to the component that integrates and controls the Automatic Speech Recognition (ASR), Machine Translation (MT), and Text-to-Speech (TTS) components as the Flexible Translation (FlexTrans) Framework, because the name 'IraqComm' usually refers to FlexTrans together with the current ASR, MT, and TTS components, as well as, on occasion, the hardware platform.

At a high level, information flow through the system is simple: speech recognition, then translation, then speech synthesis in the other language. At a more detailed level the system is more complicated. Inter-component filters are applied at various stages, the timing of the various processes needs to be coordinated, certain properties of inputs and component outputs trigger messages to the user (e.g., 'speech too loud' or 'unable to recognize speech'), the exact flow varies with system settings such as 'autotranslate', and the user can abort ASR, MT, TTS, and other audio playback at any point. The flow chart in Figure 1 shows a simplified view of part of the FlexTrans system translating a single utterance.

The complete system is far more complex than the flow chart shows, and handles a variety of user interactions, some of which are described in more detail in Section 4. It is also robust enough to cope with any combinations of GUI elements being activated simultaneously or in quick succession, as can happen with a touchscreen interface.

### *A. Speech Recognition*

The ASR components use Dynaspeak [1], SRI's high-accuracy, embeddable, speaker-independent, real-time recognition engine. The IraqComm system uses a 16 kHz sampling rate, a 10 ms frame advance rate, and Mel frequency cepstral coefficients.

During the development of the components for each language, a decision-tree state-clustered triphone model was discriminatively trained using Minimum Phone Frame Error (MPFE) training and compiled into a state graph together with the pronunciation dictionary and a heavily pruned n-gram language model. For more information on the training and use of acoustic models, see [2]. For more in-depth information on our particular approach, see [3].

T. Sobh (ed.), *Innovations and Advances in Computer Sciences and Engineering*,
DOI 10.1007/978-90-481-3658-2_92, 

During recognition, a time-synchronous Viterbi search of the state graph generates a lattice, using Gaussian shortlists to speed up the computation. A second pass of rescoring based on the SRI Language Modeling Toolkit (SRILM) [4] uses a much larger higher-order language model with several million n-grams.

The acoustic models are trained with added noise of types that can be expected in the target environment. For signal-to-noise (SNR) ratios between 5 and 15 dB, this can reduce errors by about 25%. In these low SNR cases, additional Probabilistic Optimum Filtering (POF) compensation is applied, which can reduce errors even more significantly [5].

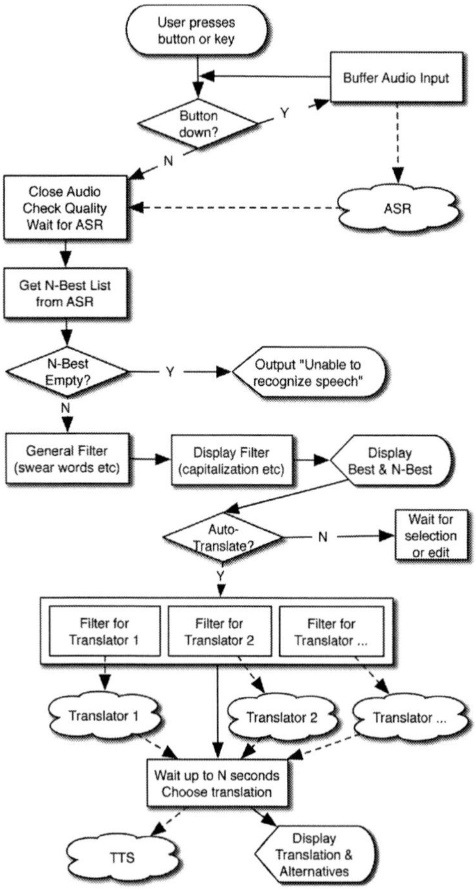


Fig. 1. Flow chart of part of the FlexTrans system

The system accommodates lists of swear words and phrases to remove before translating, in order to avoid the possibility of insulting someone because of a recognition error. This approach was chosen instead of removing these words from the ASR vocabulary, so as to avoid affecting recognition quality. In order to indicate to the user that this was done on purpose, asterisks are displayed on the screen in place of the swear words.

### *B. Translation*

In the current IraqComm system, the translation from English to Iraqi Arabic is provided by Gemini if parsing and generation succeed within a specified time period. Otherwise, a statistical translation is used. It is often the case that the rule-based translation is more accurate and more easily understandable than the statistical translation. For the translation from Iraqi Arabic to English, SRI's SRInterp statistical MT engine is applied.

The Gemini [6] based translator uses two context-free unification grammars connected by an interlingua. It parses an English utterance and converts it to a (quasi-) logical form, from which it generates an Iraqi Arabic translation (see [7] for a similar approach to Pashto translation).

In Iraqi Arabic, yes-no questions have the same word order as statements, which, depending on the intonation of the TTS, could lead to misunderstandings. One interesting property of Gemini is that it knows when an English input is a yes-no question if inverted word order is used, and can therefore mark the translation as a question. Examples of this can be found in Figure 2, questions (1), (3), and (5). The way Gemini marks these utterances as questions – by inserting *do you* – is perfectly understandable but not frequently used, so not likely to be part of the training data for a statistical system.

The SRInterp component is statistical and phrase-based. It has a probabilistic bilingual phrase table and an n-gram language model for the target language. During translation, a dynamic programming algorithm uses these resources to find the translation that is the optimal compromise between adequacy and fluency. For background on this type of approach to machine translation see [8] and [9].

### *C. Speech Synthesis*

The English and Iraqi Arabic TTS components, provided by Cepstral [10], are based on concatenative synthesis and use the Swift engine. Both voices are male and can be used at varying playback speeds. The voices can be customized for any desired lexical items via a user dictionary.

## III. EXAMPLE

The quality of speech recognition and translation is good enough to exchange information and achieve successful communication when the topic is common conversational speech or matches the training data domain well. The sample dialog in Figure 2 shows that the English translations are comprehensible although some are not perfect. For example, in (4) the *that* clause *that a person was running* is not complete. In (8), the pronoun is wrong due to a speech recognition error (*I* instead of *he*). Similar mistakes are present in the other direction. The word order of the Iraqi Arabic translation in (9) is not standard, and in (11) there is a problem with gender agreement.

| | |
|---|---|
| (1) | was anyone injured in the explosion<br>هل أي واحد كان مجروح بالإنفجار |
| (2) | لا ما كان أكو أحد بالبناية<br>no there was no one in the building |
| (3) | do you know who put the explosives there<br>هل تعرف منو يحط المتفجرات هناك |
| (4) | لا بس أنا شفت فد شخص كان ديركض قبل الإنفجار<br>no but I saw a person was running before the explosion |
| (5) | can you describe that person<br>هل تقدر توصف ذاك الشخص |
| (6) | عمره كان تقريباً أربعين سنة هو طويل كان عنده لحية<br>he was about forty years old and he's tall and he has a beard |
| (7) | which direction did he go<br>وين راح |
| (8) | اتجاه المدرسة وبعدين صعدت سيارة تويوتا حمرا بإتجاه الجنوب<br>in the direction of the school and then I caught a Toyota red towards the south |
| (9) | when was this<br>شوقت هذا كان |
| (10) | تقريباً قبل عشر دقايق<br>about ten minutes ago |
| (11) | do you remember anything else that might be helpful<br>تتذكر أي شيء ثاني ممكن تكون مفيدة |
| (12) | لا ما أعتقد<br>no I don't think so |
| (13) | thank you very much<br>شكراً جزيلاً |

Fig. 2. Sample Dialog

It is possible that these types of mistakes can lead to misunderstandings, but they do not usually cause problems because the meaning is clear in context. This is particularly true for users who have a strong incentive to try to communicate. They can take care to make sure the recognition result is correct before translating; make use of gestures and facial expressions in addition to speech; use the frequent phrases to communicate about problems (e.g., ask the Iraqi speaker to use shorter, simpler sentences when a particular long utterance turns out to have a confusing translation); look at the translation alternatives; use the 'do you understand' button or simple yes/no questions for confirmation at important points in the conversation; and know from intuition or experience what types of utterances are difficult, such as figurative speech or translating names.

It is difficult to give meaningful numbers about the accuracy of the system out of context, because accuracy varies significantly depending on the topic and type of conversation and the individual speakers. It is hard to know whether any given test is harder or easier than actual field use. Evaluating speech-to-speech translation systems is a research area in itself, with many possible approaches and tradeoffs, and no clear ideal solution. Some of the evaluation approaches currently in use in the DARPA TRANSTAC program are described in [11]. It should be noted that as development of the system continues, any evaluation is only a snapshot in time and rapidly obsolete as research progresses.

A detailed analysis of the performance of a particular set of components is not the focus of this paper, but to give at least some indication of system performance, we can mention that during the NIST evaluation in June 2008, new users – both English speaking subject matter experts and foreign language speakers – agreed that the system was usable and made it easy to have the interaction. Over 80% of the translations were considered adequate by human judges. Most of the inadequate translations were due to recognition errors.

For English, most of the recognition errors involved the unstressed reduced forms of short function words like articles, prepositions, pronouns, and auxiliary verbs, and thus could be avoided by making minor corrections to the ASR results before translating. In 40% of these cases, the correct recognition was one of the alternatives provided.

For Iraqi Arabic, most of the recognition errors involved prefixes and suffixes of otherwise correctly recognized words. Correcting these by typing is possible, but in some settings it may not be practical to have the Iraqi person do this. However, for about 20% of the utterances with Iraqi recognition errors the correct recognition was in the list of alternatives, so it is helpful to at least have the Iraqi speaker point to items in that list if possible.

The FlexTrans system provides an additional feature that can help improve the user's confidence in the adequacy of the translation. It is possible to turn on 'backtranslation' so that the English speaker can see a translation back into English of what the Iraqi speaker heard. If this backtranslation is comprehensible, the English speaker has reason to be quite confident that the original translation was correct and comprehensible to the Iraqi speaker.

The converse is not true: if the backtranslation is wrong, it is more likely that the translation into Arabic is fine and the problem is with the translation back into English. But the user cannot be sure which is the case. However, with some experience the user can learn what types of backtranslation mistakes are to be expected due to ambiguities in Iraqi Arabic (e.g., confusing *he* and *you*) or due to Gemini's marking of questions, which can result in backtranslations like *do you can describe that person* for (5).

If backtranslation becomes an important feature, it is possible to improve it by taking into account what types of problems Gemini output can pose as input for statistical MT, and for example stripping question markers before doing the backtranslation and coordinating the vocabularies carefully.

## IV. Meeting User Needs

### A. *User Interface*

The user interface was designed to be intuitive and easy for new users while providing advanced features for more experienced users. A screenshot of the IraqComm GUI can be seen in Figure 3.

Because the system needs to function in noisy environments, endpointing can be difficult, so instead it uses 'hold to talk', which is very familiar to military users of walkie talkies. The user can choose to hold down the button using the mouse, the touchscreen, or the 'E' and 'I' keys on the keyboard. The system buffers some audio in case the user starts speaking before hitting the button, and strips initial silence using start of speech (SOS) detection.

The main recognition result and translation are displayed prominently in large type in the two large text boxes; alternative recognition results and alternative translations are displayed on the right.

The user can select alternative recognition results from the n-best list, or edit the recognition result to correct minor mistakes using a regular keyboard or an on-screen keyboard.

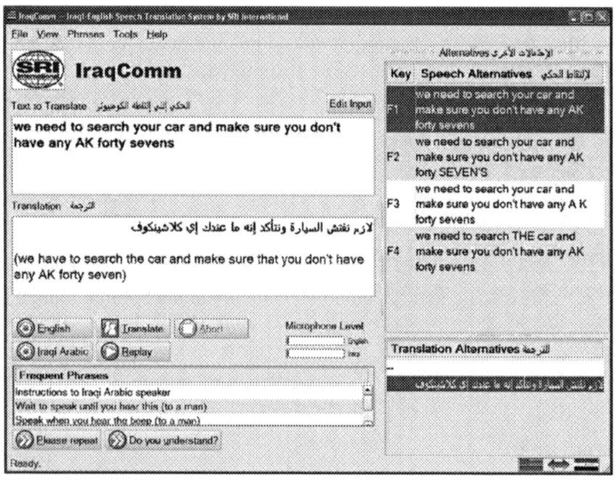


Fig. 3. IraqComm Screenshot

## V. Configurability

There are various menu items that allow users to adapt the behavior of the system to their needs. One example is "extra politeness". When consulting with a higher or equally ranked Iraqi official or elder, one may want to use more polite language than one might use in other situations.

In some situations it is desirable to be able to operate the system without needing to look at the screen or touch the computer. In eyes-free/hands-free mode, a two-button input device is used to indicate what language is recognized. There are cueing beeps for both languages, and the user can use commands like 'computer repeat'. A hierarchical dynamic grammar was developed for this purpose and merged with the regular English ASR grammar, and this command-and-control mode can be activated separately. It is also possible to play the ASR result for confirmation before translation.

The FlexTrans system can handle stereo input with one language per channel, so if a stereo audio card is available, it is possible to plug in two headsets and avoid having to pass a microphone back and forth. The audio does not get picked up from both headsets at the same time, so one speaker's breathing cannot interfere with the recognition for the other speaker, as would be the case without stereo. It is also possible to play the audio through the headsets and control the output channels for each language.

### A. *Keeping the User Informed*

The FlexTrans system was designed to be responsive to the users and keep them informed about what is happening. It is possible to abort the current operation at almost any stage. There is a status message in the bottom left corner of the screen, a working bar indicating when the system is busy, and corresponding sounds in eyes-free mode. There are also visual and/or auditory notifications when a button click is received.

### B. *Other Practical Features*

A menu item shows the conversation history, including all system inputs and outputs and associated information. The user can archive the files for future review in html format, which is is more human readable than the detailed system logs.

To speed up communication, a shortcut list of frequent phrases is provided that can be translated with one simple click, including short instructions in Iraqi Arabic explaining the basics of the system. It is easy for the user to add frequent phrases from a system translation, or by asking a trusted native speaker to record the translation.

It is also possible for the user to add custom words or phrases to the speech recognition components and the statistical translation components.

## VI. System Generality

Much of the FlexTrans system is generic and can be adapted to work on different hardware or for different languages.

### A. *Different Hardware*

The software is adaptable to a variety of hardware.

The FlexTrans system is easy to use on different screen sizes and resolutions. It is easy to include settings for new resolutions because it is using font multipliers instead of setting each font size manually. If necessary, some of the widgets can be hidden, such as the 'speech alternatives', 'translation alternatives', or 'frequent phrases', and can be made accessible from a button or menu and/or in a separate window instead. The lowest resolution we have tried so far is 640x480.

The software can be adapted to work on slower computers with less memory. It is easy to drop backup translation components. Many of the components can be adapted to be less resource-intensive, though there may be accuracy or speed tradeoffs. But it is possible to reduce vocabulary size without noticeable effect on translation quality, by removing low-frequency items.

The smallest computer we have run the system on is the Sony VAIO VGN-UX280P. The dimensions of this device are approximately 6 x 4 x 1.5 inches, with a weight of 1.2 pounds.

The FlexTrans software is written in C++ and uses Trolltech's Qt libraries. These provide both the GUI framework and other common libraries, which should minimize issues with porting the application to other platforms. The UTF-8 encoding for Unicode was used to be more compatible with old libraries and applications.

### B. *New Languages*

Where possible, the FlexTrans system was designed to facilitate supporting a new language. For example, it would be relatively easy to integrate our previous Pashto [7] work into the more generic FlexTrans architecture, and recently we have built a system to translate between English and Malay.

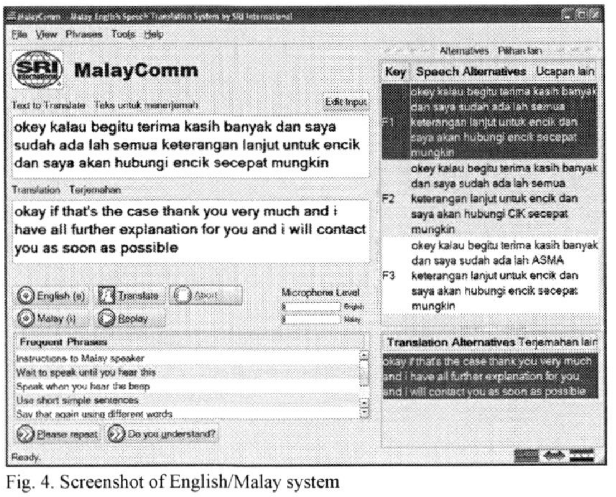

Fig. 4. Screenshot of English/Malay system

This was facilitated by several features of the FlexTrans framework.

Properties like language name, translation processes, GUI labels, etc., are set in a configuration file and can be changed easily.

Even though the current systems are intended to be primarily controlled by the English speaker, most of the FlexTrans system was designed to be symmetric, and many differences are achieved via parameters. For example, the fact that the foreign language speaker is cued with a beep is a configuration parameter rather than a hard-coded system behavior.

The FlexTrans system is built to allow for any number of translation components in each direction, and it is possible to send the input to different components based on characteristics of the input, or select which of several parallel outputs to use.

Other writing systems can be accommodated easily because Qt is Unicode-based and has bidirectional language support. Arabic right-to-left text works seamlessly.

The FlexTrans system has a notion of input and output filters so text can be appropriately prepared for a particular translator, post-processed after a particular ASR engine, preprocessed for a particular TTS engine, formatted for writing on the screen or to a log file, etc. For the IraqComm system this includes filters for converting the character set, special treatment of hyphenated words, possessive words, acronyms, and capitalization. In addition to the set of built-in filters, new ones can be created by specifying sets of regular expressions and calling other fiters, without having to recompile the software. This means that new components can be swapped in simply by installing them and specifying in the configuration file which filters to apply at various points. Some of the points where filters are applied can be seen in the flow chart in Figure 1.

## VII. LESSONS LEARNED

This fairly complex system was developed in quite a short period of time – just half a year before the first test system was sent to Iraq – and the fact that it needs to be fieldable places certain demands on its usability and reliability. Given these circumstances, we found particular value in the following procedures.

It was very helpful to have easily retrievable records of project-internal interactions, including discussions of problems and their solutions. The fact that the QA engineer was involved in all stages of the process helped avoid time-consuming loops and delays, and allowed for earlier and more focused testing. Real-time collaboration between the software and QA engineers helped constrain the search space for tracking down problems and accelerated the process of identifying their sources by ruling out potential explanations quickly.

It became clear that this type of audio-based software is subject to potential hardware and timing issues that make it particularly important to have different people test on a variety of hardware and with various configurations. In addition, the users may interact with the software in idiosyncratic ways, again emphasizing the need for a broad base of testers.

It turned out to be useful to engineer the system such that multiple versions can coexist on the same computer. This not only affected FlexTrans but also required changes to the default behavior of some of the other components, e.g. changing the default installation location for the TTS voices.

Designing the system to be easily extensible took more initial effort, but paid off very quickly.

## VIII. CONCLUSIONS

It seems clear that spoken language translation is now good enough to be useful in many situations when it is not possible to find a human translator, or to triage and help decide where the available human translators are most urgently needed. The remaining challenges in developing actual systems for a variety of languages on small portable devices all seem solvable in principle. It will be interesting to try to integrate ASR, MT, and UI more closely in a way that enables the system to benefit from feedback and to learn. It also remains to be seen what kinds of communication problems may arise as more people unfamiliar with these technologies start to use them. But miscommunication occurs even without these devices, and because they encourage more direct communication and raise awareness of the potential of misunderstandings, they might inherently counteract most of the additional sources of potential problems.

## IX. Acknowledgments

This material is based upon work supported by the Defense Advanced Research Projects Agency (DARPA) and the Department of Interior-National Business Center (DOI-NBC) under Contract Number NBCHD040058. Approved for Public Release, Distribution Unlimited. The PI for the TRANSTAC project at SRI is Kristin Precoda. The FlexTrans framework was designed and developed primarily by Michael Frandsen, with some initial design and code contributions by Shane Mason and design input by Kristin Precoda and Susanne Riehemann. Huda Jameel is an Iraqi Arabic language expert, and we would particularly like to thank her for her help with the examples in this paper. The speech recognition components were developed primarily by Dimitra Vergyri, Sachin Kajarekar, Wen Wang, Murat Akbacak, Ramana Rao Gadde, Martin Graciarena, and Arindam Mandal. Gemini translation was provided by Andreas Kathol, and SRInterp translation by Jing Zheng. Other contributions were made by Horacio Franco, Donald Kintzing, Josh Kuhn, Xin Lei, Carl Madson, Sarah Nowlin, Colleen Richey, Talia Shaham, and Julie Wong. The system also contains TTS from Cepstral, LLC.

## References

[1] H. Franco, J. Zheng, J. Butzberger, F. Cesari, M. Frandsen, J. Arnold, V.R.R. Gadde, A. Stolcke, and V. Abrash, "Dynaspeak: SRI's scalable speech recognizer for embedded and mobile systems," in *Human Language Technology*, 2002.

[2] B.H. Huang and L.R. Rabiner, "Hidden Markov Models for Speech Recognition", *Technometrics* (publ. by American Statistical Association), Vol. 33, No. 3, 1991, pp. 251–272.

[3] K. Precoda, J. Zheng, D. Vergyri, H. Franco, C. Richey, A. Kathol, and S. Kajarekar, "Iraqcomm: A next generation translation system," in *Interspeech*, 2007.

[4] A. Stolcke, "SRILM – an extensible language modeling toolkit," in *International Conference on Spoken Language Processing*, 2002, pp. 901–904.

[5] M. Graciarena, H. Franco, G. Myers, and V. Abrash, "Robust feature compensation in nonstationary and multiple noise environments," in *Eurospeech*, 2005.

[6] J. Dowding, J.M. Gawron, D. Appelt, J. Bear, L. Cherny, R. Moore, and D. Moran, "Gemini: A natural language system for spoken-language understanding," in *Human Language Technology*, 1993, pp. 43–48.

[7] A. Kathol, K. Precoda, D. Vergyri, W. Wang, and S. Riehemann, "Speech translation for low-resource languages: The case of Pashto," in *Eurospeech*, 2005.

[8] P. Brown, S. Della Pietra, V. Della Pietra, and R. Mercer, "The mathematics of statistical machine translation: parameter estimation," in *Computational Linguistics*, 19(2), 1993, pp, 263–311.

[9] P. Koehn, F.J. Och, and D. Marcu, "Statistical phrase based translation," In *Proceedings of the Joint Conference on Human Language Technologies and the Annual Meeting of the North American Chapter of the Association of Computational Linguistics* (HLT/NAACL), 2003.

[10] L. Tomokiyo, K. Peterson, A. Black, and K. Lenzo, "Intelligibility of machine translation output in speech synthesis," in *Interspeech*, 2006, pp. 2434–2437.

[11] B.A. Weiss, C. Schlenoff, G. Sanders, M.P. Steves, S. Condon, J. Phillips, and D. Parvaz, "Performance Evaluation of Speech Translation Systems," in *Proceedings of the Sixth International Language Resources and Evaluation*, 2008.

# Development of Ubiquitous Median Strip Total System in the Road

Byung-wan Jo[1], Jung-hoon Park[2], Kwang-won Yoon[2], Heoun Kim[2], Ji-Sun Choi[2]
[1] Professor, Dept. of Civil engineering, Hanyang university, Seoul, Korea
[2] Graduate Student, Dept. of civil engineering, Hanyang University, Seoul, Korea
goolss@nate.com

*Abstract* - **In order to minimize the loss of life from traffic accidents, a ubiquitous sensor network has been developed to monitor traffic accidents happening on roads real times. This will be monitoring the road conditions and the construction of the median barrier in the road has been built to intellectually notify the police station, tow car, and emergency center in.**
**In this research, an intellectual wireless sensor network system and a middleware has been developed for the purpose of recognizing and informing median barrier highway accidents. Also, sending out and receiving test from unrestricted areas, outdoor tests, methods for constructing the system, and applying functions regarding actual scenes of USN have been verified. Conclusively, a possibility of a wireless sensor network being applied to indirect facilities of society has been suggested.**

## I. Introduction

The median barrier in the road is a facility that separates the roadways according to the direction in a road that has more than four-lanes. The average width is 0.5 ~ 3m and the size of the roadways are made higher so that the traveling car will not be able to go into the opposite crossroads.

Traffic accidents that occur in mountain peaks areas, regions that deteriorate the range of vision due to bad weathers, daybreak and dark night hours where the volume of cars are low, and countryside national roads have especially been found to report the mishaps very late.
That is why this is bringing huge losses of the society, because it delays paramedics from saving lives and other material damages.

Therefore, in order to solve and amplify the problems concerning the previous methods of reporting traffic accidents, the USN/RFID basic recognition of the situation of the environment of the ubiquitous will be used with control technology to perceive the facilities condition real times.

It will be able to judge and take care of itself when a dangerous situation does occur and this will protect the driver and the passenger's life safely from the emergency.

## II. Ubiquitous & Ubiquitous Sensor Networks

### *A. Ubiquitous*

The origin of "Ubiquitous" comes from Latin, meaning omnipresent.

Developing this theory and giving rise to the field of IT for the next tech era's paradigm is the Ubiquitous Computing technology.

Ubiquitous was started out by Mark Weiser who was the chief scientist for the Palo Alto Research Center of Xerox.
He mentioned the theory of ubiquitous in a thesis titled "The Computer of the 21st Century," which was published in the 1991 September issue of Scientific American.

In this thesis, Mark Weiser justified the theory of ubiquitous as being invisible.

In other words, this means that we are able to use a computer through a network in our daily lives without being aware about it.

### *B. Ubiquitous Sensor Network*

The formation of USN has several sensor networks where the field is connected to the outer network through the gateway.

The sensor nodes that are installed on the median barrier highway will send data through a close receiver and the receiver will deliver these data to the manager.

The data that will be delivered can be sent through satellite communication, cable/wireless internet, and this Access Network will use the previous infrastructure.

The overall structure of the USN system is the same as Fig. 1. Access Network uses BcN (Broadband Convergence Network) which is the base for IPv6 and this assumes the

T. Sobh (ed.), *Innovations and Advances in Computer Sciences and Engineering*,
DOI 10.1007/978-90-481-3658-2_93, 

internet network to allow IPv6 to be applied to all sensor nodes.

Also middleware service platform is provided for the sensor network's application so that through this, the user can access the next tech era's intellectual sensor network.

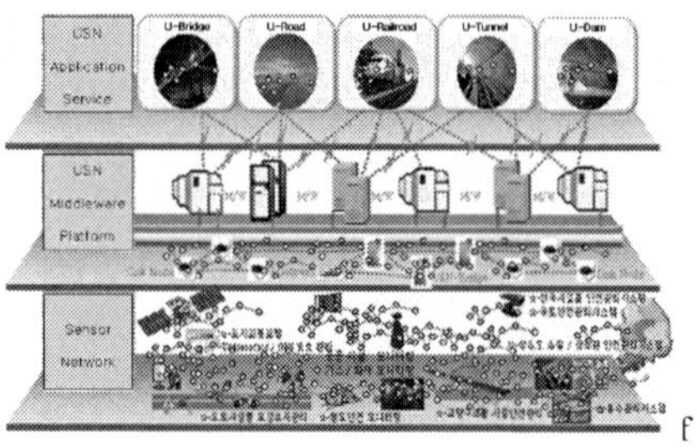

Fig. 1. USN System composition

III.. Ubiquitous median barrier System

### A. WSN(Wireless Sensor Network) System Flow

The wireless sensor network system structure that was used for the ubiquitous median barrier road can be referred to Fig. 2 and it is mainly divided into software system and hardware system.

Software system includes a structure of the process, middleware, and application. The hardware includes sensor node, mote, and sink node. The hardware system and software technology that was used on the experiment of improving the system is like Table 1.

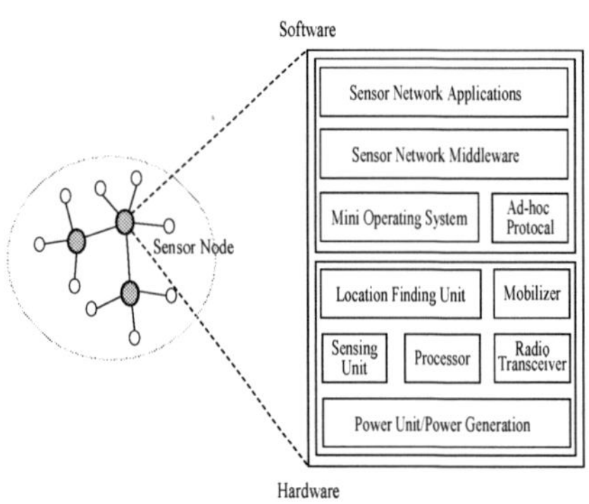


Fig. 2. Construction of Hardware

Table 1. Inner Technology of the WSN System

| Classification | | | Function |
|---|---|---|---|
| Hardware Technology | Sensor Node (Mote) | Sensor Board | · Able to attach vibrating sensor |
| | | Sensor Module | ·Microprocessor-output function<br>·Perceiving Microelecromechanical System function<br>- Put together a 3 dimensional detailed formation with the circuits, and sensor with the actuator on to the standard silicon<br>- Although super small, able to operate highly complicated operations |
| | Sink node | | · Has the function to collect data from the sensor module<br>· Has the function to connect with the outside network |
| Software Technology | Operating System (OS) | | · A sensor network that has used the event managing technique<br>·Operating system<br>- Base for the component<br>- Supports low-power mode<br>- Developed a practical sensor program using nesC<br>- Supports multi-hoop-routing function |
| | Middleware | | · Designates the address for the basic preference<br>- Designates address using ID<br>- New quality of shape following the adaptation<br>· Control the operation of cooperative network<br>- Unifies data<br>- Perceives translation and location of address<br>- Adaptable routing technique following the movement of the node<br>· Reduction in the amount of communication<br>- Reduction of traffic through unified data |

### B. Hardware System Flow

The model for the hardware system development of the system is divided into sensor and sink node.

To be more exact, there is the transmitting sensor node and receiving sink node, and the sensor node is composed of sensor board and sensor module.

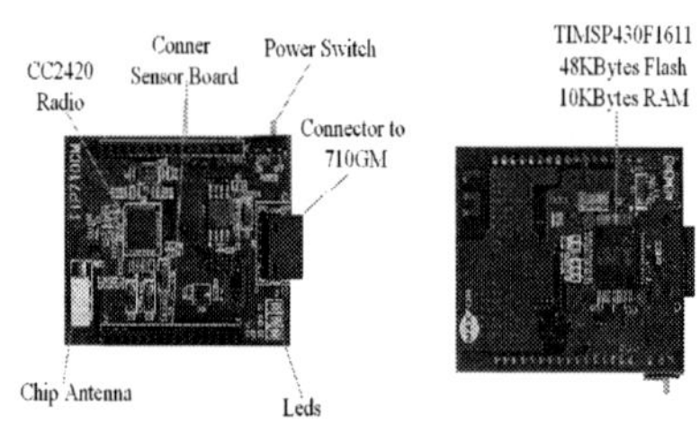


Fig. 3. Construction of Hardware

As a consequence, like Fig. 3. It is constituted by sensor module, sensor node, and sink node.

The hardware system has the sensor board and the sensor module that is able to collect and output data by perceiving traffic accidents that occur on the road.

#### a. Sensor Node (Mote)

Fig. 4. *Sensor* node is able to attach a sensor to the sensor board and module. This lets the sensor that is on top of the sensor board, perceives situations that happen on the middle barrier road and the sensor module outputs and collects all this and delivers it to the sink node. Then finally, the sink node will deliver everything to the network.

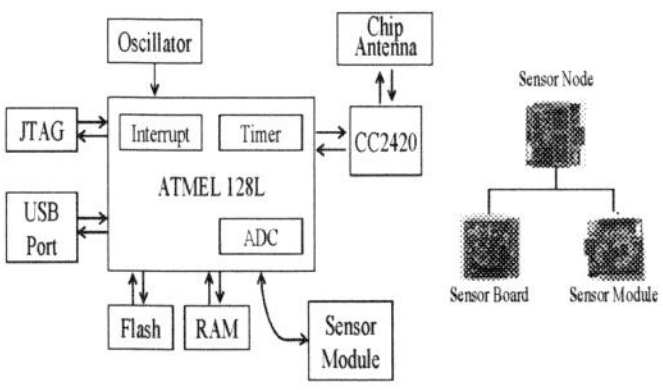


Fig. 4. Construction of Sensor Node

*b. Sensor Module*

The main features are composed of Microprocessor, RF chip, Chip Antenna. Processor uses 16 bit RISC 32 MHZ, RF Chip uses cc2420, and Operating System uses Tiny OS.

The sensor module setting that was used for development of the system in this research is the same as table. 2. and details concerning the sensor module system of the ubiquitous middle barrier road system are as follows.

The settings of the sensor modules that have been used in this research are the same as Table. 1, and the main features are composed of Microprocessor, RF chip, Chip Antenna. Processor uses 16 bit RISC 32 MHZ, RF Chip uses cc2420, and Operating System uses Tiny OS. Specific details for the system's function of the sensor module are listed below.

Table. 2 Sensor module Specifications.

| Item | Description |
|---|---|
| Processor | 16bit RISC 32Mhz (MSP430F1611) |
| RF Chip | TI (Chipcon) CC2420 |
| Internal Memory | 10KB RAM, 48KB Flash |
| External Memory | 1MB |
| Operating System | TinyOS |
| Multi-channel Radio | 2.4GHz |
| Data Rate | 250Kbyte |
| Network | Multi-hop and Ad hoc |
| Power | 3.0~3.3V |
| Range | 70m in lab |

*c. Sink Node*

Sink Node is similar to a USB type, where the transmitted information by the sensor module is received and is connected to the network. The setting for Sink Node is the same as Table 3.

Table. 3. RF Chip Specifications

| Item | Description |
|---|---|
| Internal Memory | USB |
| Data Rate | 560Kbps |

*C. Software*

*a. Middleware*

The operating system for the ubiquitous middle barrier road middleware uses TinyOS (UC Berkeley).

TinyOS has been designed as an open-source operating system for the wireless embedded sensor networks. The middle barrier middleware system has been designed to minimize the size of the codes requested from the Sensor Networks according to its limited memory.

Also, due to having a component basic structure, it is able to induce new technologies quickly and being an event-driven execution model is one of its main characteristics.

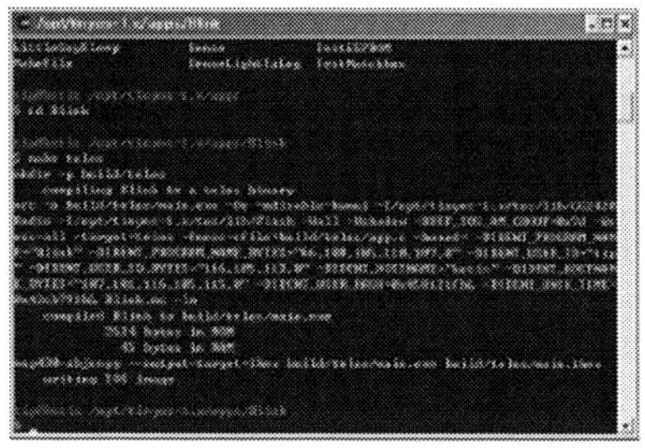

Figure 5. Developing Environment of Middleware (Cygwin Screen)

Middleware is software that allows the application to be connected with each other so that exchanging data is possible and it has the SFnet of the ubiquitous middle barrier road as well as monitoring system and data base.

SFnet, monitoring system, and data base are middleware that links applications together so they can exchange data's with each other.

*b. SFnet Program*

SFnet makes the sensor read the data sensing data from the sensor that is installed onto the middle barrier so as to deliver this to the module for an analysis. It allows the connection so that the value will be saved on the database. This takes place when the wireless sensor communication network is to be made.

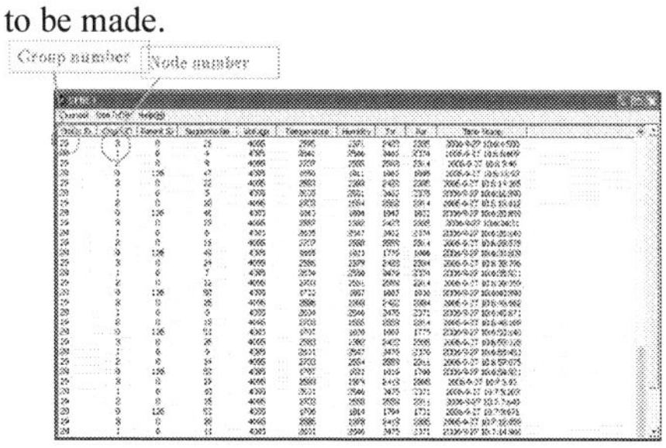


Fig. 6. Real time Sensor Sampling

First, when you start the program, input the name of the data base that has been set in advance and the connecting port number to execute it correctly.

As you can see in the figure, the nodes that are in groups of the SFnet find the most suitable route for delivering data and the corresponding state is shown through the node number.

In order to differentiate the installed groups on the middle barrier, numbers are given to mark the groups.

*c. Monitoring System*

Monitoring System is a figure that shows the actual corresponding state of each node (Fig. 7.) and it allows the monitoring of the vibrating value that occurs on the middle barrier road. When a node is selected, the data's value and the shape of the graph can be seen real-time.

Moreover, through the monitoring system, you can call the values of the data that are saved on the data base and this system also has the function to save those values.

Like Fig. 7. If the number of the node has been designated for transmission, the number of the node will appear on the screen and it will connect itself directly to the sink node to notify the transmission status.

At this point, if the transmission is interrupted, the straight line will disappear and the number of the number will not be seen on the screen and the user will know that he or she will not be receiving the information by checking the screen. The receiving status of the sensor node can be delayed or interrupted according to the given condition of its surroundings.

However, the formation of this system has suggested in using the Ad-hoc network.

The Ad-hoc network is a network that has no basic structure, because it has been made flexible due to the nodes.

Ad-hoc network does not need a base station or an access point like the basic network storage to maintain the network's formation. The nodes that have been used for the development of the system communicated with each other by using the wireless interface.

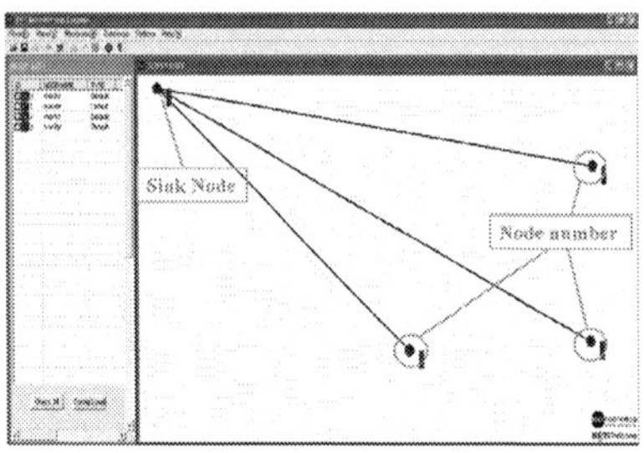


Fig. 7. Monitoring System Operation

In addition, by relying on the multi-hop-routing function, it was able to overcome the distance restriction of communication that the wireless interface had and that is why it was possible to see how the network's topology was changing dynamically, because the nodes were able to move freely.

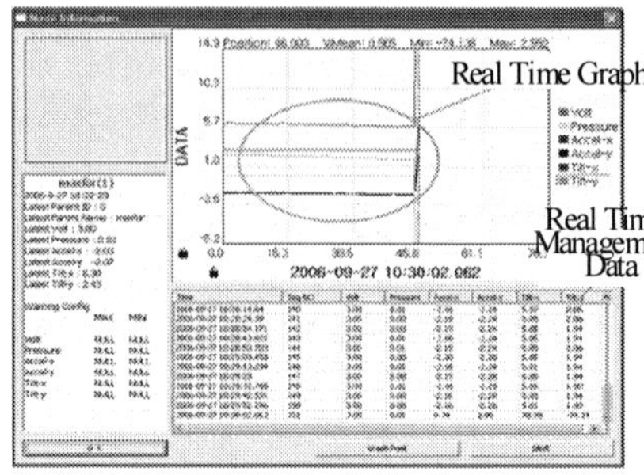


Figure. 8. Monitoring System Data Transfer

*d. Data base (DB)*

Database is a device that allows the sensor to read and the module will deliver the data to a location where it can be saved. In the database that was dealt in this research, a name was first selected before the experiment and after SFnet was launched, selecting a route through the database's name automatically saved the selected database.

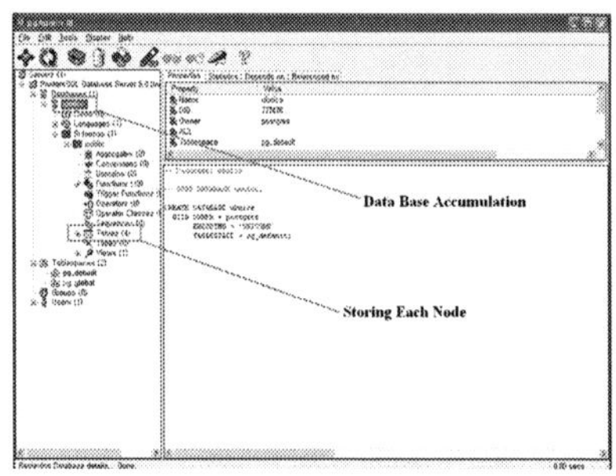


Figure. 9. Monitoring System Data Base

Like Fig. 9. Form a database after you give it a name and then each of them will automatically be saved following the designated node number.

Linking with the ubiquitous middle barrier road system monitoring system program, through the saved name of the database, the data base has the function to call out the value of the measure from real-time that the user seeks and it can also be saved for different purposes. It is believed to be possible for analyzing traffic accident impulse.

## IV. Send and receive free space distance test & Experiment the Basic USN System

In this research, the performance of the sensor was tested first and in order to compare the results, the experiment of sending out and receiving ability was done in free space.

The weather was clear and in order to minimize the effect coming from the outside environment, the test was to be done around 2 in the afternoon, which is the period of time that has the least moisture.

Fig. 10. Send and receive free space distance test

### *A. Result of the test*

According to the test results, the transmission in the free space was even possible to the end of the Track and Field area, which is 300m from beginning to end. Testing beyond this was not possible and that is why this result finalized the experiment.

### *B. Outdoor Experiments*

Tests such as the wireless acceleration sensor, 70 data can be saved per second. The usual value of the sampling also came out regularly, because the median barrier was fixated.

Fig. 11. shows the sampling value of the node at real-time. Fig. 12. shows an impulse that crossed the standard value of the median barrier has been perceived, but it succeeded in developing the system to notify the manager about the location and the shape of the node by making a pop-up come automatically on to the screen of the main sever computer.

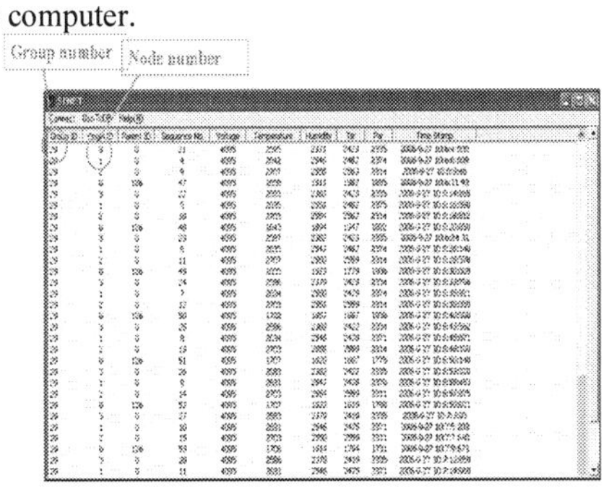


Fig. 11. Node Sampling Data Value

Fig. 12. Traffic Accidents Area & Type

## V. CONCLUSIONS

In order to perceive and quickly report traffic accidents that often occurs on roads, the following conclusion has come out by applying the wireless sensor network theory to the median barrier road.

First, creating a sink node hardware that connects the sensor node's RF Chip cc2240 with the network in USB method that perceives traffic accidents within a road, constructing a ubiquitous median barrier road system by using the open Source Tiny OS as an operating system developed at UC Berkeley, and after testing the capabilities in reality, it has been verified that wireless communication assists in perceiving and taking care of matters effectively concerning traffic accidents that occur anywhere at anytime.

Second, after outdoor tests were made, it has been confirmed that the value of node sampling can save 70 data per second and it has also succeeded in alerting the main computer with graphs when the value goes overboard or when a traffic accident occurs.

## ACKNOWLEDGMENTS

This study was financially supported by Small and Medium Business Administration under a research and developing project, and the authors would like to thank Small and Medium Business Administration and Hanyang University BK21.

## REFERENCES

[1] Kang. W.E. et al., Road Restraint Systems installment & Study on management standard : a median strip & crash cushions, moct. (1998)
[2] Choi G. D. et al., Road Restraint Systems installment & Management Guideline : A median Strip, moct. (1999).
[3]Kwan. G.S., "Ubiquitous Computing concept & aspect", ETRI,(2003) Park.S.C et al., USN Technology, jinhan M&B, (2005)
[4] David Culler, Deborah Estrin, Main Shivastava, Overview of Sensor Network, IEEE Computer, Page 41-49(2004)

# An LOD Control Interface for an OpenGL-based Softbody Simulation Framework

Miao Song
Department of Computer Science and Software Engineering
Concordia University, Montreal, Quebec, Canada
Email: m_song@cse.concordia.ca

Peter Grogono
Department of Computer Science and Software Engineering
Concordia University, Montreal, Quebec, Canada
Email: grogono@cse.concordia.ca

***Abstract*—We summarize an interactive GLUI-based interface to the real-time softbody simulation using OpenGL. This interactivity focuses not only the user being able to drag 1D-, 2D-, and 3D-deformable elastic objects selectively or all at once, but also being able to change at run-time various "knobs" and "switches" of the level-of-detail (LOD) of the objects on the scene as well as their physically-based modeling parameters. We discuss the properties of such an interface in its current iteration, advantages and disadvantages, and the main contribution of this work.**

***Index Terms*—level-of-detail (LOD), softbody simulation, OpenGL, 3D, graphics interfaces, real-time, frameworks**

## I. Introduction

### A. Problem Statement

To validate the look and feel, physical properties and realism in physical-based elastic softbody simulation visualization requires a comprehensive interface to allow "tweaking" various simulation parameters at run-time while simulation is running, instead of re-compiling and re-starting the simulation program's source code every time a single parameter is changed. The typical values in our simulation that change are various forces applied to the body, such as 4 different types of spring forces with elasticity, damping, as well as gas pressure and even user-interaction with the object by dragging it in a direction with a mouse as well as collision response forces, spring stiffness and so on. Since the simulation is real-time, another version of level-of-detail (LOD) [1] adjustments includes the number of particles, springs, subdivisions, at the geometry level. At the simulation level the variations include the complexity and sensitivity of the physical-based simulation algorithms and the time step that they can bear: the higher granularity is for the algorithm, the more computation time is required but higher accuracy of the simulation and the time step is achieved. The problem is how to visually study these properties, validate them conveniently through either expert-mode or less-than-expert-mode user interface included with the simulation program, in real-time.

### B. Proposed Solution

We propose the first iteration of the GLUI-based interface [2] to our realtime softbody simulation visualization in OpenGL [3], [4], [5] that allows "tweaking" of the simulation parameters. We introduce its visual design, as well as some details of the software design and the mapping between GLUI components and the internal state of the simulation system we are working with. We propose the current interface in its first iteration of the design and improvement.

### C. Contributions

The most important contribution of this work, aside from greatly improving the usability of the simulation system by scientists, is capture and comprehension of a hierarchy of the LOD parameters, traditional and non-traditional. For example, allowing arbitrary number of integration algorithms constitute a run-time LOD sequence, which would not normally be considered as LOD parameters, so we introduce the higher-level LOD components, such as algorithms, in addition to the more intuitive LOD parameters like the number of geometric primitives there are on the scene and so on.

### D. Softbody Simulation Framework

The system we are working with was first conceived by Song and Grogono [6], [7], that did not have much user-interactive interface to LOD and other details except the mouse drag and some function keys. For every parameter change, the simulation had to be edited and recompiled. The simulation program and its framework is written in C++ for 1D-, 2D-, and 3D- realtime physically-based two-layer elastic softbody simulation system. Originally, there was no provision to alter the simulation state at run-time or startup via configuration parameters or GUI control, so we make a contribution in improving the framework by adding the notion of state, LOD hierarchy, and its link to the GLUI.

## II. LOD Interaction Interface

### A. Visual Design

Our initial iteration of the visual design of the LOD interactivity interface is summarized in Figure 1. The LOD components are on the right-hand-side (in their initial state, not expanded). And the main simulation window is on the left (interactivity with that window constitutes for now just the mouse drag and functional keys). Following the top-down approach we bring in more details of configurable simulation parameters.

*1) Adjustable Parameters:* The adjustable LOD parameters can be categorized as dimensionality, geometry, integration algorithms, force coefficients, and particle mass.

T. Sobh (ed.), *Innovations and Advances in Computer Sciences and Engineering*,
DOI 10.1007/978-90-481-3658-2_94, © Springer Science+Business Media B.V. 2010

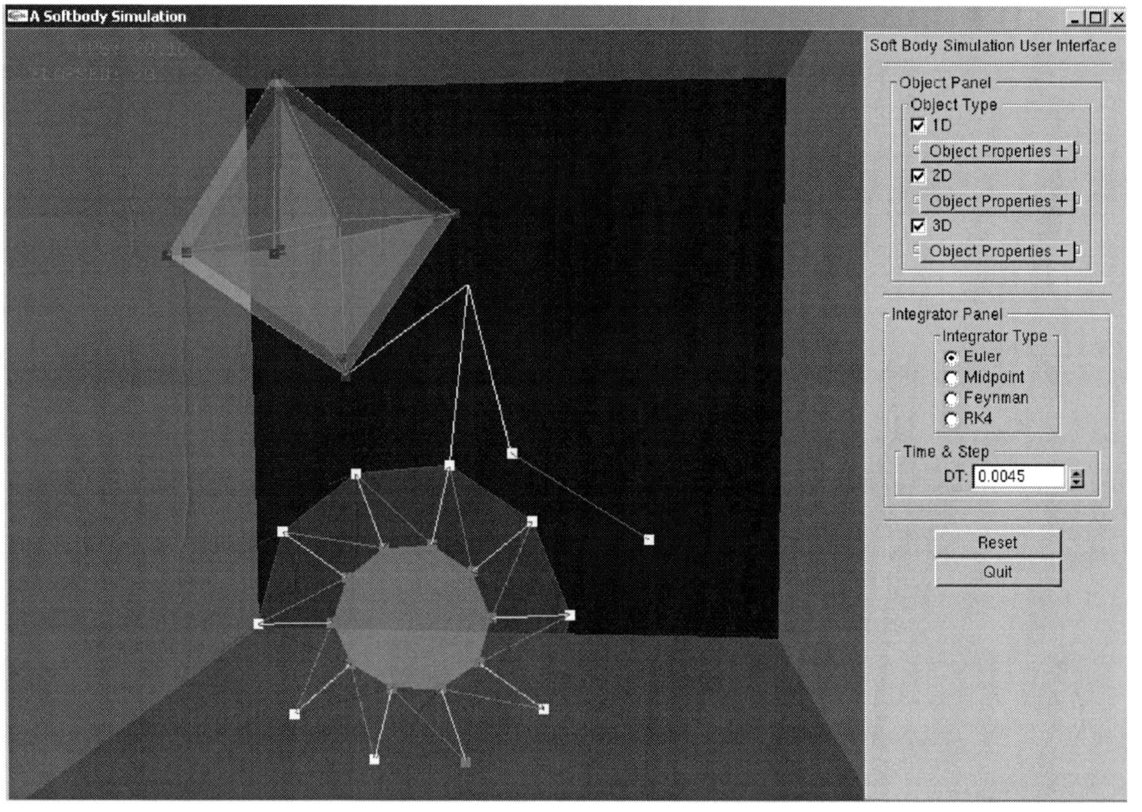


Fig. 1. Three Types of Objects Dragged by Mouse

*a) Dimensionality:* Switch of the 1-, 2-, and 3-dimensional objects present in the simulation are simply done as checkboxes as shown in Figure 2. The "Object Properties" buttons expand the higher-level representation into the finer details of the objects of each dimension type.

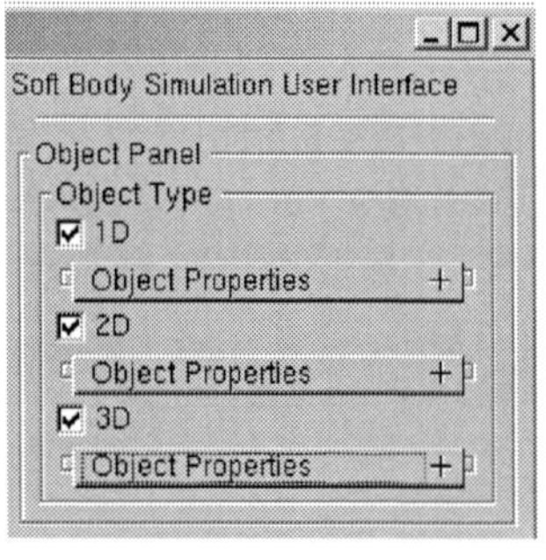


Fig. 2. Object Type Dimensionality

*b) Geometry:* In 1D there are no really any geometry changes. In 2D, the geometry means number of springs comprising the inner and outer layers of a 2D softbody. Increasing the number of springs automatically increases the number of the particles at the ends of those springs, and brings a better stability to the object, but naturally degradation in real-time performance. In Figure 3, Figure 4, Figure 5, and Figure 6 is an example of the 2D object being increased from 10 to 18 springs, from the UI and rendering of the object perspectives.

The geometry of increasing or decreasing of the number of springs in 2D, as in Figure 3, Figure 4, Figure 5, and Figure 6 is not the same approach used for the 3D objects in the original framework. The 3D object is constructed from an octahedron by an iteration of a subdivision procedure, as selectively shown in Figure 7 Figure 8, Figure 9, and Figure 10. The iteration increases dramatically the number of geometrical primitives and their interconnectivity and the corresponding data structures, so the efficiency of the real-time simulation degrades exponentially with an iteration as well as the detailed time step, which can make the users wait on a lower end hardware. Using the LOD GLUI components we designed, a researcher can fine-tune the optimal simulation parameters for the system where the simulation is running. Since it is a combination of multiple parameters, the complexity between editing and seeing the result is nearly immediate compared to the recompilation effort required earlier.

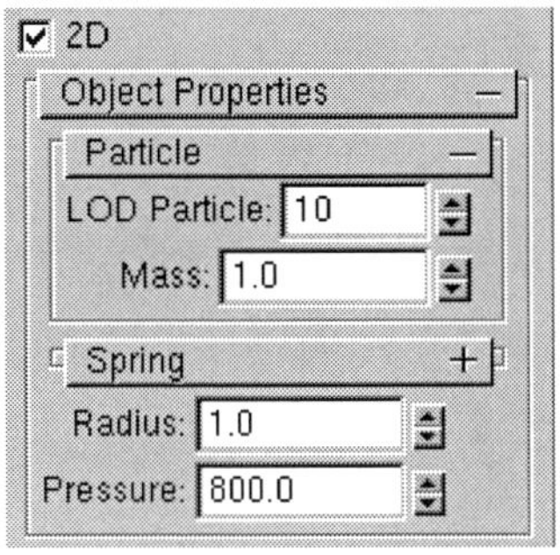


Fig. 3. LOD 2D UI with 10 Springs

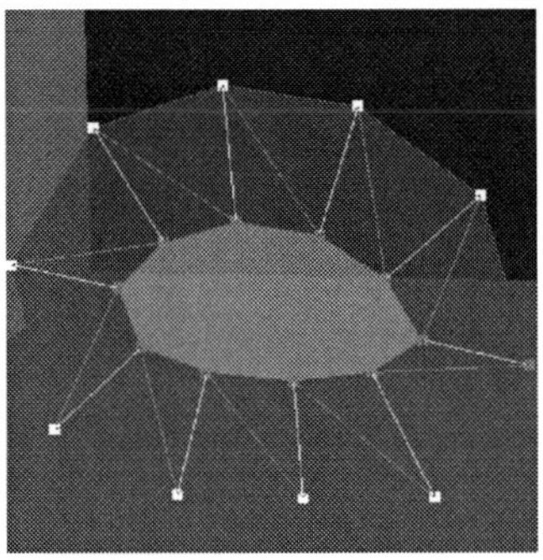

Fig. 4. LOD 2D Object with 10 Springs

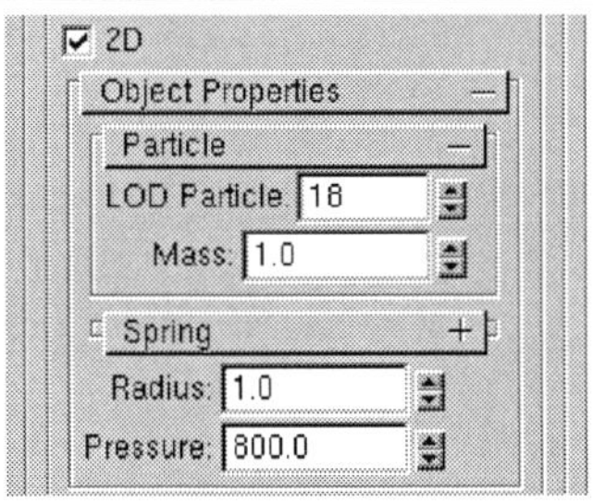


Fig. 5. LOD 2D UI with 18 Springs

*c) Integration Algorithm:* Currently available integrator types are in Figure 11, from the fastest, but least accurate and stable (Euler) to the most accurate and stable (RK4) [6].

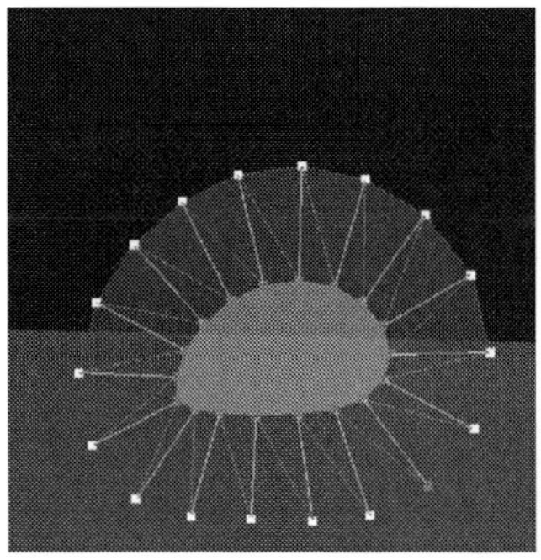

Fig. 6. LOD 2D Object with 18 Springs

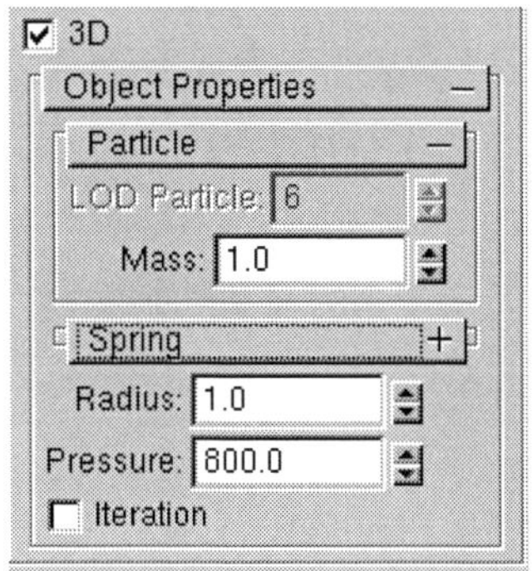


Fig. 7. LOD 3D UI without Iteration

The Softbody Simulation Framework allows addition of an arbitrary number of implementations of various integrators that can be compared in real-time with each others by switching from one to another while the simulation is running.

*d) Force Coefficients:* In Figure 12, we provide in the current form a way to adjust the coefficients used in force accumulation calculations as applied to each particle. KS and KD represent elasticity and damping factors of different types of springs. The default Hooke refers to the structural springs comprising the perimeter layers of the object, and then the radius springs, as well as the spring created by a mouse drag of a nearest point on the object and the mouse pointer.

*e) Mass:* Each object properties has sub-properties of comprising it particles, including mass, as exemplified in Figure 3, Figure 5, Figure 7, and Figure 9. We argue mass is an LOD parameter as each layer's particles of the softbody object can get different masses (but uniform across the layer), or even different masses for individual particles. The latter case however does not scale well in the present UI design, and has to be realized differently.

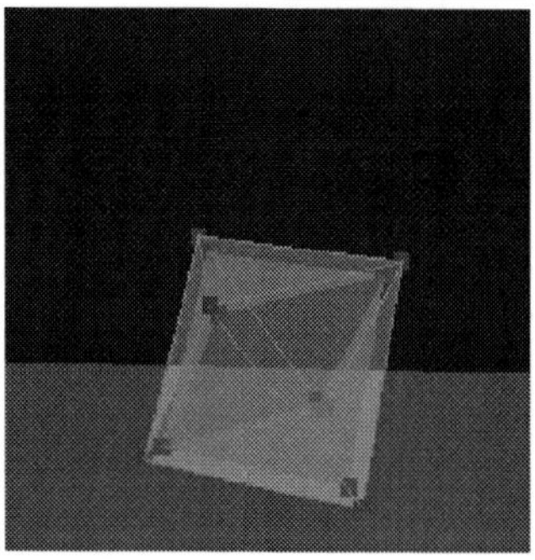

Fig. 8. LOD 3D Octahedron

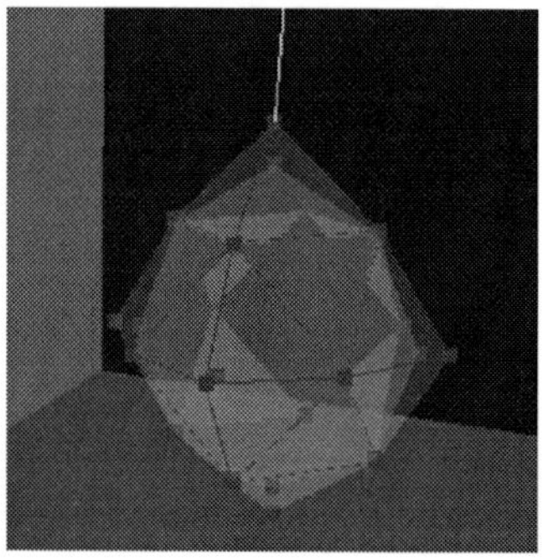

Fig. 10. LOD 3D Object with Iteration Once

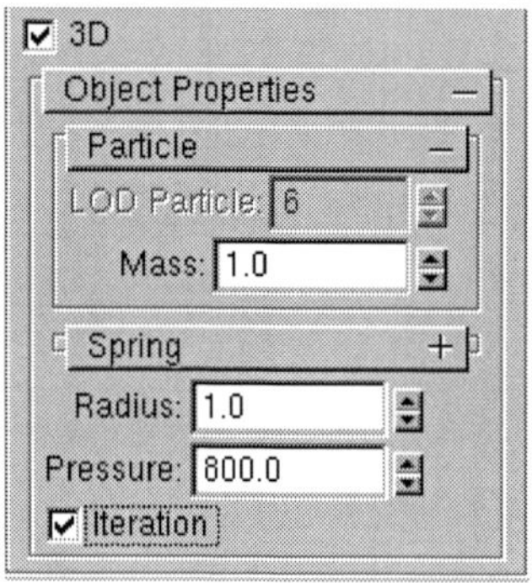


Fig. 9. LOD 3D UI with Iteration Once

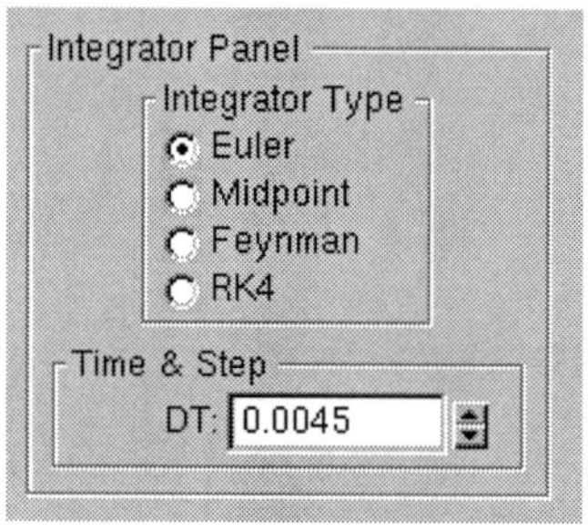


Fig. 11. Integrator Type LOD UI

### *B. Software Design*

The software design is centered around the notion of state. The state is captured by a collection of variables of different data types to hold the values of the simulation parameters. The whole architecture follows the Model-View-Controller (MVC) [8] design pattern, where the Model is the state of the geometry and LOD parameters, View is the simulation window and the GUI into the model details, and the controller that translates users actions onto the GUI into the model changes, particularly the LOD state.

*1) LOD State:* The system state has to encode a variety parameters mentioned earlier that are exposed to the interactive user at run-time. They include:

*a) Dimensionality:* The LOD on dimensionality is 1D, 2D, and 3D. All three types of objects can be rendered on the same scene at the same time. This is encoded by the instances of state variables `object1D` of type `Object1D`, `object2D` of type `Object2D`, and `object3D` of type `Object3D` as well as their corresponding Boolean flags reflecting the status of whether top render them or not. This is done to illustrate the object behavior under the same simulation environment and how the objects of different dimensionality respond to the same simulation settings.

*b) Geometry:* The 2D and 3D objects have an additional LOD parameter related to their geometry complexity. In 2D the geometry LOD specifies the number of particles the inner and outer layers have and by extension the number of structural, radial, and shear springs they are connected by. The 3D geometry LOD is based on the way the sphere is built through a number of iterations and the default set of points upon which the triangular subdivision is performed on the initial octahedron. The LOD here is encoded by the `iterations` state variable.

*c) Integration Algorithms:* The algorithmic LOD includes the selection of a physical interpolation and approximation algorithms for the run-time integration of the velocities and acceleration exhibited by the particles based on physical laws, such as Newton's, and the different types of integrators: Euler, mid-point, Feynman, and RK4. These are the four presently implemented integrators in the Integrators Framework of the system, so the corresponding state variable is `integratorType` that allows selecting any of the available integrators and witness the effects of such selection at run-time. The LOD aspect here goes about the simplicity and run-time performance of an algorithm implementation versus accuracy of approximation of the physical motion, with less accurate (but fastest) being the Euler's integrator, and most

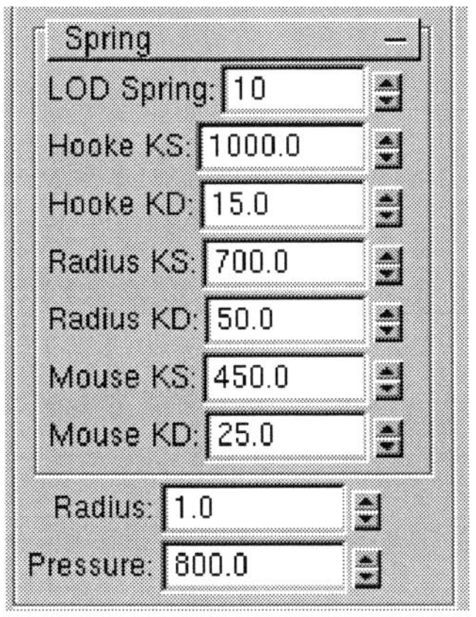


Fig. 12. Force Coefficients LOD UI

accurate (but slowest) being the RK4.

*d) Force Coefficients:* There are a number of forces that are taken into account in the simulation process. Each of which corresponds to at least one state variable supporting the simulation and the interface. The coefficients correspond to Hooke's forces on springs, gravity force, drag force, gas pressure (for 2D and 3D enclosed objects), and the collision response force. The spring forces (structural on each layer, radial, shear, and mouse drag) typically have the elasticity spring coefficient, $KS$, and the damping force coefficient, $KD$. The gravity force by default is $m \cdot g$, where both the mass $m$ of a particle and $g$ can be varied as LOD parameters.

*e) Particle Mass:* Particle mass is another LOD parameter that affects the simulation and can be tweaked through the interface. Internally, the simulation allows every particle to have its own mass (or at least two different layers in the softbody can have two distinct particle weights), but it is impractical to provide a GUI to set up each particle with such a weight, however, in most simulations we are dealing with the particles of uniform mass, so we can allow resetting the mass of all particles with a knob. Thus, the default particle mass LOD parameter is mapped to the `mass` state variable.

## III. Conclusion

We extended the original Softbody Simulation Framework with the notion of state as well as greatly enhanced the interactivity of the simulation system by exposing the state to the GLUI interface at run-time such that researchers working in the field with the framework can easily observe the effects of a vast variety of LOD parameters on the physical-based softbody simulation visualization at real-time without the need of altering the source code and recompiling prior each simulation increasing the usability of the tool. We identified a scale problem with some of the LOD parameters when mapping to a UI, such as spreading the mass over each particle or layer of particles is unmanageable and has to be specified by other means when needed.

## IV. Future Work

- Allow alteration of the Archimedean-based graphs and different types of them than a octahedron [9] not just the number of iterations as a run-time LOD parameter.
- Interactivity through haptic devices [10] with the softbody feedback (e.g. for surgeon training or interactive cinema [11]).
- Allow the state dump and reload functionality in order to display each particle and spring state (all the force contributions, velocity, and the position) at any given point in time in a text or XML file for further import into a relational database or an Excel spreadsheet for plotting and number analysis, perhaps by external tools. Reloading would enable to reproduce a simulation from some point in time, a kind of a replay, if some interesting properties or problems are found. This is useful for debugging as well.
- Continue with improvements to usability and functionality of the interface for scientific experiments.

## Acknowledgment

We thank the reviewers of this work and their suggestions. This work is sponsored in part by the Faculty of Engineering and Computer Science, Concordia University, Montreal, Quebec, Canada.

## References

[1] J. Gibbs, *Rendering Skin and Hair.* SIGGRAPH, March 2001, http://silicon-valley.siggraph.org/MeetingNotes/shrek/hairskin.pdf.

[2] P. Rademacher, "GLUI - A GLUT-based user interface library," SourceForge, Jun. 1999, http://glui.sourceforge.net/.

[3] E. Angel, *Interactive Computer Graphics: A Top-Down Approach Using OpenGL.* Addison-Wesley, 2003.

[4] OpenGL Architecture Review Board, "OpenGL," [online], 1998–2008, http://www.opengl.org.

[5] M. Woo, J. Neider, T. Davis, D. Shreiner, and OpenGL Architecture Review Board, *OpenGL Programming Guide: The Official Guide to Learning OpenGL, Version 1.2*, 3rd ed. Addison-Wesley, Oct. 1999, ISBN 0201604582.

[6] M. Song, "Dynamic deformation of uniform elastic two-layer objects," Master's thesis, Department of Computer Science and Software Engineering, Concordia University, Montreal, Canada, Aug. 2007.

[7] M. Song and P. Grogono, "A framework for dynamic deformation of uniform elastic two-layer 2D and 3D objects in OpenGL," in *Proceedings of C3S2E'08.* Montreal, Quebec, Canada: ACM, May 2008, pp. 145–158, ISBN 978-1-60558-101-9.

[8] C. Larman, *Applying UML and Patterns: An Introduction to Object-Oriented Analysis and Design and the Unified Process.* Prentice Hall PTR, 2001, ISBN 0131489062.

[9] H. Nonaka, "Operation-based notation for archimedean graph," in *Proceedings of the 12th World Multi-Conference on Systemics, Cybernetics and Informatics (WM-SCI'08)*, N. Callaos, W. Lesso, C. D. Zinn, J. Baralt, J. Boukachour, C. White, T. Marwala, and F. V. Nelwamondo, Eds. Orlando, Florida, USA: IIIS, Jun. 2008.

[10] Wikipedia, "Haptic technology — Wikipedia, The Free Encyclopedia," 2008, [accessed 20-August-2008]. [Online]. Available: http://en.wikipedia.org/w/index.php?title=Haptic_technology&oldid=232644878

[11] M. Song, "Are haptics-enabled interactive and tangible cinema, documentaries, 3D games, and specialist training applications our future?" in *Proceedings of GRAPP'09.* Lisboa, Portugal: INSTICC, Feb. 2009, to appear, grapp.org. Short position paper.

# Laboratory performance test of overload vehicles regulation system on Ubiquitous road


Byung-wan Jo[1], Kwang-won Yoon[2], Seok-won Kang[2], Jung-hoon Park[2], Heoun Kim[2]

[1] Professor, Dept. of Civil engineering, Hanyang university, Seoul, Korea

[2] Graduate Student, Dept. of civil engineering, Hanyang University, Seoul, Korea

ykwabc@nate.com



*Abstract*

**Overloaded vehicles operate damage to road, bridge, and then increasing in maintenance and repair cost because structures are reduced durability. The existing systems have many problems and need coping measure.**

**Therefore, this paper organized Ubiquitous sensor network system for development of intelligent auto overload vehicle regulation system about high speed vehicles, also axial load WIM-sensor was selected by indoor experiment through wireless protocol. And we examined possibility U-load auto overload vehicle regulation system through experiment of the transmission and reception distance.**

**If this system will apply to road and bridge, might be effective for economy and convenience through establishment of U-IT system.**


## I. INTRODUCTION

From growth of economy, most commercial traffic is using the road, overload vehicles have been increase that exert and influence upon road safety, traffic flow, social overhead capital and economical of nation.

Currently part of overloaded vehicle regulations rely on measurement of static load from installed overload checkpoint because of high accuracy. [Figure.1]

Efficiency on overloaded vehicle regulations depend on possibility regulation, need many human power and cost for raise the efficiency

Also the existing overloaded vehicles regulation system have demanded measures about problems that misconduct be possibility and incur the enmity of the people from specialized insufficiency on regulation staffs.

On this, the present research is indicating building methods of unattended U-Overloaded Vehicles Management System applied with ubiquitous technology, and practical regulating plans for to solve problems of overload vehicles regulation, and troubles that occurred by overloaded vehicles

Fig. 1. The existing overloaded vehicles regulation system

## II. Composition of Unattended U-Overloaded Vehicles Management System

### A. Ubiquitous

The origin of "Ubiquitous" comes from Latin, meaning omnipresent. Developing this theory and giving rise to the field of IT for the next tech era"s paradigm is the Ubiquitous Computing technology.

Ubiquitous was started out by Mark Weiser who was the chief scientist for the Palo Alto Research Center of Xerox.


T. Sobh (ed.), *Innovations and Advances in Computer Sciences and Engineering*,
DOI 10.1007/978-90-481-3658-2_95, © Springer Science+Business Media B.V. 2010

He mentioned the theory of ubiquitous in a thesis titled "The Computer of the 21st Century," which was published in the 1991 September issue of Scientific American.

In this thesis, Mark Weiser justified the theory of ubiquitous as being invisible. In other words, this means that we are able to use a computer through a network in our daily lives without being aware about it.

### *B. Ubiquitous Sensor Network*

The formation of USN has several sensor networks where the field is connected to the outer network through the gateway.

The sensor nodes that are installed on the median barrier highway will send data through a close receiver and the receiver will deliver these data to the manager.

The data that will be delivered can be sent through satellite communication, cable/wireless internet, and this Access Network will use the previous infrastructure.

### *C.* Formation of Whole System

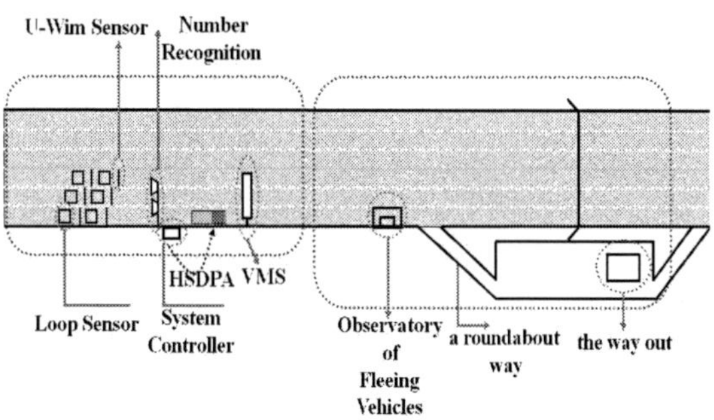


Fig. 2. Ubiquitous Intelligent Overload Regulation System

[Figure 2] indicates that the present system is made-up to be installed on regular roads for rapid judgment whether the vehicles are overloaded or not without influencing to traffic flow, send information to the overloaded vehicles, regulation offices, and related organizations.

The whole system has composed with USN Sensor Field, System Controller, and External Network. The USN Sensor Field measures the speed of vehicles, and it is formed with Loop Sensor that is for checking the passing of each vehicle, U-WIM Sensor for measuring the vehicles" axial load, and recognition equipment of license plate for cognizing-overloaded vehicles.

Likewise, system controller is composed with Sink Node that collects information that are measured by USN, Gateway that sends the information out, Middleware for program synchronization with outer network, and HSDPA/WCDMA for communication with outer network. It sends the data to outer network, enables real-time check. Overloading control office sends information of the vehicles that are subject of control to competent organization, and decides possibilities for treatment and enforcement of penalty.

### *D.* Formation of Whole System

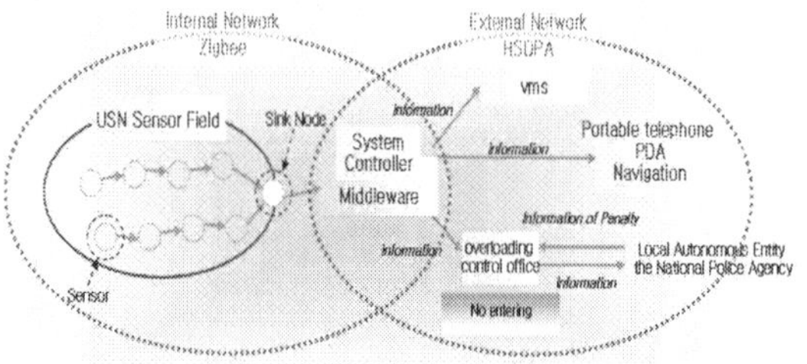


Fig. 3. Intelligent Overload Regulation System

[Figure 3] is presenting a U-Overloaded Vehicles Managreement System. It sends information that was measured from several sensors to one Sink Node through USN, and the Sink Node sends the measured information to System Controller. The System Controller sends the information to variable message sign (VMS), report the violation record to offender, send the content to the violator"s cell phone, PDA, Navigation, and so on, and inform to overloading regulation office. The overloading regulation office undertakes administration duties of infringement with competent Local Autonomous Entity and the National Police Agency.

Zigbee, one of short distance wireless communications, is used for communication in Internal Network, and 3.5-generation wireless communication HSDPA (High-Speed Downlink Packet Access) is used for External Network as it is presented [Figure 3].

### E. Overloading System Algorithm

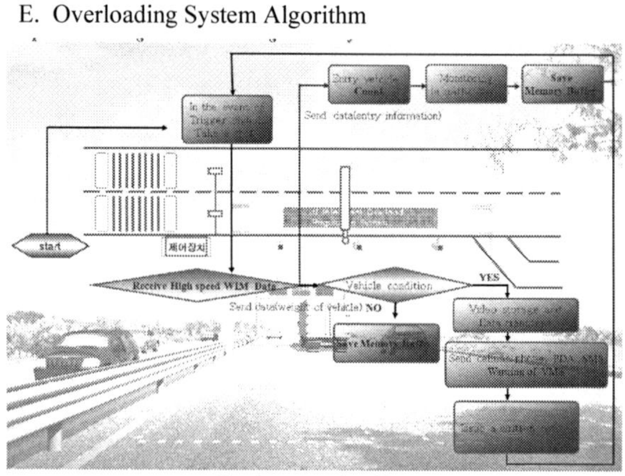


Fig. 4. system algorithm

[Figure 4] is presenting an algorithm of U-Overloaded Vehicles Management System. The judgment process order of overloaded vehicle has done as remarked above, and alternation of algorithm that enables supplement and management of operation-permitted vehicles on occasion demands is possible

## III. Performance Test of Sensor for USN Formation

To select the most suitable WIM Sensor that will be used for USN formation, we compared efficiency, cost, convenience of installation, selected two sensors, executed an inside test, and analyzed confidence of the sensors.

### A. Bending Plate Sensor Experiment and the Result

Bending Plate sensor that used a method for measuring micro transformation of Plate has used to present study. Bending Plate Sensor Experiment was done with surcharged loading on a sensor. We increased weight to five stages, did three times of tests in each stage, and calculated average value and an error. For test facilities, we used UTM strength measurement like [Figure 5(a)].

The present test was done with 23centigrade, 64 percent humidity, and 10 voltages. The test was done three times in each stage with four, six, eight, 10-ton different weight.

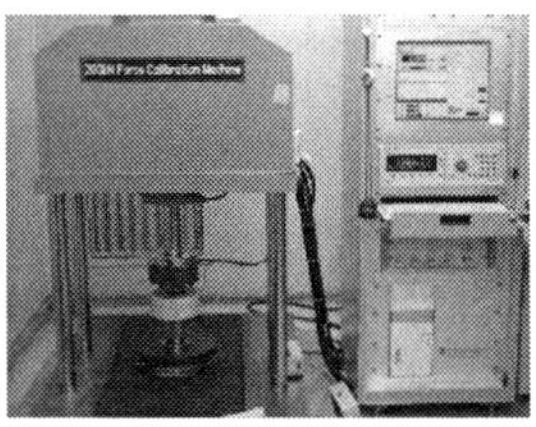

Fig. 5(a) UTM strength measurement

Fig. 5(b) Sensor data

As the result in [Table 1], when the weight was two ton, the average was 1.9867, and showed the maximum error, which is 0.67 percent. When it was six ton, the average was 3.9867, the error was 0.33 percent, and in case of six ton, the average was 5.9933, and showed the minimum error, which is 0.11 percent. In addition, in case of eight ton, and 10 ton, error rates were 0.17 percent and 0.33 percent respectively. After three times of test, the result showed that indicator"s measured weights were all agreed, and the error rate were less than from 0.11 to 0.67 percent.

Table 1 The result Bending Plate Sensor test

| **testing machine** | | **Weight (ton / KN)** | | | error | |
|---|---|---|---|---|---|---|
| Step | W(ton/KN) | first | second | third | average | error(%) |
| 1 | 2 (19.613KN) | 2.000 | 1.980 | 1.980 | 1.9867 | 0.67 |
| 2 | 4 (39.227KN) | 4.000 | 3.980 | 3.980 | 3.9867 | 0.33 |
| 3 | 6 (58.840KN) | 6.000 | 5.980 | 6.000 | 5.9933 | 0.11 |
| 4 | 8 (78.453KN) | 8.000 | 7.980 | 7.980 | 7.9867 | 0.17 |
| 5 | 10 (98.067KN) | 9.980 | 9.960 | 9.960 | 9.9667 | 0.33 |

We experiment on load variable for secure confidence and investigate application about Bending Plate Type Sensor which will be using this study also we experiment on resistance of changed temperature that temperature seasonally variable. Temperature have been changed 7step as -10□, -5□, 0□, 10□, 20□, 30□, 50□ by experiment using tester[Pic. 1] And load separated 5step from 2 Tons to 10 Tons, addition to 2 Tons each step. Laboratory humidity is 37%, have measured 3 time each temperature for experiment accuracy. [Figure 6] display load inner part and outer part on Burning&Cold Test Chamber

Fig. 6. experiment with temperature

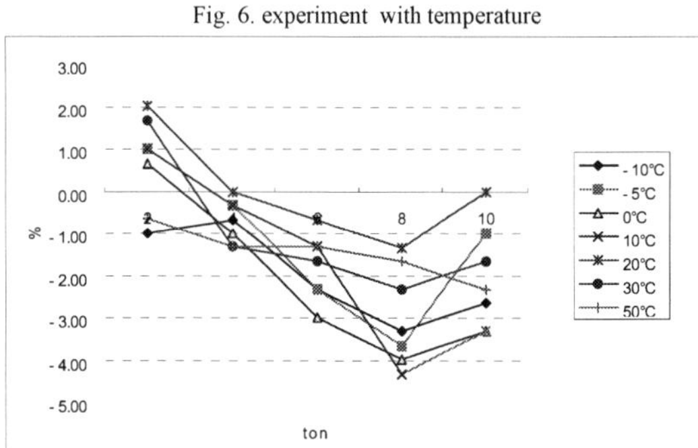


Fig. 7 experiment data with temperature

Through [Figure 7] which display diagram of measure different temperature data, as result of experiment on resistance of changed temperature on Bending Plate Type Sensor, resistance is enough using sensor on normal temperature even changed temperature from turning of the season

## IV. Designing Ubiquitous Platform

To solve the problems of existing Overloading Control System, we need to apply ubiquitous technology to Overloading Control System and construct an unattended and wireless system.

For this, we need a wireless technology and system designing technology that enables the sensor to send the measured information wirelessly and let the user and overloaded-vehicle know the violation. The present system is composed with USN Sensor Field, and System Controller.

A. USN Sensor Field

Figure.8(a) Sensor Node Figure.8(b)WIM Sensor

It is important to make wireless through USN Sensor Field composition with several sensors such as High Speed WIM Sensor that measures a high speed vehicle"s weight, Loop Sensor that sense passing of vehicles, image perception facility, and so on. In the case of Ubiquitous Sensor, it is possible to fix Sensor Node, which is [Figure 8(a)], to WIM Sensor, which is [Figure 8(b)], and send information wirelessly. Through linkage several sensors, we can compose Sensor Field with wireless communication technology such as Zigbee and Bluetooth to organize Wireless Sensor Field.

B. Wireless Sensor Node efficiency test

This experiment on distance of transmission and reception of wireless sensor node for acquire data from applied sensor node with Bending Plate Sensor.

In U-Overload Vehicles Regulation System case, we assumed suitability of communicative distance over 300m because distance of USN Sensor Field and other Internal Network is closed as within 10m.

The field experiment have been done on Seoul Hanyoung University"s playground and Salgoji Park, weather is fine, time is 2pm because of low percentage of humidity on air.

The experiment of method; when the start point, confirm the 100% transmission quality and extend distance checked transmission quality. Also, we considered sensor node based information that transmission quality is over 50%, data does not loss.

Figure.9 Test of transmission distance on free-space

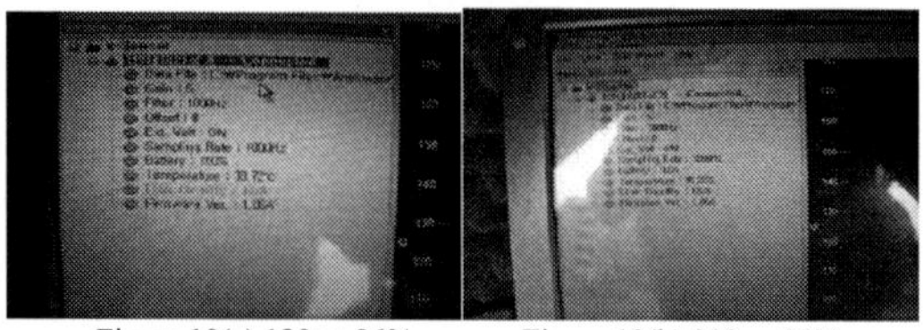

Figure.10(a) 120m: 86% Figure.10(b) 300m: 56%

As a result of [Figure 10(a)] and [Figure 10(b)], communication quality on 120m of free-space is 86%, communication quality on 300m is 56% that confirmed no problem of using wireless sensor node on this study.

C. System Controller

Gateway that performs as a Sink Node, which gathers collected information from USN, Data Logger that processes sensing data, Middleware that enables communication with the exterior, and Modem that is related to WCDMA or HSDPA, which sends direct data, are built in a System Controller.

a. Gateway

Gateway and Middleware designing technology is playing a role as a gate that information goes out when gathered information is sent from USN Sensor Field to another Sensor Field or network. Gateway is possible to function equally in Sink Node where the gathered information from Sensor Field is passing through.

b. Data Logger

Data Logger is a facility that alters analogue input of sensor measuring data to digital number, and record automatically. It convert analogue signal to digital signal through AD converter.

c. Middleware

Middleware is a programming service that arbitrates in between of more than two systems, and the present system is operating synchronization between Sensor Field and each applied service through Middleware.

d. WCDMA / HSDPA Modem

VMS, user"s cell phone, PDA, Navigation and HASPA (High-Speed Downlink Packet Access), which is for sending overloaded vehicles" information to overloading control office, is possible to send larger volume and data faster than CDMA Technology. It has merits that it does not need extra investment, and it is possible to use with a little improvement of WCDMA system.

This present system is possible to send information fast and precisely, which has been gathered and processed by HSDPA, to overloaded vehicle, control office, and related agency. In addition, it enables sending massive information, checking through wireless internet, and provides real-time confirming service.

## V. Application Plan of U-Overloaded Vehicles Management System

The U-Overloaded Vehicles Management System is able to institute free operation method according to various purposes such as restriction or regulation of overloaded vehicles.

It is a small scale USN. It transmits overloaded vehicle"s information to main control system, and it can be connected with another ITS (Intelligent Traffic System) field, and applied to national traffic system. In addition, it can be used to prevent load carrying capacity declination of structures such as bridges through controlling vehicles, which are control-objected, coming and going in several system operations.

This kind of free utilization of U-Overloaded Vehicles Management System can be used to national or personal purposes, and improve the efficiency of system application.

## VI. CONCLUSION

From this study, grasp problem of the existing overloaded vehicle regulation system, and experimented on development of system with based indoor test for unattended overloaded vehicle regulation system using Ubiquitous technology. Results of experiment are as follow.

1.) Whole system is organized USN Sensor Field, System Controller, External Network. Measured data from several sensors transmit to one Sink Node through USN and tranmit to System controller. System controller report to diver"s cell phone, PDA, Navigation and VMS (Variable Message System)

2.) This study offered standard of effective overloaded vehicle regulation from development of U-Uninhabited Overloaded vehicle regulation system algorithms.

When vehicle go through Loop Sensor, camera take a picture of vehicle and Wim Sensor measure vehicle"s load.

System controller received vehicle"s data and load data. When vehicle is not overloaded, data saved on Memory Buffer. However, vehicle is overloaded, System controller notice to diver"s cell phone as SMS Message, print contents on the VMS (Variable Message System)

3.) We experimented on static load test of Bending Plate Type Sensor. As experiment on compare with Calibration Machine"s load and Bending Plate Type Sensor"s load results. When test on load 2 tons error is 0.67%. When load 4 tons error is 0.33% and load 6 tons error is 0.11%. 8 tons and 10tons are equal two results. Error is less than 1% and that shown the accuracy for using this study.

4.) We experiment on load variable for secure confidence and investigate application about Bending Plate Type Sensor also we experiment on resistance of changed temperature that temperature seasonally variable. Temperature have been changed 7step as -10□, -5□, 0□, 10□, 20□, 30□, 50□. As a result, errors are 2%, 1.67%, 2.4%, 2.06%, 0.8%, 1.73%, 1.46% each temperature, average error is 1.73% that low error range. And resistibility is enough using sensor on the temperature change as climatic change, season change.

As result of experiment on resistance of changed temperature on Bending Plate Type Sensor, resistance is enough using sensor on normal temperature

5.)The result of transmission distance test about wireless sensor node on the free-space. Communication quality on 120m is 86%, communication quality on 300m is 56% that more than 50% quality. Considered based sensor node information that transmission quality is over 50%, data does not loss. Also acquisition data using wireless for possibility compatible with sensor node and Bending Plate Type Sensor and acquired correct data. Therefore, confirmed no problem of using wireless sensor node on U-Uninhabited Overloaded vehicle regulation system

6.) This system have applied on the road, bridge and others, will be expected economical effects because decreased labor costs through uninhabited system and improved system reliability from using high efficient sensor and fitting algorithms. Also system install is convenience from Ubiquitous Sensor Network system, installation charge is diminution through application with existing overloaded vehicle regulation system and established u-ITS

## ACKNOWLEDGMENTS

This study was financially supported by Small and Medium Business Administration under a research and developing project, and the authors would like to thank Small and Medium Business Administration and Hanyang University BK21.

## REFERENCES

[1] Eun young Kim, chung won Lee(2006) Overloading vehicles analasys and regulation , A paper of Research Seoul, 7(1), pp75~83

[2] Ho jung Kim(2003) Study use of WIM Data, graduation thesis, Myongji Univ.

[3] Tae mu Sim(1998) route trait of Overloading vehicles using Bridge Weigh-In-Motion System, a graduation thesis, Kyunghee Univ.

[4] Seunghwa ENC(2004) Overloading vehicles system development, 1, Seoul, Korea

[5] A. Emin Aktan(2002) Monitoring and Safety Evaluation of Existing Concrete Structures, State-of-the-Art Report, USA

[6] Andrew P. Nichols(2004) Quality Control Procedures For Weigh-In-Motion Data, Doctor of Philosophy, Purdue Univ.

# Design of bridge health monitoring system on Wireless Sensor Network

Byung-wan Jo[1], Heoun Kim[2], Jung-hoon Park[2], Kwang-won Yoon[2], Seok-won Kang[2], Seong-hee Han[2]

[1] Professor, Dept. of Civil engineering, Hanyang university, Seoul, Korea
[2] Graduate Student, Dept. of civil engineering, Hanyang University, Seoul, Korea
military744@nate.com

**Abstract - Recently, Advent of Ubiquitous environment expect to change paradigm on the whole society. Specially, Ubiquitous is applied the field of construction had a ripple effect on the industry to be much avail. There has been increasing interest in developing structure health monitoring system based on wireless sensor network due to recent advancement in Ubiquitous sensor network technologies. Structure health monitoring system collects data via analog sensor and then sends to analysis application through the wired network.**

**However, there are many problems such as high cost from wire-network, difficult getting location or amendment when fault occurred and others. In case introduction of Ubiquitous wireless network, transmission skill and real-time monitoring system, a control center can many infrastructures monitoring, also it curtail expenses and human power.**

**This study suggests an application model bridge maintenance system by using an ubiquitous technology for solved the problem of the structure health monitoring system that had wire-network system.**

## I. INTRODUCTION

Among infrastructures, large fabrics such as engineering works and structural materials that safety takes top priority need accurate and precise design and execution with persistently cautious maintenance management for ensuring safe usability. Currently, part of the large infrastructures established real-time monitoring system for effective maintenance management and safety control of facilities. Nonetheless, under the circumstances that sufficient visible range is not assure because of fog, such as chain-reaction collision on the Seohae Grand Bridge was occurred on February 2007 and it cannot help being a chance to confirm that large-scale mishaps can outbreak even there are minute real-time monitoring systems and about 20 stationed administrant. Therefore, bridge-monitoring system should perform more functions than only simple precision measuring. Real-time measurement and automatic judgment whether the measured value excesses the boundary line or not occurs at the same time, and when it exceeds, they need to investigate and apply the method on the scene about intelligent bridge-safety control that connects proper intelligent actuator.

This paper, we design and implement the health monitoring system based on USN(Ubiquitous Sensor Network) to solve those problems.

## II. Ubiquitous Computing and Sensor Network

### *A. Ubiquitous*

The term "ubiquitous" has its root in the Latin word "ubique,"meaning "(God) exists anywhere or in many places simultaneously," which is now being used with the meaning of "omnipresence: exists anywhere."

Introduced for the first time in 1991, the word "ubiquitous" has recently entered the stage of development as a new concept in information technology revolution. And now, nations around the globe are introducing strategies to implement ubiquitous technology in their own fashion that would sufficiently meet their specific characteristics and needs. With the United States as the central figure, improvement models have been devised and technological developments have been carried out for the advancement of ubiquitous technology. In addition, countries such as South Korea, Japan, and the European Union that rely on the IT industry as the driving force for national development are focusing on technological development and the establishment of a new-generation standard through bench-marking by specific countries. Although advanced technologies and applications of the ubiquitous concept are still presently at a rudimentary stage, we can expect to create a revolutionary U-space that we have never imagined before through the active introduction of advanced models and road maps by professionals from each associated field. As Mark Weiser suggested, ubiquitous technology shall adapt itself spontaneously within the infinitely complex world of information service calmly and

invisibly by blending itself in everyday life. Portable devices shall become super-compact due to the striking

T. Sobh (ed.), *Innovations and Advances in Computer Sciences and Engineering*,
DOI 10.1007/978-90-481-3658-2_96, © Springer Science+Business Media B.V. 2010

developments in network technology and other various forms of information technologies. These super-compact devices would become invisible tools that can freely provide services in the form of accessories in our lives. Regardless of their size, every form of objects that we frequently use in our lives today will be transformed into embedded devices based on their specific application with built-in computers. Not only that, they will develop into object-computers capable of providing network function that would allow all objects to recognize and control one another.

### *A. Ubiquitous Sensor Network (USN)*

USN possesses a special structure where multiple sensor network fields are connected to the external network through gateways. Sensor nodes transmit data to nearby sink nodes and the data received by the sensor node are transmitted to the gateway. Data transmitted to the administrator from the gateway may be transmitted through satellite communication or wired/wireless Internet, where the existing infra shall be used for the access network as such. System Internal Technology.

### *B. Ubiquitous Computing in the Field of Civil Engineering*

The concept of Ubiquitous Computing has already been introduced in the field of civil engineering and construction. In the UCLA, earthquake sensors are installed within every building around the campus to monitor the safety of the building from earthquake threats. As for roads, a real-time road weather information system can be realized by receiving data from the humidity/temperature sensors installed in multiple spots along the roads. Not only that, an effective road-based system capable of delivering real-time traffic information may be introduced in the future from the implementation of road surface sensor and video image detection network. Through the effective use of RFID and GPS technology, itis expected that the construction vehicle tracking system, construction resource management, remote operation control using wearable computers in dangerous tunnels and underground construction sites, and emergency alarm function may become available in the near future. Furthermore, the monitoring of risky conditions is expected to become significantly more efficient than that of existing wired networks with the application of a new sensor-based network, which will be connected with an automatic water gate control system capable of remote controlling the water gate opening and closing sensor and device as well as a water-surface surveillance camera installed in the water gates to monitor the water level in large-scale dams and reservoir facilities.

## III. Design and embodiment of sensor network module

Ubiquitous system which developed using Ubiquitous technology is not exist alone on Ubiquitous environment, is constantly reciprocal action with other systems.

Therefore, when design sensor network module that organized with consideration for manage reciprocal action with other systems.

Developed board on this study designed that able to transmit collected information to application software which operated on Host from sensor device. Board will be used Reference for verification and adaptation of whole system working. Sensor function and Block diagram of board are next.

Table 1. Sensor Network Module

| | |
|---|---|
| - MCU MSP430F1611 (10K RAM, 48K Flash) | - RF Transceiver module [2.4GHz] |
| - Host or console I/F Serial Interface | - User Button |
| - ISP Flash fusing JTAG port | - 16bit low power processor |
| - 3V Battery Operation | - Operating System TinyOS |

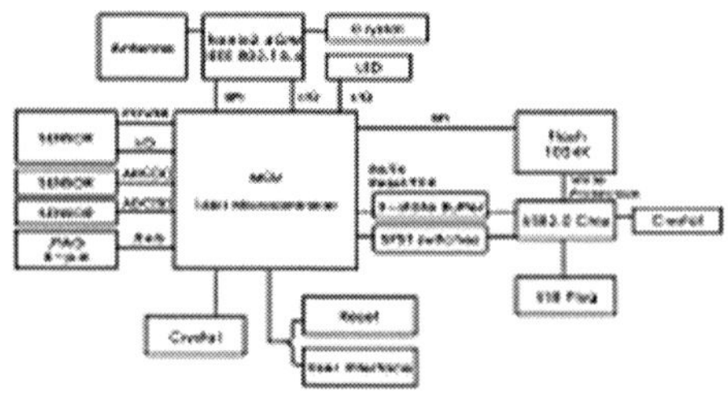

Fig 1. Block diagram

### *A. Sensor board*

Sensor board have amplifier which amplify sensor signal, filter which remove noise, 16bit A/D converter which convert ANALOG Signal into DIGITAL Signal, and Serial flash memory which remember the setting value of board.

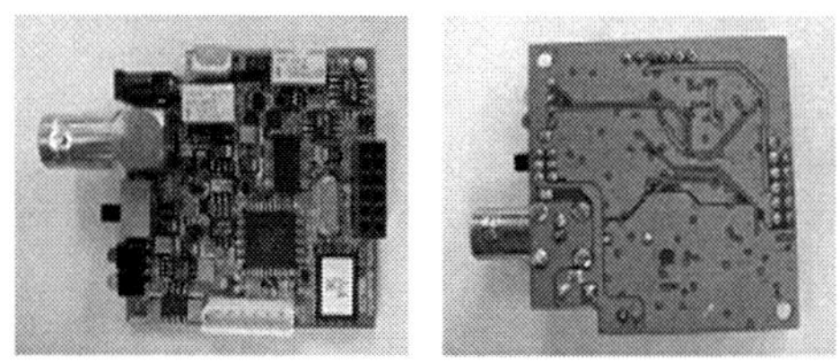

Fig 2. Sensor board

### *B. Sensor Interface Module*

SIM(Sensor Interface Module) made for connection with variety sensor and wireless communication module used civil engineering measurement and strain gage which measure strain rate and displacement.

System organization of method is various depend on kind of sensor. Case using this SIM, Interface module which can connect all Analog sensor and system without MEMS based (Attach to sensor board embody System On Board) chip size.

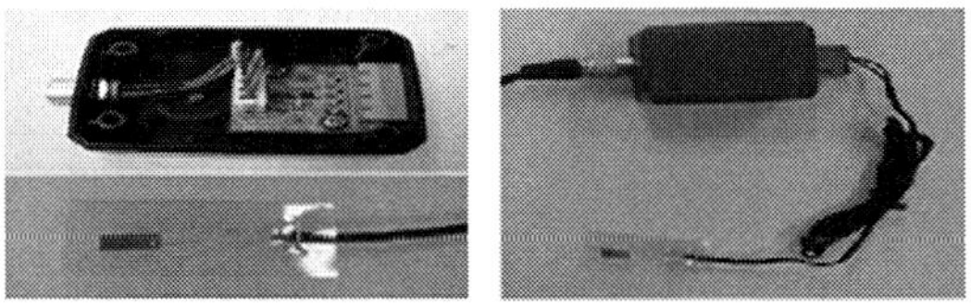

Fig 3. Strain gage and ANALOG sensor interface module

### C. Wireless Sensor Board

Wireless sensor board is made integration combine sensor board and wireless communication board.

Recently, MEMS based clinometer and accelerometer sensor are using for made One-body wireless sensor module because must use chip form on the board.

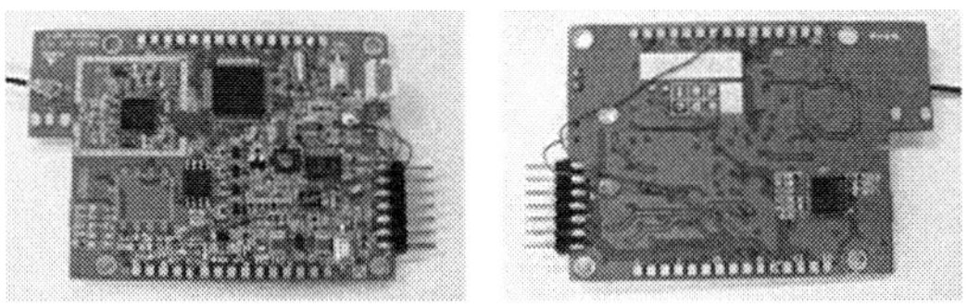

Fig 4. The wireless sensor board by the Combination of sensor and sensor board and Wireless Communication

### D. Accessary Module

When using board is related to communication and sensor interface, It is Accessory Module for construction of USN system, USB-Serial Converter is produced for support to suitable with USB Interface and Serial Interface. Program Writer is produced for written firmware and other on the board.

Fig 5. USB type Serial Program Writer

Fig 6. USB-SERIAL Converter

### E. Wireless Sensor Node for USN

IT is Wireless Sensor Node for construction USN system. One wireless sensor node is organization with more than one sensor, sensor interface module, A/D Converter, USN Module and Antenna.

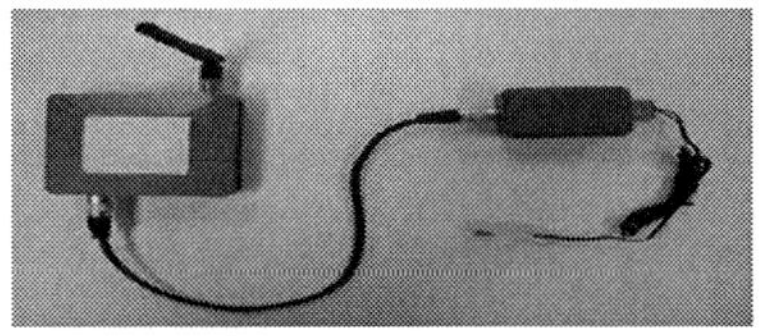

Fig 7. 1Ch, 8Ch Wireless Sensor Node for USN attached strain gage.

## IV. Field experiment

Developed sensor module on this study that have been verified accuracy field data about accelerometer, strain gauge and temperature sensor installed slab between A1-P1 Sapgyo Grand Bridge on Chungnam Dangjin City.

### A. Bridge status

| Classification | Contents | Classification | Contents |
|---|---|---|---|
| Bridge name | Sapgyo Grand Bridge | Management agency | Yesan national maintenance construction office |
| Location | Chungnam Dangjin Sin-pyung Wunjung | Route | National road No.34 |
| Length | 560.0 m | Width | 20.1 m |
| Superstructure | STEEL BOX GILDER | Design live load | DB-24 |

### B. Status of installation sensor on Sapgyo Grand Bridge

Accelerometer, strain gauge and temperature sensor installed slab between A1-P1 Sapgyo Grand Bridge on Chungnam Dangjin City.

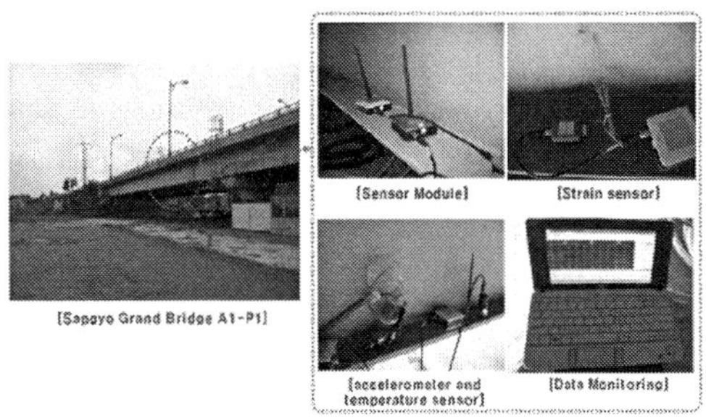


### C. Result of sensor test

Accelerometer, strain gauge and temperature sensor installed on Sapgyo Grand Bridge and display data on the graph.

In accelerometer case, acquired 100 data per sec and keep -0.001⌒.004(g). And strain gauge and temperature sensor are 1 data per sec.

Result of measured on Sapgyo Grand Bridge by accelerometer, strain gauge and temperature sensor is as follows

a. accelerometer sensor

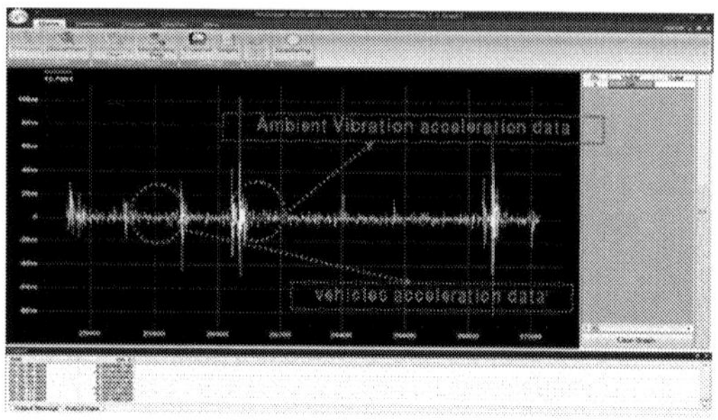


Fig 8. accelerometer data

b. strain gauge and temperature sensor

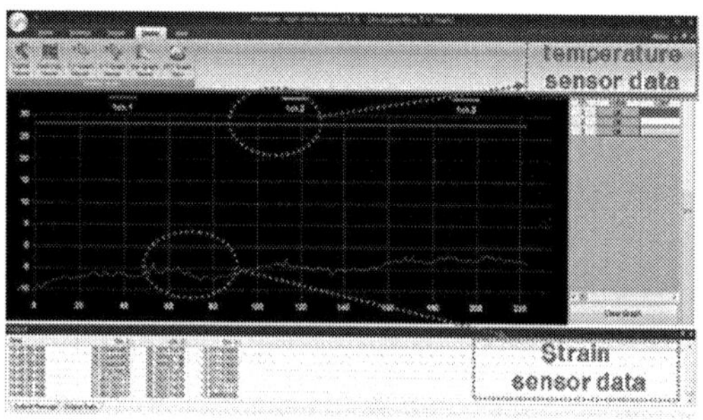


Fig 9. strain gauge and temperature data

## V. Conclusion

This study is based research for application computing skill of Ubiquitous environment on the monitoring system of bridge structure.

This paper developed sensor network module that based construction Ubiquitous environment with bridge structure monitoring. Also we have verified accuracy of data through field experiment about module.

Application of inspection system have problem that cost and difficult of cable installation even though most existing bridge without special bridge do not have monitoring system, and outworn bridges attention of inspection cause have many element of danger. However, when apply this system, we are convinced of decrease in inspection and install cost, in the addition, this system solved the problem that difficult of installation.

Finally, Ubiquitous technology that have presented new paradigm in the future will be application on the grand bridge as cable-stayed girder bridge or suspension bridge.

## ACKNOWLEDGMENTS

This study was financially supported by Small and Medium Business Administration under a research and developing project, and the authors would like to thank Small and Medium Business Administration and Hanyang University BK21.

## REFERENCES

[1] MOCT, Roadway Bridge Design Criteria , 2005

[2] Koo bong kuen, " Method for impact factor determination of bridges", Journal of the Institute of Construction Technology Vol 24, Mo. 1, 2005. 5. pp. 1~11

[3] Ryoo young dal, " NCA CIO REPORT, 2004, NIA

[4] D.H.Wnag, W.H.Liao, "Instrumentation of a Wireless Transmission System for Health Monitoring of Large Infrastructures", IEEE Instrumentation and Measure ment Technology Conference, 2001. pp. 21~23, pp. 634~639

[5] S.J. Lee, J.K Lim, "Measurement, Monitoring, and Control Systems based on Ubiquitous Computing Technology, Control System", 2005.1

[6] W.Y. Young, "Ubiquitous Computing and Civil Engineering"

[7] D.H.Wnag, W.H.Liao, "Instrumentation of a Wireless Transmission System for Health Monitoring of Large Infrastructures", IEEE Instrumentation and Measurement Technology Conference, 2001.21~23, pp.634~639

[8] Jeong-Hee Chang, Hae-Yun Choi, Sang-Ju Han, "The Study of the Health Monitoring of Bridge Systems Using Ubiquitous Computing Technology" NEXT 2007 KOREA Conference, 2007.10

# TOTALLY SENDER- AND FILE-ORDER RECOVERY TECHNIQUE FOR RELIABLE MULTICASTING SOLUTIONS USING HEARTBEAT

CHIN TECK MIN, LIM TONG MING
School of Computer Technology, Sunway University College

*Abstract* - **Guaranteed Reliable Asynchronous Unicast and Multicast (GRUM) is a middleware that allows sending and receiving of messages among peers in a group. GRUM has been designed and implemented to achieve highest successful data delivery rate to peers in a group. GRUM is a prototype that provides guaranteed, reliable and ordered multicast and unicast communication services. GRUM is an implementation of Application Layer Multicast (ALM) which makes use of heartbeat [6] in the design of the Heartbeat–driven Recovery technique to guarantee message multicasting service is ordered by sender and file as peers are sending messages to other peers in the group concurrently. Few totally ordered techniques [7][8][9] were reviewed and a Totally Sender and File Ordered algorithm is devised and implemented in the paper to provide sender and file ordered transmission in the GRUM. Experiments were carried out to test the proposed techniques and achievements and problems encountered were discussed.**

## I. INTRODUCTION

Internet has been growing very fast in the last ten years[1]. This makes Internet the preferred information super highway for all types of data transfer activities be it commercial or non-commercial transactions around the world. Internet has opened up a new way of communication among people. Media such as newspaper, magazine, video, movie and news are also taking advantage of the Internet to deliver services to its clientele. Many types of application and services are now offered through Internet. These network services use unicast/multicast [2] on top of the Inter-Process Communication model. However, multicast is useful for applications such as groupware, online conferences, interactive distance learning and applications such as online auction [2][3] and fault tolerance systems. Different applications require different communication models; for example, online conferencing system requires speed and the messages need not be in order. But for an online auction system, the order of messages is very important. GRUM middleware is designed for applications that need guaranteed multicasting services which order of senders and files is extremely important where speed of data delivery is of secondary important. A good example is a distributed retailing application that needs consistent and constant sending and receiving of transactional data coded in the form of SQL statements between retail outlets and the head office. Even though most of the national telecommunication service providers have provided few broadband Internet service options, however, these services are still riding on the existing copper-based infrastructure that does not give quality broadband network services. For large cooperates who can afford to subscribe true broadband services such as leased T1 line, they have to invest high installation cost and expensive monthly subscription fee. In view of this situation, the GRUM research project has stated several strong motivations to justify the need to design and develop a 'near' high bandwidth, ordered and reliable data sending and receiving services on the existing infrastructure to users.

## II. RESEARCH OBJECTIVES

The infrastructure of the existing public switching network has been in use for many years and the service is highly unpredictable and the bandwidth in many parts of the country is very low. The decision to adopt UDP as the communication protocol rather than TCP is due to its low overhead, connectionless and light communication properties which is suitable for the application's need and its current infrastructure. This has motivated the inception of the GRUM project. In short, GRUM engine will use UDP as the communication protocol to send and receive data among peers.

The objectives of this research are summarized as follow:

1. To design and, implement a prototype that has the following characteristics: reliable, guaranteed, ordered and auto-recovered multicasting UDP-based data delivery services.
   1.1 A reliable engine ensures that each packet sent to each receiver (or peer) is guaranteed. It takes into account sender/receiver being halted by external factor, Internet services down, and bandwidth of Internet line unstable
   1.2 The proposed GRUM will provide an auto-recovery mechanism where any missing or unreachable packets will be resent until packets are fully delivered
   1.3 The 'ordered' feature proposed in GRUM will ensure all sending services are ordered by sender and ordered by file.
2. To use GRUM with a client application as a testing system for the testing of the GRUM prototype and to collect communication data over one month period using six (6) peers as the test bed

## III. ENTITIES OF GRUM

GRUM is a group communication middleware that is not constrained by the hardware limitation. It is an implementation of ALM. A GRUM consists of the following entities in its model. Fig 1 shows the framework and entities for a typical GRUM setup.

GRUM consists of a **COORDINATOR** and few **ENDHOST**s. **COORDINATOR** is a software component which coordinates and manages all the communication activities in a group. It is responsible for the creation of a group, maintenance of

T. Sobh (ed.), *Innovations and Advances in Computer Sciences and Engineering*,
DOI 10.1007/978-90-481-3658-2_97, © Springer Science+Business Media B.V. 2010

membership. Each **ENDHOST** will connect to the **COORDINATOR** to establish a group or join a group. **ENDHOST** is a client computer which creates an class instance of **CHANNEL** to connect to **COORDINATOR** and join a group. After an **ENDHOST** joins a group, it can start sending and receiving messages either using **MULTICAST** or **UNICAST** methods which is provided by the **CHANNEL**. Since GRUM is a group communication middleware, it provides **UNICAST** and **MULTICAST** services especially for a group only. As a result, **UNICAST** service provides by GRUM is limited to communication among the group members only and could not connect to others computer that is not in the group.

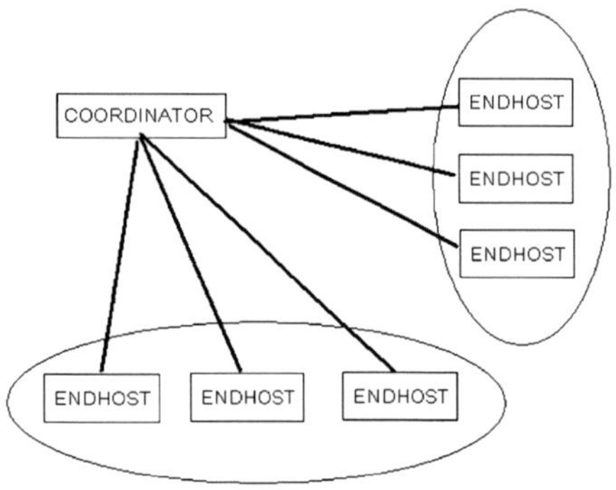


Fig1. Entities in GRUM

## IV. MULTICAST MESSAGE PASSING MODEL USING SENDER AND FILE ORDER (MSFO)

**MSFO** is a communication model used by GRUM to multicast message in a group. This model delivers the 'orderness' characteristics required by GRUM. The 'orderness' of the MSFO model achieves two (2) kind of 'orderness':

It ensures that if two senders S1 and S2 send two different files F1 and F2 respectively at time t1 and t2. Assume that S1 starts sending at time t1 and S2 starts at time t2 where t2 is t1 + 1. MSFO will guarantee that S1 finishes the sending of F1 before S2 starts its sending. This demonstrates the **Sender Orderness** of the MSFO.

If sender S1 has finished the sending of F1 at t1 but some receivers Rx where x = 1..n have not completely received all the messages as broadcasted by the **COORDINATOR**, and at time t2, sender S2 is granted a confirmed **BATCHID** to send file F2 where t2 = t2 + 1. All the messages broadcasted by S2 will be queued in the buffer until all messages Mn where n = 1 .. z for F1 are completely received by all receivers before receivers start receiving messages broadcasted by S2. This is the **File Orderness** of the MSFO.

Fig 2 Illustrates the Multicast Message Passing Model using Sender and Files Order (MSFO) in the GRUM and the working mechanism of the model.

In the MSFO model, the term message refers to the smallest unit of data (or file fragment of a predefined size) that warp with datagram to send through a network. As one **ENDHOST** sends a message, it needs to send a REQUEST message first to the **COORDINATOR** to request and lock a multicast **BATCHID**. The objective of the REQUEST message is to inform the **COORDINATOR** that a requesting **ENDHOST** intends to multicast data by using the granted **BATCHID**. The **BATCHID** is a controlled sequential number that will be added to every

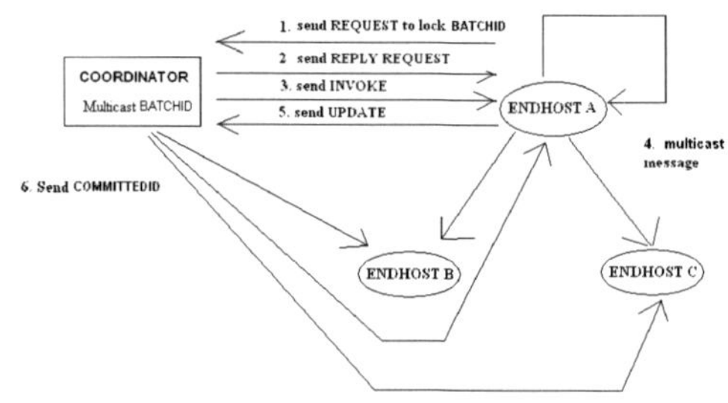


Fig 2. The proposed MSFO in GRUM multicast message passing model

message that is to be broadcasted within the group. The **BATCHID** has been designed to provide indexing order, identifier and monitoring reasons. Each message in GRUM use the **BATCHID** as an index so that every message sent is in order so that messages can be identified and controlled. When the **COORDINATOR** receives a request to lock **BATCHID** message, it will send a REPLY REQUEST message to indicate it has received the request and **ENDHOST** can stop sending of REQUEST message and wait for **COORDINATOR** to response with INVOKE message. The **COORDINATOR** will do a check to see whether the **BATCHID** is currently locked or not lock. If it is NOT locked, the **COORDINATORCOORDINATOR** will send an INVOKE message to the requesting **ENDHOST**. With this, the **ENDHOST** can start multicasting message to be sent to the members in the group. The INVOKE message is used to instruct the **ENDHOST** to start sending data to its desire receiver. The INVOKE message is always accompanied by a **BATCHID** in which the **ENDHOST** will use the **BATCHID** as starting number to be attached to the each of the message. For every subsequent message, this **BATCHID** will be incremented and the new number will be added to each of the message that is sent. After the **ENDHOST** completed multicasting the messages, the **BATCHID** will be sent back to the **COORDINATOR** using an UPDATE massage. The UPDATE message is used to inform the **COORDINATORCOORDINATOR** two things. First is **ENDHOST** has finished all the sending messages and second things is the **COMMITTEDID** which is the last value of **BATCHID** added to message plus one. After the **COORDINATOR** receives the **COMMITTEDID** from the sender, **COORDINATORCOORDINATOR** will multicast the **COMMITTEDID** to every **ENDHOST** in the group to inform all **ENDHOST** that sent message with **BATCHID** not greater than the **COMMITTEDID** are valid messages to be received and processed. In the Fig 2 an **ENDHOST** A desires to multicast a message to members of a group. In order to achieve this objective, it first sends a REQUEST message to the **COORDINATOR** to lock the multicast **BATCHID**. Then the **COORDINATOR** needs to ensure that the multicast **BATCHID** is not lock by any other **ENDHOST**. If it is not then the **COORDINATOR** send an INVOKE message to the **ENDHOST** A. The **ENDHOST** A will then start to multicast message starts from the given **BATCHID** by incrementing each message by 1. As soon as the message is sent, **ENDHOST** A will send UPDATE message to the **COORDINATOR**. After the **COORDINATOR** received the **COMMITTEDID**, **COORDINATOR** will multicast the **COMMITEDID** to every **ENDHOST** in the group. This is to inform all **ENDHOST** in the group that the valid range of running number is safe to process by **ENDHOST**s are numbers between the last **BATCHID** to the **COMMITEDID**.

The MSFO uses the Globally Totally Orderness mechanism where sending of messages can only be carried out when **ENDHOST** in a group has successfully requested a **BATCHID** before it can do any sending activity. The MSFO only allows one sending **ENDHOST** to multicast at any one time. This is because once a **BATCHID** is locked and hold by an **ENDHOST**, other **ENDHOST**s cannot start multicasting services. The **COORDINATOR** will not send an INVOKE message to the requesting **ENDHOST** until a **BATCHID** is granted.

## V. HEARTBEAT-DRIVEN RECEIVER-RELIABLE MULTICAST RECOVERY SCHEME (HRRMR)

The MSFO component anticipates a Receiver Reliable [4] model where individual receiver is responsible for detecting missing packets. [5] states that a Receiver Reliable protocol is able to facilitate the detection of lost messages, especially if a message is missing from a sequence of messages, it needs a distribution of the latest sequence number which is commonly known as heartbeats. The model requires that each **ENDHOST** receives a **COMMITTEDID** which gives the information all the messages with **BATCHID** that is less than the COMMITTEDID to be safe to receive. Then if the **ENDHOST** finds out that there is a message with a specified **BATCHID** which is less than the COMMITTEDID but is not in its repository then the missing message will be detected. The **ENDHOST** will try to recover the missing packet through the HRRMR recovery scheme. Fig 3 illustrates the HRRMR recovery scheme for multicast recovery in the GRUM.

Fig 3 shows that a sender multicasts three (3) messages to all other peers in the group by sendinga COMMITTEDID to the **COORDINATOR**. The aim is for the receiver 1, 2, and 3 to receive the multicast messages. Unfortunately receiver 3 fails to receive the message with **BATCHID** 3. The **COORDINATOR** multicasts the COMMITTEDID to all peers in the group so that each peer is able to recover any missing messages. All the **ENDHOSTS** in the group will process messages with **BATCHID** that is less than the COMMITTEDID. While receiver 3 wants to process message with **BATCHID** 3 which is less than the COMMITTEDID, it is found that the message is not in its repository. Receiver 3 will then multicast a RECOVERY message in the group asking for the lost message from other peers in the group. This **RECOVERY** message is a

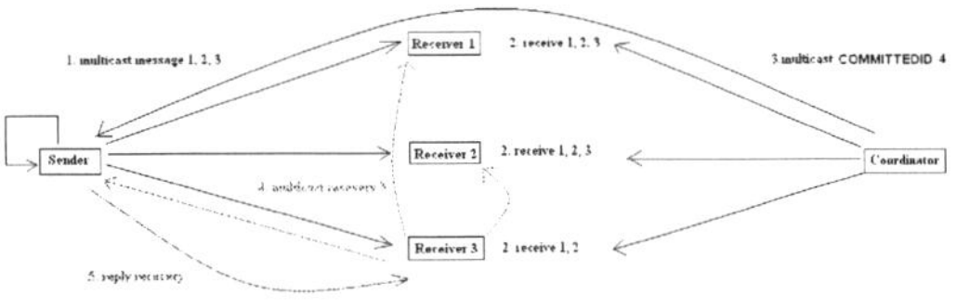


Fig 3. HRRMR recovery schema for multicast recovery in GRUM

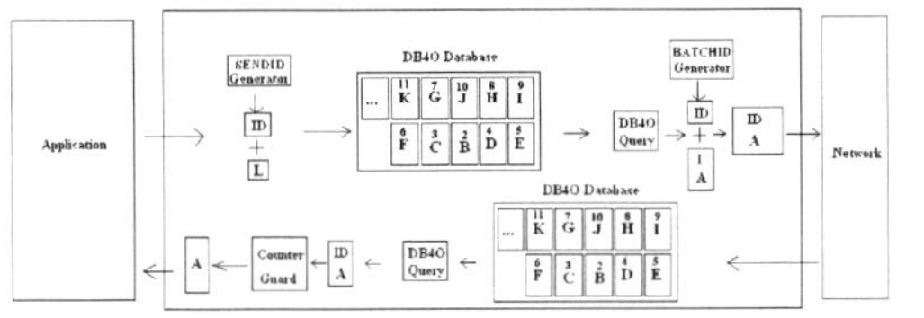


Fig 4. Illustrations of basic concept of message queuing in GRUM

message that ask from other **ENDHOSTS** in the group to find out the sender of the missing message with a specified **BATCHID**. It will then be asked to resend the specified message to the sender again using the RECOVERY message. Any peer in the group that sends the message with the **BATCHID** 3 will reply the recovery message by sending a REPLY RECOVERY message which contains the message with **BATCHID** 3. This is an implementation of Source based Recovery Scheme [6] where the source exclusively retransmits all the lost packets to the requesting receiver. This mechanism guarantees that one recovery attempt is enough for each request, and thus reduces the overhead incurred by failed recovery attempts. In this design, the receiver will send the recovery request in certain time interval until the message is recovered.

### VI. SORTING IN GRUM

The basic concept of the queuing and sorting models use in GRUM is shown in Fig 4.

An Application (Fig 4) is a program that sends and receives data among peers. All the messages that the Application sends is inserted into the DB4O database accompanied by a SENDID; a running number that keeps track of messages sent. This ID is used to maintain the order of messages as messages are sent from the DB4O database. GRUM will utilize the sorting capability provided by the SODA Query API in DB4O to query the messages sorted in descending order base on the SENDID in order to allow the data to be combined with the **BATCHID** prior to the data sending activity.

Messages received from the network are then inserted into the DB4O database. By using similar sorting capability provided by the SODA Query, querying sorted messages in descending order based **BATCHID**, each message is matched with the Counter Guard; a function that guards each message sent to applications in the correct order.

This sorting model gives GRUM the capability to sort messages by making use of the **BATCHID** information. This allows GRUM to provide a totally order and reliable multicasting capability by combining the GRUM's MULTICAST model.

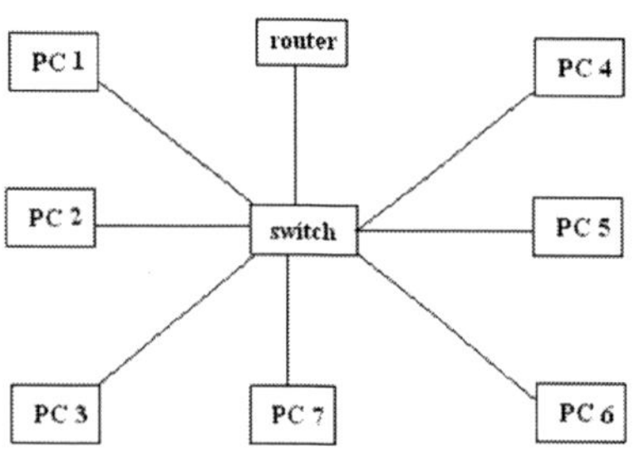


Fig 5. The testing environment

### VII. ANALYSIS OF RESULTS FROM TEST CASES

GRUM middleware was tested in a controlled environment as shown in Fig 5.

The test environment (Fig 5) consists of seven networked PCs in a local area network. They are labeled as PC1, PC2, PC3, PC4, PC5, PC6, and PC7. All the PCs are connected with a switch where the switch is connected to router. PC1 to PC6 are installed with an ENDHOST application that makes use of the ENDHOST API provided by the GRUM middleware whereas PC7 is installed with the COORDINATOR application utilizing the GRUM API.

A summary of designed test cases are shown in Table I. For Every test case we explain the result in below.

#### A. Multicast data by one sender

The first test case is to multicast data from one sender to multiple peers (or receivers). This test case uses one ENDHOST as the sender to multicast data to all the other ENDHOSTs concurrently. In this test, grum6 is the sender, whereas the rest of ENDHOSTs are the receivers. From this test case there is no data loss in transmission activities.

#### B. Multicast data by two senders.

In the second test case, the test involves a group that consists of two (2) ENDHOSTs as senders which multicast data to other receiver ENDHOSTs. In this case, grum6 and grum5 are the senders. In this case, receiver receives all the messages without losing any packets at all.

Table I
Test Cases for GRUM

| Test Cases | Purpose |
|---|---|
| 1. Multicast data by one sender | Test the multicast capability using GRUM. |
| 2. Multicast data by two senders. | Test how good the locking mechanism for **BATCHID** in GRUM. |
| 3. Multicast data with recovery by pulling out a network cable. | Test how good the recovery mechanism in GRUM |

#### C. Multicast data with recovery by pulling out a network cable.

In this test case, grum6 is the chosen sender that multicasts data to other peers in the group. In order to simulate 'lost or limited Internet service', 'some peers are down or not able to be connected to the group' or 'losing of signal from the peer in a group', this test case simply 'pull out' the connection of the peer to simulate one of the potential scenarios described. The network cable that connects to grum3 will be disconnected for about 1 minute to 'introduce' packets lost or connection lost to grum3. As a result, GRUM must demonstrate its recovery mechanism by automatically initiating the recovery algorithm to recover the missing messages From this test cast, it has shown that GRUM is very reliable and capable of recovering from unanticipated events.

## VIII. Future Enhancement and conclusion

The MSFO and HRRMR with a built-in sorting algorithm have provided a reliable multicasting model that guarantee the sender and file ***orderness*** characteristics in the core of the middleware. The **labeled** messages and pull-based recovery approach are adopted as part of the design consideration. This has made clients to responsible for the recovering activities; hence reduce the dependency of the log server for better throughput. However, tests carried out in this project have also suggested some issues that need to be addressed which may cause performance and scalability issues.

First of all, the locking of **BATCHID** dramatically reduces the performance and the tight coupling of each message causing the bottleneck to the network if the number of **ENDHOST** is large. As an Example, in **MSFO**, the send REQUEST message needs to have acknowledgment from the **COORDINATOR** else the **ENDHOST** will keep sending REQUEST message and this will cause serious communication traffic overhead in the message queue of the **COORDINATOR**. As a result, performance is very much slow down. Some messages need to have ACKNOWLEDGMENT message to ensure the message is not lost. This is because the model cannot afford to have missing messages. If there is no ACKNOWLEDGMENT message, **ENDHOST** will repetitively send REQUEST message to **COORDINATOR**; if **COORDINATOR** did not receive the REQUEST message, then the **ENDHOST** will wait infinitely with the assumption that **COORDINATOR** has not given the **ENDHOST** the permission to multicast.

ACKNOWLEDGMENT

With the opportunity here, I would like to thanks to Dr Lim Tong Meng who have help throughout the research. And I also like to thanks my friend who have supported me in the research

## REFERENCES

[1] A. M. O. K. G. Coffman. Growth of the Internet. [Online]. http://www.dtc.umn.edu/~odlyzko/doc/oft.Internet.growth.pdf

[2] R. Dasari. To Unicast Or To Multicast. [Online]. http://www.mpulsetech.com/prod/ToUcastOrMcast.pdf

[3] D. H. S. Unknown, J. Unknown, and C. d. M. C. Unknown. Establishing a Trade-Off between Unicast and Multicast Retransmission Modes for Reliable Multicast Protocol. [Online]. http://ieeexplore.ieee.org/xpl/freeabs_all.jsp?arnumber=876432

[4] D. L. Y. M. K. Wonyong Yoon. (2000) Tree-based reliable multicast in combined fixed/mobile IP networks. [Online]. http://ieeexplore.ieee.org/iel5/7152/19253/00891087.pdf

[5] D. C. S. Qixiang Sun. A Gossip-based Reliable Multicast for Large-scale High-througput Application.

[6] S. R. R. K. S. S. I. Danyang Zhang. A Recovery Algorithm for Reliable Multicasting in Reliable Networks.

[7] Gregory V. Chockler, (1997), An Adaptive Totally Ordered Multicast Protocol that Tolerates Partitions, The Hebrew University of Jerusalem

[8] X Chen, L E Moser adn P M Melliar-Smith (1997), Totally ordered gigabit multicasting, University of California, Santa Barbara

[9] Brian Whetten, Todd Montgomery and Simon Kaplan (2008), A High Performance Totally Ordered Multicast Protocol, University of California at Berkeley, West Virginia University and University of Illinois at Champaign-Urbana

# Anonymity Leakage Reduction in Network Latency

Longy O. Anyanwu, Ed.D.
Mathematics and Computer Science
Fort Hays State University, Hays, Kansas, USA
loanyanwu@fhsu.edu

Jared Keengwe, Ph.D.
Department of Teaching and Learning
University of North Dakota, ND, USA
jared.keengwe@und.nodak.edu

Gladys Arome, Ph.D.
Department of Media and Instructional Tech.
Florida Atlantic University, FL, USA
garome@fau.edu

**ABSTRACT**

Each Internet communication leaves trails here or there, that can be followed back to the user. Notably, anonymous communication schemes are purposed to hide users' identity as to personal, source and destination location and content information. Previous studies have shown that the average round trip times (RTT) leakage between network host location, $X_1$ and network destination location, $Y_1$, can be determined, [12]. Additionally, an attack from a web site with access to a network coordinate system can recover 6.8 bits/hr. of network location from one corrupt Tor router, [12]. Notably, no network capability is in existence to completely negate anonymity leakage in network latency, [12], thus, the minimization of anonymity leakage in network latency becomes critically salient. The purpose of this paper is to investigate network latency anonymity leaks, and propose practical techniques for their reduction. In this direction, we investigate the following technical question: what implementation techniques can be configured to truly reduce anonymity leaks using deployable systems. Here, an extension of the popular Tor security strategies and unique configuration of the popular network anonymity techniques (algorithms) for future implementation are presented.

**Categories and Subject Descriptors**

Network security. Network anonymity loss reduction. Secure networks and communication. Anonymous communications.

**General terms**

Network security, Reliable anonymity systems.

## 1. INTRODUCTION

The Internet promises an ever-increasing variety of available services to anyone anywhere. This social and business convenience comes with compromises to privacy. On the Internet, users have few controls, if any, over the privacy of their actions. Each communication leaves trails here or there, and often, someone can follow these trails back to the user. Notably, anonymous communication schemes are purposed to hide users' identity as to personal, source and destination location, and content information. Frankly, anonymous communication on the Internet offers new opportunities but has ill-understood risks.

Two types of anonymity are required for complete anonymization [5]. Data anonymity filters out identifying data, such as the sender field in an e-mail. Connection anonymity obscures the communication patterns. Furthermore, there are four types of connection anonymity. Sender anonymity protects the identity of the initiator. Receiver anonymity protects the identity of the responder. Mutual anonymity [11] provides both sender and receiver anonymity. Unlinkability [15] means that an attacker cannot discern sender-receiver relationships. Even if the identity of one endpoint is compromised, the identity of the other endpoint cannot be linked to it. Thus, people create special networks to protect privacy, especially the identities of the entities participating in a communication via Internet connections. Thus, the main goal of the networks is to provide anonymity for their users. Each network employs some specific anonymous communication schemes (methods) to reach the goal.

## 2. ANONYMOUS COMMUNICATION SCHEMES

Anonymous communication is not new. It has been in use for a while now. As a matter of fact, the idea was first fielded by Chaum [3]. He suggested the transmission of messages through a server which mixes message packets from different users before forwarding them to the destination, thus, concealing the identity, location and destination, and content information between senders and receivers. Consequently, many anonymity schemes have emerged, and have been used rather widely. Network latency has become a major basis to construct de-anonymization schemes for network communication. It has also become a method of comparing network transmission rates. This method has infiltrated into the anonymity scheme market. High latency anonymity schemes deliver messages at long delay rates (such as the Mixmaster and Mixminion), [6]; [13]. With high latency schemes, more bandwidth is used. On the other hand, low latency systems transmit messages at a reasonably short delay rates. Such low latency protocol systems are typified by the popularly used Tor and AN.ON systems, [7]; [16]. The benefits of using low-delay anonymity are that anonymous communications use a variety application services including remote login and web browsing, although this functionality comes at the cost of reduced anonymity guarantees. In particular, most of these services are easily defeated by a global passive adversary using relatively straightforward attacks such as packet counting, [17]. Additionally, using packet counting attacks, an attacker with control of a subset, *S*, of the nodes in the system can trace a subset, *S*, of the connections made to colluding servers and subset, $S^2$of all connections running through the system, [19].

The possibility of using latency data in traffic analysis has been mentioned several times in previous works, apparently originating in 2001, [1]. Since then some studies have investigated the amount of network latency leakage [12]. Of course, to get an upper bound on the amount of information that can be leaked under the current Internet topology, the amount of information about a host that can be gained, given a precise estimate of its RTT to a randomly chosen host, may be measured. For the general Internet, in the above study, the MIT King data set was used [9]. Then for each source host A, we computed the expected number of bits in the RTT to a random destination host B by counting, for each B, the number of hosts C such that the confidence intervals for AB and AC overlapped. Taking this count as $N_B$ and the total number of hosts as N, the information gain for AB was computed as $\log_2$ (N/NB). The cumulative distribution of expected information gain for the two data sets used in the study is shown in Figure 1. For the King data set, the average number of bits from RTT per host is 3.64, the median is 3.8.

T. Sobh (ed.), *Innovations and Advances in Computer Sciences and Engineering*,
DOI 10.1007/978-90-481-3658-2_98, 

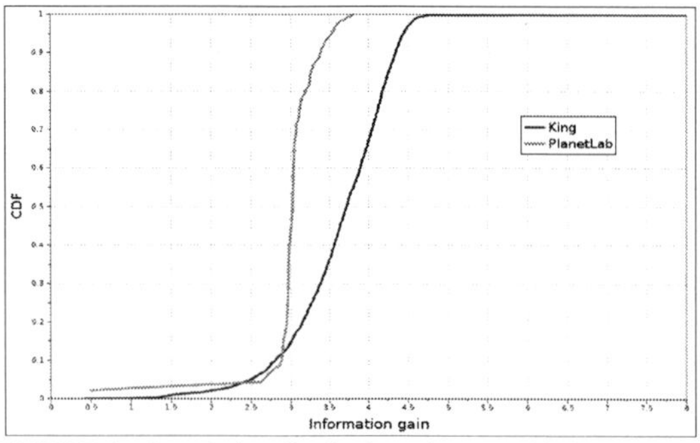


FIGURE 1: Cumulative distribution of expected information gain from RTT per host, for MIT King data set and PlanetLab nodes.

### 2.1 The Multicast Scenario

The multicast approach [2] provides a communication infrastructure that is reasonably resilient against both eavesdropping and traffic analysis. Using this protocol, entities representing applications communicate through a sequence of networked computing nodes, which is referred to as onion routers. Onion routers are generally application layer routers that realize Chaum MIXes. Onion routing connections proceed in three phases: connection setup phase, data transfer phase and connection termination phase. Over the Internet, anonymous systems [10], [18] use application level routing to provide anonymity through a fixed core set of MIXes as in the Onion Routing protocol. Each host keeps a global view of the network topology, and makes anonymous connections through a sequence of MIXes instead of making direct socket connections to other hosts. The relative percentage of malicious nodes and connectivity is shown in figure 2.

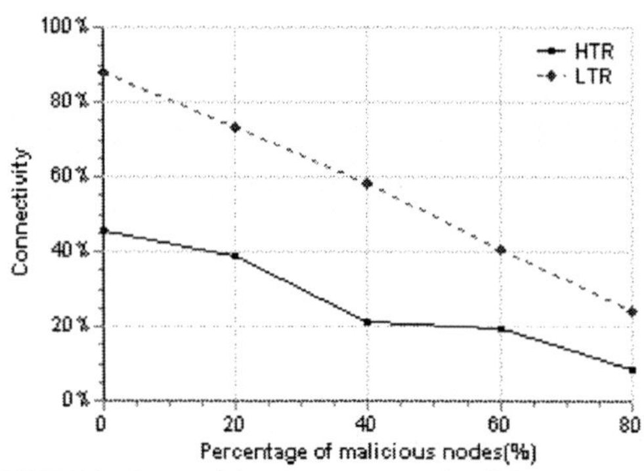


FIGURE 2: Connectivity vs. percentage of malicious nodes

Nonetheless, there are limitations to the capabilities of an attacker, simply because, access to any traffic of interest before it exits the anonymity service and arrives at his malicious servers is very rare. Some studies have investigated into the question: what information, outside of the actual bits of data packets delivered to the adversary, does a low-latency anonymity service leak, and to what extent does this leakage compromise the anonymity offered by the service? [12]). Several recent studies have explored the impact of the local attacker's access to information about the timing of events in a low-latency anonymity scheme, such as packet arrival times, using, for instance, the "circuit clogging" attack version of Murdoch and Danezis, [14], which relies on the observation that a sudden increase in the load of a Tor server will increase the latency of all connections running through it. Indeed, Murdoch and Danezis demonstrated how a corrupt Tor node and web server can exploit this property to determine the nodes in a Tor circuit, (the nodes that forward a given connection through the network).

### 2.2 The purpose of the paper.

The purpose of this paper is to investigate network latency anonymity leaks, and propose practical techniques for their reduction or even elimination. In this direction, we investigate the following technical question: what implementation techniques can be configured to truly reduce anonymity leaks? The emphasis is on deployable systems which provide strong anonymity against a strong attacker model for the Internet. The method used here, is to propose an extension of the popular Tor security strategies and to present a unique configuration of the popular network anonymity techniques (algorithms) for future implementation.

### 2.3 Typical Time-based (Latency) Attack

In such an attack, typically, the corrupt Tor node regularly sends packets on a loop through each Tor server, measuring the time the packets spend in transit. Then when the malicious server wishes to trace a connection, it modulates its throughput in a regular, on/off burst pattern. By correlating the delay at each Tor server against the timing of these burst periods, the attacker learns which nodes are in the circuit. Since the estimated number of Tor users (on the order of 105 as of April 2007) is less than the number of possible circuits (on the order of 108) seeing two connections that use the same circuit nodes is a strong indicator that the connections are from the same user. Thus at a minimum, timing information can leak the linkage between Tor connections.

Hopper, et al, in their paper titled "How Much Anonymity does Network Latency Leak?", made similar observations that typical malicious servers acting as local adversaries can observe the network latency of a connection made over a Tor circuit. They also observed that even in this scenario, if a client attempts to connect to two malicious servers (or make two connections to the same malicious server) using the same circuit, then the server-client RTTs of these connections (minus the RTT from the last node to the server) will be drawn from the same distribution, whereas other clients connecting to the server will have different RTTs. Based on this observation, they developed an attack on Tor that allows two colluding web servers to link connections traversing the same Tor circuit. The attack uses only standard HTTP, the most commonly mentioned Tor application layer, and requires no active probing of the Tor network and has very minimal bandwidth requirements. They tested this attack using several hundred randomly chosen pairs of clients and randomly chosen pairs of servers from the PlanetLab wide area testbed, [3], communicating over the deployed Tor network. Resultantly, they found suggestions that pairs of connections can have an equal error rate of roughly 17%, and the test can be tuned to support a lower false positive or false negative rate. Also, the publicly available MIT King data set, [9], a collection of pair wise RTTs between 1950 Internet

hosts, was analyzed to estimate the average amount of information that is leaked by knowing the RTT between a given host and an unknown host. The investigators found that, on average, knowing the RTT to a host from one known server yields 3.64 bits of information about the host ( and equivalently, it reduces the number of possible hosts from n to $n/2^{3.64} \approx 0.08n$). The expected bits gained per hour relative to connection, is shown in figure 3. Notably, several attack techniques exist. One widely used attack technique, the active client-identification attack, is briefly explained below.

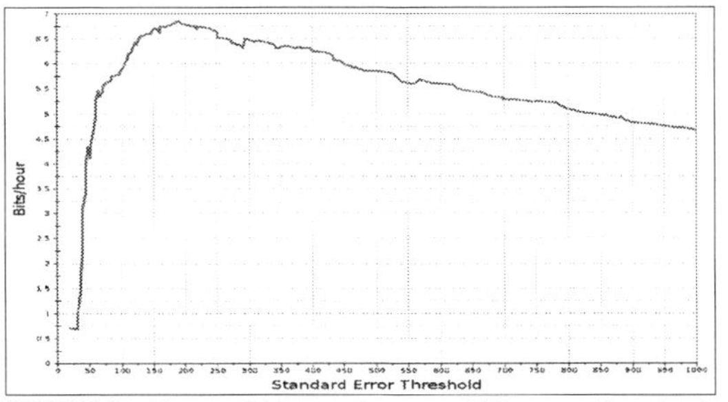


FIGURE 3: Expected bits per hour vs. 100-connection standard error threshold.

### 2.4 An Active Client-identification Attack.

Using standard protocols and taking advantage of repeated visits from a client, a malicious Tor server with access to latency oracle, can estimate the RTT between Tor servers and nodes in the RTT equivalence class of nodes of suspected client's locations. A simulated attack [12] was evaluated using over 200 runs with randomly chosen client/server pairs from the PlanetLab wide area testbed, using randomly chosen circuits among the currently deployed Tor nodes (as of Jan./Feb. 2007). The results suggest that a malicious server with a periodically reloading web page can recover, on average, about 6.8 bits of information about a client's location per hour. Thus a client's RTT equivalence class can be determined in 3 hours, on average. These results have serious implications for the design of low-latency anonymity schemes. In particular, they suggest that, without new ideas for path selection, adding delay to a connection may be unavoidable for security considerations. In turn, this has implications for design decisions: for example, if latency must be uniformly high, then TCP tunneling over such services will provide extremely low bandwidth; or if the latency of circuits can be masked with noise in the short term, then circuit lifetimes may need to be shortened. The Tor circuit constitutes the primary anonymous network system used in this study.

## 3. A BRIEF OVERVIEW OF THE TOR SYSTEM

The Tor system is a low-latency and bandwidth-efficient anonymizing layer for TCP streams. Its growing popularity and the availability of a test-bed deployment have proven to be a fertile ground for research on implementing and attacking low-delay anonymity schemes. Tor system works similarly to a circuit-switched telephone network, where a communication path, or circuit, is first established, over which all communication during a given session takes place. Anonymity is achieved by establishing that circuit through three nodes: an entry node, an intermediary (middleman), and an exit node. Only the entry node knows the identity of the client contacting it, in the form of its IP address. The middleman node knows the identities of both the entry and exit nodes, but not who the client is, or the destination he or she wishes to reach over the circuit. If the Tor server is an "exit" node, which provides a gateway between the Tor network and the Internet, it is responsible for making application-layer connections to hosts on the Internet, and serves as a relay between potentially non-encrypted Internet connections and encrypted Tor traffic. Thus, it knows the destination with whom the client wishes to communicate, but not the identity of the client. In this manner, no single node in the Tor network knows the identities of both communicating parties associated with a given circuit. All communications proceed through this encrypted tunnel.

Circuits are established iteratively by the client, who gets a list of Tor nodes and long-term keys from a directory service, selects a Tor node from that list (preferably one with high uptime and bandwidth), negotiates a communication key, and establishes an encrypted connection. To avoid statistical profiling attacks, by default each Tor client restricts its choice of entry nodes to a persistent set of three randomly chosen "entry guards". The circuit is then extended to additional nodes by tunneling through the established links. Link encryption, using ephemeral Diffie-Hellman key exchange for forward secrecy, is provided by SSL/TLS. To extend the circuit to another Tor node, the client tunnels that request over the newly-formed link. Traffic between Tor nodes is broken up into cells of 512 bytes each. Cells are padded to that size when not enough data is available. All cells from the client use layered (or "onion") encryption, in that if the client wishes for a message to be passed to example.com via Tor nodes A, B, and C (C being the exit node), the client encrypts the message with a key shared with C, then again with a key shared with B, and finally A. The message is then sent over the previously established encrypted tunnel to A (the entry node). A will peel off a layer of encryption, ending up with a message encrypted to B (note that A can not read this message, as A does not have the key shared between the client and B). A then passes on the message to B, who peels off another encryption layer, and passes the message to C. C removes the final encryption layer, ending up with a clear text message to be sent to example.com. Messages can be any communication that would normally take place over TCP. Since there is significant cryptographic overhead (such as Diffie-Hellman key exchange and SSL/TLS handshake) involved with the creation and destruction of a circuit, circuits are reused for multiple TCP streams. However, anonymity can be compromised if the same circuit is used for too long, so Tor avoids using the same circuit for prolonged periods of time, giving circuits a client-imposed maximum lifetime1.

The biggest problem with the Tor network is its vulnerability to timing attacks. If an attacker sees a packet from the user to the first Tor router and shortly afterwards a packet from the last router to the final destination, it is possible to identify the user. This is an inherent issue of low-latency anonymizers and its solution is still an open research problem. Although, it has been suggested before that this information might be a potential avenue of attack [1], it is not known to us that leaking this information had any adverse effect on the anonymity provided by schemes like Tor. An example of a typical attack on a Tor system is briefly described bellow.

### 3.1 Typical Attack Against the Tor System:

When a client, using a timing-based attack, connects to the malicious web server, that server modulates its data transmission back to the client in such a way as to make the traffic pattern easily identifiable by an observer. At least one Tor server controlled by the adversary builds "timing" circuits through each Tor server in the network (around 800 as of January/February 2007[1]). These circuits all have length one, beginning and terminating at the adversarial Tor node. By sending traffic through timing circuits to measure latency, the adversary is able to detect which Tor servers process traffic that exhibits a pattern like that which the attacker web server is generating. Since Tor does not

[1] TOR (the onion router) servers. http://proxy.org/tor.shtml, 2007.

reserve bandwidth for each connection, when one connection through a node is heavily loaded, all others experience an increase in latency. By determining which nodes in the Tor network exhibit the server-generated traffic pattern, the adversary can map the entire Tor circuit used by the client.

Recent studies have suggested that an adversary may de-anonymize any stream for which that adversary controls the entry and exit nodes [19]. The probability of this occurrence in the short term (transient client connections) is $c(c - 1)/r^2$, where c is the maximum number of nodes corruptible by the adversary in a fixed period of time, and r is the number of available Tor routers in the network. An adversary can determine if he or she controls the entry and exit node for the same stream by using a number of methods mentioned below, including fingerprinting and packet counting attacks. Indeed, it is expected that single-hop proxy services will leak more information about the client-proxy RTT, allowing fairly precise linking attacks, although the strength of the client location attack will be somewhat diminished against services that have a single proxy server location.

In summary, most existing literature on the topic focuses mainly on types of attacks and available individualized strategies for overcoming them. Notably, individualized solutions and troubleshooting leave a lot to be desired in today's ubiquitous and multi-platform communication applications. A holistic approach to network communication anonymity is critical.

### 3.2 Limitations of Existing Studies

1) The most serious of these limitations is the insufficiency of data on conditional information gain, that is, we cannot conclusively evaluate, from our data, how much additional information each run of an attack provides. This is due in part to limitations of an experimental method, which did not re-use clients; thus a "longitudinal" study may be needed to more accurately assess conditional information gain.
2) Another limitation is that the client location attack assumes that a user repeatedly accesses a server from the same network location. This assumption may sometimes be invalid in the short term due to route instability, or in the long term due to host mobility. It seems plausible that the attack can still be conducted when circuits originate from a small set of network locations, such as a user's home and office networks, but the attack would be of little use in case of more frequent changes in network location. Thus, the question remains: Will replication of the experiments, using less widely-supported tools, such as persistent HTTP over Tor, produce the same results? Even the authors of a paper titled "How Much Anonymity does Network Latency Leak?" [12], themselves acknowledged that, of course, the answer is highly dependent on both the network topology (latency in a star topology would leak no information about a host's location) and the protocol in question. This is because it is conceivable that if so much noise is added to the network latency, the signal can be undetectable. Furthermore, there is room for evaluation of alternative methods of implementing RTT oracles, and perhaps for a more sophisticated testing procedure that avoids the expense of querying the RTT oracle for every pair of Tor entry node and candidate location.

## 4. LEAKAGE REDUCTION TECHNIQUES

A number of techniques and best practices which can reduce the attacker's probability of success in client location attack have been suggested [12]. Four such combinations of configuration techniques are described below.

### 4.1 The utility of onion routers in the Tor System.

The use of onion routers in the Tor system can minimize the success probability of the Murdoch-Danezis attack by allocating a fixed amount of bandwidth to each circuit, independent of the current number of circuits, and doing "busy work" during idle time. This may undesirably compromise efficiency but certainly will hinder the success of client location attack. Additionally, by configuring Tor nodes to refuse to extend circuits to nodes which are not listed in the directory, their use of RTT oracles will be prevented. They can also drop ICMP ECHO REQUEST packets in order to raise the cost of estimating their network coordinates. Additionally, a DNS-based administrative disabling of Tor recursive lookups from "outside" will limit, if not preclude, the success of this attack. Although the security implications are not clear, making the Tor path selection algorithm latency-aware, by incorporating some notion of network coordinates into directory listings, thus, clients could construct circuits having an RTT close to one of a small number of possibilities, will reduce the high average circuit RTTs (of 5 sec[2]), reduce the effectiveness of latency-based attacks, and allow clients to explicitly trade-off some anonymity for better efficiency.

### 4.2 The Administrative Ingenuity Approach.

Tor administrative drop of ping packets and denial of other attempts to learn their network coordinates to accuracy, and the addition of sufficient delays (of forwarding data at the client) to make the RTT and timing characteristics servers independent of the underlying network topology, will hinder success probability. Furthermore, given the limited time period over which a Tor circuit is available for sampling, the introduction of high variance random delays in outgoing cells and selecting delays from an identical distribution at each Tor node would also make the timing distributions from different circuits look more alike, thus thwarting the circuit-linking attack. Of course, if the only way to thwart attacks based on latency and throughput is to add latency and restrict throughput, this would have serious implications for the design of low-latency anonymity systems and the quality of anonymity we can expect from such schemes

### 4.3 The Multicast Technique to Network Anonymity.

With the multicast technique, the source node is disallowed from gathering and storing information about the network topology, [2]. Instead, the source node initiates a path establishment process by broadcasting a *path discovery* message with some trust requirements to all of neighboring nodes. Intermediate nodes satisfying these trust requirements insert their identification (IDs) and a session key into the *path discovery* message and forward copies of this message to their selected neighbors until the message gets to its destination. The intermediate nodes encrypt this information before adding it to the message. Once the receiver node receives the message, it retrieves from the message the information about all intermediate nodes, encapsulates this information in a multilayered message, and sends it along a reverse path in the dissemination tree back to the source node. Each intermediate node along the reverse path removes one encrypted layer from the message, and forwards the message to its ancestor node until the message reaches the source node. When

2 Observed in the Hopper, et al study titled "How Much Anonymity does Network Latency Leak?", Communications of the ACM, v.24 n.2 (2007), p.84-90.

the protocol terminates, the source node ends-up with information about all the trusted intermediate nodes on the discovered route as well as the session keys to encrypt the data transmitted through each of these nodes. The multicast mechanism and the layered encryption used in the protocol ensure the anonymity of the sender and receiver nodes.

The *path discovery* phase allows a source node *S* that wants to communicate securely and privately with node *R* to discover and establish a routing path through a number of intermediate wireless nodes. An important characteristic of this phase is that none of the intermediate nodes that participated in the *path discovery* phase can discover the identity of the sending node *S* and the receiving node *R*. The source node *S* triggers the *path discovery* phase by sending a *path discovery* message to all nodes within its transmission range. The *path discovery* message has five parts. The first part is the open part. It consists of message type, *TYPE*, trust requirement, *TRUST_REQ*, and a one-time public key, *TPK*. The trust requirement indicated by *TRUST_REQ* could be *HIGH*, *MEDIUM* or *LOW*. *TPK* is generated for each *path discovery* session and used by each intermediate node to encrypt routing information appended to the *path discovery* message. The second part contains the identifier IDR of the intended receiver, the symmetric key *KS* generated by the source node and *PLS* the length of the third part, *padding*, all encrypted with the public key *PKR* of the receiver.

### 4.4 The two-pronged Approach to Attack Detection.

Venkatraman and Agrawal [21] proposed an approach for enhancing the security of AODV protocol based on public key cryptography. In this approach, two systems, EAPS (External Attack Prevention System) and IADCS (Internal Attack Detection and Correction System) were introduced. EAPS works under the assumption of having mutual trust among network nodes while IADC runs by having the mutual suspicion between the network nodes. Every route request message carries its own digest encrypted with the sender's private key hash result in order to ensure its integrity. To validate established routes, route replies are authenticated between two neighbors along them. This approach, using the Onion Routing approach and trust management system to provide trust and anonymity for the path discovery (and hence for subsequent communications using this path), prevents external attacks.

## 5. CONCLUSION

Notably, some of these anonymity loss reduction configurations and techniques are mostly hardware-based. Some more, such as the multicast strategy, are mainly software-based, and yet others, such as the dropping ping request, are mostly implemented administratively. The design considerations of these techniques will, at a minimum, certainly assure the reduction of anonymity losses. Given that the signature or pattern of these attacks is not clearly known or constant, the incorporation of any combination of the proposed techniques will, no doubt, preclude the success of existing patterns of attack.

The variety of the proposed techniques spans network communications (whether single-cast or multicast). Although, there is a detailed exploration of Tor systems for their popularity in security assurance, the techniques and methods are equally applicable to other systems. Expectedly, if and where a security breach occurs, it is immediately detected and corrected using the two-pronged approach to attack detection and correction described in the paper

## 6. REFERENCES


[1] Back, A., Moeller, U., and Stiglic, A. Traffic analysis attacks and trade-offs in anonymity providing systems. In Proc. Information Hiding Workshop (IH 2001) (April 2001), LNCS 2137, pp. 245–257

[2] Boukerche, A., El-Khatib, K., Xu, L., and Korba, L. A Novel Solution for Achieving Anonymity in Wireless Ad Hoc Networks. National Research Council of Canada and Institute for Information Technology. ACM PE-WASUN'2004, held in conjunction with the 7th ACM International Symposium on Modeling, Analysis and Simulation of Wireless and Mobile Systems. Venice, Italy. October 4-6, 2004. NRC 47402.

[3] Chaum, D. L. Untraceable electronic mail, return addresses, and digital pseudonyms. Communications of the ACM 24, 2 (1981), 84–88.

[4] Chun, B., Culler, D., Roscoe, T., Bavier, A., Peterson, L., Wawrzoniak, M., and Bowman, M. Planetlab: an overlay testbed for broad-coverage services. SIGCOMM Comput. Commun. Rev. 33, 3 (2003), 3–12.

[5] Claessens, J. Preneel, B. and Vandewalle, J. Solutions for Anonymous Communication on the Internet. In *Proceedings of the International Carnahan Conference on Security Technology*, pages 298.303. IEEE, 1999.

[6] Danezis, G., Dingledine, R., and Mathewson, N. Mixminion: Design of a Type III Anonymous Remailer Protocol. In SP '03: Proc. 2003 IEEE Symposium on Security and Privacy (Washington, DC, USA, 2003), IEEE Computer Society, p. 2.

[7] Dingledine, R., Mathewson, N., and Syverson, P. F. Tor: The second-generation onion router. In Proc. 13th USENIX Security Symposium (August 2004).

[8] Federrath, H., et al. JAP: Java anonymous proxy. http://anon.inf.tu-dresden.de/.

[9] Gil, T. M., Kaashoek, F., Li, J., Morris, R., and Stribling, J. The "King" data set. http://pdos.csail.mit.edu/p2psim/kingdata/, 2005.

[10] Goldberg, I., and Shostack, A. Freedom network 1.0 architecture, November 1999.

[11] Guan, F, Fu, X. Bettati, R. and Zhoa, M. An Optimal Strategy for Anonymous Communication Protocols. In *Proceedings of 22nd International Conference on Distributed Computing Systems*, pages 257.266. IEEE, 2002.

[12] Hopper, N., Vasserman, E. Y., Chan-Tin, E. How Much Anonymity does Network Latency Leak? Communications of the ACM, v.24 n.2 (2007), p.84-90.

[13] Moeller, U., Cottrell, L., Palfrader, P., and Sassaman, L. IETF draft: Mixmaster protocol version 2. http://www.ietf.org/internet-drafts/draft-sassaman-mixmaster-03.txt, 2005.

[14] Murdoch, S. J., and Danezis, G. Low-Cost Traffic Analysis of Tor. In SP '05: Proc. 2005 IEEE Symposium on Security and Privacy (Washington, DC, USA, 2005), IEEE Computer Society, pp. 183–195.

[15] Pfitzmann, A, and Waidner, M. Networks without User Observability. *Computers & Security*, 2(6):158.166, 1987.

[16] Reiter, M. K., and Rubin, A. D. Crowds: anonymity for web transactions. ACM Transactions on Information and System Security 1, 1 (1998), 66–92.

[17] Serjantov, A., and Sewell, P. Passive attack analysis for connection-based anonymity systems. In Proc. ESORICS 2003 (October 2003).

[18] Syverson, P. F., Goldschlag, D. M., and Reed, M. G. Anonymous connections and onion routing. In *Proceedings of the IEEE Symposium on Security and Privacy* (Oakland, California, May1997), 44–54.

[19] Syverson, P., Tsudik, G., Reed, M., and Landwehr, C. Towards an analysis of onion routing security. In Designing Privacy Enhancing Technologies: Proc. Workshop on Design Issues in Anonymity and Unobservability (July 2000), H. Federrath, Ed., Springer-Verlag, LNCS 2009, pp. 96–114.

[20] TOR (the onion router) servers. http://proxy.org/tor.shtml, 2007.

[21] Venkatraman, L., Agrawal, D.P. Strategies for enhancing routing security in protocols for mobile ad hoc networks, in *Journal of Parallel and Distributed Computing, 63.2* (February 2003), Special issue on Routing in mobile and wireless ad hoc networks, Pages: 214 – 227, Year of Publication: 2003, ISSN:0743-7315

# Enterprise 2.0 Collaboration for Collective Knowledge and Intelligence Applications

R. William Maule
Research Associate Professor
Information Sciences Department
Naval Postgraduate School
Monterey, CA 93943 USA

Shelley P. Gallup
Research Associate Professor
Information Sciences Department
Naval Postgraduate School
Monterey, CA 93943 USA

***Abstract*** - This paper describes an enterprise system implemented to support ad-hoc collaborative processes in a large-scale Department of Defense experiment that integrated civil authorities, coalition nations, and non-governmental agencies in a mock international crisis. The deployed system was a commercial service-oriented architecture that provided XML-based web services that supported user-initiated and self-managed collaborative functions, and virtual workspaces within which participants could create and secure communities of interest. Assessed is the degree to which cooperative tools and processes can be specified given the inevitability of ad-hoc information processes and emergent communications.

## I. Introduction

The U.S. Department of Defense (DoD) assists non-governmental organizations (NGOs), civil authorities, and coalition partners in emergencies and natural disasters. Humanitarian efforts in an emergency require that information systems be readily accessible and support ad-hoc collaboration. In this context the 2007 Trident Warrior experiment tested a suite of web services to provide ad-hoc collaborative support in simulated national and international crisis. The experiments took place in early 2007 with participants from United States, Canada, Australia, the United Kingdom, and NATO. Included were military, non-governmental organizations, and civilian agencies— police, fire, and hospitals.

Previous research by the project team has focused on the development of knowledge systems to support inter-agency and multinational collaboration. A more recent focus is the development of open virtual platforms to allow users to self-organize and self-administer workspaces and communities of interest (COI) for real-time computer supported cooperative work.

This paper extends previous research through comparison of a simulated, laboratory based enactment of an international crisis and disaster relief effort versus actual processes in a global experiment, specifically focusing on collaborative web services utilization. Assessed are processes for self-organization in SOA-based collaborative services to help ascertain tool selection for given tasks in specific environmental context. The user-interfaces of the tools and an example workflow are presented.

## II. Security and Context

Global disasters require a paradigm shift in national policies and unprecedented cooperation between nations. Predefined treaties and agreements, humanitarian assistance and disaster relief (HA/DR) operations, and other international crisis have changed the nature of cooperation. Partnerships between the United States and other countries are increasingly defined through strategic relations or ad-hoc cooperative agreements to support very near term needs.

The basic principle in effect is one of requisite variety. That is, the variety of potential circumstances in the world that will require collaboration with other nations is much greater than pre-defined options. Plans and systems must be put into place ahead of time to meet notional circumstances. A capacity for ad-hoc configuration of capabilities, fit to the circumstances requiring action, will be much more effective and ensure that multiple contingencies can be met simultaneously. The overall impact is a more dynamic and effective international response.

Defining the collaborative relationships between nations based upon emergent and ever changing requirements is a necessary but difficult problem for information systems providers. The assumption used by operational planners and authors of the configurations in place may not have initially included an understanding of implications inherent in emergent needs.

The United States Navy has embraced the concept of a Maritime Operations Center (MOC) that will have a "reach-back" capability for information fusion and planning. As these MOCs reach initial operational capability they will integrate with MOCs of other nations, extending the operational reach necessary in global relief or emergency operations.

How can participants reach across security boundaries between nations, and between military and civilian operations, in a way that preserves security while still allowing acceptable interaction? This has historically been a challenge in large-scale disasters or emergency relief.

T. Sobh (ed.), *Innovations and Advances in Computer Sciences and Engineering*,
DOI 10.1007/978-90-481-3658-2_99, 

Architecting information and knowledge systems, in consonance with security, environmental diversity, and national response paradigms, has a multi-dimensional challenge. Emerging capabilities in Enterprise 2.0 technologies—herein defined as Web 2.0 but with enhanced security—can enable user-defined collaborative/social spaces and contexts. This paper provides technical and operational analysis of an effort to provide potential solutions for such an implementation based on a large-scale experiment.

## III. SOA and Collaborative Services

Enterprise architectures have historically employed "system-of-systems" methodology to integrate new and legacy operations [1]. Service-Oriented Architecture (SOA) extends this paradigm through intricate weaves of XML-based web services with operations carefully orchestrated [2,3]. The result is an exponential compounding of capabilities—in the instance of the study herein for collaboration—but with a simultaneous increase in complexity for developers. International and military-civilian systems further complicate development through "need-to-know" and security classifications.

Research has addressed evolutionary design in the acquisition of complex adaptive systems and integration issues in information intensive operations [4]. Relatively new is the development of technologies sufficient to support collaboration for integrated group processes within user-defined virtual spaces—with military-grade security for those spaces [5]. Also new is a focus on capacities for "hastily formed networks" to address disasters which involve international and non-governmental coalitions [6].

This paper extends the above research by providing a means for users within a SOA-based collaborative environment to dynamically provision content, access, and collaborative capabilities. Due to rapidly changing environmental contexts in a natural disaster this must also include user defined and managed security. Thus, a requirement is that virtual spaces be sufficient to support evolving communities of interest with dynamic need-to-know classification. Yet, the system must be flexible enough to be used on an ad-hoc basis by civilian and non-government users, with very little if any training.

### A. *Cooperative Knowledge Domains*

Emergencies tend to produce large quantities of data and information, highly related in context, but difficult and time-consuming to dynamically process into actionable knowledge. Assignment of security to specific data sets or contexts is problematic. There is a need to allow users to self-organize and self-manage both knowledge and security, with the expectation that collective action on that knowledge will achieve stipulated objectives.

Within the Navy experimentation community the FORCEnet Innovation & Research Enterprise (F.I.R.E.) system supports analysis through a portal, rapid application development environment, and knowledge capability realized through various search and AI agents [7]. Tactical Applications for Collaboration in F.I.R.E. (TACFIRE) provides a suite of XML-based collaborative web services atop a SOA [8] The web services support collaboration, remote data acquisition, and high-level metadata sufficient for RSS and RDF publish/subscribe services for user-defined content [9]. Within such context lies great potential for knowledge-driven collaboration for emergency assistance operations.

User-generated content schemes, and user-defined security within virtual workspaces, can be achieved through directories that inherit security with content categorization such that key words and metadata are automatically assigned by virtue of their placement [10]. Full-text categorization can provide metadata that enables end-user search results to inherit security parameters.

Within the discussed system the workspaces include email, list services, threaded discussions in RDF format with RSS feeds, shared libraries with provisions for layered security, a task manager, shared calendar, and real-time conferencing with streaming media and VoIP. Most importantly, the generation of these capabilities, including the virtual workspaces that host the collaborative processes, are user-generated and managed without administrator involvement. Everything is searchable, with query results from across the various media repositories federated, the results organized by tab, with the output embedded into portlets.

## IV. Virtual Disaster Relief

In preparation for the experiment, and to help determine tool utilization, a class of 12 students utilized TACFIRE to support collaboration and cooperative work. The portal is presented in Fig. 1.

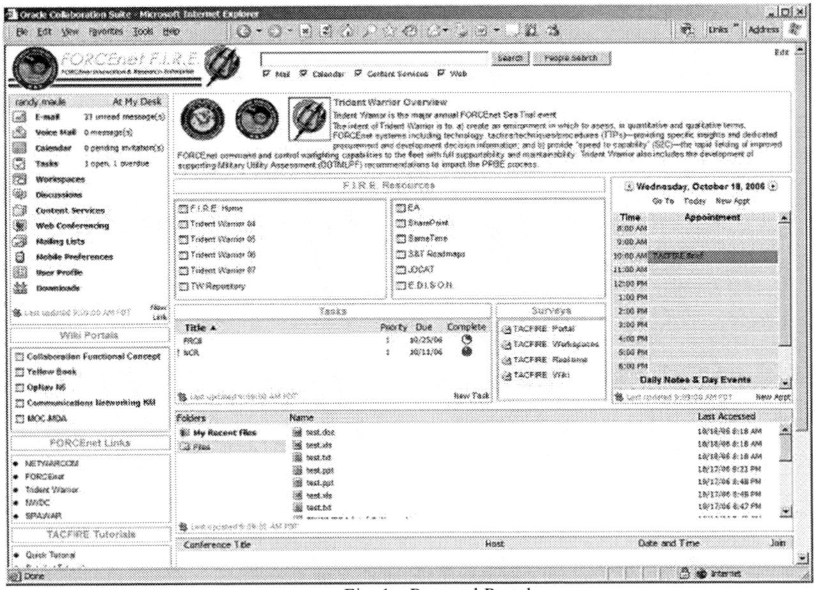


Fig. 1. Personal Portal.

Each student's personalized portal aggregated a suite of web services (Fig. 2). This included links into those services and portlets with windows into each user's personal representation of each service. Search was federated across the services, and rich media clients were available for mail and content management.

Streaming services for real-time collaboration provided instant messaging, chat, VoIP and web conferencing with shared desktops. Presence was achieved through an enhanced Jabber XMPP implementation as a separate client automatically launched on system boot. Workspaces provided the COI capability and included all of the previously discussed services plus workflow automation specific to a group or task.

A listing of the cumulative capabilities is available in Fig. 3. Of special interest in service utilization were user-selectable options for message triggering between components. Available options for email notifications included from the calendar to and from meetings and conferences; from the Enterprise Content Management (ECM) library to email and list services; from the discussion forums with email-based postings and replies and RSS output from threaded discussions; and various options for messages and email triggers from the presence server and among the real-time services.

The simulated context was a disaster relief operation that involved coordination of military, coalition, civilian, and non-government organizations (NGOs). Context closely followed a live experiment the previous year in anticipation of a similar enactment in an upcoming experiment. Student development included establishing security and management controls, and workflows contingent on completion of designated tasks.

Secure virtual workspaces were deployed, configured and managed by the students without any administrative involvement. Key to implementation of these workspaces, and their supporting web services, were workflows wherein tools were optimized for specific tasks.

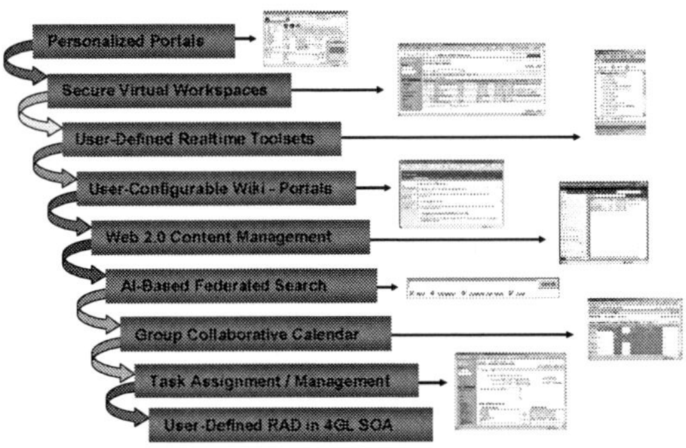


Fig. 2. High-level services.

Of special interest to the researchers was tool selection for given tasks. Students were Naval or Marine officers with experience in both collaboration and computer-supported cooperative work. New was the integration of previously separate systems into a comprehensive suite accessible through a personalized portal. Since the services were XML and integrated the ability to message and form relationships across the tools and content was new, as was the use of a workflow automation system to drive usage of the tools.

**Personalized Portal**

*Integrated Suite of Services*

- *Enterprise Content Management*
- *Mail*
- *Calendar*
- *Workspaces*
- *Discussion Threads*
- *Tasks*

*Federated Search*

*Rich Media Clients (AJAX)*

- *Mail*
- *Enterprise Content Management*

*Desktop Integration*

- *Outlook*
- *Web DAV*

**Real-Time Collaboration**

*Streaming Media*

- *Peer VoIP*
- *Chat*
- *Instant Messenger*

*Web Conferences*

- *Shared Desktops*

*Presence Server*

**Content Management**

*AI Themes and Summarizations*

*Federated Search*

*Edit-in-Place*

- *Drag-and-Drop*

*Policy-based versioning*

- *Check-in/Check-out*
- *Revision History*

**Virtual Private Database**

*Virtual Workspace*

- *Calendar*
- *Shared Files*
- *Workflow Automation*
- *Email*

*Threaded Discussion*

*RSS Feeds - XML Publish/Subscribe*

**Free-Form: Wiki-Portals**

*User-managed content*

- *User-defined relationships and security*
- *User-specified linkages*

Fig.3. Cumulative web services.

The scenario featured Navy, NGO, Homeland Security (HLS), coalition partner nations, and state, county, and local agencies (hospitals, police, etc.).

## V. Experiment and Context

Major facets of the simulation were enacted in a large-scale experiment in the spring of 2007 which involved defense forces of the United States, coalition nations, NATO, NGOs, and civilian agencies in the greater Washington DC metropolitan area.

As in the simulation, workspaces provided high-security, virtual COI (Community of Interest) capabilities, each with web services that provided communication, collaborative support and cooperative processing. Priority services included email and list services specific to the COI, threaded discussions, libraries, announcements, workflow automation within the COI, and federated search across the workspace services and content areas.

Each user was provided a personalized portal, which acted as a virtual server that aggregated web services into a single interface. The portal aggregated services as portlets, with appropriate routing and security (i.e., personal email, content areas, calendar, etc.). Within a personalized portal a user would access secure virtual workspaces to support communities of interest, with web conferences, instant messengers, email, chat, shared libraries, threaded discussion forums, and personal and group calendars.

A notification portlet provided links to new content across the workspace, including library content, discussions/RSS feeds, email and IM, calendar events, and archives from virtual real-time meetings and streamed media. Announcements and personalization features were unique to each user, such as tasks and workflow items.

Participants were instructed to not utilize broadcast VoIP since the architecture was not designed to support this service; nonetheless, broadcast VoIP was utilized and provided additional data points. Peer-peer VoIP through the instant messenger was of high quality on local networks and of fair quality between the U.S. and Europe. Across Europe the quality was said to be good but not directly measured.

Users received very little if any training. While online training materials were available this situation was not ideal since the suite of tools was quite extensive. Still, the focus of the study was ad-hoc processes for highly dynamic and rapidly evolving situations so the impact of a comprehensive and extremely evolved suite of collaborative web services with personalization was a test variable of significant interest.

Results indicated that it is feasible to implement such architecture, although tool optimization will be less than could be achieved with adequate user training. Survey results indicated favorable impressions toward the personalized portals, and their web services, and users could see a definite operational advantage to the aggregation of collaborative web services into a portal personalized for each user.

*A. Expectations and Results*

There were 40 active workspaces during the experiment, far more than was anticipated during the simulation. Each workspace/COI was initiated and administered by participants without administrative intervention. Users successfully built, secured, and deployed workspaces as needed. There were 26 active workspaces specific to distributed command operations, also far in excess of anticipated need. Several hundred users were active participants in the experiment and users of the services.

The ECM utilized a rich media interface that allowed management of all content in all workspace libraries for which a user was approved from a single interface. Participants did not fully utilize the ECM capability and this is an area that likely requires some training and may not be optimized in ad-hoc situations. Similarly, participants did not generally use a WebDAV capability that enabled files and folders to be processed via 'drag-and-drop' from personal desktops into and out of workspace libraries—a very powerful capability.

80% of survey respondents agreed that workspaces would be useful for communicating in an operational environment. 73% of respondents agreed that workspace COI improved work efficiency compared to current technologies.

| Web Conference - Detailed Metrics | Count |
|---|---|
| Web Conferences | 55 |
| Web Conference Minutes | 3779 |
| Average Conference Length(min) | 69 |
| Largest Conference (users) | 23 |
| Voice Streaming Conferences | 21 |
| PSTN | 0 |
| Computer Mic | 21 |
| Voice Streaming Minutes | 1197 |
| PSTN | 0 |
| Computer Mic | 1197 |
| Enrollment Conferences | 2 |
| Document Presentation Conferences | 3 |
| Desktop Sharing Conferences | 55 |
| Whiteboarding Conferences | 3 |
| **User Statistics - Key Metrics** | **Count** |
| Users in Web Conferences | 259 |
| Average Users per Conference | 5 |
| User Minutes | 11118 |
| Voice Users | 661 |
| Active Registered Users | 75 |
| Guest Users | 30 |

Fig. 4. Web conference utilization.

Over 200 hours were spent by command participants in web conferences (Fig. 4), with a total of 60 web conferences launched over a 'snapshot' week of the experiment by leadership from the workspace COIs, all without administrative intervention. There were a total of 719 cumulative attendances in these conferences, with the average conference lasting nearly 4 hours—a strong indicator of the robustness of the technology, and also far in excess of anticipated utilization (Fig. 5).

Civilian operations utilized fewer web conferences but the sessions lasted much longer. A total of only 5 web conferences were initiated for a cumulative 2360 minutes (39 hours 33 minutes). These conferences had a cumulative 267 user-sessions. The longest web conference was 1352 minutes (22 hours 53 minutes) and the shortest was 42 minutes. The highest number of attendees for any single web conference was 100 while the lowest was 5.

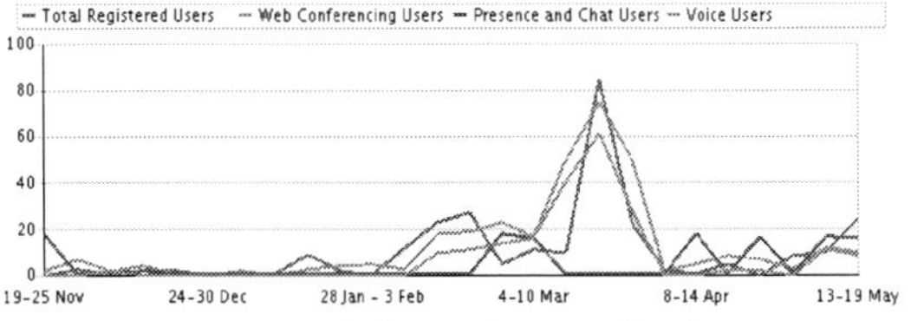


Fig. 5. Cumulative conference utilization.

The extent of utilization of the web conference and streaming media features was not anticipated. As stated earlier, the architecture was not optimized for streaming or real-time media. The heavy utilization is an indicator that when streaming media is available it will be used, whether directed to or not. The focus of the conferences was planning and operational assessment on the command side, and real-time process coordination on the civilian side. So, the focus and implementation for streaming media differed between the DoD and non-DoD groups.

95% of survey respondents agreed that the web conference tool was easy to use during real-time collaboration. 81% of respondents agreed that real-time tools could be used within an operational environment to collaborate in a timely fashion. 84% of respondents agreed that the chat tool was easy to use during real-time collaboration—despite network issues.

Network problems were prevalent during the experiment. Proper resolution would require additional servers at major centers of participation, particularly to better support real-time and streaming capabilities. Network issues were most obvious when utilizing broadcast VoIP over an extended period of time and when network fluctuations caused words to be lost in sentences.

For the one week 'snapshot' period, broadcast VoIP was utilized for 20 hours (1197 minutes)—which is substantial usage given the inevitable network issues across such a distance (United States, Atlantic Ocean, Europe) and without the benefit of a local application switch. This included both web conference and Instant Messenger (IM) VoIP.

There were 21 peer-peer VoIP sessions for a total of 192 minutes. A total of 2559 instant messages were passed during the week. 90% of survey respondents agreed that the Instant Messenger was easy to use during real-time collaboration.

Presence (ability to see who is online and immediately access that user via a 'click') was achieved through the instant messenger, chat, and web conference tools. 81% of respondents agreed that Presence was easy to use during real-time collaboration. Presence was activated automatically when the computer is turned on versus a browser-based access.

Hybrid wiki-portals were utilized to support command, civilian, and integrated operations. 64% of respondents who used the wiki-portals indicated they strongly agreed or agreed that they were easy to use. 69% found that a "YellowBook" wiki-portal developed for expertise location was easy to navigate, and 75% noted that the "YellowBook" was an easy way to find information about roles and responsibilities. 90% of respondents agreed that the wiki-portals provided accessible functionality for knowledge management.

## VI. Conclusion

Experimentation results indicate a successful utilization of XML web services atop a SOA to provide personalized portals and virtual workspaces with a comprehensive suite of collaborative capabilities. The application was DoD, coalition, NGO, and integrated civilian operations for ad-hoc utilization during a national or international disaster or emergency.

Streaming media was utilized far more than expected and a far more robust and distributed architecture would need to be employed for true operational implementation of such a comprehensive suite of XML web services. More training would be needed to optimize utilization but this was not the focus of this experiment which instead stressed ad-hoc utilization by non-trained or minimally trained personnel.

A core function was the workflow automation feature, which was not even attempted and this would be an area where training would definitely be required.

Complicating factors included: (1) logistics as various users and groups were dynamically merged into new communities; (2) technical as communications were extended from chat and portals into the personalized portals and users were first exposed to the full suite of collaborative web services (which can be intimidating); and (3) socio-technical as tools competed for utilization, with different participants favouring different tools in parallel situations.

Socio-technical factors in tool selection indicates a need for an overarching concept of operations or stated guidelines for tool optimization—which may be difficult to achieve in true ad-hoc emergency situations.

## References


1. Jamshidi, M. System-of-Systems Engineering - A Definition. *IEEE International Conference on Systems, Man, and Cybernetics*, (2005), URL: http://ieeesmc2005.unm.edu/SoSE_Defn.htm.
2. Barry, D. *Web Services and Service-Oriented Architectures*. Morgan Kaufmann, San Francisco (2003), USA.
3. Erl, T., 2005. *Service-Oriented Architecture: Concepts, Technology, and Design*. Prentice Hall, New York, USA.
4. Sage, A., and Cuppan, C. On the Systems Engineering and Management of Systems of Systems and Federations of Systems. *Information, Knowledge, Systems Management*, Vol. 2, No. 4 (2001), pp. 325-345.
5. Maule, R., and Gallup, S. TACFIRE: Enterprise Knowledge in Service-Oriented Architecture. *Proceedings of the IEEE/AFCEA 2006 Military Communications Conference (MILCOM 2006)*, ISBN 1-4244-0618-8, 23-25 October (2006), Washington, DC, USA.
6. Denning, P., and Hayes-Roth, R. Decision-Making in Very Large Networks. *Communications of the ACM*, Vol. 49, No. 11 (2006), pp. 19-23.
7. Maule, R., and Gallup, S. FORCEnet Innovation & Research Enterprise (F.I.R.E.) Architecture for Computer Supported Cooperative Work. *Proceedings of the International Conference on Software Engineering Theory and Practice (SETP-2007),* 9-12 July (2007), Orlando, Florida, USA.
8. Maule, R., and Gallup, S. Knowledge Management System with Enterprise Ontology for Military Experimentation – Case Study: F.I.R.E. and TACFIRE. *Proceedings of the International Conference on Enterprise Information Systems & Web Technologies (EISWT-2007),* 9-12 July (2007), Orlando, Florida, USA.
9. Maule, R., and Gallup, S. Fine-Grain Security for Military-Civilian-Coalition Operations Through Virtual Private Workspace Technology. *Proceedings of the 2nd International Conference on Information Warfare & Security,* 8-9 March (2007), Monterey, CA, USA.
10. Maule, R. Enterprise Knowledge Security Architecture for Military Experimentation. *Proceedings of the IEEE SMC 2005 International Conference on Systems, Man and Cybernetics,* 10-12 October (2005), Waikoloa, Hawaii.

# Knowledge Engineering Experimentation Management System for Collaboration


R. William Maule
Information Sciences Department
Naval Postgraduate School
Monterey, CA, 93943
1-831-915-2430
rwmaule@nps.edu

Shelley P. Gallup
Information Sciences Department
Naval Postgraduate School
Monterey, CA, 93943
1-831-656-1040
spgallup@nps.edu

Jack J. Jensen
Information Sciences Department
Naval Postgraduate School
Monterey, CA, 93943
1-831-656-2297
jjjensen@nps.edu



***Abstract*** - This paper presents a methodology and system for the analysis of complex experiments currently utilized by the Department of Defense. Assessed are effectiveness and efficiency of systems interactions and affected intra- and inter-organizational processes. Methodology is derived from an operational experimentation analytics and management system designed to assess next-generation military e-services and infrastructure. A focus is the expansion of systems methodology beyond traditional transaction-oriented exchanges to include management layers for complex inter-organizational processes and mechanisms to accommodate sensitive company operations.


## I. Introduction

A broad range of new technologies are being introduced to the military from the private sector and these are tested through large-scale and limited objective experiments. An enterprise information system (EIS) to manage the innovation process from introduction through experimentation to assessment was developed and is presented. Included are models for the analysis and experimentation management processes and illustrated examples of the implemented procedures as realized through the EIS. Objectives, management structures, experimentation and analysis processes, and final technology diffusion are presented as management forms and reports in the EIS.

The system is highly collaborative with security designed to support cooperative work. Management leads are responsible for information input, editing, and output such that day to day operations of the content, from innovation introduction to recommendations and diffusion, are distributed globally. Only a small team manages the database, information structure, and overall operations.

The system stresses not only engineering assessment of the base technology but also interfaces among the technologies in the "systems of systems" architecture common to the DoD. Of equal importance are the organizational structure(s) of the submitting organization and the management interface of personnel from very different specializations into a globally distributed team. This is managed through the EIS and its cooperative processes for distributed work.

Methodology establishes variables for analysis that include technical assessment of innovation capabilities along with frameworks to position the technologies within the strategic operational objectives that underlie the experiments.

## II. Management Architecture

Military systems and dynamics occur in an extremely complex environment and it is essential to accurately assess the state of operational capabilities, e.g. the associated operational activities, the ability of the process to deliver the performance expected with the people assigned, other systems needed for essential support activities, and the guidance that directs activities and interactions between all elements. Some perspective from previous studies can aid in the basic issues of content categorization for knowledge management.

### *A. Content Management*

Categorization processes can be formalized through ontology specific to the experimental context. At the highest level ontology implies theory about objects, properties of objects, and relations between objects within a specified domain of knowledge, including metadata for describing that domain [1], [2], [3]. Content can then be embedded with meaning specific to the experimental technologies. Hopefully, the architectural models can describe this meaning, and relationships between meanings can be mapped with the result presented as domain knowledge to aid analysis [4], [5], [6], [7], [8].

In application, XML can provide syntax and structure, ontology a means to define terms and relationships, and RDF (Resource Definition Framework) a method to encode ontology-defined meaning [2]. There is a basis for such practices in the DoD. Ontology has been advanced by DARPA with their DAML (DARPA Agent Markup Language), and NATO with their LC2IEDM (Land C2 Information Exchange Data Model) and JC3IEDM (Joint


T. Sobh (ed.), *Innovations and Advances in Computer Sciences and Engineering*,
DOI 10.1007/978-90-481-3658-2_100, © Springer Science+Business Media B.V. 2010

Consultation, Command and Control Information Exchange Data Model) [9], [10]. NATO and U.S. Joint Forces Command have expressed interest in experimentation around common ontology and knowledge structures [11]. Additionally the Naval Postgraduate School (NPS) has a history of testing experimental systems that use ontology and metadata to categorize knowledge specific to new technologies in the DoD [12], [13], [14], [15], [16].

### B. *Knowledge Management*

The NPS philosophy is to develop the best understanding possible of leading edge systems in a complex environment while keeping costs reasonable. A proven, rigorous process has been developed over the past 10 years to perform the assessment and this methodology is realized through an EIS. The methodology includes a proven taxonomy to direct and focus planning objectives, developments and resource requirements to ensure a comprehensive and accurate assessment of all elements under review in minimum time. The critical elements evaluated include the following:

- Processes – all aspects of the complex series of interactions that channel performance of systems to achieve certain goals.
- Systems – hardware, software, electronic, mechanical, electrical, or mechanical, all require rigorous evaluation.
- Human integration with systems and process to understand how to optimize performance with available manpower.
- Directives – complex guidance that demands, prohibits, or directs various actions associated with any process.
- Environment or context – typically highly complex in military matters, but the context must be quantified and considered when trying to assess the performance of systems.

Most important is a need to maintain a well-organized record of all aspects of the evaluation process and the eventual assessments. The archives need to be logically structured and integrated so that all evaluations can be incorporated, searched, and critical data and documents easily retrieved. Ideally, each user should have in independent view of the data.

The FORCEnet Innovative Research Enterprise (FIRE) was developed for Naval Networks Warfare Command and has evolved over the last nine years. It does not duplicate information available in other systems, but rather leverages them to track the complex dynamic between processes, systems, humans, and directives. FIRE contains taxonomy for evaluations and the data that populates that framework.

During planning, analysis, and reporting phases FIRE is a collaboration tool accessible via internet by any authorized user with no special software or hardware required. Once the military utility assessment is concluded, FIRE becomes the repository for the work just concluded, as it is for all previous work. FIRE is password protected and each user is limited to specific authorized areas.

## III. System Implementation

A depiction of the overall assessment process is shown in Fig. 1. The process begins at the top with a capabilities assessment where some form of shortfall, e.g. "gap" is identified. The program sponsor identifies potential ways to achieve the desired performance goals. These objectives are used to populate the FIRE taxonomy, an "expert process," which sets off an iterative process of defining the best way to assess the performance of the system in question, considering and measuring all the aspects listed above (processes, systems, human integration, directives, and context). Static elements are defined then operational scenarios developed to provide a match to real-world context.

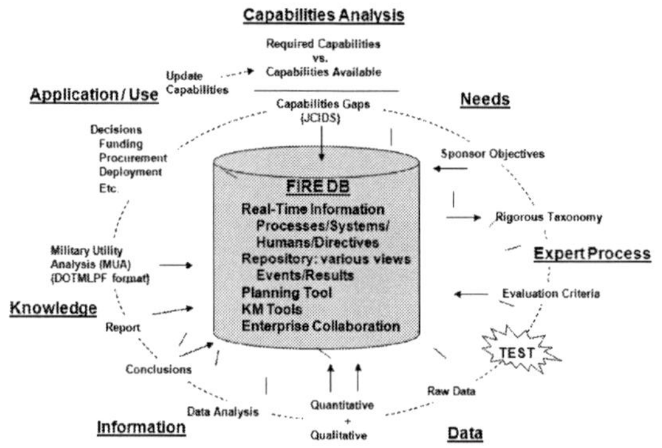


Fig. 1. Analysis flow and management processes.

Tests are run within operational scenarios and all attributes and measures defined in the taxonomy are observed and collected. The data collected are both quantitative and qualitative and both types are used in the ensuing analysis. Comprehensive reports of the conclusions and findings of the analyses are developed and the information is used to develop an assessment of military utility. This assessment looks at doctrine, organization, training, materiel, leadership, personnel, and facilities (DOTMLPF) and reports the findings and implications in each of those categories. The flow is also represented in Fig. 2 which lists the 13 step process which guided development of the FIRE management applications, forms and reports.

| Phase | 1 | 2 | 3 | 4 | 5 | 6 |
|---|---|---|---|---|---|---|
| Due Dates | Pre-CDC | CDC | Pre - IPC | Pre - IPC | Pre - MPC | Pre - MPC |
| Step | Establish Team | Concept Development | Technology / TTP Harvest | Asset Identification | Develop Experiment Objectives | IDEF / OSD / Process Action Maps |
| Required Product | Defined Names and R /R | Defines Experiment Scope and Focus Areas. Insure aligns with Naval Vision | Defines and Researches selected Tech and TTPs | Platforms ID'ed and Install Scheduled | Defines the So What and how to measure | Turns the Words into Design Diagrams |

| Phase | 7 | 8 | 9 | 10 | 11 | 12 | 13 |
|---|---|---|---|---|---|---|---|
| Due Dates | Pre - MPC | Pre-FPC | Pre - FPC | TBD | TBD | TBD | TBD |
| Step | Experiment Design | Event Definition | Data Collection Plan | Execution | Final Report | Assessment OAA | MUA |
| Required Product | Lays out the Flow and Appicable Scenarios to Meet Objectives | Defines the Detailed execution Plan | Maps the Data to be Collected to the means | Insure Plan is Flexable to changing enviornment | Must be Quick and Good | Necks down Analysis to Assessment | Necks Down Assessment to DOTMLPF Recommendations |

Fig. 2. 13-step flow for experimentation management.

The Navy can use this knowledge to make substantive decisions with respect to funding, acquisition, deployment of technology, appropriate manning, etc. Following an update to the capability database, the process repeats.

### A. *Collaborative Portals*

There are two primary portals for the FIRE services. The first was launched as TACFIRE (Tactical Applications for Collaboration in FIRE) and provides the primary user interfaces and collaborative capabilities (Fig. 3). Portals are personalized for each user with that user's email, personal and group calendar, task manager, web conferences, enterprise content manager, mailing lists, personal and group workspaces, and instant messenger.

Fig. 3. TACFIRE collaborative portal.

From the personalized portals users can navigate into the FIRE application portals specific to an ongoing or previous experiment (Fig. 4). From these portals are links to specific forms and reports that initiative leads, experiment, and analysis managers use to manage their technologies and objectives, and to coordinate data input and refinement with their distributed teams. Initiative leads serve as managers for technologies within a focus area (capability) and parse input security to technology subject matter experts (SMEs).

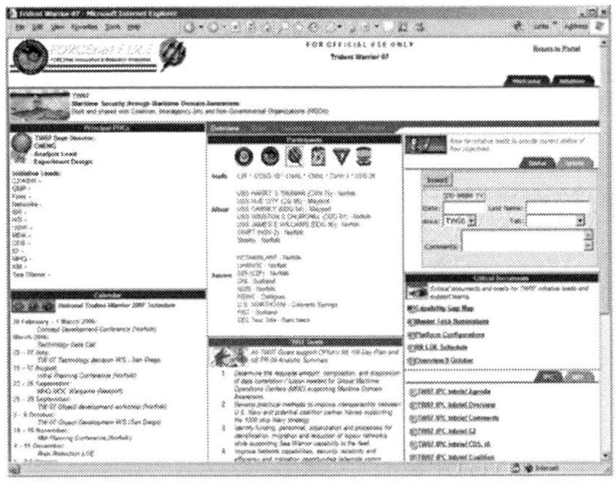

Fig. 4. FIRE experimentation portal.

## IV. Systems Integration

An important facet of DoD information systems are the processes through which systems interface. These interfaces will be aided through Service-Oriented Architecture (SOA) and there is a major push in the DoD into SOA for the ability to publish/subscribe experiment results such that program office can better use result data in their purchasing decisions.

Fig. 5 provides an example of a large-scale systems integration effort in which FIRE serves as a host for several components of an end-to-end innovation diffusion process. Working from the left side of the diagram is the needs assessment and requirements analysis. Then the EDISON process, hosted in FIRE, through which industry submits technologies into the experimentation process. Various databases are then integrated through the experimentation process until final recommendations and dissemination. To date many of the integration efforts are minimal. Evolution into SOA and service-based interfaces will enable full integration and are a work in process.

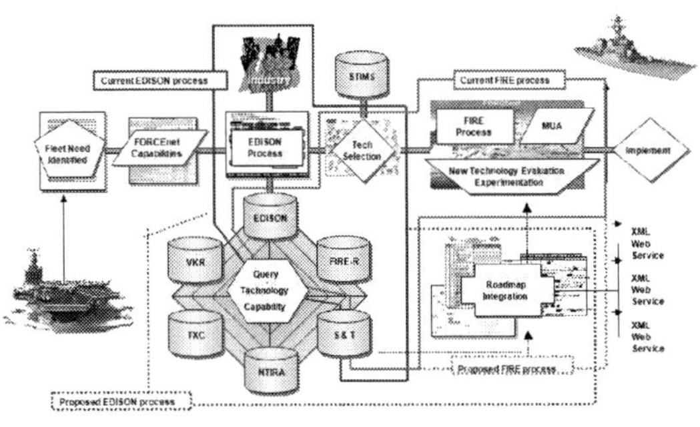


Fig. 5. High-level systems integration flow.

### *Application Interfaces*

As highlighted in Fig. 1 and 5 the management flow begins with establishing system requirements; in the Navy requirements analysis is framed as desired “capabilities”. These capabilities are then published in a Commercial Area Announcement to which industry can reply. Various means are used to collect proposals, ranging from simple white papers describing a technology to automated processes.

One of the automated processes developed as part of the FIRE configuration was named “EDISON” and addressed the “technology harvest” and “asset identification” phases of the 13-step process (Fig. 2). The EDISON process consisted of a workflow through which industry technology sponsors entered their technology information into a form in FIRE in the EDISON application in which desired capabilities (requirements) were identified. In the workflow sponsors would first fill out the specifics of their technology via a form (Fig. 6).

Submissions were automatically aggregated by capability into reports for the review committee. Another form enabled reviewers to rate each technology for its match to the requirements specification and a report was automatically generated that provided cumulative scores and reviewer comments. Finally came the approval or rejection stage and a means to provide feedback to the sponsors. A progress tab in the application provided sponsors and reviewers with current status in the review process.

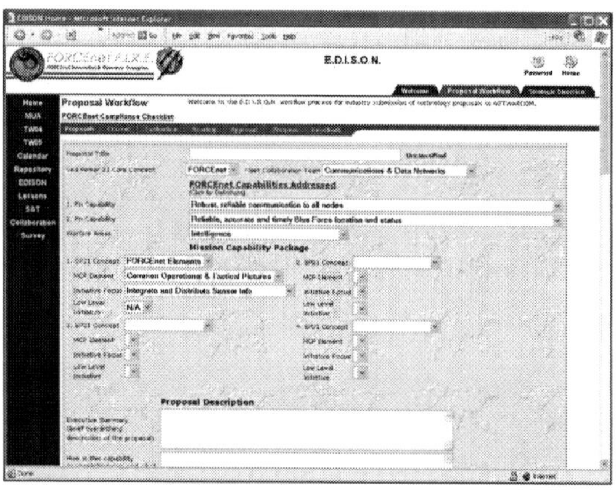


Fig. 6. EDISON technology submission process.

Those sponsors with successful submissions were then invited into the FIRE experimentation management and analysis system for the selected experiment as steps 5 and 6 of the 13-step process. This stage included a high-level screen to frame the experiment and processes (Fig. 7), followed by linked forms for objective input in a workflow from high-level to data collection specifics and analysis methodology. Various models were input to support the technology and systems integration, ranging from IDEF to Use-Case and data flow diagrams.

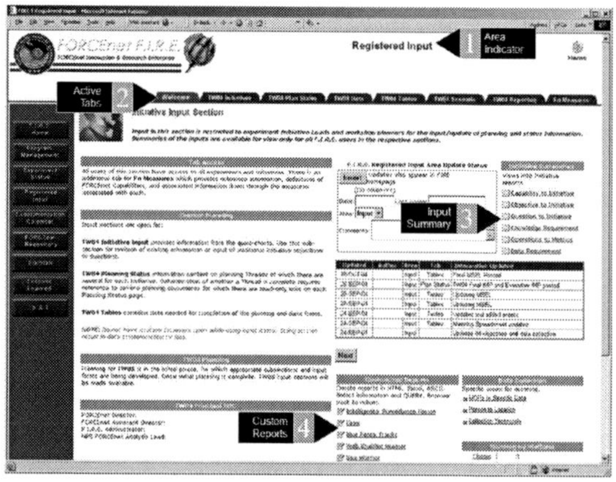


Fig. 7. Technology objectives management.

Progress through the objectives and related management processes is difficult and generally entails weekly teleconferences by the initiative lead with the technology sponsors assigned to that lead. Since the technologies may involve participants anywhere in the world a series of evaluation and progress mechanisms were devised to help manage the projects. Stoplight charts are common in the military as a means to provide the status of a system at a glance and these were adopted to monitor team progress toward completion of objectives, data sheets, models, analysis methodology, testing routines, and data collection procedures. Tests are conducted in large experiments involving several thousand DoD personnel, including several fleets, several countries, in live global operations. As such, it is important to have a clearly defined management process and analysis routine. An example of a tracking system to help ascertain status is presented in Fig. 8.

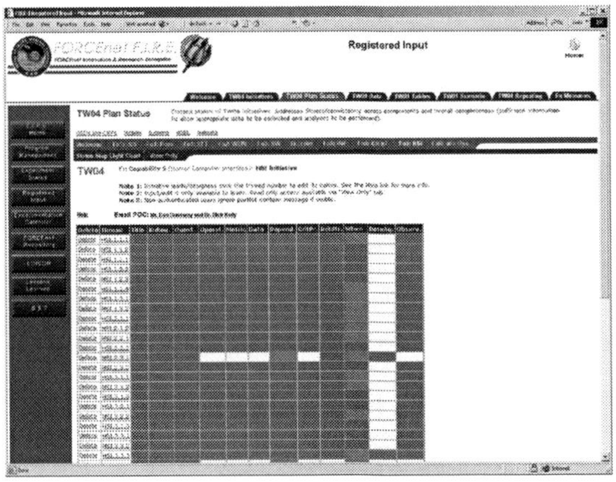

Fig. 8. Analysis management stoplights.

In process, an initiative lead would review the inputs of their technology providers and then indicate a status color. A green color would trigger an assigned member of the experiment analysis team to review all inputs. If agreed that all systems are "go" (green), indicating that all facets of the objective development process are completed, then the technology is ready for inclusion in the experiment. This includes steps 7 and 8 of the 13-step process and the manner in which the scenario tests the objective. Events must be clearly defined and a detailed execution plan presented.

A variety of instruments have been developed to help with data collection during experiment execution. These have ranged from spreadsheets to custom reports that include objectives, events, analysis methodology, and collected data.

The final step in the process is the reporting. This includes a QuickLook report that provides near immediate assessment of tested objectives and their technologies and is completed a few days after the experiment. An Operational Agent Assessment helps structure the analysis, to match the tested objectives to the requested capabilities.

A Military Utility Assessment (MUA) next provides high-level recommendations to senior decision makers. Fig. 9 provides an example of an information capability utilized within FIRE to support the MUA process. At the MUA a technology is either dismissed, recommended for adoption and integration into the funding cycle, or pushed back for further test and evaluation.

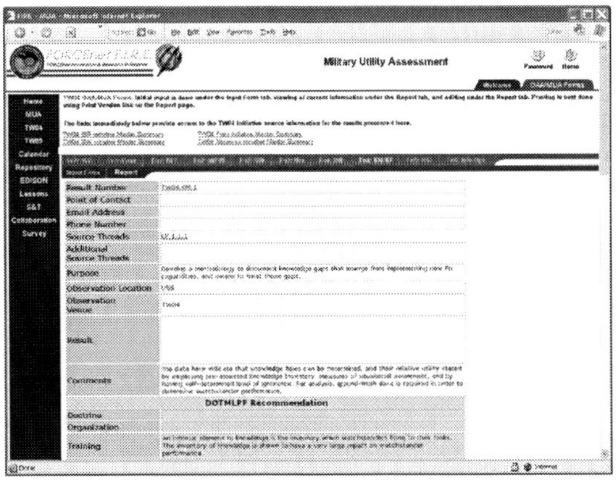

Fig. 9. Military Utility Assessment.

The Final Report is completed a few months after the end of the experiment and is generally a 1000+ page document. This includes the reports of each of the individual technologies, with experiment results from the operational scenarios, plus the integration and overall assessment from the analysis team.

## IV. CONCLUSION

This paper presented an operational enterprise information system and its supporting methodology for the analysis of complex experiments. The system is currently utilized by the Department of Defense for Navy and Joint Forces experimentation. Assessed are effectiveness and efficiency of systems interactions and affected intra- and inter- organizational processes to assess next-generation military e-services and infrastructure. A focus is the expansion of systems methodology beyond traditional transaction-oriented exchanges to include management layers for complex inter-organizational processes and mechanisms to accommodate sensitive operations.

## REFERENCES

[1] Chandrasekaran, B., Josephson, J., and Benjamins, V., "What are Ontologies, and Why Do We Need Them," *IEEE Intelligent Systems*, Vol. 14, No. 1 (1999), pp. 20-26.

[2] Berners-Lee, T. "The Semantic Web," *Scientific American*. Available http://scientificamerican.com/2001/0501issue/0501berners-lee.html, 2001.

[3] Maule, R., and Gallup, S., "Knowledge management system with enterprise ontology for military experimentation – case study: F.I.R.E. and TACFIRE. *Proceedings of the 2007 International Conference on*

*Enterprise Information Systems & Web Technologies (EISWT-07), 9-12 July, Orlando, FL, 2007.*

[4] Grimes, S., "The Semantic Web," *Intelligent Enterprise*, Vol. 5, No. 6 (2002), pp. 16, 52.

[5] Avigdor, G., and Mylopoulos, J., "Toward Web-Based Application Management Systems," *IEEE Transactions on Knowledge and Data Engineering*, Vol. 13, No. 4 (2001), pp. 683-702.

[6] Fikes, R., and Farquhar, A., "Distributed Repositories of Highly Expressive Reusable Ontologies," *IEEE Intelligent Systems*, Vol. 14, No. 2 (1999), pp. 73-79.

[7] Maule, R., Gallup, S., and Schacher, G., "Dynamic Knowledge Profiles for Organizational Metacognition," *International Journal of Knowledge, Culture and Change Management*, Vol. 4 (2005), pp. 309-317.

[8] Maule, R., "Toward a Knowledge Analytic Understanding," *International Journal of Technology, Knowledge and Society*, Vol. 1, No. 1 (2006), pp. 121-127.

[9] NATO, "The Land C2 Information Exchange Data Model. Internal Report: ADatP-32 Edition 2.0," NATO, 2000.

[10] Chamberlain, S., Boller, M., Sprung, G., Badami, V., "Establishing a Community of Interest (COI) for Global Force Management," *Paper presented at the 10th International Command & Control Research & Technology Symposium*, June 13-16, 2005, Washington, DC.

[11] Schacher, G., Maule, R., and Gallup, S., "Knowledge Management System for Joint Experimentation," *Proceedings of the NATO OSLO Symposium: Building a Vision: NATO's Future Transformation*, September 5-7, 2001, Oslo, Norway.

[12] Maule, R., Gallup, S., and McClain, B., *Knowledge Extraction for Anti-Submarine Warfare Lessons Learned from Fleet Battle Experiments*, Monterey, CA: Naval Postgraduate School, Institute for Defense Systems Engineering and Analysis, 2002.

[13] Maule, R., Schacher, G., and Gallup, S., "Knowledge Management for the Analysis of Complex Experimentation," *Journal of Internet Research*, Vol. 12, No. 5 (2002), pp. 427-435.

[14] Maule, R., Schacher, G., Gallup, S., Marachian, C., and McClain, B., *IJWA Ethnographic Qualitative Knowledge Management System*, Monterey, CA: Naval Postgraduate School, Institute for Joint Warfare Analysis, 2000.

[15] Maule, R., Gallup, S., and Schacher, G., "Broadband Internet Applications for Multi-Organizational Collaborative Processes, Workflow Automation, and Complexity Analysis in Multi-National Experimentation," *Proceedings of the 2003 Pacific Telecommunications Conference, January 18-23, 2003, Honolulu, HI.*

[16] Maule, R., Schacher, G., Gallup, S., and McClain, B., "Applied Knowledge Analytics for Military Experimentation," *Proceedings of the IEEE International Conference on Information Reuse and Integration* (IEEE IRI-2004), November 8-10, 2004, Las Vegas, NV.

# Building Information Modeling and Interoperability with Environmental Simulation Systems

Paola C. Ferrari, Neander F. Silva, Ecilamar M. Lima
Universidade de Brasília
Laboratório de Estudos Computacionais em Projeto – LECOMP
ICC Norte, Ala A, Subsolo, CEP 70910-900, Brasília, DF, BRASIL
http://lecomp.fau.unb.br/, paolaferrari@uol.com.br, neander@unb.br, ecilamar@unb.br

**Abstract – Most of the present day BIM systems do not have built-in environmental simulation tools. In this case such programs have relied so far on dedicated external simulation applications. Interoperability is therefore a major issue for BIM tools. But can data be transferred from BIM tools to simulation tools completely, understandable and ready to be used by the receiving application? In this paper we report an experiment that shows that the data transfer in the present day still requires a set of complex procedures such as preparing the files to be passed from one application to another. The data transfer from the BIM tools to environmental simulation tools seems to be quite complete, understandable and ready to use by the receiving application. However, data transferred in the opposite direction is much less customized to the receiving application and needs to be configured manually.**

## I. Research Problem

The focus of the research reported in this paper is the interoperability between BIM tools and Environmental Simulation tools with particular emphasis on natural and artificial lighting analysis of building design. Studies have already been made demonstrating the importance of natural lighting in the production of an environmentally friendly architectural design [1]. The main objective is to verify if BIM models can really be useful to design greener buildings by offering designers more accurate information about the lighting conditions in the projects.

The concept of BIM, as defined by Eastman et al in [2] refers to a modeling technology and processes that present some essential features: the building components are digitally represented as objects that include not only geometry but also data describing what they are made of and how they behave. Their data is represented coordinately in a three-dimensional (3D) space so that changes to one view are automatically updated in all the other views. The relationships between components are parametrically represented [2], so that changes to one of them result automatically in changes to all the related ones.

According to Eastman et al interoperability "depicts the need to pass data between applications, allowing multiple types of experts and applications to contribute to the work at hand. Interoperability has traditionally relied on file based exchange formats…" [3]. For the purposes of this research this means transferring data from one application to another in a complete way and can be readily used by the receiving one.

Most of the present day BIM systems do not have built-in environmental simulation tools. In this case such programs have relied so far on dedicated external simulation applications. Therefore, these systems have also relied on transferring all the necessary data to different applications in such a way that this information is complete, can be understood correctly and can be used straightaway by the receiving software for environmental simulation. The flow of data in the opposite direction must also be, in respect to the BIM paradigm, complete, understandable and ready to use by the other application.

The research questions in this paper are: 1) how is the state of the art of the present day BIM tools in relation to interoperability of such systems with external environmental simulation tools? 2) Can data be transferred from BIM tools to simulation tools completely, understandable and ready to be used by the receiving application? 3) Can data produced by the simulation tools be transferred back to the BIM tools in a way that is complete, understandable and useful?

## II. Hypothesis

We believe that interoperability between the BIM tools and environmental simulation applications in the present day still requires a set of complex procedures such as preparing the files to be passed from one application to another.

We believe that data transfer is complete, understandable and ready to use by the receiving application when the flow of information occurs from the BIM tools toward the environmental simulation tools.

We believe also that data transfer from the environmental simulation tools in the direction of the BIM tools is much more precarious. The types of data that can be actually transferred back to the BIM tools is of a much more limited nature, such as bitmapped files and spreadsheets which need

T. Sobh (ed.), *Innovations and Advances in Computer Sciences and Engineering*,
DOI 10.1007/978-90-481-3658-2_101, © Springer Science+Business Media B.V. 2010

to be interpreted, manually handled and, if possible, inserted into the original BIM system.

## III. Research Method

The experiment we carried out to verify our hypotheses consisted of the following procedures:

A model of a small building design, with all its geometry and data about materials and their behavior, was made in the BIM tool ArchiCAD (http://www.graphisoft.com).

The second step consisted of preparing a XML file and exporting it to the environmental simulation tool Ecotect (http://www.ecotect.com).

The third step consisted of running a natural and artificial lighting simulation in Ecotect for obtaining precise information about energy consumption.

This analysis allows to assess the intensity of artificial lighting and the amount of electricity spent and, therefore, to propose changes in the design project to reduce such consumption.

For this analysis to be made possible in as much as possible automated way it is vital to have a high level of interoperability between the two systems involved. A set of assessment criteria was established to perform an evaluation of the level of interoperability between the BIM tool and the environmental simulation tool. Table I shows this criteria in the next session.

## IV. Results

The first step was to produce a three-dimensional model of the building in the BIM tool ArchiCAD. Figure 1 shows the results.

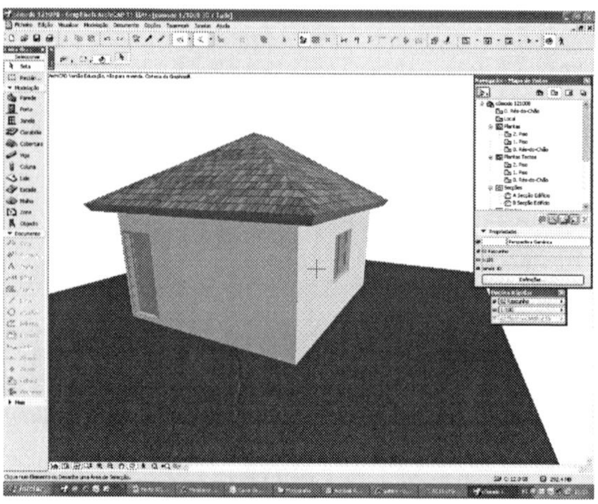

Fig.1. Virtual model in ArchiCAD.

The second step consisted of preparing the model to export from ArchiCAD into ECOTECT. This required a number of steps, which were not very well document by Graphisoft, the maker of ArchiCAD. The file had to be in the eXtensible Markup Language, XML, format, which is an extension of HTML. XML is becoming one of the main file formats for exporting data from BIM tools to environmental simulation tools [4][5]. Figure 2 shows the ArchiCAD model ready to be exported in a XML format.

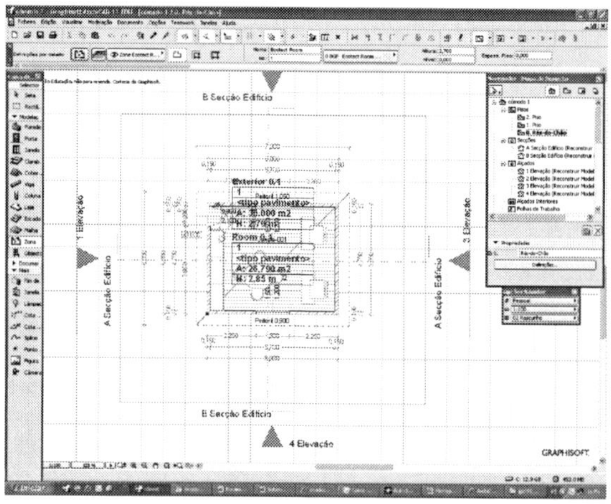

Fig.2. Model in ArchiCAD after configuration for exportation as XML file.

Figure 3 shows the environmental zone created still viewed in the ArchiCAD model.

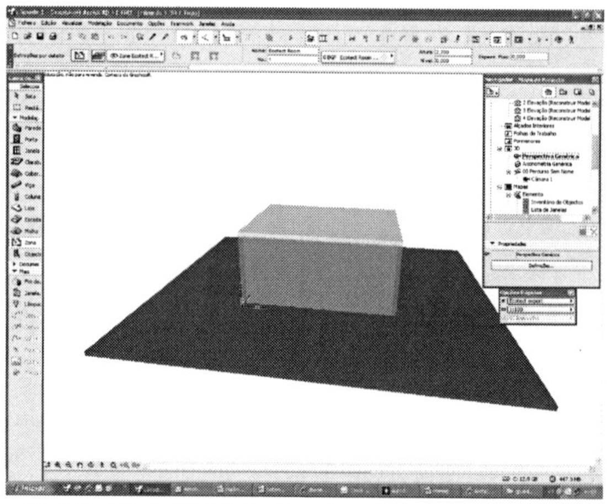

Fig.3. Model in ArchiCAD showing only environmental zone.

Figure 4 shows the model imported into ECOTECT from ArchiCAD.

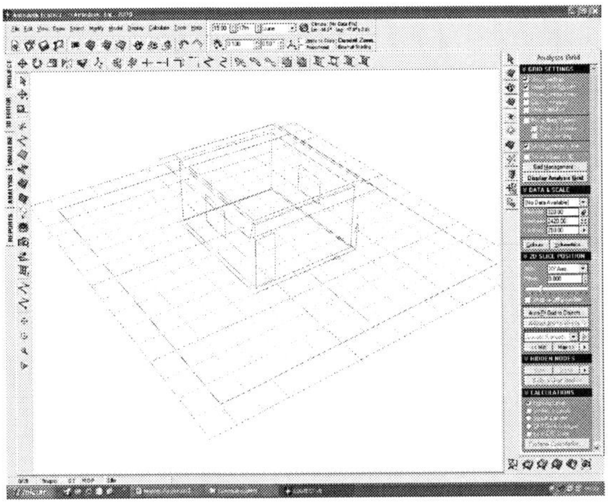

Fig.4. Model imported into ECOTECT from ArchiCAD.

It was not possible to export the roof and furniture from ArchiCAD via XML format. Therefore, we had to resort to other format in order to transfer this information to ECOTECT. Figure 5 shows the ArchiCAD model prepared to export the roof and furniture in the 3DS format. This information was not essential for the particular simulation of this research, but it was performed to test the boundaries of possible interoperability and identify possible limitations. The furniture might have been vital if we were to run an acoustics simulation, for example.

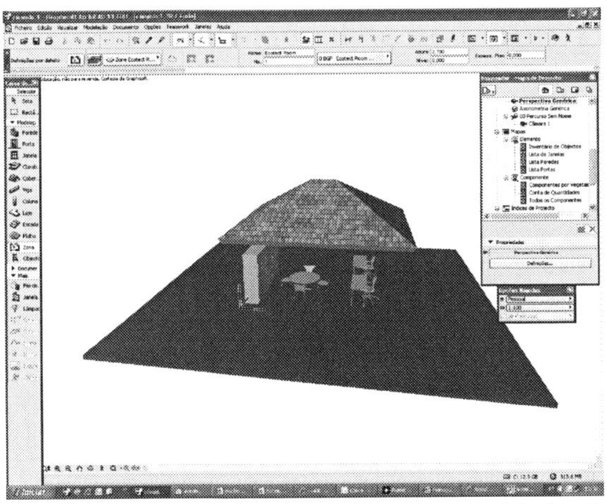

Fig.5. Model in ArchiCAD prepared to export only roof and furniture in 3DS format.

Figure 6 shows the roof and the furniture already imported into the ECOTECT model file via 3DS format.

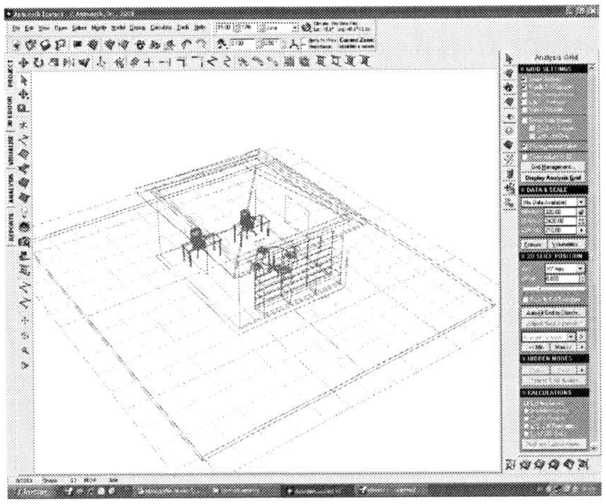

Fig.6. Model in ECOTECT including roof and furniture imported from ArchiCAD in the 3DS format.

It was them necessary to configure the geographic coordinates of our building model and we had to do this in ECOTECT. The model was located in the city of Brasilia, Brazil.

After entering the data regarding the geographic location and orientation into a ECOTECT and running a first simulation, it was possible to see the north as illustrated by Figure 7.

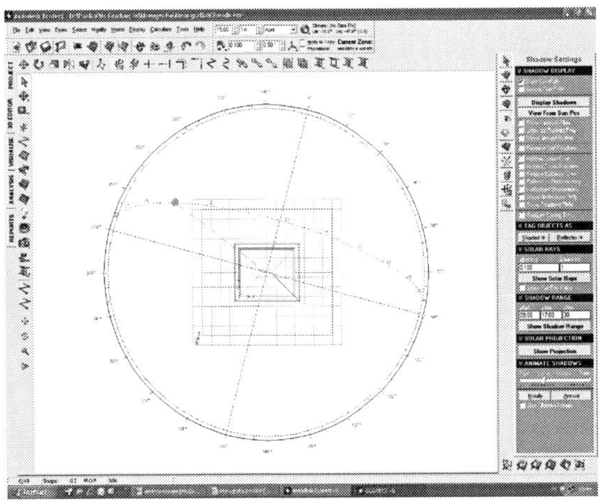

Fig.7. Setting the north in ECOTECT.

Figure 8 shows a map of the natural light intensity which is valuable information for design project evaluation that could guide the study of possible modifications and improvements. As a broader and more professional application of data of environmental concern we could mentioned the extensive use of it by world renowned architectural and engineering offices such as Arups Engineers in Norman Fosters designs [6][7].

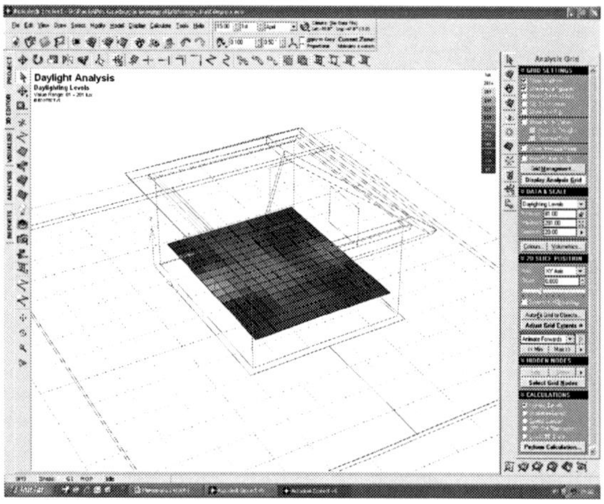

Fig.8. Natural light simulation in ECOTECT.

However, we faced some problems in importing such map into ArchiCAD from ECOTECT in the .jpg format which resulted in a blank space.

Table I below shows the assessment criteria and the resulting evaluation from our experiment.

TABLE I
ASSESSMENT CRITERIA AND EXPERIMENTATION RESULTS

| Assessment parameters of interoperability between ArchiCAD 11 & Ecotect 5.60 | | Data transfer | | |
|---|---|---|---|---|
| | | From ArchiCAD to Ecotect via 3DS | From ArchiCAD to Ecotect via XML | Ecotect outputs to ArchiCAD |
| Geometry | Does it export 2D geometry? | ▓ | ░ | ■ |
| | Does it export 3D geometry? | ░ | ▓ | ■ |
| Materials | Does it export properties of materials? | ▓ | ░ | ■ |
| | Does it export textures? | ▓ | ▓ | ■ |
| | Does is allow the specification of new materials? | ▓ | ░ | ■ |
| Elements | Does it export all constructive elements? | ░ | ▓ | ■ |
| | Does it export only some constructive elements such as doors and windows? | ▓ | ░ | ■ |
| | Does it recognize the imported constructive elements as independent lines? | ░ | ▓ | ■ |
| | Does it recognize the imported constructive elements as surfaces? | ▓ | ░ | ■ |
| | Does it export furniture? | ░ | ▓ | ■ |
| Data | Does it recognize the surfaces? | ▓ | ░ | ■ |
| | Does it export layers? | ▓ | ░ | ■ |
| Thermal Zones | Does it export thermal zones for environmental analysis? | ▓ | ░ | ■ |
| External references | Does it export as external reference? | ▓ | ▓ | ░ |

░ Yes
▓ No
■ Does not exist

## V. CONCLUSIONS

This research has contributed by showing the virtues and the difficulties of interoperability between the BIM tools and environmental simulation applications. The data transfer in the present day still requires a set of complex procedures such as preparing the files to be passed from one application to another, while the documentation of this process is vague and incomplete.

The data transfer from the BIM tool ArchiCAD to environmental simulation tool ECOTECT seems to be quite complete, understandable and ready to use by the receiving application. The XML format used in the transfer does not contain some important information such as the furniture. However, this data was not essential for the natural lighting simulation. This would become an issue if it were an acoustics the simulation. Therefore, future research should work on a more file formats which could provide a more complete range of data transfer.

The data transfer from the ECOTECT to the ArchiCAD was much more precarious. In was indeed limited in range such as bitmapped files and spreadsheets which need to be interpreted, manually handled and, if possible, inserted into the original BIM system.

On the other hand, the experiment shows a very promising interoperability between BIM systems and environmental simulation applications. Once we overcame the documentation difficulties and produced the exporting file the information read by receiving application was quite satisfactory to run the lighting simulation.

It was possible to work with the environmental resources and take the advantage of the benefits of BIM in order to make the needed modifications and improvements in the design project. As the final result this partnership allows to maximize the use of energetic resources in the built environment.

## REFERENCES

[1] Sue Roaf, Manuel Fuentes and Stephanie Thomas, "Ecohouse: A Design Guide", Architectural Press, Elsevier, Amsterdam, pp. 175-192, 2007.

[2] Chuck Eastman, Paul Teicholz, Raphael Sacks and Kathleen Liston, *"BIM Handbook – A Guide to Building Information Modeling"*, John Wiley & Sons, New Jersey, pp. 13-14, 2008.

[3] Chuck Eastman, Paul Teicholz, Raphael Sacks and Kathleen Liston, *"BIM Handbook – A Guide to Building Information Modeling"*, John Wiley & Sons, New Jersey, p. 65, 2008.

[4] Chuck Eastman, Paul Teicholz, Raphael Sacks and Kathleen Liston, *"BIM Handbook – A Guide to Building Information Modeling"*, John Wiley & Sons, New Jersey, p. 85, 2008.

[5] Eddy Krygiel and Bradley Nies, Green BIM – Successful Sustainable Design With Building Information Modeling, Sybex Wiley Publishing, Indianapolis, pp. 185-186, 2008.

[6] Hugh Whithead, "Law of Form" in Branko Kolarevic (editor) "Architecture in the Digital Age – Designing and Manufacturing", Taylor & Francis, New York and London, pp. 86-87, 2003

[7] Michael Stacey, Philip Beesley & Vincent Hui, "Digital Fabricators", University of Waterloo School of Architecture Press, p. 19, 2004.

# Index




T. Sobh (ed.), *Innovations and Advances in Computer Sciences and Engineering*,
DOI 10.1007/978-90-481-3658-2, © Springer Science+Business Media B.V. 2010

## K



## L



## M



## N